# MARINE
# GLYCOBIOLOGY
## Principles and Applications

# MARINE GLYCOBIOLOGY
## Principles and Applications

edited by
## Se-Kwon Kim

**CRC Press**
Taylor & Francis Group
Boca Raton  London  New York

CRC Press is an imprint of the
Taylor & Francis Group, an **informa** business

CRC Press
Taylor & Francis Group
6000 Broken Sound Parkway NW, Suite 300
Boca Raton, FL 33487-2742

First issued in paperback 2019

ISBN-13: 978-1-4987-0961-3 (hbk)
ISBN-13: 978-0-367-87208-3 (pbk)

---

**Library of Congress Cataloging-in-Publication Data**

---

Names: Kim, Se-Kwon, editor.
Title: Marine glycobiology : principles and applications / [edited by]
Se-Kwon Kim.
Description: Boca Raton : CRC Press/Taylor & Francis Group, 2017. | Includes
bibliographical references and index.
Identifiers: LCCN 2016010385 | ISBN 9781498709613 (hardback : alk. paper)
Subjects: | MESH: Glycomics--methods | Aquatic Organisms |
Glycoconjugates--pharmacology | Bioprospecting--methods |
Biotechnology--methods
Classification: LCC QP702.G577 | NLM QH 91.8.B5 | DDC 572/.56--dc23
LC record available at http://lccn.loc.gov/2016010385

---

# Contents

# Preface

Marine glycobiology is the study of carbohydrate and carbohydrate with molecules (protein, lipids, enzymes, or small molecules). Glycoconjugates are glycan linked with other biological molecules. The study and research on marine glycobiology is less well-known. However, the recent development on analytical instruments and chemical characterizations increases the research on glycoconjugates. The knowledge gained of the exact chemical structure of marine glyconjugates increases its use in biological and biomedical applications.

This book contains 38 chapters under different sections.

1. Section I—Chapter 1 provides a general introduction to the topics covered in this book.
2. Section II—marine glycoconjugates of reproduction and chemical communications (Chapters 3 and 4) are described.
3. Section III—Bioactivity, principles, and applications of marine glycans (Chapters 5 through 7) are explored.
4. Section IV—marine glycoproteins (Chapters 8 through 13)—deals with biomedical benefits of algal glycoproteins, marine source collagen, and detoxification in the marine environment.
5. Section V—marine glycoenzymes (Chapter 14)—discusses the sialyltransferases from the marine environment and their applications.
6. Section VI—marine carbohydrates (Chapters 15 through 29)—discusses several marine carbohydrates and their glycobiology. Marine bacterial exopolysaccharides, agar, chitin and chitosan, sulfated polysaccharides, carbohydrates from marine microbes, algal polysaccharides, and mangroves are discussed in detail. In addition, the use of these polysaccharides in pharmaceutical applications, plant growth, and various biological activities are also discussed.
7. Section VII and VIII—bioinformatics and biological role of glyconjugates (Chapters 30 and 31)—describes glycan's predictive modeling using modern algorithms and glycoconjugated bioactivity compounds and biological applications.
8. Section IX—glycoconjugates as biomedicine (Chapters 32 through 38)—presents the application of glyconjugates in biomedicine and their biotechnological applications.

I express my sincere thanks to all the authors who have contributed toward this book. Their relentless effort was the result of their strong inclination towards scientific research, and great perseverance descended from their experience. I am grateful to the experts who have contributed to this book.

I hope that fundamental as well as applied contributions to this book might serve as potential research and development leads for the benefit of humankind. Marine glycobiology will be the excellent field in the future towards the enrichment of targeted marine glycans, which further sets up a suitable for further applications. This book would be a reference book for students in academic and industrial research.

**Prof. Se-Kwon Kim**
*Busan, South Korea*

# Acknowledgments

I thank CRC Press staff for their continuous encouragement and suggestions to get this wonderful compilation published. I also extend my sincere gratitude to all the contributors for providing their help, support, and advice to accomplish this task. Further, I thank Dr. Panchanathan Manivasagan and Dr. Jayachandran Venkatesan, who worked with me throughout the course of this book project. I strongly recommend this book for marine biotechnology and glycobiology researchers/students/industrialists and hope that it helps to enhance their understanding in this field.

**Prof. Se-Kwon Kim**
*Pukyong National University*
*Busan, South Korea*

# Editor

**Se-Kwon Kim, PhD**, is a distinguished professor in the Department of Marine Bio Convergence Science and Technology and as director of Marine Bioprocess Research Center (MBPRC) at Pukyong National University, Busan, South Korea.

He received his MSc and PhD from Pukyong National University and conducted his postdoctoral studies in the Laboratory of Biochemical Engineering, University of Illinois, Urbana–Champaign, Illinois. Later, he became a visiting scientist at the Memorial University of Newfoundland and the University of British Colombia in Canada.

Dr. Kim served as president of the Korean Society of Chitin and Chitosan during 1986–1990 and the Korean Society of Marine Biotechnology during 2006–2007. In recognition of his research, he won the best paper award from the American Oil Chemists' Society. In 2002, Dr. Kim was also the chairman for the 7th Asia-Pacific Chitin and Chitosan Symposium, which was held in South Korea in 2006. He was the chief editor of the journal *Korean Society of Fisheries and Aquatic Science* during 2008–2009. Also, he is a board member of the International Marine Biotechnology Association (IMBA) and the International Society of Nutraceuticals and Functional Food (ISNFF).

His major research interests are investigation and development of bioactive substances from marine resources. His immense experience in marine bioprocessing and mass production technologies for marine bio-industry is the key asset of holding majorly funded marine bio projects in Korea. Furthermore, he expended his research fields up to the development of bioactive materials from marine organisms for their applications in oriental medicine, cosmeceuticals, and nutraceuticals. To date, he has authored around 700 research papers and 70 books and holds 130 patents.

# Contributors

**Imran Ahmad**
Chaplin School of Hospitality and Tourism
   Management
Florida International University
North Miami, Florida

**Bakrudeen Ali Ahmed**
Department of Biotechnology
Sri Shakthi Institute of Engineering and Technology
Coimbatore, Tamil Nadu, India

**Thipramalai Thankappan Ajithkumar**
National Bureau of Fish Genetic Resources
Indian Council of Agricultural Research
Lucknow, Uttar Pradesh, India

**Abdul Shirin Alijani**
Institute of Biological Sciences
and
Department of Chemical engineering
University of Malaya
Kuala Lumpur, Malaysia

**Visamsetti Amarendra**
School of Chemical and Biotechnology
SASTRA University
Thanjavur, Tamil Nadu, India

**Perumal Anantharaman**
Faculty of Marine Sciences
Annamalai University
Chidambaram, Tamil Nadu, India

**Muthuvel Arumugam**
Centre of Advanced Study in Marine Biology
Annamalai University
Chidambaram, Tamil Nadu, India

**Kazuo Azuma**
Department of Veterinary Clinical Medicine
Tottori University
Tottori, Japan

**Thangavel Balasubramanian**
Centre of Advanced Study in Marine Biology
Annamalai University
Chidambaram, Tamil Nadu, India

**Partha Pratim Bose**
Division of Molecular Medicine
Bose Institute
Kolkata, West Bengal, India

**Jaya Chakraborty**
Department of Life Science
National Institute of Technology
Rourkela, Odisha, India

**Bishnu Pada Chatterjee**
Department of Natural Science
West Bengal University of Technology
Kolkata, West Bengal, India

**Urmimala Chatterjee**
Department of Natural Science
West Bengal University of Technology
Kolkata, West Bengal, India

**Jie Chen**
School of Agriculture and Biology
Shanghai Jiao Tong University
Minhang, Shanghai, People's Republic of China

**Anong Chirapart**
Department of Fishery Biology
Kasetsart University
Bangkok, Thailand

**Katarzyna Chojnacka**
Department of Advanced Material Technologies
Wrocław University of Technology
Wrocław, Poland

**Surajit Das**
Department of Life Science
National Institute of Technology
Rourkela, Odisha, India

**Hirak R. Dash**
Department of Life Science
National Institute of Technology
Rourkela, Odisha, India

**Ariyanti Suhita Dewi**
Research and Development Centre for Marine
and Fisheries Products Competitiveness and
Biotechnology
Ministry of Marine Affairs and Fisheries
Jakarta, Indonesia

**Elangovan Dilipan**
Centre of Advanced Study in Marine Biology
Annamalai University
Chidambaram, Tamil Nadu, India

**Agnieszka Dmytryk**
Department of Advanced Material Technologies
Wrocław University of Technology
Wrocław, Poland

**Reza Farzinebrahimi**
Institute of Biological Sciences
University of Malaya
Kuala Lumpur, Malaysia

**Yuki Fujii**
Department of Pharmacy
Nagasaki International University
Nagasaki, Japan

**Alessandra Gallo**
Department of Biology and Evolution of Marine
Organisms
Stazione Zoologica Anton Dohrn
Naples, Italy

**Arijit Gandhi**
Department of Quality Assurance
Albert David Limited
Kolkata, West Bengal, India

**Katarzyna Godlewska**
Department of Advanced Material Technologies
Wrocław University of Technology
Wrocław, Poland

**Claudia Mariana Gomez-Gutierrez**
Department of Bioengineering
University of Baja California
Baja California, Mexico

**Dharmalingam Gowdhaman**
School of Chemical and Biotechnology
SASTRA University
Thanjavur, Tamil Nadu, India

**Graciela Guerra-Rivas**
Marine Pharmacology and Toxicology Laboratory
University of Baja California
Baja California, Mexico

**Imtiaj Hasan**
Department of Life and Environmental System Science
Yokohama City University
Yokohama, Japan
and
Department of Biochemistry and Molecular Biology
University of Rajshahi
Rajshahi, Bangladesh

**Yoshihiko Hayashi**
Department of Cariology
Nagasaki University
Nagasaki, Japan

**Martha Á. Hjálmarsdóttir**
Faculty of Medicine
University of Iceland
Reykjavik, Iceland

**Masahiro Hosono**
Institute of Molecular Biomembrane and Glycobiology
Tohoku Medical and Pharmaceutical University
Sendai, Japan

**Shinsuke Ifuku**
Graduate School of Engineering
Tottori University
Tottori, Japan

**Kazunari Igawa**
Department of Cariology
Nagasaki University
Nagasaki, Japan

**Takeshi Ikeda**
Department of Cariology
Nagasaki University
Nagasaki, Japan

**Dhinakarasamy Inbakandan**
Centre for Ocean Research
Sathyabama University
Chennai, Tamil Nadu, India

**Giyatmi Irianto**
Department of Food Technology
Jakarta Sahid University
Jakarta, Indonesia

**Hari Eko Irianto**
Center for Fisheries Research and Development
Agency for Marine and Fisheries Research and
    Development
Ministry Marine Affairs and Fisheries
and
Department of Food Technology
Jakarta Sahid University
Jakarta, Indonesia

**Sougata Jana**
Department of Pharmaceutics
Gupta College of Technological Sciences
Asansol, West Bengal, India

**Subrata Jana**
Department of Chemistry
V.E.C, Sarguja University
Ambikapur, Chhattisgarh, India

**Bhavanath Jha**
Division of Marine Biotechnology and Ecology
Central Salt and Marine Chemicals Research Institute
Bhavnagar, Gujarat, India

**Zedong Jiang**
College of Food and Biological Engineering
Jimei University
Xiamen, Fujian, People's Republic of China

**Kei Kaida**
Department of Oral Physiology
Nagasaki University
Nagasaki, Japan

**Kannan Kamala**
Center for Environmental Nuclear Research
SRM University
Kattankulathur, Tamil Nadu, India

**Robert A. Kanaly**
Department of Life and Environmental System Science
Yokohama City University
Yokohama, Japan

**Jeyakumar Kandasamy**
Department of Chemistry
Indian Institute of Technology—Banaras Hindu
    University
Varanasi, Uttar Pradesh, India

**Vinod K. Kannaujiya**
Center of Advanced Study in Botany
Banaras Hindu University
Varanasi, Uttar Pradesh, India

**Ramachandran Karthik**
Department of Medical Biotechnology
Chettinad Academy of Research and Education
Chennai, Tamil Nadu, India

**Kandasamy Kathiresan**
Faculty of Marine Sciences
Annamalai University
Chidambaram, Tamil Nadu, India

**Yasushi Kawakami**
School of Life and Environmental Science
Azabu University
Sagamihara, Japan

**Sarkar M.A. Kawsar**
Department of Chemistry
University of Chittagong
Chittagong, Bangladesh

**Samanta S. Khora**
Division of Medical Biotechnology
VIT University
Vellore, Tamil Nadu, India

**Se-Kwon Kim**
Department of Marine-Bio Convergence Science
and
Marine Bioprocess Research Center
Pukyong National University
Busan, South Korea

**Kirti**
Division of Medical Biotechnology
VIT University
Vellore, Tamil Nadu, India

**Yasuhiro Koide**
Department of Life and Environmental System Science
Yokohama City University
Yokohama, Japan

**Ozcan Konur**
Department of Materials Engineering
Yildirim Beyazit University
Ankara, Turkey

**Himanshu Kumar**
Department of Life Science
National Institute of Technology, Rourkela
Rourkela, Odisha, India

**Supriya Kumari**
Department of Life Science
National Institute of Technology, Rourkela
Rourkela, Odisha, India

**Francesca Maccari**
Department of Life Sciences
University of Modena and Reggio Emilia
Modena, Italy

**Neelam Mangwani**
Department of Life Science
National Institute of Technology, Rourkela
Rourkela, Odisha, India

**Thangapandi Marudhupandi**
Centre for Ocean Research
Sathyabama University
Chennai, Tamil Nadu, India

**Már Másson**
School of Health Sciences
University of Iceland
Reykjavik, Iceland

**Izabela Michalak**
Department of Advanced Material Technologies
Wrocław University of Technology
Wrocław, Poland

**Avinash Mishra**
Division of Marine Biotechnology and Ecology
Central Salt and Marine Chemicals Research Institute
Bhavnagar, Gujarat, India

**Van Duy Nguyen**
Institute of Biotechnology and Environment
Nha Trang University
Nha Trang, Vietnam

**Athapol Noomhorm**
The School of Environment, Resources and
     Development
Asian Institute of Technology
Bangkok, Thailand

**Tatsuya Oda**
Graduate School of Fisheries Science and
     Environmental Studies
Nagasaki University
Nagasaki, Japan

**Yukiko Ogawa**
Department of Pharmacy
Nagasaki International University
Sasebo, Japan

**Yoshiharu Okamoto**
Department of Veterinary Clinical Medicine
Tottori University
Tottori, Japan

**Tomohiro Osaki**
Department of Veterinary Clinical Medicine
Tottori University
Tottori, Japan

**Yasuhiro Ozeki**
Department of Life and Environmental System Science
Yokohama City University
Yokohama, Japan

**Ratih Pangestuti**
Research Center for Oceanography
Indonesian Institute of Sciences
Jakarta Utara, Republic of Indonesia

**Jainendra Pathak**
Center of Advanced Study in Botany
Banaras Hindu University
Varanasi, Uttar Pradesh, India

**Jantana Praiboon**
Department of Fishery Biology
Kasetsart University
Bangkok, Thailand

**Radhika Rajasree Santha Ravindranath**
Centre for Ocean Research
Sathyabama University
Chennai, Tamil Nadu, India

**Narayanasamy Rajendran**
Department of Zoology
Government Arts College
Chidambaram, Tamil Nadu, India

**Sultana Rajia**
Department of Life and Environmental System Science
Yokohama City University
Yokohama, Japan
and
Department of Natural Science
Varendra University
Rajshahi, Bangladesh

**Rajneesh**
Center of Advanced Study in Botany
Banaras Hindu University
Varanasi, Uttar Pradesh, India

**Richa**
Center of Advanced Study in Botany
Banaras Hindu University
Varanasi, Uttar Pradesh, India

**Idris Mohamed Saeed**
Institute of Biological Sciences
and
Department of Chemical engineering
University of Malaya
Kuala Lumpur, Malaysia

**Priyanka Sahariah**
School of Health Sciences
University of Iceland
Reykjavik, Iceland

**Hiroyuki Saimoto**
Graduate School of Engineering
Tottori University
Tottori, Japan

**Ramachandran Sarojini Santhosh**
Genetic Engineering Laboratory
and
School of Chemical and Biotechnology
SASTRA University
Thanjavur, Tamil Nadu, India

**Kandasamy Saravanakumar**
School of Agriculture and Biology
Shanghai Jiao Tong University
Minhang, Shanghai, People's Republic of China

**Ramachandran Saravanan**
Department of Marine Pharmacology
Chettinad Academy of Research and Education
Chennai, Tamil Nadu, India
and
Centre of Advanced Study in Marine Biology
Annamalai University
Chidambaram, Tamil Nadu, India

**Siswa Setyahadi**
Center for Bioindustrial Technology
Agency for the Assessment and Application of
    Technology
Puspiptek, Serpong, Banten, Indonesia

**Sabiah Shahul Hameed**
Department of Chemistry
Pondicherry University
Pondicherry, Tamil Nadu, India

**Annian Shanmugam**
Centre of Advanced Study in Marine Biology
Annamalai University
Chidambaram, Tamil Nadu, India

**Vijay Kumar Singh**
Division of Marine Biotechnology and Ecology
Central Salt and Marine Chemicals Research Institute
Bhavnagar, Gujarat, India

**Rajeshwar P. Sinha**
Center of Advanced Study in Marine Biology
Banaras Hindu University
Varanasi, Uttar Pradesh, India

**Pitchiah Sivaperumal**
Center for Environmental Nuclear Research
SRM University
Kattankulathur, Tamil Nadu, India

**Nattanun Soisarp**
Department of Fishery Biology
Kasetsart University
Bangkok, Thailand

**Arun S. Sonker**
Center of Advanced Study in Botany
Banaras Hindu University
Varanasi, Uttar Pradesh, India

**Vasuki Subramanian**
Faculty of Marine Sciences
Annamalai University
Chidambaram, Tamil Nadu, India

**Shigeki Sugawara**
Institute of Molecular Biomembrane and Glycobiology
Tohoku Medical and Pharmaceutical University
Sendai, Japan

**Syamdidi**
Research and Development Center for Marine
    and Fisheries Product Competitiveness and
    Biotechnology
Agency for Marine and Fisheries Research and
    Development
Jakarta Pusat, Indonesia

**Nallthambi Tamilkumar Varsha**
School of Chemical and Biotechnology
SASTRA University
Thanjavur, Tamil Nadu, India

**Anguchamy Veeruraj**
Centre for Ocean Research
Sathyabama University
Chennai, Tamil Nadu, India

and

Centre of Advanced Study in Marine Biology
Annamalai University
Chidambaram, Tamil Nadu, India

**Jayachandran Venkatesan**
Department of Marine-Bio Convergence Science and
    Marine Bioprocess Research Center
Pukyong National University
Busan, South Korea

**Nicola Volpi**
Department of Life Sciences
University of Modena and Reggio Emilia
Modena, Italy

**Radosław Wilk**
Department of Advanced Material Technologies
Wrocław University of Technology
Wrocław, Poland

**Shizuka Yamada**
Department of Cariology
Nagasaki University
Nagasaki, Japan

**Daiki Yamamoto**
Department of Life and Environmental System Science
Yokohama City University
Yokohama, Japan

**Takeshi Yamamoto**
Tobacco Science Research Center
Japan Tobacco Inc.
Kanagawa, Japan

**Kajiro Yanagiguchi**
Department of Cariology
Nagasaki University
Nagasaki, Japan

**Chuanjin Yu**
School of Agriculture and Biology
Shanghai Jiao Tong University
Minhang, Shanghai, People's Republic of China

# *Introduction to marine glycobiology*

*chapter one*

# Introduction to marine glycobiology

*Se-Kwon Kim and Jayachandran Venkatesan*

## Contents

Glycobiology is the study of the biology of saccharides (glycan). Glycoconjugates or glycan can be linked with other types of biological molecules. Glycoproteins and glycolipids are abundant in mammalian cells (Rademacher et al. 1988; Dwek 1996; Bertozzi and Kiessling 2001; Mengerink and Vacquier 2001; Lütteke et al. 2006; DeMarco and Woods 2008; Taylor and Drickamer 2011). Marine glycobiology is limited. However, recent developments on marine glycobiology, particularly on glycoconjugates, have resulted in several biological and biomedical applications. The research on this area is increasing (Caldwell and Pagett 2010).

## 1.1 Overview contents of the chapters in this book

In this book, we discuss marine glycoconjugates reproduction and chemical communications, marine glycans, marine glycoproteins, marine glycoenzymes, marine saccharides, bioinformatics of glycobiology, the biological role of glycoconjugates, and finally biological and biomedical applications of glycoconjugates. Glycoconjugates are extensively studied in various applications, including chromatography (Hardy et al. 1988), MRI (Fulton et al. 2006), and several medical applications (Dumitriu 1996).

Dr. Ozcan Konur (Chapter 2) presents a detailed review on glycobiology, which has been one of the most dynamic research fields in recent years with significant impact on medical research, with nearly half a million papers indexed by the Science Citation Index-Expanded (SCIE). As the topical headings of this book suggest, a number of subtopical areas have emerged during the development of this research field: glycoproteins, glycoconjugates, glycogens, glycoxidation, glycolipids, glycans, proteoglycans, glycogenomics, and glycosylation. Also, there has been a high degree of diversity among the subtopics. Although there have been nearly half a million articles and reviews related to glycoscience, it is notable that there has not been any study on the scientometric analysis of glycoscience or an overview of the citation classics in glycoscience as in other research fields. As North's new institutional theory suggests, it is important to have up-to-date information about the current public policy issues to develop a set of viable solutions to satisfy the needs of the key stakeholders. Therefore, following a scientometric overview of the research in glycoscience, brief information on a selected set of 25 citation classics in the field of the glycoscience is presented in this book to inform the key stakeholders about the influential papers in this dynamic research field as the first-ever study of its kind, complementing 10 other papers relating to the citation classics in glycoconjugates, glycans, glycosylation, glycoproteins, proteoglycans, glycolipids, glycomics, glycoenzymes, carbohydrates, and algal glycoscience. It was found that although there have been 10 research subfields among the citation classics in glycoscience, the most prolific research area has been glycoproteins. The other marginal research fields have been glycoconjugates, glycogens, glycooxidation, glycolipids, glycans, proteoglycans, glycogenomics, and glycosylation in line with the topical headings of this book. Hence, the research in glycoscience has strong public policy implications, providing strong incentives for the key stakeholders involved in the glycoscience research (Konur 2011).

## 1.2   Marine glycoconjugates

In the last three decades, the identification of glycoconjugates increased due to well-advanced analytical instruments and research activities. Synthesis and structural characterizations of marine glycojugates are important steps to know their exact functions. In this section, Dr. Alessandra Gallo et al. (Chapter 3) present marine glycoconjugates in gametes physiology and fertilization in detail. In the same section, Dr. Nicola Volpi (Chapter 4) describes heparin from marine molluscs and their occurrence, structure, and biological role in detail. In this chapter, the structure of heparin, its interaction with proteins, its anticoagulant activity, and its biological role in molluscs are explained (Volpi et al. 1995, 2012; Galeotti and Volpi 2011, 2013).

## 1.3   Marine glycans

Marine glycans perform several biological activities. *Visamsetti Amarendra and Ramachandran Sarojini Santhosh* (Chapter 5) present marine polysaccharides, glycolipids, glycoprotein, and glycoalkaloids and their respective sources. Chapter 6 describes the bioactivity of glycans, including antibacterial, antibiofilm, anticancer, anticoagulant, antifouling, antifungal, and antiviral activities (Santhosh and Suriyanarayanan 2014). To continue the biological activity of sulfated polysaccharides from seaweed, Dr. Praiboon et al. (Chapter 7) report the production and biological activity of seaweed-derived glycans (Muangmai et al. 2014; Wongprasert et al. 2014).

## 1.4   Marine glycoproteins

Glycoproteins are proteins that contain glycans covalently linked with polypeptide side chains. They have several biological applications, including antitumor and antimicrobial applications (Yamazaki 1993; Yamazaki et al. 1997). Dr. Irianto et al. (Chapter 8) present the application of glycoprotein in development of therapeutic agents. Scientists, through their investigation, have proved that bioactivities of glycoproteins extracted from both marine microalgae and macroalgae showed therapeutic benefits. Algal glycoproteins are composed of carbohydrate and protein moieties. Microalgal glycoproteins can be extracted from their culture fluid, while glycoproteins from algae are obtained from their hot water extracts. Algal glycoproteins are found to have therapeutic application as antitumor, hepatoprotective, anti-inflammatory, antioxidant, probiotic, and antidiabetic agents. Following this, Dr. Saravanan et al. (Chapter 9) present anticoagulant activity of heparin (Saravanan and Shanmugam 2011; Saravanan 2014; Manigandan et al. 2015). Dr. Veeruraj et al. (Chapter 10) present the different kinds of applications, including glycoprotein, particularly collagen (Veeruraj et al. 2012, 2013, 2015; Tamilmozhi et al. 2013; Damotharan et al. 2015). In Chapter 11, Dr. Rivas et al. discuss glycoproteins and detoxification in the marine environment. In Chapter 12, Dr. Khora et al. present the recent trends in bioprospecting of marine collagen and their applications (Khora 2013; Kim 2015b).

## 1.5   Glycolipids

Glycolipids are a structurally very heterogeneous group of membrane-bound compounds present in all living organisms, including human cells. "Glycolipid" is defined as a compound that contains one or more monosaccharides glycodically linked in to a hydrophobic moiety. Glycolipids are an essential constituent of cellular membrane and have remarkable biological functions of cell aggregation or dissociation; glycolipids act as receptors or acceptors to enable contact between biological systems. Several glycolipids have roles to play in the immune system (Warnecke and Heinz 2010). Dr. Saravanakumar et al. (Chapter 13) discuss glycolipidomics. Generally, glycolipids are classified into three major groups: glyceroglycolipids, glycosphingolipids, and glycosylphosphatidylinositols. Among these, glycosphingolipids are most common in fungal cells. Glycosphingolipids are carbohydrates containing derivatives of sphingoid, ceramide, and amino alcohol sphingosine. Other terms used to highlight the structural features of glycolipids are fucoglycosphingolipids, mannoglycosphingolipids, xyloglycosphingolipids, and cellobiose glycolipids.

## 1.6   Glycoenzymes

Dr. Takeshi Yamamoto et al. (Chapter 14) review the biological role of sialyloligosaccharides in various diseases, preparation methods of sialyloligosaccharides, screening methods of glycosyltransferases, and enzymatic properties of marine bacterial sialyltransferases.

## 1.7   Marine carbohydrates

Polysaccharides are polymeric carbohydrate molecules composed of long chains of monosaccharides linked to glycosidic linkages. There are different kinds of polymers that vary based on structure; marine carbohydrates, such as alginate, carrageenan, fucoidan, chitin, chitosan, ulvan, laminarin, are widely used for several applications (Kim 2014; Pallela 2014; Pomin and Mourão 2014; Sudha et al. 2014a,b; Kang et al. 2015).

Dr. Jana et al. and Dr. Ratih et al. (Chapters 15 and 16) present various bioactive compounds obtained from marine source; polysaccharides are attracting much attention due to their structural and functional diversity. The versatile properties of such molecules find numerous applications in various pharmaceutical and biomedical fields. Polysaccharides have been gaining interest and value in applications in the food and pharmaceutical fields. As they are derived naturally, they are easily available, nontoxic, cheap, biodegradable, and biocompatible. Chapter 15 provides an overview of the major marine polysaccharides derivatives and their applications, with a focus on drug delivery and biomedical engineering (Jana et al. 2010; Kim and Pangestuti 2011; Wijesekara et al. 2011; Kim 2012, 2015; Lin 2013; Pangestuti and Kim 2014, 2015). Other chapters in this section deal with different biological and biomedical applications.

## 1.8   Bioinformatics and glycobiology

Dr. Imran elaborates on glycans in Chapter 30. Glycans are regarded as biological materials having a high information density per fundamental unit, with structural diversity, heterogeneity, and complex nontemplate biosynthesis. Due to these facts, glycans remain a challenging part of molecular biology. With recent advances, much work has been reported on the use of modern algorithms that has made possible the characterization molecular information of complex biological systems. Machine learning algorithms are applied to glycan bioinformatics in structural analysis, biomarkers prediction, and functionality of carbohydrates.

## 1.9   Biological and biomedical application of glycoconjugates

In this section, several authors discussed about the biological and biomedical applications of glycoconjugates, including anticoagulant effects, nutraceuticals and pharmaceuticals, industrial applications, antiviral effects, and antioxidant and anticancer activities.

## 1.10   Conclusions

Marine-derived glycoconjugates show promising anticancer, antiviral, antimicrobial, anticoagulant, and anti-inflammatory properties. However, proper implementation is required to commercialize these productions. The main struggle in the commercialization of marine drug is quality. There is need to develop some organic synthesis methods to achieve the desired, application-specific quality.

## Acknowledgments

This book was supported by research funds from Pukyong National University in 2015 and a grant from the Marine Bioprocess Research Center of the Marine Biotechnology Program, funded by the Ministry of Ocean and Fisheries, Republic of Korea.

## References

Bertozzi, CR and LL Kiessling. 2001. Chemical glycobiology. *Science* 291 (5512):2357–2364.

Caldwell, GS and HE Pagett. 2010. Marine glycobiology: Current status and future perspectives. *Marine Biotechnology* 12 (3):241–252.

Damotharan, P, A Veeruraj, M Arumugam, and T Balasubramanian. 2015. *In vitro* antibacterial activity of venom protein isolated from sea snake *Enhydrina schistosa* against drug-resistant human pathogenic bacterial strains. *Journal of Coastal Life Medicine* 3 (6):453–458.

DeMarco, ML and RJ Woods. 2008. Structural glycobiology: A game of snakes and ladders. *Glycobiology* 18 (6):426–440.

Dumitriu, S. 1996. *Polysaccharides in Medicinal Applications.* CRC Press, Boca Raton, FL.

Dwek, RA. 1996. Glycobiology: Toward understanding the function of sugars. *Chemical Reviews* 96 (2):683–720.

Fulton, DA, EM Elemento, S Aime, L Chaabane, M Botta, and D Parker. 2006. Glycoconjugates of gadolinium complexes for MRI applications. *Chemical Communications* (10):1064–1066.

Galeotti, F and N Volpi. 2011. Online reverse phase-high-performance liquid chromatography-fluorescence detection-electrospray ionization-mass spectrometry separation and characterization of heparan sulfate, heparin, and low-molecular weight-heparin disaccharides derivatized with 2-aminoacridone. *Analytical Chemistry* 83 (17):6770–6777.

Galeotti, F and N Volpi. 2013. Novel reverse-phase ion pair-high performance liquid chromatography separation of heparin, heparan sulfate and low molecular weight-heparins disaccharides and oligosaccharides. *Journal of Chromatography A* 1284:141–147.

Hardy, MR, RR Townsend, and YC Lee. 1988. Monosaccharide analysis of glycoconjugates by anion exchange chromatography with pulsed amperometric detection. *Analytical Biochemistry* 170 (1):54–62.

Jana, S, D Lakshman, KK Sen, and SK Basu. 2010. Development and evaluation of epichlorohydrin cross-linked mucoadhesive patches of tamarind seed polysaccharide for buccal application. *International Journal of Pharmaceutical Sciences and Drug Research* 2:193–198.

Kang, H-K, CH Seo, and Y Park. 2015. The effects of marine carbohydrates and glycosylated compounds on human health. *International Journal of Molecular Sciences* 16 (3):6018–6056.

Khora, SS. 2013. Marine fish-derived bioactive peptides and proteins for human therapeutics. *International Journal of Pharmacy and Pharmaceutical Sciences* 5 (3):31–37.

Kim, S-K. 2012. *Marine Pharmacognosy: Trends and Applications.* CRC Press, Boca Raton, FL.

Kim, S-K. 2014. *Marine Carbohydrates: Fundamentals and Applications.* Elsevier Academic Press, Amsterdam, the Netherlands.

Kim, S-K. 2015a. *Handbook of Anticancer Drugs from Marine Origin.* Springer International Publishing, Cham, Switzerland.

Kim, S-K. 2015b. *Springer Handbook of Marine Biotechnology.* Springer, Dordrecht, the Netherlands.

Kim, S-K and R Pangestuti. 2011. Potential role of marine algae on female health, beauty, and longevity. *Advances in Food and Nutrition Research* 64:41–55.

Konur, Ozcan. 2011. The scientometric evaluation of the research on the algae and bio-energy. *Applied Energy* 88 (10):3532–3540.

Lin, S-K. 2013. *Marine Nutraceuticals: Prospects and Perspectives,* S-K Kim, (ed.), CRC Press, Boca Raton, FL; *Marine Drugs* 11 (4):1300–1303. http://www.ncbi.nlm.nih.gov/pmc/articles/PMC3705405/.

Lütteke, T, A Bohne-Lang, A Loss, T Goetz, M Frank, and C-W von der Lieth. 2006. GLYCOSCIENCES. de: An Internet portal to support glycomics and glycobiology research. *Glycobiology* 16 (5):71R–81R.

Manigandan, V, R Karthik, and R Saravanan. 2015. Marine carbohydrate based therapeutics for Alzheimer disease—Mini review. *Journal of Neurology and Neuroscience* pp. 1–6.

Mengerink, KJ and VD Vacquier. 2001. Glycobiology of sperm–egg interactions in deuterostomes. *Glycobiology* 11 (4):37R–43R.

Muangmai, N, GC Zuccarello, T Noiraksa, and K Lewmanomont. 2014. A new flat Gracilaria: *Gracilaria lantaensis* sp. nov. (Gracilariales, Rhodophyta) from the Andaman coast of Thailand. *Phycologia* 53 (2):137–145.

Pallela, R. 2014. Nutraceutical and pharmacological implications of marine carbohydrates. *Marine Carbohydrates: Fundamentals and Applications* 73:183–195.

Pangestuti, R and S-K Kim. 2014. Biological activities of carrageenan. *Marine Carbohydrates: Fundamentals and Applications* 72:113.

Pangestuti, R and S-K Kim. 2015. Seaweeds-derived bioactive materials for the prevention and treatment of female's cancer. In *Handbook of Anticancer Drugs from Marine Origin* pp. 165–176. Springer International Publishing, Cham, Switzerland.

Pomin, VH and PAS Mourão. 2014. Specific sulfation and glycosylation—A structural combination for the anticoagulation of marine carbohydrates. *Frontiers in Cellular and Infection Microbiology* 4:33.

Rademacher, TW, RB Parekh, and RA Dwek. 1988. Glycobiology. *Annual Review of Biochemistry* 57 (1):785–838.

Santhosh, RS and B Suriyanarayanan. 2014. Plants: A source for new antimycobacterial drugs. *Planta Medica* 80 (1):9–21.

Saravanan, R. 2014. Isolation of low-molecular-weight heparin/heparan sulfate from marine sources. In *Marine Carbohydrates: Fundamentals and Applications.* pp. 45–60. Elsevier Academic Press, Amsterdam, the Netherlands.

Saravanan, R and A Shanmugam. 2011. Is isolation and characterization of heparan sulfate from marine scallop *Amussium pleuronectus* (Linne.) an alternative source of heparin? *Carbohydrate Polymers* 86 (2):1082–1084.

Sudha, PN, S Aisverya, R Nithya, and K Vijayalakshmi. 2014a. Industrial applications of marine carbohydrates. In *Marine Carbohydrates: Fundamentals and Applications.* p. 145. Elsevier Academic Press, Amsterdam, the Netherlands.

Sudha, PN, T Gomathi, P Angelin Vinodhini, and K Nasreen. 2014b. Marine carbohydrates of wastewater treatment. In *Marine Carbohydrates: Fundamentals and Applications.* p. 103. Elsevier Academic Press, Amsterdam, the Netherlands.

Tamilmozhi, S, A Veeruraj, and M Arumugam. 2013. Isolation and characterization of acid and pepsin-solubilized collagen from the skin of sailfish (*Istiophorus platypterus*). *Food Research International* 54 (2):1499–1505.

Taylor, ME and K Drickamer. 2011. *Introduction to Glycobiology.* Oxford University Press, Oxford, U.K.

Veeruraj, A, M Arumugam, T Ajithkumar, and T Balasubramanian. 2012. Isolation and characterization of drug delivering potential of type-I collagen from eel fish *Evenchelys macrura*. *Journal of Materials Science: Materials in Medicine* 23 (7):1729–1738.

Veeruraj, A, M Arumugam, T Ajithkumar, and T Balasubramanian. 2015. Isolation and characterization of collagen from the outer skin of squid (*Doryteuthis singhalensis*). *Food Hydrocolloids* 43:708–716.

Veeruraj, A, M Arumugam, and T Balasubramanian. 2013. Isolation and characterization of thermostable collagen from the marine eel-fish (*Evenchelys macrura*). *Process Biochemistry* 48 (10):1592–1602.

Volpi, N, M Cusmano, and T Venturelli. 1995. Qualitative and quantitative studies of heparin and chondroitin sulfates in normal human plasma. *Biochimica et Biophysica Acta (BBA)-General Subjects* 1243 (1):49–58.

Volpi, N, F Maccari, J Suwan, and RJ Linhardt. 2012. Electrophoresis for the analysis of heparin purity and quality. *Electrophoresis* 33 (11):1531–1537.

Warnecke, D and E Heinz. 2010. Glycolipid headgroup replacement: A new approach for the analysis of specific functions of glycolipids in vivo. *European Journal of Cell Biology* 89 (1):53–61.

Wijesekara, I, R Pangestuti, and S-K Kim. 2011. Biological activities and potential health benefits of sulfated polysaccharides derived from marine algae. *Carbohydrate Polymers* 84 (1):14–21.

Wongprasert, K, T Rudtanatip, and J Praiboon. 2014. Immunostimulatory activity of sulfated galactans isolated from the red seaweed *Gracilaria fisheri* and development of resistance against white spot syndrome virus (WSSV) in shrimp. *Fish and Shellfish Immunology* 36 (1):52–60.

Yamazaki, M. 1993. Antitumor and antimicrobial glycoproteins from sea hares. *Comparative Biochemistry and Physiology Part C: Comparative Pharmacology* 105 (2):141–146.

Yamazaki, M, J Kisugi, and R Iijima. 1997. Antineoplastic glycoproteins in marine invertebrates. Gan to kagaku ryoho. *Cancer & Chemotherapy* 24 (11):1477–1485.

## chapter two

# Glycoscience
## The current state of the research

*Ozcan Konur*

## Contents

Glycobiology has been one of the most dynamic research fields in recent years with significant impact on medical research with nearly half a million papers, as indexed by the Science Citation Index Expanded (SCIE). As the topical headings of this book suggest, a number of sub-topical areas have emerged during the development of this research field: glycoproteins, glycoconjugates, glycogens, glycooxidation, glycolipids, glycans, proteoglycans, glycogenomics, and glycosylation. Since there has been a high degree of diversity among the subtopics, the term *glycoscience* rather than the term *glycobiology* was preferred for this study. Although there have been nearly half a million articles and reviews in glycoscience, it is notable that there has not been any study on the scientometric analysis of glycoscience or an overview of the citation classics in glycoscience as in other research fields. As North's New Institutional Theory suggests, it is important to have up-to-date information about the current public policy issues to develop a set of viable solutions to satisfy the needs of all the key stakeholders. Therefore, following a scientometric overview of the research in glycoscience, brief information on a selected set of 25 citation classics in the field of glycoscience is presented in this chapter to inform the key stakeholders about the influential papers in this dynamic research field as the first-ever study of its kind, complementing 10 other papers relating to the citation classics in glycoconjugates, glycans, glycosylation, glycoproteins, proteoglycans, glycolipids, glycomics, glycoenzymes, carbohydrates, and algal glycoscience. It was found that although there have been 10 research subfields among the citation classics in glycoscience, the most prolific research area has been glycoproteins. The other marginal research fields have been glycoconjugates, glycogens, glycooxidation, glycolipids, glycans, proteoglycans, glycogenomics, and glycosylation in line with the topical headings of this book. Hence, the research in glycoscience has strong public policy implications, providing strong incentives for the key stakeholders involved in glycoscience research.

## 2.1  Overview

### 2.1.1  Issues

Glycobiology has been one of the most dynamic research fields in recent years with significant impact on medical research with nearly half a million papers, as indexed by the SCIE as of September 2014 (e.g., Bernfield et al. 1999, Brown and Rose 1992, Dwek 1984).

As the topical headings of this book suggest, a number of subtopical areas have emerged during the development of this research field: glycoproteins (e.g., Simonet et al. 1997, Thiebaut et al. 1987), glycoconjugates (Domon and Costello 1988, Varki 1993), glycogens (Cross et al. 1995), glycooxidation (Baynes 1991), glycolipids (Brown and Rose 1992, Hakomori 1981), glycans (Farndale et al. 1986), proteoglycans (Bernfield et al. 1999, Kjellen and Lindahl 1991), glycogenomics (Cantarel et al. 2009), and glycosylation (Henrissat and Bairoch 1993, Kawano et al. 1997). Since there has been a high degree of diversity among the subtopics, the term *glycoscience* rather than the term *glycobiology* was preferred for this study.

Although there have been nearly half a million articles and reviews in glycoscience, it is notable that there has not been any study on the scientometric analysis of glycoscience or an overview of the citation classics in glycoscience as in the other research fields (e.g., Baltussen and Kindler 2004a,b, Dubin et al. 1993, Gehanno et al. 2007, Konur 2011, 2012a–p, 2013, 2015a–m, 2016a–b, Paladugu et al. 2002, Wrigley and Matthews 1986).

As North's New Institutional Theory suggests, it is important to have up-to-date information about the current public policy issues to develop a set of viable solutions to satisfy the needs of all the key stakeholders (Konur 2000, 2002a–c, 2006a,b, 2007a,b, 2012c,d, North 1994).

Therefore, following a scientometric overview of the research in glycoscience, brief information on a selected set of 25 citation classics in the field of glycoscience is presented in this chapter to inform the key stakeholders about the influential papers in this dynamic research field as the first-ever study of its kind, complementing 10 other papers relating to the citation classics in glycoconjugates, glycans, glycosylation, glycoproteins, proteoglycans, glycolipids, glycomics, glycoenzymes, carbohydrates, and algal glycoscience (Konur 2014a–j).

It was found that although there have been 10 research subfields among the citation classics in glycoscience, the most prolific research area has been

glycoproteins. The other marginal research fields have been glycoconjugates, glycogens, glycooxidation, glycolipids, glycans, proteoglycans, glycogenomics, and glycosylation in line with the topical headings of this book.

### 2.1.2  Methodology

A search for glycoscience was carried out in the SCIE and Social Sciences Citation Index (SSCI) databases (version 5.14) in September 2014 to locate papers relating to glycoscience using the keyword set of topic=(glyco*) in the abstract pages of the papers. Related keywords such as polysaccharides, carbohydrates, and glycans were not used in order to focus solely on the fundamental research in glycoscience.

The key bibliometric data were extracted from this search for an overview of glycoscience literature. It was necessary to focus on the key references by selecting articles and reviews.

The located highly cited 25 papers were arranged in the order of decreasing number of citations. The summary information about the located citation classics is presented in the order of decreasing number of citations for each topical area.

The information relating to the document type, affiliation of the authors, the number and gender of the authors, the country of the authors, the journal in which the paper was published, the subject area of the journal where the paper was indexed, the concise topic of the paper, and the total number of citations received for the paper for both the Web of Science and Google Scholar databases was given in tables for each paper.

### 2.1.3  Research on glycoscience: An overview

Using the keywords related to glycoscience, 489,436 references were located. A total of 448,479 of these references were articles and reviews. Meeting abstracts, notes, and editorial materials formed the remaining part of the sample. This finding suggests that the field of glycoscience was a specialized field of research with a relatively large sample size and with a specific set of shareholders such as authors, institutions, and countries.

The most prolific three authors, Y. Zhang, Y. Wang, and Y. Liu, produced 1039, 1007, and 899 papers each, respectively. The list of the most prolific authors was dominated by the Southeast Asian authors.

The most prolific country in terms of the number of publications was the United States with 145,767 papers forming 32.5% of the sample. Japan, China, and Germany followed the United States with 10.8%, 7.8%, and 7.8% of the sample, respectively. Europe dominated the most prolific country list. English was the dominant language of scientific communication in glycoscience comprising 97.5% of the sample.

The most prolific institution was National Research Centre of Scientific Research (CNRS) of France with 10,458 papers. Harvard University of the United States, National Institutes of Health (NIH) of the United States, and National Health and Medical Research Institute (INSERM) of France followed CNRS with 8163, 8033, and 7343 papers, respectively. The U.S. and European institutions dominated the most prolific institution list.

Like nanoresearch, glycoscience research has boomed after 2000 comprising 66.3% of the sample. There was a general increasing trend in the number of papers over time starting with 1183 papers in 1980 and making a peak with 26,433 papers in 2013. The research in 1980s and 1990s formed 5.2% and 26.2% of the sample with a significant rise in 1991, possibly due to the inclusion of the abstracts in the abstract pages of the indices.

The most prolific journal in terms of the number of publications was *Journal of Biological Chemistry* publishing 13,166 papers. *Journal of Virology, Proceedings of the National Academy of Sciences of the United States of America*, and *Carbohydrate Research* followed the top journal with 5990, 3977, and, 3639 papers, respectively. It is notable that the contribution of glycoscience journals such as *Glycobiology* and *Glycoconjugate Journal* to the research in glycoscience was negligible.

The most prolific subject category in terms of the number of publications was *Biochemistry and Molecular Biology* with 93,780 papers forming 20.9% of the sample. *Pharmacology Pharmacy, Cell Biology*, and *Chemistry Multidisciplinary* followed the top subject category with 8.6%, 6.6%, and 6.1% of the samples, respectively. These findings suggest that these top four subject categories share a common set of journals, where a journal was indexed under more than one subject category. It is also notable that health sciences dominated the top subject list.

The most-cited papers in glycoscience were dominated by health science applications. For example, Cross et al. (1995) discussed the inhibition of glycogen synthase kinase-3 (GSK3) by insulin mediated by protein kinase B in a paper originating from Scotland and Switzerland with 2904 citations There was a similar trend for the hottest papers published in the past 3 years between 2011 and 2013. For example, the hottest paper with 286 citations was related to Otto Warburg's contributions to current concepts of cancer metabolism (Koppenol et al. 2011).

The next section presents brief information on the most-cited papers. First, citation classics relating to glycoproteins are presented. Second, citation classics relating to nine other topics in glycoscience are presented.

## 2.2   Research on glycoproteins

### 2.2.1   Overview

The research on glycoproteins has been one of the most dynamic research areas in glycoscience with 12 citation classics in recent years. These citation classics in the field of glycoscience with more than 1481 citations were located, and the key emerging issues from these papers were presented below in the decreasing order of the number of citations (Table 2.1).

The papers were dominated by researchers from seven countries, usually through the intracountry institutional collaboration, and they were multiauthored. The number of authors for the papers fluctuated from 2 to 200. The United States was the most prolific country with eight papers, while Harvard University was the most prolific institution with three papers, showing that the United States was the global leader in this field of research.

Similarly, all these papers were published in journals indexed by the Science Citation Index (SCI) and/or SCIE. There was no paper indexed by the SSCI. The number of citations ranged from 1481 to 2730 for the Web of Science and from 1651 to 4165 for the Google Scholar databases. Eleven of the papers were articles, while only one of them was a review.

The papers were published mostly before 2000 as there were five and six papers published in the 1980s and 1990s. There was a significant gender deficit among the most-cited papers in glycoproteins as there was only 1 paper with a female first author of 12 papers.

On the one hand, most prolific journals were *Cell, Nature*, and *Proc. Natl. Acad. Sci. U. S. A.* (three papers each). It is significant that all these journals had high citation impacts. On the other hand, the most prolific subjects were biochemistry and molecular biology (five papers) and multidisciplinary sciences (six papers).

The citation classics in glycoproteins deal with the important health issues: biochemistry of P-glycoprotein (Pgp)-mediated multidrug resistance, glycoprotein hormones, bacterial transport proteins in the MDR1 (Pgp) gene, disruption of the mouse multidrug resistance protein 1a (mdr1a) Pgp gene, osteoprotegerin (OPG), use of a monoclonal antibody directed against the platelet glycoprotein IIB/IIIA receptor, structure of the hemagglutinin membrane glycoprotein of influenza virus, dimensional structure of the human class II histocompatibility antigen HLA-DR1, structure of a human immunodeficiency virus (HIV) gp120 envelope glycoprotein, MDR gene product Pgp in normal human tissues, antiphospholipid (aPL) antibodies, and multiple sequence variations and correlation of one allele with Pgp expression and activity in vivo.

Table 2.1 The research on the glycoproteins

| Number | Paper ref. | Year | Doc. | Affil. | Country | No. of authors | M/F | Journal | Subject area | Topic | Total number of citations WK | Total number of citations GS |
|---|---|---|---|---|---|---|---|---|---|---|---|---|
| 3 | Simonet et al. | 1997 | A | Amgen Inc. | The United States | 30 | M | *Cell* | Biochm. Mol. Biol., Cell Biol. | Osteoprotegerin | 2730 | 4165 |
| 6 | Thiebaut et al. | 1987 | A | Japanese Fdn. Canc. Res., NCI | Japan, the United States | 6 | M | *Proc. Natl. Acad. Sci. U. S. A.* | Mult. Sci. | Multidrug-resistance gene product P-glycoprotein in normal human tissues | 2157 | 2554 |
| 7 | Endicott and Ling | 1989 | R | Univ. Toronto | Canada | 2 | F | *Annu. Rev. Biochem.* | Biochm. Mol. Biol. | Biochemistry of P-glycoprotein-mediated multidrug resistance | 2023 | 2231 |
| 10 | Wilson et al. | 1981 | A | Natl. Inst. Med. Res., Harvard Univ. | England, the United States | 3 | M | *Nature* | Mult. Sci. | Structure of the hemagglutinin membrane glycoprotein of influenza virus | 1862 | 2209 |
| 11 | Brown et al. | 1993 | A | Harvard Univ. +1 | The United States | 7 | M | *Nature* | Mult. Sci. | Dimensional structure of the human class II histocompatibility antigen HLA-DR1 | 1822 | 2108 |
| 12 | Pierce and Parsons | 1981 | A | Univ. Calif. Los Angeles | The United States | 2 | M | *Annu. Rev. Biochem.* | Biochm. Mol. Biol. | Glycoprotein hormones | 1813 | 2010 |

(Continued)

*Table 2.1 (Continued)* The research on the glycoproteins

| Number | Paper ref. | Year | Doc. | Affil. | Country | No. of authors | M/F | Journal | Subject area | Topic | Total number of citations WK | Total number of citations GS |
|---|---|---|---|---|---|---|---|---|---|---|---|---|
| 13 | Kwong et al. | 1998 | A | Columbia Univ., Harvard Univ. +3 | The United States | 6 | M | *Nature* | Mult. Sci. | Structure of an HIV gp120 envelope glycoprotein | 1758 | 2616 |
| 14 | Chen et al. | 1986 | A | NCI, NIH | The United States | 7 | M | *Cell* | Biochm. Mol. Biol., Cell Biol. | Bacterial transport proteins in the mdr1 (P-glycoprotein) gene | 1712 | 1912 |
| 18 | Califf et al. | 1994 | A | Univ. Michigan, Johns Hopkins Univ. +40 | The United States | 200 | M | *N. Engl. J. Med.* | Med. Gen. Int. | Use of a monoclonal-antibody directed against the platelet glycoprotein IIB/IIIA receptor | 1582 | 2416 |
| 19 | Hoffmeyer et al. | 2000 | A | Humboldt Univ. +2 | Germany | 11 | M | *Proc. Natl. Acad. Sci. U. S. A.* | Mult. Sci. | Multiple sequence variations and correlation of one allele with P-glycoprotein expression and activity in vivo | 1541 | 2194 |
| 21 | Schinkel et al. | 1994 | A | Netherlands Canc. Inst. +1 | Netherlands | 11 | M | *Cell* | Biochm. Mol. Biol., Cell Biol. | Disruption of the mouse mdr1a P-glycoprotein gene | 1530 | 1922 |
| 23 | McNeil et al. | 1990 | A | Univ. New S Wales +1 | Australia | 4 | M | *Proc. Natl. Acad. Sci. U. S. A.* | Mult. Sci. | Antiphospholipid antibodies | 1481 | 1651 |

*Note:* A, Article; R, Review; M, Male; F, Female; WK, Web of Knowledge; GS, Google Scholar.

## 2.2.2 *Most-cited papers in glycoproteins*

Simonet et al. (1997) discuss about OPG, a novel secreted glycoprotein involved in the regulation of bone density in a paper originating from the United States with 2730 citations. They found that in vivo, hepatic expression of OPG in transgenic mice resulted in a profound yet nonlethal osteopetrosis, coincident with a decrease in later stages of osteoclast differentiation. They observed the same effects on administration of recombinant OPG into normal mice. They further found that in vitro, osteoclast differentiation from precursor cells was blocked in a dose-dependent manner by recombinant OPG. Furthermore, OPG blocks ovariectomy-associated bone loss in rats. They conclude that OPG can act as a soluble factor in the regulation of bone mass and imply a utility for OPG in the treatment of osteoporosis associated with an increased osteoclast activity.

Thiebaut et al. (1987) discuss the cellular localization of the multidrug-resistance gene (MRD1) product Pgp in normal human tissues in a paper originating from Japan and the United States with 2157 citations. They found that the protein was concentrated in a small number of specific sites, and most tissues examined revealed very little Pgp. However, certain cell types in the liver, pancreas, kidney, colon, and jejunum showed specific localization of Pgp. In the liver, they found Pgp exclusively on the biliary canalicular front of hepatocytes and on the apical surface of epithelial cells in small biliary ductules. In the pancreas, they find Pgp on the apical surface of epithelial cells in small ductules but not on larger pancreatic ducts. In the kidney, they found that Pgp was concentrated on the apical surface of epithelial cells of the proximal tubules. They further found that both the colon and jejunum showed high levels of Pgp on the apical surface of superficial columnar epithelial cells. The adrenal gland showed high levels of Pgp distributed diffusely on the surface of cells in both the cortex and medulla. They conclude that the glycoprotein has a role in the normal secretion of metabolites and certain anticancer drugs into bile, urine, and directly into the lumen of the gastrointestinal tract.

Endicott and Ling (1989) discuss the biochemistry of Pgp-mediated multidrug resistance in a review paper originating from the United States with 2023 citations.

Wilson et al. (1981) discuss the structure of the hemagglutinin membrane glycoprotein of influenza virus at 3-A resolution in a paper originating from England and the United States with 1862 citations in an early paper. They found that the hemagglutinin glycoprotein of influenza virus is a trimer comprising two structurally distinct regions: a triple-stranded coiled coil of α-helices that extend 76 Å from the membrane and a globular region of antiparallel β-sheet, which contains the receptor-binding site and the variable antigenic determinants, positioned on the top of this stem. They further found that each subunit had an unusual loop-like topology, starting at the membrane, extending 135 Å distally, and folding back to enter the membrane.

Brown et al. (1993) discuss the dimensional structure of the human class II histocompatibility glycoprotein HLA-DR1 from human B cell membranes in a paper originating from the United States with 1822 citations. They found that it was similar to that of class I HLA. Peptides are bound in an extended conformation that projects from both ends of an "open-ended" antigen-binding groove. They further found that a prominent nonpolar pocket into which an "anchoring" peptide side chain fits was near one end of the binding groove. A dimer of the class II αβ heterodimers was seen in crystal forms of HLA-DR1, suggesting class II HLA dimerization as a mechanism for initiating the cytoplasmic signaling events in T-cell activation.

Pierce and Parsons (1981) discuss the structure and function of glycoprotein hormones in a paper originating from the United States with 1813 citations in an early paper.

Kwong et al. (1998) discuss the structure of an HIV gp120 envelope glycoprotein in complex with the CD4 receptor and a neutralizing human antibody in a paper originating from the United States with 1758 citations. They solved the x-ray crystal structure of an HIV-1 gp120 core complexed with a two-domain fragment of human CD4 and an antigen-binding fragment of a neutralizing antibody that blocked the binding of the chemokine receptor at 2.5 Å resolution. They found that the structure revealed a cavity-laden CD4-gp120 interface, a conserved binding site for the chemokine receptor, evidence for a conformational change on CD4 binding, the nature of a CD4-induced antibody epitope, and specific mechanisms for immune evasion.

Chen et al. (1986) discuss the internal duplication and homology with bacterial transport proteins in the MDR1 (Pgp) gene from multidrug-resistant human cells in a paper originating from the United States with 1712 citations. They determined the MDR in human cells by the MDR1 gene, encoding a high-molecular-weight membrane glycoprotein (Pgp). They found that the protein, which is 1280 amino acids long, consists of two homologous parts of approximately equal length. Each half of the protein includes a hydrophobic region with six predicted transmembrane segments and a hydrophilic region. They further find that the hydrophilic regions share homology with peripheral membrane components of active transport systems in bacteria and include potential nucleotide-binding sites.

Califf et al. (1994) discuss the use of a monoclonal antibody directed against the platelet glycoprotein IIb/IIIa receptor in high-risk coronary angioplasty in a paper originating from the United States with many institutions and with 1582 citations. They found that as compared with placebo, the c7E3 Fab bolus and infusion resulted in a 35% reduction in the rate of the primary end point, whereas a 10% reduction was observed with the c7E3 Fab bolus alone. The reduction in the number of events with the c7E3 Fab bolus and infusion was consistent across the end points of unplanned revascularization procedures and nonfatal myocardial infarction. They further found that bleeding episodes and transfusions were more frequent in the group that was given the c7E3 Fab bolus and infusion than in the other two groups. They conclude that the ischemic complications of coronary angioplasty and atherectomy were reduced with a monoclonal antibody directed against the platelet IIb/IIIa glycoprotein receptor, although the risk of bleeding was increased.

Hoffmeyer et al. (2000) discuss the functional polymorphisms of the human MDR-1 gene with the emphasis on the multiple sequence variations and correlation of one allele with Pgp expression and activity in vivo in a paper originating from Germany with 1541 citations. They analyzed the MDR-1 sequence in 21 volunteers. They found a significant correlation of a polymorphism in exon 26 (C3435T) of MDR-1 with expression levels and function of MDR-1. The homozygous of individuals for this polymorphism had significantly lower duodenal MDR-1 expression and the highest digoxin plasma levels. They conclude that this polymorphism affects the absorption and tissue concentrations of numerous other substrates of MDR-1.

Schinkel et al. (1994) discuss whether the disruption of the mouse mdr1a Pgp gene leads to a deficiency in the blood–brain barrier and to increased sensitivity to drugs in a paper originating from the Netherlands with 1530 citations. They generated mice that are homozygous for a disruption of the mdr1a (also called MDR3) gene, encoding a drug-transporting Pgp. They found that the mice were viable and fertile and appeared phenotypically normal, but they displayed an increased sensitivity to the centrally neurotoxic pesticide ivermectin (100-fold) and to the carcinostatic drug vinblastine (3-fold). By comparing mdr1a (+/+) and (−/−) mice, they found that the mdr1a Pgp is the major Pgp in the blood–brain barrier and that its absence results in elevated drug levels in many tissues (especially in brain) and in decreased drug elimination. They conclude that Pgp inhibitors might be useful in selectively enhancing the access of a range of drugs to the brain.

McNeil et al. (1990) discuss whether aPL antibodies are directed against a complex antigen that includes a lipid-binding inhibitor of coagulation $\beta_2$-glycoprotein-I (apolipoprotein-H)—in a paper originating from Australia with 1481 citations. They purified a plasma/serum cofactor to homogeneity and showed that the binding of aPL antibodies to cardiolipin required the presence of the cofactor in a dose-dependent manner. They identify the cofactor as $\beta_2$-glycoprotein I ($\beta_2$GPI) (apolipoprotein H), a plasma protein that is known to bind to anionic phospholipids. They argue that the presence of $\beta_2$GPI is an absolute requirement for antibody–phospholipid interaction, suggesting that bound $\beta_2$GPI forms the antigen to which aPL antibodies are directed. They further argue that interference with the function of $\beta_2$GPI by aPL antibodies could explain the thrombotic diathesis seen in association with these antibodies.

## 2.3 Research on other topics in glycoscience

### 2.3.1 Overview

The research on other topics in glycoscience has been one of the most dynamic research areas in glycoscience with 13 citation classics in recent years. These citation classics in the field of glycoscience with more than 1430 citations were located, and the key emerging issues from these papers were presented below in decreasing order of the number of citations under a number of topical headings (Table 2.2).

The papers were grouped under nine topical headings of glycoconjugates (two papers), glycogens (one paper), glycooxidation (one paper), glycolipids (two papers), glycobiology (one paper), glycans (one paper), proteoglycans (two papers), glycogenomics (one paper), and glycosylation (two papers).

The papers were dominated by the researchers from seven countries, usually through the intracountry institutional collaboration, and they were multiauthored. The number of authors for the papers fluctuated from 1 to 12. The United States was the most prolific country with six papers.

Similarly, all these papers were published in journals indexed by the SCI and/or SCIE. There was no paper indexed by the SSCI. The number of citations ranged from 1430 to 3861 for the Web of Science and from 1688 to 4705 for the Google Scholar databases. Nine of the papers were articles, while only four of them were reviews.

The papers were published mostly before 2000 as there were four and eight papers published in the 1980s and 1990s, respectively. There was a significant gender deficit among the most-cited papers in this category as there were only 3 papers with a female first author of 13 papers.

The most prolific journal was *Annual Review of Biochemistry* with three papers. It is significant that all journals had high citation impacts. There were also no papers published in glycoscience journals such as

*Table 2.2* The research on the other topics in glycoscience

| Number. | Paper ref. | Year | Doc. | Affil. | Country | No. of authors | M/F | Journal | Subject area | Topic | Total number of citations WK | Total number of citations GS |
|---|---|---|---|---|---|---|---|---|---|---|---|---|
| 1 | Varki | 1993 | R | Univ. Calif. San Diego | The United States | 1 | M | *Glycobiology* | Biochm. Mol. Biol. | Biological roles of oligosaccharides | 3861 | 4705 |
| 2 | Cross et al. | 1995 | A | Friedrich Miescher Inst., Univ. Dundee | Scotland, Switzerland | 5 | M | *Nature* | Mult. Sci. | Inhibition of glycogen synthase kinase-3 by insulin mediated by protein kinase B | 2904 | 3953 |
| 4 | Baynes | 1991 | A | Univ. S Carolina | The United States | 1 | M | *Diabetes* | Endoc. Metabol. | Role of oxidative stress in the development of complications in diabetes | 2372 | 3480 |
| 5 | Brown and Rose | 1992 | A | Yale Univ. | The United States | 2 | F | *Cell* | Biochm. Mol. Biol, Cell Biol. | GPI-anchored proteins to glycolipid-enriched membrane subdomains | 2237 | 2807 |
| 8 | Dwek | 1984 | A | Univ. Oxford | England | 1 | M | *Chem. Rev.* | Chem. Mult. | Glycobiology | 2003 | 2440 |
| 9 | Farndale et al. | 1986 | A | Strangeways Res. Lab. | England | 3 | M | *Biochim. Biophys. Acta* | Biochm. Mol. Biol., Biophys. | Improved quantitation and discrimination of sulfated glycosaminoglycans using dimethylmethylene blue | 1975 | 2450 |

*(Continued)*

**Table 2.2 (Continued)** The research on the other topics in glycoscience

| Number. | Paper ref. | Year | Doc. | Affil. | Country | No. of authors | M/F | Journal | Subject area | Topic | Total number of citations WK | Total number of citations GS |
|---|---|---|---|---|---|---|---|---|---|---|---|---|
| 15 | Bernfield et al. | 1999 | R | Harvard Univ. | The United States | 7 | M | *Annu. Rev. Biochem.* | Biochm. Mol. Biol. | Functions of cell surface heparan sulfate proteoglycans | 1641 | 2191 |
| 16 | Cantarel et al. | 2009 | A | CNRS | France | 6 | F | *Nucleic Acids Res.* | Biochm. Mol. Biol. | Carbohydrate-Active EnZymes database (CAZy) | 1631 | 2302 |
| 17 | Hakomori | 1981 | R | Univ. Washington +1 | The United States | 1 | M | *Annu. Rev. Biochem.* | Biochm. Mol. Biol. | Glycosphingolipids in cellular interaction, differentiation, and oncogenesis | 1608 | 1737 |
| 20 | Kawano et al. | 1997 | A | Chiba Univ. +2 | Japan | 12 | M | *Science* | Mult. Sci. | Activation of $V_\alpha 14$ Natural Killer T (NKT) cells by glycosylceramides | 1530 | 1939 |
| 22 | Kjellen and Lindahl | 1991 | R | Swedish Univ. Agr. Sci. | Sweden | 2 | F | *Annu. Rev. Biochem.* | Biochm. Mol. Biol. | Structures and interactions of proteoglycans | 1529 | 1813 |
| 24 | Domon and Costello | 1988 | A | MIT | The United States | 2 | M | *Glycoconjugate J.* | Biochm. Mol. Biol. | Nomenclature for carbohydrate fragmentations in FAB-MS MS spectra of glycoconjugates | 1470 | 1688 |
| 25 | Henrissat and Bairoch | 1993 | A | Univ. Geneva, CNRS | Switzerland, France | 2 | M | *Biochem. J.* | Biochm. Mol. Biol. | New families in the classification of glycosyl hydrolases | 1430 | 1924 |

*Note:* A, Article; R, Review; M, Male; F, Female; WK, Web of Knowledge; GS, Google Scholar.

*Glycobiology* and *Glycoconjugate Journal* among the citation classics.

The most prolific subjects were biochemistry and molecular biology (nine papers) and multidisciplinary sciences (two papers).

The citation classics in this category deal with the important health and biochemical issues: structures and interactions of proteoglycans, glycosphingolipids in cellular interaction, differentiation, and oncogenesis, functions of cell surface heparan sulfate proteoglycans, new families in the classification of glycosyl hydrolases, biological roles of oligosaccharides, nomenclature for carbohydrate fragmentations in fast atom bombardment mass spectrometry (FAB-MS MS) spectra of glycoconjugates, Carbohydrate-Active EnZyme (CAZy) database, improved quantitation and discrimination of sulfated glycosaminoglycans using dimethylmethylene blue, glycosylphosphatidyl inositol (GPI)-anchored proteins to glycolipid-enriched membrane subdomains, glycobiology, role of oxidative stress in the development of complications in diabetes, the inhibition of GSK3 by insulin mediated by protein kinase B, and the activation of Vα14 Natural Killer T (NKT) cells by glycosylceramides.

## 2.3.2 Most-cited papers in glycoscience: Other topics

### 2.3.2.1 Glycoconjugates

Varki (1993) discusses the theories on the biological roles of oligosaccharide units of individual classes of glycoconjugates in a review paper originating from the United States with 3861 citations. He argues that while all of these theories are correct, exceptions to each can also be found. The biological roles of oligosaccharides span the spectrum from those that are trivial to those that are crucial for the development, growth, function, or survival of an organism. Some general principles emerge. First, it is difficult to predict a priori the functions a given oligosaccharide on a given glycoconjugate might be mediating or their relative importance to the organism. Second, the same oligosaccharide sequence may mediate different functions at different locations within the same organism or at different times in its ontogeny or life cycle. Third, the more specific and crucial biological roles of oligosaccharides are often mediated by unusual oligosaccharide sequences, unusual presentations of common terminal sequences, or by further modifications of the sugars themselves. He concludes that the only common features of the varied functions of oligosaccharides are that they either mediate "specific recognition" events or that they provide "modulation" of biological processes. In doing so, they generate much of the functional diversity required for

the development and differentiation of complex organisms and for their interactions with other organisms in the environment.

Domon and Costello (1988) discuss the systematic nomenclature for carbohydrate fragmentations in FAB-MS MS spectra of glycoconjugates in a paper originating from the United States with 1470 citations. Ai, Bi, and Ci labels are used to designate fragments containing a terminal (nonreducing end) sugar unit, whereas $X_j$, $Y_j$, and $Z_j$ represent ions still containing the aglycone (or the reducing sugar unit). Subscripts indicate the position relative to the termini analogous to the system used in peptides, and superscripts indicate cleavages within carbohydrate rings. FAB-MS/MS spectra of a native glycosphingolipid and glycopeptide, and a permethylated ganglioside, are shown as illustrations.

### 2.3.2.2 Glycogens

Cross et al. (1995) discuss the inhibition of GSK3 by insulin mediated by protein kinase B in a paper originating from Scotland and Switzerland with 2904 citations. They show that agents that prevent the activation of both MAP kinase–activated protein kinase-1 and p70$^{S6k}$ by insulin in vivo do not block the phosphorylation and inhibition of GSK3. They argue that another insulin-stimulated protein kinase inactivates GSK3 under these conditions, and they show that it is the product of the protooncogene protein kinase B (PKB, also known as Akt/RAC). Like the inhibition of GSK3, the activation of PKB is prevented by inhibitors of phosphatidylinositol 3-kinase.

### 2.3.2.3 Glycooxidation

Baynes (1991) discusses the role of oxidative stress in the development of complications in diabetes in a paper originating from the United States with 2372 citations. He argues that although glycooxidation products are present only in trace concentrations, even in diabetic collagen, studies on glycation and oxidation of protein models in vitro suggest that these products are biomarkers of more extensive underlying glycative and oxidative damage to the protein. The possible sources of oxidative stress and damage to proteins in diabetes include free radicals generated by autoxidation reactions of sugars and sugar adducts to protein and by autoxidation of unsaturated lipids in plasma and membrane proteins. He concludes that the oxidative stress may be amplified by a continuing cycle of metabolic stress, tissue damage, and cell death, leading to increased free radical production and compromised free radical inhibitory and scavenger systems, which further exacerbate the oxidative stress.

### 2.3.2.4 Glycolipids

Brown and Rose (1992) discuss the sorting of GPI-anchored proteins to glycolipid-enriched membrane subdomains during the transport to the apical cell surface in a paper originating from the United States with 2237 citations. They show that a protein with a GPI anchor can be recovered from lysates of epithelial cells in a low-density, detergent-insoluble form. Under these conditions, the protein is associated with detergent-resistant sheets and vesicles that contain other GPI-anchored proteins and are enriched in glycosphingolipids but do not contain a basolateral marker protein. They further find that the protein is recovered in this complex only after it has been transported to the Golgi complex, suggesting that protein–sphingolipid microdomains are formed in the Golgi apparatus and plasma membrane.

Hakomori (1981) discusses the glycosphingolipids in cellular interaction, differentiation, and oncogenesis in a review paper originating from the United States with 1608 citations in an early paper. He argues that the idea that glycosphingolipids (or, briefly, glycolipids) are ubiquitous components of plasma membrane and display cell type–specific patterns perhaps stemmed from the classical studies on glycolipids of erythrocyte membranes. Subsequently, plasma membranes of various animal cells were successfully isolated and analyzed; all were characterized by their much higher content of glycolipid than that was found in intracellular membranes. He concludes that it is generally assumed that glycolipids are present at the outer leaflet of the plasma membrane bilayer, although this assumption is based only on experiments with surface labeling by galactose oxidase–NaB$[^3H]_4$ of the intact and lysed erythrocyte membranes and inside-out vesicles.

### 2.3.2.5 Glycobiology

Dwek (1984) discusses glycobiology in a review paper originating from England with 2003 citations in an early paper. He focuses on the glycoproteins, protein glycosylation, oligosaccharides, and glycosylation inhibitors as antiviral agents.

### 2.3.2.6 Glycans

Farndale et al. (1986) discuss the improved quantitation and discrimination of sulfated glycosaminoglycans using dimethylmethylene blue in a paper originating from England with 1975 citations. They describe a modified form of the dimethylmethylene blue assay that has improved specificity for sulphated glycosaminoglycans, and they conclude that, in conjunction with specific polysaccharides, the dimethylmethylene blue assay can be used to quantitate individual sulphated glycosaminoglycans.

### 2.3.2.7 Proteoglycans

Bernfield et al. (1999) discuss the functions of cell surface heparan sulfate proteoglycans in a review paper originating from the United States with 1641 citations. He notes that the heparan sulfate on the surface of all adherent cells modulates the actions of a large number of extracellular ligands. The members of both cell surface heparan sulfate proteoglycan families, the transmembrane syndecans and the glycosylphosphoinositide-linked glypicans, bind these ligands and enhance the formation of their receptor-signaling complexes. These heparan sulfate proteoglycans also immobilize and regulate the turnover of ligands that act at the cell surface. They note that recent analyses of genetic defects in *Drosophila melanogaster*, mice, and humans confirm most of these activities in vivo and identify additional processes that involve cell surface heparan sulfate proteoglycans.

Kjellen and Lindahl (1991) discuss the structures and interactions of proteoglycans in a review paper originating from Sweden with 1529 citations.

### 2.3.2.8 Glycogenomics

Cantarel et al. (2009) discuss the CAZy as an expert resource for glycogenomics in a paper originating from France with 1631 citations. The CAZy database is a knowledge-based resource specialized in the enzymes that build and breakdown complex carbohydrates and glycoconjugates. They note that as of September 2008, the database describes the present knowledge on 113 glycoside hydrolase, 91 glycosyltransferase, 19 polysaccharide lyase, 15 carbohydrate esterase, and 52 carbohydrate-binding module families. These families are created based on experimentally characterized proteins and are populated by sequences from public databases with significant similarity. More than 6400 proteins have assigned EC numbers and 700 proteins have a Protein Data Bank (PDB) structure.

### 2.3.2.9 Glycosylation

Kawano et al. (1997) discuss the CD1d-restricted and TCR-mediated activation of $V_\alpha14$ NKT cells by glycosylceramides in a paper originating from Japan with 1530 citations. They identify a glycosylceramide-containing α-anomeric sugar with a longer fatty acyl chain ($C_{26}$) and sphingosine base ($C_{18}$) as a ligand for this TCR. They find that glycosylceramide-mediated proliferative responses of $V_\alpha14$ NKT cells were abrogated by treatment with chloroquine–concanamycin A or by monoclonal antibodies against CD1dN$_\beta$8, CD40/CD40L, or B7/CTLA-4/CD28 but not by interference with the function of a transporter-associated protein. They conclude that this lymphocyte shares distinct recognition systems with either T or NK cells.

Henrissat and Bairoch (1993) discuss the new families in the classification of glycosyl hydrolases based on amino acid sequence similarities in a paper originating from Switzerland and France with 1430 citations. They find that approximately half of the families were monospecific (containing only one EC number), whereas the other half were polyspecific (containing at least two EC numbers). A more than 60% increase in sequence data for glycosyl hydrolases allowed them to update the classification not only by adding more members to already identified families but also by finding 10 new families. On the basis of a comparison of 482 sequences corresponding to 52 EC entries, they then define 45 families, of which 22 are polyspecific.

## 2.4   Conclusions

The data presented on the scientometric overview of glycoscience in this study show that glycoscience has been a multidisciplinary research field where the key subject categories have been "biochemistry molecular biology," "pharmacology pharmacy," "cell biology," and "chemistry multidisciplinary." The data also show that this research field has boomed after 2000 with increasing publication rate and citations.

Although there have been 10 research subfields among the citation classics in glycoscience, the most prolific research area has been glycoproteins with 12 papers. The other marginal research fields have been glycoconjugates, glycogens, glycooxidation, glycolipids, glycans, proteoglycans, glycogenomics, and glycosylation in line with the topical headings of this book.

The citation classics in glycoscience have had common characteristics. They came from a limited number of countries, where the United States was the most prolific country. All the citation classics had more than 1400 citations in line with the definition of the citation classics.

All these citation classics were published in high-impact journals such as *Cell*, *Nature*, and *Proceedings of the National Academy of Sciences of the United States of America*. The citation classics were indexed under a number of subject categories highlighting the multidisciplinarity of the glycoscience field.

It is notable that there has been a significant gender deficit among the authors of these citation classics presented in this study.

The citation classics in glycoproteins dealt with the important health issues: biochemistry of Pgp-mediated multidrug resistance, glycoprotein hormones, bacterial transport proteins in the mdr1 (Pgp) gene, disruption of the mouse mdr1a Pgp gene, OPG, use of a monoclonal antibody directed against the platelet glycoprotein IIB/IIIA receptor, and structure of the haemagglutinin membrane glycoprotein of influenza virus. The other research issues were dimensional structure of the human class II histocompatibility antigen HLA-DR1, structure of an HIV gp120 envelope glycoprotein, MRD1-product Pgp

in normal human tissues, aPL antibodies, and multiple sequence variations and correlation of one allele with Pgp expression and activity in vivo.

Similarly, the citation classics in other fields dealt with important research issues: structures and interactions of proteoglycans, glycosphingolipids in cellular interaction, differentiation, and oncogenesis, functions of cell surface heparan sulfate proteoglycans, new families in the classification of glycosyl hydrolases, biological roles of oligosaccharides, and nomenclature for carbohydrate fragmentations in FAB-MS MS spectra of glycoconjugates. The other research issues were CAZy, improved quantitation and discrimination of sulfated glycosaminoglycans using dimethylmethylene blue, GPI-anchored proteins to glycolipid-enriched membrane subdomains, glycobiology, role of oxidative stress in the development of complications in diabetes, inhibition of GSK3 by insulin mediated by protein kinase B, and activation of Vα14 NKT cells by glycosylceramides.

The citation classics presented in this paper were helpful in highlighting important papers influencing the development of the research field in glycoscience under 10 topical areas.

Further research is recommended for the detailed studies including scientometric studies and citation classic studies for each topical area (Konur 2014 a–j).

## References

Baltussen, A. and C.H. Kindler. 2004a. Citation classics in anesthetic journals. *Anesthesia and Analgesia* 98:443–451.

Baltussen, A. and C.H. Kindler. 2004b. Citation classics in critical care medicine. *Intensive Care Medicine* 30:902–910.

Baynes, J.W. 1991. Role of oxidative stress in development of complications in diabetes. *Diabetes* 40:405–412.

Bernfield, M., M. Gotte, P.W. Park, O. Reizes, M.L. Fitzgerald, J. Lincecum et al. 1999. Functions of cell surface heparan sulfate proteoglycans. *Annual Review of Biochemistry* 68:729–777.

Brown, D.A. and J.K. Rose. 1992. Sorting of GPI-anchored proteins to glycolipid-enriched membrane subdomains during transport to the apical cell-surface. *Cell* 68:533–544.

Brown, J.H., T.S. Jardetzky, J.C. Gorga, L.J. Stern, R.G. Urban, J.L. Strominger et al. 1993. Three-dimensional structure of the human class-II histocompatibility antigen HLA-DR1. *Nature* 364:33–39.

Califf, R.M., N. Shadoff, N. Valett, E. Bates, A. Galeana, W. Knopf et al. 1994. Use of a monoclonal-antibody directed against the platelet glycoprotein IIb/IIIa receptor in high-risk coronary angioplasty. *New England Journal of Medicine* 330:956–961.

Cantarel, B.L., P.M. Coutinho, C. Rancurel, T. Bernard, V. Lombard, and B. Henrissat. 2009. The Carbohydrate-Active EnZymes database (CAZy): An expert resource for Glycogenomics. *Nucleic Acids Research* 37:D233–D238.

Chen, C.J., J.E. Chin, K. Ueda, D.P. Clark, I. Pastan, M.M. Gottesman et al. 1986. Internal duplication and homology with bacterial transport proteins in the mdr1 (P-glycoprotein) gene from multidrug-resistant human cells. *Cell* 47:381–389.

Cross, D.A.E., D.R. Alessi, P. Cohen, M. Andjelkovich, and B.A. Hemmings. 1995. Inhibition of glycogen-synthase kinase-3 by insulin-mediated by protein-kinase-B. *Nature* 378:785–789.

Domon, B. and C.E. Costello. 1988. A systematic nomenclature for carbohydrate fragmentations in FAB-MS MS spectra of glycoconjugates. *Glycoconjugate Journal* 5:397–409.

Dubin, D., A.W. Hafner, and K.A. Arndt. 1993. Citation-classics in clinical dermatological journals—Citation analysis, biomedical journals, and landmark articles, 1945–1990. *Archives of Dermatology* 129:1121–1129.

Dwek, R.A. 1984. Glycobiology: Toward understanding the function of sugars. *Chemical Reviews* 96:683–720.

Endicott, J.A. and V. Ling. 1989. The biochemistry of P-glycoprotein-mediated multidrug resistance. *Annual Review of Biochemistry* 58:137–171.

Farndale, R.W., D.J. Buttle, and A.J. Barrett. 1986. Improved quantitation and discrimination of sulfated glycosaminoglycans by use of dimethylmethylene blue. *Biochimica et Biophysica Acta* 883:173–177.

Gehanno, J.F., K. Takahashi, S. Darmoni, and J. Weber. 2007. Citation classics in occupational medicine journals. *Scandinavian Journal of Work, Environment and Health* 33:245–251.

Hakomori, S.I. 1981. Glycosphingolipids in cellular interaction, differentiation, and oncogenesis. *Annual Review of Biochemistry* 50:733–764.

Henrissat, B. and A. Bairoch. 1993. New families in the classification of glycosyl hydrolases based on amino-acid-sequence similarities. *Biochemical Journal* 293:781–788.

Hoffmeyer, S., O. Burk, O. von Richter, H.P. Arnold, J. Brockmoller, A. Johne et al. 2000. Functional polymorphisms of the human multidrug-resistance gene: Multiple sequence variations and correlation of one allele with P-glycoprotein expression and activity in vivo. *Proceedings of the National Academy of Sciences of the United States of America* 97:3473–3478.

Kawano, T., J.Q. Cui, Y. Koezuka, I. Toura, Y. Kaneko, K. Motoki et al. 1997. CD1d-restricted and TCR-mediated activation of V$_\alpha$14 NKT cells by glycosylceramides. *Science* 278:1626–1629.

Kjellen, L. and U. Lindahl. 1991. Proteoglycans: Structures and interactions. *Annual Review of Biochemistry* 60:443–475.

Konur, O. 2000. Creating enforceable civil rights for disabled students in higher education: An institutional theory perspective. *Disability and Society* 15:1041–1063.

Konur, O. 2002a. Assessment of disabled students in higher education: Current public policy issues. *Assessment and Evaluation in Higher Education* 27:131–152.

Konur, O. 2002b. Access to employment by disabled people in the UK: Is the disability discrimination act working? *International Journal of Discrimination and the Law* 5:247–279.

Konur, O. 2002c. Access to nursing education by disabled students: Rights and duties of nursing programs. *Nurse Education Today* 22:364–374.

Konur, O. 2006a. Participation of children with dyslexia in compulsory education: Current public policy issues. *Dyslexia* 12:51–67.

Konur, O. 2006b. Teaching disabled students in higher education. *Teaching in Higher Education* 11:351–363.

Konur, O. 2007a. A judicial outcome analysis of the disability discrimination act: A windfall for the employers? *Disability and Society* 22:187–204.

Konur, O. 2007b. Computer-assisted teaching and assessment of disabled students in higher education: The interface between academic standards and disability rights. *Journal of Computer Assisted Learning* 23:207–219.

Konur, O. 2011. The scientometric evaluation of the research on the algae and bio-energy. *Applied Energy* 88:3532–3540.

Konur, O. 2012a. 100 citation classics in energy and fuels. *Energy Education Science and Technology Part A-Energy Science and Research* 30(si 1):319–332.

Konur, O. 2012b. What have we learned from the citation classics in energy and fuels: A mixed study, *Energy Education Science and Technology Part A: Energy Science and Research* 30(si 1):255–268.

Konur, O. 2012c. The gradual improvement of disability rights for the disabled tenants in the UK: The promising road is still ahead. *Social Political Economic and Cultural Research* 4:71–112.

Konur, O. 2012d. The policies and practices for the academic assessment of blind students in higher education and professions. *Energy Education Science and Technology Part B: Social and Educational Studies* 4(si 1):240–244.

Konur, O. 2012e. Prof. Dr. Ayhan Demirbas' scientometric biography. *Energy Education Science and Technology Part A: Energy Science and Research* 28:727–738.

Konur, O. 2012f. The evaluation of the research on the biofuels: A scientometric approach. *Energy Education Science and Technology Part A: Energy Science and Research* 28:903–916.

Konur, O. 2012g. The evaluation of the research on the biodiesel: A scientometric approach. *Energy Education Science and Technology Part A: Energy Science and Research* 28:1003–1014.

Konur, O. 2012h. The evaluation of the research on the bioethanol: A scientometric approach. *Energy Education Science and Technology Part A: Energy Science and Research* 28:1051–1064.

Konur, O. 2012i. The evaluation of the research on the microbial fuel cells: A scientometric approach. *Energy Education Science and Technology Part A: Energy Science and Research* 29:309–322.

Konur, O. 2012j. The evaluation of the research on the biohydrogen: A scientometric approach. *Energy Education Science and Technology Part A: Energy Science and Research* 29:323–338.

Konur, O. 2012k. The evaluation of the biogas research: A scientometric approach. *Energy Education Science and Technology Part A: Energy Science and Research* 29:1277–1292.

Konur, O. 2012l. The scientometric evaluation of the research on the production of bio-energy from biomass. *Biomass and Bioenergy* 47:504–515.

Konur, O. 2012m. The evaluation of the global energy and fuels research: A scientometric approach. *Energy Education Science and Technology Part A: Energy Science and Research* 30:613–628.

Konur, O. 2012n. The evaluation of the biorefinery research: A scientometric approach, *Energy Education Science and Technology Part A: Energy Science and Research* 30(si 1):347–358.

Konur, O. 2012o. The evaluation of the bio-oil research: A scientometric approach, *Energy Education Science and Technology Part A: Energy Science and Research* 30(si 1):379–392.

Konur, O. 2012p. The evaluation of the research on the biofuels: A scientometric approach. *Energy Education Science and Technology Part A: Energy Science and Research* 28:903–916.

Konur, O. 2013. What have we learned from the research on the International Financial Reporting Standards (IFRS)? A mixed study. *Energy Education Science and Technology Part D: Social Political Economic and Cultural Research* 5:29–40.

Konur, O. 2014a. Algal glycoscience: The state of the research. Working paper. Ankara, Turkey, Yildirim Beyazit University.

Konur, O. 2014b Carbohydrates: The state of the research. Working paper. Ankara, Turkey, Yildirim Beyazit University.

Konur, O. 2014c. Glycoenzymes: The state of the research. Working paper. Ankara, Turkey, Yildirim Beyazit University.

Konur, O. 2014d. Glycomics: The state of the research. Working paper. Ankara, Turkey, Yildirim Beyazit University.

Konur, O. 2014e. Glycolipids: The state of the research. Working paper. Ankara, Turkey, Yildirim Beyazit University.

Konur, O. 2014f. Proteoglycans: The state of the research. Working paper. Ankara, Turkey, Yildirim Beyazit University.

Konur, O. 2014g. Glycoproteins: The state of the research. Working paper. Ankara, Turkey, Yildirim Beyazit University.

Konur, O. 2014h. Glycosylation: The state of the research. Working paper. Ankara, Turkey, Yildirim Beyazit University.

Konur, O. 2014i. Glycans: The state of the research. Working paper. Ankara, Turkey, Yildirim Beyazit University.

Konur, O. 2014j. Glycoconjugates: The state of the research. Working paper. Ankara, Turkey, Yildirim Beyazit University.

Konur, O. (2015a) Algal economics and optimization. In: Kim, S.K. & Lee, C.G. (Eds.) *Marine Bioenergy: Trends and Developments*, pp. 691–716. Boca Raton, FL, CRC Press.

Konur, O. (2015b) Algal high-value consumer products. In: Kim, S.K. & Lee, C.G. (Eds.) *Marine Bioenergy: Trends and Developments*, pp. 653–682. Boca Raton, FL, CRC Press.

Konur, O. (2015c) Algal photobioreactors. In: Kim, S.K. & Lee, C.G. (Eds.) *Marine Bioenergy: Trends and Developments*, pp. 81–108. Boca Raton, FL, CRC Press.

Konur, O. (2015d) Algal biosorption of heavy metals from wastes. In: Kim, S.K. & Lee, C.G. (Eds.) *Marine Bioenergy: Trends and Developments*, pp. 597–626. Boca Raton, FL, CRC Press.

Konur, O. (2015e) Current state of research on algal bioelectricity and algal microbial fuel cells. In: Kim, S.K. & Lee, C.G. (Eds.) *Marine Bioenergy: Trends and Developments*, pp. 527–556. Boca Raton, FL, CRC Press.

Konur, O. (2015f) Current state of research on algal biodiesel. In: Kim, S.K. & Lee, C.G. (Eds.) *Marine Bioenergy: Trends and Developments*, pp. 487–512, Boca Raton, FL, CRC Press.

Konur, O. (2015g) Current state of research on algal biohydrogen. In: Kim, S.K. & Lee, C.G. (Eds.) *Marine Bioenergy: Trends and Developments*, pp. 393–422. Boca Raton, FL, CRC Press. (Publication no. MB/4).

Konur, O. (2015h) Current state of research on algal biomethanol. In: Kim, S.K. & Lee, C.G. (Eds.) *Marine Bioenergy: Trends and Developments*, pp. 327–370. Boca Raton, FL, CRC Press.

Konur, O. (2015i) Current state of research on algal biomethane. In: Kim, S.K. & Lee, C.G. (Eds.) *Marine Bioenergy: Trends and Developments*, pp. 273–302. Boca Raton, FL, CRC Press.

Konur, O. (2015j) Current state of research on algal bioethanol. In: Kim, S.K. & Lee, C.G. (Eds.) *Marine Bioenergy: Trends and Developments*, pp. 217–244. Boca Raton, FL, CRC Press.

Konur, O. (2015k) Algal photosynthesis, biosorption, biotechnology, and biofuels. In: Kim, S.K. (Eds.) *Springer Handbook of Marine Biotechnology*, pp. 1131–1161. Berlin, Germany, Springer.

Konur, O. (2015l) The scientometric study of the global energy research. In: Prasad, R., Sivakumar, S., Sharma, U.C. (Eds.) *Energy Science and Technology. V. 1. Opportunities and Challenges*, pp. 475–489. Houston, TX, Studium Press LLC.

Konur, O. (2015m) The review of citation classics on the global energy research. In: Prasad, R., Sivakumar, S., Sharma, U.C. (Eds.) *Energy Science and Technology. V. 1. Opportunities and Challenges*, pp. 490–526. Houston, TX, Studium Press LLC.

Konur, O. (2016a) Scientometric overview regarding the surface chemistry of nanobiomaterials. In: Grumezescu, A. M. (Ed.) *Surface Chemistry of Nanobiomaterials, Applications of Nanobiomaterials. Volume 3*, pp. 463–486. Amsterdam, the Netherlands, Elsevier.

Konur, O. (2016b) Scientometric overview regarding the nanobiomaterials in antimicrobial therapy. In: Grumezescu, A. M. (Ed.) *NanoBioMaterials in Antimicrobial Therapy, Applications of Nanobiomaterials*, pp. 511–535. Amsterdam, the Netherlands, Elsevier.

Koppenol, W.H., P.L. Bounds, and C.V. Dang. 2011. Otto Warburg's contributions to current concepts of cancer metabolism. *Nature Reviews Cancer* 11:325–337.

Kwong, P.D., R. Wyatt, J. Robinson, R.W. Sweet, J. Sodroski, and W.A. Hendrickson. 1998. Structure of an HIV gp120 envelope glycoprotein in complex with the CD4 receptor and a neutralizing human antibody. *Nature* 393:648–659.

McNeil, H.P., R.J. Simpson, C.N. Chesterman, and S.A. Krilis. 1990. Antiphospholipid antibodies are directed against a complex antigen that includes a lipid-binding inhibitor of coagulation: $\beta_2$-glycoprotein-I (apolipoprotein-H). *Proceedings of the National Academy of Sciences of the United States of America* 87:4120–4124.

North, D. 1994. Economic-performance through time. *American Economic Review* 84:359–368.

Paladugu, R., M.S Chein, S. Gardezi, and L. Wise. 2002. One hundred citation classics in general surgical journals. *World Journal of Surgery* 26:1099–1105.

Pierce, J.G. and T.F. Parsons. 1981. Glycoprotein hormones: Structure and function. *Annual Review of Biochemistry* 50:465–495.

Schinkel, A.H., J.J.M. Smit, O. van Tellingen, J.H. Beijnen, E. Wagenaar, L. van Deemter et al. 1994. Disruption of the mouse mdr1a P-glycoprotein gene leads to a deficiency in the blood-brain-barrier and to increased sensitivity to drugs. *Cell* 77:491–502.

Simonet, W.S., D.L. Lacey, C.R. Dunstan, M. Kelley, M.S. Chang, R. Luthy et al. 1997. Osteoprotegerin: A novel secreted protein involved in the regulation of bone density. *Cell* 89:309–319.

Thiebaut, F., T. Tsuruo, H. Hamada, M.M. Gottesman, I. Pastan, and M.C. Willingham. 1987. Cellular-localization of the multidrug-resistance gene-product P-glycoprotein in normal human-tissues. *Proceedings of the National Academy of Sciences of the United States of America* 84:7735–7738.

Varki, A. 1993. Biological roles of oligosaccharides: All of the theories are correct. *Glycobiology* 3:97–130.

Wilson, I.A., J.J. Skehel, and D.C. Wiley. 1981. Structure of the haemagglutinin membrane glycoprotein of influenza-virus at 3-A resolution. *Nature* 289:366–373.

Wrigley, N. and S. Matthews. 1986. Citation-classics and citation levels in geography. *Area* 18:185–194.

# *Marine glycoconjugates of reproduction and chemical communications*

*chapter three*

# Marine glycoconjugates in gamete physiology and fertilization

*Alessandra Gallo*

## Contents

## 3.1 Introduction

The term "glycoconjugates" is the common designation for carbohydrates covalently linked to a nonsugar moiety. The major groups of glycoconjugates are glycoproteins, glycopeptides, peptidoglycans, glycolipids, and lipopolysaccharides. This class of molecules play many important and intriguing roles in biological processes such as cellular communication, protein folding, inflammation, pathogen invasion, immunology, and cancer. Moreover, there is compelling evidence that glycoconjugates play important roles in the field of reproductive biology, which represents a significant area for glycobiological investigation. Functional and structural analyses of glycoconjugates operating in the reproductive system suggest that they are required for appropriate biological activities because variations in glycosylation products contribute to reproductive functionality in male and female animals (Dell et al. 1999). In particular, they are involved in many aspects of reproductive biology, such as the coating of germ cells, the passage of sperm through the female reproductive tract, maturation of sperm, attachment of sperm to female reproductive surfaces, fertilization, embryogenesis, and implantation (Mengerink and Vacquier 2001).

Glycoconjugates were identified in many marine organisms, in which they are involved in immune response, thrust generation in nonflagellar cyanobacteria, electroreception in sharks, sperm–oocyte interaction, cell recognition in embryo development, and cell adhesion (Caldwell and Pagett 2010). Moreover, reproductive biology is one of the areas within marine science that has a pronounced glycoconjugate component (Caldwell and Pagett 2010). Nevertheless, there has been very little emphasis on glycobiological studies in reproduction, as the great majority of these studies have been focused on gametes and fertilization. Although little is known about the function of glycoconjugates in gametogenesis in marine organisms, it is better characterized in the fertilization process. This chapter aims at highlighting recent advances on the key role of glycoconjugates in gametogenesis and fertilization process of marine animals.

## 3.2 Role of glycoconjugate in oogenesis

Oogenesis is the process of formation of the female gametes, preparing them for fertilization and the following developmental events. Oogenesis can be divided into two phases: the growth phase characterized by the enlargement of the oocyte and the maturation phase during which the oocyte becomes fertilizable (Nagahama et al. 1995). During the growth phase, large amounts of glycoconjugates, in particular glycoproteins, are synthesized and represent the molecular constituents of cortical vesicles, vitelline envelope, and yolk granules of the fully grown oocyte of marine animals (Ortiz-Delgado et al. 2008; Sarasquete et al. 2002).

The major event of the growth phase is the formation and accumulation of the yolk proteins in the oocyte, a process known as vitellogenesis (Wallace 1985). In oviparous animals, yolk proteins give nutritional support to the oocyte and developing embryos, constituting 60%–90% of the total oocyte proteins. Yolk proteins in the oocyte are derived from the enzymatic cleavage of a common precursor called vitellogenin (Vg), which is a glycolipophosphoprotein showing similar characteristics in vertebrates and invertebrates (Matozzo et al. 2008).

After the uptake of Vg by the growing oocyte, it is cleaved proteolytically into a characteristic suite of yolk proteins: a high-density lipoglycoprotein called "lipovitellins," a highly phosphorylated phosvitins, and a β-component stored within membrane-bound structures called "yolk platelets" (Ortiz-Delgado et al. 2008).

The structure of the glycan moieties of Vg has been discovered only in the decapod crustacean *Cherax quadricarinatus*. In this species, it has been shown that Vg, which is produced in the hepatopancreas and secreted to the hemolymph (Abdu et al. 2002), is posttranslationally modified by N-linked oligosaccharides. The Vg sites for N-glycosylation were glycosylated with the high-mannose (Man) glycans, ranging from Glc1Man9GlcNAc2 (one glucoses, nine mannoses, and two N-acetylglucosamine (GlcNAc)) to Man5GlcNAc2 (five mannoses and two GlcNAc) and the glucose-capped oligosaccharide (Glc1Man9GlcNAc2). It has been suggested that N-glycans of Vg play an important role in the folding and subunit assembly to achieve the mature protein in the hemolymph and ovary, in keeping this hydrophopic protein in the hemolymph to improve its transport in the ovary, and in recognition and receptor-mediated endocytosis mediating the uptake of Vg into the oocytes. Once uptaken into the oocytes, N-glycans might also have a role in packaging and compacting its products in yolk bodies (Khalaila et al. 2004).

The protein and lipid components of lipovitellin have been characterized in both invertebrates and vertebrates, while, despite the structural and functional importance of the carbohydrate components of this major yolk protein (MYP), very little information is available on these units. Lipovitellin was purified only from the crustacean *Emerita asiatica* demonstrating that contained fucose (Fuc), Man, galactosamine (Gal), five different O-linked oligosaccharides, and four different N-linked oligosaccharides; however, no structural details of these glycan moieties are available (Tirumalai and Subramoniam 2001).

Different from decapods, in sea urchin species, Vg is synthesized in the intestine as a 195 kDa glycoprotein and secreted into coelomic fluid. From the coelom, this glycoprotein is absorbed by the nutritive phagocytes (accessory cells) of the ovary, where it is stored. Then, it is transported to the growing oocytes to be accumulated as a 180 kDa glycoprotein, termed major yolk protein. This decrease in molecular weight from Vg to MYPs seems to be a common phenomenon in sea urchin. It is possible that Vg is slightly modified in molecular structure after its incorporation into the gonads (Harrington and Ozaki 1986; Ozaki et al. 1986; Unuma et al. 1998).

A group of complex sialic acid containing glycosphingolipids (NGLs), the gangliosides, are found in the yolk granules of the unfertilized sea urchin oocytes. These are more complex NGLs in which oligosaccharide chains, containing sialic acid, are attached to a ceramide. In the oocytes of some sea urchin species, the chemical structures of the major gangliosides have been identified as NeuGcα2-6Glcβ1-1Cer (M5) and HSO3-NeuGcα2-6Glcβ1-1Cer (T1). The ganglioside M5 has been shown to be localized in the plasma membrane and in the yolk granule, where it is associated with yolk lipoproteins and is involved in the uptake of yolk lipoproteins into the growing oocytes during oogenesis and transported from yolk granules to other cellular components during embryogenesis. M5 ganglioside associating with yolk lipoproteins in yolk granules may be considered as a significant stored material to be utilized for early embryogenesis (Kubo and Hoshi 1990; Shogomori et al. 1993, 1997).

Sea urchin oocytes contain a variety of NGLs such as glucosylceramide, melibiosylceramide, ceramide trihexoside, and GL-5 (Kubo et al. 1990, 1992b).

The chemical structures of these glycolipids were determined, showing that ceramide moieties of these are almost identical.

Glucosylceramide (Glcβ1-1Cer) and melibiosylceramide (Galα 1-6Glcβ1-1Cer) have the same long-chain base compositions that are very characteristic, and all of them are phytosphingosines. The fatty acid compositions of these glycolsphingolipids also resemble each other (Kubo et al. 1988). The chemical structure of the ceramide trihexoside was determined (Galβl-6Galβ1-6Glcβ1-lCer) and its carbohydrate structure is a novel trisaccharide. Glucosylceramide resembles ceramide trihexoside in ceramide structure, suggesting that it is synthesized from glucosylceramide (Kubo et al. 1992a).

GL-5 chemical structure is Fucα1-3GalNAcβ1-4(Fucα1-3)GlcNAcβ1-4Glcβ1-lCer. GL-5 has a unique saccharide sequence: the reducing terminal disaccharide core is GlcNAcβ1-4Glc. The defucosylated core structure (GalNAcβ1-4GlcNAcβ1-4Glcβ1-) is a novel trisaccharide chain. Fuc directly binds to N-acetylgalactosamine (GalNAc) and the sugar structure is one of the shortest saccharide chains found so far among the difucosylated glycolipids (Inagaki et al. 1992; Kubo et al. 1992). These glycosphingolipids could participate in sperm–oocyte interaction process (Kubo et al. 1988).

Another important process of oocyte development is the assembly of an extracellular eggshell, or vitelline envelope (Dumont and Brummett 1985), which in some species is encompassed by a jelly-like layer. The vitelline envelope is an acellular structure made of glycoproteinaceous fibers (Focarelli et al. 1990; Glabe et al. 1982; Honegger and Koyanagi 2008; Wikramanayake and Clark 1994). It possesses many functions, such as prevention of polyspermy, protection of the growing oocyte and of the developing embryo, uptake of nutrients and other molecules during oogenesis, and guidance of the spermatozoa to the oocyte. The number of oocyte envelopes varies from one to several in different animal species.

Fish oocyte envelope consists of two layers. The outer layer is rich in glycolipids, while the inner one, called zona radiata, contains neutral glycoproteins (Sarasquete et al. 2002). The zona radiata consists of two main layers with different morphological characteristics. The inner layer consists mainly of proteins with few carbohydrates and its functions in fertilization and embryo development have been conserved during evolution. The outer layer has a specific macromolecular composition containing glycoproteins, carboxylated and sulfated polysaccharides, and, rarely, sialic acid, all of which contribute to the interactions between the oocyte and its aquatic environment (Shabanipour and Heidari 2004).

Zona radiata glycoconjugates of different teleost fish oocytes have been well investigated. In the flatfish *Solea senegalensis*, the presence of glycans with terminal GalNAc and/or αGal and with terminal/internal αMan is reported. As oogenesis proceeds, the glycan pattern of zona radiata decreases drastically, and in the last phase of maturation, only few βGalNAc residues are detected. Zona radiata of dusky grouper *Epinephelus marginatus* is characterized by a higher content of GlcNAc and sialic acid (Mandich et al. 2002) as in swordfish zona radiata, in which α-L-Fuc residues are also detected (Ortiz-Delgado et al. 2008). During oocyte development, the glycan pattern changes, and this may reflect the different activities of zona radiata during different phases of oogenesis (Accogli et al. 2012).

The composition of glycoprotein oligosaccharides present on the oocyte envelope of the different marine invertebrates was also detected.

In the ascidian *Phallusia mammillata*, it has been reported that α-GlcNAc and GalNAc are highly expressed on the oocyte envelope with 86% by weight of the total sugar content, while Fuc, Man, and Glc accounted for the remaining 14% (Litscher and Honegger 1991). Also in the oocyte envelope of the ascidian *Halocynthia roretzi*, GalNAc, α-GlcNAc, and Man together with Gal are the major components of the glycoprotein oligosaccharides. Moreover, in this species, a high content of arabinose, xylose, and rhamnose was also detected, but any information about their plausible biological significance is not available. Glycan analysis reveals that glycans are O-linked, and the core structures of these anionic glycans are constituted by 1,4-linked xylose and 1,3-linked Gal and GalNAc residues (Baginski et al. 1999). The vitelline coat of the mature oocyte of the ascidian *Ciona intestinalis* has been demonstrated to contain three glycoproteins, and the

most abundant carbohydrate is GalNAc, followed by Fuc and Gal (Rosati et al. 1982). Moreover, the monosaccharide rhamnose is also detected (De Santis and Pinto 1990).

Different from ascidians, α-GlcNAc is lowly expressed on the oocyte envelope of crustaceans. Glycoprotein oligosaccharides present on the oocyte envelope of different crustacean species have also been well characterized. In particular, they contain high concentration of Glc and α1-3Man and low concentration of Gal, Fuc, GlcNAc, and GalNAc. Despite its small concentration, α-GlcNAc, together with Glc and Man, has great importance during gamete interaction process. Fuc, Gal, and GalNAc exclusively have a structural function (Dupré et al. 2012; Pillai and Clark 1990).

The oocytes of echinoderms and molluscs are characterized by a second fibrous layer, the jelly coat. It is located immediately surrounding the vitelline envelope and consists primarily of a network of short peptides and sulfated fucan glycoproteins.

In the mollusk gastropode *Haliotis asinina*, the oocyte jelly coat contains two major glycoproteins of 107 kDa and 178 kDa, whereas the vitelline envelope contains a broad spectrum of protein bands. Glc is the major sugar residue of both jelly coat and vitelline envelope glycoproteins, whereas minor proportions of arabinose, fructose, Gal, and Fuc are present (Suphamungmee et al. 2010).

In different sea urchin species, chemical characterization of the oocyte jelly coat component has identified two acid glycoproteins: sialic acid–rich glycoprotein and fucose sulfate glycoconjugate (FSG) (SeGall and Lennarz 1979; Shimizu et al. 1990). FSGs are linear polysaccharides with a repeated unit composed of sulfated 1–4 hexose moieties and inducers of the acrosome reaction in sperm (Alves et al. 1997; Mulloy et al. 1994; Vacquier and Moy 1997). Among sea urchin species, FSGs differ in sugar composition (L-Fuc or L-Gal), glycosidic linkage (α1→3 or α1→4), and sulfation pattern (2- or 4-O-sulfate) (Alves et al. 1998; Vilela-Silva et al. 2002). These differences are responsible of species-specific induction of the acrosomal reaction (Alves et al. 1998; Hirohashi et al. 2008).

The sialic acid–rich glycoproteins have a low protein content, while contain many O-linked sialyglycan chains. These are attached to threonine residues on the core protein containing Neu5Gc residues and are composed of Fuc, Gal, and GalNAc residues. Most interestingly, the inner residue linkages in the polysialyglycan chains contain unique (→5-Oglycolyl-Neu5Gcα2→)n linkages (Kitazume et al. 1994). This structure represents the first naturally occurring polysialyglycan as an integral part of a glycoprotein and it potentiates the acrosome reaction induced by FSG (Hirohashi et al. 2008).

The oocyte jelly coat of starfish presents two glycoproteins: the first one is a high Man glycoprotein with a molecular mass of 80 kDa and an unique saccharide structures, but its function is not well clarified (Endo et al. 1987; Hoshi et al. 1994); the second one is a highly sulfated Fuc-rich glycoprotein named acrosome reaction–inducing substance (ARIS), which is involved in triggering the sperm acrosome reaction (Uno and Hoshi 1978). The biological activity of ARIS resides in one of the sugar fragments called fragment 1 (Hirohashi et al. 2008; Naruse et al. 2010). The molecular size of this glycan is about 10 kDa and does not contain amino acid residues. Its structure was revealed as 10 or so repeats of the following pentasaccharide unit [→4)-β-D-Xylp-(1→3)-α-D-Galp-(1→3)-α-L-Fucp-4(SO$_3$⁻)-(1→3)-α-L-Fucp4(SO$_3$⁻)-(1→4)-α-L-Fucp-(1→] (Koyota et al. 1997). This sugar chain links to the peptide part by O-glycosidic linkage trough another sugar chain with a different structure from fragment 1 (Hirohashi et al. 2008).

The second fragment isolated from ARIS was named fragment 2. It is mostly composed of sulfated glycans and retains about 10% (w/w) of the protein part. Fragment 2 has a molecular mass of 400 kDa and its glycans are O-linked. It is composed of the heptasaccharide units of   [→3)-Galp-(1→3)-Fucp-(1→3)-Galp-(1→4)-GalNAcp-(1→4)-GlcNAcp-6(SO$_3$⁻)-(1→6)-Galp4(SO$_3$⁻)-(1→4)-GalNAcp-(1→] (Gunaratne et al. 2003).

Moreover, ARIS has another acid sugar chain composed of Man, GalNAc, and GlcNAc, although it is not important for its biological activity (Hoshi et al. 2012).

## 3.3  Role of glycoconjugate in spermatogenesis

Spermatogenesis is the developmental process in which highly differentiated spermatozoa are produced from male primordial germ cells. This process begins with the mitotic proliferation of spermatogonia and formation of primary spermatocytes, passing then through meiosis into secondary spermatocytes, which differentiate into spermatids and finally through spermiogenesis into mature spermatozoa (Gallo and Costantini 2012).

It has been reported that, during spermatogenesis, the contents and profiles of different glycoproteins significantly change, suggesting the important role of such components in this process (Anakwe and Gerton 1990; Jones et al. 1988; Martinez-Menarguez et al. 1992). Moreover, it has been demonstrated that some complex germ cell–somatic cell interactions are implicated in the spermatogenesis process, and several reports suggest that glycoconjugates are involved in cell adhesion during this process (Pratt et al. 1993; Raychoudhury and Millette 1997).

Fish spermatogenesis is characterized by a cystic mode that consists of a single primary spermatogonium, which is enveloped by one or two Sertoli cells, hence forming cysts by Sertoli–Sertoli cell contacts, which formed the tubular compartment. The germ cell clone descending from this primary spermatogonium develops within the cyst so that Sertoli cells are in contact with only a single germ cell clone at a time. The clone members are formed in a synchronized pattern sequentially through the different stages of spermatogenesis and spermiogenesis. The cyst eventually opens to release spermatozoa. Sertoli cells support germ cells by providing the optimal microenvironment for germ cell development. Another important somatic cell type is Leydig cells that, together with the connective tissue, formed the interstitial compartment (Nobrega et al. 2009; Schulz and Miura 2002).

Glycoprotein oligosaccharide sequences are identified and localized in the testis of teleost fish *S. senegalensis* (Desantis et al. 2010). Glycoproteins are localized in both compartments of the testis, the interstitial and germinal compartments, but also in the basal lamina separating them. In the interstitial stroma, a very complex glycoprotein composition, including asialo- and sialoglycans in N- and O-linked oligosaccharides and N-linked glycans, was discovered. The glycosylation pattern of the interstitial compartment depending on the testicular region is analyzed. In fact, Neu5Acα2 (*N*-Acetylneuraminic acid), 3Galβ1,4GlcNAc and GalNAcα1,3(L-Fuc-α-1,2) Galβ1,3/4GlcNAcβ1 are more abundant in the medullar region than in the cortex. This different pattern is due to a differential glycoprotein compound trafficking patterns from the vascular system to each testis region. This occurs because germ cell proliferation and differentiation take place in the cortical region of the testis, while final sperm maturation in the medullar region to support distinct glycoprotein requirements in different phases of the developmental process. The interstitial compartment of fish species is characterized by the presence of melano-macrophage centers containing glycans terminating with Galβ1,3GalNAc, but their role is not known, and Leydig cells, that unlike from other teleost species, do not exhibit glycoprotein oligosaccharide sequences (Liguoro et al. 2004). The basal lamina is characterized by glycans with terminal/internal Man, internal βGlcNAc, terminal Neu5Acα2,6Gal/GalNAc, Neu5acGalβ1,3GalNAc, Galβ1,3GalNAc, Galβ1,4GlcNAc, GalNAc, αGal, and α-L-Fuc. This complex glycosylation pattern is related to the presence of meshwork composed of several glycoconjugate components responsible for biological function of basal lamina. In the germinal compartment, Sertoli cells express sialoglycans terminating with Neu5Acα2,3Galβ1,4GlcNAc that may be involved in the establishment of interaction between adjacent spermatocysts, playing a key role in the organization of spermatogenic cysts. Glycoprotein oligosaccharide sequences are absent in spermatogonia, while primary spermatocytes express, in cytoplasm and nucleus, glycans terminating with Galβ1,3GalNAc and αGalNAc, respectively (Desantis et al. 2010). Since germ cell nuclear glycoproteins are associated with the chromatin, it has been suggested that they play a role in the regulation of transcription factors and in the control of cell cycle (Varki et al. 1999). Spermatid cytoplasm exhibits N-linked glycans containing high-Man residues, as well as oligosaccharides terminating with α/βGalNAc, αGal, and α-L-Fuc. These glycoproteins were also found in fine granular structures located in the cytoplasm of spermatids.

Different from *S. senegalensis*, the Sertoli cells of other fish species, such as spotted ray and Nile tilapia, display a more complex glycosylation pattern, but sialoglycans seem to be lacking in those species (Liguoro et al. 2004; Tokalov and Gutzeit 2007).

Glycoconjugates have been also identified and localized in germ cells and somatic cells of cartilaginous fish testis. It has been observed that extensive modifications of the glycoside residue composition at the level of surface, cytoplasm, and nucleus of both type cells occur.

Cadherins, a glycoprotein superfamily containing mannosyl chains, were identified on the surface of germ and Sertoli cells. Cadherins play a key role in establishing interactions between germ and somatic cells, implied in cyst and spermatoblast formation. In addition, germ cells and Sertoli cells undergoing apoptosis change the cellular surface composition, specifically overexpressing on the cell surface GalNAc and Gal that make them susceptible to phagocytosis. During cyst maturation, Leydig cells also change their surface glycoconjugate composition due to a modification of their activity during spermatogenesis. Germ cells express new sugar residues during their differentiation, and the presence of glycoconjugates is limited to the Golgi zone in spermatocytes, to acrosome formation region in spermatids, and to the acrosome in spermatozoa (Liguoro et al. 2004). This finding suggests that the acrosome glycosylation pattern depends on stoichiometric and spatial distribution of binding sites and/or the activity of glycolytic enzymes and glycosyltransferases (Liguoro et al. 2004; Martinez-Menarguez et al. 1993). Glycoconjugates are also expressed in the nucleus of germ cell where they are involved in the control of the cell cycle (Liguoro et al. 2004).

Among the invertebrates, glycoprotein oligosaccharide sequences are identified in sea urchin testis, which expressed a heavily sialylated glycoprotein, named flagellasialin. This protein has a molecular mass ranging from 40 to 80 kDa and contains a unique polysialic acid (polySia) sulfated α2,9-linked polyNeu5Ac (Miyata et al. 2004). Flagellasialin is exclusively located to the flagellum where probably α2,9-linked polySia regulates voltage-sensitive sodium and calcium channels,

influencing the intracellular calcium and sperm motility (Kambara et al. 2011; Miyata et al. 2006). In the flagellum but also in the sperm head, it has been demonstrated the expression of a 190 kDa glycoprotein linked to α2,8-linked poly(Neu5Ac) structures. The biological function of this glycoprotein is not yet clarified. The conformational differences between α2,9- and α2,8-linked polySia structures and their co-localization in the same sperm region might reflect functional differences of these two polySia-containing glycoprotein at fertilization. The α2,8-linked polyNeu5Ac structure occurs also in glycolipids localized in the sperm head to facilitate rearrangement of the membrane proteins on the sperm surface upon sperm activation (Ijuin et al. 1996; Miyata et al. 2011). Moreover, the presence of the yolk protein precursor, Vg, in immature sea urchin male gonad, and in the female, was demonstrated (Shyu et al. 1986). It has been suggested that male Vg could be a precursor of MYP incorporated in the testicular nutritive phagocytes as a nutrient source for spermatogenesis (Unuma et al. 1998). As spermatogenesis proceeds, MYP decreases in quantity since it is utilized as a material for synthesizing new substances that participate in the formation of spermatozoa and are metabolized as an energy source during this period (Unuma et al. 2003). The immature sea urchin testis also contains a large quantity of polysaccharides, most of which are probably glycogen in the form of granules. The polysaccharide content decreases as gametogenesis proceeds in both sexes, since they are possibly used as an energy source, as suggested for *Strongylocentrotus intermedius* (Marsh and Watts 2001; Verhey and Moyer 1967).

In crustaceans, glycoproteins are identified in immature and mature germ cells. In particular, it has been demonstrated that immature germ cells and spermatozoa have a different glycoconjugate composition. In *Aristaeomorpha foliacea*, only immature germ cells express N-linked oligosaccharides, which contain terminal and internal α-D-Man, internal β-D-GlcNAc, NeuAcα2,6Gal/GalNAc, and terminal NeuAcα2,3Galβ1,4GlcNAc. On the contrary, spermatozoa express both N- and O-linked oligosaccharides in the cytoplasm. The N-linked oligosaccharides consist of terminal GlcNAc, while O-linked oligosaccharides terminated with β-Gal(1-3)-GalNAc and/or α-GalNAc, whereas they end with sialic acid linked to β-Gal(1-3)-GalNAc in the nucleus. The extracellular matrix in which spermatozoa are embedded consists of neutral and acid glycoconjugates, and differences in glycoconjugate composition between immature and mature hemispermatophores have been observed. The former expresses both N- and O-linked glycoconjugates, while mature hemispermatophores express only O-linked oligosaccharides and contain more sialyl-glycoconjugates than immature ones (Desantis et al. 2006). The role of extracellular matrix in crustaceans is not well known. It has been suggested that the acid mucopolysaccharides of spermatophores may act as a cementing agent or an antimicrobial agent or in the maintenance of spermatozoa during the storage within female spermathecae (Desantis et al. 2006; Sasikala and Subramoniam 1987; Subramonian 1991). In *A. foliacea*, it has also been demonstrated that sperm glycoprotein pattern seasonally changes and spermatozoa undergo changes in glycoconjugate composition during their transit from testis to hemispermatophore (Desantis et al. 2003).

## 3.4 Role of glycoconjugate in fertilization process

Fertilization is the process in which gametes fuse to form a new individual. This encompasses at least five steps, even if the details of the fertilization process may vary between species. In any event, spermatozoa first recognize and bind, in a species-specific manner, to the extracellular coat of the oocyte. Once bound to the oocyte coat, spermatozoa undergo the acrosome reaction, which is an exocytosis of the acrosomal vesicle located on the tip of the spermatozoa head, which releases a lytic agent. These events enable spermatozoa to cross the extracellular coat and reach the oocyte plasma membrane where the binding between the two cells occurs and causes the fusion of the genetic material of spermatozoa and oocyte, leading to the activation of oocyte metabolism and mitosis and the beginning of the development (Vacquier 1998; Yanagimachi 1994).

Glycoconjugates have been shown to play essential roles in fertilization process. The initial sperm–oocyte binding requires a specific interaction between glycoproteins of oocyte envelope and complementary sperm surface receptors (Caldwell and Pagett 2010; Tosti 1994).

The involvement of glycoconjugates in fertilization process was demonstrated for the first time in sea urchins (SeGall and Lennarz 1979). The FSG of oocyte jelly coat binds a glycoprotein receptor, named suREJ, containing one epidermal growth factor module and two C-type lectin carbohydrate-recognition modules, located exclusively in the plasma membrane just on the sperm acrosomal vesicle at the anterior apex of the spermatozoa head (Mengerink and Vacquier 2001; Moy et al. 1996). After this binding, the spermatozoa undergo consecutive morphological and biochemical changes called the acrosomal reaction, which involves both the exocytosis of the acrosomal vesicle so that the protein bindin was expelled and the polymerization of actin to form the acrosomal process (Vacquier et al. 1995).

The protein bindin, coating the outside of the acrosomal process, reacts with a matching oocyte membrane receptor (Ohlendieck et al. 1994). This receptor, named EBR1, is a 350 kDa glycoprotein containing both N- and O-linked oligosaccharide chains with different biological activities, being inactive and active, respectively. The active oligosaccharide is linked to serine or threonine via an O-glycosidic bond to GalNAc, which contains a Gal bound $\beta(1\rightarrow3)$ to the linkage sugar. The remainder of the structure consists of GlcNAc and Fuc residues, with the sulfate moieties on the Fuc residues. The O-linked oligosaccharide chains show different levels of sulfation correlated to their activity, being greatly enhanced with increasing sulfation. Two-step models involving carbohydrate and protein chains have been proposed for sperm–oocyte interaction in sea urchin. The first step is postulated to be a low-affinity ionic interaction of the sulfated O-linked oligosaccharide chains of the oocyte receptor with spermatozoa that are not species specific. This is followed by a high-affinity, species-specific interaction of one or more domains of the polypeptide chain and bindin on the acrosomal process (Dhume and Lennarz 1995). The species specificity of this interaction is determined by differences in the attachment position of the oligosaccharide chains on the protein backbone (Hirohashi and Lennarz 2001).

As in sea urchin, upon encountering the oocyte jelly coat, the starfish spermatozoa undergo the acrosome reaction. To induce the acrosome reaction, three jelly components act in concert on the spermatozoa: ARIS, a group of steroidal saponins named Co-ARIS, and an oligopeptide known as asterosap (Hoshi et al. 1988; Longo et al. 1995). When spermatozoa reach the oocyte jelly coat, the first binding occurs between ARIS and its receptor specifically located to the anterior region of the sperm head (Ushiyama et al. 1993). The tertiary structure of ARIS is important for its activity. The signal for a correct folding is derived from fragment 1. The fragment 1 has a double helix and a compact carbohydrate core with sulfates protruding in pairs away from the center of the helix. It has been demonstrated that the double helix structure of ARIS saccharide chain is important for the induction of acrosome reaction (Hirohashi et al. 2008). The first oocyte–sperm binding gates calcium channels, inducing an intracellular calcium increase. This signal causes exocytosis by which spermatozoa expose the devices essential for penetration through the oocyte coats and for fusion with the oocyte plasma membrane. To induce acrosome reaction, ARIS requires Co-ARIS (Hoshi et al. 1990). Co-ARIS is composed of a sulfated steroid and a pentasaccharide chain. Its activity requires sulfate moiety and mainly depends on the structure of steroidal side chain, while not necessarily requiring a specific structure of the saccharide chain (Nishiyama et al. 1987). Spermatozoa do not have a specific receptor for Co-ARIS. It has been suggested that steroid ring and side chain of Co-ARIS may infiltrate or be inserted into sperm plasma membrane and the sulfated group, and a sugar chain contributes to keep it in the right position and orientation to interact correctly with other components of the sperm plasma membrane (Hoshi et al. 1994). Asterosap is a glutamine-rich tetratriacontapeptide with a 25-residue ring formed by a disulfide linkage that is essential for its biological activity (Nishigaki et al. 1996). It transiently increases the intracellular pH by stimulating $Na^+/H^+$ exchange systems and calcium via the activation of asterosap receptor that is a guanylyl cyclase located in the sperm flagellar plasma membrane (Matsumoto et al. 2002, 2008; Nishigaki et al. 2000). An increase in calcium and pH is essential to trigger the acrosome reaction.

Analogous to sea urchin bindin and EBR1, lysin and its receptor are found in archaeogastropods. The abalone spermatozoon swims easily through the oocyte jelly coat and reaches the vitelline envelope. The contact between the spermatozoon and the oocyte vitelline envelope induces the polymerization of actin to generate acrosomal process and the exocytotic of acrosomal vesicle that releases protein lysin onto the surface of the vitelline envelope. Lysin is a 16 kDa nonenzymatic, cationic protein that forms a dimer, which is able to bind a giant 1000 kDa glycoprotein named vitelline envelope receptor for lysin (VERL). VERL is a long, unbranched, fibrous glycoprotein that is composed of at least 50% saccharide, containing Glc and Man residues, and comprises 30% of the vitelline envelope mass (Swanson and Vacquier 1997). Dimer binding results in monomerization and the tight, species-specific binding of lysin to VERL that involves the carbohydrate moieties of VERL linked to serine or threonine via an *O*-glycosidic bond. Upon binding lysin monomers, the fibrous VERL molecules lose cohesion and splay apart, through which the spermatozoon passes to reach the oocyte and the tip of its acrosomal process fuses with the oocyte plasma membrane (Kresge et al. 2001; Vacquier et al. 1990). Moreover, other vitelline envelope glycoproteins of 30–50 kDa, even if do not bind lysin, could be involved in mediating species selectivity based on their interaction with VERL (Swanson and Vacquier 1997; Swanson et al. 2001).

In decapod crustaceans, the first sperm–oocyte interaction is established between the apical end of a sperm's appendage, named spike, and the outermost oocyte envelope. In particular, in *Rhynchocinetes typus*, it has been demonstrated the presence of a lectin-like molecule on the tip of sperm spike that recognized specific carbohydrates on the oocyte envelope such as GlcNAc, Glc, and Man (Dupre and Barros 2011; Dupré et al. 2012). At this point, the tip of the rigid spike exerts a lytic effect upon (Dupre and Barros 2011; Dupré et al. 2012) the vitelline envelope, causing a perforation through

which the spermatozoon passes to reach the oocyte plasma membrane (Barros et al. 1986; Bustamante et al. 2001; Rios and Barros 1997).

In ascidians, the sperm–oocyte binding is mediated by an enzyme–substrate complex established between a specific sperm surface glycosidase and corresponding glycans on the surface of the vitelline coat (Hoshi 1986; Hoshi et al. 1985). In *Ascidia nigra* and *P. mamillata*, N-acetyl-glucosaminidase in sperm membranes recognizes terminal α-GlcNAc residues on the oocyte envelope (Godknecht and Honegger 1991; Xie and Honegger 1993), while in *C. intestinalis* and *H. roretzi*, the sulfated Fuc-containing glycans of vitelline coat glycoproteins are responsible for the sperm binding to the vitelline coat. In particular, the sperm glycosidase α-L-fucosidase binds to terminal L-Fuc residues of the vitelline layer (Hirohashi et al. 2008; Matsumoto et al. 2002). The degree of sulfation and the proper spacing of sulfate groups seem essential for the biological activity (Baginski et al. 1999). When spermatozoon binds the vitelline envelope, acrosome reaction occurs and the acrosomal outer membrane fuses with the plasmalemma, enclosing the acrosome, resulting in exocytosis of the acrosomal substances. After the acrosome reaction, the apical process protrudes mainly from the peripheral region of the apex of the sperm head. Although the chemical nature and the precise role of the acrosomal substances remain to be elucidated, it has been proposed that these acrosomal substances are responsible for membrane fusion between the apical process and the oocyte plasma membrane (Fukumoto 1988, 1990; Fukumoto and Numakunai 1995). The sperm–oocyte vitelline interaction activates a sperm lysine system. In particular, three proteases are involved in sperm penetration of the vitelline coat: two trypsin-like proteases, acrosin and spermosin, and one chymotrypsin-like protease. The chymotrypsin-like activity is involved in sperm penetration of the vitelline coat, but spermosin and acrosin both function to increase the fertilization rate (Lambert et al. 2002; Sawada et al. 1984, 2002).

Upon fusion of the sperm's and oocyte's cellular membranes, fertilization has been completed. At this point, the oocyte is extremely vulnerable to other sperm attempting to fertilize, so protection against the penetration of the oocyte by more than one spermatozoon, phenomenon is referred to as polyspermy, is developed. Two mechanisms are used by marine animals to prevent polyspermy, the fast and the slow block, and several studies report that glycoconjugates are implied in to block polyspermy.

In marine invertebrates, a fast block to polyspermy involves the opening of sodium channels in the oocyte plasma membrane that causes an ion flows into the oocyte, depolarizing the membrane. This depolarization prevents additional sperm from fusing to the oocyte plasma membrane. The slow block to polyspermy is due to the modifications of oocyte vitelline envelope, which transform it to a hard layer called fertilization membrane. The hardening of vitelline envelope is trigged by the contents of cortical granules that are released upon fertilization. Fertilization membrane forms a protective barrier that repels not only additional spermatozoa but also bacteria and small eukaryotic invaders (Wessel et al. 2001).

In sea urchins, fertilization membrane formation is initiated by trypsin-like proteases that cleave both a sperm-binding protein, removing supernumerary spermatozoa and preventing further sperm binding, and the proteins that connect the vitelline envelope to the plasma membrane (Foltz and Shilling 1993; Hagstrom 1956; Hirohashi and Lennarz 1998). At the same time, sugars that are released from the cortical granules attract water into the perivitelline space allowing the vitelline envelope lifts off the oocyte plasma membrane that is also due to the formation of hyaline layer (Wessel et al. 2001). Hyalin is a filamentous molecule that tends to form aggregates producing an amorphous layer (Adelson et al. 1992). In particular, its carbohydrate residues act as receptors for hyaline–hyalin and hyaline–cell interaction. Then, ovoperoxidases harden the vitelline envelope that becomes resistant to both mechanical and enzymatic modifications (Wessel et al. 2001). Ovoperoxidase released in the perivitelline space interacts with another cortical granule protein, the proteoliaisin. When this complex is associated with the nascent fertilization envelope, ovoperoxidase catalyzes the covalent cross-linking of juxtaposed tyrosine residues in adjacent polypeptide chains, including SFE 1 and SFE 9, to form a stable, macromolecular complex (Shapiro et al. 1989). Proteoliaisin targets the ovoperoxidase to the nascent fertilization envelope and protects this enzyme from proteolytic digestion (Nomura and Suzuki 1995; Somers and Shapiro 1991).

Ascidian oocytes do not have cortical granules and prevent polyspermy by first releasing a large quantity of glycosidase, followed by an electrical modification of the oocyte plasma membrane (Koyanagi and Honegger 2003; Lambert 1989; Litscher and Honegger 1991). Sperm–oocyte binding triggers oocytes to release large quantities of glycosidase that rapidly bind the vitelline-coat surface as sperm surface glycosidase blocking the binding of supernumerary spermatozoa (Honegger and Koyanagi 2008; Lambert 1986).

The spermatozoa of fish lack an acrosome, and the fertilizing spermatozoon enters the oocyte through a canal-like structure in the chorion named micropyle (Mengerink and Vacquier 2001). Following sperm–oocyte interaction, cortical alveoli fuse with the plasma membrane and discharge their contents such as polysialoglycoprotein (PSGP), hyaline, proteases, and transglutaminases into the presumptive perivitelline space.

Once released from the cortical alveoli, proteases cleave the 200 kDa PSGP (L-PSGP) into small 9 kDa PSGP (H-PSGP) (Inoue and Inoue 1986; Inoue et al. 1987; Laale 1980). Both amino acid and carbohydrate compositions of L-PSGP and H-PSGP are identical. H-PSGP is still too large to permeate the chorion. This establishes a colloid osmotic pressure, which causes an influx of mostly external water that swells the perivitelline space. The perivitelline space cushions the embryo and bathes it in a special medium containing the 9 kDa PSGP, lipids, carbohydrates, and ions as well as providing a sink for nitrogenous wastes (Eddy 1974; Peterson and Martin-Robichaud 1993; Rudy and Potts 1969). The perivitelline space also provides space for the free movement and growth of the embryo, which, due to its hypertonicity, absorbs water osmotically and slowly swells during embryogenesis (Li et al. 1989). With the establishment of the perivitelline space, the rise in hydrostatic pressure is aided by the hardening of the chorion, which contributes to the closure of the micropyle and thereby reduces the probability of polyspermy and microbial infection (Kobayashi and Yamamoto 1993; Kudo 1991). The hardening of the chorion is catalyzed by transglutaminases activated by PSGP that have proteinase activity (Kudo and Teshima 1998). Thus, activated transglutaminase forms covalent ε-(γ-glutamyl)-lysine cross-link constituents of adjacent proteins or glycoproteins of chorion (Oppen-Berntsen et al. 1990).

## 3.5 Conclusion

It has become increasingly clear over the past few years that glycoconjugates are critical to successful reproduction. They play important roles in several reproductive processes including oogenesis, spermatogenesis, fertilization, and embryogenesis.

It is clear that much remains to be learned about the roles of glycoconjugates in reproduction and that developing this information base is of the utmost importance. Due to the key role played by glycoconjugates in gametogenesis and fertilization process, disturbance of glycoconjugate expression can be the cause of impaired fertility and reproductive success. Further study of glycoconjugates on gametogenesis, as well as fertilization process, will provide a greater understanding of the molecular mechanisms that control the process of reproduction.

## References

Abdu, U., C. Davis, I. Khalaila, and A. Sagi. 2002. The vitellogenin cDNA of *Cherax quadricarinatus* encodes a lipoprotein with calcium binding ability, and its expression is induced following the removal of the androgenic gland in a sexually plastic system. *Gen. Comp. Endocrinol.* 127 (3):263–272.

Accogli, G., S. Zizza, A. Garcia-Lopez, C. Sarasquete, and S. Desantis. 2012. Lectin-binding pattern of the senegalese sole *Solea senegalensis* oogenesis. *Microsc. Res. Tech.* 75 (8):1124–1135.

Adelson, D. L., M. C. Alliegro, and D. R. McClay. 1992. On the ultrastructure of hyalin, a cell adhesion protein of the sea urchin embryo extracellular matrix. *J. Cell. Biol.* 116 (5):1283–1289.

Alves, A. P., B. Mulloy, J. A. Diniz, and P. A. Mourao. 1997. Sulfated polysaccharides from the egg jelly layer are species-specific inducers of acrosomal reaction in sperms of sea urchins. *J. Biol. Chem.* 272 (11):6965–6971.

Alves, A. P., B. Mulloy, G. W. Moy, V. D. Vacquier, and P. A. Mourao. 1998. Females of the sea urchin *Strongylocentrotus purpuratus* differ in the structures of their egg jelly sulfated fucans. *Glycobiology* 8 (9):939–946.

Anakwe, O. O. and G. L. Gerton. 1990. Acrosome biogenesis begins during meiosis: Evidence from the synthesis and distribution of an acrosomal glycoprotein, acrogranin, during guinea pig spermatogenesis. *Biol. Reprod.* 42 (2):317–328.

Baginski, T., N. Hirohashi, and M. Hoshi. 1999. Sulfated O-linked glycans of the vitelline coat as ligands in gamete interaction in the ascidian, *Halocynthia roretzi. Dev. Growth Differ.* 41 (3):357–364.

Barros, C., E. Dupré, and L. Viveros. 1986. Sperm-egg interactions in the shrimp *Rhynchocinetes typus. Gamete Res.* 14 (2):171–180.

Bustamante, E., J. Palomino, A. Amoroso, R. D. Moreno, and C. Barros. 2001. Purification and biochemical characterization of a trypsin-like enzyme present in the sperm of the rock shrimp, *Rhynchocinetes typus. Invertebr. Reprod. Dev.* 39 (3):175–181.

Caldwell, G. S. and H. E. Pagett. 2010. Marine glycobiology: Current status and future perspectives. *Mar. Biotechnol.* 12 (3):241–252.

De Santis, R. and M. R. Pinto. 1990. Gamete interaction in ascidians: Sperm binding and penetration through the vitelline coat. In *Mechanism of Fertilization*, B. Dale (ed.). Berlin, Germany: Springer-Verlag.

Dell, A., H. R. Morris, R. L. Easton, M. Patankar, and G. F Clark. 1999. The glycobiology of gametes and fertilisation. *Biochim. Biophy. Acta Gen. Subj.* 1473 (1):196–205.

Desantis, S., M. Labate, F. Cirillo, and G. M. Labate. 2003. Testicular activity and sperm glycoproteins in giant red shrimp (*Aristaeomorpha foliacea*). *J. Northwest. Atl. Fish. Sci.* 31:205–212.

Desantis, S., M. Labate, P. Maiorano, and A. Tursi. 2006. An ultrastructural and histochemical study of the germinal cells contained in hemispermatophores of males of the *Aristaeomorpha foliacea* (Risso, 1827). *Hydrobiologia* 557:41–49.

Desantis, S., S. Zizza, A. Garcia-Lopez, V. Sciscioli, E. Mananos, V. G. De Metrio, and C. Sarasquete. 2010. Lectin-binding pattern of senegalese sole *Solea senegalensis* (Kaup) testis. *Histol. Histopathol.* 25 (2):205–216.

Dhume, S. T. and W. J. Lennarz. 1995. The involvement of O-linked oligosaccharide chains of the sea urchin egg receptor for sperm in fertilization. *Glycobiology* 5 (1):11–17.

Dumont, J. N. and A. R. Brummett. 1985. Egg envelopes in vertebrates. *Dev. Biol.* 1:235–288.

Dupre, E. M. and C. Barros. 2011. In vitro fertilization of the rock shrimp, *Rhynchocinetes typus* (Decapoda, Caridea): A review. *Biol. Res.* 44 (2):125–133.

Dupré, E., D. Gòmez, A. Araya, and C. Gallardo. 2012. Role of egg surface glycoconjugate in the fertilization of the rock shrimp *Rhynchocinetes typus* (Milne-Edwards, 1837). *Lat. Am. J. Aquat. Res.* 40 (1):22–29.

Eddy, F. B. 1974. Osmotic properties of the perivitelline fluid and some properties of the chorion of Atlantic salmon eggs (*Salmo salar*). *J. Zool.* 174 (2):237–243.

Endo, T., M. Hoshi, S. Endo, Y. Arata, and A. Kobata. 1987. Structures of the sugar chains of a major glycoprotein present in the egg jelly coat of a starfish, *Asterias amurensis. Arch. Biochem. Biophys.* 252 (1):105–112.

Focarelli, R., D. Rosa, and F. Rosati. 1990. Differentiation of the vitelline coat and the polarized site of sperm entrance in the egg of *Unio elongatulus* (Mollusca, Bivalvia). *J. Exp. Zool.* 254 (1):88–96.

Foltz, K. R. and F. M. Shilling. 1993. Receptor-mediated signal transduction and egg activation. *Zygote* 1 (4):276–279.

Fukumoto, M. 1988. Fertilization in ascidians apical processes and gamete fusion in *Ciona intestinalis* spermatozoa. *J. Cell Sci.* 89 (2):189–196.

Fukumoto, M. 1990. Morphological aspects of ascidian fertilization: Acrosome reaction, apical processes and gamete fusion in *Ciona intestinalis. Invertebr. Reprod. Dev.* 17 (2):147–154.

Fukumoto, M. and T. Numakunai. 1995. Morphological aspects of fertilization in *Halocynthia roretzi* (Ascidiacea, Tunicata). *J. Struct. Biol.* 114 (3):157–166.

Gallo, A. and M. Costantini. 2012. Glycobiology of reproductive processes in marine animals: The state of the art. *Mar. Drugs* 10 (12):2861–2892.

Glabe, C. G., L. B. Grabel, V. D. Vacquier, and S. D. Rosen. 1982. Carbohydrate specificity of sea urchin sperm bindin: A cell surface lectin mediating sperm-egg adhesion. *J. Cell. Biol.* 94 (1):123–128.

Godknecht, A. and T. G. Honegger. 1991. Isolation, characterization, and localization of a sperm-bound N-acetylglucosaminidase that is indispensable for fertilization in the ascidian, *Phallusia mammillata. Dev. Biol.* 143 (2):398–407.

Gunaratne, H. M., T. Yamagaki, M. Matsumoto, and M. Hoshi. 2003. Biochemical characterization of inner sugar chains of acrosome reaction-inducing substance in jelly coat of starfish eggs. *Glycobiology* 13 (8):567–580.

Hagstrom, B. E. 1956. Further studies on cross fertilization in sea urchins. *Exp. Cell Res.* 11 (2):507–510.

Harrington, F. E. and H. Ozaki. 1986. The major yolk glycoprotein precursor in echinoids is secreted by coelomocytes into the coelomic plasma. *Cell Differ.* 19 (1):51–57.

Hirohashi, N. and W. J. Lennarz. 1998. Sperm-egg binding in the sea urchin: A high level of intracellular ATP stabilizes sperm attachment to the egg receptor. *Dev. Biol.* 201 (2):270–279.

Hirohashi, N. and W. J. Lennarz. 2001. Role of a vitelline layer-associated 350 kDa glycoprotein in controlling species-specific gamete interaction in the sea urchin. *Dev. Growth Differ.* 43 (3):247–255.

Hirohashi, N., N. Kamei, H. Kubo, H. Sawada, M. Matsumoto, and M. Hoshi. 2008. Egg and sperm recognition systems during fertilization. *Dev. Growth Differ.* 50 Suppl 1:S221–S238.

Honegger, T. G. and R. Koyanagi. 2008. The ascidian egg envelope in fertilization: Structural and molecular features. *Int. J. Dev. Biol.* 52 (5–6):527–533.

Hoshi, M. 1986. Sperm glycosidase as a plausible mediator of sperm binding to the vitelline envelope in ascidians. *Adv. Exp. Med. Biol.* 207:251–260.

Hoshi, M., T. Amano, Y. Okita, T. Okinaga, and T. Matsui. 1990. Egg signals for triggering the acrosome reaction in starfish spermatozoa. *J. Reprod. Fertil. Suppl.* 42:23–31.

Hoshi, M., R. De Santis, M. R. Pinto, F. Cotelli, and F. Rosati. 1985. Sperm glycosidases as mediators of sperm-egg binding in the ascidians. *Zool. Sci.* 2:65–69.

Hoshi, M., T. Matsui, I. Nishiyama, T. Amano, and Y. Okita. 1988. Physiological inducers of the acrosome reaction. *Cell Differ. Dev.* 25 Suppl:19–24.

Hoshi, M., H. Moriyama, and M. Matsumoto. 2012. Structure of acrosome reaction-inducing substance in the jelly coat of starfish eggs: A mini review. *Biochem. Biophys. Res. Commun.* 425 (3):595–598.

Hoshi, M., T. Nishigaki, A. Ushiyama, T. Okinaga, K. Chiba, and M. Matsumoto. 1994. Egg-jelly signal molecules for triggering the acrosome reaction in starfish spermatozoa. *Int. J. Dev. Biol.* 38 (2):167–174.

Ijuin, T., K. Kitajima, Y. Song, S. Kitazume, S. Inoue, S. M. Haslam, H. R. Morris, A. Dell, and Y. Inoue. 1996. Isolation and identification of novel sulfated and non-sulfated oligosialyl glycosphingolipids from sea urchin sperm. *Glycoconj. J.* 13 (3):401–413.

Inagaki, F., S. Tate, H. Kubo, and M. Hoshi. 1992. A novel difucosylated neutral glycosphingolipid from the eggs of the sea urchin, Hemicentrotus pulcherrimus: II. Structural determination by two-dimensional NMR. *J. Biochem.* 112 (2):286–289.

Inoue, S. and Y. Inoue. 1986. Fertilization (activation)-induced 200-to 9-kDa depolymerization of polysialoglycoprotein, a distinct component of cortical alveoli of rainbow trout eggs. *J. Biol. Chem.* 261:5256–5261.

Inoue, S., K. Kitajima, Y. Inoue, and S. Kudo. 1987. Localization of polysialoglycoprotein as a major glycoprotein component in cortical alveoli of the unfertilized eggs of Salmo gairdneri. *Dev. Biol.* 123 (2):442–454.

Jones, R., C. R. Brown, and R. T. Lancaster. 1988. Carbohydrate-binding properties of boar sperm proacrosin and assessment of its role in sperm-egg recognition and adhesion during fertilization. *Dev. Genes Evol.* 102:781–779.

Kambara, Y., K. Shiba, M. Yoshida, C. Sato, K. Kitajima, and C. Shingyoji. 2011. Mechanism regulating Ca2+-dependent mechanosensory behaviour in sea urchin spermatozoa. *Cell Struct. Funct.* 36 (1):69–82.

Khalaila, I., J. Peter-Katalinic, C. Tsang, C. M. Radcliffe, E. D. Aflalo, D. J. Harvey, R. A. Dwek, P. M. Rudd, and A. Sagi. 2004. Structural characterization of the N-glycan moiety and site of glycosylation in vitellogenin from the decapod crustacean Cherax quadricarinatus. *Glycobiology* 14 (9):767–774.

Kitazume, S., K. Kitajima, S. Inoue, F. A. Troy, II, J. W. Cho, W. J. Lennarz, and Y. Inoue. 1994. Identification of polysialic acid-containing glycoprotein in the jelly coat of sea urchin eggs. Occurrence of a novel type of polysialic acid structure. *J. Biol. Chem.* 269 (36):22712–22718.

Kobayashi, W. and T. S. Yamamoto. 1993. Factors inducing closure of the micropylar canal in the chum salmon egg. *J. Fish Biol.* 42 (3):385–394.

Koyanagi, R. and T. G. Honegger. 2003. Molecular cloning and sequence analysis of an ascidian egg beta-N-acetylhexosaminidase with a potential role in fertilization. *Dev. Growth Differ.* 45 (3):209–218.

Koyota, S., K. M. Wimalasiri, and M. Hoshi. 1997. Structure of the main saccharide chain in the acrosome reaction-inducing substance of the starfish, Asterias amurensis. *J. Biol. Chem.* 272 (16):10372–10376.

Kresge, N., V. D. Vacquier, and C. D. Stout. 2001. Abalone lysin: The dissolving and evolving sperm protein. *Bioessays* 23 (1):95–103.

Kubo, H. and M. Hoshi. 1990. Immunocytochemical study of the distribution of a ganglioside in sea urchin eggs. *J. Biochem.* 108 (2):193–199.

Kubo, H., A. Irie, F. Inagaki, and M. Hoshi. 1988. Melibiosylceramide as the sole ceramide dihexoside from the eggs of the sea urchin, Anthocidaris crassispina. *J. Biochem.* 104 (5):755–760.

Kubo, H., A. Irie, F. Inagaki, and M. Hoshi. 1990. Gangliosides from the eggs of the sea urchin, Anthocidaris crassispina. *J. Biochem.* 108 (2):185–192.

Kubo, H., G. J. Jiang, A. Irie, M. Morita, T. Matsubara, and M. Hoshi. 1992a. A novel ceramide trihexoside from the eggs of the sea urchin, Hemicentrotus pulcherrimus. *J. Biochem.* 111 (6):726–731.

Kubo, H., G. J. Jiang, A. Irie, M. Suzuki, F. Inagaki, and M. Hoshi. 1992b. A novel difucosylated neutral glycosphingolipid from the eggs of the sea urchin, Hemicentrotus pulcherrimus: I. Purification and structural determination of the glycolipid. *J. Biochem.* 112 (2):281–285.

Kudo, S. 1991. Fertilization, cortical reaction, polyspermypreventing and anti-microbial mechanisms in fish eggs. *Bull. Inst. Zool. Acad. Sin. Monogr.* 16:313–340.

Kudo, S. and C. Teshima. 1998. Assembly in vitro of vitelline envelope components induced by a cortical alveolus sialoglycoprotein of eggs of the fish Tribolodon hakonensis. *Zygote* 6 (3):193–201.

Laale, H. W. 1980. The perivitelline space and egg envelopes of bony fishes: A review. *Copeia* 1980 (2):210–226.

Lambert, C. C. 1986. Fertilization-induced modification of chorion N-acetylglucosamine groups blocks polyspermy in ascidian eggs. *Dev. Biol.* 116 (1):168–173.

Lambert, C. C. 1989. Ascidian eggs release glycosidase activity which aids in the block against polyspermy. *Development* 105 (2):415–420.

Lambert, C. C., T. Someno, and H. Sawada. 2002. Sperm surface proteases in ascidian fertilization. *J. Exp. Zool.* 292 (1):88–95.

Li, X., E. Jenssen, and H. J. Fyhn. 1989. Effects of salinity on egg swelling in Atlantic salmon (Salmo salar). *Aquaculture* 76:317–334.

Liguoro, A., M. Prisco, C. Mennella, L. Ricchiari, F. Angelini, and P. Andreuccetti. 2004. Distribution of terminal sugar residues in the testis of the spotted ray Torpedo marmorata. *Mol. Reprod. Dev.* 68 (4):524–530.

Litscher, E. and T. G. Honegger. 1991. Glycoprotein constituents of the vitelline coat of Phallusia mammillata (Ascidiacea) with fertilization inhibiting activity. *Dev. Biol.* 148 (2):536–551.

Longo, F. J., A. Ushiyama, K. Chiba, and M. Hoshi. 1995. Ultrastructural localization of acrosome reaction-inducing substance (ARIS) on sperm of the starfish *Asterias amurensis*. *Mol. Reprod. Dev.* 41 (1):91–99.

Mandich, A., A. Massari, S. Bottero, and G. Marino. 2002. Histological and histochemical study of female germ cell development in the dusky grouper *Epinephelus marginatus* (Lowe, 1834). *Eur. J. Histochem.* 46 (1):87–100.

Marsh, A. G. and S. A. Watts. 2001. Energy metabolism and gonad development. *Dev. Aquac. Fish. Sci.* 32:27–42.

Martinez-Menarguez, J. A., M. Aviles, J. F. Madrid, M. T. Castells, and J. Ballesta. 1993. Glycosylation in Golgi apparatus of early spermatids of rat. A high resolution lectin cytochemical study. *Eur. J. Cell Biol.* 61 (1):21–33.

Martinez-Menarguez, J. A., J. Ballesta, M. Aviles, M. T. Castells, and J. F. Madrid. 1992. Cytochemical characterization of glycoproteins in the developing acrosome of rats. An ultrastructural study using lectin histochemistry, enzymes and chemical deglycosylation. *Histochemistry* 97 (5):439–449.

Matozzo, V., F. Gagne, M. G. Marin, F. Ricciardi, and C. Blaise. 2008. Vitellogenin as a biomarker of exposure to estrogenic compounds in aquatic invertebrates: A review. *Environ. Int.* 34 (4):531–545.

Matsumoto, M., J. Hirata, N. Hirohashi, and M. Hoshi. 2002. Sperm-egg binding mediated by sperm alpha-L-fucosidase in the ascidian, *Halocynthia roretzi*. *Zoolog. Sci.* 19 (1):43–48.

Matsumoto, M., O. Kawase, M. S. Islam, M. Naruse, S. N. Watanabe, R. Ishikawa, and M. Hoshi. 2008. Regulation of the starfish sperm acrosome reaction by cGMP, pH, cAMP and Ca2+. *Int. J. Dev. Biol.* 52 (5–6):523–526.

Mengerink, K. J. and V. D. Vacquier. 2001. Glycobiology of sperm-egg interactions in deuterostomes. *Glycobiology* 11 (4):37R–43R.

Miyata, S., C. Sato, S. Kitamura, M. Toriyama, and K. Kitajima. 2004. A major flagellum sialoglycoprotein in sea urchin sperm contains a novel polysialic acid, an alpha2,9-linked poly-N-acetylneuraminic acid chain, capped by an 8-O-sulfated sialic acid residue. *Glycobiology* 14 (9):827–840.

Miyata, S., C. Sato, H. Kumita, M. Toriyama, V. D. Vacquier, and K. Kitajima. 2006. Flagellasialin: A novel sulfated alpha2,9-linked polysialic acid glycoprotein of sea urchin sperm flagella. *Glycobiology* 16 (12):1229–1241.

Miyata, S., N. Yamakawa, M. Toriyama, C. Sato, and K. Kitajima. 2011. Co-expression of two distinct polysialic acids, alpha2,8- and alpha2,9-linked polymers of N-acetylneuraminic acid, in distinct glycoproteins and glycolipids in sea urchin sperm. *Glycobiology* 21 (12):1596–1605.

Moy, G. W., L. M. Mendoza, J. R. Schulz, W. J. Swanson, C. G. Glabe, and V. D. Vacquier. 1996. The sea urchin sperm receptor for egg jelly is a modular protein with extensive homology to the human polycystic kidney disease protein, PKD1. *J. Cell Biol.* 133 (4):809–817.

Mulloy, B., A. C. Ribeiro, A. P. Alves, R. P. Vieira, and P. A. Mourao. 1994. Sulfated fucans from echinoderms have a regular tetrasaccharide repeating unit defined by specific patterns of sulfation at the 0–2 and 0–4 positions. *J. Biol. Chem.* 269 (35):22113–22123.

Nagahama, Y., M. Yoshikuni, M. Yamashita, T. Tokumoto, and Y. Katsu. 1995. Regulation of oocyte growth and maturation in fish. *Curr. Top. Dev. Biol.* 30:103–146.

Naruse, M., H. Suetomo, T. Matsubara, T. Sato, H. Yanagawa, M. Hoshi, and M. Matsumoto. 2010. Acrosome reaction-related steroidal saponin, Co-ARIS, from the starfish induces structural changes in microdomains. *Dev. Biol.* 347 (1):147–153.

Nishigaki, T., K. Chiba, and M. Hoshi. 2000. A 130-kDa membrane protein of sperm flagella is the receptor for asterosaps, sperm-activating peptides of starfish *Asterias amurensis*. *Dev. Biol.* 219 (1):154–162.

Nishigaki, T., K. Chiba, W. Miki, and M. Hoshi. 1996. Structure and function of asterosaps, sperm-activating peptides from the jelly coat of starfish eggs. *Zygote* 4 (3):237–245.

Nishiyama, I., T. Matsui, Y. Fujimoto, N. Ikekawa, and M. Hoshi. 1987. Correlation between the molecular structure and the biological activity of Co-ARIS, a cofactor for acrosome reaction-inducing substance. *Dev. Growth Differ.* 29 (2):171–176.

Nobrega, R. H., S. R. Batlouni, and L. R. Franca. 2009. An overview of functional and stereological evaluation of spermatogenesis and germ cell transplantation in fish. *Fish Physiol. Biochem.* 35 (1):197–206.

Nomura, K. and N. Suzuki. 1995. Sea urchin ovoperoxidase: Solubilization and isolation from the fertilization envelope, some structural and functional properties, and degradation by hatching enzyme. *Arch. Biochem. Biophys.* 319 (2):525–534.

Ohlendieck, K., J. S. Partin, R. L. Stears, and W. J. Lennarz. 1994. Developmental expression of the sea urchin egg receptor for sperm. *Dev. Biol.* 165 (1):53–62.

Oppen-Berntsen, D. O., J. V. Helvik, and B. T. Walther. 1990. The major structural proteins of cod (*Gadus morhua*) eggshells and protein crosslinking during teleost egg hardening. *Dev. Biol.* 137 (2):258–265.

Ortiz-Delgado, J. B., S. Porcelloni, C. Fossi, and C. Sarasquete. 2008. Histochemical characterisation of oocytes of the swordfish *Xiphias gladius*. *Sci. Mar.* 72 (3):549–564.

Ozaki, H., O. Moriya, and F. E. Harrington. 1986. A glycoprotein in the accessory cell of the echinoid ovary and its role in vitellogenesis. *Dev. Genes Evol.* 195 (1):74–79.

Peterson, R. H. and D. J. Martin-Robichaud. 1993. Rates of ionic diffusion across the egg chorion of Atlantic salmon (*Salmo salar*). *Physiol. Zool.* 66 (3):289–306.

Pillai, M. C. and W. H. Clark, Jr. 1990. Development of cortical vesicles in *Sicyonia ingentis* ova: Their heterogeneity and role in elaboration of the hatching envelope. *Mol. Reprod. Dev.* 26 (1):78–89.

Pratt, S. A., N. F. Scully, and B. D. Shur. 1993. Cell surface beta 1,4 galactosyltransferase on primary spermatocytes facilitates their initial adhesion to Sertoli cells in vitro. *Biol. Reprod.* 49 (3):470–482.

Raychoudhury, S. S. and C. F. Millette. 1997. Multiple fucosyltransferases and their carbohydrate ligands are involved in spermatogenic cell-Sertoli cell adhesion in vitro in rats. *Biol. Reprod.* 56 (5):1268–1273.

Rios, M. and C. Barros. 1997. Trypsin-like enzymes during fertilization in the shrimp *Rhynchocinetes typus*. *Mol. Reprod. Dev.* 46 (4):581–586.

Rosati, F., F. Cotelli, R. De Santis, A. Monroy, and M. R. Pinto. 1982. Synthesis of fucosyl-containing glycoproteins of the vitelline coat in oocytes of *Ciona intestinalis* (Ascidia). *Proc. Natl. Acad. Sci. U. S. A.* 79 (6):1908–1911.

Rudy, P. P., Jr. and W. T. Potts. 1969. Sodium balance in the eggs of the Atlantic salmon, *Salmo salar*. *J. Exp. Biol.* 50 (1):239–246.

Sarasquete, C., S. Cardenas, C. M. de Gonzalez, and E. Pascual. 2002. Oogenesis in the bluefin tuna, *Thunnus thynnus* L.: A histological and histochemical study. *Histol. Histopathol.* 17 (3):775–788.

Sasikala, S. L. and T. Subramoniam. 1987. On the occurrence of acid mucopolysaccharides in the spermatophores of two marine prawns, *Penaeus indicus* (Milne-Edwards) and *Metapenaeus monoceros* (Fabricius) (Crustacea: Macrura). *J. Exp. Marine. Biol.* 113:145–153.

Sawada, H., N. Sakai, Y. Abe, E. Tanaka, Y. Takahashi, J. Fujino, E. Kodama, S. Takizawa, and H. Yokosawa. 2002. Extracellular ubiquitination and proteasome-mediated degradation of the ascidian sperm receptor. *Proc. Natl. Acad. Sci. U. S. A.* 99 (3):1223–1228.

Sawada, H., H. Yokosawa, and S. Ishii. 1984. Purification and characterization of two types of trypsin-like enzymes from sperm of the ascidian (Prochordata) *Halocynthia roretzi*. Evidence for the presence of spermosin, a novel acrosin-like enzyme. *J. Biol. Chem.* 259 (5):2900–2904.

Schulz, R. W. and T. Miura. 2002. Spermatogenesis and its endocrine regulation. *Fish Physiol. Biochem.* 26 (1):43–56.

SeGall, G. K. and W. J. Lennarz. 1979. Chemical characterization of the component of the jelly coat from sea urchin eggs responsible for induction of the acrosome reaction. *Dev. Biol.* 71 (1):33–48.

Shabanipour, N. and B. Heidari. 2004. A histological study of the zona radiate during late oocyte developmental stages in the Caspian sea mugilid, *Liza aurata* (Risso 1810). *Braz. J. Morphol. Sci.* 21:191–195.

Shapiro, B. M., C. Somers, and P. J. Weidman. 1989. Extracellular remodeling during fertilization. In *Cell Biology of Fertilization*, H. Schatten and G. Schatten (eds.). San Diego, CA: Academic Press.

Shimizu, T., H. Kinoh, M. Yamaguchi, and N. Suzuki. 1990. Purification and characterization of the egg jelly macromolecules, sialoglycoprotein and fucose sulfate glycoconjugate, of the sea urchin *Hemicentrotus pulcherrimus*. *Dev. Growth Differ.* 32 (5):473–487.

Shogomori, H., K. Chiba, and M. Hoshi. 1997. Association of the major ganglioside in sea urchin eggs with yolk lipoproteins. *Glycobiology* 7 (3):391–398.

Shogomori, H., K. Chiba, H. Kubo, and M. Hoshi. 1993. Nonplasmalemmal localisation of the major ganglioside in sea urchin eggs. *Zygote* 1 (3):215–223.

Shyu, A. B., R. A. Raff, and T. Blumenthal. 1986. Expression of the vitellogenin gene in female and male sea urchin. *Proc. Natl. Acad. Sci. U. S. A.* 83 (11):3865–3869.

Somers, C. E. and B. M. Shapiro. 1991. Functional domains of proteoliaisin, the adhesive protein that orchestrates fertilization envelope assembly. *J. Biol. Chem.* 266 (25):16870–16875.

Subramonian, T. 1991. Chemical composition of spermatophores in decapod crustaceans. In *Crustacean Sexual Biology*, R. T. Bauer and J. A. Martin (eds.). New York: Columbia University Press.

Suphamungmee, W., P. Chansela, W. Weerachatyanukul, T. Poomtong, R. Vanichviriyakit, and P. Sobhon. 2010. Ultrastructure, composition, and possible roles of the egg coats in *Haliotis asinina*. *J. Shellfish Res.* 29 (3):687–697.

Swanson, W. J., C. F. Aquadro, and V. D. Vacquier. 2001. Polymorphism in abalone fertilization proteins is consistent with the neutral evolution of the egg's receptor for lysin (VERL) and positive darwinian selection of sperm lysin. *Mol. Biol. Evol.* 18 (3):376–383.

Swanson, W. J. and V. D. Vacquier. 1997. The abalone egg vitelline envelope receptor for sperm lysin is a giant multivalent molecule. *Proc. Natl. Acad. Sci. U. S. A.* 94 (13):6724–6729.

Tirumalai, R. and T. Subramoniam. 2001. Carbohydrate components of lipovitellin of the sand crab *Emerita asiatica*. *Mol. Reprod. Dev.* 58:54–62.

Tokalov, S. V. and H. O. Gutzeit. 2007. Lectin-binding pattern as tool to identify and enrich specific primary testis cells of the tilapia (*Oreochromis niloticus*) and medaka (*Oryzias latipes*). *J. Exp. Zool. B Mol. Dev. Evol.* 308 (2):127–138.

Tosti, E. 1994. Sperm activation in species with external fertilisation. *Zygote* 2 (4):359–361.

Uno, Y. and M. Hoshi. 1978. Separation of the sperm agglutinin and the acrosome reaction-inducing substance in egg jelly of starfish. *Science* 200 (4337):58–59.

Unuma, T., T. Suzuki, T. Kurokawa, Yamamoto T., and T. Akiyama. 1998. A protein identical to the yolk protein is stored in the testis in male red sea urchin, *Pseudocentrotus depressus*. *Biol. Bull.* 194:92–97.

Unuma, T., T. Yamamoto, T. Akiyama, M. Shiraishi, and H. Ohta. 2003. Quantitative changes in yolk protein and other components in the ovary and testis of the sea urchin *Pseudocentrotus depressus*. *J. Exp. Biol.* 206 (Pt 2):365–372.

Ushiyama, A., T. Araki, K. Chiba, and M. Hoshi. 1993. Specific binding of acrosome-reaction-inducing substance to the head of starfish spermatozoa. *Zygote* 1 (2):121–127.

Vacquier, V. D. 1998. Evolution of gamete recognition proteins. *Science* 281 (5385):1995–1998.

Vacquier, V. D., K. R. Carner, and C. D. Stout. 1990. Species-specific sequences of abalone lysin, the sperm protein that creates a hole in the egg envelope. *Proc. Natl. Acad. Sci. U. S. A.* 87 (15):5792–5796.

Vacquier, V. D. and G. W. Moy. 1997. The fucose sulfate polymer of egg jelly binds to sperm REJ and is the inducer of the sea urchin sperm acrosome reaction. *Dev. Biol.* 192 (1):125–135.

Vacquier, V. D., W. J. Swanson, and M. E. Hellberg. 1995. What have we learned about sea urchin sperm bindin? *Dev. Growth Differ.* 37 (1):1–10.

Varki, A., R. Cummings, J. Esko, H. Freeze, G. Hart, and J. Marth. 1999. Nuclear and cytoplasmic glycosylation. In *Essentials of Glycobiology*, A. Varki, R. Cummings, J. Esko, H. Freeze, G. Hart, and J. Marth (eds.). Cold Spring Harbor, NY: Cold Spring Harbor Laboratory Press.

Verhey, C. A. and F. H. Moyer. 1967. Fine structural changes during sea urchin oogenesis. *J. Exp. Zool.* 162 (2):195–225.

Vilela-Silva, A. C., M. O. Castro, A. P. Valente, C. H. Biermann, and P. A. Mourao. 2002. Sulfated fucans from the egg jellies of the closely related sea urchins *Strongylocentrotus droebachiensis* and *Strongylocentrotus pallidus* ensure species-specific fertilization. *J. Biol. Chem.* 277 (1):379–387.

Wallace, R. A. 1985. Vitellogenesis and oocyte growth in nonmammalian vertebrates. In *Oogenesis*, L. W. Browder (ed.). Berlin, Germany: Springer.

Wessel, G. M., J. M. Brooks, E. Green, S. Haley, E. Voronina, J. Wong, V. Zaydfudim, and S. Conner. 2001. The biology of cortical granules. *Int. Rev. Cytol.* 209:117–206.

Wikramanayake, A. H. and W. H. Clark. 1994. Two extracellular matrices from oocytes of the marine shrimp *Sicyonia ingentis* that independently mediate only primary or secondary sperm binding. *Dev. Growth Differ.* 36 (1):89–101.

Xie, M. and T. G. Honegger. 1993. Ultrastructural investigations on sperm penetration and gamete fusion in the ascidians *Boltenia villosa* and *Phallusia mammillata. Mar. Biol.* 116 (1):117–127.

Yanagimachi, R. 1994. Fertility of mammalian spermatozoa: Its development and relativity. *Zygote* 2 (4):371–372.

*chapter four*

# Heparin from marine mollusks
## Occurrence, structure, and biological role

*Nicola Volpi and Francesca Maccari*

## Contents

## 4.1   Introduction

Heart and vascular diseases, also including thrombosis, are the leading causes of death in the United States and Europe (Arias and Smith 2003). After the introduction of antithrombotic agents, particularly heparin and its derivatives, deadly heart diseases have decreased substantially (about 30%) when compared to malignant cancer, even if they are still the main cause of death (Arias and Smith 2003). This explains the efforts to discover and develop specific and more potent antithrombotic agents.

Commercial manufacture of heparin relies on either porcine or bovine intestinal or bovine lung tissue as raw material. The apparent link between bovine spongiform encephalopathy and the similar prion-based Creutzfeldt–Jakob disease in humans (Schonberger 1998) has limited the use of bovine heparin. Moreover, it is not easy to distinguish bovine and porcine heparins, thus making it difficult to ensure the species source of heparin (Linhardt and Gunay 1999). Furthermore, porcine heparin also has problems associated with religious restrictions on its use. Additionally, in 2007–2008, there was a heparin crisis resulting from contaminated batches of heparin (and low-molecular-weight derivatives) entering the marketplace (Liu et al. 2009), causing severe side effects and some leading to death. The contamination was traced to an adulteration of the crude heparin precursor during the process between the slaughterhouse where heparin was collected from pig intestines and the pharmaceutical manufacturing site. Finally, nonanimal sources of heparin, such as chemically synthesized, enzymatically synthesized, or recombinant heparins, are currently not available for pharmaceutical purposes (Lord and Whitelock 2014). These concerns have motivated to look for alternative, nonmammalian sources of heparin.

The presence of sulfated glycosaminoglycans (GAGs) in some taxa of invertebrates is now well documented (Cassaro and Dietrich 1977, Hovingh and Linker 1982, Nader et al. 1984, Dietrich et al. 1985, Jordan and Marcum 1986, Pejler et al. 1987, Chavante et al. 2000, Medeiros et al. 2000, Cesaretti et al. 2004, Luppi et al. 2005). A comprehensive survey of different classes of invertebrates has shown that heparan sulfate (HS)-like and/or heparin-like compounds, besides chondroitin sulfate (CS), are present in many species (Medeiros et al. 2000). Previous studies have also shown that heparin is present in several species of marine mollusks. A compound from the clam *Mercenaria mercenaria* (Jordan and Marcum 1986) exhibits several structural similarities to heparin. Heparins with high anticoagulant activity have been isolated from the mollusks *Anomalocardia brasiliana* (Dietrich et al. 1985, Pejler et al. 1987), *Tivela mactroides* (Pejler et al. 1987), *Tapes phylippinarum* (Cesaretti et al. 2004), and *Callista chione* (Luppi et al. 2005). Due to our knowledges of the sulfated polysaccharides (SPSs) in vertebrates and invertebrates, it is now possible to draw a phylogenetic tree of the distribution of sulfated GAGs in the animal kingdom.

## 4.2 Marine sulfated GAG-like molecules and their potential applications

Marine organisms are found to be very rich sources of food, feed, medicines, and energy. They have also proven to be rich sources of structurally diverse bioactive compounds with valuable pharmaceutical and biomedical potential. Marine organisms (invertebrates, vertebrates, algae, and bacteria) are sources for industrially important GAGs, complex polysaccharides, and phycocolloids like agar, carrageenan, and alginate (Cassaro and Dietrich 1977, Nader et al. 2004). Apart from industrial uses, in recent years, polysaccharides of marine origin have emerged as important classes of bioactive natural products. They have been reported to have anticoagulant, antitumor, antimutagenic, anticomplementary, immunomodulating, hypoglycemic, antiviral, hypolipidemic, and anti-inflammatory activities (Senni et al. 2011).

SPSs are a class of complex heterogeneous macromolecules containing sulfate groups in their carbohydrate backbone in various numbers and different positions. They are commonly found in marine invertebrates and algae as such as in marine vertebrates and higher animals, scarcely present in microbes, and absent in higher plants. The most studied for their biological properties are mammalian SPSs or glycoconjugates constituted by GAGs composed of negatively charged chains, most of which are covalently linked to proteins. The discovery of the biological importance of the mammalian glycoconjugates has been the beginning of a new modern research field focusing on the complex carbohydrate-based recognition phenomena, glycobiology (Shriver et al. 2004). It has been demonstrated that these particular biological properties are due to the chemical diversity of the polysaccharide chains in which the patterns of sulfate substitution can give specific biological functions. It was also noted that the chemical diversity of the SPSs is largely species specific.

Apart from the well-known presence in many marine vertebrates, the presence of SPS in some taxa of invertebrates is now well documented and they can be found in marine invertebrates. They have been isolated from marine mollusks (Volpi and Maccari 2003, 2005, 2007, 2009, Cesaretti et al. 2004, Luppi et al. 2005) or echinoderms such as sea urchins and sea cucumbers (ascidians) (Tapon-Bretaudiere et al. 2002). A comprehensive survey of different classes of invertebrates has shown that GAGs and GAG-like compounds are present in many species (Cassaro and Dietrich 1977, Nader et al. 2004).

SPSs (and GAGs), produced by the extraction and purification of different animal (and algae) tissues, have several fundamental biological activities and pharmacological properties, making them important drugs for use in clinical and pharmaceutical fields. In fact, SPSs have diverse functions in the tissues from which they originate. They are capable of binding with proteins at several levels of specificity and are involved mainly in the development, cell differentiation, cell adhesion, cell signaling, and cell–matrix interactions. These bioactive molecules present a great potential for medical, pharmaceutical, and biotechnological applications such as wound dressings, biomaterials, tissue regeneration, 3D culture scaffolds, and even drugs. Additionally, GAGs interact with a wide range of proteins involved in physiological and pathological processes showing many biological activities that can influence tissue repair and inflammatory response (Ernst and Magnani 2008).

## 4.3 Structure of heparin

Heparin is a linear natural polysaccharide consisting of 1→4 linked pyranosyluronic acid (uronic acid) and 2-amino-2-deoxyglucopyranose (D-glucosamine, GlcN) repeating units (Linhardt 2003). The uronic acid usually comprises 80%–90% L-idopyranosyluronic acid (L-iduronic acid, IdoA) (Figure 4.1a) and 10%–20% D-glucopyranosyluronic acid (D-glucuronic acid, GlcA). Heparin, with its high content of sulfo and carboxyl groups, is a polyelectrolyte, having the highest negative charge density of any known biological macromolecule. At the disaccharide level, a number of structural variations exist (Figure 4.1b), leading to sequence microheterogeneity within heparin.

Heparin is polydisperse with a molecular mass (MM) range of 5–40, an average MM of ~12 kDa, and an average negative charge of about −75, making it an extremely challenging molecule to characterize.

Heparin's complexity extends through multiple structural levels. At the proteoglycan level, different numbers of GAG chains (possibly having different

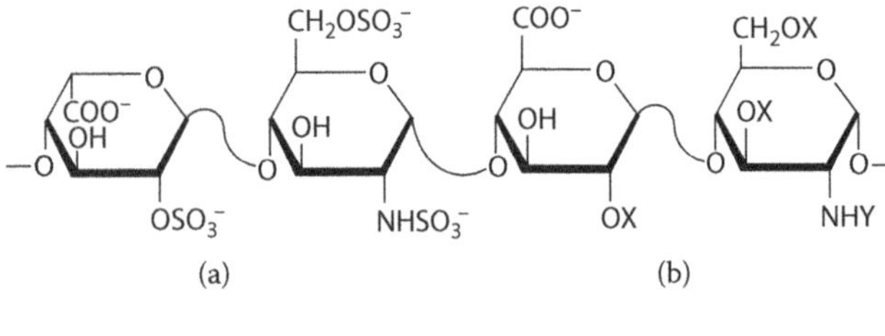

*Figure 4.1* Structure of heparin: (a) major trisulfated disaccharide repeating unit (X = sulfo or H, Y = sulfo, Ac, or H); (b) undersulfated structural variants. (Reprinted from Linhardt, R.J., *J. Med. Chem.*, 46, 2551, 2003. With permission.)

saccharide sequences) can be attached to the various serine residues present on the core protein. Heparin chains are biosynthesized and attached to a unique core protein, serglycin, found primarily in mast cells. On mast cell degranulation, proteases act on the heparin core protein to release peptidoglycan heparin, which is further processed by a β-endoglucuronidase into GAG heparin (Linhardt 2003). The chemical, physical, and biological properties of heparin are primarily ascribed to GAG structure (or sequence), saccharide conformation, chain flexibility, molecular weight, and charge density.

## 4.4 Interaction of heparin with proteins

With the discovery of increasing numbers of heparin-binding proteins (Capila and Linhardt 2002), there was a need to characterize the molecular properties, within the proteins and heparin, responsible for specific recognition (Table 4.1). The biological activities of heparin and HS primarily result from their interaction with hundreds of different proteins. By using modeling, Cardin and Weintraub (1989) demonstrated that some heparin-binding proteins had defined motifs corresponding to consensus sequences, giving the first evidence for the general structural requirements for GAG–protein interactions.

Their results suggested that if the XBBBXXBX (B is a basic and X is a hydropathic amino acid residue) sequence was contained in an α-helical domain, then the basic amino acids would be displayed on one side of the helix with the hydropathic residues pointing back into the protein core. Heparin-binding sites, commonly observed on the external surface of proteins, correspond to shallow pockets of positive charge. Thus, the topology of the heparin binding site is also an important factor in heparin-binding consensus sequences. Structural analysis of the heparin-binding sites in acidic fibroblast growth factor (FGF-1), basic FGF (FGF-2), and transforming growth factor β-1 implicated a TXXBXXTBXXXTBB motif (T defines a turn) (Hileman et al. 1998).

By screening peptide libraries, the conservation of amino acids in heparin-binding domains and the importance of spacing between basic amino acids in heparin binding were demonstrated. Peptides enriched in arginine and lysine and polar hydrogen-bonding amino acids were observed to bind heparin with highest affinity (Capila and Linhardt 2002, Gandhi and Mancera 2008). Studies on the role of the pattern and the spacing of the basic amino acids in heparin-binding domains showed that heparin interacted more tightly with peptides containing a complementary binding site of high positive charge density, while the less-sulfated HS interacted

more tightly with a complementary site on a peptide that had more widely spaced basic (Linhardt 2003).

## 4.5 Anticoagulant activity of heparin

The anticoagulant capacity of pharmaceutical heparin (120–180 USP U/mg) is the most thoroughly studied of its activities. Anticoagulation occurs when heparin binds to antithrombin III (ATIII), a serine protease inhibitor (serpin). ATIII undergoes a conformational change and becomes activated as an inhibitor of thrombin and other serine proteases in the coagulation cascade. A major breakthrough in the study of heparin-catalyzed anticoagulation resulted from the separation of distinct heparin fractions differing markedly in affinity for ATIII. Low-affinity ATIII-binding heparin comprises about two-thirds of porcine intestinal heparin and has a low anticoagulant activity (typically <20 U/mg). In contrast, high ATIII affinity heparin comprises the remaining third of porcine intestinal heparin and has a high anticoagulant activity (typically ~300 U/mg). Rosenberg and Lam (1979) and Lindahl et al. (1979) examined the ATIII-binding site by performing a partial chemical and enzymatic depolymerization of heparin. The isolation of 3-O-sulfatase from human urine, capable of desulfonating 3-O-sulfoglucosamine residues, provided the crucial clue to the structure of the ATIII-binding site (Figure 4.2).

NMR studies by several groups proved the presence of the 3-O-sulfo group within the ATIII-binding site, and chemical synthesis of a pentasaccharide containing the 3,6-di-O-sulfo group substantiated these findings (Torri et al. 1985). The ATIII-binding sequences found in certain HSs are partially responsible for the blood compatibility of the vascular endothelium (Marcum and Rosenberg 1987).

The ATIII pentasaccharide is sufficient to catalyze the ATIII-mediated inhibition of factor Xa, a critical serine protease in the coagulation cascade. To catalyze the ATIII-mediated inhibition of thrombin, 16–18 saccharide units are required (Sinay 1999). Thus, the structure–activity relationship of thrombin inhibition has been more difficult to establish because it relied on the synthesis of oligosaccharides substantially larger than ATIII for pharmacological evaluation. Recent studies show that a relatively nonspecific but highly charged thrombin-binding domain in heparin, localized on the nonreducing side of the heparin's ATIII-binding site, is required to form a ternary complex. Success in understanding the structure–activity relationship of heparin's inhibition of thrombin has resulted in a new class of potent, synthetic, but still, experimental thrombin inhibitors.

*Table 4.1* Characteristics of some selected heparin-binding proteins

| Heparin-binding protein | Physiological/pathological role | $K_d$ | Oligosaccharide size | Sequence features | Function |
|---|---|---|---|---|---|
| **Proteases/esterases** | | | | | |
| AT III | Coagulation cascade serpin | ca. 20 nM | 5-mer | GlcNS6S3S | Enhances |
| SLPI | Inhibits elastase and cathepsin G | ca. 6 nM | 12-mer to 14-mer | IS | Enhances |
| C1 INH | Inhibits C1 esterase | ca 100 nM | — | HS | Enhances |
| VCP | Protects host cell from complement | nM | — | — | Unclear |
| **Growth factors** | | | | | |
| FGF-1 | Cell proliferation, differentiation, morphogenesis, and angiogenesis | nM | 4-mer to 6-mer | IdoA2S-GlcNS6S | Activates signal transduction |
| FGF-2 | Same as FGF-1 | nM | 4-mer to 6-mer | IdoA2S-GlcNS | Same as FGF-1 |
| **Chemokines** | | | | | |
| PF-4 | Inflammation and wound healing | nM | 12-mer | HS/LS/HS | Inactivates heparin |
| IL-8 | Proinflammatory cytokine | ca 6 µM | 18-mer to 20-mer | HS/LS/HS | Promotes |
| SDF-1a | Proinflammatory mediator | ca 20 nM | 12-mer to 14-mer | HS | Localizes |
| **Lipid-binding proteins** | | | | | |
| Annexin II | Receptor for TPA and plasminogen, CMV and tenascin C | ca. 30 nM | 4-mer to 5-mer | HS | Unclear |
| Annexin V | Anticoagulant activity; influenza and hepatitis B viral entry | ca. 20 nM | 8-mer | HS | Assembles |
| ApoE | Lipid transport; AD risk factor | ca. 100 nM | 8-mer | HS | Localizes |
| **Pathogen proteins** | | | | | |
| HIV-1 gp120 | Viral entry | 0.3 µM | 10-mer | HS | Inhibits |
| CypA | Viral localization and entry | — | — | — | Inhibits |
| Tat | Transactivating factor, primes cells for HIV infection | ca. 70 nM | 6-mer | HS | Antagonizes |
| HSV gB and gC | Viral entry into cells | — | — | — | Inhibits |
| HSV gD | Viral entry and fusion | — | — | GlcNH23S | Inhibits |
| Dengue virus envelope protein | Viral localization | ca. 15 nM | 10-mer | HS | Inhibits |
| Malaria CS protein | Sporozoite attachment to hepatocytes | ca. 40 nM | 10-mer | HS | Inhibits |
| **Adhesion proteins** | | | | | |
| Selectins | Adhesion, inflammation, and metastasis | µM | >14-mer | HS with GlcNH2 | Blocks |
| Vitronectin | Cell adhesion and migration | nM | — | — | Removes |
| Fibronectin | Adhesion and traction | µM | 8-mer to 14-mer | HS with GlcNS | Reorganizes |
| HB-GAM | Neurite outgrowth in the development | ca. 10 nM | 16-mer to 18-mer | HS | Mediates |
| AP | In amyloid plaque | µM | 4-mer | HS | Assembles |

*Source:* Reprinted from Linhardt, R.J., *J. Med. Chem.*, 46, 2551, 2003. With permission.

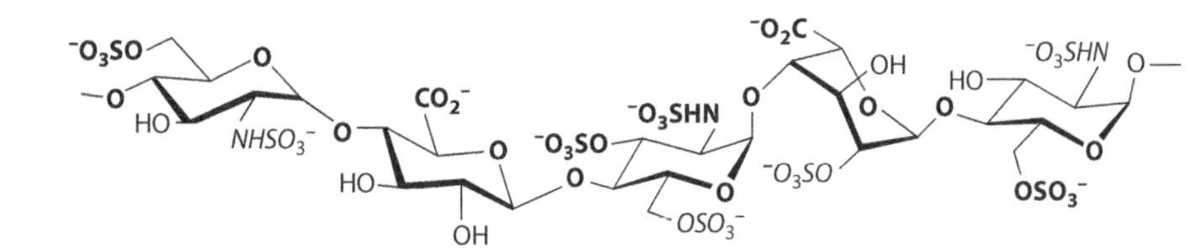

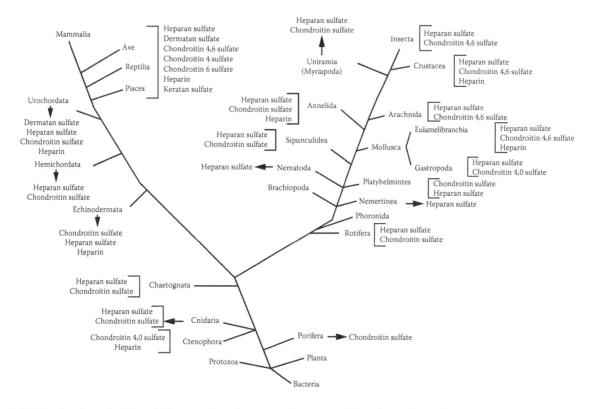

**Figure 4.2** Antithrombin III pentasaccharide-binding site. The anionic groups in bold are critical (95% loss in binding energy on removal) and those in italics are important (25%–50% loss in binding energy on removal) for interaction with ATIII. (Reprinted from Linhardt, R.J., *J. Med. Chem.*, 46, 2551, 2003. With permission.)

## 4.6 Occurrence of heparin in vertebrates and invertebrates

An updated phylogenetic tree of the distribution of sulfated GAGs in the animal kingdom is shown in Figure 4.3 (Medeiros et al. 2000). Whereas HSs are ubiquitous components of all tissue-organized metazoan, heparin has shown a very peculiar distribution in mammalian and other vertebrate tissues as well as invertebrates. CS also has a widespread distribution (Figure 4.3).

Since the earlier studies from a variety of mammals, it has been found that the lung, intestine, and liver were the organs richest in heparin (Nader et al. 2004) (Table 4.2).

Except for rabbit tissues, heparin's presence was demonstrated in lung, skin, ileum, lymph nodes, thymus, and appendix of all species studied. The absolute content of heparin varied depending on different tissues. The lack of heparin in rabbits was correlated with the absence of mast cells in the species (Nader et al. 1980). A large variation of the concentration of heparin among species is evident. Thus, bovine and dog tissues contain the highest amounts of heparin. Generally, in nonmammalian vertebrate tissues, the amount of heparin is considerably lower (Nader et al. 2004).

In invertebrates, heparin is found in few taxa: mollusks, crustacean, annelida, echinoderma, and cnidaria (Table 4.3). As observed for vertebrate heparin samples, the anticoagulant activity and MM varied according

**Figure 4.3** Distribution of sulfated glycosaminoglycans in the animal kingdom. (Data from Volpi, N., *Invertebr. Surviv. J.*, 2, 6, 2005.)

*Marine Glycobiology: Principles and Applications*

*Table 4.2* Distribution of heparin in mammalian and other vertebrates

| Tissue | Rabbit | Guinea pig | Rat | Dog | Cat | Pig | Bow | Human | Chicken | Snake | Lizard | Frog | Fish | Shark |
|---|---|---|---|---|---|---|---|---|---|---|---|---|---|---|
| | | | | | | | µg/g dry tissue | | | | | | | |
| Lung | <1 | 70 | 67 | 217 | 63 | 211 | 300 | 8 | 0.5 | 0.3 | 0.0 | 0.6 | 0.0 | |
| Liver | <1 | <1 | <1 | 141 | 1 | <1 | 50 | <1 | 0.0 | 0.0 | 0.0 | 0.8 | 1.3 | 0.0 |
| Ileum | <1 | 27 | 1 | 400 | 87 | 113 | 1015 | 32 | 0.5 | 0.9 | 0.2 | 0.0 | 0.0 | 0.0 |
| Kidney | <1 | 4 | <1 | 2 | 6 | <1 | 26 | <1 | 0.1 | 0.0 | 0.0 | 0.5 | | 0.3 |
| Aorta | <1 | <1 | 9 | 102 | <1 | 2 | 150 | <1 | | | | | | |
| Brain | <1 | <1 | <1 | <1 | <1 | <1 | <1 | <1 | 0.0 | 0,0 | 0.0 | 0.0 | 0.0 | 0.0 |
| Muscle | <1 | <1 | 36 | 9 | <1 | 5 | 2 | <1 | 0.0 | 0.0 | 0.0 | 11.4 | 0.0 | 0.0 |
| Spleen | <1 | <1 | <1 | 11 | <1 | <1 | 19 | <1 | 0.0 | | | | | 11.9 |
| Skin | <1 | <1 | 175 | 15 | 63 | 2 | 108 | 39 | 0.4 | 0.0 | 0.0 | 0.0 | 0.0 | 0.0 |
| Lymph | <1 | 11 | 5 | 160 | 74 | 242 | 180 | 41 | | | | | | |
| Thymus | <1 | 112 | 20 | 20 | | 10 | 286 | 35 | | | | | | |
| Appendix | <1 | | | 17 | 38 | | 20 | 47 | | | | | | |
| Branchia | | | | | | | | | | | | | 0.0 | 0.0 |

*Source:* Data from Volpi, N., *Invertebr. Surviv. J.*, 2, 6, 2005.

*Table 4.3* Distribution of heparin in invertebrates

| Class and species | Average MM (kDa) | Anticoagulant activity (USP) |
|---|---|---|
| Mollusks | | |
| *Cyprina islandica* | nd | 95 |
| *Mactrus pussula* | nd | 100 |
| *Mercenaria mercenaria* | 18 | 348 (Anti-IIa) |
| *Anomalocardia brasiliana* | 32 | 320 |
| *Donax striatus* | 20 | 220 |
| *Tivela mactroides* | 25 | 180 |
| *Tapes phylippinarum* | 14 | 350 |
| *Callista chione* | 11 | 97 |
| Crustacea | | |
| *Ucides cordatus* | nd | 60 |
| *Dedrocephalus brasiliensis* | 10 | 52 |
| *Penaeus brasiliensis* | 9 | 60 |
| Annelida | | |
| *Aphrodite longicornis* | nd | nd |
| *Hermodice carunculata* | nd | nd |
| Echinoderma | | |
| *Mellita quinquiesperforata* | 12 | 50 |
| Cnidaria | | |
| *Physalia* sp. | nd | nd |
| *Mnemiopsis* sp. | nd | nd |
| *Sipunculus nudus* | nd | nd |

*Source:* Data from Volpi, N., *Invertebr. Surviv. J.*, 2, 6, 2005.

*Note:* nd—not detected, the molecular mass and the anticoagulant activity are reported.

to the species analyzed. Furthermore, no correlation between MM and anticoagulant activity of the heparins is evident. All these results imply that heparins have a large structural variation depending on their origin (Nader et al. 2004).

## 4.7 Biological role of heparin in mollusks

The biological function of the clam heparins and their apparently specific ATIII-binding regions is unclear at the moment. Mollusks do not possess any blood coagulation system similar to that of mammals; yet, their heparins are capable of dramatically accelerating the inactivation of mammalian coagulation enzymes by the mammalian protease inhibitor, ATIII. It is possible that the bivalve heparin is designed to interact with an endogenous antithrombin-like protease inhibitor acting on serine protease target enzymes. The existence and function of such an enzyme system remain to be established.

In mammals, heparin is released from the mast cells in response to specific inflammatory agents such as IgE antibodies or complement fragments (anaphylatoxins). Since the discovery of mast cells by Paul Ehrlich (1879) and after the demonstration that their metachromatic properties when stained with basic dyes were related to heparin, the question of whether this compound was only confined to the mast cells or not has been a matter of controversy. Studies on the concentration of heparin and its content in different fetal and adult bovine tissues (Nader et al. 1982) have shown that a good correlation between the mast cell number and heparin concentration could be obtained in all analyzed tissues.

Studies on mast cell–deficient mice of the genotypes $W/W^v$ and $S1/S1^d$ established that mast cells originate from hematopoietic stem cells. These experiments also demonstrated that heparin is present in appreciable amounts in the skin of the breeders and of the normal progeny. On the contrary, no heparin was detected in the skin of the $W/W^v$ genotype, which is deficient in mast cells. No significant differences in the relative amounts of the other sulfated GAGs—HS, dermatan sulfate, and CS—were observed among the analyzed breeders. This suggests that heparin is not replaced by other sulfated GAGs in the animals that lack heparin. These results clearly indicate that heparin is related to the presence of mast cells (Kitamura and Go 1979; Marshall et al. 1994). Furthermore, in mammals, the heparin-containing mast cells are accumulated in lymphoid organs and in tissues exposed to the external milieu (skin, lungs, intestine), and one suggested role for this polysaccharide in mammalian is to fight external parasites (Nader et al. 2004).

Heparin and other sulfated GAGs as well as histamine were found and quantified in various organs of the mollusk *A. brasiliana*. The heparin was present in granules inside the cytoplasm of mast-like cell (Dietrich et al. 1985, Pejler et al. 1987). A good correlation between heparin and histamine content was found in the labial palp, intestine, ctenidium, mantle, and foot tissues.

Some conclusions can be derived from these studies: (1) heparin seems to be present exclusively in mast cells of vertebrates or mast-like cells in the case of mollusks; (2) its primary biological activity is not related with antithrombotic activity since mollusks, which do not possess a coagulation system, contain heparin and rabbits that possess this system are devoid of heparin; and (3) many indirect evidences suggest that in mammals and mollusks heparin and its mast cells are involved in defence mechanisms independently of the immune system and support the hypothesis that this macromolecule could function as a mechanism for the surveillance of these organisms against certain pathogens.

## 4.8 Conclusions and future perspectives

The study of heparin will certainly extend well into this new century with so many questions left still unanswered. Some focal points in the near future include (1) improved preparation and synthesis of heparins, (2) new heparin-based anticoagulants with improved properties, (3) new therapeutic applications for heparins, (4) new heparin mimetics, (5) new biomaterials, and (6) development of an improved understanding of physiology and pathophysiology through glycomics.

The preparation of heparin from mammalian tissues creates concern particularly after the recent appearance of bovine spongiform encephalopathy in Europe and its contamination by oversulfated CS. Bovine tissues are now rarely used in heparin production, and there are growing concerns about porcine tissues. Heparins prepared by alternative routes, such as defined, recombinant mammalian cell lines capable of being cultured in large-scale fermentations, or nonmammalian tissues, offer an exciting alternative to the present preparations.

New anticoagulants based on heparin's structure might offer enhanced specificity targeting one or selected groups of coagulation proteases and avoiding undesired interactions with other proteins, thus decreasing the side effects associated with heparins use. Furthermore, new therapeutic applications might include the use of heparin to treat infectious diseases, inflammation, and control of cell growth in wound healing and cancer. These new activities will require the elimination of heparin's anticoagulant activity, the engineering of appropriate pharmacokinetics and pharmacodynamics, and optimally oral bioavailability. A concern about the application of heparins to promote wound healing is that they might simultaneously promote cancer. Thus, wound healing applications will probably require localization of the drug at the site of action possibly through the application of polymers or gels. These new potential applications of heparin may be strictly associated with molecules possessing peculiar and unusual structure, as those isolated from invertebrates.

One of the most important future directions of heparin research is driven by the recent sequencing of the human genome and the field of genomics. While much attention is currently focused on the proteome encoded by the genome and the rapidly developing field of proteomics, the glycome has garnered little attention. Glycomics is the study of the structure and function of the glycome, the most important and complex of the posttranslation modifications that proteins undergo. An improved understanding of the glycome should be beneficial in better understanding genetic diseases, offering new therapeutic approaches to treat these very serious pathologies. Moreover, improved knowledge of glycomics, in vertebrate and invertebrate species, should lead to a better understanding of physiology and pathophysiology, offering new approaches to drug development.

By considering previous knowledge on the therapeutic potential of natural bioactive compounds such as SPSs, especially GAGs, it is now well documented, and this activity combined with natural biodiversity will allow the development of a new generation of therapeutics. Advances in our understanding of the biosynthesis, structure, and function of complex glycans from mammalian origin have shown the crucial role of this class of macromolecules to modulate disease processes and the importance of a deeper knowledge of structure–activity relationships. Marine environment offers a wide biodiversity, and original (sulfated) polysaccharides have been discovered presenting a great chemical diversity that is largely species specific. The study of the biological properties of the knowledges, the therapeutic potential of natural bioactive compounds such as sulfates from marine eukaryotes could provide a valid alternative to traditional polysaccharides such as GAGs. Marine (sulfated) polysaccharides present a real potential for natural product drug discovery and for the delivery of new derived products for therapeutic applications.

## 4.9　Summary

Several invertebrate species contain variable amounts of one or more types of sulfated GAGs. At present, the existence of a species-specific sulfated GAGs composition based on the relative amount and type of CSs, HS,

and heparin is well known. Heparin is an SPS belonging to the family of GAGs with numerous important biological activities, such as anticoagulant and antithrombotic properties, that derive from its interaction with diverse proteins. Unusual heparin samples for MM, fine structural organization and anticoagulant activity, are isolated and characterized from mollusks. This chapter more specifically deals with structural and biologically important aspects of heparin in invertebrates with special emphasis on the heparin from marine mollusks.

## References

Arias, E., B.L. Smith. 2003. Deaths: Preliminary data for 2001. *National Vital Statistics Reports* 51:1–44.

Capila, I., R.J. Linhardt. 2002. Heparin-protein interactions. *Angewandte Chemie International Edition* 41:390–412.

Cardin, A.D., H.J. Weintraub. 1989. Molecular modeling of protein-glycosaminoglycan interactions. *Arteriosclerosis* 9:21–32.

Cassaro, C.M., C.P. Dietrich. 1977. Distribution of sulfated mucopolysaccharides in invertebrates. *Journal of Biological Chemistry* 252:2254–2261.

Cesaretti, M., E. Luppi, M. Maccari, N. Volpi. 2004. Isolation and characterization of a heparin with high anticoagulant activity from the clam *Tapes phylippinarum*. Evidence for the presence of a high content of antithrombin III-binding site. *Glycobiology* 14:1275–1284.

Chavante, S.F., E.A. Santos, F.W. Oliveira, M. Guerrini, G. Torri, B. Casu, C.P. Dietrich, H.B. Nader. 2000. A novel heparan sulphate with high degree of N-sulphation and high heparin cofactor-II activity from the brine shrimp *Artemia franciscana*. *International Journal of Biological Macromolecules* 27:49–57.

Dietrich, C.P., J.F. de-Paiva, C.T. Moraes, H.K. Takahashi, M.A. Porcionatto, H.B. Nader. 1985. Isolation and characterization of a heparin with high anticoagulant activity from *Anomalocardia brasiliana*. *Biochimica et Biophysica Acta* 843:1–7.

Ehrlich, P. 1879. Beitrage zur kenntnis der granulierten bindegewebszellen und der eosinophilen leukocythen. *Archives of Anatomy and Physiology* 166–169.

Ernst, B., J.L. Magnani. 2008. From carbohydrate leads to glycomimetic drugs. *Nature Reviews Drug Discovery* 8:661–677.

Gandhi N.S., Mancera R.L. 2008. The structure of glycosaminoglycans and their interactions with proteins. *Chemical Biology and Drug Design* 72:455–482.

Hileman, R.E., J.R. Fromm, J.M. Weiler, R.J. Linhardt. 1998. Glycosaminoglycan-protein interactions: Definition of consensus sites in glycosaminoglycan binding proteins. *Bioessays* 20:156–167.

Hovingh, P., A. Linker. 1982. An unusual heparan sulfate isolated from lobsters (*Homarus americanus*). *Journal of Biological Chemistry* 257:9840–9844.

Jordan, R.E., J.A. Marcum. 1986. Anticoagulantly active heparin from clam (*Mercenaria mercenaria*). *Archives of Biochemistry and Biophysics* 248:690–695.

Kitamura, Y., S. Go. 1979. Decreased production of mast cells in S1/S1d anemic mice. *Blood* 53:492–497.

Lindahl, U., G. Backstrom, M. Hook, L. Thunberg, L.A. Fransson, A. Linker. 1979. Structure of the antithrombin-binding site in heparin. *Proceedings of the National Academy of Sciences of the United States of America* 76:3198–3202.

Linhardt, R.J. 2003. Heparin: Structure and activity. *The Journal of Medicinal Chemistry* 46:2551–2564.

Linhardt, R.J., N.S. Gunay. 1999. Production and chemical processing of low molecular weight heparins. *Seminars in Thrombosis and Hemostasis* 25 Suppl 3:5–16.

Liu, H., Z. Zhang, R.J. Linhardt. 2009. Lessons learned from the contamination of heparin. *Natural Product Reports* 26:313–321.

Lord, M.S., J.M. Whitelock. 2014. Bioengineered heparin: Is there a future for this form of the successful therapeutic? *Bioengineered* 5:222–226.

Luppi, E., M. Cesaretti, N. Volpi. 2005. Purification and characterization of heparin from the italian clam *Callista chione*. *Biomacromolecules* 6:1672–1678.

Marcum, J.A., R.D. Rosenberg. 1987. Anticoagulantly active heparan sulfate proteoglycan and the vascular endothelium. *Seminars in Thrombosis and Hemostasis* 13:464–474.

Marshall, J.S, S. Kawabori, L. Nielsen, J. Bienenstock. 1994. Morphological and functional characteristics of peritoneal mast cells from young rats. *Cell and Tissue Research* 276:565–570.

Medeiros, G.F., A. Mendes, R.A. Castro, E.C. Bau, H.B. Nader, C.P. Dietrich. 2000. Distribution of sulfated glycosaminoglycans in the animal kingdom: Widespread occurrence of heparin-like compounds in invertebrates. *Biochimica et Biophysica Acta* 1475:287–294.

Nader, H.B., T.M.P.C. Ferreira, J.F. Paiva, M.G.L. Medeiros, S.M.B. Jeronimo, V.M.P. Paiva, C.P. Dietrich. 1984. Isolation and structural studies of heparan sulfates and chondroitin sulfates from three species of molluscs. *Journal of Biological Chemistry* 259:1431–1435.

Nader, H.B., C.C. Lopes, H.A.O. Rocha, E.A. Santos, C.P. Dietrich. 2004. Heparins and heparinoids: Occurrence, structure and mechanism of antithrombotic and hemorrhagic activities. *Current Pharmaceutical Design* 10:951–966.

Nader, H.B., A.H. Straus, H.K. Takahashi, C.P. Dietrich. 1982. Selective appearance of heparin in mammalian tissues during development. *Biochimica et Biophysica Acta* 714:292–297.

Nader, H.B., H.K. Takahashi, A.H. Straus, C.P. Dietrich. 1980. Selective distribution of the heparin in mammals: Conspicuous presence of heparin in lymphoid tissues. *Biochimica et Biophysica Acta* 627:40–48.

Pejler, G., A. Danielsson, I. Bjork, U. Lindahl, H.B. Nader, C.P. Dietrich. 1987. Structure and antithrombin-binding properties of heparin isolated from the clams *Anomalocardia brasiliana* and *Tivela mactroides*. *Journal of Biological Chemistry* 262:11413–11421.

Rosenberg, R.D., L. Lam. 1979. Correlation between structure and function of heparin. *Proceedings of the National Academy of Sciences of the United States of America* 76:1218–1222.

Schonberger, L.B. 1998. New variant Creutzfeldt-Jakob disease and bovine spongiform encephalopathy. *Infectious Disease Clinics of North America* 12:111–121.

Senni, K., J. Pereira, F. Gueniche, C. Delbarre-Ladrat, C. Sinquin, J. Ratiskol, G. Godeau, A.M. Fischer, D. Helley, S. Colliec-Jouault. 2011. Marine polysaccharides: A source of bioactive molecules for cell therapy and tissue engineering. *Marine Drugs* 9:1664–1681.

Shriver, Z., S. Raguram, R. Sasisekharan. 2004. Glycomics: A pathway to a class of new and improved therapeutics. *Nature Reviews Drug Discovery* 3:863–873.

Sinay, P. 1999. Sugars slide into heparin activity. *Nature* 398:377–378.

Tapon-Bretaudiere, J., D. Chabut, M. Zierer, S. Matou, D. Helley, A. Bros, P.A. Mourao, A.M. Fischer. 2002. A fucosylated chondroitin sulfate from echinoderm modulates *in vitro* fibroblast growth factor 2-dependent angiogenesis. *Molecular Cancer Research* 1:96–102.

Torri, G., B. Casu, G. Gatti, M. Petitou, J. Choay, J.C. Jacquinet, P. Sinay. 1985. Mono- and bidimensional 500 MHz $^1$H-NMR spectra of a synthetic pentasaccharide corresponding to the binding sequence of heparin to antithrombin-III: Evidence for conformational peculiarity of the sulfated iduronate residue. *Biochemical and Biophysical Research Communications* 128:134–140.

Volpi, N. 2005. Occurrence and structural characterization of heparin from molluscs. *Invertebrate Survival Journal* 2:6–16.

Volpi, N., F. Maccari. 2003. Purification and characterization of hyaluronic acid from the mollusc bivalve *Mytilus galloprovincialis*. *Biochimie* 85:619–625.

Volpi, N., F. Maccari. 2005. Glycosaminoglycans composition of the large freshwater mollusc bivalve *Anodonta anodonta*. *Biomacromolecules* 6:3174–3180.

Volpi, N., F. Maccari. 2007. Characterization of a low-sulfated chondroitin sulfate isolated from the hemolymph of the freshwater snail *Planorbarius corneus*. *Marine Biology* 152:1003–1007.

Volpi, N., F. Maccari. 2009. Structural characterization and antithrombin activity of dermatan sulfate purified from marine clam *Scapharca inaequivalvis*. *Glycobiology* 19:356–367.

# *Marine glycans*

*chapter five*

# Bioactivity and mechanism of action of marine glycans

*Visamsetti Amarendra and Ramachandran Sarojini Santhosh*

## Contents

## 5.1 Introduction

Carbohydrates are one of the major macromolecules essential for life, along with deoxyribonucleic acid (DNA), ribonucleic acid (RNA), and proteins. These are ubiquitous in nature and are structural components and energy transporters in living organisms. More than 75% of the dry weight of plants is constituted by carbohydrates. The biological roles of these carbohydrates include polysaccharides as storage materials (glycogen, starch), structural materials (cellulose, lignin), and simple molecules, as well as their combination with other macromolecules as signaling factors, cellular markers, and cell membrane adhesion molecules.

The combination of carbohydrates with lipids and proteins forming glycolipids and glycoproteins is important for functional roles in cell membranes (ceramides, glycosphingolipids), hormonal synthesis (human chorionic gonadotropin, erythropoietin), transport molecules (transferrin, ceruloplasmin), structural molecules (collagens), and immunogenic molecules (immunoglobulins, histocompatibility antigens).

Increased research on pharmaceutically active agents and their applications in commercial products and screening for natural products with large resource availability are becoming major objectives. The focus of global research is shifting to natural polymers such as carbohydrates and their derivatives for applications both in health care and industry. The sources for these natural polymers from the sea include marine algae, sponges, soft corals, sea urchins, sea cucumbers, cnidarians, and ascidians. Natural polymers such as chitosan, heparin **1**, carrageenan, alginates **2**, and fucoidans **3** are major carbohydrates that form structural and storage materials for these marine organisms. The polyanionic nature and heterogeneity of these polymers endow them with various pharmaceutically important qualities, such as antioxidants (alginates **2** and hyaluronic acids **4**) (Kanchana et al. 2013; Karaki et al. 2013), antimicrobials (lectins) (Kawsar et al. 2010), anticancer agents (glycerol lipids) (Makarieva et al. 2005), anticoagulants (alginates **2**) (Karaki et al. 2013), antivirals (sulfate polysaccharides) (Wang et al. 2010), and cytoprotective agents (fucoidans **3**) (Wang et al. 2012a). Along with natural polymers, derivatives of simple sugars (mono-and oligosaccharides linked with other heterocyclic structures) such as terpene glycosides and steroidal glycosides isolated from marine microorganisms contribute a major share in the development of novel and potential therapeutic drugs with potential human health applications.

In this chapter, we describe the characteristics of marine glycans and glycosides derived from various marine sources and their applications in pharmaceutical and biotechnological industries. The chemical structures of glycans with different bioactivities are sequentially numbered in bold Arabic numerals (Figure 5.1).

## 5.2   Classification

### 5.2.1   Monosaccharides

Based on the number of carbon atoms, monosaccharides are classified into trioses (three carbons (3C); for example, glyceraldehyde), tetroses (4C; erythrose, threose), pentoses (5C; ribose, arabinose, xylose), hexoses (6C; glucose, mannose, galactose), and heptoses (7C; sedoheptulose). On the basis of their functional groups, monosaccharides are classified into aldoses and ketoses depending on the presence of an aldehyde or a ketone, respectively,

in the structure. Each sugar in the group (e.g., hexose) is differentiated by the arrangement of the hydroxyl group on the chiral carbon. These monosaccharides with polyhydroxy groups have commercial value in the food sector. Commercial preparations include high corn syrup using fructose, crystalline glucose in confectionaries and in baked foods; fructose and xylitol in frozen desserts, chewing gum, and sweeteners; as well as tagatose in antidiabetic foods and sorbitol in toothpastes.

### 5.2.2   Di- and trisaccharides

Condensation of two or three monosaccharides results in the formation of di- or trisaccharides, respectively, with the elimination of a water molecule. The commonly known sugars—sucrose, lactose, and maltose—belong to disaccharides, and maltotriose, raffinose, and melezitose belong to trisaccharides. The basic units of these di/trisaccharides are glucose, fructose, and galactose. These condensed products are commonly used in industries. Sucrose forms the basic household component as a sweetener, and at high concentrations used as a preservative, involved in the preparation of cookies, cakes, biscuits, candies, sorbets, and so on. Lactose is another important disaccharide used in the dairy industry in the preparation of artificial milk products, and it is also used in stout beer. Fructo-trisaccharides (neokestose **5**, 1-kestose **6**, and 6-kestose **7**) are commercially used as prebiotics, which have positive effects on the growth of intestinal bacteria (*Bifidobacterium* sp., and *Lactobacillus* sp.) for improved metabolism (Kilian et al. 2002).

### 5.2.3   Polysaccharides

Polysaccharides are represented with the general formula $(C_6H_{10}O_5)n$. Macromolecules are attracting major attention for their use as drug molecules because of the serious side effects of many available drugs in the market. Polysaccharides and polysaccharide–protein molecules have been found to be effective against various biological effects. These molecules possess antibiotic, antioxidant, antimutant, anticoagulant, and immunostimulant activities (Zong et al. 2012). These were characterized in bacteria, fungi, algae, plants, and animals. They serve important roles in many biological processes: as storage materials (starch, glycogen), structural components (cellulose in plants, chitin in crustaceans), signal recognition, cell–cell communication, fertilization, pathogen prevention, blood clotting, and system development (Zong et al. 2012). Alginate **2**, which is rich in mannuronic acid, an anionic polysaccharide, is largely distributed in the cell walls of brown algae (Shobharani et al. 2014). It has many uses in the paper and textile industries, pharmaceuticals, and other industrial materials.

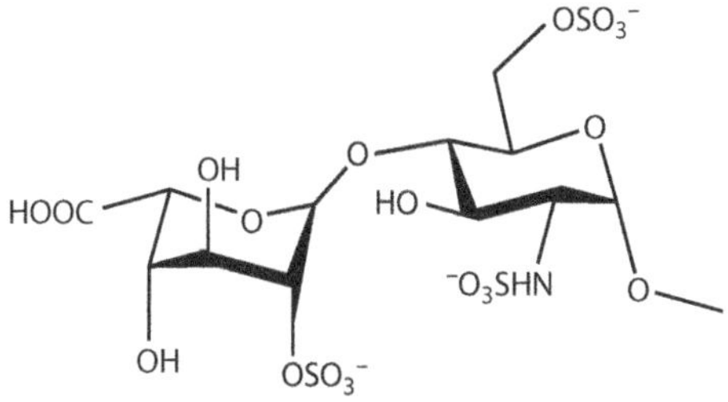

1

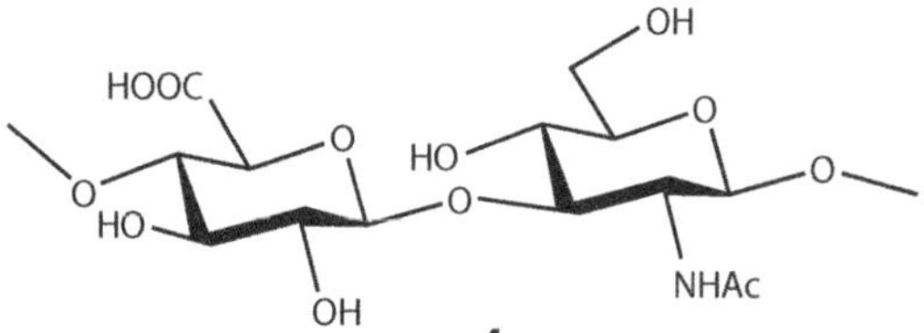

4

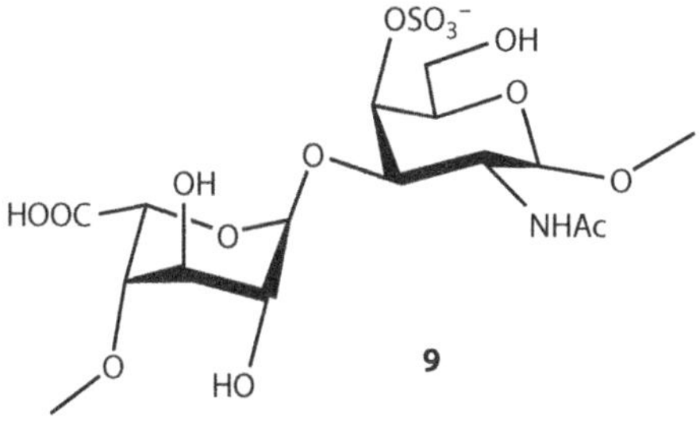

9

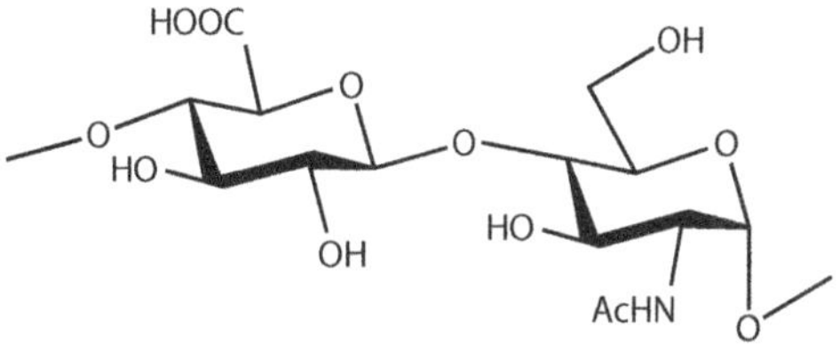

8

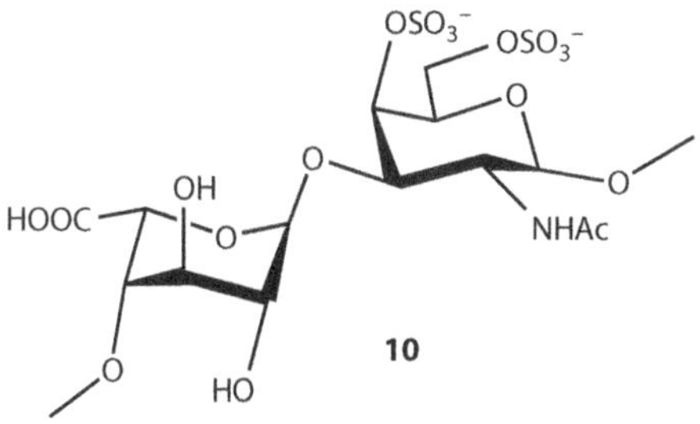

10

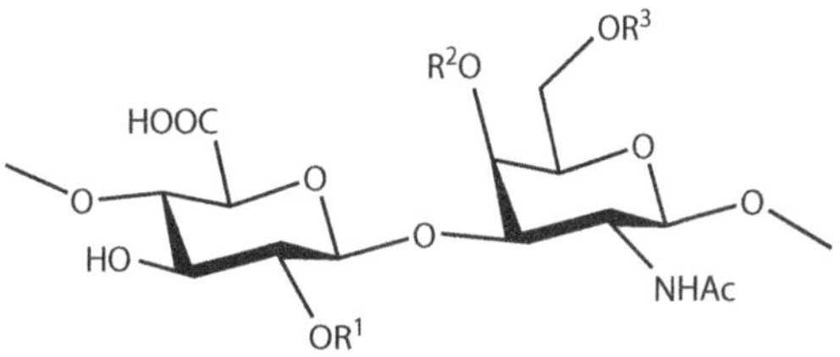

**11**: $R^{2,6} = H, R^4 = SO_3^-$

**12**: $R^{2,4} = H, R^6 = SO_3^-$

**13**: $R^{2,6} = SO_3^-, R^4 = H$

**14**: $R^{4,6} = SO_3^-, R^2 = H$

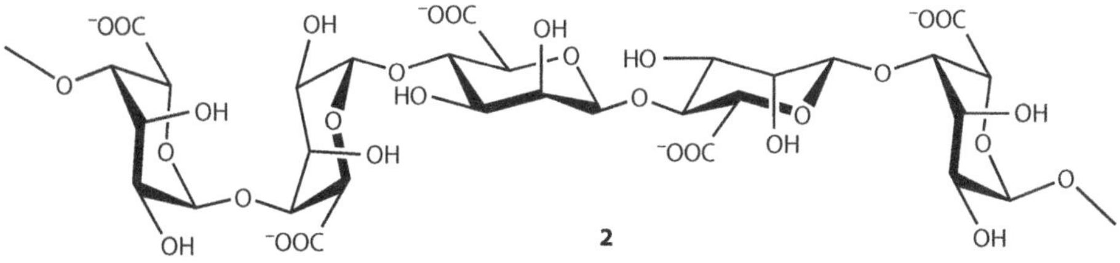

2

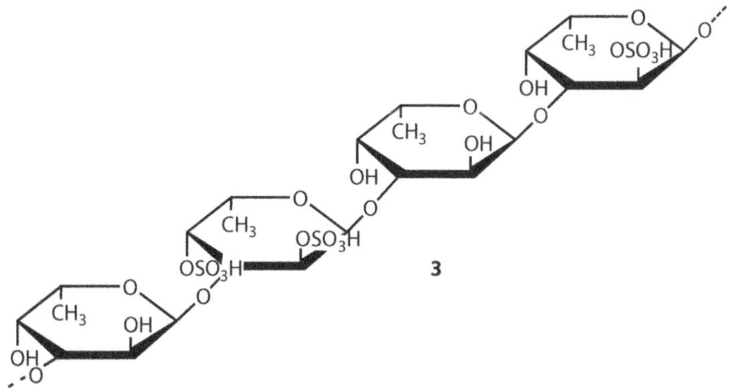

3

***Figure 5.1*** Marine glycans possessing bioactivity. *(Continued)*

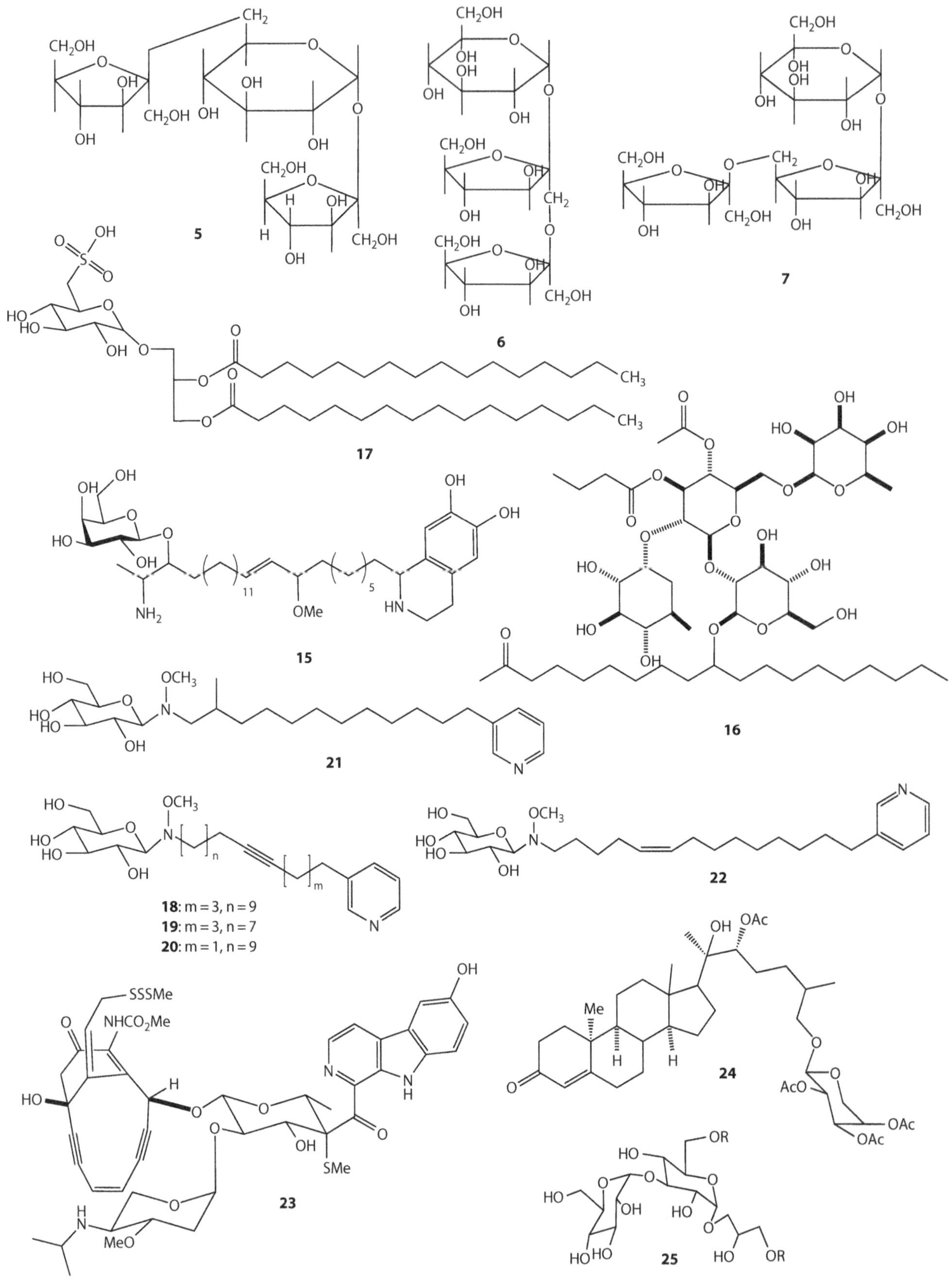

**Figure 5.1 (Continued)**  Marine glycans possessing bioactivity.

(Continued)

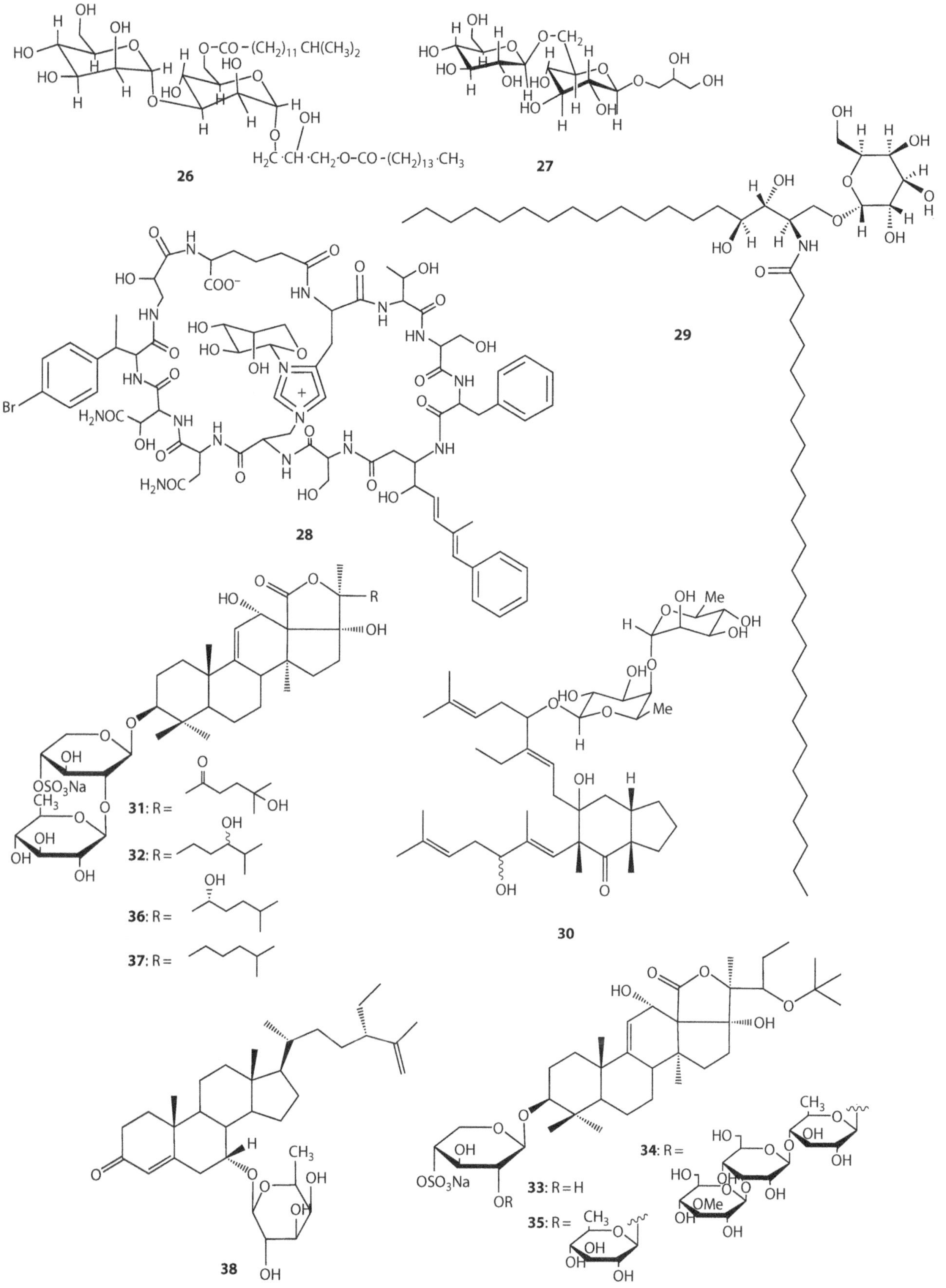

**Figure 5.1 (Continued)**  Marine glycans possessing bioactivity.

(Continued)

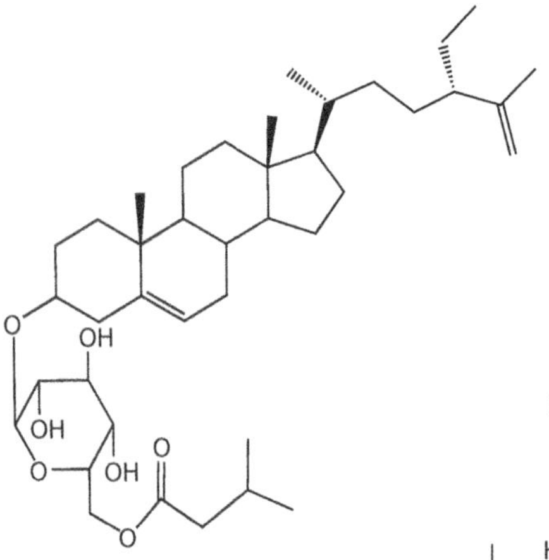

**39**

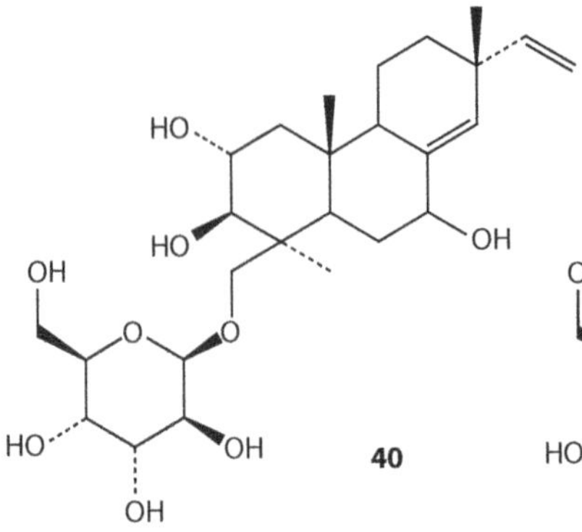

**40**

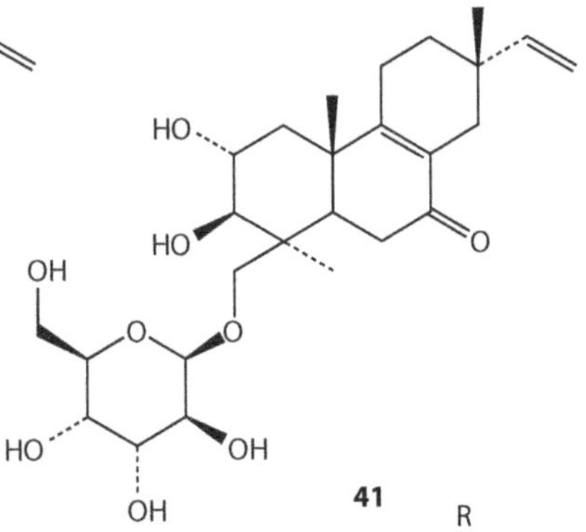

**41**

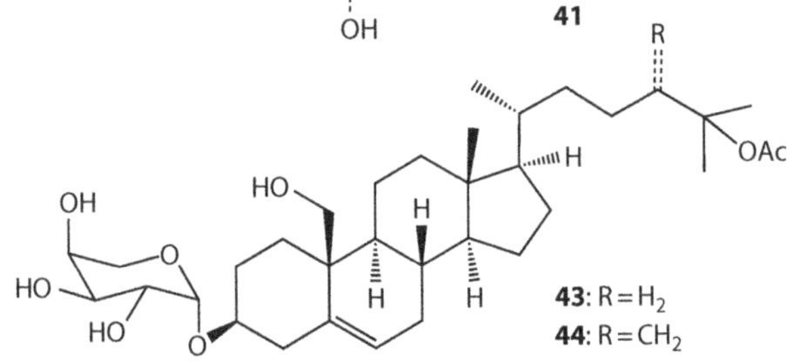

**43**: R = $H_2$
**44**: R = $CH_2$

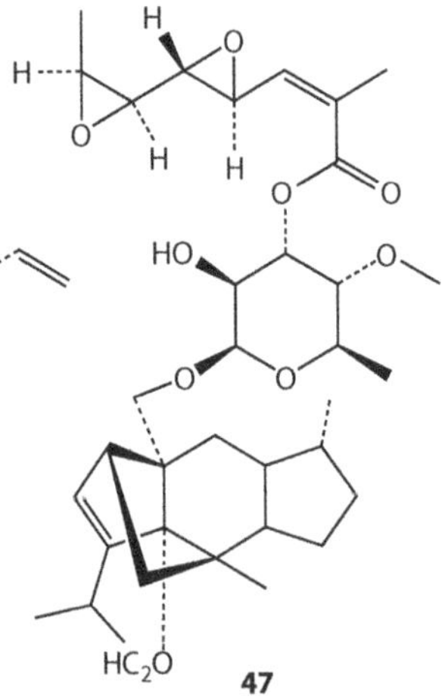

**47**

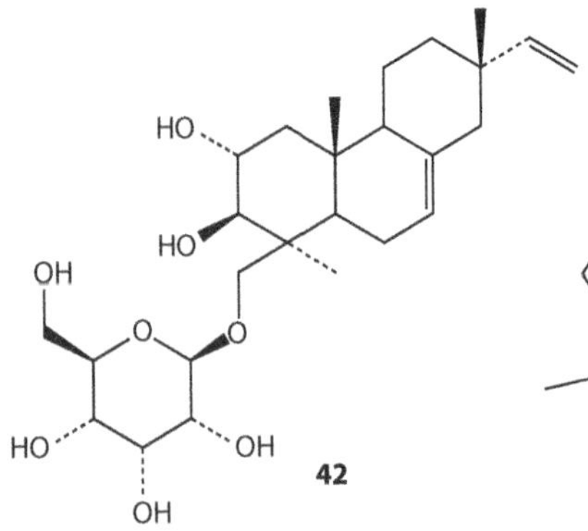

**42**

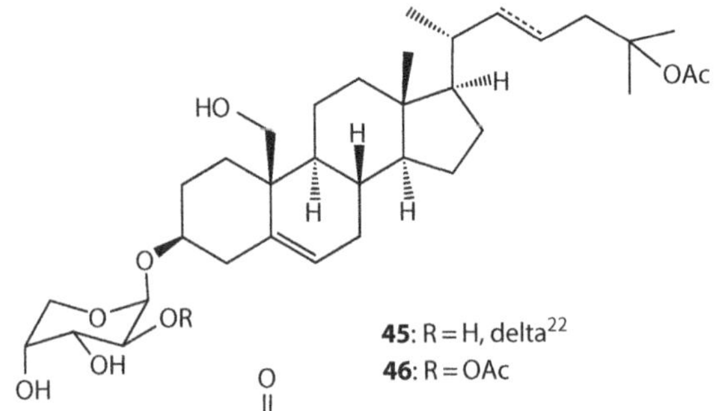

**45**: R = H, delta[22]
**46**: R = OAc

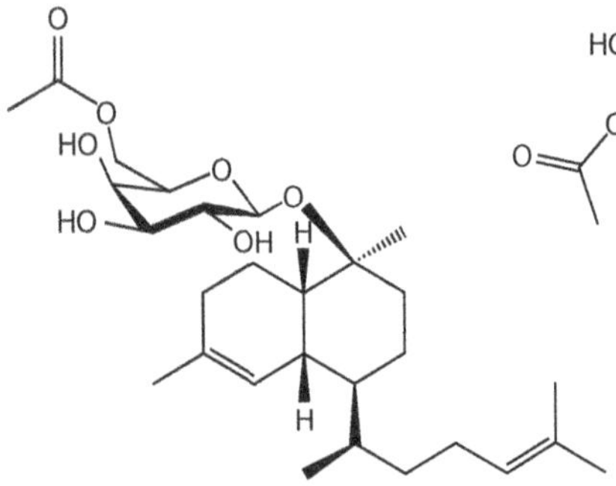

**48**

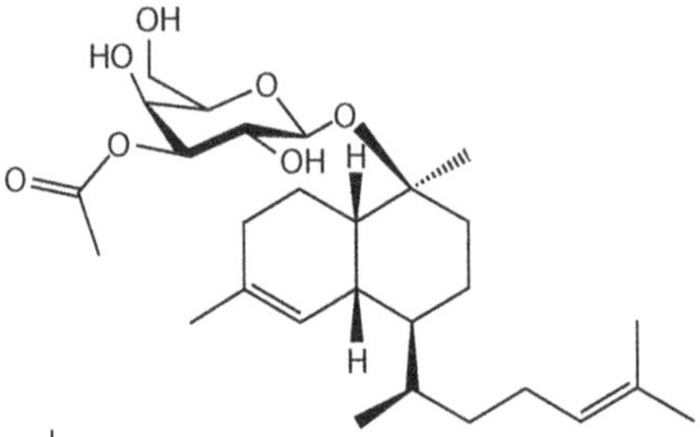

**49**

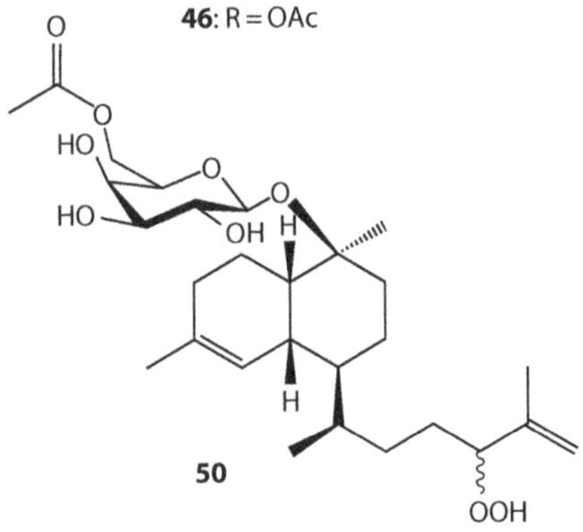

**50**

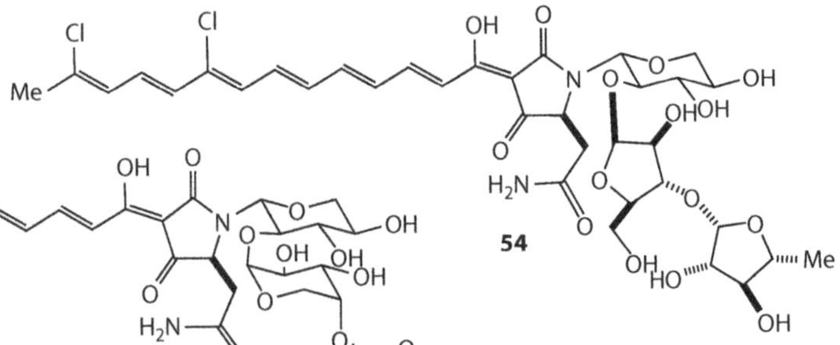

**54**

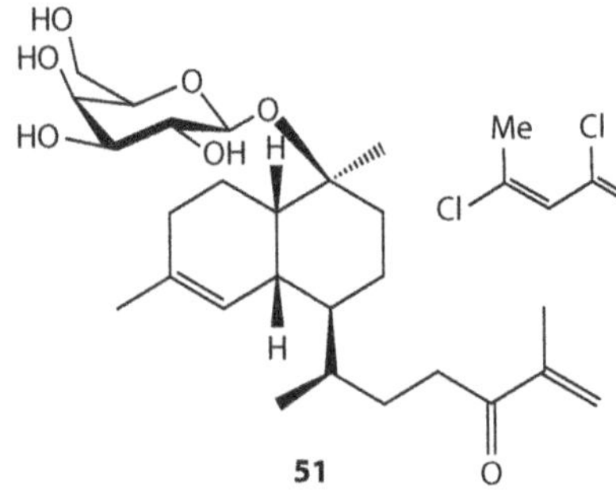

**51**

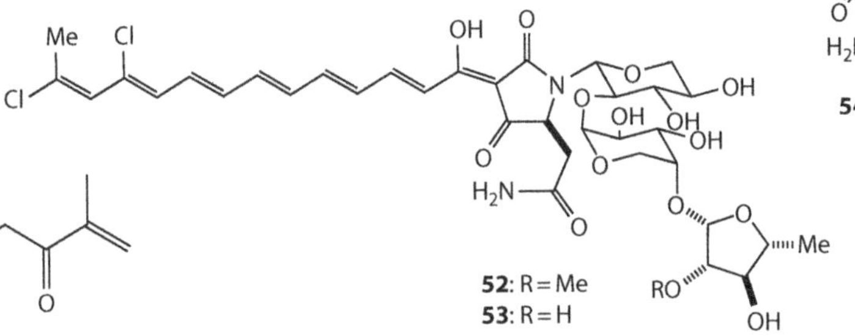

**52**: R = Me
**53**: R = H

***Figure 5.1 (Continued)*** Marine glycans possessing bioactivity. *(Continued)*

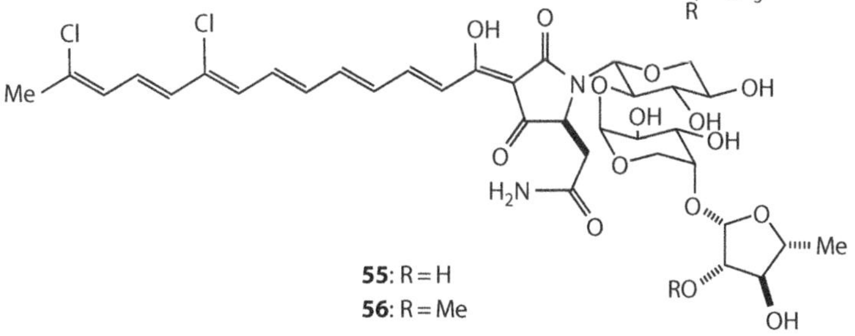

**55**: R = H
**56**: R = Me

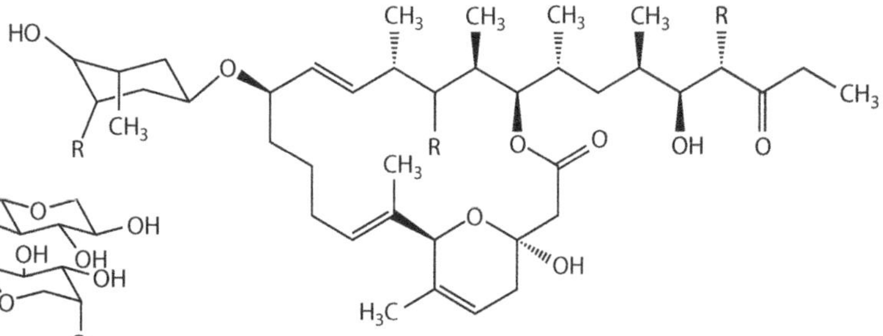

**57**

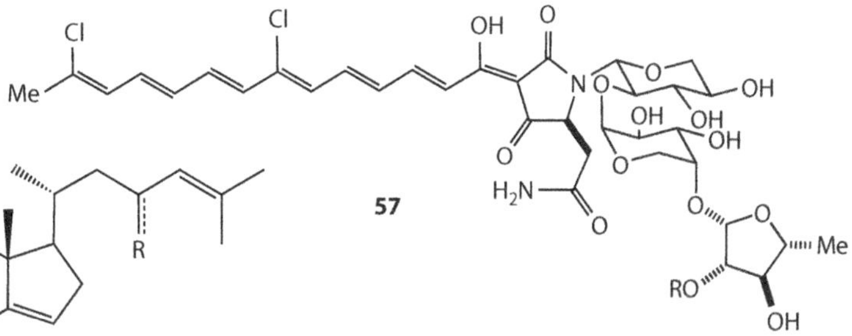

**59**: R = OH
**60**: R = = O

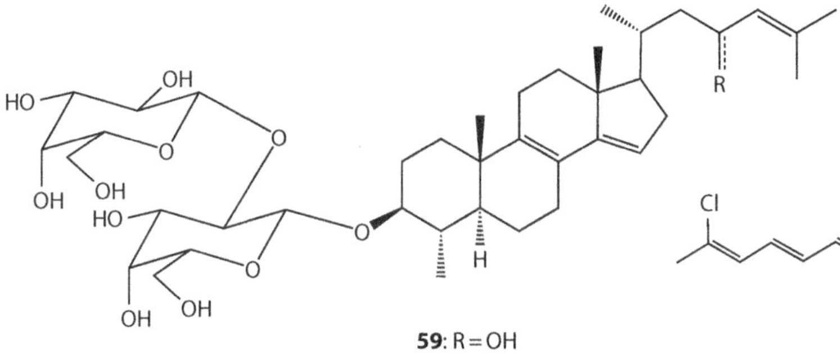

**58**

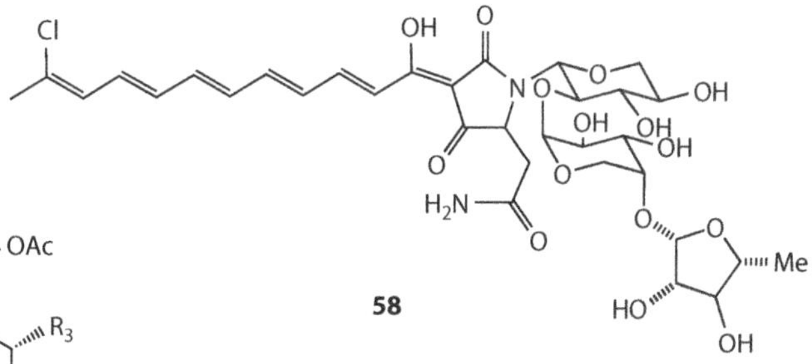

**61**

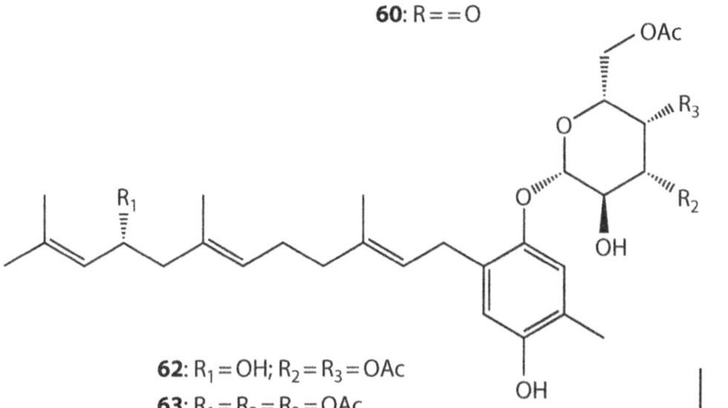

**62**: R₁ = OH; R₂ = R₃ = OAc
**63**: R₁ = R₂ = R₃ = OAc
**64**: R₁ = H; R₂ = R₃ = OAc
**67**: R₁ = H; R₂ = OH; R₃ = OAc
**68**: R₁ = H; R₂ = OAc; R₃ = OH

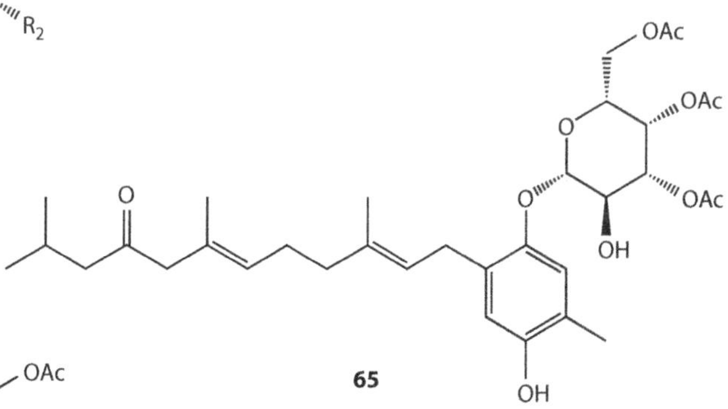

**65**

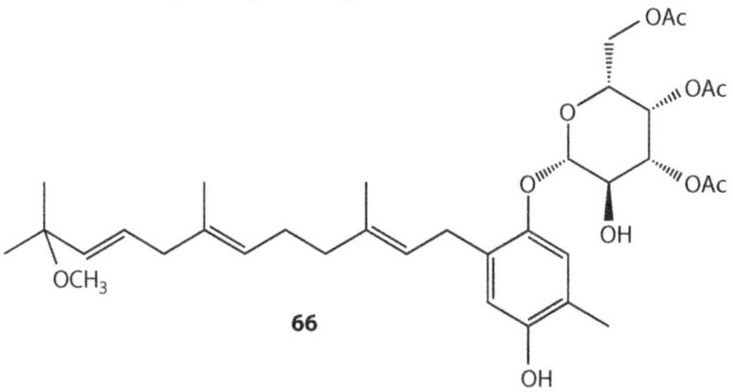

**66**

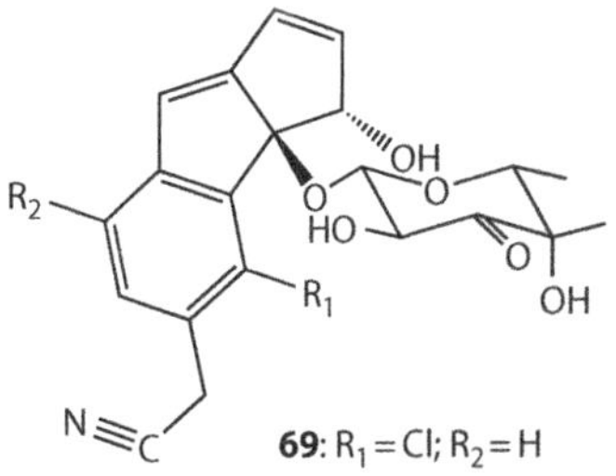

**69**: R₁ = Cl; R₂ = H
**70**: R₁ = Cl; R₂ = Cl

*Figure 5.1 (Continued)* Marine glycans possessing bioactivity.          *(Continued)*

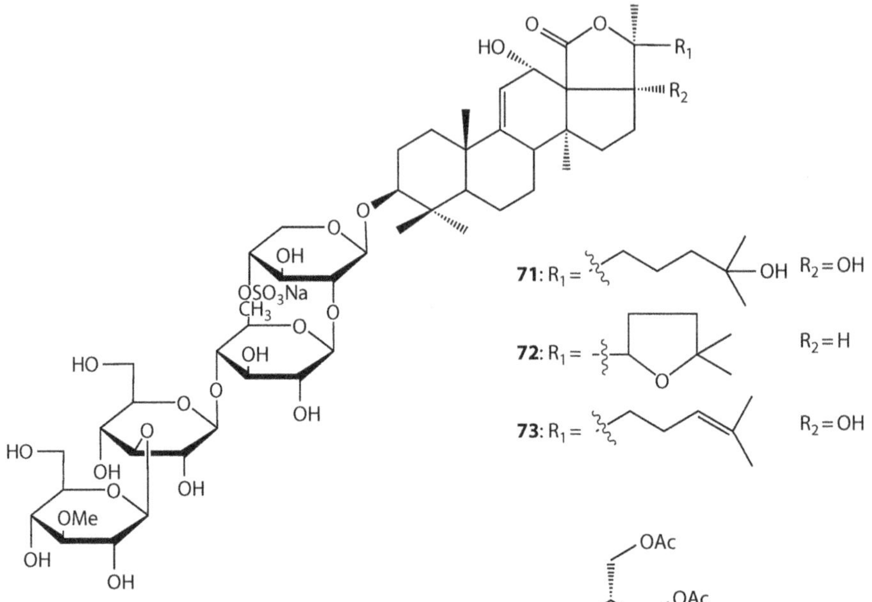

**71**: R$_1$ =                OH   R$_2$ = OH

**72**: R$_1$ =                     R$_2$ = H

**73**: R$_1$ =                R$_2$ = OH

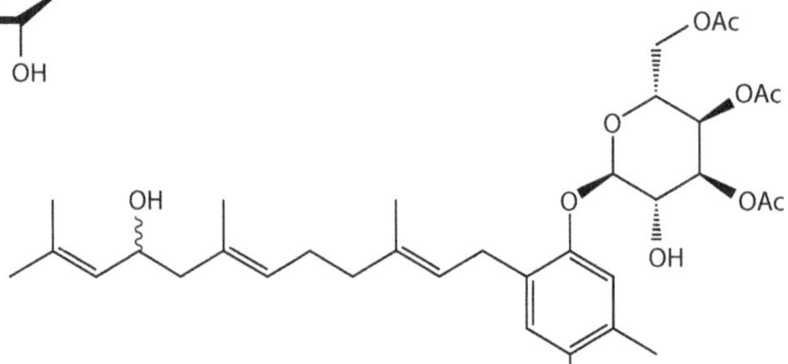

**74**

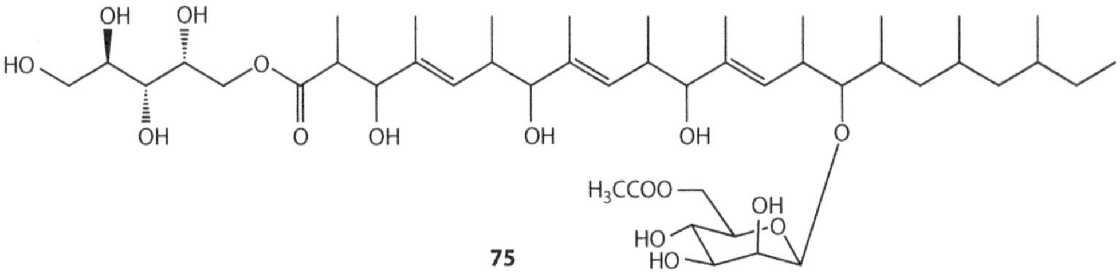

**75**

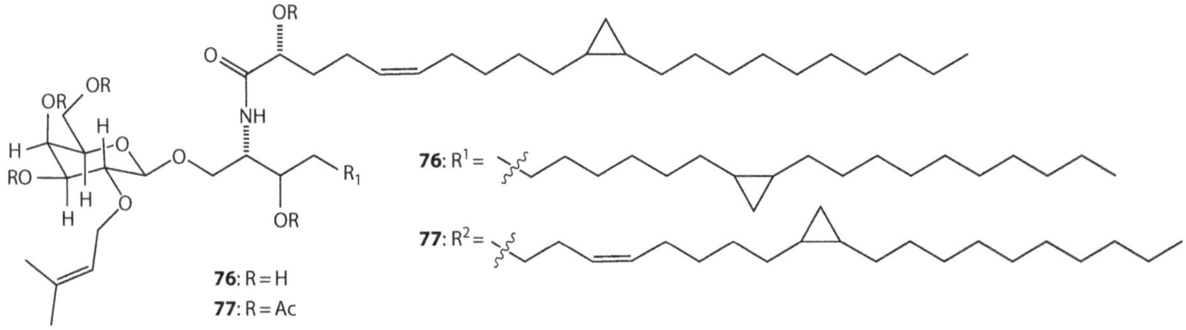

**76**: R$^1$ =

**77**: R$^2$ =

**76**: R = H
**77**: R = Ac

*Figure 5.1 (Continued)* Marine glycans possessing bioactivity.

### 5.2.3.1 Glycosaminoglycans

Glycosaminoglycans (GAGs) are natural macromolecules and are negatively charged heteropolysaccharides that are known for various biological functions. Based on the structure, these are classified into hyaluronic acids **4**, heparan sulfates **8**, dermatan sulfates **9, 10**, and chondroitin sulfates (A, C, D, and E) **11–14**. They are known for their anticoagulant property. In the past decades, many anticoagulants have been described. Hyaluronic acid (HA, also called hyaluronan **4**) is an anionic, nonsulfated GAG. It is a linear polymer consisting of repeating units of *N*-acetyl-D-glucosamine and D-glucuronic acid. It is used in the preparation of biocompatible and eco-friendly polymers for applications in drug delivery, tissue engineering, and viscosupplementation (Kanchana et al. 2013). HA has been shown to possess antiaging and antioxidant properties. Heparan sulfate **8**, which has the capacity to activate anti-thrombin that inhibits coagulation, is commercially available as anticoagulants and pro-angiogenic factors (Liu and Pedersen 2007; Fuster and Wang 2010). Dermatan sulfate **9, 10**, because of its variable sulfation patterns, is implicated in the wound-healing process, infection, and cardiovascular disease (Plichta et al. 2012). Chondroitin sulfate **11–14** in combination with glucosamine is used to treat osteoarthritis.

### 5.2.3.2 Exopolysaccharides

Microbes are well known to synthesize exopolysaccharides (EPSs). These are high-molecular-weight polymers produced by marine microbes, and are found to possess various biological activities such as antitumor, immunostimulatory, antihuman immunodeficiency virus, and anticoagulant. These have many biochemical and biomedical applications such as thickening agents, stabilizing agents, natural antioxidants, anticancer drugs, and bio-absorbents. Their role in the prevention of oxidative damage by scavenging free radicals makes them good therapeutic agents. With increase in the utilization of synthetic compounds as antioxidants for use in food, major concerns are about their toxicity, metabolism, absorption, and accumulation in the body organs and tissues. Since there is a necessity to find alternatives with similar antioxidant effect, the focus is shifting toward naturally occurring antioxidants. Another class of sulfated compounds is quite suitable for interacting with functional proteins because they are polyanionic. These are called sulfated polysaccharides (SPs). The SP–protein interactions control and regulate the activity of functional proteins in the body. The sulfate content in these polysaccharides drives their electrostatic interactions with these molecular complexes, which influence the protein functionality (Pomin et al. 2014). SPs include sulfated fucans (SFs), sulfated glycans (SGs), and GAGs. These SPs are mainly isolated from sea urchins, sea cucumbers, brown algae, green algae, red algae, and marine ascidians (tunicates). SFs, also called fucoidans, are rich in α-L-fucopyranosyl (Fuc*p*) units, with a few composed of xylose and uronic acids. The SG class of SPs is composed of α-L-, α-D-, and β-D-galactopyranosyl (Gal*p*) units. The sulfation patterns of SGs in marine algae are more complex than the regular and simpler structures in marine invertebrates. For example, in green algae, additional substitutions like pyruvates and the possibility of branches result in the complexity of SGs, whereas in red algae the backbone of repeated disaccharide units of 3-linked β-D-Gal*p* and 4-linked α-Gal units enhances the heterogeneous nature in SGs. Despite the complexity and heterogeneity of SFs and SGs in marine algae, regular and well-defined chemical structures of SFs and SGs from marine invertebrates enable us to understand the structure–function correlations that are important for the anticoagulation property (Pomin et al. 2014). Marine GAGs are unique in their structure compared to mammalian GAGs. Those include dermatan sulfate and chondroitin sulfate. GAG isolated from ascidian, *Styela plicata*, called dermatan sulfate (DS) **9, 10**, is composed of [→4)-α-L-IdoA-(2R$^1$,3R$^2$)-(1→3)-β-D-GalNAc-(4R$^3$,6R$^4$)-(1→]$_n$ (IdoA-iduronic acid; GalNAc-*N*-acetyl galactosamine). The sulfation at R$^1$, R$^2$, R$^3$, and R$^4$ is 66%, <5%, 94%, and 6%, respectively. This sulfation pattern is restricted to 2C in IdoA unit and occasionally to 4C in GalNAc unit. Chondroitin sulfates (CSs) **11–14**, another class of GAGs that are fucosylated (FucCS), are isolated from sea cucumbers. These are composed of {→3)-β-D-GalNAc-(1→4)-[α-L-FucP-(1→3)]-β-D-GlcA-(1→}$_n$. The sulfation in Fuc*p* unit is observed at 2, and/or 3, and/or 4 positions at different percentages, whereas the Fuc*p* unit is absent in mammalian CSs and sulfation is observed in GalNAc at either 4C or 6C (Pomin et al. 2014).

EPSs from fungi and bacteria are best considered as natural antioxidants. The β-type heteropolysaccharide from *Pseudomonas* PF-6 (Ye et al. 2012) and the water-soluble mannan EPSs ETW1 and ETW2 from the marine bacterium *Edwardsiella tarda* (Guo et al. 2010) are examples of EPSs as natural antioxidants. EPSs are also found in *Alteromonas macleodii* HYD657 associated with the deep-sea hydrothermal vent polychaete annelid living in extreme conditions. The EPS isolated from this isolate has found application in cosmetics because of its biological activities (Cambon-Bonavita et al. 2002). EPSs from deep-sea hydrothermal vents can be utilized as viscosifiers, gelling agents, emulsifiers, stabilizers, and texture enhancers. They have found applications in industrial products, for example, detergents, textiles, adhesives, paper, paint, food and beverages, due to their many properties such as stabilizing, suspending, thickening, gelling, coagulating, film-forming, and water retaining capability (Guezennec 2002). Chemical modifications of natural EPSs show the importance of changes in chemistry of structures, enhancing the physical and chemical properties and leading to efficient and more stable products for industrial and medical uses. As notable examples, the thickening, binding, and stabilizing properties are increased by methylation; flame-resistant properties and ion-exchange capabilities are introduced by conversion to the phosphate ester; and adhesive properties are improved with quaternization (Geresh et al. 2000). The reaction of the extracellular polysaccharide with the quaternizing agent results in increase in the elasticity of the EPS. The sulfated EPS isolated from red microalgae, *Porphyridium* sp., was reacted with 3-chloro-2-hydroxypropyltrimethyl ammonium chloride (Geresh et al. 2000). Fucoidan **3** is a complex sulfated polysaccharide distributed among marine organisms. Fucoidan from the sea cucumber *Acaudina molpadioides* possesses a protective effect against ethanol-induced gastric ulcer (Wang et al. 2012a). Fucoidans from different algae have been developed into commercial products (Morya et al. 2012; Yu et al. 2014).

## 5.3   Diversity

### 5.3.1   Glycolipids

Lipids with carbohydrate moieties are naturally present in all organisms in their cellular membranes. They are involved in various physiological functions, such as the regulation of cell proliferation and differentiation, cell recognition, signal transduction with their structures in receptors of cytotoxins, cells, viruses, transmitters, and hormones. The diverse moieties of carbohydrates in lipids result in their applications to different fields. Research on utilizing glycolipids for various applications is increasing because of their possessing various properties such as anti-adhesive, immunostimulatory, biosurfacant, antimicrobial, antifungal, and anticancer: for example, the glycerolipid A-5 from the intestine of sea urchins with anticancer activity (Sahara et al. 2002); oceanalin A **15** from the Australian sponge with antifungal activity (Makarieva et al. 2005); extracellular glycolipids from marine microbes, such as mannosyl erythritol lipids, polyol lipids, rhamno lipids, sophorose lipids, and succinoyl trehalose lipids, inducing cell differentiation of the leukemia cell line HL60 (Isoda et al. 1997); caminoside A **16** from marine sponge *Caminus sphaeroconia* with antimicrobial activity (Linington et al. 2002); and sulfoquinovosyldiacylglycerols **17** from the marine seaweed *Sargassum vulgare* with anti-herpes activity (Plouguerné et al. 2013).

### 5.3.2   Glycoproteins

Sugar moieties attached to protein molecules are referred to as glycoproteins. The glyco moieties play important roles such as signal recognition, trafficking of proteins to subcellular regions, and cell–cell signaling pathways. The major molecules involved in the interaction with proteins are glucose, galactose, mannose, inositol, and *N*-acetylglucosamine. Other than basic cellular functions, the cellular free glycoproteins possess antibacterial, antifungal, antitumor, and many other defense mechanisms. The 30 kDa D-galactose binding lectin purified from the marine sponge *Halichondria okadai* (Kawsar et al. 2010) and the 27 kDa tachylectin purified from *Suberites domuncula* (Schröder et al. 2003) show antibacterial and antifungal properties. Glycoproteins rich in xylose-rich glycans extracted from frustules of the marine diatom *Craspedostauros australis* function as cell surface adhesion molecules (Chiovitti et al. 2003).

### 5.3.3   Glycoalkaloids

Glycoalkaloids are marketed as exfoliants. Amphimedosides (A–E) **18–22** are naturally occurring glycosylated alkoxyamines. These glycosides were first isolated from the marine sponge *Amphimedon* sp., and were found to possess cytotoxic activities against P388 murine leukemia cells, with IC50 values ranging from 0.45 to 11.0 µg/mL (Takekawa et al. 2006). Shishijimicins A **23**, the enediyne antibiotic isolated from the marine ascidian *Didemnum proliferum*, inhibits the tumor cell lines 3Y1, HeLa, and P388 (Oku et al. 2003).

## 5.4   Marine sources

### 5.4.1   Marine plants

Plants in the marine ecosystem vary widely, from seaweeds, seagrasses, and mangroves to the simplest microscopic phytoplankton. Macroalgae comprising red algae, brown algae, and green algae, and microalgae comprising diatoms and dinoflagellates, come under the marine flora. Marine plants live in different habitats from salt marshes near shores to open seas. Living in diverse habitats, the marine plankton is the prime provider of nutrients to their symbiotic associates such as corals reefs, crustaceans, echinoderms, and mollusks in their food chain. The carbohydrates and related by-products from these marine plants are used in the food industry as additives, gelling agents, beverages, and other deictic foods and in many applications in biomedicine, fertilizers, bioethanol production, cosmetics, adhesives, paper industry, tissue engineering, drug delivery vehicles, and so on. Carrageenans, alignates, laminarins, SPs (fucoidans), heparin, and agar are the major carbohydrate polymers from these marine plants. Similarly, monomers of saccharides composed of heterogeneous, heterocyclic complexes form natural glycosides (terpenoid glycosides, alkaloid glycosides, polyketide glycosides, steroidal glycosides, etc.) (Véron et al. 1996; Plouguerné et al. 2013).

### 5.4.2　Marine animals

Marine invertebrates possess the capability to produce different kinds of carbohydrates such as glycans and glycosides. They include sponges, soft corals, sea urchins, sea cucumbers, cnidarians, and ascidians. Chitosans, lectins, and dextran sulfate polymers are the major polymer carbohydrates. Oligoglycosides are widely synthesized by most marine invertebrates (Kawsar et al. 2010; Badawy and Rabea 2011; Kalinin et al. 2012).

Among the marine invertebrates, the richest source for synthesizing steroid glycosides are starfish (Ivanchina et al. 2011). Bebrycoside 24 was isolated from the South China Sea gorgonian coral *Bebryceindica* (Yang et al. 2007). Being primitive, sessile, and benthic and filter feeders, associated with diverse microorganisms, and rich sources of biological compounds, sponges contain medicinally potent chemicals displaying various pharmacological properties. Examples include glycerolipids (GGL.2, {1-O-acyl-3-[α-glucopyranosyl-(1–3)-(6-O-acyl-α-mannopyranosyl)] glycerol} with 14-methyl hexadecanoic acid and 12-methyl tetradecanoic acid positioned at C-6 of the mannose unit and at glycerol moieties, respectively) 25 (antitumor agent) from the Adriatic sea sponge *Halichondria panicea* (Wicke et al. 2000), lutoside 26 (antimicrobial agent) from *Xestospongia* sp. (Bultel-Poncé et al. 1997; Bultel-Poncé et al. 1998), diglucosyl-glycerolipid (GGL11, {1, 2-O-diacyl-3-[β-glucopyranosyl-(1–6)-β-glucopyranosyl] glycerol} with 14-methyl hexadecanoic acid and 12-methyl tetradecanoic acid as the main fatty acid moieties) 27 (anticancer agent) from *Acanthella acuta* (Ramm et al. 2004), theonegramide 28 (antifungal compound) from *Entotheonella palauensis* (Bewley et al. 1996), KRN700 29 (antitumor and immunostimulatory compound) from *Agelas mauritianus* (Veldt et al. 2007; Motohashi et al. 2009), Xestovanin A 30 from *Xestospongia vanilla* (Northcote and Andersen 1989), and triterpenoid glycosides 31–37 isolated from sea cucumber *Holothuria leucospilota* (Hua et al. 2009).

## 5.5　Bioactivity of marine glycans

### 5.5.1　Antibacterial

The D-galactose-binding lectin purified from the marine sponge *H. okadai* showed inhibitory activity against gram-positive *Bacillus megaterium* and *B. subtilis* (Kawsar et al. 2010). The steroidal glycosides iyengaroside-A 38 and B 39 and the clerosterol galactoside with antibacterial properties were isolated from marine green algae *Codium iyengarii* collected from the Arabian Sea (Ali et al. 2012). These compounds were found to be active against *Corynebacterium diphtheriae*, *Escherichia coli*, *Klebsiella pneumoniae*, *Shigella dysenteriae*, and *Staphylococcus aureus* (Ali et al. 2012).

### 5.5.2　Antibiofilm

Self-synthesized biofilms of bacterial pathogens that induce chronic infections act as a barrier for the action of antibiotics. Naturally occurring EPSs have attracted the attention of researchers because of their role in inhibiting biofilm formation in many bacteria (Jiang et al. 2011). *E. coli* biofilm formation was inhibited by cell-free culture supernatants of two different bacteria associated with the marine sponge *Spongia officinalis* (Sayem et al. 2011; Rendueles et al. 2013). One of the active supernatants (SP1), which was found to be the high-molecular-weight polysaccharide (SP1 EPS), inhibited biofilm formation by *Acinetobacter* sp., *S. aureus*, *Salmonella typhimurium*, *Shigella sonnei*, *Listeria monocytogenes*, *Bacillus cereus*, *B. amyloliquefaciens*, *B. pumilus*, and *B. subtilis* without inhibiting the growth (Sayem et al. 2011).

### 5.5.3　Anticancer

The diterpene glycosides, virescenosides O, P, and Q 40–42, characterized from the marine strain *Acremonium striatisporum* KMM 4401 and associated with the holothurians *Eupentacta fraudatrix*, showed cytotoxic activities against carcinoma *in vitro* and developing eggs of the sea urchin *Strongylocentrotus intermedius* (Afiyatullov et al. 2002). The glycolipid 3′-sulfonoquinovosul-1′-monoacylglycerol isolated from the intestine of sea urchins was found to inhibit the growth of solid tumors (Sahara et al. 2002).

### 5.5.4　Anticoagulant

Alginate 2, which is rich in mannuronic acid, is synthesized by the marine lactic acid bacteria *Pediococcus acidilactici*, *Weissella paramesenteroides*, *P. pentosaceus*, and *Enterococcus faecium* by fermenting *Sargassum* sp. (Shobharani et al. 2014). Alginate is an anionic polysaccharide largely distributed in the cell walls of brown algae. It has many roles in paper and textile industries as water proofing and fire proofing material and in pharmaceutical industry as hydrophilic drug carriers and matrix materials. It has another unique feature: anticoagulant activity. According to the World Health Organization (WHO), by 2030, the main cause of deaths in the world will be cardiovascular diseases including heart diseases and strokes related to thrombosis. Heparin, a natural anticoagulant, is used in hematology and transfusion medicine. But, with difficulties in production, chemical inhomogeneity, variability in physiological activities, and so on, a natural, safe, and easy-to-use heparin substitute is highly demanded. Naturally occurring polysaccharides synthesized by seaweeds are considered better substitutes for heparin. Those include sulfated galactan from red seaweed, sulfated β-arabinan from green seaweed, and fucoidan and alginates from brown seaweed (Karaki et al. 2013; Shobharani et al. 2014). The triterpenoid glycosides leucospilotaside A **31**, leucospilotaside B **32**, leucospilotaside C **33**, holothurin A **34**, holothurin B **35**, holothurin B2 **36**, echinoside B **37** are isolated from the sea cucumber *Holothuria leucospilota* (Hua et al. 2009).

## 5.5.5    Antifouling agent

The steroidal glycosides junceellosides C, E, F, and G **43–46**, isolated from the South China Sea gorgonian *Dichotella gemmacea*, were the first reported glycosides with β-ʟ-arabinopyranose from marine sources. The sterol part of these glycosides possesses antifouling activity and shows lethality toward brain shrimp (Jiang et al. 2013).

## 5.5.6    Antifungal

Hypoxysordarin **47**, a diterpene glycoside isolated from the facultative marine fungus *Hypoxylon croceum*, was obtained from mangrove estuary driftwood. It was found to inhibit fungal protein synthesis by binding to the EF2–ribosome complex (Daferner et al. 1999; Bugni and Ireland 2004). The marine soft coral *Lemnalia* sp. was found to produce new decalin-type bicyclic diterpenoids lemnalosides A–D **48–51**, of which lemnalosides A and B possess inhibitory activity toward hyphae formation in *Streptomyces* 85E at the concentration of 20 and 2.5 µg per disk, respectively (Yao et al. 2007). New polyene tetramic acids having a trisaccharide unit, isolated from the marine sponge *Siliquariaspongia japonica*, were found to inhibit fungal pathogens such as *Aspergillus fumigatus* and *Candida albicans* (Sata et al. 1999). Aurantosides A and B **52, 53** from the sponge *Theonella swinhoei*, Aurantoside C **54** from the Philippine sponge *Homophymia conferta*, and Aurantoside D–F **55–57** from *S. japonica* are the group of tetramic acid derivatives containing xylose and arabinose units (Sata et al. 1999). Aurantoside K **58** from the sponge *Melophlus* shows a wide range of antifungal activity against *C. albicans*, *Cryptococcus neoformans*, *Aspergillus niger*, *Penicillium* sp., *Rhizopus sporangia*, and *Sordaria* sp. (Kumar et al. 2012). A yeast strain defective in double-stranded DNA repair was inhibited by the steroidal glycosides erylosides K and L **59, 60** isolated from *Erylus lendenfeldi* (Sandler et al. 2005). Venturicidin C **61**, a macrolide glycoside, possesses antifungal activity (Shaaban et al. 2014).

## 5.5.7    Antioxidant

*Pseudomonas* PF-6, the marine bacterium isolated from the sea-mud of Dalian HeiShiHi Jiao waters, produces large amounts of viscous EPSs during liquid fermentation. The EPSs, on structural characterization, were identified to belong to a β-type heteropolysaccharide (Ye et al. 2012). Two water-soluble mannan EPSs, ETW1 and ETW2, purified from the marine bacterium *Edwardsiella tarda* showed good antioxidant as well as hydroxyl and DPPH radical scavenging activities (Guo et al. 2010).

## 5.5.8    Antitussive

Research on chronic respiratory diseases such as cough, asthma, and bronchitis is important in health care for the discovery of pharmacologically active agents. Interestingly, marine glycans that possess various activities are studied for their antitussive properties. Asterosaponins possessing antifungal, antiviral, and anti-biofouling activities are chemically diverse with steroidal groups and various glycosidic linkages in their structure. *In vitro* analysis of asterosaponin isolated from the star fish *Luidia quinaria* on asthma-induced mouse showed expectorant, antitussive, and antiasthmatic properties (Guo et al. 2009).

## 5.5.9    Antiviral

A sulfated polysaccharide produced by the brown seaweed *Hydroclathrus clathratus* showed strong antiviral activity to herpes simplex virus type 1 (HSV-1) of both cyclovir-resistant strain and clinical strain (Wang et al. 2010). An anti-herpes-active glycolipid, 1,2-di-*O*-palmitoyl-3-*O*-(6-sulfo-α-ᴅ-quinovopyranosyl)-glycerol **17**, was extracted from the Brazilian brown seaweed *Sargassum vulgare* (Plouguerné et al. 2013).

## 5.5.10    Bioflocculants

Flocculants have a wide range of applications in industry because of their capability to aggregate colloids. Applications in tap-water and wastewater treatment, in fermentation, and in various food industries require biodegradable, eco-friendly, and natural flocculants. Polysaccharide-rich bioflocculant from the sea soil bacteria *Rhodococcus erythropolis* was identified to remove reactive dye from active sludge. The decolorization improved to 93.9% with the addition of copper sulfate in the treatment (Peng et al. 2014). The marine *myxobacterium* strain NU-2 isolated from a salt soil sample produced a bioflocculant rich in polysaccharide and protein content. The flocculation/bleaching property was observed to be 98.2% for acid red and 99% for emerald blue (Zhang et al. 2002).

## 5.5.11 Cytoprotective agents

Fucoidan 3 is a complex sulfated polysaccharide distributed among marine organisms. Fucoidan from the sea cucumber *Acaudina molpadioides* shows a protective effect against ethanol-induced gastric ulcer (Wang et al. 2012a). Fucoidans from different algae have been developed into commercial products (Morya et al. 2012; Yu et al. 2014).

## 5.5.12 Cytotoxic agents

Euplexides (A–G) 62–68 are farnesyl hydroquinone glycosides isolated from *Euplexaura anastomosans*. These compounds displayed moderate cytotoxicity, antioxidant activity, and inhibitory effect against phospholipase A2 (Shin et al. 1999; Seo et al. 2001; Wang et al. 2012b). Amphimedosides (A–E) 18–22 are naturally occurring glycosylated alkoxyamines. These glycosides were first discovered from the marine sponge *Amphimedon* sp. They were found to possess cytotoxic activity against P388 murine leukemia cells, with IC50 values ranging from 0.45 to 11.0 µg/mL (Takekawa et al. 2006). Two cyclopenta [*a*] indene glycosides, cyanosporasides A and B, 69, 70, possessing 3-keto-pyrohexose sugar and cyano-substituted ring systems, were isolated from the marine actinomycete *Salinispora pacifica*. Cyanosporaside A 37 was found to show weak cytotoxicity against human colon carcinoma HCT-116 cells (Oh et al. 2006). Another sea cucumber, *Holothuria scabra*, was identified to synthesize triterpene glycoside compounds that possess cytotoxic activity. Scabraside D 71, fuscocineroside C 72, and 24-dehydroechinoside A 73 inhibited the human tumor cell lines P-388, A549, MKN-28, HCT116, and MCF-7 (Hua et al. 2012). Cladionol A 74, a polyketide glycoside isolated from *Gliocladium* sp., associated with the sea grass *Syringodium isoetifolium*, was found to be moderately cytotoxic to L1210 and KB cells (Kasai et al. 2005). The sulfated polysaccharide B-1 from the marine *Pseudomonas* sp. stimulates the apoptosis of the leukemic cell line U937 (Matsuda et al. 2003). A new hydroquinone glycoside derivative, moritoside 75, isolated from the gorgonian *Euplexaura* sp., was found to inhibit the first cell division of fertilized star fish (*Asterina pectinifera*) eggs (Fusetani et al. 1985; Wang et al. 2012b).

## 5.5.13 Immunomodulatory

The development/progression of various malignancies in the body is suppressed by the inherent capacity of the immune system. The failure of the system's functional role can be overcome by the use of promising immunostimulatory molecules that are available from natural sources. Naturally occurring polysaccharides extracted from plants and microorganisms of both terrestrial and marine origin play potential roles in activating the immune system. Marine carbohydrates possessing immunostimulatory properties include simple glycans, triterpene glycosides, EPSs, and glycolipids. A triterpene glycoside, frondoside A, isolated from the sea cucumber *Cucumaria frondosa*, at its subtoxic doses shows strong immunomodulatory properties. This compound induces macrophage phagocytosis of *S. aureus* by activating lysosomal activity (Aminin et al. 2008). But organ/tissue transplantation (called *grafting* in medicine) requires immunosuppressants for a brief period for the recipient immune system to accept the transplanted organ/tissue. Natural immunosuppressants are required to have more structural similarity with the cellular-matrix molecules such as glycolipids and glycoproteins of the cell membranes. Marine sponges, which are the direct targets of researchers to screen diverse molecules, are identified to contain various immunomodulatory compounds. Plakoside A and B 76, 77, two unique glycosphingolipids of the prenylated glycolipid class from the marine sponge *Plakortis simplex*, possess strong immunosuppressive activity on activated T cells (Costantino et al. 1997).

## 5.5.14 Thickening and stabilizing agents

Amphipathic molecules that contain hydrophilic and hydrophobic domains are the major components of a wide range of products with numerous applications. To meet the global demands and to produce renewable and nontoxic products, biosurfactants and bioemulsifiers of natural origin have received increasing interest. High-molecular-weight glycans providing structural and functional heterogeneity and with a large number of reactive groups stably bind with oil droplets during emulsifications, and the resulting emulsifiers are able to meet industrial demand for new types of biopolymers with enhanced functionalities. HE39 and HE67 are glycoproteins with high protein contents, and uronic acid shows emulsification of food oils at both neutral and acidic conditions, withstanding both acid- and high-temperature treatment (Gutiérrez et al. 2007a). These bioemulsifiers are extracted from the marine proteobacterium *Halomonas* sp. (Gutiérrez et al. 2007a). The marine α-proteobacterium *Antarctobacter* sp. TG22 was detected with a high-molecular-weight glycoprotein AE22 rich in uronic acid content (Gutiérrez et al. 2007b). Compared to xanthan gum and gum arabic, AE22 polymer showed better emulsion-stabilizing properties, which would make promising stabilizing agents for biotechnological applications (Gutiérrez et al. 2007b).

## 5.6 Conclusion

Carbohydrates, a source of carbon for living organisms for building their body parts and energy, react with other biological molecules and form glycolipids, glycoproteins, and glycoalkaloids. This was established through evolution and enabled them to have biological activities. A glycan isolated from a particular organism can show activities such as antibacterial, antibiofilm, anticancer, anticoagulant, antifouling, antifungal, antioxidant, antitussive, antiviral, bioflocculants, cytoprotective, cytotoxic, immuno-modulatory, and thickening, and act as a stabilizing agent. This chapter mainly described marine glycans possessing the above bioactivities and their source organisms.

## Acknowledgments

The authors thank the Department of Science and Technology; the Department of Biotechnology; Indian council of Medical Research; Government of India, New Delhi; India and Vice Chancellor, SASTRA University, Thanjavur, India, for providing financial and infrastructural facilities.

## References

Afiyatullov, S. S., A. I. Kalinovsky, T. A. Kuznetsova, V. V. Isakov, M. V. Pivkin, P. S. Dmitrenok, and G. B. Elyakov. 2002. New diterpene glycosides of the fungus *Acremonium striatisporum* isolated from a sea cucumber. *Journal of Natural Products* 65:641–644.

Ali, M. S., M. Saleem, R. Yamdagni, and M. A. Ali. 2012. Steroid and antibacterial steroidal glycosides from marine green alga *Codium iyengarii* Borgesen. *Natural Product Letters* 16:407–413.

Aminin, D. L., I. G. Agafonova, V. I. Kalinin, A. S. Silchenko, S. A. Avilov, V. A. Stonik, P. D. Collin, and C. Woodward. 2008. Immunomodulatory properties of frondoside A, a major triterpene glycoside from the North Atlantic commercially harvested sea cucumber *Cucumaria frondosa*. *Journal of Medicinal Food* 11:443–453.

Badawy, M. E. I. and E. I. Rabea. 2011. A biopolymer chitosan and its derivatives as promising antimicrobial agents against plant pathogens and their applications in crop protection. *International Journal of Carbohydrate Chemistry* 460381.

Bewley, C. A., N. D. Holland, and D. J. Faulkner. 1996. Two classes of metabolites from *Theonella swinhoei* are localized in distinct populations of bacterial symbionts. *Experientia* 52:716–722.

Bugni, T. S. and C. M. Ireland. 2004. Marine-derived fungi: A chemically and biologically diverse group of microorganisms. *Natural Product Reports* 21:143–163.

Bultel-Poncé, V., C. Debitus, J. P. Berge, C. Cerceau, and M. Guyot. 1998. Metabolites from the sponge-associated bacterium *Micrococcus luteus*. *Journal of Marine Biotechnology* 6:233–236.

Bultel-Poncé, V., C. Debitus, A. Blond, C. Cereau, and M. Guyot. 1997. Lutoside: An Acyl-l-(Acyl-6'-Mannobiosyl)-3-Glycerol isolated from the sponge-associated bacterium *Micrococcus luteus*. *Tetrahedron Letters* 38:5805–5808.

Cambon-Bonavita, M. A., G. Raguénès, J. Jean, P. Vincent, and J. Guezennec. 2002. A novel polymer produced by a bacterium isolated from a deep-sea hydrothermal vent polychaete annelid. *Journal of Applied Microbiology* 93:310–315.

Chiovitti, A., A. Bacic, J. Burke, and R. Wetherbee. 2003. Heterogeneous xylose-rich glycans are associated with extracellular glycoproteins from the biofouling diatom *Craspedostauros australis* (Bacillariophyceae). *European Journal of Phycology* 38:351–360.

Costantino, V., E. Fattorusso, A. Mangoni, M. D. Rosa, and A. Ianaro. 1997. Glycolipids from Sponges. Plakoside A and B, two unique prenylated glycosphingolipids with immunosuppressive activity from the marine sponge *Plakortis simplex*. *Journal of American Chemical Society* 119:12465–12470.

Daferner, M., S. Mensch, T. Anke, and O. Sternerb. 1999. Hypoxysordarin, a new sordarin derivative from *Hypoxylon croceum*. *Zeitschrift für Naturforschung C* 54c:474–480.

Fusetani, N., K. Yasukawa, S. Matsunaga, and K. Hashimoto. 1985. Bioactive marine metabolites XII. Moritoside, an inhibitor of the development of starfish embryo, from the gorgonian sp. *Tetrahedron Letters* 26:6449–6452.

Fuster, M. M. and L. Wang. 2010. Endothelial heparin sulfate in angiogenesis. *Progress in Molecular Biology and Translational Science* 93:179–212.

Geresh, S., R. P. Dawadi, and S. M. Arad. 2000. Chemical modifications of biopolymers: Quaternization of the extracellular polysaccharide of the red microalga *Porphyridium* sp. *Carbohydrate Polymers* 63:75–80.

Guezennec, J. 2002. Deep-sea hydrothermal vents: A new source of innovative bacterial exopolysaccharides of biotechnological interest? *Journal of Industrial Microbiology and Biotechnology* 29:204–208.

Guo, C., X. Tang, X. Dong, and Y. Yang. 2009. Studies on the expectorant, antitussive and antiasthmatic properties of asterosaponin extracted from *Luidia quinaria*. *African Journal of Biotechnology* 8:6694–6696.

Guo, S., W. Mao, Y. Han, X. Zhang, C. Yang, Y. Chen, Y. Chen et al. 2010. Structural characteristics and antioxidant activities of the extracellular polysaccharides produced by marine bacterium *Edwardsiella tarda*. *Bioresource Technology* 101:4729–4732.

Gutiérrez, T., B. Mulloy, C. Bavington, K. Black, and D. H. Green. 2007. Partial purification and chemical characterization of a glycoprotein (putative hydrocolloid) emulsifier produced by a marine bacterium *Antarctobacter*. *Applied Microbiology and Biotechnology* 76:1017–1026.

Gutiérrez, T., B. Mulloy, K. Black, and D. H. Green. 2007. Glycoprotein emulsifiers from two marine *Halomonas* species: Chemical and physical characterization. *Journal of Applied Microbiology* 103, 1716–1727.

Hua, H. A. N., L. I. Ling, Y. I. Yang-Hua, W. A. N. G. Xiao-Hua, and P. A. N. Min-Xiang. 2012. Triterpene glycosides from sea cucumber *Holothuria scabra* with cytotoxic activity. *Chinese Herbal Medicines* 4:183–188.

Hua, H. A. N., Y. Yang-Hua, L. Ling, L. Bao-Shu, P. Min-Xiang, Y. Bing, and W. Xiao-hua. 2009. Triterpene glycosides from sea cucumber *Holothuria leucospilota*. *Chinese Journal of Natural Medicines* 7:346–350.

Isoda, H., D. Kitamoto, H. Shinmoto, M. Matsumura, and T. Nakahara. 1997. Microbial extracellular glycolipid induction of differentiation and inhibition of the protein kinase C activity of human promyelocytic leukemia cell line HL60. *Bioscience, Biotechnology, and Biochemistry* 61:609–614.

Ivanchina, N. V., A. A. Kicha, and V. A. Stonik. 2011. Steroid glycosides from marine organisms. *Steroids* 76:425–454.

Jiang, M., P. Sun, H. Tang, B. S. Liu, T. J. Li, C. Li, and W. Zhang. 2013. Steroids glycosylated with both D- and L-arabinoses from the South China Sea gorgonian *Dichotella gemmacea*. *Journal of Natural Products* 76:764–768.

Jiang, P., J. Li, F. Han, G. Duan, X. Lu, Y. Gu, and W. Yu. 2011. Antibiofilm activity of an exopolysaccharide from marine bacterium *Vibrio* sp. QY101. *PLOS ONE* 6:e18514.

Kalinin, V. I., N. V. Ivanchina, V. B. Krasokhin, T. N. Makarieva, and V. A. Stonik. 2012. Glycosides from marine sponges (Porifera, Demospongiae): Structures, taxonomical distribution, biological activities and biological roles. *Marine Drugs* 10:1671–1710.

Kanchana, S., M. Arumugam, S. Giji, and T. Balasubramanian. 2013. Isolation, characterization and antioxidant activity of hyaluronic acid from marine bivalve mollusc *Amussium pleuronectus* (Linnaeus, 1758). *Bioactive Carbohydrates and Dietary Fibre* 2:1–7.

Karaki, N., C. Scbaaly, N. Chahine, T. Faour, A. Zinchenko, S. Rachid, and H. Kanaan. 2013. The antioxidant and anticoagulant activities of polysaccharides isolated from the brown algae Dictyopteris polypodioides growing on the Lebanese coast. *Journal of Applied Pharmaceutical Science* 3:43–51.

Kasai, Y., K. Komatsu, H. Shigemori, M. Tsuda, Y. Mikami, and J. Kobayashi. 2005. Cladionol A, a polyketide glycoside from marine-derived fungus *Gliocladium* Species. *Journal of Natural Products* 68:777–779.

Kawsar, S. M. A., S. M. A. Mamun, Md. S. Rahman, H. Yasumitsu, and Y. Ozeki. 2010. In vitro antibacterial and antifungal effects of a 30 kDa D-galactoside-specific lectin from the demosponge, *Halichondria okadai*. *International Journal of Biological, Veterinary, Agricultural and Food Engineering* 4:19–25.

Kilian, S., S. Kritzinger, C. Rycroft, G. Gibson, and J. du Preez. 2002. The effects of the novel bifidogenic trisaccharide, neokestose, on the human colonic microbiota. *World Journal of Microbiology and Biotechnology* 18:637–644.

Kumar, R., R. Subramani, K. D. Feussner, and W. Aalbersberg. 2012. Aurantoside K, a new antifungal tetramic acid glycoside from a Fijian marine sponge of the genus *Melophlus*. *Marine Drugs* 10:200–208.

Linington, R. G., M. Robertson, A. Gauthier, B. B. Finlay, R. V. Soest, and R. J. Andersen. 2002. Caminoside A, an antimicrobial glycolipid isolated from the marine sponge *Caminus sphaeroconia*. *Organic Letters* 4:4089–4092.

Liu, J. and L. C. Pedersen. 2007. Anticoagulant heparan sulfate: Structural specificity and biosynthesis. *Applied Microbiology and Biotechnology* 74:263–272.

Makarieva, T. N., V. A. Denisenko, P. S. Dmitrenok, A. G. Guzii, E. A. Santalova, V. A. Stonik, J. B. MacMillan, and T. F. Molinski. 2005. Oceanalin A, a hybrid α,ω-bifunctionalized sphingoid tetrahydroisoquinoline β-glycoside from the marine sponge *Oceanapia* sp. *Organic Letters* 7:2897–2900.

Matsuda, M., T. Yamori, M. Naitoh, and K. Okutani. 2003. Structural revision of sulfated polysaccharide b-1 isolated from a marine *Pseudomonas* species and its cytotoxic activity against human cancer cell lines. *Marine Biotechnology* 5:13–19.

Morya, V. K., J. Kim, and E. K. Kim. 2012. Algal fucoidan: Structural and size-dependent bioactivities and their perspectives. *Applied Microbiology and Biotechnology* 93:71–82.

Motohashi, S., K. Nagato, N. Kunii, H. Yamamoto, K. Yamasaki, K. Okita, H. Hanaoka et al. 2009. A phase I-II study of α-galactosylceramide-pulsed IL-2/GM-CSF-cultured peripheral blood mononuclear cells in patients with advanced and recurrent non-small cell lung cancer. *The Journal of Immunology* 182:2492–2501.

Northcote, P. T. and R. J. Andersen. 1989. Xestovanin A and secoxestovanin A, triterpenoid glycosides with new carbon skeletons from the sponge *Xestospongia vanilla*. *Journal of American Chemical Society* 111:6276–6280.

Oh, D. C., P. G. Williams, C. A. Kauffman, P. R. Jensen, and W. Fenical. 2006. Cyanosporasides A and B, chloro- and cyano-cyclopenta [a] indene glycosides from the marine actinomycete "*Salinispora pacifica.*" *Organic Letters* 8:1021–1024.

Oku, N., S. Matsunaga, and N. Fusetani. 2003. Shishijimicins A–C, novel enediyne antitumor antibiotics from the ascidian *Didemnum proliferum*. *Journal of American Chemical Society* 125:2044–2045.

Peng, L., C. Yang, G. Zeng, L. Wang, C. Dal, Z. Long, H. Liu, and Y. Zhong. 2014. Characterization and application of bioflocculant prepared by *Rhodococcus erythropolis* using sludge and livestock wastewater as cheap culture media. *Applied Microbiology and Biotechnology* 98:6847–6858.

Plichta, J. K. and K. A. Radek. 2012. Sugar-coating wound repair: A review of FGF-10 and dermatan sulfate in wound healing and their potential application in burn wounds. *Journal of Burn Care Research* 33:299–310.

Plouguerné, E., L. M. de Souza, G. L. Sassaki, J. F. Cavalcanti, M. T. V. Romanos, B. A. P. da Gama, R. P. Pereira, and E. Barreto-Bergter. 2013. Antiviral sulfoquinovosyldiacylglycerols (SQDGs) from the Brazilian Brown Seaweed *Sargassum vulgare*. *Marine Drugs* 11:4628–4640.

Pomin, V. H. and P. A. S. Mourão. 2014. Specific sulfation and glycosylation—A structural combination for the anticoagulation of marine carbohydrates. *Frontiers in Cellular and Infection Microbiology* 4:33.

Ramm, W., W. Schatton, I. Wagner-Döbler, V. Wray, M. Nimtz, H. Tokuda, F. Enjyo et al. 2004. Diglucosyl-glycerolipids from the marine sponge-associated *Bacillus pumilus* strain AAS3: Their production, enzymatic modification and properties. *Applied Microbiology and Biotechnology* 64 (4):497–504.

Rendueles, O., J. B. Kaplan, and J. M. Ghigo. 2013. Antibiofilm polysaccharides. *Environmental Microbiology* 15:334–346.

Sahara, H., S. Hanashima, T. Yamazaki, S. Takahashi, F. Sugawara, S. Ohtani, M. Ishikawa et al. 2002. Anti-tumor effect of chemically synthesized sulfolipids based on sea urchin's natural sulfonoquinovosylmonoacylglycerols. *Japanese Journal of Cancer Research* 93:85–92.

Sandler, J. S., S. L. Forsburg, and D. J. Faulkner. 2005. Bioactive steroidal glycosides from the marine sponge *Erylus lendenfeldi*. *Tetrahedron* 61:1199–1206.

Sata, N. U., S. Matsunaga, N. Fusetani, and R. W. M. V. Soest. 1999. Aurantosides D, E, and F: New antifungal tetramic acid glycosides from the marine sponge *Siliquariaspongia japonica*. *Journal of Natural Products* 62:969–971.

Sayem, S. M. A., E. Manzo, L. Ciavatta, A. Tramice, A. Cordone, A. Zanfardino, M. D. Felice, and M. Varcamonti. 2011. Anti-biofilm activity of an exopolysaccharide from a sponge-associated strain of *Bacillus licheniformis*. *Microbial Cell Factories* 10:74.

Schröder, H. C., H. Ushijima, A. Krasko, V. Gamulin, N. L. Thakur, B. Diehl-Seifert, I. M. Müller, and W. E. G. Müller. 2003. Emergence and disappearance of an immune molecule, and antimicrobial lectin, in basal metazoa. *The Journal of Biological Chemistry* 278:32810–32817.

Seo, Y., J. R. Rho, K. W. Cho, and J. Shin. 2001. New farnesyl-hydroquinone glycosides from the gorgonian *Euplexaura anastomosans*. *Natural Product Letters* 15:81–87.

Shaaban, K. A., S. Singh, S. I. Elshahawi, X. Wang, L. V. Ponomareva, M. Sunkara, G. C. Copley et al. 2014. Venturicidin C, a new 20-membered macrolide produced by *Streptomyces* sp. TS-2–2. *Journal of Natural Products* 67:223–230.

Shin, J., Y. Seo, and K. W. Cho. 1999. Euplexides A-E: Novel farnesylhydroquinone glycosides from the gorgonian *Euplexaura anastomosans*. *Tetrahedron* 64:1853–1858.

Shobharani, P., V. H. Nanishankar, P. M. Halami, and N. M. Sachindra. 2014. Antioxidant and anticoagulant activity of polyphenol and polysaccharides from fermented *Sargassum* sp. *International Journal of Biological Macromolecules* 65:542–548.

Takekawa, Y., S. Matsunaga, R. W. M. van Soest, and N. Fusetani. 2006. Amphimedosides, 3-alkylpyridine glycosides from a marine sponge *Amphimedon* sp. *Journal of Natural Products* 69:1503–1505.

Veldt, B. J., H. J. J. van der Vliet, B. M. E. von Blomberg, H. van Vlierberghe, G. Gerken, N. Nishi, K. Hayashi et al. 2007. Randomized placebo controlled phase I/II trial of alpha-galactosylceramide for the treatment of chronic hepatitis C. *Journal of Hepatology* 47:356–365.

Véron, B., C. Billard, J. C. Dauguet, M. A. Hartmann. 1996. Sterol composition of *Phaeodactylum tricornutum* as influenced by growth temperature and light spectral quality. *Lipids* 9:989–994.

Wang, H., V. E. C. Ooi, and P. O. Ang Jr. 2010. Anti-herpesviral property and mode of action of a polysaccharide from brown seaweed (*Hydroclathrus clathratus*). *World Journal of Microbiology and Biotechnology* 26:1703–1713.

Wang, L. H., J. H. Sheu, S. Y. Kao, J. H. Su, Y. H. Chen, Y. H. Chen, Y. D. Su et al. 2012a. Natural product chemistry of gorgonian corals of the family Plexauridae distributed in the indo-pacific ocean. *Marine Drugs* 10:2415–2434.

Wang, Y., W. Su, C. Zhang, C. Xue, Y. Chang, X. Wu, Q. Tang, and J. Wang. 2012b. Protective effect of sea cucumber (*Acaudina molpadioides*) fucoidan against ethanol-induced gastric damage. *Food Chemistry* 133:1414–1419.

Wicke, C., M. Hüners, V. Wray, M. Nimtz, U. Bilitewski, and S. Lang. 2000. Production and structure elucidation of glycoglycerolipids from a marine sponge-associated *Microbacterium* species. *Journal of Natural Products* 63:621–626.

Yang, J., S. H. Qi, S. Zhang, Z. H. Xiao, and Q. X. Li. 2007. Bebrycoside, a new steroidal glycoside from the Chinese gorgonian coral *Bebryce indica*. *Pharmazie* 62:154–155.

Yao, G., N. B. Vidor, A. P. Foss, and L. C. Chang. 2007. Lemnalosides A-D, decalin-type bicyclic diterpene glycosides from the marine soft coral *Lemnalia* sp. *Journal of Natural Products* 70:901–905.

Ye, S., F. Liu, J. Wang, H. Wang, and M. Zhang. 2012. Antioxidant activities of an exopolysaccharide isolated and purified from marine *Pseudomonas* PF-6. *Carbohydrate Polymers* 87:764–770.

Yu, L., L. Ge, C. Xue, Y. Chang, C. Zhang, X. Xu, Y. Wang. 2014. Structural study of fucoidan from sea cucumber *Acaudina molpadioides*: A fucoidan containing novel tetrafucose repeating unit. *Food Chemistry* 142:197–200.

Zhang, J., Z. Liu, S. Wang, and P. Jiang. 2002. Characterization of a bioflocculant produced by the marine myxobacterium *Nannocystis* sp. NU-2. *Applied Microbiology and Biotechnology* 59:517–522.

Zong, A., H. Cao, and F. Wang. 2012. Anticancer polysaccharides from natural resources: A review of recent research. *Carbohydrate Polymers* 90:1395–1410.

## chapter six

# Marine glycans in relationship with probiotic microorganisms to improve human and animal health

*Van Duy Nguyen*

## Contents

## 6.1 Introduction

According to the Alimentary Pharmabiotic Centre at University College Cork in Ireland, pharmabiotics are defined as any biological entity "mined" from gastrointestinal microbiota, including probiotics, bacteriocins, bacteriophages, and bioactive molecules. They are believed to have significant impacts on the pharmaceutical, medical food, and functional food sectors in adapting with the current challenge posed by multidrug-resistant bacteria and food supply, in terms of both quantity and quality (Patterson et al., 2014). The population growth in many countries leads to an increasing demand for animal food. Besides health threats linked to climate change, heavy losses resulting from animal diseases have a huge impact on the world animal production, particularly in developing countries. Therefore, issues of food security and animal health have become a serious, widespread concern.

The use of broad-spectrum classical antibiotics may induce an antibiotic-resistant mechanism of bacteria and an accumulation of unused residues of drugs in the environment, which, in turn, results in long-term negative effects on human and animal health. Thus, scientific communities have proposed friendly alternatives such as vaccines, antibiotic substitutes, and probiotics (Nguyen et al., 2013; Cheng et al., 2014). However, the use of vaccines is often laborious, costly, and highly stressful to the animals. Therefore, pharmabiotics, or probiotics, prebiotics, bacteriocins, bacteriophages, and bioactive molecules are being introduced to be used in food, aquaculture, livestock, and clinical settings. This chapter focuses on the potential of marine glycans from seaweed and probiotic bacteria for use as prebiotics, bioactive molecules, or in combination with probiotics to develop highly effective synbiotics.

### 6.1.1 Prebiotics

Prebiotics have been defined by Gibson and Roberfroid (1995) as "a non-digestible poly or oligosaccharides food ingredient which beneficially affects the host by selectively stimulating the growth of and/or activating the metabolism of one or a limited number of health-promoting bacteria in the intestinal tract, thus improving the host's intestinal balance." There is evidence that they can help modulate the growth of gut micro-biota and stimulate bacteriocin production (Patel and Goyal, 2012). Many prebiotics are an effective addition to disease control strategies in human and terrestrial animals. Currently, prebiotics are used increasingly in aquaculture, including mannan-oligosaccharides, fructooligosaccharides, inulin, and vitamin C (Ibrahem et al., 2010; Zhou et al., 2010). Other kinds of prebiotics include galactooligosaccharides (GOSs), xylooligosac-charides (XOSs), arabinoxylooligosaccharides, and iso-maltooligosaccharides (IMOs). Ringo and coworkers (2010) summarized the roles of prebiotics in fish and shellfish as follows: effect on growth, feed conversion, gut microbiota, cell damage/morphology, resistance against pathogenic bacteria, and innate immune param-eters. Prebiotics have been commonly extracted from terrestrial plants and microbes, but data on prebiotics extracted from seaweed and marine probiotics have not been reviewed fully.

### 6.1.2  Probiotics

The term "probiotics" was introduced by Parker (1974) to describe organisms and substances that contribute to intestinal microbial balance. Probiotics are vital culture of bacteria and fungi that, when introduced through feed, have a positive effect on health. According to FAO/WHO, probiotics are defined as "live microorganisms which when consumed in adequate amounts confer a health benefit on the host." When it is being used as a treatment, probiotic bacteria are able to survive in the gastric environment and exposure to bile and pancre-atic juice in the upper small intestine to exert beneficial effects in the lower small intestine and the colon (Mottet and Michetti, 2005).

### 6.1.3  Synbiotics

Synbiotics are a combination of probiotics and prebiot-ics, which can synergistically promote the growth of beneficial bacteria or newly added species in the colon (Macfarlane et al., 2008). In the development of novel synbiotic products, it is very important to ascertain pre-biotic and probiotic interactions and influence of prebi-otics on probiotic growth and biological activity on the host human and animal.

### 6.1.4  Glycans

Glycans, which are synonymous with polysaccharides, are defined as "compounds consisting of a large num-ber of monosaccharides linked glycosidically" (IUPAC, 1997). The term "glycans" is commonly used only for those containing more than 10 monosaccharide resi-dues. Homoglycans that are composed of a single type

of monosaccharide residue are named by replacing the suffix "-ose" of the sugar by "-an," for example, man-nans, fructans, xylans, arabinans, and glucans. However, in practice, the term may also be used to refer to the carbohydrate portion of a glycoconjugate, which can be divided into glycoproteins, glycolipids, and glycos-aminoglycans (Dwek, 1996). Therefore, glycans encom-pass oligosaccharides that can exist either in free form or in large complex carbohydrate molecules attached to proteins or lipids. Glycans are mostly found on a cell surface, in an extracellular matrix, and in some other cellular compartments (Pavao, 2011).

## 6.2  Glycans as prebiotics or foods for probiotics

### 6.2.1  Concept of prebiotics as foods for probiotics

Gut microbiota of humans and animals consists of a complex set of microorganisms that live in the digestive tracts. Many of them are beneficial bacteria that utilize hard-to-digest foods, produce nutrients and energy, and protect the hosts from food-borne pathogens. Pathogens, or harmful bacteria, also reside in the digestive tracts; they release toxins and are increasingly associated with a series of diseases. Therefore, a balance of beneficial and harmful bacteria contributes to human and animal health. In fact, many types of bacteria are competing in the gut, and the winners have potential to cause health problems in the host (Gibson and Roberfroid, 2008).

There are at least three major approaches to modify or maintain the balance of intestinal microbial popula-tions. The first approach is the use of traditional broad-spectrum antibiotics, which can be effective in removing pathogenic agents from gut flora. However, these anti-biotics can destroy beneficial bacterial communities, induce a resistance mechanism in the ecosystem, accu-mulate the remaining harmful residues in foods, and so cannot be routinely used to confer animal and human health sustainably. The second approach is involved in the direct supplementation of probiotics, or live beneficial microbes, to the intestinal microbiota. They usually benefit the human and animal host by glean-ing the energy from the fermentation of undigested glycans and the subsequent absorption of short-chain fatty acids. For example, the most important of these fatty acids in humans are butyrates (metabolized by the colonic epithelium), propionates (by the liver), and ace-tates (by muscle tissue). Besides, probiotics also play a role in synthesizing vitamins B and K and in metaboliz-ing bile acids, sterols, and xenobiotics (Cummings and MacFarlane, 1997).

The third approach is related to the supply of pre-biotics or foods to microbial communities already resid-ing in human and animal intestinal tract, rather than the supplementation of an exogenous source of live microbes. Therefore, prebiotics are considered as nondigestible food ingredients that selectively stimulate the proliferation and/or activity of beneficial bacteria. This means that prebiotics are food for probiotics (Gibson and Roberfroid, 1995). In fact, most prebiotics are nondigestible, fermentable glycans.

Historically, during 1899–1906, Henry Tissier, a French pediatrician at the Pasteur Institute in Paris, isolated a bacterium characterized by a Y-shaped morphology (bifid) in the intestinal flora of breast-fed infants and named it "bifidus" and proposed the oral feeding of bifidobacteria to prevent infant diarrhea (Anukam and Reid, 2008). In 1907, Elie Metchnikoff, Deputy Director at the Pasteur Institute, proposed the theory that lactic acid bacteria are beneficial to human health and also suggested that "oral administration of cultures of fermentative bacteria would implant the beneficial bacteria in the intestinal tract" (Anukam and Reid, 2008). Thus, both Tissier and Metchnikoff are considered ancestors of probiotics and prebiotics science and technology. In 1954, Gyorgy reported that components of human milk (N-acetyl-glucosamine) promoted the growth of a *Bifidobacterium* strain. In 1957, Petuely recognized lactulose as a "bifidus" factor. In the 1970s and 1980s, Japanese researchers discovered that a number of different nondigestible oligosaccharides were "bifidus" factors. In 1995, the term "prebiotic" was coined by Gibson and Roberfroid to link the concepts of prebiotics and probiotics for promoting beneficial populations of intestinal bacteria (Gibson and Roberfroid, 1995). Currently, a prebiotic is popularly defined as a nondigestible food ingredient that beneficially affects the host by selective stimulation of a limited number of bacteria in the colon, improving host health (Gibson et al., 2004).

The prebiotic approach offers a number of advantages to modify the intestinal microbiota over using probiotics or traditional antibiotics. Prebiotics are stable to pH and heat treatment and contribute to long shelf life of a series of foods and beverages (Gibson and Roberfroid, 2008). Their physicochemical properties are useful to food taste and texture. They are also resistant to acids, proteases, and bile salts during transit within the digestive tract. Prebiotics stimulate organisms already occupied in the host, whereas the introduction of probiotics requires compatibility between the host and bacterial strains and competence with already established bacterial populations. Prebiotics also stimulate fermentative activity of gut flora and give rise to the production of short-chain fatty acids. They lower intestinal pH and provide osmotic water retention in the gut. In addition, prebiotics can be used along with probiotics to develop a complementary and synergistic drug formulation called "synbiotics".

Compared to antibiotics, prebiotics are also safer for long-term consumption and do not produce side effects, such as antibiotic-associated diarrhea, sensitivity to UV radiation, or liver damage. Also, prebiotics do not induce antimicrobial resistance genes and are not allergenic factors.

However, unlike probiotics, prebiotics when overused can cause intestinal bloating, pain, flatulence, or diarrhea. Similarly, prebiotics are not potent as antibiotics in removing specific pathogens from gut microbiota and may stimulate side effects of simple sugar malabsorption during active diarrhea (Gibson and Roberfroid, 2008).

### 6.2.2 Glycans as prebiotics

The most identified prebiotics are glycans. Although there is a wide diversity of molecular structures, these glycans share physiological characteristics related to their beneficial effects. They are nondigestible or hard to digest, nonabsorbable in the small intestine, poorly fermented by bacteria in the mouth, well fermented by purportedly beneficial bacteria in the gut, and poorly fermented by potentially pathogenic bacteria in the gut. As prebiotics, glycans usually increase the number of beneficial bacteria (probiotics) within the colon in a selective manner. This results in a wide range of physiological benefits for the host, including reduced gut infections and constipation, improved lipid metabolism, higher mineral absorption, enhanced immunomodulation, and reduced risk of carcinogenesis (Roberfroid, 2000).

To confirm prebiotic activity, glycans must display an *in vivo* effect on health of a human or an animal model. However, *in vitro* tests are useful for preliminary screening of candidate prebiotics and the prediction of the mode of action of these glycans. There are three major criteria for *in vitro* prebiotics studies: nondigestibility, fermentability, and selectivity (Kolida and Gibson, 2008). Nondigestibility means that prebiotics must neither be hydrolyzed by brush border or pancreatic enzymes nor be absorbed in the upper part of the digestive tract. This can be investigated *in vitro* by testing resistance to acidic and enzymatic hydrolysis. For fermentability, prebiotics should be fermented by colonic bacteria and this can be demonstrated *in vitro* in fecal batch culture experiments simulating pH and temperatures similar to conditions in the lower part of the digestive tract. Selectivity means that prebiotics only selectively stimulate a limited number of colon bacteria that benefit host health.

To date, several glycans, such as inulin and oligofructose, lactulose, and GOSs, have been investigated for use as potential prebiotics. Inulin occurs naturally in a range of terrestrial plants such as chicory, onion, garlic, Jerusalem artichoke, tomato, leeks, asparagus, and banana. Lactulose is manufactured from lactose via alkaline isomerization, whereas GOSs are present in low concentrations in human milk, cow's milk, and yoghurt and have also been produced biosynthetically from lactose (Kolida and Gibson, 2008). Besides, scientists suggested some tentative prebiotics such as IMOs, XOSs, and soybean oligosaccharides (Kolida and Gibson, 2008). The main commercial prebiotic agents are essentially obtained by one of the following three processes: (1) direct extraction of natural polysaccharides from plants, (2) controlled hydrolysis of such natural polysaccharides, and (3) enzymatic synthesis, using hydrolases and/or glycosyl transferases (O'Sullivan et al., 2010).

## 6.3   Seaweed glycans as prebiotics and bioactive molecules

Currently, most of these commercial prebiotics are derived from terrestrial resources, whereas marine glycans, which are very diverse and have great potential to be used as prebiotics, are still awaiting discovery. Seaweed are one of the most effective and available sources of polysaccharides. They produce about half of the organic substances in the world and possess significant potential to be utilized as prebiotics, functional food components, or nutriceuticals (Gupta and Abu-Ghannam, 2011; Holdt and Kraan, 2011). Algal polysaccharides are glycans with various kinds of glycosidic bonds of the types 1,3;1,6-β-glucans and 1,3-α-glucans, true heteroglycans and carbohydrate–protein complexes, sulfated polysaccharides (fucoidans), amino sugars and polysaccharides, glycoproteins, glycopeptides, and pectins.

Among seaweed, brown, red, and green algae are most popularly cultured for the production of biomass and bioactive molecules. Brown algae are a rich and easily renewable source of water-soluble polysaccharides (laminarans and fucoidans) and alginic acids with typical biological activity of seaweed (Gupta and Abu-Ghannam, 2011). Red algae are composed of primary carbohydrates such as floridean starch and sulfated galactans (Bouhnik et al., 2004), polysaccharides such as xylans and mannans, and especially, hydrocolloids, including agar and carrageenan. Green algae are also a diverse source of bioactive molecules with potential pharmaceutical applications, including ulvans, glucomannans, mannans, xylans, sulfated polysaccharides, and pectins (Ray and Lahaye, 1995; O'Sullivan et al., 2010).

### 6.3.1   Brown alga glycans

The prebiotic activity of laminarin and fucoidan from brown algae *Laminaria* spp. has been *in vitro* tested widely (Table 6.1). The results have shown that laminarins nondigestibile due to their resistance with hydrochloric acid, human saliva, and homogenates of the human stomach, pancreas, and small intestine (Kolida

and Gibson, 2008). They are fermentable by intestinal bacteria and degraded by the enzymes laminarases and laminarinases (Kolida and Gibson, 2008). Although selectivity of laminarins to specific bacteria has not been shown, the findings demonstrate their potential for use as prebiotics. In addition, treatment with fucoidan from *Undaria pinnatifida* positively affected on cultivable bifidobacteria and lactobacilli counts (Marzorati et al., 2010). This indicated a beneficial effect of the use of this compound because both genera were commonly associated with health benefits for the human and animal host. In fact, several species belonging to these two genera were found to enable to utilize complex carbohydrates containing fucose (Crociani et al., 1994; Topping and Clifton, 2001).

Many *in vivo* investigations confirmed the prebiotic activity of laminarins and/or fucoidan in pigs. These compounds enhanced weight gain and feed intake (Gahan et al., 2009; McDonnell et al., 2010; O'Doherty et al., 2010). Bioactive compounds from seaweed extracts of different sources have also been shown to make changes in systemic immune response in the digestive tract of pigs (Reilly et al., 2008; Smith et al., 2011; Sweeney et al., 2012). It has been noted that glycans with prebiotic properties reduced the number of *Enterobacteriaceae*, whereas increasing *Lactobacillus* and *Bifidobacterium* subsequently enhanced protective effect against intestinal diseases and disorders (Van Loo, 2004). For example, laminarin from *Laminaria* spp. extracts decreased fecal *Escherichia coli* (McDonnell et al., 2010), but fucoidan from *Laminaria hyperborea* increased in colonic *Lactobacillus* (Lynch et al., 2010), whereas a mixture of laminarin and fucoidan from *Laminaria* spp. showed both activities (O'Doherty et al., 2010) or reduced diarrhea (McDonnell et al., 2010). However, a combination of these compounds from the mixture of *L. hyperborea* and *Laminaria digitata* extracts reduced both *Enterobacterium* and *Lactobacillus* counts, which demonstrates that these compounds perhaps interact with each other to affect not only pathogens but also potential probiotics (Reilly et al., 2008). Clearly, glycans from seaweed can lead to changes in gut morphology, feed intake, immune modulation, and commensal microbial populations in animals (Leonard et al., 2010; O'Doherty et al., 2010).

The prebiotic potential of fucoidan along with alginate, another kind of glycan derived from brown alga, was also demonstrated via *in vitro* and *in vivo* experiments in mice. For example, alginate from *Lessonia* spp. was *in vitro* confirmed as a candidate prebiotic due to its selective stimulating activity on the growth of *Bifidobacterium breve* and other bifidobacteria (Akiyama et al., 1992). Alginate alone or in combination with fucoidan from *Fucus evanescens* was found to be resistant to an artificial gastric environment and selectively degraded by bifidogenic and lactogenic bacteria

*Table 6.1* Seaweed glycans as prebiotics and bioactive molecules to improve human and animal health

| | | *In vitro* test | | | *In vivo* trial or biological activity test | | | |
| --- | --- | --- | --- | --- | --- | --- | --- | --- |
| Marine glycans | Origin of glycans | Nondigestibility | Fermentability | Selectivity | Host animal | Effect on host animal | Effect on gut probiotics or microbiota | References |
| | | | | **Brown alga** | | | | |
| Laminarin | *Laminaria* spp. | Resistant to HCl, human saliva, homogenates of human upper digestive tract | Degraded by intestinal bacteria and laminarasess | Not determined | | | | Kolida and Gibson (2008) |
| Fucoidan | *Undaria pinnatifida* | Hydrolyzed by fucoidanases found only in marine bacteria and mollusks | Poorly degraded by intestinal bacteria | Selectively degraded by bifidogenic and lactogenic bacteria | | | | Marzorati et al. (2010) |
| Alginate (enzymatically depolymerized) | *Lessonia* spp. | | Confirmed | Selectively degraded by *Bifidobacterium breve* and other bifidobacteria | | | | Akiyama et al. (1992) |
| Laminarin | *Laminaria* spp. extracts | | | | Pig | Increasing daily weight gain | Reducing fecal *E. coli* | McDonnell et al. (2010) |
| Laminarin and fucoidan | *Laminaria* spp. extracts | | | | Pig | Increasing weight gain and feed intake | Not determined | Gahan et al. (2009) |
| Laminarin and fucoidan | *Laminaria* spp. extracts | | | | Pig | Reducing diarrhea | Increasing fecal lactobacilli | McDonnell et al. (2010) |
| Laminarin and fucoidan | *L. hyperborean* | | | | Pig | Not determined | Increasing in *Enterobacterium* | Lynch et al. (2010) |
| Fucoidan | *L. hyperborea* | | | | Pig | Not determined | Increasing in colonic *Lactobacillus* | Lynch et al. (2010) |
| Laminarin and fucoidan | *L. hyperborean* + *L. digitata* extracts | | | | Pig | Changing in systemic immune response | Reducing *Enterobacterium* and *Lactobacillus* | Reilly et al. (2008) |

*(Continued)*

**Table 6.1 (Continued)** Seaweed glycans as prebiotics and bioactive molecules to improve human and animal health

| Marine glycans | Origin of glycans | In vitro test | | | In vivo trial or biological activity test | | | References |
| | | Nondigestibility | Fermentability | Selectivity | Host animal | Effect on host animal | Effect on gut probiotics or microbiota | |
| --- | --- | --- | --- | --- | --- | --- | --- | --- |
| Laminarin and fucoidan | *L. digitata* extracts | | | | Pig | Increasing average daily gain and gain to feed ratio | Reducing fecal *E. coli* but increasing *Lactobacillus* | O'Doherty et al. (2010) |
| Fucoidan | *Cladosiphon okamuranus* | | | | Rat | Antiulcer effect | | Shibata et al. (1998) |
| Fucoidan (oligofucose-dodecylaniline combination [OFDA]) | *Cladosiphon okamuranus* | | | | Rat | Enhancing antiulcer activity | | Shibata et al. (2001) |
| Fucoidan | *Cladosiphon okamuranus* | | | | Rat | Antioxidant activities and provided cardioprotection against isoproterenol-induced myocardial infarction | | Thomes et al. (2010) |
| Fucoidan (high-molecular weight) | *Cladosiphon okamuranus* | | | | Rat | Antifibrogenesis activity in diethylnitrosamine or DEN-induced liver cirrhosis | | Nakazato et al. (2010) |
| Fucoidan | *Cladosiphon okamuranus* | | | | Mongolian gerbil | Increasing antimicrobial activity against *Helicobacter pylori* infection | | Shibata et al. (2003) |
| Fucoidan | *Cladosiphon okamuranus* | | | | Mouse (human T-cell leukemia virus type 1 or HTLV-1-infected T-cell) | Partially inhibiting growth of tumors | | Haneji et al. (2005) |

*(Continued)*

*Table 6.1 (Continued)* Seaweed glycans as prebiotics and bioactive molecules to improve human and animal health

| Marine glycans | Origin of glycans | In vitro test | | | In vivo trial or biological activity test | | | References |
| | | Nondigestibility | Fermentability | Selectivity | Host animal | Effect on host animal | Effect on gut probiotics or microbiota | |
|---|---|---|---|---|---|---|---|---|
| Fucoidans (intermediate molecular weight) | *Cladosiphon okamuranus* | | | | Mouse (colon 26-bearing model) | Suppressing tumor growth and increasing the number of natural killer cells in the spleen | | Azuma et al. (2012) |
| Fucoidan | *Cladosiphon okamuranus* | | | | Mouse (S-180 model) | Antitumor activity through increasing nitric oxide production | | Takeda et al. (2012) |
| Fucoidan and alginic acid (+ *B. bifidum*) | *Fucus evanescens* | Resistant to an artificial gastric environment | confirmed | Bifido- and lactobacteria | Mouse | | Increasing bifido- and lactobacteria but reducing *Staphylococcus saprophyticus* and *Proteus* | Kuznetsova et al. (2012); Zaporozhets et al. (2014) |
| Fucoidans (oversulfated) | *Cladosiphon okamuranus* | | | | Human (U937 cell) | Antitumor activity through inducing apoptosis | | Teruya et al. (2007) |
| Fucoidans | *Saccharina latissima*, *Fucus vesiculosus*, and *Cladosiphon okamuranus* | | | | | *In vitro* antithrombotic activity | | Ustyuzhanina et al. (2013) |
| Fucoidan | *Cladosiphon okamuranus* | | | | Human | Improving murine chronic colitis by repressing IL-6 synthesis in colonic epithelial cells | | Matsumoto et al. (2004) |
| Fucoidan | *Cladosiphon okamuranus* | | | | Human | Antiviral effects against Dengue virus type 2 (DEN2) | | Hidari et al. (2008) |

*(Continued)*

*Table 6.1 (Continued)* Seaweed glycans as prebiotics and bioactive molecules to improve human and animal health

| | | In vitro test | | | In vivo trial or biological activity test | | | |
| --- | --- | --- | --- | --- | --- | --- | --- | --- |
| Marine glycans | Origin of glycans | Nondigestibility | Fermentability | Selectivity | Host animal | Effect on host animal | Effect on gut probiotics or microbiota | References |
| Fucoidan | *Cladosiphon okamuranus* | | | | Human | Antiviral effects against hepatitis C virus (HCV) | | Mori et al. (2012) |
| Fucoidan | *Cladosiphon okamuranus* | | | | Bird | Antiviral effects against Newcastle disease virus (NDV) | | Elizondo-Gonzalez et al. (2012) |
| **Red alga** | | | | | | | | |
| Floridosides | *Porphyra yezoensis* | Resistant to the treatment of digestive enzymes | Confirmed | Selectively fermented by bifidobacteria | | | | Muraoka et al. (2008) |
| Floridoside, tichocarpols A and B | *Tichocarpus crinitus* | | | | Sea urchin | Feeding-deterrent activity | | Ishii et al. (2004) |
| Floridoside (0.01 mg/ mL) | *Grateloupia turuturu* | | | | Tropical barnacle | Antisettlement activity against cyprid larvae and nontoxicity | | Hellio et al. (2004) |
| Floridoside | *Laurencia undulata* | | | | Mouse (BV-2 microglia cells) | Antioxidant activities by suppressing proinflammatory responses | | Kim et al. (2013) |
| Floridoside | *Mastocarpus stellatus* | | | | Human serum | *In vitro* anticomplementary activity | | Courtois et al. (2008) |
| Floridoside and D-isofloridoside | *Laurencia undulate* | | | | Human (MRC-5, RAW264.7, HT-1080, HL-60 cells) | *In vitro* antioxidant activities | | Li et al. (2010) |

*(Continued)*

*Table 6.1 (Continued)* Seaweed glycans as prebiotics and bioactive molecules to improve human and animal health

| Marine glycans | Origin of glycans | Nondigestibility | In vitro test Fermentability | Selectivity | In vivo trial or biological activity test Host animal | Effect on host animal | Effect on gut probiotics or microbiota | References |
|---|---|---|---|---|---|---|---|---|
| Neoagaro-oligosaccharide (NAOS) | From enzymatic hydrolysis of red alga agarose | Highly resistant to digestive enzymes | Confirmed | Selectively degraded by bifidobacteria and lactobacilli | Mouse | No side effects such as eructation and bloating | Increasing probiotic bacteria but reducing putrefactive bacteria | Hu et al. (2006) |
| Agaro-oligosaccharides (AOs) | Red alga | | Degraded by human intestinal microbiota | Selectively degraded by bifidobacteria | | | | Ramnani et al. (2012) |
| AOs | Red alga | | Degraded by human intestinal microbiota | Selectively degraded by *Bacteroides uniformis* L8, and indirectly by *B. infantis, B. adolescentis, E. coli* B2 | | | | Li et al. (2014) |
| AOs | Produced from red alga (Takara Bio Inc., Shiga, Japan) | | | | Mouse | Antioxidant effect through protecting cell damage caused by reactive oxygen species | | Chen and Yan (2005); Chen et al. (2006) |
| AOs | From red alga agar hydrolysis | | | | Mouse and human | Immune modulation on RAW264.7 mouse macrophages and human monocytes | | Enoki et al. (2010) |
| AOs | Red alga | | | | Mouse | Antitumor activity on two-stage mouse skin carcinogenesis model | | Enoki et al. (2012) |

*(Continued)*

*Table 6.1 (**Continued**)* Seaweed glycans as prebiotics and bioactive molecules to improve human and animal health

| Marine glycans | Origin of glycans | In vitro test | | | In vivo trial or biological activity test | | | References |
| --- | --- | --- | --- | --- | --- | --- | --- | --- |
| | | Nondigestibility | Fermentability | Selectivity | Host animal | Effect on host animal | Effect on gut probiotics or microbiota | |
| AOs | Red alga | | | | Mouse | Antitumor activity by protecting nonsteroidal anti-inflammatory drug-induced small intestinal injury | | Higashimura et al. (2013, 2014) |
| **Green alga** | | | | | | | | |
| Heteropolysaccharide | *Ulva rigida* | | Poorly fermented by colonic bacteria | | | | | Ray and Lahaye (1995) |
| Ulvan | *Ulva* sp. | | Poorly fermented by colonic bacteria | | | | | Bobin-Dubigeon et al. (1997) |
| Ulvan | *Ulva rigida* | | | | Mouse (RAW264.7 cell) | Immune modulation in inflammatory responses related to macrophage functions | | Leiro et al. (2007) |
| Ulvan (+chitosan) | *Ulva rigida* | | | | Human (7F2 osteoblasts) | Promoting attachment and proliferation and maintaining the cell morphology and viability | | Toskas et al. (2012) |

(Kuznetsova et al., 2012). It is interesting that in an *in vivo* mice model with antibiotic-induced dysbiosis, in which the number of lactobacteria and enterococci decreased along with the increase of *Clostridium* and *Proteus*, these potential prebiotics enhanced fecal bifidobacteria and lactobacteria counts similar to those in normal gut flora but reduced the quantity of opportunistic bacteria such as *Staphylococcus saprophyticus* and *Proteus* (Kuznetsova et al., 2012).

In addition, the prebiotic activity of glycans from brown algae was tested in some rodent species and human cell lines, which demonstrated a broad spectrum of biological activities of these potential prebiotics. In 2007, a comparative study of the biological activities of nine different fucoidans from brown seaweed showed that fucoidans obtained from species different from the traditionally studied *Fucus vesiculosus* and *Ascophyllum nodosum* acted as inhibitors of inflammation, angiogenesis, and heterotypic tumor cell adhesion. Also, the results suggested the importance of 2-O-α-D-glucuronyl branch in decreasing several biological activities if weaker inhibitory potential of fucoidans from *Cladosiphon okamuranus* was not due to lower sulfate content (Cumashi et al., 2007). However, fucoidans from *C. okamuranus* were then studied commonly to demonstrate their diverse activities. They can represent potential low-toxicity antiviral compounds for the poultry industry and in the treatment of patients. For example, they expressed antiviral effects against Newcastle disease virus in birds (Elizondo-Gonzalez et al., 2012), dengue virus type 2 (Hidari et al., 2008) and hepatitis C virus in human (Mori et al., 2012). Also, assays of sulfated fucoidans from *C. okamuranus* on *Helicobacter pylori* attachment with porcine gastric mucin *in vitro* and *H. pylori*-induced gastritis *in vivo* in Mongolian gerbils demonstrated an antimicrobial effect by inhibiting the adhesion of the pathogens to the mucous membrane (Shibata et al., 2003).

Other important biological properties of fucoidans from *C. okamuranus* include antitumor, antiinflammatory, antioxidant, and antifibrotic activities. For example, the intermediate-molecular-weight fucoidans significantly suppressed tumor growth in a tumor (colon 26)-bearing mouse model, whereas the low-molecular-weight fucoidan and high-molecular-weight fucoidan groups significantly increased survival times compared to that observed in mice fed a fucoidan-free diet (Azuma et al., 2012). The results also showed that fucoidans increased the population of natural killer cells in the spleen, and these effects were related to gut immunity. Another *in vivo* assay of fucoidan resulted in partial inhibition of growth of tumors of a human T-cell leukemia virus type 1-infected T-cell line transplanted subcutaneously in severe combined immune-deficient mice (Haneji et al., 2005). The mechanism of antitumor

activity of fucoidans from brown seaweed was uncovered partially based on human and mouse cell lines. The oversulfated fucoidans induced apoptosis via the caspase-3 and caspase-7 activation-dependent pathway in human U937 cells (Teruya et al., 2007), while antitumor activity on sarcoma 180 (S-180)-bearing mice was mediated through increased nitric oxide production by fucoidan-stimulated macrophages via the nuclear factor-κB-dependent signaling pathway (Takeda et al., 2012). In addition, fucoidans improved murine chronic colitis by repressing the synthesis of interleukin 6 (IL-6) in the colonic epithelial cells (Matsumoto et al., 2004), exerted antioxidant activities, and provided cardioprotection against isoproterenol-induced myocardial infarction in rats (Thomes et al., 2010). Besides, high-molecular-weight fucoidan treatment caused antifibrogenesis in DEN-induced liver cirrhosis (Nakazato et al., 2010), while antithrombotic properties of three structurally different fucoidans from the brown seaweed *Saccharina latissima*, *F. vesiculosus*, and *C. okamuranus* were tested, revealing the influence of fucoidans on the hemostatic system (Ustyuzhanina et al., 2013). Finally, safety tests of *C. okamuranus* fucoidans showed that no significant toxicological changes were induced by fucoidan at a dose of 600 mg/kg of body weight/day and no other signs of toxicity were observed (Gideon and Rengasamy, 2008). All these results suggest that fucoidans from the brown alga *C. okamuranus* are safe and useful as a promising functional food with a wide spectrum of biological activity to correct intestinal microbiocoenosis, to arrest inflammation in the digestive tract, and to normalize immune responses and metabolic activities in humans and animals.

## 6.3.2 Red alga glycans

The prebiotic potential of floridoside, a glycan from the red alga Susabinori (*Porphyra yezoensis*), was *in vitro* demonstrated thanks to its resistance to the treatment of digestive enzymes (human saliva α-amylase, artificial gastric juice, porcine pancreatic, α-amylase, or rat intestinal acetone powder), nonabsorption using everted sacs of rat small intestine, and selective fermentation by bifidobacteria, which play an important role in human health among intestinal flora (Muraoka et al., 2008) (Table 6.1). It is worth noting that the low quality of alga Susabinori is characterized by its low degree of blackness and lower protein content than that of normalquality products, whereas the protein content of nori is inversely correlated with the carbohydrate content, and hence low-quality nori contains rich carbohydrates (Muraoka et al., 2008). Therefore, these nondigestible but selectively fermentable floridosides are useful prebiotics candidates that promote intestinal bacterial microbiota positively.

Floridosides isolated from different red alga species were also found to express diverse biological activities. Along with two phenylpropanoic acid derivatives, tichocarpol A and tichocarpol B, floridosides from *Tichocarpus crinitus* exerted feeding-deterrent activity against the sea urchin *Strongylocentrotus intermedius* (Ishii et al., 2004). In another example, both isethionic acid and floridoside were extracted from *Grateloupia turuturu* and tested for antisettlement activity against cyprid larvae of the tropical barnacle *Balanus amphitrite* and for their toxicity to nauplius larvae. The results showed that isethionic acid was active for antisettlement but had the disadvantage of being toxic to nauplius larvae, whereas floridoside was a potent inhibitor of cyprid settlement at nontoxic concentrations to nauplii (0.01 mg/mL) (Hellio et al., 2004). In addition, floridoside from *Mastocarpus stellatus*, structurally similar to the xenoantigen galactose-α-1,3-galactose, was shown to activate a complement cascade via the classical complement pathway through the recruitment and activation of natural IgM, and hence this algal molecule could be developed as a potent new anticomplementary agent for use in therapeutic complement depletion (Courtois et al., 2008). Two bioactive molecules, floridoside and D-isofloridoside, were extracted from the edible red alga *Laurencia undulata*, which exhibited antioxidant activities and inhibitory effects to matrix metallopeptidase 2 and matrix metallopeptidase 9 involved in the breakdown of extracellular matrix (Li et al., 2010). Currently, a floridoside isolated from *L. undulata* was indicated to suppress proinflammatory responses in BV-2 microglia cells by significantly inhibiting the production of nitric oxide and reactive oxygen species as well as reduce the transcription and translation of nitric oxide synthase and cyclooxygenase-2 (COX-2) by blocking the phosphorylation of mitogen-activated protein kinases (p38) and extracellular signal–regulated kinases (Kim et al., 2013). Taken together, these results indicated that floridosides from red algae could be effective candidates for application in food and pharmaceutical fields as natural marine antioxidants and active agents for neuroinflammation treatment.

In addition, agaro-oligosaccharides (AOs) derived from agarose in red algae have been shown to have potential prebiotic effects via typical *in vitro* and *in vivo* experiments. In 2006, the effects of neoagaro-oligosaccharides (NAOSs), obtained from enzymatic hydrolysis of agarose, on bacterial growth were studied. NAOSs were found to be highly resistant to digestive enzymes, which significantly stimulate the growth of bifidobacteria and lactobacilli. NAOSs were also shown to enhance the number of these probiotic bacteria but reduce putrefactive microorganisms in the mice model. Interestingly, no side effects, such as eructation and bloating, were found, which suggested the potential for AOs as prebiotics (Hu et al., 2006). Later, it was reported that AOs can be degraded and utilized by human fecal flora. Although *Bacteroides uniformis* was demonstrated to be common in human feces, only *B. uniformis* L8 enabled AOs to degrade into D-galactoses, which, in turn, were fermented by a synergistic strain called *E. coli* B2 (Li et al., 2014). On the other hand, *Bifidobacterium infantis* and *Bifidobacterium adolescentis* were found to ferment one of the intermediates of AO hydrolysis, agarotriose, as the growth substrate. It is worth noting that *B. uniformis* L8 cannot use κ-carrageenan oligosaccharides, guluronic acid oligosaccharides, and mannuronic acid oligosaccharides, and hence it is considered as a special degrader of AOs in the intestinal flora (Li et al., 2014).

Similarly, AOs produced from red alga (Takara Bio Inc., Shiga, Japan) were demonstrated to exert their *in vitro* and *in vivo* antioxidant effect in the rat model in that they could protect cell damage caused by reactive oxygen species, especially agarohexaose, exhibiting most desirable effects (Chen and Yan, 2005; Chen et al., 2006). Also, AOs from agar hydrolysis were found to have immunological effects on RAW264.7 mouse macrophages and human monocytes through reducing the elevated levels of nitric oxide, prostaglandin E$_2$ (PGE$_2$), and proinflammatory cytokines and inducing heme oxygenase-1 in activated leukocytes (Enoki et al., 2010). Currently, they have been shown to prevent tumor promotion by inhibiting PGE$_2$ elevation in chronic inflammation site on the two-stage mouse skin carcinogenesis model (Enoki et al., 2012) and prevent nonsteroidal anti-inflammatory drug-induced small intestinal injury in mice (Higashimura et al., 2013). It is therefore proposed that AOs might be good candidates in functional food to prevent reactive oxygen species–induced cell damage, inflammatory diseases, cancers, and immune disorders.

### 6.3.3 Green alga glycans

The green alga (*Ulva* sp.), which is one of the commonly consumed seaweed, contains 16.5% of water-soluble and 13.3% insoluble dietary fibers (Ray and Lahaye, 1995). Fibers can be degraded and metabolized by colonic bacteria and thus induce other physiological effects. The colonic bacterial degradability of *Ulva* sp. and its soluble and insoluble dietary fibers was tested using an *in vitro* batch system inoculated with human feces. The results showed that *Ulva* and its soluble fiber, ulvan, were poorly fermented by colonic bacteria, whereas the constitutive sugars, rhamnose and glucuronate and the aldobiouronate beta-D-glucuronosyluronate-(1,4)-L-rhamnose of the glucuronoxylorhamnan sulfate present in the soluble fiber, were highly fermented (Bobin-Dubigeon et al., 1997). Thus, ulvan is another peculiar food polysaccharide placed among soluble fibers from algae that are poorly fermented by colonic bacteria.

Also, in light of the known different effects of sulfated polysaccharides, some studies reported on the biological activities and nutritional roles of ulvans on the host. For example, the effects of *Ulva rigida* ulvans on several RAW264.7 murine macrophage activities were investigated. The results revealed that these bioactive compounds induced the expression of several chemokines, IL-6 signal transducer, and interleukin 12 (IL-12) receptor beta 1, stimulated macrophage secretion of PGE$_2$, and enhanced nitrite production, nitric oxide synthase 2 (NOS-2) and COX-2 gene expression (Table 6.1). Thus, *U. rigida* polysaccharides were suggested to be used as an experimental immunostimulant for analyzing inflammatory responses related to macrophage functions (Leiro et al., 2007). In another research, ulvans extracted from the green seaweed *U. rigida* and ulvan/chitosan membranes were found to promote the attachment and proliferation of 7F2 osteoblasts and maintain cell morphology and viability. Therefore, ulvan and chitosan were suggested for use as potential substrate materials for the cultivation of osteoblasts and for other biomedical applications (Toskas et al., 2012).

## 6.4    Glycans from marine probiotics benefiting human and animal health

Bacterial cells have a huge diversity of macromolecules such as proteins, lipids, nucleic acids, and polysaccharides. The last one includes intracellular polysaccharides acting in metabolic pathways inside the cell, peptidoglycans and lipopolysaccharides as structural polysaccharides of the cell wall and outer cell membrane, and exopolysaccharides (EPSs) linked mainly to extracellular environment through biofilm formation, stress resistance, and substrate utilization. Among them, EPSs and other glycans are used as food additives to enhance texture thanks to improving the shape, stability, and rheological behavior of food products (Patel et al., 2010). However, these glycans sometimes can lead to food spoilage, produce undesired rheological properties in products, or biofouling after biofilm formation (Azeredo and Oliveira, 2000; Badel et al., 2008).

EPS-producing lactic acid bacteria were found in the oral cavity and intestinal tract, which revealed the ecological significance of EPS biosynthesis for their survival in complex environments related to biofilm formation, stress resistance, and sucrose utilization (Iliev et al., 2006; Kenji et al., 2010). EPSs produced by probiotics have been considered to support the host strain to coaggregate and adhere to gut epithelial cells and also to interfere with pathogens adhered to gut mucus (Ruas-Madiedo et al., 2006). EPSs were also shown to

help their probiotic producers to sequester essential cations (Looijesteijn et al., 2001) in adhesion and in biofilm formation (Roberts, 1996; Badel et al., 2008). These findings demonstrate that EPSs may play a role to help probiotics to compete, colonize, and survive in the intestine of a host animal.

EPSs from terrestrial probiotics, which include lactic acid bacteria, propionibacteria, bifidobacteria, and *Bacillus* genus, have been reviewed by Patel and coworkers (2010). EPS-producing lactic acid bacteria, including *Lactobacillus, Leuconostoc, Lactococcus,* and *Streptococcus,* synthesize diverse kinds of homopolysaccharides (glucans and fructans) and heteropolysaccharides with a molecular mass from 10 to 1000 kDa (Navarini et al., 2001; Faber et al., 2002). Most lactic acid bacteria produce polysaccharides extracellularly from sucrose by glycansucrases or intracellularly from sugar nucleotide precursors by glycosyltransferases (Patel et al., 2010). It has also been shown that EPSs are produced by human intestinal bifidobacteria (Roberts et al., 1995; Amrouche et al., 2006), which act as fermentable substrates for gut flora (Salazar et al., 2008). Thus, EPSs from probiotics can be used prebiotics, which suggest a new kind of synbiotics for improving human and animal health. Similarly, EPS production from *Bacillus polymyxa* (Lee et al., 1997) and *Bacillus coagulans* (Kodali et al., 2009) was found to be associated with biomass growth, revealing that the support of adequate nutrients enhanced the biosynthesis of these glycans.

An increasing number of glycans derived from marine bacteria have been the subject of current reviews (Colliec-Jouault and Delbarre-Ladrat, 2014; Delbarre-Ladrat et al., 2014; Finore et al., 2014; Pomin, 2014); marine probiotics supporting animal health have also been summarized (Nguyen et al., 2013), whereas only a limited number of studies on glycans from marine probiotics have been reported. It is worth noting that marine EPS-producing lactic acid bacteria *Weissella cibaria, Weissella confusa, Lactobacillus plantarum,* and *Pediococcus pentosaceus* were isolated from gastrointestinal tracts of marine fish, shellfish, and shrimp. EPSs produced by these strains were resistant to stomach acid and human pancreatic amylase and selectively fermented by only *Bifidobacterium bifidum* DSM 20456 rather than by some tested *Lactobacillus* strains, which suggested the prebiotic potential of these glycans (Hongpattarakere et al., 2012). In another example, EPS-1, a novel extracellular polysaccharide produced by a strain of thermotolerant *Bacillus licheniformis* isolated from a shallow marine hot spring at Vulcano Island (Italy), was found to impair Herpes simplex virus 2 replication in human peripheral blood mononuclear cells and to induce IL-12, IFN-gamma, IFN-alpha, TNF-alpha, and IL-18, hence exerting antiviral and immunomodulatory effects. These two examples indicate that novel glycans from potential marine probiotics with prebiotic and biological activities are still waiting on the discovery. The biodiversity of marine environment and the associated chemical diversity constitute a practically unlimited resource of new bioactive glycans.

Because bacterial glycans express immense diversity in composition, they can find numerous applications in food, cosmetic, aquacultural, agricultural, and pharmaceutical settings. The production of glycans from diverse probiotics should be economically feasible so that cheaper substrates and cost-effective fermentation conditions must be considered. Probiotic strains that are facultative anaerobes would be more effective in producing energy and could overcome the problem of low oxygen level during EPS production by obligate aerobes (Patel et al., 2010).

In other points of view, because the structural and functional diversity of these marine glycans confirms the high value of the marine environment as a source of exciting drugs, bioactive molecules from nonharmful marine bacteria (not only probiotics) should be considered for use in food, feed, and pharmaceuticals. Bacterial EPSs play crucial roles in initial bacterial adhesion and biofilm formation, so they can support probiotics directly or indirectly in the intestinal environment. For example, an EPS (A101) with an average molecular weight of up to 546 kDa extracted from a marine bacterium *Vibrio* sp. QY101 not only inhibited biofilm formation of many bacteria but also disrupted established biofilm produced by *Pseudomonas aeruginosa* (but not by *Staphylococcus aureus*). Also, A101 inhibited cell aggregates of both *P. aeruginosa* and *S. aureus*, reduced cell-surface interactions in *S. aureus* only, but induced preformed cell aggregates dispersion in *P. aeruginosa* only. These findings suggest that A101 can be used to manage biofilm-associated infections and represent a new function of glycans from marine bacteria (Jiang et al., 2011).

## 6.5 Marine glycans in combination with probiotics as synbiotics

It has been suggested that prebiotics and probiotics should be used together in a complementary and synergistic manner. Food ingredients containing both probiotics and prebiotics have been termed synbiotics. Synbiotics are probiotic products, which contain living bacteria along with biologically active substances or prebiotics, which are partially or completely indigestible food ingredients that stimulate the growth and/or metabolic activity of one or several groups of bacteria in the large intestine. A number of studies comparing interventions with probiotics, prebiotics, and synbiotics showed the synbiotic combination to be the most effective. These studies used varying combination of

pro- and prebiotics and the exact nature of the synergies that lead to improved health in cultured animals (Patterson and Burkholder, 2003) and carcinogen challenge rat models (Challa et al., 1997; Femia et al., 2002; Roller et al., 2004; Le Leu et al., 2005).

Synbiotics are considered the most effective regulators of the composition and functions of gut flora (Cheng et al., 2014). However, a systematic review and meta-analysis to examine the effect of prebiotics, probiotics, and synbiotics in functional bowel disorders, including irritable bowel syndrome (IBS) and chronic idiopathic constipation (CIC) indicated little evidence for the use of prebiotics or synbiotics in these cases (Ford et al., 2014). The results showed that probiotics had beneficial effects on global IBS, abdominal pain, bloating, and flatulence scores, whereas further evidence is required before the role of prebiotics or synbiotics in IBS is confirmed. In the case of CIC, trials of probiotics and synbiotics revealed some promising results. Because there were a limited number of studies on three therapies in CIC, their effects were, however, uncertain. Thus, novel prebiotics or synbiotics from new resources are required to support diverse evidences from *in vivo* trials in human cells and animals.

Although little data were gained from marine glycans used in combination with probiotics to formulate synbiotics, some candidates were confirmed to improve human and animal health. For example, fucoidan from the marine seaweed *F. evanescens* was used along with *B. bifidum* cells to develop a new synbiotic sour milk drink called the "Bifidomarin" sour milk bioproduct (Kuznetsova et al., 2011). In this case, fucoidan not only exerted a broad spectrum of biological activities such as anti-inflammatory, antibacterial, antiviral, immunomodulating, antiproliferative, anticancer, anticoagulative, antithrombotic, and hypolipidemic activities, but it also expressed a potential prebiotic characteristic as food for probiotics in the development of synbiotic application, which led to an improvement in both immune and metabolic status in patients with digestive tract disorders and associated dysbiosis.

A current proteomic analysis of *Bifidobacterium longum* subsp. *infantis* grown on β-glucans from different sources (seaweed *L. digitata*, barley and mushroom) have demonstrated that β-glucans from different sources can be fermented by bifidobacteria via different or similar metabolic pathways. At least 17 proteins were predicted to take part in the degradation of β-glucans. Among them, intracellular glucanases were found in the cultures grown on seaweed and mushroom β-glucans. Thus, a catabolism model of β-glucans in *B. infantis* is proposed, which opens the promising use of these β-glucans as novel prebiotics in the development of synbiotics benefiting human health (Zhao and Cheung, 2013).

## 6.6  Summary

Antibiotics have played important roles in the prevention, control, and treatment of infectious diseases in humans and animals for more than 70 years. To overcome the increased rate of morbidity and mortality due to the ban on antibiotics, several alternative therapies have been suggested, which include vaccines, immunomodulatory agents, bacteriophages, antimicrobial peptides (bacteriocins included), probiotics, prebiotics, synbiotics, etc. A new term "pharmabiotics," which is defined as any biological material from gut microbiota, including probiotics, bacteriocins, bacteriophages, and bioactive molecules, has also been introduced in recent years. Synbiotics, which are a combination of prebiotics and probiotics, are now gaining scientific credibility as functional food ingredients at nutritional and therapeutic levels, and there is also an increasing evidence to show their effectiveness to benefit human and animal health.

The most known prebiotics are glycans. As discussed in this chapter, glycans from seaweed and marine probiotics are increasingly being expected to become novel prebiotics or synbiotics. Typical examples of seaweed glycans include laminarans, fucoidans, and alginates from brown algae; floridosides and AOs from red algae; and ulvans from green algae. They function not only as prebiotics to support the growth of probiotics but also exert a broad spectrum of biological properties such as antiviral, antimicrobial, antitumor, anti-inflammatory, antioxidant, antifibrotic, and immunomodulatory activities.

In addition, EPSs from bacteria, which perhaps play an ecological role for the survival of the host in complex environments related to biofilm formation, stress resistance, and substrate utilization, are potential candidates for the development of drugs. EPSs produced by probiotics can help the host strain to compete pathogens for an adhesion site in the gut mucus, to sequester essential metals, and to support the growth of other probiotics. Although these glycans derived from marine probiotics have still been studied sparingly, surely, they are expected to formulate into novel prebiotics or synbiotics for future therapies in humans and animals.

## References

Akiyama, H., Endo, T., Nakakita, R. et al. 1992. Effect of depolymerized alginates on the growth of bifidobacteria. *Biosci Biotechnol Biochem* 56:355–356.

Amrouche, T., Boutin, Y., Prioult, G., and Fliss, I. 2006. Effects of bifidobacterial cytoplasm, cell wall and exopolysaccharide on mouse lymphocyte proliferation and cytokine production. *Int Dairy J* 16:70–80.

Anukam, K.C. and Reid, G. 2008. Probiotics: 100 years (1907–2007) after Elie Metchnikoff's observations. In *Communicating Current Research and Educational Topics and Trends in Applied Microbiology* (2007 Edition), ed. A. Mendez-Vilas. Badajoz, Spain: Formatex.org, pp. 466–474.

Azeredo, J. and Oliveira, R. 2000. The role of exopolymers produced by *Sphingomonas paucimobilis* in biofilm formation and composition. *Biofouling* 16:17–27.

Azuma, K., Ishihara, T., Nakamoto, H. et al. 2012. Effects of oral administration of fucoidan extracted from *Cladosiphon okamuranus* on tumor growth and survival time in a tumor-bearing mouse model. *Mar Drugs* 10(10):2337–2348.

Badel, S., Laroche, C., Gardarin, C., Bernardi, T., and Michaud, P. 2008. New method showing the influence of matrix components in *Leuconostoc mesenteroides* biofilm formation. *Appl Biochem Biotechnol* 151:364–370.

Bobin-Dubigeon, C., Lahaye, M., and Barry, J.L. 1997. Human colonic bacterial degradability of dietary fibres from sea-lettuce (*Ulva* sp.). *J Sci Food Agric* 73:149–159.

Bouhnik, Y., Raskine, L., Simoneau, G. et al. 2004. The capacity of nondigestible carbohydrates to stimulate fecal bifidobacteria in healthy humans: A double-blind, randomized, placebo controlled, parallel group, dose response relation study. *Am J Clin Nutr* 80:1658–1664.

Challa, A., Rao, D.R., Chawan, C.B., and Shackelford, L. 1997. *Bifidobacterium longum* and lactulose suppress azoxymethane-induced colonic aberrant crypt foci in rats. *Carcinogenesis* 18:517–521.

Chen, H., Yan, X., Zhu, P., and Lin, J. 2006. Antioxidant activity and hepatoprotective potential of agaro-oligosaccharides in vitro and in vivo. *Nutr J* 5:31.

Chen, H.M. and Yan, X.J. 2005. Antioxidant activities of agaro-oligosaccharides with different degrees of polymerization in cell-based system. *Biochim Biophys Acta* 1722(1):103–111.

Cheng, G., Hao, H., Xie, S. et al. 2014. Antibiotic alternatives: The substitution of antibiotics in animal husbandry? *Front Microbiol* 5:217.

Colliec-Jouault, S. and Delbarre-Ladrat, C. 2014. Marine-derived bioactive polysaccharides from microorganisms. In *Natural Bioactive Molecules. Impacts and Prospects*, ed. G. Brahmachari. New Delhi, India: Narosa Publishing House, pp. 5.1–5.21.

Courtois, A., Simon-Colin, C., Boisset, C., Berthou, C., Deslandes, E., Guézennec, J., and Bordron, A. 2008. Floridoside extracted from the red alga *Mastocarpus stellatus* is a potent activator of the classical complement pathway. *Mar Drugs* 6(3):407–417.

Crociani, F., Alessandrini, A., Mucci, M.M., and Biavati, B. 1994. Degradation of complex carbohydrates by *Bifidobacterium* spp. *Int J Food Microbiol* 24:199–210.

Cumashi, A., Ushakova, N.A., Preobrazhenskaya, M.E. et al. 2007. A comparative study of the anti-inflammatory, anticoagulant, antiangiogenic, and antiadhesive activities of nine different fucoidans from brown seaweeds. *Glycobiology* 17(5):541–552.

Cummings, J.H. and MacFarlane, G.T. 1997. Role of intestinal bacteria in nutrient metabolism. *Clin Nutr* 16:3–9.

Delbarre-Ladrat, C., Sinquin, C., Lebellenger, L., Zykwinska, A., and Colliec-Jouault, S. 2014. Exopolysaccharides produced by marine bacteria and their applications as glycosaminoglycan-like molecules. *Front Chem* 2:85.

Dwek, R.A. 1996. Glycobiology: Toward understanding the function of sugars. *Chem Rev* 96(2):683–720.

Elizondo-Gonzalez, R., Cruz-Suarez, L.E., Ricque-Marie, D., Mendoza-Gamboa, E., Rodriguez-Padilla, C., and Trejo-Avila, L.M. 2012. In vitro characterization of the antiviral activity of fucoidan from *Cladosiphon okamuranus* against Newcastle Disease Virus. *Virol J* 9:307.

Enoki, T., Okuda, S., Kudo, Y., Takashima, F., Sagawa, H., and Kato, I. 2010. Oligosaccharides from agar inhibit pro-inflammatory mediator release by inducing heme oxygenase 1. *Biosci Biotechnol Biochem* 74(4):766–770.

Enoki, T., Tominaga, T., Takashima, F., Ohnogi, H., Sagawa, H., and Kato, I. 2012. Anti-tumor-promoting activities of agaro-oligosaccharides on two-stage mouse skin carcinogenesis. *Biol Pharm Bull* 35(7):1145–1149.

Faber, E.J., van Haaster, D.J., Kamerling, J.P., and Vliegenthart, J.F. 2002. Characterization of the exopolysaccharide produced by *Streptococcus thermophilus* 8S containing an open chain nonionic acid. *Eur J Biochem* 269:5590–5598.

Femia, A.P., Luceri, C., Dolara, P. et al. 2002. Antitumorigenic activity of the prebiotic inulin enriched with oligofructose in combination with the probiotics *Lactobacillus rhamnosus* and *Bifidobacterium lactis* on azoxymethane-induced colon carcinogenesis in rats. *Carcinogenesis* 23:1953–1960.

Finore, I., Di Donato, P., Mastascusa, V., Nicolaus, B., and Poli, A. 2014. Fermentation technologies for the optimization of marine microbial exopolysaccharide production. *Mar Drugs* 12:3005–3024.

Ford, A.C., Quigley, E.M.M., Lacy, B.E. et al. 2014. Efficacy of prebiotics, probiotics, and synbiotics in irritable bowel syndrome and chronic idiopathic constipation: Systematic review and meta-analysis. *Am J Gastroenterol* 109:1547–1561.

Gahan, D.A., Lynch, M.B., Callan, J.J., O'Sullivan, J.T., and O'Doherty, J.V. 2009. Performance of weanling piglets offered low-, medium- or high-lactose diets supplemented with a seaweed extract from *Laminaria* spp. *Animal* 3:24–31.

Gibson, G.R., Probert, H.M., Van Loo, J., Rastall, R.A., and Roberfroid, M. 2004. Dietary modulation of the human colonic microbiotia: Updating the concept of prebiotics. *Nutr Res Rev* 17:259–275.

Gibson, G.R. and Roberfroid, M. 1995. Dietary modulation of the human colonic microbiota: Introducing the concept of prebiotics. *J Nutr* 125:1401–1412.

Gibson, G.R. and Roberfroid, M.B. 2008. *Handbook of Prebiotics*. Boca Raton, FL: CRC Press.

Gideon, T.P. and Rengasamy, R. 2008. Toxicological evaluation of fucoidan from *Cladosiphon okamuranus*. *J Med Food* 11(4):638–642.

Gupta, S. and Abu-Ghannam, N. 2011. Bioactive potential and possible health effects of edible brown seaweeds. *Trends Food Sci Technol* 22(6):315–326.

Haneji, K., Matsuda, T., Tomita, M. et al. 2005. Fucoidan extracted from *Cladosiphon okamuranus* Tokida induces apoptosis of human T-cell leukemia virus type 1-infected T-cell lines and primary adult T-cell leukemia cells. *Nutr Cancer* 52(2):189–201.

Hellio, C., Simon-Colin, C., Clare, A.S., and Deslandes, E. 2004. Isethionic acid and floridoside isolated from the red alga, *Grateloupia turuturu*, inhibit settlement of *Balanus amphitrite* cyprid larvae. *Biofouling* 20(3):139–145.

Hidari, K.I., Takahashi, N., Arihara, M., Nagaoka, M., Morita, K., and Suzuki, T. 2008. Structure and anti-dengue virus activity of sulfated polysaccharide from a marine alga. *Biochem Biophys Res Commun* 376(1):91–95.

Higashimura, Y., Naito, Y., Takagi, T. et al. 2013. Oligosaccharides from agar inhibit murine intestinal inflammation through the induction of heme oxygenase-1 expression. *J Gastroenterol* 48(8):897–909.

Higashimura, Y., Naito, Y., Takagi, T. et al. 2014. Preventive effect of agaro-oligosaccharides on non-steroidal anti-inflammatory drug-induced small intestinal injury in mice. *J Gastroenterol Hepatol* 29(2):310–317.

Holdt, S.L. and Kraan, S. 2011. Bioactive compounds in seaweed: Functional food applications and legislation. *J Appl Phycol* 23:543–597.

Hongpattarakere, T., Cherntong, N., Wichienchot, S., Kolida, S., and Rastall, R.A. 2012. In vitro prebiotic evaluation of exopolysaccharides produced by marine isolated lactic acid bacteria. *Carbohydr Polym* 87(1):846–852.

Hu, B., Gong, Q., Wang, Y. et al. 2006. Prebiotic effects of neo-agaro-oligosaccharides prepared by enzymatic hydrolysis of agarose. *Anaerobe* 12:260–266.

Ibrahem, M.D., Fathi, M., Mesalthy, S., and Abd El-Aty, A.M. 2010. Effect of dietary supplementation of inulin and vitamin C on the growth, hematology, innate immunity, and resistance of Nile tilapia (*Oreochromis niloticus*). *Fish Shellfish Immunol* 29:241–246.

Iliev, I., Ivanova, I., and Ignatova, C. 2006. Glucansucrases from lactic acid bacteria (LAB). *Biotechnol Biotechnol Equip* 20:5–20.

Ishii, T., Okino, T., Suzuki, M., and Machiguchi, Y. 2004. Tichocarpols A and B, two novel phenylpropanoids with feeding-deterrent activity from the red alga *Tichocarpus crinitus*. *J Nat Prod* 67(10):1764–1766.

IUPAC. 1997. *Compendium of Chemical Terminology*, 2nd edn. (the *"Gold Book"*). Compiled by A.D. McNaught and A. Wilkinson. Oxford, U.K.: Blackwell Scientific Publications.

Jiang, P., Li, J., Han, F. et al. 2011. Antibiofilm activity of an exopolysaccharide from marine bacterium *Vibrio* sp. QY101. *PLoS One* 6(4):e18514.

Kenji, F., Tala, S., Kentaro, N. et al. 2010. Effects of carbohydrate source on physicochemical properties of the exopolysaccharide produced by *Lactobacillus fermentum* TDS030603 in a chemically defined medium. *Carbohydr Polym* 79:1040–1045.

Kim, M., Li, Y.X., Dewapriya, P., Ryu, B., and Kim, S.K. 2013. Floridoside suppresses pro-inflammatory responses by blocking MAPK signaling in activated microglia. *BMB Rep* 46(8):398–403.

Kodali, V.P., Das, S., and Sen, R. 2009. An exopolysaccharide from a probiotic: Biosynthesis dynamic, composition and emulsifying activity. *Food Res Int* 42:695–699.

Kolida, S. and Gibson, G.R. 2008. The prebiotic effect: Review of experimental and human data. In *Handbook of Prebiotics*, eds. Gibson, G.R. and Roberfroid, M. Boca Raton, FL: CRC Press, pp. 69–93.

Kuznetsova, T.A., Zaporozhets, T.S., Besednova, N.N. et al. 2011. The study of the prebiotic potential of biologically active substances from marine hydrobionts and the development of new functional nutrition products. *Vestn Dal'nevost Fil Ross Akad Nauk* 2:147–150.

Kuznetsova, T.A., Zaporozhets, T.S., Makarenkova, I.D. et al. 2012. The prebiotic potential of polysaccharides from the brown alga *Fucus evanescens* and significance for the clinical use. *Tikhookean Med J* 1:37–40.

Le Leu, R.K., Brown, I.L., Hu, Y. et al. 2005. A synbiotic combination of resistant starch and *Bifidobacterium lactis* facilitates apoptotic deletion of carcinogen-damaged cells in rat colon. *J Nutr* 135:996–1001.

Lee, I.Y., Seo, W.T., Kim, G.J. et al. 1997. Optimization of fermentation conditions for production of exopolysaccharide by *Bacillus polymyxa*. *Bioprocess Biosyst Eng* 16:71–75.

Leiro, J.M., Castro, R., Arranz, J.A., and Lamas, J. 2007. Immunomodulating activities of acidic sulphated polysaccharides obtained from the seaweed Ulva rigida C. Agardh. *Int Immunopharmacol* 7(7):879–88.

Leonard, S.G., Sweeney, T., Bahar, B., Lynch, B.P., and O'Doherty, J.V. 2010. Effect of maternal fish oil and seaweed extract supplementation on colostrum and milk composition, humoral immune response, and performance of suckled piglets. *J Anim Sci* 88(9):2988–97.

Li, M., Li, G., Zhu, L. et al. 2014. Isolation and characterization of an agaro-oligosaccharide (AO)-hydrolyzing bacterium from the gut microflora of Chinese individuals. *PLoS One* 9(3):e91106.

Li, Y.X., Li, Y., Lee, S.H., Qian, Z.J., and Kim, S.K. 2010. Inhibitors of oxidation and matrix metalloproteinases, floridoside, and D-isofloridoside from marine red alga *Laurencia undulata*. *J Agric Food Chem* 58(1):578–586.

Looijesteijn, P.J., Trapet, L., de Vries, E., Abee, T., and Hugenholtz, J. 2001. Physiological function of exopolysaccharides produced by *Lactococcus lactis*. *Int J Food Microbiol* 64:71–80.

Lynch, M.B., Sweeney, T., Callan, J.J., O'Sullivan, J.T., and O'Doherty, J.V. 2010. The effect of dietary *Laminaria*-derived laminarin and fucoidan on nutrient digestibility, nitrogen utilization, intestinal microflora and volatile fatty acid concentration in pigs. *J Sci Food Agric* 90:430–437.

Macfarlane, G.T., Steed, H., and Macfarlane, S. 2008. Bacterial metabolism and health-related effects of galacto-oligosaccharides and other prebiotics. *J Appl Microbiol* 104:305–344.

Marzorati, M., Verhelst, A., and Luta, G. 2010. *In vitro* modulation of the human gastrointestinal microbial community by plant-derived polysaccharide rich dietary supplements. *Int J Food Microbiol* 139:168–176.

Matsumoto, S., Nagaoka, M., Hara, T., Kimura-Takagi, I., Mistuyama, K., and Ueyama, S. 2004. Fucoidan derived from *Cladosiphon okamuranus* Tokida ameliorates murine chronic colitis through the down-regulation of interleukin-6 production on colonic epithelial cells. *Clin Exp Immunol* 136(3):432–439.

McDonnell, P., Figat, S., and O'Doherty, J.V. 2010. The effect of dietary laminarin and fucoidan in the diet of the weanling piglet on performance, selected fecal microbial populations and volatile fatty acid concentrations. *Animal* 4:579–585.

Mori, N., Nakasone, K., Tomimori, K., and Ishikawa, C. 2012. Beneficial effects of fucoidan in patients with chronic hepatitis C virus infection. *World J Gastroenterol* 8(18):2225–2230.

Mottet, C. and Michetti, P. 2005. Probiotic: Wanted dead or alive. *Dig Liver Dis* 37:3–6.

Muraoka, T., Ishihara, K., Oyamada, C. et al. 2008. Fermentation properties of low quality red alga Susabinori *Porphyra yezoensis* by intestinal bacteria. *Biosci Biotechnol Biochem* 72(7):1731–1739.

Nakazato, K., Takada, H., Iha, M., and Nagamine, T. 2010. Attenuation of N-nitrosodiethylamine-induced liver fibrosis by high-molecular-weight fucoidan derived from *Cladosiphon okamuranus*. *J Gastroenterol Hepatol* 25(10):1692–1701.

Navarini, L., Abatangelo, A., Bertocchi, C., Conti, E., Bosco, M., and Picotti, F. 2001. Isolation and characterization of the exopolysaccharide produced by *Streptococcus thermophilus* SFi20. *Int J Biol Macromol* 28:219–226.

Nguyen, V.D., Le, M.H., and Trang, S.T. 2013. Application of probiotics from marine microbes for sustainable marine aquaculture development. In *Marine Microbiology: Bioactive Compounds and Biotechnological Applications*, ed. Kim, S.K. Weinheim, Germany: Wiley-VCH, pp. 307–349.

O'Doherty, J.V., Dillon, S., Figat, S., Callan, J.J., and Sweeney, T. 2010. The effects of lactose inclusion and seaweed extract derived from *Laminaria* spp. on performance, digestibility of diet components and microbial populations in newly weaned pigs. *Anim Feed Sci Technol* 157:173–180.

O'Sullivan, L., Murphy, B., McLoughlin, P. et al. 2010. Prebiotics from marine macroalgae for human and animal health applications. *Mar Drugs* 8(7):2038–2064.

Parker, G.A. 1974. Assessment strategy and the evolution of animal conflicts. *J Theor Biol* 47:223–243.

Patel, A.K., Singhania, R.R., Pandey, A., and Chincholkar S.B. 2010. Probiotic bile salt hydrolase: current developments and perspectives. *Appl Biochem Biotechnol* 162(1):166–80.

Patel, S. and Goyal, A. 2012. The current trends and future perspectives of prebiotics research: A review. *3 Biotech* 2:115–125.

Patterson, E., Cryan, J.F., Fitzgerald, G.F., Ross, R.P., Dinan, T.G., Stanton C. 2014. Gut microbiota, the pharmabiotics they produce and host health. *Proc Nutr Soc* 73(4):477–489.

Patterson, J.A. and Burkholder, K.M. 2003. Application of prebiotics and probiotics in poultry production. *Poult Sci* 82:627–631.

Pavao, M.S.G. (ed.) 2011. *Glycans in Diseases and Therapeutics*. Berlin, Germany: Springer.

Pomin, V.H. 2014. Marine medicinal glycomics. *Front Cell Infect Microbiol* 4:5.

Ramnani, P., Chitarrari, R., Tuohy, K. et al. 2012. In vitro fermentation and prebiotic potential of novel low molecular weight polysaccharides derived from agar and alginate seaweeds. *Anaerobe* 18:1–6.

Ray, B. and Lahaye, M. 1995. Cell wall polysaccharides from the marine green alga *Ulva "rigida"* (Ulvales, Chlorophyta). Chemical structure of ulvan. *Carbohydr Res* 274:313–318.

Reilly, P., O'Doherty, J.V., Pierce, K.M., Callan, J.J., O'Sullivan, J.T., and Sweeney, T. 2008. The effects of seaweed extract inclusion on gut morphology, selected intestinal microbiota, nutrient digestibility, volatile fatty acid concentrations and the immune status of the weaned pig. *Animal* 2:1465–1473.

Ringo, E., Olsen, R.E., Gifstad, T.O., Dalmo, R.A., Amlund, H., Hemre, G. I., and Bakke, A.M. 2010. Prebiotics in aquaculture: A review. *Aquacult Nutr* 16:117–136.

Roberfroid, M.B. 2000. Prebiotics and probiotics: Are they functional foods? *Am J Clin Nutr* 71(6):1682s–1687s.

Roberts, C.M., Fett, W.F., Osman, S.F., Wijey, C., O'Connor, J.V., and Hoover, D.G. 1995. Exopolysaccharide production by *Bifidobacterium longum* BB-79. *J Appl Bacteriol* 78:463–468.

Roberts, I.S. 1996. The biochemistry and genetics of capsular polysaccharide production in bacteria. *Annu Rev Microbiol* 50:285–315.

Roller, M., Pietro Femia, A., Caderni, G., Rechkemmer, G., and Watzl, B. 2004. Intestinal immunity of rats with colon cancer is modulated by oligofructose-enriched inulin combined with *Lactobacillus rhamnosus* and *Bifidobacterium lactis*. *Br J Nutr* 92:931–938.

Ruas-Madiedo, P., Gueimonde, M., Margolles, A., de los Reyes-Gavilán, C., and Salminen, S. 2006. Exopolysaccharides produced by probiotic strains modify the adhesion of probiotics and enteropathogens to human intestinal mucus. *J. Food Prot* 69:2011–2015.

Salazar, N., Gueimonde, M., Hernández-Barranco, A.M., Ruas-Madiedo, P., and de los Reyes-Gavilán, C.G. 2008. Exopolysaccharides produced by intestinal *Bifidobacterium* strains act as fermentable substrates for human intestinal bacteria. *Appl Environ Microbiol* 74:4737–4745.

Shibata, H., Iimuro, M., Uchiya, N. et al. 2003. Preventive effects of *Cladosiphon* fucoidan against *Helicobacter pylori* infection in Mongolian gerbils. *Helicobacter* 8:59–65.

Shibata, H., Nagaoka, M., Takagi, I.K., Hashimoto, S., Aiyama, R., and Yokokura, T. 2001. Effect of oligofucose derivatives on acetic acid-induced gastric ulcer in rats. *Biomed Mater Eng* 11(1):55–61.

Shibata, H., Nagaoka, M., Takeuchi, I.K. et al. 1998. Antiulcer effect of fucoidan from brown seaweed, *Cladosiphon okamuranus* TOKIDA. *Jpn Pharmacol Ther* 26:1211–1215.

Smith, A.G., O'Doherty, J.V., Reilly, P., Ryan, M.T., Bahar, B., and Sweeney, T. 2011. The effects of laminarin derived from *Laminaria digitata* on measurements of gut health: Selected bacterial populations, intestinal fermentation, mucin gene expression and cytokine gene expression in the pig. *Br J Nutr* 105:669–677.

Sweeney, T., Collins, C.B., Reilly, P., Pierce, K.M., Ryan, M., and O'Doherty, J.V. 2012. Effect of purified beta-glucans derived from *Laminaria digitata*, *Laminaria hyperborea* and *Saccharomyces cerevisiae* on piglet performance, selected bacterial populations, volatile fatty acids and pro-inflammatory cytokines in the gastrointestinal tract of pigs. *Br J Nutr* 108:1226–1234.

Takeda, K., Tomimori, K., Kimura, R., Ishikawa, C., Nowling, T.K., and Mori, N. 2012. Anti-tumor activity of fucoidan is mediated by nitric oxide released from macrophages. *Int J Oncol* 40(1):251–260.

Teruya, T., Konishi, T., Uechi, S., Tamaki, H., and Tako, M. 2007. Anti-proliferative activity of oversulfated fucoidan from commercially cultured *Cladosiphon okamuranus* TOKIDA in U937 cells. *Int J Biol Macromol* 41(3):221–226.

Thomes, P., Rajendran, M., Pasanban, B., and Rengasamy, R. 2010. Cardioprotective activity of *Cladosiphon okamuranus* fucoidan against isoproterenol induced myocardial infarction in rats. *Phytomedicine* 18(1):52–57.

Topping, D.L. and Clifton, P.M. 2001. Short-chain fatty acids and human colonic function: Roles of resistant starch and nonstarch polysaccharides. *Physiol Rev* 81:1031–1064.

Toskas, G., Heinemann, S., Heinemann, C., Cherif, C., Hund, R.D., Roussis, V., and Hanke, T. 2012. Ulvan and ulvan/chitosan polyelectrolyte nanofibrous membranes as a potential substrate material for the cultivation of osteoblasts. *Carbohydr Polym* 89(3):997–1002.

Ustyuzhanina, N.E., Ushakova, N.A., Zyuzina, K.A. et al. 2013. Influence of fucoidans on hemostatic system. *Mar Drugs* 11(7):2444–2458.

Van Loo, J.A. 2004. Prebiotics promote good health: the basis, the potential, and the emerging evidence. *J Clin Gastroenterol* 38(6 Suppl):S70–5.

Zaporozhets, T.S., Besednova, N.N., Kuznetsova, T.A. et al. 2014. The prebiotic potential of polysaccharides and extracts of seaweeds. *Russ J Mar Biol* 40(1):1–9.

Zhao, J. and Cheung, P.C. 2013. Comparative proteome analysis of *Bifidobacterium longum* subsp. infantis grown on β-glucans from different sources and a model for their utilization. *J Agric Food Chem* 61(18):4360–4370.

Zhou, Q.C., Buentello, J.A., and Gatlin III, D.M. 2010. Effect of dietary prebiotics on growth performance, immune response and intestinal morphology of red drum (*Scianops ocellatus*). *Aquaculture* 309:253–257.

*chapter seven*

# Principle and biological properties of sulfated polysaccharides from seaweed

**Jantana Praiboon, Anong Chirapart, and Nattanun Soisarp**

## Contents

## 7.1 Introduction

Polysaccharides with high sulfate functionalization are generally known as sulfated polysaccharides. Most of them are highly polyanionic in nature due to the attachment of negatively charged functional groups to the central sugar backbone of linear or branched polymers (Raveendran et al., 2013). These anionic polymers are widespread in nature, occurring in a great variety of organisms such as mammals, marine invertebrates, microorganisms, and seaweed (Costa et al., 2010).

Seaweed are a rich source of polysaccharides, and the type of polysaccharide is different depending on the taxonomic group. Different carbohydrates, including agar, carrageenan, and alginate, are extracted from seaweed.

These carbohydrates are used widely in the food and pharmaceutical industries as functional ingredients such as stabilizers. Moreover, sulfated polysaccharides derived from seaweed have been drawing much attention as important bioactive natural products. These sulfated polysaccharides exhibit many beneficial biological functions such as anticoagulant, antiviral, antioxidant, anticancer, anti-inflammatory, anti-hyperlipidemia, antimicrobial, and immonuomodulatory activities. In general, the biological activities of sulfated polysaccharides are related to the molecular size, the type of sugar and sulfate content, and the sulfate position. The type of linkage and molecular geometry are also known to play a role in their activity. In this chapter, we describe and summarize the principle and biological properties of seaweed sulfated polysaccharides.

## 7.2 Principle of seaweed polysaccharides

Seaweed contain large amounts of polysaccharides, notably cell wall structural, but also mucopolysaccharides and storage polysaccharides (Murata and Nakazoe, 2001). Polysaccharides are polymers of simple sugars (monosaccharides) linked together by glycosidic bonds (Lahaye, 2001). The total polysaccharide concentrations in the seaweed species of interest range from 4% to 76% of their dry weight, and the highest contents are found in species such as *Ascophyllum*, *Porphyra*, and *Palmaria*; however, green seaweed species such as *Ulva* also have a high content, up to 65% of dry their weight (Kraan, 2012).

Polysaccharides can be homopolymers or heteropolymers of neutral (pentoses and hexose) and/or anionic sugars (hexoses), substituted or not by nonsugar compounds, linear, or ramified (Delattre et al., 2011). A part, but not all, of them have a regular chemical structure based on 2–8 sugar repeat units. They are stereoregular and may adopt an ordered conformation. These features lead to their specific behaviors in solution. The resulting conformations are spirals, sheets, and single, double, and triple helices (Rinaudo, 2004). The principal cell wall polysaccharides in green seaweed are ulvans, while those in red seaweed are agarans and carrageenans and those in brown seaweed are alginates and fucans as well as the storage polysaccharide laminarin (Rioux et al., 2007; Jiao et al., 2011).

### 7.2.1 Green seaweed

#### 7.2.1.1 Ulvans and oligo-ulvans
The name "ulvan" is derived from the original terms "ulvin" and "ulvacin," but it is now being used to refer to polysaccharides from members of Ulvales (Lahaye and Robic, 2007). Ulvan extraction can be achieved sequentially with hot sodium oxalate solution, hot water, and 1 and 4 M KOH solution at 20°C (Ray and Lahaye, 1995). Recovery of ulvan is generally done by precipitation by adding an alcohol or a quaternary ammonium salt (Lahaye and Robic, 2007).

Ulvans are the major constituents of green seaweed cell walls, representing 8%–29% of the algal dry weight (Lahaye and Robic, 2007). They are ramified acidic and sulfated polysaccharides (Vera et al., 2011) constituted by a central backbone of disaccharide units formed by an L-rhamnose 3-sulfate linked to (1) D-guluronic acid residue (ulvabiouronic acid unit A), (2) L-iduronic acid residue (ulvabiuronic acid unit B), (3) D-xylose 4-sulfate residue (ulvabiose unit A), or (4) D-xylose residue (ulvabiose unit B), as shown in Figure 7.1. In addition, ulvans

show ramifications in position O-2 of the rhamnose 3-sulfate residue. Other oligo-ulvans have been obtained by depolymerization of cell wall polysaccharides from *Ulva armoricana*, *U. rigida*, *U. lactuca*, *U. compressa*, and *U. intestinalis* using 2 M HCl at 100°C for 45 min, which produces mainly monosaccharide and disaccharide units (Briand et al., 2010, 2011). The oligo-ulvans with a molecular weight of 50–60 kDa have been obtained by ultrasound fragmentation and further purification by size exclusion chromatography (Jaulneau et al., 2010).

#### 7.2.1.2 Rhamnan sulfate
Some green seaweed produce specific sulfated polysaccharides, which are mainly composed of the α-L-rhamnose moiety (Lee et al., 2010). A water-soluble polysaccharide extracted from *Ulva lactuca* was found to be a sulfated glucurono-xylo-rhamnoglycan, in which the sulfate groups are linked to rhamnose (Percival and Wold, 1963). A similar sulfated polysaccharide was isolated from *U. rigida* (Ray and Lahaye, 1995). The rhamnan sulfate polysaccharide from cell walls of *Monostroma latissimum* consists of large amounts of rhamnose residues and appears to be an entire homopolysaccharide, with structural properties different from those of ulvan, and heterogeneity of cell wall polysaccharides different from those of *U. lactuca* and *U. rigida* from the Ulvaceae (Lee et al., 1998). Another work reported the structure of the rhamnan sulfate isolated from commercially cultured *Monostroma nitidum*, which consisted of (1 → 3)-linked α-L-rhamnosyl residues, a part of which at C-2 was substituted with (1 → 2)-linked trisaccharide side chains, the nonreducing end of which was α-D-glucuronic acid (Figure 7.2; Nakamura et al., 2011). The authors also reported that the sulfate groups substituted at C4 and C2 position of the L-rhamnosyl residues on main and side chain, respectively.

#### 7.2.1.3 Others
*Codium* species biosynthesize sulfated galactans constituted by 3-linked β-D-galactopyranose residues partially sulfated on C-4 and/or C-6, with ramifications of the backbone on C-6 and important amounts of pyruvate (Fernández et al., 2014). The highly pyruvated of 1,3-β-D-galactan sulfate from *Codium yezoense* and the similar polysaccharide from *Codium isthmocladium* represent another type of polysaccharide found in green seaweed. Sulfated β-D-mannans was isolated from *Codium vermilara* (Fedorov et al., 2013). Besides, highly sulfated pyranosic arabinans were detected and characterized as part of the sulfated polymers from *Codium decorticatum* (Fernández et al., 2014).

### 7.2.2 Brown seaweed

The cell wall of brown algae is mainly composed of fucoidan, alginate, and laminarin (3:1:1), as well as their

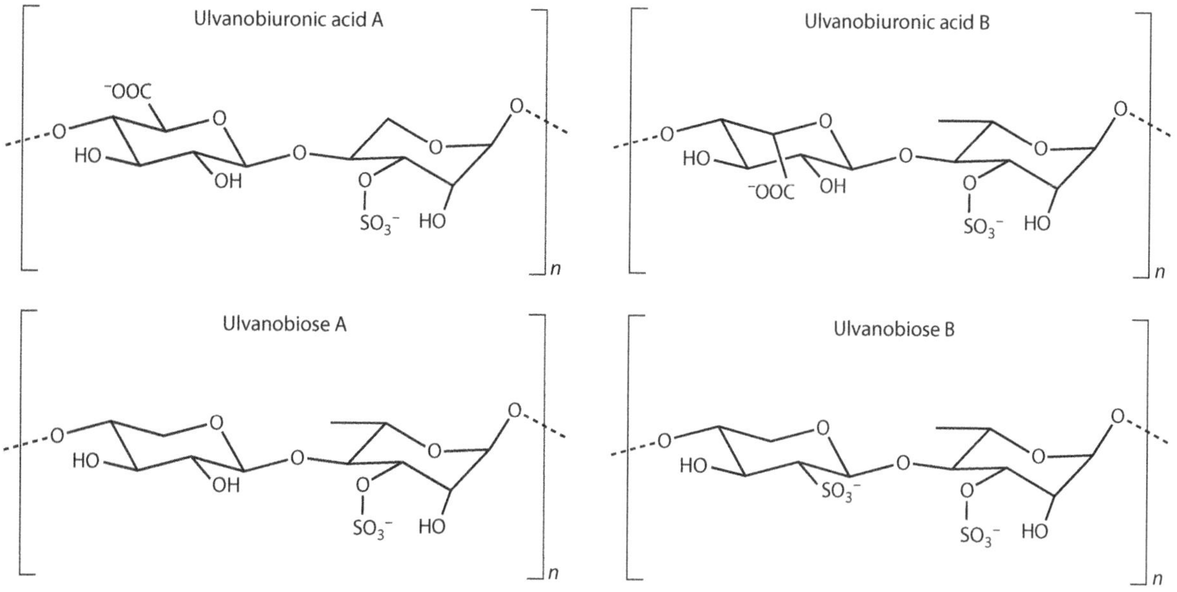

**Ulvanobiuronic acid A**

**Ulvanobiuronic acid B**

**Ulvanobiose A**

**Ulvanobiose B**

*Figure 7.1* Structure of ulvans isolated from green seaweed.

derivatives. These reserve polysaccharides provide strength and flexibility, maintain ionic equilibrium, and prevent desiccation. They are suitable as thickeners and gelation agents.

### 7.2.2.1 Alginate

The source of alginate or algin is mainly brown macroalgae, where it is the major structural component of the cell wall and intercellular matrix. Brown seaweed that grow in more turbulent conditions usually have higher alginate content than those growing in calmer waters (McHugh, 2003). While any brown seaweed can be used as a source of alginate, the actual chemical structure of the alginate varies from one genus to another, and similar variability is found in the properties of the alginate that is extracted from the seaweed. Since the main applications of alginate are in thickening aqueous solutions and forming gels, its quality is judged by how well it performs in these uses. Alginate

can be produced from a variety of the brown seaweed species such as *Ascophyllum nodosum, Durvillaea potatorum, Ecklonia* sp., *Laminaria digitata, Lessonia nigrescens, Macrocystis pyrifera,* and *Sargassum sinicola* (McHugh, 2003; Yabur et al., 2007; Vauchel et al., 2008; Fertah et al., 2014). However, *Sargassum* is used only when nothing else is available: its alginate is usually borderline quality and the yield usually low (McHugh, 2003). Commercially available alginate is typically extracted from *Laminaria hyperborea, L. digitata, L. japonica, Ascophyllum nodosum,* and *Macrocystis pyrifera* by treatment with aqueous alkali solutions, typically NaOH. The alginate produced from the giant kelp *Macrocystis pyrifera,* which is most frequently used for cell immobilization, yields gels with lower strength and stability than gels produced from other alginates (Skjåk-Bræk and Martinsen, 1991). The alginate content is 22%–30% for *Ascophyllum nodosum* and 25%–44% for *Laminaria digitata* (Qin, 2008).

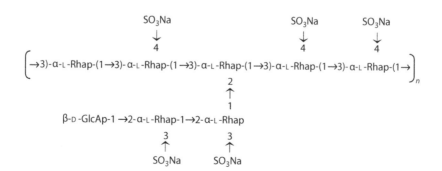

*Figure 7.2* Structure of rhamnan sulfate isolated from commercially cultured *Monostroma nitidum* sulfated galactan.

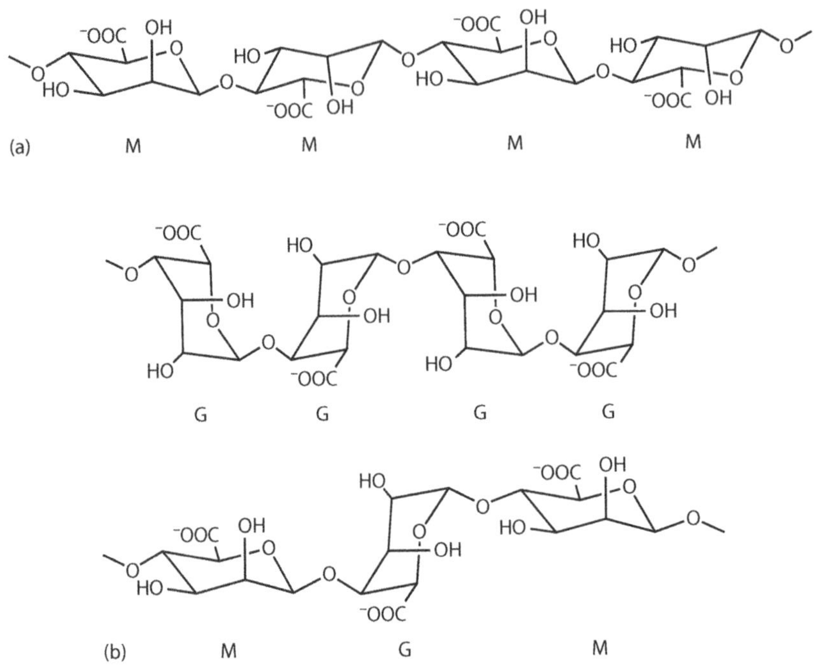

*Figure 7.3* Alginate block types: (a) poly-D-mannuronate and (b) poly-L-guluronate. G, guluronic acid; M, mannuronic acid.

Alginic acid is a linear anionic polysaccharide containing 1,4-linked β-D-mannuronic acid (M) and α-L-guluronic acid (G) residues (Balboa et al., 2013). The monomers can be linked by calcium ions through binding consecutive blocks of guluronic acid to form gels. Figure 7.3 illustrates the blocks that are composed of consecutive G residues (GGGGGG), consecutive M residues (MMMMMM), and alternating M and G residues (GMGMGM). Alginates extracted from different sources differ in M and G contents as well as in the length of each block, and more than 200 different alginates are currently being manufactured (Tønnesen and Karlsen, 2002). The G-block content of *Laminaria hyperborean* stems is 60%, and for other commercially available alginates it is in the range of 14.0%–31.0% (Qin, 2008).

The relative amounts and occurrence of both monomers and their sequential arrangement in the polymeric chain are related to the origin of the alginate, both environmental and genetic.

### 7.2.2.2 Fucans

In general, sulfated polysaccharides of brown seaweed are called fucans. These include the compounds fucoidan, ascophyllan, sargassan, and glucuronoxylofucan (Shanmugam and Mody, 2000). Fucans are a heterogeneous family of polysaccharides based on L-fucose, ranging from fucose- and sulfate-rich molecules to molecules less sulfated and richer in uronic residues (Table 7.1). Fucans are extracted from the matrix phase of the cell wall of brown algae by organic solvents or

*Table 7.1* Average chemical composition (g/100 g) of the different fucan species

|                | Fucoidans | Ascophyllans | Sargssans |
|----------------|-----------|--------------|-----------|
| L-fucose       | 50–90     | ~25          | 25–45     |
| Uronic acid    | <8        | ~25          | ~12       |
| Other sugars   | 5–40      | ~25          | ~36       |
| Sulfate groups | 35–45     | ~13          | 15–21     |
| Proteins       | <4        | ~12          | 4         |

*Source:* Modified from Chaubet, F. et al., Relationships between chemical characteristics and anticoaglulant activity of low molecular weight fucans from marine algae, in Paulsen, B.S., ed., *Bioactive Carbohydrate Polymers*, Kluwer Academic Publishers, Dordrecht, the Netherlands, 2000, pp. 59–84.

water followed with aqueous species probably coexist in the thallus of the algae (Chaubet et al., 2000). Fucans are very soluble and do not develop highly viscous solutions. The technologies to obtain them include (1) extraction of low-molecular-weight compounds; (2) extraction with water, acid, or calcium chloride, eventually aided by either hydrolytic enzymes, or ultrasound, or microwaves; (3) precipitation; and (4) purification (Balboa et al., 2013). The composition of fucans in algae varies according to the species (Table 7.2). Moreover, for a given species the overall proportions of the constituents depend upon the tissue and numerous other factors such as the age, habitat, or season, and these variation can have an effect upon the properties and qualities of the various extracts (Chaubet et al., 2000).

*7.2.2.2.1 Fucoidan*   Fucoidan is a class of sulfated, fucose-rich polysaccharides found in the fibrillar cell walls and intercellular spaces of brown seaweed of the class Phaeophyceae (Berteau and Mulloy, 2003).

*Table 7.2* Diversity and relative abundance of the sulfated fucans from phaeophyta

| Order | Genus | Fucoidans | Ascophyllans | Sargassans |
|-------|-------|-----------|--------------|------------|
| Estocarpales | *Ectocarpus* | | + (traces) | |
| | *Sorocarpus* | + | + | + |
| Chordariales | *Chorda* | ++ | | |
| | *Heterochordia* | ++ | | |
| | *Leathesis* | ++ | | |
| | *Nemacystus* | ++ | | |
| | *Sphaerotrichia* | ++ | | |
| | *Tinocladia* | ++ | | |
| Desmarestiales | *Desmarestia* | + | + | ++ |
| Dictyosiphonales | *Asperococcus* | | | + |
| Scytosiphonales | *Colpomenia* | | | ++ |
| | *Scytosiphon* | + | + | + |
| Dictyotales | *Dictyopteris* | + | ++ | |
| | *Dictyota* | + | ++ | + |
| | *Padina* | | ++ | + |
| | *Taonia* | | | + |
| Laminariales | *Alaria* (thalles) | | + | + |
| | *Alaria* (sporophylls) | + | + | ++ |
| | *Ecklonia* | ++ | ++ | + |
| | *Eisenia* | ++ | + | ++ |
| | *Kjellmaniella* | ++ | + | |
| | *Laminaria* | ++ | | |
| | *Lessonia* | ++ | | |
| | *Macrocystis* | ++ | | |
| | *Nereocystis* | | | |
| | *Undaria* (thalles) | | | |
| | *Undaria* (sporophylls) | | | |
| Fucales | *Ascophyllum* | + | ++ | + |
| | *Bifurcaria* | + | + | ++ |
| | *Fucus* | + | + | |
| | *Halidrys* | + | + | |
| | *Himanthalia* | + | ++ | |
| | *Hizikia* | ++ | + | |
| | *Pelvetia* | + | + | |
| | *Sargassum* | | ++ | |
| | *Turbinaria* | | | |

*Source:*   Modified from Chaubet, F. et al., Relationships between chemical characteristics and anticoagulant activity of low molecular weight fucans from marine algae, in Paulsen, B.S., ed., *Bioactive Carbohydrate Polymers*, Kluwer Academic Publishers, Dordrecht, the Netherlands, 2000, pp. 59–84.

*Table 7.3* Brown seaweed containing fucoidan

| Species | Order | %dry wt | References |
|---------|-------|---------|------------|
| *Dictyota dichotoma* | Dictyotales | 7.2 | Rabanal et al. (2014) |
| *Dictyota dichotoma* | Dictyotales | 3.9 | Kantachumpoo and Chirapart (2010) |
| *Dictyota caribaea* | Dictyotales | 22.2 | García-Ríos et al. (2012) |
| *Padina perindusiata* | Dictyotales | 21.6 | García-Ríos et al. (2012) |
| *Padina australis* | Dictyotales | 11.8 | Kantachumpoo and Chirapart (2010) |
| *Colpomenia sinuosa* | Scytosiphonales | 4.33 | Kantachumpoo and Chirapart (2010) |
| *Hydroclathrus clathratus* | Scytosiphonales | 19.7 | Kantachumpoo and Chirapart (2010) |
| *Sargassum filipendula* | Fucales | 26 | García-Ríos et al. (2012) |
| *Sargassum polycystum* | Fucales | 5.1 | Kantachumpoo and Chirapart (2010) |
| *Turbinaria turbinata* | Fucales | 32.1 | García-Ríos et al. (2012) |
| *Turbinaria conoides* | Fucales | 10.4 | Kantachumpoo and Chirapart (2010) |
| *Undaria pinnatifida* | Laminariales | 3.2–16.0 | Skriptsova et al. (2010) |
| *Punctaria plantaginea* | Dictyosiphonales | 19.2 | Bilan et al. (2002) |
| *Adenocystis utricularis* | Ectocarpales | 19.1 | Ponce et al. (2003) |
| *Cladosiphon okamuranus* | Chordariales | 2 | Sakai et al. (2003) |

They are present in several orders, mainly Fucales and Laminariales but also in Chordariales, Dictyotales, Dictyosiphonales, Ectocarpales, and Scytosiphonales (Table 7.3). The yield of fucoidan varies depending on algal species and season (Skriptsova et al., 2010). They account for 10%–20% dry weight, consisting mainly of sulfated L-fucose and small proportions of galactose, mannose, xylose, glucose, rhamnose, and uronic acids (Balboa et al., 2013). The fucoidan polymers are easily extracted from the cell wall of brown algae with hot water or acid solution (Ale et al., 2011). Almost all biological studies use a commercially available crude preparation of sulfated polysaccharides from *Fucus vesiculosus* rather than a purified fucoidan (Mulloy et al., 1994). Fucoidan from different algal species differ from each other and vary not only in the position and level of sulfation and molecular mass but sometimes also in the structures of the main carbohydrate chain, for example, the fucoidan from the brown algae *Fucus serratus*, *Ascophyllum nodosum*, *F. vesiculosus*, *F. evanescens*, *Laminaria saccharina*, *Cladosiphon okamuranus*, and *Chorda filum* (Figure 7.4).

*7.2.2.2.2 Ascophyllan* Ascophyllans are intrafibrillar matrix polysaccharides that play a role in the structure of the cell wall in the adult thallus by crosslinking alginates and cellulose microfibrils of the skeleton, whereas fucoidans are components of the intercellular matrix. They are xylofucoglycuronans with large proportions of 4-sulfated L-fucose ($\alpha$-1 → 2 linkgaes), D-xylose, and uronic residues. Their structure is highly heteropolymeric, with no long sequence of the same monosaccharide; xylose and glucuronic acids would then appear as 1 → 4 linked or terminal units (Chaubet et al., 2000). Ascophyllan

is distinguishable from fucoidan in that it contains a much higher amount of uronic acids than fucoidan. Moreover, the level of uronic acid in ascophyllan is nearly equal to that of fucose (Table 7.1). Ascophyllan isolated from *Ascophyllum nodosum* as a sulfated fucan preparation is distinguished from fucoidan by the presence of a backbone of uronic acid with fucose-containing branches (3-*O*-D-xylosyl-L-fucose-4-sulfate). Specifically, ascophyllan has fucose and xylose in nearly equimolecular proportion, whereas fucoidans have a much higher ratio of fucose than xylose (Jiang et al., 2010).

*7.2.2.2.3 Sargassan* Sargassans are found in *Sargassum* but also in the mature sporophyll of *Undaria* (Table 7.2). Their structure remains controversial, but many authors agree on linear chains of 1 → 4 linked D-galactose or glucose branced at C6 with L-fucosyl-3-sufate or, occasionally, a uronic acid, probably D-glucuronic acid (Figure 7.5). In general, galactose is found in the terminal position or 1 → 3 linked with other hexose, and hence does not form real homogalactan sequences (Chaubet et al., 2000).

*7.2.2.2.4 Laminaran or laminarin* Laminaran is a polysaccharide of brown algae, composed of (1,3)-$\beta$-D-glucopyranose residues with an occasional branching point at C6 and with a variable proportion of the glucose chains terminated at the potential reducing end with a molecule of mannitol. This polysaccharide is not part of the cell wall and does not naturally contain sulfate groups. It is of relatively low molecular weight (~3–6 kDa) and its structural features are species dependent. The solubility of laminarans depends on the branching level: the higher the branching degree,

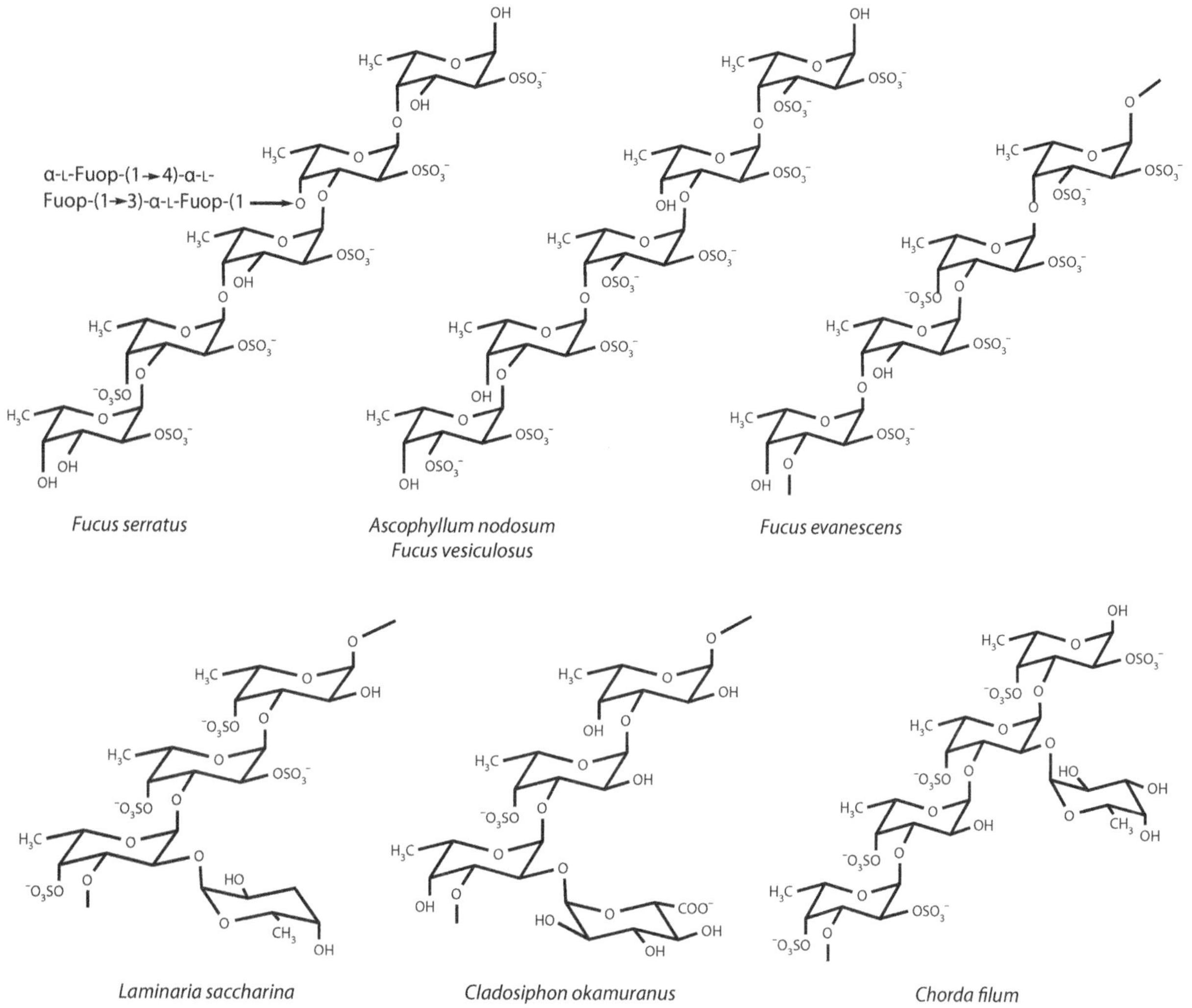

α-L-Fuop-(1→4)-α-L-
Fuop-(1→3)-α-L-Fuop-(1 →

**Fucus serratus**

**Ascophyllum nodosum
Fucus vesiculosus**

**Fucus evanescens**

**Laminaria saccharina**

**Cladosiphon okamuranus**

**Chorda filum**

*Figure 7.4* Chemical structure of fucoidan from brown seaweed.

the higher the solubility; thus, low-branched laminarans are soluble only in warm water (60°C–80°C). Moreover, they are not thickening nor gelling agents, nor do they have any current commercial use on an industrial scale (Rupérez et al., 2002; Balboa et al., 2013).

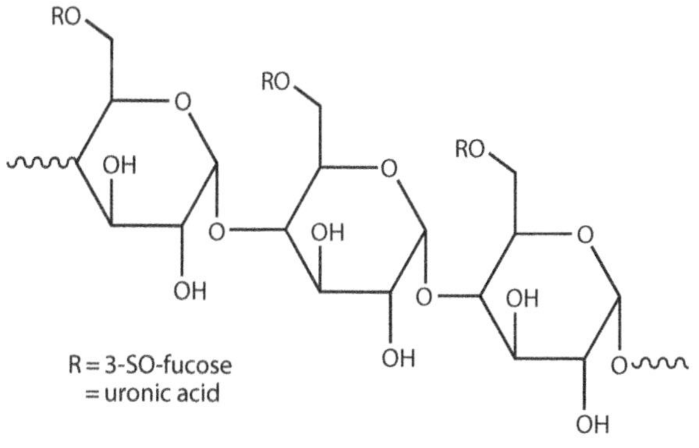

R = 3-SO-fucose
= uronic acid

*Figure 7.5* Chemical structure of sargassan.

## 7.2.3   Red seaweed

### 7.2.3.1   Agar

Most agars are extracted from the species *Gelidium* and *Gracilaria* with the related genera of *Gracilariopsis* and *Hydropuntia*. Closely related to *Gelidium* are the species of *Pterocladia*, and small quantities of these are collected, mainly in Azores (Portugal) and New Zealand. *Gelidiella acerosa* is the main source of agar in India (McHugh, 2003). *Ahnfeltia* species have been used in both Russia and Japan, one source being the island of Sakhalin (Russia).

Pioneering studies on the structure of agar polysaccharides have been conducted in Japan (Araki, 1966). Agar polymers synthesized by species of the genus *Gracilaria* constitute a complex mixture of molecules. Agar polysaccharides are usually composed of a repeating agarobiose unit of alternating 1,3-linked-D-galactose and 1,4-linked 3,6-anhydro-L-galactose residues (Araki, 1966). This repeating sequence can be

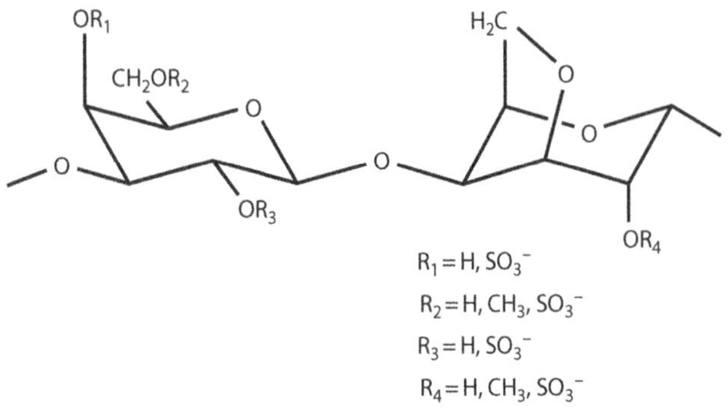

$R_1 = H, SO_3^-$

$R_2 = H, CH_3, SO_3^-$

$R_3 = H, SO_3^-$

$R_4 = H, CH_3, SO_3^-$

*Figure 7.6* Chemical structure of agar polysaccharide.

substituted by methoxyl, sulfate esters, and pyruvate ketal groups (Figure 7.6) and can also occur in its biological precursor form, where L-galactose-6-sulfate replaces the 3,6-anhydro-L-galactose residue (Duckworth and Yaphe, 1971; Duckworth et al., 1971; Lahaye et al., 1986; Chirapart et al., 1995). In comparison with agars from *Gelidium* and *Pterocladia*, agar from *Gracilaria* can have higher degrees of sulfation, methoxylation, and pyruvylation (Murano, 1995). An agar extracted from *Gracilariopsis lemaneiformis* has been reported as composed of the biological precursor to agarobiose repeating units and agarobiose containing 6-*O*-methyl agarobiose and a small amount of 2-*O*-methyl-α-L-galactopyranose residues (Chirapart et al., 1995).

According to the substitution, agar polysaccharides can be classified into three groups (Yaphe and Duckworth, 1971): (1) sulfated agar, (2) methylated agar, and (3) pyruvated agar. The sulfate group is found more in *Gelidium* and *Gracilaria* agars, the methylate group is found in *Gracilaria arcuata* (Tako et al., 1999) and *Hydropuntia edulis* (as *Gracilaria edulis*, Villanueva and Montaño, 1999; Praiboon et al., 2006) agars, and rarely found with pyruvic acid in some *Gracilaria caudate* (Barros et al., 2013).

### 7.2.3.2 Carrageenan

Carrageenan is a naturally occurring anionic sulfated linear polysaccharide, extracted from certain red seaweed of the Rhodophyceae family (Rochas et al., 1989; Chen et al., 2002). Several carrageenophytes have been reported particularly from *Chondrus crispus*, *Eucheuma* spp., *Iridaea undulosa*, *Hypnea musciformis*, *Solieria filiformis*, and *Agardhiella ramosissima* (McHugh, 2003; Mtolera and Buriyo, 2004; Araújo et al., 2012; Batista et al., 2014).

There are several different carrageenans with slightly different chemical structures and properties (McHugh, 2003). Carrageenans are formed by alternate units of D-galactose and 3,6-anhydro-galactose

(3,6-AG) joined by α-1,3 and β-1,4-glycosidic linkages. Carrageenans can be classified based on the amount and position of sulfate groups, based on their family, or based on their properties (Prajapati et al., 2014). Depending on the amount and position of the $SO_3^-$ groups, carrageenan are classified into λ (lambda), κ (kappa), ι (iota), ν (nu), μ (mu), θ (theta), ξ (ksi), γ (gamma), β (beta), and ω (omega) (Figure 7.7), all containing about 22%–35% of sulfate groups (Stanley, 1987; Barabanova and Yermak, 2012).

Based on the family, the first class involves the kappa (κ) family, which contains subclasses like kappa, iota, mu, and nu carrageenan. The second class involves lambda carrageenan, which contains subclasses like lambda, xi, and pi. The third class involves the beta family containing subclasses like beta and gamma (Prajapati et al., 2014). Mollion et al. (1986) introduced another family named the omega family in which sulfate groups are on the C6 of the 1,3-linked galactopyranosyl units. Alpha carrageenan was included as a subclass in the beta family because 1,3-linked galactopyranosyl units are not sulfated at C4 but sulfation was observed at C2 of 3,6-anhydrogalactose groups. According to their properties, carrageenans can split into two groups: gelling agent (kappa, iota) and thickening agent (lambda). It is now known that kappa and lambda carrageenans do not occur together in the same plant but are incorporated at different stages of the reproductive cycle (Stanley, 1987). Kappa carrageenan occurs in the haploid gametophytic plants and the lambda type in the diploid tetrasporophytes (McCandless et al., 1973; Jackson and McCandless, 1979; Carrington et al., 2001).

Complex hybrid galactans, with both agar- and carrageenan-like structures, have also been reported (Sen et al., 2002). Several species produce complex heteropolysaccharides containing uronic acids together with neutral or sulfated monosaccharides or galaclan.

### 7.2.3.3 Porphyran

Porphyran is a sulfated polysaccharide isolated from seaweed of the order Bangiales especially from the genus *Porphyra*. It is obtained from red algae of the kingdom Rhodophyta. Chemically, porphyran is related to agarose, consisting of a linear backbone of alternating 3-linked β-D-galactose and 4-linked 3,6-anhydro-α-L-galactose units. The L-residues are mainly composed of α-L-galactosyl 6-sulfate units, and the 3,6-anhydrogalactosyl units are minor (Figure 7.8) (Vo et al., 2015).

Studies of sulfated polysaccharide from 40 species of *Porphyra* revealed that the 3,6-anhydrogalactose content varies between 5% and 19% and sulfate content between 6% and 11%. However, Zhang et al. (2003) reported that porphyran obtained from *Porphyra*

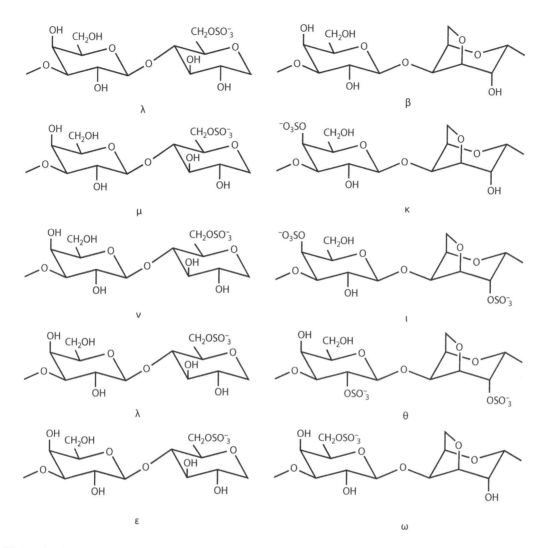

*Figure 7.7* Units of sulfated D-galactose and anhydrogalactose in kappa, lamda, and iota carrageenans from red seaweed.

*haitanensis* contained a higher sulfate content (17.4%–33.5%) than other *Porphyra* species. It is possible that the sulfate content of porphyran is related to the environmental temperature. *Porphyra haitanensis* grows in the subtropical, high-temperature environment, and thus has high sulfate content prophyran. Several investigations of the structure and biological activity of porphyrans isolated from different species have been undertaken (Zhang et al., 2003, 2004, 2009a; Bhatia et al., 2013).

## 7.3   Biological properties of seaweed polysaccharides

Sulfated polysaccharides of many seaweed species have been reported for their biological attributes, such as anticoagulant, antiviral, antioxidant, anticancer, anti-inflammatory, anti-hyperlipidemic, antimicrobial, and immunomodulatory activities. Data on the biological activity of sulfated polysaccharides from seaweed are summarized in Tables 7.4 through 7.6.

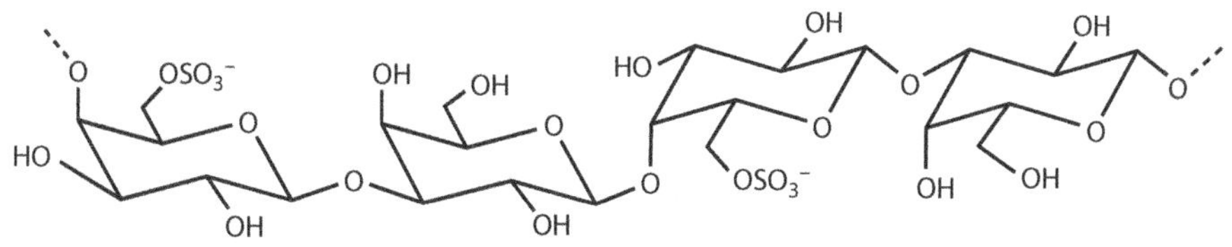

*Figure 7.8* Chemical structure of porphyran.

*Table 7.4* Bioactive polysaccharides from green seaweed

| Species | Compounds | Biological activity | In vivo/ in vitro | Model | Active concentration/ remark | Molecular weight | References |
|---|---|---|---|---|---|---|---|
| *Codium cylindricum* | Sulfated galactan | Anticoagulant activity | *In vitro* | Prolonged activated partial thromboplastin (APTT), prothrombin (PT), and thrombin time (TT) | Showed weaker activity than heparin | nd | Matsubara et al. (2001) |
| *Codium vermilara* | Sulfated arabinan | Anticoagulant activity | *In vitro* | Prolonged APTT, PT, and TT | Showed weaker activity than heparin and somewhat higher than that of dermatan sulfate | 180 kDa | Fernández et al. (2013) |
| *Enteromorpha prolifera* | Six different MW sulfated polysaccharides (DEP1–6) | Antioxidant activity | *In vitro* | (a) $O_2^-$ | (a) $IC_{50}$ of DEP1–4 were 16, 16.5, 16.7, and 28 mg/mL, respectively | 446.5, 247.0, 76.1, 19.0, 5.0, and 3.1 kDa | Li et al. (2013) |
| | | | | (b) HO· | (b) $IC_{50}$ of MW 3.1 kDa was 0.39 mg/mL | | |
| | | | | (c) Metal chelating | (c) The chelating effect of MW 3.1 kDa was 77.3% at 5 mg/mL | | |
| | | | | (d) Reducing power | (d) The reducing power DEP1–6 was 0.281, 0.285, 0.301, 0.324, 0.339, and 0.391 at 2.5 mg/mL, respectively | | |
| *Gayralia oxysperma* | Sulfated heterorhamnan | Antiviral | *In vitro* | Against HSV-1 and HSV-2 | $IC_{50}$: 0.27 to 0.3 µg/mL for HSV-1 and values 10-fold lower from 0.036 to 0.054 µg/mL for HSV-2 | 1519 kDa and 91–820 kDa | Cassolato et al. (2008) |
| *Gayralia oxysperma* | Sulfated heterorhamnan | Antitumor activity | *In vitro* | Inhibited the growth of human glioblastoma cells (U87MG) | Promoted 48.4%, 46.1%, 26.6%, and 28% of reduction of cell viability at 100 µg/mL | 109 and 251 kDa | Ropellato et al. (2015) |
| *Monostroma nitidum* | Sulfate polysaccharide | Anticoagulant activity | *In vitro* | Prolonged activated partial thromboplastin (APTT), prothrombin (PT), and thrombin (TT) time | Showed weaker activity than heparin | nd | Mao et al. (2008) |

(Continued)

*Table 7.4 (Continued)* Bioactive polysaccharides from green seaweed

| Species | Compounds | Biological activity | In vivo/ in vitro | Model | Active concentration/ remark | Molecular weight | References |
|---------|-----------|---------------------|-------------------|-------|------------------------------|------------------|------------|
| *Ulva conglobata* | Ulvan | Anticoagulant activity | *In vitro* | Prolonged activated partial thromboplastin (APTT), | Showed weaker activity than heparin | nd | Mao et al. (2006) |
| *Ulva fasciata* | Ulvan with different molecular (UFP1, UFP2, UFP3, and UFP4) | Antioxidant activity | *In vitro* | (a) $O_2^-$ | (a) Scavenging effect: UFP3 > UFP2 > UFP1 > UFP4 were 66.17%, 51.60%, 45.60%, and 42.54% at 10 mg/mL, respectively | 264, 75, 29, and 2 kDa | Shao et al. (2013) |
| | | | | (b) HO· | (b) The $EC_{50}$ was 6.03, 5.48, and 3.23 mg/mL for UFP1–UFP3, respectively | | |
| | | | | (c) ABTS | (c) $EC_{50}$ of UFP2, UFP3, and UFP4 were 0.86, 0.87, and 5.80 mg/mL, respectively | | |
| | | | | (d) Reducing power | (d) The reducing power of four UFP fractions increased in the order UFP2 > UFP3 > UFP4 > UFP1 | | |
| *Ulva fasciata* | Ulvan with different molecular and its partially desulfation | Antioxidant activity | *In vitro* | (a) $O_2^-$ | (a) The best scavenging activity was UFP2 (73.7% at 8 mg/mL) | 1.9–259 kDa | Shao et al. (2014b) |
| | | | | (b) HO· | (b) The best scavenging activity was UFP2 (76.98% at 8 mg/mL) | | |
| | | | | (c) ABTS | (c) The best scavenging activity was UFP2 (94.05% at 4 mg/mL) | | |
| | | | | (d) DPPH | (d) UFP2 > UFP3 > UFP1 and DS-UFP2 > DS-UFP3 > DS-UFP1 | | |
| | | Anticancer activity | *In vitro* | (e) Inhibited the growth of DLD intestinal cancer cells | (e) Inhibitory activity of S-UFP3 was 74.70% at the concentration of 4.0 mg/mL | | |

(Continued)

**Table 7.4 (Continued)** Bioactive polysaccharides from green seaweed

| Species | Compounds | Biological activity | In vivo/ in vitro | Model | Active concentration/ remark | Molecular weight | References |
|---|---|---|---|---|---|---|---|
| *Ulva pertusa* | Ulvan | Anti-hyperlipidemia activity | *In vivo* | Decreased total cholesterol, low-density lipoprotein cholesterol, and triglyceride; increased high-density lipoprotein cholesterol; and decreased the atherogenic index | 120, 250, and 500 mg/kg | — | Yu et al. (2003a) |
| *Ulva pertusa* | Ulvan (U) and degraded fractions (U1, U2) | Anti-hyperlipidemic activity | *In vivo* | U: lowered the level of serum total cholesterol and low-density lipoprotein; U1-U1: reduced triglyceride and increased high-density lipoprptein | 0.5 g/100 g diet | 151.6, 64.5, and 28.2 | Yu et al. (2003b) |
| *Ulva pertusa* | Ulvan (U) and degraded fractions (U1, U2, and U3) | Antioxidant activity | *In vitro* | (a) $O_2^-$ <br> (b) HO· <br> (c) Reducing power <br> (d) Metal chelating | (a) $IC_{50}$ of U3 was 22.1 µg/mL <br> (b) $IC_{50}$ of U3 was 2.8 µg/mL. The order of hydroxyl scavenging activity was (U3 > U1 (U) > U2) <br> (c) The order of reducing power was U3 > U1 (U) > U2 <br> (d) Chelating ability of U3 was 23.9%, 50.2%, and 94.3% at 0.6, 1.3, and 3.4 mg/mL | 151.7, 64.5, 58.0, and 28.2 | Qi et al. (2005a) |

(Continued)

*Table 7.4 (Continued)* Bioactive polysaccharides from green seaweed

| Species | Compounds | Biological activity | *In vivo/ in vitro* | Model | Active concentration/ remark | Molecular weight | References |
|---|---|---|---|---|---|---|---|
| *Ulva pertusa* | Ulvan (U) and their derivatives (HU1–HU5) | Antioxidant activity | *In vitro* | (a) $O_2^-$ | (a) $EC_{50}$ of U and HU1–HU5 was 20, 5.8, 6.1, 11.4, 11.5, and 10.9 µg/mL | nd | Qi et al. (2005b) |
| | | | | (b) HO· | (b) $EC_{50}$ of HU1, HU3, HU4, and HU5 was 1.43, 0.82, 0.53, and 0.46 mg/mL, respectively; U and HU2 could not be read. | | |
| | | | | (c) Reducing power | (c) Reducing ability of all samples were lower than ascorbic acid, α-tocopherol, and BHA | | |
| | | | | (d) Metal chelating | (d) HU4 and HU5 have higher chelating ability than other samples at high concentration | | |

*Note:*  nd, no data.

*Table 7.5* Bioactive polysaccharides from brown seaweed

| Species | Compound | Biological activity | In vivo/ in vitro | Model | Active concentration/ remark | Molecular weight | References |
|---|---|---|---|---|---|---|---|
| *Ascophyllum nodosum* | Asscophyllan | Anticancer | *In vivo* | Stimulated respiratory burst in RAW264.7 macrophages | 10–1000 µg/mL | 390 kDa | Wang et al. (2013) |
| *Capsosiphon fulvescens* | Sulated polysaccharide | Anticancer activity | *In vitro* | Inhibited the proliferation of AGS gastric cancer cells by inducing apoptosis via an IGF-IR-mediated PI3K/Akt pathway | $IC_{50}$ was 2.5 mg/mL | nd | Kwon and Nam (2007) |
| *Cladosiphon novae-caledoniae kylin* | Fucoidan | Anti-angiogenic | *In vitro* | Suppress expression and secretion of VEFG of human uterine carcinoma HeLa cell | 20% (final concentration) | nd | Ye et al. (2005) |
| *Colpomenia sinuosa* | Sulfated fucan | Antimicrobial activity | *In vitro* | Inhibited growth of *Candida albicans* | 2 mg/mL | nd | Kantachumpoo and Chirapart (2010) |
| *Dictyota cervicornis* | Sulfated polysaccharide | Anticoagulant activity | *In vitro* | Prolonged activated partial thromboplastin time (APTT) | Anticoagulant activity 1.4-fold less than heparin | nd | Costa et al. (2010) |
| *Ecklonia kurome* | Fucoidan | Anticoagulant activity | *In vitro* | Inhibited thrombin by blocking the formation of prothombinase and prevention the generation of factor Xa | (a) The concentration required for 50% inhibition of the activity of generated thrombin was ~11 (intrinsic) and ~15 µg/mL (extrinsic) (b) The concentration required for 50% inhibition of the generation of thrombin by activation of the intrinsic and extrinsic pathway were ~1.5 and ~5.5 µg/mL | nd | Nishino et al. (1999) |

(Continued)

*Table 7.5 (Continued)* Bioactive polysaccharides from brown seaweed

| Species | Compound | Biological activity | In vivo/ in vitro | Model | Active concentration/ remark | Molecular weight | References |
|---|---|---|---|---|---|---|---|
| *Eisenia bicyclis* | Laminarin and fucoidan | Antitumor | *In vitro* | Inhibited colony formation of SK-Mel-28 and DLD-1 cell | Inhibitory effect of laminaran on SK-Mel-28 was 32%–34% at 100–200 µg/mL and on DLD-1 was 44%–49% at 100–200 µg/mL Inhibitory effect of fucoidan on SK-Mel-28 was 12%–16% at 100–200 µg/mL and on DLD-1 was 46%–45% at 100–200 µg/mL | nd | Ermakova et al. (2013) |
| *Fucus vesiculosus* | Fucoidan | Antioxidant activity | *In vitro* | Ferric reducing ability (FRAP) | FRAP values at 4 and 30 min were 209.8 and 263.9 µmol Fe (II) Trolox/g sample dry weight, respectively | $1600 \times 10^{-3}$ kDa | Rupérez et al. (2002) |
| *Fucus vesiculosus* | Nature fucoidan (NF) and Oversulfated fucoidan (OSF) | Anti-angiogenic and antitumor | *In vivo* | Suppressed the neovascularization induced by Sarcoma 180 cell Inhibited the growth of Lewis lung carcinoma or B16 melanoma cells | 5 mg/5 kg | nd | Koyanagi et al. (2003) |
| *Hydroclathrus clathratus* | Sulfated polysaccharide | Antiviral activity | *In vitro* | Antiviral against HSV-2 | $EC_{50}$ of H3-a1 and H3-b1 were 1.7 and 1.67 µg/mL, respectively | nd | Hui et al. (2007) |
| *Laminaria japonica* | Fucoidan | Antioxidant activity | *In vitro* | Inhibited low-density lipoprotein oxidation | 0.4 mg/mL | >200 kDa | Xue et al. (2001) |

*(Continued)*

*Table 7.5 (**Continued**)* Bioactive polysaccharides from brown seaweed

| Species | Compound | Biological activity | In vivo/ in vitro | Model | | Active concentration/ remark | Molecular weight | References |
|---|---|---|---|---|---|---|---|---|
| *Laminaria japonica* | Three sulfated polysaccharide fraction (F1, F3, and F3) | Antioxidant activity | *In vitro* | (a) $O_2^-$ | | (a) $EC_{50}$ of F1–F3 were 1.7, 3.8, and 14.5 µg/mL, respectively | nd | Wang et al. (2008) |
| | | | | (b) HO· | | (b) $EC_{50}$ of F1 and F3 were 2.30 and 1.42 mg/mL | | |
| | | | | (c) Metal chelating | | (c) The highest activity was 18.97%, 29.48%, and 28.87% at the concentration of 0.76, 1.17, and 0.76 mg/mL | | |
| | | | | (d) Reducing power | | (d) The reducing power of F1–F3 at 1 mg/mL were 0.094, 0.079, and 0.084, respectively | | |
| *Laminaria japonica* | Fucoidan and derivative | Antioxidant activity | *In vitro* | (a) $O_2^-$ | | PHDF exhibited the highest scavenging ability with $EC_{50}$ was ~4 µg/mL | ~5.3–6.8 kDa | Wang et al. (2009) |
| | | | | (b) HO· | | PHDF exhibited the highest scavenging ability with $EC_{50}$ was 1.99 mg/mL | | |
| | | | | (c) DPPH | | PHDF exhibited the highest scavenging ability | | |
| | | | | (d) Reducing power | | DFPS exhibited the highest scavenging ability | | |
| *Laminaria japonica* | Low-molecular-weight sulfated polysaccharide | Antioxidant activity | *In vitro* | (a) $O_2^-$ | | (a) $IC_{50}$ was 0.14 mg/mL | 8,000–10,000 | Zhao et al. (2004) |
| | | Hepatoprotective activity | *In vivo* | (b) HO· | | (b) $IC_{50}$ was 0.2 mg/mL | | |
| | | | | (c) Killing *E. coli* by hypochlorous acid | | (c) The maximum protection (100%) at 3.0 mg/mL | | |
| | | | | (d) Liver injury induced by $CCl_4$ and D-GalN of ICR mice | | (d) 50–100 mg/kg | | |

*(Continued)*

Table 7.5 (*Continued*) Bioactive polysaccharides from brown seaweed

| Species | Compound | Biological activity | In vivo/in vitro | Model | Active concentration/remark | Molecular weight | References |
|---|---|---|---|---|---|---|---|
| *Lessonia vadosa* | Fucoidan | Anticoagulant activity | In vitro | Prolonged thrombin time (TT) | Native polysaccharide presented a good inhibition of coagulation in relation to heparin activity | 320,000 (Native) | Chandia and Matsuhiro (2008) |
| | | Elicitor properties | | Activated defense enzyme activities in tobacco plant | Native polysaccharide presented comparable activity as the depolymerized fucoidan | 32,000 (depolymerized fucoidan) | |
| *Lobophora variegata* | Heterofucan | Anti-inflammatory activity | In vivo | Inhibition of an inflammatory symptom of mouse ear edema and reduction in serum TNF-α | 50 mg/kg | nd | Paiva et al. (2011) |
| *Padina gymnospora* | Heterofucan | Anticoagulant activity | In vitro | Prolonged activated partial thromboplastin time (APTT) | PF1 showed anticoagulant activity 2.5-fold less than heparin | 18 kDa | Silva et al. (2005) |
| *Sargassum fulvellum* | Sulfated fucan | Anticoagulant activity | In vivo | Prolonged activated partial thromboplastin (APTT), prothrombin (PT), and thrombin (TT) time | Showed weaker activity than heparin | nd | Zoysa et al. (2008) |
| *Sargassum hemiphyllum* | Fucoidan | Anti-inflammatory activity | In vitro | Reduced IL-1β, IL-6, TNF-R, and NO in LPS-stimulated RAW264.7 cell Inhibited mRNA expressions of IL-β, iNOS, and COX-2; downregulated NF-κB in nucleus | 1–5 mg/mL | nd | Hwang et al. (2011) |

(*Continued*)

*Table 7.5 (Continued)* Bioactive polysaccharides from brown seaweed

| Species | Compound | Biological activity | In vivo/ in vitro | Model | Active concentration/ remark | Molecular weight | References |
|---|---|---|---|---|---|---|---|
| *Sargassum horneri* | Sulfated polysaccharide (SHP30, SHP60, and SHP80) | Antioxidant activity | *In vitro* | (a) O₂⁻<br>(b) HO·<br>(c) ABTS<br>(d) DPPH<br>(e) FRAP | SHP30 < SHP60 < SHP80<br>SHP30 > SHP60 > SHP80<br>SHP30 < SHP60 < SHP80<br>SHP30 < SHP60 < SHP80<br>SHP30 > SHP60 > SHP80 | $1.58 \times 10^{-3}$ kDa,<br>$1.92 \times 10^{-3}$ kDa,<br>11.2 kDa | Shao et al. (2014a) |
| | | Antitumor activity | *In vitro* | Inhibited MKN45 and DLD cancer cells | SHP30 exhibited highest activity against MKN45 cells (51.92% at 2 mg/mL) SHP30, and SHP60 exhibited inhibition activity against DLD cells were 85.3% and 76.37% at 4.0 mg/mL and 44% for SHP80 at 6.0 mg/mL | | |
| *Sargassum pallidum* | Purified polysaccharide | Antitumor activity | *In vitro* | Inhibited growth of HepG2, A549, and MGC-803 cell | Inhibition (%) at 1 mg/mL of HepG2 was 35%–81% A549 was 15%–67% MGC803 was 11%–79% | nd | Ye et al. (2008) |
| | | Antioxidant activity | | DPPH | Scavenging activity was lower than 10% at 3.8 mg/mL | | |
| *Sargassum polycystum* | Sulfated fucan | Antimicrobial activity | *In vivo* | Inhibited growth of *Candida albicans* | 2 mg/mL | nd | Kantachumpoo and Chirapart (2010) |
| *Sargassum stenophyllum* | Fucoidan | Anti-angiogenic and antitumor activity | *In vivo* | Inhibited angiogenesis in fertilized chicken eggs and Inhibited the growth of B16F10 murine melanoma cell | 1500 µg/disk, decreased the vessel number in CAM by 64%<br><br>150 µg/animal/day, suppressed the tumor growth | nd | Dias et al. (2005) |
| *Sargassum swartzii* | Crude polysaccharide | Antibacterial activity | *In vitro* | Inhibited growth of *E. coli* | 100 µg | 50 kDa | Vijayabaskar et al. (2012) |

*(Continued)*

Table 7.5 (Continued) Bioactive polysaccharides from brown seaweed

| Species | Compound | Biological activity | In vivo/in vitro | Model | Active concentration/remark | Molecular weight | References |
|---|---|---|---|---|---|---|---|
| Sargassum vulgare | Alginate | Antioxidant activity | | TAC, reducing power (RW), DPPH, ABTS, $H_2O_2$ | Inhibition (%) at 1 mg/mL; TAC: 32.34% RW: increase with increasing concentration DPPH: 25.33% ABTS: 55% $H_2O_2$: 47.23% | | Sousa et al. (2007) |
| Sargassum vulgare | Alginate | Anticancer activity | In vivo | Inhibited growth of sarcoma 180 in mice | 50 and 100 m/m²/day | 194,330 kDa | Dore et al. (2013a) |
| Sargassum vulgare | Fucan | Anti-angiogenic and antitumor activities | In vivo | Inhibited VEGF secretion in endothelial cells and antiproliferative activity on HeLa cell line | VEGF inhibition of fractionated (SV1) was 75.9% at 100 µg/µL and 71.4% at 50 µg/µL for purified (PSV1) | 160 kDa | Dore et al. (2013a) |
| Sargassum vulgare | Fucan | Anticoagulant activity | In vitro | (a) Prolonged APTT and stimulated the enzyme activity of FXa | (a) 50 and 100 µg/mL (APTT) 12.5–25 µg/10 µL (FXa) | nd | Dore et al. (2013b) |
| | | Anti-inflammatory activity | In vivo | (b) Anti-thrombotic trail in rat Carrageenan-induced paw edema model | (b) Ten times higher than heparin 10, 30, and 50 mg/kg | | |
| | | Antioxidant activity | In vitro | (c) DPPH | (c) Scavenging ability was 22.2% at 2.5 mg/mL | | |
| Stoechospermum marginatum | Sulfated fucan | Antiviral activity | In vitro | Inhibited HSV-1 and HSV-2 | $EC_{50}$ was 0.63–10 µg/mL | 40 kDa | Adhikari et al. (2006) |
| Undaria pinnatifida | galactofucan Sulfate (GFS) and its fraction (F1M, F2M, F4M) | Antiviral activity | In vitro | Antiviral against HSV-1, HSV-2, and HCMV | $IC_{50}$ of GSF was 1.1, 0.2, and 0.5 µg/mL, respectively; $IC_{50}$ of F1M was 4.6, 1.0, and 4.0 µg/mL, respectively; $IC_{50}$ of F2M was 1.1, 0.2, and 0.5 µg/mL, respectively; $IC_{50}$ of F4M were 3.1, 1.0, and 2.0 µg/mL, respectively | GFS = 710 kDa F1M = 150 kDa F2M = 290 kDa F4M = ND | Hemmingson et al. (2006) |
| Undaria pinnatifida | Fucoidan | Antiviral activity | In vivo | Antiviral against HSV-1 | 5 mg/day | nd | Hayashi et al. (2008) |

Note: nd, no data.

*Table 7.6* Bioactive polysaccharides from red seaweed

| Species | Compound | Biological activity | In vivo/in vitro | Model | Active concentration/remark | Molecular weight | References |
|---|---|---|---|---|---|---|---|
| *Acanthophora spicifera* | Sulfated galactan | Antiviral activity | *In vitro* | Inhibited HSV-1 and HSV-2 | $IC_{50}$ of C (25°C) was 1.0– >50 µg/mL $IC_{50}$ of H (first hot water extraction) was 0.8– >50 µg/mL | nd | Duarte et al. (2004) |
| *Bostrychia montagnei* | Sulfated galactan and its fraction (B1–B6) | Antiviral activity | *In vitro* | Inhibited HSV-1 (F), TK-HSV-1 (B2006), TK-HSV-1 (Field), and HSV-2 (G) | The most active fractions were BCW, B4, and BHW. $IC_{50}$ of BCW was 12.9, 11.2, 1.9, and 1.2 µg/mL, respectively. $IC_{50}$ of B4 was 15.4, 12.4, 4.1, and 5.8 µg/mL, respectively. $IC_{50}$ of BHW was 20.5, 20.7, 1.5, and 3.3 µg/mL, respectively | BCW = 27,500 B1 = 31,300 B2 = 5,600 B3 = 11,900 B4 = 43,700 B5 = 42,500 B6 = 34,000 BHW = 45,700 | Duarte et al. (2001) |
| *Chondrus ocellatus* | λ-Carrageenan (PC1–5) | Antitumor and Immunomodulation activities | *In vivo* | Inhibited growth and immune response in ICR mice of transplanted S180 and H22 tumor | 200 mg/kg/day Low molecular weight showed higher activity than high molecular weight | 650, 240, 140, 15, and 9.3 kDa | Zhou et al. (2004) |
| *Chondrus ocellatus* | λ-Carrageenan | Antitumor activity | *In vivo* | Inhibited growth of S180 tumor by mixture of 5-Fu on transplanted H-22 tumor mice | 200 mg/kg/day | 15 kDa (PC-4) | Zhou et al. (2006) |
| *Cryptonemia seminervis* | Sulfated galactan | Antiviral activity | *In vitro* | Inhibited HMPV replication | $ED_{50}$ was 1.2 µg/mL for S fraction | 51.6–63.8 kDa | Mendes et al. (2014) |
| *Gigartina skottsbergii* | Carrageenan | Antiviral activity | *In vitro* | Inhibited HSV-1 and HSV-2 | $IC_{50}$ of µ/ν-carrageenan was lower than 1µg/mL $IC_{50}$ of κ/ι-carrageenan was in the range 1.6–4.1 µg/mL | µ/ν-carrageenan = 198,000 κ/ι-carrageenan = 75,000–124,000 | Carlucci et al. (1977) |
| *Gracilaria corticata* | Sulfated galactan | Antiviral activity | *In vitro* | Inhibited HSV-1 and HSV-2 | $IC_{50}$ of HSV-1 and HSV-2 was 0.19–50.0 and 0.24–45.9 µg/mL | 52,404–165,197 | Muzumder et al. (2002) |

*(Continued)*

*Table 7.6 (Continued)* Bioactive polysaccharides from red seaweed

| Species | Compound | Biological activity | In vivo/in vitro | Model | Active concentration/remark | Molecular weight | References |
|---|---|---|---|---|---|---|---|
| *Mastocarpus stellatus* | Carrageenan | Antioxidant activity | *In vitro* | (a) FRAP<br>(b) ABTS<br>(c) $O_2^-$ | (a) The highest reducing power at 4 and 30 min was 40.2 and 44.9 µmol TE/g sample dw, respectively<br>(b) Positive correlation between sulfate content and ABTS<br>(c) Negligible (0.01–0.2 µmol VitC equivalent/g for all samples) | 8–1425 kDa | Gómez-Ordóñez et al. (2014) |
|  |  | Anticoagulant activity | *In vitro* | (d) Prolonged activated partial thromboplastin (APTT) and prothrombin (PT) | (d) The activity was lower than heparin |  |  |
| *Nemalion helminthoides* | Sulfated xyloman | Immunomodulatory activity | *In vitro* | (a) Cytotoxicity on macrophage: RAW264.7<br>(b) Induced proliferation of macrophage of the murine cell line RAW 264.7<br>(c) Stimulated the production of NO, IL-6, and TNFα of the murine cell line RAW 264.7<br>(d) Immunoprotection against HSV-2 infection | (a) 28% and 32% at 8 µg/mL<br>(b) 0.5 and 10 µg/mL<br>(c) 30 mg/kg<br>(d) Reduced morbidity by 75% and 100% | 13.6 and 11.7 kDa | Pérez-Recalde et al. (2014) |
|  |  |  | *In vivo* |  |  |  |  |
| *Porphyra* sp. | Porphyran | Anticancer activity | *In vitro* | Inhibited cell proliferation and induced apoptosis in AGS human gastric cancer cell | 0.25%–5.0% | nd | Kwon and Nam (2006) |
| *Porphyra haitanensis* | Porphyran | Antioxidant activity | *In vitro* | (a) $O_2^-$<br>(b) HO· | (a) $IC_{50}$ of fractions F1–F3 were 0.14, 1.6, and 0.06 mg/mL<br>(b) $IC_{50}$ of all fractions (F1–F3) were >1 mg/mL | nd | Zhang et al. (2003) |

(Continued)

**Table 7.6 (Continued)** Bioactive polysaccharides from red seaweed

| Species | Compound | Biological activity | In vivo/in vitro | Model | Active concentration/remark | Molecular weight | References |
|---|---|---|---|---|---|---|---|
| *Porphyra haitanensis* | Porphyran | Antioxidant activity | *In vivo* | (c) $H_2O_2$ induced hemolysis in rat erythrocytes | (c) F1 and F2 exhibited weak protective effect at high concentration while F3 had no effect | 850 kDa | Zhang et al. (2004) |
| | | | | (d) Lipid peroxide of rat liver mosome | (d) $IC_{50}$ of fraction F1-F3 was 1.6, 0.8, and 1.8 mg/mL | | |
| | | | | (a) Inhibited lipid peroxidation in aging mice | (a) At high dose (100–200 mg/kg) exhibited similar or stronger effect than vitamin C (200 mg/kg) | | |
| | | | | (b) Increased SOD activity | (b) Dose-dependent manner | | |
| | | | | (c) Increased GSH-Px activity | (c) 50 mg/kg in liver and brain, 100 mg/kg in heart | | |
| | | | | (d) Total antioxidant capacity (TAOC) | (d) 100 mg/kg | | |
| *Porphyra haitanensis* | Porphyran | Antioxidant activity | *In vitro* | (a) DPPH | (a) P3 > P2 > P1 > P (low to high MW) $IC_{50}$ of P3 was 2.01 mg/mL | MW 5,876–29,695 | Zhao et al. (2006) |
| | | | | (b) Reducing power | (b) P3 showed highest reducing power | | |
| *Porphyra haitanensis* | Porphyran (P) and their derivatives (AP, PP, and BP) | Antioxidant activity | *In vitro* | (a) $O_2^-$ | (a) $IC_{50}$ of AP, PP, and BP were 1.14, 4.06, and 3.54 µg/mL | nd | Zhang et al. (2009b) |
| | | | | (b) HO· | (b) $IC_{50}$ of P, PP, and BP were 6.55, 0.31, and 1.04 µg/mL | | |
| | | | | (c) Reducing power | (c) P was 0.42 at 6.17 mg/mL AP was 1.13 at 5.95 mg/mL BP was 0.94 at 5.71 mg/mL | | |
| *Porphyra yezoensis* | Porphyran | Anti-inflammatory activity | *In vitro* | Inhibited NO production in LPS-stimulated RAW264.7 cell | 1000 µg/mL | 30 kDa | Isaka et al. (2015) |
| *Porphyra vietnamensis* | Porphyran | Immunomodulatory activity | *In vivo* | Cellular and humoral immune responses in Wistar albino rats | 200–500 mg/kg | nd | Bhatia et al. (2013) |
| *Schizymenia binderi* | Sulfated galactan | Antiviral activity | *In vitro* | Inhibited HSV-1 (F), TK-HSV-1 (B2006), TK-HSV-1 (Field), and HSV-2 (G) | $IC_{50}$ were 0.76, 0.18, 0.21, and 0.63 µg/mL, respectively | nd | Matushiro et al. (2005) |

*Note:* nd, no data.

## 7.3.1 Anticoagulant activity

Anticoagulant activity is the most widely studied property of sulfated polysaccharides (Costa et al., 2010). Seaweed polysaccharides are sometimes used as modulators of a coagulant and as alternatives to heparin, which is the popular anticoagulant used as medicine. The anticoagulant activity of sulfated polysaccharide has been identified from many species of seaweed, especially from brown and red seaweed. Fewer anticoagulants have been found in green seaweed. The proposed mechanisms of action of sulfate polysaccharides on anticoagulant activity involve predominantly HC-II-mediated antithrombin activities, direct antithrombin action (thrombin–fibrinogen complex), AT-III involvement, anti-factor Xa, and fibrinolytic activities.

Fucoidan extracted from *Ecklonia kurome* exhibited antithrombin activity by the inhibition of the generation of thrombin by blocking the formation of prothrombinase and preventing the generation of the intrinsic factor Xa (Nishino et al., 1999). It was found that the concentration required for 50% inhibition of the activity of generated thrombin was ~11 µg/mL for the intrinsic pathway and ~15 µg/mL for the extrinsic pathway; the concentrations required for 50% inhibition of the generation of thrombin by activation of the intrinsic and extrinsic pathway were ~1.5 and ~5.5 µg/mL. This concentration was about 1/10 to 1/7 that of the activity of the generated thrombin in plasma. The sulfate polysaccharide isolated from *Sargassum fulvellum* exhibited anticoagulant activity by inhibiting both intrinsic and extrinsic blood coagulation pathways (Zoysa et al., 2008). The heterofucan isolated from *Sargassum vulgare* exhibited anticoagulant activity via its direct action on the enzymatic activity of thrombin and stimulation of FXa enzymatic activity (Dore et al., 2013b). Moreover, the heterofucan extracted from *Padina gymnospora* with a molecular weight of 18 kDa exhibited anticoagulant activity 2.5-fold less than that of heparin, and 3-*O*-sulfation at C-3 of 4-α-L-fucose 1 → 4 units of this polymer was responsible for their activity (Silva et al., 2005).

Among the carrageenans, λ-type was the most potent anticoagulant at low concentration. The anticoagulant activity of λ-carrageenan was nearly twice that of unfractionated carrageenan and four times that of κ-carrageenan. However, the most active carrageenan showed ~1/15 of the activity of heparin (Shanmugam and Mody, 2000). Liang et al. (2014) reported that the anticoagulant activity of agarose and κ-carrageenan was largely dependent on the position of sulfate substitution of the polysaccharides. The existence of G4S and DA2S, rather than G-6S, is helpful in increasing the anticoagulant activity. The positional influence on the anticoagulant activity is supposed to be in the order A-2, G-2 > G-4 > G-6.

Sulfated polysaccharides from several *Codium* species have been reported regarding their anticoagulant activity. The high-molecular-weight xylo-arabinogalactans from *Codium fragile* ssp. *atalnticum* have been isolated, and their anticoagulant activity was attributed to the potentiation of heparin cofactor (HC II) and antithrombin III (AT III) (Rogers et al., 1990; Jurd et al., 1995). However, the highly sulfated galactoarabionglucan extracted from *Codium pugniformis* exhibits its anticoagulant activity by potentiated AT III but not HC II (Matsubara et al., 2000). The sulfated galactan isolated from *Codium cylindricum* did not potentiate both AT III and HC II. These results indicate that the anticoagulant mechanism of this polysaccharide might be different from that of other anticoagulants isolated from other species of *Codium* (Matsubara et al., 2001). Fernández et al. (2013) reported that a highly sulfated 3-linked β-arabinan with arabinose in the pyranose form obtained from *Codium vermilara* exhibited direct thrombin inhibition, and the sulfate groups on C-2 were shown to interact more intensely with the thrombin structure. This sulfated arabinan show anticoagulant activity by a mechanism different from those found previously for other sulfated polysaccharides and glycosaminoglycans.

Ulvan isolated from *U. conglobata* has been reported with regard to its anticoaglulant activity, which is by direct inhibition of thrombin and the potentiation of HC II. Polysaccharides extracted from various species of *Caulerpa* (*C. racemosa, C. taxifolia, C. scalpelliformis, C. veravalensis*, and *C. peltata*) showed high anticoagulant activity, which was associated with the sugar, sulfate content, and molecular weight of the extract (Shanmugam et al., 2001). Arata et al. (2015) reported that pyruvated and sulfated galactan of green seaweed the in order Bryopsidales *Penicillus capitatus, Udotea flabellum*, and *Halimeda opuntia* exhibited anticoagulant activity by potentiating AT III. Mao et al. (2008) reported that the sulfated polysaccharide of *Monostroma nitidum* has potent thrombin inhibitors mediated by HCII and coagulation factor Xa inhibition. However, the activity of these green seaweed polysaccharides was lower than that of heparin.

The relationship between structure and anticoagulant activity has been investigated. Gómez-Ordóñez et al. (2014) reported that the higher sulfate content, together with very high MW, might have a positive influence on anticoagulant capacity. The inhibitory activity of fucoidan extracted from *Pelvetia canaliculata* was reduced with decreased molecular weight. Fucoidan with a molecular weight of 58 kDa was the most effective, and a low molecular weight of 21 kDa was moderately effective for anticoagulant activity (Colliec et al., 1994).

## 7.3.2  Antiviral activity

In recent years, screening assays of the antiviral activity of the extract from a number of marine algae have led to the identification of many carbohydrate polymers having potent inhibitory effects against herpes simplex virus type 1 (HSV-1) and 2 (HSV-2), human cytomegalovirus (HCMV), human immunodeficiency virus type 1 (HIV-1), respiratory syncytial virus, and influenza virus. These polysaccharides include fucans, sulfated galactans (Mazumder et al., 2002; Matsuhiro et al., 2005), ulvan, and mannans (Lee et al., 1999; Cassolato et al., 2008).

The sulfated fucan fraction of *Stoechospermum marginatum* was found to have antiviral activity against HSV-1 and HSV-2 by inhibiting virus replication at concentrations that do not have any effect on the cell viability (Adhikari et al., 2006). Moreover, the sulfated galactofucan of *Undaria pinnatifida* exhibited *in vitro* antiviral activity against HSV-1, HSV-2, and HCMC (Hemmingson et al., 2006). Focoidan from *U. pinnatifida* exhibited *in vivo* antivirus activity against HSV-1 by inhibition of virus replication and stimulation of both innate and adaptive immune defense functions (Hayashi et al., 2008). Fucoidans from *Sargassum* were found to have antiviral activity against HSV-1), HSV-2, hepatitis A virus (HAV), coxsackie virus (CVB 3), human cytomegalovirus (HCMV), and HIV-1. Fucoidan can inhibit the initial stages of viral infection, attachment to and penetration into host cells, and replication stages after virus penetration; however, the exact mechanism remains unknown. In addition, fucoidans isolated from *Sargassum* have been found to have low cytotoxicity (>1000 µg/mL) to virus host cell lines such as normal Vero cells (African green monkey kidney cells), suggesting the potential for development of safe antiviral drugs based on these fucoidans (Liu et al., 2012). Sulfated heteropolysaccharides from *Hydroclathrus clathratus* showed antiviral activity against HSV-2 with low cytotoxicity to Vero and HEp-2 cells (Hui et al., 2008).

Sulfated polysaccharides from red seaweed are active against retroviruses. Carrageenan, a common cell wall polysaccharide from red algae, is co-internalized into infected cell with HSV, inhibiting the virus. Carrageenan also interferes with the fusion (syncytium formation) between cells infected with HIV and inhibits the specific retroviral enzyme reverse transcriptase (Neushyul, 1990). Matsuhiro et al. (2005) found that sulfated galactan extracted from *Schizymenia binderi* exhibited antiviral activity against HSV-1 and HSV-2 by interfering with the initial adsorption of the virus to the cells. High molecular weight sulfated galactan extracted from *Gracilaria corticata* has been reported to have antiviral activity against HSV-1 and HSV-2 by inhibition of the initial virus attachment to the host cell (Mazumder et al., 2002). Mendes et al. (2014) reported

that the anti-metapneumovirus (HMPV) activity of the sulfated and pyruvylated DL-hybrid galactans obtained from *Cryptonemia seminervis* and its depolymerized products occurred early in the replicative cycle. These polysaccharideds inhibited HMPV replication by binding to the viral particle. Moreover, the depolymerized galactans inhibited the recognition of cell receptor by HMPV and penetration to the host cell.

A rhamnan sulfate isolated from the green seaweed *Monostroma latissimum* showed potent antiviral effect against HCMV and HIV-1. The antiviral action of rhamnan sulfate was not only due to the inhibition of virus adsorption but also might involve the later replication of virus in the host cells (Lee et al., 1999). Cassolato et al. (2008) reported that the sulfated heterorhamnan isolated from *Gayralia oxysperma* exhibited antiviral activity against HSV-1 and HSV-2. The hydrophobic character of the methyl group at C-5 of the rhamnosyl units is an additional factor that could contribute to the high and specific activity of this the sulfated heterorhamnan.

The correlation between the structure (molecular weight and sulfate content) of polysaccharide and is antiviral activity against HSV-1 and HSV-2 virus has been reported for sulfated galactans of *Acanthophora spicifera* (Duarte et al., 2004) and *Bostrychia montagnei* (Duarte et al., 2001). Moreover, the distribution of sulfate groups and a backbone able to adopt a definite shape in compelling circumstance would be other factors involved in the formation of a polysaccharide–virus complex. Carlucci et al. (1997) found that cyclized µ/ν-carrageenan (IC3) isolated from *Gigartina skottsbergii* exhibited antiviral activity against HSV-1 and HSV-2 higher than λ-carrageenan and κ/ι-carrageenan. The difference between these polysaccharides is in the position of the sulfate group in the β-D-galactose units. This result indicated that the position of sulfate in polymer also had an effect on their antiviral activity.

## 7.3.3  Antioxidant activity

In recent years, seaweed polysaccharides have been shown to play an important role as free-radical scavengers and antioxidants for prevention of oxidative damage in living organisms. Several compounds with antioxidative activity have been isolated from seaweed.

The reducing power of crude polysaccharide and fraction from seaweed has been reported for species *Ecklonia* (Athukorala et al., 2006), *Fucus vesiculosus* (Rupérez et al., 2002), *Laminaria japonica* (Xue et al., 2001; Wang et al., 2008, 2009), *Sargassum horneri* (Shao et al., 2014a), *S. pallidum* (Ye et al., 2008), and *S. swartzii* (Vijayabaskar et al., 2012), etc. In brown seaweed, fucoidan has shown higher reducing ability than other polysaccharide fractions, that is, alginate and laminarian (Balboa et al., 2013). It has been reported that the acetylated and benzoylated derivatives of porphyran extracted from the red alga *Porphyra haitanensis* showed excellent reducing power by their high hydrogen-donation ability (Zhang et al., 2009b). Gómez-Ordóñez et al. (2014) reported that the degree of sulfation could play a more determining role than MW in the reducing ability of sulfated polysaccharides from the red seaweed *Mastocarpus stellatus*. High-sulfate-content ulvan exhibited stronger reducing power than natural ulvan (Qi et al., 2005b).

α-α-Diphenylpicrylhydrazyl (DPPH) is a stable free radical widely used for evaluating natural antioxidants, algae, or algal products due to its stability, simplicity, and reproducibility. Some antioxidants may react slowly or be inert to DPPH. The water-soluble polysaccharide from *Petalonia binghamiae, Scytosiphon lomentaria*, and *Turbinaria ornata* was a less potent DPPH scavenger than the phenolic fraction. Concentration-dependent radical scavenging was reported for polysaccharides from *Sargassum palladium* (Ye et al., 2008) and *S. swartzii* (Vijayabaskar et al., 2012). The polysaccharide fractions of *S. palladium* purified by DEAE-52 anion-exchange chromatography showed lower scavenging activity than the crude polysaccharide (Ye et al., 2008). Agaro-oligosaccharides (agarohexose) have been reported to have the ability of DPPH radical scavenging (Chen and Yan, 2005). Zhao et al. (2006) reported that low-molecular-weight porphyran exhibited scavenging activity higher than that of its high-molecular-weight counterpart. The benzoylated derivative of low-molecular-weight fucoidan showed higher scavenging activity than other derivatives and native polysaccharide due to the presence of benzoyl group in the fucoidan molecule, which could have activated the hydrogen atom of the anomeric carbon. The higher the activation capacity of the group, the stronger the hydrogen atom donating capacity. Moreover, this benzoylated derivative showed higher scavenging activity than that of butylated hydroxyanisole (BHA), a standard antioxidant (Wang et al., 2009).

ABTS (2,2′-azino-bis(3-ethylbenzothiazoline-6-sulfonic acid)) radical is oxidized to the radical cation (ABTS$^{.+}$) and its decolorization is used to measure the antioxidant capacity in both water-soluble and lipid-soluble food samples, and is expressed as TEAC (Trolox equivalent antioxidant capacity). This activity was reported for water-soluble crude polysaccharide from *Sargassum swartzii* (Vijayabaskar et al., 2012), *S. horneri* (Shao et al., 2014a), and *Ulva fasciata* (Shao et al., 2013, 2014b). A positive correlation was found between the sulfate content and ABTS of carrageenan isolated from *M. stellatus* (Gómez-Ordóñez et al., 2014).

Inhibition of superoxide radical formation was reported for sulfated polysaccharides from *Enteromorpha prolifera* (Li et al., 2013), *Ulva fasciata* (Shao et al., 2013a, b), *U. pertusa* (Qi et al., 2005a), *Laminaria japonica* (Wang

et al., 2008, 2009), *Sargassum horneri* (Shao et al., 2014a), *Mastocarpus stellatus* (Gómez-Ordóñez et al., 2014), and *P. haitanensis* (Zhang et al., 2003, 2009a). Three fractions of sulfated galactan (F1–F3) were obtained by ion-exchange chromatography from *P. haitanesis*. Fraction F3 with highest sulfate content showed the strongest superoxide radical scavenging activity. This result suggested that sulfate content is important for the antioxidant activities of porphyran (Zhang et al., 2003a). Moreover, the F1 fraction of this seaweed was characterized, and the *in vivo* antioxidant was also determined. The results found that this polysaccharide increased the total antioxidant capacity and the activity of superoxide dismutase (SOD) and glutathione peroxidase (GSH-Px) in aging mice (Zhang et al., 2004). The acetylated derivative of porphyran obtained from this species showed higher activity than its native porphyrin as well as phosphorylated and benzoylated derivatives (Zhang et al., 2009). Zhang et al. (2004) reported that the antioxidant activity of porphyran depends on their molecular weight, with the low-molecular-weight compound showing higher activity than the high-molecular-weight one. Shao et al. (2013b) reported that the higher sulfate content of polysaccharides resulted in greater scavenging effect on the superoxide radical.

The effects of chemical modification of fucoidan on the superoxide scavenging activity were also reported. The oversulfated, benzoylated, acetylated, and phosphorylated derivative of low-molecular-weight fucoidan prepared from *Laminaria japonica* showed stronger scavenging activity than vitamin C and κ-carrageenan oligosaccharide derivative (Wang et al., 2009). The acetylation and benzoylation of ulvan, a heteropolysaccharide extracted from *U. pertusa*, also could enhance the scavenging activity (Qi et al., 2005b). The sulfated polysaccharide of *Enteromorpha prolifera* showed great inhibitory effects on superoxide radical at a low concentration compared to vitamin C, and samples with high molecular weight exhibited higher inhibitory effects (Li et al., 2013). Yuan et al. (2005) synthesized oversulfated, acetylated, and phosphorylated derivatives of κ-carrageenan. They found that all derivatives exhibited antioxidant activity higher than that of native κ-carrageenan.

It had been found that high-sulfate-content ulvan prepared from *U. pertusa* had more effective scavenging activity on the hydroxyl radical than natural ulvan. Moreover, their activity was higher than that of vitamin C (Qi et al., 2005b). Li et al. (2013) reported that the sulfated polysaccharide of *Enteromorpha prolifera* with low molecular weight showed stronger inhibitory effect on the hydroxyl radical than the high-molecular-weight counterpart. Phosphorylated derivative of porphyran obtained from *P. haitanensis* showed higher hydroxyl radical scavenging activity than native porphyran and its benzoylated derivatives (Zhang et al., 2009).

Secondary or preventive antioxidative action by metal-ion chelation reflects the inhibition of the interaction between metals and lipids through the formation of insoluble metal complexes or generation of steric resistance without directly interacting with the oxidative species. The presence of transition-metal ions, such as $Cu^{2+}$, $Fe^{2+}$, $Fe^{3+}$, in a system can induce the initiation and accelerate the rate of oxidation, either by promoting the decomposition of hydroperoxides or through the production of hydroxyl radicals by the Fenton reaction (Balboa et al., 2013). This activity was reported for the water extracts of polysaccharides from many species of seaweed. The low-molecular-weight (3.1 kDa) sulfated polysaccharide from *Enteromorpha prolifera* exhibited higher chelating ability than the initial polysaccharide (high MW) (Li et al., 2013). It was found that high-sulfate-content ulvan prepared from *U. pertusa* had more effective chelating ability than natural ulvan (Qi et al., 2005b).

Several extracts have proven efficient against model lipid peroxidation, such as the crude polysaccharide of *E. cava* (Athukorala et al., 2006). Other systems include the liposome as a model of the cell membrane for *in vitro* biochemical research. Alginate fractions from *L. japonica* inhibited yolk homogenate lipid peroxidation (Zhao et al., 2012), and *T. ornata* crude polysaccharides (Ananthi et al., 2010) were used to protect rat liver homogenates. Moreover, low-molecular-weight sulfated polysaccharide (LA) and crude fucoidan (FA) of *Laminaria japonica* inhibited the AAPH-induced low-density lipoprotein oxidation (Xue et al., 2001).

It has been found that the sulfate content and molecular weight have a significant effect on the antioxidant activity of sulfated polysaccharides (Qi et al., 2005a; Shao et al., 2013, 2014). Li et al. (2013) reported that molecular weight was the most significant factor that influences the antioxidant activities when the sulfate content and monosaccharide composition of the samples were similar.

### 7.3.4 Anticancer activity

Many of seaweed sulfated polysaccharide and/or their derivatives, such as degraded and oversufated products obtained by chemical modification, show anticancer properties. They either can have a direct inhibitory action on cancer cells and tumor or influence different stages of carcinogenesis and tumor development such as angiogenesis, and recover the broken balance between proliferation and programmed cell death (apoptosis).

Angiogenesis is a multistep process whereby the new blood vessels develop from the preexisting vasculature. It involves migration, proliferation, and differentiation of mature endothelial cells, and is regulated by the interaction of endothelial cell with angiogenesis-inducing factors and extracellular matrix components (Fedorov et al., 2013). Any imbalance in the control of this complex system may promote numerous angiogenesis-dependent diseases such as cancer, rheumatoid arthritis, or diabetic retinopathy. One of the most specific and important factors involved in angiogenesis is the vascular endothelial growth factor (VEGF), which links specifically to different membrane receptors of endothelial cells. In addition to stimulating angiogenesis, VEGF is also important for maintaining the integrity and permeability of blood vessels (Dore et al., 2013a).

The anticancer properties of sulfated polysaccharides have been established many times by *in vitro* and *in vivo* experiments. Fucoidan has been reported to have antitumor activity by inhibiting the invasion/angiogenesis of tumor cells. Fucoidan isolated from the enzymatic digest of *Cladosiphon novae-caledoniae kylin* showed anti-angiogenic activity on human uterine carcinoma HeLa cells by suppressed expression and secretion of VEGF (Ye et al., 2005). The fucoidan from *Sargassum horneri*, *Ecklonia cava*, *Eisenia bicyclis*, and *Costaria costata* showed anticancer activity against the human skin melanoma cell line SK-MEL-28 and the human colon cancer cell line DLD-1 (Ermakova et al., 2011, 2013). The purified fraction of *Sargassum palladium* showed anticancer activities against HepG2 (human hepatoma cell line), A549 (human lung cancer cell line), and MGC-803 (human gastric cancer cell line) (Ye et al., 2008). The heterofucan purified from *Sargassum vulgare* has been reported to have anti-angiogenic activity by inhibiting tubulogenesis in rabbit aorta endothelial cells (RAECs) and VEGF secretion. Moreover, this polysaccharide exhibited antiproliferative activity against HeLa cells (Dore et al., 2013b). Liu et al. (2012) reported that fucoidan from *Sargassum* had considerable *in vivo* anticancer activity and could significantly reduce tumor weight and prolong survival time of tumor-bearing animals. However, *in vitro* experiments revealed that *Sargassum* fucoidans had very limited cytotoxicity ($IC_{50} \approx 0.1–1$ mg/mL). Ascophyllan from *Ascophyllum nodosum* showed potent cytotoxic effects on Vero (African green monkey kidney) and XC (rat sarcoma) cells in a similar concentration-dependent manner. Moreover, a significant growth-promoting effect of ascophyllan on MDCK (Madin-Darby canine kidney) cells was observed under normal growth conditions (Jiang et al., 2010).

The differences in the mechanism of apoptosis probably depend on the structural characteristic of sulfated polysaccharides and the type of cell lines. Kwon and Nam (2006) reported the antitumor activity of porphyran on AGS gastric cancer cell. They found that porphyran induced apoptosis via activation of pro-apoptotic molecule, including Bax and caspase-3, and suppression of anti-apoptotic Bcl-2. Porphyran showed inhibition of

IGF-I receptor phsphorylation as well as Akt activation, which are known to be important for apoptosis-related signaling; in addition, porphyran decreased the level of Akt activation and stimulated caspases-3 activation. A similar result was obtained from the sulfated polysaccharide of the brown seaweed *Capsosiphon fulvescens*. This polysaccharide exhibited anticancer activity against AGS gastric cancer cells by inducing apoptosis via an IGF-IR-mediated PI3K/Akt pathway (Kwon and Nam, 2007). The SHP30 fraction prepared from *Sargassum horneri* exhibited antiproliferative activity against MKN45 gastric cancer cells and DLD intestinal cancer cells. This polysaccharide could inhibit the growth of DLD cells by inducing apoptosis through the regulation of apoptosis-associated gene expressions such as Bcl-2 and Bax (Shao et al., 2014a).

Chemical moidifications, such as sulfation and oversulfation, of fucoidan fraction and their anti-angiogenic activity have also been reported. Natural and oversulfated fucoidans (NF and OSF) obtained from *Fucus vesiculosus* showed anti-angiogenic properties by suppressing the mitogenic and chemotactic actions of the vascular endothelial growth factor 165 ($VEGF_{165}$) on human umbilical vein endothelial cells (HUVECs) by preventing the binding of $VEGF_{165}$ to its cell surface receptor. These polysaccharides suppressed the neovascularization induced by Sarcoma 180 cells that had been implanted in mice and inhibited the growth of Lewis lung carcinoma and B16 melanoma in mice. Moreover, the suppressive effect of OSF was higher than that of NF, indicating that increasing the number of sulfate groups in the fucoidan molecule contributes to the effectiveness of its anti-angiogenic and antitumor activities (Koyanagi et al., 2003).

Liang et al. (2014) demonstrated that the cytotoxicity of the red seaweed polysaccharide (agarose and κ-carrageenan) against HUVECs was dependent on the position of sulfate substitution of the polysaccharides. The substitution of sulfate groups on G-6 will cause the strongest cytotoxicity. The cytotoxicity of sulfate groups is supposed to be in the order G-6 > G-4 > G-2 > A-2. Moreover, the molecular weight of λ-carrageenan had a notable effect on tumor-inhibiting activities, and the low-molecular-weight samples were more effective than the high-molecular-weight sample extracted from the red alga *Chondrous ocellatus* (Zhou et al., 2004).

Among the seaweed, green seaweed has been less studied in comparison with brown and red seaweed as a source of polysaccharide with anticancer and cancer-preventive properties. However, their anticancer properties have been reported, mainly for Ulvan. The polysaccharide DS-UFP3 fraction, prepared from *U. fasciata*, with the lowest sulfate content but highest uronic acid content and molecular weight, exhibited the best antitumor activity against DLD intestinal cancer cells, indicating that both the sulfate content and uronic acid may have contributed to the antitumor activities of the polysaccharides from *U. fasciata* (Shao et al., 2013). The sulfated heterorhamnans and their Smith-degraded products of *Gayralia oxysperma* exhibited antitumor activity in human glioblastoma cells (U87MG). These sulfated polysaccharides induced an increase in the number of cells in the G1 phase with concomitant increase of the mRNA levels of p53 and p21. Moreover, the presence of 2-linked disulfated rhamnose residues together with the molecular weight could be important factors to be correlated with the inhibitory effect on human glioblastoma cells (Ropellato et al., 2015).

### 7.3.5 Anti-inflammatory activity

The heterofucan from the brown algae *Lobophora variegata* has potential anti-inflammatory activity by reducing articular edema and serum TNF-α cytokine in acute zymosan-induced arthritis in rats (Paiva et al., 2011). Dore et al. (2013b) reported that the heterofucan extracted from *S. vulgare* was able to reduce the edema by decrease in cell infiltration in the plantar region of the animals tested. Similar results were reported by Ananthi et al. (2010) when testing a polysaccharide from *Turbinaria ornata* in carrageenan-induced paw edema.

Nitric oxide (NO) is a gaseous free-radical involved in various physiological processes, such as vasodilatation, smooth muscle regulation, neurotransmission, innate immune responses, and apoptosis. However, an excessive amount of NO produced by nitric oxide synthase (iNOS) in activated macrophages contributes to numerous severe inflammatory diseases, including sepsis and arthritis. Therefore, the effective inhibition of nitric oxid production in activated macrophages can lead to an anti-inflammatory effect (Isaka et al., 2015). Hwang et al. (2011) reported that the fucoidan isolated from *Sargassum hemiphyllum* could reduce interleukin (IL)-1b, IL-6, TNF-a, and NO; inhibit mRNA expressions of IL-b, iNOS, and COX-2; and downregulate of NF-kB (nuclear factor kappa-light-chain-enhancer of activated B cells) in the nucleus for the mouse macrophage cells (RAW264.7) induced by lipopolysaccharide (LPS). Porphyran extracted from discolored nori (*Porphyra yezoensis*) showed anti-inflammatory activity by inhibited NO production in LPS-stimulated RAW264.7 cells by preventing the expression of inducible iNOS, and molecular weight was an important factor for its activity (Isaka et al., 2015).

### 7.3.6 Anti-hyperlipidemia activity

Fucoidans isolated from *Sargassum wightii* and *S. henslowianum* were found to have hypolipidemic effects, reduce the serum total cholesterol, triglyceride, and low-density lipoprotein, and increase high-density lipoprotein in mice and rats with diet-induced hyperlipidemia. Interestingly, fucodians introduced into the animals by subcutaneous injection or gastric perfusion both exerted the hypolipidemic effect (Liu et al., 2012).

Ulvan isolated from *Ulva pertusa* was found to have anti-hyperilipidemia activity, by lowering the content of plasma total cholesterol, low-density lipoprotein cholesterol, and triglyceride and increasing the content of serum high-density lipoprotein cholesterol. Moreover, it decreased the atherogenic index (Yu et al., 2003a). It has been found that the molecular weight of ulvan has an effect on their anti-hyperlipidemia activity. High-molecular-weight ulvan lowered the level of serum total cholesterol and low density lipoprotein, whereas low-molecular-weight fractions reduced triglyceride and increased high-density lipoprotein (Yu et al., 2003b).

### 7.3.7 Antimicrobial activity

The lyase depolymerized product of alginate with MW 4.235 kDa exhibited a broad spectrum of antibacterial activity against *Eshcerichia coli*, *Staphylococcus aureus*, *Salmonella paratyphi* B, and *Bacillus subtilis* with minimal inhibitory concentration (MIC) of 0.312, 0.225, 0.016, and 0.325 µg/mL, respectively (Hu et al., 2005). Sulfated fucan extracted from *Sargassum polycystum* and *Colpomenia sinuosa* could inhibited the growth of the yeast *Candida albican* at a concentration 2 mg/mL (Kantachumpoo and Chirapart, 2010). Crude polysaccharide from *Sargassum swartzii* could inhibit the growth of *E. coli* better than antibiotic drug amplicillin (Vijayabaskar et al., 2012).

### 7.3.8 Immunomodulatory activity

Sulfated polysaccharides from seaweed are known to have modulated immune functions, which are more dependent on the structure and molecular weight. It has various effects on innate immune and complement systems, which may further reduce the pro-inflammatory and several allergic reactions. The interest in sulfated polysaccharides stems from complement activation, nitric oxide synthase, pro-inflammatory cytokine induction, interference with the migration of leukocytes, elevation of primary antibody response, binding to pattern recognition receptors, influencing macrophage function, and modifying intestinal cecal flora.

Ascophyllan was capable of inducing the secretion of TNF-α and granulocyte colony stimulating factor (G-CSF) in the mouse macrophage cell line RAW264.7 (Nakayasu et al., 2009). Wang et al. (2013) investigated the effects of ascophyllan on macrophage activation. They found that ascophyllan is capable of inducing reactive oxygen species (ROS) generation in RAW264.7 cells through the activation of NADPH oxidase. Moreover,

the ROS level induced by ascophyllan was comparable to that induced by phorbol myristate acetate (PMA), a potent stimulator of the respiratory burst activity in macrophages. Similar results were reported by Dore et al. (2013b), who found that the heterofucan isolated from *Sargassum vulgare* was able to stimulate the proliferation of macrophages, cells of the innate immune system, and is a potential target for new studies related to the role of this compound in the fight against cancer via stimulation of the innate immune system.

Porphyran extracted from *Porphyra vietnamensis* showed immune-modulation effect by increased weight of the thymus, spleen, and lymphoid organ cellularity of Wistar albino rats. This polysaccharide is capable of increasing neutrophil adhesion to nylon fibers as well as a dose-dependent increase in antibody titer values. Moreover, a decreased response to delayed-type hypersensitive (DTH) reaction induced by sheep red blood cells (SRBCs) was observed. It also prevented myelosuppression in cyclophosphamide-treated rats (Bhatia et al., 2013). Pérez-Recalde et al. (2014) reported that sulfated xylomannans from the red seaweed *Nemalion helminthoides* could stimulate macrophage cells and induce the production of cytokines (both IL-6 and TNFα) in RAW 264.7 cells and in mice. Besides, they could act as immunoprotectors against herpetic infection. The molecular mechanism of the action of these compounds in immune cells could involve pattern recognition receptors (PRR; mannose receptor, DC-sign, dectin, and C lectin, among others) as well, as was demonstrated with polysaccharides from other sources.

## 7.4 Conclusion

Sulfated polysaccharides from seaweed comprise a wide range of chemical compositions. These include sulfated fucan in brown seaweed and sulfated galactan in red seaweed and different sulfated polysaccharides in green seaweed. The bioactive polysaccharides discussed here are obtained from different seaweed exhibiting different chemical structures and displaying a large variety of biological effects on specific targets. These sulfated polysaccharides seem to be very useful and promising for biological research to clarify the mechanism of action in the human body as well as in the design of very specific and potent new functional ingredients for a wide variety of industrial applications such as functional pharmaceuticals and foods, functional cosmetics, and new medicines.

In general, the biological activities of sulfated polysaccharides are related to the molecular size, the type of sugar, and the sulfate content and sulfate position. The type of linkage and molecular geometry are also known to play a role in their activity. The difference in molecular weight, sulfate content/position, sugar composition, and their potent bioactivities of seaweed polysaccharides could be due to the extraction method or to seasonal or geographical variations. The low-molecular-weight fractions are usually more biocompatible and active than the crude polysaccharide, but there is no consensus on the optimal molecular size; nor is there a definitive correlation between their structure and activity. Chemical modifications, such as sulfation, oversulfation, acetylation, benzylation, and phosphorylation, are also used to improve the activity of native polysaccharides. However, little information is available about the mechanisms and relationships between their biological activities and chemical structure. Therefore, further evaluation of the polysaccharides' bioactivities is needed for the utilization of these marine resources.

## References

Adhikari, U., C.G. Mateu, K. Chattopadhyay, C.A. Pujol, E.B. Damonte, and B. Ray. 2006. Structure and antiviral activity of sulfated fucans from *Stoechospermum marginatum*. *Phytochemistry* 67: 2474–2482.

Ale, M.T., J.D. Mikkelsen, and A.S. Meyer. 2011. Important determinants for fucoidan bioactivity: A critical review of structure-function relations and extraction methods for fucose-containing sulfated polysaccharides from brown seaweeds. *Mar. Drugs.* 9: 2106–2130.

Ananthi, S., H.R.B. Raghavendran, A.G. Sunil, V. Gayathri, G. Ramakhrisnan, and H.R. Vasanthi. 2010. In vitro antioxidant and in vivo anti-inflammatory potential of crude polysaccharide from *Turbinaria ornata* (Marine brown alga). *Food Chem. Toxicol.* 48: 187–192.

Araki, C. 1966. Some recent studies on the polysaccharides of agarophytes. In E.G. Young and J.L. McLachlan (eds.), *Proceedings of the Fifth International Seaweed Symposium*. Pergamon Press, Oxford, U.K., pp. 3–17.

Arata, P.X., I. Quintana, D.J. Canelón, B.E. Vera, R.S. Compagnone, and M. Ciancia. 2015. Chemical structure and anticoagulant activity of highly pyruvylated sulfated galactans from tropical green seaweeds of the order Bryopsidales. *Carbohydr. Polym.* 122: 376–386.

Araújo, I.W.F., J.A.G. Rodrigues, E.S.O. Vanderlei, G.A. Paula, T.B. Lima, and N.M.B. Benevides. 2012. *Iota*-carrageenans from *Solieria filiformis* (Rhodophyta) and their effects in the inflammation and coagulation. *Acta Sci. Technol.* 34: 127–135.

Athukorala, Y., Kim, K.N., and Y.J. Jeon. 2006. Antiproliferative and antioxidant properties of an enzymatic hydrolysate from brown alga, *Ecklonia cava*. *Food Chem. Toxicol.* 44: 1065–1074.

Balboa, E.M., E. Conde, A. Moure, E. Falqué, and H. Domímgiez. 2013. *In vitro* antioxidant properties of crude extracts and compound from brown algae. *Food Chem.* 138: 1764–1785.

Barabanova, A.O. and I.M. Yermak. 2012. Structural peculiarities of sulfated polysaccharides from red algae *Tichocarpus crinitus* (Tichocarpaceae) and *Chondrus pinnulatus* (Gigartinaceac) collected at the Russian Pacific Coast. In S.-K. Kim (ed.), *Handbook of Marine Macroalgae*. Wiley-Blackwell, Hoboken, NJ, pp. 193–204.

Barros, F.C.N., D.C. Silva, V.G. Sombra, J.S. Maciel, J.P.A. Feitosa, A.L.P. Freitas, and R.C.M. Paula. 2013. Structural characterization of polysaccharide obtained from red seaweed *Gracilaria caudata* (J Agardh). *Carbohydr. Polym.* 92: 598–603.

Batista, J.A., E.G.N. Dias, T.V. Brito et al. 2014. Polysaccharide isolated from *Agardhiella ramosissima*: Chemical structure and anti-inflammation activity. *Carbohydr. Polym.* 99: 59–67.

Berteau, O. and B. Mulloy. 2003. Sulfated fucans, fresh perspectives: Structures, functions, and biological properties of sulfated fucans and an overview of enzymes active toward this class of polysaccharide. *Glycobiology* 13(6): 29R–40R.

Bhatia, S., P. Rathee, K. Sharma, B.B. Chaugule, N. Kar, and T. Bera. 2013. Immuno-modulation effect of sulphated polysaccharide (prophyran) from *Porphyra vietnamensis*. *Int. J. Biol. Macromol.* 57: 50–56.

Bilan, M.I., A.A. Grachev, N.E. Ustuzhanina, A.S. Shashkov, N.E. Nifantiev, and A.I. Usov. 2002. Structure of a fucoidan from the brown seaweed *Fucus evanescens* C.Ag. *Carbohydr. Res.* 337: 719–730.

Briand, X., Cluzet, S., Dumas, B., Esquerré-Tugayé, M.T., and Salamagne, S. 2010. Ulvans as activators of plant defense and resistance reactions against biotic and abiotic stresses. US Patent 7,820,176, October 26, 2010.

Briand, X., Cluzet, S., Dumas, B., Esquerré-Tugayé, M.-T., and Salamagne, S. 2011. Use of ulvans as elicitors of mechanisms for nitrogen absorption and protein synthesis. US Patent 7,892,311, February 22, 2011.

Carlucci, M.J., C.A. Pujol, M. Ciancia, M.D. Noseda, M.C. Matulewicz, W.B. Domonte, and A.S. Cerezo. 1997. Antiherpetic and anticoagulant properties of carrageenans from the red seaweed *Gigartina skottsbergii* and their cyclized derivatives: Correlation between structure and biological activity. *Int. J. Biol. Marcromol.* 20: 97–105.

Carrington, E., S.P. Grace, and T. Chopin. 2001. Life history phases and the biomechanical properties of the red alga *Chondrus crispus* (Rhodophyta). *J. Phycol.* 37: 699–704.

Cassolato, J.E.F., M.D. Noseda, C.A. Pujol, F.M. Pellizzari, E.B. Damonte, and M.E.R. Duarte. 2008. Chemical structure and antiviral activity of the sulfated heterorhamnan isolated from the green seaweed *Gayralia oxysperma*. *Carbohydr. Res.* 343: 3085–3095.

Chandía, N.P. and B. Matsuhiro. 2008. Characterization of a fucoidan from *Lessonia vadosa* (Phaeophyta) and its anticoagulant and elicitor properties. *Int. J. Biol. Macromol.* 42: 235–240.

Chaubet, F., L. Chevolot, J. Jozefonvicz, P. Durand, and C. Boisson-vidal. 2000. Relationships between chemical characteristics and anticoaglulant activity of low molecular weight fucans from marine algae. In B.S. Paulsen (ed.), *Bioactive Carbohydrate Polymers*. Kluwer Academic Publishers, Dordrecht, the Netherlands, pp. 59–84.

Chen, H.M. and X.J. Yan. 2005. Antioxidant activities of agaro-oligosaccharide with degree of polymerization in cell-base system. *Biochem. Biophys. Acta* 1722: 103–111.

Chen, Y., M.L. Liao, and D.E. Dustan. 2002. The rheology of K+-κ-carrageenan as a weak gel. *Carbohydr. Polym.* 50: 109–116.

Chirapart, A., M. Ohno, H. Ukeda, M. Sawamura, and H. Kusunose. 1995. Chemical composition of agars from a newly reported Japanese agarophyte, *Gracilariopsis lemaneiformis*. *J. Appl. Phycol.* 7: 359–365.

Colliec, S., V.C. Boissan, and J. Jozefonvicz. 1994. A low molecular weight fucoidan from the brown seaweed *Plevetia canaliculata*. *Phytochemistry* 35: 697–700.

Costa, L.S., G.P. Fidelis, S.L. Cordeiro et al. 2010. Biological activities of sulfated polysaccharides from tropical seaweeds. *Biomed. Pharmacother*. 64: 21–28.

Delattre, C., T.A. Fenoradosoa, and P. Michaud. 2011. Galactans: An overview of their most important sourcing and applications as natural polysaccharides. *Braz. Arch. Biol. Technol*. 54(6): 1075–1092.

Dias, P.F., J.M. Siqueira Jr., L.F. Vendruscolo, T.J. Neiva, A.R. Gagliardi, M. Maraschin, and R.M. Ribeiro-do-Valle. 2005. Anti-angiogenic and antitumoral properties of a polysaccharide isolated from the seaweed *Sargassum stenophyllum*. *Cancer Chemother. Pharmocol*. 56: 436–446.

Dore, C.M.P.G., M.G.C.F. Alves, N.D. Santos, A.K.M. Cruz, R.B.G. Câmara, A.J.G. Castro, L.G. Alves, H.B. Nader, and E.L. Leite. 2013a. Antiangiogenic activity and direct antitumor effect from a sulfated polysaccharide isolated from seaweed. *Microvasc. Res*. 88: 12–18.

Dore, C.M.P.G., M.G. das C.F. Alves, L.S. Will, T.G. Costa, D.A. Sabry, L.A. de S. Rêgo, C.M. Accardo, H.A.O. Rocha, L.G.A. Filgueira, and E.L. Leite. 2013ba. A sulfated polysaccharide, fucans, isolated from brown algae *Sargassum vulgare* with anticoagulant, antithrombotic, antioxidant and anti-inflammatory effects. *Carbohydr. Polym*. 91: 467–475.

Duarte, M.E.R., J.P. Cauduro, D.G. Noseda, M.D. Noseda, A.G. Gonçalves, C.A. Pujol, E.B. Damonte, and A.S. Cerezo. 2004. The structure of the agaran sulfate from *Acanthophora spicifera* (Rhodomelaceae, Ceramiales) and its antiviral activity. Relation between structure and antiviral activity in agarans. *Carbohydr. Res*. 339: 335–347.

Duarte, M.E.R., D.G. Noseda, M.D. Noseda, S. Tulio, C.A. Pujol, and E.B. Damonte. 2001. Inhibitory effect of sulfated galactans from the marine alga *Bostrychia montagnei* on herpes simplex virus replication in vitro. *Phytomedicine* 8(1): 53–58.

Duckworth, M., K.C. Hong, and W. Yaphe. 1971. The agar polysaccharides of *Gracilaria* species. *Carbohydr. Res*. 18: 1–9.

Duckworth, M. and W. Yaphe. 1971. The structure of agar. Part II. The use of a bacterial agarose to elucidate structural features of the charged polysaccharides in agar. *Carbohydr. Res*. 16: 435–445.

Ermakova, S., R. Men'shova, O. Vishchuk, S.M. Kim, B.H. Um, V. Isakov, and T. Zvyagintseva. 2013. Water-soluble polysaccharides from the brown alga *Eisenia bicyclis*: Structural characteristics and antitumor activity. *Algal Res*. 2: 51–58.

Ermakova, S., R. Sokolova, S.M. Kim, B.H. Um, V. Isakov, and T. Zvyagintseva. 2011. Fucoidan from brown seaweeds *Sargassum hornery, Eclonia cava, Costaria costata*: Structural characteristics and anticancer activity. *Appl. Biochem. Biotechnol*. 164: 841–850.

Fedorov, S.N., S.P. Ermakova, T.N. Zkvyagintseva, and V.A. Stonik. 2013. Anticancer and cancer preventives properties of marine polysaccharide: Some results and prospects. *Mar. Drugs*. 11: 4876–4901.

Fernández, P.V., I. Quintana, A.S. Cerezo, J.J. Caramelo, L. Pol-Fachin, H. Verli, J.M. Estevez, and M. Ciancia. 2013. Anticoagulant activity of a unique sulfated pyranosic (1 → 3)-β-L-arabinan through direct interaction with thrombin. *J. Biol. Chem*. 288: 223–233.

Fernández, P.V., M.P. Raffo, J. Alberghina, and M. Ciancia. 2014. Polysaccharides from the green seaweed *Codium decorticatum*. Structure and cell wall distribution. *Carbohydr. Polym*. 117: 836–844.

Fertah, M., A. Belfkira, El montassir Dahmane, M. Taourirte, and F. Brouillette. 2014. Extraction and characterization of sodium alginate from Moroccan *Laminaria digitata* brown seaweed. *Arab. J. Chem*. http://dx.doi.org/10.1016/j.arabjc.2014.05.003.

García-Ríos, V., E. Ríos-Leal, D. Robledo, and Y. Freile-Pelegrin. 2012. Polysaccharides composition from tropical brown seaweeds. *Phycol. Res*. 60: 305–315.

Gómez-Ordóñez, E., A. Jiménez-Escrig, and P. Rupérez. 2014. Bioactivity of sulfated polysaccharides from the edible red seaweed *Mastocarpus stellatus*. *Bioactive Carbohydr. Dietary Fibre* 3: 29–40.

Hayashi, K., T. Nakano, M. Hashimoto, K. Kanekiyo, and T. Hayashi. 2008. Defensive effects of a fucoidan from brown alga *Undaria pinnatifida* against herpes simplex virus infection. *Int. Immunopharmacol*. 8(1): 109–116.

Hemmingson, J.A., R. Falshaw, and R.H. Furneaux. 2006. Structure and antiviral activity of the galactufucan sulfates extracted from *Undaria pinnatifida* (Phaeophyta). *J. Appl. Phycol*. 18: 185–193.

Hu, X., X. Jiang, J. Gong, H. Hwang, Y. Liu, and H. Guan. 2005. Antibacterial activity of lyase-depolymerized products of alginate. *J. Appl. Phycol*. 17: 57–60.

Hui, W., O.E. Vincent, and A. Put. 2007. Antiviral polysaccharide isolated from Hong Kong brown seaweed *Hydroclathrus clathratus*. *Sci. China C Life Sci*. 50(5): 611–618.

Hwang, P.A., S.Y. Chien, Y.L. Chan, M.K. Lu, C.H. Wu, Z.L. Kong, and C.J. Wu. 2011. Inhibition of lipopolysaccharide (LPS)-induced inflammatory responses by sargassum hemiphyllum sulfated polysaccharide extract in RAW 264.7 macrophage cells. *J. Agric. Food Chem*. 59: 2062–2068.

Isaka, S., K. Cho, S. Nakazono, R. Abu, M. Ueno, D. Kim, and T. Oda. 2015. Antioxidant and anti-inflammatory activities of porphyran isolatedfrom discolored nori (*Porphyra yezoensis*). *Int. J. Biol. Macromol*. 74: 68–75.

Jackson, S.G. and E.L. McCandless. 1979. Incorporation of [35S]sulfate and [14C]bicarbonate into karyotype specific polysaccharides of *Chondrus crispus*. *Plant Physiol*. 64: 585–589.

Jaulneau, V., C. Lafitte, C. Jacquet, S. Fournier, S. Salamagne, X. Briand, M.-T. Esquerré-Tugayé, and B. Dumas. 2010. Ulvan, a sulphated polysaccharide from green algae, activates plant immunity through the jasmonic acid signaling pathway. *J. Biomed. Biotechnol*. 2010: 525291.

Jiang, A., T. Okimura, T. Kokose, Y. Yamasaki, K. Yamaguchi, and T. Oda. 2010. Effect of sulfated fucan, ascophyllan, from the brown algae *Ascophyllum nodosum* on various cell lines: A comparative study on ascophyllan and fucoidan. *J. Biosci. Bioeng*. 110(1): 113–117.

Jiao, G., G. Yu, J. Zhang, and S.H. Ewart. 2011. Chemical structures and bioactivities of sulphated polysaccharides from marine algae. *Mar. Drugs* 9: 196–223.

Jurd, K.M., D.J. Roger, G. Blunden et al. 1995. Anticoagulant properties of sulfated polysaccharides and proteoglycan from *Codium fragile* spp. *atlanticum*. *J. Appl. Phycol*. 7(4): 339–345.

Kantachumpoo, A. and A. Chirapart. 2010. Components and antimicrobial activity of polysaccharides extracted from Thai brown seaweeds. *Kasetsart J. (Nat. Sci.)* 44: 220–233.

Koyanagi, S., N. Tanigawa, H. Nakagawa, S. Soeda, and H. Shimeno. 2003. Oversulfation of fucoidan enhances its anti-angiogenic and antitumor activities. *Biochem. Pharmacol.* 65: 173–179.

Kraan, S. 2012. Chapter 2: Algal polysaccharides, novel applications and outlook. In C.-F. Chang (ed.), *Carbohydrates: Comprehensive Studies on Glycobiology and Glycotechnology*. InTech Publisher, Rijeka, Croatia, 570pp. http://cdn.intechopen.com/pdfs-wm/41116.pdf, accessed November 15, 2014.

Kwon, M.J. and T.J. Nam. 2006. Porphyran induce apopotosis related signal pathway in AGS gastric cancer cell lines. *Life Sci.* 79: 1956–1962.

Kwon, M.J. and T.J. Nam. 2007. A polysaccharide of the marine alga *Capsosiphon fulvescens* induces apoptosis in AGS gastric cancer cells via an IGF-IR-mediated PI3K/Akt pathway. *Cell Biol. Int.* 31: 768–775.

Lahaye, M. 2001. Chemistry and physico-chemistry of phyco-colloids. *Cahiers Biol. Mar.* 42: 137–157.

Lahaye, M. and A. Robic. 2007. Structure and functional properties of ulvan, a polysaccharide from green seaweeds. *Biomacromolecules* 8: 1765–1774.

Lahaye, M., C. Rochas, and W. Yaphe. 1986. A new procedure for determining the heterogeneity of agar polymers in the cell walls of *Gracilaria* spp. (Gracilariaceae, Rhodophyta). *Can. J. Bot.* 64: 579–585.

Lee, J.B., K. Hayashi, T. Hayashi, U. Sankawa, and M. Maeda. 1999. Antiviral activities against HSV-1, HCMV, and HIV-1 of rhamnan sulfate from *Monostroma latissimum*. *Planta Med.* 65(5): 439–441.

Lee, J.B., S. Koizumi, K. Hayashi, and T. Hayashi. 2010. Structure of rhamnan sulfate from the green alga *Monostroma nitidum* and its anti-herpetic effect. *Carbohydr. Polym.* 81: 572–577.

Lee, J.B., T. Yamagaki, M. Maeda, and H. Nakanishi. 1998. Rhamnan sulfate from cell walls of *Monostroma latissimum*. *Phytochemistry* 37: 810–814.

Li, B., S. Liu, R. Xing, K. Li, R. Li, Y. Qin, X. Wang, Z. Wei, and P. Li. 2013. Degradation of sulfated polysaccharides from *Enteromorpha prolifera* and their antioxidant activities. *Carbohydr. Polym.* 92: 1991–1996.

Liang, W., X. Mao, X. Peng, and S. Tang. 2014. Effects of sulfate group in red seaweed polysaccharides on anticoagulant activity and cytotoxicity. *Carbohydr. Polym.* 101: 776–785.

Liu, L., M. Heinrich, S. Myers, and S.A. Dworjanyn. 2012. Toward a better understanding of medicinal uses of the brown seaweed *Sargassum* in traditional Chinese medicine: A phytochemical and pharmacological review. *J. Ethnopharmacol.* 142: 591–619.

Mao, W., X. Zang, Y. Li, and H. Zhang. 2006. Sulfated polysaccharide from marine green algae *Ulva conglobata* and their anticoagulant activity. *J. Appl. Phycol.* 18: 9–14.

Mao, W.J., F. Fang, H.Y. Li, X.H. Qi, H.H. Sun, Y. Chen, and S.D. Guo. 2008. Heparinoid-active two sulfated polysaccharides isolated from marine green algae *Monostroma nitidum*. *Carbohydr. Polym.* 74: 834–839.

Matsubara, K., Y. Matsuura, A. Bacic, M.L. Liao, K. Hori, and K. Miyazawa. 2001. Anticoagulant properties of a sulfated galactan preparation from a marine green alga, *Codium cylindricum*. *Int. J. Biol. Macromol.* 28: 395–399.

Matsubara, K., Y. Matsuura, K. Hori, and K. Miyazawa. 2000. An anticoagulant proteoglycan from the marine green alga, *Codium pugniformis*. *J. Appl. Phycol.* 12: 9–14.

Matsuhiro, B., A.F. Conte, E.B. Domonte, A.A. Kolender, M.C. Matulewicz, E.G. Mejías, C.A. Pujol, and E.A. Zúñiga. 2005. Structural analysis and antiviral activity of a sulfated galactan from red seaweed *Schizymenia binderi* (Gigartinales, Rhodophyta). *Carbohydr. Res.* 340: 2392–2402.

Mazumder, S., P.K. Ghosal, C.A. Pujol, M.J. Carlucci, E.B. Damonte, and B. Raya. 2002. Isolation, chemical investigation and antiviral activity of polysaccharides from *Gracilaria corticata* (Gracilariaceae, Rhodophyta). *Int. J. Biol. Macromol.* 31: 87–95.

McCandless, E.L., J.S. Craigie, and J.A. Walter. 1973. Carrageenans in the gametophytic and sporophytic stages of *Chondrus crispus*. *Planta* 112: 201–212.

Mendes, G.S., M.E.R. Duarte, F.G. Colodi, M.D. Noseda, L.G. Ferreira, S.D. Berté, J.F. Cavalcanti, N. Santos, and M.T.V. Romanos. 2014. Structure and anti-metapneumovirus activity of sulfated galactans from the red seaweed *Cryptonemia seminervis*. *Carbohydr. Polym.* 101: 313–323.

Mollion, M.J., S. Moreau, and D. Christian. 1986. Isolation of a new type of carrageenan from *Rissoeulla verruculosa* (Bert) J. Agardh (Rhodophyta, Gigartinales). *Bot. Mar.* 29: 549–552.

Mtolera, M.S.P. and A.S. Buriyo. 2004. Studies on Tanzanian hypneaceae: Seasonal variation in content and quality of kappa-carrageenan from *Hypnea musciformis* (Gigartinales: Rhodophyta). *Western Indian Ocean J. Mar. Sci.* 3: 43–49.

Mulloy, B., A.C. Ribeiro, A.P. Alves, R.P. Vieira, and P.A.S. Mourao. 1994. Sulfated fucans from echinoderms have a regular tetrasaccharide repeating unit defined by specific patterns of sulfation at the O-2 and O-4 positions. *J. Biol. Chem.* 269: 22113–22123.

Murano, E. 1995. Chemical structure and quality of agars from *Gracilaria*. *J. Appl. Phycol.* 7:245–254.

Murata, M. and J. Nakazoe. 2001. Production and use of marine algae in Japan. *Jarq–Jpn. Agric. Res. Q.* 35(4): 281–290.

Muzumder, S., P.K. Ghosal, C.A. Pujol, M.J. Carlucci, E.B. Damonte, and B. Ray. 2002. Isolation, chemical investigation and antiviral activity of polysaccharides from *Gracilaria corticata* (Gracilariaceae, Rhodophyta). *Int. J. Biol. Macromol.* 31: 87–95.

Nakamura, M., Y. Yamashiro, T. Konishi, I. Hanasiro, and M. Tako. 2011. Structural characterization of rhamnan sulfate isolated from commercially cultured *Monostroma nitidum* (Hitoegusa). *Nippon Shokuhin Kagaku Kogaku Kaishi* 58: 245–251.

Nakayasu, S., R. Soegima, K. Yamaguchi, and T. Oda. 2009. Biological activity of fucose-containing polysaccharide ascophyllan isolated from the brown alga *Ascophyllum nodosum*. *Biosci. Biotechnol. Biochem.* 73(4): 961–964.

Neushul, M. 1990. Antiviral carbohydrates from marine red algae. *Hydrobiologia* 204(1):99–104.

Nishino, T., A. Fukuda, T. Nagumo, M. Fujihara, and E. Kaji. 1999. Inhibition of the generation of thrombin and factor Xa by a fucoidan from the brown seaweed *Ecklonia kurome*. *Thromb. Res.* 96: 37–49.

Paiva, A.A.O., A.J.G. Castro, M.S. Nascimento, L.S. Will, N.D. Santos, R.M. Araújo, C.A.C. Xavier, F.A. Rocha, and

E.L. Leite. 2011. Antioxidant and anti-inflammatory effect of polysaccharides from *Lobophora variegata* on zymosan-induced arthritis in rats. *Int. Immunopharmacol.* 11: 1241–1250.

Percival, E. and J.K. Wold. 1963. The acid polysaccharide from the green seaweed *Ulva lactuca*. Part II. The site of the ester sulphate. *J. Chem. Soc.* 0: 5459–5468.

Pérez-Recalde, M., M.C. Matulewicza, C.A. Pujolb, and M.J. Carlucci. 2014. *In vitro* and *in vivo* immunomodulatory activity of sulfatedpolysaccharides from red seaweed *Nemalion helminthoides. Int. J. Biol. Macromol.* 63: 38–42.

Ponce, N.M.A., C.A. Pujol, E.B. Damonte, M.L. Flores, and C.A. Stortz. 2003. Fucoidans from the brown seaweed *Adenocystis utricularis*: Extraction methods, antiviral activity and structural studies. *Carbohydr. Res.* 338: 153–165.

Praiboon, J., A. Chirapart, Y. Akakabe, O. Bhumibhamon, and T. Kajiwara. 2006. Physical and chemical characterization of agar polysaccharides extracted from the Thai and Japanese species of *Gracilaria. Science Asia* 32(Suppl. 1): 11–17.

Prajapati, V.D., P.M. Maheriya, G.K. Jani, and H.K. Solanki. 2014. Carrageenan: A natural seaweed polysaccharide and its applications. *Carbohydr. Polym.* 105: 97–112.

Qi, H., Q. Zhang, T. Zhao, R. Chen, H. Zhang, X. Niu, and Z. Li. 2005b. Antioxidant activity of different sulfate content derivatives of polysaccharide extracted from *Ulva pertusa* (Chlorophyta) *in vitro. Int. J. Biol. Macromol.* 37: 195–199.

Qi, H., T. Zhao, Q. Zhang, Z. Li, and Z. Zhao. 2005a. Antioxidant activity of different molecular weight sulfated polysaccharides from *Ulva pertusa* Kjellm (Chlorophyta). *J. Appl. Phycol.* 17: 527–534.

Qin, Y. 2008. Alginate fibres: An overview of the production processes and applications in wound management. *Polym. Int.* 57: 171–180.

Rabanal, M., N.M.A. Ponce, D.A. Navarro, R.M. Gómez, and C.A. Stortz. 2014. The system of fucoidans from the brown seaweed *Dictyota dichotoma*: Chemical analysis and antiviral activity. *Carbohydr. Polym.* 101: 804–811.

Raveendran, S., Y. Yoshida, T. Maekawa, and D.S. Kumar. 2013. Pharmaceutically versatile sulfated polysaccharide based bionano platforms. *Nanomed. Nanotechnol., Biol. Med.* 9: 605–626.

Ray, B. and M. Lahaye. 1995. Cell-wall polysaccharides from the marine green alga *Ulva "rigida"* (Ulvales, Chlorophyta). Extraction and chemical composition. *Carbohydr. Res.* 274: 251–261.

Rinaudo, M. 2004. Role of substituents on the properties of some polysaccharides. *Biomacromolecules* 5: 1155–1165.

Rioux, L.E., S.L. Turgeon, and M. Beaulieu. 2007. Characterization of polysaccharides extracted from brown seaweeds. *Carbohydr. Polym.* 69: 530–537.

Rochas, C., M. Rinaudo, and S. Landry. 1989. Relation between the molecular structure and mechanical properties of carrageenan gels. *Carbohydr. Polym.* 10: 115–127.

Rogers, D.J., K.M. Jurd, G. Blundsen, S. Paoletti, and F. Zanetii. 1990. Anticoagulant activity of proteoglycan in extracts of *Codium fragile* spp. *atlanticum. J. Appl. Phycol.* 2: 357–361.

Ropellato, J., M.M. Carvalho, L.G. Ferreira et al. 2015. Sulfated heterorhamnans from the green seaweed *Gayralia oxysperma*: Partial depolymerization, chemical structure and antitumor activity. *Carbohydr. Polym.* 117: 476–485.

Rupérez, P., O. Ahrazem, and J.A. Leal. 2002. Potential antioxidant capacity of sulfated polysaccharides from the edible marine brown seaweed *Fucus vesiculosus. J. Agric. Food Chem.* 50: 840–845.

Sakai, T., K. Ishizuka, and I. Kato. 2003. Isolation and characterization of a fucoidan-degrading marine bacterium. *Mar. Biotechnol.* 5: 409–416.

Sen, A.K., A.K. Das, K.K. Sarkar, A.K. Siddhanta, R. Takano, K. Kamei, and S. Hara. 2002. An agaroid-carrageenan hybrid type backbone structure for the antithrombotic sulfated polysaccharide from *Grateloupia indica* boergensen (Halymeniales, Rhodophyta). *Bot. Mar.* 45: 331–338.

Shanmugam, M. and K.H. Mody. 2000. Heparinoid-active sulphate polysaccharides from marine algae as potential blood anticoagulant agents. *Curr. Sci.* 79: 1672–1683.

Shanmugam, M., B.K. Pamavat, K.H. Mody, R.M. Oza, and A. Tewari. 2001. Distribution of heparinoid-active sulphated polysaccharides in some Indian marine green algae. *Indian J. Mar. Sci.* 30(4): 222–227.

Shao, P., M. Chen, Y. Pei, and P. Sun. 2013a. In vitro antioxidant activities of different sulfated polysaccharides from chlorophytan seaweeds *Ulva fasciata. Int. J. Biol. Macromol.* 59: 295–300.

Shao, P., X. Chen, and P. Sun. 2013b. In vitro antioxidant and antitumor activities of different sulfated polysaccharides isolated from three algae. *Int. J. Biol. Macromol.* 62:155–161.

Shao, P., X. Chen, and P. Sun. 2014a. Chemical characterization, antioxidant and antitumor activity ofsulfated polysaccharide from *Sargassum horneri. Carbohydr. Polym.* 105: 260–269.

Shao, P., Y. Pei, Z. Fang, and P. Sun. 2014b. Effects of partial desulfation on antioxidant and inhibition of DLD cancer cell of *Ulva fasciata* polysaccharide. *Int. J. Biol. Macromol.* 65: 307–313.

Silva, T.M.A., L.G. Alves, K.C.S. de Queiroz, M.G.I. Santos, C.T. Marques, S.F. Chavante, H.A.O. Rocha, and E.L. Leite. 2005. Pratial characterization and anticoagulant activity of a heteroglycan from the brown seaweed *Padina gymnospora. Braz. J. Med. Biol. Res.* 38: 532–533.

Skjåk-Bræk, G. and A. Martinsen. 1991. Application of some algal polysaccharides in biotechnology. In M.D. Guiry and G. Blunden (eds.), *Seaweed Resources in Europe: Uses and Potentials*. Wiley, New York, pp. 219–257.

Skriptsova, A.V., N.M. Shevchenko, T.N. Zvyagintseva, and T.I. Imbs. 2010. Monthly changes in the content and monosaccharide composition of fucoidan from *Undaria pinnatifida* (Laminariales, Phaeophyta). *J. Appl. Phycol.* 22: 79–86.

Sousa, A.d.P.A., M.R. Torres, C. Pessoa, M.O.d. Moraes, F.D.R. Filho, A.P.N.N. Alves, and L.V. Costa-Lotufo. 2007. *In vivo* growth-inhibition of Sarcoma 180 tumor by alginate from brown seaweed *Sargassum vulgare. Carbohydr. Polym.* 69: 7–13.

Stanley, N.F. 1987. Carrageenans. In D.J. McHugh (ed.), *Production and Utilization of Products from Commercial Seaweeds*. FAO Fisheries Technical Papers, Rome, Italy. ISBN: 9251026122.

Tako, M., M. Higa, K. Medoruma, and Y. Nakasone. 1999. A highly methylated agar from red seaweed, *Gracilaria arcuata*. *Bot. Mar.* 42: 513–517.

Tønnesen, H.H. and J. Karlsen. 2002. Alginate in drug delivery systems. *Drug Dev. Ind. Pharm.* 28(6): 621–630.

Vauchel, P., R. Kaas, A. Arhaliass, R. Baron, and J. Legrand. 2008. A new process for extracting alginates from *Laminaria digitata*: Reactive extrusion. *Food and Bioprocess Technol.* 1: 297–300.

Vera, J., J. Castro, A. Gonzalez, and A. Moenne. 2011. Seaweed polysaccharides and derived oligosaccharides stimulate defense responses and protection against pathogens in plants. *Mar. Drugs.* 9: 2514–2525.

Vijayabaskar, P., N. Vaseela, N., and G. Thirumaran. 2012. Potential antibacterial and antioxidant properties of a sulfated polysaccharide from the brown marine algae *Sargassum swartzii*. *Chin. J. Nat. Med.* 10(6): 0421–0428.

Villanueva, R. and N. Montaño. 1999. Highly methylated agar from *Gracilaria edulis* (Gracilariales, Rhodophyta). *J. Appl. Phycol.* 11: 225–227.

Vo, T.S., D.H. Ngo, K.H. Kang, W.K. Jung, and S.K. Kim. 2015. The beneficial properties of marine polysaccharide in alleviation of allergic response. *Mol. Nutr. Food Res.* 59: 129–138.

Wang, J., F Wang, Q. Zhang, Z. Zhang, X. Shi, and P. Li. 2009. Synthesized different derivaties of low molecular fucoidan extracted from *Laminaria japonica* and their potential antioxidant activity in vitro. *Int. J. Biol. Macromol.* 44: 379–384.

Wang, J., Q. Zhang, Z. Zhang, and Z. Li. 2008. Antioxidant activity of sulfated polysaccharide fractions extracted from *Laminaria japonica*. *Int. J. Biol. Macromol.* 42: 127–132.

Wang, Y., Z. Jiang, D. Kim, M. Ueno, T. Okimura, K. Yamaguchi, T. Oda. 2013. Stimulatory effect of the sulfated polysaccharide ascophyllan on the respiratory burst in RAW264.7 macrophages. *Int. J. Biol. Macromol.* 52: 164–169.

Xue, C.H., Y. Fang, H. Lin, L. Chen, Z.J. Li, D. Deng, and C.X. Lu. 2001. Chemical characters and antioxidative properties of sulfated polysaccharides from *Laminaria japonica*. *J. Appl. Phycol.* 13: 67–70.

Yabur, R., Y. Bashan, and G. Hernández-Carmona. 2007. Alginate from the macroalgae *Sargassum sinicola* as a novel source for microbial immobilization material in wastewater treatment and plant growth promotion. *J. Appl. Phycol.* 19: 43–53.

Yaphe, W. and M. Duckworth. 1971. The relationship between structures and biological properties of agars. In Nisizawa, K. (ed.), *Proceedings of the Seventh International Seaweed Symposium*. University of Tokyo Press, Tokyo, Japan.

Ye, H., K. Wang, C. Zhou, J. Liu, and X. Zeng. 2008. Purification, antitumor and antioxidant activities *in vitro* of polysaccharides from the brown seaweed *Sargassum pallidum*. *Food Chem.* 111: 428–432.

Ye, J., Y. Li, K. Teruya, Y. Katakura, A. Ichikawa, H. Eto, M. Hosoi, S. Nishimoto, and S. Shirahata. 2005. Enzyme-digested fucoidan extracts derived from seaweed

Mozuku of *Cladosiphon novae-caledoniae kylin* inhibit invasion and angiogenesis of tumor cells. *Cytotechnology* 47: 117–126.

Yu, P., N. Li, X. Liu, G. Zhou, Q. Zhang, and P. Li. 2003b. Antihyperlipidemic effect of different molecular weight sulfated polysaccharides from *Ulva pertusa* (Chlorophyta). *Pharmacol. Res.* 48: 543–549.

Yu, P., Q. Zhang, N. Li, Z. Xu, Y. Wang, and Z. Li. 2003a. Polysaccharides from *Ulva pertusa* (Chlorophyta) and preliminary studies on their antihyperlipidemia activity. *J. Appl. Phycol.* 15: 21–27.

Yuan, H., W. Zhang, X. Li, X. Lü, N. Li, X. Gao, and J. Songa. 2005. Preparation and in vitro antioxidant activity of κ-carrageenan oligosaccharides and their oversulfated, acetylated, and phosphorylated derivatives. *Carbohydr. Res.* 340: 685–692.

Zhang, Q., N. Li, X. Liu, Z. Zhao, Z. Li, and Z. Xu. 2004. The structure of a sulfated galactan from *Porphyra haitanensis* and its in vivo antioxidant activity. *Carbohydr. Res.* 339: 105–111.

Zhang, Q., P. Yu, Z. Li, H. Zhang, Z. Xu, and P. Li. 2003. Antioxidant activities of sulfated polysaccharide fractions from *Porphyra haitanesis*. *J. Appl. Phycol.* 15: 305–310.

Zhang, Z., Q. Zhang, J. Wang, X. Shi, H. Song, and J. Zhang. 2009b. *In vitro* antioxidant activities of acetylated, phosphorylated and benzoylated derivatives of porphyran extracted from *Porphyra haitanensis*. *Carbohydr. Polym.* 78: 449–453.

Zhang, Z., Q. Zhang, J. Wang, H. Zhang, X. Niu, and P. Li. 2009a. Preparation of the different derivatives of the low-molecular-weight porphyran from *Porphyra haitanensis* and their antioxidant activities *in vitro*. *Int. J. Biol. Macromol.* 45(1): 22–26.

Zhao, G.F., Y.P. Sun, H. Xin, Y.N. Zhang, Z.E. Li, and Z.H. Xu. 2004a. In vivo antitumor and immunonmodulation activities of different molecular weight lambda-carrgeenan from *Chondrus ocellatus*. *Pharmacol. Res.* 50: 47–53.

Zhao, T., Q. Zhang, H. Qi, H. Zhang, X. Niu, Z. Xu, and Z. Li. 2006. Degradation of porphyran from *Porphyra haitanensis* and the antioxidant activities of the degraded porphyrans with different molecular weight. *Int. J. Biol. Macromol.* 38: 45–50.

Zhao, X., B. Li, C. Xue, and L. Sun. 2012. Effect of molecular weight on the antioxidant property of low molecular weight alginate from *Laminara japonica*. *J. Appl. Phycol.* 24: 295–300.

Zhao, X., C.H. Xue, Z.J. Li, T.P. Cai, H.Y. Liu, and H.T. Qi. 2004b. Antioxidant and hepatoprotectivie activities of low molecular weight sulfated polysaccharide from *Laminaria japonica*. *J. Appl. Phycol.* 16: 111–115.

Zhou G., Sheng W., Yao W., and C. Wang. 2006. Effect of low molecular lambda-carrageenan from *Chondrus ocellatus* on antitumor H-22 activity of 5-Fu. *Pharmacol. Res.* 53(2): 129–134.

Zoysa, M.D., C. Nikapitiya, Y.J. Jeon, Y. Jee, and J.J. Lee. 2008. Anticoagulant activity of sulfated polysaccharide isolated from fermented brown seaweed *Sargassum fulvellum*. *J. Appl. Phycol.* 20: 67–74.

*section four*

---

*Marine glycoproteins*

*chapter eight*

# Biomedical benefits of algal glycoproteins

*Hari Eko Irianto and Ariyanti Suhita Dewi*

## Contents

## 8.1 Introduction

Marine natural products have attracted multidisciplinary interests due to their unique pharmaceutical properties. Supported by the advancing technology in SCUBA diving, the exploration of marine resources has become more accessible nowadays. However, for the past six decades, marine natural products research focused mainly on small molecules, and research on marine macromolecules is surprisingly constricted, despite their biomedical potential. Research on marine biopolymers is predominantly converged to their functions in biological systems, such as reproduction, chemical communication, and self-defense (Caldwell and Pagett, 2010).

Among other marine resources, marine algae have long been used as nutritional food source. Marine algae have been consumed as part of daily diet in the Asia Pacific region due to their health benefits. The biological activities of marine algae extracts are influenced by the presence of active compounds, such as polysaccharides, phlorotannins, carotenoids, minerals, peptides, and sulfolipids (Thangam et al., 2014). Polysaccharides from brown algae, such as *Laminaria japonica*, *Saccharina japonica*, and *Undaria pinnatifida*, for example, have been reported to exhibit antioxidant, anti-inflammatory, anticoagulant, antiviral, and anticancer properties (Han et al., 2011; Kim et al., 2012b; Rafiquzzaman et al., 2013). Polysaccharides from green algae, *Capsosiphon fulvescens*, were reported to stimulate immune system and also exhibited anticancer activities (Kim et al., 2013a).

The benefits of algae toward improving health have enforced their increasing demand as food source.

The production of marine algae worldwide has reached 19 million tons in 2010 (FAO, 2012), while farmed algae have overshadowed the production of algae collected from the wild, which accounted for only 4.5% of the total algae production in 2010. Thus, algae aquaculture is recently expanded to compensate the lack of natural availability. Despite their biological activities, most studies on algae focused on their carbohydrate content, whereas investigations on their proteins were rarely found.

The protein content in algae is different among species. Brown algae, excluding *U. pinnatifida*, have significantly lower protein content (3%–10% dry wt.) than green and red algae, which contain 10%–47% (dry wt.) of protein. Algal proteins have several potential uses such as a protein source in functional foods, food dye, and even fish feed (Fleurence, 1999). Recently, various researches have been emphasizing on algal glycoproteins (GPs) and their bioactivities.

Marine GPs are composed of carbohydrate and protein moieties. Their functions in aquatic animals are crucial to mediate sperm–egg binding, to promote larval settlement and bioadhesion, to compose chemoattractants, and to support immune response and pathogenesis (Cadwell and Pagett, 2010). Due to their pivotal roles, GPs from marine animals have attracted interests for pharmacological applications.

Algal GPs can be obtained from both micro- and macroalgae. Microalgal GPs are mostly reported from *Chlorella vulgaris*, owing to its immunostimulant activity *in vitro* and *in vivo* (Tanaka et al., 1998). Meanwhile, macroalgal GPs have also been isolated from green and read algae, such as *L. japonica*, *S. japonica*, *Capsosiphon fulvescens, Codium decorticatum, Hizikia fusiformis, Ulva lactuca,* and *Palmira palmata.* The bioactivities of their GP have been reported to exhibit antitumor, antioxidant, immunomodulator, probiotic, hypoglycemic, anti-inflammatory, antibacterial, and hepatoprotective properties. Algal GPs are rarely isolated from brown algae due to their low protein content, with the exception of *U. pinnatifida.* It is also reported that algal protein content and their digestibility are season dependent (Fleurence, 1999). This becomes limitation when algal proteins are to be developed for functional foods.

This chapter will discuss the isolation of algal GPs, their biological properties, and their potential therapeutic applications.

## 8.2 Algae GPs as potential therapeutic agents

### 8.2.1 Extraction and isolation of GPs

Algal GPs are distinguished between those of micro- and macroalgae. Microalgal GPs are prepared from their culture fluid, prior to purification by column chromatography. The molecular mass of the isolated GP is then determined by gel filtration by comparison with the protein standard (Konishi et al., 1996; Tanaka et al., 1998; Noda et al., 2002; see Figure 8.1).

Meanwhile, GPs from algae are obtained from their hot water extracts. Polysaccharides are separated by centrifugation, filtration, and precipitation with ethanol.

GP-containing filtrate is then purified by precipitation with ammonium sulfate. Dialysis against distilled water and centrifugation are used to purify GP prior to use (Go et al., 2009).

### 8.2.2 Biological properties of GPs

GP from *C. vulgaris* is acidic and composed of galactose-containing carbohydrate (67%) and protein (33%) that contribute to a molecular weight of 218 kDa (Konishi et al., 1996). Surprisingly, another strain of *C. vulgaris* contain smaller GP, known as ARS-2, with a molecular weight of 6.1 kDa, and consists of 6 linked-β1-6-galactopyranose carbohydrate and protein with a similar composition. The protein moiety of *C. vulgaris* supernatant (CVS) plays a vital role in its antitumor activity, in which its terminal amino acid sequence has been decoded as DVGEAFPTVVDALVA (Tanaka et al., 1998).

High-level protein among macroalgae are usually found in green algae, such as *C. fulvescens* (Kim et al., 2013a) and *C. decorticatum* (Thangam et al., 2014). The latter contains GP with a mass of 48 kDa that is composed of more than 60% of protein (Thangam et al., 2014). Furthermore, GP from *U. lactuca* contained 12.5% of protein and residues of glucuronic acid, xylose, rhamnose and glucose (Abdel-Fattah and Sary, 1987). On the other hand, GP from *L. japonica* (LjGP) only had 10% protein content (Go et al., 2009, 2010; Han et al., 2011; Park et al., 2011).

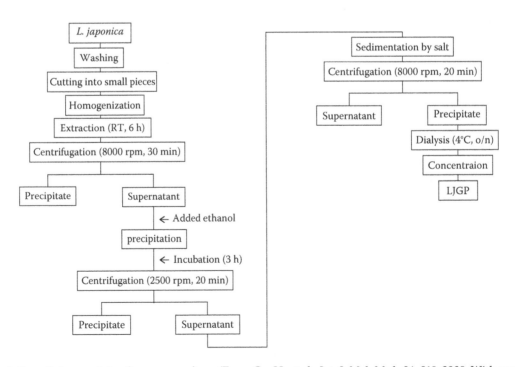

*Figure 8.1* Isolation of glycoproteins from macroalgae. (From Go, H. et al., *Int. J. Mol. Med.,* 24, 819, 2009. With permission.)

## 8.2.3   Chlorella vulgaris *extract and CVS as biological response modifiers*

The use of *C. vulgaris* as biological response modifier (BRM) was initially detected when its hot water extract augmented resistance against Meth-A tumor in mice (Tanaka et al., 1984). Investigation on its mechanism of action then revealed the involvement of polymorpho-nuclear leukocytes (PMNs) (Konishi et al., 1985). PMN is crucial in protecting the host against infection by various bacteria and fungi. *C. vulgaris* extract (CVE) amplified the resistance of mice toward intraperitoneal infection of *E. coli* (Tanaka et al., 1986). Oral administration of CVE significantly decreased the amount of bacteria to an undetectable level within 24 h compared to control by increasing the number of PMN cells. This was considered as a crucial advantage of CVE in terms of practical use, since BRM agents mostly lose their effectiveness when given orally (Hasegawa et al., 1989). Furthermore, CVE also increased the number of leukocytes in peripheral blood, bone marrow, and spleen in cyclophosphamide (CY)-treated rats, which indicated its role as an immunostimulant. CVE improved host resistance against immunosuppressive conditions and susceptibility to infection due to the side effects of chemotherapeutic drugs (Hasegawa et al., 1990). When administered subcutaneously, CVE enhanced the bacteriocidal activity of PMN in the CY-treated mice (Konishi et al., 1990). The potential of the culture supernatant of *C. vulgaris* to prevent stress-induced immunosuppression was reiterated by reducing the apoptosis of thymocyte in mice undergoing psychological stress in the communication box (Hasegawa et al., 2000).

CVE was also found to prevent infection by *Listeria monocytogenes* in mice via modulation of αβ-T cells (Hasegawa et al., 1994). Furthermore, CVE reduced the infection of *L. monocytogenes* in LP-BM5 murine leukemia–infected mice. This result confirmed the activity of CVE to protect immunodeficient host toward opportunistic infection (Hasegawa et al., 1995). Dantas et al. (1999) reported that CVE protected *L. monocytogenes*-infected mice by increasing the amount of NK (natural killer) cells that led to the increase in IFN-γ (gamma interferon) levels. CVE stimulated the hematopoietic response in *L. monocytogenes*-infected mice and prolonged their survival upon exposure to a lethal dose of bacteria (Dantas and Queiroz, 1999). CVE improved cell-mediated immunity by activating the macrophages to produce interleukin-12 (IL-12) and finally enhanced the response of Th1 against bacteria (Hasegawa et al., 1997). Administration of CVE in mice also increased the production of IL-2 at 48 and 72 h after bacterial inoculation (Queiroz et al., 2002). However, in protein-deficient mice, CVE improved immune function by stimulating the production of ILs, IL-2, and IL-4 but not IFN-γ (An et al., 2010). Moreover, CVE was effective for the prophylactic treatment of

poststress myelosuppression in stress-induced rats when exposed to *L. monocytogenes* (Queiroz et al., 2004, 2008).

The ability of CVE to improve Th-1 response via the production of IFN-γ also proved useful for allergic prevention (Hasegawa et al., 1999). This was later confirmed by Bae and coworkers (2013), who observed that CVE suppressed histamine release in mice via Th1 enhancement.

The protective effect of CVE treatment was also found when tested against lead-exposed mice. CVE chelating activity occurred in association with its immunoprotective effect during simultaneous exposure to lead (Queiroz et al., 2003) via modulation of cytokine production (Queiroz et al., 2011). Similarly, CVE stimulated the expression of metallothioneins in cadmium-exposed mice to protect liver damage (Shim et al., 2008).

### 8.2.4   Antitumor activities

CVE, which mainly contains GP, was initially recognized to contain bioactive compounds from their positive results on several immunostimulant assays. The antitumor activities of CVS have been tested against sarcoma (Tanaka et al., 1990), fibrosarcoma, lymphoma (Tanaka et al., 1998), carcinoma (Justo et al., 2001), and HepG2 cells (Yusof et al., 2010).

The acetone extract of CV that was given orally to BALB/c (inbred strain of mice) or CDF1 (hybrid mice) with Meth-A tumor significantly suppressed the growth of tumor. Antigen-specific concomitant immunity of mice against tumor was mediated by cytostatic T cells but not by cytotoxic T cells (Tanaka et al., 1990). Tumor-bearing mice were treated with CVS to reveal its role in preventing metastasis or tumor growth by enhancing thymocytes (Tanaka et al., 1998). The protein moiety of CVS also showed protective effect against myelosuppression by 5-fluorouracil (5FU), which prolonged survival without affecting the antitumor activity of the drug (Konishi et al., 1996). Additionally, an immunopotentiator agent composed of 56% of galactose-rich carbohydrate and 36% of protein was also isolated from CVS. It was reported to exhibit even greater antitumor activity against rechallenged tumors and was claimed to be comparable to that of a streptococcal preparation, a widely used OK-432 (Noda et al., 1998).

CVE exhibited antitumor activity toward mice that were inoculated with Ehrlich ascites tumor. The administration of CVE was proven to increase the bone marrow and spleen granulocyte–macrophage progenitor cells (Justo et al., 2001). CVE also modulated immunomyelopoietic activity and thus delayed tumor outgrowth through immunotherapeutic mechanisms in tumor-bearing mice (Ramos et al., 2010). When tested against Hep2 cells, CVS induced apoptosis signaling cascades by increasing the expression of p53, Bax, caspase-3, and Bcl-2 proteins, leading to DNA damage and apoptosis

(Yusof et al., 2010). Additionally, CVE promoted proliferation of (non-transformed rat small intestinal/epithelium cell lines) IEC-6 cells via mitogen-activated protein kinase (MAPK), (phosphatidylinositol-3-kinases) PI3K/ (serine/threonine-specific protein kinase) Akt, and canonical Wnt pathways (Song et al., 2012).

Tanaka and coworkers (1998) proposed that the antitumor mechanism of action of CVS was through immunopotentiation. However, in case of another water-soluble GP, ARS-2, Hasegawa et al. (2002) suggested the involvement of toll-like receptor 2 in the antitumor activity as well.

In terms of clinical application, CVS is comparable to OK-432 that has been used widely for cancer patients. The administration of CVS may be differentiated into IT and IV to inhibit spontaneous and recurring tumor, whereas SC injection can be employed to enhance metastasis inhibition. The latter suggested that the regional lymph node might play an important role in developing immunity against tumor (Tanaka et al., 1998).

A study by Konishi and coworkers (1996) showed that *C. vulgaris* GP not only exhibited antitumor effect but also prolonged the survival of tumor-induced mice that were treated with 5FU without interfering in its activity. CVS also proved to protective against bacteria-induced infection following gut transplantation by promoting the recovery of hematopoiesis (Konishi et al., 1996; see Figure 8.2).

An LjGP was found to inhibit proliferation of HT-29 colon cancer in a dose-dependent manner via apoptosis induction. It was proposed that the apoptosis was triggered via several mechanisms, such as DNA fragmentation, sub-G1 arrest, caspase-3 activation, and PARP (poly ADP ribose polymerase) degradation. Furthermore, cell

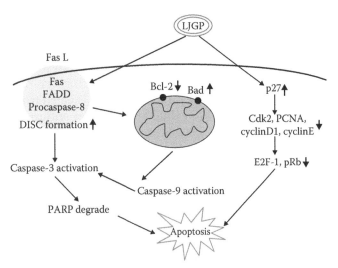

*Figure 8.2* Proposed mechanism of action of apoptosis by LjGP in HT-29 cells. (Reprinted from *Toxicol. In Vitro*, 21, Go, H., Hwang, H.J., and Nam, T.J., A glycoprotein from Laminaria japonica induces apoptosis in HT-29 colon cancer cells, 1546. Copyright 2010, with permission from Elsevier.)

lines treated with LjGP also carried out programmed cell death by activating Fas, FADD (Fas-associated protein with death domain), and procaspase (Go et al., 2010). LjGP also affected cell membrane by inducing anoikis or extracellular matrix disruption. LjGP activated the matrix metalloproteinase (MMP), integrin, and epithelial cathedrin and eventually downregulate PI3K, MAPK, and Wnt signaling pathways (Go et al., 2011). LjGP not only exhibited antitumor activity, it also promoted cell proliferation in normal colon cell lines via the EFGR signaling pathway. LjGP stimulated EGFR, Akt, Shc/Grb2 binding, and ERK (extracellular signal regulated kinases) activation but inhibited JNK phosphorylation (Go et al., 2009).

Apoptosis-induced activity of LjGP against AGS human gastric cell lines was also proposed through different mechanisms, such as upregulation of Bax expression, downregulation of Bcl-2 and IAP family members, activation of caspase 3 and 9, and degradation of PARP. The treatment of LjGP also inhibited tumor growth by downregulating telomerase activity and expression of human telomerase reverse transcriptase, and decreasing levels of cyclooxygenase (COX-2) that correspond to the synthesis of prostglandin $E_2$ (Han et al., 2011).

Similar to LjGP, the apoptotic activity of a GP from *C. fulvescens* (CfGP) toward AGS cell lines was also reported (Kim et al., 2012a). Expression of CfGP promoted apoptosis through the caspase–cascade pathway. It also triggered the activation of Bcl-2 family proteins that play important roles in CfGP-induced apoptosis. CfGP also stimulated the sub-G1 phase arrest to inhibit AGS cell growth. Additionally, AGS cell proliferation was prevented through the downregulation of the expression of growth-related proteins such as TGF-β1, FAK, PI3K, Akt, and small GTPase, excluding RhoB. The activation of the latter was important to promote cell apoptosis. AGS cell migration was restrained by suppressing the expression of integrin $\alpha_v$, $\beta_1$, $\beta_2$, and $\beta_5$. CfGP also decreased MMP expression in a dose-dependent manner and thus inhibited AGS growth and invasion (Kim et al., 2013b).

Investigation on the GP from *C. decorticatum* (CdGP), green algae from India, revealed its antitumor activities against breast cancer cell lines MDA-MB-231 via apoptosis induction. Induction of mitochondria-mediated intrinsic apoptotic pathway was observed by the loss of mitochondrial membrane potential, downregulation of Bax/Bcl-2, cytochrome-c release, and activation of caspase 3 and 9. CdGP also regulated cell apoptosis by causing G2/M phase cell cycle arrest (Thangam et al., 2014).

### 8.2.5  Other biological activities

In a quest to explore the potential use of algal GPs, various researches were conducted to test them for other biological activities, such as hepatoprotective, antiinflammatory, antioxidant, probiotic, and antidiabetic.

Liver injury may occur as a result of the long-term use of acetaminophen. Thus, a study was conducted on the hepatoprotective activity of the GP from *H. fusiformis* (HFGP) to screen its potential use for the treatment of AAP-induced liver injury. HFGP blocks AAP toxicity by inhibiting the activation of caspase 3 and 9. It also activates ERK, which plays an important role in the regulation of oxidative stress (Hwang et al., 2008). Hepatoprotective effect of CVE was also observed in $CCl_4$-induced mice by stimulating the antioxidant activity and nitric oxide (NO) production in liver and by inhibiting the cytochrome P450 2B1/2B2 (CYP2B1/2) (Kim et al., 2009).

Aside from exhibiting hepatoprotective activities, algal GP has also been investigated for its anti-inflammatory activities. The use of LjGP for the treatment of neurological diseases was studied by Park and coworkers (2011). LjGP inhibited the production of NO and prostaglandin $E_2$ in LPS-stimulated BV-2 cells. Additionally, it downregulated the expression of NO synthase, COX-2, and pro-inflammatory cytokines. LjGP suppressed pro-inflammatory agents through the regulation of nuclear factor κβ and extracellular signal–regulated kinase, as well as inactivation of the p-38 and ERK of MAPK and PI3K/Akt pathways in activated microglia. Similarly, *P. yezoensis* GP regulated nuclear factor κβ and MAP kinase via inhibition of TLR4 signaling (Shin et al., 2011).

Reactive oxygen species (ROS) is a well-known cause for oxidative stress that may be linked to the development of degenerative diseases. Therefore, an antioxidant is usually employed to prevent the damaging effects of ROS. The hot water extracts of both *Spirulina* and *Chlorella* were tested for their antioxidant and antiproliferative activities. It was shown that *Spirullina*'s extract had higher antioxidant and antiproliferative activity against HSC (hematopoietic stem cells) and HepG2 (liver hepatocellular carcinoma) cells than that of *Chlorella*'s (Wu et al., 2005). Both GPs from *S. japonica* (SjGP) and *U. pinnatifida* (UpGP) were reported to exhibit antioxidant properties (Kim et al., 2012b; Rafiquzzaman et al., 2013). In a series of antioxidant assays, SjGP showed significant antioxidant activities at a concentration of 5 mg/mL with optimal activity at pH 7–9 and 40°C (Kim et al., 2012b). In similar assays, UpGP exhibited potent activities at even smaller concentration with $IC_{50}$ between 0.08 and 0.25 mg/mL (Rafiquzzaman et al., 2013). Additionally, both GPs prevented DNA cleavage under oxidative stress. Both activities were found to be pH dependent. Further study on the relationship between GPs' structure and their antioxidant activities suggested the necessity of carbohydrate and protein in the structure for their optimum activity (Kim et al., 2012b; Rafiquzzaman et al., 2013).

UpGP was found to exhibit antidiabetic activity by inhibiting α-glucosidase, when used in combination with acarbose, and increasing glucose transport in yeast cell. This finding suggested that UpGP could be taken along with antidiabetic drugs to reduce the adverse side effects in patients. Furthermore, the glycoside moiety in UpGP was found critical to the inhibitory effect against α-glucosidase (Rafiquzzaman et al., 2015).

Exploration of algal GP from *S. japonica* as functional foods was expanded by testing its potential as probiotic. SjGP was shown to increase the probiotic properties of *L. plantarum*. The pretreatment of probiotic bacteria with SjGP increased cell adhesion, autoaggregation, and growth in Caco-2 cells (Kim et al., 2015); (Table 8.1).

***Table 8.1*** Therapeutic potential of algal glycoproteins

| Algae | Bioactivity | Mechanism | References |
|---|---|---|---|
| *C. vulgaris* | Antitumor | Apoptosis and immunopotentiation | Tanaka et al. (1990, 1998); Yusof et al. (2010) |
| | Immunostimulant | Activation of polymorphonuclear (PMN) leukocytes | Konishi et al. (1985); Tanaka et al. (1986); Hasegawa et al. (1989, 1990, 2000); Konishi et al. (1990) |
| | Antibacteria | Activation of adaptive immune system | Hasegawa et al. (1997); Queiroz et al. (2008) |
| | Antiallergic | Activation of adaptive immune system | Hasegawa et al. (1999) |
| *L. japonica* | Antitumor | Apoptosis | Go et al. (2010, 2011); Han et al. (2011) |
| | Anti-inflammatory | Activation of nuclear factor κβ and extracellular signal–regulated kinase | Park et al. (2011) |
| | Probiotic | Unknown | Kim et al. (2015) |
| *S. japonica* | Antioxidant | Unknown | Kim et al. (2012b) |
| *C. decorticatum* | Antitumor | Apoptosis | Thangam et al. (2014) |
| *U. pinnatifida* | Antioxidant | Unknown | Rafiquzzaman et al. (2013) |
| | Antidiabetic | Inhibition of alpha-glucosidase | Rafiquzzaman et al. (2015) |
| *C. fulvescens* | Antitumor | Apoptosis | Kim et al. (2012a) |
| *H. fusiformis* | Hepatoprotective | Activation of caspase 3 and 9 | Hwang et al. (2008) |
| *P. yezoensis* | Anti-inflammatory | Activation of nuclear factor κβ and extracellular signal–regulated kinase | Shin et al. (2011) |

## 8.3  Concluding remarks

The potential of marine algae has motivated many countries to conventionally develop farming and processing activities due to various reasons, such as for poverty alleviation and phycocolloid industry establishment. The recent findings discussed here inform that GPs extracted from marine algae were demonstrated to have therapeutic benefits. This fact may change the direction of algal utilization in the future. Probably, uses of algae by taking advantage of the bioactivities of GP will bring more profit compared to existing conventional utilization. Therefore, effective and efficient extraction techniques of algal GP should be developed to provide a more feasible technology.

## References

Abdel-Fattah, A. F., Sary, H. H., *Phytochemistry* 1987, 26 (5), 1447–1448.

An, H. J., Rim, H. K., Jeong, H. J., Hong, S. H., Um, J. Y., Kim, H. M., *Immunopharmacology and Immunotoxicology* 2010, 32 (4), 585–592.

Bae, M. J., Shin, H. S., Chai, O. H., Han, J. G., Shon, D. H., *Journal of the Science of Food and Agriculture* 2013, 93 (12), 3133–3136.

Caldwell, G. S., Pagett, H. E., *Marine Biotechnology* 2010, 12 (3), 241–252.

Dantas, D. C. M., Kaneno, R., Queiroz, M. L. S., *Immunopharmacology and Immunotoxicology* 1999, 21 (3), 609–619.

Dantas, D. C. M., Queiroz, M. L. S., *International Journal of Immunopharmacology* 1999, 21 (8), 499–508.

FAO, *The State of World Fisheries and Aquaculture 2012*. Food and Aquaculture Organization of the United Nations, Rome, Italy, 2012.

Fleurence, J., *Trends in Food Science and Technology* 1999, 10 (1), 25–28.

Go, H., Hwang, H. J., Kim, I. H., Nam, T. J., *Open Journal of Preventive Medicine* 2011, 1 (02), 49.

Go, H., Hwang, H. J., Nam, T. J., *International Journal of Molecular Medicine* 2009, 24 (6), 819–824.

Go, H., Hwang, H. J., Nam, T. J., *Toxicology In Vitro* 2010, 24 (6), 1546–1553.

Han, M. H., Kim, G. Y., Moon, S. K., Kim, W. J., Nam, T. J., Choi, Y. H., *International Journal of Oncology* 2011, 38 (2), 577–584.

Hasegawa, T., Ito, K., Ueno, S., Kumamoto, S., Ando, Y., Yamada, A., Nomoto, K., Yasunobu, Y., *International Journal of Immunopharmacology* 1999, 21 (5), 311–323.

Hasegawa, T., Kimura, Y., Hiromatsu, K., Kobayashi, N., Yamada, A., Makino, M., Okuda, M., Sano, T., Nomoto, K., Yoshikai, Y., *Immunopharmacology* 1997, 35 (3), 273–282.

Hasegawa, T., Matsuguchi, T., Noda, K., Tanaka, K., Kumamoto, S., Shoyama, Y., Yoshikai, Y., *International Immunopharmacology* 2002, 2 (4), 579–589.

Hasegawa, T., Noda, K., Kumamoto, S., Ando, Y., Yamada, A., Yoshikai, Y., *International Journal of Immunopharmacology* 2000, 22 (11), 877–885.

Hasegawa, T., Okuda, M., Makino, M., Hiromatsu, K., Nomoto, K., Yoshikai, Y., *International Journal of Immunopharmacology* 1995, 17 (6), 505–512.

Hasegawa, T., Okuda, M., Nomoto, K., Yoshikai, Y., *Immunopharmacology and Immunotoxicology* 1994, 16 (2), 191–202.

Hasegawa, T., Tanaka, K., Ueno, K., Ueno, S., Okuda, M., Yoshikai, Y., Nomoto, K., *International Journal of Immunopharmacology* 1989, 11 (8), 971–976.

Hasegawa, T., Yoshikai, Y., Okuda, M., Nomoto, K., *International Journal of Immunopharmacology* 1990, 12 (8), 883–891.

Hwang, H. J., Kim, I. H., Nam, T. J., *Food and Chemical Toxicology* 2008, 46 (11), 3475–3481.

Justo, G. Z., Silva, M. R., Queiroz, M. L. S., *Immunopharmacology and Immunotoxicology* 2001, 23 (1), 119–132.

Kim, E. Y., Kim, Y. R., Nam, T. J., Kong, I. S., *International Journal of Food Science and Technology* 2012a, 47 (5), 1020–1027.

Kim, H. K., Li, L., Lee, H. S., Park, M. O., Bilehal, D., Li, W., Kim, Y. H., *Food Science and Biotechnology* 2009, 18 (5), 1186–1192.

Kim, Y. M., Kim, I. H., Nam, T. J., *Nutrition and Cancer—An International Journal* 2012b, 64 (5), 761–769.

Kim, Y. M., Kim, I. H., Nam, T. J., *International Journal of Oncology* 2013a, 43 (5), 1395–1401.

Kim, Y. M., Kim, I. H., Nam, T. J., *Molecular Medicine Reports* 2013b, 8 (1), 11–16.

Kim, E. Y., Rafiquzzaman, S. M., Lee, J. M., Noh, G., Jo, G. A., Lee, J. H., Kong, I. S., *Journal of Applied Phycology* 2015, 27(2), 965–973.

Konishi, F., Mitsuyama, M., Okuda, M., Tanaka, K., Hasegawa, T., Nomoto, K., *Cancer Immunology Immunotherapy* 1996, 42 (5), 268–274.

Konishi, F., Tanaka, K., Himeno, K., Taniguchi, K., Nomoto, K., *Cancer Immunology Immunotherapy* 1985, 19 (2), 73–78.

Konishi, F., Tanaka, K., Kumamoto, S., Hasegawa, T., Okuda, M., Yano, I., Yoshikai, Y., Nomoto, K., *Cancer Immunology Immunotherapy* 1990, 32 (1), 1–7.

Noda, K., Ohno, N., Tanaka, K., Okuda, M., Yadomae, T., Nomoto, K., Shoyama, Y., *Phytotherapy Research* 1998, 12 (5), 309–319.

Noda, K., Tanaka, K., Yamada, A., Ogata, J., Tanaka, H., Shoyama, Y., *Phytotherapy Research* 2002, 16 (6), 581–585.

Park, H. Y., Han, M. H., Kim, G. Y., Kim, N. D., Nam, T. J., Choi, Y. H., *Journal of Food Science* 2011, 76 (7), T156–T162.

Queiroz, J. D., Malacrida, S. A., Justo, G. Z., Queiroz, M. L. S., *Immunopharmacology and Immunotoxicology* 2004, 26 (3), 455–467.

Queiroz, J. D., Torello, C. O., Neto, J. P., Valadares, M. C., Queiroz, M. L. S., *Brain, Behavior, and Immunity* 2008a, 22 (7), 1056–1065.

Queiroz, M. L. S., Bincoletto, C., Valadares, M. C., Dantas, D. C. M., Santos, L. M. B., *Immunopharmacology and Immunotoxicology* 2002, 24 (3), 483–496.

Queiroz, M. L. S., da Rocha, M. C., Torello, C. O., Queiroz, J. D., Bincoletto, C., Morgano, M. A., Romano, M. R., Paredes-Gamero, E. J., Barbosa, C. M. V., Calgarotto, A. K., *Food and Chemical Toxicology* 2011, 49 (11), 2934–2941.

Queiroz, M. L. S., Rodrigues, A. P. O., Bincoletto, C., Figueiredo, C. A. V., Malacrida, S., *International Immunopharmacology* 2003, 3 (6), 889–900.

Queiroz, M. L. S., Torello, C. O., Perhs, S. M. C., Rocha, M. C., Bechara, E. J. H., Morgano, M. A., Valadares, M. C., Rodrigues, A. P. O., Ramos, A. L., Soares, C. O., *Food and Chemical Toxicology* 2008, 46 (9), 3147–3154.

Rafiquzzaman, S. M., Kim, E. Y., Kim, Y. R., Nam, T. J., Kong, I. S., *International Journal of Biological Macromolecules* 2013, 62, 265–272.

Rafiquzzaman, S. M., Min Lee, J., Ahmed, R., Lee, J. H., Kim, J. M., Kong, I. S., *International Journal of Food Science and Technology* 2015, 50(1), 143–150.

Ramos, A. L., Torello, C. O., Queiroz, M. L. S., *Nutrition and Cancer—An International Journal* 2010, 62 (8), 1170–1180.

Shim, J. Y., Shin, H. S., Han, J. G., Park, H. S., Lim, B. L., Chung, K. W., Om, A. S., *Journal of Medicinal Food* 2008, 11 (3), 479–485.

Shin, E. S., Hwang, H. J., Kim, N. H., Nam, T. J., *International Journal of Molecular Medicine* 2011, 28 (5), 809–815.

Song, S. H., Kim, I. H., Nam, T. J., *International Journal of Molecular Medicine* 2012, 29 (5), 741–746.

Tanaka, K., Koga, T., Konishi, F., Nakamura, M., Mitsuyama, M., Himeno, K., Nomoto, K., *Infection and Immunity* 1986, 53 (2), 267–271.

Tanaka, K., Konishi, F., Himeno, K., Taniguchi, K., Nomoto, K., *Cancer Immunology Immunotherapy* 1984, 17 (2), 90–94.

Tanaka, K., Tomita, Y., Tsuruta, M., Konishi, F., Okuda, M., Himeno, K., Nomoto, K., *Immunopharmacology and Immunotoxicology* 1990, 12 (2), 277–291.

Tanaka, K., Yamada, A., Noda, K., Hasegawa, T., Okuda, M., Shoyama, Y., Nomoto, K., *Cancer Immunology Immunotherapy* 1998, 45 (6), 313–320.

Thangam, R., Senthilkumar, D., Suresh, V., Sathuvan, M., Sivasubramanian, S., Pazhanichamy, K., Gorlagunta, P. K. et al., *Journal of Agricultural and Food Chemistry* 2014, 62 (15), 3410–3421.

Wu, L. C., Ho, J. A. A., Shieh, M. C., Lu, I. W., *Journal of Agricultural and Food Chemistry* 2005, 53 (10), 4207–4212.

Yusof, Y. A. M., Saad, S. M., Makpol, S., Shamaan, N. A., Ngah, W. Z. W., *Clinics* 2010, 65 (12), 1371–1377.

*chapter nine*

# Partial sequencing, structural characterization, and anticoagulant activity of heparan sulfate and sulfated chitosan from selected Indian marine mollusks

*Ramachandran Saravanan, Ramachandran Karthik, and Annian Shanmugam*

## Contents

## 9.1 Introduction

Glycobiology is a branch of carbohydrate chemistry that deals with the structure, biosynthesis, and biological activity of carbohydrates that are generally distributed in natural sources. Sugars or saccharides are fundamental constituents of all living things, and because of their characteristics of the assorted functions, they play a vital role in biology, medicine, and biomedical engineering. The term "glycobiology" was coined by Prof. Raymond Dwekin in 1988 to recognize the coming together of the traditional disciplines of carbohydrate chemistry and biochemistry (Rademacher et al., 1988). It is an amalgamation of carbohydrate chemistry, enzymology, and biochemistry with a modern understanding of the cell and molecular biology of glycosaminoglycans (GAGs) and, in particular, they are linked to proteins and lipids in the cells (Saravanan, 2014).

## 9.2 History of glycobiology

The structural investigation of proteoglycans (PG) was started in the beginning of the twentieth century with the research on "chondromucoid" of cartilage and the preparation of anticoagulants, for example, heparin from liver/kidney of porcine/bovine sources (Saravanan and Shanmugam, 2011). Between 1930 and 1960, great strides were made in the preparations and structural analyses of mucopolysaccharides chemistry, especially in hyaluronan, dermatan sulfate (DS), keratan sulfate (KS), different isomeric forms of chondroitin sulfate (CS), heparin, and heparan sulfate (HS). In concert, these mucopolysaccharides are also called "glycosaminoglycans" to point to the existence of amino sugars and other sugars in polymeric forms. Consequent analysis afforded insights into the association of the sequences to core proteins, and these structural studies paved the path for the biosynthetic studies that pursued (Saravanan et al., 2010).

The improved chromatographic procedure of isolation and fractionation of PG and GAGs dates back to the 1970s (Volpi, 1994). PG from cartilage is obtained by density gradient ultracentrifugation, enlightening a complex of PG, hyaluronan, and a core protein. In addition, it is realized that the production of PG is a general property of animal cells and that PG and GAGs are present on the cell surface, inside the cell, and in the extracellular matrix. This assessment guides to a speedy growth of the field and the ultimate positive reception of PG functions in cell adhesion and cell signaling, in addition to a host of other biological activities. At present, investigation with somatic cell mutants in addition

to research by means of gene knockout and silencing methods in a array of replica creatures, counting nematode worms (*Caenorhabditis elegans*), fruit flies (*Drosophila melanogaster*), African clawed frogs (*Xenopus laevis*), zebrafish (*Danio rerio*), and mice (*Mus musculus*), are aspired at improving our knowledge of the role of PG in progress and physiology. Consecutively, human diseases correlated with abnormal biosynthesis or dilapidation of PG has been identified, and some are named congenital disorders of glycosylation. Various analytical methods have been developed, including mass spectroscopic methods and glycan (GLY) array relevancies that offer novel tools for understanding PG structure and function (Esko et al., 2009).

## 9.3  Biological significance of glycan

The biological significance of GLYs is mainly classified into three types: (1) health, (2) energy, and (3) materials science (Figure 9.1). GLYs are ubiquitous throughout the natural world. Each living cell is covered with a GLYs layer that is endowed with a cell/tissue-specific identity, and several proteins entertain GLYs tags that can act like a dimmer switch to adapt to protein activity.

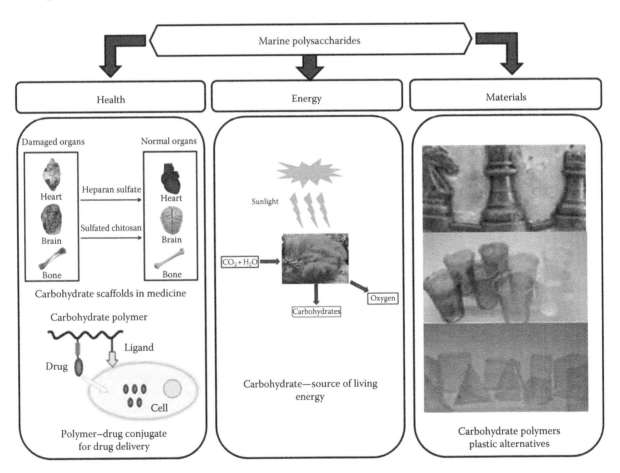

*Figure 9.1* (**See color insert.**) Role of marine polysaccharides in health, energy, and materials.

## 9.3.1   Health

GLYs play vital functions in approximately every biological reaction and are riveted in all major diseases in clinics. They are fundamental to cell linkage and movement, for example, when white blood cells move around to a site of damage or infectivity. GLYs surrounded by cells support strain the expression of genes and proteins, shaping ingredient of a cell's reaction to biological signals. Moreover, GLYs have the ability to recognize pathogens like bacteria, virus, and fungi and track the immune response against a particular antigen. These immunological properties of GLYs are used as therapeutic drugs/components of drugs to assist extravagant infectious and transinfectious diseases and as biomarkers to detect diseases like cancer.

Presently, glycoscience also contributes to personalized medicine against disease/disorder. Nowadays, a lot of investment is made to study the human genome analysis to develop tailored drugs for treatments. A major problem in human genome sequencing is binding of GLYs to genes. Since a moratorium presently prohibits genomics could help develop tailored medical care.

## 9.3.2   Energy

The biological reactions of photosynthesis incarcerate sunlight and convert it to starch that fuels the growth of plants and the animals that feed on them. Some of these oligosaccharides are used to make plant cell wall structures composed of GLYs, cell wall proteins, and lignin. Large-scale cohort of biofuels from biomass requires a resourceful process to exchange the energy stored in plant cell walls into liquid fuel. Nevertheless, a phenomenon called insubordination—the confrontation of plant cell walls to the degradation that is essential to convert biomass into fuel—makes existing techniques for producing biofuel costly. Glycoscience can perhaps help enzymes and other catalysts to improve the speed of reactions. As cell wall structures can differ from plant to plant species and even within different cell/tissues of a single plant, an improved perceptive of plant cell wall creation could also help researchers opt for or maneuver crops to exploit biomass production, while minimizing the use of fertilizers and water.

## 9.3.3   Materials

The use of plastics in the day-to-day life of humans is not completely avoidable. Many of the most widely used plastics are made from polymers found in petroleum. Due to their cost, degradation, supply, and pollution problems, scientist are looking at alternative sources of plastics. Following this, there is increasing pressure to find alternative sources of polymer materials. Recently, scientists have turned their attention to develop GLYs-based polymers, which are extracted/isolated from natural resources. GLYs-based polymers are of pure form that are modified chemically to provide a wide range of new properties like water insolubility, or are still broken down to their component sugar moieties to formulate chemical forerunners for use in chemical, medical, and engineering industries (NAS, 2012).

## 9.4   Marine resources

In general, over 60% of the active components in pharmaceutical formulations are obtained from natural products or are synthetic derivatives or mimetics (Cragg and Newman, 2013). Serious investigation of all agents endorsed by the Food and Drug Administration (FDA) in the past decades evidently point towards those natural components as resources of novel entity structures of drugs, but the final drug leads are not essentially functional. Secondary metabolites obtained from marine organisms are the result of millions of years of development and natural assortment: even a particular species comprises a library of secondary metabolites that is authenticated for bioactivity. In consequence of the different enzymatic reactions, normal innovations have an inherent aptitude to distinguish and bind macromolecules, hinder their activity, and modulate biological progressions (Kiuru et al., 2014).

The maritime territory of India is the world's largest with an economic zone covering 2,305,143 km². The Indian Ocean, the Arabian Sea, and the Bay of Bengal can be regarded as its most important bioactive resources (Law of the Sea, 2011). Marine polysaccharides and oligosaccharides have been deemed recently as astonishing resources for the innovation of handy foods, biochemical, pharmaceuticals, cosmetics, and biomaterials. Marine oligosaccharides are commonly achieved by the hydrolysis of polysaccharides, which are primarily extracted from seaweeds and other sea creatures. Based on their sources, marine polysaccharides are classified into (1) plant polysaccharides, (2) animal polysaccharides, and (3) microbial polysaccharides. The chemical components and structures of most marine polysaccharides and oligosaccharides are complex and heterogeneous. Therefore, more research should be focused on the determination of their structure and activity in order to exploit the use of carbohydrates in humans (Lang et al., 2014).

## 9.4.1   GAGs analogs from marine invertebrates

Among the sulfated GAGs, heparin/HS performs various biological roles in cells after binding to protein, which regulates anticoagulation, homeostasis, anti-inflammations, fibrinolysis, angiogenesis, defense mechanisms, endocytosis, apoptosis, cell matrix assembly, cell recognition, cell proliferation, cell invasion, cell migration, and cell adhesion. Sulfated GAGs are ubiquitous to the animal kingdom of the eukarya domain, whereas only nonsulfated chains of GAGs are found in the bacteria kingdom. Except for hyaluronan (also known as hyaluronic acid), all GAGs occur in the tissues as PG, where the polysaccharide chains are covalently associated to a core protein. The GAGs are polydisperse linear polysaccharides composed of alternate residues of hexosamine and uronic acid connected by glycosidic linkages (Sampaio et al., 2006).

### 9.4.1.1  Heparin/HS

Anionic anticoagulant heparin is a naturally occurring sulfated GAG arranged by disaccharide residues containing a hexuronic acid (α-L-iduronicacid or β-D-glucuronic acid) connected at 1,4 positrons to α-D-glucosamine. The heparin/HS molecules consist of a heterogeneous mixture of polymers with a similar backbone, which results from distinctions of sulfation on the D-glucosamine (N-acetylated, N-sulfated, O-sulfated at C6 and/or C3) and on the uronic acid residue (O-sulfated at C2) (Lindahl et al., 1989). Due to its unique binding to antithrombin (blood clotting protein), connecting a specific pentasaccharide unit surrounding a 3-O-sulfated glucosamine, heparin is considered to possess powerful anticoagulant activity (Lindahl et al., 1980, 1989). In fact, based on its capability to inhibit fluid phase coagulation, unfractionated heparin (UH) isolated from porcine/bovine intestinal mucosa has been used clinically for several decades. On the other hand, the therapeutic use of UH is limited mostly by its potent hemorrhagic outcome; hence, patients under heparin remedy have to be closely examined (Hirsh, 1984). The bioavailability of UH is also poor, requiring several daily dosing, and has various adverse effects like heparin-induced thrombocytopenia (Hirsh and Raschke, 2004; Hirsh et al., 2004). To avoid these problems, different low-molecular-weight heparins (LMWHs) have been produced by degrading UH by using chemical/enzyme hydrolysis (Hirsh and Raschke, 2004).

### 9.4.1.2  Global historical developmental status of heparin

Bjork and Lindhal (1982) and Casu (1985) from Italy reviewed the mechanism of the anticoagulant activity of heparin. Later Dietrich et al. (1985) described the characterization of heparin obtained from mollusks (*Anomalocardia brasiliana*, *Donax striatus*, and *Tivela mactroides*) by using nuclear magnetic resonance (NMR) spectroscopy and they explained the chemical structure, disaccharide composition, and anticoagulant activity

of heparin. This finding is a big milestone in heparin research. After years of study, Albano and Maurao (1986) from Brazil isolated large amounts of the novel class of sulfated polysaccharides from the tunic of *Styela plicata* (Chordata Tunicata) and they found that this tissue does not contain GAGs, as is the case for mammal connective tissues, but a unique sulfated galactan. At the same time, Hovingh et al. (1986) from the United States showed that heparin-like polysaccharides occur not only in mammals but also in lower vertebrate species. The very next year, Pejler et al. (1987) from Brazil found heparin-like polysaccharides in certain invertebrates.

Linhardt et al. (1988) from the United States developed a new single HPLC technique of quantifying the oligosaccharide composition of commercial heparin obtained from bovine and porcine sources and they observed 86% of HMWH and only 5% of LMWH oligosaccharides. Two other scientists from the United States found the molecular interaction of protein with GAGs and they have reported around 49 regions in 21 proteins are identified as potential heparin-binding sites based on the sequence organizations of their basic and nonbasic residues (Cardin and Weintraub, 1989). Linhardt et al. (1990) explained the difference between heparin and HS and the substrate specificity of heparin and HS by using enzyme hydrolysis with heparin lyases I–III. In the year 1992, Linhardt et al. have isolated and characterized the human heparin. Volpi (1993) from Italy investigated the fast and slow moving heparins by agarose gel electrophoresis from mixtures of GAGs such as DS and CS.

The structure and vibrational spectra of heparins, physiochemical/biochemical properties of low molecular weight HS and anticoagulant activity of heparin were reported in various countries like Japan, the United States, the Netherlands, France, Brazil, and Italy (Sushko et al., 1994; Atha et al., 1995, 1996; Paiva et al., 1995; Jin et al., 1997; Nader et al., 1999; Medeiros et al., 2000; Volpi, 2005). Only bovine lung or porcine intestine tissues are presently used as raw materials to prepare commercial, pharmaceutical heparins. But the facade of bovine spongiform encephalopathy, "mad cow disease," and its apparent link to the similar prion-based Creutzfeldt–Jakob disease in humans, has limited the use of bovine heparin. Furthermore, it is not easy to discriminate bovine and porcine heparins, making it tricky to guarantee the species source of heparin (Linhardt and Gunay, 1999). Porcine heparin also has problems with its use, associated with religious restrictions among members of the Muslim and Jewish faiths (Warda et al., 2003). So at this juncture, there is need for an alternative non-mammalian source of heparin for remedial actions.

### 9.4.1.3  Chitin/chitosan

Chitin is the second most abundant natural biopolymer in the world, with an estimated 106–107 tons in marine biomass alone (Park and Kim, 2010). Chitin is made up of long-chain homopolymer of *N*-acetyl-D-glucosamine (NGlcAc), (1–4)-linked 2-acetamido-2-deoxy-β-D-glucan, a plagiaristic unit of glucose. Strong acids can split the chitin into simple glucosamine units, water-insoluble homopolymer, acetic acid, and chitosan. Chitins are processed further into two major derivatives of chitosan and amino glucose through many nonspecific enzymes, for instance, cellulases, lipases, proteases, and chitosanases (Lee et al., 2008; Lin et al., 2009). Chitin is one of the major structural carbohydrates of shells of insects/invertibrates, such as crab, shrimp, cuttlefish, etc. Seafood industries process the meat from marine organisms and throw away the cuttlebone as a waste, which is a potent sources for the production of chitin/chitosan derivatives, both of which have various applications in pharma/biotech industry. These shell wastes are potential in commercial-scale chitin production and the production of chitosan and its oligomers (Kim and Mendis, 2006). These chitins have a α-crystallographic structure where the main chains arrange in an antiparallel fashion with strong intermolecular hydrogen bonding (Minke and Blackwell, 1978). Chitin may also be obtained from squid pens (Kurita et al., 1994), and so presents a β-crystallographic structure where chitin chains arrange in a parallel fashion with relatively weak intermolecular forces.

Particularly, chitosan is a risk-free biopolymer obtained by the deacetylation of chitin, and presently chitosan and its oligosaccharides have entertained considerable concentration due to their biological roles and properties in viable applications. In the past few decades, as a resource of bioactive material, chitosan and chito-oligosaccharides, which are deprivation products of chitin or chitosan isolated by enzymatic or acidic hydrolysis, are considered as an array of biomedicals, including wound dressings and drug delivery systems in nanoscience (Agnihotri et al., 2004; Kumar et al., 2004) and in food processing and chemical industries (Knorr, 1984; Razdan and Pettersson, 1994; Kurita, 1998; Xia et al., 2011). Chitin and its derivatives have delivered biological potential for a broad range of applications, for example, in the food and medical fields (Ngo et al., 2009; Ribeiro et al., 2009; Li et al., 2010; Lim et al., 2010), agriculture (Kulikov et al., 2006) and aquaculture (Wang and Chen, 2005; Rajesh Kumar et al., 2008), dental (Khor and Lim, 2003; Arnaud et al., 2010) and cosmetics (Mark et al., 1986; Morganti and Morganti, 2008), wastewater treatment (Mckay et al., 1982; Covas et al., 1992), and membranes (Chatelet et al., 2001; Kim et al., 2007).

### 9.4.1.4 Global historical developmental status of chitin

In the period 1780–1855, the director of the botanical garden in Nancy, Henri Braconnot, placed the establishments of the polymer science of carbohydrate; following the discovery of chitin, the first mucopolysaccharide discovered 30 years prior to cellulose, he prolonged with his initiative extracting carbohydrate from edible fungi *Agaricus bisporus* and extracted inulin from the tubers of *Heliantus tuberous*. The discovery of chitin became the basis for some chemical experiments carried out on raw fabric isolated from *Agaricus volvaceus*, *A. acris*, *A. cantharellus*, *A. piperatus*, *Hydnum repandum*, *H. hybridum*, and *Boletus viscidus*. The fungal material is, to a degree, purified by boiling in dilute KOH, which eliminates proteins and pigments to produce the chitin–glucan complex (Muzzarelli et al., 2012).

The structure and the physicochemical/biochemical characteristics of chitosan were determined using infrared (IR) spectroscopy, $^1$H-NMR, and $^{13}$C-NMR spectrum by Brugnerotto et al. (2001); Tolaimate et al. (2000) from France. Xing et al. (2005) from China reported preparation of chitosan with high molecular weight and high sulfur content by treating dimethyl formamide and formic acid instead of dichloro acetic acid. Yen and Maus (2004,2007) from Taiwan investigated crude crab chitosan, which exhibited distinctly arranged microfibrillar crystalline structure in scanning electron microscope (SEM). Aimoli et al. (2006) from Brazil examined the early stages of *in vitro* calcification on chitosan films using three different techniques, such as x-ray fluorescence (XRF), atomic force microscopy (AFM), and x-ray diffraction (XRD), by immersing chitosan membranes in stimulated body fluid. In 2005, Vikhoreva et al. from Russia also reported the preparation and anticoagulant activity of low-molecular-weight sulfated chitosan shown by Fourier transform IR and NMR methods and that elemental analysis also demonstrated the anticoagulant activity of chitosan sulfates and anti-Xa activity like heparins. Except India, all the other countries like China, Japan, Taiwan, Brazil, the United Kingdom, and the United States have contributed to the sequencing units of polymer of heparin and chitin. In consideration of this, polymer type of HS from Indian marine scallop (*Amussium pleuronectus*) and β-chitosan from Indian cuttlefish (*Sepia pharaonis*) are dealt with in this chapter.

## 9.5 Methods for preparations

### 9.5.1 Extraction of molluskan HS

The extraction of molluskan HS is carried out by the method proposed by Saravanan (2014). The whole body wet tissues of selected marine animals homogenized in 5 L of ice-cold acetone and after decantation overnight at 4°C are cut into small pieces, the resulting supernatant is discarded. The pellet is suspended in 5 L of ice-cold chloroform/methanol (2:1). The defatted material is finally washed with cold ethyl ether, dried, and the defatted tissues are used for further study. The protein and nucleic acid contaminants are removed by treating the defatted tissues with pronase and endonuclease digestion. Finally the deproteinized sample is treated with chondroitin lyase to remove CS, dialyzed against distilled water, freeze-dried, and treated with ion-exchange chromatography and gel filtration chromatography. Active fractions are pooled, dialyzed, and again freeze-dried. The white powder is called marine scallop HS, which is used for further study.

### 9.5.2 Chitin extraction

Cuttlebones are removed from the selected cuttlefish and washed with tap water, distilled water, and then millipore water. They are then air-dried for a few days and powdered in a motor and pestle, and finally that powder is used for the extraction of chitin and chitosan (Subhapradha et al., 2013). The processing of chitin from cuttlebone involves removing mineral and protein matrix in which the chitin is implanted. Normally, it has been attained by washing the dried powder shells after pulverizing in concentrated acids (HCl 1.82% for 3–4 h) and alkali (1% NaOH for 24 h), but the order of the processing steps is inconsistent, which is clearly explained in Figure 9.2. Chitin is present within copious taxonomic clusters. However, viable chitins are usually isolated from marine crustaceans, such as crab, shrimp, and cuttlebone, mostly because a huge quantity of waste is a by-product of seafood processing industries. Herein,

α-chitin is manufactured while squid pens are used to create β-chitin. The purified structure of α-chitin has been scrutinized more broadly than that of either α- or β-form, because it is the frequently used polymorphic form. Only a few studies have been carried out on the structure γ-chitin extracted from natural sources. It has been implied that γ-chitin may be a damaged configuration of either α- or β-chitin relatively than a true third polymorphic form (Roberts, 1992).

Chains are organized in sheets or stacks, the chains in any one sheet having the same direction or "sense" are called as α-chitin. In β-chitin, neighboring sheets beside the common c axis have identical directions; the sheets are similar, while in α-chitin, adjacent sheets along the c axis have the opposite direction; they are antiparallel. In γ-chitin, every third sheet has the opposite direction to the two preceding sheets (Roberts, 1992). A schematic representation of the three structures is shown in Figure 9.3.

### 9.5.3 MALDI-TOF MS

Structural analyses of the marine scallop HS and cuttlefish chitosan were carried out using an UltrafleXtreme matrix-assisted laser desorption/ionization time-of-flight mass spectrometry or MALDI-TOF MS (Bruker Daltonics, Bremen, Germany) in reflectron (positive ion) mode operated at 26 kV (extraction lens 19.95 kV, focus lens 16 kV) accelerating voltage in the positive ion mode with a time delay of 90 ns. The spectrometer emits at 337 nm using a 50 Hz pulsed nitrogen laser. A set of

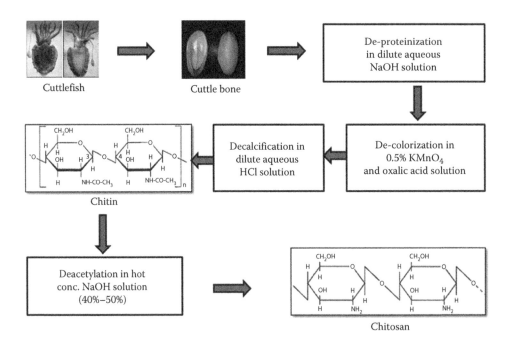

*Figure 9.2* Flowchart describing preparation of chitin and chitosan from cuttlebone.

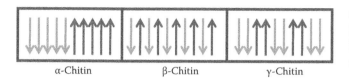

| α-Chitin | β-Chitin | γ-Chitin |

*Figure 9.3* Different forms of chitin.

turbo molecular pumps are operated to maintain the ion source and the flight tube at a pressure of about $7 \times 10^{-7}$ mbar. A 1:1 (v/v) saturated solution of matrix (α-cyano-4-hydroxycinnamic acid) in the purified HS and chitosan were used as the sample, whereas standards solutions of HS and shrimp chitosan were served for external calibration. For MALDI-MS analyses, 1 volume part of a sample solution is mixed with 1 volume part of the matrix solution, and 0.25 µL of this mixture is applied to a stainless steel target, allowed to dry, the steel target plate inserted, passed through a laser, and the TOF is counted.

### 9.5.4   One-dimensional NMR analysis

One-dimensional (1D)-NMR ($^1$H and $^{13}$C) spectroscopy is performed using conditions described by Sudo et al. (2001). Consequently, around 1–2 mg of marine scallop was freeze-dried from 0.5 mL portions of $^2$H$_2$O (99.6%, Sigma) to eliminate exchange protons with D$_2$O. The dried sample is re-dissolved in 0.5 mL of $^2$H$_2$O (99.6%) and transferred to an NMR tube. All spectra are determined on a Varian BRUKER-500 MHz spectrometer equipped with 5 mm triple resonance tunable probe with standard Varian software at 294.9 K on 700 µL marine scallop HS at 0.5–1.0 mM. The HOD signal is suppressed by presaturation for 3 s.

### 9.5.5   Two-dimensional NMR nuclear Overhauser enhancement spectroscopy analysis

Two-dimensional nuclear Overhauser enhancement spectroscopy (NOESY) is achieved using the method described by Mulloy (1996). Approximately 1–2 mg of marine scallop HS is freeze-dried from 0.5 mL portions of $^2$H$_2$O (99.6%, Sigma) to obtain exchange protons with D$_2$O. The dried marine scallop HS is redissolved in 0.5 mL of $^2$H$_2$O (99.6%) and transferred to the NMR tube. NOESY spectra are recorded using the phase-sensitive mode and run with spin-lock field of about 10 kHz at 60°C. The HOD signal is suppressed by presaturation for 3 s.

### 9.5.6   Anticoagulant activity assay

The activated partial thromboplastin time (APTT) and prothrombin time (PT) of HS of marine scallop and sulfated chitosan from cuttlebone are assayed by adopting the method of Mauray et al. (1995) using human blood as a source. The activity was expressed as the activity was expressed as IU/mg.

## 9.6   Results and discussion

Marine homo/heteropolysaccharides play vital roles in biochemical/cancer biology with variations in glycosylation products contributing toward disease and immunology, molecular and cellular communication, and molecular developmental biology. GAG-originated oligosaccharides are generally observed by MS in the negative-ion detection mode, but under certain investigational situations, positive-ion detection authorizes the acquisition of high-quality MS. GAGs, and more predominantly HS, have been involved in the regulation of a wide range of biological commitments because of their definite interactions with a large number of proteins present in the cell. These interactions appear to depend on the number and position of sulfate content in individual sugar moieties and on the way sulfated glycosidic units are rearranged to form functional domains. The fundamental unit of heparin or heparan-like GAGs is a disaccharide encompassing an uronic acid residue linked 1–4 to a N-acetyl-glucosamine. It is replicated several times to form heterogeneous combinations of chains of different lengths. Various adjustments inside the basic disaccharide unit, including two potential epimers of the uronic acid (L-iduronic and D-glucuronic), sulfation at the N, 3-O, 6-O position of the glucosamine and 2-O position of the uronic acid, create a significant amount of diversity (Bultel et al., 2010). Table 9.1 shows the list of partial sequence structures of HS disaccharides from marine scallop and their molecular weights. From the table, the base peak of marine scallop HS is 656.01 and the molecular weight is 698.28 Da.

MALDI-TOF MS and sequencing is one of the cardinal techniques of structural characterization and sequencing of carbohydrates due to its high sensitivity, precision, and speed. It produces definitive information on the structure of homogenous and heterogeneous complex oligosaccharides, which are otherwise difficult to characterize (Lang et al., 2014). The MALDI-TOF MS spectra of two isomeric standards ΔUA-NGlcS6S and ΔUA-NGlcS (Figure 9.4) show nearly identical dissociation patterns with the predominant exception being the ion at $m/z$ 293.88 and 460.91, which is unique to ΔUA-NGlcAc; 316.92 and 649.98 are unique to ΔUA-NGlcS6S. The chemical structures of natural/synthesized polymers are characterized usually by determining the composition of repeat units, that is, monomers, end-groups that cap the polymer chains and the molecular mass distribution. Polymer classification

*Table 9.1* Partial sequence structures of the disaccharides of HS from marine scallop

| | | | Marine scallop HS | | | |
|---|---|---|---|---|---|---|
| S. no. | Expected, *m/z* | Observed, *m/z* | Oligomers | R2 | R6 | R |
| 1. | 285.07 | 293.88 | NGlcAc6S | H | H | $OCH_3$ |
| 2. | 315 | 316.92 | ΔUA-NGlcAc | H | H | $CO_2H$ |
| 3. | 339.1 | 334.94 | ΔUA-NGlcAc | H | H | Ac |
| 4. | — | 364.04 | ND | — | — | — |
| 5. | 379.11 | 378.95 | ΔUA-NGlcAc | H | H | $CH_3CO$ |
| 6. | 401.09 | 400.94 | NGlcAcOS | H | H | $SO_3H$ |
| 7. | 419.06 | 422.94 | ΔUA-NGlcS | H | H | $SO_3H$ |
| 8. | 447.18 | 444.93 | ΔUA-NGlcAc6S | H | $SO_3^-$ | Ac |
| 9. | 459.07 | 460.91 | ΔUA-NGlcAc6S | H | $SO_3H$ | $CH_3CO$ |
| 10. | 479.02 | 479.03 | ΔUA-NGlcAc1S | $SO_3^-$ | H | Ac |
| 11. | — | 489.54 | ND | — | — | — |
| 12. | 503.02 | 505.91 | ΔUA(1S)NGlc | $SO_3^-$ | H | H |
| 13. | — | 523.03 | ND | — | — | — |
| 14. | 539.03 | 538.08 | ΔUA2S-NGlcAc6S | $SO_3^-$ | $SO_3^-$ | Ac |
| 15. | 544.95 | 550.06 | ΔUA2S-NGlcAc | $SO_3^-$ | Ac | H |
| 16. | | 568.07 | ND | | | |
| 17. | 585.80 | 587.08 | ΔUA-NGlcAc6S | H | $SO_3H$ | $CH_3CO$ |
| 18. | — | 649.98 | ND | — | — | — |
| 19. | — | 656.01 | ND | — | — | — |
| 20. | — | 671.99 | ND | — | — | — |
| 21. | — | 698.28 | ND | — | — | — |

with MS is not effective as MS necessitates gas phase ions for a flourishing investigation and polymers are arranged in large, ensnared chains that are not easily exchanged to gas phase ions. Traditional MS techniques developed for the structural analysis of polymers, similar to pyrolysis gas chromatography, use thermal energy to vaporize nonvolatile samples like polymers. Although thermal energy also dissociates polymers into constituent parts leading to simple fragmentation and the full chemical structure information employed is lost during vaporization.

A new analytical technique has been developed to determine this aspect of polymer sample isolated from marine resources. This new technique is known as MALDI-MS, it is possibly the most important MS technique used to analyze polymer systems of GAGs. It is a special technique of MS using specific sample preparation methods and low flounce laser desorption to create the analyte ions. Totally 21 ionized fragments are collected; among the 21 fragments 17 are sequenced (Zaia, 2005; Tissot et al., 2007) and 4 fragments are not determined (ND), which is clearly mentioned in Table 9.1.

Molecular mass and polydispersity information can be employed to verify artificial pathways, investigation poverty methods, look for preservatives and impurities, and evaluate result formulations. End-group and

chemical structure data are significant to perceptive structure–possessions relationships of polymer hexose amine and uronic acid formulations in marine scallop HS. MALDI-MS affords a direct technique of determining end-group mass and composition in addition to the mass and composition of disaccharide oligomer repeat units. MALDI-MS was used to determine the base peak at 656.01 Da and molecular weight of marine scallop HS having a mass of 698.28 Da by ionization and departure devoid of any poverty of the polymer.

Figure 9.5 represents the $^1$H-NMR of marine scallop, and the chemical shifts in protons in marine scallop are detailed in Table 9.2, which are compared with standard HS.

$^{13}$C-NMR analysis revealed the presence of signals corresponding to $O$-$CH_2$ at 16.78 ppm, ΔUA-NGlcAc6S at 57.42 ppm representing carbon-6 of hexosamine, ΔUA-NGlcAc at 69.20 ppm representing β-glucuronic acid, and ΔUA2S-NGlcS6S at 75.96 ppm representing carbon-6 of hexosamine. $^{13}$C-NMR spectrum confirmed that the $N$-acetylglcosamine is sulfated at the 6-position and not at the 4-position and also contains signals attributed to the nonsulfated uronic acid residues (Figure 9.6).

The NOESY spectrum of marine scallop HS is shown in Figure 9.7. The cross-peak showed nonreducing end residues of $N$-AcGlcA and UA. The chemical

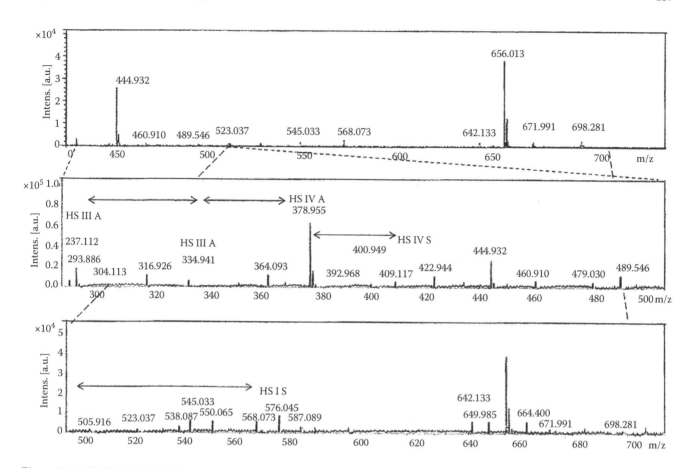

*Figure 9.4* MALDI-TOF MS of heparan sulfate from marine scallop.

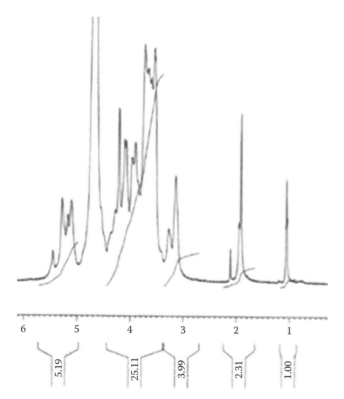

*Figure 9.5* ¹H-NMR spectrum of HS of marine scallop.

shift of the anomeric proton of GlcpNAp is linked to GlcA (4.71 ppm) and UA (4.62 ppm) in HS. Figure 9.8 represents the disaccharide sequences of HS from marine scallop. The anisotropic effect of uronic residues, such as a carboxyl group, based on the different conformation might strongly affect the magnetic environment of the anomeric protons of GlcpNAp residue. One way is to determine the relationship between the structure and the biological activity of sulfated polysaccharides isolated from marine sources, with the nature of the polysaccharide backbone and the extent and position of sulfate in the GAGs. Consequently, the HS extracted from marine scallop are characterized using 1 and 2 D-NMR spectrum. The ¹H-NMR chemical shifts (Figure 9.5) revealed the presence of signals corresponding to methyl of NGlcAp at 2.107 ppm, H-2 of GlcAp at 3.143 ppm, and H-5 and H-2 signals of GlcAp at 4.364 and 3.538 ppm (all the ring protons are merged) in marine scallop GAG from marine molluskan. This strongly suggests that GAG obtained from marine scallop is HS.

In addition, the presence of H-1 and H-2 signals of NGlcs at 5.450 and 3.274 ppm, respectively, and the lack of *L*-IdoAρ signal at 4.997 ppm shows that GAG isolated from marine scallop is HS. The same type of ¹H chemical shifts were found in turkey intestinal HS by

**Table 9.2** Chemical shifts of proton of standard HS and marine scallop

| S. no. | Chemical shift (ppm) | H1-GlcAp | H1-NGlcAp | H1-IdoAp | H2-GlcAp | H5-IdoAp | H4-IdoAp | H2-GlcAp | Methyl of NGlcAp |
|---|---|---|---|---|---|---|---|---|---|
| 1. | Standard HS | 5.61 | 5.15 | 4.15 | 3.25 | 4.25 | 4.05 | 3.22 | 2.02 |
| 2. | Marine scallop | 5.45 | 5.28 | 4.21 | 3.27 | 4.36 | 4.10 | 3.14 | 2.10 |

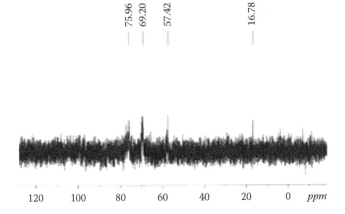

**Figure 9.6** $^{13}$C-NMR spectrum of HS of marine scallop.

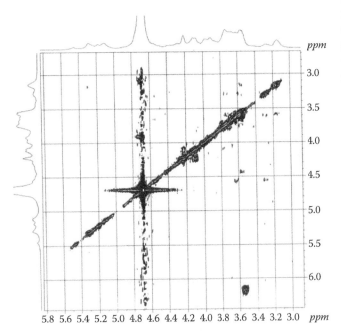

**Figure 9.7** 2D-NMR spectrum of HS of *A. pleuronectus*.

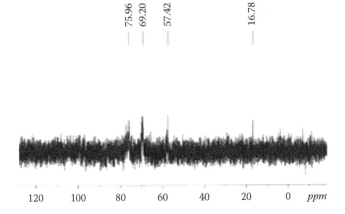

**Figure 9.8** Disaccharide sequences of marine molluskan heparan sulfate, R = H of SO$_3$.

Warda et al. (2003). From this we confirmed the isolated GAG from marine scallop is HS and not heparin.

The MALDI-TOF MS of the sulfated chitosan resulted in uniform isomeric subunits of 207 Da corresponding to the chitosan monomers NGlc and NGlcAc (Figure 9.9). Totally 26 ionized fragments of sulfated chitosan are collected; among the 26, 10 oligomers of chitosan are sequenced and 16 are ND. The sulfated chitosan oligosaccharide is observed to be a mixture of 12-mers, with individual units coupled to the homogenous component of NGlc and heterogeneous unit of NGlc-NAcGlu. The fragmentation pattern (Table 9.3) envisaged the symmetrical breakdown of the oligomers imparting a combined molecular weight of 1277 with a base peak of 649 *m/z*. Park et al. in 2010 reported a similar fragmentation framework of 15-mer chitosan oligosaccharides predominantly composed of NAcGlc and relatively less fraction of NGlc. The presence of NAcGlc at the reducing end of the chitosan oligosaccharide is deemed responsible for the hydrolysis of chitosan by chitosonase to result in low-molecular-weight chitosan.

Acid hydrolysis is thus found to lower the cleavage rate of NGlcAc-NGlcAc and NGlcAc-NGlc. The degree of deacetylation determines the extent of ionized fragmentation and breakdown of NGlc and NAcGlc oligomeric subunits. Enzymatic hydrolysis of chitosan oligosaccharides resulted in the formation of NGlcAc at the nonreducing ends (Kumar et al., 2005).

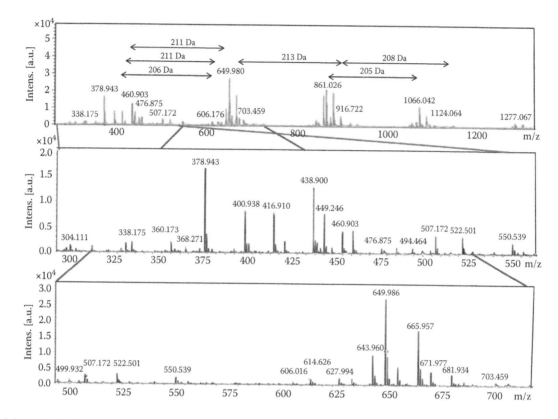

*Figure 9.9* MALDI-TOF MS of sulfated chitosan from cuttlebone of cuttlefish.

The MS of the chitooligosaccharides dissolved in 0.5% acetic acid isolated from crab shell was reported to contain highly intensive quasimolecular [M+Na]⁺ ions and less intensive [M+K]⁺. The ionized fragmentation pattern showed the subunit fractions range from 447.5 to 1478.2 *m/z* (Lopatin et al. 1995). Shells from shrimps and crustaceans are sources of chitin and chitosan from the seafood industry. Positive-ion mode fragmentation of chitooligosaccharides from crustaceans has been reported to result in higher abundance of NAcGlc when it was paired with NAcGlc than with NGlc at reducing ends (Lang et al., 2014).

The *in vitro* anticoagulant activities of APTT and PT of purified HS from marine scallop are found to be 13.7 and 24.2 s, respectively (Table 9.4).

## 9.7  Summary

This chapter draws attention to the glycobiology of the two foremost carbohydrates of marine mollusks. The study reveals details about the isolation, purification, and partial sequencing of HS PG from the marine scallop (*A. pleuronectes*) and sulfated chitosan from cuttlebone (*S. pharaonis*) and their structural intricacies using high-end characterization techniques such as MALDI-TOF MS and 1D and 2D NMR. The 1D and 2D NMR

signals showed the presence of UA and NAcGlc corresponding to the HS at a low molecular weight of 698 Da determined by the fragmentation of NGlc and NAcGlc. The molecular weight of sulfated chitosan of cuttlebone obtained from cuttlefish is found to be 1277.06 Da. Both HS and sulfated chitosan exhibited commendable anticoagulant activities in comparison with standard bovine source HS, which is evident from the clotting times of PT and APTT. Evidently, from these results, HS from marine scallops and sulfated chitosan from cuttlebone are found to be oligosaccharides of low molecular weight with intriguing anticoagulant activities. Thus the HS and the sulfated chitosan could be considered alternative anticoagulant compounds of the future.

## Acknowledgments

Author RK thanks Chettinad Academy of Research and Education (CARE) for providing the fellowship for performing the study. The authors acknowledge Dr. Rajaian Pushpabai Rajesh, Molecular Biophysics Unit, Indian Institute of Science, Bangalore, India, for MALDI-TOF experiments. The authors also acknowledge Professor Ramesh Rao, Head, Department of Pathology, CARE, for providing the facilities to conduct anticoagulant activity experiments.

*Table 9.3* Partial sequence of the cuttlebone sulfated chitosan from cuttlefish

| | Sulfated chitosan | | | |
|---|---|---|---|---|
| S. no. | Expected, *m/z* | Observed, *m/z* | Oligomer | Types |
| 1. | 485.1 | 494 | $(NGlc)_3$ | $[M+Na]^+$ |
| 2. | 509.7 | 507 | $(NGlc)_2$-NGlcAc | $[M+Na]^+$ |
| 3. | 664 | 665 | $(NGlc)_4$ | $[M+Na]^+$ |
| 4. | 685.9 | 681 | $(NGlc)_3$-NGlcAc | $[M+K]^+$ |
| 5. | 749.7 | 703 | $(NGlc)_2$-$(NGlcAc)_2$ | — |
| 6. | 846.9 | 861 | $(NGlc)_4$-$(NGlcAc)$ | $[M+K]^+$ |
| 7. | 985.9 | 916 | $(NGlc)_6$ | $[M+Na]^+$ |
| 8. | 1023.9 | 1066 | $(NGlc)_5$-$(NGlcAc)$ | $[M+K]^+$ |
| 9. | 1131.9 | 1124 | $(NGlc)_7$ | $[M+Na]^+$ |
| 10. | 1308 | 1277 | $(NGlc)_7$-$(NGlcAc)$ | $[M+Na]^+$ |
| 11. | 751.8 | ND | $(NGlc)$-$(NGlcAc)_3$ | — |
| 12. | 824.8 | ND | $(NGlc)_5$ | $[M+Na]^+$ |
| 13. | 829.8 | ND | $(NGlc)_5$ | $[M+Na]^+$ |
| 14. | 1007.9 | ND | $(NGlc)_5$-$(NGlcAc)$ | — |
| 15. | 1147 | ND | $(NGlc)_7$ | $[M+Na]^+$ |
| 16. | 1169 | ND | $(NGlc)_6$-$(NGlcAc)$ | $[M+K]^+$ |
| 17. | 1185 | ND | $(NGlc)_5$-$(NGlcAc)_2$ | $[M+Na]^+$ |
| 18. | 1211 | ND | $(NGlc)_5$-$(NGlcAc)_2$ | $[M+Na]^+$ |
| 19. | 1330.1 | ND | $(NGlc)_7$-$(NGlcAc)$ | $[M+K]^+$ |
| 20. | 1346 | ND | $(NGlc)_6$-$(NGlcAc)_2$ | $[M+Na]^+$ |
| 21. | 1350 | ND | $(NGlc)_6$-$(NGlcAc)_2$ | $[M+Na]^+$ |
| 22. | 1372.1 | ND | $(NGlc)_6$-$(NGlcAc)_2$ | $[M+Na]^+$ |
| 23. | 1512.1 | ND | $(NGlc)_8$-$(NGlcAc)$ | $[M+K]^+$ |
| 24. | 1672.2 | ND | $(NGlc)_9$-$(NGlcAc)$ | $[M+K]^+$ |
| 25. | 1835.3 | ND | $(NGlc)_{10}$-$(NGlcAc)$ | $[M+K]^+$ |
| 26. | 1875.4 | ND | $(NGlc)_9$-$(NGlcAc)_2$ | $[M+Na]^+$ |

*Table 9.4* Anticoagulant activity marine scallop HS and sulfated chitosan from cuttlefish

| S. no. | Oligosaccharide | PT (s) | APTT (s) |
|---|---|---|---|
| 1. | Marine scallop HS | $13.7 \pm 0.5$ | $24.2 \pm 0.5$ |
| 2. | Sulfated chitosan | $13.9 \pm 0.3$ | $25.9 \pm 0.3$ |
| 3. | Shrimp chitosan | $14.4 \pm 0.5$ | $26.2 \pm 0.4$ |
| 5. | Standard heparin | $13.7 \pm 0.2$ | $23.5 \pm 0.5$ |

# References

Agnihotri, S., N. Mallikarjuna, and T. Aminabhavi. 2004. Recent advances on chitosan-based micro-and nanoparticles in drug delivery. *Journal of Controlled Release* 100: 5–28.

Aimoli, C. G., M. A. Torres, and M. M. Beppu. 2006. Investigations into the early stages of "in vitro" calcification on chitosan films. *Materials Science and Engineering C* 26: 78–86.

Albano, R. M. and P. A. S. Maurao. 1986. Isolation, fractionation and preliminary characterization of a novel class of sulfated glycans from the tunic of *Styela plicata* (Chordata tunicata). *Journal of Biological Chemistry* 261: 758–765.

Arnaud, T., B. Neto, and F. Diniz. 2010. Chitosan effect on dental enamel deremineralization: An *in vitro* evaluation. *Journal of Dentistry* 38: 848–852.

Atha, D. H., B. Coxon, V. Reipa, and A. K. Gaigalas. 1995. Physicochemical characterization of low molecular weight heparin. *Journal of Pharmaceutical Sciences* 84: 360–364.

Atha, D. H., A. K. Gaigalas, and V. Reipa. 1996. Structural analysis of heparin by Raman spectroscopy. *Journal of Pharmaceutical Sciences* 85: 52–56.

Bjork, I. and U. Lindahl. 1982. Mechanism of the anticoagulant action of heparin. *Molecular and Cellular Biochemistry* 48: 161–182.

Brugnerottoa, J., J. Lizardib, F. M. Goycooleab, W. ArguÈelles-Monalc, J. Desbrie Aresa, and M. Rinaudo. 2001. An infrared investigation in relation with chitin and chitosan characterization. *Polymer* 42: 3569–3580.

Bultel, L., M. Landoni, E. Grand, A. S. Couto, and J. Kovenskya. 2010. UV-MALDI-TOF mass spectrometry analysis of heparin oligosaccharides obtained by nitrous acid controlled degradation and high performance anion exchange chromatography. *Journal of the American Chemical Society for Mass Spectrometry* 21: 178–190.

Cardin, A. D. and H. J. Weintraub. 1989. Molecular modeling of protein-glycosaminoglycan interactions. *Arteriosclerosis* 9: 21–32.

Casu, B. 1985. Structure and biological activity of heparin. *Advances in Carbohydrate Chemistry and Biochemistry* 43: 51–134.

Chatelet, C., O. Damour, and A. Domard. 2001. Influence of the degree of acetylation on some biological properties of chitosan films. *Biomaterials* 22: 261–268.

Covas, P. C., L. Alvarez, and M. W. Arguelles. 1992. The adsorption of mercuric ions by chitosan. *Journal of Applied Polymer Science* 46: 1147–1150.

Cragg, G. M. and D. J. Newman. 2013. Natural products: A continuing source of novel drug leads. *Biochimica et Biophysica Acta* 1830: 3670–3695.

Dietrich, C. P., J. F. de Paiva, C.T. Moraes, H. K. Takahashi, M. A. Porcionatto, and H. B. Nader. 1985. Isolation and characterization of a heparin with high anticoagulant activity from *Anomalocardia brasiliana*. *Biochimica et Biophysica Acta* 843: 1–7.

Esko, J. K., K. Kimata, and U. Lindahl. 2009. Proteoglycans and sulfated glycosaminoglycans. In *Essentials of Glycobiology*, Varki, A. (ed.). 2nd edn. Cold Spring Harbor, NY: Cold Spring Harbor Laboratory Press, 2009: 229–248.

Esteves, A. I. S., M. Nicolai, M. Humanes, and J. Goncalves. 2011. Sulfated polysaccharides in marine sponges: Extraction methods and anti-HIV activity. *Marine Drugs* 9: 139–153.

Hirsh, J. 1984. Heparin induced bleeding. *Nouvelle Revue Francaise D'hematologie* 26: 261–266.

Hirsh, J., N. Heddle, and J. G. Kelton. 2004. Treatment of heparin-induced thrombocytopenia. *Archives of Internal Medicine* 164(4): 361–369.

Hirsh, J. and R. Raschke. 2004. Heparin and low-molecular-weight heparin: The seventh ACCP conference on antithrombotic and thrombolytic therapy. *Chest* 126: 188S–203S.

Hovingh, P., M. Piepkorn, and A. Linker. 1986. Biological implications of the structural, antithrombin affinity and anticoagulant activity relationships among vertebrate heparins and heparan sulphates. *Biochemical Journal* 237: 573–581.

Jin, J. L., J. P. Abrahams, R. Skinner, M. Petitou, R. N. Pike, and R. W. Carrell. 1997. The anticoagulant activation of antithrombin by heparin. *Proceedings of the National Academy of Sciences of the United States of America* 94: 14683–14688.

Khor, E. and L. Lim. 2003. Implantable applications of chitin and chitosan. *Biomaterials* 24: 2339–2349.

Kim, C., Y. Choi, S. Lee, Y. Gin, and Y. Son. 2007. Development of chitosan dermal scaffold and its characterization. *Key Engineering Materials* 342: 181–184.

Kim, S. K. and E. Mendis. 2006. Bioactive compounds from marine processing byproducts—A review. *Food Research International* 39(4): 383–393.

Kiuru P., M. V. D. Auria, C. D. Muller, P. Tammela, H. Vuorela, and J. Yli-Kauhaluoma. 2014. Exploring marine resources for bioactive compounds—Review. *Planta Medica* 80: 1234–1246.

Knorr, D. 1984. Use of chitinous polymers in food: A challenge for food research and development. *Food Technology (USA)* 38: 85–89.

Kulikov, S., S. Chirkov, A. Ilina, S. Lopatin, and V. Varlamov. 2006. Effect of the molecular weight of chitosan on its antiviral activity in plants. *Applied Biochemistry and Microbiology* 42: 200–203.

Kumar, A. B., M. C. Varadaraj, L. R. Gowda, and R. N. Tharanathan. 2005. Characterization of chito-oligosaccharides prepared by chitosanolysis with the aid of papain and pronase, and their bactericidal action against *Bacillus cereus* and *Escherichia coli*. *Biochemical Journal* 391: 167–175.

Kumar, M., R. Muzzarelli, C. Muzzarelli, H. Sashiwa, and A. Domb. 2004. Chitosan chemistry and pharmaceutical perspectives. *Chemical Reviews* 104: 6017–6084.

Kurita, K. 1998. Chemistry and application of chitin and chitosan. *Polymer Degradation and Stability* 59: 117–120.

Kurita, K., S. Ishii, K. Tomita, S. Nishimura, and K. Shimoda. 1994. Reactivity characteristics of squid β-chitin as compared with those of shrimp chitin: High potentials of squid chitin as a starting material for facile chemical modifications. *Journal of Polymer Science Part A: Polymer Chemistry* 32: 1027–1032.

Lang, Y., X. Zhao, L. Liu, and G. Yu. 2014. Applications of mass spectrometry to structural analysis of marine oligosaccharides. *Marine Drugs* 12: 4005–4030.

Law of the Sea. 2011. Part V—Exclusive economic zone, Article 56, United Nations. http://www.un.org/depts/los/convention_agreements/texts/unclos/part5.htm. Retrieved August 28, 2011.

Lee, D., W. Xia, and J. Zhang. 2008. Enzymatic preparation of chitooligosaccharides by commercial lipase. *Food Chemistry* 111: 291–295.

Li, Z., L. Yubao, Z. Yi, W. Lan, and J. Jansen. 2010. *In vitro* and *in vivo* evaluation on the bioactivity of ZnO containing nano-hydroxyapatite/chitosan cement. *Journal of Biomedical Materials Research A* 93: 269–279.

Lim, C., N. Yaacob, Z. Ismail, and A. Halim. 2010. *In vitro* biocompatibility of chitosan porous skin regenerating templates (PSRTs) using primary human skin keratinocytes. *Toxicology In Vitro* 24: 721–727.

Lin, S., Y. Lin, and H. Chen. 2009. Low molecular weight chitosan prepared with the aid of cellulase, lysozyme and chitinase: Characterisation and antibacterial activity. *Food Chemistry* 116: 47–53.

Lindahl, U., G. Backstrom, L. Thunberg, and I. G. Leder. 1980. Evidence for a 3-O-sulfated D-glucosamine residue in the antithrombin-binding sequence of heparin. *Proceedings of the National Academy of Sciences of the United States of America* 77: 6551–6555.

Linhardt, R. J. and N. S. Gunay. 1999. Production and chemical processing of low molecular weight heparins. *Seminars in Thrombosis and Haemostasis* 25: 5–6.

Linhardt, R. J., K. G. Rice, Y. S. Kim, D. L. Lohse, H. M. Wang, and D. Loganathan. 1988. Mapping and quantification of the major oligosaccharide components of heparin. *Biochemical Journal* 254: 781–787.

Linhardt, R. J., J. E. Turnbull, H. M. Wang, D. Loganathan, and J. T. Gallagher. 1990. Examination of the substrate specificity of heparin and heparan sulfate lyases. *Biochemistry* 29: 2611–2617.

Linhardt, R. J., H. M. Wang, D. Loganathan, and J. H. Bae. 1992. Search for the heparin antithrombin-III binding site precursor. *Journal of Biological Chemistry* 267: 2380–2387.

Liu, C., L. Q. Lin, Y. Gao, L. Ye, Y. Xing, and T. Xi. 2007. Characterization and antitumor activity of a polysaccharide from *Strongylocentrotus nudus* eggs. *Carbohydrate Polymers* 67: 313–318.

Lopatin, S. A., M. M. Ilyin, V. N. Pusatobaev, Z. A. Bezchetnikova, V. P. Varlamov, and V. A. Davankov. 1995. Mass spectrometric analysis of N-acetyl chitooligosaccharides prepared through enzymatic hydrolysis of chitosan. *Analytical Biochemistry* 227: 285–288.

Mark, H., N. Bikales, C. Overberger, and G. Menges. 1986. *Encyclopedia of Polymer Science and Engineering.* New York: Wiley-Interscience.

Mauray, S., C. Sternberg, J. Theveniaux, J. Millet, C. Sinquin, J. T. Bretaudiere, and A. M. Fischer. 1995. Venous antithrombotic and anticoagulant activities of a fucoidan fraction. *Thrombosis Haemostasis* 74(5): 1280–1285.

Mckay, G., H. Blair, and J. Gardner. 1982. Adsorption of dyes on chitin. I. Equilibrium studies. *Journal of Applied Polymer Science* 27: 3043–3057.

Medeiros, G. F., A. Mendes, R. A. Castro, E. C. Bau, H. B. Nader, and C. P. Dietrich. 2000. Distribution of sulfated glycosaminoglycans in the animal kingdom: Widespread occurrence of heparin-like compounds in invertebrates. *Biochimica et Biophysica Acta* 1475: 287–294.

Minke, R. and J. Blackwell. 1978. The structure of α-chitin. *Journal of Molecular Biology* 120: 167–181.

Morganti, P. and G. Morganti. 2008. Chitin nanofibrils for advanced cosmeceuticals. *Clinical Dermatology* 26: 334–340.

Mulloy, B. 1996. High-field NMR as a technique for the determination of polysaccharide structures. *Molecular biotechnology*, 6(3): 241–265.

Muzzarelli, R. A. A., J. Boudrant, D. Meyer, N. Manno, M. DeMarchis, and M. G. Paoletti. 2012. Current views on fungal chitin/chitosan, human chitinases, food preservation, glucans, pectins and inulin: A tribute to Henri Braconnot, precursor of the carbohydrate polymers science, on the chitin bicentennial—Review. *Carbohydrate Polymers* 87: 995–1012.

Nader, H. B., S. F. Chavante, E. A. dos-Santos, F. W. Oliveira, J. F. de-Paiva, S. M. B. Jeronimo et al. 1999. Heparan sulfates and heparins: Similar compounds performing the same functions in vertebrates and invertebrates? *Brazilian Journal of Medical and Biological Research* 32: 529–538.

Ngo, D., S. Lee, M. Kim, and S. Kim. 2009. Production of chitin oligosaccharides with different molecular weights and their antioxidant effect in RAW 264.7 cells. *Journal of Functional Foods* 1: 188–198.

Paiva, J. F., E. A. Santos, W. Jeske, J. Fareed, H. B. Nader, and C. P. Dietrich. 1995. A comparative study on the mechanism of the anticoagulant action of mollusc and mammalian heparins. *Comparative Biochemistry and Physiology* 111: 495–499.

Park, B. K. and M. M. Kim. 2010. Applications of chitin and its derivatives in biological medicine. *International Journal of Molecular Sciences* 11: 5152–5164.

Park, J. K., H. S. Jo, C. G. Lee, and Y. I. Park. 2010. Significance of the molecular weight and degree of N-deacetylation of chitosan-oligosaccharides on the biological applications. *Journal of Chitin and Chitosan* 15: 195–202.

Pejler, G., A. Danielsson, I. Bjork, U. Lindahl, C. P. Dietrich, and H. B. Nader. 1987. Structure and antithrombin—Binding properties of heparin isolated from the clams *Anomalocardia brasiliana* and *Tivela mactroides*. *Journal of Biological Chemistry* 262: 1413–1421.

Rademacher, T. W., R. B. Parekh, and R. A. Dwek. 1988. Glycobiology. *Annual Review of Biochemistry* 57: 785–838.

Rajesh Kumar, S., V. I. Ahmed, V. Parameswaran, R. Sudhakaran, V. S. Babu, and A. S. Hameed. 2008. Potential use of chitosan nanoparticles for oral delivery of DNA vaccine in Asian sea bass (*Lates calcarifer*) to protect from *Vibrio* (Listonella) *anguillarum. Fish Shellfish Immunology* 25: 47–56.

Razdan, A. and D. Pettersson. 1994. Effect of chitin and chitosan on nutrient digestibility and plasma lipid concentrations in broiler chickens. *British Journal of Nutrition* 72: 277–288.

Ribeiro, M., A. Espiga, D. Silva, P. Baptista, J. Henriques, C. Ferreira et al. 2009. Development of a new chitosan hydrogel for wound dressing. *Wound Repair and Regeneration* 17: 817–824.

Roberts, G. A. F. 1992. *Chitin Chemistry.* London, U.K.: Macmillan.

Sampaio, L. O., I. L. S. Tersariol, C. C. Lopes, R. I. Boucas, F. D. Nascimento, H. A. O. Rocha, and H. B. Nader. 2006. Heparins and heparan sulfates. Structure, distribution and protein interactions. In *Insights into Carbohydrate Structure and Biological Function*, Verli, H. (ed.). Transworld Research Network, Kerala, pp. 1–85.

Saravanan, R. 2014. Isolation of low molecular weight heparin/heparan sulfate from marine sources. In *Food and Nutrition Research*, Kim, S.-K. (ed.). Academic Press, Elsevier Publications, 72: 45–60.

Saravanan, R. and A. Shanmugam. 2011. Is isolation and characterization of heparan sulfate from marine scallop *Amussium pleuronectus* (Linne) an alternative source of heparin? *Carbohydrate Polymers* 86(2): 1082–1084.

Saravanan, R., S. Vairamani, and A. Shanmugam. 2010. Glycosaminoglycans from marine clam *Meretrix meretrix* (Linne.) are an anticoagulant. *Preparative Biochemistry and Biotechnology* 40: 305–315.

Subhapradha, N., P. Ramasamy, S. Vairamani, P. Madeswaran, A. Srinivasan, and A. Shanmugam. 2013. Physicochemical characterization of β-chitosan from *Sepioteuthis lessoniana* gladius. *Food Chemistry* 141: 907–913.

Sudo, M., K. Sato, A. Chaidedgumjorn., H. Toyoda., T. Toida., and T. Imanari. 2001. 1-H nuclear magnetic resonance spectroscopic analysis for determination of glucuronic and iduronic acids in dermatan sulfate, heparin, and heparan sulfate. *Analytical biochemistry*, 297(1): 42–51.

Sushko, N. I., S. P. Firsov, R. G. Zhbankov, V. M. Tsarenkov, M. Marchewka, and C. Ratajczak. 1994. Vibrational spectra of heparins. *Journal of Applied Spectroscopy* 61: 704–707.

Tissot, B., N. Gasiunas, A. K. Powell, Y. Ahmed, Z. Zhi, S. M. Haslam et al. 2007. Towards GAG glycomics: Analysis of highly sulfated heparins by MALDI-TOF mass spectrometry. *Glycobiology* 17(9): 972–982.

The National Academy of Sciences (NAS). 2012. *Transforming Glycoscience: A Roadmap for the Future.* Washington, DC: National Academies Press.

Tolaimate, A., J. Desbrieres, M. Rhazi, A. Alagui, M. Vincendon, and P. Vottero. 2000. On the influence of deacetylation process on the physiochemical characteristics of chitosan from squid chitin. *Polymer* 41: 2463–2469.

Vikhoreva, G., G. Bannikova, P. Stolbushkina, A. Panov, N. Drozd, V. Makarov et al. 2005. Preparation and anticoagulant activity of a low-molecular-weight sulfated chitosan. *Carbohydrate Polymers* 62(4): 327–332.

Volpi, N. 1993. "Fast moving" and "slow moving" heparins, dermatan sulfate, and chondroitin sulfate: Qualitative and quantitative analysis by agarose-gel electrophoresis. *Carbohydrate Research* 247: 263–278.

Volpi, N. 1994. Fractionation of heparin, dermatan sulfate and chondroitin sulfate by sequential precipitation: A method to purify a single glycosaminoglycans species from a mixture. *Analytical Biochemistry* 218: 382–391.

Volpi, N. 2005. Occurrence and structural characterization of heparin from molluscs—Review. *Invertebrate Survival Journal* 2: 6–16.

Wang, S. and J. Chen. 2005. The protective effect of chitin and chitosan against *Vibrio alginolyticus* in white shrimp *Litopenaeus vannamei*. *Fish Shellfish Immunology* 19: 191–204.

Warda, M., W. Mao, T. Toida, and R.J. Linhardt. 2003. Turkey intestine as a commercial source of heparin? Comparative structural studies of intestinal avian and mammalian glycosaminoglycans. *Comparative Biochemistry and Physiology B* 134: 189–197.

Xia, W., P. Liu, J. Zhang, and J. Chen. 2011. Biological activities of chitosan and chitooligosaccharides. *Food Hydrocolloids* 25: 170–179.

Xing, R., H. Yu, S. Liu, W. Zhang, Q. Zhang, Z. Li et al. 2005. Antioxidant activity of differently region selective chitosan sulfates *in vitro*. *Bioorganic and Medicinal Chemistry* 13:1387–1392.

Yen, M.T., and J.L. Mau. 2004. Physico-chemical properties of chitin from shiitake stipes and crab shells. *Annual of Tainan Woman's College of Arts and Technology* 23: 229–240.

Yen, M. T. and J. L. Mau. 2007. Selected physical properties of chitin prepared from shiitake stripes. *Food Science and Technology* 40: 558–563.

Zaia, J. 2005. Principles of mass spectrometry of glycosaminoglycans. *Journal of Biomacromolecular Mass Spectrometry* 1(1): 3–36.

*chapter ten*

# Biomedical potential of natural glycoproteins with special reference to marine collagen

**Anguchamy Veeruraj, Muthuvel Arumugam, Thipramalai Thankappan Ajithkumar, and Thangavel Balasubramanian**

## Contents

## 10.1 Introduction

Glycoproteins are proteins having covalently bound carbohydrates, and they are found in all living animals in both soluble and insoluble forms with diverse functions and properties (Sharon and Lis, 1981; Lennarz, 1983). Certainly, there are more proteins that contain covalently bound carbohydrates in their molecule than those that are devoid of them. The carbohydrate content of glycoproteins ranges from less than 1% to over 80% of the molecule. These carbohydrate units are involved in various biological activities. Glycoproteins can be broadly classified into three types depending on the nature and function of their carbohydrate units: typical glycoproteins (numerous glycoproteins), glycosaminoglycans (GAGs, a group of compounds often classified as glycoproteins), and collagen (a unique type of glycosylated proteins). Collagen represents one of the major groups of proteins found in the animal kingdom, occurring in multicellular organisms ranging from sponges to mammals. The presence of carbohydrate in collagens from extensive sources suggests that it could be playing an essential biological role.

Biologically active analogs of the extracellular matrix (ECM) are synthesized by grafting GAG chains onto type I collagen and by controlling the physicochemical properties of the resulting graft copolymer.

When multicellular organisms evolved, a material was needed between the cells to give the tissue shape, resistance to pressure, torsion, and tension and to provide the structural framework for some specialized tissues such as bone and cartilage. The superfamily of collagen can generally be divided into two divergent subfamilies, one of which includes the vertebrate interstitial collagen genes and the invertebrate collagen genes, for example, the vertebrate type IV and type IX collagen genes (Fields, 1988). The evolution of genes of fibrillar collagens has been assumed to develop from six Gly-X-Y triplets coding an early gene, since the collagen genes often have 54 base-pair-long exons (Kivirikko and Myllylä, 1983).

The most important types of collagen are found in connective tissue and comprise types I, II, III, V, and XI. However, type I is the main collagen of skin and bone, and therefore is the most abundant form, accounting for 90% of the body's total collagen (Alberts et al., 2002; Hing, 2004). Collagen molecules consist of carefully arranged arrays of tropocollagen molecules, which are long, rigid molecules, composed of three left-handed helices of peptides and associated laterally to form collagen fibrils with a characteristic banded structure (Hing, 2004). Finally, fibrils associate to form collagen fibers.

Collagen is extracted from various sources, considering it is one of the most abundant proteins on earth. Nonetheless, common sources of collagen for tissue engineering applications include bovine skin and tendons, porcine skin, and rat tail, among others. Conversely, biopolymer-based films are a possible answer to the demanded environmentally friendly, biodegradable packaging materials, and they are often derived from abundant renewable sources (Ötles and Ötles, 2004). In this case, collagen is produced from seafood waste, which is available in large quantities and at feasible cost. On the one hand, collagen is used as a raw material for the production of gelatin, cosmetics, and foods, and as an alternative for edible and/or biodegradable film manufacture (Kittiphattanabawon et al., 2005). On the other hand, its physicochemical properties and potential use in biodegradable films for food packaging are still under exploration (Wolf et al., 2006). Nowadays, the primary sources of industrial collagen are calf skin and bones. However, these sources have a high risk of bovine spongiform encephalopathy (BSE) or transmissible SE (TSE). In addition, collagen extracted from bovine sources are prohibited for the Sikhs and Hindus, while porcine collagen cannot be consumed by Muslims and Jews, both of whom require bovine meat to be religiously prepared. Therefore, currently new alternatives are being sought for collagen sources, mainly from seafood by-products, such as the skin, bones, and fins of fish, as well as marine organisms such as sponges, jellyfish, sea anemones, and squids. From this point of view, collagen from marine sources can reap economic and environmental benefits. Several forms of collagen are currently being investigated as scaffolds for cardiovascular, musculoskeletal, and nervous tissue engineering (Duan et al., 2007). As a result, alternative sources of collagen, especially from aquatic animals including mollusks, have received much attention (Shen et al., 2007). Marine life forms are also a considerable source of collagen, being extracted from sponges (Exposito et al., 1991, 2002), fish (Sugiura et al., 2009; Veeruraj et al., 2012; Tamilmozhi et al., 2013), squid species like Illex and squids (Sikorski and Kolodziejska, 1986; Veeruraj et al., 2015), and jellyfish (Song et al., 2006). These collagens are widely used in the industry but less for research and clinical use. Collagen has been attracting attention as a biomaterial due to its unique characteristics such as high tensile strength, lower antigenicity, bioresorbability, induction of coagulation of blood platelets, effect on cell differentiation and wound healing, and the possibility of the control of various characteristics through physical and chemical modifications (Stenzel et al., 1974). Numerous innovations have taken place in the field of collagen-based biomaterials, ranging from injectable collagen solutions to bone regeneration scaffolds. Multiple cross-linking patterns within collagen

as well as with other biopolymers were explored in order to improve the generation of tissue and its function. Hence, the purpose of this review is to summarize available information on the biomedical and pharmaceutical potential of marine-sourced natural polymers of collagen-based biomaterials, their processing, physicochemical properties, and structural characteristics, the dosage forms of natural polymers, and the application of collagen products in medicine, and to be discuss recent developments with special focus on its use for wound healing, drug delivery, and tissue engineering.

## 10.2 Structural and functional properties of collagen

Collagens constitute nearly 30% of all proteins in living animals. Among the 29 collagen types, type IV collagen is a major and important component of basement membranes (Khoshnoodi et al., 2008; Bächinger et al., 2010). Different collagen types assemble into different structures. It is a right-handed triple super helical rod structure, and has the unique ability to form insoluble fibers that have high tensile strength (Gelse et al., 2003). Collagen plays an important role in the structural maintenance of tissues and is widely used in the food, pharmaceutical, biomedical, and cosmetic industries due to its excellent biocompatibility and biodegradability, and weak antigenicity compared with other natural polymers. It is widely distributed in ocular tissues such as the sclera, stroma, and cornea. Collagen type I is commonly found in connective tissues, including tendons, bones, and skins (Muyonga et al., 2004) and forms triple-helical conformations, whereas polypeptide chains, the so-called α chains, coil into a left-handed helix with about 18 amino acids per turn. Type I collagen is the most abundant protein present in mammals and is the most thoroughly studied protein because if its diverse functions due to its physical and chemical properties. Collagen type I usually forms fibrils with a length of 300 nm and a fibrillar diameter of up to 1000 nm. The three polypeptide subunits of type I collagen have similar amino acid compositions, and each polypeptide is composed of about 1050 amino acids, containing approximately 33% glycine, 25% proline, and 25% hydroxyproline, with a relative abundance of lysine. Native collagen is water insoluble, and for many pharmaceutical applications collagen is modified to improve its water solubility. The molecules have high hydroxyproline and hydroxylysine content and glycine as every third amino acid, which is a prerequisite for triple-helical folding. Type I collagen is trimeric $[(\alpha1)_2\beta2]$ and exists as a triple helix. However, glycine occupies the restricted space where the three helical α chains come together in the center of the triple-helical structure (Prockop et al., 1979). Therefore, each of the three chains has the repeating structure Gly-X-Y, in which X and Y can be any amino acid, but they are frequently the imino acids proline (about 100 of the X positions) and hydroxyproline (about 100 of the Y positions). Because both proline and hydroxyproline are rigid, cyclic amino acids, they limit the rotation of the polypeptide backbone and thus contribute to the stability of the triple helix (Prockop et al., 1979). Hydroxyproline plays an essential role in stabilizing the triple helix structure of collagen by hydrogen bonding between the hydroxyl group and water. Collagen polypeptides that lack hydroxyproline can fold into a triple-helical conformation at low temperatures, but the triple helix formed is not stable at mammalian body temperature (Prockop et al., 1979). The amount of Gly-Pro-Hyp sequences is the main, but not exclusive, factor for varying collagen thermostability ranging from Antarctic fish to animals living at very high temperatures near thermal vents at the bottom of oceans (Burjanadze, 2000).

The fundamental feature of collagen is an elegant structural motif in which three parallel polypeptide strands in a left-handed, polyproline type II (PPII) helical conformation coil about each other with one residue stagger to form a right-handed triple helix. The tight packing of PPII helices within the triple helix mandates that every third residue be Gly, resulting in a repeating XaaYaaGly sequence, where Xaa and Yaa can be any amino acid. This repetition occurs in all types of collagen, although it is disrupted at certain locations within the triple-helical domain of nonfibrillar collagens (Brazel et al., 1987). The amino acids in the Xaa and Yaa positions of collagen are often (2S)-proline (Pro, 28%) and (2S,4R)-4-hydroxyproline (Hyp, 38%), respectively. The stability of the collagen molecules is ensured by interstrand hydrogen-bonding forces. Their structure of collagen is a right-handed triple helix of three staggered, left-handed PPII helices, with all peptide bonds in the trans conformation and two hydrogen bonds within each triplet. In 1955, this structure was distinguished by Rich and Crick (1955, 1961) and Cowan et al. (1955) to be the triple-helical structure accepted today, which has a single interstrand N-H$_{(Gly)}$···O=C$_{(Xaa)}$ hydrogen bond per triplet and a 10-fold helical symmetry with a 28.6 Å axial repeat (10/3 helical pitch) (Shoulders and Raines, 2009). Fibril-forming collagens are the most commonly used in the production of collagen-based biomaterials, and type I collagen is presently the gold standard in the field of tissue engineering and nano-biomaterial matrices.

## 10.3   Collagen from marine source

The diversity in the marine environment signifies the development and application of novel chemico-biological molecules used in various biomedical and pharmaceutical applications. The importance of these molecules is increasing day by day because of their increased need in solving various health problems. Apart from protein molecules like collagen, chemical molecules of marine origin are gaining importance because of their potential in the development of various cosmeceutical, pharmaceutical, and biomedical materials. Hence, there is a great potential in marine bioprocess industry to convert and utilize most of these by-products as valuable products. With the rapid development of seafood processing industries, huge quantities of by-products are discarded, which may cause pollution and emit offensive odors. Optimal use of these by-products is a promising way to protect the environment, to produce value-added products to increase revenue to the fish processors, and to create new job/business opportunities. Currently, enormous amounts of protein-rich by-products, accounting for 60%–75% of the raw material, are produced during aquatic product processing in the world. Fish skin is an important source of collagen, which can be used as a replacement for mammalian sources. Therefore, many studies have been conducted to extract and screen for potential industrial applications of collagens from marine food by-products, such as skins of squid (*Doryteuthis singhalensis*) (Veeruraj et al., 2015), eel (*Evenchelys macrura*) (Veeruraj et al., 2013), sail fish (*Istiophorus platypterus*) (Tamilmozhi et al., 2013), Amur sturgeon (*Acipenser schrenckii*) (Wang et al., 2014), hammerhead shark (*Sphyrna lewini*) (Chi et al., 2014), balloon fish (*Diodon holocanthus*) (Huang et al., 2011), and bighead carp (*Hypophthalmichthys nobilis*) (Liu et al., 2012).

Marine foods are the richest source of collagens in terms of production and application in various bioprocesses and biomedical engineering. Several species of fish have been well investigated for their tissue collagen and gelatin for various applications in functional foods, nutraceuticals, and as important food ingredients of medical and biomedical importance (Hayashi et al., 2012). Other marine mammals (e.g., baleen whales, sea otters, and walruses) have significant mean bone collagen, which is well explored for various biophysicochemical analyses (Schoeninger and DeNiro, 1984). Moreover, type I collagen from aquatic animals may provide an alternative collagen source, with shark skin collagen in particular being potentially important (Kimura et al., 1981). The major content of imino acids and the type of cross-linkages vary from mammalian collagen, which reflects the low denaturation temperature of shark skin collagen (Nomura et al., 2000). However, the low denaturation temperature can be improved by exogenous cross-linking for biomedical applications. Easy extractability and better yield of collagen are advantageous for its applications. Hence, shark collagen is a promising source of type I collagen for various biomedical applications.

## 10.4   *Physical, biological, and their mechanisms of collagen with copolymer composite*

ECM is a complex organization of structural proteins such as collagens and a wide variety of proteoglycans that are found within tissues and organs in all animals. In humans, collagen consists of one-third of the total protein, accounts for three-quarters of the dry weight of skin, and is the most prevalent component of the ECM. With the understanding that *in situ* cells exist within a complex three-dimensional structure, a wide variety of tissue engineering scaffolds have been created for a multitude of applications.

Collagen is the most widely used tissue-derived natural polymer, and it is a main component of the ECMs of mammalian tissues, including the skin, bone, cartilage, tendon, and ligament. Physically formed collagen gels are thermally reversible and offer a limited range of mechanical properties, and chemical cross-linking of collagen using glutaraldehyde or diphenylphosphoryl azide can improve the physical properties (Muyonga et al., 2004). Various sources of collagens are used in different applications depending on their thermal stability, which may have a direct correlation with the living environment and body temperature of those species (Rigby, 1968; Veeruraj et al., 2013). The lower thermal stability of fish collagen than that of mammalian collagen limits its wide application (Nagai and Suzuki, 2000). Moreover, the thermal stability of collagen has a direct correlation with the relative abundance of imino acids (hydroxyproline and proline), and higher contents of these imino acids increase the thermal stability of collagen due to the higher cross-link density (Wong, 1989). Hydroxyproline may stabilize the triple helix by hydrogen-bonded water bridges, as originally proposed by Ramachandran et al. (1973). The difference in imino acid content among the animals is associated with the different living environments of their sources, particularly the habitat temperature (Regenstein and Zhou, 2007).

The stability of collagen is found to be proportional to the total content of pyrrolidine imino acids. A recent study by Veeruraj et al. (2013) showed that the Pro + Hyp rich zones of the molecules are most likely involved in the formation of junction zones stabilized by hydrogen bonding in marine fish collagen. However, these results show that the hydroxylysine range of values is between 4 and 6 per 1000 residues, as found in the acid-soluble collagen (ASC) and pepsin-soluble collagen (PSC) from the marine eel. Therefore, this isolated type I collagen may find wide application due to its denaturation temperature which is close to that of mammalian collagen. According to Asghar and Henrickson (1982), hydroxylysine contributes to the formation and stabilization of cross-links in the collagen. There are also interesting consequences of the variation in $T_d$ with variation in temperature of their living environment.

Type I collagen hydrogels have been widely used as scaffolds in tissue engineering for various tissues and organs (Yannas, 1972, 1992; Badylak et al., 2002). Conventional gels of type I collagen are prepared by neutralization of an acidic pH solution of collagen molecules, which self-assemble via noncovalent bonds to form an entangled network of thick collagen fibrils. The mechanical properties of a collagen-based gel depend on the thickness, length, and three-dimensional density of the collagen fibrils. The presence or absence of covalent cross-linking can also affect the mechanical properties (Badylak et al., 2002). As some collagen gels may not have the desired mechanical strength, they may be cross-linked by both chemical and physical methods to increase their strength or to decrease their absorptivity (degradability *in vivo*) (Badylak et al., 2002).

Heat stability and cross-linking of collagen molecules are important features for their use as biomaterials (Everaerts et al., 2008). Song et al. (2006) investigated the cross-linking agent 1-ethyl-3-(3-dimethylaminopropyl) carbodiimide hydrochloride (EDC), and they were able to decrease the enzymatic degradation of jellyfish collagen scaffolds *in vitro*. Thus, this cross-linking method seems to be suitable to modulate the biodegradability of jellyfish collagen. Although the melting temperature of jellyfish collagen is below human body temperature, it could be used in combination with other polymers such as chitosan to make resorbable biomaterials, as reported by Wang et al. (2010), who developed an injectable chitosan/marine collagen composite gel. The self-assembled protein complexes of collagen fibril have high tensile strength, and collagen gels consisting of an entangled network of thick collagen fibrils have some mechanical strength and elasticity. However, they often are not mechanically strong enough nor are they as elastic as rubber. Recently, Yunoki et al. (2007) prepared an elastic hydrogel of salmon collagen by a combination of cross-linking with EDC and partial heat denaturation.

Collagen-based scaffolds used in cell culture containing 30% chitosan enhanced fibroblast proliferation compared to pure collagen scaffolds, and the scaffold of collagen and chitosan has promising properties of mechanical strength, biodegradation rate, and cell proliferation stimulating ability, which are crucial for tissue engineering applications. Glutaraldehyde (GA)-treated collagen/chitosan scaffold is a potential candidate as a dermal equivalent with improved biostability and good biocompatibility (Maa et al., 2003; Tangsadthakun et al., 2006).

Collagen–chitosan scaffolds supplemented with different concentrations (0.1%–0.5%) of aloe vera (AV) were prepared and tested *in vitro* for their possible application in tissue engineering. The results suggested that the AV gel-blended collagen–chitosan scaffolds could be a promising candidate for tissue engineering applications (Jithendra et al., 2013). Therefore, in the future, technical developments in collagen processing as well as combination of collagen with other biopolymeric materials should allow specific tailoring of the release rate to the desired kinetics for local delivery.

## 10.5 Collagen-based biomaterials for biomedical application

Collagen-based biomaterials can be classified into two categories based on the extent of their purification from natural sources: decellularized collagen matrices that maintain the original tissue properties and ECM structure, and functional scaffolds prepared via extraction, purification, and collagen polymerization. Recently, many techniques have been developed for the production of an acellular collagen matrix or ECM. The first technique has elegantly reviewed the three methods used for tissue decellularization: physical, chemical, and enzymatic (Gilbert et al., 2006). Physical methods include snap-freezing that disrupts the cells by forming ice crystals, high pressure that bursts cells, and agitation that induces cell lysis, and they are used most often in combination with chemical methods to make possible penetration of active molecules into the tissue. Chemical methods of decellularization include a variety of reagents that can be used to remove the cellular content of the ECM. In order to produce collagen-based biomaterials, different approaches have been developed to extract collagen from biological tissues. As Table 10.1 shows, based on their application, the products are gels, sponges, tubes, spheres, and membranes.

Generally, a wide range of materials (both synthetic and natural) are available for developing controlled release systems. Therefore, natural polymers have gained special interest over synthetic polymers mainly because of the former's better biocompatibility, biocompatible degradation products (in contrast to some synthetic monomers), cost effectiveness, and availability in bulk. The role of natural polymers in biomedical application is pivotal, especially as hydrogels, carrier systems, wound dressings, oral dosage forms, and implants. Biopolymers from marine sources have been studied and utilized in pharmaceutical and biotechnological product development for a number of years, particularly collagen and chitosan as biodegradable biopolymers in drug-delivery systems.

*Table 10.1* Biomedical applications of collagen biomaterials

| S. no. | Composition | Biomaterial form | Applications |
| --- | --- | --- | --- |
| 1. | Collagen | Gel | Cosmetic skin defects |
| | | | Drug delivery |
| | | | Vitreous replacement surgery |
| | | | Coating of bioprostheses |
| | | Films/membrane | Wound dressing |
| | | | Drug delivery |
| | | | Dialysis |
| | | | Tissue regeneration |
| | | | Corneal shields |
| | | | Skin patches |
| | | Sponge | 3D cell culture |
| | | | Wound dressing |
| | | | Hemostatic agent |
| | | | Skin replacement |
| | | | Drug delivery |
| | | Hollow fiber tubing | Cell culture |
| | | | Nerve regeneration |
| | | Sphere | Micro-carrier for cell culture |
| | | | Drug delivery |
| | | Rigid form | Bone repair |
| 2. | Collagen + GAG | Membrane | Tissue regeneration |
| | | | Skin patches |
| 3. | Collagen + hydroxyapatite | Power sponge | Bond-filling and repair drug delivery (BMP) |

### 10.5.1 Collagen for tissue engineering applications

Tissue engineering is a multidisciplinary field which involves the "application of the principles and methods of engineering and life sciences toward the fundamental understanding of structure–function relationships in normal and pathological mammalian tissues and the development of biological substitutes that restore, maintain or improve tissue function" (Shalak and Fox, 1988). The goal of tissue engineering is to overcome the limitations of conventional treatments based on organ transplantation and biomaterial implantation (Langer and Vacanti, 1993; Sachlos and Czernuszka, 2003). In this technique, tissue loss or organ failure may be treated either by implantation of an engineered biological substitute or, alternatively, with *ex vivo* perfusion systems. Tissue engineering has attracted a great deal of attention in science, engineering, medicine, and society in general. As a

**Figure 10.1** Chemical structure of (a) poly(glycolic acid) and (b) poly(lactic acid), and typical structures of (c) porous scaffolds of poly(lactide-*co*-glycolide) and (d) nonwoven fabrics of poly(glycolic acid). The latter materials have been widely used for tissue engineering applications. (Reprinted from *Trends Biotechnol.*, 16(5), Kim, B.-S. And Mooney, D.J., Development of biocompatible synthetic extracellular matrices for tissue engineering, 7. Copyright 1998, with permission from Elsevier.)

result, highly porous scaffolds are needed that play a critical role in cell seeding, proliferation, and new tissue formation in three dimensions (Liu and Ma, 2004). Scaffolds are three-dimensional (3D) substrates for cells, and their main goal is to act as a template for tissue regeneration (Bolgen et al., 2008). Meanwhile, the scaffold should also have adequate mechanical properties to enable the new tissue to perform normal mechanical functions.

One potential substitute is type 1 collagen/glycosaminoglycan (CG)-based scaffolds, which have been successfully used in clinical settings such as viable treatments for conjunctiva and epithelial regeneration (Compton et al., 1998; Hsu et al., 2000). Much of this success is due to a number of useful properties, including low antigenicity, biodegradability, high porosity, and a high ligand density (O'Brien et al., 2005). CG scaffolds are typically composed of type 1 collagen, an abundant connective protein in the bone, and the GAG chondroitin 6-sulfate, a proteoglycan commonly found in the bone matrix. Both components have been shown to be important for cell attachment, proliferation, and differentiation (Pieper et al., 1999).

The polymer potentially mimics many roles of ECMs found in tissues and comprises various amino acids and sugar-based macromolecules. It brings the cells together and controls the tissue structure, regulates the function of the cells, and allows the diffusion of nutrients, metabolites, and growth factors (Alberts et al., 1994). Various types of polymers have been studied and utilized to date in tissue engineering, and aliphatic polyesters including poly(glycolic acid) (PGA), poly(lactic acid) (PLA), and copolymers

(PLGA) are the most widely used synthetic polymers (Figure 10.1; Kim and Mooney 1998; Thomson et al., 1999). Therefore, these types of polymers have a long history of use in medical applications and are considered as safe in many situations by the Food and Drug Administration (FDA). However, the use of these types of polymer scaffolds requires the surgeon to make sufficiently large incisions (cuts) to enable the placement of the polymer/cell constructs. An exciting alternative approach to cell delivery for tissue engineering is the use of polymers (i.e., hydrogels) that can be injected into the body. This approach enables the clinician to transplant the cell and polymer combination in a minimally invasive manner. There are structural similarities between hydrogels and the macromolecular-based components in the body, and they are considered biocompatible (Jhon and Andrade, 1973). Hydrogels have found numerous applications in tissue engineering as well as drug delivery. Tissue engineering is the most recent application of hydrogels, in which they are used as scaffolds to engineer new tissues (Figure 10.2; Langer and Vacanti, 1993; Jen et al. 1996; Vacanti, 2003).

## 10.5.2 Collagen for bone tissue engineering

Investigations into synthetic and natural inorganic ceramic materials (e.g., hydroxyapatite and tricalcium phosphate) as candidate scaffold materials have been aimed mostly at bone tissue engineering (Burg et al., 2000). Synthetic and natural polymers are an attractive alternative and versatile in their applications to the

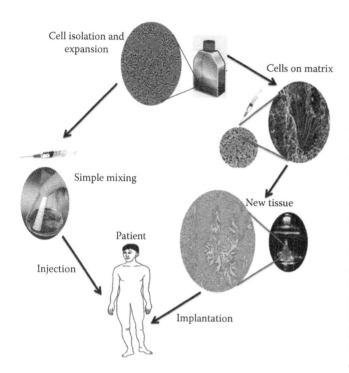

*Figure 10.2* Schematic illustration of typical tissue engineering approaches. Cells are obtained from a small biopsy from a patient, expanded *in vitro*, and transplanted into the patient either by injection using a needle or through any other minimally invasive delivery approach, or by implantation at the site following an incision (cut) by the surgeon to allow placement.

growth of most tissues. Bone tissue repair accounts for approximately 500,000 surgical procedures per year in the United States alone (Geiger et al., 2003). The major solid components of human bone are collagen (a natural polymer, also found in skin and tendons) and a substituted hydroxyapatite (HA) (a natural ceramic, also found in teeth). Although these two components even when used separately provide a relatively successful means of augmenting bone growth, the composite of the two natural materials exceeds this success. The ECM of bone tissue is a composite composed of type I collagen fibers reinforced with calcium-deficient hydroxyapatite (HA) crystals which are elongated and plate-like (Rho et al., 1998). Collagen–HA scaffolds are typically prepared by freeze-drying to achieve the high levels of interconnected porosity (~80%–99%) required for cell infiltration, nutrient transport, vascularization, and tissue regeneration (Wahl et al., 2007; Cunniffe et al., 2010; Harley et al., 2010; Jones et al., 2010; Lyons et al., 2010). Moreover, the porosity, pore size, and pore morphology of freeze-dried collagen scaffolds can be tailored by controlling the freezing rate, freezing temperature, and collagen concentration (O'Brien et al., 2005; Harley et al., 2010; Yunoki et al., 2010).

Bone engineering typically uses an artificial ECM (or scaffold), osteoblasts or cells that can become osteoblasts (i.e., bone marrow mesenchymal stem cells), and regulating factors that promote cell adhesion, differentiation, proliferation, and bone formation. Figure 10.3

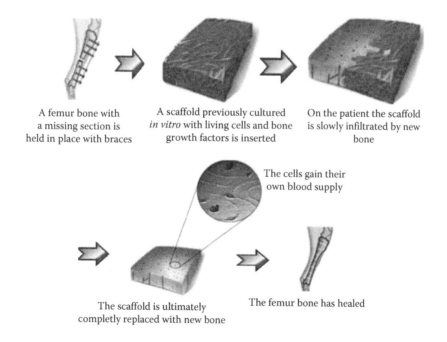

*Figure 10.3* Schematic diagram of bone tissue engineering concept. (From Rodrigues, S.C., Salgado, C.L., Sahu, A., Garcia, M.P., Fernandes, M.H., Monteiro, F.J.: Preparation and characterization of collagen-nanohydroxyapatite biocomposite scaffolds by cryogelation method for bone tissue engineering applications. *J. Biomed. Mater. Res.* 2012. 101A. 1080–1094. Copyright Wiley-VCH Verlag GmbH & Co. KGaA. Reproduced with permission.)

shows the bone tissue engineering concept using a hypothetical example of a femur. As can be seen, a tissue substitute is constructed in the laboratory by combining a scaffold with living cells and growth factors. When the construct is mature enough, it is implanted in the patient to repair the bone femur. Currently, the scientific challenge of bone tissue engineering is developing suitable 3D scaffolds that act as a template for cell adhesion and proliferation in favored 3D orientations. The scaffolds provide the necessary support for the cells to proliferate, and their architectures define the ultimate shapes of the new bones (Webster, 2007). Over the past decade, one of the main goals of bone tissue engineering has been to develop biodegradable materials as bone substitutes for filling large bone defects.

Current strategies of bone repair have limitations because of the synthetic graft materials or because of scaffold that support only *ex vivo* bone tissue engineering. Hence, biomimetic strategies have been considered important for the investigation of naturally occurring porous structures for their use as templates for bone growth (Clarke et al., 2011). Marine organisms such as corals, algae (diatoms), and sponges are rich in mineralized porous structures, some of which are currently being used as bone graft materials (Biocoral, Pro Osteon 200R, Pro Osreon 500R Algipore, etc.), and others are in their early stages of development. In addition to their use as structural mineralized materials, collagen from marine organisms has also been considered as a functionally important bone grafting material. The isolation, characterization, and design of collagen and its composite scaffolds for *in vitro* and *in vivo* applications as biomedical materials for bone tissue engineering involve a continuous process of search from marine sources (Rao et al., 2011). The structure and chemical composition of marine sponges, possessing calcareous or siliceous spicules and/or collagen/sponges, mimic the cancellous architecture of bone tissue and the complex canal system to create a porous biomimetic environment that is ideal for cellular integration within the bone matrix and promote osteogenesis *in vitro* (Green et al., 2002; Lin et al., 2011). Recent studies on collagen scaffolds from marine sponge species that belong to Callyspongiidae have proven to serve as potent bone tissue engineering scaffolds (Lin et al., 2011). Moreover, a tricomponent system of marine sponge collagen, Hap, and chitosan has become a promising bioscaffold for the assessment of bone tissue engineering applications (Pallela et al., 2012). Jellyfish collagen scaffolds are gaining much importance, as there is not much impurity in their body parts. Therefore, the extraction of jellyfish collagen from their umbrella or arms has been very interesting for applications in biomaterial sciences. Although the integration of these collagens with other biomacromolecules is at the initial stage, attempts are

being made to generate bone tissue engineering scaffolds or devices.

Wahl and Czernuszka (2006) have reported the most common routes to the fabrication of collagen (Col) and HA composites for bone analogs. The regeneration of diseased or fractured bones is a challenge faced by current technologies in tissue engineering. HA/collagen composites (Col-HA) have the potential of mimicking and replacing skeletal bones. Both *in vivo* and *in vitro* studies show the importance of the collagen type, mineralization conditions, porosity, manufacturing conditions, and cross-linking.

Nanohydroxyapatite (nHA)/collagen composites more closely mimic native bone composition and structure and significantly improve bone cell function and host integration leading to a faster recovery. A nanocomposite system composed of nHA/collagen/osteoblasts was developed in conjunction with PLA (Liao and Cui, 2004; Liao et al., 2004). The novel graft substitute (nHAC/PLA) supported cell adhesion, proliferation, and migration into the matrix to a depth of ~200–400 μm in a period of 12 days. *In vivo*, these nanocomposites were evaluated in a rabbit segmental defect model, in which the integration of the defect and evidence of bone formation were reported. Pek et al. (2008) prepared a porous nanocomposite bone scaffold containing collagen fibers and apatite nanocrystals to match the chemical composition and crystalline structure of natural bones. Fish collagen is produced from the skins of wild deep-sea fish such as cod, haddock, and pollock. This collagen hydrolysate acts a therapeutic agent of potential utility in the treatment of osteoarthritis and osteoporosis (Moskowitz, 2000). Also, collagen hydrolysate was used in patients with osteoarthritis (Bello, 2006).

Recently, collagen-apatite (Col-Ap) scaffolds were widely employed for bone tissue engineering. Very recently, Xia et al. (2014) fabricated Col-Ap scaffolds with a unique multilevel lamellar structure consisting of co-aligned micro and macropores. The basic building blocks of this scaffold are bone-like mineralized collagen fibers developed via a biomimetic self-assembly process in a collagen-containing modified simulated body fluid (m-SBF). This biomimetic method preserves the structural integrity and the great tensile strength of collagen by reinforcing the collagen hydrogel with apatite nanoparticles. Therefore, owing to the biomimetic composition, tunable structure, improved mechanical strength, and good biocompatibility of this novel scaffold, it has great potential to be used in bone tissue engineering applications.

### 10.5.3  Collagen for wound-healing processes

Wound healing is generally assumed to occur in three phases: inflammation, proliferation, and maturation.

The first phase of wound healing, such as the inflammatory phase, is characterized by two parts: the hemostatic part and the cellular part. The second phase of wound healing, or the proliferative phase, is characterized by three parts, the re-epithelialization part, the neovascularization part, and the collagen deposition part. The third phase of wound healing, or the maturation phase, deals with remodeling of the scar, and it is sometimes called the remodeling phase. Among the recent advancements in wound care management, slow release of drugs exhibited by hydrogels provides scope for developing novel wound dressings (Kim et al., 1992). The ideal drug-delivery system should be inert, biocompatible, mechanically strong, comfortable for the patients, capable of achieving optimum drug loading, simple to administer and remove, safe from accidental release, and easy to formulate and sterilize.

Collagen fibers, sponges, and fleeces have long been used in medicine as hemostatic agents. Collagen-based implants have been used as vehicles for the delivery of cultured keratinocytes and drugs for skin replacement and burn wounds (Leipziger et al., 1985). Implanted collagen sponges are infiltrated by amorphous connective tissue containing GAG, fibronectin, and new collagen, followed by various cells – primarily fibroblasts and macrophages. Fleischmajer et al. (1993) used a CG composite as a scaffold in their research. It has been shown that the isolated bone marrow stem cells (BMSCs) are potent in the healing, keratinization, and vascularization of full-thickness skin wounds. Other researchers have also reported the healing potential of BMSCs in skin regeneration by using the CG scaffold (Liu et al., 2008). Significant wound healing was reported for BMSCs as compared to allogenic neonatal dermal fibroblasts (Kobayashi and Spector, 2009). Healing was evaluated in terms of the re-epithelialization speed, the thickness of the regenerated epidermis, and the number of fibroblasts (Wu et al., 2007).

In the form of sponges, collagen is useful in the treatment of different wounds, such as pressure sores, donor sites, leg ulcers, and decubitus ulcers, as they adhere well to wet wounds, absorb large quantities of tissue exudates, preserve a moist environment, and encourage the formation of new granulation tissue and epithelium on the wound (Gorham, 1991). Collagen films have been employed to improve the cicatricial repair of mechanical and chemical damages (Gopinath et al., 2005). In severe wound infection, systemic administration of drugs may lead to insufficient drug concentration reaching the site of infection, drug-associated side effects, and systemic toxicity. These has been overcome successfully by topical application of drugs; collagen dressings with antibiotics have also been developed to control infection (Stemberger et al., 1997; Sripriya et al., 2007). Furthermore, the biocompatibility

of collagen has been established, and collagen matrices stimulate the biological phenomena involved in the success of wound healing, such as myofibroblastic differentiation and fibroblastic proliferation (Helary et al., 2006). Moreover, Sampath Kumar et al. (2012) reported the extraction of type I collagen from the bone of two marine fishes, *Megalaspis cordyla* and *Otolithes ruber*, and studied its efficacy in wound healing. These results suggested that the characterized collagen was cross-linked with glutaraldehyde and possessed 3D pores, and acted efficiently in healing the excision wounds made on Wistar rats.

## 10.5.4 Collagen for drug-delivery properties

Collagen has unique applications in drug-delivery systems, for example, collagen shields in ophthalmology, sponges for burns/wounds, mini-pellets and tablets for protein delivery, gel formulation in combination with liposomes for sustained drug delivery, as controlling material for transdermal delivery, as nanoparticles for gene delivery, and as basic matrices for various cell culture systems (Lee et al., 2001). Liposomes are widely used in drug delivery because of their biodegradability and flexibility in terms of composition and size. It was found that coupling liposomes to the gel matrices enhanced the stability of the system because of the antioxidant effect of collagen molecules when they were immobilized (Weiner et al., 1985). Cross-linking the functional liposomes to a collagen gel matrix further maintained the release rate of the entrapped marker. A novel drug-delivery system comprising liposomes sequestered in a collagen gel has been demonstrated, which controlled the release profiles of insulin and growth hormone into the circulation. Collagen-based systems soaked in liposomes were found to deliver significantly higher levels of the immunosuppressive agent cyclosporin A to the cornea, anterior sclera, aqueous humor, and vitreous in rabbit eyes than collagen shields without liposomes.

The importance of drug delivery is applicable not only to the physiological response but also to the mechanism of drug release, especially as a carrier for on-site delivery of the potent antibacterial agent. The scaffolds used for this purposes are thin films, gels, sponges, and bilayered dressings with or without chemical modifications (Ruszczak and Friess, 2003). Fibrillar collagen matrices used alone are capable of only moderating the release of drugs, and hence a significant nonfibrillar collagen is necessary to modulate the diffusive parameters. Further chemical modifications such as succinylation and electrostatic charge interactions are some of the strategies adopted to improve the release pattern of drugs. In this category, several wound-healing materials such as composite collagen–chondroitin-4 and -chondroitin-6-sulfate (Berthod et al., 1998), polyurethane-coated atelocollagen–rifampicin composite (Suh et al., 1994) materials, bilayered dressing with carboxymethyl-chitin hydrogel materials, and chitosan acetate thin films were able to sustain antibiotic delivery. Cheng et al. (2013) reported a novel electrochemical encapsulation method to encapsulate various nanomaterials and biomolecules into collagen. The electrochemical encapsulation process involves assembling collagen along with nano/biomaterials using an isoelectric focusing mechanism. It has been shown that a wide range of nanomaterials such as carbon nanotubes, polymeric nanoparticles, magnetic calcium phosphate nanoparticles, and biomolecules can be encapsulated into collagen. These novel collagen-based composite materials possess improved electrical, mechanical, antimicrobial, magnetic, and bioactive properties. This novel electrochemical encapsulation process offers a means to fabricate novel biomaterials for various biomedical applications such as tendon/ligament, nerve, skin tissue engineering, tendon/ligament to bone grafts, sutures, and so on.

The major advantages using collagen as a biomaterial for the development of delivery systems are the availability in abundance and easy purification, nonantigenicity, biodegradability and bioresorbability, nontoxicity, biocompatibility, synergy with bioactive components, high tensile strength, and easy modifiability as desired by utilizing functional groups compatible with synthetic polymers (Werkmeister and Ramshaw, 1991). Collagen are widely utilized in cosmetic surgery as a healing aid for burn patients, for the reconstruction of bone, and for a wide variety of dental, orthopedic, and surgical purposes. Although fish are the major resources for collagen from the marine environment, research on their biomedical implications is not well

explored (Ikoma et al., 2003). In this regard, investigations should focus on more effective uses of underutilized resources, such as type I collagen, which can be prepared from the skin, bone, and fins of marine fish and fish wastes. The use of collagen as a drug-delivery system is very comprehensive and diverse (Friess, 1998). Recently, Veeruraj et al. (2012) reported the drug-delivery potential of type I collagen isolated from the marine eel *Evenchelys macrura*. This collagen was successfully tested for its drug-delivery capacity through standard drugs embedded in gels and films of ASCs and PSCs. Wound dressings that act also as delivery systems represent an interesting application for collagen-based applications. Recent studies have shown the feasibility and, more importantly, the benefits of implants delivering antibiotics (Shanmugasundaram et al., 2006; Sripriya et al., 2007; Adhirajan et al., 2009). The drug-delivery property of collagen-based biomaterials also display great potential for ulcer treatment (Sun et al., 2009) and abdominal wall defect reconstruction (Liyanage et al., 2006; Ansaloni et al., 2007; Bellows et al., 2008). Collagen scaffolds have also shown to accurately deliver cells, proteins, drugs, and nucleic acids on a predictable and long-term basis (Sano et al., 2003; Takeshita and Ochiya, 2006; Takeshita et al., 2009; Glattauer et al., 2010). Finally, a recent clinical trial using adenovirus in collagen gel has cleared the path for future clinical studies on gene therapy delivered by a collagen matrix (Mulder et al., 2009). The biodegradability of collagen and its low immunogenicity make it a substrate of choice for internal and topical pharmacogenomical applications.

Table 10.2 lists the types of collagen drug-delivery applications and preparations that use collagen as a

*Table 10.2* Applications of collagen-based drug delivery system

| Application mode | Drug | Indication |
|---|---|---|
| Inserts | Pilocarpine, erythromycin, gentamycin | Infection, glaucoma |
| Shields | Gentamycin, vancomycin, tobramycin, polymyxin B, trimethoprim, amphotericin B, 5-FU, pilocarpine, steroids, flurbiprofene | Infection, glaucoma Glaucoma, inflammation |
| Particles | Cyclosporine | Allograft implication |
| Gels | Keterolac | Inflammation |
| Aqueous injection | Vinblastine, cisplatin, 5-FU, [111]In, [90]Y-labeled monoclonal antibodies, TGF-β, FGF, insulin, growth hormone | Cancer treatment Wound repair |
| Sponges | Gentamicin, cefotaxime, fusidic acid, clindamycin, all-trans retinoic acid, growth factors, bone morphogenic proteins | Infection, cervical dysplasia Tissue regeneration |
| Films | Medroxyprogesterone acetate, human growth hormone, immunostimulants, tetracycline, growth factors | Tissue repair and regeneration |
| Microparticles | Retinol, tretinoin, tetracaine, lidocaine, ethacridine lactate | Local anesthetic, dermal |
| Monolithic devices | Minocycline, lysozyme, intereukin-2, interferon | Infection and wound repair |

*Source:* Reprinted from *Adv. Drug Deliv. Rev.*, 55(12), Ruszczak, Z., Effect of collagen matrices on dermal wound healing, 17. Copyright 2003, with permission from Elsevier.

matrix. It may (1) act as a reservoir to increase the contact time with the host, (2) reversibly bind a drug such that it is released in a delayed mode, and (3) reduce the likelihood of systemic toxicity. Most of the delivery devices mentioned have been designed to prolong the release of therapeutics. Controlled delivery is essential for drugs that have narrow line of demarcation between therapeutic and toxic concentration. An extensive compilation on different types of collagen-based delivery systems for varied applications, including antibacterial, recombinant proteins, growth factors, and gene therapy, has been presented by Ruszczak and Friess (2003). Very recently, Nithya (2013) reported the efficiency of composite collagen nanofiber scaffolds in *in vitro* culture studies on cell adhesion, cell proliferation, and cell spreading and also *in vivo* studies on the wound site of a diabetic animal model. For local antibiotic delivery, the goal should be to maintain the highest possible, but not toxic, local drug concentration without producing systemic effects. This can be achieved by physical and possibly also chemical incorporation of the drug into a collagen matrix in the course of manufacturing to ensure drug immobilization. In future, drugs may be complexed with collagen through direct binding of the drug to free amino or carboxylic groups of the collagen molecule. Therefore, drug delivery or release will occur by diffusion from a collagen matrix implanted or injected as such or polymerized after intra-tissue injection (Chvapil, 1979; Ramshaw et al., 1995; Stemberger et al., 1997).

## 10.6   Summary

Collagen is the most widely used tissue-derived natural protein and is a main component of the ECMs (natural polymer) of all living animal tissues including the skin, bone, cartilage, tendon, and ligament. Marine collagens present an enormous variety of structures, providing a real potential for natural product drug discovery and for the delivery of new marine-derived products for therapeutic applications. Pharmaceutical applications of marine collagen have been widely realized due to the fulfillment of many requirements of drug-delivery systems, such as good biocompatibility, low antigenicity, and degradability upon implantation. Collagen-based biomaterials are of utmost importance for tissue engineering and regenerative medicine. The next generation of collagen drug-delivery system will be focused on both drug combination and different release profiles, which will lead to better infection control. Moreover, together with this development and better understanding of benefits coming from local drug delivery, some new collagen-based system may, in selected indications, even replace the current standard of systemic antibiotic

treatment. The collagen gel is one of the first natural polymers to be used as a promising matrix for drug delivery and tissue engineering. Collagen can also be used in biomedical applications as a decellularized ECM serving as a scaffolding material for tissue regeneration. We hope that the information given in this chapter will be helpful to new marine researchers who are interested in biomedicine, especially bone tissue engineering. In summary, marine-derived, collagen-based biomaterials are expected to be useful for a variety of structures, presenting real potential for natural, biodegradable new products for drug discovery and for the delivery of poly-matrix substances for various therapeutic applications in the future.

## References

Adhirajan, N., N. Shanmugasundaram, S. Shanmuganathan, and M. Babu. 2009. Collagen-based wound dressing for doxycycline delivery: In-vivo evaluation in an infected excisional wound model in rats. *The Journal of Pharmacology and Pharmacotherapeutics* 61: 1617–1623.

Alberts, B., D. Bray, J. Lewis, M. Raff, K. Roberts, and J.D. Watson. 1994. *Molecular Biology of the Cell*. Garland Publishing, New York, p. 971.

Alberts, B.J.A., J. Lewis, M. Raff, K. Roberts, and P. Walter. 2002. *Molecular Biology of the Cell*, 4th edn. Garland Science, New York, 1643pp.

Ansaloni, L., F. Catena, S. Gagliardi, F. Gazzotti, L. D'Alessandro, and A.D. Pinna. 2007. Hernia repair with porcine small-intestinal submucosa. *Hernia* 11: 321–326.

Asghar, A. and R.L. Henrickson. 1982. Chemical, biochemical, functional and nutritional characteristics of collagen in food systems. *Advances in Food Research* 28: 237–273.

Bächinger, H.P., K. Mizuno, J. Vranka, and S. Boudko. 2010. Collagen formation and structure. In *Comprehensive Natural Products II: Chemistry and Biology*, L. Mander and H.W. Liu (eds.). Elsevier, Oxford, U.K., pp. 469–530.

Badylak, S., K. Kokini, B. Tullius, A. Simmons-Byrd, and R. Morff. 2002. Morphological study of small intestinal submucosa as a body wall repair device. *Journal of surgical Research* 103(2): 190–202.

Beck, K., and B. Brodsky. 1998. Supercoiled protein motifs: the collagen triple-helix and the alpha-helical coiled coil. *Journal of structural biology* 122(1-2):17–29.

Bella, J., M. Eaton, B. Brodsky, and H.M. Berman. 1994. Crystal and molecular structure of a collagen-like peptide at 1.9 Å resolutions. *Science* 266:75–81.

Bello, A.E. 2006. Collagen hydrolysate for the treatment of osteoarthritis and other joint disorders: A review of literature. *Current Medical Research Opinions* 22: 2221–2232.

Bellows, C.F., W. Jian, M.K. McHale, D. Cardenas, J.L. West, S.P. Lerner, and G.E. Amiel. 2008. Blood vessel matrix: A new alternative for abdominal wall reconstruction. *Hernia* 12: 351–358.

Berisio, R., L. Vitagliano, L. Mazzarella, and A. Zagari. 2002. Crystal structure of the collagen triple helix model [(Pro-Pro-Gly)$_{10}$]$_3$. *Protein Science* 11: 262–270.

Berthold, A., K. Cremer, and J. Kreuter. 1998. Collagen microparticles: Carriers for glucocorticosteroids. *European Journal of Pharmaceutics and Biopharmaceutics* 45: 23–29.

Bolgen, N., Y. Yang, P. Korkusuz, E. Guzel, A.J. El Haj, and E. Piskin. 2008. Three-dimensional ingrowth of bone cells within biodegradable cryogel scaffolds in bioreactors at different regimes. *Tissue Engineering Part A* 14: 1743–1750.

Brazel, D., I. Oberbäumer, H. Dieringer et al. 1987. Completion of the amino acid sequence of the α1 chain of human basement membrane collagen (type IV) reveals 21 nontriplet interruptions located within the collagenous domain. *European Journal of Biochemistry* 168: 529–536.

Burg, K.J.L., S. Porter, and J.F. Kellam. 2000. Biomaterial developments for bone tissue engineering. *Biomaterials* 21: 2347–2359.

Burjanadze, T.V. 2000. New analysis of the phylogenetic change of collagen thermostability. *Biopolymers* 53: 523–528.

Cheng, X., V. Poenitzsch, L. Cornell, C. Tsao, and T. Potter. 2013. Electrochemical bioencapsulation of nanomaterials into collagen for biomedical applications. *Journal of Encapsulation and Adsorption Sciences* 3: 16–23.

Chi, C.F., Z.H. Cao, B. Wang, F.Y. Hu, Z.R. Li, and B. Zhang. 2014. Antioxidant and functional properties of collagen hydrolysates from spanish mackerel skin as influenced by average molecular weight. *Molecules* 19: 11211–11230.

Chvapil, M. 1979. Industrial uses of collagen. In *Fibrous Proteins: Scientific, Industrial and Medical Aspects*, D.A.D. Parry and L.K. Creamer (eds.). Academic Press, London, U.K., pp. 247–269.

Clarke, A.W., F. Alyas, T. Morris, C.J. Robertson, J. Bell, and D.A. Connell. 2011. Skin-derived tenocyte-like cells for the treatment of patellar tendinopathy. *The American Journal of Sports Medicine* 39: 614–623.

Compton, C.C., C.E. Butler, I.V. Yannas, G. Warland, and D.P. Orgill. 1998. Organized skin structure is regenerated in vivo from collagen-GAG matrices seeded with autologous keratinocytes. *Journal of Investigative Dermatology* 110: 908–916.

Cowan, P.M., S. McGavin, and A.C.T. North. 1955. The polypeptide chain configuration of collagen. *Nature* 176: 1062–1064.

Cunniffe, G.M., G.R. Dickson, S. Partap, K.T. Stanton, and F.J. O'Brien. 2010. Development and characterisation of a collagen nano-hydroxyapatite composite scaffold for bone tissue engineering. *Journal of Materials Science: Materials in Medicine* 21: 2293–2298.

Duan, X.D., C. McLaughlin, M. Griffith, and H. Sheardown. 2007. Biofunctionalization of collagen for improved biological response: Scoffolds for corneal tissue engineering. *Biomaterials* 28: 78–88.

Everaerts, F., M. Torrianni, M. Hendriks, and J. Feijen. 2008. Biomechanical properties of carbodiimide crosslinked collagen: Influence of the formation of ester crosslinks. *Journal of Biomedical Materials Research Part A* 85: 547–555.

Exposito, J.Y., C. Cluzel, R. Garrone, and C. Lethias. 2002. Evolution of collagens. *Anatomical Record* 268: 302–316.

Exposito, J.Y., D. Le Guellec, Q. Lu, and R. Garrone. 1991. Short chain collagens in sponges are encoded by a family of closely related genes. *Journal of Biological Chemistry* 266: 21923–21928.

Fields, C. 1988. Domain organization and intron positions in *Caenorhabditis elegans* collagen genes: The 54-bp module hypothesis revisited. *Journal of Molecular Evolution* 28: 55–63.

Fleischmajer, R., E.D. MacDonald, P. Contard, and J.S. Perlish. 1993. Immunochemistry of a keratinocyte-fibroblast co-culture model for reconstruction of human skin. *Journal of Histochemistry and Cytochemistry* 41: 1359–1366.

Friess, W. 1998. Collagen-biomaterial for drug delivery. *European Journal of Pharmaceutics and Biopharmaceutics* 45: 113–136.

Geiger, M., R.H. Li, and W. Friess. 2003. Collagen sponges for bone regeneration with rhBMP-2. *Advanced Drug Delivery Review* 55: 1613–1629.

Gelse, K., E. Poschi, and T. Aigner. 2003. Collagens, structure, function, and biosynthesis. *Advanced Drug Delivery* 55: 1531–1546.

Gilbert, T.W., T.L. Sellaro, and S.F. Badylak. 2006. Decellularization of tissues and organs. *Biomaterials* 27: 3675–3683.

Glattauer, V., J.F. White, W.B. Tsai, C.C. Tsai, T.A. Tebb, S.J. Danon, J.A. Werkmeister, and J.A. Ramshaw. 2010. Preparation of resorbable collagen-based beads for direct use in tissue engineering and cell therapy applications. *Journal of Biomedical Materials Research A* 92: 1301–1309.

Gopinath, D., M.S. Kumar, D. Selvaraj, and R. Jayakumar. 2005. Pexiganan-incorporated collagen matrices for infected wound-healing processes in rat. *Journal of Biomedical Materials Research Part A* 73: 320–331.

Gorham, S.D. 1991. Collagen. In *Biomaterials*, D. Byrom (ed.). Stockton Press, New York, pp. 55–122.

Green, D., D. Walsh, S. Mann, and R.O. Oreffo. 2002. The potential of biomimesis in bone tissue engineering: Lessons from the design and synthesis of invertebrate skeletons. *Bone* 30: 810–815.

Harley, B.A., A.K. Lynn, Z. Wissner-Gross, W. Bonfield, I.V. Yannas, and L.J. Gibson. 2010. Design of a multiphase osteochondral scaffold. II. Fabrication of a mineralized collagenglycosaminoglycan scaffold. *Journal of Biomedical Materials Research Part* 92A: 1066–1077.

Hayashi, Y., Y. Shizuka, K.Y. Guchi et al. 2012. Chitosan and fish collagen as biomaterials for regenerative medicine. In *Advances in Food and Nutrition Research*, S.-K. Kim (ed.). Academic Press, New York, pp. 105–120.

Helary, C., L. Ovtracht, B. Coulomb, G. Godeau, and M.M. Giraud-Guille. 2006. Dense fibrillar collagen matrices: A model to study myofibroblast behaviour during wound healing. *Biomaterials* 27: 4443–4452.

Hing, K.A. 2004. Bone repair in the twenty-first century: Biology, chemistry or engineering? *Philosophical Transactions of the Royal Society of London, Series A: Mathematical Physical and Engineering Sciences* 362: 2821–2850.

Hsu, W.C., M.H. Spilker, I.V. Yannas, and P.A. Rubin. 2000. Inhibition of conjunctival scarring and contraction by a porous collagen-glycosaminoglycan implant. *Investigative Ophthalmology Visual Science* 41: 2404–2411.

Huang, Y.R., C.Y. Shiau, H.H. Chen, and B.C. Huang. 2011. Isolation and characterization of acid and pepsin-solubilized collagens from the skin of balloon fish (*Diodon holocanthus*). *Food Hydrocolloids* 25: 1507–1513.

Ikoma, T., H. Kobayashi, J. Tanaka, D. Walsh, and S. Mann. 2003. Physical properties of type I collagen extracted from fish scales of *Pagrus major* and *Oreochromis niloticus*. *International Journal of Biological Macromolecules* 32: 199–204.

Jen, A.C., M.C. Wake, and A.G. Mikos. 1996. Review: Hydrogels for cell immobilization. *Biotechnology and Bioengineering* 50: 357–364.

Jhon, M.S. and J.D. Andrade. 1973. Water and hydrogels. *Journal of Biomedical Materials Research Part A* 7: 509–522.

Jithendra, P., A.M. Rajam, T. Kalaivani, A.B. Mandal, and C. Rose. 2013. Preparation and characterization of aloe vera blended collagen-chitosan composite scaffold for tissue engineering applications. *ACS Applied Materials Interfaces* 5: 7291–7298.

Jones, G.L., R. Walton, J. Czernuszka, S.L. Griffiths, A.J.E. Haj, and S.H. Cartmell. 2010. Primary human osteoblast culture on 3D porous collagen hydroxyapatite scaffolds. *Journal of Biomedical Materials Research* 94A: 1244–1250.

Khoshnoodi, J., V. Pedchenko, and B.G. Hudson. 2008. Mammalian collagen IV. *Microscopy Research and Technique* 71: 357–370.

Kim, B.S. and D.J. Mooney. 1998. Development of biocompatible synthetic extracellular matrices for tissue engineering. *Trends in Biotechnology* 16: 224–230.

Kim, S.R., Y. Kim, M.A. Costa, and G. An. 1992. Inhibition of sucrose enhancer effect of the potato proteinase inhibitor II promoter by salicylic acid. *Plant Physiology* 98: 1479–1483.

Kimura, S., Y. Takema, and M. Kubota. 1981. Octopus skin collagen: Isolation and characterization of collagen comprising two distinct a chains. *Journal of Biological Chemistry* 256: 13230–13234.

Kittiphattanabawon, P., S. Benjakul, W. Visessanguan, T. Nagai, and M. Tanaka. 2005. Characterisation of acid-soluble collagen from skin and bone of bigeye snapper (*Priacanthus tayenus*). *Food Chemistry* 89: 363–372.

Kivirikko, K. and R. Myllylä. 1983. Kollageenien geeniperhe. *Duodecim* 99: 1373–1382.

Kobayashi, M. and M. Spector. 2009. *In vitro* response of the bone marrow-derived mesenchymal stem cells seeded in a type-I collagen-glycosaminoglycan scaffold for skin wound repair under the mechanical loading condition. *Molecular & Cellular Biomechanics* 6: 217–227.

Langer, R. and J.P. Vacanti. 1993. Tissue engineering. *Science* 260: 920–926.

Lee, C.H., A. Singla, and Y. Lee. 2001. Biomedical applications of collagen. *International Journal of Pharmaceutics* 221: 1–22.

Leipziger, L.S., V. Glushko, B. Dibernado, F. Shafaie, J. Noble, J. Nichols, and O.M. Alvarez. 1985. Dermal wound repair: Role of collagen matrix implants and synthetic polymer dressings. *Journal of the American Academy of Dermatology* 12: 409–419.

Lennarz, W.J. 1983. Glycoprotein synthesis and embryonic development. *CRC Critical Review Biochemistry* 14: 257–272.

Liao, S.S. and F.Z. Cui. 2004. *In vitro* and *in vivo* degradation of mineralized collagen-based composite scaffold: Nanohydroxyapatite/collagen/poly(L-lactide). *Tissue Engineering* 10: 73–80.

Liao, S.S., F.Z. Cui, W. Zhang et al. 2004. Hierarchically biomimetic bone scaffold materials: Nano-HA/collagen/PLA composite. *Journal of Biomedical Materials Research* 69B: 158–165.

Lin, Z., K.L. Solomon, X. Zhang, N.J. Pavlos, T. Abel, C. Willers, K. Dai, J. Xu, Q. Zheng, and M. Zheng. 2011. *In vitro* evaluation of natural marine sponge collagen as a scaffold for bone tissue engineering. *International Journal of Biological Sciences* 7(7): 968–977.

Liu, D.S., L. Liang, J.M. Regenstein, and P. Zhou. 2012. Extraction and characterisation of pepsin-solubilised collagen from fins, scales, skins, bones and swim bladders of bighead carp (*Hypophthalmichthys nobilis*). *Food Chemistry* 133: 1441–1448.

Liu, X.H. and P.X. Ma. 2004. Polymeric scaffolds for bone tissue engineering. *Annals of Biomedical Engineering* 32: 477–486.

Liu, Y., H.S. Ramanath, and D.A. Wang. 2008. Tendon tissue engineering using scaffold enhancing strategies. *Trends in Biotechnology* 26: 201–209.

Liyanage, S.H., G.S. Purohit, J.N. Frye, and P. Giordano. 2006. Anterior abdominal wall reconstruction with a Permacol implant. *Journal of Plastic, Reconstructive & Aesthetic Surgery* 59: 553–555.

Lyons, F.G., A.A. Al-Munajjed, S.M. Kieran, M.E. Toner, C.M. Murphy, G.P. Duffy, and F.J. O'Brien. 2010. The healing of bony defects by cell-free collagen-based scaffolds compared to stem cell-seeded tissue engineered constructs. *Biomaterials* 31: 9232–9243.

Maa, L., C. Gaoa, Z. Maoa, J. Zhoua, J. Shena, X. Hub, and C. Han. 2003. Collagen/chitosan porous scaffolds with improved biostability for skin tissue engineering. *Biomaterials* 24: 4833–4841.

Moskowitz, R.W. 2000. Role of collagen hydrolysate in bone and joint disease. *Seminars in Arthritis and Rheumatism* 30(2): 87–99.

Mulder, G., A.J. Tallis, V.T. Marshall, D. Mozingo, L. Phillips, G.F. Pierce, L.A. Chandler, and B.K. Sosnowski. 2009. Treatment of nonhealing diabetic foot ulcers with a platelet-derived growth factor gene-activated matrix (GAM501): Results of a phase 1/2 trial. *Wound Repair Regeneration* 17: 772–779.

Muyonga, J.H., C.G.B. Cole, and K.G. Duodu. 2004. Characterization of acid soluble collagen from skin of young and adult Nile perch (*Lates nilotics*). *Food Chemistry* 85: 81–89.

Nagai, T. and N. Suzuki. 2000. Preparation and characterization of several fish bone collagens. *Journal of Food Biochemistry* 24: 427–436.

Nithya, T.G. 2013. Efficacy of mesenchymal stem cell incorporation in different composite electrospun collagen nanofibers for diabetic wound healing. *Asian Journal of Pharmaceutical and Clinical Research* 6: 149–152.

Nomura, Y., S. Toki, Y. Ishii, and K. Shirai. 2000. Improvement of material property of shark type I collagen by composing with porcine type I collagen. *Journal of Agricultural and Food Chemistry* 48: 6332–6336.

O'Brien, F.J., B.A. Harley, I.V. Yannas, and L.J. Gibson. 2005. The effect of pore size on cell adhesion in collagen-GAG scaffolds. *Biomaterials* 26: 433–441.

Ötles, S. and S. Ötles. 2004. Biobased packaging materials for the food industry-types of biobased packaging materials (Természetes alapú élelmiszeripari csomagoló anyagok-Természetes csomagolóanyagok típusai). *Évofylam* 53: 116–119.

Pallela, R., J. Venkatesan, V.R. Janapala et al. 2012. Biophysicochemical evaluation of chitosan-hydroxyapatite-marine sponge collagen composite for bone tissue engineering. *Journal of Biomedical Materials Research Part A* 100A(2): 486–495.

Pek, Y.S., S. Gao, M.S.M. Arshad, K.J. Leck, and Y.Y. Ying. 2008. Porous collagen-apatite nanocomposite foams as bone regeneration scaffolds. *Biomaterials* 29: 4300–4305.

Pieper, J.S., A. Oosterhof, P.J. Dijkstra, J.H. Veerkamp, and T.H. van Kuppevelt. 1999. Preparation and characterization of porous crosslinked collagenous matrices containing bioavailable chondroitin sulphate. *Biomaterials* 20(9): 847–858.

Prockop, D.J., K.I. Kivirikko, L. Tudermann, and N.A. Guzman. 1979. The biosynthesis of collagen and its disorders. *New England Journal of Medicine* 301: 13–23, 77–78.

Ramachandran, G.N., M. Bansal, and R.S. Bhastnagar. 1973. A hypothesis on the role of hydroxyproline in stabilizing collagen structure. *Biochimica et Biophysica Acta* 323: 166–171.

Ramshaw, J.A.M., J.A. Werkmeister, and V. Glattauer. 1995. Collagen based biomaterials. *Biotechnology Review* 13: 336–382.

Rao, J.V., R. Pallela, and G.V.S.B. Prakash. 2011. Prospects of marine sponge collagen and its applications in cosmetology. In *Marine Cosmeceuticals: Trends and Prospects*, S.K. Kim (ed.). CRC–Taylor & Francis Group, Boca Raton, FL, pp. 77–103.

Regenstein, J.M. and P. Zhou. 2007. Collagen and gelatin from marine by-products. In *Maximising the Value of Marine By-Products*, F. Shahidi (ed.). Wood Head Publishing Limited, Cambridge, U.K., pp. 279–303.

Rho, J.Y., L. Kuhn-Spearing, and P. Zioupos. 1998. Mechanical properties and the hierarchical structure of bone. *Medical Engineering and Physics* 20: 92–102.

Rich, A. and F.H.C. Crick. 1955. The structure of collagen. *Nature* 176: 915–916.

Rich, A. and F.H.C. Crick. 1961. The molecular structure of collagen. *Journal of Molecular Biology* 3: 483–506.

Rigby, B.J. 1968. Amino-acid composition and thermal stability of the skin collagen of the antarctic ice-fish. *Nature* 219: 166–167.

Rodrigues, S.C., Salgado, C.L., Sahu, A., Garcia, M.P., Fernandes, M.H., and F.J. Monteiro. 2012. Preparation and characterization of collagen-nanohydroxyapatite biocomposite scaffolds by cryogelation method for bone tissue engineering applications. *Journal of Biomedical Materials Research A* 101:1080–1094.

Ruszczak, Z. and W. Friess. 2003. Collagen as a carrier for on-site delivery of antibacterial drugs. *Advanced Drug Delivery Reviews* 55: 1679–1698.

Sachlos, E. and J.T. Czernuszka. 2003. Making tissue engineering scaffolds work. Review on the application of solid free-form fabrication technology to the production of tissue engineering scaffolds. *European Cells and Materials* 5: 29–40.

Sampath Kumar, N.S., R.A. Nazeer, and R. Jaiganesh. 2012. Wound healing properties of collagen from the bone of two marine fishes. *International Journal of Peptide Research and Therapeutics* 18: 185–192.

Sano, A., M. Maeda, S. Nagahara, T. Ochiya, K. Honma, H. Itoh, T. Miyata, and K. Fujioka. 2003. Atelocollagen for protein and gene delivery. *Advanced Drug Delivery Review* 55: 1651–1677.

Schoeninger, M.J. and M.J. DeNiro. 1984. Nitrogen and carbon isotopic composition of bone collagen from marine and terrestrial animals. *Geochimica et Cosmochimica Acta* 48: 625–639.

Shalak, R. and C.F. Fox. 1988. Preface. In *Tissue Engineering*, R. Shalak and C.F. Fox (eds.). Alan R. Liss, New York, pp. 26–29.

Shanmugasundaram, N., J. Sundaraseelan, S. Uma, D. Selvaraj, and M. Babu. 2006. Design and delivery of silver sulfadiazine from alginate microspheres-impregnated collagen scaffold. *Journal of Biomedical Materials Research Part B: Applied Biomaterials* 77: 378–388.

Sharon, N. and H. Lis. 1981. Glycoproteins: Research booming on long-ignored, ubiquitous compounds. *Chemical Engineering News* 59: 21–44.

Shen, X.R., H. Kurihara, and K. Takahashi. 2007. Characterization of molecular species of collagen in scallop mantle. *Food Chemistry* 102: 1187–1191.

Shoulders, M.D. and R.T. Raines. 2009. Collagen structure and stability. *Annual Review of Biochemistry* 78: 929–958.

Sikorski, Z.E. and I. Kolodziejska. 1986. The composition and properties of squid meat. *Food Chemistry* 20: 213–224.

Song, E., S.Y. Kim, T. Chun, H.J. Byun, and Y.M. Lee. 2006. Collagen scaffolds derived from a marine source and their biocompatibility. *Biomaterials* 27: 2951–2961.

Sripriya, R., M.S. Kumar, M.R. Ahmed, and P.K. Sehgal. 2007. Collagen bilayer dressing with ciprofloxacin, an effective system for infected wound healing. *Journal of Biomaterials Science, Polymer Edition* 18: 335–351.

Stemberger, A., H. Grimm, F. Bader, H.D. Rahn, and R. Ascherl. 1997. Local treatment of bone and soft tissue infections with the collagen-gentamicin sponge. *European Journal of Surgery* 163(S578): 17–26.

Stenzel, K.H., T. Miyata, and A.L. Rubin. 1974. Collagen as a biomaterial. *Annual Review of Biophysics and Bioengineering* 3: 231–252.

Sugiura, H., S. Yunoki, E. Kondo, T. Ikoma, J. Tanaka, and K. Yasuda. 2009. *In vivo* biological responses and bioresorption of tilapia scale collagen as a potential biomaterial. *Journal of Biomaterials Science, Polymer Edition* 20: 1353–1368.

Suh, H., S.W. Suh, and B.G. Min. 1994. Anti-infection treatement of a treanscutaneous device by a collagen-rifampicine composite. *ASAIO Journal* 40: M406–M411.

Sun, W., H. Lin, B. Chen, W. Zhao, Y. Zhao, Z. Xiao, and J. Dai. 2009. Collagen scaffolds loaded with collagen-binding NGF-beta accelerate ulcer healing. *Journal of Biomedical Materials Research A* 92A: 887–895.

Takeshita, F. and T. Ochiya. 2006. Therapeutic potential of RNA interference against cancer. *Cancer Sciences* 97: 689–696.

Takeshita, F., N. Hokaiwado, K. Honma, A. Banas, and T. Ochiya. 2009. Local and systemic delivery of siRNAs for oligonucleotide therapy. *Methods in Molecular Biology* 487: 83–92.

Tamilmozhi, S., A. Veeruraj, and M. Arumugam. 2013. Isolation and characterization of acid and pepsin-solubilized collagen from the skin of sailfish (*Istiophorus platypterus*). *Food Research International* 54: 1499–1505.

Tangsadthakun, C., S. Kanokpanont, N. Sanchavanakit, T. Banaprasert, and S. Damrongsakkul. 2006. Properties of Collagen/Chitosan Scaffolds for Skin Tissue Engineering. *Journal of Metals, Materials and Minerals* 16(1): 37–44.

Thomson, R.C., A.G. Mikos, E. Beahm, J.C. Lemon, W.C. Satterfield, T.B. Aufdemorte, and M.J. Miller. 1999. Guided tissue fabrication from periosteum using preformed biodegradable polymer scaffolds. *Biomaterials* 20: 2007–2018.

Vacanti, J.P. 2003. Tissue and organ engineering: Can we build intestine and vital organs? *Gastrointestinal Surgery* 7: 831–835.

Veeruraj, A., M. Arumugam, and T. Balasubramanian. 2013. Isolation and characterization of thermostable collagen from the marine eel-fish (*Evenchelys macrura*). *Process Biochemistry* 48: 1592–1602.

Veeruraj, A., M. Arumugam, T. Ajithkumar, and T. Balasubramanian. 2012. Isolation and characterization of drug delivering potential of type-I collagen from eel fish *Evenchelys macrura. Journal of Materials Science: Materials in Medicine* 23: 1729–1738.

Veeruraj, A., M. Arumugam, T. Ajithkumar, and T. Balasubramanian. 2015. Isolation and characterization of collagen from the outer skin of squid (*Doryteuthis singhalensis*). *Food Hydrocolloids* 43: 708–716.

Wahl, D.A. and J.T. Czernuszka. 2006. Collagen-hydroxyapatite composites for hard tissue repair. *European Cells and Materials* 11: 43–56.

Wahl, D.A., E. Sachlos, C. Liu, and J.T. Czernuszka. 2007. Controlling the processing of collagen-hydroxyapatite scaffolds for bone tissue engineering. *Journal of Materials Science and Materials Medicine* 18: 201–209.

Wang, L., Q. Liang, T. Chen, Z. Wang, J. Xu, and H. Ma. 2014. Characterization of collagen from the skin of Amur sturgeon (*Acipenser schrenckii*). *Food Hydrocolloids* 38: 104–109.

Wang, W., S. Itoh, T. Aizawa, A. Okawa, K. Sakai, T. Ohkuma, and M. Demura. 2010. Development of an injectable chitosan/marine collagen composite gel. *Biomedical Materials* 5: 065009.

Webster, T.J. (ed.). 2007. *Nanotechnology for the Regeneration of Hard and Soft Tissues*, 1st edn. World Scientific Publishing Company, Hakensack, NJ, p. 260.

Weiner, A.L., S.S. Carpenter-Gren, E.C. Soehngen, R.P. Lenk, and M.C. Popescu. 1985. Liposomecollagen gel matrix: A novel sustained drug delivery system. *Journal of Pharmaceutical Sciences* 74: 922–925.

Werkmeister, J.A. and J.A.M. Ramshaw. 1991. Monoclonal antibodies to type V collagen as markers for new tissue deposition associated with biomaterial implants. *Journal of Histochemistry Cytochemistry* 39: 1215–1220.

Wolf, K.L., P.J.A. Sobral, and V.R.N. Telis. 2006. Characterizations of collagen fibers for biodegradable films production. In *IUFoST World Congress, 13th World Congress of Food Science and Technology*, Brazil, pp. 801–802. http://dx.doi.org/10.1051/IUFoST:20060929.

Wong, D.W.S. 1989. *Mechanism and Theory in Food Chemistry.* Van Nostrand Reinhold, New York, p. 428.

Wu, X., L. Black, G. Santacana-Laffitte, and C.W. Patrick. 2007. Preparation and assessment of glutaraldehyde-cross-linked collagen-chitosan hydrogels for adipose tissue engineering. *Journal of Biomedical Materials Research A* 81: 59–65.

Xia, Z., M.M. Villa, and M. Wei. 2014. A biomimetic collagen–apatite scaffold with a multi-level lamellar structure for bone tissue engineering. *Journal of Material Chemistry B* 2: 1998–2007.

Yannas, I.V. 1972. Collagen gelatine in the solid state. *Journal of Macromolecular Science—Reviews in Macromolecular Chemistry and Physics* C7(1): 49–104.

Yannas, I.V. 1992. Tissue regeneration by use of collagen-glycosaminoglycan copolymers. *Clinical Materials* 9: 179–187.

Yunoki, S., K. Mori, T. Suzuki, N. Nagai, and M. Munekata. 2007. Novel elastic material from collagen for tissue engineering. *Journal of Materials Science: Materials in Medicine* 18: 1369–1375.

Yunoki, S., T. Ikoma, and J. Tanaka. 2010. Development of collagen condensation method to improve mechanical strength of tissue engineering scaffolds. *Materials Characterization* 61: 907–911.

*chapter eleven*

# Glycoproteins and detoxification in the marine environment

*Graciela Guerra-Rivas and Claudia Mariana Gomez-Gutierrez*

## Contents

## 11.1 Introduction

Imagine a suddenly changing world where everything turns different in a few seconds. Picture yourself traveling through a medium that makes you face strong gradients of temperature, humidity, and oxygen levels. Forming this image will make easier to have a notion of the kind of difficulties of marine life. In this environment, living beings confront problems that terrestrials do not have to face due to more stable conditions over the crust of the earth. For this reason, microorganisms, plants, and animals from the sea have been forced by the environment to develop specialized mechanisms to cope harsh and changeable conditions, such as high salt concentration, extreme temperatures and pH, high pressures, estivation, low oxygen availability, and low nutrient concentration. On top of this, cities of the world have undergone steadily an impressive development increasing human pressures on the sea including waste water discharges, land reclamation, and oil exploitation. As a consequence, marine organisms are constantly exposed to mixtures of chemical contaminants or xenobiotics that they have to fight to survive. One of the results of this complex marine environment on their bodies is the possession of unique metabolic routes with characteristic metabolites, proteins, and enzymes, some of which are not found in their terrestrial counterparts. Also, they possess their own special defense mechanisms to deal with xenobiotics. Thus, marine habitats represent a huge reservoir of specialized molecules and a wide range of amazing molecular arrangements to be discovered. Glycoproteins are one of the diverse groups of molecules that play an important role in the struggle of marine living beings to cope xenobiotics.

## 11.2 Mechanisms for detoxification: Biotransformation, phase I and phase II

Xenobiotic elimination is a complex process that is governed by the chemical flux into an organ, followed by efflux back to the circulation, cellular metabolism, and excretion (Pang et al., 2009). Regarding the physiological and molecular mechanisms for this process, much information has been gathered in the past five decades. New breakthroughs encompassing biochemistry, molecular biology, genetics, and the steady growth in the "omics" sciences have been supported by modern technologies for a better insight into the way organisms face harmful molecules.

When natural and man-made organic foreign compounds contaminate aquatic ecosystems, they are dispersed in the water as a function of polarity. The more hydrophilic they are, the more they dissolve in this solvent and the bigger the opportunity they have to "wrap" the inhabitants. However, low-polarity compounds have a higher probability to get inside the organism due to their similarity to biological membranes, which permit the entrance to the cells. Once inside, the contaminant will elicit a defense response from the host, aiming to remove the xenobiotic and its deleterious effects. The final result will be a combination of the effects of the xenobiotic at its site of action (toxicodynamics) and the effects of the body on xenobiotic delivery to its site of action (toxicokinetics). In other words, the toxic action of a chemical will be a consequence of the interaction between the chemical and a molecular target within the living organism. On the one hand, toxicodynamic aspects are determined by processes such as xenobiotic–receptor interaction and are specific to the class of the chemical. On the other hand, toxicokinetic aspects are determined by general processes, such as absorption, distribution, and body elimination, which apply irrespective of the toxicodynamic properties (Figure 11.1).

For all three stages, physicochemical properties of the xenobiotic play an important role in the membrane transport involved since lipophilicity enables it to penetrate lipid membranes (Figure 11.2).

Lipophilic compounds are also well suited to be transported by lipoproteins in the blood, which makes them to be expeditiously absorbed by the target organ. However, elimination requires a certain degree of hydrophilic character for an efficient excretion. Therefore, lipophilic compounds must be modified and converted to more polar compounds.

Elimination as a whole is achieved by the xenobiotic detoxification system, which is mainly composed by metabolizing enzymes and transporters, and proceeds in three phases called phase I, phase II, and phase III (Wang et al., 2012). Phase I and phase II biotransformation processes convert the hydrophobic xenobiotic molecules into polar compounds by various biochemical reactions. Phase I is characterized by the following functions: highly hydrophobic moieties are converted through oxidation, reduction,

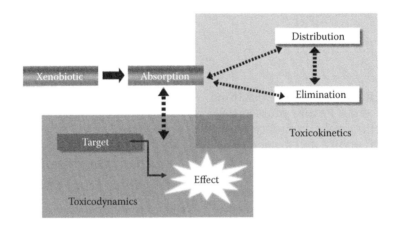

*Figure 11.1* The general route of a xenobiotic. Interplay between toxicokinetic and toxicodynamic aspects during the toxicological process. The toxic effects of the xenobiotic depends on the concentration of altered molecular targets.

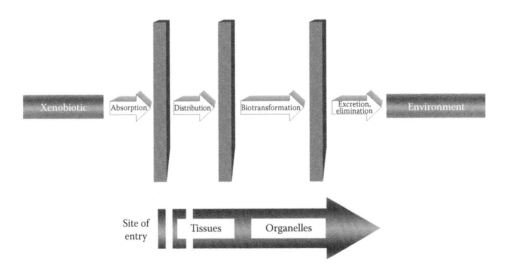

*Figure 11.2* Xenobiotics pass through membranes. The absorption, distribution, and excretion of xenobiotics involve passing through various cell and organ membranes, which occurs through various transport mechanisms.

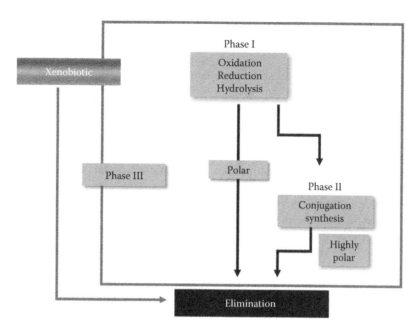

***Figure 11.3*** Elimination stage of a xenobiotic. Phases I and II are structural modifications of the foreign compound to make it more eliminable (polar) through natural mechanisms. Phase III refers to the role of active membrane transporters that function to shuttle drugs and other xenobiotics across cellular membranes.

and hydrolysis reactions to carboxylic acids, alcohols, amines, or any compound more polar than the original xenobiotic (Torres et al., 2008). Among all the enzymes participating in phase I metabolism, cytochromes P450 (CYPs) are the best known in the field of xenobiotic and drug-metabolizing agents. CYPs belong to a monooxygenase superfamily that are widely distributed in nature, from bacteria to man. For marine organisms, members of the CYP superfamily have been reported in mollusks (Livingstone et al., 1985; Solé et al., 1994), fishes (Shahraki et al., 2013; Uno et al., 2012), mammals (Teramitsu et al., 2000; Tilley et al., 2003), crustaceans (James, 1990; James and Boyle, 1998), and microalgae (Petroutsos et al., 2007; Warshawsky et al., 1995; Pflugmacher and Sandermann, 1998) among others (Torres et al., 2008). The P450 enzymes catalyze a myriad of reactions including N- and O-dealkylation, aliphatic and aromatic hydroxylation, N- and S-oxidation, and deamination, which are performed on a wide variety of foreign and endogenous compounds (Pflugmacher and Sandermann, 1998; Guengerich, 2001; Torres et al., 2008; Yoshimoto and Auchus, 2014). Phase II enzymes have also been well characterized and its conduction of conjugation reactions is recognized. It is known that they conduct conjugation reactions by introducing a large polar moiety such as glutathione, sulfate, glucuronide, and amino acid groups in xenobiotic compounds or their metabolites with groups such as –COOH, –OH, or –NH₂ (Torres et al., 2008; Omiecinski et al., 2011). All the products, typically more hydrophilic

than parent compounds, are ready to be excreted using the available elimination pathway in the organisms (Figure 11.3).

Important phase II enzymes include UDP-glucuronosyl transferases, sulfatases, acetyltransferases, and glutathione *S*-transferases (GST). Compared with phase I reactions, there have been few studies of phase II reactions in aquatic animals, and this is especially true for invertebrates and critical for microalgae. For some marine fishes, a variety of conjugation enzymes have been reported and some xenobiotic metabolism products have been detected (Gonzalez et al., 2009). These findings suggest that they are able to generate glycosides, glucuronides, sulfates, and mercapturic acid as excretable forms (James, 1987; Di Bello et al., 2007; Ikenaka et al., 2013). There are only a few studies for marine invertebrates reporting GST activity, for example, in polychaetes, mollusks, echinoderms, sea urchins, crustaceans, and ascidians (Livingstone, 1998; Jørgensen, 2008; Trisciani et al., 2012). Also, there is some evidence that glycosides, glucuronides, and sulfates are synthesized during elimination process by mollusks, crustaceans, and echinoderms (James, 1987). Fewer studies have been done for marine microalgae. Although these organisms represent the largest biomass component of aquatic systems, there is little information available on the activities on these enzymes, and more detailed knowledge is required (Pflugmacher and Sandermann, 1998; Torres et al., 2008).

## 11.3 Phase III: The "defensome" in marine habitats

The capability of a species to modify chemical structure is not paired to its ability to distinguish between "good" and "bad" products. It is well known that biotransformation generally results in harmless metabolites, but as a double-edged sword, phase I and phase II enzyme function can lead to harmful results, as an increased toxicity or a new activity against the organism (Shimada et al., 1996; Kondraganti et al., 2003; Goldstone et al., 2006). Underlying and complementary to biotransformation process, there is a complex arrangement of elements needed to support metabolite processing to avoid deleterious effects of nocive molecules. One of such mechanisms, functioning as a first line of defense in aquatic species, is mediated by the activity of transporters that prevent the accumulation of harmful chemicals on the cells. This cellular mechanism, called multixenobiotic resistance (MXR) system, was first demonstrated by Kurelec in marine organisms (Kurelec, 1992) and is similar to the surface carbohydrate-containing protein observed in *Chinese hamster* ovary cells earlier by Juliano and Ling (1976). These researchers termed this transporter as the P-glycoprotein (P-gp) (permeability protein), and it is recognized as a member of a family of membrane transporters that function in the resistance of cancer cells to chemotherapeutics, currently referred to as the multiple drug resistance or multidrug-resistant (MDR) phenotype (Omiecinski et al., 2011). P-gps belong to ATP-binding cassette (ABC) proteins or multidrug efflux transporters, the largest family of transmembrane (TM) proteins (Dean, 2002). These proteins bind ATP and use the energy to drive the transport of molecules across all cell membranes (Dean, 2002).

The MXR P-gp, according to his discoverer, pumps out of cells and organisms both endogenous chemicals and xenobiotics, including also some man-made chemicals (Kurele, 1997).

Besides a wide distribution among terrestrial organisms including man, P-gp-like proteins have been described in a variety of marine organisms. Sponges, mussels, oysters, worms, corals, clams, diatoms, rotifers, sea urchins, and fish have been reported as possessing ABC transporters, although there is not enough knowledge regarding molecular features for all of them (Kroeger et al., 1994; Bard, 2000; Achard et al., 2004; Goldstone et al., 2006; Kingtong et al., 2007; Ivanina and Sokolova, 2008; Venn et al., 2009; Kim et al., 2013; Franzellitti et al., 2014). These transporters, together with phase I and II enzymes and others, orchestrate the xenobiotic detoxification process. Along with phase I and II enzymes and ABC proteins in charge of biotransformation and efflux of xenobiotics, respectively, there are other factors involved in the protection of marine

living beings against chemical threatens. First, there are soluble receptors and ligand-activated transcription factors that function as cellular sensors of harmful molecules or damage. Second, there are also antioxidant enzymes such as superoxide dismutases, catalases, and others performing protection against ROS particles and other noxious radicals from external/internal sources. Hence, in an integral context, the defensome concept refers to an "integrated network of genes and pathways that allow an organism to mount an orchestrated defense against toxic chemicals" (Goldstone et al., 2006).

## 11.4 Glycoproteins in marine organisms detoxification

Proteins are classified as ABC transporters based on the sequence and organization of their ATP-binding domains, also known as nucleotide-binding folds or NBFs (Goldstone et al., 2006). The characteristic motifs in all ATP-binding proteins, Walker A and Walker B, are separated by 90–120 amino acids. A functional protein is recognized by two NBFs and two TM domains, with 6–11 membrane-spanning α-helices that provide the substrate specificity. NBFs, located in the cytoplasm, transfer the energy to transport specific substrates across the membranes, mostly in a unidirectional fashion. This implies the movement of compounds from the cytoplasm to the outside of the cell or into an extracellular compartment, such as endoplasmic reticulum, mitochondria, or peroxisome. As mentioned before, ABC transporters can function as a shuttle of hydrophobic or slightly hydrophilic compounds either as a part of a metabolic process or in the transport of foreign compounds in the elimination process. ABC genes are widely distributed on eukaryotic genomes and are highly conserved between species. In mammals, there are seven ABC gene subfamilies and the last report indicates a total of 48 known human genes, the best known in nature. From these reports, three well-known examples are the P-gps, members of the ABCB subfamily; the MDR proteins or ABCC proteins; and the MXR or ABCG proteins (Dean, 2002).

MXR, in the context of marine environments, was proposed for the first time as a cellular defense mechanism more than 20 years ago (Kurelec, 1992). Since then, the question of how cells and marine organisms thrive in contaminated waters has gradually found an answer. Although there are much less studies on P-gps related to humans and cancer biology, MXR has attracted interest from researchers due to the central role of these proteins in marine species survival. At present, there are several lines of evidence that have shown conclusively the involvement of P-gps in the detoxification process. For both, mollusks and fish, excellent reviews have been published to account for the facts that make aquatic organisms to adapt, survive, and reproduce in polluted waters.

## 11.4.1   Mussel ABC proteins as a model of a barrier with MXR

Oysters, clams, and mussels are frequently used as biomonitors of marine contaminants and as test organisms in myriads of toxicological studies due to their filter-feeding characteristic. On taking their food from the surroundings, they are capable of separating seawater discarding it and, at the same time, keeping solutes. As a result, filter feeders concentrate on any contaminant possibly present in their environment. Taking advantage of this biological feature, researchers have been able to measure any accumulated chemical or biomarker searching for answers about the state of health of both marine organisms and the environment. Mussels (in particular, *Mytilus edulis* L.) have been useful in the field of ecotoxicology because they have a higher resistance to pollutants than oysters and clams, and up to now, they are not related to episodes of massive mortality or to pathogens infection (Costa et al., 2009).

Soon after the evidence that a multixenobiotic transport was presented in some marine organisms, a mussel, and several sponges (Kurelec, 1992), Cornwall and his colleagues characterized a likely candidate to be responsible for the phenomenon: they found a 170 kDa protein in the gills of the mussel *Mytilus californianus* related to MDR transport protein in various ways (Cornwall et al., 1995). These similarities included binding membrane capacity, verapamil inhibition, and cross-reactivity with an antibody to a conserved region of the mammalian MDR protein. Also, this protein, as MDR transport protein, acted on efflux of moderately hydrophobic compounds. These findings provided some added insight into the resistance of invertebrates to xenobiotics. A new breakthrough for characterization came with the analysis of processes in gill tissue. Once again, a mussel, *M. californianus*, was examined to have an insight into the nature of transporter function, gene expression, and localization of transport protein (Luckenbach and Epel, 2008). In this work, Lukenbach and Epel cloned two complete sequence coding for ABC with high degrees of identity to mammalian MDR transporters of the type of ABCB and ABCC subfamilies. Overall identity of mussel ABCB with ABCB1, ABCB4, and ABCB11 P-gps in vertebrates accounts for the resistance of mussels against a broad array of xenobiotics and the expression pattern in many tissues. All these findings were the support to demonstrate that ABC transporters form an active, physiological barrier at the tissue–environment interface in mussel gills, protecting the animal against environmental xenotoxicants.

As P-gp expression in mussels indicates the possibility of the MXR induction as a response to organic pollution, studies with inducers and inhibitors have been performed. Results have confirmed that P-gp activity is induced according to the level of seawater pollution (Smital et al., 2003). It is also induced by the concentration of various chemical stressors including substrates for P-gp, such as the pesticides pentachlorophenol and chlorthal, as well as nonsubstrates, such as 1,1-dichloro-2,2-bis(*p*-chlorophenyl) ethylene or DDE, a product of the DDT breakdown and sodium arsenite (Eufemia and Epel, 2000). Also, it has been found that multixenobiotic defense systems are compromised by synthetic musk fragrances, which contain compounds widely used in pharmaceutical and personal care products, thus enabling toxic MXR substrates to accumulate in cells of marine organisms (Luckenbach et al., 2004).

## 11.5   Summary

Microorganisms, plants, and animals from the sea have developed specialized mechanisms to deal with harsh conditions in their habitat, including harmful xenobiotics. Glycoproteins play an important role in xenobiotics elimination by marine organisms, since they are part of the chemical defensome in marine habitats. The multixenobiotic response is a cellular mechanism against harmful compounds mediated by the P-gp, an efflux transporter belonging to the ABC proteins. P-gp in mussel was first discovered more than 20 years ago, and it has been characterized as a 170 kDa protein possessing the structural motifs of ABC proteins previously described for terrestrial mammals. Also, it has similar patterns of induction, stimulation, and inhibition.

## References

Achard, M., Baudrimont, M., Boudou, A., and Bourdineaud, J. P. Induction of a multixenobiotic resistance protein (MXR) in the Asiatic clam *Corbicula fluminea* after heavy metals exposure. *Aquatic Toxicology* 67 (2004): 347–357.

Bard, S. Multixenobiotic resistance as a cellular defense mechanism in aquatic organisms. *Aquatic Toxicology* 48(4) (2000): 357–389.

Cornwall, R., Toomey, B. H., Bard, S., Bacon, C., Jarman, W. M., and Epel, D. Characterization of multixenobiotic/multidrug transport in the gills of the mussel *Mytilus californianus* and identification of environmental substrates. *Aquatic Toxicology* 31: 277–296.

Costa, M. M., Prado-Alvarez, M., Gestal, C., Li, H., Roch, P., Novoa, B., and Figueras, A. Functional and molecular immune response of Mediterranean mussel (*Mytilus galloprovincialis*) haemocytes against pathogen-associated molecular patterns and bacteria. *Fish & Shellfish Immunology* 26(3) (2009): 515–523.

Dean, M. C. *The Human ATP-Binding Cassette (ABC) Transporter Superfamily*. Bethesda, MD: NCBI, 2002.

Di Bello, D., Vaccaro, E., Longo, V., Regoli, F., Nigro, M., Benedetti, M., Gervasi, P. G., and Pretti, C. Presence and inducibility by β-naphthoflavone of CYP1A1, CYP1B1 and phase II enzymes in *Trematomus bernacchii*, an Antarctic fish. *Aquatic Toxicology* 84 (2007): 19–26.

Eufemia, N. A. and Epel, D. Induction of the multixenobiotic defense mechanism (MXR), P-glycoprotein, in the mussel *Mytilus californianus* as a general cellular response to environmental stresses. *Aquatic Toxicology* 49 (2000): 89–100.

Franzellitti, S., Buratti, S., Capolupo, M., Du, B., Haddad, S. P., Chambliss, C. K., and Fabbri, E. An exploratory investigation of various modes of action and potential adverse outcomes of fluoxetine in marine mussels. *Aquatic Toxicology* 151 (2014): 14–26.

Goldstone, J. V., Hamdoun, A., Cole, B. J., Howard-Ashby, M., Nebert, D. W., Scally, M., and Stegeman, J. J. The chemical defensome: Environmental sensing and response genes in the *Strongylocentrotus purpuratus* genome. *Developmental Biology* 300(1) (2006): 366–384.

González, J. F., Reimschuessel, R., Shaikh, B., and Kane, A. S. Kinetics of hepatic phase I and II biotransformation reactions in eight finfish species. *Marine Environmental Research* 67(4–5) (2009): 183–188.

Guengerich, F. P. Common and uncommon cytochrome P450 reactions related to metabolism and chemical toxicity. *Chemical Research in Toxicology* 14(6) (2001): 611–650.

Ikenaka, Y., Oguri, M., Saengtienchai, A., Nakayama, S. M. M., Ijiri, S., and Ishizuka, M. Characterization of phase-II conjugation reaction of polycyclic aromatic hydrocarbons in fish species: Unique pyrene metabolism and species specificity observed in fish species. *Environmental Toxicology and Pharmacology* 36(2) (2013): 567–578.

Ivanina, A. V. and Sokolova, I. M. Effects of cadmium exposure on expression and activity of P-glycoprotein in eastern oysters, *Crassostrea virginica* Gmelin. *Aquatic Toxicology* 88(1) (2008): 19–28.

James, M. O. Conjugation of organic pollutants in aquatic species. *Environmental Health Perspectives* 71 (1987): 97.

James, M. O. Isolation of cytochrome P450 from hepatopancreas microsomes of the spiny lobster, *Panulirus argus*, and determination of catalytic activity with NADPH cytochrome P450 reductase from vertebrate liver. *Archives of Biochemistry and Biophysics* 228(1) (1990): 8–17.

James, M. O. and Boyle, S. M. Cytochromes P450 in crustacea. *Comparative Biochemistry and Physiology. Part C, Pharmacology, Toxicology & Endocrinology* 121 (1998): 1–3.

Jørgensen, A., Giessing, A. M. B., Rasmussen, L. J., and Andersen, O. Biotransformation of polycyclic aromatic hydrocarbons in marine polychaetes. *Marine Environmental Research* 65(2) (2008): 171–186.

Juliano, R. L. and Ling, V. A surface glycoprotein modulating drug permeability in *Chinese hamster* ovary cell mutants. *Biochimica et Biophysica Acta* 455(1) (1976): 152–162.

Kim, R. O., Kim, B. M., Jeong, C. B., Nelson, D. R., Lee, J. S., and Rhee, J. S. Expression pattern of entire cytochrome P450 genes and response of defensomes in the benzo[a]pyrene-exposed monogonont rotifer *Brachionus koreanus*. *Environmental Science and Technology* 47 (2013): 13804–13812.

Kingtong, S., Chitramvong, Y., and Janvilisri, T. ATP-binding cassette multidrug transporters in Indian-rock oyster *Saccostrea forskali* and their role in the export of an environmental organic pollutant tributyltin. *Aquatic Toxicology* 85(2) (2007): 124–132.

Kondraganti, S. R., Fernandez-Salguero, P., Gonzalez, F. J., Ramos, K. S., Jiang, W., and Moorthy, B. Polycyclic aromatic hydrocarbon-inducible DNA adducts: Evidence by 32P-postlabeling and use of knockout mice for Ah receptor-independent mechanisms of metabolic activation in vivo. *International Journal of Cancer. Journal International Du Cancer* 103(1) (2003): 5–11.

Kroeger, N., Bergsdorf, C., and Sumper, M. A new calcium binding glycoprotein family constitutes a major diatom cell wall component. *The EMBO Journal* 13(19) (1994): 4676.

Kurelec, B. The multixenobiotic resistance mechanism in aquatic organisms. *Critical Reviews in Toxicology* 22(1) (1992): 23–43.

Kurelec, B. A new type of hazardous chemical: The chemosensitizers of multixenobiotic resistance. *Environmental Health Perspectives* 105 (1997): 855–860.

Livingstone, D. R. The fate of organic xenobiotics in aquatic ecosystems: Quantitative and qualitative differences in biotransformation by invertebrates and fish. *Comparative Biochemistry and Physiology: A Molecular and Integrative Physiology* 120 (1998): 43–49.

Livingstone, D. R., Moore, M. N., Lowe, D. M., Nasci, C., and Farrar, S. V. CYP2 responses of microsomal NADPH-cytochrome c reductase activity and cytochrome P450 in digestive glands of *Mytilus edulis* and *Littorina littorea* to environmental and experimental exposure to pollutants. *Aquatic Toxicology* 7 (1985): 79–91.

Luckenbach, T., Corsi, I., and Epel, D. Fatal attraction: Synthetic musk fragrances compromise multixenobiotic defense systems in mussels. *Marine Environmental Research* 58 (2004): 215–219.

Luckenbach, T. and Epel, D. ABCB- and ABCC-type transporters confer multixenobiotic resistance and form an environment-tissue barrier in bivalve gills. *American Journal of Physiology. Regulatory, Integrative and Comparative Physiology* 294(6) (2008): R1919–R1929.

Omiecinski, C. J., Vanden Heuvel, J. P., Perdew, G. H., and Peters, J. M. Xenobiotic metabolism, disposition, and regulation by receptors: From biochemical phenomenon to predictors of major toxicities. *Toxicological Sciences: An Official Journal of the Society of Toxicology* 120(Suppl. 1) (2011): S49–S75.

Pang, K. S., Maeng, H. J., and Fan, J. Interplay of transporters and enzymes in drug and metabolite processing. *Molecular Pharmaceutics* 6 (2009): 6.

Petroutsos, D., Wang, J., Katapodis, P., Kekos, D., Sommerfeld, M., and Hu, Q. Toxicity and metabolism of p-chlorophenol in the marine microalga *Tetraselmis* marina. *Aquatic Toxicology* 85(3) (2007): 192–201.

Pflugmacher, S. and Sandermann, H. J. Cytochrome P450 monooxygenases for fatty acids and xenobiotics in marine macroalgae. *Plant Physiology* 117(1) (1998): 123–128.

Shahraki, J., Motallebi, A., and Pourahmad, J. Oxidative mechanisms of fish hepatocyte toxicity by the harmful dinoflagellate *Cochlodinium polykrikoides*. *Marine Environmental Research* 87–88 (2013): 52–60.

Shimada, T., Hayes, C. L., Yamazaki, H., Amin, S., Hecht, S. S., Guengerich, F. P., and Sutter, T. R. Activation of chemically diverse procarcinogens by human cytochrome P-450 1B1. *Cancer Research* 56(13) (1996): 2979–2984.

Smital, T., Sauerborn, R., and Hackenberger, B. K. Inducibility of the P-glycoprotein transport activity in the marine mussel *Mytilus galloprovincialis* and the freshwater mussel *Dreissena polymorpha*. *Aquatic Toxicology* 65(4) (2003): 443–465.

Solé, M., Porte, C., and Albaigés, J. Mixed-function oxygenase system components and antioxidant enzymes in different marine bivalves: Its relation with contaminant body burdens. *Aquatic Toxicology* 30(3) (1994): 271–283.

Teramitsu, I., Yamamoto, Y., Chiba, I., and Iwata, H. Identification of novel cytochrome P450 1A genes from five marine mammal species. *Aquatic Toxicology* 51 (2000): 145–153.

Tilley, R. E., Kemp, G. D., and Hall, A. J. Cryostorage of hepatic microsomes from two marine mammal species: Effects on cytochrome P450-monooxygenase activities and content. *Marine Pollution Bulletin* 46(5) (2003): 654–658.

Torres, M. A., Barros, M. P., Campos, S. C. G., Pinto, E., Rajamani, S., Sayre, R. T., and Colepicolo, P. Biochemical biomarkers in algae and marine pollution: A review. *Ecotoxicology and Environmental Safety* 71(1) (2008): 1–15.

Trisciani, A., Perra, G., Caruso, T., Focardi, S., and Corsi, I. Phase I and II biotransformation enzymes and polycyclic aromatic hydrocarbons in the Mediterranean mussel (*Mytilus galloprovincialis*, Lamarck, 1819) collected in front of an oil refinery. *Marine Environmental Research* 79 (2012): 29–36.

Uno, T., Ishizuka, M., and Itakura, T. Cytochrome P450 (CYP) in fish. *Environmental Toxicology and Pharmacology* 34(1) (2012): 1–13.

Venn, A. A, Quinn, J., Jones, R., and Bodnar, A. P-glycoprotein (multi-xenobiotic resistance) and heat shock protein gene expression in the reef coral *Montastraea franksi* in response to environmental toxicants. *Aquatic Toxicology* 93(4) (2009): 188–195.

Wang, Y. M., Ong, S. S., Chai, S. C., and Chen, T. Role of CAR and PXR in xenobiotic sensing and metabolism. *Expert Opinion on Drug Metabolism & Toxicology* 8(7) (2012): 803–817.

Warshawsky, D., Cody, T., Radike, M., Reilman, R., Schumann, B., LaDow, K., and Schneider, J. Biotransformation of benzo[a]pyrene and other polycyclic aromatic hydrocarbons and heterocyclic analogs by several green algae and other algal species under gold and white light. *Chemico-Biological Interactions* 97(2) (1995): 131–148.

Yoshimoto, F. K. and Auchus, R. J. The diverse chemistry of cytochrome P450 17A1 (P450c17, CYP17A1). *The Journal of Steroid Biochemistry and Molecular Biology* 1 (2014): 1–14.

*chapter twelve*

# Recent trends in bioprospecting of marine collagen

*Kirti and Samanta S. Khora*

## Contents

## 12.1  Introduction

Marine bioprospecting is the search for novel compounds from natural sources in the marine environment. Marine ecosystem is a rich source of a variety of organisms that are source of immeasurable value for obtaining compounds that can be used in drug development, cosmetics, and other value-added products. Marine ecosystem is generally preferred for bioprospecting because of its great diversity, but much of the marine biome has not been studied and is underinvestigated. There is a high probability of finding new bioactive compounds from the marine environment that will be of great benefit. The increase in discoveries from the marine environment is basically due to technological advancements in exploring the ocean and the genetic diversity it contains.

Collagen is the most abundant structural protein in animals (Muyonga et al., 2004). It is fibrous in nature and makes up about one-third of the total amount of proteins in the body. It is an essential component of all body tissues such as the skin, bone, tendons, muscles, cartilage, and blood vessels. Gelatin is a colorless, translucent protein fraction derived from parent protein collagen, which is present in the tendons, ligaments, and tissues of mammals, by partially hydrolyzing the parent protein. Collagen's molecular structure consists of three polypeptide α-chains that are twisted together to form a triple helix. The distinctive feature of collagen is the regular arrangement of amino acids in each of the three chains of the collagen subunit. The sequence of amino acids is characterized by a repetitive unit of glycine (Gly)-proline (Pro)-X, or Gly-X-hydroxyproline (Hyp), where Gly accounts for one-third of the sequence while X and Y can be any other amino acid residue (Gorham, 1991).

Most of the industries generally use cows and pigs as the source of collagen. However, concerns related to the transmission of highly infectious diseases such as bovine spongiform encephalopathy, transmissible

*Table 12.1* Different sources of marine collagen

| Organisms | Phylum | Tissue/organ | References |
|---|---|---|---|
| Invertebrates | | | |
| *Chondrosia reniformis* | Porifera | Sponge material | Swatschek et al. (2002) |
| *Spirastrella inconstans* | Porifera | Sponge material | Sudharsan et al. (2013) |
| *Cyanea nozakii* | Cnidaria | Umbrella | Zhang et al. (2014) |
| *Asterias amurensis* | Echinodermata | Body tissues | Lee et al. (2009) |
| *Holothuria parva* | Echinodermata | Skin | Adibzadeh et al. (2014) |
| *Penaeus chinensis* | Arthropoda | Muscle | Minamisako and Kimura (1989) |
| Fish | | | |
| *Lagocephalus gloveri* | Chordata | Skin | Senaratne et al. (2006) |
| *Parupeneus heptacanthus* | Chordata | Scales | Matmaroh et al. (2011) |
| *Katsuwonus pelamis* | Chordata | Bones | Nagai and Suzuki (2000) |
| *Thunnus obesus* | Chordata | Bones | Jeong et al. (2013) |
| *Diodon holocanthus* | Chordata | Skin | Huang et al. (2011) |
| Mammals | | | |
| *Balaenoptera acutorostrata* | Chordata | Body pieces | Nagai et al. (2008) |

spongiform encephalopathy, and foot and mouth disease from pigs and cattle have limited the extraction of collagen from these sources for industrial purposes (Kittiphattanabawon et al., 2005). The major advantage of collagen extracted from marine sources is that it is free of these risks; also, it is not associated with religious issues. Fisheries and the aquaculture industry produce million tons of fish wastes per annum all over the world. Marine processing by-products or fish wastes such as skin, scales, bones, and fins have high collagen content. This waste, if utilized in the manufacture of value-added products such as collagen, could contribute significantly to the economic value of fish. Fish skin is a major by-product of the fish processing industry, causing wastage and pollution, but could provide a valuable source of collagen. Collagen is widely used in cosmetic surgery, anti-wrinkle creams, skin care products, shampoos, wound-healing aids, treating burns, and in the food industry. Collagen has excellent biocompatibility, low antigenicity, high biodegradability, and good mechanical and cell-binding properties, which has enabled the use of collagen as a biomaterial in a variety of connective tissue applications (Nimni, 1988; Friess, 1998; Lee et al. 2001).

## 12.2   Sources of marine collagen

The marine species available as the source of collagen can be roughly divided into three categories: marine invertebrates such as cuttlefish, jellyfish, star fish, and sea urchin; sea mammals such as seal and whale; and fishes such as jawless fish, cartilaginous fish, and bony

fish (Eastoe and Leach, 1977). Scientists have found that the skin, bone, scale, fin and cartilage of freshwater and marine fish, scallop mantle (Shen et al., 2007), the adductor of pearl oyster (Mizuta et al., 2002a,b), the muscle layer of ascidian (Mizuta et al., 2002a,b), and the marine sponge *Spirastrella inconstans* can be used as new sources of collagen. The different sources of marine collagen are listed in Table 12.1. Marine fishes are the prominent source of collagen. Based on their living environment, fishes are usually subdivided into four groups: hot-water fish, warm-water fish, cold-water fish, and ice-water fish (Eastoe and Leach, 1977). Collagen has been isolated and characterized from various marine sources such as the scales of marine fishes (Thuy et al., 2014), the skin of flatfish (Heu et al., 2010), scales of the spotted golden goatfish (Matmaroh et al., 2011), skin and bone of bigeye snapper (Kittiphattanabawon, 2005), and the skin of *Lagocephalus gloveri* (Senaratne et al., 2006).

## 12.3   Conventional methods of collagen isolation

Collagen is the major component of the extracellular matrix, and more than 27 genetic isoforms have been identified. Collagen type I, II, and III are the most abundant and well investigated for biomedical applications (Mocan et al., 2011). There are several methods of collagen isolation. Extraction with neutral salt as well as acid and enzymatic solutions are the main methods used for collagen type I isolation (Wohllebe and Carmichael, 1978; Kuznetsova and Leikin, 1999; Xong, 2008).

### 12.3.1 Salt precipitation method

This method involves the treatment of tendon pieces with neutral salt solutions such as 0.05 M $Na_2HPO_4$, pH, 8.7–9.1. Then the concentration of sodium chloride is gradually increased to 4 M, which facilitates collagen isolation. The supernatant containing the salt-soluble fraction of collagen is dialyzed against 0.01 M $Na_2HPO_4$ (Wohllebe and Carmichael, 1978).

### 12.3.2 Acid-soluble collagen isolation method

Collagen extraction is often carried out by direct extraction with organic acids such as acetic acid, citric acid, or lactic acid (Skierka and Sadowska, 2007). The solubility of collagen in acid solution plays the main role in the extraction efficiency. Among all acids, acetic acid is the most promising solvent in extracting collagen from different sources (Muyonga et al., 2004; Kittiphattanabawon et al., 2005; Wang et al., 2009; Singh et al., 2011). This method involves the treatment of tendon pieces with 0.5 M acetic acid in the presence of 5 mM EDTA for 2–4 days; pH, 2.5–3. Centrifugation is done and the supernatant thus collected containing acid-soluble fractions of collagen is dialyzed against 0.05 M $Na_2HPO_4$ (Brodsky, 1982). Acid-solubilized collagen has been extracted from different sources such as the scales of marine fishes from Japan and Vietnam (Thuy et al., 2014), skin of flat fish (Heu et al., 2010), scales of the golden goatfish (Matmaroh et al., 2011), from skin and bone of bigeye snapper (Kittiphattanabawon et al., 2005), skins of underutilized fishes (Bae et al., 2008), umbrellas of jellyfish (Zhang et al., 2014), and the skin wastes of the marine eel (Veeruraj et al., 2013).

The yield of collagen by acid extraction may be influenced by many factors, such as the acid concentration, the ratio of raw material to the acid solution, incubation temperature, and incubation time. The yield of collagen not only varies with the processing conditions but also depends on raw materials based on species and tissues. Many researchers have obtained high collagen yield from fish skin and bone by the acid extraction method. However, for some marine species, particularly marine invertebrates like jellyfish, the yield of acid-solubilized collagen was low. Hence, an extraction process involving partial enzymatic digestion is often used to increase the yield. The different procedures used for the extraction of collagen from different sources are listed in Table 12.2.

### 12.3.3 Enzyme-soluble collagen isolation method

To increase the yield of extracted collagen, various enzymes, such as trypsin and α-amylase, have been used to facilitate the solubilization of collagen (Johns and Courts, 1977). Pepsin proteolysis is widely used for marine collagen extraction. This method involves the treatment of tendon pieces in acetic acid in the presence of 5 mM EDTA and pepsin with concentration 0.05 g/100 g of tissue at 4°C for 2–4 days (Mocan et al., 2011). The collagen obtained by this method is called "pepsin-soluble collagen." This method of collagen extraction is also highly influenced by the incubation temperature and time (Nomura et al., 1996). Increase in temperature and time also influence the yield significantly. Generally, most of the preparation steps are done at low temperature, that is, 4°C–6°C. Many researchers have studied the extraction of pepsin-solubilized collagen from different sources such as the skin of flat fish (Heu et al., 2010), scales of golden goatfish (Matmaroh et al., 2011), skin of brownstripe red snapper (Jongjareonrak et al., 2005), and umbrellas of jellyfish (Zhang et al., 2014).

## 12.4 Yields of marine collagen

The yields of collagen extracted from different marine sources are influenced by the source of the raw material, the extraction method, the nature and concentration of the acid or alkali used, incubation temperature, and the incubation time. The collagen content can decrease as a result of denaturation of the protein during the extraction process or due to differences in the environmental temperature (Nagai et al., 2002). The yield of acid-solubilized collagen from the skin of flatfish was found to be 57.3% and pepsin-solubilized collagen was 85.5% (Heu et al., 2010). The yield was found to increase by the use of pepsin. Similarly, the yields of acid-soluble collagen (ASC) and pepsin-soluble collagen (PSC) from the skin of striped catfish were found to be 5.1% and 7.7%, respectively, based on the wet weight of the skin (Singh et al., 2011). The yields of collagens from the skin of Japanese sea bass, Chub mackerel, and Bullhead shark were found to be 51.4%, 49.8%, and 50.1%, respectively, on the basis of lyophilized dry weight (Kolodziejska et al., 1999). The collagen content in cod skins was found to be dependent on the fishing season. The collagen content in the skin of Baltic cod (*Gadus morhua*) was 21.5% on wet weight basis and ~71.2% on dry weight basis (Sadowska et al., 2003). The yield of ASC from the umbrella of jellyfish (*Cyanea nozakii*) was 13.0% (dry weight), which was more than that of PSC, that is, 5.5% (dry weight). The ASC yields on a dry weight basis for fresh water young and adult Nile perch (*Lates niloticus*) were 63.1% and 58.7%, respectively (Muyonga et al., 2004). The yield percentage of ASC and PSC from different sources is listed in Table 12.3.

*Table 12.2* Procedures for the isolation of collagen from different marine sources

| Organism | Tissue/organ | Procedure | References |
|---|---|---|---|
| Lizard fish, gray mullet, yellowback seabream | Scales | Treatment of scales with 0.1 M NaOH for 6 h, then demineralization with 0.5 M Na$_2$EDTA for 24 h, wash with cold distilled water, extraction with acetic acid for 4 days, centrifugation of the extract at 20,000$g$ for 1 h, salting out with NaCl, collection of precipitate by centrifugation, dissolution in acetic acid, dialysis, and lyophilization. | Thuy et al. (2014) |
| Jellyfish | Umbrella | Extraction with 0.5 M acetic acid for 3 days, centrifugation at 20,000$g$ for 1 h, salting out of the supernatant by NaCl, precipitate collection by centrifugation at 20,000$g$ for 30 min, dissolution of the pellet in 0.5 M acetic acid, dialysis against 0.1 M acetic acid, and lyophilization. | Zhang et al. (2014) |
| Catfish | Skin | Removal of non-collagenous protein of the skin by mixing with 0.1 M NaOH for 6 h, washing with cold water, defatting with 10% butyl alcohol for 48 h, washing with cold water, soaking in 0.5 M acetic acid for 24 h, filtration, precipitation of collagen by adding NaCl, collection of the precipitate by centrifugation, dissolving of the pellet in acetic acid, and dialysis. | Singh et al. (2011) |
| Red sea cucumbers | Body wall | Washing the body wall with distilled water, treating with a disaggregating solution (0.5 M NaCl + 50 mM EDTA + 0.2 M β-mercaptoethanol + 0.1 M Tris–HCl), stirring for 3 days, filtering the suspension and the centrifugation of the filtrate, then again treating with 0.1 M NaOH and stirring, suspending the collagen fibrils in acetic acid with porcine pepsin for 2 days, salting out by adding NaCl to final 0.8 M, centrifugation, dialysis, and lyophilization. | Park et al. (2012) |
| Silvertip shark | Skeletal and head bone | Homogenizing the bone in phosphate buffer, treating the homogenate with double-distilled water, decalcification with 0.5 M EDTA, then soaking the shark bone in 0.5 M acetic acid containing 1% pepsin for 4 days, centrifugation at 15,400$g$ for 30 min, salting out the supernatant by adding 2 M NaCl, redissolving the precipitate in a minimum volume of 0.5 M acetic acid, dialysis, and lyophilization. | Jeevithan et al. (2014) |

*Table 12.3* Yields of acid-soluble collagen (ASC) and pepsin-soluble collagen (PSC) from different marine sources

| Organisms | Tissue/organ | Type of collagen | Yield, % | References |
|---|---|---|---|---|
| Flatfish | Skin | ASC | 57.3 | Heu et al. (2010) |
| | | PSC | 85.5 | |
| Striped catfish | Skin | ASC | 5.1 | Singh et al. (2011) |
| | | PSC | 7.7 | |
| Japanese sea bass | Caudal fin | ASC | 5.2 | Nagai (2004) |
| Baltic cod | Skin | ASC | 21.5 (wet weight of skin) | Sadowska et al. (2003) |
| | | | 71.2 (dry weight of skin) | |
| Jellyfish | Umbrella | ASC | 13.0 | Zhang et al. (2014) |
| | | PSC | 5.5 | |
| Brown-backed toadfish | Skin | PSC | 54.3 | Senaratne et al. (2006) |
| Eel | Skin | ASC | 80 | Veeruraj et al. (2013) |
| | | PSC | 7.10 | |

## 12.5 Structure of marine collagen

Collagen is the predominant fibrous protein in animals. It is a rod-like structure that resists stretching of the cell and provides high tensile strength. All collagens are composed of three polypeptide α-chains coiled around each other to form the triple-helix configuration. Tropocollagen is a subunit of larger collagen fibril aggregates. The diameter of the triple helix is ~1.5 nm and the length of each subunit is ~300 nm. These three left-handed α-chains are twisted together into a right-handed coiled-coil triple helix, which represents a quaternary structure of collagen that is stabilized by numerous hydrogen-bonds and van der Waals interaction (Brinckmann et al., 2005) as well as some covalent bonds (Harkness, 1966), and further associated into right-handed microfibrils and fibrils, which are further assembled into collagen fibers (He et al., 2011). Depending on the type of collagen, the molecule may be made up of either three identical α-chains (homotrimers) or two or three different α-chains (heterotrimers). The distinctive feature of collagen is the regular arrangement of amino acids in each of the three chains of collagen subunits, and the primary structure of collagen shows sequence homology across genus and adjacent family line. Collagen has the most unusual amino acid composition in which glycine (Gly), proline (Pro), hydroxyproline (Hyp), and hydroxylysine (Hyl) are dominant. Glycine accounts for one-third of the sequence. These amino acids are arranged in a repetitive tripeptide sequence, Gly-X-Y, where X can be any amino acid but is frequently hydroxyproline or hydroxylysine (Brinckmann et al., 2005). X-ray diffraction, electron microscopy, hydrodynamic methods, and a variety of spectroscopy techniques are used extensively in characterizing marine collagen. The fibrillar structure of collagen extracted from marine eel was observed under scanning electron microphotography. Collagen fibrils were found to form bundles and were not parallel. These collagen bundles varied in width and thickness, often gave branches, and intertwined with each other (Veeruraj et al., 2013). Collagen from the skin and bone of Spanish mackerel displayed a uniform and regular network ultrastructure as observed by SEM microphotography (Li et al., 2013). The fish collagens have a characteristic amino acid pattern of that of calf collagen; however, there are differences in the contents of imino acids and amino acids with a hydroxyl group in the side chain. The stability of the α-chain helix in the secondary structure of protein is believed to be through restricted rotation imposed by the imino acids in the peptide linkage. Both hydrogen bonds and intramolecular covalent cross-links are responsible for providing stability to tertiary structure of collagen. The number and nature of bonds and cross-links influences the physical properties and the stability of collagen (Nagai, 2004). Fourier transform infrared (FTIR) spectra of collagen describe the regions of amide I, amide II, and amide III, which are related to the configuration of the polypeptide. The molecular structure of marine collagen is shown in Figure 12.1.

## 12.6 Physiochemical properties of marine collagen

Marine collagen and gelatin, especially those from fish, have lower denaturation temperature than mammalian gelatins, and care is needed for their preparation and storage, as there are high chances of microbiological attack and thermal denaturation (Jones, 1977). The lower hydroxyproline content of fish collagen compared to bovine collagen contributes to its low stability (Winter and Page, 2000). The thermal denaturation temperature of eel skin collagen was found to 38.5°C (ASC) and 35.0°C (PSC). This can be due to the higher cross-linkage of marine eel skin collagen (Veeruraj et al., 2013). The thermal denaturation temperature ($T_d$) of skin collagen from pig and calf is 37°C and 40.8°C, respectively (Ikoma et al., 2003). The higher imino acid content facilitates intra and intermolecular cross-linking, which in turn provides more stability to the triple-helical structure of the collagen molecule. Generally, cold-water fish collagen has low melting temperature because its imino acid content is very low (Sadowska et al., 2003). Marine collagen with a high denaturation temperature finds wide application because its close denaturation temperature with mammalian collagen. The denaturation temperature of marine collagen is lower than that of land-animal collagen. Thermal stability of collagen not only depends on the imino acid content but is also related to the environmental and body temperature of the fish species (Rigby, 1968). Viscosity is the other important property of collagen and gelatin for their commercial applications. Denaturation of collagen structure caused by heat treatment is associated with the changes in viscosity (Nagai et al., 1999). The $T_d$ of *Lagocephalus gloveri* was found to be 28°C (Senaratne et al., 2006), that of ocellate pufferfish 28°C (Nagai et al., 2002), that of the skin of tiger puffer 28.4°C, and that of red stingray 32.2°C (Bae et al., 2008). Physiochemical properties of collagen from different marine sources are listed in Table 12.4.

## 12.7 Synthesis of gelatin from collagen

Gelatin is derived from collagen by partial thermal hydrolysis, and it has broad applications in the food and

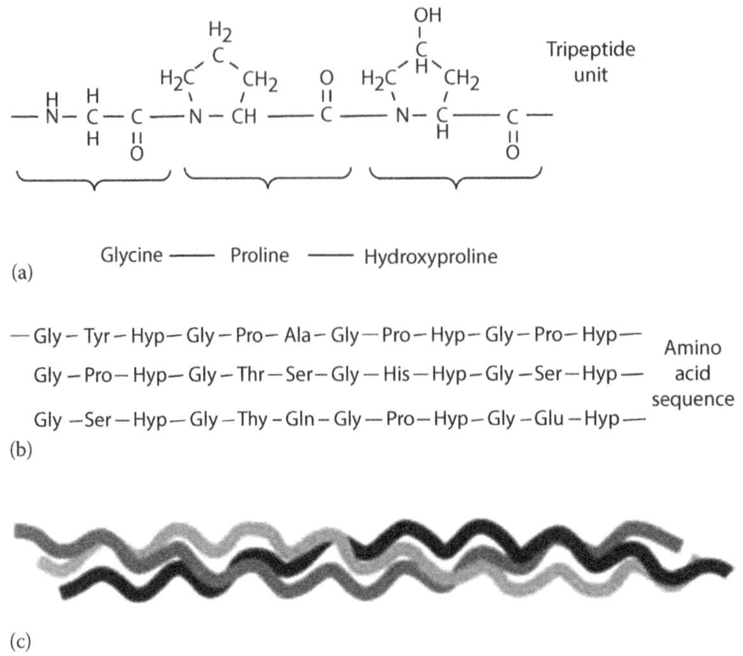

Figure 12.1 (a–c) Structure of collagen. (Adapted from Brodsky, B. et al., Collagen and gelatins, in *Biopolymers Online*, 2005, pp. 119–129.)

*Table 12.4* Physiochemical properties of collagen from different marine organisms

| Source | Proximate composition (%) | | Denaturation temperature | pH solubility | NaCl concentration | References |
|---|---|---|---|---|---|---|
| Scale of spotted golden goatfish | Moisture: 23.79 ± 1.57 Protein: 34.46 ± 0.1 Fat: 0.83 ± 0.1 Ash: 42.31 ± 0.56 | | ASC: 41.58°C PSC: 41.01°C | ASC: 4.96 PSC: 5.39 | ASC: 20 g/L PSC: 30 g/L | Matmorah et al. (2011) |
| Skin of brown-backed toadfish | Moisture: 73.4 Protein: 90.3 Fat: 1.3 Ash: 8.4 | | 28°C | — | — | Senaratne et al. (2006) |
| Skin of balloon fish | Moisture: 62.23 Protein: 21.95 Fat: 0.73 Ash: 15.87 | | ASC: 29.64°C PSC: 30.30°C | pH 1–5 | ASC: 1% PSC: 2% | Huang et al. (2011) |
| Umbrella of jellyfish | Moisture: 97.41 ± 0.09 Protein: 1.99 ± 0.06 Fat: 0.21 ± 0.05 Ash: 0.40 ± 0.05 | | ASC and PSC: 23.8°C | — | — | Zhang et al. (2014) |
| Skin and bone of big eye snapper | Skin Moisture: 64.08 ± 0.05 Protein: 32.0 ± 0.19 Fat: 0.98 ± 0.23 Ash: 3.23 ± 1.41 | Bone 62.27 ± 0.29 13.3 ± 0.43 8.77 ± 0.46 14.40 ± 0.68 | Skin: 31.0°C Bone: 31.5°C | Skin: 2 Bone: 5 | 3% NaCl | Kittiphattanabawon et al. (2005) |

*Note:* Data deficient.

pharmaceutical industries (Hao et al., 2009). A number of seafood and seafood-processing wastes such as the skin, scales, and bones have been studied as sources of gelatin. The yield and quality of gelatin are influenced not only by the species or tissue from which it is extracted but also by the synthesis process. The ultimate aim in gelatin production is the conversion of collagen into gelatin with maximum yield and good physiochemical properties. The acid extraction process of gelatin involves pretreatment of the raw materials with an acid solution followed by an extraction that is carried out in an acid medium (Gomez-Guillen and Montero, 2001). Acid treatment leads to the breakage of some interchain cross-linkages, which facilitates the extraction process. The period of acid pretreatment varies with the type of raw materials, and the pretreatment temperature influences the extraction process. The pretreatment temperature can be room temperature for sea mammals and warm-water fish, but the optimal temperature of gelatin extraction from cold-water fish is less than 10°C (Zhou and Regenstein, 2005). The alkaline extraction process of gelatin involves the pretreatment of the raw material with an alkaline solution, which can remove considerable amounts of non-collagenous materials (Zhou and Regenstein, 2005) and also break some interchain cross-linkages. Temperature plays a critical role in all pretreatment processes. Gelatin was extracted from the skin of dog shark (*Scoliodon sorrakowah*) and skipjack tuna (*Katsuwonus pelamis*), and their physiochemical properties were determined. There were considerable differences between the yield and functional properties of gelatin from the skin of these sources (Shyni et al., 2014).

## 12.8   Improving the quality of collagen from marine sources

Collagen and gelatin obtained from cold-water fish have low melting points and gelling temperature. However, many applications require them to be gels at room temperature. This property of gelatin can be enhanced by the use of transglutaminase (TGase). This enzyme catalyzes the reaction between a lysine residue and a glutamine residue, which helps in making covalent chemical cross-linkages within the gelatin network (Babin and Dickinson, 2001). Many studies have been carried out on the effects of TGase on fish gelatins from megrim, cod, and hake (Fernandez-Diaz et al., 2001; Gomez-Guillen et al., 2001; Kolodziejska et al., 2004).

The quality and yield of collagen and gelatin are influenced by the procedures used for extraction. Many studies have reported an increase in the yield of collagen with the use of pepsin during the process of extraction. It is reported that the pepsin cleaves the cross-linked

molecules at the telopeptide region, which results in increased collagen extraction efficacy (Nalinanon et al., 2010). Zhou and Regenstein (2004) reported that to obtain a high-quality gelatin extract, the alkaline concentration should be high enough to remove the non-collagenous protein and the acid concentration should be in the proper range to get an optimal weak acid extraction medium. The extraction temperature also affects the extraction process and yield.

Impurities, such as insoluble particles, inorganic salts, and pigments, present in extracted collagen and gelatin can be removed by chromatographic techniques and also by ultrafiltration (Simon et al., 2002). Activated charcoal can also be used to remove the compounds responsible for unpleasant fish flavors.

## 12.9   Bioactivity and applications of marine collagen

Collagen and gelatin are the source of biologically active peptides with favorable health benefits for nutritional or pharmaceutical applications. Both contain bioactive peptides that are inactive within their sequence and can be released during gastrointestinal digestion or by controlled enzymatic hydrolysis.

### 12.9.1   Antioxidant activity

Oxidation of biomolecules mediated by free radicals leads to unfavorable impacts on food and the biological system (Zhang et al., 2011). These free radicals, generated as the result of oxidation, cause disruption in the biological system by damaging the major biomolecules (DNA, protein, lipids, and small cellular molecules), which in turn leads to cardiovascular and neurodegenerative diseases (Esmaeili et al., 2011). Antioxidants play a vital role in both food systems as well as in the human body to reduce oxidative stress. At present, many synthetic antioxidants such as butylated hydroxytoluene (BHT), butylated hydroxyanisole (BHA), and *tert*-butylhydroquinone (TBHQ) are extensively used in order to reduce the damage caused by free radicals. However, the possible toxicity associated with these synthetic antioxidants, as well as possible consumer rejection, has led to decreased use of these products (Zhuang et al., 2012). Therefore, there is growing interest in the use of natural antioxidants. Because of the enormous volume of fish processing waste generated annually, a great deal of attention has been paid in the procurement of collagen antioxidant peptides from these marine sources. Recently, a number of researches have demonstrated that peptides derived from different marine collagen and gelatin hydrolysates act as potential antioxidants, which have been isolated from various marine sources such as brown-striped red snapper skin (Khantaphant and Benjakul, 2008), jelly fish umbrella (Zhuang et al., 2010), jumbo squid (Mendis et al., 2005; Rajapakse et al., 2005), oyster (Qian et al., 2008), blue mussel (Rajapakse et al., 2005), Alaska Pollock skin (Je et al., 2007), mackerel muscle protein (Wu et al., 2003), fish skin gelatin (Mendis et al., 2005), and tuna backbone (Je et al., 2007). Some of these collagen/gelatin-derived antioxidant peptides have been further purified and sequenced. The exact mechanism behind the antioxidant activity of these peptides is not fully known. Many researchers have reported the beneficial effects of peptides by scavenging free radicals and preventing oxidative damage by interrupting the radical chain reaction of lipid peroxidation (Mendis et al., 2005; Qian et al., 2008). It has been reported that hydrophobic amino acids play an important role in these antioxidant activities (Mendis et al., 2005). It is observed that of the different oxidative systems, the peptides derived from marine fish collagen and gelatin have greater antioxidant activities than α-tocopherol (Jung et al., 2005).

## 12.9.2   Antimicrobial properties

Antimicrobial peptides are basically involved in host defense and are widely distributed in nature. They are mostly cationic peptides and show the ability to kill target cells rapidly. The molecular mechanism of action of cationic peptides is not fully known. They generally adopt a strongly amphipathic or amphiphilic three-dimensional structure on interaction with bacterial membranes. The difference in membrane composition of Gram-positive and Gram-negative bacteria influences the mode of action and bacterial specificity of these antibacterial compounds (Floris et al., 2003). There is not enough information on the antimicrobial properties of collagen- or gelatin-derived peptides. Gomez-Guillen et al. (2010) reported the antimicrobial properties of peptide fractions obtained from tuna and squid skin gelatin. These gelatin hydrolysates were found to be highly active against different strains of bacteria (both Gram-positive and Gram-negative), mainly *Lactobacillus acidophilus* and *Bifidobacterium animalis* subsp. *Lactis*, *Shewanella putrefaciens*, and *Photobacterium phosphoreum*. Fractions obtained from squid were found inhibiting the growth of *Aeromona hydrophila*, whereas tuna hydrolysate fractions were found to be more active than squid ones for *Lactobacillus acidophilus*, *Pseudomonas aeruginosa*, and *Salmonella choleraesuis*. Fish gelatin peptides have a repeated motif of Gly-Pro-Ala triplets in their structure (Kim and Mendis, 2006), and this hydrophobic nature allows the peptide to enter the membrane and helps in pore formation as the positive charge would initiate the

peptide's interaction with the negatively charged bacterial surface (Wieprecht et al., 1997).

## 12.9.3   Anti-hypertensive properties

Hypertension or high blood pressure is one of the most common lifestyle-related diseases. It is one of the major risk factor for myocardial infarction, congestive heart failure, arteriosclerosis, stroke, and end-stage renal diseases (Kearney et al., 2005). Angiotensin-converting enzyme (ACE) is a vasoactive enzyme that regulates the blood pressure by means of the rennin–angiotensin system, and the inhibition of catalytic action of this enzyme suppress blood pressure elevation (Chen et al., 2007). Potent ACE inhibitory hydrolysates and peptides have been obtained from collagenous materials not only from terrestrial sources but also from sources such as fish skins (Byun and Kim, 2001; Nagai et al., 2006), fish cartilage (Nagai et al., 2006), scales (Fahmi et al., 2004), and sea cucumbers (Zhao et al., 2007). The structure–activity relationship of ACE inhibitory peptides has not yet been established. It has been observed that ACE-inhibitory peptides compete with biologically important ACE substrates by interaction of their C-terminal tripeptide residues (Ondetti and Cushman, 1982). The high concentration of hydrophobic amino acids and high proline contents also contribute to the ACE inhibitory activity of collagen and gelatin hydrolysates (Vermeirssen et al., 2004; Contreras et al., 2009).

## 12.9.4   Pharmaceutical applications of collagen and gelatin

Gelatin, a hydrolyzed product of parent protein collagen, is used in the manufacture of tablets, capsules, and pastilles. Gelatin is used for coating drugs, in the preparation of hard and soft capsules, as an antioxidant hydrolysate, in wound healing, and as a component in dietry health supplements. Other pharmaceutical applications include its use in tablets, emulsions, ointments, syrups, as a matrix for implants, and in intravenous infusions (Saddler and Horsey, 1987; Pollack, 1990; Rao, 1995). Fish gelatin, particularly from cold-water fish sources in the nongelling and nonhydrolyzed form, was used as a carrier in a pharmaceutical composition designed in such a way that releases the active ingredient in the oral cavity on contact with saliva (Murray et al., 2004).

It is reported that the oral consumption of collagen peptide provides beneficial effects to the body in improving joint health and as a food supplement, which improves low bone mineral density in people suffering from malnutrition and degenerative joint diseases (Wu et al., 2004). Other than this, collagen peptides can thicken hair (Scala et al., 1976), treat nail disorders such as brittle nails (Minaguchi et al., 2005), and stimulate the formation of collagen fibrils in the dermis, and so on (Matsuda et al., 2006).

## 12.9.5   Role of collagen in bone and joint disorders

Osteoarthritis and osteoporosis are major health concerns related to musculoskeletal disorders. Their treatment includes analgesics, anti-inflammatory agents, and lubricating and nutritional supplements. Collagen hydrolysates are safe compounds with low toxicity, and they provide greater symptomatic relief than pharmaceutical drugs (Aleman and Alvarez, 2013). It was found in clinical studies that in patients suffering from joint diseases, gelatin hydrolysates seemed to exert a direct effect on the cartilage (Moskowitz, 2000). It has also been observed that some gelatin-derived peptides are absorbed in the intestine when orally administered and accumulate preferentially in the cartilage, where they finally may stimulate cartilage metabolism (Oesser et al., 1999). Fish collagen hydrolysates have shown marked effect on chondrogenic differentiation of equine adipose tissue-derived stromal cells (Raabe et al., 2010).

## 12.9.6   Application of marine collagen in tissue engineering and regeneration

Collagen is ubiquitous in the mammalian body. There are various biomedical applications of marine fish collagens. Collagen is the major constituent of the extracellular matrix (Lin et al., 2009). Langer and colleagues (Langer, 1990; Langer and Vacanti, 1993, 1999) first elucidated the basic principles of tissue engineering; that is, cells, genes, and proteins are delivered via a degradable material, termed a scaffold, in order to generate tissue. Green et al. (2002) first reported the fiber skeleton of natural marine sponge and suggested its application for tissue-engineered bone. Lin et al. (2011) identified and characterized natural marine sponges as potential bioscaffolds for osteogenesis. Their study indicated that the natural marine sponge skeleton is favorable as a new source of bioscaffold for the repair of bone defects. Jellyfish collagen is also an important source as a matrix component for tissue engineering because of the presence of low impurity. On a dry-weight basis, edible jellyfish contains more than 40% collagen (Ehrlich, 2010). Fibrillized collagen scaffolds were reported from the jellyfish *Rhopilema esculentum* (Hoyer et al., 2012). Different parameters influencing fibril formation of jellyfish collagen were also analyzed. It was observed that collagen fibrils producing a matrix mimic the cartilage extracellular matrix.

## 12.9.7   Applications of collagen and gelatin in food industry

The nutritive value of a protein can be assessed by determining its amino acid content and comparing it with essential amino acid content of an ideal reference protein (Fennema, 1996). Collagen and gelatin are usually used as functional ingredients rather than nutritional ingredients in food application, as collagen and gelatin are deficient in several essential amino acids, and it will be very difficult to achieve satisfactory improvement. Collagen can be used as a clarifying agent to remove colloidal suspensions during the production of alcoholic drinks and fruit juices (Courts, 1977). Collagen casings are mainly produced from the collagen extracted from cattle, sheep, and pig hides. It can also be derived from fish. Because of the high demand for sausage products, collagen casings were developed, which had some advantages over the natural gut casings (Hood, 1987). Gelatin obtained from the degradation of collagen can be applied in wine fining and juice clarification. Gelatin in food provides a melt-in-the-mouth function. It is also used in meat products to absorb meat juices and to give them form (Gelatin Manufacturers Institute of America, 1993).

## 12.10   Conclusion

Collagen and gelatin from marine sources may not be able to completely replace those from terrestrial sources, but they still offer novel and viable properties and fulfill the demands of the food and pharmaceutical industries. Collagen can be obtained effectively from various marine sources such as marine invertebrates, sea mammals, and fish, and have potential for commercial applications. Marine collagen and gelatin can be extracted from the skin, scales, and bones of warm- or cold-water fish. This provides a lot of variation, particularly in the amino acid content, which leads to diverse applications in the food and pharmaceutical industries. There are several methods to improve the quality and yield of collagen and gelatin from marine sources. The quality of collagen and gelatin can be improved by optimization of the extraction process, by enzymatic modification, and by the removal of impurities. Fish wastes show wide variability in their composition and deficient microbiological quality, and these could limit the successful application of these wastes in health-related practices on a large scale. Much work required to be done to explore safe sources of marine collagen and to find their applicability in various fields. Development of advanced processes to enhance the quality and to remove the impurities would certainly make marine collagen a versatile ingredient for the food, pharmaceutical, and cosmetic industries.

# References

Adibzadeh, N., S. Aminzadeh, S. Jamilli, A. A. Karkhane, and N. Farrokhi. 2014. Purification and characterization of pepsin-solubilized collagen from skin of sea cucumber *Holothuria parva*. *Applied Biochemistry and Biotechnology* 173:143–154.

Aleman, A. and O. M. Alvarez. 2013. Marine collagen as a source of bioactive molecules: A review. *The Natural Products Journal* 3:105–114.

Babin, H. and E. Dickinson. 2001. Influence of transglutaminase treatment on the thermoreversible gelation of gelatin. *Food Hydrocolloids* 15:271–276.

Bae, I., K. Osatomi, A. Yoshida, K. Osako, A. Yamaguchi, and K. Hara. 2008. Biochemical properties of acid—Soluble collagen extracted from the skins of underutilised fishes. *Food Chemistry* 108:49–54.

Brinckmann, J., H. Notbohm, P. K. Mueller, and Editors. 2005. *Collagen: Primer in Structure, Processing and Assembly*. Heidelberg, Germany: Springer-Verlag.

Brodsky, B. 1982. Variations in collagen fibril structure in tendons. *Biopolymers* 21:935–951.

Brodsky, B., J. A. Werkmeister, and J. A. M. Ramshaw. 2005. Collagen and gelatins. In *Biopolymers Online*, pp. 119–129.

Byun, H. G. and S. K. Kim. 2001. Purification and characterization of angiotensin I converting enzyme (ACE) inhibitory peptides from Alaska pollack (*Theragra chalcogramma*) skin. *Process Biochemistry* 36:1155–1162.

Chen, Q., G. Xuan, M. Fu et al. 2007. Effect of angiotensin converting enzyme inhibitory peptide from rice dregs protein on antihypertensive activity in spontaneously hypertensive rats. *Asian Pacific Journal of Clinical Nutrition* 16:281–285.

Contreras, M., R. Carrón, M. J. Montero, M. Ramos, and I. Recio. 2009. Novel casein-derived peptides with antihypertensive activity. *International Dairy Journal* 19:566–573.

Courts, A. 1977. Uses of collagen in edible products. In *The Science and Technology of Gelatin*, Ward, A. G. and Courts, A. (eds.). New York: Academic Press, pp. 395–412.

Eastoe, J. E. and A. A. Leach. 1977. Chemical constitution of gelatin. In *The Science and Technology of Gelatin*, Ward, A. G. and Courts, A. (eds.). New York: Academic Press, pp. 73–107.

Ehrlich, H. 2010. Chitin and collagen as universal and alternative templates in biomineralization. *International Geology Review* 52:661–699.

Esmaeili, N., H. Ebrahimzadeh, K. Abdi, and S. Safarian. 2011. Determination of some phenolic compounds in *Crocus sativus* and its antioxidant activities study. *Pharmacognosy Magazine* 7:74–80.

Fahmi, A., S. Morimura, H. C. Guo, T. Shigematsu, K. Kida, and Y. Uemura. 2004. Production of angiotensin I converting enzyme inhibitory peptides from sea bream scales. *Process Biochemistry* 39:1195–1200.

Fennema, O. R. 1996. *Food Chemistry*. New York: Marcel Dekker.

Fernandez-Diaz, M. D., P. Monter, and M. C. Gomez-Guillen. 2001. Gel properties of collagens from skins of cod (*Gadus morhua*) and hake (*Merluccius merluccius*) and their modification by the coenhancers magnesium sulphate, glycerol and transglutaminase. *Food Chemistry* 74:161–167.

Floris, R., I. Recio, B. Berkhout, and S. Visser. 2003. Antibacterial and antiviral effects of milk proteins and derivatives thereof. *Current Pharmaceutical Design* 9:1257–1275.

Friess, W. 1998. Collagen biomaterial for drug delivery. *European Journal of Pharmaceutics and Biopharmaceutics* 45:113–136.

Gelatin Manufacturers Institute of America (GMIA). 1993. *Gelatin*. New York: GMIA.

Gómez-Guillén, M. C., M. E. López-Caballero, A. López de Lacey, A. Alemán, B. Giménez, and P. Montero. 2010. Antioxidant and antimicrobial peptide fractions from squid and tuna skin gelatin. In *Sea By-Products as a Real Material: New Ways of Application*, Le Bihan, E. and Koueta, N. (eds.). Kerala, India: Transworld Research Network Signpost, pp. 89–115.

Gomez-Guillen, M. C. and P. Montero. 2001. Method for the production of gelatin of marine origin and product thus obtained. International Patent PCT/S01/00275.

Gomez-Guillen, M. C., A. I. Sarabia, M. T. Solas, and P. Montero. 2001. Effect of microbial transglutaminase on the functional properties of megrim (*Lepidorhombus boscii*) skin gelatin. *Journal of the Science of Food and Agriculture* 81:665–673.

Gorham, S. D. 1991. Collagen as a biomaterial. In *Biomaterials*, Byron, D. (ed.). New York: Stockton Press, pp. 55–122.

Green, D., D. Walsh, S. Mann, and R. O. Oreffo. 2002. The potential of biomimesis in bone tissue engineering: Lessons from the design and synthesis of in-vertebrate skeletons. *Bone* 30:810–815.

Hao, S., L. Li, X. Yang et al. 2009. The characteristics of gelatin extracted from sturgeon (*Acipenser baeri*) skin using various pretreatments. *Food Chemistry* 115:124–128.

Harkness, R. D. 1966. Collagen. *Science Progress* 54:257–274.

He, L., C. Mu, J. Ski, Q. Zhang, B. Shi, and Q. Lin. 2011. Modification of collagen with a natural cross-linker, procyanidin. *International Journal of Biological Macromolecules* 48:354–359.

Heu, M. S., J. H. Lee, H. J. Kim, S. J. Jee, J. S. Lee, Y. J. Jeon, F. Shahidi, and J. S. Kim. 2010. Characterization of acid- and pepsin-soluble collagens from flatfish skin. *Food Science and Biotechnology* 19:27–33.

Hood, L. L. 1987. Collagen in sausage casings. In *Advances in Meat Research, Collagen as a Food*, Pearson, A. M., Dutson T. R., and Bailey, A. J. (eds.). New York: Van Nostrand Reinhold, pp. 109–129.

Hoyer, B., A. Bernhardt, S. Heinemann, I. Stachel, M. Meyer, and M. Gelinsky. 2012. Biomimetically mineralized salmon collagen scaffolds for application in bone tissue engineering. *Biomacromolecules* 13:1059–1066.

Huang, Y. R., Shiau, C. Y., Chen, H. H., Huang, B. C. 2011. Isolation and characterization of acid and pepsin-solubilized collagen from the skin of balloon fish (*Diodon holocanthus*). *Food Hydrocolloids* 25:1507–1513.

Ikoma, T., H. Kobayashi, J. Tanaka, D. Walash, and S. Mann. 2003. Physical properties of type I collagen extracted from fish scales of *Pagrus major* and *Oreochromis niloticus*. *International Journal of Biological Macromolecules* 32:199–204.

Je, J., Z. Qian, H. Byun, and S. Kim. 2007. Purification and characterization of an antioxidant peptide obtained from tuna backbone protein by enzymatic hydrolysis. *Process Biochemistry* 42:840–846.

Jeevithan, E., B. Bao, Y. Bu, Y. Zhou, Q. Zhao, and W. Wu. 2014. Type II collagen and gelatin from silvertip shark (*Carcharhinus albimargunatus*) cartlilage: Isolation, purification, physiochemical and antioxidant properties. *Marine Drugs* 27:3852–3873.

Jeong, H. S., J. Venkatesan, and S. K. Kim. 2013. Isolation and characterization of collagen from marine fish (*Thunnus obesus*). *Biotechnology and Bioprocess Engineering* 18:1185–1191.

Johns, P. and A. Courts. 1977. Relationship between collagen and gelatin. In *The Science and Technology of Gelatin*, Ward, A. G. and Courts, A. (eds.). New York: Academic Press, pp. 164–165.

Jones, N. R. 1977. Uses of gelatin in edible products. In *The Science and Technology of Gelatin*, Ward, A. G. and Courts, A. (eds.). New York: Academic Press, pp. 365–394.

Jongjareonrak, A., S. Benjakul, W. Visessaguan, and M. Tanaka. 2005. Isolation and characterisation of collagen from bigeye snapper (*Priacanthus macracanthus*) skin. *Journal of the Science of Food and Agriculture* 85:1203–1210.

Jung, W. K., P. J. Park, H. G. Byun, S. H. Moon, and S. K. Kim. 2005. Preparation of hoki (*Johnius belengerii*) bone oligophosphopeptide with a high affinity to calcium by carnivorous instestine crude proteinase. *Food Chemistry* 91:333–340.

Kearney, P. M., M. Whelton, M. K. Reynolds, P. Muntner, P. K. Whelton, and J. He. 2005. Global burden of hypertension. *Lancet* 365:217–223.

Khantaphant, S. and S. Benjakul. 2008. Comparative study on the proteases from fish pyloric caeca and the use for production of gelatine hydrolysate with antioxidative activity. *Comparative Biochemistry and Physiology* 151:410–419.

Kim, S. K. and E. Mendis. 2006. Bioactive compounds from marine processing byproducts—A review. *Food Research International* 39:383–393.

Kittiphattanabawon, P., S. Benjakul, W. Visesssangun, T. Nagai, and M. Tanaka. 2005. Characterisation of acid-soluble collagen from skin and bone of bigeye snapper (*Priacanthus tayenus*). *Food Chemistry* 89:363–372.

Kolodziejska, I., Z. E. Sikorski, and C. Niecikowska. 1999. Parameters affecting the isolation of collagen from squid (*Illex argentinus*) skins. *Food Chemistry* 66:153–157.

Kuznetsova, N. and S. Leikin. 1999. Does the triple helical domain of Type I collagen encode molecular recognition and fiber assembly while telopeptides serve as catalytic domains? *Journal of Biological Chemistry* 274:36083–36088.

Langer, R. 1990. New methods of drug delivery. *Science* 249:1527–1533.

Langer, R. S. and J. P. Vacanti. 1999. Tissue engineering: The challenges ahead. *Scientific American* 280:8689.

Lee, C. H., A. Singla, and Y. Lee. 2001. Biomedical applications of collagen. *International Journal of Pharmaceutics* 221:1–22.

Lee, K. J., Park, H. Y., Kim, Y. K., Park, J., Yoon H. D. 2009. Biochemical characterization of collagen from the starfish *Asterias amurensis*. *Journal of the Korean Society for Applied Biological Chemistry* 52:221–226.

Lin, Y., C., F. J. Tan, K. G. Marra, S. S. Jan, and D. C. Liu. 2009. Synthesis and characterization of collagen/hyaluronan/chitosan composite sponges for potential biomedical applications. *Acta Biomaterialia* 5:2591–2600.

Liu, D., L. Liang, J. M. Regenstein, and P. Zhou. 2012. Extraction and characterization of pepsin-solubilised collagen from fins, scales, skin, bones, and swim bladders of bighead carp (*Hypophthalmichthys nobilis*). *Food Chemistry* 133:1441–1448.

Matmaroh K., S. Benjakul, T. Prodpran, A. Encarnacio, and H. Kishimura. 2011. Characteristics of acid soluble collagen and pepsin soluble collagen from scale of spotted golden goatfish (*Parupeneus heptacanthus*). *Food Chemistry* 129:1179–1186.

Matsuda, N., Y. Koyama, Y. Hosaka et al. 2006. Effects of ingestion of collagen peptides on collagen fibrils and glycosaminoglycans in dermis. *Journal of Nutritional Science and Vitaminology* 52:211–215.

Mendis, E., N. Rajapakse, H. G. Byun, and S. K. Kim. 2005. Investigation of jumbo squid (*Dosidicus gigas*) skin gelatin peptides for their *in vitro* antioxidant effects. *Life Sciences* 77:2166–2178.

Minaguchi, J., Koyama, Y. I., Meguri, N., Hosaka, Y., Ueda, H., Kusubata, M., Hirota, A., Irie, S., Mafune, N., Takehana, K. 2005. Effects of ingestion of collagen peptide on collagen fibrils and glycosaaminoglycans in Achilles tendon. *Journal of Nutritional Science and Vitaminology* 51:69–174.

Minamisako, K., Kimura, S. 1989. Characterization of muscle collagen from fleshy prawn *Penaeus chinensis*. *Comparative Biochemistry and Physiology Part B: Comparative Biochemistry* 94:349–353.

Mizuta, S., S. Isobe, and R. Yoshinaka. 2002a. Existence of two molecular species of collagen in the muscle layer of the ascidian (*Halocynthia roretzi*). *Food Chemistry* 79:9–13.

Mizuta, S., T. Miyagi, T. Nishimiya, and R. Yoshinaka. 2002b. Partial characterization of collagen in mantle and adductor of pearl oyster (*Pinctada fucata*). *Food Chemistry* 79:319–325.

Mocan, E., O. Tagadiuc, and V. Nacu. 2011. Aspects of collagen isolation procedure. *Clinical Research Studies* 320:1–5.

Moskowitz, R. W. 2000. Role of collagen hydrolysate in bone and joint disease. *Seminars in Arthritis and Rheumatism* 30:87–99.

Murray, O., M. Hall, P. Kearney, and R. Green. 2004. Fast-dispersing dosage forms containing fish gelatin. Patent number: US6709669B1.

Muyonga, J. H., C. G. B. Cole, and K. G. Duodu. 2004. Extraction and physico-chemical characterization of Nile perch (*Lates niloticus*) skin and bone gelatin. *Food Hydrocolloids* 18:581–592.

Nagai, T. 2004. Collagen from diamondback squid (*Thysanoteuthis rhombus*) outer skin. *Zeitschrift fur Naturforschung* 59:271–275.

Nagai, T., Y. Araki, and N. Suzuki. 2002. Collagen of the skin of ocellate puffer fish (*Takifugu rubripes*). *Food Chemistry* 78:173–177.

Nagai, T., T. Nagashima, A. Abe, and N. Suzuki. 2006. Antioxidative activities and angiotensin I-converting enzyme inhibition of extracts prepared from chum salmon (*Oncorhynchus keta*) cartilage and skin. *International Journal of Food Properties* 9:813–822.

Nagai, T., T. Ogawa, T. Nakamura, T. Ito, H. Nagakawa, K. Fujiki, M. Nakao, and T. Yano. 1999. Collagen from edible jellyfish exumbrella. *Journal of the Science of Food and Agriculture* 79:855–858.

Nagai, T., Suzuki, N. 2000. Isolation of collagen from fish waste material-skin, bone and fins. *Food Chemistry* 68:277–281.

Nagai, T., Suzuki, N., Nagashima T. 2008. Collagen from common minke whale (*Balaenoptera acutorostrata*) unesu. *Food Chemistry* 111:296–301.

Nalinanon, S., S. Benjakul, and H. Kishimura. 2010. Collagens from the skin of arabesque greenling (*Pleurogrammus azonus*) solubilized with the aid of acetic acid and pepsin from albacore tuna (*Thunnus alalunga*) stomach. *Journal of the Science of Food and Agriculture* 90:1492–1500.

Nimni, M. E. 1988. *Collagen Biochemistry*, Vol. I. Boca Raton, FL: CRC Press.

Nomura, Y., H. Sakai, Y. Ishii, and K. Shirai. 1996. Preparation and some properties of type I collagen from fish scales. *Biosciences Biotechnology and Biochemistry* 60:2092–2094.

Oesser, S., M. Adam, W. Babel, and J. Seifert. 1999. Oral administration of (14) C labeled gelatin hydrolysate leads to an accumulation of radioactivity in cartilage of mice (C57/BL). *Journal of Nutrition* 129:1891–1895.

Ondetti, M. A. and D. W. Cushman. 1982. Enzymes of the renin-angiotensin system and their inhibitors. *Annual Review of Biochemistry* 51:283–308.

Park, S. Y., Lim, H. K., Lee, S., Cho, M. 2012. Pepsin-solubilised collagen (PSC) from Red Sea cucumber (*Stichopus japonicus*) regulates cell cycle and the fibronectin synthesis in HaCaT cell migration. *Food Chemistry* 132:487–492.

Pollack, S. V. 1990. Silicone, fibre, and collagen implantation for facial lines and wrinkles. *The Journal of Dermatologic Surgery and Oncology* 16:957–961.

Qian, Z., W. Jung, and S. Kim. 2008. Free radical scavenging activity of a novel antioxidative peptide purified from hydrolysate of bullfrog skin, *Rana catesbeiana Shaw*. *Bioresource Technology* 99:1690–1698.

Raabe, O., C. Reich, S. Wenisch, A. Hild, M. Burg-Roderfeld, H. C. Siebert, and S. Arnhold. 2010. Hydrolyzed fish collagen induced chondrogenic differentiation of equine adipose tissue-derived stromal cells. *Histochemistry and Cell Biology* 134:545–554.

Rajapakse, N., E. Mendis, H. G. Byun, and S. K. Kim. 2005. Purification and *in vitro* antioxidative effects of giant squid muscle peptides on free radical-mediated oxidative systems. *Journal of Nutritional Biochemistry* 16:562–569.

Rao, K. P. 1995. Recent developments of collagen based materials for medical applications and drug delivery systems. *Journal of Biomaterials Science, Polymer Edition* 7:623–645.

Rigby, B. J. 1968. Amino-acid composition and thermal stability of the skin collagen of the Antarctic ice-fish. *Nature* 219:166–167.

Saddler, J. M. and P. J. Horsey. 1987. The new generation gelatins: A review of their history, manufacture and properties. *Anesthesia* 42:998–1004.

Sadowska, M., I. Kolodziejska, and C. Niecikowska. 2003. Isolation of collagen from the skins of Baltic cod (*Gadus morhua*). *Food Chemistry* 81:257–262.

Scala, N., R. S. Hollies, and K. P. Sucher. 1976. Effect of daily gelatin ingestion on human scalp hair. *Nutrition Reports International* 13:579–592.

Senaratne, L. S., P. J. Park, and S. K. Kim. 2006. Isolation and characterization of collagen from brown backed toadfish (*Lagocephalus gloveri*) skin. *Bioresource Technology* 97:191–197.

Shen, X. R., H. Kurihara, and K. Takahashi. 2007. Characterization of molecular species of collagen in scallop mantle. *Food Chemistry* 102:1187–1191.

Shyni, K., G. S. Hema, G. Ninan, S. Mathew, C. G. Joshy, and P. T. Lakshmanan. 2014. Isolation and characterization of gelatins from the skin of skipjack tuna (*Katsuwonus pelamis*), dog shark (*Scoliodon sorrakowah*), and rohu (*Labeo rohita*). *Food Hydrocolloids* 39:68–76.

Simon, A., L. Vandanjon, G. Levesque, and P. Bourseau. 2002. Concentration and desalination of fish gelatin by ultrafiltration and continuous diafiltration processes. *Desalination* 144:313–318.

Singh, P., S. Benjakul, S. Maqsood, and H. Kishimura. 2011. Isolation and characterisation of collagen extracted from the skin of striped Catfish (*Pangasianodon hypophthalmus*). *Food Chemistry* 124:97–105.

Skierka, E. and M. Sadowska. 2007. The influence of different acids and pepsin on the extractability of collagen from the skin of Baltic cod (*Gadus morhua*). *Food Chemistry* 105:1302–1306.

Sudharsan, S., P. Seedevi, R. Saravanan, P. Ramasamy, S. V. Kumar, S. Vairamani, A. Srinivasan, and A. Shanmugam. 2013. Isolation, characterization and molecular weight determination of collagen from marine sponge *Spirastrella inconstans* (Dendy). *African Journal of Biotechnology* 12:504–511.

Swatschek, D., W. Schatton, J. Kellermann, W. E. G. Muller, and J. Kreuter. 2002. Marine sponge collagen: Isolation, characterization and effects on the skin parameters surface-pH, moisture and sebum. *European Journal of Pharmaceutics and Biopharmaceutics* 53:107–113.

Thuy, L. T. M., E. Okazaki, and K. Osako. 2014. Isolation and characterization of acid-soluble collagen from the scales of marine fishes from Japan and Vietnam. *Food Chemistry* 149:264–270.

Veeruraj, A., M. Arumugam, and T. Balasubramanian. 2013. Isolation and characterization of thermostable collagen from the marine eel-fish (*Evenchelys macrura*). *Process Biochemistry* 48:1592–1602.

Vermeirssen, V., J. Van Camp, and W. Verstraete. 2004. Bioavailability of angiotensin I converting enzyme inhibitory peptides. *British Journal of Nutrition* 92:357–366.

Wang, L., B. Yang, and X. Du. 2009. Extraction of acid-soluble collagen from Grass carp (*Ctenopharyngodon idella*) skin. *Journal of Food Process Engineering* 32:743–751.

Wieprecht, T., M. Dathe, R. M. Epand, M. Beyermann, E. Krause, W. L. Maloy, D. L. McDonald, and M. Bienert. 1997. Influence of the angle subtended by the positively charged helix face on the membrane activity of amphipathic, antibacterial peptides. *Biochemistry* 36:12869–12880.

Winter, A. D. and A. P. Page. 2000. Prolyl 4-hydroxylase is an essential procollagen-modifying enzyme required for exoskeleton formation and the maintenance of body shape in the nematode *Caenorhabditis elegans*. *Molecular Cell Biology* 20:4084–4093.

Wohllebe, M. and D. Carmichael. 1978. Type-I trimer and type-I collagen in neutral-salt-soluble lathyritic-rat dentine. *European Journal of Biochemistry* 92:183–188.

Wu, H. C., H. M. Chen, and C. Y. Shiau. 2003. Free amino acids and peptides as related to antioxidant properties in protein hydrolysates of mackerel (*Scomber austriasicus*). *Food Research International* 36:949–957.

Wu, J., M. Fujioka, K. Sugimoto, G. Mu, and Y. Ishimi. 2004. Assessment of effectiveness of oral administration of collagen peptide on bone metabolism in growing and mature rats. *Journal of Bone and Mineral Metabolism* 22:547–553.

Xiong, X. 2008. *New Insights into Structure and Function of Type I Collagen*, Dissertation, Institut fur Grenzflachenverfahrenstechnik der Universitat, Stuttgart, Germany, 111pp.

Zhang, J., R. Duan, L. Huang, S. Yujie, and J. M. Regenstein. 2014. Characterisation of acid-soluble and pepsin-solubilised collagen from jellyfish (*Cyanea nozakii* Kishinouye). *Food Chemistry* 150:20–26.

Zhang, Z., J. Wang, Y. Ding, X. Dai, and Y. Li. 2011. Oral administration of marine collagen peptides from Chum salmon skin enhances cutaneous wound healing and angiogenesis in rats. *Journal of the Science of Food and Agriculture* 91:2173–2179.

Zhao, Y., B. Li, Z. Liu, S. Dong, X. Zhao, and M. Zeng. 2007. Antihypertensive effect and purification of an ACE inhibitory peptide from sea cucumber gelatine hydrolysate. *Process Biochemistry* 42:1586–1591.

Zhou, P. and J. M. Regenstein. 2004. Optimization of extraction condition for Pollock skin gelatin. *Journal of Food Science* 69:393–398.

Zhou, P. and J. M. Regenstein. 2005. Effects of alkaline and acid pretreatment on *Alaska pollock* skin gelatin extraction. *Journal of Food Science* 70:392–396.

Zhuang, Y., L. Sun, Y. Zhang, and G. Liu. 2012. Antihypertensive effect of long-term oral administration of jellyfish (*Rhopilema esculentum*) collagen peptides on renovascular hypertension. *Marine Drugs* 10:417–426.

Zhuang, Y. L., L. P. Sun, X. Zhao, H. Hou, and B. F. Li. 2010. Investigation of gelatine polypeptides of jellyfish (*Rhopilema esculentum*) for their antioxidant activity *in vitro*. *Food Technology and Biotechnology* 48:222–228.

## chapter thirteen

# Marine fungi

## Glycolipidomics

Kandasamy Saravanakumar, Kandasamy Kathiresan,
Narayanasamy Rajendran, Chuanjin Yu, and Jie Chen

### Contents

## 13.1 Introduction

Marine fungi are those that grow and sporulate in the marine environment. They may be of obligate or facultative form. Obligate marine fungi exclusively exist in estuarine or marine environment, while facultative forms are terrestrial ones that can also grow in estuarine or marine environs (Kohlmeyer and Kohlmeyer, 1979). Marine fungi are mostly of saprophytic or heterotrophic filamentous forms and spore-forming eukaryotic microorganisms. Their morphological structure is studied by direct observation and phylogenetic identity by molecular sequencing. In general, they have been studied for biotechnological applications as industrial enzymes, natural products, and for agriculture biocontrol, as well as in bioremediation, fuel conservation, and waste management.

Marine fungi are taxonomically distinct (Jones et al., 2009), saline-tolerant (Jennings, 1986), and special in their biochemical properties (Damare et al., 2006). By virtue of their novelty, marine fungi have potential applications in omic studies (Damare et al., 2008). Generally, marine fungi can be isolated from nutrient-rich substrates such as decaying wood (harbor), coral reef (Le Campion-Alsumard et al., 1995), sea grass (Thirunavukkarasu, 2012), mangroves (Saravanakumar 2012), and deep-sea sediments (Damare, 2007). Marine fungi are distinct and diverse (Sridhar, 2005). Among the marine substrates, mangroves constitute the second largest source for the isolation of obligate marine fungi (Raghukumar, 2004), which reportedly produce many structurally and pharmaceutically novel metabolites. This chapter focuses on the glycolipids of marine fungi and their properties, biological functions, and applications.

## 13.2 Microbial glycolipids

Glycolipids are a structurally very heterogeneous group of membrane-bound compounds present in all living organisms, including humans. The term "glycolipid" refers to a compound that contains one or more monosaccharides glycodically linked to a hydrophobic moiety (Warnecke and Heinz, 2010) such as an acylglycerol, a sphingoid, a ceramide (*N*-acylsphingoid), or a prenyl phosphate (Lang, 2002; Brandenburg and Holst, 2005). Glycolipids are essential constituents of the cellular

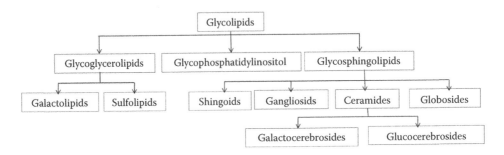

***Figure 13.1*** Classification of glycolipids.

membrane and have the remarkable biological functions of cell aggregation or dissociation; glycolipids act as receptor of accepter to provide the contact. Several glycolipids have a role to play in the immune system.

Generally, glycolipids are classified as belonging to three major groups: glyceroglycolipids, glycosphingolipids, and glycosylphosphatidylinositols. The detailed classification of glycosphingolipids is given in Figure 13.1. Among the major groups of glycolipids, glycosphingolipids are most common in fungal cells. Glycosphingolipids are carbohydrates containing derivatives of sphingoid, ceramide, and the amino alcohol sphingosine. Some other terms are also used to highlight the structural features of glycolipids, such as fucoglycosphingolipid, mannoglycosphingolipid, xyloglycosphingolipid, and cellbiose glycolipid (IUPAC-IUBMB, 1998).

## 13.3   Fungal glycolipidomics

Glycolipids are an interesting group of compounds occurring in the cell wall of animals, microbes, and plants (Pinto et al., 2008). Fungal glycolipids are composed of sugar units – usually glucose and galactose–hydrophobic ceramides, $C_{19}$ sphingoid, C-9 methyl branches, and unsaturated linkages with hydroxy-hexadecanoic acids of low molecular weight (Pinto et al., 2008). However, fungi are predominantly known to be producers of cellbiose glycolipids as secondary metabolites with the biological function as a cell surface hydrophobic factor. The GATA-like transcriptome gene expression of the fungus *Ustilago maydis* for the production of glycolipids has been studied with the two genes *emt* 1 and *cyp* 1, and it was proved that these genes are essential for the production of fungal glycolipids (Hewald et al., 2005). The synthesis of glycolipids from fungi requires the expression of the enzyme acetyltransferase and its gene *Ahd1* and the acyl/acetyl transferase genes *Uat2*, and *Fat2* in *U. maydis* and *P. flocculosa* (Teichmann et al., 2011). Biosynthetic pathways and gene cluster analysis of glycolipids from *U. maydis* have revealed that the gene expressions of cytochrome 450 (*Cyp*1 and *Cyp*2), fatty acid synthase (*fas2*), hydroxylases

(*uhd*1 and *ahd*1), glycosyl transferase (*ugt*1), and ABC transporter gene (*atr 1*) are necessary for the synthesis of glycolipids (Teichmann et al., 2011).

### 13.3.1   Glycoconjugates in fungal cell wall

Glycoconjugates are composed of glycoproteins, peptides, glucons, polysaccharides, phosphoric acid, phospholipids, nitrogen, and glycolipid molecules. They are found in the cell wall. Fungal cell walls are substantially thicker than bacterial cell walls and normally make up to 10%–30% of the biomass. They are freely accessible to small molecules, the solute transport system, and the signaling receptors in the cell membrane. The fungal cell wall plays a role in osmotic support, selective permeability, and interaction with the environment (Gonzalez et al., 2009). Fungal cell walls consist of covalently cross-linked polysaccharides of β-glycans and chitin and several polysaccharides covalently cross-linked through glycosidic bonds (Pinto et al., 2008).

### 13.3.2   Fungal exo-glycolipids

Generally, the glycolipid molecules are found in the cell membrane of all eukaryotic cells. All types of glycolipids are biosurfactants, but not vice versa (Mukherjee et al., 2006; Khopade et al., 2012). Simple glycolipids contain one or more sugars, while complex glycolipids such as gangliosides contain a branch chain with several sugars. Cell membranes of the fungi are of many types, and are assembled from four compounds: (1) phospholipids molecules, (2) transmembrane proteins, (3) inerter protein network, and (4) cell surface markers, which are not identical. Glycolipids are formed in the fungal cell wall by the process of glycosylation in the endoplasmic reticulum membrane sections and transfer of the Golgi complex followed by the plasma membrane. These add the sugar molecule chain to lipids called sugar-coating lipids, which extend to the outside of fungal cells. The difference between fungal species is in the glycolipids, which are used as the cell surface layer or marker.

Glycolipids are also compounds of fatty acids that contain carbohydrates and nitrogen, but not phosphoric acid, and they include certain compounds of gangliosides, sulfolipids, and salfatids (Pinto et al., 2008). The glycolipids are markers for the identification of cell surface changes, and serve as fundamental building blocks of fungi, energy molecule, or store, component of membrane constituents, signal molecule to interact with the environmental compounds through the outer matrix, lectins, growth factors, and a potential factor of pathogenesis and immune responses (Hakomori, 1990; Springer and Lasky, 1991; Pinto et al., 2008). However, the properties and mechanisms of action of fungal glycolipids remain unclear.

### 13.3.3 Marine fungal glycolipids

There has been a growth in the research on glycolipids from marine resources due to their potential novelty in biotechnological applications. Muralidhar et al. (2003) have reviewed glycolipids from marine resources such as algae (Lo et al., 2001), microorganisms, bacteria (Batrakov et al., 1998), fungi (Abraham et al., 1994), yeasts (Zinjarde and Pant, 2002), actionbacteria (Kokare et al., 2007), sponges (Pettit et al., 1999), gorgonians (Shin and Seo, 1995), sea anemones (Sugita et al., 1994), bryozoans (Ojika et al., 1997), tunicates (Loukaci et al., 2000), marine annelids (Noda et al., 1992), star fish (Sugiyama et al., 1988), sea cucumber (Higuchi et al., 1994), sea urchin (Babu et al., 1997), crinoids (Arao et al., 1999), molluscs (Yamaguchi et al., 1992), and marine crabs (Asai et al., 2000).

In terrestrial fungi, yeasts have glycolipids as major constituents, while in many fungal species glycolipids are not the major component. However, the content of glycolipids was high (11%–16% of total lipids) in *Blastocladiella emersonii*, and the major compound of glycolipid present in the fungus is GalDAG and Gal$_2$DAG (Mills and Cantino, 1974). The content of glycolipids is also high (61%–48%) in the mycelia of *Macrophomina phaseoline*, whereas it is low in sclerotia (14%–62%). The glycolipid content varies according to the constituents of the fermentation medium. The major compound of the fungal glycolipids has been identified as GalDAG and Gal$_2$DAG, based on structural characterization. Further, the major glycolipids of fungi are glycosphingolipids and D-glucosylceramides. Galactocerebrosides are reportedly present in the fungal species *Aspergillus niger*, *Candida utilis*, and *Saccharomyces cerevisiae* (Wagner and Zofcsik, 1966). Besides, the fungal species *Fusarium lini* and *Phycomycetes blakesleeanus*, the mushrooms, are also known to produce glycolipids (Weiss et al., 1973). Glycolipids have been widely studied in fungi such as *Torulaspora delbrueckii*, *Saccharomyces cerevisiae*,

*Candida glabrata*, *Kluyveromyces yarrowii*, *F. pedrosoi*, and *K. polyporus* (Saito et al., 2006; Pinto et al., 2008). The long-chain sphingadinene was first reported from *Aspergillus oryzae* (Fujino and Ohishi, 1976) and subsequently from *Schizophyllum commune* (Ballio et al., 1979), *Fusicoccum amygdali* (Ballio et al., 1979), *Clitocybe geotrope*, *Aspergillus fumigatus* (Villas-Boas et al., 1994), *C. nebularis* (Fodegal et al., 1986), *A. niger* (Levery et al., 2000), *A. versicolor* (Walenkamp et al., 1999), *Candida albicans* (Matsubara et al., 1987), *Acremonium chrysogenum* (Sakaki et al., 2001), *Cryptococcus neoformans* (Rodrigues et al., 2000), *Colletotrichum gloeosporioides* (da Silva et al., 2004), *Fonsecaea pedrosoi* (Nimrichter et al., 2005), *Hansenula anomala* (Ng and Laneelle, 1977), *Fusarium* sp. (Duarte et al., 1998), *Histoplasma capsulatum* (Toledo et al., 2001), *Kluyveromyces waltii* (Takakuwa et al., 2002), *Paracoccidioides brasiliensis* (Takahashi et al., 1996), *Magnaporthe grisea* (Koga et al., 1998), *Pichia pastoris* (Sakaki et al., 2001), *Saccharomyces kluyveri* (Takakuwa et al., 2002), *Pseudallescheria boydii* (Pinto et al., 2002), *Termitomyces albuminosus* (Qi et al., 2001), *Sporothrix schenckii* (Toledo et al., 2001), *Candida bombicola* (Ito and Inoue, 1982), *C. antarctica*, *Schizonella melanogramma*, *Geotrichum candidum* (Deml et al., 1980; Kurz et al., 2003), and *Ustilago maydis* (Haskins, 1950; Hewald et al., 2006). A smut fungus, *Ustilago scitaminea* NBRC 132730, and yeasts *Pseudozyma crassa* are producers of the glycolipid biosurfactant mannosylerythritol (Fukuoka et al., 2008; Morita et al., 2009). Recently, the fungus *Pseudozyma* species was shown to convert cellulosic materials into commercially important biosurfactant glycolipids (Faria et al., 2014).

Very few studies are available on glycolipids of marine fungi (Table 13.1). The white rot marine fungus *Nia vibrissae* is a producer of glycolipids with inhibitory activity on the binding of endotoxin lipopolysaccharide (LPS) to human endotoxin receptor (Helmholz et al., 1999). The marine fungus *Gliocladium roseum* KF-1040 is a producer of roselipins, which can inhibit the enzyme diacylglycerol acyl transferase (Matsumoto et al., 1999; Tobata et al., 1999; Tomada et al., 1999). Glycolipids derived from the marine yeasts *Calyptogena soyoae* and *Yarrowia lipolytica* are effective in the degradation of hydrocarbons (Zinjarde and Pant, 2002; Konishi et al., 2010). Glycolipids synthesized from filamentous endosymbiotic *Aspergillus ustus* show significant antimicrobial activity (Kiran et al., 2009). Several marine fungi such as *Penicillum* sp. F23-2 (Sun et al., 2009), *Linincola laevis* (Abraham et al., 1994), *Fusarium* sp. (Li et al., 2002), and *Microsphaeropsis olivacea* (Keusgen et al., 1996) significantly produce glycolipids with unknown applications. The glycolipids derived from the sponge-associated marine fungus *Aspergillus ustus* MSF3 are efficient in antimicrobial activity (Seghal Kiran et al., 2009).

*Table 13.1* Biological activity of glycolipids from marine fungi

| Species | Compound | Biological activity | References |
|---|---|---|---|
| *Nia vibrissa* (White rot marine fungi) | Glycolipid | Inhibit the binding of endotoxin lipopolysaccharide (LPS) to human endotoxin receptor | Helmholz et al. (1999) |
| Marine fungi *Gliocladium roseum KF-1040* | Roselipins | Inhibit the enzyme diacylglycerol acyl transferase | Omura et al. (1999); Tomada et al. (1999); Tobata et al. (1999) |
| Yeasts: *Yarrowia lipolytica* | Lipid carbohydrate protein complex of glycolipid | Hydrocarbon degradation | Zinjarde and Pant (2002) |
| Yeasts: *Calyptogena soyoae* | Glycolipids | Hydrocarbon degradation | Konishi et al. (2010) |
| Endosymbiotic *Aspergillus ustus (MSF3)* | Biosufactant (glycolipids) | Antimicrobial activity | Kiran et al. (2009) |
| *Penicillum* sp. F23-2 | Glycolipids | — | Sun et al. (2009) |
| *Linincola laevis* | Ceramide | — | Abraham et al. (1994) |
| *Fusarium* sp. | Sphingosine glycoside (3′*E*, 4*E*)-1,3-dihytoxy 2-[(2′-hytorxy octadecanoyloxy) amino]-10-methyl-3′,4,9-octadecatriene | — | Li et al. (2002) |
| *Microsphaeropsis olivacea* | Cerebroside | — | Keusgen et al. (1996) |
| Sponge-associated marine fungi *Aspergillus ustus* MSF3 | Glycolipid | Antimicrobial activity | Seghal Kiran et al. (2009) |
| *Thraustochytrium globosum* | Glycosphingolipid, thrustochytrosides A-C | — | Jenkins et al. (1999) |

## 13.3.4 Synthesis, isolation, and purification of fungal glycolipids

Generally, glycolipids are predominantly synthesized from bacteria, algae, and actinobacteria by a series of methods in different cell compartments, catalyzed by enzymes, carbon sources, polarity, and so on (Kitamoto et al., 2002; Zeidan and Hannun 2007). However, the fungal isolates are known for the synthesis of cellobiose glycolipids. The method of the synthesis (Figure 13.2), isolation, and purification of cellobiose glycolipids (Frautz et al., 1986) are now described. In brief, the inoculums of the glycolipid-producing strains are cultured in potato dextrose broth at 28°C in a rotary shaker at 180 rpm for 4 days. Following this, the inoculum is transferred to a yeast and nitrogen-based medium without ammonium sulfate and 5% glucose as carbon source. The glycolipids are harvested for the further purification after incubation for 4 days at 28°C in a rotary shaker of 180 rpm. The extracellular glycolipids are extracted from the supernatant of the fungal culture using ethyl acetate, then dissolved in methanol, and purified by vacuum column chromatography and thin layer chromatography with the eluting solution chloroform/methanol/distilled water (65:25:4 v/v). The sugar-containing compound of glycolipids is obtained by spraying a solution of glacial acid/sulfuric acid/p-anisaldehyde (50:1:0.5 v/v). Further, the glycolipids are characterized by HPLC-MS, spectrometric analysis, MALDI analysis, and $^1$H and $^{13}$C NMR. Besides, Hubert et al. (2012) have described the new method of preparation of microbial glycolipids and purification from complex microbial crude extracts.

## 13.4 Application of fungal glycolipids

Microbially synthesized glycolipids have more advantages than the chemically synthesized ones due to their properties such as environmental friendliness, economic viability, low toxicity, high selectivity, as well as specificity at extreme temperature, pH, and salinity (Sanchez et al., 2013). Thus, glycolipids find applications in various industries such as bioremediation of hydrocarbons, food, cosmetics, biocontrol, pharmaceuticals. In fungi, glycolipids are well-known bioactive metabolites with unique properties of biodegradation, mild production conditions, and antimicrobial activity (Banat et al., 2000; Morita et al., 2009; Naing et al., 2014). Among the glycolipids, mannosylerythritol lipids are promising microbial surfactants, richly produced by microorganisms (Fukuoka et al., 2008; Morita et al., 2009). This glycolipid is composed of

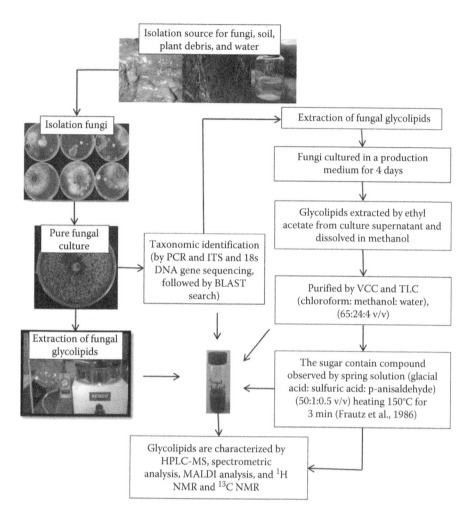

*Figure 13.2* Method of synthesis, isolation, and purification cellobiose glycolipids.

4-*O*-β-ᴅ-mannopyranosyl-erythritol, fatty acids, and a hydrophobic moiety (Kitamoto et al., 2002).

## 13.4.1 Biocontrol

Several fungal molecules and enzymes are reported to be present in the biocontrol fungus *Trichoderma*. Recently, it was shown that the glycolipid derived from the fungus *Paenibacillus ehimensis* MA2012 could control plant fungal and bacterial pathogens, such as the bacteria *Erwininia carotovota* spp. *Carotovora, Ralstonia solanacearum, Pseudomonas syringae* pv. *tabaci, Xanthomonas campestris* pv. *Vesicatoria* and the fungi *Colletotrichum gloeosporioides, Fusarium oxysporum* f. sp. *lycopersici, Phytophthora capsici,* and *Rhizoctonia solani* (Naing et al., 2014). The biocontrol fungus *Pseudozyma flocculosa* produces an antifungal glycolipid called flocculosin with anti-phytopathogenic properties as *O*-glycosidically linked 3,15,16-trihydroxypalmitic acid (Mimee et al., 2005; Hammami et al., 2010; Teichmann et al., 2011). A type of microbial glycolipids, rhamnolipid, is a very

promising for agricultural applications with enormous anti-phytofungal properties (Nitschke et al., 2005; Banat et al., 2010).

## 13.4.2 Bioremediation

Environmental pollution has led to enormous impacts that cause disorders in humans and animals. This has given rise to research interest in the bioremediation of environmental pollution by using biological or chemical methods. However, biological methods are highly environmentally friendly compared to chemical methods of bioremediation. Among the environmental pollutants, hydrocarbons in the environment pose the biggest challenge, resulting in environmental destruction. In chemical degradation, hydrocarbon can produce toxic by-products, which are dangerous to humans and animals (Ron and Rosenberg, 2002). Hence, biologically derived glycolipids can be used for degrading hydrocarbons and have environmental biocompatibility with the human skin (Wang et al., 2005; Pornsunthorntawee et al., 2009).

These lipids can be used for the soil surface bioremediation because they can degrade toxic aromatic and polycyclic compounds in the soil (Lai et al., 2009). Several kinds of microbial glycolipids are widely used in hydrocarbon degradation. Among the glycolipids, rhamnolipids are highly potent (Megharaj et al., 2011; Souza et al., 2014). The glycolipid mannosylerythritol, derived from the *Pseudozyma* sp. NII08165, is highly stable against alkaline pH and temperature. It is efficient in the removal of stains and can be used as a detergent addictive or formulation (Sajna et al., 2013).

### 13.4.3 Food and cosmetics

Microbial glycolipids have remarkable properties, such as biocompatibility and nontoxicity, which make them useful to the food and cosmetics industries (Rieger, 1997). Glycolipids are used as emulsifiers, foaming and wetting agents, solubilizers, antimicrobial agents, and adhesive agent in foods (Banat et al., 2000; Singh and Cameotra, 2004; Shoeb et al., 2013). Glycolipid surfactants such as sophorolipids, rhamnolipids, and mannosylerythritol lipids are regularly used in anti-dandruff and anti-wrinkle formulations, deodorants, nail care products, toothpastes, and skin roughness removers and antiaging compounds in the cosmetics industry, due to their low irritancy, anti-irritating effects, moisturizing property, and compatibility (Piljac and Piljac, 1999).

### 13.4.4 Medicinal value

Currently, research on discovering novel microbial molecules with high bioactivity are of increased interest to scientists in natural products research. Several secondary metabolites of medicinal value are reported from microbial sources. However, very little research has been done on the hydrophobic and hydrophilic nature of glycolipids derived from microbial sources. Sanchez et al. (2013) has reviewed the biological activity of glycolipids derived from microorganisms. Interestingly, microbial extracellular glycolipids derived from *Candida antarctica* and *Aureobasidium* sp. could inhibit the human promyelocytic leukemia cell line HL60 (Isoda et al., 1997). Mannosylerythritol is a type of glycolipid that has wide biomedical applications, such as gene transfection between liposomes and the plasma membrane and cell differentiation (Rodrigues et al., 2006; Liu et al., 2010). Glycolipids isolated from the fungus *Candida antarctica* have remarkable antimicrobial activity against clinical pathogens (Rodrigues et al., 2006). Glycolipids can also act against cancer and other degenerative diseases (Ruckhäberle et al., 2008). Hence the elucidation and functional analysis glycolipids, especially sphingolipids, from fungal cell membranes have received due attention because of their pharmaceutical utilization (Sanchez et al., 2013). Besides some enzymes, such as serine palmitoyl transferase, ceramide synthase, and inositol phosphoceramide synthase, the glycolipids of fungal cells are reported to have significant antifungal activity (Vicente et al., 2003). The structural phytosphingosines derived from fungi and plants have considerable anti-inflammatory activity and are antagonists for a number of cytokines expressed in epidermal keratinocytes. Besides, these compounds induce the human cancer cell death via apoptosis through the activation of caspase and cytochrome C (Park et al., 2003).

## 13.5 Conclusion and future prospects

Microbial glycolipids have become promising compounds for a variety of biotechnological applications during the past few decades. While fungal glycolipids have been studied, marine fungi are the least studied for glycolipids. Glycolipids are the most important constituents of the cell membrane in marine fungi, and these compounds may be structurally unique with efficient biological activities compared to those from terrestrial fungi. Marine fungi may produce chemical entities to survive against environmental stresses. The properties and mechanisms of action of the glycolipids in fungi are largely unclear, and these deserve much research.

## Acknowledgments

The authors are thankful to the authorities of Shanghai Jiao Tong University, P.R. China[1], Annamalai University[2], Tamil Nadu, India, and the Government Arts College[3], Chidambaram, Tamil Nadu, India, for providing the necessary facilities. This work was supported by the Special Project of Basic Work Project for Science and Technology of China (No. 2014FY120900).

## References

Abraham, S. P., T. D. Hoang, M. Alam, and E. B. G. Jones, 1994. Chemistry of the cytotoxicity principals of the marine fungus *Lignincola laevis*. *Pure Appl. Chem.*, 66: 2391–2394.

Arao, K., M. Inagaki, and R. Higuchi, 1999. Constituents of crinoidea.1. Isolation and structure of inositolphosphoceramide from the feather star *Comanthus japonica*. *Chem. Pharm. Bull.*, 47: 687–689.

Asai, N., N. Fusetani, S. Matsunaga, and J. Sasaki, 2000. Sex pheromones of the hair crab *Erimacrus isenbeckii*. Part 1: Isolation and structures of novel ceramides. *Tetrahedron*, 56: 9895–9899.

Babu, U. V., S. P. S. Bhandari, and H. S., Garg, 1997. Temnosides A and B, two new glycosphingolipids from the sea urchin *Temnopleurus toreumaticus* of the Indian Coast. *J. Nat. Prod.*, 60: 732–734.

Ballio, A., C. G. Casinovi, M. Framondino, G. Marino, G. Nota, and B. Santurbano, 1979. A new cerebroside from *Fusicoccum amygdale* Del. *Biochem. Biophys. Acta*, 573: 51–60.

Banat, I. M., A. Franzetti, I. Gandolfi, G. Bestetti, M. G. Martinotti, and L. Fracchia, 2010. Microbial biosurfactants production, applications and future potential. *Appl. Microbiol. Biotechnol.*, 87: 427–444.

Banat, I. M., R. S. Makkar, and S. S. Cameotra, 2000. Potential commercial applications of microbial surfactants. *Appl. Microbiol. Biotechnol.*, 53: 495–508.

Batrakov, S. G., D. I. Nikitin, V. I. Sheichenko, and A. O. A. Ruzhitsky, 1998. Novel sulfonic-acid analogue of ceramide is the major extractable lipid of the gram-negative marine bacterium *Cyclobacterium marinus* WH. *Biochim. Biophys. Acta*, 1391: 79–91.

Brandenburg, K. and O. Holst, 2005. Glycolipids: Distribution and biological function. *eLS*. doi: 10.1038/npg.els.0003912.

da Silva, A. F., M. L. Rodrigues, S. E. Farias, I. C. Almeida, M. R. Pinto, and E. Barreto-Bergter, 2004. Glucosylceramides in *Colletotrichum gloeosporioides* are involved in the differentiation of conidia into mycelial cells. *FEBS Lett.*, 561(1–3): 137–143.

Damare, S., 2007. Deep sea Fungi: occurrence and adaptations. PhD, Goa University, India.

Damare, S., M, Nagarajan and C. Raghukumar, 2008. Spore germination of fungi belonging to *Aspergillus* species under deep-sea conditions. *Deep-Sea Res I.* 55: 670–678.

Damare, S., C. Raghukumar, and S. Raghukumar, 2006. Fungi in deep-sea sediments of the Central Indian Basin. *Deep-Sea Res. I*, 53: 14–27.

Deml, G., T. Anke, F. OberWinkler, B. M. Giannetti, and W. Steglich, 1980. *Phytochemistry*, 19: 83–87.

Duarte, R. S., C. R. Polycarpo, R. Wait, R. Hartmann, and E. Barreto-Bergter, 1998. Structural characterization of neutral glycosphingolipids from *Fusarium* species. *Biochim. Biophys. Acta*, 1390: 186–196.

Faria, N. T., M. Santos, C. Ferreira, S. Marques, F. C. Ferreira, and C. Fonseca, 2014. Conversion of cellulosic materials into glycolipid biosurfactants, mannosylerythritol lipids, by *Pseudozyma* spp. under SHF and SSF processes. *Microb. Cell Fact.*, 13: 155.

Fodegal, M., H. Mickos, and T. Norberg, 1986. Isolation of *N*-2'-hydroxydecanoyl-1-*O*-β-D-glucopyranosil-9-methyl-4,8-D-erythro-sphingadienine from fruiting bodies of two Basidiomycetes fungi. *Glycoconj. J.*, 3: 233–237.

Frautz, B., S. Lang, and F. Wagner, 1986. Formation of cellobios lipids by growing and resting cells of *Ustilago maydis*. *Biotechnol. Lett.*, 8: 757–762.

Fujino, Y. and M. Ohnishi, 1976. Structure of cerebroside in *Aspergillus oryzae. Biochim. Biophys. Acta*, 486(1): 161–171.

Fukuoka, T., T. Morita, M. Konishi, T. Imura, and D. Kitamoto, 2008. A basidiomycetous yeast, *Pseudozyma tsukubaensis*, efficiently produces a novel glycolipid biosurfactant. The identification of a new diastereomer of mannosylerythritol lipid-B. *Carbohydr. Res.*, 343: 555–560.

Gonzalez, M., P. W. J. de Groot, F. M. Klis, and P. N. Lipke, 2009. Glycoconjugate structure and function in fungal cell walls. In *Microbial Glycobiology Structures, Relevance and Applications*, Moran, A. P. (ed.). Academic Press, Amsterdam, the Netherlands, pp. 169–183.

Hakomori, S., 1990. Bifunctional role of glycosphingolipids. Modulators for transmembrane signaling and mediators for cellular interactions. *J. Biol. Chem.*, 265(31): 18713–18716.

Hammami, W., F. Chain, D. Michaud, and R. R. Bélanger, 2010. Proteomic analysis of the metabolic adaptation of the biocontrol agent *Pseudozyma flocculosa* leading to glycolipid production. *Proteome Sci.*, 8: 7.

Haskins, R. H., 1950. Biochemistry of the Ustilaginales. I. Preliminary cultural studies of *Ustilago zeae. Can. J. Res. C*, 28: 213–223.

Helmholz, H., P. Etoundi, and U. Lindequist, 1999. Cultivation of the marine basidiomycete *Nia vibrissa* (Moore & Meyers). *J. Biotechnol.*, 70: 203–206.

Hewald, S., K. Josephs, and M. Bölker, 2005. Genetic analysis of biosurfactant production in *Ustilago maydis. Appl. Environ. Microbiol.*, 71: 3033–3040.

Hewald, S., U. Linne, M. Scherer, M. A. Marahiel, J. Kämper, and M. Bölker, 2006. Identification of a gene cluster for biosynthesis of mannosylerythritol lipids in the basidiomycetous fungus *Ustilago maydis. Appl. Environ. Microbiol.*, 72: 5469–5477.

Higuchi, R., M. Inagaki, K. Tokogawa, T. Mayamoto, and T. Komori, 1994. Constituents of Holothuroideae, IV— Isolation and structure of three new cerebrosides, CE-2b, CE-2c and CE-2d, from the sea cucumber *Cucumaria echinata. Liebigs Ann. Chem.*, 79–81.

Hubert, J., K. Ple, M. Hamzaoui, G. Nuissier, I. Hadef, R. Reynaud, A. Guilleret, and J. H. Renault, 2012. New perspectives for microbial glycolipid fractionation and purification processes. *C. R. Chimie*, 15: 18–28.

Isoda, H., D. Kitamoto, H. Shinmoto, M. Matsumura, and T. Nakahara, 1997. Microbial extracellular glycolipid induction of differentiation and inhibition of the protein kinase C activity of human promyelocytic leukemia cell line HL60. *Biosci. Biotechnol. Biochem.*, 61: 609–614.

Ito, S. and S. Inoue, 1982. Sophorolipids from *Torulopsis bombicola* as microbial surfactants in alkane fermentation. *Biotechnol. Lett.*, 4: 3.

IUPAC-IUBMB, Joint Commission on Biochemical Nomenclature (JCBN), whose members are A. Cornish-Bowden, A. J. Barrett, R. Cammack, M. A. Chester, D. Horton, C. LieÂbecq, K. F. Tipton, and B. J. Whyte, 1998. Nomenclature of glycolipids. *Carbohydr. Res.*, 312: 167–175.

Jenkins, K.M., P. R. Jensen, and W. Fenical, 1999. Thraustochytrosides A-C: new glycosphingolipids from a unique marine protist, *Thraustochytrium globosum. Tetrahedron Lett.* 40: 7637–7640.

Jennings, D. H., 1986. Fungal growth in the sea. In *The Biology of Marine Fungi*, Moss, S. T. (ed.). Cambridge University Press, Cambridge, U.K., pp. 1–10.

Jones, E. B. G., J. Sakayaroj, S. Suetrong, S. Somrithipol, and K. L. Pang, 2009. Classification of marine ascomycota, anamorphic taxa and basidiomycota. *Fungal Divers.*, 35: 1–187.

Keusgen, M., C. M. Yu, J. M. Curtis, D. Brewer, and S. W. Ayer, 1996. A cerebroside from the marine fungus *microsphaeropsis olivacea* (Bonord.) Höhn. *Biochem. Syst. Ecol.*, 24: 465–468.

Khopade, A., R. Biao, X. Liu, K. Mahadik, L. Zhang, and C. Kokare, 2012. Production and stability studies of the biosurfactant isolated from marine *Nocardiopsis* sp. B4. *Desalination*, 285: 198–204.

Kiran, G. S., T. A. Hema, R. Gandhimathi, J. Selvin, A. Manilal, S. Sujith, and K. Natarajaseenivasan, 2009. Optimization and production of a biosurfactant from the sponge-associated marine fungus *Aspergillus ustus* MSF3. *Colloid Surf. B*, 73: 250–256.

Kitamoto, D., H. Isoda, and T. Nakahara, 2002. Functions and potential applications of glycolipids biosurfactants— From energy–saving materials to gene delivery carriers. *J. Biosci. Bioeng.*, 94: 187–201.

Koga, J., T. Yamauchi, M. Shimura, N. Ogawa, K. Oshima, K. Umemura, M. Kikuchi, and N. Ogasawara, 1998. Cerebrosides A and C, sphingolipid elicitors of hypersensitive cell death and phytoalexin accumulation in rice plants. *J. Biol. Chem.*, 273: 31985–31991.

Kohlmeyer, J. and E. Kohlmeyer, 1979. *Marine Mycology. The Higher Fungi*. Academic Press, New York.

Kokare, C. R., S. S. Kadam, K. R. Mahadik, and B. A. Chopade, 2007. Studies on bioemulsifier production from marine *Streptomyces* sp. S1. *Indian J. Biotechnol.*, 6: 78–84.

Konishi, M., T. Fukuoka, T. Nagahama, and T. Morita, 2010. Biosurfactant-producing yeast isolated from *Calyptogena soyoae* (deep-sea cold-seep clam) in the deep sea. *J. Biosci. Bioeng.*, 110: 169–175.

Kurz, M., C. Eder, and D. Isert, 2003. Ustilipids, acylated β-D-mannopyranosyl D-erythritols from *Ustilago maydis* and *Geotrichum candidum*. *J. Antibiot.*, 56: 91–101.

Lai, C., Y. Huang, Y. Wei, and J. Chang, 2009. Biosurfactant-enhanced removal of total petroleum hydrocarbons from contaminated soil. *J. Hazard. Mater.*, 167: 609–614.

Lang, S., 2002. Biological amphiphiles (microbial biosurfactants). *Curr. Opin. Colloids Interface Sci.*, 7: 12–20.

Le Campion-Alsumard, T., S. Golubic, and K. Priss, 1995. Fungi in corals: Symbiosis or disease? Interactions between polyps and fungi causes pearl-like skeleton biomineralization. *Mar. Ecol. Prog. Ser.*, 117: 137–147.

Levery, S. B., M. S. Toledo, R. L. Doong, A. H. Straus, and H. K. Takahashi, 2000. Comparative analysis of ceramide structural modification found in fungal cerebrosides by electrospray tandem mass spectrometry with low energy collision-induced dissociation of Li+ adduct ions. *Rapid Commun. Mass Spectrom.*, 14(7): 551–563.

Li, H., Y. Lin, X. Liu, S. Zhou, and L. L. P. Vrijmoed, 2002. Study on secondary metabolites of marine fungus *Fusarium* sp. from South China Sea. *Haiyang Kexue*, 26: 57–59.

Liu, J., A. Zou, and B. Mu, 2010. Surfactin effect on the physiochemical property of PC liposome. *Colloids Surf. A: Physicochem. Eng. Aspects*, 361(1): 90–95. doi:10.1016/j.colsurfa.2010.03.021.

Lo, J. M., W. L. Wang, Y. M. Chiang, and C. M. Chen, 2001. Ceramides from the Taiwan red alga *Ceratodictyon spongiosum* and symbiotic sponge *Sigmadocia symbiotica*. *J. Chin. Chem. Soc.*, 48: 821–826.

Loukaci, A., V. Bultel-Ponce, A. Longeon, and M. Guyot, 2000. New lipids from the tunicate Cystodytes cf. dellechiajei, as PLA2 inhibitors. *J. Nat. Prod.*, 63: 799–802.

Matsubara, T., A. Hayashi, Y. Banno, T. Morita, and Y. Nozawa, 1987. Cerebroside of the dimorphic human pathogen, *Candida albicans*. *Chem. Phys. Lipids*, 43(1): 1–12.

Matsumoto, T., A. Ishiyama, Y. Yamaguchi, R. Masuma, H. Ui, K. Shiomi, H. Yamada, and S. Omura, 1999. Novel cyclopentanone derivatives pentenocins A and B, with interleukin-1 beta converting enzyme inhibitory activity, produced by *Trichoderma hamatum* FO-6903. *J. Antibiot.*, 52: 754–757.

Megharaj, M., B. Ramakrishnan, K. Venkateswarlu, N. Sethunathan, and R. Naidu, 2011. Bioremediation approaches for organic pollutants: A critical perspective. *Environ. Int.*, 37: 1362–1375.

Mills, G. L. and E. C. Cantino, 1974. Lipid composition of the zoospores of *Blastocladiella emersonii*. *J. Bacteriol.*, 118: 192.

Mimee, B., C. Labbé, R. Pelletier, and R. R. Bélanger, 2005. Antifungal activity of flocculosin, a novel glycolipid isolated from *Pseudozyma flocculosa*. *Antimicrob. Agents Chemother.*, 49: 1597–1599.

Morita, T., T. Fukuoka, T. Imura, and D. Kitamoto, 2009. Production of glycolipid biosurfactants by basidiomycetous yeasts. *Biotechnol. Appl. Biochem.*, 53: 39–49.

Mukherjee, S., P. Das, and R. Sen, 2006. Towards commercial production of microbial surfactants. *Trends Biotechnol.*, 24: 509–515.

Muralidhar, P., P. Radhika, N. Krishna, D. Venkata Rao, and Ch. Bheemasankara Rao, 2003. Sphingolipids from marine organisms: A reviews. *Nat. Prod. Sci.*, 9 (3): 117–142.

Naing, K. W., Anees, M., Kim, S.J., Nam, Y., Kim, Y.C., and K.Y. Kim, 2014. Characterization of antifungal activity of Paenibacillus ehimensis KWN38 against soil borne phytopathogenic fungi belonging to various taxonomic groups. Ann. Microbiol. 64: 55–63.

Ng, K. H. and M. A. Laneelle, 1977. Lipids of the yeast *Hansenula anomala*. *Biochimie*, 59(1): 97–104.

Nimrichter, L., M. L. Rodrigues, E. G. Rodrigues, and L. R. Travassos, 2005. The multitude of targets for the immune system and drug therapy in the fungal cell wall. *Microbes Infect.*, 7(4): 789–798.

Nitschke, M. and S. Costa, and J. Contiero, 2005. Rhamnolipid surfactants: an update on the general aspects of these remarkable biomolecules. *Biotechnol. Prog.*, 21: 593.

Noda, N., R. Tanaka, K. Miyahara, and T. Kawasaki, 1992. Two novel galactosylceramides from *Marphysa sanguinea*. *Tetrahedron Lett.*, 33: 7527–7530.

Ojika, M., G. Yoshino, and Y. Sakagami, 1997. Novel ceramide 1-sulfates, potent DNA topoisomerase I inhibitors isolated from the Bryozoa *Watersipora cucullata*. *Tetrahedron Lett.*, 38: 4235–4238.

Omura, S., H. Tomada, N. Tobata, Y. Ohyama, T. Abe, and M. Namikoshi, 1999. Roselipins, novel fungal metabolites having a highly methylated fatty acid modified with a mannose and an arabinitol. *J. Antibiot.*, 52: 586.

Park, E. J., S. Y. Park, and E. H. Joe, 2003. 15d-PGJ2 and rosiglitazone suppress Janus kinase-STAT inflammatory signaling through induction of suppressor of cytokine signaling 1 (SOCS1) and SOCS3 in glia. *J. Biol. Chem.*, 278: 14747–14752.

Pettit, G. R., J. Xu, D. E. Gingrich, M. D. Williams, D. L. Doubek, J. C. Chapuis, and J. M. Schimdt, 1999. Antineoplastic agents. Part 395. Isolation and structure of agelagalastatin from the Papua New Guinea marine sponge *Agelas* sp. *Chem. Commun.*, 915–916.

Piljac, T. and G. Piljac, 1999. Use of rhamnolipids in wound healing, treating burn shock, atherosclerosis, organ transplants, depression, schizophrenia and cosmetics. European Patent 1,889,623. Paradigm Biomedical Inc., New York.

Pinto, M. R., E. Barreto-Bergter, and C. P. Taborda, 2008. Glycoconjugates and polysaccharides of fungal cell wall and activation of immune system. *Braz. J. Microbiol.*, 39: 195–208.

Pinto, M. R., M. L. Rodrigues, L. R. Travassos, R. M. Haido, R. Wait, and E. Barreto-Bergter, 2002. Characterization of glucosylceramides in *Pseudallescheria boydii* and their involvement in fungal differentiation. *Glycobiology*, 12(4): 251–260.

Pornsunthorntawee, O., S. Maksung, O. Huayyai, R. Rujiravanit, and S. Chavadej, 2009. Biosurfactant production by *Pseudomonas aeruginosa* SP4 using sequencing batch reactors: Effects of oil loading rate and cycle time. *Bioresour. Technol.*, 100: 812–818.

Qi, J., M. Ojika, and Y. Sakagami, 2001. Neuritogenic cerebrosides from an edible Chinese mushroom. Part 2: Structures of two additional termitomycesphins and activity enhancement of an inactive cerebroside by hydroxylation. *Bioorg. Med. Chem.*, 9: 2171–2177.

Raghukumar, S. 2004. The role of fungi in marine detrital processes. In *Marine Microbiology: Facets & Opportunities*, Ramaiah, N. (ed.). National Institute of Oceanography, Goa, India, pp. 91–101.

Rieger, M. N. 1997. Surfactant chemistry and classification. In *Surfactants in Cosmetics*, 2nd edn., Rieger, M. N. and Rhein, L. D. (eds.). Mercel Dekker, New York.

Rodrigues, L. R., J. A. Teixeira, H. C. V. Meib, and R. Oliveira, 2006. Isolation and partial characterization of a biosurfactant produced by *Streptococcus thermophilus* A. *Colloids Surf. B: Biointerfaces*, 53: 105–112.

Rodrigues, M. L., L. R. Travassos, K. R. Miranda, A. J. Franzen, S. Rozental, W. de Souza, C. S. Alviano, and E. Barreto-Bergter, 2000. Human antibodies against a purified glucosylceramide from *Cryptococcus neoformans* inhibit cell budding and fungal growth. *Infect. Immun.*, 68: 7049–7060.

Ron, E. Z. and E. Rosenberg, 2002. Biosurfactants and oil bioremediation. *Curr. Opin. Biotechnol.*, 13: 249–252.

Ruckhäberle, E., T. Kam, L. Hanker, R. Gätje, D. Metzler, and U. Holtrich, 2008. Prognostic relevance of glucosylceramide synthase (GCS) expression in breast cancer. *J. Cancer. Res. Clin. Oncol.*, 10: 1–10.

Saito, K., N. Takakuwa, M. Ohnishi, and Y. Oda, 2006. Presence of glucosylceramide in yeast and its relation to alkali tolerance of yeast. *Appl. Microbiol. Biotechnol.*, 71(4): 515–521.

Sajna, K. V., R. K. Sukumaran, H. Jayamurthy, K. Konda Reddy, S. Kanjilal, R. B. N. Prasad, and A. Pandey, 2013. Studies on biosurfactant from *Pseudozyma* sp. NII 08165 and their potential applications as laundry detergent additives. *Biochem. Eng. J.*, 78: 85–92.

Sakaki, T., U. Zähringer, D. C. Warnecke, A. Fahl, W. Knogge, and E. Heinz, 2001. Sterol glycosides and cerebrosides accumulate in *Pichia pastoris*, *Rhynchosporium secalis* and other fungi under normal conditions or under heat shock and ethanol stress. *Yeast*, 18(8): 679–695.

Sánchez, A.J.C., H.H. Sanchez, and M.E.J. Flores, 2013. Biological activity of glycolipids produced by microorganisms: New trends and possible therapeutic alternatives. *Microbiol. Res.* 168: 22–32.

Saravanakumar, K., 2012. Studies on mangroves derived *Trichoderma* and their biotechnological applications. PhD thesis, Annamalai University, Tamil Nadu, India, 290pp.

Seghal Kiran, G., T. A. Hema, R. Gandhimathia, J. Selvina, T. Anto Thomasa, T. Rajeetha Ravji, and K. Natarajaseenivasan, 2009. Optimization and production of a biosurfactant from the sponge-associated marine fungus *Aspergillus ustus* MSF3. *Colloids Surf. B*, 3: 250–256.

Shin, J. and Y. Seo, 1995. Isolation of new ceramides from the Gorgonian *Acabaria undulata*. *J. Nat. Prod.*, 58: 948–953.

Shoeb, E., F. Akhlaq, U. Badar, J. Akhter, and S. Imtiaz, 2013. Classification and industrial applications of biosurfactants. *Acad. Res. Int.*, 4: 3.

Singh, P. and S. S. Cameotra, 2004. Potential applications of microbial surfactants in biomedical sciences. *Trends Biotechnol.*, 22(3): 142–146.

Souza, E.C, T.C.V. Penna, and R.P.S. Oliveira, 2014. Biosurfactant-enhanced hydrocarbon bioremediation: An overview. *International Biodeterioration & Biodegradation* 89:88–94.

Springer, T. A. and L. A. Lasky, 1991. Sticky sugars for selectins. *Nature*, 349: 196–197.

Sridhar, K. R., 2005. Diversity of fungi in mangrove ecosystem. In *Microbial Diversity: Current Perspectives and Potential Applications*, Satyanarayanan, T. and Johri, B. N. (eds.). I.K. International Pvt. Ltd., New Delhi, India, pp. 129–148.

Sugita, M., K. Aoki, A. Sakata, and T. Hori, 1994. Glycosphingolipids in Coelenterata. Characterization of cerebrosides from the sea anemone, *Actinogeton* sp. *Shiga Daigku Kyoikugakubu Kiyo, Shizen Kagaku, Kyoiku Kagaku*, 44: 25–30.

Sugiyama, S., M. Honda, and T. Komori, 1988. Biologically active glycosides from Asteroidea, XV. Asymmetric synthesis of phytosphingosine anhydro base: Assignment of the absolute stereochemistry. *Liebigs Ann. Chem.*, 619–625.

Sun, H. H., W. J. Mao, Y. Chen, S. D. Guo, H. Y. Li, X. H. Qi, Y. L. Chen, and J. Xu, 2009. Isolation, chemical characteristics and antioxidant properties of the polysaccharides from marine fungus *Penicillium* sp. F23-2. *Carbohydr. Polym.*, 78: 117–124.

Takahashi, H. K., S. B. Levery, M. S. Toledo, E. Suzuki, M. E. Salyan, S. Hakomori, and A. H. Straus, 1996. Isolation and possible composition of glucosylceramides from *Paracoccidioides brasiliensis*. *Braz. J. Med. Biol. Res.*, 29(11): 1441–1444.

Takakuwa, N., M. Kinoshita, Y. Oda, and M. Ohnishi, 2002. Existence of cerebroside in *Saccharomyces kluyveri* and its related species. *FEMS Yeast Res.*, 2(4): 533–538.

Teichmann, B., C. Labbe, F. Lefebvre, M. Bolker, U. Linne, and R. R. Be Langer, 2011. Identification of a biosynthesis gene cluster for flocculosin a cellobiose lipid produced by the biocontrol agent *Pseudozyma flocculosa*. *Mol. Microbiol.*, 79: 1483–1495.

Thirunavukkarasu, N., T. S. Suryanarayanan, K. P. Girivasan, A. Venkatachalam, V. Geetha, J. P. Ravishankar, and M. Doble, 2012. Fungal symbionts of marine sponges from Rameswaram, southern India: Species composition and bioactive metabolites. *Fungal Divers.*, 55: 37–46.

Tobata, N., Y. Ohyama, H. Tomada, T. Abe, M. Namikoshi, and S. Omura. 1999. Structure elucidation of roselipins, inhibitors of diacylglycerol acyltransferase produced by *Gliocladium roseum* KF-1040. *J. Antibiot.*, 52: 815.

Toledo, M.S., S.B. Levery, E. Suzuki, A.H. Straus, and H. K. Takahashi, 2001. Characterization of cerebrosides from the thermally dimorphic mycopathogen Histoplasma capsulatum: expression of 2- hydroxy fatty N-acyl (E)-Delta(3)- unsaturation correlates with the yeast-mycelium phase transition. *Glycobiology* 11: 113–124.

Tomada, H., Y. Ohyama, T. Abe, N. Tobata, M. Namikoshi, and S. Omura, 1999. Roselipins, inhibitors of diacylglycerol acyltransferase, produced by Gliocladium roseum KF-1040. *J. Antibiot.*, 52: 689.

Vicente, F., A. Basilio, A. Cabello, and F. Peláez, 2003. Natural products as a source of antifungals. *Clin. Microbiol. Infect.*, 9: 15–32.

Villas-Boas, M. H. S., H. Egge, G. Pohlentz, R. Hartmann, and E. Barreto-Bergter, 1994. Structural determination of *N*-2′-hydroxyoctadecanoyl-1-*O*-beta-d-glucopyranosil-9-methyl-4,8-Dsphingadienine from species of *Aspergillus. Chem. Phys. Lipids* 70: 11–1910.

Wagner, H. and W. Zofcsik, 1966. On new sphingolipids of yeast. *Biochem. Z.*, 344: 314–316.

Walenkamp, A. M., W. S. Chaka, A. F. Verheul, V. V. Vaishnav, R. Cherniak, F. E. Coenjaerts, and I. M. Hoepelman, 1999. *Cryptococcus neoformans* and its cell wall components induce similar cytokine profiles in human peripheral blood mononuclear cells despite differences in structure. *FEMS Immunol. Med. Microbiol.*, 26: 309–318.

Wang, X., L. Gong, S. Liang, X. Han, C. Zhu, and Y. Li, 2005. Algicidal activity of rhamnolipid biosurfactants produced by *Pseudomonas aeruginosa. Harmful Algae*, 4: 433–443.

Warnecke, D. and E. Heinz, 2010. Glycolipid head group replacement: A new approach for the analysis of specific functions of glycolipids in vivo. *Eur. J. Cell Biol.*, 89: 53–61.

Weiss, B., R. L. Stiller, and R. C. Jack, 1973. Sphingolipids of the fungi *Phycomycetes blakesleeanus* and *Fusarium lini. Lipids* 8: 25–30.

Yamaguchi, Y., K. Konda, and A. Hayashi, 1992. Studies on the chemical structure of neutral glycosphingolipids in eggs of the sea hare, *Aplysia juliana. Biochim. Biophys. Acta*, 1165: 110–118.

Zeidan, Y. H. and Y. A. Hannun, 2007. Translational aspects of sphingolipid metabolism. *Trends Mol. Med.*, 13: 327–336.

Zinjarde, S. S. and A. Pant, 2002. Emulsifier from a tropical marine yeast, *Yarrowia lipolytica* NCIM 3589. *J. Basic Microbiol.*, 42: 67–73.

# *Marine glycoenzymes*

## chapter fourteen

# Sialyltransferases from marine environments

## Preparation of sialyloligosaccharides and its application

*Takeshi Yamamoto*

### Contents

## 14.1 Introduction

All mammalian cells have various carbohydrates on their surfaces in the form of sugar chains covalently bound to membrane proteins and lipids (Varki 1993, Schauer 2004). The majority of sugar chains fall into three groups. One is those attached to lipids and the others are those attached to the amide nitrogen of asparagine side chains in proteins (N-linked sugar chains) and attached to an oxygen atom of the side chains of serine or threonine in proteins (O-linked sugar chains). These sugar chains are usually composed of a limited number of monosaccharides including galactose, mannose, fucose, N-acetylgalactosamine, N-acetylglucosamine, and N-acetylneuraminic acid (Neu5Ac) with several glycosidic linkage patterns (Gagneux and Varki 1999, Angata and Varki 2002). The structures of these monosaccharides are shown in Figure 14.1. These monosaccharides are transferred into sugar chains by specific glycosyltransferases from corresponding sugar nucleotides, donor

substrates of glycosyltransferases, such as UDP-GlcNAc, UDP-Gal, GDP-Fuc, cytidine 5′-monophosphate (CMP)-Neu5Ac, and so on (Stults et al. 1988, Burda and Aebi 1999). Generally, mammalian glycosyltransferases show very strict acceptor and donor substrates specificity. To date, it has been demonstrated that the sugar chains of these glycoproteins and glycolipids play important roles in a variety of biochemical phenomena (Paulson 1989, Hakomori 1990, Kannagi 2002).

Glycosylation of proteins was considered to be specific to eukaryotes around the mid-1990s. However, recent progress has revealed that glycoproteins are present in prokaryotes as well (Erickson and Herzberg 1993, Young et al. 2002), although bacterial sugar chains differ structurally from the eukaryotic counterparts (Wacker et al. 2002). Furthermore, evidence is provided that protein glycosylation pathways similar to those in eukaryotes exist in Gram-negative bacteria, and many bacterial glycosyltransferases have also been reported (Wacker et al. 2002, Yamamoto 2006). In case of Gram-negative

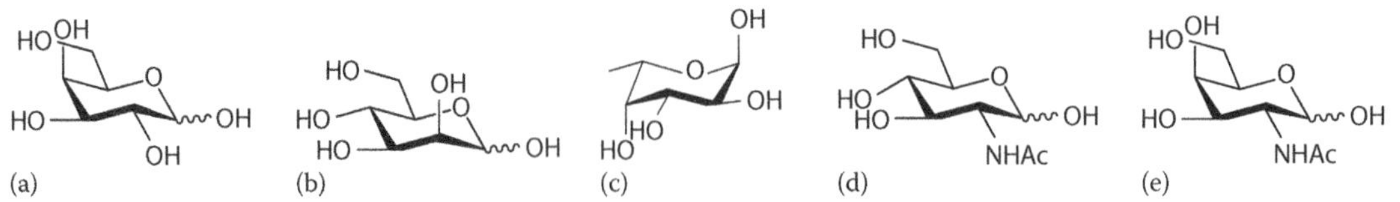

**Figure 14.1** Structure of major monosaccharides consisting sugar chains: (a) D-galactose, (b) D-mannose, (c) L-fucose, (d) *N*-acetyl-D-glucosamine, and (e) *N*-acetyl-D-galactosamine.

bacteria, it has been revealed that glycoconjugates, such as lipooligosaccharides (LOS), lipopolysaccharides, and glycoproteins, are present on their surface (Wakarchuk et al. 1998, Muldon et al. 2002). In particular, the sugar chain structures of glycoconjugates in human pathogens have well been studied. It has been demonstrated that some pathogenic bacteria have evolved to escape immune systems of hosts by mimicking surface sugar chains of the host cells, which are crucial for self/nonself recognition and that the glycoconjugates of the pathogens are involved in the virulence and adhesion to the host cells (Guerry et al. 2002).

Among the monosaccharides consisting of sugar chains described earlier, Neu5Ac, so-called sialic acid or Sia, often occurs at nonreducing termini of sugar chains of glycoconjugates and is thought to be one of the most important monosaccharides (Schauer 2004). For example, protein sialylation is reported to make half-lives of protein much longer in mammals. To be concrete, the introduction of two more sialylated sugar chains by genetic engineering to erythropoietin, which has three of the sialylated sugar chains in the native form, extended considerably the half-life in blood compared to the native protein (Elliott et al. 2003). Another example is insulin with sugar chains. Insulin is composed of two polypeptide chains and not naturally glycosylated. Several types of sialylated insulin were prepared by chemoenzymatic approach and tested for the blood glucose–lowering activity, and the modified insulin showed prolonged effects compared to the native form (Sato et al. 2004). Thus, sialylated sugar chains are thought to be very important in several biological phenomena. Furthermore, sialyltransferases are considered to be key enzymes in the synthesis of sialylated sugar chains and in the sialylation of glycoconjugates.

## 14.2  Sias: Structure and biological roles of sialylated sugar chains

### 14.2.1  Structure of Sias

Sias are a family of monosaccharides comprising over 50 naturally occurring derivatives of neuraminic acid (Neu), 5-amino-3,5-dideoxy-D-glycero-D-galacto-non-2-ulopyranonic acid (Angata and Varki 2002, Schauer 2004). Structurally, the Sia derivatives carry a variety of substitutions at the amino and/or hydroxyl groups. To be concrete, the amino acid group is often acetylated, glycosylated, or deaminated. The hydroxyl groups can be acetylated at O7, O8, or O9, singly or in combination (Paulson 1989, Hakomori 1990), and can also be lactylated, phosphorylated, methylated, or sulfurylated. The three major members of the Sia group are Neu5Ac, *N*-glycolylneuraminic acid (Neu5Gc), and 2-keto-3-deoxy-D-glycero-D-galacto-noninic acid (KDN) (Paulson 1989, Schauer 2004). The structures of these typical Sias are shown in Figure 14.2. Although Sias are widely distributed in higher animals and some microorganisms, only Neu5Ac is ubiquitous, and Neu5Gc is not found in bacteria (Paulson 1989, Angata and Varki 2002). Usually, Sias exist in the sugar chain of glycoconjugates including glycoproteins and glycolipids and are linked to the terminal positions of the sugar chains of the glycoconjugates (Lasky 1992). Sias are usually transferred from donor substrate of sialyltransferases, CMP-Neu5Ac, to sugar chains by sialyltransferases and attached to sugar chains through four main linkage patterns—Neu5Acα2-6Gal, Neu5Acα2-3Gal, Neu5Acα2-6GalNAc, and Neu5Acα2-8(9)Neu5Ac. These glycosidic linkage patterns are formed by specific sialyltransferases (Taniguchi et al. 2002).

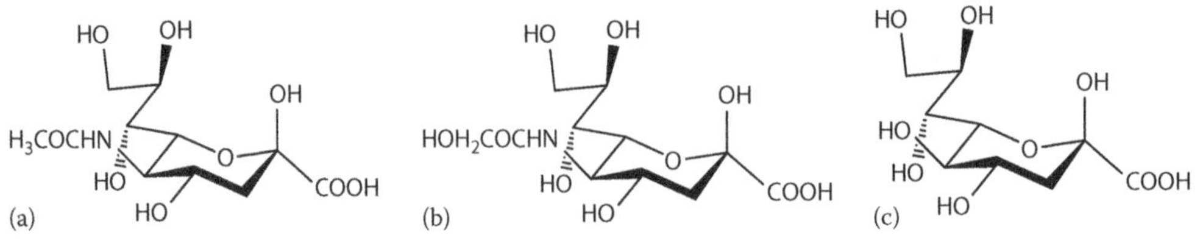

**Figure 14.2** Structure of sialic acids: (a) *N*-acetylneuraminic acid (Neu5Ac), (b) *N*-glycolylneuraminic acid (Neu5Gc), and (c) 2-keto-3-deoxy-D-glycero-D-galacto-noninic acid (KDN).

Till date, it has been clarified that these sialylated sugar chains play important roles in many biological processes, including immunological responses, viral infections, cell–cell recognition, inflammation, and so on (Angata and Varki 2002, Suzuki 2005).

## 14.2.2 Examples of the relationship between sugar chains and diseases

### 14.2.2.1 Influenza

The relationship between the sugar chain structure of the host cell and host cell recognition by influenza virus is a well-investigated biological phenomenon (Weis et al. 1988, Suzuki 2005). Many reports have shown that influenza A and B viruses bind via viral hemagglutinin to host cell surface receptors, Neu5Ac- or Neu5Gc-linked glycoproteins or glycolipids (Connor et al. 1994, Suzuki 2005). Furthermore, it has been clearly demonstrated that these influenza viruses also recognize the sugar chain structure of the host cell (Connor et al. 1994). Confirming evidence has shown that the avian influenza viruses recognize Neu5Acα2-3Galβ1-3/4GlcNAc structures, and the human influenza viruses recognize Neu5Acα2-6Galβ1-3/4GlcNAc structures (Connor et al. 1994). The host cell specificities of the influenza A and B viruses are determined mainly by the linkage of Neu5Ac or Neu5Gc to the penultimate galactose residues and core structure of the host glycoproteins or glycolipids. For this reason, the distribution of Neu5Ac or Neu5Gc and their linkage patterns on the host cell surface are important determinants of host tropism.

### 14.2.2.2 Guillain–Barré syndrome

Guillain–Barré syndrome (GBS) is the prototypic postinfectious autoimmune disease (Yuki 2001). Epidemiological studies clearly demonstrate that many GBS patients develop the syndrome following infection by the Gram-negative bacterium *Campylobacter jejuni* (Yuki et al. 2004). GBS patients develop neuropathy, which is the most common cause of generalized paralysis (Ropper 1992). The serotypes of *C. jejuni* associated with GBS-derived neuropathies express ganglioside-like LOS structures on the cell surface. Detailed investigations of LOS structures isolated from GBS patients reveal that the core oligosaccharides mimic gangliosides located in the neural tissue. Terminal oligosaccharide moieties identical to those of the gangliosides GM1, GD3, and GT1a have been found in *C. jejuni* O:19 strains (Aspinall et al. 1992, 1994a,b, Gilbert et al. 2000). The structures of GM1 and LOS are shown in Figure 14.3. The development of GBS in patients following infection with *C. jejuni* is thought to be related to the molecular mimicry of gangliosides and the cross-reaction of antibodies against LOS, leading to neuropathy. The molecular mimicry basis of GBS is consistent with the finding that most GBS patients develop autoantibodies that react with GM1 subsequent to *C. jejuni* enteritis (Ho et al. 1999, Ogawara et al. 2000). Molecular mimicry has been proposed as a pathogenic mechanism underlying autoimmune disease (Yuki 2005). It has also been demonstrated in the development of Fisher syndrome after infection with *Haemophilus influenzae* in the production of anti-GQ1b autoantibodies mediated by the GQ1b-mimicking LOS on the bacterial surface (Koga et al. 2005).

## 14.3 Major methods for preparing sugar chains

A large number of oligosaccharides exist in nature. For example, many kinds of sialylated oligosaccharides, such as 3′-sialyllactose, 6′-sialyllactose, and sialyllacto-N-neotetraose, are contained in milk of various animals (Haeuw-Fievre et al. 1993, Kunz et al. 2000). However, the purification and isolation of sialylated oligosaccharides from natural sources is very difficult due to their structural complexity. On the other hand, an abundant supply of sugar chains is essential for a detailed investigation of the biological role and function of the sugar

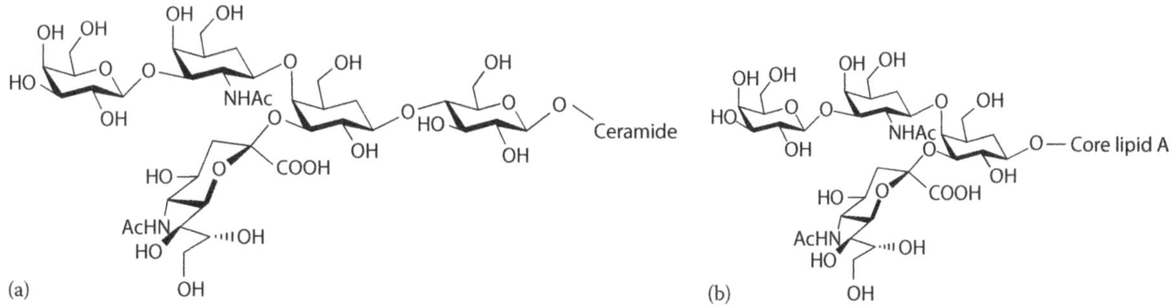

*Figure 14.3* Structures of GM1 and typical lipopolysaccharides of *Campylobacter jejuni*: (a) GM1 and (b) typical lipopolysaccharides identified in *C. jejuni*.

chains. Major methods for preparing sialylated oligosaccharides are described in the following sections.

## 14.3.1 Chemical synthesis

There is a large body of knowledge for the chemical synthesis of oligosaccharides including sialyloligosaccharides (Schmidt 1986, Kanie and Hindsgaul 1992, Wang et al. 1996). In general, however, chemical synthesis of oligosaccharides requires multiple protection and deprotection steps, so production of these compounds is very complex and expensive. Furthermore, in the case of sialyloligosaccharide synthesis, there are some points to be considered. Usually, Sias naturally exist in carbohydrate chains only as the α-linkage form. Placement of the carboxyl group mounted in the anomeric position and 3-deoxy structure, both of which are thought to be negative structural features of Sia to form α-linked sialyloligosaccharides from the viewpoint of chemical synthesis, have to be considered for the chemical synthesis of α-linkage form sialyloligosaccharides (Ando and Imamura 2004). Placement of the carboxyl group in the anomeric position of sialic acid makes the corresponding oxocarbenium ion unstable due to its electron-withdrawing nature. Moreover, due to the 3-deoxy structure of sialic acid, the neighboring effect is not able to be exploited so that the formation of the stereochemically disfavored equatorial glycoside becomes hard and the corresponding 2,3-en-derivatives are concomitantly formed as a major by-product in the course of the coupling reaction. Therefore, chemical synthesis of α-linkage form sialyloligosaccharides is considered to be one of the most difficult modes of glycosylation. However, several excellent chemical synthesis methods with respect to high stereoselectivity and coupling yield have been developed (Ando et al. 2003, Ito et al. 2003). Using these methods, kilogram-scale quantities of several sialyloligosaccharides (protected form) have been already produced.

## 14.3.2 Enzymatic synthesis

In comparison with chemical synthesis of oligosaccharides, enzymatic synthesis of oligosaccharides using glycosyltransferases is thought to be superior for reaction yield, ease of reaction (Izumi and Wong 2001). But this method involved two problems that had to be solved: One problem was the limited number of available enzymes for oligosaccharides production. The other was the difficulty of an abundant supply of donor substrates of glycosyltransferases, such as UDP-galactose for galactosyltransferases and CMP-Neu5Ac for sialyltransferase. However, on the one hand recent progress in the understanding of bacterial glycosyltransferases has made available, in large quantities, glycosyltransferases, including sialyltransferases, galactosyltransferases, and fucosyltransferases (Koizumi 2003). In particular, the number of sialyltransferases available for a large-scale production of sialyloligosaccharides is increasing. For example, α2,3-sialyltransferases were cloned from *Photobacterium* sp. (Tsukamoto et al. 2008), *Photobacterium phosphoreum* (Tsukamoto et al. 2007), and *Vibrio* sp. (Takakura et al. 2007) and α2,6-sialyltransferases were cloned from *Photobacterium damselae* (Yamamoto et al. 1998a), *Photobacterium leiognathi* (Yamamoto et al. 2007), and so on. On the other hand, recent progress in sugar nucleotide production using the bacterial enzymes also has made sugar nucleotides available in large quantities. Specifically, CMP-Neu5Ac was produced efficiently by the combination of recombinant *Escherichia coli* overexpressing CMP-Neu5Ac biosynthetic genes and *Corynebacterium ammoniagenes* contributing cytidine 5′-triphosphate (CTP) (Endo et al. 2000): in this case, it was reported that 17 g of CMP-Neu5Ac was obtained. As described earlier, it is possible to obtain some kinds of glycosyltransferases and sugar nucleotides in large quantities. Therefore, gram-scale production of some kinds of sialyloligosaccharides by enzymatic methods is now available. Actually, gram-scale productions of sialyloligosaccharides, such as 6′-sialyllactose, 3′-sialyllactose, and 6′-sialyl-N-acetyllactosamine, are carried out in our laboratory using bacterial sialyltransferases, routinely.

## 14.3.3 Fermentation method

On the one hand, a unique strategy for oligosaccharide production has been developed using a combination of metabolically engineered bacteria (Koizumi et al. 1998, Endo et al. 1999, Endo et al. 2001). This method is based on the utilization of whole cells that are involved in the synthesis of oligosaccharides and on systems that allow sugar–nucleotide recycling by these whole cells. Kyowa Hakko Kogyo developed this technology by engineering a *C. ammoniagenes* strain to produce and regenerate UTP from an inexpensive substrate (Koizumi et al. 1998). This strain has been used in combination with two recombinant *E. coli* strains, one that overexpresses the genes of sugar–nucleotide biosynthesis and the other that overexpresses a glycosyltransferase gene. This technology efficiently produces several oligosaccharides, such as globotriose, lactosamine, and sialyllactose. However, this method requires permeabilization of the bacteria with xylene, which kills the cells, to allow a passive circulation of the substrate between different types of cells used; the reactions occur in nongrowing cells.

On the other hand, a remarkable and excellent oligosaccharide production system has also been developed using a single growing metabolically engineered bacterium by Dr. Eric Samain at the Centre de Recherches sur les Macromolécules Végétales (Preim et al. 2002). In this system, oligosaccharides are produced in a bacterium that overexpresses the recombinant glycosyltransferase genes and that maintains the pool level of the sugar nucleotides by its enzymatic cellular machinery (Antoine et al. 2003, Drouillard et al. 2006). To date, it has been reported that several oligosaccharides, such as lacto-*N*-neotetraose, lacto-*N*-neohexaose, and sialyllactose, are efficiently produced by this system (Antoine et al. 2003, Drouillard et al. 2006, Fierfort and Samain 2008).

Very recently, it has been reported that more than 25 g/L of both 6′-sialyllactose and 3′-sialyllactose has been produced by a high-density cell culture of *E. coli* strains overexpressing bacterial sialyltransferases of α2,6-sialyltransferase from *Photobacterium* sp. and α2,3-sialyltransferase from *Neisseria meningitidis*, respectively (Fierfort and Samain 2008, Drouillard et al. 2010). In the case of the sialyllactose production system using the transformed bacterium by *Photobacterium* sp. JT-ISH-224 α2,6-sialyltransferase gene, formation of KDO-lactose and 6,6′-disialyllactose has been observed as well as 6′-sialyllactose production (Drouillard et al. 2010). On the contrary, no formation of by-product has been observed in case of sialyllactose production system using the transformed bacterium by the *N. meningitidis* α2,3-sialyltransferase gene. These results indicate that the 6′-sialyllactose production system with marine bacterial sialyltransferase has the possibility to produce various kinds of sialylated carbohydrate chains, but examination of the culture conditions is necessary to efficiently produce the aimed compounds (Drouillard et al. 2010).

## 14.4 Sialyltransferases from marine environment

Glycoproteins, such as erythropoietin, granulocyte colony-stimulating factor, tissue-type plasminogen activator, and so on, have been produced by biotechnology in the pharmaceutical industry, and these recombinant glycoproteins have already been used as drugs clinically (Walsh 2010, Modjtahedi et al. 2012, Elvin et al. 2013). Regarding the quality of these therapeutic glycoproteins, the glycosylation is critical because the glycosylation is known to give a drastic impact on the therapeutic efficacy, serum half-life, immunogenicity, and so on (Kontermann 2011, Jedrzejewsk et al. 2013).

Nowadays, these recombinant therapeutic glycoproteins are produced not only in Chinese hamster ovary cell lines but also in cell lines from baby hamster kidney, murine tissues, and so on (Ghaderi et al. 2012). Although the N-linked sugar chains of proteins produced in these cells very well resemble the native human glycan structure, some different points are observed (Bork et al. 2009, Hossler et al. 2009). For example, the terminal of N-linked sugar chains is capped with the Neu5Ac in human native glycoproteins. However, glycoproteins expressed in the cell lines described earlier are incompletely capped by Neu5Ac (Tanemura et al. 2013). It has been shown that these sialylated N-linked sugar chains strongly affect the solubility, stability, and immunogenic properties of the respective glycoproteins by many studies. Thus, sialyltransferases for *in vitro* sialylation of recombinant therapeutic glycoproteins are very important in pharmaceutical industry to control the glycosylation of therapeutic proteins to a desired, homogenous, and bioactive glycoform (Raju et al. 2001, Khmelnitsky 2004, Ribitsch et al. 2014).

Till date, we had screened marine bacteria for sialyltransferases and found many marine bacteria that produce sialyltransferases. So, I introduce two screening methods of marine bacterial sialyltransferases that we had used in our research activities.

### 14.4.1 Screening bacteria for sialyltransferases by measuring enzyme activity

Here, I introduce this screening method in brief. The detailed procedure of this screening method was described in our previous report (Yamamoto et al. 1996).

Seawater, sea sand, mud, seaweed, and small animals including various kinds of fishes and shells were collected from various coastal locations in Japan, and these samples were used as bacterial sources. Bacteria that grew on marine agar 2216 were isolated from the samples. After the cultivation with liquid medium, bacteria were harvested from the culture broth by centrifugation, and lysed by sonication on ice, and then measured immediately for sialyltransferase activity. Sialyltransferase activity was confirmed as follows: the reaction mixture contained the bacterial lysate as the sample of enzyme, lactose (acceptor substrate), CMP-Neu5Ac containing $^{14}$C-labeled CMP-[4,5,6,7,8,9-$^{14}$C]-Neu5Ac (donor substrate), 100 mM Bis–Tris buffer (pH 6.0), 0.5 M NaCl, and 0.03% Triton X-100. The reaction was carried out at 25°C for 2 h. After the reaction, the radioactivity of [4, 5, 6, 7, 8, 9-$^{14}$C]-Neu5Ac that had transferred to the acceptor substrate was measured using a liquid scintillation counter, and the amount of Neu5Ac transferred was calculated. Using this procedure, we have isolated many bacteria that produce sialyltransferase. For instance, *P. phosphoreum* JT-ISH-467 that showed α2,3-sialyltransferase activity was isolated from the outer skin of the Japanese common squid, *Todarodes pacificus* (Tsukamoto et al. 2007); *P. damselae* JT0160 that expressed α2,6-sialyltransferase activity was isolated from seawater (Yamamoto et al. 1996); *Photobacterium* sp. JT-ISH-224 that contained both α2,3- and α2,6-sialyltransferase activities was isolated from the gut of the Japanese barracuda, *Sphyraena pinguis* (Tsukamoto et al. 2008); and *P. leiognathi* JT-SHIZ-145 that expressed α2,6-sialyltransferase activity was isolated from the outer skin of the Japanese squid, *Loliolus japonica* (Yamamoto et al. 2007).

### 14.4.2   Screening bacteria for sialyltransferases by lectin staining

Lectins are sugar-binding proteins that are highly specific for their sugar moieties. For instance, *Sambucus sieboldiana* agglutinin (SSA) recognizes both Neu5Acα2-6Gal and Neu5Acα2-6GalNAc structures of sialyloligosaccharides in glycoconjugates (Shibuya et al. 1989). In case of SSA, this lectin recognizes monosaccharides at nonreducing end (Neu5Ac) and also the glycosidic linkage that they formed (α2-6 linkage). The detail procedure of this screening method was described in our previous report (Kajiwara et al. 2010). Thus, I introduce this screening method in brief.

Using biotin-labeled SSA, we carried out lectin staining of the cells of *P. damselae* JT0160, *P. leiognathi* JT-SHIZ-145, *P. phosphoreum* JT-ISH-467, and *Photobacterium* sp. JT-ISH-224 and observed the cells using differential interference contrast (DIC) and fluorescence microscopy (Kajiwara et al. 2010). To be concrete, after cultivation of isolated bacteria with liquid medium, the bacterial cells were collected by centrifugation and collected cells were spotted onto glass slides. After spotting the cells, cells were fixed with paraformaldehyde solution and blocked with bovine serum albumin phosphate-buffered saline (PBS) solution. After the glass slides were washed, biotin-labeled SSA was added. After the reaction, the cells were washed again with PBS solution and Alexa 594-labeled streptavidin solution was added. After washing the cells with PBS solution, the cells were mounted using Prolong Gold antifade reagent and observed using DIC and fluorescence microscopy. The SSA bound to *Photobacterium* sp. JT-ISH-224, *P. damselae* JT0160, and *P. leiognathi* JT-SHIZ-145. These *Photobacterium* strains produce α2,6-sialyltransferases, so the lectin staining indirectly detected α2,6-sialyltransferase-producing bacteria (Figure 14.4). SSA did not bind to *P. phosphoreum* JT-ISH-467, which produces only α2,3-sialyltransferase. Therefore, the SSA lectin might be useful to screen for not only Neu5Acα2-6Gal and/or Neu5Acα2-6GalNAc structures on the bacterial cell surface but also the production of α2,6-sialyltransferase.

We consider that this efficient method would be applicable to the screening of other glycosyltransferases by changing the type of lectin used. As a result, we have isolated marine bacteria that produce fucosyltransferases by this lectin staining method using biotin-labeled

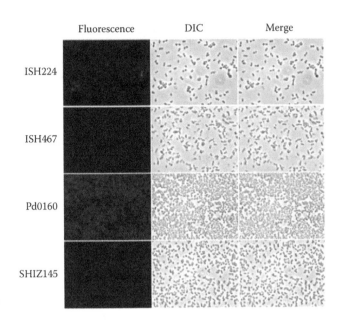

**Figure 14.4** Detection of Neu5Ac residues on marine bacterial cells by biotin-labeled SSA lectin staining. ISH224, *Photobacterium* sp. JT-ISH-224; ISH467, *P. phosphoreum* JT-ISH-467; Pd0160, *P. damselae* JT0160; SHIZ145, *P. leiognathi* JT-SHIZ-145. The SSA lectin specific for Neu5Acα2–6Gal/GalNAc bound to cell surfaces of marine bacteria producing α2,6-sialyltransferases (red staining in SSA column, detected with fluorescein Alexa 594-labeled streptavidin). DIC: Results of observation by differential interference contrast microscopy, Fluorescence: Results of SSA lectin staining observed by fluorescence microscopy, Merge: Merged results DIC and Fluorescence. (Data from Kajiwara, H. et al., *Microbes Environ.*, 25, 152, 2010. With permission.)

*Aleuria aurantia* lectin, which recognizes the fucose residue in carbohydrate chains (Kajiwara et al. 2012).

## 14.5   Enzymatic characterization of marine bacterial sialyltransferases

We have isolated marine bacteria that produce sialyltransferases during the course of our study. Among the bacteria, the identified bacteria were classified in the genus *Photobacterium* or closely related to genus *Vibrio*. From these marine bacteria, my research team had purified sialyltransferases and clarified their enzymatic properties.

### 14.5.1   An α2,6-sialyltransferase produced by *Photobacterium damselae* JT0160

It has been demonstrated that the α2,6-sialyltransferase produced by *P. damselae* JT0160 has unique acceptor substrate specificity compared with the mammalian counterparts. First, it has been revealed that

*P. damselae* α2,6-sialyltransferase transfers Neu5Ac to both glycoproteins including asialo-N-linked and asialo-O-linked glycoproteins and glycolipids (Yamamoto et al. 1998b, Kushi et al. 2010). In addition, *P. damselae* α2,6-sialyltransferase recognizes lactose and *N*-acetyllactosaminide as acceptor substrates with almost equal *Km* values, although rat liver α2,6-sialyltransferase has a *Km* value approximately 33 times higher for lactose than for *N*-acetyllactosaminide (Yamamoto et al. 1996). From these results, it was indicated that *P. damselae* α2,6-sialyltransferase did not recognize the 2-acetamido group in the *N*-acetylglucosaminyl residue, unlike the mammalian enzymes (Yamamoto et al. 1996). The acceptor substrate specificity of mammalian sialyltransferases is generally high, and mammalian enzymes are specific for the type of the sugars and linkages. For example, mammalian sialyltransferases did not recognize fucosylated carbohydrate chains as an acceptor substrate. In eukaryotes, the carbohydrate chain moieties of glycoconjugate are synthesized by a series of glycosyltransferases, and the order of glycosylation is determined strictly. Usually, fucosylation is performed by fucosyltransferases after sialylation in mammalian cell. Thus, mammalian sialyltransferases did not recognize fucosylated carbohydrate chain as acceptor substrates and cannot transfer Neu5Ac to them. However, *P. damselae* α2,6-sialyltransferase does recognizes fucosylated carbohydrate chains as an acceptor substrate and transfers Neu5Ac to the galactose residue of carbohydrate chains at position 6 in 2'-fucosyllactose (Kajihara et al. 1996). In addition, ã2,6-sialyltransferase also recognizes 3'-sialyllactose and gives 3',6'-disialyllactose as the reaction product (Kajihara et al. 1996).

The *P. damselae* α2,6-sialyltransferase is less specific, which may allow various sialylated glycans to be prepared, as described earlier. Furthermore, it has been revealed that α2,6-sialyltransferase uses both Galβ1,3GlcNAc and Galβ1,6GlcNAc as acceptor substrates and gives the corresponding enzymatic reaction products, respectively (Mine et al. 2010a). These results indicate that this α2,6-sialyltransferase is not sensitive to the nature of the second sugar from the nonreducing terminus and to the linkage between the terminal two sugars. Furthermore, it has also been revealed that α2,6-sialyltransferase has a unique donor substrate specificity. The sialyltransferase recognizes CMP-Neu5Ac, CMP-KDN as donor substrates and many CMP–Sia derivatives with the nonnatural modification of an azido or acetylene group at positions C5, C7, C8, and/or C9 (Kajihara et al. 1999, Yu et al. 2006).

## 14.5.2 *An α2,3-sialyltransferase produced by* Photobacterium phosphoreum *JT-ISH-467*

To date, three genes encoding the marine bacterial α2,3-sialyltransferase have been cloned (Takakura et al. 2007, Tsukamoto et al. 2007, 2008), and all these genes have been expressed as recombinant active form enzymes in an *E. coli* protein expression system. During the course of 3'-sialyllactose production using α2,3-sialyltransferase cloned from *Photobacterium* sp. JT-ISH-224, it has been revealed that the α2,3-sialyltransferase can transfer Neu5Ac to both O-2 and O-3' hydroxyl groups of lactose simultaneously, giving 2,3'-disialyllactose as an enzymatic reaction product (Mine et al. 2010b). By nuclear magnetic resonance (NMR) spectroscopy analysis, it has been confirmed that the reaction product contains only α-form. The relative configuration between C-1 and C-3 of the α-glucopyranose residue is superimposable with that between C-4 and C-2 of galactopyranose (Figure 14.5). Therefore, it is expected that the enzyme recognizes the α-glucopyranose residue as acceptor substrate and transfers Neu5Ac to O-2 hydroxyl group of the α-glucopyranose. In addition, it has also been demonstrated that the α2,3-sialyltransferase can transfer Neu5Ac to the β-anomeric hydroxyl groups of mannose and give the corresponding reaction product (Mine et al. 2010a). In this case, the stereochemistry of β-mannose residue from C-2 to O-5 in the pyranose ring is superimposable on that from C-4 to C-2 of galactopyranose (Figure 14.6), except for the difference between the C-2 carbon atom in the Gal and O-5 oxygen atom in the Man (Mine et al. 2010a). Thus, it is strongly expected that the α2,3-sialyltransferase recognizes the acceptor substrate mainly through the stereochemical structure of the C-4 to C-3 of Gal (Mine et al. 2010a).

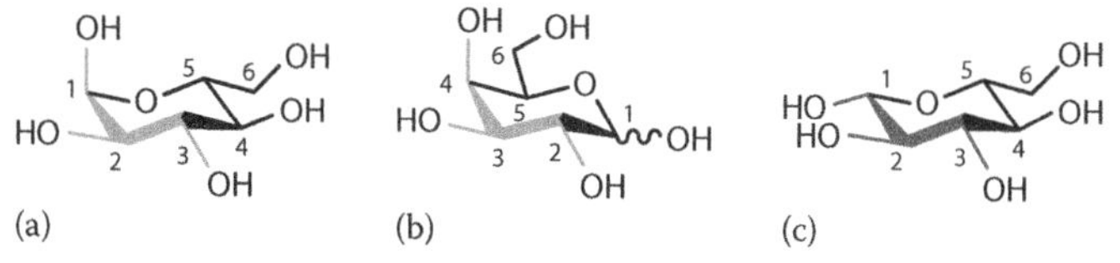

(a)                                  (b)                                  (c)

*Figure 14.5* Structures of D-galactopyranose, α-D-glucopyranose, and β-D-glucopyranose: relative configuration between C-1 and C-3 of the α-D-glucopyranose (a) is superimposable on that between C-4 and C-2 of D-galactopyranose (b). However, the relative configuration between C-1 and C-3 of the β-D-glucopyranose (c) is not identical with that between C-4 and C-2 of D-galactopyranose (b).

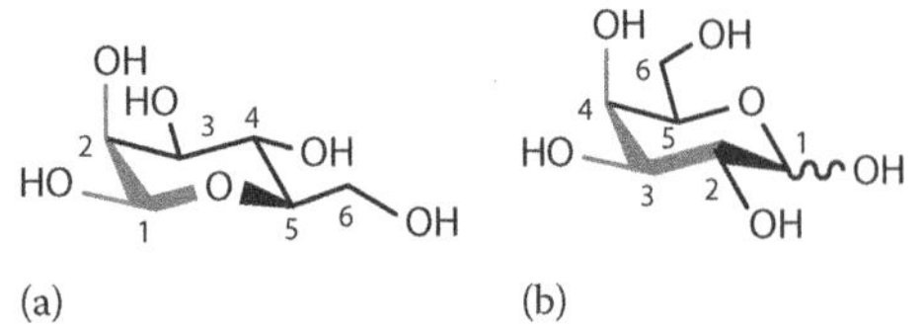

*Figure 14.6* Structures of D-galactopyranose and β-D-mannopyranose: relative configuration between C-2 and O-5 of β-D-mannopyranose (a) is superimposable on that from C-4 to C-2 of D-galactopyranose (b).

Very recently, it has also been reported that this α2,3-sialyltransferase can catalyze the transfer of Neu5Ac residue to inositols, noncarbohydrates, and carbohydrates with a diol structure corresponding to the C-3 and C-4 of the galactopyranose moiety. The α2,3-sialyltransferase does recognize both *epi*-inositol and 1D-*chiro*-inositol as acceptor substrates and gives the corresponding reaction products (Mine et al. 2010c). As described earlier, marine bacterial sialyltransferases have very broad acceptor substrate specificities compared to those of mammalian counterparts, and this indicate that marine bacterial sialyltransferases are thought to be good tools for preparing several kinds of sialyloligosaccharides enzymatically.

## 14.6 Sialyloligosaccharides as a drug candidate for disease

As described earlier, many reports have demonstrated that sialyloligosaccharides play an important role in various biological phenomena and are related to some diseases developed following infection by some bacteria and viruses.

Additionally, some diseases related to defects in protein glycosylation are reported. One is caused by defects in glycosyltransferase activities and the other is caused by defects in sugar biosynthesis. An example of the former is α-dystroglycanopathies, which are a group of congenital muscular dystrophies, such as Walker–Warburg syndrome (Muntoni et al. 2002, Endo 2004) and muscle–eye–brain disease (Santavuori et al. 1989). It has been demonstrated that the pathogenic mechanism of these congenital muscular dystrophies involve the dystrophin–glycoprotein complex (DGC). DGC is composed of α-dystroglycan (α-DG), β-dystroglycan (β-DG), dystrophin, and some other molecules (Michele and Campbell 2003). DGC acts as a transmembrane linker between the extracellular matrix and intracellular cytoskeleton. The α-DG binds to laminin, the intracellular domain of β-DG interacts with dystrophin in the skeletal muscle (Winder 2001, Michele and Campbell 2003),

and sugar chains of α-DG have a role in binding to laminin, neurexin, and agrin (Michele et al. 2002, Michele and Campbell 2003). As Endo et al. demonstrated, the sugar chains of α-DG include sialyl *O*-mannosyl oligosaccharides and Siaα2–3Galβ1–4GlcNAcβ1–2Man and are laminin-binding ligands of α-DG (Chiba et al. 1997). To date, muscle–eye–brain disease is caused by failure in the formation of GlcNAcβ1–2Man linkage of *O*-mannosyl sugar chains on α-DG (Yoshida et al. 2001, Manya et al. 2003).

An example of the latter is distal myopathy with rimmed vacuoles (DMRVs)—hereditary inclusion body myopathy (hIBM) or UDP-N-acetylglucosamine 2-epimerase/N-acetylmannosamine kinase (GNE) myopathy—which is a moderately progressive autosomal recessive myopathy (Ikeuchi et al. 1997). The disease gene underlying DMRV–hIBM has been predicted to be GNE, encoding UDP-*N*-acetylglucosamine-2-epimerase and *N*-acetylmannosamine kinase (Eisenberg et al. 2001, Nishino et al. 2002). These enzymes are essential for Sia biosynthesis. Thus, the patients with this disease show hyposialylation in various organs (Malicdan et al. 2009). A prophylactic treatment with sialyllactose, Neu5Ac, and *N*-acetylmannosamine was tested to confirm effectiveness in a mouse model of DMRV-hIBM. Oral treatment with the Sia metabolites described earlier completely precluded the development of the myopathic phenotype in the model mice (Malicdan et al. 2009). In addition, the therapeutic effect of 6′-sialyllactose oral administration had been examined with 50-week-old GNE myopathy mice for 30 weeks. As a result, spontaneous locomotion activity was recovered in 6′-sialyllactose-treated mice, while Neu5Ac-treated mice slowed the disease progression. Treatment with 6′-sialyllactose led to marked restoration of hyposialylation in the muscle and consequently to robust improvement in muscle size, contractile parameters, and pathology as compared to Neu5Ac. These results indicated that GNE myopathy can be treated even at a progressive stage and 6′-sialyllactose had a remarkable advantage than free Sia (Yonekawa et al. 2014).

## 14.7 Summary

In this chapter, the biological role of sialyloligosaccharides in various diseases, preparation methods of sialyloligosaccharides, screening methods of glycosyltransferases, and enzymatic properties of marine bacterial sialyltransferases have been reviewed.

As written in this chapter, some enzymes produced by marine bacteria are thought to be powerful tools for the preparation of biologically important molecules. Thus, basic studies of the marine bacterial enzymes need to be performed for future application.

# References

Ando, H. and A. Imamura. 2004. Proceeding in synthetic chemistry of sialo-glycoside. *Trends in Glycoscience and Glycotechnology* 16:293–303.

Ando, T., H. Ishida, and M. Kiso. 2003. First total synthesis of alpha-(2-3)/alpha-(2-6)-disialyl lactotetraosyl ceramide and disialyl Lewis A ganglioside as cancer-associated carbohydrate antigens. *Carbohydrate Research* 338:503–514.

Angata, T. and A. Varki. 2002. Chemical diversity in the sialic acids and related α-keto acids: An evolutionary perspective. *Chemical Reviews* 102:439–469.

Antoine, T., B. Priem, A. Heyraud, L. Greffe, M. Gilbert, W. W. Wakarchuk, J. S. Lam, and E. Samain. 2003. Large-scale in vivo synthesis of the carbohydrate moieties of gangliosides GM1 and GM2 by metabolically engineered *Escherichia coli*. *Chembiochem* 4:406–412.

Aspinall, G. O., A. G. McDonald, and H. Pang. 1994a. Lipopolysaccharides of *Campylobacter jejuni* serotype O:19: Structures of O antigen chains from the serostrain and two bacterial isolates from patients with the Guillain-Barré syndrome. *Biochemistry* 33:250–255.

Aspinall, G. O., A. G. McDonald, H. Pang, L. A. Kurjanezyk, and J. L. Penner. 1994b. Lipopolysaccharides of *Campylobacter jejuni* serotype O:19: Structures of core oligosaccharide regions from the serostrain and two bacterial isolates from patients with the Guillain-Barré syndrome. *Biochemistry* 33:241–249.

Aspinall, G. O., A. G. McDonald, T. S. Raju, H. Pang, S. D. Mills, L. A. Kurjanezyk, and J. L. Penner. 1992. Serological diversity and chemical structures of *Campylobacter jejuni* low-molecular-weight lipopolysaccharides. *Journal of Bacteriology* 174:1324–1332.

Bork, K., R. Horstkorte, and W. Weidemann. 2009. Increasing the sialylation of therapeutic glycoproteins: The potential of the sialic acid biosynthetic pathway. *Journal of Pharmaceutical Sciences* 98:3499–3508.

Burda, P. and M. Aebi. 1999. The dolichol pathway of N-linked glycosylation. *Biochimica et Biophysica Acta* 1426:230–257.

Chiba, A., K. Matsumura, H. Yamada, T. Inazu, T. Shimizu, S. Kusunoki, I. Kanazawa et al. 1997. Structures of sialylated O-linked oligosaccharides of bovine peripheral nerve adystroglycan. *Journal of Biological Chemistry* 272:2156–2162.

Connor, R. J., Y. Kawaoka, R. G. Webster, and J. C. Paulson. 1994. Receptor specificity in human, avian, and equine H2 and H3 influenza virus isolates. *Virology* 205:17–23.

Drouillard, S., H. Driguez, and E. Samain. 2006. Large-scale synthesis of H-antigen oligosaccharides by expressing *Helicobacter pylori* alpha1,2-fucosyltransferase in metabolically engineered *Escherichia coli* cells. *Angewandte Chemie International Edition in English* 45:1778–1780.

Drouillard, S., T. Mine, H. Kajiwara, T. Yamamoto, and E. Samain. 2010. Efficient synthesis of 6'-sialyllactose, 6',6-disialyllactose and 6'-KDO-lactose by metabolically engineered *Escherichia coli* expressing a multifunctional sialyltransferase from the *Photobacterium* sp. JT-ISH-224. *Carbohydrate Research* 345:1394–1399.

Eisenberg, I., N. Avidan, T. Potikha, H. Hochner, M. Chen, T. Olender, M. Barash et al. 2001. The UDP-N-acetylglucosamine 2-epimerase/N-acetylmannosamine kinase gene is mutated in recessive hereditary inclusion body myopathy. *Nature Genetics* 29:83–87.

Elliott, S., T. Lorenzini, S. Asher, K. Aoki, D. Brankow, L. Buck, L. Buese et al. 2003. Enhancement of therapeutic protein in vivo activities through glycoengineering. *Nature Biotechnology* 21:414–421.

Elvin J. G., R. G. Couston, and C. F. Van der Walle. 2013. Therapeutic antibodies: Market considerations, disease targets and bioprocessing. *International Journal of Pharmaceutics* 440:83–98.

Endo, T. 2004. Structure, function and pathology of O-mannosyl glycans. *Glycoconjugate Journal* 21:3–7.

Endo, T., S. Koizumi, K. Tabata, and A. Ozaki. 1999. Large-scale production of N-acetyllactosamine through bacterial coupling. *Carbohydrate Research* 316:179–183.

Endo, T., S. Koizumi, K. Tabata, and A. Ozaki. 2000. Large-scale production of CMP-NeuAc and sialylated oligosaccharides through bacterial coupling. *Applied Microbiology and Biotechnology* 53:257–261.

Endo, T., S. Koizumi, K. Tabata, and A. Ozaki. 2001. Large-scale production of the carbohydrate portion of the sialyl–Tn epitope, α-NeupAc-(2 → 6)-D-GalpNAc, through bacterial coupling. *Carbohydrate Research* 330:439–443.

Erickson, P. R. and M. C. Herzberg. 1993. Evidence for the covalent linkage of carbohydrate polymers to a glycoprotein from *Streptococcus sanguis*. *Journal of Biological Chemistry* 268:23780–23783.

Fierfort, N. and E. Samain. 2008. Genetic engineering of *Escherichia coli* for the economical production of sialylated oligosaccharides. *Journal of Biotechnology* 134:261–265.

Gagneux, P. and A. Varki. 1999. Evolutionary considerations in relating oligosaccharide diversity to biological function. *Glycobiology* 9:747–755.

Ghaderi, D., M. Zhang, N. Hurtado-Ziola, and A. Varki. 2012. Production platforms for biotherapeutic glycoproteins. Occurrence, impact, and challenges of non-human sialylation. *Biotechnology and Genetic Engineering Reviews* 28:147–176.

Gilbert, M., J. R. Brisson, M. F. Karwaski, J. Michniewicz, A. M. Cunningham, Y. Wu, N. M. Young, and W. W. Wakarchuk. 2000. Biosynthesis of ganglioside mimics in *Campylobacter jejuni* OH4384. *Journal of Biological Chemistry* 275:3896–3906.

Guerry, P., C. M. Szymanski, M. M. Prendergast, T. E. Hickey, C. P. Ewing, D. L. Pattarini, and A. P. Moran. 2002. Phase variation of *Campylobacter jejuni* 81-176 lipooligosaccharide affects ganglioside mimicry and invasiveness in vitro. *Infection and Immunity* 70:787–793.

Haeuw-Fievre, S., J. M. Wieruszeski, Y. Plancke, J. C. Michalski, J. Montreuil, and G. Streker. 1993. Primary structure of ¹H-NMR spectroscopy and fast-atom-bombardment mass spectrometry. Evidence for a new core structure, the para-lacto-N-octanose. *European Journal of Biochemistry* 215:361–371.

Hakomori, S. 1990. Bifunctional role of glycosphingolipids. *Journal of Biological Chemistry* 265:18713–18716.

Ho, T. W., H. J. Willison, I. Nachamkin, C. Y. Li, J. Veitch, H. Ung, G. R. Wang et al. 1999. AntiGD1a antibody is associated with axonal but not demyelinating forms of Guillain-Barré syndrome. *Annals of Neurology* 45:168–173.

Hossler, P., S. F. Khattak, and Z. J. Li. 2009. Optimal and consistent protein glycosylation in mammalian cell culture. *Glycobiology* 19:936–949.

Ikeuchi, T., T. Asaka, M. Saito, H. Tanaka, S. Higuchi, K. Tanaka, K. Saida et al. 1997. Gene locus autosomal recessive distal myopathy with rimmed vacuoles maps to chromosome 9. *Annals of Neurology* 41:432–437.

Ito, H., H. Ishida, B. E. Collins, S. E. Fromholt, R. L. Schnaar, and M. Kiso. 2003. Systematic synthesis and MAG-binding activity of novel sulfated GM1b analogues as the mimics of Chol-1 (alpha-series) gangliosides: Highly active ligands for neural siglecs. *Carbohydrate Research* 338:1621–1639.

Izumi, M. and C.-H. Wong. 2001. Microbial sialyltransferases for carbohydrate synthesis. *Trends in Glycoscience and Glycotechnology* 13:345–360.

Jedrzejewsk, P. M., I. J. D. Val, K. M. Polizzi, and C. Kontoravdi. 2013. Applying quality by design to glycoprotein therapeutics: Experimental and computational efforts of process control. *Pharmaceutical Bioprocessing* 1:51–69.

Kajihara, Y., S. Akai, T. Nakagawa, R. Sato, T. Ebata, H. Kodama, and K. Sato. 1999. Enzymatic synthesis of Kdn oligosaccharides by a bacterial α-(2 → 6)-sialyltransferase. *Carbohydrate Research* 315:137–141.

Kajihara, Y., T. Yamamoto, H. Nagae, M. Nakashizuka, T. Sakakibara, and I. Terada. 1996. A novel α-2,6-sialyltransferase: Transfer of sialic acid to fucosyl and sialyl trisaccharides. *Journal of Organic Chemistry* 61:8632–8635.

Kajiwara, H., M. Toda, T. Mine, H. Nakada, H. Wariishi, and T. Yamamoto. 2010. Visualization of sialic acid produced on bacterial cell surfaces by lectin staining. *Microbes and Environments* 25:152–155.

Kajiwara, H., M. Toda, T. Mine, H. Nakada, and T. Yamamoto. 2012. Isolation of fucosyltransferase-producing bacteria from marine environments. *Microbes and Environments* 27:515–518.

Kanie, O. and O. Hindsgaul. 1992. Synthesis of oligosaccharides, glycolipids and glycopeptides. *Current Opinion in Structural Biology* 2:674–681.

Kannagi, R. 2002. Regulatory roles of carbohydrate ligands for selectins in homing of lymphocytes. *Current Opinion in Structural Biology* 12:599–608.

Khmelnitsky, Y. L. 2004. Current strategies for in vitro protein glycosylation. *Journal of Molecular Catalysis B: Enzymatic* 31:73–81.

Koga, M., M. Gilbert, J. Li, S. Koike, M. Takahashi, K. Furukawa, K. Hirata, and N. Yuki. 2005. Antecedent infections in Fisher syndrome: A common pathogenesis of molecular mimicry. *Neurology* 64:1605–1611.

Koizumi, S. 2003. Large-scale production of oligosaccharides using bacterial functions. *Trends in Glycoscience and Glycotechnology* 15:65–74.

Koizumi, S., T. Endo, K. Tabata, and A. Ozaki. 1998. Large-scale production of UDP-galactose and globotriose by coupling metabolically engineered bacteria. *Nature Biotechnology* 16:847–850.

Kontermann, R. E. 2011. Strategies for extended serum half-life of protein therapeutics. *Current Opinion in Biotechnology* 22:868–876.

Kunz, C., S. Rudloff, W. Baier, N. Klein, and S. Strobel. 2000. Oligosaccharides in human milk: Structural, functional and metabolic aspects. *Annual Review of Nutrition* 20:699–722.

Kushi, Y., H. Kamimiya, H. Hiratsuka, H. Nozaki, H. Fukui, M. Yanagida, M. Hashimoto, K. Nakamura, S. Watarai, T. Kasama, H. Kajiwara, and T. Yamamoto. 2010. Sialyltransferases of marine bacteria efficiently utilize glycosphingolipid substrates. *Glycobiology* 20:187–198.

Lasky, L. A. 1992. Selectins: Interpreters of cell-specific carbohydrate information during inflammation. *Science* 258:964–969.

Malicdan, M. C. V., S. Noguchi, Y. K. Hayashi, I. Nonaka, and I. Nishino. 2009. Prophylactic treatment with sialic acid metabolites precludes the development of the myopathic phenotype in the DMRV-hIBM mouse model. *Nature Medicine* 15:690–695.

Manya, H., K. Sakai, K. Kobayashi, K. Taniguchi, M. Kawakita, T. Toda, T. Endo. 2003. Loss-of-function of an N-acetylglucosaminyltransferase, POMGnT1, in muscle–eye–brain disease. *Biochemical and Biophysical Research Communications* 306:93–97.

Michele, D. E., R. Barresi, M. Kanagawa, F. Saito, R. D. Cohn, J. S. Satz, J. Dollar et al. 2002. Post-translational disruption of dystroglycan-ligand interactions in congenital muscular dystrophies. *Nature* 418:417–422.

Michele, D. E. and K. P. Campbell. 2003. Dystrophin–glycoprotein complex: Post-translational processing and dystroglycan function. *Journal of Biological Chemistry* 278:15457–15460.

Mine, T., H. Kajiwara, T. Murase, Y. Kajihara, and T. Yamamoto. 2010b. An α2,3-sialyltransferase cloned from *Photobacterium* sp. JT-ISH-224 transfers N-acetylneuraminic acid to both O-2 and O-3′ hydroxyl groups of lactose. *Journal of Carbohydrate Chemistry* 29:51–60.

Mine, T., T. Miyazaki, H. Kajiwara, K. Naito, K. Ajisaka, and T. Yamamoto. 2010a. Enzymatic synthesis of unique sialyloligosaccharides using marine bacterial α-(2 → 3)- and α-(2 → 6)-sialyltransferases. *Carbohydrate Research* 345:1417–1421.

Mine, T., T. Miyazaki, H. Kajiwara, N. Tateda, K. Ajisaka, and T. Yamamoto. 2010c. A recombinant α-(2 → 3)-sialyltransferase with an extremely broad acceptor substrate specificity from *Photobacterium* sp. JT-ISH-224 can transfer N-acetylneuraminic acid to inositols. *Carbohydrate Research* 345:2485–2490.

Modjtahedi, H., S. Ali, and S. Essapen. 2012. Therapeutic application of monoclonal antibodies in cancer: Advances and challenges. *British Medical Bulletin* 104:41–59.

Muldon, J., A. S. Shashkov, A. P. Moran, J. A. Ferris, S. N. Senchenkova, and A. V. Savage. 2002. Structures of two polysaccharides of *Campylobacter jejuni* 81116. *Carbohydrate Research* 337:2223–2229.

Muntoni, F., M. Brockington, D. J. Blake, S. Torelli, and S. C. Brown. 2002. Defective glycosylation in muscular dystrophy. *Lancet* 360:1419–1421.

Nishino, I., S. Noguchi, K. Murayama, A. Driss, K. Sugie, Y. Oya, T. Nagata et al. 2002. Distal myopathy with rimmed vacuoles is allelic to hereditary inclusion body myopathy. *Neurology* 59:1689–1693.

Ogawara, K., S. Kuwabara, M. Mori, T. Hattori, M. Koga, and N. Yuki. 2000. Axonal Guillain-Barré syndrome: Relation to anti-ganglioside antibodies and *Campylobacter jejuni* infection in Japan. *Annals of Neurology* 48:624–631.

Paulson, J. C. 1989. Glycoproteins: What are the sugar chains for? *Trends in Biochemical Sciences* 14:272–276.

Preim, B., M. Gilber, W. W. Wakarchuk, A. Heyraud, and E. Samain. 2002. A new fermentation process allows large scale production of human milk oligosaccharides by metabolically engineered bacteria. *Glycobiology* 12:235–240.

Raju, T. S., J. B. Briggs, S. M. Chamow, M. E. Winkler, and A. J. S. Jones. 2001. Glycoengineering of therapeutic glycoproteins: In vitro galactosylation and sialylation of glycoproteins with terminal *N*-acetylglucosamine and galactose residues. *Biochemistry* 40:8868–8876.

Ribitsch, D., S. Zitzenbacher, P. Augustin, K. Schmölzer, T. Czabany, C. Luley-Goedl, M. Thomann et al. 2014. High-quality production of human α-2,6-sialyltransferase in Pichia pastoris requires control over N-terminal truncations by host-inherent protease activities. *Microbial Cell Factories* 13:138–150.

Ropper, A. H. 1992. The Guillain-Barré-syndrome. *The New England Journal of Medicine* 326:1130–1136.

Santavuori, P., H. Somer, K. Sainio, J. Rapola, S. Kruus, T. Nikitin, L. Ketonen, and J. Leisti. 1989. Muscle–eye–brain disease (MEB). *Brain and Development* 11:147–153.

Sato, M., T. Furuike, R. Sadamoto, N. Fujitani, T. Nakahara, K. Nikura, K. Monda et al. 2004. Glycoinsulins: Dendritic sialyloligosaccharide-displaying insulins showing a prolonged blood-sugar-lowering activity. *Journal of the American Chemical Society* 126:14013–14022.

Schauer, R. 2004. Sialic acids: Fascinating sugars in higher animals and man. *Zoology* 107:49–64.

Schmidt, R. R. 1986. New methods for the synthesis of glycosides and oligosaccharides—Are there alternatives to the koenigs-knorr method? New synthetic methods. *Angewandte Chemie International Edition in English* 25:212–235.

Shibuya, N., K. Tazaki, Z. Song, G. E. Tarr, I. J. Goldstein, and W. J. A. Peumans. 1989. Comparative study of bark lectins from three elderberry (*Sambucus*) species. *Journal of Biochemistry* 106:1098–1103.

Stults, C. L. M., C. C. Sweeley, and B. A. Macher. 1988. Glycosphingolipids: Structure, biological source, and properties. *Methods in Enzymology* 179:167–214.

Suzuki, Y. 2005. Sialobiology of influenza: Molecular mechanism of host range variation of influenza virus. *Biological and Pharmaceutical Bulletin* 28:399–408.

Takakura, Y., H. Tsukamoto, and T. Yamamoto. 2007. Molecular cloning, expression and properties of an α/β-galactoside α2,3-sialyltransferase from *Vibrio* sp. JT-FAJ-16. *Journal of Biochemistry* 142:403–412.

Tanemura, M., E. Miyoshi, H. Nagano, H. Eguchi, K. Taniyama, W. Kamiike, M. Mori, and Y. Doki. 2013. Role of α-gal epitope/anti-Gal antibody reaction in immunotherapy and its clinical application in pancreatic cancer. *Cancer Science* 104:282–290.

Taniguchi, N., K. Honke, and M. Fukuda. 2002. *Handbook of Glycosyltransferases and Related Genes*, 1st edn. Springer-Verlag: Tokyo, Japan, pp. 267–356.

Tsukamoto, H., Y. Takakura, T. Mine, and T. Yamamoto. 2008. *Photobacterium* sp. JT-ISH-224 produces two sialyltransferases, α-/β-galactoside α2,3-sialyltransferase and β-galactoside α2,6-sialyltransferase. *Journal of Biochemistry* 143:187–197.

Tsukamoto, H., Y. Takakura, and T. Yamamoto. 2007. Purification, cloning and expression of an α-/β-galactoside α2,3-sialyltransferase from a luminous marine bacterium, *Photobacterium phosphoreum. Journal of Biological Chemistry* 282:29794–29802.

Varki, A. 1993. Biological roles of oligosaccharides: All of the theories are correct. *Glycobiology* 3:97–130.

Wacker, M., D. Linton, P. G. Hitchen, M. Nita-Lazar, S. T. Haslam, S. J. North, M. Panico et al. 2002. *N*-linked glycosylation in *Campylobacter jejuni* and its functional transfer into *E. coli. Science* 298:1790–1793.

Wakarchuk, W. W., M. Gilbert, A. Martin, Y. Wu, J. R. Brisson, P. Thibault, and J. C. Richards. 1998. Structure of an α-2,6-sialylated lipooligosaccharide from *Neisseria meningitidis* immunotype L1. *European Journal of Biochemistry* 54:626–633.

Walsh, G. 2010. Biopharmaceutical benchmarks. *Nature Biotechnology* 28:917–924.

Wang, Z., X.-F. Zhang, Y. Ito, Y. Nakahara, and T. Ogawa. 1996. A new strategy for stereoselective synthesis of sialic acid-containing glycopeptide fragment. *Bioorganic and Medicinal Chemistry* 4:1901–1908.

Weis, W., J. H. Brown, S. Cusack, J. C. Paulson, J. J. Skehel, and D. C. Wiley. 1988. Structure of influenza virus haemagglutinin complexed with its receptor, sialic acid. *Nature* 333:426–431.

Winder, S. J. 2001. The complexities of dystroglycan. *Trends in Biochemical Sciences* 26:118–124.

Yamamoto, T., Y. Hamada, M. Ichikawa, H. Kajiwara, T. Mine, H. Tsukamoto, and Y. Takakura. 2007. A β-galactoside α2,6-sialyltransferase produced by a marine bacterium, *Photobacterium leiognathi* JTSHIZ-145, is active at pH 8. *Glycobiology* 17:1167–1174.

Yamamoto, T., H. Nagae, Y. Kajihara, and I. Terada. 1998b Mass production of bacterial α2,6-sialyltransferase and enzymatic syntheses of sialyloligosaccharides. *Bioscience Biotechnology and Biochemistry* 62:210–214.

Yamamoto, T., M. Nakashizuka, H. Kodama, Y. Kajihara, and I. Terada. 1996. Purification and characterization of a marine bacterial β-galactoside α2,6-sialyltransferase from *Photobacterium damsela* JT0160. *Journal of Biochemistry* 120: 104–110.

Yamamoto, T., M. Nakashizuka, and I. Terada. 1998a. Cloning and expression of a marine bacterial beta-galactoside alpha 2,6-sialyltransferase gene from *Photobacterium damsela* JT0160. *Journal of Biochemistry* 123:94–100.

Yamamoto, T., Y. Takakura, and H. Tsukamoto. 2006. Bacterial sialyltransferase. *Trends in Glycoscience and Glycotechnology* 18:253–265.

Yonekawa, T., M. C. V. Malicdan, A. Cho, Y. K. Hayashi, I. Nonaka, T. Mine, T. Yamamoto et al. 2014. Sialyllactose ameliorates myopathic phenotypes in symptomatic GNE myopathy model mice. *Brain* 137:2670–2679.

Yoshida, A., K. Kobayashi, H. Manya, K. Taniguchi, H. Kano, M. Mizuno, T. Inazu et al. 2001. Muscular dystrophy and neuronal migration disorder caused by mutations in a glycosyltransferase, POMGnT1. *Developmental Cell* 1:717–724.

Young, N. M., J. R. Brisson, J. Kelly, D. C. Watson, L. Tessier, P. H. Lanthier, H. C. Jarrell et al. 2002. Structure of the *N*-linked glycan present on multiple glycoproteins in the Gram-negative bacterium, *Campylobacter jejuni*. *Journal of Biological Chemistry* 277:42530–42539.

Yu, H., S. Huang, H. Chokhawala, M. Sun, H. Zheng, and X. Chen. 2006. Highly efficient chemoenzymatic synthesis of naturally occurring and non-natural alpha-2,6-linked sialosides: A *P. damsela* alpha-2,6-sialyltransferase with extremely flexible donor-substrate specificity. *Angewandte Chemie International Edition in English* 45:3938–3944.

Yuki, N. 2001. Infectious origins of and molecular mimicry in Guillain-Barré and Fisher syndromes. *The Lancet Infectious Diseases* 1:29–37.

Yuki, N. 2005. Carbohydrate mimicry: A new paradigm of autoimmune diseases. *Current Opinion in Immunology* 17:577–582.

Yuki, N., K. Suzuki, M. Koga, Y. Nishimoto, M. Odaka, K. Hirata, K. Taguchi et al. 2004. Carbohydrate mimicry between human ganglioside GM1 and *Campylobacter jejuni* lipo-oligosaccharide causes Guillain-Barré syndrome. *Proceeding of the National Academy of Sciences of the United States of America* 101:11404–11409.

*section six*

---

*Marine carbohydrates*

*chapter fifteen*

# Polysaccharides from marine sources and their pharmaceutical approaches

*Sougata Jana, Arijit Gandhi, and Subrata Jana*

## Contents

## 15.1  Introduction

Marine resources are nowadays widely studied because the oceans cover more than 70% of the world's surface. Among the 36 known living phyla, 34 are found in marine environments with more than 300,000 known species of fauna and flora. The rationale of searching for bioactive compounds from the marine environment stems from the fact that marine plants and animals have adapted to different types of marine environment and these creatures are constantly under tremendous selection pressure including space competition, predation, surface fouling, and reproduction Kijou and Swangwong 2004, Yan 2004, Jana et al. 2013b). The marine environment covers a wide thermal, pressure, and nutrient range, and it has extensive photic and non-photic zones. This wide variability has facilitated extensive specification at all phylogenetic levels, from microorganisms to mammals. Despite the fact that the biodiversity in the marine environment far exceeds that of the terrestrial environment, research into the use of marine natural products as pharmaceutical agents is still in its infancy. This may be due to the lack of ethnomedical history and the difficulties involved in the collection of marine organisms. But with the development of new diving techniques, remotely operated machines, and so on, it is possible to collect marine samples; in fact, during the past decade, over 4200 novel compounds have been isolated from various marine animals such as tunicates, sponges, soft corals, bryozoans, sea slugs, and marine organisms (Harvey 2000, Proksch et al. 2002, Jha and Zi-Rong 2004, Jirge and Chaudhari 2010).

Among the different bioactive compounds obtained from the sea, polysaccharides have been gaining interesting and valuable applications in the food and pharmaceutical fields. As they are derived from natural sources, they are easily available, nontoxic, cheap, biodegradable, and biocompatible. Polysaccharides can be obtained from a number of sources including seaweeds, plants, bacteria, fungi, insects, crustaceans, and animals, and can be structurally tuned through genetic engineering. The term "polysaccharide" encompasses very diverse and large carbohydrates that may be composed of only one kind of repeating monosaccharide (termed homo-polysaccharides or homoglycans; e.g., cellulose) or formed by two or more different monomeric units (hetero-polysaccharides or heteroglycans; e.g., agar, alginate, carrageenan). The conformation of the polysaccharide chains is markedly dependent not only on the pH and ionic strength of the medium, particularly in the case of polyelectrolytes, but also on the temperature and the concentration of certain molecules. Polysaccharides are divided into two subtypes: anionic and cationic. Several anionic and cationic polysaccharides are widely available in nature and have gained keen interest in the food and pharmaceutical fields (Colquhoun et al. 2001, Guezennec 2002, Coviello et al. 2007, D'Ayala et al. 2008). The biological activity of naturally occurring polysaccharides has been increasingly utilized for human applications and creating a strong position in the biomedical field. Because of their different chemical structures and physical properties, these natural sources can be used in the different applications, from tissue engineering to the preparation of drug vehicles for controlled release. This chapter focuses on the present use and the diverse applications of marine polysaccharides in the pharmaceutical sector.

## 15.2 Production of polysaccharides by marine sources

The polysaccharide content of marine sources varies greatly in brown and red algae. Polysaccharides may comprise up to 74% of the total organic matter (Romankevich 1984), while they range from 20% to 40% in planktonic algae (Parsons et al. 1961), and the content in zooplanktons is two to four times lower than that in phytoplanktons. Phytoplanktons serve as the principal source of polysaccharides found in seawater, particles, and sediments; the photosynthetic conversion of carbon dioxide to biomass is the basis of polysaccharide production. Phytoplankton polysaccharides' composition has been surveyed in both field samples and laboratory monocultures. In general, glucose has been found to be the most common monosaccharide, probably due to the fact that phytoplanktons store polysaccharides primarily in the form of glucose. The major source of polysaccharides in marine sediments is sinking particles. Methods of analysis of polysaccharides

in sediments usually involve extraction, separation, and, finally, quantitation of the sugars. Polysaccharides in sediments comprise ~10% of the total organic carbon. Studies of anoxic surface sediments show that amino acids and polysaccharides are the major constituents of organic matter (Burdige and Zheng 1998). Polysaccharides are remineralized rapidly during epigenetic and digenetic bacterial activity, and only less than 10% of these compounds are stable enough to be incorporated into the sediment a few centimetres below the sediment/water interface (Degens et al. 1964, Seifert et al. 1990).

## 15.3 Major marine polysaccharides

### 15.3.1 Chitosan

Chitin and chitosan have wide applications in the medical field, such as wound dressing, hypocholesterolemic agents, blood anticoagulant, antithrombogenic agents, and drug delivery systems, in addition to other uses such as wound-healing materials and cosmetic preparations. Chitosan and its derivatives have been studied extensively as carriers for drug delivery, cancer therapy, and in the biomedical field, and is generally regarded as safe materials. Chitin is a known biodegradable natural polymer based on polysaccharides, and is obtained from crustacean shell (e.g., crabs, shrimps, and lobsters) and fungi such as yeasts and plants. It principally occurs in animals of the phylum Arthopoda (Knaul et al. 1999, Zheng et al. 2001). In 1823, Ojer named it "Chitin" from Greek word "khiton," meaning "envelope," present in certain insects. In 1894, Hope Seyle named it as "chitosan."

#### 15.3.1.1 Chemical structure

Chitosan is a linear copolymer consisting of B (1-4)-linked 2-amino-2-deoxy-D-glucose (D-glucosamine) and 2-acetamido-2-deoxy-D-glucose (N-acetyl-D-glucosamine) units (Figure 15.1). Chitosan is the second most abundant natural polymer after cellulose, and is obtained by alkaline N-deacetylation of chitin, which is the primary structural component of the outer skeletons of many marine creatures such as crustaceans, crabs, shrimp shells, and many other species such as insects and fungi (Addo et al. 2010, Jana et al. 2013b). Chitosan is degraded by lysozyme present in the various mammalian tissues, which leads to the production of N-acetyl-D-glucosamine and D-glucosamine, which also play an important physiological role in the in vivo biochemical processes.

#### 15.3.1.2 Method of manufacture

Chitosan is manufactured by chemically treating the shells of crustaceans such as shrimps and crabs (Figure 15.1).

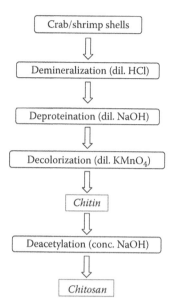

*Figure 15.1* Chitosan production flow diagram.

The process involves the separation of proteins by treating with alkali and of minerals such as calcium carbonate and calcium phosphate by treatment with acid. Initially, the shells are deproteinized by treatment with 3%–5% aqueous sodium hydroxide solution. The resulting product is neutralized, and calcium is removed by treatment with 3%–5% aqueous hydrochloric acid solution at room temperature to precipitate chitin. The chitin is dried and deacetylated to give chitosan. This can be achieved by treatment with 40%–45% aqueous sodium hydroxide solution at moderate temperature (110°C), and the precipitate is washed with water. The crude sample is dissolved in 2% acetic acid and the insoluble material is removed. The resulting clear supernatant solution is neutralized with aqueous sodium hydroxide to give a white precipitate of chitosan. It can be further purified and ground to a fine, uniform powder or granules (Gupta and Ravikumar 1986, Niederhofer and Maller 2004, Jana et al. 2013b).

### 15.3.1.3   Properties of chitosan
The degree of N-deacetylation (40%–98%) and molecular weight (50,000–2,000,000 Da) play very important roles

in the physicochemical properties of chitosan. Hence, they have a major effect on the biological properties.

Chemical properties of chitosan are as follows (Gupta and Ravikumar 1986, Errington et al. 1993, Singla and Chawla 2001, Niederhofer and Maller 2004, Rowe et al. 2006):

- Cationic polyamine.
- High charge density at pH 6.5.
- Adheres to negatively charged surfaces.
- Forms gels with polyanions.
- High-molecular-weight linear polyelectrolyte
- Viscosity is <5 cPs.
- Chelates with certain transitional metals.
- Amiable to chemical modification.
- Reactive amino/hydroxyl groups.
- The molecular weight of chitosan ranges from $1 \times 10^5$ to $3 \times 10^5$; average molecular weight ranges from $3.8 \times 10^3$ to $2000 \times 10^3$.
- Degree of acetylation of chitosan ranges from 66% to 99.8%.
- Moisture content is >10%.

Biological properties of chitosan are as follows (Gebelein and Dunn 1990, Skaugrud 1991, RaviKumar 2000):

- Biocompatible
- Natural polymer
- Biodegradable to normal body constituents
- Safe and nontoxic
- Hemostatic, bacteriostatic, and fungistatic
- Spermicidal
- Anticarcinogenic
- Anticholesteremic
- Reasonable cost
- Versatile

### 15.3.1.4   Chitosan derivatives
Chitosan (Figure 15.2) has two free hydroxyl groups at C3 (primary) and C6 (secondary) and a primary amino group at C2.

The amino and hydroxyl groups provide the opportunity for chemical modification. Chitosan can be depolymerized to reduce the molecular weight and

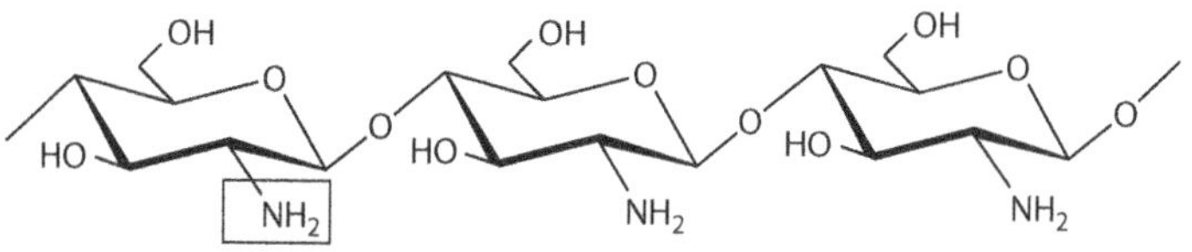

*Figure 15.2* Chitosan is a linear polysaccharide composed of randomly distributed β-(1–4)-linked D-glucosamine (deacetylated unit) and N-acetyl-D-glucosamine (acetylated unit).

*Table 15.1* Different chitosan derivatives

| Purpose | Reaction | Product | References |
|---|---|---|---|
| Reduce molecular weight (low solubility, better solubility) | Depolymerization | Low-molecular-weight chitosan | Elsayed et al. (2009), Wu et al. (2008), Kasaai et al. (2003), Ren et al. (2005), Zhao et al. (2009) |
| | | Chitosan oligomers | Einbu et al. (2007) |
| Improve cationic properties | Deacetylation | Chitosan with different degrees of deacetylation | Knill et al. (2009), Baxter et al. (1992) |
| | Quaternization | Quaternized chitosan | Younessi et al. (2004), Cafaggi et al. (2007), Zheng et al. (2009) |
| Improve chitosan's water solubility | Acylation, alkylation | N-acyl chitosan, N-alkyl chitosan | Naoji et al. (2000), Hirano et al. (1981) |
| | Pegylation | PEG-chitosan | Opanasopit et al. (2006, 2007), Yang et al. (2008) |
| | Carboxyalkylation | Carboxyalkyl-chitosan | Chen and Tan (2006), You et al. (2007) |
| Amphoteric polyelectrolytes | Phosphorylation | Phosphorylated chitosan, N-alkyl-N-methylene phosphonic chitosan | Jayakumar and Tamura (2006), Jayakumar et al. (2006, 2008) |
| | Sulfatation | Sulfonic chitosan | Zhang et al. (2003, 2004, 2008) |
| Cell-targeting | Alkylation, cross-linking | Sugar-modified chitosan | Park et al. (2003) |
| Photosensitive derivatives | Azidation | Azidated-chitosan | Ishihara et al. (2002), Aiedeh and Taha (2001) |
| Amphiphilic derivatives | Acylation, alkylation, grafting, cross-linking | N-acyl chitosan | Tien et al. (2003) |
| | | N-alkyl chitosan | Naoji et al. (2000) |
| | | Crown-ether-bound chitosan | Ravi Kumar et al. (2004) |
| | | Cyclodextrin-linked chitosan | He et al. (2011) |
| | | Graft derivatives | Jiang et al. (2006), Jun et al. (2004) |
| Miscellaneous | Several reactions (thiolation, Michael addition, succinylation, etc.) | Thiolated chitosan | Bernkop-Schnurch et al. (2003), Jayakumar et al. (2007) |
| | | Chitosan–dendrimer hybrid | Ravi Kumar et al. (2004) |
| | | Polyurethane-type chitosans | Ravi Kumar et al. (2004) |
| | | Oxychitin and fluorinated chitins | Ravi Kumar et al. (2004) |
| | | Tyrosine glucan | Ravi Kumar et al. (2004) |
| | | Lauryl succinyl chitosan | Rekha and Sharma (2009) |

viscosity, which results in better solubility in aqueous medium. Besides, chitosan is modified by introducing non-pH-dependent positive charges on the chitosan backbone, so that its properties such as mucoadhesive and absorption enhancer can be retained without the presence of positive charge on the polymer. Hydrophobic moieties are introduced into the backbone of chitosan through different methods such as alkylation, acylation, graft polymerization, and some other methods, as shown in Table 15.1.

### 15.3.1.5   Applications of chitosan and its derivatives

*15.3.1.5.1 Colon-specific drug delivery* Chitosan has been used for the specific delivery of insulin to the colon. Chitosan capsules were coated with an enteric coating (hydroxy propyl methyl cellulose phthalate) and contained various additional absorption enhancers and

enzyme inhibitors apart from insulin. It was found that the capsules specifically disintegrated in the colon. This disintegration was suggested as due either to the lower pH in the ascending colon as compared to the terminal ileum or to the presence bacterial enzymes that can degrade chitosan (Tozaki et al. 1997).

*15.3.1.5.2 Buccal drug delivery* Chitosan is an excellent polymer to be used for buccal delivery because it has muco/bioadhesive properties and can act as an absorption enhancer (Bernkop-Schnürch et al. 2003). Directly compressible bioadhesive tablets of ketoprofen containing chitosan and sodium alginate in the weight ratio of 1:4 showed sustained release for 3 h after intra-oral (into sublingual site of rabbits) drug administration (Sawayangi et al. 1982). Buccal tablets based on chitosan microspheres containing chlorhexidine diacetate provided prolonged release of the drug in the buccal cavity, improving the antimicrobial activity of the drug (Ritthidej et al. 1994).

*15.3.1.5.3 Nasal drug delivery* Compared to oral or subcutaneous administration, nasal administration enhances the bioavailability and improves safety and efficacy. Chitosan effectively enhances the absorption of hydrophilic compounds such as proteins and peptide drugs across nasal and intestinal epithelia. Nasal administration of diphtheria toxoid (DT) incorporated into chitosan microparticles resulted in a protective systemic and local immune response against DT with enhanced IgG production. Nasal formulations induce significant serum IgG responses similar to secretory IgA levels, which are superior to parenteral administration of the vaccine (Hamman et al. 2003). Studies have shown that bioadhesive chitosan microspheres of pentazocine for intranasal systemic delivery significantly improved the bioavailability with sustained and controlled blood level profiles compared to intravenous and oral administration (Kean et al. 2005).

*15.3.1.5.4 Transdermal drug delivery* A membrane of chitosan was developed by Dureja et al. (2001) using a model drug diclofenac sodium. According to He et al. (2008), *N*-trimethyl chitosan can be used for transdermal drug delivery.

*15.3.1.5.5 Vaginal drug delivery* Some mucoadhesive vaginal gels were prepared using hydroxyethylcellulose (HEC) mixed with chitosan or its derivative 5-methylpyrrolidinone-chitosan (MPCS) loaded with the model antibacterial drug metronidazole (Perioli et al. 2008). Sandri et al. (2004) evaluated the mucoadhesive and permeation enhancing properties of four different chitosan derivatives: 5-methyl-pyrrolidinone chitosan, two low molecular mass chitosans, and a partially re-acetylated chitosan, via vaginal and buccal mucosa using acyclovir as model drug.

*15.3.1.5.6 Absorption promotion activity* Chitosan acts as a permeability enhancer to increase the transmucosal absorption of the drugs that normally do not pass the tight junctional barrier. The enhancing properties of chitosan and its derivatives have been attributed to their interactions with the tight junctions and cellular membrane components to reversibly open the tight junctions and hence increase the paracellular permeability of hydrophilic compounds. The permeation-enhancing properties of chitosan were suggested to be due to a combination of mucoadhesion and a transient opening of the tight junctions in the mucosal cell membrane (Dodane et al. 1999, Elsayed et al. 2009). Majumdar et al. (2008) studied the effect of chitosan, benzalkonium chloride (BAK), and disodium ethylendiaminetetraacetic acid

(EDTA), alone and in combination, on the permeation of acyclovir (ACV) across excised rabbit cornea. It was found BAK at 0.005% in combination with 0.01% EDTA and 0.1% chitosan increased transcorneal ACV permeation by 5.5-fold (Majumdar et al. 2008). Mei et al. (2008) assessed and compared the absorption-promoting effect of chitosans of different molecular weights, trimethyl chitosans, and thiolated chitosans for intranasal absorption of 2,3,5,6-tetramethylpyrazine phosphate (TMPP). An in situ nasal perfusion technique in rats was utilized to test the rate and extent of TMPP absorption. Trimethylated chitosan (TMC) exhibited stronger absorption-enhancing ability than the homopolymer chitosan (Mei et al. 2008). Jonker et al. (2002) prepared different degrees of quaternization of TMC (22.1%–48.8%). Everted intestinal sac (rats) and the single-pass intestinal perfusion methods were used to determine the effect of the polymers (0.0625%–0.5% w/v) on the permeation of the hydrophilic model compound. It was shown that the permeation-enhancing effects depended on the degree of quaternization of TMC. In both models, the best permeation-enhancing results were obtained with the highest degree of quaternization of TMC (48.8%) at a concentration of 0.5% w/v (Jonker et al. 2002). Lidong et al. (2013) studied the potential of *N*-succinyl chitosan as a novel permeation enhancer for the intranasal absorption of isosorbide dinitrate (ISDN). A series of *N*-succinyl chitosans (NSCS) with different degrees of succinylation and molecular weight were synthesized. An in situ nasal perfusion technique in rats was utilized to investigate the effect of the degree of NSCS substitutions, NSCS molecular weight, and concentration on the intranasal absorption of ISDN. Better promoting effect was observed for 0.1% NSCS compared to 0.5% chitosan (Lidong et al. 2013).

*15.3.1.5.7 Protein drug delivery* Novel polyampholyte hydrogels based on carboxymethyl chitosans (CMCs) can be used in oral delivery system for protein drugs (Chen et al. 2004). Chitosan–TBA (4-thiobutylamidine) conjugates can be considered as a vehicle for the nasal delivery of peptide drugs (Kraulan et al. 2006).

### 15.3.2 Carrageenan

Carrageenan (CG) is the generic name for a family of high molecular weight, sulfated polysaccharides (SPs) obtained by extraction from certain species of red seaweeds. It is composed of galactose and anhydrogalactose units linked by glycosidic unions (Coviello et al. 2007, Jiao et al. 2011). CG is widely used because of its excellent physical functional properties, such as gelling, thickening, emulsifying, and stabilizing abilities (Campo et al. 2009, Chen et al. 2010, Kavitha Reddy et al. 2011).

CGs are mainly obtained from different species of Rhodophyta: Chondrus, Eucheuma, Gigartina, and Hypnea (Campo et al. 2009, Jiao et al. 2011). They are mainly composed of d-galactose residues linked alternately in 3-linked β-d-galactopyranose and 4-linked α-d-galactopyranose units and are classified according to the degree of substitution that occurs on their free hydroxyl groups. Substitutions are generally either addition of ester sulfate or the presence of the 3,6-anhydride on the 4-linked residue (Nanaki et al. 2010). In addition to d-galactose and 3,6-anhydro-d-galactose as the main sugar residues and sulfate as the main substituent, other carbohydrate residues may be present in CG preparations, such as glucose, xylose, and uronic acids, as well as some substituents, such as methyl ethers and pyruvate groups (De Ruiter and Rudolph 1997). These polysaccharides can also be traditionally categorized into six basic forms: kappa (κ), iota (ι), lambda (λ), mu (μ), nu (ν), and theta (θ) CG (Knutsen et al. 1994). Three most important types of CG (important from the commercial point of view) are κ, ι, and λ, and their structures are presented in Figure 15.3.

### 15.3.2.1 Sources and production of carrageenan

The main species of seaweed from which CG is manufactured are Chondrus, Eucheuma, Gigartina, and Hypnea. Specific details of the extraction processes are regarded as trade secrets by the several manufacturers

of CG, but broadly these follow a similar pattern. The seaweed is dried quickly to prevent degradation, and is then baled for shipment to processing facilities. The seaweed is repeatedly washed to remove gross impurities such as sand, salt, and marine life, and then subjected to a hot alkali extraction process, which releases the CG from the cell. Once the CG is in a hot solution, it undergoes clarification, which is then converted into powder (Hilliou et al. 2006).

In general, several methods have been used to remove CG from the solution. The first method is the "freeze–thaw" technique. The solution is gelled with various salts and then frozen. Upon thawing, water is removed and the resultant mass, primarily CG and its salt, is ground to the desired particle size. The second method, referred to as "alcohol precipitation method," is to place the concentrated solution of CG in 2-propanol or other alcohols, leading to CG precipitation out of the solution. The solvents are then evaporated, and the precipitated CG is dried and ground to the desired particle size. Another method is the "KCl precipitation" process, where, after hot extraction, the filtrate is evaporated to reduce the volume. The filtrate is then extruded through spinnerets into a cold 1.0%–1.5% solution of KCl. The resulting gel threads are further washed with KCl solution and are pressed, dried, and milled to CG powder (McHugh 1987). The economics of extraction processes is strongly dependent the cost of the energy required to bring the CG into solution and its subsequent recovery in the dry form (McHugh 1987). Commercial CG is usually standardized by blending different batches of CG and adding salt or sugar to obtain the desired gelling or thickening properties.

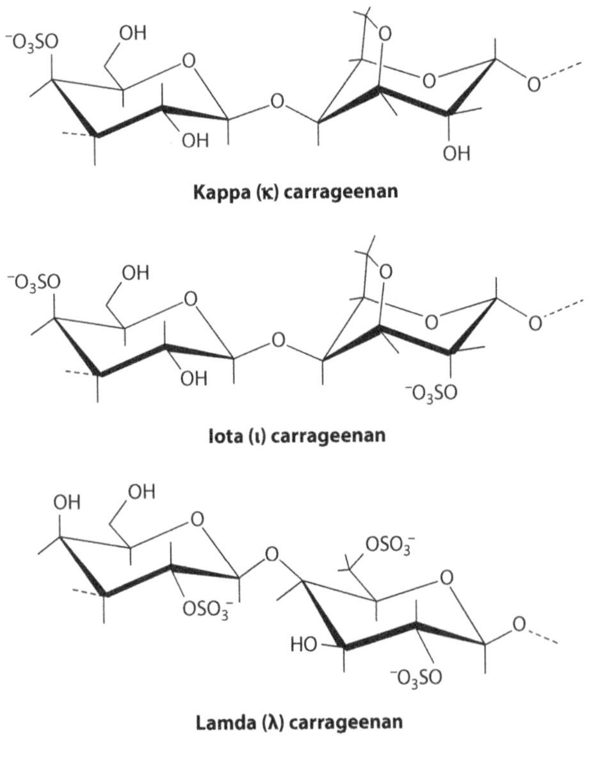

### 15.3.2.2 Properties of carrageenan

The chemical reactivity of CGs is primarily due to their half-ester sulfate groups which are strongly anionic, being comparable to inorganic sulfate in this respect. The free acid is unstable, and commercial CGs are available as stable sodium potassium and calcium salts or, most commonly, as a mixture of these. The associated cations together with the conformation of the sugar units in the polymer chain determine the physical properties of CGs (Michel et al. 1997).

κ- and τ-CG are known to undergo a thermally induced disorder–order transition, where at elevated temperatures both chains exist as random coils with a large amount of conformational entropy. Upon cooling, entropy is reduced, and chains reorient into a more ordered conformation, which is believed to consist of a double helix, aggregated mono-helices, or aggregated helical dimmers (Stone and Nickerson 2012).

The function of CGs in various applications depends largely on their rheological properties. CGs, as linear, water-soluble polymers, typically form highly viscous

**Kappa (κ) carrageenan**

**Iota (ι) carrageenan**

**Lamda (λ) carrageenan**

*Figure 15.3* Chemical structure of carrageenan.

aqueous solutions. The viscosity depends on the concentration, temperature, the presence of other solutes, as well as the type of CG and its molecular weight (Lai et al. 2000). Viscosity increases nearly exponentially with concentration and decreases with temperature. CGs are susceptible to depolymerization through acid-catalyzed hydrolysis. At high temperatures and low pH, this may rapidly lead to complete loss of function (Stanley 2011).

CGs have also been shown to possess several potential pharmaceutical properties including anticoagulant, anticancer, anti-hyperlipidemic, and immunomodulatory activities. In vitro studies indicate that CGs may also have antiviral effects, inhibiting the replication of hepatitis A viruses (Buck et al. 2006, Wijesekara et al. 2011). CG also exhibits antioxidant and free-radical scavenging activities. Additionally, a comparison of a variety of compounds reveals that CG is an extremely potent infection inhibitor for a broad range of sexually transmitted human papilloma virus.

### 15.3.2.3 Applications of carrageenan

*15.3.2.3.1 Anticoagulant and antithrombotic activity* Many studies have reported the anticoagulant activity of carrageenan. Among the CG types, λ-CG has approximately twice the activity of unfractionated CG and four times the activity of κ-CG. However, the most active CG has approximately one-fifteenth of the activity of heparin. The principal basis of the anticoagulant activity of CG appears to be an antithrombic property. λ-CG shows greater antithrombic activity than κ-CG probably due to its higher sulfate content (Wunderwald et al. 1979, Shanmugam and Mody 2000).

*15.3.2.3.2 Antiviral activity* CG is a selective inhibitor of several enveloped viruses, including human pathogens such as human immunodeficiency virus (HIV), herpes simplex virus (HSV), human cytomegalovirus, human rhinoviruses, and others (Girond et al. 1991, Carlucci et al. 1999, Caceres et al. 2000, Stiles et al. 2008). CG acts primarily by preventing the binding or the entry of virions into cells (Grassauer et al. 2008). Carlucci et al. (2004) reported that λ-CG and partially cyclicized μ-CG from *Gigartina skottsbergii* showed potent antiviral effect against different strains of HSV type 1 and type 2. Similar results were also reported by Zacharopoulos and Phillips (1997), who described the ability of CG solutions (λ, κ, or τ) to prevent HSV-2 infection.

*15.3.2.3.3 Antitumor and immunomodulatory activities* Several studies have reported that CGs have antiproliferative activity in cancer cell lines in vitro, as well as inhibitory activity of tumor growth in mice (Yuan et al. 2004, Zhou et al. 2004, 2006). In addition, they have antimetastatic activity by blocking the interactions between

cancer cells and the basement membrane, and inhibit tumor cell proliferation and tumor cell adhesion to various substrates, but their exact mechanisms of action are not completely understood so far. It was reported that the oral administration of several seaweeds can cause a significant decrease in the incidence of carcinogenesis in vivo (Yamamoto et al. 1986). Hagiwara et al. (2001) investigated the modifying effects of CG on colonic carcinogenesis in male rats.

*15.3.2.3.4 Use of carrageenan as a drug-delivery vehicle*

*15.3.2.3.4.1 Carrageenan-based microparticle delivery systems* For the preparation of microcapsules based on the negative charge, CG is commonly used in combination with cationic charged polymers or cationic polyelectrolytes. For example, τ-CG was used in combination with spermine to prepare trypsin-loaded microcapsules, with functional integrity of trypsin, a hydrolytic enzyme that retained at τ-CG concentrations below 10.5 mM (Patil and Bergisadi 1998). CG was also used together with chitosan (CS) to prepare glucose oxidase-loaded microcapsules for oral controlled release (Briones and Sato 2010). Among the different types of CGs, the order of encapsulation efficiency of CS/CG complexes decreased in the order κ > τ > λ. The CS/κ-CG complex had the lowest release rate and was able to preserve 80.2% of glucose oxidase activity in pH 1.2 solution, 73.3% in chitosanase solution, and 66.4% in pepsin solution.

*15.3.2.3.4.2 Carrageenan-based beads* Compared with microcapsules and microspheres, CGs are more extensively reported as the excipients to prepare beads due to their easy gelling, thermoreversibility of the gel network, and appropriate viscoelastic properties (Hezaveh and Muhamad 2012). In an early study, κ-CG (Gelcarin®GP-812NF) was used to control the release of acetaminophen from beads prepared by a cross-linking technique (Garcia et al. 1996). Sustained-release mefenamic acid beads based on κ-CG were also employed to reduce the daily dose and minimize gastrointestinal disturbance caused by the drug (Ozsoy and Bergisadi 2000). Polyelectrolyte complex hydrogel beads based on CS and CG have been studied as controlled-release devices for colon-specific delivery of sodium diclofenac (Piyakulawat et al. 2007). Sipahigil and Dortunc (2001) prepared CG (κ, τ) beads using the ionotropic gelation method as a controlled release system for a slightly water soluble drug ibuprofen and a freely water soluble drug verapamil hydrochloride. The results showed that about 30% of ibuprofen and about 70% of verapamil. HCl were released in 6 and 5 h, respectively, from the CG beads prepared (Sipahigil and Dortunc 2001). CG-based hydrogel beads have been used for the controlled delivery of platelet derived growth factor in bone tissue engineering, which satisfied the requirement for a fully functional vascular network (Santo et al. 2009), and alginate–CG hydrogel beads have been employed as cell delivery systems for applications in regenerative medicine (Popa et al. 2011). Ghanam and Kleinebudde (2011) studied the suitability of κ-CG pellets for the formulation of multiparticulate tablets with modified drug-releasing efficiency. Sufficiently long release properties were obtained with κ-CG pellets containing theophylline as a model drug and coated with Kollicoat® SR 30 D (Ghanam and Kleinebudde 2011).

*15.3.2.3.4.3 Carrageenan-based nanoparticles* Nanoparticles based on CS–CG complex can be obtained in an aqueous environment under very mild conditions based on electrostatic interaction, thus avoiding the use of organic solvents or other aggressive processes (Pinheiro et al. 2012). Thus, CS–CG nanoparticles have potential applications not only in drug delivery but also in tissue engineering and regenerative medicine (Bulmer et al. 2012). It was reported that κ-CG nanoparticles exhibited noncytotoxic behavior in in vitro tests using L929 fibroblasts, and provided controlled release for up to 3 weeks with ovalbumin as a model protein, and CS/CG ratio had a significant effect on the properties of the nanoparticles (Grenha et al. 2010).

*15.3.2.3.4.4 Carrageenan-based oral extended-release tablets* CG alone or in combination with other polymers has been widely used for decades in order to modulate drug release in a satisfactory way. A recent study was also conducted to investigate the effect of polymer blends containing CGs (τ, λ) and cellulose ethers (HPMC, sodium carboxymethyl cellulose, methyl cellulose, hydroxypropyl cellulose) on ibuprofen release from tablets prepared by direct compression. Most of the formulations showed linear release profiles, and the release of ibuprofen was sustained over 12–16 h (Nerurkar et al. 2005). In addition, CG in combination with different swollen polymers can also be used to prepare three-layered matrix tablets. Compared to HPMC, pectin, guar gum, xanthan gum, CS, and ethyl cellulose, CG has been regarded a superior polymer for the delivery of metoprolol tartrate in three-layered matrix tablets due to its better accordance to the target release profile; CG-based formulations also exhibited a super case II release mechanism (Baloglu and Senyigit 2010). Recent studies showed that when CS/CG-based tablets were transferred from simulated gastric fluid to simulated intestinal fluid, in situ polyelectrolyte film could be formed on the surface of tablets, and the polyelectrolyte film could further control drug release (Li et al. 2013a,b).

*15.3.2.3.5 As a gelling agent/viscosity-enhancing agent for controlled release and prolonged retention* Because of on their physicochemical properties, κ-CG and τ-CG can form gels easily, especially in the presence of monovalent and divalent ions. The gel formed by τ-CG has suitable rheological and sustained-release characteristics, with potential use as vehicles for oral delivery of drugs to dysphagic patients (Miyazaki et al. 2011). Many studies have focused on adding CG to other polymers in order to utilize the gelling properties of CG to achieve good controlled-release profiles. For instance, after κ-CG is added to agarose hydrogels containing a micellar drug solution of cetyltrimethylammonium bromide–camptothecin, release rate of the drug decreased significantly (Liu and Li 2007). In another study, pre-gelatinized starch gels containing various ratios of κ-CG could offer a wide range of bioadhesive strength, leading to improved physical properties and prolonged buccal delivery of miconazole (Lefnaoui and Moulai-Mostefa 2011). In a recent report, the permeation-enhancing effect of AT1002, a biologically active fragment of *Zonula occludens* toxin that could reversibly open intercellular tight junction safter binding to the zonulin receptor, was significantly promoted by the bioadhesive agent τ-CG (Song and Eddington 2012).

### 15.3.3 *Alginate*

Alginate is found in the cell walls of marine organisms. Alginate is a water-soluble, linear polysaccharide extracted from brown seaweed and is composed of alternating blocks of 1-4-linked α-ʟ-guluronic and β-ᴅ-mannuronic acid residues.

Figure 15.4 shows the structures of mannuronic and guluronic acid residues and the binding between these residues in alginate. Because of the particular shapes of the monomers and their modes of linkage in the polymer, the geometries of the G-block regions, M-block regions, and alternating regions are substantially different. Specifically, the G-blocks are buckled, while the M-blocks have a shape referred to as an extended ribbon. If two G-block regions are aligned side by side, a diamond shaped hole is formed. This hole has dimensions that are ideal for the cooperative binding of calcium ions. The homopolymeric regions of β-ᴅ-mannuronic acid blocks and β-ʟ-guluronic acid blocks are interspersed with regions of alternating structure (β-ᴅ-mannuronic acid–β-ʟ-guluronic acid blocks) (Haug and Larsen 1962, Haug et al. 1967, Jana et al. 2011).

#### 15.3.3.1 *Properties of alginate*

As a natural biomaterial, alginate is highly stable, safe, nontoxic, hydrophilic and biodegradable. In addition, it has abundant resources in nature and low cost of processing (Yang et al. 2011a, Pawar and Edgar 2012).

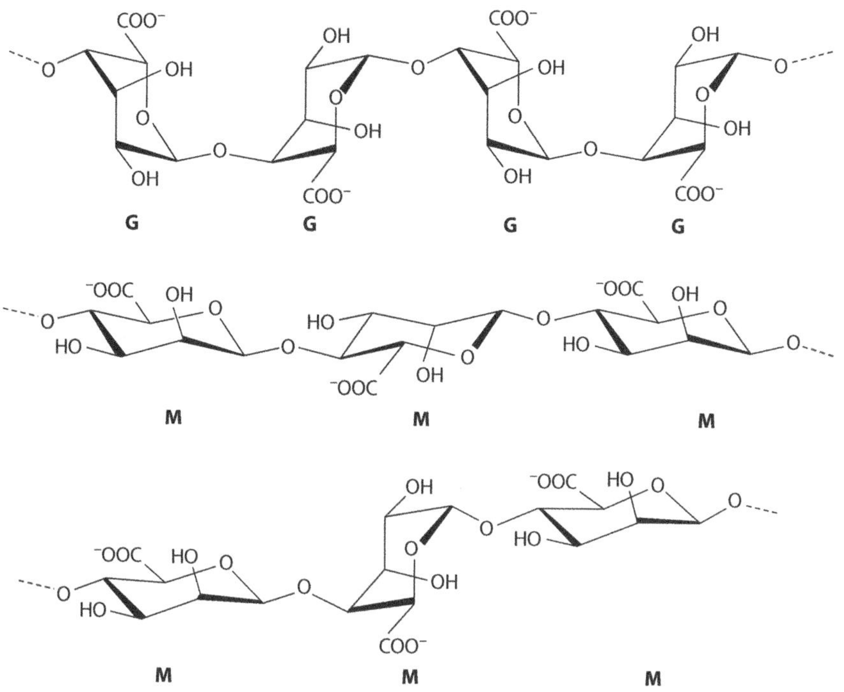

*Figure 15.4* Chemical structures of G-block, M-block, and alternating block in alginate.

Alginate, an algal polysaccharide, is widely used in the food industry as a stabilizer, or as a thickening or an emulsifying agent. Alginate when mixed with calcium ions can produce a gel structure, which finds use as a thickening agent in the food industry, and in drug release systems during pharmaceutical applications. It is one of the important biomaterials used in wound healing and cell culture (George and Abraham 2006, Martinez et al. 2011). Alginates were mainly used in the manufacture of paper, pharmaceuticals, cosmetic creams, and processed foods. The graft copolymers of this alginate polysaccharide find applications in diverse fields such as pharmaceutical, biomedical, agriculture, and environmental sciences. Polymers with promising applications in the biomedical field as delivery systems of therapeutic agents, and the bioseparation devices have been attracting much attention due to nontoxic nature of alginate. Because of their gelling ability in the presence of divalent cations, such as calcium and barium, stabilizing properties and high viscosity in aqueous solutions, alginate and its derivatives have been, extensively utilized in biomedical applications of cell transplantation, drug delivery, and bulk agent gels. However, as no hydrolytic or enzymatic chain breakages occur within alginate chains, high-molecular-weight alginate polymers cannot be easily degraded and may be very slow to clear from the body (Davidovich-Pinhas et al. 2009, Moebus et al. 2009, El-Sherbiny et al. 2011, Jana et al. 2013b). It was noted that the biological activity of alginates depends on its molecular weight, sulfate content, and anionic group that endows antioxidant activity (Xue et al. 1998).

### 15.3.3.2   Applications

*15.3.3.2.1 Controlled drug delivery*   Chemical reaction between sodium alginate and calcium chloride to form calcium alginate was utilized for the preparation of microspheres. Calcium alginate microbeads represent a useful tool for oral sustained/controlled drug delivery but suffer from several drawbacks, mainly related to the stability, and rapid drug release at higher pH, which in most cases is too fast due to increased porosity. To overcome such drawbacks, a different approach was adopted to develop calcium alginate microbeads coated with guar gum (GG) and locust bean gum (LBG) as drug release modifiers to improve stability and prolong the release of encapsulated drug. Deshmukh et al. (2009) investigated in vitro release of diclofenac sodium from microspheres made by combining sodium alginate, locust bean gum, and xanthan gum. The drug entrapment efficacy of all the formulations was in the range of 90.6%–98.9%. The drug entrapment efficacy of microspheres increased with increase in concentration of hydrophilic gums. The microspheres were found to be effective in sustaining drug release for more than 12 h. The drug release was diffusion-controlled and followed first-order kinetics. Stability studies showed that there was no significant change in drug content and dissolution profile of the microspheres (Deshmukh et al. 2009).

Jana et al. (2015) prepared alginate–locust bean gum interpenetrating microspheres by the calcium ion-induced ionotropic gelation technique for prolonged release of aceclofenac. The microspheres provided sustained release of aceclofenac over a period of 8 h. Pharmacodynamic study of the microspheres showed prolonged anti-inflammatory activity in CG-induced rat paw model following oral administration (Jana et al. 2015). Moebus et al. (2009) used hydrogel-forming polymers (e.g., alginates and poloxamers) as encapsulation materials for controlled drug delivery to mucosal tissue.

*15.3.3.2.2 Antimicrobial activity* Very recently, Zahran et al. (2014) investigated the antimicrobial activity of alginate/silver (Ag) nanocomposite on cotton fabric, using a simple one-step, rapid synthetic route by the reduction of silver nitrate using alkali-hydrolyzed alginate solution, which acts as both a reducing and capping agent. Fourier transform infrared spectra, color coordinates, silver content, silver release percent, and SEM images of treated fabric samples confirmed the successful physical deposition of AgNP/alginate composite on the fabric. The treated fabrics demonstrated excellent antibacterial activity against the tested bacteria *Escherichia coli*, *Staphylococcus aureus*, and *Pseudomonas aeruginosa*. A slight decrease in the antibacterial feature of the cotton fabrics was observed after successive washings (Zahran et al. 2014).

*15.3.3.2.3 Mucoadhesive drug delivery* Davidovich-Pinhas et al. (2009) prepared mucoadhesive drug delivery systems based on hydrated thiolated alginate. Extensive studies have shown that dry, non-cross-linked, compressed tablets made from thiolated polymers adhere better to the mucus layer compared to the native polymers (Davidovich-Pinhas et al. 2009). In another study, Motwani et al. (2008) prepared mucoadhesive chitosan (CS)-sodium alginate (ALG) nanoparticles as a new vehicle for the prolonged topical ophthalmic delivery of the antibiotic gatifloxacin. A modified co-acervation or ionotropic gelation method was used to produce gatifloxacin-loaded submicroscopic nanoreservoir systems. Nanoparticles showed fast release during the first hour followed by more gradual drug release during the next 24 h period following a non-Fickian diffusion process (Motwani et al. 2008).

*15.3.3.2.4 Antitumor activity* Martínez et al. (2012) studied the antitumor activity of tamoxifen (TMX)-loaded nanoparticles using different mixtures of alginate and cysteine, where the disulfide bond reduced bovine serum albumin. These systems showed an enhancement of the TMX antitumor activity, since lower tumor evolutions and lower tumor growth rates were observed in mice treated with them (Martínez et al. 2012).

Yang et al. (2011a) reported a high-performance nanoparticle for photodynamic detection of colorectal cancer, where alginate is physically complexed with folic acid-modified chitosan to form nanoparticles with improved drug release in the cellular lysosome. The nanoparticles displayed no differences in particle size or zeta potential when it was loaded with 5-aminolevulinic acid (5-ALA; 27% loading efficiency). Moreover, the nanoparticles were readily taken up by colorectal cancer cells via folate receptor-mediated endocytosis. Subsequently, the loaded 5-ALA was release in the lysosome, and this was promoted by the reduced attraction intensity between chitosan and 5-ALA via the deprotonated alginate, resulting in a higher intracellular protoporphyrin-IX (PpIX) accumulation for photodynamic detection. These studies demonstrate that the alginate-incorporated and folic acid-conjugated chitosan nanoparticles are excellent vectors for colorectal-specific delivery of 5-ALA for fluorescent endoscopic detection (Yang et al. 2011a).

## 15.4    Other marine polysaccharides

### 15.4.1    Sulfated polysaccharides

In recent years, various sulfated polysaccharides (SPs), isolated from marine algae, with a wide range of important biological activities have attracted much attention in the fields of food, pharmaceutical, and cosmetic industries. Marine algae are the most important source of these polymers, along with different animals such as mammals and invertebrates (Mourao 2007). These SPs demonstrate various health-beneficial biological activities such as anti-HIV-1 (Schaeffer and Krylov 2000), immunomodulation (Leiro et al. 2007), and anticancer (Rocha et al. 2005) activities. SPs are by-products in the preparation of alginates from edible brown seaweeds and could be used as a rich source of natural antioxidants with potential application in the food industry. Porphyran is an SP isolated from seaweeds of order Bangiales especially from the genera Porphyra. It is obtained from red algae of Kingdom Rhodophyta. Chemically, porphyran is related to agarose, and consists of a linear backbone of alternating 3-linked β-D-galactose and 4-linked 3,6-anhydro-α-L-galactose units. The L residues are mainly composed of α-L-galactosyl 6-sulfate units; the 3,6-anhydrogalactosyl units are minor. Porphyran has been reported to possess various pharmaceutical properties such as antioxidant, antitumor, immunostimulant, hypotensive, antifatigue, antibacterial, anticoagulant, anticancer, antiviral, antihyperlipidemic, and hepatoprotective activity (Bhatia et al. 2008). Fucoidans are a complex series of (SPs) found widely in the cell walls of brown seaweeds. In recent years, different brown algae were analyzed for

**Table 15.2** Application summary of marine polysaccharides

| Polysaccharide(s) | Active agent | Purpose | Reference |
|---|---|---|---|
| Alginate + hydrolyzed chitosan | — | Preparation of a novel fiber for wound care application | Sweeney et al. (2014) |
| Fucoidan | — | Anti-HIV activity | Thuy et al. (2015) |
| Chitosan | Adenosine triphosphate (ATP) | To restore depletion of ATP in macrophages | Giacalone et al. (2014) |
| Chitosan | Epirubicin | Anticancer therapy | Wang et al. (2007) |
| Chitosan | Bis-demethoxy curcumin | Slow and sustained diffusion-controlled release of the drug | Subramanian et al. (2014) |
| Chitosan + carageenan | Nisin | Antibacterial activity | Chopra et al. (2014) |
| Carrageenan | Methylene blue | Intestinal targeting | Hezaveh and Muhamad (2012) |
| Carrageenan | Sodium fluorescein | Trans-scleral delivery of macromolecules | Thrimawithana et al. (2011) |
| Chitosan | Insulin | Oral delivery | Makhlof et al. (2011) |
| Alginate + chitosan | 5-Aminosalicylic acid | Colon-specific delivery | Mladenovska et al. (2007) |
| Chitosan + carageenan | — | Gastroprotective and anti-ulcerogenic effect | Volod'ko et al. (2014) |
| Chitosan | Tetracaine hydrochloride | Ocular drug delivery | Addo et al. (2010) |
| Agar | Ciprofloxacin | Antibacterial activity | Blanco-Fernandez et al. (2011) |
| Glycosaminoglycan-like marine exopolysaccharide | — | Bone and cartilage tissue engineering | Rederstorff et al. (2011) |
| Sulfated polysaccharide, p-KG03 | — | Inhibition of influenza A virus infection | Kim et al. (2012) |
| IPSI-A, IPSI-B, and IPSII | — | Antioxidant activity | Yingying et al. (2014) |
| Sulfated galactan | — | Anticoagulant activity | Ngoa and Kim (2013) |
| Fucoidan | — | Superoxide radical scavenging property | Ngoa and Kim (2013) |
| Chitosan | Chlorpheniramine | Oral controlled release | Kumari and Kundu (2007) |
| Porphyran | — | Antiallergic activity | Ishihara et al. (2005) |

their fucoidan content. The low molecular weight fractions of algal fucoidans (less than 30 kDa) obtained by depolymerization have been shown to exhibit some heparin-like properties, with fewer side effects (Karim et al. 2011). Such polysaccharides do not occur in other divisions of algae and in land plants. However, the related biopolymers were found in marine invertebrates such as sea cucumbers or sea urchins (Cumashi et al. 2007).

The applications of marine polysaccharides are summarized in Table 15.2.

## 15.5   Future directions

In future, efforts would be made to improve these novel biomaterials, for further enhancement in results in different applications. Moreover, designing new scaffolds from natural polymers such as alginate and gelatin would be possible for soft tissue engineering. Chitin/chitosan has great potential in a variety of biomedical and industrial applications. The physicochemical and mechanical properties of chitosan would be utilized in fabricating particles and films that could be modulated for specific purposes. Efforts should also be made to prepare nanofibrous scaffolds from other natural polymers, including silk, for hard and soft tissue engineering. And the best use of these marine sources would be in the fields of food and cosmetics industries, effluent treatment, and medicine.

## 15.6   Conclusion

The application of marine polysaccharides in pharmaceutical field has been increasing very fast in the past decade. Marine polysaccharides are employed not only for the delivery of small chemical drugs and proteins but also for tissue regeneration with therapeutic biomacromolecules and cell delivery. At present, some of these are mainly used as polymer matrix in oral extended-release tablets, as a novel extrusion aid for the production of pellets, and as a carrier/stabilizer in micro/nanoparticle systems. Moreover, based on the special characteristics of marine polysaccharides, such as charge, gelling, and viscosity properties, these have been used as gelling agents/viscosity enhancing agents for controlled drug release and prolonged retention. Other potential applications are being explored. The increasing knowledge of the physicochemical properties and structural analysis of marine polysaccharides enable better utilization of marine polysaccharides as drug carriers. They have great importance for optimal drug delivery. Meanwhile, polymer–drug and polymer–polymer interactions also play a significant role in pharmaceutical formulation development. Although marine polysaccharides have already been employed as the carrier of drug delivery systems, further analysis of the drug delivery mechanisms is required. At present, many reports still focus on in vitro drug release from marine polysaccharide-based formulations, but in vivo and in vitro–in vivo correlation studies are essential in the near future. Importantly, taking into account MPs' biological properties such as antitumor activities, immunomodulatory effects, anticoagulant activities, and inflammatory responses, safety evaluation of these compounds needs to be carried out. Moreover, furthering our current understanding of the fundamental properties of marine polysaccharides may provide strong theoretical bases for future application of these compounds in advanced drug delivery systems.

## References

Addo, R.T., A. Siddig, and R. Siwale. 2010. Formulation, characterization and testing of tetracaine hydrochloride-loaded albumin-chitosan microparticles for ocular drug delivery. *J. Microencapsul.* 27:95–104.

Aiedeh, K. and M.O. Taha. 2001. Synthesis of iron-crosslinked chitosan succinate and iron-crosslinked hydroxamated chitosan succinate and their in vitro evaluation as potential matrix materials for oral theophylline sustained-release beads. *Eur. J. Pharm. Sci.* 13:159–168.

Baloglu, E. and T. Senyigit. 2010. A design and evaluation of layered matrix tabletformulations of metoprolol tartrate. *AAPS PharmSciTech* 11:563–573.

Baxter, A., M. Dillon, I.C.D.A. Taylor, and G.A.F. Roberts. 1992. Improved method for IR determination for the degree of deacetylation of chitosan. *Int. J. Biol Macromol.* 14:166–169.

Bernkop-Schnurch, A., M. Hornof, and T. Zoidl. 2003. Thiolated polymers-thiomers: Synthesis and in vitro evaluation of chitosan-2-iminothiolane conjugates. *Int. J. Pharm.* 260:229–237.

Bernkop-Schnürch, A., C.E. Kast, and D. Guggi. 2003. Permeation enhancing polymers in oral delivery of hydrophilic macromolecules: thiomer/GSH systems. *J. Control. Release* 93:95–103.

Bhatia, S., A. Sharma, K. Sharma, M. Kavale, B.B. Chaugule, and K. Dhalwal. 2008. Novel algal polysaccharides from marine source: Porphyran. *Pharmacogn. Rev.* 2:271–276.

Blanco-Fernandez, B., A. Lopez-Viota, A. Concheiro, and C. Alvarez-Lorenzo. 2011. Synergistic performance of cyclodextrin–agar hydrogels for ciprofloxacin delivery and antimicrobial effect. *Carbohydr. Polym.* 85:765–774.

Briones, A.V. and T. Sato. 2010. Encapsulation of glucose oxidase (GOD) in polyelec-trolyte complexes of chitosan–carrageenan. *React. Funct. Polym.* 70:19–27.

Buck, C.B., C.D. Thompson, J.N. Roberts, M. Mueller, D.R. Lowy, and J.T. Schiller. 2006. Carrageenan is a potent inhibitor of papillomavirus infection. *PLoS Pathog.* 2:671–680.

Bulmer, C., A. Margaritis, and A. Xenocostas. 2012. Encapsulation and controlled release of recombinant human erythropoietin from chitosan–carrageenan nanoparticles. *Curr. Drug Deliv.* 9:527–537.

Burdige, D.J. and S.L. Zheng. 1998. The biogeochemical cycling of dissolved organic nitrogen in estuarine sediments. *Limnol. Oceanogr.* 43:1796–1813.

Caceres, P.J., M.J. Carlucci, E.B. Damonte, B. Matsuhiro, and E.A. Zuniga. 2000. Carrageenans from chilean samples of *Stenogramme interrupta* (Phyllophoraceae): Structural analysis and biological activity. *Phytochemistry* 53:81–86.

Cafaggi, S., E. Russo, and R. Stefani. 2007. Preparation and evaluation of nanoparticles made of chitosan or N-trimethyl chitosan and a cisplatin-alginate complex. *J. Control. Release* 121:110–123.

Campo, V.L., D.F. Kawano, D.B. Silva, and D.I. Carvalho. 2009. Carrageenans: Biological properties, chemical modifications and structural analysis—A review. *Carbohydr. Polym.* 77:167–180.

Carlucci, M.J., L.A. Scolaro, and E.B. Damonte. 1999. Inhibitory action of natural carrageenans on herpes simplex virus infection of mouse astrocytes. *Chemotherapy* 45:429–436.

Carlucci, M.J., L.A. Scolaro, M.D. Noseda, A.S. Cerezo, and E.B. Damonte. 2004. Protective effect of a natural carrageenan on genital herpes simplex virus infection in mice. *Antiviral Res.* 64:137–141.

Chen, H.M., X.J. Yan, F. Wang, W.F. Xu, and L. Zhang. 2010. Assessment of the oxidative cellular toxicity of carrageenan oxidative degradation product towards Caco-2 cells. *Food Res. Int.* 43:2390–2401.

Chen, L., Z. Tian, and Y. Du. 2004. Synthesis and pH sensitivity of carboxymethyl chitosan-based polyampholyte hydrogels for protein carrier matrices. *Biomaterials* 25:3725–3732.

Chen, Y., and H. Tan. 2006. Crosslinked carboxymethylchitosan-g-poly(-acrylic acid) copolymer as a novel super absorbent polymer. *Carbohydr. Res.* 341:887–896.

Chopra, M., P. Kaur, M. Bernela, and R. Thakur. 2014. Surfactant assisted nisin loaded chitosan-caragenan nanocapsule synthesis for controlling food pathogens. *Food Control* 37:158–164.

Colquhoun, I.J., A. Jay, J. Eagles, V.J. Morris, K.J. Edwards, and A.M. Griffin. 2001. Structure and conformation of a novel genetically engineered polysac-charide P2. *Carbohydr. Res.* 330:325–333.

Coviello, T., P. Matricardi, C. Marianecci, and F. Alhaique. 2007. Polysaccharidehydrogels for modified release formulations. *J. Control. Release* 119:5–24.

Cumashi, A., N.A. Ushakova, M.E. Preobrazhenskaya, A. D'Incecco, A. Piccoli, and L. Totani. 2007. A comparative study of the anti-inflammatory, anticoagulant, antiangiogenic, and antiadhesive activities of nine different fucoidans from brown seaweeds. *Glycobiology* 17:541–552.

D'Ayala, G.G., M. Malinconico, and P. Laurienzo. 2008. Marine derived polysaccharides for biomedical applications: Chemical modification approaches. *Molecules* 13:2069–2106.

Davidovich-Pinhas, M., O. Harari, and H. Bianco-Peled. 2009. Evaluating the mucoadhesive properties of drug delivery systems based on hydrated thiolated alginate. *J. Control. Release* 136:38–44.

De Ruiter, G.A. and B. Rudolph. 1997. Carrageenan biotechnology. *Trends Food Sci. Technol.* 8:389–395.

Degens, E.T., J.R. Reuter, and K.N.F. Shaw. 1964. Biochemical compounds in offshore California sediments and sea waters. *Geochimica et Cosmochimica Acta* 28:45–66.

Deshmukh, V.K., J.K. Jadhav, V.J. Masirkar, and D.M. Sakarkar. 2009. Formulation, optimization and evaluation of controlled release alginate microspheres using synergy gum blends. *Res. J. Pharm. Technol.* 2:324–327.

Dodane, V., M. Amin Khan, and J.R. Mervin. 1999. Effect of chitosan on epithelial permeability and structure. *Int. J. Pharm.* 182:21–32.

D'Souza, G., A. R. Kreimer, R. Viscidi, M. Pawlita, C. Fakhry, W. M. Koch, W. H. Westra, and M. L. Gillison, 2007. Case–control study of human papillomavirus and oropharyngeal Cancer. *N. Engl. J. Med.* 356:1944–1956.

Dureja, H., A.K. Tiwary, and S. Gupta. 2001. Simulation of skin permeability in chitosan membranes. *Int. J. Pharm.* 213:193–198.

Einbu, A., H. Garsdalen, and I.C.M. Varum. 2007. Kinetics and hydrolysis of chitin/chitosan oligomers in concentrated hydrochloric acid. *Carbohydr. Res.* 3421:1055–1062.

Elsayed, A., M.A. Remawi, N. Qinna, A. Farouk, and A. Badwan. 2009. Formulation and characterization of an oily-based system for oral delivery of insulin. *Eur. J. Pharm. Biopharm.* 73:269–279.

El-Sherbiny, I., M. Abdel-Mogib, A. Elsayed, and D. Hugh. 2011. Synthetic Biodegradable pH-responsive alginate-poly(lactic-co-glycolic acid) nano/micro hydrogel matrices for oral delivery of silymarin. *Carbohydr. Polym.* 83:1345–1354.

Errington, N., S.E. Harding, K.M. Varum, and L. Illum. 1993. Hydrodynamic characterization of chitosan varying in degree of acetylation. *Int. J. Biol. Macromol.* 15:113–117.

Garcia, A. M. and E. S. Ghaly. 1996. Preliminary spherical agglomerates of water soluble drug using natural polymer and cross-linking technique. *J. Control. Release.* 40:179–186.

Gebelein, C.G. and R.L. Dunn (eds.). 1990. *Progress in Biomedical Polymers.* Plenum Press, New York, pp. 283–290.

George, M. and T. Abraham. 2006. Polyionic hydrocolloids for the intestinal delivery of protein drugs: Alginate and chitosan—A review. *J. Control. Release* 114:1–14.

Ghanam, D. and P. Kleinebudde. 2011. Suitability of κ-carrageenan pellets for the formulation of multiparticulate tablets with modified release. *Int. J. Pharm.* 16:9–18.

Giacalone, G., H. Hillaireau, P. Capiau, H. Chacun, F. Reynaud, and E. Fattal. 2014. Stabilization and cellular delivery of chitosan–polyphosphate nanoparticles by incorporation of iron. *J. Control. Release* 194:211–219.

Girond, S., J.M. Crance, H. Van Cuyck-Gandre, J. Renaudet, and R. Deloince. 1991. Antiviral activity of carrageenan on hepatitis A virus replication in cell culture. *Res. Virol.* 142:261–270.

Grassauer, A., R. Weinmuellner, C. Meier, A. Pretsch, E. Prieschl-Grassauer, and H. Unger. 2008. Iota-carrageenan is a potent inhibitor of rhinovirus infection. *Virol. J.* 5: 1–13.

Grenha, A., M.E. Gomes, M. Rodrigues, V.E. Santo, J.F. Mano, and N.M. Neves. 2010. Development of new chitosan/carrageenan nanoparticles for drug delivery applications. *J. Biomed. Mater. Res. A* 92A:1265–1272.

Guezennec, J. 2002. Deep-sea hydrothermal vents: A new source of innovative bacterial exopolysaccharides of biotechnological interest. *J. Ind. Microbiol. Biotechnol.* 29:204–208.

Gupta, K.C. and M.N.V. Ravikumar. 1986. An overview on chitin and chitosan applications with an emphasis on controlled drug release formulations. *Sci. Rev. Macromol. Chem. Phys.* 4:273.

Hagiwara, A., K. Miyashita, T. Nakanishi, M. Sano, S. Tamano, I Asai, M Nakamura, K. Imaida, N. Ito, and T. Shirai. 2001. Lack of tumor promoting effects of carrageenan on 1,2-dimethylhydrazine-induced colorectal carcinogenesis in male F344 rats. *J. Toxicol. Pathol.* 14:37–43.

Hamman, J.H., C.M. Schultz, and A.F. Kotze. 2003. N-trimethyl chitosan chloride: Optimum degree of quaternization for drug absorption enhancement across epithelial cells. *Drug Dev. Ind. Pharm.* 29:161–172.

Harvey, A. 2000. Strategies for discovering drug from previously unexplored natural products. *Drug Discov. Today* 5:294–300.

Haug, A. and B. Larsen. 1962. Quantitative determination of the uronic acid composition of alginates. *Acta Chem. Scand.* 16:1908–1918.

Haug, A., B. Larsen, and O. Smidsrod. 1967. Studies on the sequence of uronic acid residues in alginic acid. *Acta Chem. Scand.* 21:691–704.

He, W., X. Guo, and M. Zhang. 2008. Transdermal permeation enhancement of N-trimethyl chitosan for testosterone. *Int. J. Pharm.* 356:82–87.

He, Z.X., Z.H. Wang, H.H. Zhang, X. Pan, W.R. Su, and D. Liang. 2011. Doxycycline and hydroxypropyl-β-cyclodextrin complex in poloxamer thermal sensitive hydrogel for ophthalmic delivery. *Acta Pharm. Sin. B* 1:254–260.

Hezaveh, H. and I.I. Muhamad. 2012. The effect of nanoparticles on gastrointestinal release from modified κ-carrageenan nanocomposite hydrogels. *Carbohydr. Polym.* 89:138–145.

Hilliou, L., F.D.S. Larotonda, P. Abreu, A.M. Ramos, A.M. Sereno, and M.P. Goncalves. 2006. Effect of extraction parameters on the chemical structure andgel properties of hybrid carrageenans obtained from *Mastocarpus stellatus*. *Biomol. Eng.* 23:201–208.

Hirano, S., K. Tobetto, and Y. Noishiki. 1981. SEM ultrastructure studies of N-acyl- and N-benzylidene–chitosan and chitosan membranes. *J. Biomed. Mater. Res.* 15:903–911.

Ishihara, K., C. Oyamada, R. Matsushima, M. Murata, and Muraoka. 2005. Inhibitory effect of porphyran, prepared from dried Nori, on contact hypersensitivity in mice. *Biosci. Biotechnol. Biochem.* 69:1824–1830.

Ishihara, M., K. Nakanishi, K. Ono, M. Sato, M. Kikuchi, Y. Saito, H. Yura et al. 2002. Photocrosslinkable chitosan as a dressing for wound occlusion and accelerator in healing process. *Biomaterials* 23:833–840.

Jana, S., A. Das, A.K. Nayak, K.K. Sen, and S.K. Basu. 2013a. Aceclofenac-loaded unsaturated esterified alginate/gellan gum microspheres: In vitro and in vivo assessment. *Int. J. Biol. Macromol.* 57:129–137.

Jana, S., A. Gandhi, S. Chakraborty, K.K. Sen, and S.K. Basu. 2013b. Marine microorganisms and their versatile applications in bioactive compounds, Chapter 22. In Kim, S.-K. (ed.), *Marine Microbiology: Bioactive Compounds and Biotechnological Applications*, 1st edn. Wiley-VCH Verlag GmbH & Co. KGaA, Weinheim, Germany, pp. 379–391.

Jana, S., A. Gandhi, K.K. Sen, and S.K. Basu. 2011. Natural polymers and their application in drug delivery and biomedical field. *J. PharmaSciTech* 1:16–27.

Jana, S., Gandhi, A., Sen, K.K., and Basu, S.K. 2013c. Biomedical applications of chitin and chitosan derivatives, Chapter 18. In Kim, S.K. (ed.), *Chitin and Chitosan Derivatives: Advances in Drug Discovery and Developments*, 1st edn. Taylor & Francis Group, LLC, Boca Raton, FL, pp. 337–360.

Jana, S., A. Gandhi, S. Sheet, and K.K. Sen. 2015. Metal ion-induced alginate–locust bean gum IPN microspheres for sustained oral delivery of aceclofenac. *Int. J. Biol. Macromol.* 72:47–53.

Jayakumar, R., H. Nagahama, T. Furuike, and H. Tamura. 2008. Synthesis of phosphorylated chitosan by novel method and its characterization. *Int. J. Biol. Macromol.* 42:335–339.

Jayakumar, R., R.L. Reis, and J.F. Mano. 2006. Phosphorous containing chitosan beads for controlled oral drug delivery. *J. Bioact. Compat. Polym.* 21:327–340.

Jayakumar, R., R.L. Reis, and J.F. Mano. 2007. Synthesis and characterization of pH-sensitive thiol-containing chitosan beads for controlled drug delivery applications. *Drug Deliv.* 14:9–17.

Jayakumar, R. and H. Tamura. 2006. Preparative methods of phosphorylated chitin and characterization. *Polym. Prepr.* 55:2101.

Jha, R.K. and X. Zi-rong. 2004. Biomedical compounds from marine organisms. *Mar. Drugs* 2:123–146.

Jiang, G.B., D. Quan, K. Liao, and H. Wang. 2006. Novel polymer micelles prepared from chitosan grafted hydrophobic palmitoyl groups for drug delivery. *Mol. Pharm.* 3:152–160.

Jiao, G., G. Yu, J. Zhang, and H.S. Ewart. 2011. Chemical structures and bioactivities of sulfated polysaccharides from marine algae. *Mar. Drugs* 9:196–223.

Jirge, S.S. and S.Y. Chaudhari. 2010. Marine: The ultimate source of bioactives and drug metabolites. *Int. J. Res. Ayurveda Pharm.* 1:55–62.

Jonker, C., J.H Hamman, and A.F Kotzé. 2002. Intestinal paracellular permeation enhancement with quaternised chitosan: in situ and in vitro evaluation. *Int. J. Pharm.* 238:205–213.

Jun, L, N. Peihong, Z. Mingzu, and Y. Zhangqing. 2004. Graft copolymerization of (dimethylamino)ethyl methacrylate on to chitosan initiated by ceric ammonium nitrate. *J. Macromol. Sci., Pure Appl. Chem.* 41:685–696.

Karim, S., P. Jessica, G. Farida, D. Christine, S. Corinne, and R. Jacqueline. 2011. Marine polysaccharides: A source of bioactive molecules for cell therapy and tissue engineering. *Mar. Drugs* 9:1664–1681.

Kasaai, M.R., J. Arul, and G. Charlet. 2003. Fragmentation of chitosan by ultrasonic irradiation. *Ultrason. Sonochem.* 15:1001–1008.

Kavitha Reddy, G.K.M., S. Shobharani, and G. Switi. 2011. Natural polysac-charides: Versatile excipients for controlled drug delivery systems. *Asian J. Pharm. Sci.* 6:275–286.

Kean, T., S. Roth, and M. Thanou. 2005. Trimethylated chitosans as non-viral gene delivery vectors: Cytotoxicity and transfection efficiency. *J. Control. Release* 103:643–653.

Kijoa, A. and P. Swangwong. 2004. Drugs and cosmetics from the sea. *Mar. Drugs* 2:73–82.

Kim, M., J.H. Yim, S. Kim, H.S. Kim, W.G. Lee, S.J. Kim, P. Kang, and C. Lee. 2012. In vitro inhibition of influenza A virus infection by marine microalga-derived sulfated polysaccharide p-KG03. *Antiviral Res.* 93:253–259.

Knaul, J.Z., S.M. Hudson, and A.M. Creber. 1999. Improved mechanical properties of chitosan fibers. *J. Appl. Polym. Sci.* 72:1721–1732.

Knill, C.J., J.F. Kennedy, J. Mistry, M. Miraftab, G. smart, M.R. Groocock, and H.J. Williams. 2009. Acid hydrolysis of commercial chitosan. *J. Chem. Technol. Biotechnol.* 80:1291–1296.

Knutsen, S.H., D.E. Myslabodski, B. Larsen, and A.I. Usov. 1994. A modified system of nomenclature for red algal galactans. *Bot. Mar.* 37:163–169.

Kraulan, A.H., D. Guggi, and A. Bernkop-Schnürch. 2006. Thiolated chitosan microparticles: A vehicle for nasal peptide drug delivery. *Int. J. Pharm.* 307:270–277.

Kumari, K. and P.P. Kundu. 2007. Semi-interpenetrating polymer networks (IPNs) of chitosan and L-alanine for monitoring the release of chlorpheniramine maleate. *J. Appl. Polym. Sci.* 103:3751.

Lai, V.M.F., P.A.L. Wong, and C.Y. Lii. 2000. Effects of cation properties on sol-gel transition and gel properties of κ-carrageenan. *J. Food Sci.* 65:1332–1337.

Lefnaoui, S. and N. Moulai-Mostefa. 2011. Formulation and in vitro evaluation of kappa-carrageenan-pregelatinized starch-based mucoadhesive gels containing miconazole. *Starch-Starke* 63:512–521.

Leiro, J., R. Castro, J. Arranz, and J. Lamas. 2007. Immunomodulating activities of acidic sulphated polysaccharides obtained from the seaweed Ulva rigida C Agardh. *Int. Immunopharmacol.* 7:879–888.

Li, L., L. Wang, Y. Shao, R. Ni, T. Zhang, and S. Mao. 2013a. Drug release characteristics from chitosan-alginate matrix tablets based on the theory of self-assembled film. *Int. J. Pharm.* 450:197–207.

Li, L., L. Wang, Y. Shao, Y. Tian, C. Li, and Y. Li. 2013b. Elucidation of release characteristics of highly soluble drug trimetazidine hydrochloride from chitosan-carrageenan matrix tablets. *J. Pharm. Sci.* 102:2644–2654.

Lidong, N., J. Wang, L. Wang, and S. Mao. 2013. A novel permeation enhancer: N-succinyl chitosan on the intranasal absorption of isosorbide dinitrate in rats. *Eur. J. Pharm. Sci.* 16:301–306.

Liu, J. and L. Li. 2007. Diffusion of camptothecin immobilized with cationic surfactant into agarose hydrogel containing anionic carrageenan. *J. Biomed. Mater. Res. A* 83A:1103–1109.

Majumdar, S., K. Hippalgaonkar, and Repka, M.A. 2008. Effect of chitosan, benzalkonium chloride and ethylenediaminetetraacetic acid on permeation of acyclovir across isolated rabbit cornea. *Int. J. Pharm.* 348:175–178.

Makhlof, A.Y., Y. Tozuka, and H. Takeuchi. 2011. Design and evaluation of novel pH-sensitive chitosan nanoparticles for oral insulin delivery. *Eur. J. Pharm. Sci.* 42:445–451.

Martinez, A., R. Lozanob, J. Teijon, and M. Blanco. 2011. Synthesis and characterization of thiolated alginate-albumin nanoparticles stabilized by disulfide bonds: Evaluation as drug delivery systems. *Carbohydr. Polym.* 83:1311–1321.

Martínez, A., E. Muñiz, I. Iglesias, J.M. Teijón, and M.D. Blanco. 2012. Enhanced preclinical efficacy of tamoxifen developed as alginate–cysteine/disulfide bond reduced albumin nanoparticles. *Int. J. Pharm.* 436:574–581.

McHugh, D.J. 1987. In McHugh, D.J. (ed.), *Production and Utilization of Products from Commercial Seaweeds.* Food and Agriculture Organization of the United Nations, Rome, Italy.

Mei, D., S. Mao, W. Sun, Y. Wang, and T. Kissel. 2008. Effect of chitosan structure properties and molecular weight on the intranasal absorption of tetramethylpyrazine phosphate in rats. *Eur. J. Pharm. Biopharm.* 70:874–881.

Michel, A.S., M.M. Mestdagh, and M.A.V. Axelos. 1997. Physico-chemical properties of carrageenan gels in presence of various cations. *Int. J. Biol. Macromol.* 21:195–200.

Miyazaki, S., M. Ishitani, A. Takahashi, T. Shimoyama, K. Itho, and D. Attwood, D. 2011. Carrageenan gels for oral sustained delivery of acetaminophen to dysphagic patients. *Biol. Pharm. Bull.* 34:164–166.

Mladenovska, K., R.S. Raicki, E.I. Janevik, T. Ristoski, M.J. Pavlova, and Z. Kavrakovski. 2007. Colon-specific delivery of 5-aminosalicylic acid from chitosan-Ca-alginate microparticles. *Int. J. Pharm.* 342:124–136.

Moebus, K., J. Siepmann, and R. Bodmeier. 2009. Alginate–poloxamer microparticles for controlled drug delivery to mucosal tissue. *Eur. J. Pharm. Biopharm.* 72:42–53.

Motwani, S.K., S. Chopra, S. Talegaonkar, K. Kohli, F.J. Ahmad, and R.K. Khar. 2008. Chitosan–sodium alginate nanoparticles as submicroscopic reservoirs for ocular delivery: Formulation, optimisation and in vitro characterisation. *Eur. J. Pharm. Biopharm.* 68:513–525.

Mourao, P.A. 2007. A carbohydrate-based mechanism of species recognition in sea urchin fertilization. *Braz. J. Med. Biol. Res.* 40:5–17.

Nanaki, S., E. Karavas, L. Kalantzi, and D. Bikiaris. 2010. Miscibility study of car-rageenan blends and evaluation of their effectiveness as sustained releasecarriers. *Carbohydr. Polym.* 79:1157–1167.

Naoji, K.N., N. Tatsumoto, T. Sano, and K. Toya. 2000. A simple preparation of half *N*-acetylated chitosan highly soluble in water and aqueous organic solvents. *Carbohydr. Res.* 324:268–274.

Nerurkar, J., H.W. Jun, J.C. Price, and M.O. Park. 2005. Controlled-release matrixtablets of ibuprofen using cellulose ethers and carrageenans: Effect of for-mulation factors on dissolution rates. *Eur. J. Pharm. Biopharm.* 61:56–68.

Ngoa, D. and S.K. Kim. 2013. Sulfated polysaccharides as bioactive agents from marine algae. *Int. J. Biol. Macromol.* 62:70–75.

Niederhofer, A. and B.W. Maller. 2004. A method for direct preparation of chitosan with low molecular weight from fungi. *Eur. J. Pharm. Biopharm.* 57:101–105.

Opanasopit, P., T. Ngawhirunpat, A. Chaidedgumjorn, T. Rojanarata, A. Apirakaramwong, S. Phongying, C. Choochottiros, and S. Chirachanchai. 2006. Incorporation of camptothecin into *N*-phthaloyl chitosan-g-mPEG self-assembly micellar system. *Eur. J. Pharm. Biopharm.* 64:269–276.

Opanasopit, P., T. Ngawhirunpat, T. Rojanarata, C. Choochottiros, and S. Chirachanchai. 2007. *N*-phthaloylchitosan-g-mPEG design for all-trans retinoic acid-loaded polymeric micelles. *Eur. J. Pharm. Sci.* 30:424–431.

Ozsoy, Y. and N. Bergisadi. 2000. Preparation of mefenamic acid sustained release beads based on kappa-carrageenan. *Boll. Chim. Farm.* 139:120–123.

Park, J.H., Y.W. Cho, H. Chung, I.C. Kwon, and S.Y. Jeong. 2003. Synthesis and characterization of sugar-bearing chitosan derivatives: Aqueous solubility and biodegradability. *Biomacromolecules* 4:1087–1091.

Parsons, T.R., K. Stephens, and J.D.H. Stickland. 1961. On the chemical composition of eleven species of marine phytoplankton. *J. Fish. Res. Board Can.* 18:1001–1016.

Patil, R.T. and T.J. Speaker. 1998. Carrageenan as an anionic polymer for aqueous microencapsulation. *Drug Deliv.* 5:179–182.

Pawar, S.N. and K.J. Edgar. 2012. Alginate derivatization: A review of chemistry, properties and applications. *Biomaterials* 33:3279–3305.

Perioli, L., V. Ambrogi, L. Venezia, C. Pagano, M. Ricci, and C. Rossi. 2008. Chitosan and a modified chitosan as agents to improve performances of mucoadhesive vaginal gels. *Colloids Surf. B: Biointerfaces* 66:141–145.

Pinheiro, A.C., A.I. Bourbon, B.G.D. Medeiros, L.H.M. da Silva, M.C.H. da Silva, and M.G. Carneiro-da-Cunha. 2012. Interactions between kappa-carrageenan and chitosan in nanolayered coatings—Structural and transport properties. *Carbohydr. Polym.* 87:1081–1090.

Piyakulawat, P., N. Praphairaksit, N. Chantarasiri, and N. Muangsin. 2007. Preparation and evaluation of chitosan/carrageenan beads for controlled release of sodium diclofenac. *AAPS PharmSciTech* 8:E1–E11.

Popa, E.G., M.E. Gomes, and R.L. Reis. 2011. Cell delivery systems using alginate–carrageenan hydrogel beads and fibers for regenerative medicine applications. *Biomacromolecules* 12:3952–3961.

Proksch, P., R.A. Edrada, and R. Ebel. 2002. Drugs from the seas current status and microbiological implications. *Appl. Microbiol. Biotechnol.* 59:125–134.

Ravi Kumar, M.N.V. 2000. A review of chitin and chitosan applications. *React. Funct. Polym.* 46:1–27.

Ravi Kumar, M.N.V., R.A.A. Muzzarelli, C. Muzzarelli, H. Sashiwa, and A.J. Domb. 2004. Chitosan chemistry and pharmaceutical perspectives. *Chem. Rev.* 104:6017–6084.

Rederstorff, E., P. Weiss, S. Sourice, P. Pilet, F. Xie, C. Sinquin, S. Colliec-Jouault, J. Guicheux, and S. Laïb. 2011. An in vitro study of two GAG-like marine polysaccharides incorporated into injectable hydrogels for bone and cartilage tissue engineering. *Acta Biomater.* 7:2119–2130.

Rekha, M.P. and Sharma, C.P. 2009. Synthesis and evaluation of lauryl succinyl chitosan particles towards oral insulin delivery and absorption. *J. Control. Release* 135:144–151.

Ren, D., H. Yi, W. Wang, and X. Ma. 2005. The enzymatic degradation and swelling properties of chitosan matrices with different degrees of *N*-acetylation. *Carbohydrates* 340:2403–2410.

Ritthidej, G.C., P. Chemto, S. Pummangura, and P. Menasveta. 1994. Chitin and chitosan as disintigrant in paracetamol tablets. *Drug Dev. Ind. Pharm.* 20:2109–2134.

Rocha, H.A., C.R. Franco, E.S. Trindade, S.S. Veiga, E.L. Leite, and H.B. Nader. 2005. Fucan inhibits Chinese hamster ovary cell (CHO) adhesion to fibronectin by binding to the extracellular matrix. *Planta Med.* 71:628–633.

Romankevich, E.A. 1984. *Geochemistry of Organic Matter in the Ocean*. Springer, Berlin, Germany, pp. 32–34.

Rowe, R.C., P.J. Sheskey, and S.C. Owen (eds.). 2006. *Handbook of Pharmaceutical Excipients*, 5th edn. The Pharmaceutical Press, London, U.K., pp. 132–135.

Sandri, G., S. Rossi, F. Ferrari, M.C. Bonferoni, C. Muzzarelli, and C. Caramella. 2004. Assessment of chitosan derivatives as buccal and vaginal penetration enhancers. *Eur. J. Pharm. Sci.* 21:351–359.

Santo, V.E., A.M. Frias, M. Carida, R. Cancedda, M.E. Gomes, and J.F. Mano. 2009. Carrageenan-based hydrogels for the controlled deliveryof PDGF-BB in bone tissue engineering applications. *Biomacromolecules* 10:1392–1401.

Sawayangi, C.Y., N. Nambu, and T. Nagai. 1982. Directly compressed tablets of chitin or chitosan in addition to lactose or potato starch. *Chem. Pharm. Bull.* 30: 2935–2940.

Schaeffer, D.J. and V.S. Krylov. 2000. Anti-HIV activity of extracts and compounds from algae and cyanobacteria. *Ecotoxicol. Environ. Saf.* 45:208–227.

Seifert, R., K.C. Emeis, A. Spitzy, K. Strahlendorf, W. Michaelis, and E.T. Degens. 1990. Geochemistry of labile organic matter in sediments and interstitial waters recovered from sites 651 and 653, Leg 107 in the Tyrrhenian Sea. *Proc. Ocean Drill. Program Sci. Results* 10:591–602.

Shanmugam, M. and K.H. Mody. 2000. Heparinoid-active sulphated polysaccharides from marine algae as potential blood anticoagulant agents. *Curr. Sci.* 79:1672–1683.

Singla, A.K. and M. Chawla. 2001. Chitosan: Some pharmaceutical and biological aspects—An update. *J. Pharm. Pharmacol.* 53:1047–1067.

Sipahigil, O. and B. Dortunc. 2001. Preparation and in vitro evaluation of verapamil HCl and ibuprofen containing carrageenan beads. *Int. J. Pharm.* 228:119–128.

Skaugrud, O. 1991. Chitosan—New biopolymer for cosmetics and drugs. *Drug Cosmet. Ind.* 148:24–29.

Song, K.H. and N.D. Eddington. 2012. The impact of AT1002 on the delivery of ritonavir in the presence of bioadhesive polymer, carrageenan. *Arch. Pharm. Res.* 35:937–943.

Stanley, N. 2011. FAO Corporate Document Repository—Chapter 3: Production, properties and uses of carrageenan. FMC Corporation, Marine Colloids Division, Rockland, ME.

Stiles, J., L. Guptill-Yoran, G.E. Moore, and R.M. Pogranichniy. 2008. Effects of κ-carrageenan on in vitro replication of feline herpesvirus and on experimentally induced herpetic conjunctivitis in cats. *Invest. Ophthalmol. Vis. Sci.* 49:1496–1501.

Stone, A.K. and M.T. Nickerson. 2012. Formation and functionality of whey proteinisolate-(kappa-, iota-, and lambda-type) carrageenan electrostatic complexes. *Food Hydrocoll.* 27:271–277.

Subramanian, S.B., A.P. Francis, and T. Devasena. 2014. Chitosan–starch nanocomposite particles as a drug carrier for the delivery of bis-desmethoxy curcumin analog. *Carbohydr. Polym.* 114:170–178.

Sweeneya, R., G. Miraftaba, and G. Collyer. 2014. Absorbent alginate fibres modified with hydrolysed chitosan forwound care dressings—II. Pilot scale development. *Carbohydr. Polym.* 102:920–927.

Thrimawithana, T.R., S.A. Young, C.R. Bunt, C.R. Green, and R.G. Alany. 2011. In-vitro and in-vivo evaluation of carrageenan/methylcellulose polymeric systems for transscleral delivery of macromolecules. *Eur. J. Pharm. Sci.* 44:399–409.

Thuy, T.T.T., B.M. Ly, T.T.T. Van, N.V. Quanga, H.C. Tuc, Y. Zheng, C. Seguin-Devauxc, B. Mid, and U. Ai. 2015. Anti-HIV activity of fucoidans from three brown seaweed species. *Carbohydr. Polym.* 115:122–128.

Tien, C., M. Lacroix, P. Ispas-Szabo, and M. Mateescu. 2003. N-acylated chitosan: Hydrophobic matrices for controlled drug release. *J. Control. Release* 93:1–13.

Tozaki, H., J. Komoike, C. Tada, T. Maruyama, A. Terabe, T. Suzuki, A. Yamamoto, and S. Muranishi. 1997. Chitosan capsules for colon-specific drug delivery: Improvement of insulin absorption from the rat colon. *J. Pharm. Sci.* 86:1016–1021.

Volod'ko , A.V., V.N. Davydova, E. Chusovitinb, I.V. Sorokina, M.P. Dolgikh, T.G. Tolstikova, S.A. Balagan, N.G. Galkin, and I.M. Yermak. 2014. Soluble chitosan–carrageenan polyelectrolyte complexes and their gastroprotective activity, *Carbohydr. Polym.* 10:1087–1093.

Wang, Y.S., L.R. Liu, Q. Jiang, and Q.Q. Zhang. 2007. Self aggregated nanoparticles of cholesterol-modified chitosan conjugate as a novel carrier of epirubicin. *Eur. Polym. J.* 43:43–51.

Wijesekara, I., R. Pangestuti, and S.K. Kim. 2011. Biological activities and potential health benefits of sulfated polysaccharides derived from marine algae. *Carbohydr. Polym.* 84:14–21.

Wu, T., S. Zivanovic, D.G. Hayes, and J. Weiss. 2008. Efficient reduction of chitosan molecular weight by high-intensity ultrasound: Underlying mechanism and effect of process parameters. *J. Agric. Food Chem.* 56:5112–5119.

Wunderwald, P., W.J. Schrenk, and H. Port. 1979. Antithrombin BM from human plasma: An antithrombin binding moderately to heparin. *Thromb. Res.* 15:49–60.

Xue, C., G. Yu, T. Hirata, J. Terao, and H. Lin. 1998. Antioxidative activities of several marine polysaccharides evaluated in a phosphatidylcholine-liposomal suspension and organic solvents. *Biosci. Biotechnol. Biochem.* 62:206–209.

Yamamoto, I., H. Maruyama, M. Takahashi, and K. Komiyama. 1986. The effect of dietary or intraperitoneally injected seaweed preparations on the growth of sarcoma-180 cells subcutaneously implanted into mice. *Cancer Lett.* 30:125–131.

Yan, H.Y. 2004. Harvesting drugs from the seas and how Taiwan could contribute to this effort. *Changhua J. Med.* 9:1–6.

Yang, J.S., Y.J. Xie, and W. He. 2011a. Research progress on chemical modification of alginate: A review. *Carbohydr. Polym.* 84:33–39.

Yang, S., F. Lin, H. Tsai, C. Lin, H. Chin, J. Wong, and M. Shieh. 2011b. Alginate-folic acid-modified chitosan nanoparticles for photodynamic detection of intestinal neoplasms. *Biomaterials* 32:2174–2182.

Yang, X., Q. Zhang, Y. Wang, H. Chen, H. Zhang, F. Gao, and L. Liu. 2008. Self-aggregated nanoparticles from methoxy poly(ethylene glycol)-modified chitosan: Synthesis; characterization; aggregation and methotrexate release in vitro. *Colloids Surf. B, Biointerfaces* 61:125–131.

Yingyinga, S., W. Hui, G. Ganlina, P. Yinfanga, and Y. Binlun. 2014. The isolation and antioxidant activity of polysaccharides from themarine microalgae *Isochrysis galbana*. *Carbohydr. Polym.* 113:22–31.

You, I., F.Q. Hu, Y.Z. Du, and H. Yuan. 2007. Polymeric micelles with glycolipid like structure and multiple hydrophobic domains for mediating molecular target delivery of paclitaxel. *Biomacromolecules* 8:2450–2456.

Younessi, P., R.A. Mohammad, and K. Shamimi. 2004. Preparation and ex vivo evaluation of TEC as an absorption enhancer for poorly absorbable compounds in colon specific drug delivery. *Acta Pharm.* 54:339–345.

Yuan, H., J. Song, X. Li, N. Li, and J. Dai. 2004. Immunomodulation and antitumor activity of κ-carrageenan oligosaccharides. *Pharmacol. Res.* 50:47–53.

Yuan, H., J. Song, X. Li, N. Li, and J. Dai. 2006. Immunomodulation and antitumor activity of κ-carrageenan oligosaccharides. *Cancer Lett.* 243:228–234.

Zacharopoulos, V.R. and D.M. Phillips. 1997. Vaginal formulations of carrageenan protect mice from herpes simplex virus infection. *Clin. Diagn. Lab. Immunol.* 4:465–468.

Zahrana, M.K., H.B. Ahmeda, and M.H. El-Rafie. 2014. Surface modification of cotton fabrics for antibacterial application bycoating with AgNPs–alginate composite. *Carbohydr. Polym.* 108:145–152.

Zhang, C., Q. Ping, H. Zhang, and J. Shen. 2003. Preparation of *N*-alkyl-O-sulfate chitosan derivatives and micellar solubilization of taxol. *Carbohydr. Polym.* 54:137–141.

Zhang, C., P. Qineng, and H. Zhang. 2004. Self-assembly and characterization of paclitaxel loaded *N*-octyl-O-sulfate chitosan micellar system. *Colloids Surf. B, Biointerfaces* 39:69–75.

Zhang, C., G. Qu, Y. Sun, T. Yang, Z. Yao, W. Shen, Z. Shen, Q. Ding, H. Zhou, and Q. Ping. 2008. Biological evaluation of *N*-octyl-O-sulfate chitosan as a new nano-carrier of intravenous drugs. *Eur. J. Pharm. Sci.* 33:415–423.

Zhao, X., A. Kong, Y. Hou, C. Shan, H. Ding, and Y. Shan. 2009. An innovative method for oxidative degradation of chitosan with molecular oxygen catalyzed by metal phthalocyanine in neutral ionic liquid. *Carbohydr. Res.* 344:2010–2013.

Zheng, H., Y. Du, J. Yu, R. Huang, and L. Zhang. 2001. Preparation and characterization of chitosan/poly(vinyl alcohol) blend fibers. *J. Appl. Polym. Sci.* 80:2558–2565.

Zheng, Y., Z. Cai, and X. Song. 2009. Preparation and characterization of folate conjugated *N*-trimethyl chitosan nanoparticles as protein carrier targeting folate receptor: In vitro studies. *J. Drug Target.* 17:294–303.

Zhou, G., W. Sheng, W. Yao, and C. Wang. 2006. Effect of low molecular-carrageenan from *Chondrus ocellatus* on antitumor H-22 activity of 5-Fu. *Pharmacol. Res.* 53:129–134.

Zhou, G., Y.P. Sun, X. Xin, Y. Zhang, L. Li, and Z. Xu. 2004. In vivo antitumor and immunomodulation activities of different molecular weight lambda-carrageenans from *Chondrus ocellatus*. *Pharmacol. Res.* 50:47–53.

*chapter sixteen*

# Pharmaceutical importance of marine algal-derived carbohydrates

*Ratih Pangestuti, Se-Kwon Kim*

## Contents

## 16.1 Introduction

Today, research is being focused on discovering pharmaceutical agents with selective pharmaceutical effects. As a consequence of increase in demand, serious research is under way for finding therapeutic agents from natural sources. Marine organisms are known to have a rich source of structurally diverse bioactive compounds with valuable pharmaceutical potentials. The earliest biologically active substance of marine origin was a toxin named "holothurin," which was extracted from sea cucumber (*Actinopyga agassizi*). Since then, the search for drugs and natural products of interest from marine organisms has continued (Laurienzo 2010).

The ocean is rich in biodiversity, and marine flora alone constitute more than 90% of oceanic biomass. Marine flora include microflora (bacteria, actinobacteria, cyanobacteria, and fungi), microalgae, macroalgae, and flowering plants (mangroves and other halophytes). Algae are simple chlorophyll-containing organisms composed of one cell or grouped together in colonies or as organisms with many cells and sometimes collaborating together as simple tissues. Algae are found everywhere on earth: in the sea, rivers, lakes, on soil and walls, in animal and plants, in fact, just about everywhere where there is a light to carry out photosynthesis. They vary greatly in size from unicellular (microns) to giant kelps up to 70 m long and growing at up to 0.5 m/day (El Gamal 2010). The term "algae" in this chapter is used to refer to marine macroalgae.

The field of natural polysaccharides of marine origin is already large and expanding, and marine algae are the most abundant source of polysaccharides. The principal cell wall polysaccharides in red algae are agars and carrageenan, those in green algae are ulvans, and those in brown algae are alginates and fucans, as well as the storage polysaccharide laminarin (Vera et al. 2011). Polysaccharides of algal origin are unbranched polysaccharides obtained from the cell membranes of some species of algae, largely used as gelatin and thickener in food industry and as a gel for electrophoresis in molecular biology and biochemistry. Chemically, it is constituted by galactose sugar molecules; it is the primary structural support for algae's cell walls. The field of natural polysaccharides of marine algal is already large and expanding. Over the past few years, medical and pharmacological industries have shown an increased interest in algae-derived polysaccharides. Due to the wide variations in their molecular weights, structural parameters, and physiological characteristics, algal polysaccharides show diverse pharmacological activities.

This chapter focuses on the progress of marine algal polysaccharides of interest in pharmaceuticals. The more innovative and appealing fields of application and strategies for their modification are reported. Finally, an update on the recent literature on the more common marine algal polysaccharides is given.

## 16.2  Application of marine algal polysaccharides in pharmaceuticals

### 16.2.1  Alginate

Alginate is a linear polysaccharide composed of 1,4-linked β-D-mannuronic acid and α-L-guluronic acid. Alginate, which exists in brown algae as the most abundant polysaccharide and is located in the intercellular matrix, was first described by the British chemist Stanford in 1881. Its main function is believed to be skeletal, giving both strength and flexibility to the algal tissue. Because of their ability to retain water and gelling, viscosifying, and stabilizing properties, alginates are widely used industrially (Stephen and Phillips 2010). The annual production of alginates is estimated to be approximately 38,000 tons worldwide.

In the past few decades, medical and pharmaceutical industries have shown an increased interest in biopolymers in general and alginates in particular. The ability of alginates to form soft hydrogels in the presence of calcium ions forms the basis for a wide variety of industrial applications, including foods and pharmaceuticals (Christensena 2011). Presently, research on alginates has largely shifted toward biomedical applications. Alginates can easily be formulated into a variety of soft, elastic gels, fibers, foams, nanoparticles, multilayers, etc., at physiological conditions ensuring the preservation of cell viability and function.

#### 16.2.1.1  Encapsulation

Alginates have been widely used to encapsulate enzymes, proteins, and live cells. Encapsulation represents one of the current leading methodologies aimed at the delivery of biological products to patients for the treatment of multiple diseases. Alginate is the most frequently employed material for the elaboration of the polymer matrix and outer biocompatible membrane because of its mild gelling and biocompatibility and biodegradability properties. However, its successful exploitation requires knowledge of microencapsulation technology and of the main properties of alginates, including their composition, purity, and viscosity.

#### 16.2.1.2  Disintegrants

Disintegrants are substances or mixture of substances that promote the breakup of an orally ingested tablet into smaller fragments (disintegrate) in an aqueous environment to release the active ingredient from the tablet and cause a physiological effect in the body. The more efficient the disintegrant, the faster the disintegration of the tablet, and thus the faster the active ingredient can be available for absorption into the gastrointestinal tract. Alginate is an effective disintegrant tablet, which has low sodium compared to other commonly used disintegrants. Alginates are insoluble in water, slightly acidic in reaction, and should be used in acidic or neutral granulation. Alginates do not retard flow and can be successfully used with ascorbic acid, multivitamin formulations, and acid salts of organic bases. Furthermore, as alginates are obtained from brown algae, they are inherently starch free and can be considered a natural material. Hence, alginate is a well-suited disintegrant for dietary supplements when there is a need to prepare natural disintegrants with low sodium contents.

#### 16.2.1.3  Biofilms

Biofilms are formed from natural polymers of faunal or floral origin (Silva et al. 2009, Seixas et al. 2013). The production of edible films from natural polymers has received much attention due to the excellent bio-degradability, biocompatibility, and edibility of the films (Fazilah et al. 2011). Some properties of alginates, which are natural polymers, are as follows: low cost, high stability, good gelling properties, biocompatibility, low toxicity, and easy modification chemistry and biochemistry, great potential for use in the preparation of biofilms, and are used in coating of food and drugs. Sodium alginate films are typically cast from aqueous solutions containing sodium alginate (5%–10%). The selection of sodium alginate will depend on the desired solid level and viscosity of the solution. The resulting films are water soluble, clear, and transparent, and the addition of a plasticizer such as glycerine provides greater film flexibility. Moreover, alginate biofilms are appropriate to load additives and antibacterial compounds. A mixture of starch and alginate to form edible film has been reported to improve the mechanical properties of biofilm.

### 16.2.1.4 Gastric reflux control

Gastroesophageal reflux disease is a common digestive system dysfunction associated with a spectrum of various symptoms of indigestion in patients of all backgrounds (Wang et al. 2013). There are several commercial products that suppress gastric reflux by the physical effect of creating a gelatinous raft that floats in the stomach. As an example, the use of oral sodium alginate solutions may provide an alternative method of relief by creating a protective pH-neutral layer over gastric contents without absorption into the bloodstream. This type of medication is classified as a reflux suppressant because the effect is not on the gastric acid itself, but rather it results in the formation of a viscous mechanical block of the reflux episode. This protective layer in the stomach, formed by the floating and foaming mechanism, may improve incidence, duration, and severity of postprandial symptoms from gastroesophageal reflux disease in adults.

## 16.2.2 Agar

Agar is an unbranched polysaccharide obtained from the cell membranes of some species of marine red algae, primarily from the genera *Gelidium* and *Gracilaria*. Agar is a linear polysaccharide composed of the repeating disaccharide unit of $(1 \rightarrow 3)$-linked β-D-galactose and $(1 \rightarrow 4)$-linked 3,6-anhydro-α-L-galactose residues. Agar helix is more compact compared to carrageenan due to the smaller amount of sulfate groups. Agar is a thermoreversible gelling polysaccharide, which sets at 30°C–40°C. Being less sulfated than furcellaran and carrageenans, agar can form strong gels, which are subject to pronounced syneresis, attributed to strong

aggregation of double helices (not weakened by the sulfate groups). The ability to form reversible gels in high-temperature aqueous solutions is the most important property of agar.

Agar is largely used as gelatin and thickener in food industry and as a gel for electrophoresis in microbiology (d'Ayala et al. 2008). Mid-quality agars are used as the gel substrate in biological culture media. The use of agar in solid culture media for bacteriological research was first described in 1881. Agar is also used in silk and paper industries, as a substitute for isinglass and in dying fabrics. In pharmaceutical industries, agar is used as biofilms, bulking agents or laxatives, suppositories, anticoagulants, and as an ingredient in tablets and capsules, as well as in different types of emulsions.

### 16.2.2.1 Biofilms

Agar has been few used as biofilm due to poor aging. Both photodegradation and fluctuations in ambient temperature and humidity alter agar crystallinity, leading to the formation of microfractures and polymer embrittlement. However, it was reported that agar-based films display better moisture barrier properties than cassava starch films. Gelled agar chains can stabilize film-forming emulsion to create a macronetwork. This macronetwork entraps flattened lipid particles, improving barrier performance by increasing tortuosity. The mechanical properties of agar-hydrogenated vegetable oil–emulsified films are comparable with some proteins and low-density polyethylene films.

### 16.2.2.2 Laxatives

Agar has been used as a laxative in daily treatment for chronic constipation for decades since it forms a smooth nonirritant bulk in the digestive tract (Rawat et al. 2012). Such agar is in the form of flakes that absorb from 12 to 15 times its weight of fluids. Agar is used as an ingredient in the preparation of capsules and suppositories, in surgical lubricants, in the preparation of emulsions, and as a suspending agent for barium sulfate in radiology. Agar is also used as a disintegrating and excipient in tablets. Sulfated agar has antilipemic activity similar to heparin. It is believed to inhibit the aerobic oxidation of ascorbic acid.

### 16.2.2.3 Suppositories

Suppositories are a unit solid dosage form of medicament intended for insertion into body cavity (Hanan and Durgin 2008). Suppositories are the alternate dosage forms for drugs with less bioavailability when taken orally. An advantage of suppositories over other dosage forms is reduction of side effects—gastrointestinal irritation and avoidance of both the unpleasant taste and first-pass metabolism—as the rectal route can deliver more than 60% of the administered drug directly into the systemic circulation, thus avoiding loss of drug due to the first-pass effect. It is also highly recommended for treating unconscious patients and children. The suppositories made with agar have the following advantages: they easily leave the molds and are not affected to the same extent by exposure as are those containing sodium stearate.

### 16.2.2.4 Disintegrants

Agars have been used as disintegrants because they absorb water and swell significantly but do not become gelatinous in water at physiological temperature. Their pharmaceutical dosage form, which disintegrates rapidly in water within 10 s, having an open-matrix network structure comprised mannitol and natural gum, can be used particularly for oral administration to pediatric and geriatric patient. High gel strength of agar makes it a potential candidate as a disintegrant. The use of agar powder as a disintegrating agent for the development of rapidly disintegrating oral tablets has also been investigated (Hayase et al. 2011).

### 16.2.2.5 Prosthetic dentistry

In dentistry, agar is used to make accurate negative casts of teeth, sockets, and entire edentulous gums to form accurate artifacts.

## 16.2.3 Carageenan

Carrageenan is a sulfated polygalactan extracted from marine red algae with 15%–40% of ester sulfate content and an average relative molecular mass well above 100 kDa (Necas and Bartosikova 2013). Carrageenan is formed by the linear backbone built up by β-D-galactose and 3,6-anhydro-α-D-galactose with variable density in the sulfated group. Based on the sulfate group content, carrageenan can be classified into various types such as λ, κ, ι, ε, and μ. These classifications do not reflect definitive chemical structures but only general differences in the composition and degree of sulfation at specific locations in the polymer. *Chondrus crispus*, *Gigartina stellata*, *Iridaea* spp., *Eucheuma* spp., and *Kappaphycus* spp. are the principal raw materials used for carrageenan extraction.

### 16.2.3.1 Antibiotic production

Antibiotics have been widely produced since the pioneering efforts a century ago, and the importance of antibiotics to medicine has led to much research into their discovery and production. The κ-carrageenan has been used in an attempt to improve the production of tetracycline and chlortetracycline. Tetracyclines represent one of the most important groups of antibiotics.

### 16.2.3.2 Substance for the elaboration of the drugs with highly selective metal-binding properties

Carrageenans possess metal-binding properties, whereby they can be successfully used in various fields of pharmacy (Khotimchenko et al. 2010). Carrageenans can strongly bind and hold metal ions; this property can be used for the creation of new drugs for the elimination of metals from the body or targeted delivery of these metal ions for healing purposes. Besides, such compounds may be the base for elaboration of drugs purposed for targeted radioactive therapy of malignant tumors. Therefore, it can be concluded that carrageenan is the more appropriate substance for elaboration of the drugs with highly selective metal-binding properties.

### 16.2.3.3 Microbicides for sexually transmitted diseases

The polyanion category of microbicides includes carrageenans. Carrageenans are chemically related to heparan sulfate, which many microbes utilize as a biochemical receptor for initial attachment to the cell membranes (Kim and Pangestuti 2011). Thus, carrageenan and other microbicides of its class act as decoy receptors for viral binding. Buck et al. reported that carrageenan, particularly ι-carrageenan, inhibits human papilloma virus (HPV) infection three orders magnitude more potent than heparin, a highly effective model for HPV inhibitor (Roberts et al. 2007). Carrageenan acts primarily by preventing the binding of HPV virions to cells and blocks HPV infection through a second, postattachment heparin sulfate–independent effect. These mechanisms are consistent as carrageenan resembles heparan sulfate. Furthermore, some milk-based products block HPV infectivity *in vitro*, even when diluted millionfold (Buck et al. 2006). In another study, carrageenan was reported to inhibit genital transmission of HPV in female mouse model cervicovaginally (Schiller and Davies 2004). Carrageenan was able to generate antigen-specific immune responses and antitumor effects in female mice vaccinated with HPV-16 E7 peptide vaccine (Zhang et al. 2010). Previous work indicates that carrageenan-based microbicides are safe for vaginal use, as shown by the Carraguard trial in 165 women.

## 16.2.4 Fucoidan

Fucoidans are sulfated polysaccharides containing substantial percentages of L-fucose and sulfate ester groups, constituents of brown algae, and some marine invertebrates (i.e., sea urchins and sea cucumbers). Fucoidans have been extensively studied due to their varied biological activities, including anticoagulant and antithrombotic activities, antivirus activities, antitumor and immunomodulatory activities, antiadhesive activities, antiangiogenic activities, anti-inflammatory activities, blood lipids reducing activities, antioxidant and anticomplementary properties, activity against hepatopathy activities, uropathy and renalpathy activities, neuroprotective activities, gastric protective effects, and therapeutic potential in surgery (Li et al. 2008).

### 16.2.4.1  Dietary supplements

Due to their wide biological activities, fucoidans were used as dietary supplements. Fucoidan from brown algae is partially hydrolyzed and then mixed with other ingredients for use as a dietary supplement in beverage, capsule, or tablet form. The fucoidan is partially hydrolyzed with acid and heat. The main effective ingredient in fucoidan is the fucose, one of the eight essential biological sugars. Fucoidan provides special saccharides and biological sugars, which have recently been identified as being absolutely essential for cell-to-cell communication through glycoproteins and glycolipids.

### 16.2.4.2  Biomaterials

The structure and bioactivity of fucoidan are similar to those of heparin, a highly sulfated glycosaminoglycan composed of D-glucosamine, L-iduronic acid, and D-glucuronic acid. Heparin can stabilize growth factor, enhance binding with cell surface receptors, and serve as an antithrombotic inhibitor for clinical therapy. However, heparin has been reported to possess side effects. By contrast, fucoidan comes from natural nontoxic algae and could have great potential for biomaterials.

### 16.2.4.3  Oral delivery system

A rationally designed oral delivery carrier must overcome various significant barriers in the gastrointestinal tract prior to delivery to the bloodstream. One of these barriers is the low pH of the gastric medium in the stomach. Chitosan/fucoidan nanoparticles are effective pH-sensitive carriers, and they have great potential for application in oral delivery systems (Huang and Lam 2011).

## 16.2.5  Ulvan

Ulvan is the major water-soluble polysaccharide extracted from marine green algae genus *Ulva*. Ulvan polysaccharides possess unique structural properties where the main constituents of ulvan are sulfated rhamnose residues linked to uronic acids, resulting in a repeated disaccharide unit D-glucuronosyl-(1,4)-L-rhamnose

3-sulfate, called aldobiouronic acid (Jaulneau et al. 2010). These polysaccharides present several potentially valuable biological properties for agricultural, food, and pharmaceutical applications, among which are anticoagulant (heparin like), antioxidant, antithrombotic, immunomodulating, strain-specific anti-influenza, anticoagulant, antihyperlipidemic, and antitumoral activities.

The material suitable for biomedical and pharmaceutical applications—tissue engineering, regenerative medicine, and drug delivery—must be biocompatible and biodegradable, and its products of degradation must be safe and easily cleared from the host organisms. Most of the materials obtained from natural resources are able to fulfill these strict requirements, but particular attention is focused on biopolymers of plant and marine algal origins due to their abundance and minor concerns for purification. Ulvan represents an advantageous versatile platform of "unique" marine algal polysaccharides that, along with their abundance and renewability, would potentially display the properties that match the criteria for biomedical applications. The presence of both rhamnose and iduronic acid confers a particularly peculiar character to ulvan, as both these monosaccharides are rarely found in algal polysaccharides. However, it is the presence of both an iduronic acid and a sulfated sugar moiety that is the basis for the comparison of ulvan with mammalian sulfated glycosaminoglycans, especially sulfated chondroitin or dermatan. These results position ulvan structures as prospective blocks that can be further functionalized to acquire the desired stability and biological interactivity to be used as tissue-engineered structures (Alves et al. 2013).

Other pharmaceutical applications of ulvan are based on rhamnosylated saccharides, since L-rhamnose is a fundamental component of the surface antigens of many microorganisms and is specifically recognized by a number of mammalians lectins. Ulvan is also a potential source of iduronic acid, another rare sugar found in mammalian glycosaminoglycans, which is required in the synthesis of heparin analogs with antithrombotic activities (Lahaye and Robic 2007).

## 16.2.6  Laminarin

Laminarin is the principal storage glucan found in brown algae, representing up to 35% of the dry weight. It is mainly a linear polysaccharide constituted by 25–50 glucose units linked by β-1,3-glycosidic bonds and in some cases having β-1,6-glycosidic bonds and ramifications in the O-6 position. Laminarins have an average molecular weight of 5 kDa, but they can differ in the terminal reducing end corresponding to a glucose residue in G-type laminarin and to a mannitol residue in M-type laminarin.

Laminarins were used in pharmaceutical compositions for mimicking heparin activity and for therapeutic use. instead of heparin, in preventing restenosis by the inhibition of vascular smooth muscle cell proliferation; in accelerating wound healing by activating the release of active growth factors stored in the extracellular matrix; and for inhibiting tumor cell metastasis by the inhibition of heparanase activity. Laminarins were also used as active ingredients in cosmetic or pharmaceutical, particularly dermatological, compositions. It has been discovered that laminarins possess stimulating, regenerating, conditioning, and energizing effects on human dermis fibroblasts and human epidermis keratinocytes.

## 16.3   Summary

Marine organisms, particularly marine algae, are known to have a rich source of structurally diverse bioactive compounds with valuable pharmaceutical potentials. Medical and pharmacological industries have shown an increased interest in algae-derived polysaccharides. Marine algal polysaccharides such as agar, alginates, carrageenan, fucoidan, ulvan, and laminarin possess potentially valuable biological properties. Marine algal polysaccharides have been applied in pharmaceuticals for many purposes such as encapsulation, disintegrants, biofilms, gastric reflux control, suppositories, laxatives, microbicides for sexually transmitted diseases, antibiotic production, prosthetic dentistry, biomaterials, oral delivery system, dietary supplements, and substance for the elaboration of the drugs with highly selective metal-binding properties. Marine algal polysaccharides are suitable for biomedical and pharmaceutical applications, due to their biocompatible and biodegradable properties, and their degradation products are safe and easily cleared from host organisms.

## References

Alves, A., Sousa, R.A., and Reis, R.L. 2013. Processing of degradable ulvan 3D porous structures for biomedical applications. *Journal of Biomedical Materials Research Part A*, 101: 998–1006.

Buck, C.B., Thompson, C.D., Roberts, J.N., Müller, M., Lowy, D.R., and Schiller, J.T. 2006. Carrageenan is a potent inhibitor of papillomavirus infection. *PLoS Pathogens*, 2: e69.

Christensena, B.E. 2011. Alginates as biomaterials in tissue engineering. *Carbohydrate Chemistry: Chemical and Biological Approaches*, 37: 227–258.

D'ayala, G.G., Malinconico, M., and Laurienzo, P. 2008. Marine derived polysaccharides for biomedical applications: Chemical modification approaches. *Molecules*, 13: 2069–2106.

El Gamal, A.A. 2010. Biological importance of marine algae. *Saudi Pharmaceutical Journal*, 18: 1–25.

Fazilah, A., Maizura, M., Abd Karim, A., Bhupinder, K., Rajeev, B., Uthumporn, U., and Chew, S. 2011. Physical and mechanical properties of sago starch–alginate films incorporated with calcium chloride. *International Food Research Journal*, 18: 1027–1033.

Hanan, Z. and Durgin, J. 2008. *Pharmacy Practice for Technicians*. Delmar Publishers and Cengage Learning, New York, 656pp.

Hayase, N., Iwayama, K., Ohtaki, K.-I., Yamashita, Y., Awaya, T., and Matsubara, K. 2011. Dissolution behaviors of tablet and capsule covered with oblate or agar jelly for taking medicine easily. *Yakugaku zasshi: Journal of the Pharmaceutical Society of Japan*, 131: 161–168.

Huang, Y.C. and Lam, U.I. 2011. Chitosan/fucoidan pH sensitive nanoparticles for oral delivery system. *Journal of the Chinese Chemical Society*, 58: 779–785.

Jaulneau, V., Lafitte, C., Jacquet, C., Fournier, S., Salamagne, S., Briand, X., Esquerré-Tugayé, M.T., and Dumas, B. 2010. Ulvan, a sulfated polysaccharide from green algae, activates plant immunity through the jasmonic acid signaling pathway. *Journal of Biomedicine & Biotechnology*, 2010: 525291.

Khotimchenko, Y.S., Khozhaenko, E.V., Khotimchenko, M.Y., Kolenchenko, E.A., and Kovalev, V.V. 2010. Carrageenans as a new source of drugs with metal binding properties. *Marine Drugs*, 8: 1106–1121.

Kim, S.-K. and Pangestuti, R. 2011. Potential role of marine algae on female health, beauty, and longevity. *Advances in Food and Nutrition Research*, 64: 41–55.

Lahaye, M. and Robic, A. 2007. Structure and functional properties of ulvan, a polysaccharide from green seaweeds. *Biomacromolecules*, 8: 1765–1774.

Laurienzo, P. 2010. Marine polysaccharides in pharmaceutical applications: An overview. *Marine Drugs*, 8: 2435–2465.

Li, B., Lu, F., Wei, X., and Zhao, R. 2008. Fucoidan: Structure and bioactivity. *Molecules*, 13: 1671–1695.

Necas, J. and Bartosikova, L. 2013. Carrageenan: A review. *Veterinarni Medicina*, 58: 187–205.

Rawat, A., Srivastava, S., and Ojha, S.K. 2012. Herbal remedies for management of constipation and its ayurvedic perspectives. *Journal of International Medical Science Academy*, 25: 19–22.

Roberts, J.N., Buck, C.B., Thompson, C.D., Kines, R., Bernardo, M., Choyke, P.L., Lowy, D.R., and Schiller, J.T. 2007. Genital transmission of HPV in a mouse model is potentiated by nonoxynol-9 and inhibited by carrageenan. *Nature Medicine*, 13: 857–861.

Schiller, J.T. and Davies, P. 2004. Delivering on the promise: HPV vaccines and cervical cancer. *Nature Reviews Microbiology*, 2: 343–347.

Seixas, F., Turbiani, F., Salomão, P., Souza, R., and Gimenes, M. 2013. Biofilms composed of alginate and pectin: Effect of concentration of crosslinker and plasticizer agents. *Chemical Engineering Transactions*, 32: 1693–1698.

Silva, M.a.D., Bierhalz, A.C.K., and Kieckbusch, T.G. 2009. Alginate and pectin composite films crosslinked with Ca²⁺ ions: Effect of the plasticizer concentration. *Carbohydrate Polymers*, 77: 736–742.

Stephen, A.M. and Phillips, G.O. 2010. *Food Polysaccharides and Their Applications*. CRC Press: New York.

Vera, J., Castro, J., Gonzalez, A., and Moenne, A. 2011. Seaweed polysaccharides and derived oligosaccharides stimulate defense responses and protection against pathogens in plants. *Marine Drugs*, 9: 2514–2525.

Wang, Y.K., Hsu, W.H., Wang, S.S., Lu, C.Y., Kuo, F.C., Su, Y.C., Yang, S.F., Chen, C.Y., Wu, D.C., and Kuo, C.H. 2013. Current pharmacological management of gastroesophageal reflux disease. *Gastroenterology Research and Practice*, 2013: 1–13.

Zhang, Y.Q., Tsai, Y.C., Monie, A., Hung, C.F. and Wu, T.C. 2010. Carrageenan as an adjuvant to enhance peptide-based vaccine potency. *Vaccine*, 28: 5212–5219.

*chapter seventeen*

# Marine bacterial exopolysaccharides
## Functional diversity and prospects in environmental restoration

**Jaya Chakraborty, Neelam Mangwani, Hirak R. Dash,**
**Supriya Kumari, Himanshu Kumar, and Surajit Das**

## Contents

## 17.1 Introduction

Marine ecosystems are among Earth's principal ecosystem encasing vast bacterial diversity. These bacteria play a central role in each tropical level and are vital mediators involved in the maintenance of several changing environmental conditions (de Carvalho and Fernandes 2010). As a survival strategy to unfavorable conditions (or as secondary metabolites), they secrete several products. Some of these secondary products help in protection and are also important from biotechnological to environmental perspectives (Jensen and Fenical 1994). One of these products is exopolysaccharides (EPSs), which constitute the major fraction of the dissolved organic matter reservoir in the ocean (Gutierrez et al. 2013; Eronen-Rasimus et al. 2014). Structurally, EPSs are high-molecular-weight polymers of carbohydrates. The term "EPS" was coined by Sutherland in 1972 to describe high-molecular-weight carbohydrate polymers produced by bacteria. The synthesis of EPS by bacteria is one of the strategies for growth under the natural environment (Costerton 1999). Bacteria present in aquatic systems are usually enclosed in a matrix, which is rich in EPS (Decho 1990; Costerton 1999). Most of the bacteria in a moist environment survive as adherent communities (known as biofilms), attached to the substratum and enclosed in a milieu of biomolecules that are rich in protein, nucleic acid, and EPS. Thus, EPSs also constitute structural framework of biofilms (White 1986). The formation of biofilm or production of EPS supports the survival of bacteria under extreme temperature, salinity, and nutrient availability (Poli et al. 2010).

The EPS-producing bacteria have been isolated from different zones of the marine environment, such as surface water, sediments, mangroves, hydrothermal vents, and deep sea (Poli et al. 2010, 2011). Many marine bacteria from genera *Halomonas, Bacillus, Pseudomonas, Pseudoalteromonas, Marinobacter, Alteromonas, Alcanivorax,* and *Rhodococcus* are known to produce EPS (Calvo et al. 2002; Bhaskar and Bhosle 2006; Iyer et al. 2006; Dash et al. 2014; Mangwani et al. 2014). EPSs from bacteria have a very prospective role in bioremediation of hydrocarbons and heavy metals in the marine environment. Chemically, EPSs from marine bacteria are polyionic and rich in uronic acid. Ionic nature assists in the binding with cations (metal cations), whereas uronic acid interfaces with hydrophobic organic compounds such as hydrocarbons (Iyer et al. 2006). A modest quantity of amino acids and peptides are also present in EPS, which confer amphiphilic properties to the EPS. EPSs have been reported to emulsify hydrocarbons, crude oil, and refined petroleum products (Abbasi and Amiri 2008; Abari et al. 2012). Apart from that, EPSs produced by some marine bacteria are also helpful in controlling biofouling (Kim et al. 2011). Bacterial EPSs serve diverse roles in the marine environment. It acts as a source of carbon, provides protection, helps in trace metal entrapment, cell immobilization, formation of mixed communities, and degradation of pollutants (Poli et al. 2010). This chapter illustrates the diverse aspects of bacterial EPS, their structural and functional diversity, genetic regulations, and environmental applications.

## 17.2   Functional diversity of EPSs

Carbohydrate polymers have enormous structural diversity to serve a wide variety of functions. Their structures vary from linear to highly branched, in which monosaccharides (sugars) constitute the major structural frame. The sugar composition of polysaccharides reflects their functional versatility. Based upon their structural unit, EPSs are having different functions in the bacterial life cycle. They are classified as structural polysaccharides that provide support and function, capsular polysaccharides (CPSs) that help in cellular interaction and virulence, and biofilm-associated polysaccharides that help in the formation and stabilization of biofilm matrix, cell-to-cell, and cell-to-surface interactions.

### 17.2.1   Structural polysaccharides

Structural polysaccharides are the polymeric carbohydrate molecules that provide shape, structure, and integrity to the living entities like plants, animals, bacteria, fungi, and algae. High structural variety and complexity of the polysaccharides draw much attention due to potential applications in diverse technological spheres (Dumitriu 2005).

Polysaccharides produced by soil bacteria form an important communication factor, which allows them to build various forms of interactions. Surface polysaccharides provide the primary framework of all bacteria. There exists structural diversity in the cell wall

makeup of both Gram-negative and Gram-positive bacteria. In both, the cell wall is constructed from the polymer peptidoglycan. Peptidoglycan is a composite of long glycan strands that are cross-linked by stretchable peptides (Figure 17.1). This structural elastic network protects the shape and integrity of the cells from lysis (Matias et al. 2003). The glycan strands are made up of alternating N-acetylglucosamine and N-acetylmuramic acid (MurNAc) residues linked by β-1,4 glycosidic bonds (Rogers et al. 1980). The D-lactoyl group of each MurNAc residue is substituted by a peptide stem, whose composition is most often L-Ala-g-D-Glu-meso-$A_2$pm (or L-Lys)-D-Ala-D-Ala ($A_2$pm, 2,6-diaminopimelic acid) in a nascent peptidoglycan. The glycan strands are cross-linked between the carboxyl group of D-Ala at position 4 and the amino group of the diamino acid at position 3, through direct linkage or through a short peptide bridge. Thus, the heteropolymer has the presence of an unusual sugar (MurNAc), of ϒ-bonded D-Glu, of L–D (and even D–D) bonds, and of nonprotein amino acids (e.g., $A_2$pm). As an example, the glycan strands in *Escherichia coli* are broad with a mean of 20–30 disaccharide units of which some ranges up to 80 units (Harz et al. 1990). The disaccharide units in the glycan strand are synthesized with a covalently linked peptide, which is cross-linked to the peptide originating from another glycan strand (Koch 2000). 3D structural determination by NMR study of a relaxed dimeric peptidoglycan segment suggested that subsequent peptides are spaced by 120° (Meroueh et al. 2006).

The Gram-negative bacterial cell wall is three layers thick (Yao et al. 1999). Lipopolysaccharides (LPSs) are amphiphilic macromolecules that form the external layer of the outer membrane of Gram-negative bacteria. They constitute of an asymmetric lipid bilayer with phospholipids and glycolipids in the inner leaflet (Muhlradt and Golecki 1975). Primarily, the LPS layer in the outer membrane bestows Gram-negative bacteria with a strong permeability barrier against toxic compounds such as antibiotics, allowing survival in an adverse environment (Nikaido 2003). In addition, the innate immune response is potentially activated by bacterial LPS. It also acts as a conserved pathogen-associated molecular pattern recognized by innate immune receptors (Miller et al. 2005). LPS is composed of three moieties, lipid A, a core polysaccharide, and an O-specific polysaccharide (OPS). Lipid A is the domain that is highly hydrophobic in nature, which anchors the LPS molecule into the membrane and helps in the regulation of biological activities of the LPS. The core oligosaccharide and an OPS protrude into the environment carrying their antigenic determinants.

The Gram-positive cell wall is a complex arrangement of macromolecules consisting of a peptidoglycan that surrounds the cytoplasmic membrane decorated with glycopolymers (teichoic acid) or polysaccharides and proteins. The composition of the peptide chains in

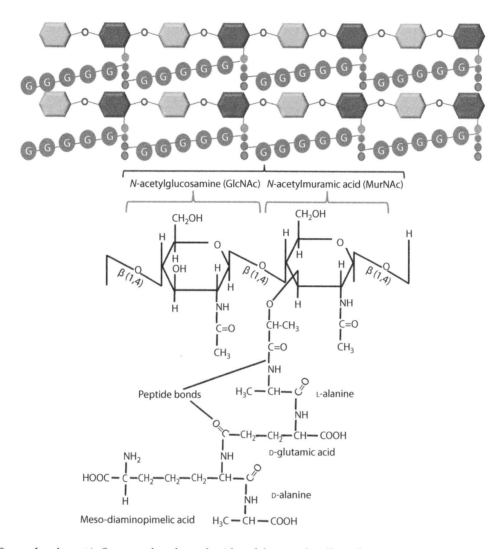

*Figure 17.1* (**See color insert.**) Structural polysaccharide of bacterial cell wall; Peptidoglycan structure composed of *N*-acetylglucosamine (GlcNAc) and *N*-acetylmuramic acid (MurNAc).

the peptidogylcan differs in different species directly or indirectly through cross-linked short chains of one or more amino acids. This three-dimensional arrangement around the cell results in bacterial integrity and rigidity (Chapot-Chartier and Kulakauskas 2014). The basic structure of peptidoglycan is modified partially in many of the bacterial species. The glycan chains undergo N-acetylation or O-acetylation or amidation of the free carboxyl groups of the amino acids in the peptide chains (Vollmer 2008). Post O-acetylation of MurNAc in Gram-positive pathogenic bacteria leads to resistance against lysozyme (Moynihan and Clarke 2011). In addition to the maintenance of bacterial cell integrity, the cell wall also helps in the growth of bacteria and resists the internal turgor pressure. Acting as the interface between bacterial cell and the external environment, it aids in bacterial interactions with other surfaces and host organisms.

The Gram-positive bacterial cell wall withstands high internal osmotic pressure of 20–25 atm, for example, *Staphylococcus aureus*, whereas the Gram-negative bacterial cell wall withstands osmotic pressure of about 2–3 atm, for example, *E. coli* (Mitchell and Moyle 1956). Recently, 61 signature proteins have been discovered from the genome of α-proteobacteria, which might be functionally encoded for the adaptation mechanism by marine bacteria (Kainth and Gupta 2005; Dash et al. 2013). Studies on the cell wall fraction of a marine pseudomonad showed the presence of two distinct protein components. One of the protein content was designated as the "labile" component, which is the substrate of a lytic enzyme system from the cell wall fraction. The second protein component was designated as the "resistant" component, which is not affected by the lytic enzyme. The labile protein located in the outer membrane of the double membrane structure varies over the range of 0%–50% based on the medium used for growth, whereas the resistant protein accounted for 35% of autolysed cell wall residue, and the value was independent of the growth medium (Brown 1961).

Apart from structural polysaccharides in bacteria, primary and secondary cell walls of all terrestrial and aquatic plants and seaweeds are composed of hemicellulosic fractions, which are a large group of well-characterized polysaccharides. These fractions are made up of a relatively limited number of sugar residues, principally D-xylose, D-mannose, D-glucose, D-galactose, L-arabinose, D-glucuronic acid, 4-*O*-methyl-D-glucuronic acid, D-galacturonic acid, L-rhamnose, L-fucose, and various O-methylated neutral sugars to a lesser extent. Dry weight of 20%–35% of hemicellulose polymers constitutes the wood and annual plants. The hemicellulosic fraction of wheat straw is composed of β-1,4-linked D-xylopyranose units with side chains of various lengths containing L-arabinose, D-glucuronic acid or 4-*O*-methyl ether, D-galactose, and possibly D-glucose. Structural polysaccharides also include arabinoxylans, chitin, cellulose, and pectins.

## 17.2.2 Capsular polysaccharides

CPSs is present in a wide range of bacterial species that constitute repeating units of long-chain polysaccharides. They can be either homo- or heteropolymers composed of repeating monosaccharides joined by glycosidic linkages (Roberts 1996). These act as the outermost layer of the cell, which is involved in direct interactions between the bacteria and their surrounding environment. These interactions implicate capsules as the major virulence factor existing in many bacterial pathogens.

Usually, the bacterial extracellular polysaccharides are classified as CPSs, where they are intimately associated with the cell surface or as slime polysaccharides, which are loosely attached to the cell (Whitfield 1988). Due to the association of CPS with O-antigenic moieties of LPS, the differentiation of CPS with other cell surface polysaccharides is quite complex. Capsules are widely distributed in diverse pathogens like *Actinobacillus pleuropneumoniae*, *E. coli*, *Neisseria meningitidis*, *Sinorhizobium meliloti*, *S. aureus*, and *Streptococcus pneumoniae* (Kroncke et al. 1990). A wide array of structurally diverse CPSs are produced by a single species, which are the basis for different serotyping schemes in the bacteria (Moxon and Kroll 1990).

The bacterial capsules have a wide range of functions, which include resistance to specific and nonspecific host community, desiccation, and adherence prevention (Roberts 1996). By the formation of a hydrated gel by the capsule around the bacterial cell surface, the harmful effects of desiccation can be relieved. This property is an advantage in the survival and transmission of

pathogenicity by encapsulated bacteria from one host to another (Roberson and Firestone 1992). Furthermore, CPS promotes adherence of bacteria in many other surfaces, thus facilitating colonization of a particular niche forming biofilm and persistence of the microbes during colonization (Costerton et al. 1987; Jenkinson 1994). Complement-mediated killing is also resisted by bacterial capsules that provide a permeability barrier to the complement components masking the cell surface structure–activating complement (Howard and Glynn 1971). Thus, capsules are responsible for resistance to complement-mediated opsonophagocytosis.

Bacterial species demonstrate a great structural diversity in synthesizing capsules. Furthermore, chemically identical CPSs may also be synthesized by different bacterial species. The production of CPSs in Gram-negative bacteria occurs by means of one of the two widespread assembly systems, Wzy-dependent or ATP-binding cassette (ABC) transporter–dependent pathways (Cuthbertson et al. 2009, 2010). As a model bacterium, *E. coli* has been studied for providing an overview of different proteins required for CPS synthesis (Figure 17.2). The capsule system has been differentiated into four groups based on the organization of the key genes, assembly system, and regulatory features (Whitfield 2006). The isolates causing gastrointestinal disease have groups 1 and 4 capsules synthesized by the Wzy-dependent pathway. This pathway involves the synthesis of polyprenol-linked CPS repeat units in the cytoplasm flipped across the inner membrane by a Wzx protein and polymerized into the full-length CPS by a Wzy protein forming a capsule structure on the cell surface. Groups 2 and 3 capsules require an ABC transporter–dependent assembly system that assembles a phospholipid at the reducing end of the polysaccharide chain, which helps in the attachment of CPS to the cell surface (Jann and Jann 1990). Groups 1 and 4 capsules are highly predominant in *Erwinia* and *Klebsiella*, whereas capsules belonging to groups 2 and 3 are found in mucosal pathogens like *Campylobacter jejuni*, *Haemophilus influenzae*, *Mannheimia haemolytica*, *N. meningitidis*, *Pasteurella multocida*, etc. (Wilkie et al. 2012).

## 17.2.3 Biofilm-associated polysaccharides

Biofilm is an aggregation of bacteria enclosed in a matrix consisting of a mixture of extracellular polymeric compounds primarily consisting of EPSs. Biofilms comprise the normal environment for most microbial cells in many natural and artificial habitats. It consists of complex associations of cells, extracellular products, and detritus trapped within the biofilm or released from cells (Christensen 1989). Biofilm secreting the mixture of polysaccharides forms the major cementing material within the matrix having diverse composition with

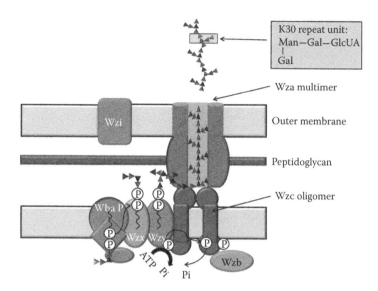

*Figure 17.2* Overview of different proteins required for capsular polysaccharide synthesis. Capsular polymer of serotype K30 is formed by a sequential process initiated by the galactose-1-P transferase (WbaP). Wzx flips them across the membrane and becomes substrates for Wzy-dependent blockwise polymerization. Transphosphorylation of Wzc and its dephosphorylation by Wzb are required to maintain high-level polymerization and capsule export. Wza, the outer-membrane lipoprotein, is vital for the export of the capsule to the surface, and Wzi is involved in the determination of the extent of surface attachment of the capsule.

different physical and chemical properties. Most microbial EPSs are polyanionic due to the presence of uronic acids or ketal-linked pyruvate, whereas minority of them are neutral macromolecules. Inorganic residues, such as phosphate and sulfate, also confer polyanionic status (Sutherland 1990). Mack et al. (1996) exemplified the polycationic EPS, an adhesive polymer obtained from *Staphylococcus epidermidis* associated with biofilms. Backbone composition of sequences of 1,4-β- or 1,3-β-linkages confers high rigidity, for example, cellulosic backbone of xanthan from *Xanthomonas campestris*. The polysaccharides can thus form diverse types of structures within a biofilm. Mayer et al. (1999) suggested that electrostatic and hydrogen bonds are the governing forces during biofilm formation. Several ionic interactions may be involved, but more delicate chain complex formation in which one macromolecule fits into the other may result in either floc formation or networks very poorly soluble in aqueous solvents. In biofilm-associated EPSs, the interaction with a wide range of other molecular species includes lectins, proteins, lipids, and other polysaccharides (Sutherland 2001).

Traditionally, microorganisms are characterized and identified as single-cell planktons (Donlan 2002). However, meticulous studies of sessile communities in different environments have led to the conclusion that planktonic microbial growth rarely exists in nature. The investigation of microbial aggregates on tooth surfaces by Antonie van Leeuwenhoek resulted in the identification of microbial biofilms (Socransky et al.

1998; Costerton et al. 1999). During a study of natural marine bacteria populations, Zobell also discovered many microbes that are found attached to the solid surfaces than found in the surrounding medium. It has become evident that bacterial function and growth within a population is a fundamental aspect of bacterial survival and a typical life style of microorganisms (Davey and O'Toole 2000). The term "biofilm" was coined and described in 1978 (Costerton et al. 1978). Since then, it has been well documented that biofilm-associated microbes differ from their planktonic relatives in terms of their genes that are transcribed (Donlan 2002). The development of biofilms by the bacteria occurs on a number of different surfaces, such as natural aquatic, soil environments, living tissues, medical devices, or industrial or potable water piping systems (Flemming and Wingender 2001). Clusters of different microbial populations are found in almost all moist environments where the flow of nutrients is available and surface attachment is possible (Singh et al. 2006). Biofilms have been found to protect the microbial community from environmental stresses (Ahimou et al. 2007). Therefore, the formation of biofilms in natural and industrial environments allows bacteria to develop resistance to bacteriophages, amoebae, chemically diverse biocides, host immune responses, and antibiotics (Costerton 1999). These important characteristics have resulted in emerging biofilm science and biofilm engineering as an intensively developing area of research (Costerton 2003).

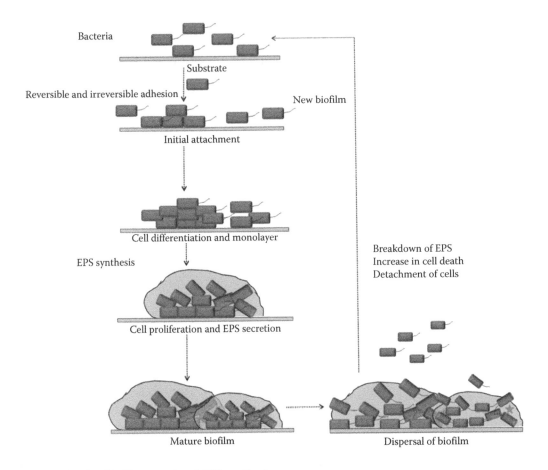

*Figure 17.3* Steps involved in biofilm-associated EPS synthesis.

Over 99% of microorganisms on the Earth live within the biopolymers consisting of a mixture of polymeric polysaccharide compounds produced by the aggregation of bacteria, algae, fungi, and protozoa. The formation of biofilms is a prerequisite for the existence of all microbial aggregates, and it is as an essential step in the survival of bacterial populations (Figure 17.3). The proportion of EPS in biofilms comprises between approximately 50% and 90% of the total organic matter (Donlan 2002). The composition and quantity of EPS production are dependent on the type of microorganisms, age of the biofilms, and the different environmental conditions under which the biofilms exist (Mayer et al. 1999). These include different levels of oxygen and nitrogen, extent of desiccation, temperature, pH, and availability of nutrients. The changing environment and existence of microorganisms in such diverse conditions suggest that they are able to respond to their environments and can change their EPS and adhesion abilities, depending on the properties of the surface onto which they attach. It has been reported that microbial colonization on solid surfaces can be affected by a diverse range of parameters. For example, the degree of colonization

on certain surfaces has been found to increase with surface roughness because the "valleys" present can allow the microbes to reside in a protected area with reduced shear forces and the surface roughness provides a surface with increased surface area for the bacterial attachment. Further, microorganisms have been found to attach more rapidly to hydrophobic and nonpolar surfaces than hydrophilic surfaces. Cell surface hydrophobicity, the presence of fimbriae and flagella, and the degree of EPS production are the other essential factors that have been shown to profoundly influence the rate and degree of attachment of microbial cells to different surfaces. Three types of forces involved in this process are electrostatic interactions, hydrogen bonds, and London dispersion forces (Flemming and Wingender 2001). These binding forces are likely to contribute to the overall stability of biofilm matrices. Different components of EPS have also been found to influence the extent to which microorganisms can adhere to both hydrophilic and hydrophobic surfaces. It has also been shown that the formation of EPS leads to irreversible attachment with different environmental surfaces (van Hullebusch et al. 2003; Romani et al. 2008).

## 17.3  Biosynthesis and genetic regulation of EPSs

The EPS synthesis by bacteria depends on the growth phase, mode of growth, nutrients, pH, temperature, salt, etc. It synthesizes EPS intracellularly throughout growth or in different growth phases. The synthesis of EPS involves a large number of genes, which code for enzymes and regulatory proteins (Mishra and Jha 2013). The biosynthesis of EPS occurs in four steps: transport of sugar inside the cell, phosphorylation of sugar, polymerization of sugar units, and export of EPS outside the cytoplasm (Madhuri and Prabhakar 2014). The EPS synthesis starts with the transport of monosaccharide (e.g., glucose) inside a bacterial cell by active transport, diffusion, or group translocation (Figure 17.4a). Inside the cytoplasm, sugar molecules are phosphorylated into sugar-6-phosphates, which are further converted into sugar-1-phosphates by phosphoglucomutases (Jolly et al. 2002). Thereafter, sugar-1-phosphate is converted

into a sugar nucleotide by the enzyme pyrophosphorylase. For example, glucose-1-phosphate is converted into uridine diphosphate glucose (UDP-glucose) by UDP-glucose pyrophosphorylase (Lieberman and Markovitz 1970; Jolly et al. 2002). UDP-glucose serves as the fundamental intermediate for interconversion of other sugars involved in polysaccharide assembly and synthesis. The bacterial polysaccharides consist of repeating units of sugar moieties that are synthesized by glycosyltransferases, which transfer a sugar moiety to a glycosyl carrier lipid (located in cytoplasm membrane). A lipid carrier is linked to a monosaccharide component via a pyrophosphate bridge of the terminal alcohol group of the carrier lipid (Campbell et al. 1997). In the last step, monosaccharide units of assembled polysaccharides are further modified by different enzymatic activities like acetylation, acylation, methylation, and sulfation and exuded from the cell in the form of loose slime or a capsule with the help of flippase, permease, or ABC transporters (Liu et al. 1996; Daniels et al. 1998; Sutherland 2001).

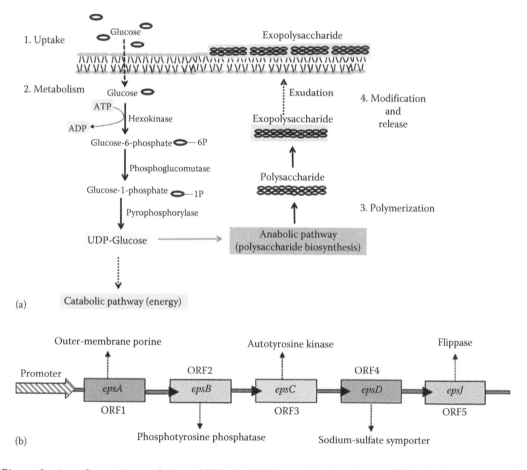

*Figure 17.4* Biosynthesis and genetic regulation of EPS: (a) EPS biosynthesis pathway. (With kind permission from **Springer Science+Business Media**: *The Prokaryotes,* Rosenberg, E., DeLong, E., Lory, S., Stackebrandt, E., and Thompson, F., eds., Microbial exopolysaccharides, 2013, pp. 179–192, Mishra, A. and Jha, B.) (b) Organization of *eps* gene clusters of *Halomonas maura.* (From Arco, Y. et al., *Microbiology,* 151, 2841, 2005. With permission.)

In general, EPS synthesis is regulated by gene(s), which are part of chromosomes. However, in some bacteria like *S. meliloti*, EPS synthesis by two-component regulatory systems (i.e., megaplasmids and chromosome) has been reported (Cheng and Walker 1998).

The genetic mechanism for the synthesis of EPS has been studied in a marine bacterium *Halomonas maura*. In the bacterium, *epsABCDJ* genes are part of *eps* gene cluster that regulates the biosynthesis of EPS known as the mauran (Figure 17.4b). They reported that the mauran is synthesized via a Wzy-like biosynthesis system and *epsA*, *epsB*, *epsC*, *epsD*, and *epsJ* codes for outer membrane porin, phosphotyrosine phosphatase, autotyrosine kinase, sodium–sulfate symporter, and flippase, respectively, which are collectively involved in the assembly and polymerization of the mauran (Arco et al. 2005). *epsA* gene product is a Wzy-like protein, which is a member of outer membrane auxiliary protein family. It is involved in the translocation of high-molecular-mass polysaccharides across the outer membrane. *epsB* gene product, phosphotyrosine phosphatase, dephosphorylates the consumed undecaprenyl pyrophosphate carrier so that it can again participate in the polymerization cycle. EpsC is a membrane-localized protein, which is involved in translocation, surface assembly, and maintenance of the polysaccharide chain length. In the bacterium, EpsA and EpsC contribute toward cell-surface assembly of the mauran. EpsD is a sodium–sulfate symporter, particularly sodium sulfate/carboxylate cotransporters (Markovich and Murer 2004). EpsJ acts as flippase, which transfers assembled lipid-linked repeating units of polysaccharides across the cytoplasmic membrane (Arco et al. 2005; Llamas et al. 2006).

In *Pseudoalteromonas* sp. SM9913, Yu et al. (2014) identified *epsT* gene that codes for UDP-glucose lipid carrier transferase. Deletion of *epsT* resulted in 73% decrease in EPS synthesis, indicating its importance in the EPS production in the bacterium. Bacteria can have multiple gene clusters, which regulate the biosynthesis of different types of EPS. For instance, in *Pseudomonas putida*, EPS synthesis is regulated by *pea* and *peb* gene clusters. These EPSs constitute a structural frame for the biofilm. Other gene clusters *alg* and *bcs* are also present in *P. putida*, which regulate the synthesis of alginate and cellulose, respectively (Nilsson et al. 2011). Similarly, EPS biosynthesis in *Pseudomonas aeruginosa* is controlled by *pel*, *psi*, and *alg* gene clusters (Darzins and Chakrabarty 1984; Jackson et al. 2004; Vasseur et al. 2005). *Vibrio fischeri* is a marine bacterium that lives in symbiotic association (in light organ) with the squid *Euprymna scolopes*. In the bacterium, polysaccharide synthesis is controlled by *syp* gene cluster, which consists of 18 genes *sypA* to *sypR*. *syp locus* codes for EPSs involved in colonization and biofilm formation (Shibata et al. 2012).

## 17.4  Distribution of EPS-producing marine bacteria

EPSs are the substantial component of the extracellular polymers, surrounding most microbial cells in the marine environment. With due advancement of time, increasing concern is laid upon the isolation of new EPS-producing bacteria from extreme environments like deep-sea hydrothermal vents, mangroves, hypersaline environments, cold water, and sediments.

### 17.4.1  Hydrothermal vents

Marine bacteria dwelling as microbial biofilms have various macromolecules, such as polysaccharides, proteins, nucleic acids, and lipids. These form the architectural matrix of intracellular space within microbial biofilms and unattached aggregates in the marine environment (Wingender et al. 1999). The primary components of the macromolecules are polysaccharides that generally represent 40%–95% of the extracellular polymeric substances (Flemming and Wingender 2001). EPS constitutes a large fraction of the reduced carbon reservoir in the ocean and enhances the survival of marine bacteria, influencing the physicochemical environment in proximity to the bacterial cell (Poli et al. 2010). Marine bacterial EPSs are involved in response to environmental stress, in recognition processes and cell–cell interactions, and in adherence of biofilms to surfaces (Weiner et al. 1995). Moreover, EPS helps in creating microenvironments for cell function, metabolism, and reproduction in case of host cell and microbial symbiosis. Thermophilic bacteria from marine habitats growing in hot and salty environments with huge temperature fluctuations have lipids, enzymes, and biopolymers with unique adaptation properties (Nicolaus et al. 2000). EPSs from thermophilic bacteria *Bacillus thermantarticus*, *Thermococcus litoralis*, and *Thermotoga maritima* were reported to have the ability to grow beyond normal boundaries of life and utilized for industrial purposes (Manca et al. 1996; Rinker and Kelly 2000). Nicolaus et al. (2002) isolated thermophilic bacteria from hot marine springs around Ischia (Italy) and from hydrothermal vents in the Gulf of Naples (Italy), which were able to produce EPSs. Additionally, EPSs from the extreme environments, especially Antarctic marine ecosystems, have been explored due to promising biotechnological applications (Nichols et al. 2005a). Nichols et al. (2005b) isolated a Gram-negative, aerobic, gliding, orange–yellow marine bacterium *Olleya marilimosa* from a particulate material sampled from the Southern Ocean, producing EPS in liquid culture.

Unfavorable conditions of high pressure, extreme temperatures, and heavy metals constitute the deep-sea hydrothermal vent environments. Hydrothermal deep-sea vents endeavor as a source of novel EPS-producing mesophilic and thermophilic heterotrophic bacteria. Deep-sea hydrothermal vents are areas of active tectonics with diverse physicochemical characteristics. The thermal gradients render the development of psychrophilic, mesophilic, and thermophilic microorganisms. Hydrothermal environment exposes the microorganisms to high hydrostatic pressure and high concentrations of toxic metals and sulfides. EPS synthesized from this origin will have novel chemical composition with a wide range of biological and metal-binding properties (Guezennec 2002). From a hydrothermal vent at a depth of 2600 m in North Fiji Basin, *Alteromonas macleodii*, an aerobic mesophilic bacterium, was isolated. It consisted of repeating units of hexasaccharide comprising of three uronosyl residues with a branch point at a galacturonosyl residue and a side chain terminated by a 4,6-*O*-(1-carboxyethylidene)-β-D-Man*p* having wide application in water treatment and metal removal. Furthermore, the xanthan-like EPS produced had applications as a food-thickening agent and had bone-healing properties (Raguénès et al. 1996; Rougeaux et al. 1998; Zanchetta and Guezennec 2001; Table 17.1).

A heterotrophic facultative sulfur-dependent hyperthermophilic archaeon from a shallow submarine thermal spring with an optimal growth temperature of 88°C was isolated. Its EPS contained mannose as the only monosaccharide constituent, which is a peculiar feature for the prokaryotes. In addition to that, sulfated EPS and EPSs such as sulphoevernan,

chondroitin sulfate, dextran sulfate, heparin, and mannan sulfate facilitated in the protection of eukaryotic cells from viruses, including human immunodeficiency virus type 1, by inhibiting virus particle adsorption to host cells (Ito et al. 1989). EPSs having unique properties are synthesized from a variety of species like *Bacillus, Haloferax, Haloarcula, Methanosarcina, Sulfolobus, T. maritima*, and *T. litoralis* (Maugeri et al. 2002). These bacterial EPSs have enhanced toxic metal chelation capability and are utilized as biosorbents in wastewater treatment alternative to physical and chemical methods. Usually, EPSs from marine bacteria like *Pseudomonas* and *Vibrio* are reported to have antitumor, antiviral, and immune-stimulating activities. The organisms obtained from deep-sea hydrothermal vents secrete a low-molecular-weight heparin like EPS that exhibits anticoagulant property. Polysaccharides constituting L-fucose are exhibited to have a potential role in the prevention of tumor cell colonization of lung, regulation of white blood cells formation, production of cosmeceuticals as skin moisturizing agents, and in synthesis of antigens for antibody production (de Jesús Paniagua-Michel et al. 2014). Free-radical depolymerization and sulfation of marine bacterial EPS from the extreme environment of hydrothermal vents activated the complement system, primarily GY 785 DROS. This can help in opening an outlook for the treatment of diseases caused by deregulation of the immune system and overactivation of the complement system (Courtois et al. 2014).

*Table 17.1* EPSs produced by organisms of hydrothermal vents and their applications in environmental biotechnology

| Organisms | EPS | Application | References |
|---|---|---|---|
| *Pseudoalteromonas* | Octasaccharide repeating unit with two side chains | Gelling properties | Rougeaux et al. (1999) |
| *Alteromonas macleodii* | Sulfated heteropolysaccharide, high uronic acids with pyruvate. The repeating unit is a branched hexasaccharide containing Glc, Man, Gal, GlcA, GalA, pyruvated mannose | Water treatment and metal removal | Rougeaux et al. (1998) |
| Facultative sulfur-dependent hyperthermophilic archaeon | Mannose sulphoevernan, chondroitin sulfate, dextran sulfate, heparin, and mannan sulfate | Inhibits virus particle adsorption to host cells | Ito et al. (1989) |
| *Thermococcus litoralis* | Mannose | Helps in biofilm formation | Rinker and Kelly (1996) |
| *Geobacillus* sp. | A pentasaccharide repeating unit (two of them with a gluco–galacto configuration and three with a *manno* configuration) | Pharmaceutical application | Nicolaus et al. (2002) |
| *Bacillus thermodenitrificans* | Trisaccharide repeating unit and a *mannopyranoside* configuration | Immunomodulation and antiviral activities | Arena et al. (2009) |
| *Bacillus thermantarticus, Thermococcus litoralis, Thermotoga maritima* | Thermophilic EPS | Industrial purposes | Manca et al. (1996); Rinker and Kelly (2000) |

### 17.4.2 Mangroves

The complex and dynamic ecosystems varying in salinity, water level, and nutrient availability constitute the mangrove environment with diverse microbial communities. They are dominant intertidal wetlands situated along coastlines of tropical and subtropical regions and form the significant sinks for pollution from freshwater and contaminated tidal water discharges (Bernard et al. 1996). The major group present in the mangrove regions includes Rhizobiales, Campylobacterales, Methylococcaceae, and Vibrionales (Gomes et al. 2010). The tropical mangroves constitute 91% of bacteria and fungi of the total microbial biomass, whereas algae and protozoa represent only 7% and 2%, respectively. *Pantoea agglomerans* strain KFS-9 produced a water-soluble EPS from mangrove forests, which is a protein-bound polysaccharide composed of arabinose, glucose, galactose, and gulcuronic acid showing free radical–scavenging ability. It established a high antioxidant activity by effectively quenching the free radicals, hydroxyl and superoxide. This demonstrated the protective effect of water-soluble EPS against UV radiation (Wang et al. 2007). *Pseudomonas* sp. and *Azotobacter* sp. were isolated from Pichavaram mangrove sediment, which synthesized alginate EPSs (Lakshmipriya and Sivakumar 2012). An EPS with an unusual sugar composition was produced from the bacteria *Vibrio neocaledonicus* NC470 screened from New Caledonia with high N-acetyl-hexosamines and uronic acid content and low amount of neutral sugar. This NC470 strain showed a high metal-binding capacity of 370 mg Cu(II) g$^{-1}$ EPS and up to 70 mg Ni(II) g$^{-1}$ EPS (Chalkiadakis et al. 2013). Another marine bacterium isolated from a microbial mat in Polynesian atoll synthesized a hexosamine-rich EPS, which could effectively bind 400 mg Cu(II) g$^{-1}$ EPS and 65 mg Ni(II) g$^{-1}$ EPS, respectively (Guezennec et al. 2011).

### 17.4.3 Hypersaline marine environment

The extreme habitats with high salt concentration, low oxygen concentration, high or low temperature, and sometimes very high pH constitute the hypersaline environment. The factors affecting biodiversity include pressure, low nutrient availability, solar radiation, and the presence of heavy metals and toxic compounds (Rodriguez-Valera 1988). Usually, the hypersaline environments have high salinity of alkaline pH within 9–11.5. They are inhabited by halotolerant microorganisms as well as halophilic microorganisms (0.5 and 2.5 M NaCl) and extreme halophiles (above 2.5 M NaCl). Moderate and extreme halophiles have been isolated not only from hypersaline ecosystems (salt lakes, marine salterns, and saline soils) but also from alkaline ecosystems (alkaline lakes). Various biochemical strategies to

adapt high saline conditions include compatible solutes that are used for the maintenance of cellular structure and function, like ectoine, bacteriorhodopsins, EPSs, hydrolases, and biosurfactants (Poli et al. 2010). The members of Halobacteriales, for example, *Haloarcula, Haloferax, Halobacterium, Halorubrum, Natronobacterium, Natronococcus,* and the species *Salinibacter ruber* are the EPS-producing bacteria. These bacteria play an important ecological role in the extreme environments.

Halophilic bacteria distributed in hypersaline environment require 3%–15% w/v NaCl for the proper growth (Kushner and Kamekura 1988). *Haloferax mediterranei* isolated from Mediterranean Sea synthesizes EPS constituting (1→4)-β-D-GlcpNAcA-(1→6)-α-D-Manp-(1→4)-β-D-GlcpNAcA-3-O-SO$_3^-$, which showed application in oil recovery, especially in oil deposits with high salinity (Parolis et al. 1996). *Hahella chejuensis* isolated from the marine sediment sample collected from Marado, Cheju Island, Republic of Korea, synthesized an EPS rich in glucose and galactose and effectively synthesized biosurfactant for the detoxification of polluted areas from petrochemical oils (Lee et al. 2001). Thus, the bioproducts such as EPSs, halophilic enzymes, and compatible solutes have enormous applications in biotechnology (Ventosa et al. 2008).

Many hypersaline EPS producers were discovered from Spain and Morocco of which *H. maura* and *Halomonas eurihalina* were the commonest (Bouchotroch et al. 2001; Quesada et al. 2004), whereas *Halomonas ventosae* (Martínez-Cánovas et al. 2004a) and *Halomonas anticariensis* (Martínez-Cánovas et al. 2004b) were also present. *Salipiger mucescens* was isolated from a saline soil bordering a saltern on the Mediterranean coast at Calblanque, which synthesized EPS on its cell surface (Martínez-Cánovas et al. 2004c). Ortega-Morales et al. (2007) reported the production of exopolymers from tropical intertidal biofilm bacteria. *Microbacterium* sp. and *Bacillus* sp. synthesized an EPS that was a glycoprotein and a true polysaccharide dominated by neutral sugars but with significant concentrations of uronic acids and hexosamines. EPS from *Microbacterium* sp. possesses a higher surfactant activity than that of commercial surfactants. Due to its anionic nature, it chelates cations helping in bioremediation. EPSs from *Bacillus* sp. have potential biomedical applications in tissue regeneration (Ortega-Morales et al. 2005).

A halophilic EPS-producing chemoheterotrophic bacterium *Palleronia marisminoris* was isolated from hypersaline soil bordering a saline saltern on the Mediterranean seaboard in Murcia (Spain) (Martínez-Checa et al. 2005). Three novel halophilic species *Idiomarina fontislapidosi, Idiomarina ramblicola,* and *Alteromonas hispanica* belonging to the family Alteromonadaceae were isolated, which effectively synthesized EPSs. The EPS of *A. hispanica* was composed of glucose, mannose, and xylose and was less viscous and pseudoplastic in behavior. Additionally, EPS of all three species showed emulsification activity and capacity to bind metals (Mata et al. 2008). *Halomonas alkaliantarctica* strain CRSS from Salt Lake in Cape Russell in Antarctica synthesized EPS of high viscosity in composition and it primarily constitutes glucose and fructose (Poli et al. 2007). *Halomonas almeriensis*–synthesized EPSs composed of mannose, glucose, and rhamnose. The EPS showed the presence of sulfates and rendered its use as a biodetoxifier and emulsifier (Llamas et al. 2012). Three *Halomonas* strains were isolated from the Deepwater Horizon site, which produced EPSs that exhibit amphiphilic properties. Amphiphilic EPSs aid in hydrocarbon degradation, and it was reported that *Halomonas* sp. TG39 increased the solubilization of aromatic hydrocarbons present in oil-contaminated surface waters (Gutierrez et al. 2013).

### 17.4.4   Cold marine environment

The deep sea is a major component of our planet's biosphere and represents 75% of the total volume of the oceans. These environments are influenced by high pressure, low temperature, and low nutrient concentration. Psychrotrophic bacteria possess the capability of growing above 20°C besides tolerating low temperature, whereas psychrophilic bacteria prefer to grow below 20°C (0°C–20°C) (Gounot 1986). In recent times, the deep-sea psychrotolerant bacteria living in abyssal ecological community have been investigated. These bacteria adapt to the changing deep-sea ecosystem by altering their morphology, physiology, and metabolism (Maugeri et al. 2002). *Pseudoalteromonas* strain SM9913 isolated from deep-sea sediment in the Bohai Gulf, Yellow Sea, China, synthesized linear arrangement of $\alpha$-(1 $\rightarrow$ 6) linkage of glucose with a high degree of acetylation, showing flocculation behavior and high biosorption capacity. The EPS yield was inversely proportional to temperature. The major sugar unit of the EPS was 6-linked glucose (61.8%); other sugar units present were terminal arabinofuranosyl (11.0%) and glucopyranosyl (11.2%) residues and a small amount of other sugar derivatives with a different structure as compared to the EPSs reported from other marine bacteria. Additionally, EPS enhanced the stability of the cold-adapted protease MCP-01 secreted by the same strain by preventing its autolysis. It also helped in binding with various metal ions like $Fe^{2+}$, $Zn^{2+}$, $Cu^{2+}$, and $Co^{2+}$. The EPS acts as a very good flocculating agent to conglomerate colloidal and suspended particles (Qin et al. 2007; Li et al. 2008). Two strains *Pseudoalteromonas* strain CAM025 and *Pseudoalteromonas* strain CAM036 were also isolated from particles collected in melted Antarctic sea and particles captured by plankton net towed through the Southern Ocean, respectively. *Pseudoalteromonas* strain CAM025 synthesized sulfated heteropolysaccharides with high levels of uronic acid with acetyl groups, whereas *Pseudoalteromonas* strain CAM036 also synthesized sulfated heteropolysaccharides with high levels of uronic acids with acetyl and succinyl groups. These effectively had cryoprotection ability and trace metal-binding activity (Mancuso Nichols et al. 2004). *Colwellia psychrerythraea* strain 34H isolated from Arctic marine sediments could tolerate the temperature from 8°C to 14°C at high pressures of 400–600 atm, and this extreme environmental conditions stimulated EPS production in the isolate. EPS production is not only a survival strategy but also a source of potent biotechnological applications (Marx et al. 2009). EPS from the deep-sea psychrophilic bacterium *Pseudoalteromonas* sp. SM9913 was investigated and found to effectively absorb $Pb^{2+}$ and $Cu^{2+}$. The maximum adsorption capacity of the EPS determined by Dubinin–Radushkevich isotherm equation for $Pb^{2+}$ and $Cu^{2+}$ was 243.3 and 36.7 mg/g, respectively. The primary functional groups in the bacteria for binding metal ions were C–O–C, acetyl, and hydroxyl groups of polysaccharide (Zhou et al. 2009).

### 17.4.5   Sediments and surface water

The water from stream, river, lake, wetland, and ocean constitutes the surface water. The surface water is enriched with the primary microflora such as *Alteromonas*, *Colwellia*, and *Pseudoalteromonas*. EPS-producing *Exiguobacterium* sp. was isolated and characterized from coastal (Mandapam) area of Pak Bay, Tamil Nadu, India, which showed the presence of amino sugars. Sucrose in the medium increased the production of EPSs (Karuppiah et al. 2014). Different species of marine bacteria dwelling in water and sediments produced a variety of EPSs, which represent a class of important products of growing interest for many industrial sectors. *Vibrio parahaemolyticus* was isolated from the marine water (Su and Liu 2007), and *Pseudomonas* sp. strain NCMB 2021 was isolated from Halifax, Nova Scotia (Wrangstadh et al. 1986), which synthesized EPS. *Pseudoalteromonas* sp. strain S9 was isolated from the marine sediment, and *Pseudoalteromonas atlantica* and *Hyphomonas* strain MHS-3 were isolated from shallow water sediments in Puget Sound (Quintero and Weiner 1995). A marine bacterium *Enterobacter cloacae* produced an EPS, which emulsified hexane, benzene, xylene, kerosene, paraffin oil, cottonseed oil, coconut oil, jojoba oil, castor oil, groundnut oil, and sunflower oil (Iyer et al. 2006). Another bioflocculant-producing marine bacterium *Virgibacillus* sp. was isolated from the marine sediment of Algoa Bay, which was a polysaccharide (Cosa et al. 2011). *Halomonas* sp. were isolated from Arabian

Sea as a potent biofilm former synthesizing EPS (Nisha and Thangavel 2014). Thus, various EPS-producing marine bacteria have been traced from various habitats and ecosystems.

## 17.5 Environmental applications

Marine ecosystems having huge microbial biodiversity, which produces EPS of various structures and compositions, are biotechnologically very relevant. The isolation and identification of new microorganisms provide wider opportunities for new industrial applications such as helping in hydrocarbon remediation, metal remediation, and also in biofouling control.

### 17.5.1 Hydrocarbon remediation

Marine environments are polluted by a variety of organic compounds, from both terrestrial and atmospheric sources. The contamination of ocean principally with crude oil remains a major threat. Every year, about 1.3 million tons of petroleum enters into the marine environment (NRC 2003; McGenity et al. 2012). The components of crude oil (particularly aromatic hydrocarbons fractions) are toxic to marine organisms (Carls et al. 1999; Heintz 2007). However, the microbial community residing nearby polluted sites is a vital player involved in the degradation of these chemicals. Microbes have adapted several strategies such as increase of population size, genetic modification, biofilm formation, and secretion of biosurfactant to degrade organic chemicals (Singh et al. 2006). Many marine microorganisms also produce EPS for the enhanced utilization of organic compounds as a sole carbon source. EPS extracted from marine bacteria can significantly influence the overall performance of bioremediation. It also helps in the attachment of microbial cells on the surface of xenobiotics and subsequently helps in their degradation (Ta-Chen et al. 2008). EPSs from marine bacteria have higher level of uronic acids, and they have the ability to interface with hydrophobic organic chemicals (Kennedy and Sutherland 1987; Janecka et al. 2002; Gutierrez et al. 2008, 2009).

*Alcanivorax borkumensis* is one of the leading oil-degrading marine bacterium as it can form biofilm at oil–water interface to increase bioavailability and degradation of oil (Schneiker et al. 2006). Bacteria under genus *Colwellia* secreted EPS over oil and formed flocs. Baelum et al. (2012) reported that *Colwellia* can synthesize EPS and degrade components of crude oil at extremely low temperature (5°C). EPSs are natural emulsifying agents and are important for remediation of hydrocarbons (Ta-Chen et al. 2008). Bacteria such as *H. eurihalina* and *Enterobacter cloacear* produced EPS with emulsifying properties (Calvo et al. 2002; Iyer et al. 2006).

EPS can enhance the solubility of hydrophobic substrates (e.g., PAHs, biphenyl) by numerous hydrophobic interactions (Pan et al. 2010). EPS produced by marine bacterium *Halomonas* sp. exhibited amphiphilic properties and high emulsifying qualities. The EPS produced by the bacterium increases the solubilization and degradation of many hydrophobic organic pollutants such as phenanthrene, fluorene, pyrene, and biphenyl (Gutierrez et al. 2013). Table 17.2 illustrates the list of EPS-producing and hydrocarbon-degrading marine bacteria. Extracellular enzymes immobilized in EPS play a beneficial role in biodegradation. Abari et al. (2012) reported that EPS associated with toluene-utilizing marine bacterium *Sporosarcina halophila* retains peroxidase enzymes such as laccase and catalase (present in EPS), which considerably affected the degradation of toluene by the bacterium.

EPS produced by a bacterium can also increase the degradation potential of other bacteria. Iwabuchi et al. (2002) reported that EPS produced by marine bacterium *Rhodococcus rhodochrous* (named as S-2 EPS) increased the degradation of multiple aromatic components in crude oil by native consortia of seawater. They observed 50% increase in crude oil degradation rate in the presence of S-2 EPS. EPS can also enhance microbial hydrocarbon

*Table 17.2* List of EPS-producing marine bacteria associated with hydrocarbon degradation

| Class | Genus | Source | Hydrocarbon degradation | References |
|---|---|---|---|---|
| Gammaproteobacteria | *Enterobacter* | Sediment | *n*-Hexadecane | Hua et al. (2010); Iyer et al. (2006) |
| | *Halomonas* | Seawater | PAHs, alkane, biphenyl | Calvo et al. (2002); Gutierrez et al. (2007, 2013) |
| | *Pseudomonas* | Seawater | PAHs | Mangwani et al. (2014) |
| | *Colwellia* | Deep seawater | Crude oil | Marx et al. (2009) |
| | *Alcanivorax* | Seawater | Alkanes | Schneiker et al. (2006) |
| | *Pseudoalteromonas* | Deep seawater | PAHs, alkane | Chronopoulou et al. (2015) |
| Actinobacteria | *Rhodococcus* | Seawater | *n*-Hexadecane, PAHs | Iwabuchi et al. (2000, 2002) |

degradation in soil microcosm and provide a prospective tool for *in situ* bioremediation (Ron and Rosenberg 2002; Venosa and Zhu 2003; Cappello et al. 2012).

## 17.5.2   Metal remediation

EPSs are composed of various negatively charged functional groups and vary in their interaction with ionic compounds (Zhang et al. 2010). Thus, EPSs are able to form organometallic complex by electrostatic attraction with heavy metals (Beech and Sunner 2004). This adds a promising application of EPS for toxic heavy metals and radionuclide remediation (Lloyd and Renshaw 2005). Occasionally, EPS also includes small fractions of nonsugar components such as proteins and nucleic acids, which are important for tertiary structure and physical properties of the EPS (Iyer et al. 2005). The occurrence of these nonsugar residues imparts acidic nature to the EPS and the presence of uronic acid confers negative surface charge (Decho and Lopez 1993). Table 17.3 describes the report of bacterial EPS for potential application in metal biosorption.

Qin et al. (2007) reported the structure of EPS from a psychrotolerant bacterium *Pseudoalteromonas* sp. SM9913 isolated from deep sea. The EPS from bacterium was rich in acidic polysaccharides with net negative charge. The EPSs from bacterium have high level of acetyl group (polyionic nature) and were able to bind a range of metal ions such as $Zn^{2+}$, $Cu^{2+}$, and $Co^{2+}$. In another study, Gutierrez et al. (2012) reported the metal-binding properties of the EPS produced by *Halomonas* sp. TG39. The EPS from the bacterium was rich in uronic acid which efficiently bound trace metals (e.g., Fe, Zn, Mn, and Si) and toxic metals (e.g., Cd and Pb). Thus, in the marine environment, EPS also helps in trace metal sequestration necessary for cellular functioning (Gutierrez et al. 2012). Maugeri et al. (2002) characterized

EPS from *Bacillus licheniformis* isolated from marine hot spring. The EPS from bacterium has mannopyranoside configuration, and it could immobilize (or resist) $Cd^{2+}$, $Zn^{2+}$, $As^{2+}$, and $Hg^{2+}$.

EPS produced by marine sulfate-reducing bacteria (SRB) can bind with heavy metals such as Cr, Ni, and Mo. The binding efficiency of Mo to SRB-EPS is considerably greater as compared to Cr and Ni (Beech and Cheung 1995). Binding of heavy metals (Pb and Cd) to EPS produced by purple nonsulfur bacterium *Rhodobium marinum* was also reported (Panwichian et al. 2011). A photosynthetic marine bacterium *Rhodovulum* sp. PS88 has been reported to remove Cd even at high NaCl concentration (Watanabe et al. 2003). EPSs from this strain consist of RNA polymers. Thus, phosphoryl and hydroxyl groups of EPS are involved in Cd biosorption by PS88. The binding of divalent metal ions to EPS is affected by many factors such as pH, presence of other ions, and composition of polysaccharides. The presence of sulfate esters and pyruvate in EPS also increases the metal-binding capacity (Osman and Fett 1989; Loaec et al. 1997, 1998). The ionic radius of metals is also critical for complexation of metals to EPS. The 100–110 pm of ionic radius is suitable for EPS–metal complex (Loaec et al. 1998). The use of EPS for heavy metal removal through biosorption is an economical alternative to chemical methods such as precipitation, coagulation, and ion exchange (Pal and Paul 2008). Thus, a huge diversity of marine bacterial EPS in terms of structure and chemical composition provides a prospective way for toxic metal remediation.

## 17.5.3   Biofouling control

Biological fouling is the attachment of organism(s) to the submerged or wetted artificial surfaces. Biofouling can occur by two different ways: microbial fouling

*Table 17.3* List of EPS-producing marine bacteria associated with heavy metal bioremediation

| Class | Genus | Source | Metal biosorption | References |
|---|---|---|---|---|
| Gammaproteobacteria | *Enterobacter* | Sediment | Cd, Cu, Co, Cr | Iyer et al. (2005, 2006) |
| | *Alteromonas* | Deep-sea hydrothermal vent | Zn, Cd, Pb | Loaec et al. (1997) |
| | *Marinobacter* | | Pb and Cu | Bhaskar and Bhosle (2006) |
| | *Pseudoalteromonas* | Deep-sea sediments | Zn, Cu, Co | Qin et al. (2007) |
| | *Pseudomonas* | Sediments | Cd | Chakraborty and Das (2014) |
| | *Halomonas* | Estuary | Cu, Zn, Pb, Cd, U | Gutierrez et al. (2012) |
| | Sulfate-reducing bacteria | Corroded surface | Cr, Ni, Mo, Cu, Zn | Beech and Cheung (1995); Pal and Paul (2008) |
| Alphaproteobacteria | *Rhodobium* | Seawater | Cd, Cu, Pb, Zn | Panwichian et al. (2011) |
| | *Rhodovulum* | Sediments | Cd | Watanabe et al. (2003) |
| Bacilli | *Bacillus* | Seawater | Hg, Zn | Maugeri et al. (2002) |

(adhesion of microorganisms to the surface) and macrofouling (adhesion of seaweed, mollusks, mussels, or other organisms) (Delauney et al. 2010). It is an ongoing and crucial problem for surfaces in contact with water, creating huge economic losses (Yebra et al. 2004; Cao et al. 2011). The succession of fouling starts with the adsorption of organic and inorganic chemicals to the immersed surfaces (primary film). After that, microbial cells adhere to the primary film, accumulate, and produce polymeric matrix forming a microbial film over the surface. In the next step, multicellular species, microalgae, and debris get attached to the microbial film. At the last stage, large marine organisms such as macroalgae, barnacles, and mussels attach to the preexisting microbial film (Delauney et al. 2010).

There are many techniques used to prevent fouling of artificial surfaces such as the use of biocide-coated surfaces, photocatalytic material coating, and energy methods (Nakayama et al. 1998; Morris and Walsh 2000; Bearinger et al. 2003). However, natural antifouling chemicals produced by aquatic organisms or plants are the most promising methods for biofouling control. Coating of surfaces with biological polymers or secondary metabolites can also prevent the formation of biofilm and subsequent biofouling. Guezennec et al. (2012) used EPS from *Alteromonas*, *Pseudomonas*, and *Vibrio* spp. as antibiofouling coating. They reported that EPS can inhibit the primary colonization of bacteria, thereby minimizing successive biofouling. The EPSs used as antibiofouling coating have 28%–41% of uronic acid and are rich in neutral sugars such as fucose, glucose, galactose, rhamnose, and mannose. The presence of polysaccharide film changes the hydrophobic/hydrophilic balance, which is important for adhesion of cells to surfaces (Yaskovich 1998; Guezennec et al. 2012). *Vibrio alginolyticus*, *V. proteolyticus*, and *V. vulnificus* also produce antibiofouling EPSs (Qian et al. 2006; Kim et al. 2011). EPS extracted from *V. alginolyticus* prevented larval attachment and metamorphosis of marine invertebrates (e.g., *Hydroides elegans*, *Balanus amphitrite*, and *Bugula neritina*) causing fouling (Qian et al. 2006). EPS produced by marine bacteria are nontoxic and do not have biocidal activity. Their application will not cause any harm to native organisms residing nearby the EPS-coated surface. Thus, EPS from marine bacteria can act as antibiofouling mediators in a nontoxic manner.

## 17.6   Summary

The marine environment is a rich source of microbial diversity, which can be used for the production of novel compounds that are useful for industrial and environmental applications. In addition to environmental aspects, they also have antitumor, antiulcer, immunomodulatory, antiviral, and cholesterol-lowering activities. Extremes of temperature, salinity, and water activity flourish diverse microbial life. Thus, biochemical products (i.e., EPS) from these polyextremophilic microbes are also versatile in terms of structure and functionality. Diverse assemblages of marine bacteria that are synthesizing EPS have potential functions, and their production is not limited to taxa. Primarily, diversity is seen in the monomeric compositions, linkage bonds, and associated conjugates. The intrinsic functions include morphological, structural, and protective functions, while applied functions are seen in human usage, medical, cosmetics, pharmaceutical, dairy products, and other forms of industrial and environmental applications. Although a myriad of applications are available for the EPS produced by bacteria, it is vital with respect to human usage that the EPS meets generally regarded as safe status or at least has a cost-effective means of neutralizing toxic constituents in cases of environmental applications such as in water (waste and municipal) treatment, metal remediation, and biofouling control. Production cost, largely, has been a limiting factor in the nonindustrial application of several prospective EPSs, and this is exemplified in the bioflocculation process. However, the search for bacteria with high EPS and EPS conjugate bioflocculant yields is still ongoing. Genetic and metabolic engineering applied together supported by statistical tools and predictive studies can lead to promising production of functional EPS with environment restoration ability.

## *Acknowledgments*

The authors are thankful to the authorities of NIT, Rourkela, for providing facilities and to the Department of Biotechnology, Government of India, for research grants on marine bacterial biofilm-based bioremediation.

## *References*

Abari, A.H., G. Emtiazi, and S.M. Ghasemi. 2012. The role of exopolysaccharide, biosurfactant and peroxidase enzymes on toluene degradation by bacteria isolated from marine and wastewater environments. *Jundishapur Journal of Microbiology* 5:479–485.

Abbasi, A. and S. Amiri. 2008. Emulsifying behaviour of an exopolysaccharide produced by *Enterobacter cloacae*. *African Journal of Biotechnology* 7:1574–1576.

Ahimou, F., M.J. Semmens, G. Haugstad, and P.J. Novak. 2007. Effect of protein, polysaccharide, and oxygen concentration profiles on biofilm cohesiveness. *Applied and Environmental Microbiology* 73:2905–2910.

Arco, Y., I. Llamas, F. Martínez-Checa, M. Argandoña, E. Quesada, and A. del Moral. 2005. *epsABCJ* genes are involved in the biosynthesis of the exopolysaccharide mauran produced by *Halomonas Maura*. *Microbiology* 151:2841–2851.

Arena, A., C. Gugliandolo, G. Stassi, B. Pavone, D. Iannello, G. Bisignano, and T.L. Maugeri. 2009. An exopolysaccharide produced by *Geobacillus thermodenitrificans* strain B3-72: Antiviral activity on immunocompetent cells. *Immunology Letters* 123:132–137.

Baelum, J., S. Borglin, R. Chakraborty, J.L. Fortney, R. Lamendella et al. 2012. Deep-sea bacteria enriched by oil and dispersant from the Deepwater Horizon spill. *Environmental Microbiology* 14:2405–2416.

Bearinger, J.P., S. Terrettaz, R. Michel, N. Tirelli, H. Vogel, M. Textor, and J.A. Hubbell. 2003. Chemisorbed poly(propylene sulphide)-based copolymers resist biomolecular interactions. *Nature Materials* 2:259–264.

Beech, I. and C.S. Cheung. 1995. Interactions of exopolymers produced by sulphate-reducing bacteria with metal ions. *International Biodeterioration and Biodegradation* 35:59–72.

Beech, I.B. and J. Sunner. 2004. Biocorrosion: Towards understanding interactions between biofilms and metals. *Current Opinion in Biotechnology* 15:181–186.

Bernard, D., H. Pascaline, and J.J. Jeremie. 1996. Distribution and origin of hydrocarbons in sediments from lagoons with fringing mangrove communities. *Marine Pollution Bulletin* 32:734–739.

Bhaskar, P. and N.B. Bhosle. 2006. Bacterial extracellular polymeric substance (EPS): A carrier of heavy metals in the marine food-chain. *Environment International* 32:191–198.

Bouchotroch, S., E. Quesada, A. del Moral, I. Llamas, and V. Béjar. 2001. *Halomonas maura* sp. nov., a novel moderately halophilic, exopolysaccharide-producing bacterium. *International Journal of Systematic and Evolutionary Microbiology* 51:1625–1632.

Brown, A.D. 1961. The peripheral structures of gram-negative bacteria: I. Cell wall protein and the action of a lytic enzyme system of a marine pseudomonad. *Biochimica et Biophysica Acta* 48(2):352–361.

Calvo, C., F. Martínez-Checa, F. Toledo, J. Porcel, and E. Quesada. 2002. Characteristics of bioemulsifiers synthesised in crude oil media by *Halomonas eurihalina* and their effectiveness in the isolation of bacteria able to grow in the presence of hydrocarbons. *Applied Microbiology and Biotechnology* 60:347–351.

Campbell, J.A., G.J. Davies, V. Bulone, and B. Henrissat. 1997. A classification of nucleotide-diphospho-sugar glycosyltransferases based on amino acid sequence similarities. *Biochemical Journal* 326:929.

Cao, S., J. Wang, H. Chen, and D. Chen. 2011. Progress of marine biofouling and antifouling technologies. *Chinese Science Bulletin* 56:598–612.

Cappello, S., M. Genovese, C. Della Torre, A. Crisari, M. Hassanshahian, S. Santisi, R. Calogero, and M.M. Yakimov. 2012. Effect of bioemulsificant exopolysaccharide (EPS2003) on microbial community dynamics during assays of oil spill bioremediation: A microcosm study. *Marine Pollution Bulletin* 64:2820–2828.

Carls, M.G., S.D. Rice, and J.E. Hose. 1999. Sensitivity of fish embryos to weathered crude oil: Part I. Low-level exposure during incubation causes malformations, genetic damage, and mortality in larval pacific herring (*Clupea pallasi*). *Environmental Toxicology and Chemistry* 18:481–493.

Chakraborty, J. and S. Das. 2014. Characterization and cadmium-resistant gene expression of biofilm-forming marine bacterium *Pseudomonas aeruginosa* JP-11. *Environmental Science and Pollution Research* 21:14188–14201.

Chalkiadakis, E., R. Dufourcq, S. Schmitt, C. Brandily, N. Kervarec, D Coatanea, and C. Simon-Colin. 2013. Partial characterization of an exopolysaccharide secreted by a marine bacterium, *Vibrio neocaledonicus* sp. nov., from New Caledonia. *Journal of Applied Microbiology* 114(6):1702–1712.

Chapot-Chartier, M.P. and S. Kulakauskas. 2014. Cell wall structure and function in lactic acid bacteria. *Microbial Cell Factories* 13(1):S9.

Cheng, H.P. and G.C. Walker. 1998. Succinoglycan is required for initiation and elongation of infection threads during nodulation of alfalfa by *Rhizobium meliloti*. *Journal of Bacteriology* 180(19):5183–5191.

Christensen, B.E. 1989. The role of extracellular polysaccharides in biofilms. *Journal of Biotechnology* 10:181–202.

Chronopoulou, P.M., G.O. Sanni, D.I. Silas-Olu, J.R. Meer, K.N. Timmis, C.P. Brussaard, and T.J. McGenity. 2015. Generalist hydrocarbon-degrading bacterial communities in the oil-polluted water column of the North Sea. *Microbial Biotechnology* 8(3):434–447.

Cosa, S., L.V. Mabinya, A.O. Olaniran, O.O. Okoh, K. Bernard, S. Deyzel, and A.I. Okoh. 2011. Bioflocculant production by *Virgibacillus* sp. Rob isolated from the bottom sediment of Algoa Bay in the Eastern Cape, South Africa. *Molecules* 16(3):2431–2442.

Costerton, J.W., K.J. Cheng, G.G. Geesy, T.I. Ladd, J.C. Nickel et al. 1987. Bacterial biofilms in nature and disease. *Annual Review of Microbiology* 41:435–464.

Costerton, J.W., G.G. Geesey, and K.J. Cheng. 1978. How bacteria stick. *Scientific American* 238:86–95.

Costerton, J.W., P.S. Stewart, and E.P. Greenberg. 1999. Bacterial biofilms: A common cause of persistent infections. *Science* 284:1318–1322.

Costerton, W., R. Veeh, M. Shirtliff, M. Pasmore, C. Post, and G. Ehrlich. 2003. The application of biofilm science to the study and control of chronic bacterial infections. *Journal of Clinical Investigation* 112:1466–1477.

Courtois, A., C. Berthou, J. Guézennec, C. Boisset, and A. Bordron. 2014. Exopolysaccharides isolated from hydrothermal vent bacteria can modulate the complement system. *PloS One* 9(4):e94965.

Cuthbertson, L., V. Kos, and C. Whitfield. 2010. ABC transporters involved in export of cell surface glycoconjugates. *Microbiology and Molecular Biology Reviews* 74:341–362.

Cuthbertson, L., I.L. Mainprize, J.H. Naismith, and C. Whitfield. 2009. Export of extracellular polysaccharides in gram-negative bacteria; the pivotal roles of OPX and PCP protein families. *Microbiology and Molecular Biology Reviews* 73:155–177.

Daniels, C., C. Vindurampulle, and R. Morona. 1998. Overexpression and topology of the *Shigella flexneri* O-antigen polymerase (Rfc/Wzy). *Molecular Microbiology* 28:1211–1222.

Darzins, A. and A.M. Chakrabarty. 1984. Cloning of genes controlling alginate biosynthesis from a mucoid cystic fibrosis isolate of *Pseudomonas aeruginosa*. *Journal of Bacteriology* 159:9–18.

Dash, H.R., N. Mangwani, J. Chakraborty, S. Kumari, and S. Das. 2013. Marine bacteria: Potential candidates for enhanced bioremediation. *Applied Microbiology and Biotechnology* 97(2):561–571.

Dash, H.R., N. Mangwani, and S. Das. 2014. Characterization and potential application in mercury bioremediation of highly mercury-resistant marine bacterium *Bacillus thuringiensis* PW-05. *Environmental Science and Pollution Research* 21:2642–2653.

Davey, M.E. and G.A. O'Toole. 2000. Microbial biofilms: From ecology to molecular genetics. *Microbiology and Molecular Biology Reviews* 64:847–867.

de Carvalho, C.C. and P. Fernandes. 2010. Production of metabolites as bacterial responses to the marine environment. *Marine Drugs* 8:705–727.

de Jesús Paniagua-Michel, J., J. Olmos-Soto, and E.R. Morales-Guerrero. 2014. Algal and microbial exopolysaccharides: New insights as biosurfactants and bioemulsifiers. *Marine Carbohydrates: Fundamentals and Applications* 73:221.

Decho, A.W. 1990. Microbial exopolymer secretions in ocean environments: Their role (s) in food webs and marine processes. *Oceanography and Marine Biology: An Annual Review* 28:73–153.

Decho, A.W. and G.R. Lopez. 1993. Exopolymer microenvironments of microbial flora: Multiple and interactive effects on trophic relationships. *Limnology and Oceanography* 38:1633–1645.

Delauney, L., C. Compere, and M. Lehaitre. 2010. Biofouling protection for marine environmental sensors. *Ocean Science* 6:503–511.

Donlan, R.M. 2002. Biofilms: Microbial life on surfaces. *Emerging Infectious Diseases* 8:881–890.

Dumitriu, S. 2005. *Polysaccharides: Structural Diversity and Functional Versatility*. CRC Press, Marcel Dekker, New York, pp. 1–40.

Eronen-Rasimus, E., H. Kaartokallio, C. Lyra, R. Autio, H. Kuosa, G.S. Dieckmann, and D.N. Thomas. 2014. Bacterial community dynamics and activity in relation to dissolved organic matter availability during sea-ice formation in a mesocosm experiment. *Microbiology Open* 3:139–156.

Flemming, H.C. and J. Wingender. 2001. Relevance of microbial extracellular polymeric substances (EPSs)—Part I: Structural and ecological aspects. *Water Science and Technology* 43:1–8.

Gomes, N.C.M., D.F.R. Cleary, F.N. Pinto, C. Egas, A. Almeida et al. 2010. Taking root: Enduring effect of rhizosphere bacterial colonization in mangroves. *PLoS One* 5:14065.

Gounot, A.M. 1986. Psychrophilic and psychrotrophic microorganisms. *Experientia* 42:1192–1197.

Guezennec, J. 2002. Deep-sea hydrothermal vents: A new source of innovative bacterial exopolysaccharides of biotechnological interest? *Journal of Industrial Microbiology and Biotechnology* 29(4):204–208.

Guezennec, J., J.M. Herry, A. Kouzayha, E. Bachere, M.W. Mittelman, and M.N. Bellon-Fontaine. 2012. Exopolysaccharides from unusual marine environments inhibit early stages of biofouling. *International Biodeterioration and Biodegradation* 66:1–7.

Guezennec, J., X. Moppert, G. Raguenes, L. Richert, B. Costa, and C. Simon-Colin. 2011. Microbial mats in French Polynesia and their biotechnological applications. *Process Biochemistry* 46:16–22.

Gutierrez, T., D. Berry, T. Yang, S. Mishamandani, L. McKay, A. Teske, and M.D. Aitken. 2013. Role of bacterial exopolysaccharides (EPS) in the fate of the oil released during the deepwater horizon oil spill. *PloS One* 8(6):e67717.

Gutierrez, T., D.V. Biller, T. Shimmield, and D.H. Green. 2012. Metal binding properties of the EPS produced by *Halomonas* sp. TG39 and its potential in enhancing trace element bioavailability to eukaryotic phytoplankton. *BioMetals* 25:1185–1194.

Gutierrez, T., G. Morris, and D.H. Green. 2009. Yield and physicochemical properties of EPS from *Halomonas* sp. strain TG39 identifies a role for protein and anionic residues (sulfate and phosphate) in emulsification of n-hexadecane. *Biotechnology and Bioengineering* 103:207–216.

Gutierrez, T., B. Mulloy, K. Black, and D.H. Green. 2007. Glycoprotein emulsifiers from two marine Halomonas species: chemical and physical characterization. *Journal of Applied Microbiology* 103(5):1716–1727.

Gutierrez, T., T. Shimmield, C. Haidon, K. Black, and D.H. Green. 2008. Emulsifying and metal ion binding activity of a glycoprotein exopolymer produced by *Pseudoalteromonas* sp. strain TG12. *Applied and Environmental Microbiology* 74:4867–4876.

Harz, H., K. Burgdorf, and J.V. Holtje. 1990. Isolation and separation of the glycan strands from murein of *Escherichia coli* by reversed-phase high-performance liquid chromatography. *Analytical Biochemistry* 190:120–128.

Heintz, R.A. 2007. Chronic exposure to polynuclear aromatic hydrocarbons in natal habitats leads to decreased equilibrium size, growth, and stability of pink salmon populations. *Integrated Environmental Assessment and Management* 3:351–363.

Howard, C.J. and A.A. Glynn. 1971. The virulence for mice of strains of *Escherichia coli* related to the effects of K antigens on their resistance to phagocytosis and killing by complement. *Immunology* 20:767–777.

Hua, X., Z. Wu, H. Zhang, D. Lu, M. Wang, Y. Liu, and Z. Liu. 2010. Degradation of hexadecane by *Enterobacter cloacae* strain TU that secretes an exopolysaccharide as a bioemulsifier. *Chemosphere* 80:951–956.

Ito, M., M. Baba, K. Hirabayashi, T. Matsumoto, M. Suzuki, S. Suzuki, S. Shigeta, E. de Clercq. 1989. *In vitro* activity of mannan sulfate, a novel sulfated polysaccharide, against human immunodeficiency virus type I and other enveloped viruses. *European Journal of Clinical Microbiology* 8:171–173.

Iwabuchi, N., M. Sunairi, H. Anzai, M. Nakajima, and S. Harayama. 2000. Relationships between colony morphotypes and oil tolerance in *Rhodococcus rhodochrous*. *Applied and Environmental Microbiology* 66:5073–5077.

Iwabuchi, N., M. Sunairi, M. Urai, C. Itoh, H. Anzai, M. Nakajima, S. Harayama. 2002. Extracellular polysaccharides of *Rhodococcus rhodochrous* S-2 stimulate the degradation of aromatic components in crude oil by indigenous marine bacteria. *Applied and Environmental Microbiology* 68:2337–2343.

Iyer, A., K. Mody, and B. Jha. 2005. Biosorption of heavy metals by a marine bacterium. *Marine Pollution Bulletin* 50:340–343.

Iyer, A., K. Mody, and B. Jha. 2006. Emulsifying properties of a marine bacterial exopolysaccharide. *Enzyme and Microbial Technology* 38:220–222.

Jackson, K.D., M. Starkey, S. Kremer, M.R. Parsek, and D.J. Wozniak. 2004. Identification of *psl*, a locus encoding a potential exopolysaccharide that is essential for *Pseudomonas aeruginosa* PAO1 biofilm formation. *Journal of Bacteriology* 186:4466–4475.

Janecka, J., M.B. Jenkins, N.S. Brackett, L.W. Lion, and W.C. Ghiorse. 2002. Characterization of a *Sinorhizobium* isolate and its extracellular polymer implicated in pollutant transport in soil. *Applied and Environmental Microbiology* 68:423–426.

Jann, B. and K. Jann. 1990. Structure and biosynthesis of the capsular antigens of *Escherichia coli*. In *Bacterial Capsules*. Springer, Berlin, Germany, pp. 19–42.

Jenkinson, F.F. 1994. Adherence and accumulation of oral *Streptococci*. *Trends in Microbiology* 2:209–212.

Jensen, P.R. and W. Fenical. 1994. Strategies for the discovery of secondary metabolites from marine bacteria: Ecological perspectives. *Annual Review of Microbiology* 48:559–584.

Jolly, L., S.J. Vincent, P. Duboc, and J.R. Neeser. 2002. Exploiting exopolysaccharides from lactic acid bacteria. *Antonie Van Leeuwenhoek* 82:367–374.

Kainth, P. and R.S. Gupta. 2005. Signature proteins that are distinctive of alpha proteobacteria. *BMC Genomics* 6(1):94.

Karuppiah, P., V. Venkatasamy, and T. Ramasamy. 2014. Isolation and characterization of exopolysaccharide producing bacteria from Pak Bay (Mandapam). *International Journal of Oceanography and Marine Ecological System* 3:1–8.

Kennedy, A.F. and I.W. Sutherland. 1987. Analysis of bacterial exopolysaccharides. *Biotechnology and Applied Biochemistry* 9:12–19.

Kim, M., J.M. Park, H.J. Um, K.H. Lee, H. Kim, J. Min, and Y.H. Kim. 2011. The antifouling potentiality of galactosamine characterized from *Vibrio vulnificus* exopolysaccharide. *Biofouling* 27:851–857.

Koch, A.L. 2000. Simulation of the conformation of the murein fabric: The oligoglycan, penta-muropeptide, and cross-linked nona-muropeptide. *Archives of Microbiology* 174:429–439.

Kroncke, K.D., J.R. Golecki, and K. Jann. 1990. Further electron microscopic studies on the expression of *Escherichia coli* group II capsules. *Journal of Bacteriology* 172(6):3469–3472.

Kushner, D.J. and M. Kamekura. 1988. Physiology of halophilic bacteria. In *Halophilic Bacteria*, Rodríguez-Valera, F. (ed.). CRC Press, Boca Raton, FL, pp. 109–138.

Lakshmipriya, V.P. and P.K. Sivakumar. 2012. Isolation and characterization of total heterotrophic bacteria and exopolysaccharide produced from mangrove ecosystem. *International Journal of Pharmaceutical and Biological Archives* 3(3):679–684.

Lee, H.K.; J. Chun, E.J. Moon, S.H. Ko, D.S. Lee, H.S. Lee, and K.S. Bae. 2001. *Hahella chejuensis* gen. nov., sp. nov., an extracellular-polysaccharide-producing marine bacterium. *International Journal of Systematic and Evolutionary Microbiology* 51:661–666.

Li, W.W., W.Z. Zhou, Y.Z. Zhang, J. Wang, and X.B. Zhu. 2008. Flocculation behavior and mechanism of an exopolysaccharide from the deep-sea psychrophilic bacterium *Pseudoalteromonas* sp. SM9913. *Bioresource Technology* 99(15):6893–6899.

Lieberman, M.M. and A. Markovitz. 1970. Depression of guanosine diphosphate-mannose pyrophosphorylase by mutations in two different regulator genes involved in capsular polysaccharide synthesis in *Escherichia coli* K-12. *Journal of Bacteriology* 101:965–972.

Liu, D., R.A. Cole, and P.R. Reeves. 1996. An O-antigen processing function for Wzx (RfbX): A promising candidate for O-unit flippase. *Journal of Bacteriology* 178:2102–2107.

Llamas, I., H. Amjres, J.A. Mata, E. Quesada, and V. Béjar. 2012. The potential biotechnological applications of the exopolysaccharide produced by the halophilic bacterium *Halomonas almeriensis*. *Molecules* 17(6):7103–7120.

Llamas, I., A. Del Moral, F. Martínez-Checa, Y. Arco, S. Arias, and E. Quesada. 2006. *Halomonas maura* is a physiologically versatile bacterium of both ecological and biotechnological interest. *Antonie Van Leeuwenhoek* 89:395–403.

Lloyd, J.R. and J.C. Renshaw. 2005. Bioremediation of radioactive waste: Radionuclide–microbe interactions in laboratory and field-scale studies. *Current Opinion in Biotechnology* 16:254–260.

Loaec, M., R. Olier, and J. Guezennec. 1997. Uptake of lead, cadmium and zinc by a novel bacterial exopolysaccharide. *Water Research* 31:1171–1179.

Loaec, M., R. Olier, and J. Guezennec. 1998. Chelating properties of bacterial exopolysaccharides from deep-sea hydrothermal vents. *Carbohydrate Polymers* 35:65–70.

Mack, D., W. Fischer, A. Krokotsch, K. Leopold, R. Hartmann, H. Egge, and R. Laufs. 1996. The intercellular adhesin involved in biofilm accumulation of *Staphylococcus epidermidis* is a linear β-1,6-linked glucosaminoglycan: Purification and structural analysis. *Journal of Bacteriology* 178:175–183.

Madhuri, K. and K.V. Prabhakar. 2014. Microbial exopolysaccharides: Biosynthesis and potential applications. *Oriental Journal of Chemistry* 30:1401–1410.

Manca, M.C., L. Lama, R. Improta, E. Esposito, A. Gambacorta, and B. Nicolaus. 1996. Chemical composition of two exopolysaccharides from *Bacillus thermoantarcticus*. *Applied and Environmental Microbiology* 62:3265–3269.

Mancuso Nichols, C.A., S. Garon, J.P. Bowman, G. Raguénès, and J. Guézennec. 2004. Production of exopolysaccharides by Antarctic marine bacterial isolates. *Journal of Applied Microbiology* 96(5):1057–1066.

Martínez-Cánovas, M.J., V. Béjar, F. Martínez-Checa, and E. Quesada. 2004b. *Halomonas anticariensis* sp. nov., from Fuente de Piedra, a saline-wetland wildfowl reserve in Málaga, southern Spain. *International Journal of Systematic and Evolutionary Microbiology* 54:1329–1332.

Mangwani, N., S.K. Shukla, T.S. Rao, and S. Das. 2014. Calcium-mediated modulation of *Pseudomonas mendocina* NR802 biofilm influences the phenanthrene degradation. *Colloids and Surfaces B: Biointerfaces* 114:301–309.

Markovich, D. and H. Murer. 2004. The SLC13 gene family of sodium sulphate/carboxylate cotransporters. *Pflügers Archiv* 447(5):594–602.

Martínez-Cánovas, M.J., E. Quesada, I. Llamas, and V. Béjar. 2004a. *Halomonas ventosae* sp. nov., a moderately halophilic, denitrifying, exopolysaccharide-producing bacterium. *International Journal of Systematic and Evolutionary Microbiology* 54:733–737.

Martínez-Cánovas, M.J., E. Quesada, F. Martınez-Checa, A. Del Moral, and V. Béjar. 2004c. *Salipiger mucescens* gen. nov., sp. nov., a moderately halophilic, exopolysaccharide-producing bacterium isolated from hypersaline soil, belonging to the α-Proteobacteria. *International Journal of Systematic and Evolutionary Microbiology* 54:1735–1740.

Martínez-Checa, F., E. Quesada, M.J. Martínez-Cánovas, I. Llamas, and V. Béjar. 2005. *Palleronia marisminoris* gen. nov., sp. nov., a moderately halophilic, exopolysaccharide-producing bacterium belonging to the 'Alphaproteobacteria', isolated from a saline soil. *International Journal of Systematic and Evolutionary Microbiology* 55(6):2525–2530.

Marx, J.G., S.D. Carpenter, and J.W. Deming. 2009. Production of cryoprotectant extracellular polysaccharide substances (EPS) by the marine psychrophilic bacterium *Colwellia psychrerythraea* strain 34H under extreme conditions. *Canadian Journal of Microbiology* 55(1):63–72.

Mata, J.A., V. Béjar, P. Bressollier, R. Tallon, M.C. Urdaci, E. Quesada, and I. Llamas. 2008. Characterization of exopolysaccharides produced by three moderately halophilic bacteria belonging to the family Alteromonadaceae. *Journal of Applied Microbiology* 105(2):521–528.

Matias, V.R.F., A. Al-Amoudi, J. Dubochet, and T.J. Beveridge. 2003. Cryo-transmission electron microscopy of frozen-hydrated sections of *Escherichia coli* and *Pseudomonas aeruginosa*. *Journal of Bacteriology* 185:6112–6118.

Maugeri, T.L., C. Gugliandolo, D. Caccamo, A. Panico, L. Lama, A. Gambacorta, and B. Nicolaus. 2002. A halophilic thermotolerant *Bacillus* isolated from a marine hot spring able to produce a new exopolysaccharide. *Biotechnology Letters* 24:515–519.

Mayer, C., R. Moritz, C. Kirschner, W. Borchard, R. Maibaum et al. 1999. The role of intermolecular interactions: Studies on model systems for bacterial biofilms. *International Journal of Biological Macromolecules* 26(1):3–16.

McGenity, T.J., B.D. Folwell, B.A. McKew, and G.O. Sanni. 2012. Marine crude-oil biodegradation: A central role for interspecies interactions. *Aquatic Biosystems* 8:10.

Meroueh, S.O., K.Z. Bencze, D. Hesek, M. Lee, J.F. Fisher, T.L. Stemmler, and S. Mobashery. 2006. Three-dimensional structure of the bacterial cell wall peptidoglycan. *Proceedings of the National Academy of Sciences of the United States of America* 103(12):4404–4409.

Miller, S.I., R.K. Ernst, and M.W. Bader. 2005. LPS, TLR4 and infectious disease diversity. *Nature Reviews Microbiology* 3:36–46.

Mishra, A. and B. Jha. 2013. Microbial exopolysaccharides. In *The Prokaryotes*, Rosenberg, E., DeLong, E., Lory, S., Stackebrandt, E., and Thompson, F. (eds.). Springer, Berlin, Germany, pp. 179–192.

Mitchell, P. and J. Moyle. 1956. Osmotic function and structure in bacteria. In *Bacterial Anatomy, Sixth Symposium of the Society for General Microbiology*. Cambridge University Press, Cambridge, U.K.

Morris, R.S. and M.A. Walsh. 2000. Zinc oxide photoactive material. US Patent US6063849 A.

Moxon, E.R. and J.S. Kroll. 1990. The role of bacterial polysaccharide capsules as virulence factors. In *Bacterial Capsules*, Jann, K. and Jann, B. (eds.). Springer, Berlin, Germany, pp. 65–85.

Moynihan, P.J. and A.J. Clarke. 2011. O-Acetylated peptidoglycan: Controlling the activity of bacterial autolysins and lytic enzymes of innate immune systems. *The International Journal of Biochemistry and Cell Biology* 43:1655–1659.

Muhlradt, P.F. and J.R. Golecki. 1975. Asymmetrical distribution and artifactua reorientation of lipopolysaccharide in the outer membrane bilayer of *Salmonella typhimurium*. *European Journal of Biochemistry* 51:343–352.

Nakayama, T., H. Wake, K. Ozawa, H. Kodama, N. Nakamura, and T. Matsunaga. 1998. Use of a titanium nitride for electrochemical inactivation of marine bacteria. *Environmental Science and Technology* 32:798–801.

National Research Council (NRC). 2003. *Oil in the Sea III: Inputs, Fates, and Effects*. National Academy Press, Washington, DC.

Nichols, C.M., J.P. Bowman, and J. Guezennec. 2005b. *Olleya marilimosa* gen. nov., sp. nov., an exopolysaccharide-producing marine bacterium from the family Flavobacteriaceae, isolated from the Southern Ocean. *International Journal of Systematic and Evolutionary Microbiology* 55(4):1557–1561.

Nichols, C.M., J. Guezennec, and J.P. Bowman. 2005a. Bacterial exopolysaccharides from extreme marine environments with special consideration of the southern ocean, sea ice, and deep-sea hydrothermal vents: A review. *Marine Biotechnology* 7(4):253–271.

Nicolaus, B., L. Lama, A. Panico, V.S. Moriello, I. Romano, and A. Gambacorta. 2002. Production and characterization of exopolysaccharides excreted by thermophilic bacteria from shallow, marine hydrothermal vents of flegrean ares (Italy). *Systematic and Applied Microbiology* 25(3):319–325.

Nicolaus, B., A. Panico, M.C. Manca, L. Lama, A. Gambacorta, T. Maugeri, C. Gugliandolo, and D. Caccamo. 2000. A thermophilic *Bacillus* isolated from eolian shallow hydrothermal vent, able to produce exopolysaccharide. *Systematic and Applied Microbiology* 23:426–432.

Nikaido, H. 2003. Molecular basis of bacterial outer membrane permeability revisited. *Microbiology and Molecular Biology Reviews* 67:593–656.

Nilsson, M., W.C. Chiang, M. Fazli, M. Gjermansen, M. Givskov, and T. Tolker-Nielsen. 2011. Influence of putative exopolysaccharide genes on *Pseudomonas putida* KT2440 biofilm stability. *Environmental Microbiology* 13:1357–1369.

Nisha, P. and M. Thangavel. 2014. Isolation and Characterization of Biofilm Producing Bacteria from Arabian Sea. *Research Journal of Recent Sciences* 3:132–136.

Ortega-Morales, B.O., C.C. Gaylarde, G.E. Englert, and P.M. Gaylarde. 2005. Analysis of salt-containing biofilms on limestone buildings of the Mayan culture at Edzna, Mexico. *Geomicrobiology Journal* 22(6):261–268.

Ortega-Morales, B.O., J.L. Santiago-García, M.J. Chan-Bacab, X. Moppert, E. Miranda-Tello, M.L. Fardeau, and J. Guezennec. 2007. Characterization of extracellular polymers synthesized by tropical intertidal biofilm bacteria. *Journal of Applied Microbiology* 102(1):254–264.

Osman, S.F. and W.F. Fett. 1989. Structure of an acidic exopoly-saccharide of *Pseudomonas marginalis* HT041B. *Journal of Bacteriology* 171:1760–1762.

Pal, A. and A. Paul. 2008. Microbial extracellular polymeric substances: Central elements in heavy metal bioreme-diation. *Indian Journal of Microbiology* 48:49–64.

Pan, X., J. Liu, and D. Zhang. 2010. Binding of phenanthrene to extracellular polymeric substances (EPS) from aero-bic activated sludge: A fluorescence study. *Colloids and Surfaces. B, Biointerfaces* 80:103–106.

Panwichian, S., D. Kantachote, B. Wittayaweerasak, and M. Mallavarapu. 2011. Removal of heavy metals by exopolymeric substances produced by resistant purple nonsulfur bacteria isolated from contaminated shrimp ponds. *Electronic Journal of Biotechnology* 14:2.

Parolis, H., L.A.S. Parolis, I.F. Boán, F. Rodríguez-Valera, G. Widmalm, Mc. Manca, P.E Jansson, and I.W. Sutherland. 1996. The structure of the exopolysaccharide produced by the halophilic Archaeon *Haloferax medi-terranei* strain R4 (ATCC 33500). *Carbohydrate Research* 295:147–156.

Poli, A., G. Anzelmo, and B. Nicolaus. 2010. Bacterial exopoly-saccharides from extreme marine habitats: Production, characterization and biological activities. *Marine Drugs* 8(6):1779–1802.

Poli, A., P. Di Donato, G.R. Abbamondi, and B. Nicolaus. 2011. Synthesis, production, and biotechnologi-cal applications of exopolysaccharides and polyhy-droxyalkanoates by archaea. *Archaea* 2011:693253. doi:10.1155/2011/693253.

Poli, A., E. Esposito, P. Orlando, L. Lama, A. Giordano, F. de Appolonia, B. Nicolaus, and A. Gambacorta. 2007. *Halomonas alkaliantarctica* sp. nov., isolated from saline lake Cape Russell in Antarctica, an alkalophilic mod-erately halophilic, exopolysaccharide-producing bacte-rium. *Systematic and Applied Microbiology* 30:31–38.

Qian, P.Y., S. Dobretsov, T. Harder, and C.K. Lau. 2006. Anti-fouling exopolysaccharides isolated from cultures of *Vibrio alginolyticus* and *Vibrio proteolyticus*. US Patent US7090856 B2.

Qin, G., L. Zhu, X. Chen, P.G. Wang, and Y. Zhang. 2007. Structural characterization and ecological roles of a novel exopolysaccharide from the deep-sea psy-chrotolerant bacterium *Pseudoalteromonas* sp. SM9913. *Microbiology* 153:1566–1572.

Quesada, E., V. Bejar, M.R. Ferrer, C. Calvo, I. Llamas et al. 2004. Moderately halophilic, exopolysaccharide-producing bacteria. In *Halophilic Microorganisms*, Ventosa, A. (ed.). Springer, Heidelberg, Germany, pp. 297–314.

Quintero, E.J. and R.M Weiner. 1995. Physical and chemical characterization of the polysaccharide capsule of the marine bacterium, *Hyphomonas* strain MHS-3. *Journal of Industrial Microbiology* 15(4):347–351.

Raguénès, G., P. Pignet, G. Gauthier, A. Peres, R. Christen, H. Rougeaux, G. Barbier, and J. Guezennec. 1996. Description of a new polymer-secreting bacterium from a deep-sea hydrothermal vent, *Alteromonas macleo-dii* subsp. *fijiensis*, and preliminary characterization of the polymer. *Applied and Environmental Microbiology* 62(1):67–73.

Rinker, K.D. and R.M. Kelly. 1996. Growth physiology of the hyperthermophilic archaeon *Thermococcus litoralis*: Development of a sulfur-free defined medium, charac-terization of an exopolysaccharide, and evidence of bio-film formation. *Applied and Environmental Microbiology* 12:4478–4485.

Rinker, K.D. and R.M. Kelly. 2000. Effect of carbon and nitro-gen sources on growth dynamics and exopolysaccha-rides production for the hyperthermophilic archaeon *Thermococcus litoralis* and bacterium *Thermotoga mari-tima*. *Biotechnology and Bioengineering* 5:537–547.

Roberson, E. and M. Firestone. 1992. Relationship between desiccation and exopolysaccharide production in soil *Pseudomonas* sp. *Applied and Environmental Microbiology* 58:1284–1291.

Roberts, I.S. 1996. The biochemistry and genetics of capsular polysaccharide production in bacteria. *Annual Review of Microbiology* 50:285–315.

Rodriguez-Valera, F. 1988. Characteristics and microbial ecol-ogy of hypersaline environments. In *Halophilic Bacteria*, vol. 1, Rodriguez-Valera, F. (ed.). CRC Press, Boca Raton, FL, pp. 3–30.

Rogers, H.J., H.R. Perkins, and J.B. Ward. 1980. *Microbial Cell Walls and Membranes*. Chapman and Hall, London, U.K.

Romani, A.M., K. Fund, J. Artigas, T. Schwartz, S. Sabater, and U. Obst. 2008. Relevance of polymeric matrix enzymes during biofilm formation. *Microbial Ecology* 56:427–436.

Ron, E. and E. Rosenberg. 2002. Biosurfactants and oil biore-mediation. *Current Opinion in Biotechnology* 13:249–252.

Rougeaux, H., J. Guezennec, R.W. Carlson, N. Kervarec, R. Pichon, and P. Talaga. 1999. Structural determina-tion of the exopolysaccharide of *Pseudoalteromonas* strain HYD 721 isolated from a deep-sea hydrothermal vent. *Carbohydrate Research* 315:273–285.

Rougeaux, H., P. Talaga, R.W. Carlson, and J. Guerzennec. 1998. Structural studies of an exopolysaccharide pro-duced by *Alteromonas macleodii* subsp. *fijiensis* originat-ing from a deep-sea hydrothermal vent. *Carbohydrate Research* 312:53–59.

Schneiker, S., V.A.P.M. dos Santos, D. Bartels, T. Bekel, M. Brecht et al. 2006. Genome sequence of the ubiquitous hydrocarbon-degrading marine bacterium *Alcanivorax borkumensis*. *Nature Biotechnology* 24:997–1004.

Shibata, S., E.S. Yip, K.P. Quirke, J.M. Ondrey, and K.L. Visick. 2012. Roles of the structural symbiosis polysaccharide (*syp*) genes in host colonization, biofilm formation, and polysaccharide biosynthesis in *Vibrio fischeri*. *Journal of Bacteriology* 194:6736–6747.

Singh, R., D. Paul, and R.K. Jain. 2006. Biofilms: Implications in bioremediation. *Trends in Microbiology* 14:389–397.

Socransky, S.S., A.D. Haffajee, M.A. Cugini, C. Smith, and R.L. Kent Jr. 1998. Microbial complexes in subgingival plaque. *Journal of Clinical Periodontology* 25:134–144.

Su, Y.C., and C. Liu. 2007. *Vibrio parahaemolyticus* : a concern of seafood safety. *Food Microbiology* 24(6):549–558.

Sutherland, I.W. 1972. Bacterial exopolysaccharides. *Advances in Microbial Physiology* 8:143–213.

Sutherland, I.W. 1990. *Biotechnology of Exopolysaccharides*. Cambridge University Press, Cambridge, U.K.

Sutherland, I.W. 2001. Microbial polysaccharides from gram-negative bacteria. *International Dairy Journal* 11:663–674.

Ta-Chen, L., J.S. Chang, and C.C. Young. 2008. Exopolysaccharides produced by *Gordonia alkanivorans* enhance bacterial degradation activity for diesel. *Biotechnology Letters* 30:1201–1206.

van Hullebusch, E.D., M.H. Zandvoort, and P.N.L. Lens. 2003. Metal immobilisation by biofilms: Mechanisms and analytical tools. *Reviews in Environmental Science and Biotechnology* 2:9–33.

Vasseur, P., I. Vallet-Gely, C. Soscia, S. Genin, and A. Filloux. 2005. The *pel* genes of the *Pseudomonas aeruginosa* PAK strain are involved at early and late stages of biofilm formation. *Microbiology* 151:985–997.

Venosa, A.D. and X. Zhu. 2003. Biodegradation of crude oil contaminating marine shorelines and freshwater wetlands. *Spill Science and Technology Bulletin* 8:163–178.

Ventosa, A., E. Mellado, C. Sanchez-Porro, and M.C. Marquez. 2008. Halophilic and halotolerant micro-organisms from soils. In *Microbiology of extreme soils*, Dion, P. and Nautiyal, C.S. (eds.). Springer, Berlin, Germany, pp. 87–115.

Vollmer, W. 2008. Structural variation in the glycan strands of bacterial peptidoglycan. *FEMS Microbiology Reviews* 32:287–306.

Wang, H., X. Jiang, H. Mu, X. Liang, and H. Guan. 2007. Structure and protective effect of exopolysaccharide from *P. agglomerans* strain KFS-9 against UV radiation. *Microbiological Research* 162(2):124–129.

Watanabe, M., K. Kawahara, K. Sasaki, and N. Noparatnaraporn. 2003. Biosorption of cadmium ions using a photosynthetic bacterium, *Rhodobacter sphaeroides* S and a marine photosynthetic bacterium, *Rhodovulum* sp. and their biosorption kinetics. *Journal of Bioscience and Bioengineering* 95:374–378.

Weiner, R., Langille, S., and Quintero, E. 1995. Structure, function and immunochemistry of bacterial exopolysaccharides. *Journal of Industrial Microbiology and Biotechnology* 15:339–346.

White, D.C. 1986. Non-destructive biofilm analysis by Fourier transform spectroscopy (FT/IR). *Presented at the Perspectives in Mcrobial Ecology, Proceedings of the Fourth International Symposium on Microbial Ecology*, Ljubljana, Yugoslavia, pp. 442–446.

Whitfield, C. 1988. Bacterial Extracellular Polysaccharides. *Canadian Journal of Microbiology* 34:415–420.

Whitfield, C. 2006. Biosynthesis and assembly of capsular polysaccharides in *Escherichia coli*. *Annual Review of Biochemistry* 75:39–68.

Wilkie, I.W., M. Harper, J.D. Boyce, and B. Adler. 2012. *Pasteurella multocida*: Diseases and pathogenesis. *Current Topics in Microbiology and Immunology* 361:1–22.

Wingender, J., T.R. Neu, and H.C. Flemming. 1999. What are bacterial extracellular polymeric substances? In *Microbial Extracellular Polymeric Substances*. Springer, Berlin, Germany, pp. 1–19.

Wrangstadh, M., P.L. Conway, and S. Kjelleberg. 1986. The production and release of an extracellular polysaccharide during starvation of a marine Pseudomonas sp. and the effect there of on adhesion. *Archives of Microbiology* 145(3):220–227.

Yao, X., M. Jericho, D. Pink, and T. Beveridge. 1999. Thickness and elasticity of gram-negative murein sacculi measured by atomic force microscopy. *Journal of Bacteriology* 181:6865–6875.

Yaskovich, G.A. 1998. The role of cell surface hydrophobicity in adsorption immobilization of bacterial strains. *Applied Biochemistry and Biotechnology* 4:373–376.

Yebra, D.M., S. Kiil, and K. Dam-Johansen. 2004. Antifouling technology—Past, present and future steps towards efficient and environmentally friendly antifouling coatings. *Progress in Organic Coatings* 50:75–104.

Yu, Z., N. Zhou, H. Qiao, and J. Qiu. 2014. Identification, cloning, and expression of L-amino acid oxidase from marine *Pseudoalteromonas* sp. B3. *The Scientific World Journal* 2014: Article ID 979858.

Zanchetta, P. and J. Guezennec. 2001. Surface thermodynamics of osteoblasts: Relation between hydrophobicity and bone active biomaterials. *Colloids and Surfaces B: Biointerfaces* 22:301–307.

Zhang, D., X. Pan, K.M. Mostofa, X. Chen, G. Mu, F. Wu, Q. Fu. 2010. Complexation between Hg (II) and biofilm extracellular polymeric substances: an application of fluorescence spectroscopy. *Journal of Hazardous Materials* 175(1):359–365.

Zhou, W.Z., W.W. Li, Y.Z. Zhang, B.Y. Gao, and J. Wang. 2009. Biosorption of Pb²⁺ and Cu²⁺ by an exopolysaccharide from the deep-sea psychrophilic bacterium *Pseudoalteromonas* sp. SM9913. *Huan Jing Ke Xue* 30(1):200–205.

*chapter eighteen*

# Agar-abundant marine carbohydrate from seaweeds in Indonesia

*Production, bioactivity, and utilization*

*Syamdidi, Hari Eko Irianto and Giyatmi Irianto*

## Contents

## 18.1 Introduction

Agar is one of the marine carbohydrates and, as a polysaccharide complex, obtained through bleaching and hot water extraction of agarocytes from the red alga Rhodophyceae. There are two genera of seaweeds used as agar sources for agar industries in Indonesia: *Gelidium* and *Gracilaria*. In general, *Gelidium, Acanthopeltis, Ceramium, Pterocladia,* and *Gracilaria* are predominant raw materials in agar production in the world. Agar consists of about 70% agarose and 30% agaropectin (Scott and Eagleson 1998). Agarose is a neutral gelling fraction, which consists of a linear polymer of alternating D-galactose and 3,6-anhydrogalactose units. Agaropectin is a non-gelling fraction, which consists of 1,3 glycosidically linked D-galactose units, some of which are sulfated at position 6. Chemical structures of agarose and agaropectin are shown in Figure 18.1.

Seaweed is abundant in Indonesia, of which around 555 species have been already identified. About 21 species have been utilized as raw material by seaweed processing industries (Aslan 1991). Tremendous development of seaweed farming during the past 5 years has brought about the situation in which the seaweed production of Indonesia (in volume) is higher than in the Philippines; the production volume in 2013 reached 9,298,474 tons (Ministry of Marine Affairs and Fisheries 2014). Most of them are mainly *Eucheuma* sp. and *Gracilaria* sp.

Red algae are the most popular seaweeds that are used for food, pharmaceutical, and other industries. From 17 genera of red algae, 34 species have already been used for various purposes. Moreover, 23 species are able to be cultured: 6 species from the genus *Eucheuma*, 3 species from *Gelidium*, 10 species from *Gracillaria*, and 4 species from *Hypnea*. The genus *Eucheuma, Gracilaria,* and *Gelidium* are commonly found and cultured in Indonesian waters. However, only the genera producing agar (agarophytes) and carrageenan (carrageenophytes) are commercially cultured to support the seaweed industry in Indonesia and also to fulfill worldwide demand (Kordi 2011).

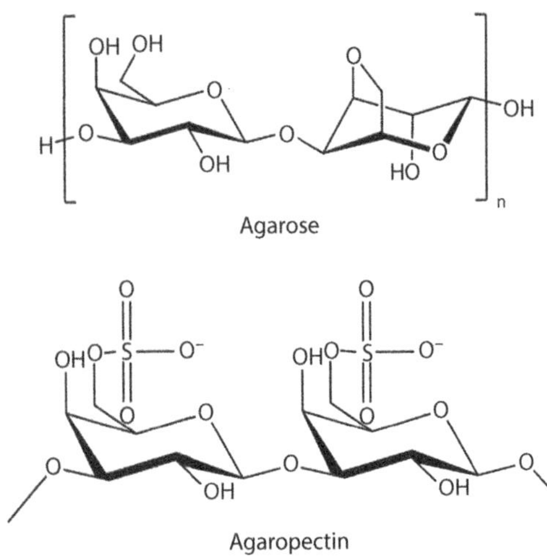

***Figure 18.1*** Structure of agarose (1,4)-3,6 anhydro L-galactose, (1,3) D-galactose, and agaropectin. (From Ramadhan, W., Utilization of agar powder as texturizer in guava (*Psidium guajava* L.) spread sheet and shelf-life prediction, Faculty of Fisheries and Marine Science, Bogor Agricultural University, Bogor, Indonesia, 2011. With permission.)

Many countries have developed methods of agar processing, but only two methods are widely used: the freeze-thaw method, and the pressing method. Indonesia is currently the largest agar producer in the world. Agar produced by Indonesian processing factories is not only used for domestic consumption but also export to several countries, for example, China, Japan, and European countries.

Several investigations have been conducted by researchers all over the world to reveal the health benefits of agar, the findings of which are expected to guide into new prospective utilization. Some research has shown that the bioprospecting substances of agar have the potential to alleviate health problems. Therefore, new value-added products based on those findings are expected to be developed in the future, and Indonesia is poised to play an important role due to the abundance resources of agar-producing seaweeds as raw material.

## 18.2   Agar industries in Indonesia

World trade of agar in terms of raw material of agar and finished products shows an increasing trend. The world demand was estimated to be 10,000 tons/year of raw material and 3,500 tons/year of finished products (Directorate General of Aquaculture 2005). Indonesia can develop the agar industry in three directions: as a raw material supplier, an agar producer, or both. Agar production looks more strategic for Indonesia, because the industry can contribute to the welfare of people by providing them job opportunities.

Agar industry is growing rapidly in Indonesia. Production of agar in Indonesia started before the World War II. In 1930, the first agar factory was established in Kudus, Central Java. This was followed by others in Jakarta, Surabaya, and Makassar (Sulistyo 2002). PT Sinar Kencana, an agar factory in Surabaya, East Java, was founded in 1947. In 1955, there were five companies operating with a total production around 13.7 tons per annum (tpa). After four decades, 10 new agar companies were established with the total production reaching 108.7 tpa. Thus the number of agar producing companies became 15 in 1993 (Zatnika 1997) but currently only 11 companies are existing. One of them is the largest agar company in the world with total production of over 3000 tpa.

### 18.2.1   Agar source

Red algae genera such as *Gelidium, Gracilaria, Grateloupia, Halymenia, Hypnea,* and *Porphyra* (Table 18.1) are used as raw material in agar production in Indonesia. They are naturally available throughout the Indonesian waters. Currently, the demand of *Gracilaria* has grown to supply raw material for agar processing because of the increasing demand of agar and for establishing new agar factories. Seaweed from the wild cannot fulfill raw material need of the agar industry, which has led to the exploration of another source of raw material, farmed seaweeds. Culturing *Gracilaria* was developed later on around the year 1980 to ensure continuous supply of raw material for the agar industry. *Gracilaria* is now widely cultured by the farmers along the coast, particularly in Sulawesi and Java. One of the biggest *Gracilaria* culture ponds is located in Tangerang, Banten Province.

### 18.2.2   Seaweed harvesting

Seaweed is normally harvested after 45 days if cultured in the beach but after 60–75 days if cultured in ponds (Kordi 2011). The Directorate General of Aquaculture (2005) suggested that *Gracilaria* is harvested after 90 days of cultivation, and the next harvesting can be performed after 60 days. Harvesting should be performed at the right age to obtain the optimum yield of agar during processing. Harvesting age also affects the quality of the extracted agar, in which underage seaweed will result in a lower agar quality.

Harvested *Gracilaria* is then washed with fresh water to free it from unwanted materials including other seaweed species. Seaweeds are subsequently sun-dried by placing them on drying racks, mats, or floors. The drying step should not be delayed so as to avoid deterioration due to fermentation. Drying can take 3–4 days till achieving a moisture content of about 25% as required by the market. While waiting for marketing, dried

***Table 18.1*** Seaweeds used as agar source in Indonesia

| Seaweed genera | Location in Indonesia |
|---|---|
| *Gelidium amansii* | Alor Islands, Tanimbar Islands, Maluku Islands |
| *G. rigidium* | Scattered across Indonesian coast |
| *G. latifolium* | Bengkulu, Lampung, South part of Java Island, West Nusa Tenggara Islands |
| *Gracilaria confervoides* | Scattered across Indonesian coast |
| *G. verrucosa* | West Sumbawa, Sawu Island, South Sulawesi, Southeast Sulawesi |
| *G. eucheumoides* | South Lampung, South Java Island, Southeast Celebes, South and Southeast Moluccas |
| *G. lichenoides* | Scattered across Indonesian coast |
| *G. gigas* | Scattered across Indonesian coast |
| *G. taenoides* | Riau Islands, Bangka and Belitung Islands, Lampung |
| *Grateloupia filicina* | West and South Java Island, South Lampung, Seribu Islands |
| *Halymenia durvillei* | South and Southeast Celebes, Ambon and Seram Islands, Papua, East Nusa Tenggara, Lombok, Sumbawa and Halmahera |
| *Hypnea cervicornis* | Riau Islands, Bali, Tawi-tawi |
| *H. divacirata* | Riau Islands, Moluccas Islands |
| *H. musciformis* | Scattered across Indonesian coast |
| *Porphyra atropurpurea* | Halmahera Island and Kei Island |

*Source:*  Anggadiredja, J.T. et al., *Seaweed: Culturing, Processing and Marketing Potential Fisheries Commodity*, Penebar Swadaya, Jakarta, Indonesia, 2006.

seaweed is stored for a certain period. According to Rodarte et al. (2010), the agar content of *Gracilaria* from the tropics decreases in a few months because of hydrolisis. The hydrolysis in *Gracilaria* could be due to the presence of agarolytic bacteria, of which the most important is *Bacillus cereus*, and to the presence of the algae's own agarolytic enzyme. Storage of *Gracilaria cornea* for 2 years and *Gracilaria eucheumatoides* for 1 year resulted in a reduction of a gel strength of 17% and 35%, respectively.

### 18.2.3   Agar extraction

In Indonesia, there are three types of agar products in the market: agar sheet, agar bar, and agar powder. Basically, the extraction methods used to make those products are similar.

Agar is extracted from *Gracilaria* by means of two steps of cooking using water, in which the ratio of dried seaweed and water is approximately 1:20. The first cooking is carried out with the dried seaweed and an agar/water ratio of 1:14 at 85°C–95°C and pH 6–7 for 2 h. The agar extract and seaweed pulp are separated using fabrics. The seaweed pulp obtained is boiled again for the second cooking with the dried seaweed and with a water ratio of 1:6 for 1 h. The agar extract is then added with 2%–3% KOH or KCl for gel formation (Directorate General of Aquaculture 2005). The gel undergoes further processing for producing agar sheet, agar bar, or agar powder by applying specific treatments for each product.

Mostly, farmed *Gracilaria* produces a soft-textured gel resulting in difficulties for further processing,

particularly for industries operated on a small scale. To improve the gel properties, the farmed *Gracilaria* is mixed with that harvested from the nature before agar extraction (Aji et al. 2003).

Some authors (Rao and Bekheet 1976; Chapman and Chapman 1980; Robello et al. 1997; Montano et al. 1999) have found that soaking the seaweed in an alkali solution can improve the gel strength of agar. The chemical structures of the agar precursor in seaweed and of the idealized agar after alkali treatment are shown in Figure 18.2. Kusuma et al. (2013) have shown that the concentration of NaOH affected the gel strength, sulfate content, ash content, moisture content, and yield

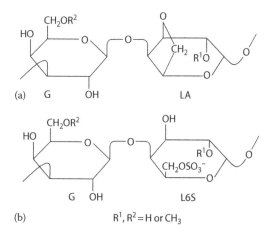

***Figure 18.2*** Chemical structure of agar. (a) Idealized agar. (b) Agar precursor prior to extraction (From Distantina, S., et al., *Jurnal Rekayasa Proses*, 2, 11, 2008. With permission.)

of agar *G. verrucosa*. The higher the concentration of NaOH, the higher the gel strength, the ash content, the water content, and the yield, but the lower the sulfate levels. The best agar quality was obtained using 6% NaOH solution. Distantina et al. (2008) and Van et al. (2008) noted that soaking seaweeds in an alkali solution resulted in lower extraction rate and lower yield, but higher gel strength compared to soaking in an acid solution.

### 18.2.4 Agar quality

The quality of Indonesian agar is regulated under National Standardization Board. This institution regulates the specification of agar for both safety and wholesomeness (Table 18.2). Most agar industries in Indonesia have no problem meeting the standards required by the Board. Moreover, they usually sell the products above requirement standard, especially for moisture content, which ranges from 10% to 15%. International standards

*Table 18.2* Indonesian agar specification

| Parameter | Unit | Requirement |
|---|---|---|
| *Sensory* | — | Min. 7 (1–9 scale)[a] |
| *Chemical quality* | | |
| Moisture content | % | Max. 22 |
| Ash content[b] | % | Max. 6.5 |
| Acid-insoluble ash[b] | % | Max. 0.5 |
| Starch[b] | — | Negative |
| Gelatin and protein[b] | — | Negative |
| *Microbial quality* | | |
| TPC | Colony/g | Max. 5000 |
| *Escherichia coli* | APM/g | <3 |
| Salmonella | Per 25 g | Negative |
| Yeast and moulds | Colony/g | Max. 300 |
| *Heavy metals[b]* | | |
| Arsenic (As) | mg/kg | Max. 3 |
| Cadmium (Cd) | mg/kg | Max. 1 |
| Mercury (Hg) | mg/kg | Max. 1 |
| Lead (Pb) | mg/kg | Max. 3 |
| Zinc (Zn) | mg/kg | Max. 40 |
| *Physic[b]* | | |
| Water absorption | — | Min. five times |
| Foreign insoluble matter | % | Max. 1 |
| Particle size (pass 60 mesh size) | % | Min. 80 |

*Source:* National Standardization Board, *SNI Agar Powder*, BSN, Jakarta, Indonesia, 2014.

[a] For each sensory attribute.
[b] If only requested.

published by, for example, by the EU (European Union) and JECFA (The Joint FAO/WHO Expert Committee on Food Additives) are quite similar to the Indonesian standard. However, in industrial trading, moisture content, color, and odor are the most demanding parameters. Other parameters are conditional, depending on the buyer's requirements.

## 18.3 Bioactivity of agar

Studies have been carried out to reveal the bioactivity of agar-producing seaweeds, not specifically agar itself. Some species of red algae were reported to exhibit a broad spectrum of biological activities. The bioactivity of the most common agar seaweeds in Indonesia—from genus *Gracilaria* and *Gelidium*—has been examined. Almeida et al. (2011) have excellently reviewed the bioactivity of *Gracilaria* from some studies that highlight the potential pharmaceutical uses, particularly in terms of toxic, cytotoxic, spermicidal, anti-implantation, antibacterial, antiviral, antifungal, antiprotozoa, antihypertensive, antioxidant, anti-inflammatory, analgesic, and spasmolytic effects in the gastrointestinal tract.

### 18.3.1 Anti-inflammatory

Anti-inflammatory activities have been reported for a sulfated polysaccharide fraction from *Gracilaria caudate* (Chavez et al. 2013), a galactan from *Gelidium crinale* (Sousa et al. 2013), and polysaccharide fractions from *G. verrucosa* (Yoshizawa et al. 1996). Sajiki (1997) reported that *G. verrucosa*, *G. asiatica*, *G. lichenoides* and other species contain PGE2, which produce physiological effects including hyperthermia, hypotension, smooth-muscle dilatation, hyperalgesia, and gastric secretion inhibition (Minghetti and Levi 1998).

### 18.3.2 Contraception activity

A methanol/methylene chloride (1:1) extract from *G. corticata* exhibited post-coital contraceptive activity due to enhanced pre-implantation without any marked side effects when introduced orally to female rats. This shows that red marine algae are a potential source for post-coital contraceptive drugs. Extracts from *G. edulis* demonstrated 100% inhibition of sperm motility, and this effect was related to the disruption of the plasma membrane by spermicidal compounds (Almeida et al. 2011).

### 18.3.3 Antioxidant

The antioxidant activity of *Gracilaria* is related to the anti-inflammatory effects. Swantara and Parwata (2011) found that *G. coronopifolia* has 90.27% antioxidant activity consisting of 1-nonadecane, hexadecanoic acid,

9-octadecanoic acid, cholest-4,6-dien-3-ol, cholest-5-en-3β-ol (cholesterol), and cholest-4-en-3-one. Murugan and Iyer (2012) noted that *G. edulis* possesses antioxidant activity.

### 18.3.4  Gastrointestinal and cardivascular effects

Aqueous extract from dried *G. verrucosa* or fresh *G. chorda* at a dose of 0.5 mg per animal was able to control gastrointestinal disorders in mice, the active compounds being zeaxanthin, antheraxanthin, carotenoids, pyrimidine 2-amino-4-carboxy, non-alkaloid nitrogen heterocycle, steroids, 5-α-poriferastane, 3-β-6-α-diol poriferastane, 5-α-3-β-6-β-diol, and gigatinine. Meanwhile, a 90% ethanol extract from *G. edulis* showed diuretic activity, and an aqueous extract from *G. lichenoides* administered intravenously showed antihypertensive effect in rats (Almeida et al. 2011).

### 18.3.5  Antibiotic activity

*Gracilaria corticata* were found to be active against gram-positive cultures of *Bacillus* (Bhakuni and Rawat 2005). The authors also found that an extract of *Gelidium cartilagineum* and *Chondrus crispus* were active against influenza B and mumps virus. This activity has been attributed to the presence of a galactan unit in the polysaccharides, agar, and carrageenan present in both species.

Another genus of *Gracilaria*—*Gracilaria* Greville—contains a gelatinous nontoxic colloidal carbohydrate in the cell wall and intercellular spaces of the algae and has wide use in the preparation of food, ice creams, jellies, soups, bacteriological samples, and cosmetics (Gosh et al. 2012; Kerton et al. 2013). They are sources of important bioactive metabolites with antibiotic activity (Smit 2004).

A chloroform extract of *G. edulis* (Gmelin) Silva was found to have antibacterial activity against some bacterial strains, in which the isolated steroids (carotenoids, β-cryptoxanthin, and β-carotene) and carbohydrates are suspected as the active compounds. An ethanol extract from *G. debilis* showed antibacterial activity against *Staphylococcus aureus*. A 95% ethanol extract from whole dried *G. cervicornisalgae* was active against *S. aureus* at a concentration of 5.0 mg/mL. A methanol extract from fresh *G. corticata* was active against *Bacillus subtilis*, *B. megaterium*, *S. aureus*, and *Streptococcus viridans*. Ethanol extracts from *G. domigensis* and *G. sjoestedii* showed antibacterial activity against *E. coli* and *S. aureus*. Ethanol extracts from *G. debilis*, *G. domingensis*, and *G. sjoestedti* were active against *Candida albicans*. An ethanol extract from *G. domigensis* was active against *Mycobacterium smegmatis* and *Neurospora crassa*. *G. domigensis* has as chemical

constituents polysaccharide CT-1, palmitic acid, and steroids (stigmasterol, sitosterol, campesterol, cholest-7-en-3-β-ol, and brassicasterol) (Almeida et al. 2011).

### 18.3.6  Antiviral activity

Extracts from *G. bursa-pastoris* and *Gracilaria* sp. were inactive against the herpes simplex 1 virus (HSV) and the human immunodeficiency virus (HIV) when evaluated in cell cultures. Granin BP and citrullinyl-arginine proteins were isolated from these extracts. A methanol extract from dried *G. pacifica* at a concentration of 200.0 µg/mL was active against Sindbis virus, but was not effective against HSV when tested at a concentration of 400 µg/mL. Extracts and compounds obtained from *Gracilaria* sp. with anti-HIV activity are also active against other retroviruses such as HSV (Almeida et al. 2011).

### 18.4  Agar utilization

The many uses of agar and agarose are related to the formation of thermoreversible gels at low concentration in water with large hysteresis. The physicochemical and rheological properties of these algal polysaccharides are linked to their chemical structure (Lahaye and Rochas 1991). Agar has been in use as a source of food in some Asian countries for a long time. Most of agar utilization is for food (~80%–90%) and the rest is used for biotechnological applications (FAO 2003; Armisen and Galatas 2009). Based on their types, agar was divided into natural agar and industrial agar. Natural agar was produced by traditional method while the industrial agar using freezing-defreezing techniques by artificial freezing, or accelerated synaeresis through pressure (Armisen 1995) (Table 18.3).

Agar is used as a thickening agent and also to improve the texture of processed food such as jellies, dairy products, fruit pastilles, chewing gum, canned meats, soups, confectionery, and baked goods and icings, as well as frozen and salted fish (Armisen and Galatas 2009). Its neutral taste and high fiber content make agar superior to other hydrocolloid additives. These properties have led to broadening the application of agar to many food products. Agar is generally safe for human consumption. According to the U.S. Food and Drug Administration (FDA), agar is classified as GRAS (Generally Recognized as Safe), as an E406 additive by the European Commission, and is registered as 9002-18-0 in the Register Service of the Chemical Abstracts (Armisen and Galatas 2009).

In future, agar is expected to be widely used in biotechnology application because of its benefits. About 10% of the production is classified as too low for a GRAS product. Agar consists of agarose, which is a

*Table 18.3* Agar grades, their application and plant origins

| | Agar type applications | Type of seaweed |
| --- | --- | --- |
| Natural agar | "Strip" | Produced mostly with *Gelidium* by traditional methods |
| | "Square": accustomed only in Far East traditional kitchen | |
| Industrial agar | Food-grade agar used for industrial food production | *Gelidium, Gracilaria, Gracilariopsis, Pterocladia, Ahnfeltia, Gelidiella* |
| | Pharmacological agar | *Gelidium* |
| | Clonic plants production grade | *Gelidium* or *Pterocladia* |
| | Bacteriological grade used for bacteriological culture media formulation | *Gelidium* or *Pterocladia* |
| | Purified agar used in biochemistry and in culture media for very difficult bacteria | *Gelidium* |

*Source:* Armisen, R., *J. Appl. Phycol.*, 7, 231, 1995. With permission.

neutral polysaccharide, and agaropectin, which is polysaccharide sulfate. As a gelling agent, agar is widely used in the food, pharmaceutical, and cosmetics industries. Agarose is widely used in the field of biotechnology, both as a culture medium and an electrophoresis medium. Pure agarose is one of main ingredients of microbial and plant culture media (Istini et al. 2001).

The possible use of agar in food application is related to its properties: its gel-forming capability and its unique, reversible gelling performance. In the food industry, agar mostly used as gelling agent for bread, jelly, candies, dairy products, and ice cream. The food industry has been growing fast during the last decades in line with the increasing world population. Food trend nowadays is changing to healthy products that are low in fat but high in fiber. Agar contains high fiber, which can be used in food fortification to increase health benefit due to its indigestibility by the human metabolic system. Compared to carrageenan, agar is composed of soluble fibers since its content is above 94%, which is higher than that of carrageenan (67%–80%) (Armisen and Galatas 2009). Murata and Nakazoe (2001) found that agar consumption led to a decrease in the concentration of blood glucose and caused an anti-aggregation effect on red blood cells. In the beverage industry, agar is used in beer, coffee, and wine purifying process and also as an emulsifier for chocolate products. Agar with a concentration 0.1%–1% can be used as an emulsifier for some products such as yoghurt, cheese, jelly, and bakery products.

Agar consists of two fractions: agarose and agaropectin (Armisen and Galatas 1987). Agarose has low sulfate content and hence mostly used for biotechnology application (Wahyuni 2003). Agarose is used as an ingredient of making gels for electrophoresis of protein

isolation and purification. In addition, agarose is used in chromatography columns, which has been commercialized production with brand Sepharose (Pharmacia) and Bio-Gel A (Bio-Rad). Agarose also has been widely used in up to 250,000 Da for particle separation as well as for virus, protein, and chromosome separation (Wahyuni 2003).

## 18.5   Conclusions

Indonesia has abundance resources for marine carbohydrate production, especially red algae as a source of agar. However, the existing utilization of those resources is mainly for the production of conventional food products, namely agar powder. Recent exploration has revealed other benefits of agar, which is actually not only to fulfill human consumption needs but also to overcome some health problems based on the bioactive substances contained in agar-producing seaweeds. A new use for agar by taking advantage of its bioactive property is to generate products with significantly higher added value instead of producing conventional products. It is expected that the high added value can be realized by all stakeholders involved in the agar industry, including seaweed farmers whose role is often neglected. This way, the goals of the Indonesian government of improving the national welfare and alleviating poverty through seaweed development can be achieved as planned.

## References

Aji, N., Ariyani, F., and Suryaningrum, T.D. 2003. Technology of seaweed utilization. Research Center for Marine and Fisheries Product Processing and Socioeconomic: Jakarta, Indonesia (in Indonesian).

Almeida, C.L.F., Falcão, H., Lima, G.R.M., Montenegro, C.A, Lira, N.S., and Athayde-Filho, P.F. 2011. Bioactivities from marine algae of the genus *Gracilaria*. *International Journal of Molecular Science*, 12: 4550–4573.

Anggadiredja, J.T., Zatnika, A., Purwoto, H., and Istini, S. 2006. *Seaweed: Culturing, Processing and Marketing Potential Fisheries Commodity*. Penebar Swadaya: Jakarta, Indonesia (in Indonesian).

Armisen, R.1995. Worldwide use and importance of Gracilaria. *Journal of Applied Phycology*, 7: 231–243.

Armisen, R. and Galatas, F. 1987. Production, properties and uses of agar. In: *Production and Utilization of Products from Commercial Seaweeds*, McHugh, D.J. (ed.). FAO Fisheries Technical Paper. FAO: Rome, Italy.

Armisen, R. and Galatas, F. 2009. Agar. In: *Handbook of Hydrocolloids*, Williams, P.A. and Phillips, G.O. (eds.). Woodhead Publishing: Cambridge, U.K., pp. 807–828.

Aslan, M.L. 1991. *Seaweed Cultures*. Kanisius Publishing, Yogyakarta, Indonesia, pp. 11–34 (in Indonesian).

Bhakuni, D.S. and Rawat, D.S. 2005. *Bioactive Marine Natural Products*. Anamaya Publishers: New Delhi, India.

Chapman, V.J. and Chapman, C.J. 1980. *Seaweed and Their Uses*, 3rd edn. Chapman and Hall Ltd.: London, U.K., pp. 148–193.

Chaves Lde, S., Nicolau, L.A., Silva, R.O., Barros, F.C., Freitas, A.L., Aragao, K.S., Ribeiro Rde, A., Souza, M.H., Barbosa, A.L., and Medeiros, J.V. 2013. Anti-inflammatory and anti-nociceptive effects in mice of a sulfated polysaccharide fraction extracted from the marine red algae *Gracilaria caudata*. *Immunopharmacology and Immunotoxicology*, 35: 93–100.

Directorate General of Aquaculture. 2005. Seaweed profiles of Indonesia. Directorate of Aquaculture, Ministry for Marine Affairs and Fisheries: Jakarta, Indonesia (in Indonesian).

Distantina, S., Anggraeni, D.R., and Fitri, L.E. 2008. Effects of concentration and kind of soaking solution on extraction rate and gel properties of agar from *Gracilaria verrucosa*. *Jurnal Rekayasa Proses*, 2(1): 11–16 (in Indonesian).

FAO. 2003. A guide to the seaweed industry. FAO Fisheries Technical Paper 441. Retrieved September 24, 2014 from http://www.fao.org/docrep/006/y4765e/y4765 e00.htm# Contents.

Ghosh, R., Banerjee, K., and Mitra, A. 2012. Eco-biochemical studies of common seaweeds in the lower Gangetic Delta. In: *Handbook of Marine Macroalgae: Biotechnology and Applied Phycology*, Se-Kwon, K. (ed.), 1st edn. John Wiley & Sons, Ltd.: New Delhi, India, pp. 45–57.

Istini, S., Abraham, S., and Zatnika, A. 2001. Purification process of agar from *Gracilaria* sp. *Jurnal Sains dan Teknologi Indonesia*, 3(9): 89–93 (in Indonesian).

Kerton, F.M., Liu, Y., Omaria, K.W., and Hawboldt, K. 2013. Green chemistry and the ocean-based biorefinery. *Green Chemistry*, 15: 860–871.

Kordi, M.G.H. 2011. *The Secret of Seaweed Culture in Bay and Pond*. Lily Publisher: Jakarta, Indonesia (in Indonesian).

Kusuma, W.I., Santosa, G.W., and Pramesti, R. 2013. Pengaruh Konsentrasi NaOH yang Berbeda Terhadaap Mutu Agar Rumput Laut *Gracilaria verrucosa*. *Journal of Marine Research*, 2(2): 120–129 (in Indonesian).

Lahaye, M. and Rochas, C. 1991. Chemical structure and physico-chemical properties of agar. *Hydrobiologia*, 221: 137–148.

Minghetti, L. and Levi, G. 1998. Microglia as effector cells in brain damage and repair: Focus on prostanoids and nitric oxide. *Progress in Neurobiology*, 54: 99–125.

Ministry of Marine Affairs and Fisheries. 2014. Marine and fisheries statistics 2013. Center of Data Statistics and Information: Jakarta, Indonesia (in Indonesian).

Montano, N.E., Villanueva, R.D., and Romero, J.B. 1999. Chemical characteristic and gelling properties of agar from two Philippine *Glacilaria* spp. (Glacilariales, Rhodophyta), *Journal of Applied Phycology*, 11: 27–34.

Murata, M. and Nakazoe, J. 2001. Production and use of marine algae in Japan. *Japan Agriculture Research Quarterly*, 35: 281–290.

Murugan, K. and Iyer, V.V. 2012. Antioxidant and antiproliferative activities of marine algae, *Gracilaria edulis* and *Enteromorpha lingulata*, from Chennai Coast. *International Journal of Cancer Research*, 8: 15–26.

National Standardization Board. 2014. *SNI Agar Powder*. BSN, Jakarta, Indonesia (in Indonesian).

Ramadhan, W. 2011. Utilization of agar powder as texturizer in guava (*Psidium guajava* L.) spread sheet and shelflife prediction. Faculty of Fisheries and Marine Science, Bogor Agricultural University: Bogor, Indonesia (in Indonesian).

Rao, A.V. and Bekheet, I.A. 1976. Preparation of agar—Agar from the red seaweed *Pterocladia capillacea* of the Coast of Alexandria, Egypt. *Applied Environmental Microbiology*, 32(4): 479–482.

Robello, J., Ohno, M., Ukeda, H., and Sawamura, M. 1997. Agar quality of commercial agarophytes from different geographical origins: 1. Physical and rheological properties. *Journal of Applied Phycology*, 8: 517–521.

Rodarte, M.A.V., Carmona, G.H., Montesinos, Y.E.R., Higuera, D.L.A., Rodríguez, R.R., and Álvarez, J.I.M. 2010. Seasonal variation of agar from *Gracilaria vermiculophylla*, effect of alkali treatment time, and stability of its Colagar. *Journal of Applied Phycology*, 22: 753–759.

Sajiki, J. 1997. Effect of acetic acid treatment on the concentrations of arachidonic acid and prostaglandin E2 in the red algae, *Gracilaria asiatica* and *G. rhodocaudata*. *Fish Science*, 63: 128–131.

Scott, T. and Eagleson, M. 1988. *Concise Encyclopedia: Biochemistry*, 2nd edn. Walter de Gruyter: New York p. 18.

Smit, A.J. 2004. Medicinal and pharmaceutical uses of seaweed natural products: A review. *Journal of Applied Phycology*, 16: 245–262.

Sousa, A.A., Benevides, N.M., de Freitas Pires, A., Fiuza, F.P., Queiroz, M.G., Morais, T.M., Pereira, M.G., and Assreuy, A.M. 2013. A report of a galactan from marine alga *Gelidium crinale* with in vivo anti-inflammatory and antinociceptive effects. *Fundamental and Clinical Pharmacology*, 27: 173–180.

Sulistijo. 2002. Research on seaweed culture in Indonesia. Paper presented at *the Speech on Research Professor Inauguration*. LIPI: Jakarta, Indonesia (in Indonesian).

Swantara, I.M.D and Parwata, I.M.O.A. 2011. *Study on Antioxidant Compound of Seaweed from Bali Coast*. The excellent research University of Udayana, Denpasar, Indonesia, pp. 89–96 (in Indonesian).

Van, T.T.T., Ly, B.M., Buu, N.Q., and Kinh, C.D. 2008. Structural characterization of agar extracted from six red seaweed species growing in the coast of Vietnam. *Asean Journal on Science and Technology for Development*, 25(2): 395–403.

Wahyuni, M. 2003. *Biotechnology in Fisheries Product Processing*. Institut Pertanian Bogor: Bogor, Indonesia (in Indonesian).

Yoshizawa, Y., Tsunehiro, J., Nomura, K., Itoh, M., Fukui, F., Ametani, A., and Kaminogawa, S. 1996. In vivo macrophage-stimulation activity of the enzyme-degraded water-soluble polysaccharide fraction from a marine alga (*Gracilaria verrucosa*). *Bioscience Biotechnology Biochemistry*, 60: 1667–1671.

Zatnika A. 1997. *The Profile of Indonesian Seaweed Industry*. Seaweed team of BPPT, BPPT: Jakarta, Indonesia (in Indonesian).

*chapter nineteen*

# Bioprospecting potential of marine natural polymers of chitin and chitosan

*Anguchamy Veeruraj, Muthuvel Arumugam, Thipramalai Thankappan Ajithkumar, and Thangavel Balasubramanian*

## Contents

## 19.1 Introduction

Natural products have great economic and ecological importance, and many natural products are yet to be discovered. The marine environment is a rich source for the production of natural bioactive metabolites, which are used in various clinical trials (Skropeta 2008). Over 60% of natural products can be considered as drugs in the pharmaceutical industry (Newman et al. 2003). Efforts on the discovery of drugs from the oceans during the past 40 years have yielded only a few marine-derived compounds for the treatment of human diseases (Hill and Fenical 2010), encompassing only vidarabine (Ara-A), cytarabine (Ara-C), both used clinically since 1970s, and, most recently, ziconitide (Prialt®), trabectedin (Yondelis®), and the halichondrin B synthetic derivative eribulin mesylate (Halaven™). Of course, this number is disappointing given the high expectations for the extraction of drugs from the sea. The realization of very few marine natural products to be found on the shelves of pharmacies certainly is not due to the limited biological activity or low chemical diversity produced by marine organisms. Polysaccharides, as a class of natural macromolecules, have a tendency to be extremely bioactive, and are generally derived from crustacean shell wastes.

Heteropolysaccharides of glycosaminoglycans (GAGs) are characterized by a repeating disaccharide unit without branched chains in which one of the two monosaccharides is always an amino sugar (*N*-acetylgalactosamine or *N*-acetylglucosamine) and the other is a uronic acid. They are present on all animal cell surfaces and in the extracellular matrix where they are known to bind and regulate different proteins (e.g., growth factors, enzymes and cytokines). After purification, they are used in numerous contexts such as foods, cosmetics, and in the clinic (Kogan et al. 2007; Zou et al. 2009). Polysaccharides are extremely common in nature, and cellulose is the most common organic compound on the planet. It is said that the second most common polysaccharide in the world after cellulose is chitin. Most naturally derived polysaccharides are emerging as ingredients in biomedical applications because they are biodegradable, water soluble, and functionally active.

The current status of polysaccharide-based materials for biomedical use can also be attributed to new possibilities for chemical modifications that enhance their functional activities for specific purposes. These strategies involve combinations of polysaccharides with other polymers and applications of nanotechnology. Some of the major recent applications of polysaccharides in the biomedical field include controlled drug delivery, tissue regeneration, wound dressing, dental implants, blood plasma expanders, vaccines, and nonviral gene delivery. Polysaccharide-based release systems that carry molecules of interest within their networks have been developed for the biomedical and pharmaceutical sectors to transport drugs and other bioactive compounds to targeted sites. Therefore, the development of new technologies to find novel bioactive compounds from by-products of marine processing will bring more value out of what is today considered a waste and represents unique challenges and opportunities for the seafood industry. Moreover, the marine invertebrate species are getting much attention due to the extraction of these polysaccharide groups of chitin and chitosan derived from chitin for safety reasons as well as due to the notion that "life originated from the marine." However, exploration of chitin and chitosan from marine animals is still at the initial stages. In this chapter, we will present brief information on the biological properties of marine natural polymers such as chitin and chitosan connected with their physicochemical, structural, and functional properties and also several biomedical applications of both polymers, emphasizing the effect of the polymer characteristics for these applications.

## 19.2 Physicochemical parameters of chitin and chitosan

The properties of chitin and chitosan are highly dependent on the physicochemical nature of the polymers. Chitosan is poly($\beta$-1,4-D-glucosamine) derived from the N-deacetylation of chitin, and it may be fully or partially N-deacetylated. The degree of acetylation (% DD) is typically less than 0.35 and is an important parameter associated with the physicochemical properties of chitosan, because it is linked directly to chitosan's cationic properties (Pochanavanich and Suntornsuk 2002). Recently, some researchers have shown that chitosan obtained from *M. circinelloides* grown on yam bean medium presented 83% DD. This result is in accordance with those of Amorim et al. (2001), Pochanavanich and Suntornsuk (2002), and Chatterjee et al. (2005), who all reported that the deacetylation degree of chitosan from fungi is between 80% and 90%. The only difference between the structure of chitosan and that of cellulose is the amine ($NH_2$) group at the C-2 position of chitosan instead of the hydroxyl (–OH) group found in cellulose.

Chitosan is insoluble in most solvents, but it dissolves well in dilute organic acids such as acetic acid, formic acid, succinic acid, lactic acid, and malic acid below pH 6.0. Because of that, chitosan can be considered a strong base, as it possesses primary amino groups with a $pK_a$ value of 6.3. The presence of the amino groups indicates that the pH substantially alters the charged state and properties of chitosan (Cho et al. 2000). At low pH, these amines get protonated and become positively charged, which makes chitosan a water-soluble cationic polyelectrolyte. The solubility of chitosan is dependent on the degree of deacetylation, distribution of the acetyl groups along the main chain, molecular weight (MW), and the method of deacetylation employed (Yi et al. 2005; Rinaudo 2006). The degree of ionization depends on the pH and $pK_a$ based on studies on the role of the protonation of chitosan. The best solvent for chitosan was found to be formic acid, where solutions are obtained in an aqueous systems containing 0.2%–100% formic acid (Kienzle-Sterzer et al. 1982). The most commonly used solvent is 1% acetic acid (as a reference) at about pH 4.0. Chitosan is also soluble in 1% hydrochloric acid and dilute nitric acid but insoluble in sulfuric and phosphoric acids. But concentrated acetic acid solutions at high temperature can cause depolymerization of chitosan (Rinaudo et al. 1999a,b). The concentration of the acid plays an important role in imparting the desired functionality (Mima et al. 1983). Chitosan, as stated above, is soluble at pH <6, and the amount of acid needed depends on the quantity of chitosan to be dissolved (Yi et al. 2005).

Commercially, almost all chitosan is obtained from chitin, previously isolated from crustacean exoskeletons (Rinaudo 2006). Basically, chitin isolation from these sources consists of three important steps: demineralization (acid removal of calcium carbonate), deproteinization (removal of proteins), and depigmentation (removal of pigments), which involve the use of hydrochloric acid baths such as 2.5% HCl solution (Shahidi and Synowiecki 1991), alkaline treatment using typically 2% NaOH solutions (Shahidi and Synowiecki 1991), and a solid–liquid extraction with acetone or other solvents or a mild oxidizing treatment (Tolaimate et al. 2003; Hayes et al. 2008). Depending on the severity of these treatments, such as temperature, reaction time, concentration of the chemicals, and concentration and size of the crushed shells, the physicochemical characteristics of the extracted chitin will vary (Kurita 2006). Then, chitosan is obtained from chitin by a deacetylation reaction.

According to the literature (Rhazi et al. 2000; Tolaimate et al. 2003; Chandumpai et al. 2004), two methods are used for the deacetylation of chitin: the Broussignac process and the Kurita process. According to Broussignac (1968), a mixture of solid potassium hydroxide (50% w/w), 96% ethanol (25% w/w), and monoethylene glycol (25% w/w), which is nearly an anhydrous reaction medium, is used as the deacetylation reagent. In the Kurita process (Kurita et al. 1993), a suspension of chitin in aqueous sodium hydroxide solution (50% w/v) is heated up to a certain temperature under a nitrogen stream with stirring. In both processes, after the desired reaction time, the solid is filtered off and washed with distilled water to neutral pH. Derivative processes using different reaction conditions (reaction time, temperature, concentration, and nature of alkaline reagent) and multiple repeating steps have been also studied (Tolaimate et al. 2003). The Broussignac process presents the advantage of giving better quality chitosan (higher MW and higher degree of deacetylation) and can be used with stainless steel reactors (industrially relevant), while the Kurita process does not allow accomplishment of very high deacetylation without larger reaction time and higher temperature, thus resulting in higher degradation of the polysaccharide chain (Tolaimate et al. 2003).

The average viscosimetric molecular weight ($M_V$) of chitosan from *M. circinelloides* obtained in this study was $2.70 \times 10^4$ g/mol, which is considered relatively low (Pochanavanich and Suntornsuk 2002), but agreed with reports of molar weights ranging between $1.0 \times 10^4$ and $9.0 \times 10^5$ g/mol (Nadarajah et al. 2001; Santos et al. 2003; Chatterjee et al. 2005). Pochanavanich and Suntornsuk (2002) reported that chitosan with a low MW can be used to reduce the tensile strength and elongation of the chitosan membrane, but it increases its permeability. Thus, fungal chitosan could have potential application in medicine and agriculture.

Most commercial chitosans have a degree of deacetylation >70% and a MW ranging between 100,000 and 1.2 million Da (Li et al. 1997). Chitosans are of commercial importance due to their high percentage of nitrogen compared to synthetically substituted cellulose, rendering them useful for metal chelation and polyoxysalt and film formulations. Recently, several hydrolytic enzymes, such as lysozyme (Vårum et al. 1996), hemicellulose, and papain (Pantaleone et al. 1992), were found to catalyze the cleavage of the glycosidic linkage in chitosan. An understanding of their physicochemical properties is essential for better application in the entire biomedical field. Currently, applications have been reported in agriculture, food, and cosmetics, and their properties are being investigated for both medical and pharmaceutical applications. Therefore, many biochemists have investigated chitin and chitosan and found them biocompatible, biodegradable, and nontoxic, due to which they have wide applicability in conventional pharmaceutical formulations; however, few researchers have analyzed their physiochemical properties such as gram MW, viscosity, effects of NaCl and pH, extent of deacetylaltion, reminisce after burnt, water log content, stability (thermogravimetric), Fourier transform infrared (FTIR) spectra, presence of heavy metals, and so on (Qina et al. 2004; Cheba 2011; Puvvada et al. 2012).

The synthesis of chitosan involves various chemical steps, such as the preparation of chitin from the crude shells, removal of the proteins in the shells, demineralization for the removal of the carbon and other salts present in the crude form, and deacetylation of the chitin that would result in chitosan (Puvvada et al. 2012). Although chitin is biosynthesized by more than 106 species in three polymorphic configurations (α, β, and γ; Figure 19.1), in laboratories or on an industrial scale, it is usually isolated from the exoskeletons of crustaceans, particularly shrimps and crabs (Tolaimate et al. 2003). α-Chitin is obtained in this way and β-chitin can be extracted from squid pens, while γ-chitin is obtained from fungi and yeast (Campana-Filho et al. 2007; Beckham and Crowley 2011). α-Chitin is the most common form, whereas β-chitin is more reactive and shows a higher affinity for solvents (Kurita et al. 1994). Finally, β-chitin is easily converted into α-chitin by alkaline treatment followed by flushing with water (Noishiki et al. 2003).

Chitosan is the only natural cationic gum that becomes viscous on being neutralized with acid, and this material is used in creams, lotions, as well as permanent waving lotions and nail lacquers (Kumar 2000; Krajewska 2004). Ramos et al. (2003) have suggested chitosan as emulsifiers in cosmetics and pharmaceuticals, and modifying chitosan by introducing phosphoric and alkyl groups into its structure results in the presence of hydrophobic and hydrophilic groups that control the solubility properties. In many cases, emulsion stabilization is achieved by the addition of specially designed polymers that have hydrophilic and hydrophobic segments. For this reason, chitosan has been suggested as an emulsifier, because it is absorbed at the interfacial surface, thus stabilizing the emulsion. From the biological point of view, these parameters affect the solubility of chitin and chitosan in water and organic solvents. The physicochemical properties of chitosan solutions can be controlled by manipulating the solution conditions (temperature, pH, ionic strength and concentration, solvent). The degree of substitution of the hydroxyl and amino groups, or the degree of quaternization of the amino groups, also affects the mechanical and biological properties of chitosans. Finally, considerable research effort has been directed toward developing safe and efficient chitin/chitosan-based products because many factors, such as the size of nanoparticles, can determine the biomedical characteristics of medicinal products (Kumirska et al. 2011).

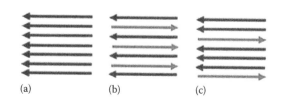

(a)   (b)   (c)

*Figure 19.1* Schematic representation of the three polymorphic forms of chitin: (a) α-chitin, (b) β-chitin, and (c) γ-chitin.

## 19.3 Structural and functional properties of chitin and chitosan

Chitin is the second most abundant natural polysaccharide after cellulose on earth, and it is a high MW linear homopolymer of β-(1,4) linked by N-acetyl glucosamine (N-acetyl-2-amino-2-deoxy-D-glucopyranose) units (Figure 19.2). Chitin is a naturally occurring mucopolysaccharide, and usually occurs in animals, particularly in crustaceans, molluscs, and insects, where it is a major constituent of the exoskeleton, and in certain diatoms, nematodes, and fungi, where it is the principal fibrillar polymer in the cell wall. Chitin is structurally similar to cellulose; however, less consideration has been paid to chitin than cellulose, primarily due to its inertness. Hence, it remains an essentially underutilized resource. Chitin has a crystalline structure, and it constitutes a network of organized fibers, which confers rigidity and resistance to organisms. Deacetylation of chitin yields chitosan, which is a relatively reactive compound and is produced in numerous forms, such as powder, paste, film, and fiber. Chitosan, a copolymer of glucosamine and N-acetyl glucosamine units linked by 1–4 glucosidic bonds, is a cationic polysaccharide and its chemical formula is $C_4H_{11}NO_4$ (Tomihata and Ikada 1997). Chitin and chitosan with beneficial biological and antimicrobial properties are attractive materials for wound care.

Chitin polymers can be found in three forms, α, β, and γ. α-Chitin is known to have a parallel sheet conformation and is the most abundant form in nature. This form can be found in the shells of crabs and shrimps. β-Chitin is found in the spines of diatoms, squid pens, and pogonophoran tubes, and the polymer is made of antiparallel sheets. γ-Chitin, which occurs in fungi and yeast, is comprised both the α and β forms, and thus is a mixture of both antiparallel and parallel sheets. The chains are associated with each other by very strong hydrogen bonds between the amide groups and carbonyl groups of adjacent chains. These hydrogen linkages are responsible for the high insolubility of the chains in water and for the formation of fibrils. Chitin, which has a compact conformation made of highly acetylated regions and sheet-rich three-dimensional structures, is poorly water soluble. These properties make the industrial and commercial exploitation of this structure difficult. To enhance water solubility, chemically modified or hydrolyzed derivatives are usually produced. For example, alkaline hydrolysis removes the acetyl groups and leaves just the amino groups, allowing the polymer to be converted from a poorly water soluble molecule into a highly water soluble one.

According to Santos et al. (2003), deacetylation and regeneration processes cause disturbance to the initial crystalline reticulum of chitin, inducing a reordering of the hydrogen linkage of chitosan. This can be observed in the IR central band at ~3483 and 3305 $cm^{-1}$, in the region of (1) the axial deformation of OH, which appears as an overlapping band of the axial deformation of NH, indicating an intermolecular hydrogen-bond formation; and (2) the displacement of the higher frequency band, indicating an increase in the structural order. These data are in accordance with those reported in the literature when comparing the infrared spectra of chitin and chitosan obtained by microbiological methods (Andrade et al. 2000; Amorim et al. 2001; Pochanavanich and Suntornsuk 2002; Santos et al. 2003).

Chitosan is a cationic polysaccharide made up of the same units and glycosidic linkage as chitin. However, low amounts of N-acetyl D-glucosamine (GlcNAc) are found in chitosan, usually <30%. Physicochemical characteristics such as hydrophobicity and interchain

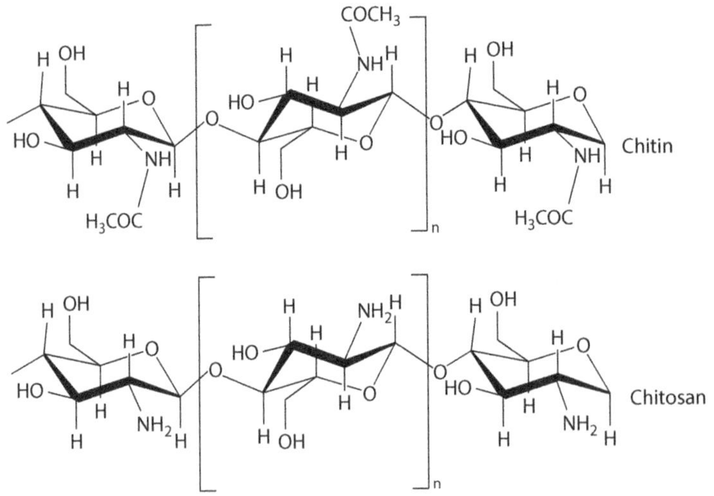

*Figure 19.2* Chemical structure of chitin and chitosan.

interactions depend on the amount and distribution of the acetyl groups (Figure 19.2). Another physicochemical characteristic that varies naturally among different chitosan samples is the MW. Based on this characteristic, three categories of chitosan exist. These categories are named, according to their different MWs, as high MW chitosans (HMWCs), medium MWCs (MMWCs), and low MWCs (LMWCs). The MW ranges between 10 and 100 kDa for LMWC, between 100 and 300 kDa for MMWC, and >300 kDa for HMWC. In aqueous solution, HMWC samples are more viscous than LMWC or MMWC samples. Although LMWC materials can be obtained by size exclusion chromatography of unmodified chitosans, enzymatic methods can be additionally employed to produce LMWC derivatives. Although chitosan and its derivatives are cationic by nature, structural differences among them account for the differences in their biological activities and physicochemical properties (Zhang et al. 2009; Ozhan Aytekin et al. 2012).

## 19.4   Biomedical applications of chitin and chitosan and their derivatives

Chitin and chitosan possess especially interesting biological and biomedical properties; therefore, they have been used in many applications, mainly in the medical and pharmaceutical fields. Not all chitin/chitosan samples show the same biological activities (Xia et al. 2011). The most important properties of chitosan and their potential biomedical and other applications are summarized in Table 19.1.

Potential applications of chitosan can be exploited only if its usable forms are properly developed and prepared. In solution and gel, it can be used as a bacteriostatic, fungistatic, and coating agent. Gels and suspensions might play roles as carriers for the slow release or controlled action of drugs, as an immobilizing medium, and as an encapsulation material. Films and membranes are used in dialysis, contact lenses, dressings, and the encapsulation of mammal cells, including cell cultures. Chitosan sponges are used in dressings and in preventing bleeding in mucous membranes; they are also used as resorbable sutures, nonwovens for dressings, and drug carriers in the form of hollow fibers.

The drug-delivery potential of chitin and chitosan finds various applications, including oral, nasal, parenteral, and transdermal administration, in implants, and in gene delivery. Chitin and chitosan-based particulate systems are attracting biological, pharmaceutical, and biomedical applications as potential drug-delivery devices. The most promising developments at present are in pharmaceutical and biological areas, followed by cosmetics. The most important aspects of the biological and biomedical applications of chitin and chitosan are due to their following properties.

### 19.4.1   Antimicrobial activity of chitin and chitosan

The antimicrobial potential of chitin and chitosan is often discussed in the recent reports, and depends on their physical properties. These natural biopolymers have been used against a broad spectrum of target organisms such as bacteria, fungi, viruses, and algae (Vinsova and Vavrikova 2008). Kong et al. (2010) have investigated the ability of chitosan to inhibit a wide variety of bacteria, fungi, yeasts, and viruses, as well as its application in a broad range or variety of antimicrobial agents, in experiments involving *in vivo* and *in vitro* interactions in different forms (solutions, films, composites). Chitosan possesses unique properties, which make it an ideal ingredient for the development of antimicrobial edible films and nontoxic biopolymers, and it has been confirmed to serve as a matrix to obtain edible films containing essential oils (Bonilla et al. 2012). The interaction (binding or chelation) of chitosan with endotoxins of gram-negative bacteria decreased their acute toxicity. Because of the strong chelating ability of chitosan, external chelating agents such as ethylenediaminetetraacetic acid (EDTA) may not be required when antimicrobial agents such as nisin are supplemented with chitosan to control gram-negative bacteria.

Sudharshan et al. (1992) studied the antimicrobial effect of water-soluble chitosan, such as chitosan lactate, chitosan hydroglutamate, and chitosan derived from *Absidia coerulea* fungi, on different bacterial cultures. They observed that chitosan glutamate and chitosan lactate were also bactericidal against both gram-positive and gram-negative bacteria in the range of 1–5 log cycle reduction within 1 h. In the same study, the authors reported that chitosan was no longer bactericidal at pH 7 because of two major reasons: the presence of a significant proportion of uncharged amino groups, and the poor solubility of chitosan.

Studying the antimicrobial activity of chitin, chitosan, and their derivatives against bacteria, Wang (1992) observed that a much higher concentration of chitosan (1% ± 1.5%) was required for the complete inactivation of *Staphylococcus aureus* after 2 days of incubation at pH 5.5 or 6.5 in the medium. Furthermore, Chang et al. (1989) established that chitosan concentrations ≥0.005 were sufficient to elicit complete inactivation of *S. aureus*. Accordingly, Darmadji and Izumimoto (1994) found the effect of chitosan in meat preservation. Simpson et al. (1997) reported the antimicrobial effect of different

***Table 19.1*** Antimicrobial effects of chitosan preparations: a summary of *in vitro* studies

| Chitosan preparations | Microorganisms major | Results/conclusions | References |
|---|---|---|---|
| Chitin and chitosan powder | *Escherichia coli, Pseudomonas aeruginosa, Enterococcus faecalis,* and *Staphylococcus saprophyticus* | Effective antimicrobial effect | Andres et al. (2007) |
| Chitosan solution | *Staphylococcus aureus* | Chitosan led to multiple changes in the bacterial gene expression; binding of chitosan to teichoic acid led to death of bacteria | Raafat et al. (2008) |
| Chitosan and chitosan oligomer | Gram-positive and gram-negative bacteria | Chitosan generally showed stronger bactericidal effects with gram-positive bacteria than with gram-negative bacteria | No et al. (2002) |
| *N*-carboxybutyl chitosan | 298 strains gram-positive bacteria, gram-negative bacteria and *Candida* spp. | Particularly active against *Candida* and gram-positive bacteria; no bactericidal activity was observed with streptococci and enterococci | Muzzarelli et al. (1990) |
| Solutions of chitosan hydrochloride, carboxymethyl chitosan, chitosan oligosaccharide, and *N*-acetyl-ᴅ-glucosamine | *Candida albicans, Candida krusei,* and *Candida glabrata* | Antifungal activity decreases with declining molecular mass and increasing masking of the protonated amino groups with functional groups | Seyfarth et al. (2008) |
| Thiazolidinone derivatives of chitosan | *E. coli, Shigella dysenteriae, P. aeruginosa,* and *Bacillus subtilis.* | Better antimicrobial activity than chitosan Schiff bases | Kulkarni et al. (2005) |
| Chitosan hydrolysate | *Bacillus cereus, E. coli, S. aureus, P. aeruginosa, Salmonella enterica* serovar *typhi,* and *Saccharomyces cerevisiae* | Strong activity at 100 ppm against many pathogens and yeast species | Tsai et al. (2004) |
| Chitosan film loaded with thyme oil | *E. coli, Klebsiella pneumoniae, P. aeruginosa,* and *S. aureus* | Chitosan films with 1.2% (v/v) showed antimicrobial activity on all microorganisms tested | Altiok et al. (2010) |
| Chitosan hydrochloride, 5-methyl-pyrrolidinone chitosan | *S. aureus, Staphylococcus epidermidis, P. aeruginosa, C. albicans,* and *Aspergillus niger* | Antimicrobial activity against bacteria and *C. albicans* is shown by the dressing | Rossi et al. (2007) |
| Chitosan incorporated with polyphosphate and silver | *S. aureus* and *P. aeruginosa* | Complete kill of *P. aeruginosa* and >99.99% kill of *S. aureus* | Ong et al. (2008) |
| Argininely functionalized chitosan | *Pseudomonas fluorescens* and *E. coli* | At the concentration of 5000 mg/L, 6% and 30% substituted chitosan-arginine killed 2.7 logs and 4.5 logs of *P. fluorescens,* and 4.8 logs and 4.6 logs of *E. coli* in 4 h, respectively | Tang et al. (2010) |
| Chitosan hydrogel | *P. aeruginosa, E. coli, S. aureus,* and *Fusarium solani* | Excellent antimicrobial/ antifungal activities | Li et al. (2011) |

cultures of bacteria on raw shrimp, with different concentrations of chitosan, and observed the variations in their degree of susceptibility to chitosan. Liu et al. (2006) determined the effect of MW and concentration of chitosan on the antibacterial activity of *Escherichia coli*. The results showed that chitosans with different MW but with the same degree of deacetylation could be obtained by acetic acid hydrolysis.

Moreover, Muzzarelli et al. (1990) reported the antimicrobial efficacy of *N*-carboxybutyl chitosan, which was prepared from crustacean chitosan (DDA = 73%), against 298 strains of gram-positive and

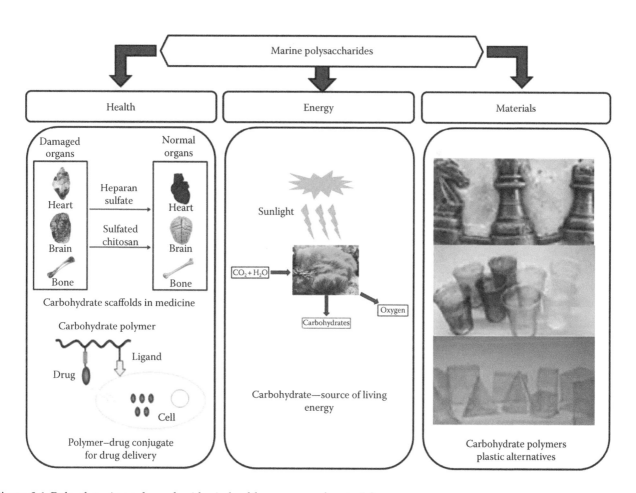

*Figure 9.1* Role of marine polysaccharides in health, energy, and materials.

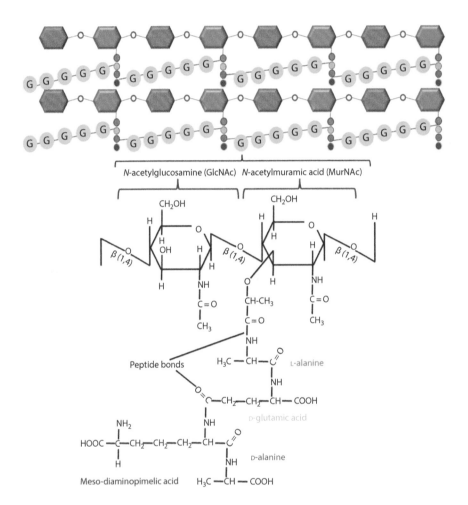

**Figure 17.1** Structural polysaccharide of bacterial cell wall; Peptidoglycan structure composed of *N*-acetylglucosamine (GlcNAc) and *N*-acetylmuramic acid (MurNAc).

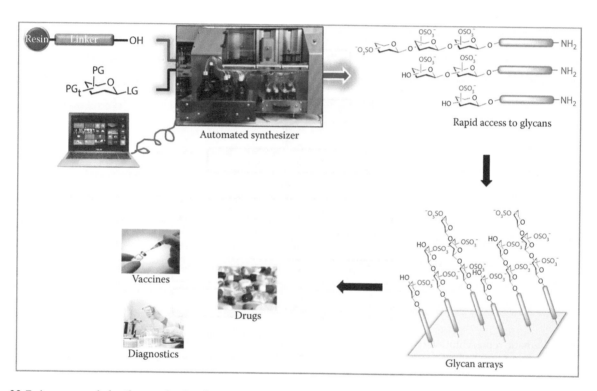

*Figure 22.7* An approach for the synthesis of marine polysaccharides.

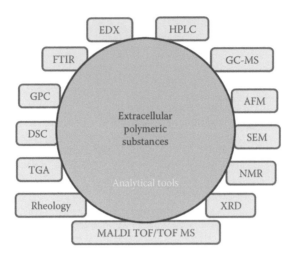

*Figure 26.1* Analytical tools commonly used for EPSs analysis.

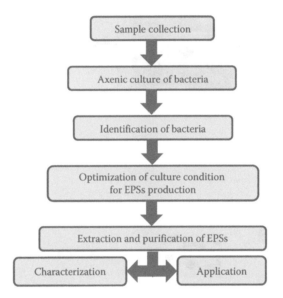

*Figure 26.2* Schematic representation of the commonly used methodology for EPSs studies.

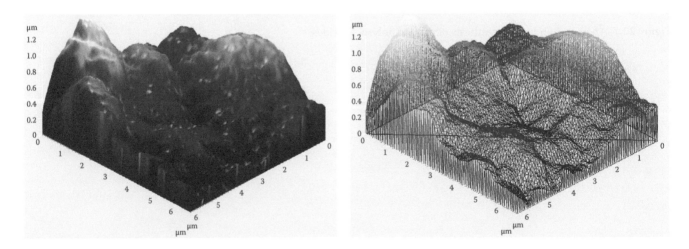

*Figure 26.5* Representative AFM image of EPSs.

gram-negative pathogens and *Candida* spp. They found that *N*-carboxybutyl chitosan was particularly active against *Candida* and gram-positive bacteria. Kulkarni et al. (2005) described the antibacterial activity of chitosan after its conversion into thiazolidinone derivatives (TDCs). TDCs were prepared by converting chitosan into chitosan's Schiff's bases, followed by treatment with mercaptoacetic acid. Polymer samples (both original chitosan and chemically modified chitosan TDCs) at a concentration of 100 ppm were tested for antimicrobial activity against *E. coli*, *Shigella dysenteriae*, *P. aeruginosa*, and *Bacillus subtilis* using a disk diffusion method by measuring the zone of inhibition. It was observed that the antibacterial activity of chitosan increases by ~10-fold in the corresponding TDC. The increase in antibacterial activity in chemically modified chitosan was proposed to be due to the newly introduced groups, leading to increased interaction and the formation of polyelectrolyte complexes between the polymer and the bacterial cell wall. The diffusive permeability of the polymer was also an important parameter for antibacterial activity. Similar *in vitro* studies on the antimicrobial effects of chitosan as well as its derivatives and complexes were also carried out by Tsai et al. (2004), Rossi et al. (2007), Ong et al. (2008), and Altiok et al. (2010). Table 19.1 gives a summary of the literature on the *in vitro* studies.

The mechanism of the antibacterial activity of chitosan works by making the bacteria flocculate and killing them. The exact mechanism of the antimicrobial action of chitin, chitosan, and their derivatives is still imperfectly known, but different mechanisms have been proposed. One of the reasons for the antimicrobial characteristic of chitosan, as discussed in previous sections, is its positively charged amino groups that interact with negatively charged microbial cell membranes, leading to the leakage of proteinaceous and other intracellular constituents of the microorganisms. It has been reported that quaternized chitosan with a high degree of substitution of the quaternary ammonium exhibits a strong interaction with negative charges on the bacterial cell surface and shows better antibacterial activity than chitosan (Young et al. 1982; Papineau et al. 1991; Seo et al. 1992; Sudharshan et al. 1992; Fang et al. 1994; Chen et al. 1998).

## 19.4.2 Antioxidant activities of chitin and chitosan

The antioxidant activities of chitosan have been extensively studied both *in vitro* and *in vivo* using different methodologies, and recent studies have shown that these are related to its structural characteristics, including MW, degree of deacetylation (DD), and its sources. MW and DD may also present some synergistic effects on the biological activities of chitosan. The antioxidant

potency of the chitin derivative-glucosamine HCl was investigated by employing various established *in vitro* systems, such as super oxide ($O_2^-$)/hydroxyl (ox)–radical scavenging, reducing power, and ferrous ion chelating potency (Xing et al. 2005). Yen and Coworkers investigated the antioxidant properties of chitosan prepared from crab shells (Yen et al. 2008) and shiitake stipes (Yen et al. 2007). Both studies showed that the antioxidant activities of chitosan increased with the increase of DD during preparation. A longer *N*-deacetylation time results in more amino groups on the C-2 positions, which contribute significantly to the antioxidant activities. Chitosan from both sources showed the greatest antioxidant activities in hydroxyl radical scavenging, conjugated diene formation, and reducing power assays, showing $EC_{50}$ <1.5 mg/mL, whereas they showed less satisfactory activities in (1,1-diphenyl-2-picrylhydrazyl) DPPH radicals and ferrous ion chelating ability tests, showing $EC_{50}$ as high as 16.3 mg/mL. $EC_{50}$ is defined as the effective concentration at which the antioxidant activity is reduced to 50%.

Xing et al. (2005) reported a major effect on the antioxidant activities of chitosan, with lower MW having more pronounced scavenging effects on superoxide and hydroxyl radicals than those with higher MW. Additionally, Sun et al. (2007) investigated the antioxidant properties of chitosan oligomers obtained by $H_2O_2$ degradation, and confirmed that MW is negatively related to the antioxidant activities. Chitosan oligomers with MW as small as 2300 Da have the greatest antioxidant activities against the superoxide anion and hydroxyl radicals. Chitosan with higher MW was considered to be more compact in structure due the stronger intramolecular hydrogen bonds, such as $N_2 \cdots O_6$ and $O_3 \cdots O_5$, than that with lower MW. Therefore, the hydroxyl and amino groups in chitosan with lower MW are more flexible to react with free radicals, and hence exhibit higher free-radical scavenging effects.

Many other modifications have been also reported to improve the free-radical scavenging activities of chitosan, such as the Schiff base reaction (Guo et al. 2005), quaternization (Guo et al. 2006; Xing et al. 2008), carboxymethylation (Guo et al. 2005; Sun et al. 2008), and acylation (Sun et al. 2011). Chemical cross-linking methods are generally adopted to graft antioxidants to chitosan molecules, including essential oils (Jung et al. 2006; Chen et al. 2009) and phenolics (Pasanphan et al. 2010; Cho et al. 2011). The antioxidant properties of grafted chitosan are significantly improved, especially for DPPH scavenging and metal ion chelating activities, whereas native chitosan has very little activity. However, because of the toxic and irritating reagents that are needed in the chemical modification process, such modification is less popular. Another novel and attractive method has been recently developed and aroused increasing interest as an alternative to chemical cross-linking modification, and enzymatic modification.

The antioxidant properties of chitosan have been shown not only by *in vitro* antioxidant assays but also in many *in vivo* studies using various animal models as well as clinical trials. Anraku et al. (2012) investigated the effect of high-MW chitosan on antioxidant stress and chronic renal failure using nephrectomized rats. These result showed that the ingestion of chitosan over a 4-week period not only resulted in a significant decrease in the ratio of oxidized to reduced albumin and in an increase in the biological antioxidant potential but also alleviated renal failure. Another recent study also revealed the antiaging effect of high-MW chitosan in glutathione-dependent antioxidant system in rats (Anandan et al. 2013). The oral administration of chitosan was reported to significantly attenuate oxidative stress in the heart tissue of aged rats by maintaining antioxidant enzymes. The antioxidant effect of high-MW chitosan has been recently demonstrated in a clinical trial (Anraku et al. 2011). Besides those of high-MW chitosan, the antioxidant properties of low-MW chitosan, especially oligosaccharides, have also been widely studied in many *in vivo* experiments. Similar results as those of chitosan oligosaccharides have been observed in diabetic rats induced by streptozotocin (Yuan et al. 2009). These results suggested that low-MW chitosan possessed greater antioxidant activities than high-MW chitosan, as observed by *in vitro* assays; however, MW-dependent antioxidant activities have not been well understood yet. Furthermore, the *in vivo* evaluation of chitosan derivatives has not been well established so far because some toxic chemicals are usually used in the modification process, which may pose potential toxicity to animals or humans.

### 19.4.3 Anticancer activity of chitin and chitosan

Approximately 60% of the drugs approved for cancer treatment are derived from natural sources; for example, vincristine, etoposide, irinotecan, taxane, and camptothecin are plant-derived compounds. One of the important compounds in cancer treatment is chitin, $(C_8H_{13}O_5N)_n$, which was first derived from a fungus and is the first known polysaccharide (Domard and Domard 2002). Potent acids can dissociate chitin, a water-insoluble polymer, into acetic acid and chitosan. Chitosan is a nontoxic polymer, and it is the main component of arthropod exoskeletons, tendons, and respiratory tract lining, as well as the cell walls of the majority of fungi.

The antitumor activity of chitin/chitosan is effected by the stimulation of the immune system (production of lymphokines, including interleukins 1 and 2, stimulation of NK, etc.) (Jeon and Kim 2002; Qin et al. 2002). Maeda and Kimura (2004) investigated the antitumor

effect of three water-soluble low-MW chitosans (21, 46, and 130 kDa) and various doses of 650 kDa chitosan in sarcoma 180 bearing mice. They found that LMWC (21 and 46 kDa) and smaller oligosaccharides could activate the intestinal immune system of animals, thus preventing tumor growth. But no antitumor effect was observed after the oral administration of chitosan samples, even of low MW (46 kDa). The same authors confirmed that high-MW chitosan (650 kDa) prevents the adverse reactions of some cancer chemotherapeutic drugs. Based on experimental results concerning the inhibition of angiogenesis and the induction of apoptosis, it was confirmed that chitooligosaccharides (COSs) are more potent angio-inhibitory and antitumor compounds. Moreover, Wang et al. (2006) reported that chitin oligosaccharides (DP 1–6) also reduced the number of K562 cells (human erythromyeloblastoid leukemia cell line).

Previously, Jeon and Kim (2002) had found that oligomers of chitosan exhibited antitumor activities *in vitro* and *in vivo* and low-MW of chitosan had remarkable antimetastatic effects against lung cancer in mice. The viability of cancer cells was amazingly decreased in some cell lines, whether or not the COS derivatives were positively or negatively charged (Huang et al. 2006). Moreover, an antioxidant and antiproliferative effect on a human bladder cancer cell line ($T_{24}$) was reported after 48 h, depending on chitosan concentration (Kuppusamy and Karuppaiah 2012). Lim et al. (2011) investigated the effect of one of the potential compounds from chitin, astaxanthin, on a human esophagus cancer cell line (TE-4) over 24, 48, and 72 h, and found a dose-dependent inhibition. At the same time, Kuppusamy and Karuppaiah (2013) analyzed the cytotoxic efficacy of chitosan and cyclophosphamide by evaluating cell viability in $T_{24}$ human urinary bladder cancer cell line. Treatment of $T_{24}$ cells with increasing concentration of chitosan and cyclophosphamide led to a concentration-dependent decrease in cell migration. It was demonstrated that chitosan had a cytotoxic effect on $T_{24}$ human bladder cancer cell line comparable to that of the drug cyclophosphamide.

## 19.5 Chitin and chitosan based biomaterials using biomedical aspects

Chitosan, a derivative of the biopolymer chitin, has been extensively applied in biomedical and pharmaceutical research because of its low toxicity and good biocompatibility. Olteanu (2007) has reported the current applications of various types of chitosan derivatives in the fields of drug delivery, tissue engineering, wound dressing, antimicrobial preparations, biotechnology, pharmaceuticals, and cosmetics. Wound is defined as "loss or breaking of cellular and anatomic or functional continuity of living tissues" (Ramzi et al. 1994). Wound healing is a biological process that is initiated by trauma and often terminated by scar formation. Wound dressing is one of the most promising medical applications of chitin and chitosan. The adhesive nature of chitin and chitosan, together with their antifungal and bactericidal character, as well as their permeability to oxygen, is a very important property associated with the treatment of wounds and burns. Different derivatives of chitin and chitosan have been prepared for this purpose in the form of hydrogels, fibers, membranes, scaffolds, and sponges (Jayakumar et al. 2011). The analysis of the kinetics of this biological process in response to different forms of dermal substitution is important for the development of efficient therapeutic products capable of stimulating wound healing (Alborova et al. 2008). Chitosan is able to accelerate re-epithelialization and normal skin regeneration, and to confer considerable antibacterial activity against a broad spectrum of bacteria (Mi et al. 2001; Azad et al. 2004).

The carboxymethyl derivatives of chitin and chitosan have been shown to be promising for adsorbing metal ions, as drug delivery systems, in wound healing, as antimicrobial agents, in tissue engineering, as components in cosmetics and food, and for antitumor activities. The focus has been on the preparative methods and applications of carboxymethyl and succinyl derivatives of chitin and chitosan, with particular emphasis on their uses as materials for biomedical applications (Jayakumar et al. 2010). Ribeiro et al. (2009) evaluated the applicability of a chitosan hydrogel in wound dressing. In their study, fibroblast cells isolated from rat skin were used to assess the cytotoxicity of the hydrogel. The results showed that chitosan hydrogel was able to promote cell adhesion and proliferation. Chitosan's positive charge allows electrostatic interactions with glycosaminoglycans, which are growth factors that enhance cell growth and proliferation (Lee et al. 2010). Cell viability studies have shown that the hydrogel and its degradation by-products are non-cytotoxic.

Moreover, some researchers have reported the use of chitosan and its derivatives for the delivery of antimicrobials (Aoyagi et al. 2007; Rossi et al. 2007; Vimala et al. 2010), growth factors (Obara et al. 2005; Alemdaroglu et al. 2006; Park et al. 2009), and other drugs (Jayakumar et al. 2007; Dai et al. 2009). Chitosan will gradually depolymerize to release N-acetyl-β-D-glucosamine, which initiates fibroblast proliferation, helps in ordered collagen deposition, and stimulates increased levels of natural hyaluronic acid synthesis at the wound site, which helps in faster wound healing and scar prevention (Paul and Sharma 2004). The best biomaterials for wound dressing should be biocompatible and promote the growth of the dermis and epidermis layers. Chen et al. (2008) reported the composite nanofibrous membrane of chitosan/collagen, which is known for its beneficial effects on wound healing. The various forms of wound dressing materials based on chitin and chitosan derivatives that are commercially available are given in Table 19.2 (Brown et al. 2009).

Recently, Sung et al. (2010) developed a minocycline-loaded wound dressing with enhanced healing effect. The cross-linked hydrogel films were prepared with ploy(vinyl alcohol) (PVA) and chitosan using freeze-drying. Their gel properties, *in vitro* protein adsorption, release, *in vivo* wound healing effect, and histopathology were then evaluated. Niyas Ahamed and Sastry (2011) showed that gentamycin-loaded chitosan–cellulose–silver nanoparticle composite (C-Ch-Ag) films had more wound-healing properties than blank C-Ch-Ag film. These films were evaluated for their water absorption capacity, antibacterial activity, tensile strength, and *in vivo* wound healing property using the excision wound model using albino rats. The drug-loaded films showed significant difference in water absorption capacity and antibacterial activity when compared to an optimized blank composite film.

*Table 19.2* Commercially available chitin and chitosan and its derivative-based wound dressing materials

| S. no. | Materials with references | Trade name | Manufacturer |
|---|---|---|---|
| 1. | Chitin and its derivatives (Brown et al. 2009) | Syvek-Patch® | Marine Polymer Technologies |
| | | Chitopack C® | Eisai Co. Japan |
| | | Chitopack S® | Eisai Co. Japan |
| | | Beschitin® | Unitika Co. Japan |
| 2. | Chitosan and its derivatives (Brown et al. 2009) | Tegasorb® | 3M |
| | | Tegaderm® | 3M |
| | | Hemcon Bandage® | HemCon |
| | | Chitodine® | IMS |
| | | Trauma DEX® | Medafor |

Recently, Nguyen et al. (2013) carried out *in vivo* studies and found greater wound closure in wounds treated with a curcumin-composite sponge than those with a composite sponge without curcumin and untreated sponge. These sponges were evaluated for their water absorption capacity, antibacterial activity, *in vitro* drug release, and *in vivo* wound healing by the excision wound model using rabbits. The obtained results revealed that the combination of curcumin, chitosan, and gelatin could improve the wound-healing activity in comparison to chitosan and gelatin without curcumin. The electrospinning technique is a versatile method to spin polymers into continuous fibers with diameters ranging from a few micrometers to a few nanometers. Electrospinning creates seemingly endless ultrafine fibers that collect in a random pattern. Because of their intrinsic features, polymeric nanofibers are attractive materials for biomedical and biotechnological applications such as tissue engineering, nanocomposites for dental application, controlled drug delivery, medical implants, wound dressings, biosensors, and filtration. Pillai and Sharma (2009) suggested that polymeric nanofibers have inherent biodegradability, biofunctionality, and biocompatibility of the biopolymer, and electrospun chitin and CS fibers have special advantages whereby properties such as cytocompatibility, tissue responses, and so on, could be controlled for critical applications.

Wu et al. (2012) have formulated nano-carriers for drug and gene delivery, hydrogels for tissue engineering, and membranes for hemodialysis. There are still many potential applications of chitosan and its derivatives that need to be developed, indicating that they are to reach their full potential. Although chitosan could be used as potential gene carriers due to its biodegradability, biocompatibility, and strong DNA-binding ability, it has several disadvantages such as poor solubility in physiological pH, low transfection efficiency, and low cell specificity (Park et al. 2004). It can be molded into various forms (gels, membranes, sponges, beads, and scaffolds) and has exceptional pore-forming ability for potential applications in tissue engineering, drug delivery, and wound healing (Muzzarelli 2009; Jayakumar et al. 2010).

Chitosan is a potentially useful pharmaceutical material owing to its good biocompatibility and low toxicity, and it has been combined with a variety of biopolymers and bioceramic systems, such as alginate, hyaluronic acid, amylopectin, carbon nanotubes, poly(methyl methacrylate), poly(lactic acid), growth factors, HA, and calcium phosphate (Muzzarelli 1997; Di Martino et al. 2005; Venkatesan and Kim 2010; Dash et al. 2011; Pallela et al. 2012; Venkatesan et al. 2012; Croisier and Jérôme 2013). Therefore, evaluation of the biocompatibility of various biomedical-grade chitosan derivatives is necessary to engineer materials that are of high quality and biocompatibility for human wound management. Moreover, Tran et al. (2013) have presented some useful chemical routes to functionalize hybrid chitosan-based nanomaterials and demonstrated their use in some prospective biomedical applications (antibacterial and antiproliferative activities, drug delivery, biodetection). The ability of these materials has been clearly demonstrated by many physicochemical methods. The results have suggested material design/synthesis for enhanced therapeutic and diagnostic effects and the most important biomedical applications of chitosan-based nanoparticles and nanocomposites.

## 19.6   Marine chitin, chitosan and their biomedical importance

Chitin and chitosan are unique and typical marine polysaccharides waiting for prospective applications and have been attracting the attention of many researchers from different disciplines. Chitin and its derivatives exhibit a variety of physicochemical and biological properties, resulting in numerous applications in areas ranging from wastewater treatment to agrochemical, environmental, and industrial uses. In addition, its lack of toxicity and allergic reactions, as well as its biocompatibility, biodegradability, and bioactivity, make it a very attractive substance for diverse applications as a biomaterial in the pharmaceutical and medical fields. Chakraborty and Ghosh (2010) have reported the production of chitosan from marine trash crustaceans and its application in mercury removal from seawater. They have concluded that its bioadsorption of heavier metals is waiting to be explored, which will open green avenues in the field of pollution abatement. Recently, the isolation and identification of chitin and chitosan from the cuttle bone of *Sepia prashadi* Winckworth have been reported by Jothi and Nachiyar (2013). Chitin and chitosan were successfully extracted from the cuttle bone, with a chitin yield of 24% from the cuttlebone of *S. prashadi* and a chitosan yield of 62.6% from chitin. Moreover, the degree of acetylation of chitosan was calculated as 55.95% using the FTIR stretching bands, which encourages extending the study to other potential sources for the extraction of chitin and chitosan. Cuttlebones are widely used as a homemade remedy for ear ache and skin diseases in India and China (Lane 1962). Further, cuttlebone has also reported antibacterial and antifungal activity against some human pathogenic microorganisms (Rajaganapathi 2001). Chitin and chitosan are used in the preparation of the materials used in wound healing, as antibacterial and antifungal agents, and in dialysis membranes, biomedical beads, fabrics, and gauzes (Subasinghe 1999). Nair and Madavan (1989) and Das et al. (1996) demonstrated that the crabs *Scylla serrata* and *Portunus pelagicus* contain 16.07% and

20.19% of chitin, respectively, in their body shell waste. The cuttlebone of *S. officinalis* was found to possess 20% chitin (Tolaimate et al. 2003), whereas, in general, squid/octopus was estimated to have 3%–20% of chitin (Patil and Satam 2002).

Recently, Vázquez et al. (2013) discussed the recent progress in eco-friendly processes to extract and purify the biomacromolecules such as chondroitin sulfate (CS), hyaluronic acid (HA), and chitin/chitosan (CH/CHs) from marine by-products with the aim of proposing environmentally friendly processes by combining various microbial, chemical, enzymatic, and membranes strategies and technologies. It was suggested that CS, HA, and CH/CHs have attracted increasing attention because of their beneficial effects on several aspects of the human health, in the formulation of cosmeceuticals and anti-aging products, nutraceuticals, and food ingredients, as well as their application in bio- and nano technological processes. For a long time, extensive studies have been conducted on the clarification of the general aspects of the chemical structures, features, novel applications, and more sustainable processes for their production.

Sulfated chitin and chitosan have a variety of biological functions, including anti-HIV-1, antioxidant, antimicrobial, blood anticoagulation, and hemagglutination inhibition activities. Moreover, some of these sulfated chitin and chitosan derivatives function in drug delivery, adsorption of metal ions, prevention of cancer metastasis, or as elicitors of conflict to late blight in potato (Vasyukova et al. 2000; Xing et al. 2004; Jayakumar et al. 2007). Therefore, many studies have been made on amino-derivatized chitosans, which possess numerous biological activities such as antioxidant, antihypertensive, enzyme inhibition, and antimicrobial (Lee et al. 2010). Among the amino-derivatized chitosans, aminoethyl-chitosan, prepared from 50% deacetylated chitosan, showed activity against HIV-1 with an $IC_{50}$ value of 17 µg/mL (Artan et al. 2008). Thus, aminoethyl-chitosan has been used as a new-generation drug candidate against HIV. The possibilities of designing new drug candidates and pharmaceuticals to support reducing or regulating HIV infection-related chronic malfunctions are promising.

Chemical modification of chitin and chitosan can generate novel compounds that possess good pharmacological properties such as antiviral activity. Nishimura et al. (1998) found that chitin sulfate had good inhibitory effect on HIV-1 infection and that its inhibition action on HIV-1 depended significantly on the sites of sulfation. In addition, it was reported that aminoethyl-chitosan, prepared from 50% deacetylated chitosan, also shows good inhibitory activity against HIV-1 *in vitro* (Vo and Kim 2010). A great deal of interest has thus been generated regarding marine-derived anti-HIV agents such as phlorotannins, sulfated chitooligosaccharides, sulfated polysaccharides, lectins, and bioactive peptides. Moreover, these evidences suggested that, because of their valuable biological functions and beneficial health effects, marine-derived anti-human immunodeficiency (HIV virus) agents have the potential as active ingredients for the preparation of novel pharmaceutical products. However, until now, most studies on the anti-HIV activity of marine-derived HIV inhibitors have been carried out *in vitro* or in animal model systems, and further research is needed in order to understand their activities in human subjects.

## 19.7   Summary

Oceans not only consist of water, but they are also an abundant source of various materials. There is a need to develop several bioactive biomaterials from marine sources. Marine organisms have interesting applications as medicines, dietary supplements, and cosmetics, and also have research and industrial uses. The importance of the biopolymers chitin and chitosan-based particulate systems is due to their biological, physicochemical, pharmaceutical, and biomedical potential, for example, in wound-healing and as drug-delivery devices. Many approaches are currently available to deliver drugs to the specific sites of action. For the development of targeted delivery systems, chitosan and its derivatives possess many advantages such as biocompatibility, biodegradability, mucoadhesivity, and other unique biological properties. All these intrinsic properties vary with MW and DD, which are the most important characteristics of chitosan. Because of the abundant amino groups, chitosan carries many positive charges in acidic media, and has become a popular biopolymer to develop encapsulation and delivery systems for the food industry, such as nano/microparticles, hydrogel beads, and nanocomposites. Overall, it is clear that chitosan and its derivatives are useful carriers for low-MW drugs requiring targeted delivery, and proper modification of chitosan, its functional properties, and biological activities can be further enhanced as more applications are developed. Major efforts will be required to identify interesting bioprospecting and highly demanding processes (in terms of skills and expertise, funding, and time) and to develop finished products in the future.

## References

Alborova, A., J. Lademann, L. Meyer, A. Kramer, H. Richter, A. Patzelt, W. Sterry, and S. Koch. 2008. *In vivo* analysis of wound healing by optical methods. *GMS Krankenhaushyg Interdiszip* 3: 1–4.

Alemdaroglu, C., Z. Degim, N. Celebi et al. 2006. An investigation on burn wound healing in rats with chitosan gel formulation containing epidermal growth factor. *Burns* 32: 319–327.

Altiok, D., E. Altiok, and F. Tihminlioglu. 2010. Physical, antibacterial and antioxidant properties of chitosan films incorporated with thyme oil for potential wound healing applications. *Journal of Materials Science: Materials in Medicine* 21: 2227–2236.

Amorim, R.V.S., W. Souza, K. Fukushima, and G.M. Campos-Takaki. 2001. Faster chitosan production by mucoralean strains in submerged culture. *Brazilian Journal of Microbiology* 32: 20–23.

Anandan, R., B. Ganesan, T. Obulesu et al. 2013. Antiaging effect of dietary chitosan supplementation on glutathione-dependent antioxidant system in young and aged rats. *Cell Stress Chaperones* 18: 121–125.

Andrade, V.S., B.B. Neto, W. Souza, and G.M. Campos-Takaki. 2000. A factorial designs analysis of chitin production by *Cunninghamella elegans. Canadian Journal of Microbiology* 46: 1042–1045.

Andres, Y., L. Giraud, C. Gerente, and P. Le Cloiree. 2007. Antibacterial effects of chitosan powder mechanisms of action. *Environmental Technology* 28: 1357–1363.

Anraku, M., T. Fujii, Y. Kondo et al. 2011. Antioxidant properties of high molecular weight dietary chitosan *in vitro* and *in vivo. Carbohydrate Polymers* 83: 501–505.

Anraku, M., H. Tomida, A. Michihara et al. 2012. Antioxidant and renoprotective activity of chitosan in nephrectomized rats. *Carbohydrate Polymers* 89: 302–304.

Aoyagi, S., H. Onishi, and Y. Machida. 2007. Novel chitosan wound dressing loaded with minocycline for the treatment of severe burn wounds. *International Journal of Pharmaceutics* 330: 138–145.

Artan, M., F. Karadeniz, M.M. Kim, and S.K. Kim. 2008. Chitosan derivatives as HIV-1 inhibitors. *Journal of Biotechnology* 136S: S527–S540.

Azad, A.K., N. Sermsintham, S. Chandrkrachang, and W.F. Stevens. 2004. Chitosan membranes as a wound-healing dressing: Characterization and clinical application. *Journal of Biomedical Materials Research Part B: Applied Biomaterials* 69: 216–222.

Beckham, G.T. and M.F. Crowley. 2011. Examination of the α-chitin structure and decrystallization thermodynamics at the nanoscale. *Journal of Physical Chemistry B* 115: 4516–4522.

Bonilla, J., L. Atares, M. Vargas, and A. Chiralt. 2012. Effect of essential oils and homogenization conditions on properties of chitosan-based films. *Food Hydrocolloids* 26: 9–16.

Broussignac, P. 1968. Chitosan: A natural polymer not well known by the industry. *Chimie et Industrie Genie Chimque* 99: 1241–1247.

Brown, M.A., M.R. Daya, and J.A. Worley. 2009. Experience with chitosan dressings in a civilian EMS system. *The Journal of Emergency Medicine* 37: 1–7.

Campana-Filho, S.P., D. de Britto, E. Curti, F.R. Abreu, M.B. Cardoso, M.V. Battisti, P.C. Sim, R.C. Goy, R.L. Signini, and R. Lavall. 2007. Extraction, structures and properties of α- and β-chitin. *Quimica Nova Journal* 30: 644–650.

Chakraborty, S. and U. Ghosh. 2010. Production of chitosan from marine trash crustaceans and its application in mercury removal from sea water. *International Journal of Chemical and Analytical Science* 1: 186–192.

Chandumpai, A., N. Singhpibulporn, D. Faroongsarng, and P. Sornprasit. 2004. Preparation and physico-chemical characterization of chitin and chitosan from the pens of the squid species, *Loligo lessoniana* and *Loligo formosana. Carbohydrate Polymers* 58: 467–474.

Chang, D.S., H.R. Cho, H.Y. Goo, and W.K. Choe. 1989. A development of food preservation with the waste of crab processing. *Bulletin in Korean Fish Society* 22: 70–78.

Chatterjee, S., M. Adhya, A.K. Guah, and B.P. Chatterjee. 2005. Chitosan from *Mucor rouxii*: Production and physico-chemical characterization. *Process Biochemistry* 40: 395–400.

Cheba, B.A. 2011. Chitin and chitosan: Marine biopolymers with unique properties and versatile applications. *Global Journal of Biotechnology and Biochemistry* 6(3): 149–153.

Chen, C., W. Liau, and G. Tsai. 1998. Antibacterial effects of N-sulfonated and N-sulfobenzoyl chitosan and application to oyster preservation. *Journal of Food Protection* 61: 1124–1128.

Chen, F., Z. Shi, K.G. Neoh, and E.T. Kang. 2009. Antioxidant and antibacterial activities of eugenol and carvacrol-grafted chitosan nanoparticles. *Biotechnology and Bioengineering* 104: 30–39.

Chen, J.P., G.Y. Chang, and J.K. Chen. 2008. Electrospun collagen/chitosan nanofibrous membrane as wound dressing. *Colloids and Surfaces A: Physicochemical Engineering Aspects* 313–314: 183–188.

Cho, Y.S., S.K. Kim, and J.Y. Je. 2011. Chitosan gallate as potential antioxidant biomaterial. *Bioorganic and Medicinal Chemistry Letters* 21: 3070–3073.

Cho, Y.W., J. Jang, C.R. Park, and S.W. Ko. 2000. Preparation and solubility in acid and water of partially deacetylated chitins. *Biomacromolecules* 1: 609–614.

Croisier, F. and C. Jérôme. 2013. Chitosan-based biomaterials for tissue engineering. *European Polymer Journal* 49: 780–792.

Dai, M., X. Zheng, X. Xu et al. 2009. Chitosan-alginate sponge: Preparation and application in curcumin delivery for dermal wound healing in rat. *Journal of Biomedicine and Biotechnology* 595126: 1–8.

Darmadji, P. and M. Izumimoto. 1994. Effect of chitosan in meat preservation. *Meat Science* 38: 243–254.

Das, N.G., P.A. Khan, and Z. Hossain. 1996. Chitin from the shell of two coastal portunid crabs of Bangladesh. *Indian Journal of Fish* 43: 413–415.

Dash, M., F. Chiellini, R.M. Ottenbrite, and E. Chiellini. 2011. Chitosan—A versatile semi-synthetic polymer in biomedical applications. *Progress in Polymer Science* 36: 981–1014.

Di Martino, A., M. Sittinger, and M.V. Risbud. 2005. Chitosan: A versatile biopolymer for orthopaedic tissue-engineering. *Biomaterials* 26: 5983–5990.

Domard, A. and M. Domard. 2002. Chitosan: Structure-properties relationship and biomedical applications. In *Polymeric Biomaterials*, S. Dumitriu (ed.). Marcel Dekker Inc., New York, Vol. 2, pp. 187–212.

Fang, S.W., C.F. Li, and D.Y.C. Shih. 1994. Antifungal activity of chitosan and its preservative effect on low-sugar Candied Kumquat. *Journal of Food Protection* 56: 136–140.

Guo, Z., R. Xing, S. Liu et al. 2005. The synthesis and antioxidant activity of the Schiff bases of chitosan and carboxymethyl chitosan. *Bioorganic & Medicinal Chemistry Letters* 15: 4600–4603.

Guo, Z.Y., H.Y. Liu, X.L. Chen, X. Ji, and P. Li. 2006. Hydroxyl radicals scavenging activity of N-substituted chitosan and quaternized chitosan. *Bioorganic & Medicinal Chemistry Letters* 16: 6348–6350.

Hayes, M., B. Carney, J. Slater, and W. Bruck. 2008. Mining marine shellfish wastes for bioactive molecules: Chitin and chitosan—Part A: Extraction methods. *Journal of Biotechnology* 3: 871–877.

Hill, R.T. and W. Fenical. 2010. Pharmaceuticals from marine natural products: Surge or ebb? *Current Opinion in Biotechnology* 21: 777–779.

Huang, R., E. Mendis, N. Rajapakse, and S.K. Kim. 2006. Strong electronic charge as an important factor for anticancer activity of chitooligosaccharides (COS). *Life Science* 78: 2399–2408.

Jayakumar, R., M. Prabaharan, S.V. Nair, S. Tokura, H. Tamura, and N. Selvamurugan. 2010. Novel carboxymethyl derivatives of chitin and chitosan materials and their biomedical applications. *Progress in Materials Science* 55: 675–709.

Jayakumar, R., M. Prabaharan, P.T. Sudheesh Kumar, S.V. Nair, and H. Tamura. 2011. Biomaterials based on chitin and chitosan in wound dressing applications. *Biotechnology Advances* 29: 322–337.

Jayakumar, R., R.L. Reis, and J.F. Mano. 2007. Synthesis and characterization of pH-sensitive thiol-containing chitosan beads for controlled drug delivery applications. *Drug Delivery* 14: 9–17.

Jeon, Y.J. and S.K. Kim. 2002. Antitumor activity of chitosan oligosaccharides produced in ultrafiltration membrane reactor system. *Journal of Microbiological Biotechnology* 12: 503–507.

Jothi, N. and R.K. Nachiyar. 2013. Identification and isolation of chitin and chitosan from cuttle bone of *Sepia prashadi* Winckworth, 1936. *Global Journal of Biotechnology & Biochemistry* 8: 33–39.

Jung, B.O., S.J. Chung, and S.B. Lee. 2006. Preparation and characterization of eugenol-grafted chitosan hydrogels and their antioxidant activities. *Journal of Applied Polymer Science* 99: 3500–3506.

Kienzle-Sterzer, C.A., D. Rodriguez-Sanchez, and Ch. Rha. 1982. Dilute solution behavior of a cationic polyelectrolyte. *Journal of Applied Polymer Science* 27: 4467–4470.

Kogan, G., L. Soltés, R. Stern, and P. Gemeiner. 2007. Hyaluronic acid: A natural biopolymer with a broad range of biomedical and industrial applications. *Biotechnology Letters* 29: 17–25.

Kong, M., X.G. Chen, K. Xing, and H.J. Park. 2010. Antimicrobial properties of chitosan and mode of action: A state of the art review. *International Journal of Food Microbiology* 144: 51–63.

Krajewska, B. 2004. Application of chitin- and chitosan-based materials for enzyme immobilizations: A review. *Enzyme and Microbial Technology* 35: 126–139.

Kulkarni, A.R., V.H. Kulkarni, J. Keshavayya et al. 2005. Antimicrobial activity and film characterization of thiazolidinone derivatives of chitosan. *Macromolecular Bioscience* 5: 490–493.

Kumar, M.N.V.R. 2000. A review of chitin and chitosan applications. *Reactive and Functional Polymers* 46: 1–27.

Kumirska, J., M.X. Weinhold, J. Thöming, and P. Stepnowski. 2011. Biomedical activity of chitin/chitosan based materials-influence of physicochemical properties apart from molecular weight and degree of N-acetylation. *Polymers* 3: 1875–1901.

Kuppusamy, S. and J. Karuppaiah. 2012. Antioxidant and cytotoxic efficacy of chitosan on bladder cancer. *Asian Pacific Journal of Tropical Medicine* 2: 769–773.

Kuppusamy, S. and J. Karuppaiah. 2013. Comparative study of cytotoxic efficacy of chitosan and cyclophosphamide against T24 human urinary bladder cancer cell line. *IJAPBC* 2: 21–24.

Kurita, K. 2006. Chitin and chitosan: Functional biopolymers from marine crustaceans. *Marine Biotechnology* 8: 203–226.

Kurita, K., S. Ishii, K. Tomita, S.I. Nishimura, and K. Shimoda. 1994. Reactivity characteristics of squid β-chitin as compared with those of shrimp chitin: High potentials of squid chitin as a starting material for facile chemical modifications. *Journal of Polymer Science Part A: Polymer Chemistry* 32: 1027–1032.

Kurita, K., K. Tomita, T. Tada, S. Ishii, S.I. Nishimura, and K. Shimoda. 1993. Squid chitin as a potential alternative chitin source deacetylation behavior and characteristic properties. *Journal of Polymer Science Part A: Polymer Chemistry* 31A: 485–491.

Lee, D.S., Y.M. Kim, M.S. Lee, C.B. Ahn, W.K. Jung, and J.Y. Je. 2010. Synergistic effects between aminoethyl-chitosans and β-lactams against methicillin-resistant *Staphylococcus aureus* (MRSA). *Bioorganic and Medicinal Chemistry Letters* 20: 975–978.

Li, Q., E. T. Dunn, E. W. Grandmaison, and M. F. A. Goosen. 1997. Applications and Properties of Chitin and Chitosan. In *Applications of Chitin and Chitosan*, M. F. A. Goosen (ed.).Technomic Publishing Company, Lancaster, PA. *Process Biochemistry* 37: 1359–1366.

Li, Q., E.T. Lunn, E.W. Grandmason, and M.F. Goosen. 1997. Applications and properties of chitosan. In *Applications and Properties of Chitosan*, M.F.A. Goosen (ed.). Technomic Publishing Co., Inc. Lancaster, U.K., pp. 1–3.

Li, Z., X. Guo, S. Matsushita, and J.Guna.2011. Differentiation of cardiosphere-derived cells into a mature cardiac lineage using biodegradable poly (N-isopropylacrylamide) hydrogels. *Biomaterials* 32: 3220–3232.

Lim, S.A., J.Y. Lee, W.H. Jung et al. 2011. Anticancer effects of astaxanthin and alpha-tocopherol in esophageal cancer cell lines. *Korean Journal of Helicobacter and Upper Gastrointestinal Research* 11: 170–175.

Liu, N., X.G. Chen, H.J. Park, C.G. Liu, C.S. Liu, X.H. Meng, and L.J. Yu. 2006. Effect of MW and concentration of chitosan on antibacterial activity of *Escherichia coli*. *Carbohydrate Polymers* 64: 60–65.

Maeda, Y. and Y. Kimura. 2004. Antitumor effects of various low-molecular-weight chitosans are due to increased natural killer activity of intestinal intraepithelial lymphocytes in sarcoma 180-bearing mice. *Journal of Nutrition* 134: 945–950.

Mi, F.L., S.S. Shyu, Y.B. Wu, S.T. Lee, J.Y. Shyong, and R.N. Huang. 2001. Fabrication and characterization of sponge-like asymmetric chitosan membranes as a wound dressing. *Biomaterials* 22: 165–173.

Mima, S., M. Miya, R. Iwamoto, and S. Yoshikawa. 1983. Highly deacetylated chitosan and its properties. *Journal of Applied Polymer Science* 28: 1909–1917.

Muzzarelli, R., R. Tarsi, O. Filippini et al. 1990. Antimicrobial properties of N-carboxybutyl chitosan. *Antimicrobial Agents and Chemotherapy* 34: 2019–2023.

Muzzarelli, R.A.A. 1997. Human enzymatic activities related to the therapeutic administration of chitin derivatives. *Cellular and Molecular Life Sciences* 53: 131–140.

Muzzarelli, R.A. 2009. Chitins and chitosans for the repair of wounded skin, nerve, cartilage and bone. *Carbohydrate Polymers* 76: 167–182.

Nadarajah, K., J. Kader, M. Mohd, and D.C. Paul. 2001. Production of chitosan by fungi. *International Journal of Biological Science* 4: 263–265.

Nair, K.G.R. and P. Madhavan. 1989. Advances in chitin research. In *Recent Trends in Processing Low Cost Fish*, K. Balachandran et al. (eds.). Society of Fisheries Technologists, Kochi, India, p. 174.

Newman, D.J., G.M. Cragg, and K.M. Snader. 2003. Natural products as sources of new drugs over the period 1981–2002. *Journal of Natural Products* 66: 1022.

Nguyen, V.C., V.B. Nguyen, and M. Hsieh. 2013. Curcumin-loaded chitosan/gelatin composite sponge for wound healing application. *International Journal of Polymer Science* 106570: 1–7.

Nishimura, S.I., H. Kai, K. Shinada, T. Yoshida, S. Tokura, K. Kurita, H. Nakashima, N. Yamamoto, and T. Uryu. 1998. Regioselective syntheses of sulfated polysaccharides: Specific anti-HIV-1 activity of novel chitin sulfates. *Carbohydrate Research* 306: 427–433.

Niyas Ahamed, M.I. and T.P. Sastry. 2011. Wound dressing application of chitosan based bioactive compounds. *International Journal of Pharmacy and Life Sciences* 2: 991–996.

Noishiki, Y., H. Takami, Y. Nishiyama, M. Wada, S. Okada, and S. Kuga. 2003. Alkali-induced conversion of β-chitin to α-chitin. *Biomacromolecules* 4: 896–899.

Obara, K., M. Ishihara, M. Fujita et al. 2005. Acceleration of wound healing in healing-impaired db/db mice with a photocrosslinkable chitosan hydrogel containing fibroblast growth factor-2. *Wound Repair and Regeneration* 13: 390–397.

Olteanu, C.E. 2007. Applications of functionalized chitosan. *Scientific Study and Research* 8: 227–256.

Ong, S.Y., J. Wu, S.M. Moochhala et al. 2008. Development of a chitosan-based wound dressing with improved hemostatic and antimicrobial properties. *Biomaterials* 29(32): 4323–4332.

Ozhan Aytekin, A., S. Morimura, and K. Kida. 2012. Physiological activities of chitosan and N-trimethyl chitosan chloride in U937 and 3T3-L1 cells. *Polymers for Advanced Technologies* 23: 228–235.

Pallela, R., J. Venkatesan, V.R. Janapala, and S.K. Kim. 2012. Biophysicochemical evaluation of chitosan-hydroxyapatite-marine sponge collagen composite for bone tissue engineering. *Journal of Biomedical Materials Research Part A* 100: 486–495.

Pantaleone, D., M. Yalpani, and M. Scollar. 1992. Unusual susceptibility of chitosan to enzymic hydrolysis. *Carbohydrate Research* 237: 325–332.

Papineau, A.M., D.G. Hoover, D. Knorr, and D.F. Farkas. 1991. Antimicrobial effect of water-soluble chitosans with high hydrostatic pressure. *Food Biotechnology* 5: 45–57.

Park, C.J., S.G. Clark, C.A. Lichtensteiger et al. 2009. Accelerated wound closure of pressure ulcers in aged mice by chitosan scaffolds with and without bFGF. *Acta Biomaterialia* 5: 1926–1936.

Park, J.H., S. Kwon, J.O. Nam et al. 2004. Self-assembled nanoparticles based on glycol chitosan bearing 5β-cholanic acid for RGD peptide delivery. *Journal of Controlled Release* 95: 579–588.

Pasanphan, W., G.R. Buettner, and S. Chirachanchai. 2010. Chitosan gallate as a novel potential polysaccharide antioxidant: An EPR study. *Carbohydrate Polymers* 345: 132–140.

Patil, Y.T. and S.B. Satam. 2002. Chitin and chitosan. Treasure from crustacean shell waste. *Sea Food Export Journal* XXXIII: 31–38.

Paul, W. and C.P. Sharma. 2004. Chitosan and alginate wound dressing: A short review. *Trends in Biomaterials and Artificial Organs* 18: 18–24.

Pillai, C.K.S. and C.P. Sharma. 2009. Electrospinning of chitin and chitosan nanofibres. *Trends Biomaterials and Artificial Organs* 22: 179–201.

Pochanavanich, P. and W. Suntornsuk. 2002. Fungal chitosan production and its characterization. *Letters in Applied Microbiology* 35: 17–21.

Puvvada, Y.S., S. Vankayalapati, and S. Sukhavasi. 2012. Extraction of chitin from chitosan from exoskeleton of shrimp for application in the pharmaceutical industry. *International Current Pharmaceutical Journal* 1: 258–263.

Qin, C.Q., Y.M. Du, L. Xiao, Z. Li, and X.H. Gao. 2002. Enzymic preparation of water soluble chitosan and their antitumor activity. *International Journal of Biological Macromolecules* 31: 111–117.

Qina, C., B. Zhoub, L. Zenga, Z. Zhanga, Y. Liub, Y. Dub, and L. Xiaob. 2004. The physicochemical properties and antitumor activity of cellulase-treated chitosan. *Food Chemistry* 84: 107–115.

Raafat, D., K., von Bargen, A. Haas, and H.G. Sahl. 2008. Insights into the mode of action of chitosan as an antibacterial compound. *Applied and Environmental Microbiology* 74:3764–3773.

Rajaganapathi, J. 2001. Antimicrobial activities of marine molluscs and purification of anti-HIV protein. PhD, thesis, Annamalai University, Tamil Nadu, India, pp. 1–130.

Ramos, V.M., N.M. Rodriguez, M.F. Diaz, M.S. Rodriguez, A. Heras, and E. Agullo. 2003. N-methylene phosphonic chitosan. Effect of preparation methods on its properties. *Carbohydrate Polymers* 52: 39–46.

Ramzi, S.C., K. Vinay, and R. Stanley. 1994. Pathologic basis of diseases. WB Saunders Company, Philadelphia, 5: p 86.

Rhazi, M., J. Desbrieres, A. Tolaimate, A. Alagui, and P. Vottero. 2000. Investigation of different natural sources of chitin: Influence of the source and deacetylation process on the physicochemical characteristics of chitosan. *Polymer International* 49: 337–344.

Ribeiro, M.P., A. Espiga, D. Silva et al. 2009. Development of a new chitosan hydrogel for wound dressing. *Wound Repair and Regeneration* 17: 817–824.

Rinaudo, M. 2006. Chitin and chitosan: Properties and application. *Progress in Polymer Science* 31: 603–632.

Rinaudo, M., G. Pavlov, and J. Desbrieres. 1999a. Influence of acetic acid concentration on the solubilization of chitosan. *Polymer* 40: 7029–7032.

Rinaudo, M., G. Pavlov, and J. Desbrieres. 1999b. Solubilization of chitosan in strong acid medium. *International Journal of Polymer Analysis and Characterization* 5: 267–276.

Rossi, S., M. Marciello, G. Sandri et al. 2007. Wound dressings based on chitosans and hyaluronic acid for the release of chlorhexidine diacetate in skin ulcer therapy. *Pharmaceutical Development and Technology* 12: 415–422.

Santos, J.E., J.P. Soares, E.R. Dockal, S.P. Campana Filho, E.T.G. Cavalheiro. 2003. Caracterização de quitosanas comerciais de diferentes origens. *Polím. Ciência e Tecnol.* 13: 242–249.

Seo, H.J., K. Mitsuhashi, and H. Tanibe. 1992. *Advances in Chitin and Chitosan.* Elsevier Applied Science, New York, pp. 34–40.

Seyfarth F., S. Schliemann, P. Elsner, and U. C. Hipler. 2008. Antifungal effect of high- and low-molecular-weight chitosan hydrochloride, carboxymethyl chitosan, chitosan oligosaccharide and N-acetyl-d-glucosamine against *Candida albicans, Candida krusei and Candida glabrata.* International Journal of Pharmaceutics 353(1-2): 139–148.

Shahidi, F. and J. Synowiecki. 1991. Isolation and characterization of nutrients and value-added products from snow crab (*Chinoecetes opilio*) and shrimp (*Pandalus borealis*) processing discards. *Journal of Agricultural and Food Chemistry* 39: 1527–1532.

Simpson, B.K., N. Gagne, I.N.A. Ashie, and E. Noroozi. 1997. Utilization of chitosan for preservation of raw shrimp (*Pandalus borealis*). *Food Biotechnology* 11: 25–44.

Skropeta, D. 2008. Deep sea natural products, *Natural Product Report* 25: 989–1216.

Subasinghe, S. 1999. Chitin from shellfish waste-health benefits over-shadowing industrial areas. *Infofish International* 3/99: 58–65.

Sudharshan, N.R., D.G. Hoover, and D. Knorr. 1992. Antibacterial action of chitosan. *Food Biotechnology* 6: 257–272.

Sun, T., Q. Yao, D.X. Zhou, and F. Mao. 2008. Antioxidant activity of N-carboxymethyl chitosan oligosaccharides. *Bioorganic and Medicinal Chemistry Letters* 18: 5774–5776.

Sun, T., D.X. Zhou, J.L. Xie, and F. Mao. 2007. Preparation of chitosan oligomers and their antioxidant activity. *European Food Research Technology* 225: 451–456.

Sun, T., Y. Zhu, J. Xie, and X.H. Yin. 2011. Antioxidant activity of N-acyl chitosan oligosaccharide with same substituting degree. *Bioorganic and Medicinal Chemistry Letters* 21: 798–800.

Sung, J.H., M.R. Hwang, J.O. Kim, J.H. Lee, Y.I. Kim, J.H. Sun, W. Chang, S.G. Jin, J.A. Kim, W.S. Lyoo, S.S. Han, S.K. Ku, C.S. Yong, and H.G. Choi. 2010. Gel characterization and in vivo evaluation of minocycline-loaded wound dressing with enhanced wound healing using polyvinyl alcohol and chitosan. *International Journal of Pharmaceutics* 392: 232–240.

Tolaimate, A., J. Desbrieres, M. Rhazi, and A. Alagui. 2003. Contribution to the preparation of chitins and chitosans with controlled physic-chemical properties. *Polymer* 44: 7939–7952.

Tomihata, K. and Y. Ikada. 1997. *In vitro* and *in vivo* degradation of films of chitin and its deacetylated derivatives. *Biomaterials* 18: 567–575.

Tran, Q.H., V.Q. Nguyen, and A.T. Le. 2013. Silver nanoparticles: Synthesis, properties, toxicology, applications and perspectives. *Advances in Natural Sciences: Nanoscience and Nanotechnology* 4 (033001): 1–20.

Tsai, G.J., S.L. Zhang, and P.L. Shieh. 2004. Antimicrobial activity of a low-molecular-weight chitosan obtained from cellulase digestion of chitosan. *Journal of Food Protection* 67: 396–398.

Vårum, K.M., H.K. Holme, M. Izume, B.T. Stokke, and X. Smidsrød. 1996. Determination of enzymatic hydrolysis specificity of partially N-acetylated chitosans. *Biochimica et Biophysica Acta* 1291: 5–15.

Vasyukova, N.I., G.I. Chalenko, N.G. Gerasimova, E.A. Perekhod, O.L. Ozeretskovskaya, A.V. Irina, V.P. Varlamov, and A.I. Albulov. 2000. Chitin and chitosan derivatives as elicitors of potato resistance to late blight. *Applied Biochemistry and Microbiology* 36: 372–376.

Vázquez, J.A., I.R. Amado, M.I. Montemayor, J. Fraguas, M.P. González, and M.A. Murado. 2013. Chondroitin sulfate, hyaluronic Acid and chitin/chitosan production using marine waste sources: characteristics, applications and eco-friendly processes: A review. *Marine Drugs* 11(3): 747–774.

Venkatesan, J. and S.K. Kim, 2010. Chitosan Composites for Bone Tissue Engineering-An Overview. *Marine Drugs* 8(8): 2252–2266.

Venkatesan, J., R. Pallela, I. Bhatnagar, and S.K. Kim. 2012. Chitosan amylopectin/hydroxyapatite and chitosan chondroitin sulphate/hydroxyapatite composite scaffolds for bone tissue engineering. *International Journal of Biological Macromolecules* 51: 1033–1042.

Vimala, K., Y.M. Mohan, K.S. Sivudu et al. 2010. Fabrication of porous chitosan films impregnated with silver nanoparticles: A facile approach for superior antibacterial application. *Colloids and Surfaces B Biointerfaces* 76: 248–258.

Vinsova, J. and E. Vavrikova. 2008. Recent advances in drugs and prodrugs design of chitosan. *Current Pharmaceutical Design* 14: 1311–1326.

Vo, T.S. and S.K. Kim. 2010. Potential anti-HIV agents from marine resources: An overview. *Marine Drugs* 8: 2871–2892.

Wang, G. 1992. Inhibition and inactivation of five species of food borne pathogens by chitosan. *Journal of Food Protection* 55: 916–919.

Wang, S.L., T.Y. Lin, Y.H. Yen, H.F. Liao, and Y.J. Chen. 2006. Bioconversion of shellfish chitin wastes for the production of *Bacillus subtilis* W-118 chitinase. *Carbohydrate Research* 341: 2507–2515.

Wu, G., H. Gao, and J. Ma. 2012. Chitosan-based biomaterials. *Material Matters* 7: 1–4.

Xia, W., P. Liu, J. Zhang, and J. Chen. 2011. Biological activities of chitosan and chitooligosaccharides. *Food Hydrocolloids* 25: 170–179.

Xing, R., S. Liu, Z. Guo et al. 2008. Relevance of molecular weight of chitosan-N-2-hydroxypropyl trimethyl ammonium chloride and their antioxidant activities. *European Journal of Medicinal Chemistry* 43: 336–340.

Xing, R., S. Liu, H. Yu, Q. Zhang, Z. Li, and P. Li. 2004. Preparation of low-molecular-weight and high-sulfate-content chitosans under microwave radiation and their potential antioxidant activity *in vitro*. *Carbohydrate Research* 339: 2515–2519.

Xing, R., S. Liu, Z.Y. Guo et al. 2005 Relevance of molecular weight of chitosan and its derivatives and their antioxidant activities *in vitro*. *Bioorganic and Medicinal Chemistry* 13: 1573–1577.

Yen, M.T., Y.H. Tseng, R.C. Li, and J.L. Mau. 2007. Antioxidant properties of fungal chitosan from shiitake stipes. *LWT-Food Science Technology* 40: 255–261.

Yen, M.T., J.H. Yang, and J.L. Mau. 2008. Antioxidant properties of chitosan from crab shells. *Carbohydrate Polymer* 74: 840–844.

Yi, H., L.Q. Wu, W.E. Bentley et al. 2005. Biofabrication with chitosan. *Biomacromolecules* 6: 2881–2894.

Young, D.H., H. Kohle, and H. Kauss. 1982. Effect of chitosan on membrane permeability of suspension cultured *Glycine max* and *Phaseolus vulgaris* cells. *Plant Physiology* 70: 1449–1454.

Yuan, W.P., B. Liu, C.H. Liu et al. 2009. Antioxidant activity of chito-oligosaccharides on pancreatic islet cells in streptozotocin-induced diabetes in rats. *World Journal of Gastroenterology* 15: 1339–1345.

Zhang, L., M. Wang, X. Kang et al. 2009. Oxidative stress and asthma: Proteome analysis of chitinase-like proteins and FIZZ1 in lung tissue and bronchoalveolar lavage fluid. *Journal of Proteome Research* 8: 1631–1638.

Zou, X.H., Y.Z. Jiang, G.R. Zhang, H.M. Jin, T.M. Nguyen, and H.W. Ouyang. 2009. Specific interactions between human fibroblasts and particular chondroitin sulphate molecules for wound healing. *Acta Biomaterials* 5: 1588–1595.

## chapter twenty

# Bioactivities of sulfated polysaccharide porphyran isolated from edible red alga Porphyra yezoensis

Zedong Jiang and Tatsuya Oda

## Contents

## 20.1 Introduction

Marine algae are rich sources of polysaccharides in nature. Some marine algal polysaccharides such as agaran, carrageenan, and alginate have already been well explored and widely utilized as additives in the food industry as gelling, thickening, and emulsifying agents due to their rheological properties. In the recent years, many marine algal polysaccharides, especially fucoidan, ascophyllan, carrageenan, and agaran, have drawn much attention from diverse research fields to develop new drugs, cosmeceutical and nutraceutical food, or other supplements to keep the human body healthy, because of their wide variety of biological activities including anticoagulant, antioxidant, antitumor/anticancer, antiviral, antibacterial, immunomodulatory, anti-hypolipidemic, anti-hypotensive, antiallergic, and anti-angiogenic (Ren et al. 1994; Berteau and Mulloy 2003; Zhang et al. 2003a,b; Tsuge et al. 2004; Ishihara et al. 2005; Nakamura et al. 2006; Cumashi et al. 2007; Inoue et al. 2009; Cui et al. 2010; Do et al. 2010; Sinha et al. 2010; Jiao et al. 2011; Wijesekara et al. 2011; Abu et al. 2013; Ngo and Kim 2013; Jiang et al. 2014). Some sulfated polysaccharides from marine algae are known to exhibit immunomodulatory activities through activating or suppressing the immune system (Jiao et al. 2011; Bhatia et al. 2013). Among the marine algal polysaccharides, fucoidans, which are fucose-containing sulfated polysaccharides from brown seaweeds, have been extensively studied with regard to their structural characteristics and biological activities with particular focus on its immunomodulatory actions (Li et al. 2008; Jiao et al. 2011).

The red algae *Porphyra* species such as *P. yezoensis*, *P. haitanensis*, and *P. tenera* are abundantly cultivated in East and Southeast Asia including Japan, Korea, and China as an important food source (Zhang et al. 2004; Ishihara et al. 2005). *Porphyra* species are extremely rich in nutrients and contain abundant polysaccharides, proteins, vitamins, and minerals. *P. yezoensis* and *P. tenera* are commonly known as "nori," and are popularly consumed worldwide as a component of the traditional Japanese food such as "sushi" (Ishihara et al. 2005; Hatada et al. 2006). Besides being widely utilized in food manufacture, *Porphyra* species have actually been used as a traditional Chinese herbal medicine to cure some diseases for more than 1000 years (Zhang et al. 2004).

Porphran, a kind of agaran (agar polysaccharide), is one of the main constituents of *Porphyra* species, and usually occupies nearly 40% of the total mass in dried algae (Ishihara et al. 2005). It is a linear sulfated polysaccharide existing in the cell wall and intracellular matrix of the laver of *Porphyra* species (Morrice et al. 1983; Takahashi et al. 2000;

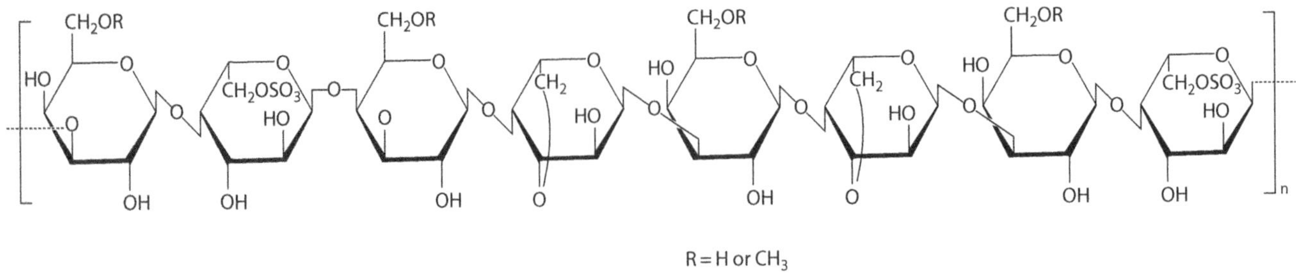

R = H or CH$_3$

*Figure 20.1* Basic structure of porphyran isolated from *P. yezoensis*.

Yoshimura et al. 2006). Porphyran is similar to agarose in its basic structure, and is mainly constituted of 1,3-linked β-D-galactopyranosyl residues alternating with 1,4-linked 3,6-anhydro-α-L-galactopyranosyl residues, but differs in some residues occurring as α-L-galactopyranosyl-6-sulfate and 6-O-methyl derivatives (Figure 20.1) (Mackie and Preston 1974; Morrice et al. 1983; Hama et al. 1999, 2010). Chemical compositions and entire the structures of porhyrans are known to be different, depending on species, harvest season, and growth environment.

Porphyrans show a relatively low viscosity and no gelling property due to the relatively high amount of sulfate ester as compared to agar and agarose (Takahashi et al. 2000; Jiao et al. 2011). In recent decades, several investigations on the nutritional and physiological functions of polysaccharides isolated from various *Porphyra* species have been undertaken (Tsuge et al. 2004; Zhang et al. 2004; Ishihara et al. 2005; Kwon and Nam 2006; Inoue et al. 2009; Kitano et al. 2012). In addition to the health benefit as a major dietary fiber in "Nori," porphyrans are found to exhibit several beneficial biological activities, such as antitumor, immunomodulatory, anti-hyperlipidemic, anti-hypercholesterolemic, antioxidant, antiallergic, and glucose-metabolism improving (Ren et al. 1994; Yoshizawa et al. 1993, 1995; Tsuge et al. 2004; Zhang et al. 2004; Ishihara et al. 2005; Kwon and Nam 2006; Inoue et al. 2009; Kitano et al. 2012). Hence, porphyran seems to be a valuable polysaccharide with health benefits, and can be used in nutraceutical food, cosmetic, and pharmaceutical industries.

Although several studies on the biological activities of porphyran have been conducted, the underlying mechanisms of action of porphyran, especially the effects on the immune system, are not fully understood. Thus, in this chapter, we describe the biological activities of porphyran isolated from *P. yezoensis*, mainly focusing on the action toward macrophages in terms of the effects on the secretion of nitric oxide (NO) and

tumor necrosis factor-α (TNF-α) from lipopolysaccharide (LPS)-stimulated macrophages.

## 20.2 Inhibitory effect of sulfated polysaccharide porphyran on nitric oxide production in LPS-stimulated RAW264.7 macrophages

### 20.2.1 Source and preparation of porphyran

Porphyran was prepared from *P. yezoensis* ("Nori") as reported previously (Hama et al. 1998; Hama et al. 1999). In brief, dry "Nori" sheets (3 g) were homogenized with 85% ethanol (500 mL) and heated at 75°C with constant stirring for 1 h, and then filtered to remove 85% ethanol-soluble substances. This extraction was repeated again. The residue was washed with methanol and dried to obtain 2 g of partially decolorized powder. The decolorized powder was extracted with 1000 mL of distilled water at 95°C with constant stirring for 1.5 h and centrifuged at 8000$g$ for 20 min, and then the supernatant was concentrated to 600 mL by using a rotary evaporator, and then digested with 50 units of nuclease P1 at 37°C for 24 h. After that, the mixture was heated for 5 min in a boiling water bath, and centrifuged to remove the precipitates. To the supernatant, sodium acetate and acetic acid were added to make a 0.5 M sodium acetate solution, pH 5.0, and then ethanol was added to a concentration of 42% (v/v). After 1 h, the precipitates in the solution were removed by centrifugation, and ethanol was added to the supernatant to a concentration of 60%. The precipitate was dissolved in distilled water, dialyzed against distilled water, and lyophilized to obtain 0.4 g of porphyran. Before use, porphyran solution was filtered through an endotoxin-removing filter (Zetapor Dispo filter) purchased from Wako Pure Chemicals Industries, Ltd. (Osaka, Japan).

## 20.2.2   Cytotoxic effect of porphyran

It has been reported that sulfated polysaccharides, including fucoidan and ascophyllan, are capable of inducing apoptotic cell death of several cancer cells (Nakayasu et al. 2009; Zhang et al. 2013). Since porphyran has been reported to inhibit cell proliferation and induce apoptosis in AGS human gastric cancer cells (Kwon and Nam 2006), we first examined the cytotoxicity of porphyran on the mouse macrophage cell line RAW264.7, which were obtained from American Type Culture Collection (Rockville, MD) and cultured in a $CO_2$ incubator at 37°C in Dulbecco's Modified Eagle's Minimum Essential Medium (DMEM) supplemented with 10% fetal bovine serum (FBS), penicillin (100 IU/mL), and streptomycin (100 mg/mL). The cytotoxicity of polysaccharide samples was measured by the Alamar blue assay, as described by Jiang et al. (2010). Briefly, adherent RAW264.7 cells in 96-well plates (3 × 10⁴ cells per well) were treated with varying concentrations of porphyran or fucoidan (positive control) for 24 h in the

growth medium, and then Alamar blue reagent was added to the cells at a final concentration of 10%. After 2 h incubation at 37°C, the absorbance of each well was measured at 570 and 600 nm using a multiwell scanning spectrophotometer (Thermo Electron Co., Yokohama, Japan). As seen in Figure 20.2a, porphyran showed no cytotoxic effect on RAW264.7 cells up to the concentration of 500 μg/mL. In contrast to the lack of significant cytotoxic effect of porphyran to RAW264.7 cells, fucoidan, a sulfated polysaccharide, showed cytotoxicity to RAW264.7 cells in a concentration-dependent manner at the same concentration range tested.

In order to further study the cytotoxic effect of porphyran and fucoidan on RAW264.7 cells, DNA fragmentation assay was performed as described by Kuramoto et al. (2005). In brief, adherent RAW264.7 cells in 35 mm diameter culture dishes (2 × 10⁶ cells per dish) were treated with 500 μg/mL of porphyran or fucoidan in the growth medium at 37°C for 24 h. After removal of the medium, the cells were lysed in 300 μL of ice-cold lysis

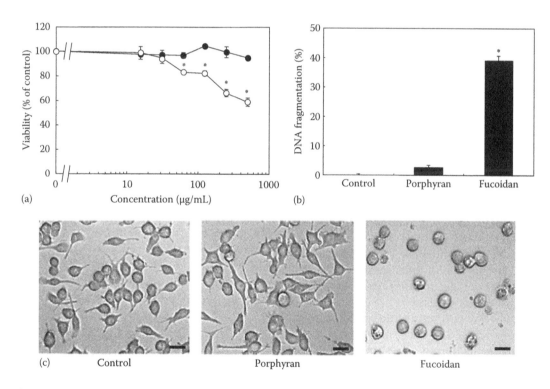

*Figure 20.2* Cytotoxic effects of porphyran and fucoidan on RAW264.7 cells. (a) Adherent cells (3 × 10⁴ cells per well in 96-well plates) in the growth medium were treated with varying concentrations of porphyran (●) or fucoidan (○) at 37°C. After 24 h, the cell viabilities were estimated by the Alamar blue assay. Data represent means ± SD of triplicate measurements. Asterisks indicate significant differences between with and without fucoidan ($p < 0.05$). (b) Adherent RAW264.7 cells in 35 mm diameter culture dishes (2 × 10⁶ cells per dish) were treated with 500 μg/mL of fucoidan or porphyran in the growth medium at 37°C. After 24 h, DNA fragmentations of the treated cells were examined by diphenylamine assay. Asterisk indicates significant difference between with and without fucoidan ($p < 0.05$). (c) Adherent RAW264.7 cells in 35 mm diameter culture dishes equipped with quartz glass were treated with 500 μg/mL of fucoidan or porphyran in the growth medium at 37°C. After 24 h, the cells were observed under a phase-contrast microscope (BIOREVO BZ9000, KEYENCE Co., Osaka, Japan). The bar indicates 20 μm.

buffer containing 0.5% Triton X-100, 10 mM Tris-HCl, pH 8.0, and 20 mM EDTA. The cell lysates were centrifuged for 20 min at 15,000$g$ to separate the DNA fragments (supernatant) from intact DNA (pellet). The DNA contents in the supernatant and pellet fractions were measured using the diphenylamine reagent, and the extent of DNA fragmentation was estimated. DNA fragmentation (Figure 20.2b) and morphological changes (Figure 20.2c) were induced in the cells treated with 500 µg/mL fucoidan, suggesting that the cytotoxicity of fucoidan was accompanied by induction of apoptosis, while no such significant apoptosis-related changes were observed in porphyran-treated RAW264.7 cells (Figure 20.2). Therefore, it seems likely that porphyran and fucoidan are quite different in terms of cytotoxicity to RAW264.7 cells. These results suggest that cellular toxicity of porphyran on RAW264.7 may not be a main reason for any observed effects of porphyran on RAW264.7 cells in the following studies.

## 20.2.3 Effect of porphyran on NO production in LPS-treated RAW264.7 cells

NO is a gaseous free radical involved in various physiological processes such as vasodilation, smooth muscle regulation, neurotransmission, and apoptosis (Lowenstein et al. 1994; Huang et al. 1995; Nelson et al. 1995; Dimmeler and Zeiher 1997). NO also plays an important role in anti-infectious immune responses as an important modulator in both innate and adaptive immunity (Bogdan et al. 2000). On the contrary the excess amount of NO released from activated macrophages contributes to numerous severe inflammatory diseases including sepsis and arthritis (Szabâo 1998; Weinberg 2000). Since NO particularly induces impaired vascular reactivity and causes pathological changes (Symeonides and Balk 1999), the selective inhibition of NO production in inflammatory cells (e.g., macrophages) may provide an important therapeutic strategy for inflammatory diseases.

To examine whether porphyran affects LPS-induced NO production in RAW264.7 cells, nitrite as a stable biomarker of NO production in LPS-stimulated mouse macrophage RAW264.7 cells was measured, because of the extremely short half-life of NO. Our preliminary study demonstrated that LPS potently induced NO production in RAW264.7 cells, even at a very low concentration (1 ng/mL). Time-course analysis showed that NO production was initiated at 6 h, and reached a stable and maximum level at 18 h, and the level was maintained 24 h after stimulation with LPS (data not shown). Thus, in this study, we evaluated nitrite concentration at 18 h after the addition of LPS. Monolayers of RAW264.7 cells in 96-well plates (3 × 10⁴ cells per well) were pretreated with varying concentrations of porphyran for 1 h in

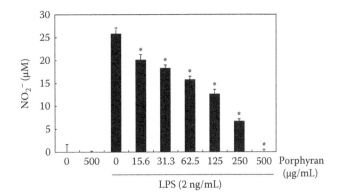

**Figure 20.3** Effects of porphyran on NO production in LPS-treated RAW264.7 cells. Adherent cells (3 × 10⁴ cells per well in 96-well plates) were preincubated with various concentrations of porphyran (0–500 µg/mL) in DMEM supplemented with 10% FBS at 37°C for 1 h, followed by the addition of LPS (final 2 ng/mL). After 18 h of incubation, the NO levels in the supernatants of the treated cells were estimated by Griess assay. Data represent means ± SD of triplicate measurements. Asterisks indicate significant differences between with and without porphyran ($p < 0.05$).

the growth medium, followed by stimulation with LPS (2 ng/mL). After 18 h of incubation, nitrite level in the medium of each treated RAW264.7 cells was determined by Griess assay, as described by Yamanishi et al. (2009). As shown in Figure 20.3, porphyran inhibited NO production in LPS-stimulated RAW264.7 cells in a concentration-dependent manner, and almost complete inhibition was attained at 500 µg/mL. Porphyran itself had no activity in inducing NO production from RAW264.7 cells at least at 500 µg/mL.

Considering that 500 µg/mL porphyran alone did not affect NO production and was not cytotoxic to RAW264.7 cells (Figure 20.2), it seems obvious that inhibition of LPS-induced NO production in RAW264.7 cells by porphyran is unrelated to any of its cytotoxic effects.

## 20.2.4 Effect of porphyran on the expression of inducible NO synthase in LPS-treated RAW264.7 cells

It is known that inducible NO synthase (iNOS) is a main enzyme responsible for NO production in activated macrophages (Yamanishi et al. 2009). To study whether the inhibition of NO production by porphyran is due to the inhibition of the expression of iNOS, which catalyzes the generation of NO from L-arginine, reserve transcription-polymerase chain reaction (RT-PCR) and immunoblot analysis were performed to detect the levels of iNOS mRNA and protein expression, respectively.

Adherent RAW264.7 cells in 24-well plates (5 × 10⁵ cells per well) were pretreated with porphyran at the concentrations of 0, 250, and 500 µg/mL for 1 h in the growth medium, and then incubated with or without LPS at the final concentration of 2 ng/mL. After 4 h of incubation, total RNA of each treated cells was isolated using Sepasol-RNA I Super (Nacalai tesque Co., Kyoto, Japan). Total RNA (2.5 µg) was reverse transcribed with an oligo dT primer in 10 µL using the PrimeScript® 1st strand cDNA synthesis kit (TaKaRa Bio Inc., Otsu, Shiga, Japan) according to the manufacturer's instructions. PCR was performed with 1 cycle of 70 s at 95°C, 25 cycles of 55 s at 93°C, 45 s at 61°C, 40 s at 72°C, and 1 cycle of 100 s at 72°C, in a 25 µL reaction mixture containing 12.5 µL of GoTag Green Master Mix (Promega KK., Tokyo, Japan), 0.5 µL of forward and reverse iNOS primers (1 µM each) or β-actin primers (10 pM each), 0.5 µL of first strand cDNA, and 11 µL nuclease-free water. The primer sequences were as follows: 5'-CAACCAGTATTATGGCTCCT-3' (forward) and 5'-GTGACAGCCCGGTCTTTCCA-3' (reverse) for mouse iNOS (Chen et al. 2007); and 5'-GGAGAAGATCTGGCACCACACC-3' (forward) and 5'-CCTGCTTGCTGATCCACATCTGCTGG-3' (reverse) for mouse β-actin (Chang et al. 2005). The β-actin primer was used as an internal control. Each PCR reaction (10 µL) product was run on 2% agarose gels containing 0.1 µg/mL ethidium bromide, and the amplified products (835 bp for iNOS and 840 bp for β-actin) were detected by a light capture instrument (ATTO Co., Tokyo, Japan). Western blot analysis of iNOS protein was performed on whole-cell extracts. Adherent RAW264.7 cells in 35 mm diameter culture dishes (2 × 10^6 cells per dish) were pretreated with 0, 250, or 500 µg/mL porphyran for 1 h in the growth medium. After 5 h of incubation with LPS (final 2 ng/mL), the cells were washed three times with ice-cold PBS and lysed with 100 µL extraction buffer (10 mM HEPES, 150 mM NaCl, 1 mM EGTA, 1% CHAPS, and 1% Triton X-100) containing 1% (v/v) protease inhibitor cocktail. After shaking for 30 min at 4 °C, the extracts were obtained from the supernatant after centrifugation at 15,000g for 10 min, and the protein concentrations were measured with the BCA protein assay (Bio-Rad Lab., Hercules, CA, USA) using bovine serum albumin as standard. The extract was mixed with equal volume of 2× SDS-sample buffer and incubated at 100 °C for 5 min. Samples containing 20 µg of protein were applied on 8% SDS-PAGE, and then electrically transferred to a poly(vinylidene difluoride) (PVDF) membrane. The membrane was blocked with 1% skimmed milk in TBS-0.1% Tween 20 (TBST). Immunostaining of the blot was performed with the specific antibody against mouse iNOS (Upstate Biotechnology, Lake Placid, NY) and β-actin (Abcam Inc., Cambridge, MA). Horseradish peroxidase conjugated-goat anti-rabbit IgG (Upstate Biotechnology) was used as a secondary antibody. The blots were detected using an enhanced chemiluminescence kit (Amersham

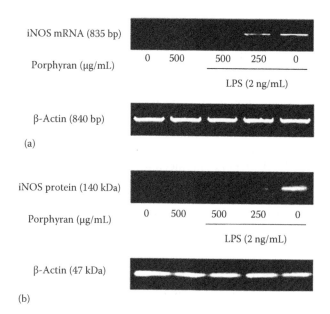

*Figure 20.4* Effects of porphyran on LPS-induced iNOS mRNA and iNOS protein levels in RAW264.7 cells. (a) Adherent cells (5 × 10^5 cells per well in 24-well plates) were preincubated with various concentrations of porphyran (0, 250, or 500 µg/mL) in DMEM supplemented with 10% FBS at 37°C for 1 h, followed by the addition of LPS (final 2 ng/mL). After 4 h incubation, the total RNA of each treated cells was subjected to RT-PCR analysis. (b) Adherent cells in 35 mm diameter culture dishes (2 × 10^6 cells per dish) were preincubated with various concentrations of porphyran (0, 250, or 500 µg/mL) in DMEM supplemented with 10% FBS at 37°C for 1 h, followed by the addition of LPS (final 2 ng/mL). After 5 h of incubation, the whole-cell lysates were analyzed by western blot analysis.

Biosciences UK Ltd., Buckinghamshire, UK). As shown in Figure 20.4, treatment of cells with 2 ng/mL LPS resulted in increased expression of both iNOS mRNA and protein, while iNOS expression was not detected in RAW264.7 cells treated with control normal medium or 500 µg/mL porphyran alone. Pretreatment of cells with porphyran, especially at 500 µg/mL, inhibited iNOS expression by nearly 100% at the transcription and translation levels in LPS-stimulated RAW264.7 cells. Since the iNOS mRNA level was significantly reduced by porphyran in LPS-treated RAW264.7 cells, while β-actin mRNA levels remained unchanged, the possibility of the nonspecific actions of porphyran on gene expression may be excluded. Thus, the inhibition of NO production is due to the inhibition of iNOS mRNA transcription.

Similar to our present results on porphyran, it was recently reported that fucoidan, a sulfated polysaccharide derived from brown algae, inhibits NO production and the expression of iNOS in LPS-activated macrophages (Yang et al. 2006). The inhibitory effect of fucoidan on NO production and iNOS expression in other cell types was also reported (Cui et al. 2010; Do et al. 2010). Fucoidan shows various biological activities. Particularly, its anti-inflammatory and anti-complement actions have drawn significant attention (Blondin et al. 1996; Berteau and Mulloy 2003). The suppressive effects on iNOS expression in activated macrophages might partly explain its anti-inflammatory actions. In contrast, it was also reported that fucoidan induces NO production from macrophages via a p38 MAP kinase and NF-κB-dependent mechanism (Nakamura et al. 2006). Fucoidan has a highly complicated structure, and the chemical structure can vary depending on the algal sources. Hence, one possible explanation for this discrepancy is that the fucoidan used in each study may be different. In fact, a comparative study on the fucoidans prepared from nine species of brown algae showed that the source and composition of fucoidan affect the biological activities (Cumashi et al. 2007). In the case of porphyran, further studies are also necessary to clarify the structure–activity relationship, especially its effects on NO production in activated macrophages.

## 20.2.5   *Effect of porphyran on TNF-α production in LPS-treated RAW264.7 cells*

LPS activates various intracellular signaling cascades leading to the secretion of various inflammatory cytokines, such as TNF-α and IL-6, in addition to the induction of NO production via iNOS expression (Kim et al. 2014). In fact, increased level of TNF-α was detected in LPS-treated RAW264.7 cells in our experiments (Figure 20.5). We next investigated whether porphyran could also affect the secretion of TNF-α from LPS-stimulated RAW264.7 cells. Adherent RAW264.7 cells in 24-well plates ($5 \times 10^5$ cells per well) were treated with porphyran at the concentrations of 0, 250, and 500 µg/mL for 1 h at 37°C in the growth medium. After a further 4 h incubation with LPS (final 2 ng/mL), the levels of TNF-α in culture supernatants of treated cells were measured by sandwich ELISA with two antibodies to two different epitopes on TNF-α molecule by similar method as described by Yamanishi et al. (2007). The TNF-α concentrations were estimated from reference to a standard curve for serial twofold dilution of murine recombinant TNF-α. Interestingly, porphyran showed only a marginal inhibitory effect on the TNF-α production in LPS-treated RAW264.7 cells even at 500 µg/mL, at which, however, NO production was completely inhibited (Figure 20.5). This observation suggests that porphyran may not interrupt the extracellular interaction between LPS and the specific receptors, but affect the intracellular signaling pathways linked with NO production such as the activation process of transcription factor.

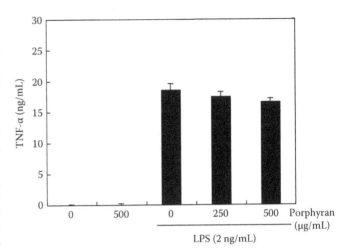

**Figure 20.5** Effect of porphyran on LPS-induced TNF per α secretion from RAW264.7 cells. Adherent cells ($5 \times 10^5$ cells per well in 24-well plates) were preincubated with various concentrations of porphyran (0, 250, or 500 µg/mL) in DMEM supplemented with 10% FBS at 37°C for 1 h, followed by the addition of LPS (final 2 ng/mL). After 4 h of incubation, the TNF-α levels in the culture medium of the treated cells were measured by ELISA. Data represent means ± SD of triplicate measurements.

## 20.2.6   *Effect of porphyran on NF-κB activation in LPS-stimulated RAW264.7 cells*

NF-κB is an essential transcription factor for the induction of several inflammatory mediators including iNOS (Muller et al. 1993; Guha and Mackman 2001). Therefore, an electrophoretic mobility shift assay (EMSA) on the nuclear extracts from LPS-stimulated RAW264.7 cells was performed using IRDye® 700 infrared dye-labeled oligonucleotide with NF-κB consensus sequence (LI-COR Bioscience, Lincoln, NE) to examine whether porphyran-caused inhibition of iNOS induction was due to the suppression of NF-κB activation. Adherent RAW264.7 cells in 35 mm diameter culture dishes ($2 \times 10^6$ cells per dish) were incubated in serum-free DMEM for 2 h at 37°C, and then porphyran (0, 250, or 500 µg/mL) was added to the cells and incubated for 1 h. After further 30 min of incubation with LPS (final 2 ng/mL), the nuclear proteins were extracted from the cells as described (Zingarelli et al. 2002). EMSA was carried out with Odyssey® IRDye® 700 infrared dye-labeled double-stranded oligonucleotides coupled with the EMSA buffer kit (LI-COR Bioscience) according to the manufacturer's instructions. Briefly, 5 µg of the nuclear extract was incubated with 1 µL of IRDye 700 infrared dye-labeled double-stranded oligonucleotides, 2 µL of 10× binding buffer, 2.5 mM DTT, 0.25% Tween-20, and 1 µg of poly (dI-dC) in a total volume of 20 µL for 20 min at room temperature in the dark. Sample proteins were separated on 4% polyacrylamide gel in 0.25× Tris–borate–EDTA running buffer for 60 min at 100 V. The gel was scanned by direct infrared fluorescence detection on the Odyssey imaging system (LI-COR Bioscience). NF-κB IRDye 700 infrared dye-labeled oligonucleotide's sequences of the double-stranded DNA probes used were as follows: 5′-AGTTGA<u>GGGGACTTTCCC</u>AGGC-3′ and 3′-TCAACT<u>CCCCTGAAAGGG</u>TCCG-5′ (LI-COR Bioscience) (Kim and Kim 2005; Pokharel et al. 2007). The underlined nucleotides indicate the binding sites for NF-κB. The specificity of the binding was examined using competition experiments, in which 100-fold excess of the unlabeled oligonucleotides was added to the reaction mixture before adding the infrared dye-labeled oligonucleotide. As shown in Figure 20.6a, an increase in the band thickness of the slow-migrating nucleoprotein complex was observed after the LPS treatment for 30 min, suggesting that NF-κB was activated by LPS. Pretreatment with porphyran (250 or 500 µg/mL) resulted in the inhibition of the LPS-inducible NF-κB DNA binding in a concentration-dependent manner. The addition of 100-fold excess of the unlabeled oligonucleotides with same sequences to the nuclear extract resulted in almost complete abolishment of the binding activity, which confirms that the binding is specific to NF-κB (Figure 20.6b). To examine whether porphyran directly inhibits NF-κB binding to DNA, the nuclear extracts prepared from the LPS-treated RAW264.7 cells were treated with porphyran and labeled NF-κB oligonucleotides. The *in vitro* exposure of the LPS-activated nuclear extracts to porphyran caused no decrease in the NF-κB DNA binding (Figure 20.6c). Hence, it seems likely that porphyran itself does not inhibit the NF-κB DNA binding activity.

It is well known that NF-κB exists in the cytosol as an inactive trimeric complex in which the p50/p65 protein dimer is associated with IκB, known as an inhibitory subunit (Romics et al. 2004). The process of the activation of NF-κB proceeds through phosphorylation and degradation of the IκB-α inhibitory subunit and subsequent translocation of p65/p50 complex into the nucleus (Wang et al. 2002). Thus, the effect of porphyran on nuclear translocation of p65 was also examined by immunoblotting. First, the analysis of IκB-α and phospho-IκB-α in cytosolic extracts and NF-κB p65 in nuclear extracts were performed, respectively. In brief, adherent RAW264.7 cells (2 × 10$^6$ cells per dish) were incubated in serum-free DMEM for 2 h at 37°C, and then porphyran (0, 250, or 500 µg/mL) was added to the cells and incubated for 1 h. After 30 min of incubation with LPS (final 2 ng/mL), the cells were washed three times with ice-cold PBS and incubated with 100 µL ice-cold cytosol extraction buffer (10 mM HEPES, pH 7.9, 1.5 mM MgCl$_2$, 10 mM KCl, 0.2% Igepal CA-630, 1 mM dithiothreitol, 20 mM β-glycerophosphate, 1 mM sodium orthoranadate, 0.5 mM phenylmethysulfonyl fluoride,

1 µg/mL leupeptin, and 1 µg/mL aprotinin) for 25 min on ice. The cytosolic extracts were collected after centrifugation at 7000$g$ for 5 min at 4°C. The nuclear pellets were resuspended in 30 µL ice-cold nuclear extraction buffer (20 mM HEPES, pH 7.9, 1.5 mM MgCl$_2$, 0.45 M NaCl, 25% glycerol, 0.2 mM EDTA, 1 mM dithiothreitol, 0.5 mM phenylmethysulfonyl fluoride, 1 µg/mL leupeptin and 1 µg/mL aprotinin) and incubated on ice for 25 min, and the nuclear extracts were obtained after centrifugation at 15,000$g$ for 10 min at 4°C. After measuring the protein concentrations, the extract was mixed with equal volume of 2× SDS-sample buffer and incubated at 100°C for 5 min. Samples containing 20 µg of protein were applied on 12.5% SDS-PAGE, and then electrically transferred to a PVDF membrane. Western blot analyses of the cytosolic and nuclear extracts were conducted by the similar way as described before except for using specific anti-NF-κB p65, anti-IκB-α, and anti-phospho-IκB-α antibody (Cell Signaling Thechnology, Inc., Beverly, MA, USA). Our results revealed that the nuclear levels of the p65 protein increased in LPS-treated RAW264.7 cells, and the LPS-induced nuclear translocation of p65 was blocked by porphyran in a concentration-dependent manner (Figure 20.6d). In addition, immuno blot analysis using specific antibodies indicated that both the phosphorylation and degradation of IκB-α were also inhibited by a pretreatment with porphyran (Figure 20.6e). These results suggest that porphyran blocks the nuclear translocation of NF-κB through the inhibition of the phosphorylation and subsequent degradation of IκB-α. A specific NF-κB inhibitor peptide (SN50) (Calbiochem, La Jolla, CA) blocked LPS-induced NO production in a concentration-dependent manner, while the inactive control peptide SN50M had no effect (Figure 20.6f). These results indicate that NF-κB activation is required for LPS-induced NO production, and may support the idea that inhibition of NF-κB activation by porphyran is its main action mechanism to inhibit NO production in LPS-stimulated RAW264.7 cells.

In this study, we found that LPS-induced NF-κB activation was inhibited by porphyran. When the cells are stimulated with extracellular stimulus, phosphorylation and subsequent degradation of IκB occur. Such changes lead to the dissociation of IκB from the complex and to the production of activated NF-κB, which translocates into the nucleus and activates the target gene expression. Immunoblot analysis using the specific antibodies demonstrated that the LPS-induced phosphorylation and degradation of IκB-α, which is prerequisite for p65 activation, was suppressed in the porphyran-pretreated cells. In addition, the translocation of p65 into nucleus in response to LPS was also inhibited by porphyran. Thus, these results suggest that the inhibition of NF-κB activation through the inhibition of the phosphorylation and degradation of IκB-α may be an

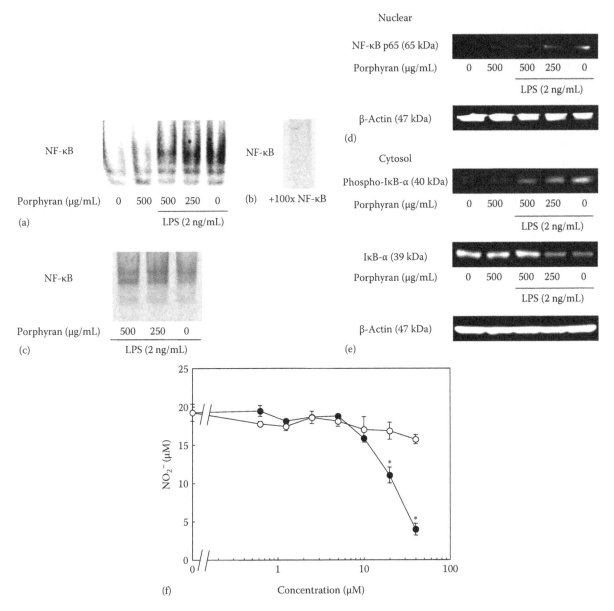

*Figure 20.6* (a–c) Effects of porphyran on LPS-induced NF-κB activation in RAW264.7 cells. (a) *In vivo* effect of porphyran on DNA binding of NF-κB. Adherent cells in 35 mm diameter culture dish (2 × 10⁶ cells per dish) were incubated in serum-free DMEM at 37°C for 2 h, and then the cells were incubated with various concentrations of porphyran (0, 250, or 500 μg/mL) in serum-free DMEM at 37°C for 1 h, followed by the addition of LPS (final 2 ng/mL). After 30 min of incubation, gel shift assay of NF-κB was performed with nuclear extracts prepared from the treated cells as described in the text. (b) EMSA was performed on the nuclear extract prepared from LPS-stimulated RAW264.7 cells in the presence of 100-fold excess unlabeled oligonucleotides with the same NF-κB consensus sequence. (c) *In vitro* effect of porphyran on DNA binding of NF-κB. Porphyran (0, 250, or 500 μg/mL) was added to the assay mixture containing the nuclear extracts prepared from the LPS-stimulated RAW264.7 cells and infrared dye-labeled oligonucleotides with NF-κB consensus sequence, incubated for 20 min, and then subjected to EMSA. (d, e) Effects of porphyran on the nuclear translocation of NF-κB p65 and phosphorylation and degradation of IκB-α in LPS-treated RAW264.7 cells. (d) Western blot analysis was conducted on NF-κB p65 in the nuclear extracts prepared from RAW264.7 cells treated with porphyran and subsequent stimulation with LPS as described in the text. (e) Western blot analyses were conducted on the IκB-α and phosphorylated IκB-α in the cytosolic extracts prepared from RAW264.7 cells treated with porphyran and subsequent stimulation with LPS as described in the text. (f) Effects of the specific NF-κB inhibitor peptide (SN50) and its inactive control peptide (SN50M) on NO production in LPS-treated RAW264.7 cells. Adherent cells (3 × 10⁴ cells per well in 96-well plates) were preincubated with various concentrations of SN50 (●) or SN50M (○) in DMEM supplemented with 10% FBS at 37°C for 2 h, followed by the addition of LPS (final 2 ng/mL). After 18 h of incubation, the NO levels in the supernatants of the treated cells were estimated as described in the text. Data represent means ± SD of triplicate measurements. Asterisks indicate significant differences between with and without SN50 ($p < 0.05$).

action mechanism of porphyran to inhibit iNOS expression in LPS-stimulated RAW264.7 cells.

In the present study, we found that porphyran inhibited the production of NO through the inhibition of iNOS expression in LPS-stimulated RAW264.7 cells. These results differ from those obtained using murine peritoneal macrophages, which demonstrated that porphyran fraction prepared from *P. yezoensis* induced NO and TNF-α production (Yoshizawa et al. 1993, 1995). Although the exact reason for these differences is uncertain now, it may be due to the different cell types used in the studies, as well as differences in sources and preparation procedures of the porphyrans used. In fact, it has been known that porphyran has a quite complicated and heterogeneous structure depending on its source, and the even the manner of cultivation or harvest time also influences the structure of porphyran even when derived from same algal species (Rees 1961; Araki et al. 1977). Although further studies are required for the clarification of the mechanism of porphyran in inhibiting NO production in LPS-stimulated macrophages as well as for its structure–activity relationship, it seems that our results can exclude the possibility that porphyran directly interferes with the interaction between NF-κB protein and DNA, since no significant inhibitory effect of porphyran on NF-κB DNA binding was observed in *in vitro* EMSA (Figure 20.6c).

### 20.2.7  Superoxide radical scavenging activity of porphyran

It has been reported that sulfated polysaccharides from marine algae, including fucoidan and porphyrin, have antioxidant activities (Lee et al. 2003; Zhang et al. 2003a,b, 2009a,b; Leung et al. 2006) and that naturally occurring antioxidant compounds such as β-carotene have suppressive effect on NF-κB activation in macrophages stimulated with LPS (Bai et al. 2005). Thus, there is a possibility that the antioxidant property of porphyran might be partly responsible for the inhibitory effects on the inflammatory response of LPS-stimulated RAW264.7 cells. To study this point, the antioxidant activity of porphyran was assessed by the surperoxide radical scavenging activity (Nishimiki et al. 1972). For this assay, 0.5 μL of 30 mM nitro blue tetrazolium (NBT) and 0.5 μL of 46.8 mM nicotinamide adenine dinucleotide (NADH) in 16 mM Tris–HCl buffer (pH 8.0) were added to 100 μL of the sample solution in 16 mM Tris–HCl buffer (pH 8.0) (final 10–1000 μg/mL). The reaction was initiated by adding 0.5 μL of 6 mM 5-methylphenazium-methyl sulfate (PMS) to the mixture. The reaction mixture was incubated at room temperature for 5 min, and the absorbance was measured at 560 nm with a spectrophotometer. The

decrease in absorbance of the reaction mixture reflected increased superoxide radical scavenging activity. The superoxide radical scavenging activity was calculated using the following equation:

$$O_2^- \text{ scavenging activity}(\%) = (1 - A_{sample}/A_{control}) \times 100$$

where

$A_{sample}$ is the absorbance in the presence of the sample

$A_{control}$ is the absorbance of Tris–HCl buffer alone without the sample.

As shown in Figure 20.7, porphyran significantly scavenged the surperoxide radical in a concentration-dependent manner. Porphyran at the final concentration of 500 μg/mL scavenged surperoxide anion radical by more than 50%.

The activation status of various transcription factors are influenced by the intracellular redox condition and controlled by the generation of ROS (Lee et al. 2003). Regarding naturally occurring antioxidants, which have inhibitory activities on inflammatory responses of macrophages, it has recently been reported that lycopene, a major carotenoid in tomato, inhibited LPS-induced production of NO through the suppression of NF-κB activation in RAW264.7 cells but had no effect on TNF-α (Feng et al. 2010). The similarity between porphyran and lycopene in terms of the effect on LPS-stimulated RAW264.7 cells suggests that the antioxidant property may be one of the important factors responsible for the anti-inflammatory activities

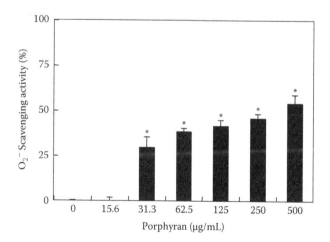

*Figure 20.7* Scavenging effects of porphyran on the superoxide anion radical. Scavenging activity of porphyran toward superoxide anion radical was measured as described in the text. Experiments were performed in triplicate, and values shown are the means ± SD. Asterisks indicate significant differences between with and without porphyran ($p < 0.05$).

toward LPS-stimulated macrophages. In addition to lycopene, diosgenin, a steroid saponin found in several plants, which has antioxidant activity, has also been reported to inhibit iNOS expression in LPS/interferon γ-activated murine macrophages without affecting the secretion of TNF-α (Jung et al. 2010). Since diosgenin inhibited LPS/interferon γ-induced NF-κB and AP-1 activation, the inhibition of iNOS expression by diosgenin was considered at the transcriptional level. As another example, prodigiosin, isolated from the marine bacterium *Hahella chejuensis*, has also been reported to suppress LPS-induced NO production by inhibiting p38 mitogen-activated protein kinase (MAP kinase), JNK MAP kinase, and NF-κB activation in mouse peritoneal macrophages, but had no effect on the production of cytokines including TNF-α (Huh et al. 2007). Although porphyran, lycopene, diosgenin, and prodigiosin are chemically unrelated compounds, they may have a similar action mechanism on activated macrophages. These compounds inhibit NO production through the inhibition of NF-κB activation without affecting the production of TNF-α, suggesting that NF-κB specially regulates iNOS expression but not the secretion of TNF-α. Other several natural compounds with antioxidant or free-radical-scavenging activities have been reported to inhibit NO production by activated macrophages through the inhibition of the NF-κB activation (Han et al. 2001; Pergola et al. 2006; Yang et al. 2006; Tang et al. 2011). Thus, it seems likely that the specific inhibition of NO production via suppression of NF-κB activation may be a common action mechanism of anti-inflammatory agents with antioxidant activity. Further studies are required to clarify this point as well as exact action mechanism of porphyran on the inhibition of NO production in activated macrophages.

## 20.3  Summary

Porphyran, extracted from an edible red alga, is a sulfated polysaccharide with a wide variety of biological activities including antitumor, antioxidant, and immunomodulatory. In this study, we examined the effect of porphyran isolated from *P. yezoensis* on NO production in the mouse macrophage cell line RAW264.7. Although no significant activity of porphyran to induce NO or TNF-α production in RAW264.7 cells was observed at the concentration range tested (10–500 µg/mL), it was found, for the first time, that porphyran inhibited NO production and expression of iNOS in RAW264.7 cells stimulated with LPS. In the presence of 500 µg/mL porphyran, NO production and expression of iNOS were completely suppressed. On the contrary, porphyran showed only a marginal effect on the secretion of TNF-α from LPS-stimulated

RAW264.7 cells. EMSA using an infrared dye-labeled oligonucleotide with NF-κB consensus sequence suggested that porphyran inhibited LPS-induced NF-κB activation. The LPS-inducible nuclear translocation of p65 and the phosphorylation and degradation of IκB-α were also inhibited by the pretreatment with porphyran. Our results obtained in *in vitro* analysis suggest that porphyran suppresses NO production in LPS-stimulated macrophages by blocking NF-κB activation. This inhibition may explain some of the anti-inflammatory effects of porphyran. Based on the findings described in this chapter, we would like to insist that porphyran can be applicable as a promising therapeutic agent for the treatment of inflammatory diseases.

## References

Abu, R., Jiang, Z., Ueno, M., Okimura, T., Yamaguchi, K., and Oda, T. (2013) *In vitro* antioxidant activities of sulfated polysaccharide ascophyllan isolated from *Ascophyllum nodosum*. *International Journal of Biological Macromolecules*, 59, 305–312.

Araki, S., Oohusa, T., Saitoh, M., and Sakurai, T. (1977) The quality of "nori", dried laver, in special reference to the contents of 3,6-anhydro-galactose in porphyran. *The Japanese Journal of Phycology*, 25, 19–23.

Bai, S.K., Lee, S.J., Na, H.J., Ha, K.S., Han, J.A., Lee, H., Kwon, Y.G., Chung, C.K., and Kim, Y.M. (2005) Beta-carotene inhibits inflammatory gene expression in lipopolysaccharide-stimulated macrophages by suppressing redox-based NF-kappaB activation. *Experimental and Molecular Medicine*, 37, 323–334.

Berteau, O. and Mulloy, B. (2003) Sulfated fucans, fresh perspectives: Structures, functions, and biological properties of sulfated fucans and an overview of enzymes active toward this class of polysaccharide. *Glycobiology*, 13, 29R–40R.

Bhatia, S., Rathee, P., Sharma, K., Chaugule, B.B., Kar, N., and Bera, T. (2013) Immuno-modulation effect of sulphated polysaccharide (porphyran) from Porphyra vienamensis. *International Journal of Biological Macromolecules*, 57, 50–56.

Blondin, C., Chaubet, F., Nardella, A., Sinquin, C., and Jozefonvicz, J. (1996) Relationships between chemical characteristics and anticomplementary activity of fucans. *Biomaterials*, 17, 597–603.

Bogdan, C., Röllinghoff, M., and Diefenbach, A. (2000) Reactive oxygen and reactive nitrogen intermediates in innate and specific immunity. *Current Opinion in Immunology*, 12, 64–76.

Chang, S.H., Mun, S.H., Ko, N.Y., Lee, J.H., Jun, M.H., Seo, J.Y., Kim, Y.M., Choi, W.S., and Her, E. (2005) The synergistic effect of phytohemagglutinin and interferon-γ on the expression of tumor necrosis factor-α from RAW 264.7 cells. *Immunology Letters*, 98, 137–143.

Chen, C.Y., Peng, W.H., Tsai, K.D., and Hsu, S.L. (2007) Luteolin suppresses inflammation-associated gene expression by blocking NF-κB and AP-1 activation pathway in mouse alveolar macrophages. *Life Science*, 81, 1602–1614.

Cui, Y., Zhang, L., Zhang, T., Luo, D., Jia, Y., Guo, Z., Zhang, Q., Wang, X., and Wang, X. (2010) Inhibitory effect of fucoidan on nitric oxide production in lipopolysaccharide-activated primary microglia. *Clinical and Experimental Pharmacology and Physiology*, 37, 422–428.

Cumashi, A., Ushakova, N.A., Preobrazhenskaya, M.E., D'Incecco, A., Piccoli, A., Totani, L., Tinari, N. et al. (2007) An comparative study of the anti-inflammatory, anticoagulant, antiangiogenic, and antiadhesive activities of nine different fucoidans from brown seaweeds. *Glycobiology*, 17, 541–552.

Dimmeler, S. and Zeiher, A.M. (1997) Nitric oxide and apoptosis: Another paradigm for the double-edged role of nitric oxide. *Nitric Oxide*, 1, 275–281.

Do, H., Pyo, S., and Sohn, E.H. (2010) Suppression of iNOS expression by fucoidan is mediated by regulation of p38 MAPK, JAK/STAT, AP-1 and IRF-1, and depends on up-regulation of scavenger receptor B1 expression in TNF-α- and IFN-γ-stimulated C6 glioma cells. *Journal of Nutritional Biochemistry*, 21, 671–679.

Feng, D., Ling, W.-H., and Duan, R.-D. (2010) Lycopene suppresses LPS-induced NO and IL-6 production by inhibiting the activation of ERK, p38 MAPK, and NF-κB in macrophages. *Inflammation Research*, 59, 115–121.

Guha, M. and Mackman, N. (2001) LPS induction of gene expression in human monocytes. *Cellular Signalling*, 13, 85–94.

Hama, Y., Nakagawa, H., Kurosawa, M., Sumi, T., Xia, X., and Yamaguchi, K. (1998) A gas chromatographic method for the sugar analysis of 3, 6-anhydrogalactose-containing algal galactans. *Analytical Biochemistry*, 265, 42–48.

Hama, Y., Nakagawa, H., Mochizuki, K., Sumi, T., and Hatate, H. (1999) Quantitative anhydrous mercaptolysis of algal galactans followed by HPLC of component sugars. *The Journal of Biochemistry*, 125, 160–165.

Hama, Y., Tsuneoka, A., Morita, R., Nomoto, O., Yoshinaga, K., Hatate, H., Sumi, T., and Nakagawa, H. (2010) Selective hydrolysis of the 3,6-anhydrogalacotosidic linkage in red algal galactan: A combination of reductive acid hydrolysis and anhydrous mercaptolysis. *Bioscience, Biotechnology, and Biochemistry*, 74, 1895–1900.

Han, Y.J., Kwon, Y.G., Chung, H.T., Lee, S.K., Simmons, R.L., Billiar, T.R., and Kim, Y.M. (2001) Antioxidant enzymes suppress nitric oxide production through the inhibition of NF-κB activation: Role of $H_2O_2$ and nitric oxide in inducible nitric oxide synthase expression in macrophages. *Nitric Oxide*, 5, 504–513.

Hatada, Y., Ohta, Y., and Horikoshi, K. (2006). Hyperproduction and application of α-agarase to enzymatic enhancement of antioxidant activity of porphyran. *Journal of Agricultural and Food Chemistry*, 54, 9895–9900.

Huang, P.L., Huang, Z., Mashimo, H., Bloch, K.D., Moskowitz, M.A., Bevan, J.A., and Fishman, M.C. (1995) Hypertension in mice lacking the gene for endothelial nitric oxide synthase. *Nature*, 377, 239–242.

Huh, J.E., Yim, J.H., Lee, H.K., Moon, E.Y., Rhee, D.K., and Pyo, S. (2007) Prodigiosin isolated from *Hahella chejuensis* suppresses lipopolysaccharide-induced NO production by inhibiting p38 MAPK, JNK and NF-κB activation in murine peritoneal macrophages. *International Immunopharmacology*, 7, 1825–1833.

Inoue, N., Yamano, N., Sakata, K., Nagao, K., Hama, Y., and Yanagita, T. (2009) The sulfated polysaccharide porphyran reduces apolipoprotein B100 secretion and lipid synthesis in HepG2 cells. *Bioscience, Biotechnology, and Biochemistry*, 73, 447–449.

Ishihara, K., Oyamada, C., Matsushima, R., Murata, M., and Muraoka, T. (2005) Inhibitory effect of porphyran, prepared from dried "Nori," on contact hypersensitivity in mice. *Bioscience, Biotechnology, and Biochemistry*, 69, 1824–1830.

Jiang, Z., Abu, R., Isaka, S., Nakazono, S., Ueno, M., Okimura T., Yamaguchi, K., and Oda, T. (2014) Inhibitory effect of orally-administered sulfated polysaccharide ascophyllan isolated from *Ascophyllum nodosum* on the growth of sarcoma-180 solid tumor in mice. *Anticancer Research*, 34, 1663–1671.

Jiang, Z., Okimura, T., Yokose, T., Yamasaki, Y., Yamaguchi, K., and Oda, T. (2010) Effects of sulfated fucan, ascophyllan, from the brown Alga *Ascophyllum nodosum* on various cell lines: A comparative study on ascophyllan and fucoidan. *Journal of Bioscience Bioengineering*, 110, 113–117.

Jiao, G., Yu, G., Zhang, J., and Ewart, H.S. (2011) Chemical structures and bioactivities of sulfated polysaccharides from marine algae. *Marine Drugs*, 9, 196–223.

Jung, D.H., Park, H.J., Byun, H.E., Park, Y.M., Kim, T.W., Kim, B.O., Um, S.H., and Pyo, S. (2010) Diosgenin inhibits macrophage-derived inflammatory mediators through downregulation of CK2, JNK, NF-κB and AP-1 activation. *International Immunopharmacology*, 10, 1047–1054.

Kim, J.H., Kim, Y.S., Hwang, J.W., Han, Y.K., Lee, J.S., Kim, S.K., Jeon, Y.J. et al. (2014) Sulfated chitosan oligosaccharides suppress LPS-induced NO production via JNK and NF-κB inactivation. *Molecules*, 19, 18232–18247.

Kim, J.W. and Kim, C. (2005) Inhibition of LPS-induced NO production by taurine chloramine in macrophages is mediated though Ras-ERK-NF-κB. *Biochemical Pharmacology*, 70, 1352–1360.

Kitano, Y., Murazumi, K., Duan, J., Kurose, K., Kobayashi, S., Sugawara, T., and Hirata, T. (2012) Effect of dietary porphyran from the red alga, *Porphyra yezoensis*, on glucose metabolism in Diabetic KK-Ay mice. *Journal of Nutritional Science and Vitaminology*, 58, 14–19.

Kuramoto, T., Uzuyama, H., Hatakeyama, T., Tamura, T., Nakashima, T., Yamaguchi, K., and Oda, T. (2005) Cytotoxicity of a GalNAc-specific C-type lectin CEL-I toward various cell lines. *The Journal of Biochemistry*, 137, 41–50.

Kwon, M.J. and Nam, T.J. (2006) Porphyran induces apoptosis related signal pathway in AGS gastric cancer cell lines. *Life Science*, 79, 1956–1962.

Lee, A.K., Sung, Y.C., Kim, Y.C., and Kim, S.G. (2003) Inhibition of lipopolysaccharide-inducible nitric oxide synthase, TNF-α and COX-2 expression by sauchinone effects on I-κBα phosphorylation, C/EBP and AP-1 activation. *Britain Journal of Pharmacology*, 139, 11–20.

Leung, M.Y.K., Liu, C., Koon, J.C.M., and Fung, K.P. (2006) Polysaccharide biological response modifiers. *Immunology Letters*, 105, 101–114.

Li, B., Lu, F., Wei, X., and Zhao, R. (2008) Fucoidan: Structure and bioactivity. *Molecules*, 13, 1671–1695.

Lowenstein, C.J., Dinerman, J.L., and Snyder, S.H. (1994) Nitric oxide: A physiologic messenger. *Annals of Internal Medicine*, 120, 227–237.

Mackie, W. and Preston, R.D. (1974) Cellwall and intercellular region polysaccharides. In *Algal Physiology and Biochemistry*, Stewart, W.P.D. (Ed.). Balckwell: London, U.K., pp. 65–66.

Morrice, L.M., McLean, M.W., Long, W.F., and Williamson, F.B. (1983) Porphyran primary structure. *European Journal of Biochemistry*, 133, 673–684.

Muller, J.M., Ziegler-Heitbrock, H.W., and Baeuerle, P.A. (1993) Nuclear factor kappa B, a mediator of lipopolysaccharide effects. *Immunobiology*, 187, 233–256.

Nakamura, T., Suzuki, H., Wada, Y., Kodama, T., and Doi, T. (2006) Fucoidan induces nitric oxide production *via* p38 mitogen-activated protein kinase and NF-κB-dependent signaling pathways through macrophage scavenger receptors. *Biochemical and Biophysical Research Communications*, 343, 286–294.

Nakayasu, S., Soegima, R., Yamaguchi, K., and Oda, T. (2009) Biological activities of fucose-containing polysaccharide ascophyllan isolated from the brown alga *Ascophyllum nodosum*. *Bioscience, Biotechnology, and Biochemistry*, 73, 961–964.

Nelson, R.J., Demas, G.E., Huang, P.L., Fishman, M.C., Dawson, V.L., Dawson, T.M., and Snyder, S.H. (1995) Behavioral abnormalities in male mice lacking neuronal nitric oxide synthase. *Nature*, 378, 383–386.

Ngo, D.H. and Kim, S.K. (2013) Sulfated polysaccharides as bioactive agents from marine algae. *International Journal of Biological Macromolecules*, 62, 70–75.

Nishimiki, M., Rao, N.A., and Yagi, K. (1972) The occurrence of superoxide anion in the reaction of reduced phenazine methosulfate and molecular oxygen. *Biochemical and Biophysical Research Communications*, 46, 849–864.

Pergola, C., Rossi, A., Dugo, P., Cuzzocrea, S., and Sautebin, L. (2006) Inhibition of nitric oxide biosynthesis by anthocyanin fraction of blackberry extract. *Nitric Oxide*, 15, 30–39.

Pokharel, Y.R., Liu, Q.H., Aryal, D.K., Kim, Y.G., Woo, E.R., and Kang, K.W. (2007) 7,7'-Dihydroxy bursehernin inhibits the expression of inducible nitric oxide synthase through NF-κB DNA binding suppression. *Nitric Oxide*, 16, 274–285.

Rees, D.A. (1961) Estimation of the relative amounts of isomeric sulfate esters in some sulfated polysaccharides. *Journal of the Chemical Society*, 5168–5171.

Ren, D., Noda, H., Amano, H., Nishino, T., and Nishizawa, K. (1994) Study on antihypertensive antihyperlipidemic effects of marine algae. *Fisheries Science*, 60, 33–40.

Romics, L., Kodys, K., Dolganiuc, A., Graham, L., Velayudham, A., Mandrekar P., and Szabo, G. (2004) Diverse regulation of NF-κB and peroxisome proliferator-activated receptors in murine nonalcoholic fatty liver. *Hepatology*, 40, 376–385.

Sinha, S., Astani, A., Ghosh, T, Schnitzler, P., and Ray, B. (2010) Polysaccharides from *Sargassum tenerrimum*: Structural features, chemical modification and anti-viral activity. *Phytochemistry*, 71, 235–242.

Symeonides, S. and Balk, R.A. (1999) Nitric oxide in the pathogenesis of sepsis. *Infectious Disease Clinics of North America*, 13, 449–463.

Szabâo, C. (1998) Role of nitric oxide in endotoxic shock. An overview of recent advances. *Annals of the New York Academy of Sciences*, 851, 422–425.

Takahashi, K., Hirano, Y., Araki, S., and Hattori, M. (2000) Emulsifying ability of porphyran prepared from dried nori, *Porphyra yezoensis*, a red alga. *Journal of Agricultural and Food Chemistry*, 48, 2721–2725.

Tang, S., Shen, X.Y., Huang, H.Q., Xu, S.W., Yu, Y., Zhou, C.H., Chen, S.R., Le, K., Wang, Y.H., and Liu, P.Q. (2011) Cryptotanshinone suppressed inflammatory cytokines secretion in RAW264.7 macrophages through inhibition of the NF-κB and MAPK signaling pathways. *Inflammation*, 34, 111–118.

Tsuge, K., Okabe, M., Yoshimura, T., Sumi, T., Tachibana, H., and Yamada, K. (2004) Dietary effect of porphyran from *Porphyra yezoensis* on growth and lipid metabolism of Sprague-Dawley rats. *Food Science and Technology Research*, 10, 147–151.

Wang, T., Zhang, X., and Li, J.J. (2002) The role of NF-κB in the regulation of cell stress responses. *International Immunopharmacology*, 2, 1509–1520.

Weinberg, J.B. (2000) Nitric oxide synthase 2 and cyclooxygenase 2 interactions in inflammation. *Immunology Research*, 22, 319–341.

Wijesekara, I., Pangestuti, R., and Kim, S.K. (2011) Biological activities and potential health benefits of sulfated polysaccharides derived from marine algae. *Carbohydrate Polymer*, 84, 14–21.

Yamanishi, T., Hatakeyama, T., Yamaguchi, K., and Oda, T. (2009) CEL-I, an *N*-acetylgalactosamine (GalNAc)-specific C-type lectin, induces nitric oxide production in RAW264.7 mouse macrophage cell line. *The Journal of Biochemistry*, 146, 209–217.

Yamanishi, T., Yamamoto, Y., Hatakeyama, T., Yamaguchi, K., and Oda, T. (2007) CEL-I, an invertebrate *N*-acetylgalactosamine-specific C-type lectin, induces TNF-alpha and G-CSF production by mouse macrophage cell line RAW264.7 cells. *The Journal of Biochemistry*, 142, 587–595.

Yang, J.W., Yoon, S.Y., Oh, S.J., Kim, S.K., and Kang, K.W. (2006) Bifunctional effects of fucoidan on the expression of inducible nitric oxide synthase. *Biochemical and Biophysical Research Communications*, 346, 345–350.

Yoshimura, T., Tsuge, K., Sumi, T., Yoshiki, M., Tsuruta, Y., Abe, S., Nishino, S., Sanematsu, S., and Koganemaru, K. (2006) Isolation of porphyran-degrading marine microorganisms from the surface of red alga, *Porphyra yezoensis*. *Bioscience, Biotechnology, and Biochemistry*, 70, 1026–1028.

Yoshizawa, Y., Ametani, A., Tsunehiro, J., Nomura, K., Itoh, M., Fukui, F., and Kaminogawa, S. (1995) Macrophage stimulation activity of the polysaccharide fraction from a marine alga (*Porphyra yezoensis*): Structure-function relationships and improved solubility. *Bioscience, Biotechnology, and Biochemistry*, 59, 1933–1937.

Yoshizawa, Y., Enomoto, A., Todoh, H., Ametani, A., and Kaminogawa, S. (1993) Activation of murine macrophage by polysaccharide fractions from a marine alga (*Porphyra yezoensis*). *Bioscience, Biotechnology, and Biochemistry*, 57, 1862–1866.

Zhang, Q., Li, L., Liu, X., Zhao, Z., Li, Z., and Xu, Z. (2004) The structure of a sulfated galactan from *Porphyra haitanesis* and its in vivo antioxidant activity. *Carbohydrate Research*, 339, 105–111.

Zhang, Q., Li, N., Zhou, G., Lu, X., Xu, Z., and Li, Z. (2003a) *In vivo* antioxidant activity of polysaccharide fraction from *Porphyra haitanesis* (Rhodephyta) in aging mice. *Pharmacology Research*, 48, 151–155.

Zhang, Q., Yu, P., Li, Z., Zhang, H., Xu, Z., and Li, P. (2003b) Antioxidant activities of sulfated polysaccharide fractions from *Porphyra haitanesis*. *The Journal of Applied Phycology*, 15, 305–310.

Zhang, Z., Teruya, K., Eto, H., and Shirahata, S. (2013) Induction of apoptosis by low-molecular-weight fucoidan through calcium- and caspase-dependent mitochondrial pathways in MDA-MB-231 breast cancer cells. *Bioscience, Biotechnology, and Biochemistry*, 77, 235–242.

Zhang, Z., Zhang, Q., Wang, J., Shi, X., Song, H., and Zhang, J. (2009a) *In vitro* antioxidant activities of acetylated, phosphorylated and benzoylated derivatives of porphyran extracted from *Porphyra haitanensis*. *Carbohydrate Polymer*, 78, 449–453.

Zhang, Z., Zhang, Q., Wang, J., Zhang, H., Niu, X., and Li, P. (2009b) Preparation of the different derivatives of the low-molecular-weight porphyran from *Porphyra haitanensis* and their antioxidant activities *in vitro*. *International Journal of Biological Macromolecules*, 45, 22–26.

Zingarelli, B., Hake, P.W., Yang, Z., O'Coonor, M., Denenberg, A., and Wong, H.R. (2002) Absence of inducible nitric oxide synthase modulates early reperfusion-induced NF-κB and AP-1 activation and enhances myocardial damage. *The FASEB Journal*, 16, 327–342.

*chapter twenty-one*

# Marine polysaccharides as biostimulants of plant growth

*Izabela Michalak, Agnieszka Dmytryk, Katarzyna Godlewska,*
*Radosław Wilk, and Katarzyna Chojnacka*

## Contents

## 21.1 Introduction

The high yield potential of modern cultivars is often limited by various environmental stresses (biotic and abiotic) that affect the crop yield. The plants endogenous resistance or tolerance to the biotic and abiotic stress can be efficiently enhanced by some natural active products, which are classified as biostimulants. For the first time, biologically derived activators were defined under general name "biostimulants" in 1997 by Zhang and Schmidt, who limited this group to "materials that, in minute quantities, promote plant growth" (Zhang and Schmidt 1997). The definition enabled to distinguish biostimulants from nutrients and soil amendments; however, further research showed oversimplification of Zhang and Schmidt description in relation to the actual capacity of biostimulants, and thus more adequate specification had to be introduced. Currently, plant growth biostimulants (PGBs) are considered as substances and materials, excluding nutrients, fertilizers, and pesticides, the application of which affects plant physiological processes, responsible not just for the growth but for the development and/or stress response. Moreover, biostimulants can be applied to plants, seeds, or growing substrates, but always at low concentration, which is provided by specific formulations (Kauffmann et al. 2007, du Jurdain 2012). The European Biostimulant Industry Council (EBIC) has extended the definition of PGB by involving the activity of living microorganisms and preparations designed for application to rhizosphere.

Biostimulants are compounds of biological origin that act by increasing natural capabilities of plants to cope with stress (Dąbrowski 2008, Calvo et al. 2014). Some act as elicitors—compounds that are inducers of resistance against pests and damages— and others provide beneficial inorganic ingredients and organic materials (e.g., amino acids) (Dąbrowski 2008). Benefits of active ingredients contained in natural materials have been known and used in plant cultivation (Khan et al. 2009).

***Table 21.1*** Polysaccharides against abiotic stress occurred in plant

| Stress factor | Polysaccharide property responsible for activity against this factor |
|---|---|
| General | Induction of activity of ribonuclease (Briand et al. 2007) |
| | Induction of signaling pathways, mediated by jasmonic acid (JA) and salicylic acid (SA) (Forcat et al. 2008, Jaulneau et al. 2010) |
| Drought | Hygroscopic properties |
| | Reduction of water outflow by membrane stabilization (Valluru and Van den Ende 2008, Livingston et al. 2009) |
| | Induction of signaling pathway, mediated by abscisic acid (ABA) (Forcat et al. 2008, Jaulneau et al. 2010) |
| | Regulation of membrane permeability by enhanced production of proline (Rai 2002, González et al. 2013) |
| Deficiency of nutrients | Gelling properties |
| | In case of micronutrients—capability to chelate |
| Frost | Gelling properties |
| | Prevention or reduction of membrane-freezing water adhesion (Livingston et al. 2009) |
| | Membrane stabilization (Valluru and Van den Ende 2008, Livingston et al. 2009) |
| Heat | Prevention of both leakage of liposome aqueous content and membrane fusion after rehydration (Hincha et al. 2003, Valluru and Van den Ende 2008) |
| Light intensity | UV-absorbing properties (photoprotective compounds) |
| Oxidative stress | Presence of sulfate groups in polysaccharide molecule (sulfated polysaccharides), capable of reacting with oxidative agent |
| | Induction of the activity of the following enzymes: peroxidase, ascorbate peroxidase, glutathione peroxidase, glutathione-$S$-transferase, and superoxide dismutase (Briand et al. 2007) |
| Salinity | Chelating properties |
| | Regulation of membrane permeability by enhanced production of proline (Rai 2002, González et al. 2013) |

Despite the long history of application, plant growth–promoting biomolecules still need to be investigated due to their diversity (Karnok 2000).

Biostimulants of plant growth can be polysaccharides that are derived from marine organisms. Marine polysaccharides can be divided into different types such as plant polysaccharides (macroalgae and microalgae), marine animal polysaccharides of vertebrates (fish) and invertebrates (mollusks, sponge, arthropods–crustaceans), and microbial polysaccharides (bacteria, fungi) according to their different sources (Senni et al. 2011, Wang et al. 2012, Trincone 2014). Marine environment offers enormous species biodiversity. Natural polysaccharides extracted from marine organisms present a great chemical diversity that depends on the species (Senni et al. 2011).

In this chapter, the application of marine organisms as a source of polysaccharides useful in agriculture was discussed. The properties of marine polysaccharides that protect the plant from abiotic stress are shown in Table 21.1. The second part was devoted to the isolation of the polysaccharides from the raw material (extraction and homogenization). The comparison of these two methods is presented in Figure 21.1. Finally, the effect of extracted polysaccharides on the growth of plants was determined.

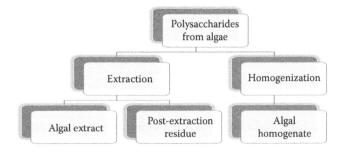

***Figure 21.1*** The methods of isolation of the polysaccharides from the marine biomass.

## 21.2  Marine sources of polysaccharides

Marine polysaccharides are important biological compounds that widely exist in marine organisms (Wang et al. 2012). Examples of aquatic organisms that are able to produce polysaccharides of agricultural significance are presented in Table 21.2. Among them proved action as biostimulants have mainly polysaccharides from seaweeds and chitin mainly from crustaceans.

***Table 21.2*** Examples of marine organisms as a source of polysaccharides of agricultural significance

| Marine organisms | Polysaccharides | References |
|---|---|---|
| | **Marine plant polysaccharides** | |
| *Macroalgae* | | |
| *Ascophyllum nodosum, Fucus serratus, Fucus vesiculosus, Laminaria hyperborean, Sargassum muticum* (B) | Laminarin, mannitol, fucoidan, alginate (alginic acid) | Sharma et al. (2012) |
| *Laminaria digitata* (B) | Laminarin | Mercier et al. (2001) |
| *Ascophyllum nodosum* (B) | Laminaran | Patier et al. (1993) |
| *Laminaria digitata* (B) | Laminarin | Aziz et al. (2003) |
| *Pelvetia canaliculata* (B) | Sulfated fucan | Klarzynski et al. (2003) |
| *Lessonia trabeculata* (B) | Oligosaccharide–polyglucuronic acid fraction from sodium alginate | Laporte et al. (2007) |
| *Lessonia vadosa* (B) | Oligosaccharide–polymannuronic acid fraction from sodium alginate | Laporte et al. (2007) |
| *Lessonia vadosa* (B) | Alginate–polymannuronic acid fraction | Chandía et al. (2004) |
| *Ulva* sp., *Enteromorpha* sp. (G) | Ulvan | Lahaye and Robic (2007) |
| *Ulva* spp. (G) | Ulvan | Jaulneau et al. (2010) |
| *Eucheuma cottonii, Eucheuma spinosum, Gigartina acicularis, Gigartina pistillata* (R) | Carrageenan | Mercier et al. (2001) |
| *Hypnea musciformis* (R) | Carrageenan | Bi et al. (2011) |
| *Schizymenia binderi* (J. Agardh) (R) | Oligosaccharide–sulfated galactan fraction | Laporte et al. (2007) |
| *Microalgae* | | |
| *Anabaena PCC 7120, Oscillatoria angustissima, Scenedesmus obliquus, Chlorella vulgaris* (G) | Intracellular (IPS) and extracellular polysaccharide - exopolysaccharide (EPS) | El-Sheekh et al. (2012) |
| | **Marine animal polysaccharides of vertebrates** | |
| Fish: Atlantic mackerel (*Scomber scombrus*), Japanese jack mackerel (*Trachurus japonicus*), Pacific bluefin tuna (*Thunnus orientalis*), yellowfin sole (*Limanda aspera*), broadbanded thornyhead (*Sebastolobus macrochir*), golden threadfin bream (*Nemipterus virgatus*), Nile tilapia (*Oreochromis niloticus*) | Glycosaminoglycans | Arima et al. (2013) |
| | **Marine animal polysaccharides of invertebrates** | |
| **Molluscs (phylum** *Mollusca***)** | | |
| Clam: *Tridacna maxima, Perna viridis* (bivalves) | Glycosaminoglycans | Arumugam et al. (2009) |
| Sea snails: *Babylonia spirata, Phalium glaucum* (gastropods) | Glycosaminoglycans | Periyasamy et al. (2013) |
| Squid: *Watasenia scintillans, Enoploteuthis chunii, Berryteuthis magister* (cephalopods) | Glycosaminoglycans | Arima et al. (2013) |
| **Sponges (phylum** *Porifera***)** | | |
| *Aplysina fulva, Chondrilla nucula, Dysidea fragilis, Hymeniacidon heliophila* | Sulfated polysaccharides | Zierer and Mourao (2000) |
| *Aplysina fulva, Dysidea robusta, Hymeniacidon heliophila* | Sulfated polysaccharides | Vilanova et al. (2009) |
| Skeletons of marine glass sponges—*Farrea occa* (Hexactinellida) | Chitin | Ehrlich et al. (2007) |

(*Continued*)

***Table 21.2 (Continued)*** Examples of marine organisms as a source of polysaccharides of agricultural significance

| Marine organisms | Polysaccharides | References |
| --- | --- | --- |
| **Phylum: Arthropods (*Arthropoda*); Subphylum: Crustaceans (*Crustacea*)** | | |
| Crab | Chitosan | Kulikov et al. (2006) |
| Shrimp | Chitosan | Nge et al. (2006) |
| Crab shell | Chitin | Burrows et al. (2007) |
| Crawfish shell | Chitin | No et al. (1989) |
| White shrimp (*Litopenaeus vannamei*) | Chitin, glycosaminoglycans | Cahú et al. (2012) |
| **Microbial polysaccharides** | | |
| Thraustochytrid protists | Exopolysaccharide | Jain et al. (2005) |
| *Xanthomonas campestris* | Exopolysaccharide | Boyle and Reade (1983) |
| *Cyanothece* sp. BH68K (C) | Exopolysaccharide | Reddy et al. (1996) |
| *Phormidium* J-1, *Anabaenopsis circularis* PCC 6720 (C) | Emulcyan—sulfated Heteropolysaccharide | Bar-Or and Shilo (1987) |

*Note:* B, brown algae; G, green algae; R, red algae; C, cyanobacteria.

## 21.2.1 Marine plant polysaccharides

A well-known source of polysaccharides is marine seaweeds. The characteristics of algal polysaccharides from brown algae (*Phaeophyta*), for example, alginate, fucoidan, and laminarin; from red algae (*Rhodophyta*), such as carrageenan, agar, and porphyran; and green algae (*Chlorophyta*), for example, ulvan were presented in many review papers (e.g., Laurienzo 2010, Jiao et al. 2011, Senni et al. 2011, Vera et al. 2011, Kraan 2012, Stadnik and de Freitas 2014, Tuhy et al. 2015). Among them of particular importance in agriculture are sulfated polysaccharides, the naturally occurring glycosaminoglycans. These compounds contain hemi-ester sulfate groups in their sugar residues. They are found in marine algae and higher animals, rarely in microbes, and absent in higher plants (Shanmugam and Mody 2000). The major sulfated polysaccharides synthesized by seaweeds include galactans (e.g., agarans and carrageenans), ulvans, and fucoidans (Jiao et al. 2011, Wijesekara et al. 2011).

In the literature, a main attention is paid to the beneficial role of polysaccharides in nutraceuticals, pharmaceuticals, and cosmeceuticals. The biological role of polysaccharides involves anticoagulant, anticancer, antiallergic, antiviral, antioxidative, and immunomodulating activities (Wijesekara et al. 2011, Ngo and Kim 2013). Little information is available on their application in the agriculture.

Some literature data show that aqueous extracts of some seaweed species that are rich in polysaccharides can act as plant biostimulants (Sharma et al. 2012). Seaweed extracts are being increasingly used in agriculture to induce plant resistance to abiotic and biotic stresses (Craigie 2010). Marine macroalgae are considered as a novel source of plant defense elicitors (Khan et al. 2009). Various algal polysaccharides (e.g.,

laminarin, carrageenan, ulvans, alginates, and fucans) have the potential to induce disease resistance in plants (Mercier et al. 2001, Vera et al. 2011). In addition, oligosaccharides obtained by depolymerization of seaweed polysaccharides also induce protection against viral, fungal, and bacterial infections in plants (Vera et al. 2011). For example, ulvan, a sulfated polysaccharide from green algae, activates plant immunity (Jaulneau et al. 2010). Seaweed polysaccharides and derived oligosaccharides can also influence plant growth by enhancing carbon and nitrogen assimilation, basal metabolism, and cell division (González et al. 2013). Moreover, seaweed extracts are able to act as chelators of metal cations from the marine environment due to high content of polysaccharides that are rich in functional groups capable of binding micronutrient ions in a reversible process (Tuhy et al. 2015).

The interest in marine microalgae is growing increasingly, especially due to sulfated exopolysaccharides (EPS)—compounds released by many species of microalgae (Raposo et al. 2013). Production of exopolysaccharides by algae is a defense system developed to cope with oxidative stress induced by toxins (e.g., microcystin) (El-Sheekh et al. 2012). Microalgae, as members of the soil microbial communities (e.g., *Chlamydomonas mexicana* and *C. sajao*), are also known to contribute to soil fertility. They participate in nitrogen fixation and in the stabilization of soils largely through the production of extracellular polymers, composed of rhamnose, fucose, ribose, arabinose, xylose, mannose, galactose, glucose, and uronic acid (Barclay and Lewin 1985). The interest in microalgae is increasing not only because of their biochemical diversity but also due to possibility to control their growth conditions in a bioreactor (Laurienzo 2010).

## 21.2.2 Marine animal polysaccharides

The marine animal polysaccharides include chitosan that can be derived from crustaceans (e.g., crayfish, crab, shrimp, and lobster) (No et al. 1989, Singh and Ray 2000, Kulikov et al. 2006, Burrows et al. 2007, Cahú et al. 2012) and sponge (Ehrlich et al. 2007). Chitin is one of the most abundant polysaccharides and is usually produced from the shells of crabs and shrimps (Burrows et al. 2007). Chitin/chitosan and their derivatives are applied in the agriculture as seed, leaf, vegetable, and fruit coating, growth enhancers, and stimulator for the plant hormones responsible for root formation, stem growth, and fruit development (Nge et al. 2006, Cheba 2011). Moreover, in agriculture, chitosan is used as a fertilizer and for the controlled release of agrochemicals. Its function is to increase the crop yield, to stimulate the immunity of plants, to protect plants against microorganisms, and to stimulate their growth (Nge et al. 2006).

Important groups of marine polysaccharides are glycosaminoglycans isolated from bivalve mollusks (mussel, oyster, scallops, clam), gastropod (abalone, sea snail), cephalopods (squid, cuttlefish, octopus) (Arumugam et al. 2009, Arima et al. 2013, Periyasamy et al. 2013), and fish (Arima et al. 2013). Glycosaminoglycans have been widely used in medicine because of their anticoagulant, antithrombotic, and antilipemic activities (Periyasamy et al. 2013).

Marine animals can also be a source of sulfated polysaccharides isolated from sponge (Zierer and Mourao 2000, Vilanova et al. 2009), for example, sulfated glucan in the species *Aplysina fulva* (Zierer and Mourao 2000).

## 21.2.3 Marine microbial polysaccharides

Marine microbial polysaccharides mainly involve the exopolysaccharides (EPSs). Microbial extracellular polymeric substances produced by both prokaryotes (eubacteria and archaebacteria) and eukaryotes (phytoplankton, fungi, and algae) are a topic of current research interest (Bhaskar and Bhosle 2005). Among all the marine microorganisms, bacteria and phytoplankton, including diatoms, cyanobacteria, and dinoflagellates, are considered to be the major sources of EPS in marine waters (De Philippis and Vincenzini 1998, Bhaskar and Bhosle 2005, Jain et al. 2005).

The bacteria that live in extreme marine environments developed special metabolic pathways and protective mechanisms to survive in these unfavorable environments. Thus, they evolved the ability to produce bioactive compounds including EPSs (Chi and Fang 2005). An overview of knowledge on EPSs produced by marine bacteria is presented by Chi and Fang

(2005), Bhaskar and Bhosle (2005), and De Philippis and Vincenzini (1998).

Many new marine microbial EPSs possess novel chemical composition, with unique properties and structures. Therefore, they can be potentially applied in products such as pharmaceuticals (e.g., anticancer), food additives, adhesives, and textiles. They can also be used in oil recovery, metal removal in mining, and industrial waste treatments (Bhaskar and Bhosle 2005, Chi and Fang 2005). The unique gelling properties of microbial EPS are important in the transport and transformation of organic matter, complexation of dissolved metal ions, and in biogeochemical cycling of elements. EPSs are rich in organic carbon, and therefore are important source for different organisms in the food chain (Bhaskar and Bhosle 2005). EPS mediates also the initial attachment of cells to different substrata and protection against environmental stress and dehydration (Vu et al. 2009). Reddy et al. (1996) suggested that the polysaccharides excreted by *Cyanothece* BH68 act as a physical barrier to the atmospheric oxygen and also serve as a chelator for iron and calcium ions, which are both essential for nitrogen fixation. In the future, the biggest attention will be paid to microbial exopolysaccharides characterized by high specific viscosity, tolerance to extreme conditions of temperature, salinity, pH, and appropriate rheological behavior (Boyle and Reade 1983). Among all marine organisms, microorganisms have the greatest advantage due to recent advances in biological techniques, and their polysaccharides can be produced *in vitro* (Laurienzo 2010).

## 21.3 Technology of production of polysaccharides

Polysaccharides for commercial use are produced by many different methods such as enzymatic, chemical, mechanical, and physical (Hahn et al. 2012). All techniques are still being investigated to improve their effectiveness. Soxhlet extraction and other traditional methods such as maceration, micronization, or sonication are modified to increase the effectiveness and to eliminate the residues (Kaufmann and Christen 2002). Modern methods such as supercritical fluid extraction (SFE), microwave-assisted extraction (MAE), and pressurized solvent extraction allow to limit or even to avoid the contamination and to accelerate the extraction process (Kaufmann and Christen 2002). Another option for extraction of polysaccharides is the application of hot and cold distilled water (Zvyagintseva et al. 2003). Extraction with hot water is the most widely used technology for the extraction of polysaccharides. The downside of hot water polysaccharide extraction results from lower yields, high temperatures, and long extraction times (Huang et al. 2010).

A special attention should be paid to the pretreatment of raw material technology that has a significant impact on the efficiency of extraction. To perform extraction of polysaccharides from the marine sources, the biomass is usually harvested from the beaches, coastal area, or directly from the sea. According to the raw material and future use of polysaccharide, specific methods of preparing material (including pretreatment methods)—extraction, separation, and purification up to the required quality—are selected. Different technologies used for polysaccharides extraction from marine resources are shown in Figure 21.2.

### 21.3.1 Preparation and treatment

Preparation of marine biomass begins with washing to remove undesirable contamination (sand, stones, salt, etc.). Next step is to reduce the water content. The seaweed used in commercial process of alginate production must contain 83% of dried matter (17% moisture) (Hernández-Carmona et al. 2013). For supercritical extraction, the raw material should contain less than 10% of moisture. After drying at low temperatures, marine biomass is milled to obtain a higher surface-to-volume ratio (Hahn et al. 2012). Before appropriate extraction process in order to increase the effectiveness of polysaccharide extraction, different pretreatment methods are used, for example, rehydration by formaldehyde solution (Hernández-Carmona et al. 2013), hydrochloric acid pretreatment in high temperature (pH 3, 15 min, 100°C) (Nagaoka et al. 1999), ethanol pretreatment (24 h, 70°C) (Ponce et al. 2003), mixture of methanol/chloroform/water 4/2/1 pretreatment (Whyte and Southcot 1970), or freezing.

### 21.3.2 Production of polysaccharides by microorganisms

Marine environment is a very rich source of microorganisms. Different marine microorganisms have the ability to produce extracellular polymeric substances, which include many polysaccharides (Vu et al. 2009). For example, gram-positive bacteria, *Lactobacillales*, *Leuconostocaceae*, and *Streptococcaceae*, synthesize large amounts of high-molecular-weight homopolysaccharide—dextran (Duboc and Mollet 2001, Naessens et al. 2005, Kumar and Mody 2009). Umezawa et al. (1983) described marine bacteria *Flavobacterium uliginosum*, which produces an active polysaccharide named marinactan. The technology of microbiological production of marine polysaccharides starts from the isolation and screening of potential strains capable of producing polysaccharides. Marine bacteria isolated from the environment (e.g., seawater, marine biomass, algae, and sea mud) are then incubated in a suitable sterile medium in optimum growth condition (pH, temperature). Microorganisms can produce polysaccharides both as consortium of many bacterial strains and as single colonies. After reaching a satisfactorily high concentration of polysaccharides, the medium in which the microbial production process was carried out is centrifuged/filtered. Then, the obtained suspension is subjected to processes of precipitation, isolation, purification, and extraction of polysaccharides. The understanding of the requirements of microorganisms, the selection of growth conditions, and the ability to isolate individual strains cause the diverse bacteria commercially used for the production of some polysaccharides (Umezawa et al. 1983).

### 21.3.3 Organic solvent extraction

Classical extraction procedure consists of extraction of earlier prepared and dry marine biomass with different organic (acetone, methanol, ethanol, chloroform, ammonium oxalate–oxalic acid) and inorganic (acid/alkaline solution, water) solvents in a Soxhlet apparatus from a few to tens of hours in different temperatures (Nagaoka et al. 1999, Ponce et al. 2003, Cumashi et al. 2007). For example, to obtain a mixture of fucose, mannose, galactose, xylose, glucose, rhamnose, and arabinose from algae *Hizikia fusiforme* (Li et al. 2006), extraction with $H_2O$ (1:10) in 2 h at 70°C was conducted three times, then precipitated with EtOH and $CaCl_2$, and finally dried. Cumashi et al. (2007) used different kinds of algal species such as *Laminaria saccharina*, *Laminaria digitata*, *Fucus vesiculosus*, *Fucus spiralis*, and *Ascophyllum nodosum* to obtain fucose, xylose, mannose, glucose, and galactose but with the use of the same extraction procedure (extraction with 2% $CaCl_2$ for 5 h at 85°C, precipitation with cetavlon, transformation of cetavlon salts into calcium salts, and an alkaline treatment to remove acetyl groups and to transform fucoidan into sodium salts [Usov et al. 1998]). Commercially developed technology of alginate production from seaweed includes acid pretreatment by hydrochloric acid in pH 4, extraction using $Na_2CO_3$ in pH 10 at 80°C, filtration to eliminate the residues using filter press, precipitation and concentration of calcium alginate, conversion of calcium alginate to alginic acid, pressing and conversion of alginic acid to sodium alginate, drying, milling, and blending to obtain the final product such as sodium alginate (Hernández-Carmona et al. 2013).

### 21.3.4 Enzymatic extraction process

In this process, pretreatment methods based on microwaves and ultrasounds are widely used to increase

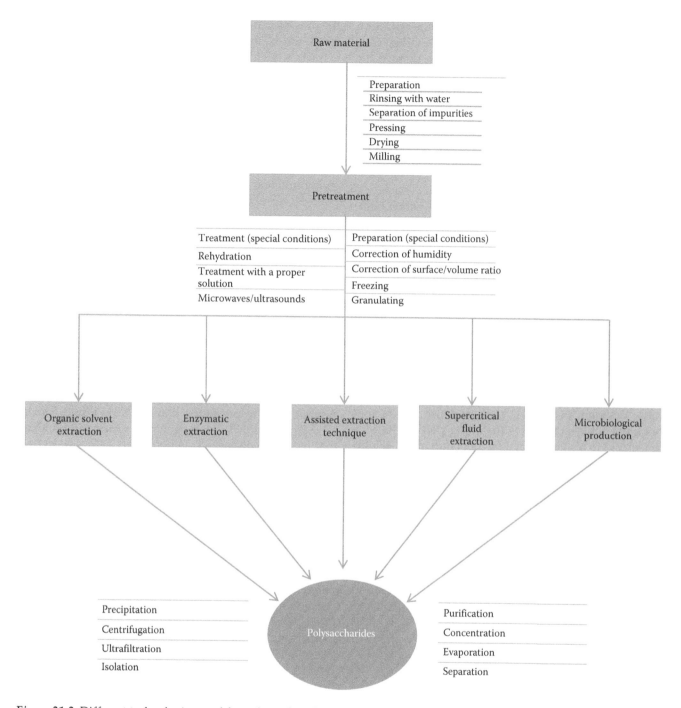

*Figure 21.2* Different technologies used for polysaccharides extraction from marine resources.

effectiveness of polysaccharides extraction. Application of enzymes such as carbohydrases, cellulase, proteases, β-glucanase, hemicellulase, xylanase to marine biomass requires the creation of special conditions such as optimal temperature, pH, amount of enzyme, and time. Widely used in enzymatic extraction are commercially available enzymes such as cellulase, kojizyme, termamyl, xylanase, agarase, carrageenans, neutrase, alcalase, and viscozyme (Heo et al. 2003, Ahn et al. 2004,

Park et al. 2004). An exemplary enzymatic extraction process is as follows: the seaweed samples are milled into powder using a grinder or mills. Suitable mixture of enzymes solution is chosen depending on the kind of the raw material that is subjected to the process and prepared at the appropriate concentration. The next step consists of mixing previously prepared solution of enzymes with dry algae biomass. The enzymatic hydrolysis reaction is conducted from a few to several hours at a suitable temperature and pH to achieve the optimum hydrolysis degree. Obtained hydrolysates are then clarified by ultrafiltration or centrifugation to remove the residue of unhydrolyzed material. The degree of enzymatic hydrolysis is determined in relation to the weight loss of biomass samples, which is expressed in percentage (Chen et al. 1999, Li et al. 2012).

### 21.3.5 Microwave- and ultrasound-assisted extraction

Plant cell wall consists mainly of cellulose, which hampers the process of isolating polysaccharides from marine biomass. The use of physical methods, such as microwave and ultrasound, can accelerate and facilitate the extraction of polysaccharides and even increase the performance (Li et al. 2012). MAE by electromagnetic radiation interacts with the water content in plant cells. The effectiveness of the application of microwave energy is dependent on the solvent used for the process and the raw material (Kaufmann and Christen 2002). Vibrations that occur within the cell during MAE increase the temperature that activates the process of evaporation of the water and exerts pressure on the cell walls. Violation of the cell wall structure makes that contained active substances including polysaccharides transferred to a solution in a facilitated manner (Hahn et al. 2012). As an example, MAE of sulfated polysaccharides (fucoidan) from brown seaweed was described in the work of Rodriguez-Jasso et al. (2011) and Wang et al. (2010). The disadvantage of MAE is related to inhomogeneous heating (Zhao et al. 2011). In ultrasound-assisted extraction, the cheapest and the most available source of ultrasound irradiation is the ultrasonic bath (Tadeo et al. 2010). The drawbacks of the sonication treatment are compositional changes and degradation of the polysaccharide preparations (Ebringerová and Hromádková 2009).

### 21.3.6 Supercritical fluid extraction

The technology using carbon dioxide in SFE is relatively a novel technology but very safe for humans and the environment and allows the isolation of mixture of many bioactive compounds including polysaccharides without residues of unwanted compounds. The effects of SFE process depend largely on a properly prepared raw material. Dried and ground marine biomass is subjected to a pressure between 100 and 1000 bar generally at a temperature of 20°C–45°C. The process should not be carried out at too high temperatures, which can result in the loss of biological activity of extracted compounds. SFE allows to obtain high yield in relatively short time. Because of nonpolar character of carbon dioxide in SFE,

the process can also be carried out with the use of cosolvents, for example, ethanol that allows to increase the extraction yield and extract more compounds, in particular of polar character. In SFE process, as a solvent next to $CO_2$, water is also used. Extract sought composed of individual components is possible to obtain by changing the technological parameters as temperature or pressure is possible (Hauthal 2001, Hamburger et al. 2004). King (1989) in his study showed that using water in supercritical state allows to extract more polar compounds at lower temperatures (less polar compounds are extracted at higher temperature). Further research on novel extraction techniques is needed since the extraction of polysaccharides is considered to be of great importance for the development of their applications (Huang et al. 2010).

## 21.4 Homogenates as a source of polysaccharides

Another source of polysaccharides can be algal homogenates (Dillon et al. 2007, Avila et al. 2012, Brooks and Franklin 2013). The discovery of the homogenization goes back to the beginning of the century when Auguste Gaulin presented it at the Paris World Fair in 1900 (Paquin 1999). "Homogenate" means cell biomass that has been disrupted (Dillon et al. 2007, Avila et al. 2012, Brooks and Franklin 2013). It is quite common that a homogenate is not entirely homogenous. It can refer to a composition of cells that have been disrupted to the point where a majority of the cells in the preparation have been ruptured (Dillon et al. 2007). The process of homogenizing is related to blending two or more substances into a homogenous or uniform mixture.

There are embodiments where a homogenate is successfully created, but it is also possible that the biomass is predominantly intact but homogeneously distributed throughout the mixture (Brooks and Franklin 2013). To extract intracellular products, it is necessary to disrupt the cell. The method used may vary depending on the type of cell and its cell wall composition. The disruption must be effective, and the method should not be too harsh so that the product recovered remains in its active form (Burden 2008). Disruption of the cells may be accomplished by the use of blending, the French press, and also even centrifugation (Coragliotti et al. 2010). The mechanical–physical techniques used for cells disruption are presented in Figure 21.3.

Homogenization of the algae can be done by a conventional homogenizer in water or preferably in a buffer (e.g., 0.05 M phosphate buffer [pH 7.0]) (Shinmen et al. 1988). It has been notified that applying mechanical forces causes an increase in surface area of splitted solid materials (Markert 1995). In cases of intact cells

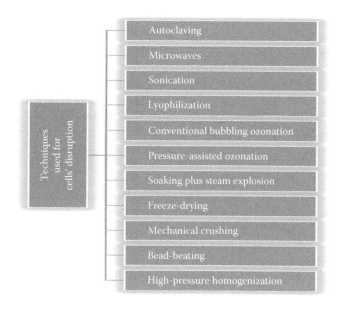

*Figure 21.3* The mechanical–physical techniques used for cell disruption. (Data from Michalak, I. and Chojnacka, K., *Eng. Life Sci.*, 14, 581, 2014.)

or homogenized cell materials, they may be resuspended in water or other suitable liquid to allow the release of polysaccharides. Higher temperature or vortexing could be used to enhance the release of polysaccharides that may be separated from the cells or insoluble cellular materials (Dillon et al. 2007). In some embodiments,

the invention comprises polysaccharides from a species of the genus *Porphyridium*, optionally sulfated at a level of 4.7 or higher, that have been reduced in molecular weight through milling or shearing (e.g., sonication or ball milling). Sonication is a method of high-frequency sound that is produced electronically and transported through a metallic tip to an appropriately concentrated cellular suspension. During the mentioned process, the creation of cavities in cell suspension occurs (Dillon et al. 2007).

Thiruvengadam and Hochrein (1978) have presented a method and an apparatus for the homogenization of multicomponent stream including a liquid and an insoluble component (either a liquid or a substantially divided solid). The multicomponent fluid stream passes through a turbulent shear layer designed such that a cavitating flow regime occurs. Premixing the fluid and the insoluble component may improve the operating efficiency of the apparatus. It was pointed out that the usage of conduit to recirculate separated components of the emulsion through the homogenization apparatus to effect intimate mixing may be desirable. The preferable upstream pressure should be maintained at 25 times the downstream pressure to develop the necessary flow turbulent velocity shear layer, thereby providing the most efficient homogenization of the multicomponent stream (Thiruvengadam and Hochrein 1978). The schematic illustration of the homogenization technology is presented in Figure 21.4.

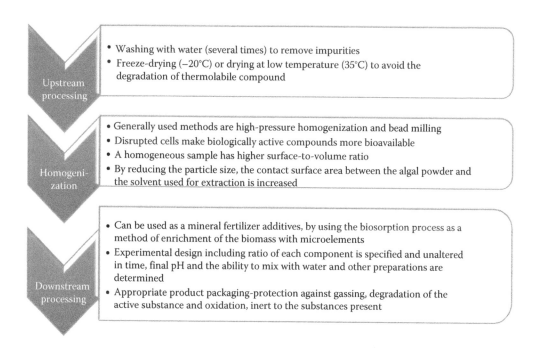

*Figure 21.4* Schematic illustration of the processing of algal biomass. (Data from Michalak, I. and Chojnacka, K., *Eng. Life Sci.*, 14, 581, 2014; Wilk, R. and Chojnacka, K., *Mar. Algae Extracts Proc. Prod. Appl.*, 1, 161, 2015; Ho, C.W. et al., *Biotechnol. Bioproc. Eng.*, 13, 577, 2008.)

The most common mechanical methods used for cell disruption are high-pressure homogenization (HPH) and bead milling. Both methods involve the application of high shear forces to deform and rupture the cell wall (Ho et al. 2008). Dynamic HPH (Paquin 1999) was adapted for cell disruption (Middelberg 1995). High pressure (up to 1500 bar) is applied, followed by an instant expansion through a nozzle. Cell disruption is accomplished by three different mechanisms: impingement on the valve, high liquid shear in the orifice, and sudden pressure drop upon discharge, causing an explosion of the cell. This method is mainly used for the release of intracellular molecules (Brooks et al. 2010).

In ball or bead mills, cells are agitated in suspension with grinding media, which are small abrasive particles (e.g., glass or ceramic beads). The cells break due to shear forces, grinding between beads, and collisions with the beads (Clayton et al. 2013). Biomolecules are released by the beads that disrupt the cells (Avila et al. 2012). The advantages and disadvantages of these two methods are presented in Table 21.3.

Microalgal cell homogenate usually contains less than 20% of purified or isolated polysaccharide compositions and less than 10% proteins, nucleic acids, and intracellular components such as other polysaccharides. Homogenization augments significantly increase the amount of solvent-accessible polysaccharides by breaking apart cell walls that are largely composed of polysaccharides (Dillon et al. 2007, Coragliotti et al. 2010, Avila et al. 2012).

Produced algal homogenates can be used as mineral fertilizer additives, after the application of biosorption process as a method of biomass enrichment with microelements. Biosorption is a well-known process, discussed in many papers. It is a passive, reversible single-stage process (Michalak et al. 2013). Marine algae, because of their availability both from fresh and saltwater, relatively high surface area, and high binding affinity, are suitable as biosorbents. Biosorption by algae has been mainly attributed to the cell wall structure containing functional groups such as amino, hydroxyl, carboxyl, and sulfate, which can act as binding sites for metal ions via both electrostatic attraction and complexation (Sari and Tuzen 2008). The biomass constitutes a source of additional nutrients important for plants since they increase crop yields and improve crops quality because of their influence on soil conditions (Michalak et al. 2013).

An example of a commercially available homogenate is Göemar BM 86, which is produced based on PhysioActivator™ technology produced by French company Goëmar. The active ingredients of BM 86 are GA 142 seaweed (*A. nodosum*) filtrates obtained by an exclusive Goëmar process that preserves and concentrates the active ingredients of seaweeds. The product also includes boron (2.1% w/w), molybdenum (0.02% w/w), and magnesium (4.8% w/w). The manufacturer of BM 86 states that it is a physiological activator of fruit setting that activates trees' nutritional pathways and the biosynthesis of flowering hormones. Fruit setting is improved. The product is recommended for fruits (e.g., apple, pear, cherry nectarine, peach, plum, apricot, strawberry, blackcurrant, raspberry, and kiwi fruit) (Goëmar 2014). Another well-known commercial products containing polysaccharides isolated from seaweed biomass are Iodus® and Vacciplant® (Goëmar). Both these preparations owe their activity to laminarin from *L. digitata*. The content of this polysaccharide enables to implement the vaccination concept, according to which

*Table 21.3* Advantages and disadvantages of high-pressure homogenization and ball mills

| Homogenization technique | Advantages | Disadvantages |
|---|---|---|
| **High-pressure homogenization** | Good reproducibility | Unable to homogenize tissue without prior dissociation (clog the device) |
| | Well-established homogenization technology on large scale | High-temperature process |
| | Quick mechanical process | High energy input |
| | Organic solvent-free method | Possible degradation of the components caused by high-pressure homogenization |
| **Ball mills** | Use of different sizes of beads | Nonselective release of the product |
| | Versatile (varying the bead size and thoroughness of the homogenization) | Micronization of the cell debris |
| | Extremely high reproducibility | |
| | No risk of cross-contamination | |

*Sources:* Data from Hakeem, K.R. et al., *Biomass and Bioenergy: Application*, Springer, Cham, Cambodia, 2014; Balasundaram, B. et al., *Trends Biotechnol.*, 8, 477, 2009.

plants are provided with capacity to protect themselves from pathogen attack and disease recurrence (Klarzynski 2006; Goëmar 2014).

## 21.5 Application of polysaccharides as biostimulants of plant growth

Biostimulants are usually complex materials, composed of various constituents, and thus indication of bioactive compounds is not a simple task (Karnok 2000). Nevertheless, among PGB, polysaccharides might be distinguished, since they were proved to act as elicitors (former name) or pathogen-associated molecular patterns (in current use), inducing proper signaling pathways (Jaulneau et al. 2010).

Polysaccharides (otherwise known as glycans) are structural and energy storage biopolymers, found in living organisms of all kinds, including especially plants. These macromolecules are also considered as the major class of compounds isolated from the marine environments (Venugopal 2011). It has been estimated that marine biomes cover about 70% of the biosphere and are inhabited by at least one million of species, except for microbes, the number of which might be equally high (Miller and Spoolman 2012). Aquaculture is therefore a rich source of polysaccharides, characterized by different structures, biochemical features, and hence activities.

As mentioned before, among marine organisms, algae and crustaceans are the most common to produce polysaccharides of PGB function. In general, polysaccharide content in seaweeds ranges between 4% and 76% of the dry weight depending on species. The highest concentration of glycans was reported for brown algae *Ascophyllum* and red algae *Porphyra* and *Palmaria*; nevertheless, some species of green seaweeds—such as *Ulva*—contain a great amount of polysaccharides (Kraan 2012). Considering crustacean, exoskeletons of various species contain 13%–42% of chitin (Nakkarike et al. 2011). Both mentioned groups of organisms generate a million tonnes of biomass annually, a part of which are by-products or wastes (Burrows et al. 2007). Thereby, proper management of marine resources is a great goal to reach.

Algae are considered as one of the major class of biostimulants (du Jurdain 2012) and their first applications to plants dates back to the prehistoric period (Craigie 2010). Most of the used algal preparations contain whole cells, homogenates, or extracts, thus growth-promoting effect of components has not been understood well enough. Several successful works on potential of algal polysaccharides to stimulate suitable response to controlled environmental changes were reported. In the recent work, Stadnik and de Freitas (2014) reviewed a number of investigations on elicitor activity of macroalgae polysaccharides in the presence of plant stress agents. Beside glycans, seaweed oligosaccharides were proved to play PGB role successfully (González et al. 2013). A summary of the stimulating and protecting capacity of the major groups of algal poly- and oligosaccharides on plants is shown in Table 21.4.

Considering applicability of biostimulants from marine animals, both chitin and chitosan, as well as chitooligosaccharides, were proven to elicit activation of host-defense genes and immune response against diseases of various origins (Bautista-Baños et al. 2006, Eckardt 2008, du Jurdain 2012). It was discovered that chitooligosaccharides induce plant defense with the capacity of mimicking pathogens. As a result, glycans are bound with proper receptors and trigger the same cascade of reactions (du Jurdain 2012).

Despite the known activity of chitin, its deacetylated form—chitosan—is more soluble in diluted acids and, unlike most of glycans, shows cationic-like characteristics, and thus has been examined more frequently (No et al. 1989, Kulikov et al. 2006, Nge et al. 2006, Burrows et al. 2007, Ehrlich et al. 2007). Nevertheless, chitin was chosen as the main compound of the soil amendment—SolActif® (France Chitine 2014), designed for restoring chitinolytic organisms that attack, repel, or otherwise antagonize disease-causing pathogens. On the other hand, the activity of KaitoSol® foliar biostimulant (AGENT; a spin-off company from The University of Cambridge and registered in Malaysia, KaitoSol® 2014) is related to chitosan, contained in active chitosan component, which hastens growth, enhances yield and general plant strength, as well as improves resistance to diseases.

Due to increasing interest of industry in PGBs, chitosan has been successfully tested as a suitable component of formulations for seed treatment. Chitosan oligomers were shown to promote germination of coated zucchini (*Cucurbita pepo*) seeds (Velásquez et al. 2012). Such promising results suggest expected benefits of using chito(oligo)saccharides in future products.

## 21.6 Summary

In this chapter, polysaccharides, isolated from marine organisms by extraction and homogenization, as biostimulants of plant growth were discussed. It was shown that these organisms are a rich source of polysaccharides that influence positively the plant growth and yield. These compounds applied in the plant cultivation allow to cope with the biotic and particularly with abiotic stress. The application of novel extraction techniques

*Table 21.4* Algal poly- and oligosaccharides as stimulators of plant growth, development, and stress response

| Polysaccharides | Effect on plants |
|---|---|
| Alginates | Increase the production of compounds related to plant immune response, such as anthocyanin, phenolic acids, glyceollins, and chitinase |
| | *The fraction of polymannuronic acid (obtained by alginate hydrolysis):* |
| | Induces an activation of phenylalanine ammonia lyase and peroxidase, the former of which is a key enzyme of the phenylpropanoid pathway (Bagal et al. 2012), while the latter participates in the biosynthesis of lignin and stress response (Vicuna 2005) |
| Oligoalginates | *Polymannuronic acid:* |
| | Affect growth and plant resistance to pathogens |
| | Enhance height and foliar biomass of plants |
| | Stimulate plant growth by the activation of ascorbate peroxidase (antioxidant enzyme) |
| | *Both polymannuronic acid and polyglucuronic acid:* |
| | Induce seed germination and increase seed yield |
| | Stimulate growth, in general, as well as enhance growth and length of roots and increase shoot length and biomass (e.g., weight of tubers) |
| | Increase content of proline, total chlorophylls and carotenoids, and essential oil(s) |
| | Enhance activity of nitrate reductase, which participates in nitrogen assimilation |
| | Elicit defense against pathogen infections (tobacco mosaic virus) by activating phenylalanine ammonia lyase |
| | *Both polymannuronic acid and polyglucuronic acid:* |
| | Stimulate root elongation (in case of carrot and rice, such effect is not observed using oligosaccharide of one type only) |
| | Enhance length and dry weight of shoots and increase root length |
| | Increase content of total chlorophylls, carotenoids, and alkaloids (e.g., morphine and codeine) |
| | Elicit nitrate reductase activity |
| Carrageenans | κ-*Carrageenans:* |
| | Act as both efficient growth-promoting agent and potent plant protectant |
| | Increase plant height, diameter of stem, and number of branches, leaves, and fruits (e.g., pods) and hasten flowering |
| | Enhance plant resistance and immune response by |
| | Inducing the activity of β-1,3-glucanase |
| | Enhancing the production of induced second metabolites |
| | λ-*Carrageenans:* |
| | Might elicit plant immune response through various signaling pathways, mediated by ethylene, jasmonic acid, and salicylic acid depending on plant species, as well as activate the production of proteinase inhibitor and chitinase, which are involved in defense against pathogens |
| | κ-, ι-, *and* λ-*Oligo-carrageenans:* |
| | Improve plant growth and development, including |
| | Stimulation of microspore embryogenesis, plant height, root and shoot growth, foliar biomass, and the thickness of a tree trunk |
| | Enhancement of efficiency of photosynthesis and basal metabolism, as well as activity of RuBisCO (enzyme involved in $CO_2$ fixation), glutamine synthase and glutamate dehydrogenase (ammonium-fixating enzymes), and pyruvate dehydrogenase (the first enzyme in the Krebs cycle) |
| | Increase in the production of both cyclins (A and D) and cyclin-dependent protein kinases (A and B), which improves cell cycle activity and hence cell division |
| | Stimulate the production of holocellulose and α-cellulose in the wood, decreasing the lignin content at the same time |
| | Increase content of essential oils (terpenes), part of which shows antimicrobial, or even antiviral, activities |
| | Elicit the response against plant disease through the systematic effect, including increase in the activity of phenylalanine ammonia lyase and accumulation of phenolics |
| | Enhance resistance to viral, fungal, and bacterial infections and elicit partial suppression of these pathogens by the accumulation of phenylpropanoid compounds |

*(Continued)*

**Table 21.4 (Continued)** Algal poly- and oligosaccharides as stimulators of plant growth, development, and stress response

| Polysaccharides | Effect on plants |
|---|---|
| Fucans | *Induce plant resistance to stress agents through* |
| | Early defense response, including alkalinization of extracellular media and production of hydrogen peroxide, |
| | Late defense response, such as enhancing activity of phenylalanine ammonia lyase and lipoxygenase |
| | Increase the production of polyamine, glutathione-*S*-transferase, and pathogenesis-related proteins |
| | Participate in red pigment formation |
| Laminarans (laminarins) | Mimic pathogen attack |
| | Participate in cellular recognizing mechanisms and plant–pathogen interactions (exogenously, being released by fungi cell wall degradation, or endogenously, as a result of fragmented callose) |
| | *Elicit plant defense response by* |
| | Inducing ethylene-dependent defense pathway |
| | Inducing the production and activity of phenylalanine ammonia lyase, stimulating the production of lipoxygenase, activating protein kinase, and enhancing activity of chitinase and β-1,3-glucanase |
| | Stimulating $Ca^{2+}$ influx, oxidative burst, and alkalinization of extracellular media |
| | Increasing the production of phytoalexins and inducing their accumulation and stimulating the production of pathogenesis-related proteins, at the same time, avoiding hypersensitive reaction |
| | Enhancing the production of $H_2O_2$ and antifungal compounds |
| | Are applicable in the laminarins or sulfated laminarins Elicit plant defense (against tobacco mosaic virus) by changing patterns of gene expression and inducing pathogenesis-related signaling pathways |
| | Are applicable in the sulfated laminarins |
| | Stimulate plant resistance to stress by increasing the production of $H_2O_2$, phytoalexins, and phenolics, as well as by activating salicylic acid–dependent defense pathway |
| Ulvans | Reduce aggressiveness of infection, up to 50% and protect plant for several days after the treatment |
| | Activate defense reactions, which are considered as plant genotype dependent; in case of fungal diseases, efficiency of ulvan activity is multiple and depends also on the species of pathogens |
| | *Elicit plant immune response by* |
| | Activating the expression of many defense-related genes, e.g., β-glucuronidase, salicylic acid– and jasmonic acid–responsive genes (jasmonic acid–dependent pathway aims to respond against necrotrophic pathogens) |
| | Enhancing the activity of glucanase and peroxidase and inducing phenylalanine ammonia lyase |
| | Increasing the production of phenolics, phytoalexins, and proteins related with the pathogenesis |
| | Inducing more frequent hypersensitive response in epidermal cells |
| | Controlling biotrophic and hemibiotrophic fungi |
| | Enhance the postinfection activity of glucanase and/or peroxidase (however, this effect might be notable not earlier than after a few days) |
| | May activate plant defense response indirectly by priming effect, providing faster and more efficient immune reactions, e.g., ulvan enhance the chitin-elicited oxidative burst and thus the production of hydrogen peroxide increases |

*Sources:* Data from Sharma, S.H.S. et al., *J. Appl. Phycol.*, 24, 1081, 2012; González, A. et al., *J. Plant Growth Regul.*, 32, 443, 2013; Stadnik, M.J. and de Freitas, M.B., *Trop. Plant Pathol.*, 39, 111, 2014.

and homogenization enables to obtain polysaccharides without loss of their biological activity. The enormous variety of polysaccharides that can be extracted from marine plants and animals or produced by marine bacteria means that the field of marine polysaccharides is constantly evolving. It is difficult to overvalue the importance of polysaccharides for the great number of applicative fields in which they appear. Marine polysaccharides present a real potential for natural product discovery and for the delivery of new marine derived products for agricultural applications. When reduction of the use of synthetic agrochemicals is expected, the use of biostimulants becomes a particularly promising option. The potential in the bioactivity of marine polysaccharides is still considered underexploited, and these compounds, including the derived oligosaccharides, are an extraordinary source of chemical diversity.

## Acknowledgments

This work was supported by the grant entitled "Biologically active compounds in extracts from Baltic seaweeds (2012/05/D/ST5/03379)" attributed by the National Science Centre in Poland and another grant entitled "Innovative technology of seaweed extracts—Components of fertilizers, feed, and cosmetics (PBS/1/A1/2/2012)" attributed by the National Centre for Research and Development in Poland.

## References

Ahn, C. B., D. S. Kang, T. S. Shin, and B. M. Jung. 2004. Free radical scavenging activity of enzymatic extracts from a brown seaweed *Scytosiphon lomentaria* by electron spin resonance spectrometry. *Food Research International* 37:253–258.

Arima, K., H. Fujita, R. Toita, A. Imazu-Okada, N. Tsutsumishita-Nakai, N. Takeda, Y. Nakao et al. 2013. Amounts and compositional analysis of glycosaminoglycans in the tissue of fish. *Carbohydrate Research* 366:25–32.

Arumugam, M., T. Balasubramanian, M. Warda, and R. J. Linhardt. 2009. Studies on glycosaminoglycans isolated from bivalves molluscs *Tridacna maxima* and *Perna viridis*. *Our Nature* 7:10–17.

Avila, J., G. Brooks, A. G. Day, A. Somanchi, and A. Coragliotti. 2012. Compositions for improving the health and appearance of skin, Patent no. US8298548 B2.

Aziz, A., B. Poinssot, X. Daire, M. Adrian, A. Bézier, B. Lambert, J.-M. Joubert, and A. Pugin. 2003. Laminarin elicits defense responses in grapevine and induces protection against *Botrytis cinerea* and *Plasmopara viticola*. *Molecular Plant-Microbe Interactions* 16:1118–1128.

Bagal, U. R., J. H. Leebens-Mack, W. W. Lorenz, and J. F. D. Dean. 2012. The phenylalanine ammonia lyase (PAL) gene family shows a gymnosperm-specific lineage. *BMC Genomics* 13:S1. doi:10.1186/1471-2164-13-S3-S1.

Balasundaram, B., S. Harrison, and D. G. Bracewell. 2009. Advances in product release strategies and impact on bioprocess design. *Trends in Biotechnology* 8:477–485.

Barclay, W. R. and R. A. Lewin. 1985. Microalgal polysaccharide production for the conditioning of agricultural soils. *Plant and Soil* 88:159–169.

Bar-Or, Y. and M. Shilo. 1987. Characterization of macromolecular flocculants produced by *Phormidium* sp. strain J-1 and by *Anabaenopsis circularis* PCC 6720. *Applied and Environmental Microbiology* 53:2226–2230.

Bautista-Baños, S., A. N. Hernández-Lauzardo, M. G. Velázquez-del Valle, M. Hernández-López, E. Ait Barka, E. Bosquez-Molina, and C. L. Wilson. 2006. Chitosan as a potential natural compound to control pre and postharvest diseases of horticultural commodities. *Crop Protection* 25:108–118.

Bhaskar, P. V. and N. B. Bhosle. 2005. Microbial extracellular polymeric substances in marine biogeochemical processes. *Current Science* 88:45–53.

Bi, F., S. Iqbal, M. Arman, A. Ali, and M.-Ul Hassan. 2011. Carrageenan as an elicitor of induced secondary metabolites and its effects on various growth characters of chickpea and maize plants. *Journal of Saudi Chemical Society* 15:269–273.

Boyle, C. D. and A. E. Reade. 1983. Characterization of two extracellular polysaccharides from marine bacteria. *Applied and Environmental Microbiology* 46:392–399.

Briand, X., S. Cluzet, B. Dumas, M.-T. Esquerre-Tugaye, and S. Salamagne. 2007. Use of ulvans as activators of plant defence and resistance reactions against biotic or abiotic stresses, US Patent no. 2007/0232494 A1.

Brooks, G. and S. Franklin. 2013. Cosmetic compositions comprising microalgal components, Patent no. US8557249 B2.

Brooks, G., S. Franklin, J. Avila, S. M. Decker, E. Baliu, W. Rakitsky, J. Piechocki, and D. Zdanis. 2010. Food compositions of microalgal biomass, Patent no. WO2010045368 A2.

Burden, D. W. 2008. Guide to the homogenization of biological samples. *Random Primers* 7:1–14.

Burrows, F., C. Louime, M. Abazinge, and O. Onokpise. 2007. Extraction and evaluation of chitin from crab exoskeleton as a seed fungicide and plant growth enhancer. *American-Eurasian Journal of Agricultural and Environmental Sciences* 2:103–111.

Cahú, T. B., S. D. Santos, A. Mendes, C. R. Córdula, S. F. Chavante, L. B. Carvalho Jr., H. B. Nader, and R. S. Bezerra. 2012. Recovery of protein, chitin, carotenoids and glycosaminoglycans from Pacific white shrimp (*Litopenaeus vannamei*) processing waste. *Process Biochemistry* 47:570–577.

Calvo, P., L. Nelson, and J. W. Kloepper. 2014. Agricultural uses of plant biostimulants. *Plant Soil* 383:3–41. doi:10.1007/s11104-014-2131-8.

Chandía, N. P., B. Matsuhiro, E. Mejías, and A. Moenne. 2004. Alginic acids in *Lessonia vadosa*: Partial hydrolysis and elicitor properties of the polymannuronic acid fraction. *Journal of Applied Phycology* 16:127–133.

Cheba, B. A. 2011. Chitin and chitosan: Marine biopolymers with unique properties and versatile applications. *Global Journal of Biotechnology and Biochemistry* 6:149–153.

Chen, X. G., Z. M. Wang, W. Zheng, and C. Lin. 1999. Extraction of *Flammulina velutipes* polysaccharides using enzymatic technique and its application. *Journal of Sichuan University of Science and Technology* 4:38–41.

Chi, Z. and Y. Fang. 2005. Exopolysaccharides from marine bacteria. *Journal of Ocean University of China* 4:67–74.

Clayton, R. L., S. N. Falling, and J. S. Kanel. 2013. Continuous process for recovering and concentrating valuable components from microalgae such as algal oils, for use as biofuel feedstocks; Agitation with solid particles, then adsorptive bubble separation, Patent no. US8512998 B2.

Coragliotti, A., S. Franklin, A. G. Day, and S. M. Decker. 2010. Microalgal polysaccharide compositions, Patent no. WO2010111710 A1.

Craigie, J. S. 2010. Seaweed extract stimuli in plant science and agriculture. *Journal of Applied Phycology* 23:371–393.

Cumashi, A., N. A. Ushakova, M. E. Preobrazhenskaya, A. D'Incecco, A. Piccoli, L. Totani, N. Tinari et al. 2007. A comparative study of the anti-inflammatory, anticoagulant, antiangiogenic, and antiadhesive activities of nine different fucoidans from brown seaweeds. *Glycobiology* 5:541–552.

Dąbrowski, Z. T. 2008. *Biostimulators in Modern Agriculture. Solanaceous Crops*, Z. T. Dąbrowski (ed.). Warsaw, Poland: Editorial House Wieś Jutra, Limited.

De Philippis, R. and M. Vincenzini. 1998. Exocellular polysaccharides from cyanobacteria and their possible applications. *FEMS Microbiology Reviews* 22:151–175.

Dillon, H. F., A. Somanchi, K. Rao, and P. J. H. Jones. 2007. Nutraceutical compositions from microalgae and related methods of production and administration, Patent no. WO2007136428 A2.

du Jurdain, P. 2012. The science of plant biostimulants—A bibliographic analysis. CONTRACT 30-CE0455515/00-96. ec.europa.eu/enterprise/sectors/chemicals/files/fertilizers/final_report_bio_2012_en.pdf. Accessed on November 10, 2014.

Duboc, P. and B. Mollet. 2001. Applications of exopolysaccharides in the dairy industry. *International Dairy Journal* 11:759–768.

Ebringerová, A. and Z. Hromádková. 2009. An overview on the application of ultrasound in extraction, separation and purification of plant polysaccharides. *Central European Journal of Chemistry* 8:243–257.

Eckardt, N. A. 2008. Chitin signaling in plants: Insights into the perception of fungal pathogens and rhizobacterial symbionts. *The Plant Cell* 20:241–243.

Ehrlich, H., M. Krautter, T. Hanke, P. Simon, C. Knieb, S. Heinemann, and H. Worch. 2007. First evidence of the presence of chitin in skeletons of marine sponges. Part II. Glass sponges (*Hexactinellida: Porifera*). *Journal of Experimental Zoology (Molecular and Developmental Evolution)* 308B:473–483.

El-Sheekh, M. M., H. M. Khairy, and R. El-Shenody. 2012. Algal production of extra and intra-cellular polysaccharides as an adaptive response to the toxin crude extract of *Microcystis aeruginosa*. *Iranian Journal of Environmental Health Sciences and Engineering* 9:10. doi:10.1186/1735-2746-9-10.

European Biostimulant Industry Council (EBIC). http://www.biostimulants.eu/. Accessed November 6, 2014.

Forcat, S., M. H. Bennett, J. W. Mansfield, and M. R. Grant. 2008. A rapid and robust method for simultaneously measuring changes in the phytohormones ABA, JA and SA in plants following biotic and abiotic stress. *Plant Methods* 4:16–23.

France Chitine. SolActif®, http://www.france-chitine.com. Accessed November 6, 2016.

Goëmar. Arysta Life Science, BM 86™, http://www.arysta-na.com/biosolutions/products/bm-86/general-information.html. Accessed November 6, 2014.

Goëmar. Arysta Life Science, Vacciplant®, http://www.arysta-na.com/biosolutions/products/vacciplant/general-information.html. Accessed November 6, 2014.

González, A., J. Castro, J. Vera, and A. Moenne. 2013. Seaweed oligosaccharides stimulate plant growth by enhancing carbon and nitrogen assimilation, basal metabolism, and cell division. *Journal of Plant Growth Regulation* 32:443–448.

Hahn, T., S. Lang, R. Ulber, and K. Muffler. 2012. Novel procedures for the extraction of fucoidan from brown algae. *Process Biochemistry* 47:1691–1698.

Hakeem, K. R., M. Jawaid, and U. Rashid. 2014. *Biomass and Bioenergy: Application*. Cham, Cambodia: Springer.

Hamburger, M., D. Baumann, and S. Adler. 2004. Supercritical carbon dioxide extraction of selected medicinal plants—Effects of high pressure and added ethanol on yield of extracted substances. *Phytochemical Analysis* 15:46–54.

Hauthal, W. H. 2001. Advances with supercritical fluids—Review. *Chemosphere* 43:123–135.

Heo, S. J., K. W. Lee, C. B. Song, and Y. J. Jeon. 2003. Antioxidant activity of enzymatic extracts from brown seaweeds. *Algae* 18:71–81.

Hernández-Carmona, G., Y. Freile-Pelegrín, and E. Hernández-Garibay. 2013. Conventional and alternative technologies for the extraction of algal polysaccharides. *Functional Ingredients from Algae for Foods and Nutraceuticals* 475–516.

Hincha, D. K., E. Zuther, and A. G. Heyer. 2003. The preservation of liposomes by raffinose family oligosaccharides during drying is mediated by effects on fusion and lipid phase transitions. *Biochimica et Biophysica Acta* 1612:172–177.

Ho, C. W., W. S. Tan, W. B. Yap, T. C. Ling, and B. T. Tey. 2008. Comparative evaluation of different cell disruption methods for release of recombinant hepatitis b core antigen from *Escherichia coli*. *Biotechnology and Bioprocess Engineering* 13:577–583.

Huang, S.-Q., J.-W. Li, Z. Wang, H.-X. Pan, J.-X. Chen, and Z.-X. Ning. 2010. Optimization of alkaline extraction of polysaccharides from *Ganoderma lucidum* and their effect on immune function in mice. *Molecules* 15:3694–3708.

Jain, R., S. Raghukumar, R. Tharanathan, and N. B. Bhosle. 2005. Extracellular polysaccharide production by *Thraustochytrid* protists. *Marine Biotechnology* 7:184–192.

Jaulneau, V., C. Lafitte, C. Jacquet, S. Fournier, S. Salamagne, X. Briand, M.-T. Esquerré-Tugayé, and B. Dumas. 2010. Ulvan, a sulfated polysaccharide from green algae, activates plant immunity through the jasmonic acid signaling pathway. *Journal of Biomedicine and Biotechnology* 2010:Article ID 525291, 11pp. doi:10.1155/2010/525291.

Jiao, G., G. Yu, J. Zhang, and H. S. Ewart. 2011. Chemical structures and bioactivities of sulfated polysaccharides from marine algae. *Marine Drugs* 9:196–223.

KaitoSol®. http://kaitosol.com/. Accessed November 6, 2016.

Karnok, K. J. 2000. Promises, promises: Can biostimulants deliver? *Golf Course Management* 68:67–71.

Kaufmann, B. and P. Christen. 2002. Recent extraction techniques for natural products: Microwave-assisted extraction and pressurized solvent extraction. *Phytochemical Analysis* 13:105–113.

Kauffman, G. L., D. P. Kneivel, and T. L. Watschke. 2007. Effects of a biostimulant on the heat tolerance associated with photosynthetic capacity, membrane thermostability, and polyphenol production of perennial ryegrass. *Crop Science* 47:261–267.

Khan, W., U. P. Rayirath, S. Subramanian, M. N. Jithesh, P. Rayorath, D. M. Hodges, A. T. Critchley, J. S. Craigie, J. Norrie, and B. Prithiviraj. 2009. Seaweed extracts as biostimulants of plant growth and development. *Journal of Plant Growth Regulation* 28:386–339.

King, J. W. 1989. Fundamentals and applications of supercritical fluid extraction in chromatographic science. *Journal of Chromatographic Science* 27:355–364.

Klarzynski, O. 2006. Stimulation of natural defenses: A growing technique for plant protection against diseases. http://www.abim.ch/fileadmin/documents-abim/presentations2006/session5/3_klarzynski_ABIM-Lucerne_2006.pdf. Accessed on November 6, 2014.

Klarzynski, O., V. Descamps, B. Plesse, J.-C. Yvin, B. Kloareg, and B. Fritig. 2003. Sulfated fucan oligosaccharides elicit defense responses in tobacco and local and systemic resistance against tobacco mosaic virus. *Molecular Plant-Microbe Interactions* 16:115–122.

Kraan, S. 2012. Algal polysaccharides, novel applications and outlook. In: *Carbohydrates—Comprehensive Studies on Glycobiology and Glycotechnology*, C.-F. Chang (ed.). InTech Rijeka, Croatia. doi:10.5772/51572. http://www.intechopen.com/books/carbohydrates-comprehensive-studies-on-glycobiology-and-glycotechnology/algal-polysaccharides-novel-applications-and-outlook. Accessed on November 12, 2014.

Kulikov, S. N., S. N. Chirkov, A. V. Il'ina, S. A. Lopatin, and V. P. Varlamov. 2006. Effect of the molecular weight of chitosan on its antiviral activity in plants. *Prikladnaya Biokhimiya i Mikrobiologiya* 42:224–228.

Kumar, A.S. and Mody, K. 2009. Microbial exopolysaccharides: Variety and potential applications. In: *Microbial Production of Biopolymers and Polymer Precursors: Applications and Perspectives* Bernd H.A. Rehm (ed.). Norfolk, U.K.: Caister Academic, pp. 229–254.

Lahaye, M. and A. Robic. 2007. Structure and functional properties of ulvan, a polysaccharide from green seaweeds. *Biomacromolecules* 8:1765–1774.

Laporte, D., J. Vera, N. P. Chandía, E. A. Zúñiga, B. Matsuhiro, and A. Moenne. 2007. Structurally unrelated algal oligosaccharides differentially stimulate growth and defense against tobacco mosaic virus in tobacco plants. *Journal of Applied Phycology* 19:79–88.

Laurienzo, P. 2010. Marine polysaccharides in pharmaceutical applications: An overview. *Marine Drugs* 8:2435–2465.

Li, S. N., H. Zhang, D. D. Han, and K. H. Row. 2012. Optimization of enzymatic extraction of polysaccharides from some marine algae by response surface methodology. *Korean Journal of Chemical Engineer* 29:650–656.

Li, B., Wei, X. J., Sun, J. L., Xu, S. Y. Structural investigation of a fucoidan containing a fucose-free core from the brown seaweed, Hizikia fusiforme. *Carbohydrate Research* 341:1135–1146.

Livingston, D. P., K. H. Dirk, and A. G. Heyer. 2009. Fructan and its relationship to abiotic stress tolerance in plants. *Cellular and Molecular Life Sciences* 66:2007–2023.

Markert, B. 1995. Sample preparation (cleaning, drying, homogenization) for trace element analysis in plant matrices. *Science of the Total Environment* 176:45–61.

Mercier, L., C. Lafitte, G. Borderies, X. Briand, M. T. Esquerré-Tugayé, and J. Fournier. 2001. The algal polysaccharide carrageenans can act as an elicitor of plant defense. *New Phytologist* 149:43–51.

Michalak, I. and K. Chojnacka. 2014. Algal extracts: Technology and advances. *Engineering in Life Sciences* 14:581–591.

Michalak, I., Ł. Tuhy, A. Saeid, and K. Chojnacka. 2013. Bioavailability of Zn (II) to plants from new fertilizer components produced by biosorption. *International Journal of Agronomy and Plant Production* 4:3522–3536.

Middelberg, A. P. J. 1995. Process-scale disruption of microorganisms. *Biotechnology Advances* 13:491–551.

Miller Jr., G. T. and S. Spoolman. 2012. *Living in the Environment*, International Edition, 17th edn. Cengage Learning Asia, Nelson Education Ltd., Canada.

Naessens, M., A. Cerdobbel, W. Soetaert, and E. J. Vandamme. 2005. Leuconostoc dextransucrase and dextran: Production, properties and applications. *Journal of Chemical Technology and Biotechnology* 80:845–860.

Nagaoka, M., H. Shibata, I. Kimura-Takagi, S. Hashimoto, K. Kimura, T. Makino, R. Aiyama, S. Ueyama, and T. Yokokura. 1999. Structural study of fucoidan from *Cladosiphon okamuranus* TOKIDA. *Glycoconjugate Journal* 16:19–26.

Nakkarike, S. M., B. Narayan, M. Hosokawa, and K. Miyashita. 2011. Value addition to seafood processing discards. In: *Handbook of Seafood Quality, Safety and Health Applications*, C. Alasalvar, K. Miyashita, F. Shahidi, and U. Wanasundara (eds.). Chichester, U.K.: Wiley-Blackwell Publishing Ltd.

Nge, K. L., N. Nwe, S. Chandrkrachang, and W. F. Stevens. 2006. Chitosan as a growth stimulator in orchid tissue culture. *Plant Science* 170:1185–1190.

Ngo, D.-H. and S.-K. Kim. 2013. Sulfated polysaccharides as bioactive agents from marine algae. *International Journal of Biological Macromolecules* 62:70–75.

No, H. K., S. P. Meyers, and K. S. Lee. 1989. Isolation and characterization of chitin from crawfish shell waste. *Journal of Agricultural and Food Chemistry* 37:575–579.

Paquin, P. 1999. Technological properties of high pressure homogenizers: The effect of fat globules, milk proteins, and polysaccharides. *International Dairy Journal* 9:329–335.

Park, P. J., F. Shahidi, and Y.-J. Jeon. 2004. Antioxidant activities of enzymatic extracts from and edible seaweed *Sargassum horneri* using ESR spectroscopy. *Journal of Food Lipids* 11:15–27.

Patier, P., J.-C. Yvin, B. Kloareg, Y. Liénart, and C. Rochas. 1993. Seaweed liquid fertilizer from *Ascophyllum nodosum* contains elicitors of plant D-glycanases. *Journal of Applied Phycology* 5:343–349.

Periyasamy, N., S. Murugan, and P. Bharadhirajan. 2013. Anticoagulant activity of marine gastropods *Babylonia spirata* LIN, 1758 and *Phalium glaucum* LIN, 1758 collected from Cuddalore, southeast cost of India. *International Journal of Pharmacy and Pharmaceutical Sciences* 5:117–121.

Ponce, N. M., C. A. Pujol, E. B. Damonte, M. L. Flores, and C. A. Stortz. 2003. Fucoidans from the brown seaweed *Adenocystis utricularis*: Extraction methods, antiviral activity and structural studies. *Carbohydrate Research* 338:153–165.

Rai, V. K. 2002. Role of amino acids in plant responses to stresses. *Biologia Plantarum* 45:481–487.

Reddy, K. J., B. W. Soper, J. Tang, and R. L. Bradley. 1996. Phenotypic variation in exopolysaccharide production in the marine, aerobic nitrogen-fixing unicellular cyanobacterium *Cyanothece* sp. *World Journal of Microbiology and Biotechnology* 12:311–318.

Rodriguez-Jasso, R. M., S. I. Mussatto, L. Pastrana, C. N. Aguilar, and J. A. Teixeira. 2011. Microwave-assisted extraction of sulfated polysaccharides (fucoidan) from brown seaweed. *Carbohydrate Polymers* 86:1137–1144.

Sari, A. and M. Tuzen. 2008. Biosorption of cadmium (II) from aqueous solution by red algae (*Ceramium virgatum*): Equilibrium, kinetic and thermodynamic studies. *Journal of Hazardous Materials* 157:448–454.

Senni, K., J. Pereira, F. Gueniche, C. Delbarre-Ladrat, C. Sinquin, J. Ratiskol, G. Godeau, A. M. Fischer, D. Helley, and S. Colliec-Jouault. 2011. Marine polysaccharides: A source of bioactive molecules for cell therapy and tissue engineering. *Marine Drugs* 9:1664–1681.

Shanmugam, M. and K. H. Mody. 2000. Heparinoid-active sulfated polysaccharides from marine algae as potential blood anticoagulant agents. *Current Science* 79:1672–1682.

Sharma, S. H. S., G. Lyons, C. McRoberts, D. McCall, E. Carmichael, F. Andrews, R. Swan, R. McCormack, and R. Mellon. 2012. Biostimulant activity of brown seaweed species from Strangford Lough: Compositional analyses of polysaccharides and bioassay of extracts using mung bean (*Vigno mungo* L.) and pak choi (*Brassica rapa chinensis* L.). *Journal of Applied Phycology* 24:1081–1091.

Shinmen, Y., H. Yamada, and S. Shimizu. 1988. Process for production of eicosapentaeonic acid from algae, Patent no. EP027774 A2.

Singh, K. and A. R. Ray. 2000. Biomedical applications of chitin, chitosan, and their derivatives. *Journal of Macromolecular Science* 40:69–83.

Stadnik, M. J. and M. B. de Freitas. 2014. Algal polysaccharides as source of plant resistance inducers. *Tropical Plant Pathology* 39:111–118.

Tadeo, J. L., C. Sánchez-Brunete, B. Albero, and A. I. García-Valcárcel. 2010. Application of ultrasound-assisted extraction to the determination of contaminants in food and soil samples. *Journal of Chromatography A* 1217:2415–2440.

Thiruvengadam, A. P. and A. A. Hochrein Jr. 1978. Homogenizing method and apparatus, Patent no. US4127332 A.

Trincone, A. 2014. Molecular fishing: Marine oligosaccharides. *Frontiers in Marine Science* 1:1–5.

Tuhy, Ł., K. Chojnacka, I. Michalak, and A. Witek-Krowiak. 2015. Algal extracts as a carrier of micronutrients—Utilitarian properties of new formulations. In: *Marine Algae Extracts: Processes, Products, Applications*, S.-K. Kim and K. Chojnacka (eds.). Weinheim, Germany: Wiley-VCH Verlag GmbH & Co. KGaA 28:467–488.

Umezawa, H., Y. Okami, S. Kurasawa, T. Ohnuki, M. Ishizuka, T. Takeuchi, T. Shiio, and Y. Yugari. 1983. Marinactan, antitumor polysaccharide produced by marine bacteria. *Journal of Antibiotics* 36:471–477.

Usov, A. I., G. P. Smirnova, M. I. Bilan, and A. S. Shashkov. 1998. Polysaccharides of algae: 53. Brown alga *Laminaria saccharina* (L.) Lam. as a source of fucoidan. *Bioorganicheskaia khimiia* 24:382–389.

Valluru, R. and W. Van den Ende. 2008. Plant fructans in stress environments: Emerging concepts and future prospects. *Journal of Experimental Botany* 59:2905–2916.

Velásquez, C. L., A. Chirinos, M. Tacoronte, and A. Mora. 2012. Chitosan oligomers as biostimulants to zucchini (*Cucurbita pepo*) seeds germination. Short communication. *Agriculture (Poľnohospodárstvo)* 58:113–119.

Venugopal, V. 2011. Polysaccharides: Their characteristics and marine sources. In: *Marine Polysaccharides. Food Applications*, V. Venugopal (ed.). Boca Raton, FL: Taylor & Francis Group, LLC.

Vera, J., J. Castro, A. Gonzalez, and A. Moenne. 2011. Seaweed polysaccharides and derived oligosaccharides stimulate defense responses and protection against pathogens in plants. *Marine Drugs* 9:2514–2525.

Vicuna, D. 2005. The role of peroxidases in the development of plants and their responses to abiotic stresses. PhD thesis, Dublin Institute of Technology, Dublin, Ireland.

Vilanova, E., C. C. Coutinho, and P. A. S. Mourão. 2009. Sulfated polysaccharides from marine sponges (Porifera): An ancestor cell–cell adhesion event based on the carbohydrate–carbohydrate interaction. *Glycobiology* 19:860–867.

Vu, B., M. Chen, R. J. Crawford, and E. P. Ivanova. 2009. Bacterial extracellular polysaccharides involved in biofilm formation. *Molecules* 14:2535–2554.

Wang, J., J. Zhang, B. Zhao, X. Wang, Y. Wu, and J. Yao. 2010. A comparison study on microwave-assisted extraction of *Potentilla anserina* L. polysaccharides with conventional method: Molecule weight and antioxidant activities evaluation. *Carbohydrate Polymers* 80:84–93.

Wang, W., S.-X. Wang, and H.-S. Guan. 2012. The antiviral activities and mechanisms of marine polysaccharides: An overview. *Marine Drugs* 10:2795–2816.

Whyte, J. N. C. and B. A. Southcot. 1970. An extraction procedure for plants extracts from red alga rhodomela-larix. *Phytochemistry* 9:1159–1161.

Wijesekara, I., R. Pangestuti, and S.-K. Kim. 2011. Biological activities and potential health benefits of sulfated polysaccharides derived from marine algae. *Carbohydrate Polymers* 84:14–21.

Wilk, R. and K. Chojnacka. 2015. Downstream processing in the technology of algal extracts—From the component to the final formulation. *Marine Algae Extracts: Processes, Products, and Applications* 1:161–179.

Zhang, X. and R. E. Schmidt. 1997. The impact of growth regulators on the a-tocopherol status in water-stressed *Poa pratensis*. *International Turfgrass Society Research Journal* 8:1364–1371.

Zhao, W., Z. Yu, J. Liu, Y. Yu, Y. Yin, S. Lin, and F. Chen. 2011. Optimized extraction of polysaccharides from corn silk by pulsed electric field and response surface quadratic design. *Journal of the Science of Food and Agriculture* 91:2201–2209.

Zierer, M. S. and P. S. Mourao. 2000. A wide diversity of sulfated polysaccharides are synthesized by different species of marine sponges. *Carbohydrate Research* 328:209–216.

Zvyagintseva, T. N., N. M. Shevchenko, A. O. Chizhov, T. N. Krupnova, E. V. Sundukova, and V. V. Isakov. 2003. Water-soluble polysaccharides of some far-eastern brown seaweeds. Distribution, structure, and their dependence on the developmental condition. *Journal of Experimental Marine Biology and Ecology* 294:1–13.

*chapter twenty-two*

# Carbohydrates in drug discovery
## Insights into sulfated marine polysaccharides

*Jeyakumar Kandasamy and Sabiah Shahul Hameed*

### Contents

## 22.1 Introduction to carbohydrates

Nucleic acids, proteins, and carbohydrates are important biopolymers responsible for most of the signal transduction processes in living organisms (Figure 22.1).[1] Carbohydrates mainly exist in the form of glycoproteins and glycolipids on cell surfaces and are involved in many biological processes, including fertilization, cell growth, bacterial and viral recognition, metastasis, inflammation, immune response, and many other cell–cell communications.[2] The structure and nature of cell surface carbohydrates significantly differ between diseased and normal ones. Pathogens including viruses, bacteria, and parasites infect their host cells by recognizing and binding to these specific carbohydrate epitopes. Soluble proteins such as lectins, toxins, and antibodies also bind to cell surface carbohydrates and play a fundamental role in many diseases.[3]

Similar to other biomolecules such as proteins and nucleic acids, glycans also had a rich history in therapeutic applications.[4-6] However, due to their high complexity and structural diversity, glycans were omitted in the last century as DNA- and protein-focused treatments became readily accessible. Nevertheless, the recent development in carbohydrate synthesis and analysis fuelled the interest in glycan-based drug discovery.[7] This chapter is mainly focused on two different aspects: (1) a brief description of carbohydrate-based drugs and vaccines currently available on the market for the treatment of various diseases and (2) the possibilities of sulfated marine carbohydrates in drug discovery.

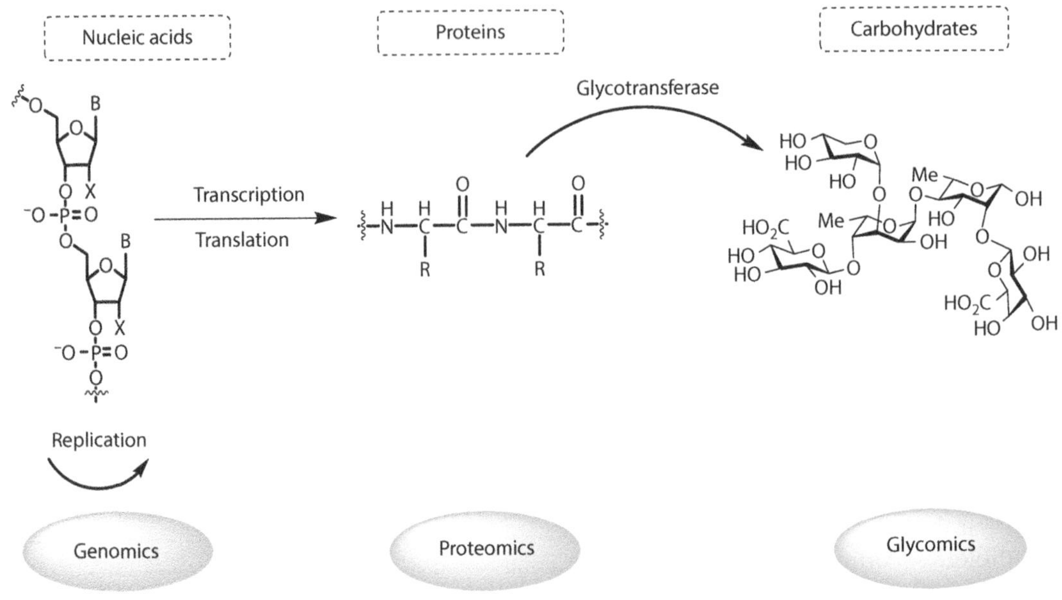

***Figure 22.1*** A fundamental biological role of nucleic acids, proteins, and carbohydrates.

## 22.2 Carbohydrates as therapeutic agents: Drugs and vaccines

Carbohydrates make an ideal platform for drug discovery because they are extensively involved in human health and disease. Carbohydrates are used as therapeutic molecules (e.g., drugs and vaccines) in many treatments and also as a target for drugs.[4–9] Currently, there are several monosaccharides, oligosaccharides, and polysaccharides that are clinically used as antiviral, antibiotic, anticoagulant, antidiabetic, antirheumatic, and anticonvulsant agents. This section focuses on the carbohydrate drugs and vaccines that provided major breakthroughs in modern medicine (Figure 22.2).

### 22.2.1 Antiviral drugs: Tamiflu and Zanamivir

Tamiflu (**1**) and Zanamivir (**2**) are the monosaccharides frequently used for preventing and treating viral infection caused by the Influenza A virus.[10,11] Both of these drugs are known as reversible competitive inhibitors of influenza neuraminidase. Tamiflu and Zanamivir mimic the transition state of the hydrolysis of terminal *N*-acetylneuraminic acid by neuraminidase. By blocking the activity of neuraminidase, it prevents new viral particles' release through the cleaving of terminal sialic acid on glycosylated hemagglutinin. Tamiflu is taken orally as a pill while Zanamivir is taken as an inhaler. In fact, Tamiflu is a prodrug that contains additional ethyl ester to increase hydrophobicity, cell permittivity, and oral bioavailability.

### 22.2.2 Antibiotics: Aminoglycosides

Aminoglycosides (Figure 22.3) are highly potent, broad-spectrum antibiotics that exert antibacterial activity by binding to the ribosomal decoding site (A-Site) and reducing the fidelity of protein synthesis.[12] The first drug from this family, streptomycin (isolated from *Streptomyces griseus*), was discovered by Selman Waksman in 1944. In fact, streptomycin was the first effective antibiotic drug against *Mycobacterium tuberculosis*. Few years later, in 1949, neomycin was isolated from *Streptomyces fradiae* that found better activity than streptomycin against aerobic gram-negative bacilli. In the following decades, several aminoglycoside drugs such as paromomycin, kanamycin, tobramycin, and gentamicin were isolated from soil bacteria by intense search for natural products with antibacterial activity. Due to high toxicity (e.g., nephrotoxicity, cytotoxicity, and ototoxicity) and bacterial resistance of natural aminoglycosides, several semisynthetic analogs such as amikacin, dibekacin, netilmicin, and isepamicin were introduced into clinical use in the 1970s and 1980s.[13]

In addition to the aminoglycoside antibiotics, there are several antibiotics found in nature with glycosidic linkages (e.g., macrolide antibiotics, erythromycin A, and moenomycin A). Recently, aminoglycosides have also emerged as frontline pharmacogenetic agents in treating human genetic disorders due to their unique ability to suppress nonsense mutations and produce functional proteins.[14]

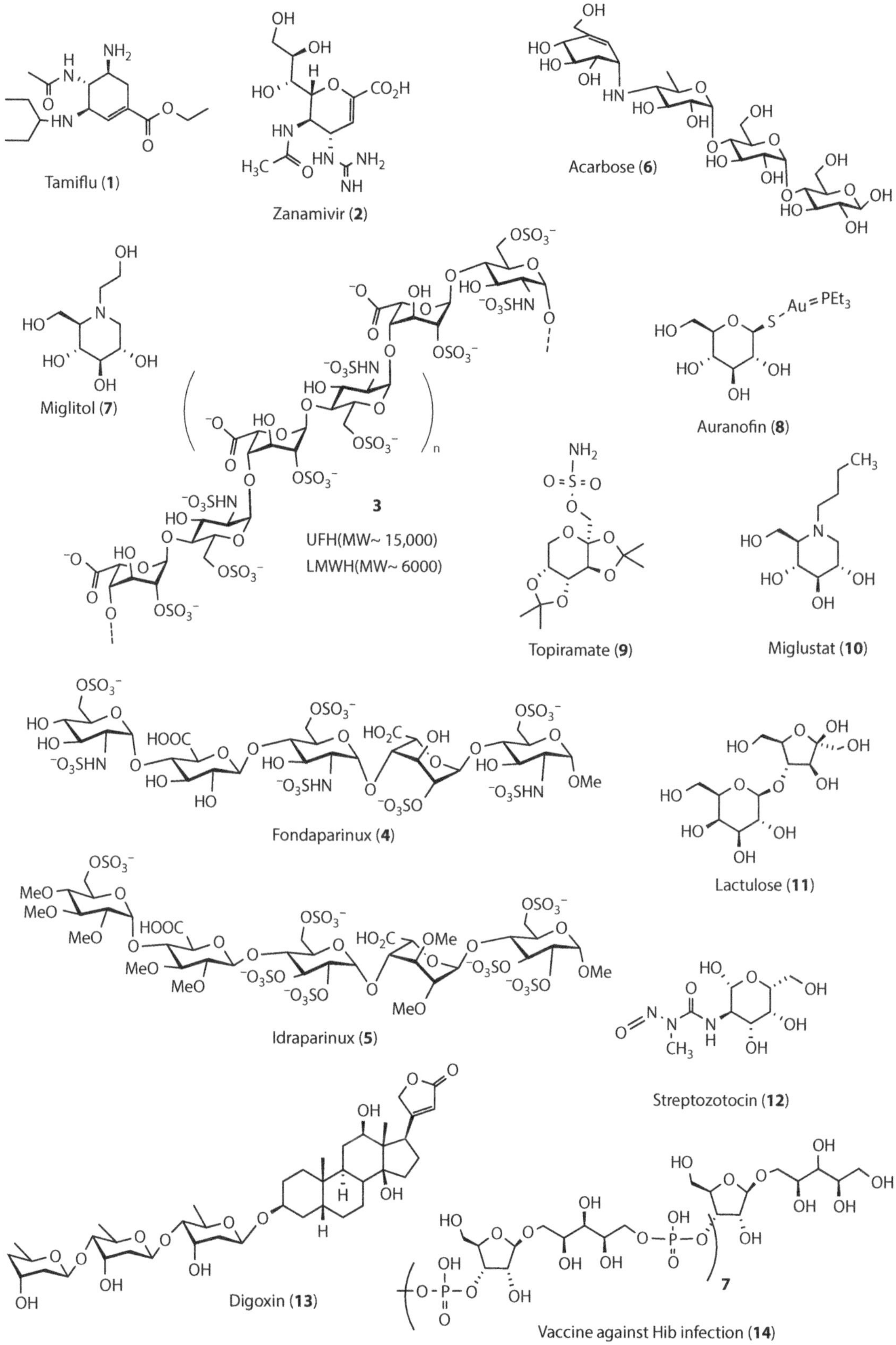

*Figure 22.2* Carbohydrates as therapeutic agents.

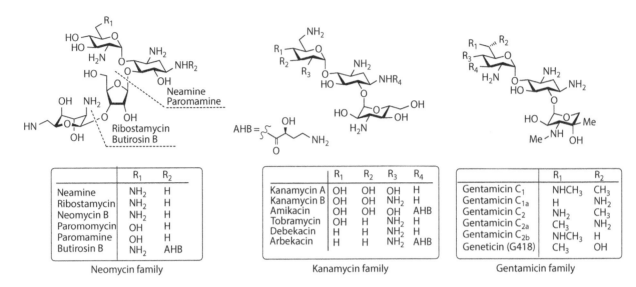

| | R₁ | R₂ |
|---|---|---|

| | $R_1$ | $R_2$ |
|---|---|---|
| Neamine | $NH_2$ | H |
| Ribostamycin | $NH_2$ | H |
| Neomycin B | $NH_2$ | H |
| Paromomycin | OH | H |
| Paromamine | OH | H |
| Butirosin B | $NH_2$ | AHB |

Neomycin family

| | $R_1$ | $R_2$ | $R_3$ | $R_4$ |
|---|---|---|---|---|
| Kanamycin A | OH | OH | OH | H |
| Kanamycin B | OH | OH | $NH_2$ | H |
| Amikacin | OH | OH | OH | AHB |
| Tobramycin | OH | H | $NH_2$ | H |
| Debekacin | H | H | $NH_2$ | H |
| Arbekacin | H | H | $NH_2$ | AHB |

Kanamycin family

| | $R_1$ | $R_2$ |
|---|---|---|
| Gentamicin $C_1$ | $NHCH_3$ | $CH_3$ |
| Gentamicin $C_{1a}$ | H | $NH_2$ |
| Gentamicin $C_2$ | $NH_2$ | $CH_3$ |
| Gentamicin $C_{2a}$ | $CH_3$ | $NH_2$ |
| Gentamicin $C_{2b}$ | $NHCH_3$ | H |
| Geneticin (G418) | $CH_3$ | OH |

Gentamicin family

**Figure 22.3** Structures of aminoglycoside antibiotics.

### 22.2.3 Anticoagulant drug: Heparin

Heparin belongs to the family of glycosaminoglycans (GAGs) and has been widely used as an anticoagulant drug over the past 50 years.[15] It is a sulfated linear polymer composed of disaccharide repeating units that predominantly contain alternative iduronic acid and glucosamine moieties (**3**). In addition to the anticoagulant and antithrombotic activities, heparin also possesses the inhibitory effect on smooth muscle cell proliferation, anti-inflammatory effect, antitumor effect, and the antiviral activities. There are three forms of heparin that currently exist in the market: (1) unfractionated heparin (UFH) (molecular weight (MW) ~15,000), (2) low MW heparin (LMWH) (MW ~6,000), and (3) synthetic heparins (SHs) (MW < 2,000). UFHs and LMWHs (**3**) have high risk of viral and bacterial impurities since they are extracted from animal sources.[16] SH such as fondaparinux (**4**) and idraparinux (**5**) is structurally well defined and completely free of chemical as well as biological contaminations.[17] These synthetic analogs showed better activity, bioavailability, and lower side effects than natural polymer, and therefore, nowadays natural heparins are slowly replaced by the synthetic constructs.

### 22.2.4 Antidiabetics drugs: Acarbose and miglitol

Diabetes describes a group of metabolic diseases in which the person has high level of blood glucose over a prolonged period. It is due to either insufficient insulin production in our body (diabetic type 1) or the body's cells do not respond properly to the produced insulin (diabetic type 2). Acarbose (**6**) is a pseudooligosaccharide of microbial origin and has become a blockbuster diabetes drug in the past decades.[18] Acarbose inhibits alpha-glucosidase and alpha-amylase enzymes that are needed for digesting carbohydrates into glucose. A recent study concludes that acarbose is an effective, safe, and well-tolerated drug for type-2 diabetes.[19] Miglitol is another sugar-based oral antidiabetic drug that is also primarily used in the treatment of diabetes type 2. In contrast to acarbose, miglitol is systemically absorbed; however, it is not metabolized and is filtered through kidneys.

### 22.2.5 Antirheumatoid arthritis drug: Auranofin

Rheumatoid arthritis (RA) is an autoimmune disease that causes a chronic, systemic inflammatory disorder that affects many tissues, organs, and flexible joints. Auranofin (**8**) is a gold-incorporated glucose monosaccharide, widely used in the treatment of RA. It is an oral drug that improves arthritis symptoms including joint pains and morning stiffness. Auranofin mode of action is not well understood; however, it has been proposed that auranofin may act as an inhibitor of kappa B kinase and thioredoxin reductase enzymes, which would lead to a decreased immune response and decreased free radical production, respectively. This compound is also under investigation for the treatment of HIV antiretroviral therapy.[20]

### 22.2.6 Anticonvulsant drug: Topiramate

Topiramate (**9**) is an anticonvulsant drug used in the treatment of epilepsy in children and adults.[21] Topiramate is also very effective in treating various psychological problems, including migraines, bipolar disorders, weight loss, and cluster headaches. Although the mechanism of action is not known, preclinical studies have revealed four properties that may contribute to topiramate activity as anticonvulsant and antimigraine agents.[22,23] Electrophysiological and biochemical evidence suggests that topiramate (1) blocks voltage-dependent sodium channels, (2) enhances the activity of the neurotransmitter gamma-aminobutyrate at some subtypes of the GABA-A receptor, (3) antagonizes the AMPA/kainate subtype of the glutamate receptor, and (4) inhibits the carbonic anhydrase enzyme, particularly isozymes II and IV.

### 22.2.7   Gaucher disease: Miglustat

Miglustat (**10**) is an iminosugar, a synthetic analog of D-glucose that is primarily used in the treatment of type I Gaucher disease.[24,25] Gaucher disease is an autosomal recessive disorder where people have a defect in the enzyme called "glucocerebrosidase" that acts on a fatty substance glucosylceramide. Accumulation of glucosylceramide can cause liver and spleen enlargement, bone disease, etc. Miglustat exerts the activity by selectively inhibiting glucosylceramide synthase enzyme. It is also being investigated for the treatment of other genetic disorders such as Tay–Sachs and cystic fibrosis.

### 22.2.8   Lactulose for chronic constipation and hepatic encephalopathy

Lactulose (**11**) is a nondigestible synthetic disaccharide composed of fructose and galactose moieties, primarily used in the treatment of chronic constipation in patients of all ages.[26] Lactulose is neither absorbed in the intestine nor broken down by human enzymes and thus stays in the digestive bolus through most of its course, causing retention of water through osmosis leading to softer, easier to pass stool. Lactulose is also recommended for hepatic encephalopathy, which is a liver disease caused by hyperammonemia (high blood ammonia ($NH_3$)). Lactulose helps to trap the $NH_3$ in the colon.

### 22.2.9   Anticancer drug: Streptozotocin

Streptozotocin (**12**) is a natural glucose-substituted N-methyl-N-nitrosourea, a broad-spectrum antibiotic and an experimental anticancer agent that has shown diabetogenic activity in animals and clinical activity in the treatment of malignant insulinomas in man.[27,28] Streptozotocin mainly targets beta cells by entering through the glucose transporter GLUT2 and causing alkylation of DNA. DNA damage induces activation of poly ADP-ribosylation, depletion of cellular $NAD^+$ and ATP, and formation of superoxide radicals, leading to the destruction of beta cells.

### 22.2.10   Cardiac glycosides: Digoxin

Digoxin (**13**), also known as digitalis, a purified cardiac glycoside extracted from the foxglove plant, is widely used for the treatment of such heart conditions as atrial fibrillation, atrial flutter, and sometimes heart failure.[29] Digoxin's primary mechanism of action is the ability to inhibit sodium–potassium ATPase (sodium pump), mainly in the myocardium. This inhibition increases the intracellular sodium content, which in turn increases the intracellular calcium content that leads to increased cardiac contractility.

### 22.2.11   Carbohydrate-based vaccines

Vaccination is considered to be one of the most powerful approaches to save lives from infectious diseases. Vaccines are usually administered to healthy individuals to prevent a serious infection. Carbohydrate epitopes or glycotopes present on the surface of various diseased cells (e.g., bacteria, parasites, and viruses) are often structurally different from the normal cell surface sugars.[30,31] Such cell surface carbohydrate markers are the basis for carbohydrate-based vaccines. The first carbohydrate conjugate vaccine against *Haemophilus influenzae* type b (**14**) was developed and licensed in the late 1980s. Later on, many other glycoconjugate vaccines have been developed against bacterial pathogens such as *Neisseria meningitidis*, *Streptococcus pneumoniae*, and group B *Streptococcus*. In fact, glycoconjugate vaccines are among the safest and most efficacious vaccines developed during the past 30 years, and they are currently used in the immunization process in various countries. Recently, intense efforts have been focused on the use of synthetic carbohydrate antigens to develop vaccines against parasites, malaria, leishmaniasis, and cancer infections.[32]

## 22.3   Marine: A source of biologically active natural products

Oceans occupy maximum part of the earth's surface and hosts nearly 80% of living species. Marine has been considered a promising source of various biologically important natural products, and so far, more than 20,000 natural products have been isolated. Ocean life forms such as sponges, ascidian, aplysia, algae, corals, bryozoa, worm, sea squirts, sea hares, sea cucumbers, and fish species are the potential source of various alkaloids, terpenoids, steroids, polypeptides, polyethers, macrolides, and polysaccharides. During the past decade, dozens of products from marine organisms were identified with promising biological and therapeutic activities.[33–37] The antiviral drugs vidarabine (Ara-A, **15**) and zidovudine

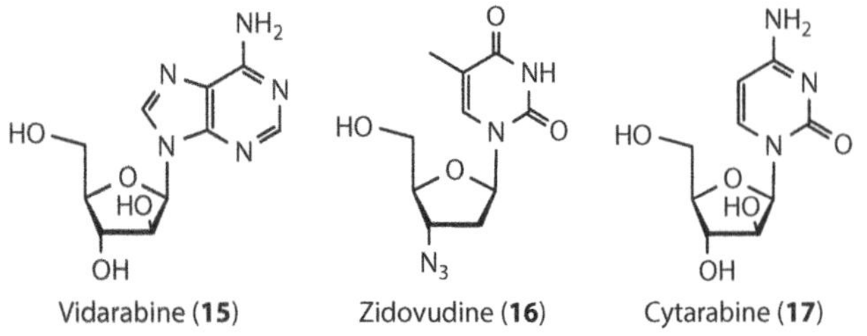

**Figure 22.4** Structures of clinically used antiviral and anticancer drugs from marine origin.

(AZT, **16**) and the anticancer agent cytarabine (Ara-C, **17**) were developed from the extracts of sponges found on a Caribbean reef (Figure 22.4). The painkiller Ziconotide is derived from the toxin of the cone snail species *Conus magus*. Many other marine natural products, including dolastatin 10, ecteinascidin 743, bryostatin 1, etc., are under clinical trials for the treatment of various cancers including breast and liver cancers, tumors, and leukemia.[38]

## 22.4   Marine carbohydrates

Marine polysaccharides are important macromolecules widely present in various marine organisms.[39–42] The field of natural polysaccharides of marine origin is already large and expanding. Marine polysaccharides present an enormous variety of structures and are still underexploited in drug discovery. Marine polysaccharides can be divided into three different types based on their origin such as marine animal polysaccharides, plant polysaccharides, and microbial polysaccharides. All these marine polysaccharides have been shown to have a variety of biological activities such as

antitumor, antiviral, anticoagulant, antioxidant, immunoinflammatory effects, and other medicinal applications. Seaweeds are one of the most abundant sources of polysaccharides. Alginates, agar and agarose, and carrageenans were obtained in high quantity from various seaweeds. Cellulose and amylose have been extracted from the macroalga *Ulva*, while chitin and chitosan are derived from the exoskeleton of marine crustaceans. Recently, microalgae have become particularly interesting in drug development, because it is easy to control their growth in a bioreactor. An effort for screening and identification of biologically active compounds from marine is the current focus of many research groups and pharmaceutical industries.

### 22.4.1   Sulfated polysaccharides from marine sources

Sulfated polysaccharides as well as their lower MW oligosaccharides are known to pose various biological key functions by interacting with a wide range of proteins at different levels of specificity.[43] These anionic polymers are widespread in nature and are found in many different organisms. Sulfated polysaccharides are preliminarily involved in cell differentiation, cell adhesion, cell signaling, and cell–matrix interactions and exert therapeutic activities like anticoagulant, antiviral, and immunoinflammatory activities. GAGs are the best examples of sulfated polysaccharides and are mainly composed of disaccharide repeating units consisting of an alternative uronic acid and amino sugar moieties (Figure 22.5).[44] GAGs are usually present in all animals; some of them such as heparin and dermatan sulfate are extracted from mammalian mucosa for therapeutic usage. The mammalian GAGs are better studied for their biological

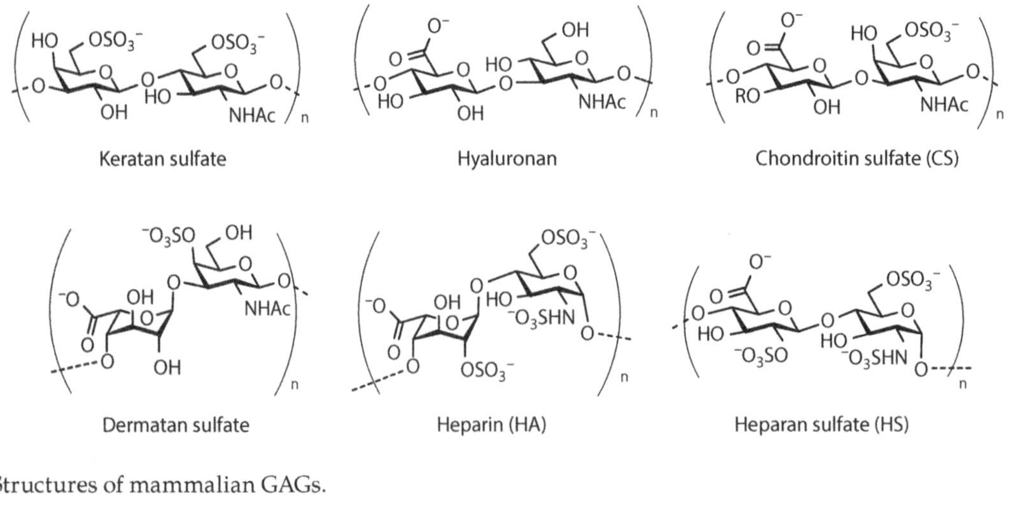

**Figure 22.5** Structures of mammalian GAGs.

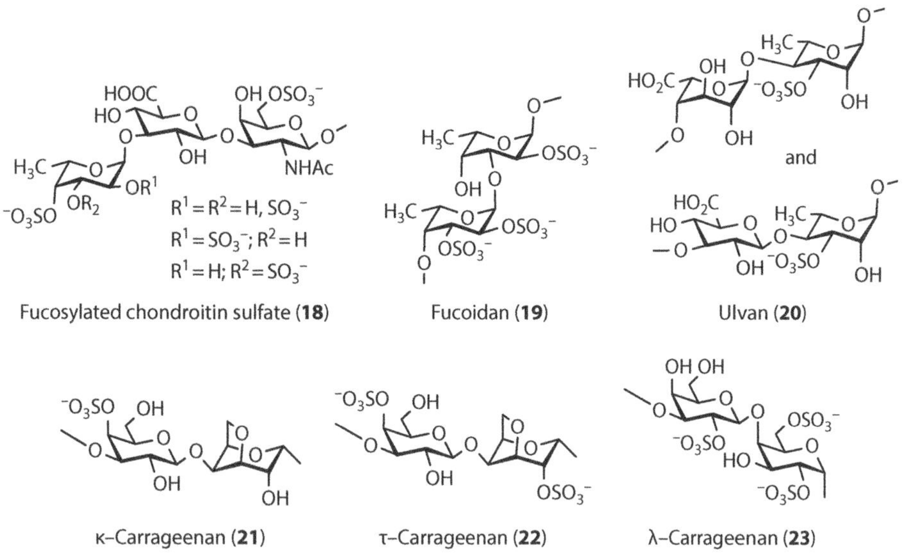

**Figure 22.6** Structures of some sulfated marine carbohydrates.

function in humans. Among the various sulfated mammalian GAGs, heparin is one the most studied for its biological role and has been clinically used as an anticoagulant in the past several decades (see Section 22.2.3). It has been demonstrated that nature of the sugar (disaccharide repeating units), length of the polysaccharides, and sulfation patterns can effectively influence the specific biological functions of these sulfated GAGs.[45,46]

Besides mammalian sources, marine organisms such as macro- and microalgae, cyanobacteria, invertebrates, and chordata also produce varieties of sulfated carbohydrates including carrageenans, ulvan, fucoidans, etc., while some of them have become valuable additives in the food and pharmaceutical industry.[47–55] Anticoagulant and antithrombotic activities are among the most widely studied properties of these sulfated polysaccharides. In the following section, we will discuss the biological importance of some marine sulfated polysaccharides.

### 22.4.1.1  Marine glycosaminoglycans

Hyaluronic acid, chondroitin sulfate, dermatan sulfate, and heparan sulfate have been isolated from marine invertebrates such as sea urchins and sea cucumbers.[56] Most of these marine-GAGs have the same backbone structures as the mammalian GAGs but have different lengths and sulfation patterns. Since animal source of heparin suffers from various disadvantages such as severe side effects, chemical and biological contaminations, and poor availability, marine-originated polysaccharides could be an alternative source of heparin.[57] Other than marine GAGs, there are several

other sulfated polysaccharides that have been isolated from marine origin with various therapeutic activities (Figure 22.6).[47–55]

### 22.4.1.2  Fucosylated chondroitin sulfate: From marine invertebrates

Fucosylated chondroitin sulfate (**18**) is a structurally distinct marine GAG found in sea cucumber species.[58] It has the same backbone composition within disaccharide repeating units (i.e., alternative glucuronic acid and galactosamine) as usually found in mammalian chondroitin sulfates; however, in addition to that, a sulfated fucosyl unit is linked to the glucuronic acid residues at the third (3-OH) position. The sulfation pattern on the fucose moiety varies accordingly with different sources from which they are isolated. Depending on this, different biological responses such as anticoagulant, antithrombotic, anti-inflammatory, anticancer, antiviral, and pro-angiogenic activities were observed.

### 22.4.1.3  Fucans: Sulfated polysaccharides from brown algae and marine invertebrates

Fucans (**19**), also called fucoidans, are sulfated polysaccharides that are mainly composed of L-fucose backbone.[59] Fucoidans can be isolated from both marine algae (e.g., brown seaweeds) and marine invertebrates (e.g., sea urchins or sea cucumbers). Fucoidans derived from seaweeds are all highly branched, while invertebrates produce linear structures. Biological properties of sulfated fucoidans extracted from marine invertebrates have been extensively studied. Similar to mammalian GAGs, fucans present anticoagulant and antithrombotic activities. LMW fucoidans (<30 kDa) have been shown to exhibit some heparin-like properties, with less side effects. In addition, they are also active on cell growth, migration, and adhesion.

### 22.4.1.4 Ulvan: Sulfated polysaccharides from green algae

Ulvan (**20**) is the major water-soluble sulfated polysaccharide found in green seaweed and is composed of disaccharide repeating units containing rhamnose with either glucuronic or iduronic acid moieties.[60] In addition, there are some minor units that have been found with the xylose moiety. Ulvan presents various applications in agricultural, food, and pharmaceutical industries. Ulvan also shows various therapeutic activities such as anticoagulant, antioxidant, antihyperlipidemic, and antitumoral activities.

### 22.4.1.5 Carrageenans: Sulfated polysaccharides from red algae

Carrageenans are high MW sulfated D-galactans found in red seaweed and is composed of repeating disaccharide units with alternating 3-linked β-D-galactose and 4-linked α-D-galactose.[61] There are at least 15 different carrageenan structures found in the literature. However, the most industrially relevant carrageenans

are κ, ι, and λ forms, the structures of which are illustrated in Figure 22.6, **21–23**. Carrageenans have been widely used in the food industry as thickening, gelling, and protein-suspending agents. Besides the well-known biological activities related to inflammatory and immune responses, carrageenans are shown to be potent inhibitors of herpes and HPV viruses. Recent finding indicates that these polysaccharides may offer some protection against HIV infection as well.

## 22.5 Summary, conclusion, and future perspective

Marine organisms offer a great diversity of natural products including sulfated polysaccharides with interesting therapeutic values. Marine algae and invertebrates are found rich in sulfated polysaccharides. Carrageenans, fucoidans, and ulvans are isolated in high quantity from red, brown, and green algae, respectively. Sulfated fucans, galactans, and GAGs are also isolated from various marine invertebrates. In addition, some derivatives of marine sulfated polysaccharides can be obtained by chemical modifications of existing polysaccharides to optimize the biological properties.[53] All these sulfated polysaccharides are highly anionic nature capable of interacting with certain cationic proteins and

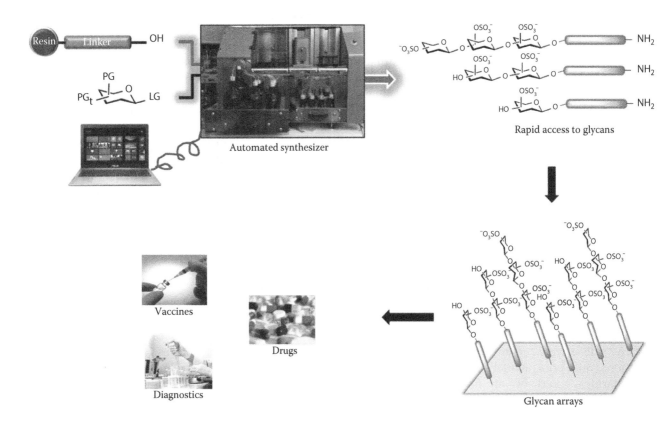

*Figure 22.7* (**See color insert.**) An approach for the synthesis of marine polysaccharides.

modify various biological processes. Also, these sulfated polysaccharides possess antioxidant, antiallergic, antihuman immunodeficiency virus, anticancer, and anticoagulant activities. Therefore, this marine origin should offer potentially safer and cheaper compounds than mammalian polysaccharides for drug discovery.

Although marine sulfated polysaccharides are known to have therapeutic values, most of them remain unexplored in the area of drug development. It is largely due to their complexity and heterogeneity with respect to size and sulfation pattern that troubled the evaluation of biological activities and the determination of structure–activity relationship. Therefore, it is essential to have structurally well-defined marine carbohydrates in sufficient purity and quantity for initial biological evaluations. There are many different approaches currently practiced in carbohydrate synthesis, which includes a target-oriented stepwise chemical synthesis, relative reactivity based on pot synthesis, automated solid-phase synthesis, and enzymatic synthesis.[62–66] Among these approaches, automated solid-phase synthesis has received much attention recently as it provides a desired product with functional linker in a rapid manner by eliminating the multiple purification steps (Figure 22.7). Recent advancements in carbohydrate analysis using glycoarray technology provide a better understanding of the biological processes involving complex carbohydrates. Therefore, the conjugation-ready oligosaccharides produced by automated synthesis can be subsequently printed on microarrays or conjugated to proteins for immunological evaluation.

Automated synthesis would be a more suitable approach for producing marine sulfated carbohydrates of interest as it majorly consists of monosaccharide or disaccharide repeating units. In addition, synthesis of mammalian GAGs such as hyaluronic acid, chondroitin sulfate, and dermatan sulfate has already been demonstrated using automated technique.[67–69] Biological evaluations of synthetic marine sulfated carbohydrates would undoubtedly lead to new findings in therapeutic applications and drug discovery.

## References

1. Cooper, G. M. *The Cell: A Molecular Approach,* 2nd edn. Sunderland, MA: Sinauer Associates, 2000.
2. Varki, A. and J. B. Lowe. Biological roles of glycans. In: *Essentials of Glycobiology,* A. Varki, R. D. Cummings, J. D. Esko et al. (eds.), 2nd edn. Cold Spring Harbor, NY: Cold Spring Harbor Laboratory Press, 2009.
3. Varki, A., R. D. Cummings, J. D. Esko, H. H. Freeze, P. Stanley, C. R. Bertozzi, G. W. Hart et al. *Essentials of Glycobiology,* 2nd edn. Cold Spring Harbor, NY: Cold Spring Harbor Laboratory Press, 2009.
4. Wong, C.-H. *Carbohydrate-Based Drug Discovery,* Vols. 1 and 2. Weinheim, Germany: Wiley VCH, 2003.
5. Hodgson, J. Carbohydrate-based therapeutics. *Nat. Biotechnol.* 9(7) (1991): 609–613.
6. Michelle, K. and L. Joshi. Carbohydrates in therapeutics. *Cardiovasc. Hematol. Agents Med. Chem.* 5(3) (2007): 186–197.
7. Hudak, J. E. and C. R. Bertozzi. Glycotherapy: New advances inspire a reemergence of glycans in medicine. *Chem. Biol.* 21(1) (2014): 16–37.
8. Seeberger, P. H. and D. B. Werz. Synthesis and medical applications of oligosaccharides. *Nature* 44(7139) (2007): 1046–1051.
9. Tiwari, V. K., R. C. Mishra, A. Sharma, and R. P. Tripathi. Carbohydrate based potential chemotherapeutic agents: Recent developments and their scope in future drug discovery. *Mini-Rev. Med. Chem.* 12(14) (2012): 1497–1519.
10. Asano, N. Glycosidase inhibitors: Update and perspectives on practical use. *Glycobiology* 13(10) (2003): 93R–104R.
11. von Itzstein, M., W.-Y. Wu, G. B. Kok, M. S. Pegg, J. C. Dyason, B. Jin, T. Van Phan et al. Rational design of potent sialidase-based inhibitors of influenza virus replication. *Nature* 363(6428) (1993): 418–423.
12. Arya, D. P. *Aminoglycoside Antibiotics.* Hoboken, NJ: Wiley-Interscience, 2007.
13. Hainrichson, M., I. Nudelman, and T. Baasov. Designer aminoglycosides: The race to develop improved antibiotics and compounds for the treatment of human genetic diseases. *Org. Biomol. Chem.* 21(2) (2008): 227–239.
14. Zingman, L. V., S. Park, T. M. Olson, Alekseev, A. E., and A. Terzic. Aminoglycoside-induced translational readthrough in disease: Overcoming nonsense mutations by pharmacogenetic therapy. *Clin. Pharmacol. Ther.* 81(1) (2007): 99–103.
15. Ofosu, F. A., I. Danishefsky, and J. Hirsh. *Heparin and Related Polysaccharides.* New York: New York Academy of Sciences, 1989.
16. Liu, H., Z. Zhang, and R. J. Linhardt. Lessons learned from the contamination of heparin. *Nat. Prod. Rep.* 26(3) (2009): 313.
17. de Kort, M., R. C. Buijsman, and C. A. A. van Boeckel. Synthetic heparin derivatives as new anticoagulant drugs. *Drug Discov. Today* 10(11) (2005): 769–779. doi:10.1016/s1359-6446(05)03457-4.
18. Truscheit, E., W. Frommer, B. Junge, L. Muller, D. D. Schmidt, and W. Wingender. Chemistry and biochemistry of microbial α-glucosidase inhibitors. *Angew. Chem. Int. Ed. Engl.* 20(9) (1981): 744–761.
19. Mikhailets, G. A. and M. S. Polyak. Acarbose (review). *Pharm. Chem. J.* 23(1) (1989): 82–87.
20. Lewis, M. G., S. DaFonseca, N. Chomont, A. T. Palamara, M. Tardugno, A. Mai, M. Collins et al. Gold drug auranofin restricts the viral reservoir in the monkey AIDS model and induces containment of viral load following ART suspension. *AIDS* 25(11) (2011): 1347–1356.
21. Maryanoff, B. E., S. O. Nortey, J. F. Gardocki, R. P. Shank, and S. P. Dodgson. Anticonvulsant O-alkyl sulfamates. 2,3:4,5-Bis-O-(1-Methylethylidene)-β-D-fructopyranose sulfamate and related compounds. *J. Med. Chem.* 30(5) (1987): 880–887.
22. Perucca, E. A. 1997. Pharmacological and clinical review on topiramate, a new antiepileptic drug. *Pharmacol. Res.* 35(4) (1997): 241–256.

23. Mula, M., A. E. Cavanna, and F. Monaco. Psychopharmacology of topiramate: From epilepsy to bipolar disorder. *Neuropsychiatr. Dis. Treat.* 2(4) (2006): 475–488.

24. Kuter, D. J., A. Mehta, C. E. M. Hollak, P. Giraldo, D. Hughes, N. Belmatoug, M. Brand et al. Miglustat therapy in type 1 Gaucher disease: Clinical and safety outcomes in a multicenter retrospective cohort study. *Blood Cells Mol. Dis.* 51(2) (2013): 116–124.

25. Pastores, G. M., D. Elstein, M. HrebÃcek, and A. Zimran. Effect of Miglustat on bone disease in adults with type 1 Gaucher disease: A pooled analysis of three multinational, open-label studies. *Clin. Ther.* 29(8) (2007): 1645–1654.

26. Weber, Jr., F. L. Lactulose and combination therapy of hepatic encephalopathy: The role of the intestinal microflora. *Digest. Dis.* 14(1) (1996): 53–63.

27. Agarwal, M. K. Streptozotocin: Mechanisms of action. *FEBS Lett.* 120(1) (1980): 1–3.

28. Piroska, E. Streptozotocin for insulinoma. *Ann. Intern. Med.* 75(3) (1971): 477.

29. Gheorghiade, M. Digoxin in the management of cardiovascular disorders. *Circulation* 109(24) (2004): 2959–2964.

30. Vliegenthart, J. F. G. Carbohydrate based vaccines. *FEBS Lett.* 580(12) (2006): 2945–2950.

31. Huang, Y.-L. and C.-Yi Wu. Carbohydrate-based vaccines: Challenges and opportunities. *Expert Rev. Vaccines* 9(11) (2010): 1257–1274.

32. Ouerfelli, O., J. D. Warren, R. M. Wilson, and S. J. Danishefsky. Synthetic carbohydrate-based antitumor vaccines: Challenges and opportunities. *Expert Rev. Vaccines* 4(5) (2005): 677–685.

33. Uemura, D., K. Nakamura, and M. Kitamura. Biologically active marine natural products. *Heterocycles* 78(1) (2009): 1.

34. Scheuer, P. Some marine ecological phenomena: Chemical basis and biomedical potential. *Science* 248(4952) (1990): 173–177.

35. Hentschel, U. Natural products from marine microorganisms. *Chembiochem* 3(11) (2002): 1151–1154.

36. Bharate, S. B., S. D. Sawant, P. P. Singh, and R. A. Vishwakarma. Kinase inhibitors of marine origin. *Chem. Rev.* 113(8) (2013): 6761–6815.

37. Proksch, P., R. A. Edrada-Ebel, and R. Ebel. Drugs from the sea—Opportunities and obstacles. *Mar. Drugs* 1(1) (2003): 5–17.

38. Simmons, T. L., E. Andrianasolo, K. McPhail, P. Flatt, and W. H. Gerwick. Marine natural products as anticancer drugs. *Mol. Cancer. Ther.* 4 (2005): 333–342.

39. Kim, S. *Marine Carbohydrates.* Burlington, VT: Elsevier Science, 2014.

40. Sudha, P. N., S. Aisverya, R. Nithya, and K. Vijayalakshmi. Industrial applications of marine carbohydrates. *Adv. Food Nutr. Res.* 73(2014): 145–181.

41. Laurienzo, P. Marine polysaccharides in pharmaceutical applications: An overview. *Mar. Drugs* 8(9) (2010): 2435–2465.

42. Wang, W., S.-X. Wang, and H.-S. Guan. The antiviral activities and mechanisms of marine polysaccharides: An overview. *Mar. Drugs* 10(12) (2012): 2795–2816.

43. José, K. Sulfated oligosaccharides: New targets for drug development? *Curr. Med. Chem.* 16(18) (2009): 2338–2344.

44. Capila, I. and R. J. Linhardt. Heparin-protein interactions. *Angew. Chem. Int. Ed. Engl.* 41(3) (2002): 390–412.

45. Lever, R. and C. P. Page. Novel drug development opportunities for heparin. *Nat. Rev. Drug Discov.* 1(2) (2002): 140–148.

46. Casu, B. and J. E. Scott. Heparin and heparin-like substances. New concepts on structure and activity. *Pharmacol. Res.* 13(7) (1981): 717–719.

47. Raposo, M., R. de Morais, and A. B. de Morais. Bioactivity and applications of sulfated polysaccharides from marine microalgae. *Mar. Drugs* 11(1) (2013): 233–252.

48. Jiao, G., G. Yu, J. Zhang, and H. Ewart. Chemical structures and bioactivities of sulfated polysaccharides from marine algae. *Mar. Drugs* 9(12) (2011): 196–223.

49. Wijesekara, I., R. Pangestuti, and S.-K. Kim. Biological activities and potential health benefits of sulfated polysaccharides derived from marine algae. *Carbohydr. Polym.* 84(1) (2011): 14–21.

50. Esteves, A. I. S., M. Nicolai, M. Humanes, and J. Goncalves. Sulfated polysaccharides in marine sponges: Extraction methods and anti-HIV activity. *Mar. Drugs* 9(2) (2011): 139–153.

51. Wang, L., X. Wang., H. Wu, and R. Liu. Overview on biological activities and molecular characteristics of sulfated polysaccharides from marine green algae in recent years. *Mar. Drugs* 12(9) (2014): 4984–5020.

52. Mao, W., X. Zang, Y. Li, and H. Zhang. Sulfated polysaccharides from marine green algae *Ulva conglobata* and their anticoagulant activity. *J. Appl. Phycol.* 18(1) (2006): 9–14.

53. d'Ayala, G. G., M. Malinconico, and P. Laurienzo. Marine derived polysaccharides for biomedical applications: Chemical modification approaches. *Molecules* 13(9) (2008): 2069–2106.

54. Zaporozhets, T. S., S. V. Ermakova, T. N. Zvyagintseva, and N. N. Besednova. Antitumor effects of sulfated polysaccharides produced from marine algae. *Biology Bulletin Reviews* 4(2) (2014): 122–132.

55. Ale, M. T., H. Maruyama, H. Tamauchi, J. D. Mikkelsen, and A. S. Meyer. Fucose-containing sulfated polysaccharides from brown seaweeds inhibit proliferation of melanoma cells and induce apoptosis by activation of caspase-3 in vitro. *Mar. Drugs* 9(12) (2011): 2605–2621.

56. Colliec-Jouaulta, S., C. Bavingtonb, and C. Delbarre-Ladrata. Heparin-like entities from marine organisms. *Handb. Exp. Pharmacol.* 207 (2012): 423–449.

57. Pavão, M. S. G. Glycosaminoglycans analogs from marine invertebrates: Structure, biological effects, and potential as new therapeutics. *Front Cell Infect. Microbiol.* 4 (2014): 123.

58. Pomin, V. H. Holothurian fucosylated chondroitin sulfate. *Mar. Drugs* 12(1) (2014): 232–254.

59. Fitton, J. H. Therapies from fucoidan; Multifunctional marine polymers. *Mar. Drugs* 9 (2011): 1731–1760.

60. Lahaye, M. and A. Robic. Structure and functional properties of ulvan, a polysaccharide from green seaweeds. *Biomacromolecules* 8(2007): 1765–1774.

61. Necas, J. and L. Bartosikova. Carrageenan: A review. *Veterinarni Medicina* 58(4) (2013): 187–205.

62. Hanessian, S. *Preparative Carbohydrate Chemistry.* New York: Marcel Dekker Inc., 1997.

63. Khan, S. H. and R. A. O'Neill. 1996. *Modern Methods in Carbohydrate Synthesis*. Amsterdam, the Netherlands: Harwood Academic, 1996.

64. Plante, O. J., E. R. Palmacci, and P. H. Seeberger. Automated solid-phase synthesis of oligosaccharides. *Science* 291 (2001): 1523–1527.

65. Seeberger, P. H. and D. B. Werz. Automated synthesis of oligosaccharides as a basis for drug discovery. *Nat. Rev. Drug Discov.* 4(9) (2005): 751–763.

66. Hsu, C.-H., S.-C. Hung, C.-Y. Wu, and C.-H. Wong. Toward automated oligosaccharide synthesis. *Angew. Chem. Int. Ed. Engl.* 50(50) (2007): 11872–11923.

67. Eller, S., M. Collot, J. Yin, H. S. Hahm, and P. H. Seeberger. Automated solid-phase synthesis of chondroitin sulfate glycosaminoglycans. *Angew. Chem. Int. Ed.* 52(22) (2013): 5858–5861.

68. Kandasamy, J., F. Schuhmacher, H. S. Hahm, J. C. Klein, and P. H. Seeberger. Modular automated solid phase synthesis of dermatan sulfate oligosaccharides. *Chem. Commun.* 50(15) (2014): 1875–1877.

69. Walvoort, M. T. C., A. G. Volbeda, N. R. M. Reintjens, H. van den Elst, O. J. Plante, H. S. Overkleeft, G. A. van der Marel, and J. D. C. Codée. Automated solid-phase synthesis of hyaluronan oligosaccharides. *Org. Lett.* 14(14) (2012): 3776–3779.

# chapter twenty-three

# Cyanobacterial extracellular polysaccharide sheath pigment, scytonemin

## A novel multipurpose pharmacophore

Jainendra Pathak, Rajneesh, Richa, Arun S. Sonker,
Vinod K. Kannaujiya, and Rajeshwar P. Sinha

## Contents

## 23.1 Introduction

Cyanobacteria are Gram-negative photosynthetic prokaryotes inhabiting a wide range of habitats, ranging from oceans to fresh water, hot springs to the Antarctic and Arctic regions, soil to bare rocks and stones, deserts to ice shelves, and are also found in the form of endosymbionts in plants, lichens, and several protists (Quesada and Vincent 1997, Baracaldo et al. 2005, Thajuddin and Subramanian 2005). These photolysis-mediated oxygen-evolving cosmopolitan prokaryotes have survived and flourished on Earth for over two billion years (Sergeev et al. 2002). Nitrogen fixation ability of cyanobacteria and their role in maintaining fertility of rice fields are well established (Sinha and Häder 1996, Vaishampayan et al. 1998, 2001, Tirkey and Adhikary 2005). Cyanobacteria tend to dominate the microbial populations of many stressed habitats by producing a huge amount of extracellular polymeric substances

(EPSs) in the form of sheaths, slime, and capsules (Chakraborty and Pal 2014). The structural composition and productivity of EPS vary with cyanobacterial species and types of stress exposure. Absorption of solar radiation to perform photosynthesis and nitrogen fixation exposes cyanobacteria to lethal ultraviolet-B (UV-B; 280–315 nm) radiation due to depletion of the stratospheric ozone layer, owing to atmospheric pollutants such as chlorofluorocarbons, chlorocarbons, and organobromides that are released anthropogenically (Crutzen 1992, Kerr and McElroy 1993, Lubin and Jensen 1995, Weatherhead and Andersen 2006). Depletion of ozone layer has been reported in Antarctic (Hodgson et al. 2005) and Arctic regions, where more than 70% decrease in ozone layer has been reported during late winter and early spring in the polar vortex (Smith et al. 1992, Von der Gathen et al. 1995). Aquatic ecosystems are also affected by harmful doses of UV-B and especially UV-A radiation that penetrate deep into the water column (up to several hundred meters) (Häder et al. 2007). Ultraviolet radiation (UVR) adversely affects living systems such as bacteria (Peak et al. 1984), cyanobacteria, phytoplankton, macroalgae (Sinha et al. 2001), plants (Quaite et al. 1992), animals (Kripke et al. 1992), and humans (Stein et al. 1989). The highly energetic UV-B affects DNA and proteins directly as well as indirectly (Castenholz and Garcia-Pichel 2000, Sinha and Häder 2002, Castenholz 2004, Cadoret et al. 2005, Häder and Sinha 2005, Mazur-Marzec 2006, Solheim et al. 2006, Björn 2007, Rath and Adhikary 2007). Direct effect involves the denaturation of DNA and RNA, whereas indirect effects include the production of reactive oxygen species (ROS) (Häder and Sinha 2005). UV-A radiation is not absorbed directly by DNA but induces DNA damage either by producing a secondary photoreaction of existing DNA photoproducts or via indirect photosensitizing reactions (Hargreaves et al. 2007). UV-B and UV-A radiations affect a number of physiological and biochemical processes in cyanobacteria such as survival, growth, pigmentation, photosynthetic oxygen production, motility, $N_2$ uptake, phycobiliprotein composition, and $CO_2$ uptake (Sinha et al. 2005, Wright et al. 2005, Xue et al. 2005, Bhandari and Sharma 2006, Helbling 2006, Cheng et al. 2007, Wulff et al. 2007, Vincent and Roy 1993). A wide range of cyanobacteria are able to synthesize and secrete EPS mainly of polysaccharide in nature, which may be secreted into the surrounding environment or remain attached to the cell surface (Neu and Marshall 1990). It is believed that EPSs in cyanobacteria play a major role in protecting cells from various stresses in extreme habitats. However, certain autotrophs including cyanobacteria have evolved mitigation strategies to counteract the toxicity of UVR such as light-dependent repair of UV-induced damage of DNA (Britt 1995, Kim and Sancar 1995, Pakker et al. 2000, Sinha and Häder 2002, Häder and Sinha 2005), accumulation of carotenoids (Davis 1976) and detoxifying enzymes or radical quenchers and antioxidants (Rastogi et al. 2013), and synthesis of photoprotective compounds such as mycosporine-like amino acids (MAAs) and scytonemin (Sinha et al. 1998, 1999, Sinha and Häder 2008, Rastogi and Sinha 2009, Rastogi et al. 2013) to counteract the damaging effects of UV radiation. Synthesis and accumulation of photoprotective compounds like scytonemin and MAAs in sheath of several cyanobacterial species confirm the crucial role of sheath in harboring UV-absorbing substances, and thus protecting the cyanobacterial cells from deleterious effects of UV radiation (Chakraborty and Pal 2014). Thus, this extracellular polysaccharide sheath pigment, scytonemin, which is produced by several species of cyanobacteria, plays a significant role in screening of UVR (Sinha et al. 1998, Sinha and Häder 2008, Rastogi et al. 2013).

## 23.2 Cyanobacterial EPSs

### 23.2.1 Structural composition of EPSs

Cyanobacterial EPSs are mainly composed of high-molecular-mass heteropolysaccharides, with variable compositions and roles according to the microorganism and the environmental conditions (Pereira et al. 2009). Cyanobacteria possess a unique cell wall that combines the presence of an outer membrane and lipopolysaccharides, as in Gram-negative bacteria, with a thick and highly cross-linked peptidoglycan layer similar to Gram-positive bacteria (Hoiczyck and Hansel 2000, Stewart et al. 2006). The EPS associated with the cell surface can be referred to as sheaths, capsules, and slimes, according to their thickness, consistency, and appearance (De Philippis and Vincenzini 1998, 2003). The sheath is defined as a thin, dense layer loosely surrounding cells or cell groups usually visible in light microscopy without staining. The sheath of the cyanobacterium *Lyngbya aestuarii* has been shown to have a sulfated proteoglycan (Robbins et al. 1998). The polypeptide comprises 12.9% of the sheath dry weight and sulfate esters account for 2.0%. Aspartic acid and alanine represent 32.5% of the polypeptide component. The dominant monosaccharide is glucose, averaging 18.0% of the dry weight. At least 13 different monosaccharide linkages have been identified. This morphologically rigid sheath is a single sulfated proteoglycan (Robbins et al. 1998). Both the sulfate groups and the uronic acids contribute to the anionic nature of the EPS, conferring a negative charge and a "sticky" behavior to the overall macromolecule (Decho 1990, Leppard et al. 1996, Arias et al. 2003, De Philippis and Vincenzini 2003, Mancuso Nichols et al. 2005). To date, up to 12 different monosaccharides have been identified in cyanobacterial EPS: the hexoses (glucose, galactose, mannose, and fructose); the pentoses (ribose, xylose, and arabinose); the deoxyhexoses (fucose, rhamnose, and methyl rhamnose); and the acidic hexoses (glucuronic and galacturonic acid) (De Philippis and Vincenzini 1998, 2003, De Philippis et al. 2001). In a few cases, the presence of additional types of monosaccharides such as *N*-acetyl glucosamine, 2, 3-*O*-methyl rhamnose, 3-*O*-methyl rhamnose, 4-*O*-methyl rhamnose, and 3-*O*-methyl glucose has been reported (Hu et al. 2003). The monosaccharide most frequently found at the highest concentration in cyanobacterial EPS is glucose, although there are polymers where other sugars, such as xylose, arabinose, galactose, or fucose, are present at higher concentrations than glucose (Tease et al. 1991, Bender et al. 1994, Gloaguen et al. 1995, Fischer et al. 1997, De Philippis and Vincenzini 1998, 2003, Parikh and Madamwar 2006). It has been reported that cyanobacterial EPSs are composed not only composed of carbohydrates but also of other macromolecules such as polypeptides (Kawaguchi and Decho 2000). Polypeptides enriched with glycine, alanine, valine, leucine, isoleucine, and phenylalanine have been reported in the EPS of *Cyanospira capsulata* and *Nucula calcicola* (Flaibani et al. 1989, Marra et al. 1990), and in *Schizothrix* sp., small proteins specifically enriched with aspartic and glutamic acid have been observed (Kawaguchi and Decho 2002). In general, the chemical composition, the type, and the amount of the exopolysaccharides produced by a given cyanobacterial strain are stable features, mostly depending on the species and the cultivation conditions (Nicolaus et al. 1999). However, the sugar composition of the EPS produced by a certain strain may, qualitatively and quantitatively, vary slightly, especially with the age of the culture (Gloaguen et al. 1995, De Philippis and Vincenzini 1998).

### 23.2.2   Putative roles of EPS in cyanobacteria

The capability of cyanobacteria to survive in severe habitats has been related to the protective mechanisms that they have developed. Among these mechanisms, one of the most diffused within the cyanobacteria is the ability to synthesize external polysaccharide layers that protect the cells from unfavorable environmental conditions such as dehydration, phagocytosis, antibody recognition, and even lysis by viruses (Figure 23.1) (Dudman 1977, Tease and Walker 1987, Hill et al. 1994, Scott et al. 1996, Hoiczyk 1998, De Vuyst and Degeest 1999, Sutherland 1999, Ruas-Madiedo et al. 2002). Owing to their hydrophilic/hydrophobic characteristics, EPSs are able to trap and accumulate water, creating a gelatinous layer around the cells that regulates water uptake and loss and stabilizes the cell membrane during periods of desiccation (Grilli Caiola et al. 1993, 1996, Tamaru et al. 2005). Upon rehydration, cyanobacteria can rapidly

recover metabolic activities and repair cellular components (Scherer et al. 1984, 1986, Satoh et al. 2002, Fleming and Castenholz 2007). Another possible role that has been thoroughly investigated is the capacity of the sheath, capsules, and slime to protect the cyanobacterial cells from the harmful effects of UV radiations. It was demonstrated that UV irradiation induces the synthesis of the extracellular polysaccharide matrix in *Nostoc commune* (Ehling-Schulz et al. 1997, Wright et al. 2005) and also that the UV-screen pigments are accumulated in the sheath and in the extracellular matrix, constituting a barrier against the penetration of the harmful UV radiations (Böhm et al. 1995, Ehling-Schulz and Scherer 1999, Dillon et al. 2002, Fleming and Castenholz 2007, 2008). The UV-absorbing pigment scytonemin was found in the sheath of a number of cyanobacteria living in environments characterized by a high level of solar irradiation (Garcia-Pichel and Castenholz 1991, Ehling-Schulz et al. 1997, Ehling-Schulz and Scherer 1999). The ultraviolet shielding pigment scytonemin, while located in the sheath, is not an integral part of the sheath. Scytonemin is easily removed by solvent extraction, and there are no significant differences in the composition of pigmented and nonpigmented sheaths (Robbins et al. 1998). Pigmented sheath contained 10.7% protein, while nonpigmented sheath contained 14.5% protein. Scytonemin appears to play no role in sheath integrity. Its ease of removal indicates that it is not covalently bound to the carbohydrate. The unchanged sheath composition and integrity, with and without scytonemin, suggest that it plays no role in sheath structural chemistry. The sheath appears to play the role of a support matrix, allowing the pigment to be retained on the filament surface and thus to serve as a UV shield (Garcia-Pichel and Castenholz 1991). Structural analysis of purified pigment by nuclear magnetic resonance

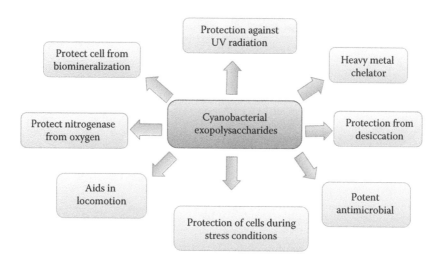

*Figure 23.1* Putative roles of EPS in cyanobacteria.

spectroscopy (NMR) identified two components: a monosubstituted phenol fragment and a disubstituted phenyl ring fragment (Robbins et al. 1998). These data were in agreement with the full structure of scytonemin as reported by Proteau et al. (1993).

## 23.3 Cyanobacterial extracellular polysaccharide sheath pigment: Scytonemin

### 23.3.1 Structure of scytonemin

Nägeli (1849) first reported this pigment in some terrestrial cyanobacteria and later termed it scytonemin (Nägeli and Schwenderer 1877). It is a highly stable, lipid-soluble, yellow-brown hydrophobic pigment, reported from extracellular polysaccharide sheaths (Figure 23.2) of about 300 cyanobacteria. It is a dimer of indole and phenolic subunits and has a molecular mass of 544 Da. The two subunits in scytonemin are linked through an olefinic carbon atom that is unique among natural products. This linkage provides scytonemin a new ring system and has been termed "the scytoneman skeleton" (Proteau et al. 1993). Scytonemin exists in two forms—oxidized (green) and reduced (red) (Garcia-Pichel and Castenholz, 1991)—which were named fuscochlorin and fuscorhodin, respectively (Kylim 1927, 1937) (Figure 23.3a, b). During the process of extraction, redox and

acid–base conditions determine the existence of the two forms. On the basis of $^1H$ and $^{13}C$ NMR and mass spectrometry (MS) experiments, three new pigments, dimethoxyscytonemin, tetramethoxyscytonemin, and scytonin (Figure 23.3c through e), have been reported (Bultel-Poncé et al. 2004) from the organic extracts of *Scytonema* sp., and all these forms are derived from the "scytoneman skeleton" of the scytonemin. Recently, Grant and Louda (2013) have reported a new member of "scytoneman skeleton," scytonemin-3a-imine (Figure 23.3f) from the cyanobacterium *Scytonema hofmanni*, growing under intense light conditions.

### 23.3.2 Scytonemin gene cluster

Gene cluster associated with scytonemin biosynthesis has been studied in *Nostoc punctiforme* by Soule et al. (2007). The genomic region flanking the mutation revealed an 18-gene cluster (NpR1276 to NpR1259) (Figure 23.4) among which the gene clusters NpR1274-NpR1271 play central role in scytonemin biosynthesis. NpR1273 was found to be directly involved in scytonemin biosynthesis in *N. punctiforme*. All 18 genes are induced by UV-A irradiation, and relative transcription levels generally peak after 48 h of continuous UV-A exposure. All the genes in the 18-gene region are cotranscribed as part of a single transcriptional unit (Soule et al. 2009a). This 18-gene cluster consists of two

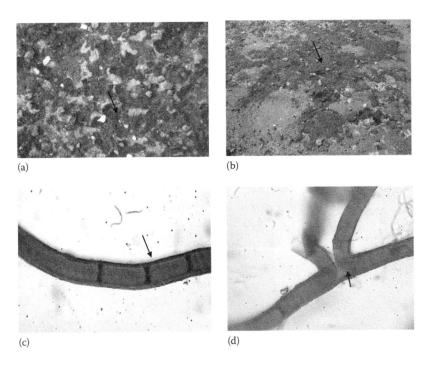

(a)

(b)

(c)

(d)

*Figure 23.2* (a, b) Cyanobacteria harboring the rooftop of the Department of Botany, Banaras Hindu University, Varanasi, as black-brown leathery mats. (c, d) Photographs showing the filaments of *Scytonema* sp. with sheaths typically colored by the presence of a yellow-brown UV-protective pigment, scytonemin.

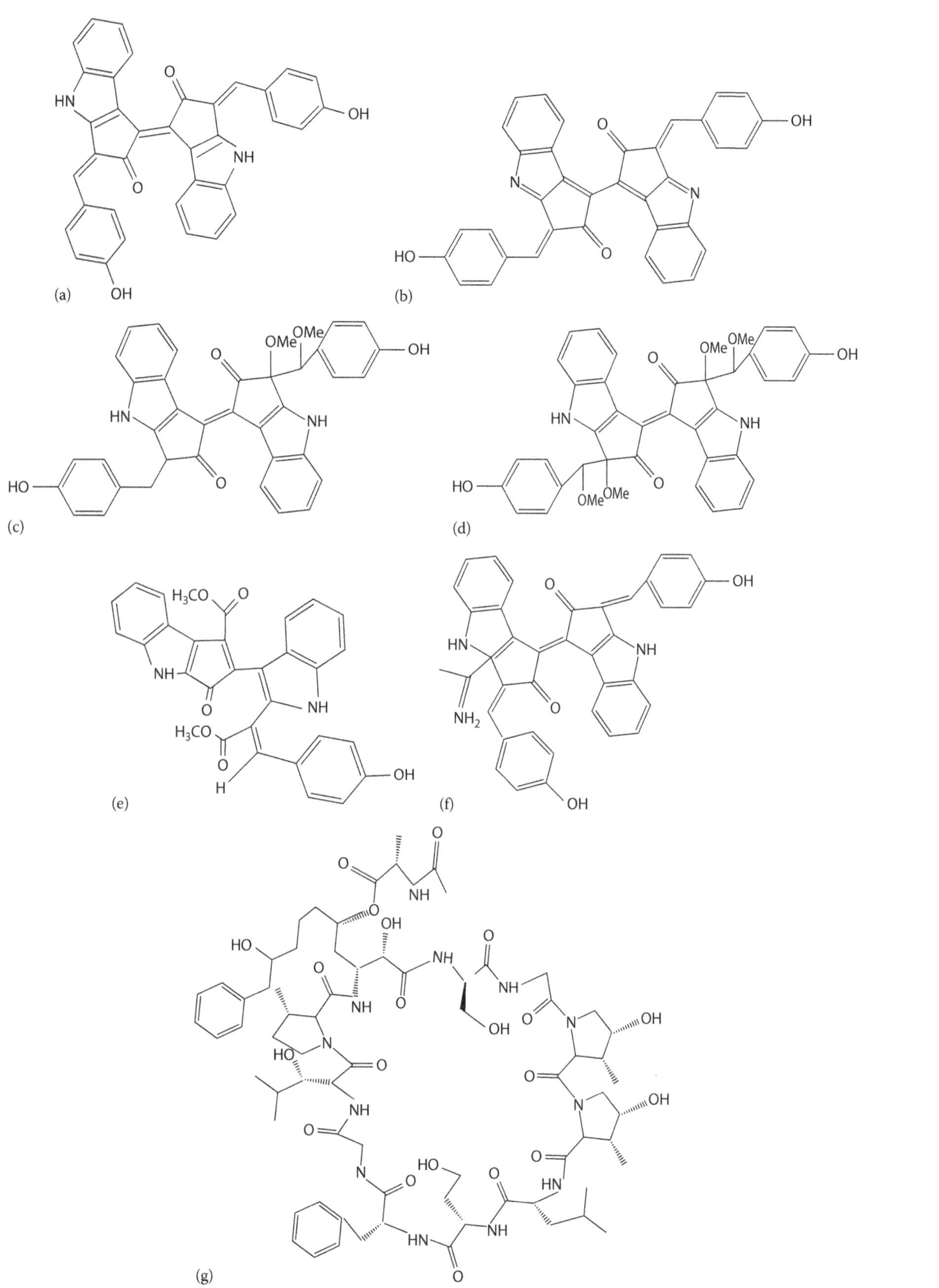

*Figure 23.3* Chemical structures of different forms of scytonemin: (a) oxidized scytonemin, (b) reduced scytonemin, (c) dimethoxyscytonemin, (d) tetramethoxyscytonemin, (e) scytonin, (f) scytonemin-3a-imine, and (g) scytonemin A.

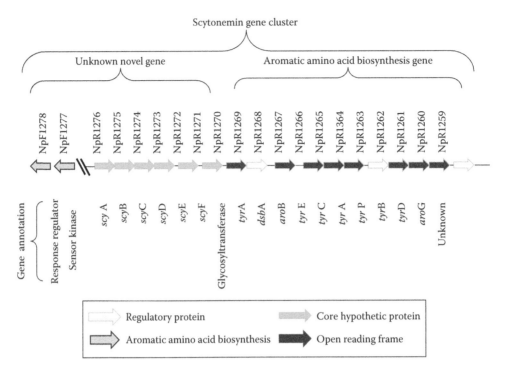

*Figure 23.4* Structure of genome associated with the biosynthesis of scytonemin in *N. punctiforme*. The ORFs from NpR1276 to NpR1259 are labeled with particular proteins. The ORFs in the downstream region of these clusters (NpR1269 to NpR1260) are predicted to encode a set of enzymes involved in the shikimic acid and aromatic amino acid biosynthesis pathways. Most of the upstream ORFs within these clusters (NpR1276 to NpR1270) encode products annotated as hypothetical proteins. The hatch marks indicate a break in the distance scale. (Modified from Soule, T. et al. *J. Bacteriol.*, 189, 4465, 2007.)

major cluster architectures that are thought to evolve from rearrangements of large sections, such as genes responsible for aromatic amino acid biosynthesis and through the insertion of genes that potentially confer additional biosynthetic capabilities. Differential transcriptional expression analysis demonstrated increased transcription of entire gene cluster after exposure to UV radiation. The knowledge from an evolutionary phylogenetic analysis in combination with the fact that the scytonemin gene cluster is distributed across several cyanobacterial lineages led to conclusion that the distribution of this gene cluster is best explained through an ancient evolutionary origin (Sorrels et al. 2010). Balskus and Walsh (2008) have proposed possible biosynthetic route (Figure 23.5) for the scytonemin biosynthesis, and the acyloin reaction was found to be the key step in constructing the carbon framework of this ecologically and evolutionary important pigment. Two enzymes encoded by open reading frame (ORF) NpR1275 and NpR1276 from the gene cluster have been functionally characterized and are involved in the initial step of scytonemin biosynthesis (Singh et al. 2010). However, the products of NpR1263 and NpR1269 ORFs are still to be functionally characterized. On the basis of conservation and location, two additional conserved clusters, NpF5232 to NpF5236 and a putative two-component

regulatory system (NpF1277 and NpF1278), are likely involved in scytonemin biosynthesis and regulation, respectively (Soule et al. 2009b). The production of the oxidized scytonemin from *N. commune* has been investigated in different culture conditions. High culture temperature, strong illumination intensity, and light–dark cycle (12:12 h) were found to be good for producing the oxidized scytonemin by *N. commune*.

### 23.3.3 Induction and synthesis of scytonemin

Metabolites of aromatic amino acid biosynthesis are thought to be the precursors of scytonemin biosynthesis, which is induced under high photon fluence rate (Garcia-Pichel and Castenholz 1991). Formaldehyde, chloramphenicol, and low temperature (4°C) counteract the effect of bright light in enhancing scytonemin production. Scytonemin synthesis involves the protein synthesis pathway. UV-A treatment is very efficient in inducing the synthesis of scytonemin (Garcia-Pichel and Castenholz 1991), whereas blue, green, or red light at the same fluence rates do not cause any significant increase in scytonemin. Biosynthesis of scytonemin is also greatly affected under different abiotic stresses such as heat, osmotic, and oxidative stress (Dillon et al. 2002). Elevation in temperature and oxidative stress,

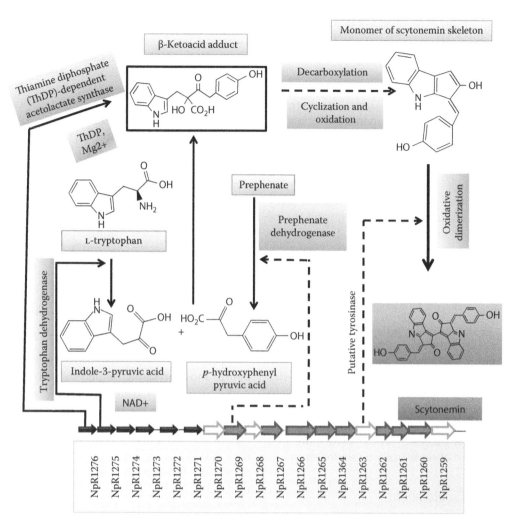

*Figure 23.5* Biosynthetic route for the scytonemin and corresponding gene products involved in each step. Continuous arrow represents functionally characterized gene products, while gene products indicated by broken arrow are still to be functionally characterized for their involvement in the corresponding step. (Modified from Singh et al., 2010.)

in combination with UV-A, has a synergistic effect on increased production of scytonemin, whereas osmotic stress causes decrease in scytonemin synthesis in the presence and in the absence of scytonemin-inducing irradiance. Solar radiation is not always required for the production of scytonemin because this pigment has been reported to be synthesized in the cyanobacterium *Calothrix* deficient in Fe or Mg and grown under low illumination (Sinclair and Whitton 1977). Fleming and Castenholz (2007) showed that cells hydrated for 2 days, in between desiccation periods, had high scytonemin synthesis compared to the cells that were hydrated for 1 day in *N. punctiforme*, but in *Chroococcidiopsis*, periodic desiccation inhibits scytonemin synthesis. Fleming and Castenholz (2008) suggested that the greater the restriction in nitrogen accessibility, the greater is the production of scytonemin. As multiple environmental signals influence scytonemin synthesis, it can be said that the regulation of scytonemin induction is a part of the complex stress response pathway. Balskus et al. (2011) examined biosynthesis and accumulation of cyanobacterial sunscreening pigment scytonemin within intertidal microbial mat communities using a combination of chemical, molecular, and phylogenetic approaches. Both laminated (layered) and nonlaminated mats contained scytonemin, with morphologically distinct mats having different cyanobacterial community compositions. Within laminated microbial mats, regions with and without scytonemin had different dominant oxygenic phototrophs, with scytonemin-producing areas consisting primarily of *L. aestuarii* and scytonemin-deficient areas dominated by a eukaryotic alga. The nonlaminated mat was populated by a diverse group of cyanobacteria and did not contain algae. The amplification and phylogenetic assignment of scytonemin biosynthetic gene *scyC* from laminated mat samples confirmed that the dominant cyanobacterium in these areas, *L. aestuarii*, is likely responsible for sunscreen production. Hence, understanding of the molecular basis of scytonemin assembly could be utilized to explore its synthesis and function within natural microbial communities. The productivity of the existing synthesis systems restrains its applications in medicine and cosmetics.

Recently, the generation of the monomer moiety of scytonemin from tryptophan and tyrosine in *Escherichia coli* has been reported (Malla and Sommer 2014). The biosynthetic pathway of scytonemin was heterologously expressed from *N. punctiforme*, and it was found that only three enzymes from *N. punctiforme* were required for the *in vivo* production of monomer moiety of scytonemin in *E. coli*. It was also found that the constructed recombinant *E. coli* strains were capable of producing novel alkaloids as shunt products. The recombinant *E. coli* strain expressing putative scytonemin biosynthetic gene cluster produced 4.2 mg L$^{-1}$ (2.46 μg mg$^{-1}$ dry cell weight) of the monomer moiety of scytonemin without supplementation of extracellular substrates, whereas upon supplementation with 1 mM of the substrates to the *E. coli* strain harboring *scyABC* genes, 8.9 mg L$^{-1}$ (4.56 μg mg$^{-1}$ dry cell weight) of the monomer moiety of scytonemin was produced in 5 days. Promising results from this study—combining this cell factory with the previously described chemical dimerization process—would contribute to a sustainable production of semisynthetic scytonemin.

### 23.3.4 Stability of scytonemin

The stability of UV-absorbing compounds may be critical for their ultimate utility as a photoprotective compound. Stability of scytonemin was investigated under three different abiotic stress conditions such as UV-B radiation, heat, and strong oxidizing agent H$_2$O$_2$ (0.25%) (Rastogi et al. 2013). The partially purified scytonemin was treated with the aforementioned different stress conditions for an hour after which their remaining contents were analyzed in the high performance liquid chromatography (HPLC) system. The effect of heat was assayed by incubating scytonemin in an incubator shaker at 60°C. The content of scytonemin was lower under 0.25% H$_2$O$_2$ stress than that under UV and heat stress. High stability of scytonemin can be evident by the fact that about 84% of the scytonemin screen in desiccated *N. punctiforme* cells remained intact even after almost 2 months of constant exposure to UV-A radiation (Fleming and Castenholz 2007).

## 23.4 Scytonemin: A novel multipurpose pharmacophore

### 23.4.1 Role of scytonemin in photoprotection

The presence of a UV-absorbing compound like scytonemin most probably helped cyanobacteria to survive and protect them from lethal effects of UV radiation when there was no stratospheric ozone layer in the Precambrian era. This assumption is supported by the fact that scytonemin has an *in vivo* maximum absorption at 370 nm, whereas purified scytonemin shows a maximum absorption at 386 nm, but it also absorb significantly at 252, 278, and 300 nm (Figure 23.6). HPLC chromatograms and absorption spectra of *Rivularia* sp. extracted with 1:1 (v/v) methanol to ethyl acetate ratio revealed the presence of a single peak of scytonemin with a retention time of 2.8 min (Figure 23.7).

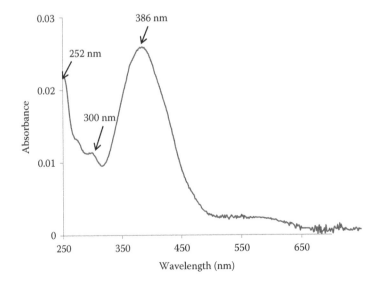

*Figure 23.6* Absorption spectrum of scytonemin extracted from *Scytonema* sp. having peaks at 252, 300, and 386 nm.

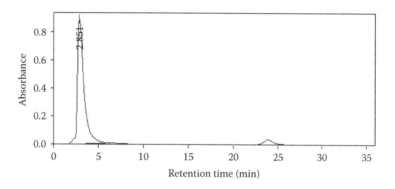

*Figure 23.7* HPLC chromatogram of partially purified scytonemin from *Scytonema* sp.

The evidence for the photoprotective role of scytonemin has been shown in a number of cyanobacteria from various harsh habitats (Garcia-Pichel and Castenholz 1991, Garcia-Pichel et al. 1992, Sinha et al. 1999, Gröniger et al. 2000, Hunsucker et al. 2001). The relevance of UV sunscreens such as scytonemin for protection has also been reported from cyanobacterial lichens such as *Collema*, *Gonohymenia*, and *Petulla*, growing in high light intensity habitats, and it is shown that scytonemin is located extracellularly in the sheath of the outer thallus part (Büdel et al. 1997). The UV sunscreen role of scytonemin has been demonstrated in the terrestrial cyanobacterium *Chlorogloeopsis* sp. (Garcia-Pichel et al. 1992). In cyanobacterial cultures, 5% of the cellular dry weight is contributed by the scytonemin, while it may be higher in naturally occurring cyanobacteria (Castenholz 1997). Pentecost (1993) has examined a correlation between UV flux and scytonemin content in a population of *Scytonema* and *Rivularia* sp. A variable and even negative correlation was found between UV and scytonemin in *Scytonema*, but a positive correlation was found in *Rivularia*. A high amount of scytonemin is required for uninhibited photosynthesis under high UV flux in a monospecific population of *Calothrix* sp. (Brenowitz and Castenholz 1997). The absorption spectra of methanolic extracts of the terrestrial cyanobacterium *Tolypothrix byssoidea* showed a prominent absorption at 260 and 384 nm, corresponding to the presence of scytonemin (Adhikary and Sahu 1998). The presence of scytonemin in cyanobacterial sheath has been reported to reduce the entry of UV-A radiation in the cell by 90% (Garcia-Pichel et al. 1992, Proteau et al. 1993). The scytonemin is highly stable and performs its screening activity without any further metabolic investment even under prolonged physiological inactivity (e.g., desiccation) when other ultraviolet protective mechanisms such as active repair of biosynthesis-damaged cellular component would be ineffective (Brenowitz and Castenholz 1997, Ehling-Schulz etal. 1997, Sinha et al. 1999).

### 23.4.2   Radical scavenging property of scytonemin

The photoprotective function of scytonemin has been studied in terms of its radical scavenging capacity (Rastogi et al. 2013). The total antioxidant activities of different concentrations of scytonemin have been determined by the decolorization of 2,2-diphenyl-1-picryl-hydrazyl (DPPH) radicals. The sunscreening pigment scytonemin exhibited the radical scavenging activity. The dose-dependent antioxidant activity of scytonemin was 12%, 33%, and 57% at concentrations of 0.5, 1.0, and 2.0 mg mL$^{-1}$, respectively. Antioxidant activity of the cyanobacterial pigment scytonemin was found to be relatively moderate in the DPPH assay. Scytonemin purified from the terrestrial cyanobacterium *N. commune* quenched an organic radical *in vitro* and accounted for up to 10% of the total activity of an ethanol extract of *N. commune* (Matsui et al. 2012). Hence, it can be concluded that the presence of scytonemin might help cyanobacteria to cope with oxidative stress under intense solar radiation.

### 23.4.3   Anti-inflammatory and antiproliferative properties of scytonemin

Scytonemin may have application in sunscreens as it has been found to act as ultraviolet sunscreen, with the greatest absorption in the spectral range of UV-A (Proteau et al. 1993, Rastogi and Sinha 2009). Scytonemin production in certain cyanobacteria is believed to be the earliest mechanism evolved for UV protection, more ancient than the flavonoids or melanins. Its ring structure, the "scytoneman skeleton," is unique among natural products, is believed to be a condensation product of tryptophan- and phenylpropanoid-derived subunits, and is also closely related to nostodione A (Proteau et al. 1993). Lack of chirality, multiple dissection points and phenolic groups that could be easily modified, and its relation to other antiproliferative agents make scytonemin a prime candidate having potential to be used as a pharmacophore that can be used for developing new therapies targeting hyperproliferative disorders. In the recent past, many pharmaceutical companies have put their efforts in developing a wide range of Chk inhibitors; however, a majority of them are weak in terms of inhibition of the kinase or its specificity toward other kinases. The majority of these inhibitors including scytonemin have been designed and tested against Chk1, although many also show inhibition of Chk2. Targeting Plk1 *in vitro* and *in vivo* through multiple mechanisms has shown promising results. Various small-molecule inhibitors have also shown activity against Plk1 (Schmit and Ahmad 2007). Scytonemin, wortmannin, and LY294002 are compounds that inhibit Plk1 activity nonselectively. Scytonemin A (Figure 23.3g), a cyclic peptide characterized from a *Scytonema* sp., has also been shown to be a strong calcium agonist (Helms et al. 1988). Scytonemin inhibits human *polo*-like kinase, a serine/threonine kinase, that plays an integral role in regulating the G2/M transition in the cell cycle. Scytonemin was shown to be active, with an $IC_{50}$ of 2 mM (Stevenson et al. 2002a). It also inhibits other cell cycle kinases with similar potency. Scytonemin inhibited cell proliferation ($IC_{50} = 7.8$ mM) in human T-cell leukemia Jurkat cells and induced apoptosis in 24% of the cells (at 3 mM). Scytonemin inhibited *polo*-like kinase 1 activity in a concentration-dependent manner with an $IC_{50}$ of 2 mM against the recombinant enzyme. Biochemical analysis showed that scytonemin reduced glutathione-S-transferase (GST)*polo*-like kinase 1 activity in a time-independent fashion, suggesting reversibility and with a mixed-competition mechanism with respect to ATP. Although scytonemin was less potent against protein kinase A and Tie2, a tyrosine kinase, it did inhibit other cell cycle–regulatory kinases like Myt1, checkpoint kinase 1, cyclin-dependent kinase 1/cyclin B, and protein kinase C-2 with $IC_{50}$ values similar to that seen for *polo*-like kinase 1. Consistent with these effects, scytonemin effectively attenuated, without chemical toxicity, the growth factor- or mitogen-induced proliferation of three cell types commonly implicated in inflammatory hyperproliferation. Similarly, scytonemin (up to 10 mM) was not cytotoxic to nonproliferating endotoxin-stimulated human monocytes. In addition, Jurkat T cells treated with scytonemin were induced to undergo apoptosis in a non-cell cycle–dependent manner consistent with its activities on multiple kinases. Scytonemin possesses both anti-inflammatory and antiproliferative properties. The dual kinase inhibitory activity may be of value therapeutically in acute and possibly chronic disorders where both inflammation and proliferation are prevalent. Limiting both neovascularization and the presence of inflammatory mediators at the affected site offers a broader spectrum of activity, which could hypothetically be more efficacious in the management

of complex inflammatory disorders. Whether or not the ability of scytonemin to inhibit both inflammation and proliferation is due solely to its effects on these two kinases, scytonemin offers a novel pharmacophore, which may serve as a template for synthesizing more potent and selective inhibitors (Stevenson et al. 2002b).

### 23.4.4 Role of scytonemin in apoptosis

Itoh et al. (2013) recently reported the induction of autophagic cell death in human T-lymphoid cell line Jurkat cells by reduced scytonemin isolated from *N. commune*. Reduced scytonemin (R-scy), isolated from *N. commune* Vaucher, has been shown to suppress the human T-lymphoid Jurkat cell growth. To reveal the mechanisms underlying the R-scy-mediated inhibition of Jurkat cell growth, cell morphology, DNA fragmentation, and microtubule-associated-protein light chain 3 (LC3) modifications were examined in these cells. Multiple vacuoles and the conversion of LC3-I to LC3-II in R-scy-treated cells were observed, suggesting that the R-scy-induced Jurkat cell growth inhibition is attributable to the induction of type II programmed cell death (PCDII; autophagic cell death or autophagy). The cells treated with R-scy produced large amounts of ROS, leading to the induction of mitochondrial dysfunction. However, the elimination of R-scy-induced ROS by treatment with *N*-acetyl-L-cysteine markedly opposed R-scy-induced PCDII. Based on these results, it can be concluded that ROS formation plays a critical role in R-scy-induced PCDII. Itoh et al. (2014) demonstrated the inhibitory effect of R-scy isolated from *N. commune* on Lipopolysaccharide/ Interferon (LPS/IFN)γ-stimulated nitric oxide (NO) production in murine macrophage RAW264 cells and examined the molecular mechanisms underlying the inhibitory effect of R-scy on LPS/IFNγ-costimulated inflammation. It was observed that R-scy induced anti-inflammatory effects in LPS/IFNγ-stimulated RAW264 cells by activating nuclear factor erythroid 2-related factor 2 (Nrf2)/antioxidant response element (ARE) signaling by generating ROS. They demonstrated that R-scy suppresses LPS/IFNγ-induced inflammation via the upregulation of the Keap1/Nrf2/ARE pathway. ROS levels increase by R-scy pretreatment suggests that ROS act as a second messenger in intracellular signal transduction. In cell redox signaling, ROS modify the sulfhydryl groups of Keap1, thereby activating the Nrf2/ARE signaling cascade (Itoh et al. 2004, Nguyen et al. 2004). ROS generated after R-scy pretreatment act upstream of the modification of Keap1 sulfhydryl groups and activation of p38 MAPK and PI3K/Akt in a cascade that results in the promotion of HO-1 expression via Nrf2/ARE signaling. These findings in R-scy-treated RAW264 cells were consistent with those of previous reports showing that protein kinases, such as MAPKs and PI3K/Akt phosphorylate-released-Nrf2, exert anti-inflammatory effects (Bang et al. 2012, Lee et al. 2012, Park et al. 2013). Increased HO-1 expression in cells contributes to the suppression of inflammatory responses, thereby providing a feasible mechanism of anti-inflammatory properties of R-Scy. Hence, it could be concluded that R-scy isolated from *N. commune* suppresses LPS/IFNγ-induced inflammation by inducing HO-1 expression via the activation of Nrf2/ARE signaling.

## 23.5  Summary

Scytonemin is not only an efficient photoprotective compound synthesized by cyanobacteria for providing protection against lethal UV radiation but also a multipurpose biological compound with numerous applications. In addition to its important role in cyanobacterial adaptation, this pigment certainly plays a vital role in the survival of microbial communities exposed to high solar radiation. The high concentration of scytonemin in many cyanobacterial sheaths might be providing significant protection to other microorganisms living within and beneath the upper layer of sheathed cyanobacteria (Powell et al. 2015). Scytonemin has also been identified and characterized as an antiproliferative pharmacophore that inhibits cell cycle kinases (Stevenson et al. 2002a). Reduced scytonemin isolated from *N. commune* plays a role in induction of autophagic cell death in human T-lymphoid cell line Jurkat cells (Itoh et al. 2013). ROS generated by R-scy treatment play a critical role as second messengers, regulating R-scy-induced anti-inflammatory effects. Extensive work is going on to explore the ecological, industrial, and pharmaceutical importance of scytonemin, which shows potent anti-inflammatory and antiproliferative properties, in addition to UV protection properties, but further extensive studies are needed to establish the potential of scytonemin as a multipurpose pharmacophore.

## Acknowledgments

Jainendra Pathak and Vinod K. Kannaujiya are thankful to CSIR, New Delhi, India, for the financial support in the form of fellowships. The work was also partially supported by a project grant (No. SR/WOS-A/LS-140/2011) sanctioned to Richa from the Department of Science and Technology. Arun S. Sonker and Rajneesh are also thankful to UGC-CSIR and DBT, respectively, for the fellowship grants.

## References

Adhikary, S. P. and J. K. Sahu. 1998. UV protecting pigments of the terrestrial cyanobacterium *Tolypothrix byssoidea*. *Journal of Plant Physiology* 153: 770–773.

Arias, S., A. del Moral, M. R. Ferrer, R. Tallon, E. Quesada, and V. Béjar. 2003. Mauran, an exopolysaccharide produced by the halophilic bacterium *Halomonas maura*, with a novel composition and interesting properties for biotechnology. *Extremophiles* 7: 319–326.

Balskus, E. P., R. J. Case, and C. T. Walsh. 2011. The biosynthesis of cyanobacterial sunscreen scytonemin in intertidal microbial mat communities. *FEMS Microbiology Ecology* 77(2): 322–332.

Balskus, E. P. and C. T. Walsh. 2008. Investigating the initial steps in the biosynthesis of cyanobacterial sunscreen scytonemin. *Journal of American Chemical Society (ACS Publications)* 130: 15260–15261.

Bang, S. Y., J. H. Kim, H. Y. Kim, Y. J. Lee, S. Y. Park, S. J. Lee, and Y. Kim. 2012. *Achranthes japonica* exhibits anti-inflammatory effect via NF-κB suppression and HO-1 induction in macrophages. *Journal of Ethnopharmacology* 144: 109–117.

Baracaldo, P. S., P. K. Hayes, and C. E. Blank. 2005. Morphological and habitat evolution in the cyanobacteria using a compartmentalization approach. *Geobiology* 3: 145–165.

Bender, J., S. Rodriguez-Eaton, U. M. Ekanemesang, and P. Phillips. 1994. Characterization of metal-binding bioflocculants produced by the cyanobacterial component of mixed microbial mats. *Applied and Environmental Microbiology* 60: 2311–2315.

Bhandari, R. and P. K. Sharma. 2006. High-light-induced changes on photosynthesis, pigments, sugars, lipids and antioxidant enzymes in freshwater (*Nostoc spongiaeforme*) and marine (*Phormidium corium*) cyanobacteria. *Photochemistry and Photobiology* 82: 702–710.

Björn, L. O. 2007. Stratospheric ozone, ultraviolet radiation and cryptogams. *Biological Conservation* 135: 326–333.

Böhm, G. A., W. Pfleiderer, P. Boger, and S. Scherer. 1995. Structure of a novel oligosaccharide-mycosporine-amino acid ultraviolet A/B sunscreen pigment from the terrestrial cyanobacterium *Nostoc commune*. *Journal of Biological Chemistry* 270: 8536–8539.

Brenowitz, S. and R. W. Castenholz. 1997. Long-term effects of UV and visible irradiance on natural populations of a scytonemin-containing cyanobacterium (*Calothrix* sp.). *FEMS Microbiology Ecology* 24: 343–352.

Britt, A. B. 1995. Repair of DNA damage induced by ultraviolet radiation. *Plant Physiology* 108: 891–896.

Büdel, B., U. Karsten, and F. Garcia-Pichel. 1997. Ultraviolet-absorbing scytonemin and mycosporine-like amino acid derivatives in exposed, rock-inhabiting cyanobacterial lichens. *Oecologia* 112: 165–172.

Bultel-Poncé, V., F. Felix-Theodose, C. Sarthou, J. F. Ponge, and B. Bodo. 2004. New pigment from the terrestrial cyanobacterium *Scytonema* sp. collected on the MitrakaInselberg, French Guyana. *Journal of Natural Products* 67: 678–681.

Cadoret, J. C., B. Rousseau, I. Perewoska, C. Sicora, O. Cheregi, I. Vass, and J. Houmard. 2005. Cyclic nucleotides, the photosynthetic apparatus and response to a UV-B stress in the cyanobacterium *Synechocystis* sp. PCC 6803. *Journal of Biological Chemistry* 280: 33935–33944.

Castenholz, R. W. 1997. Multiple strategies for UV tolerance in cyanobacteria. *Spectrum* 10: 10–16.

Castenholz, R. W. 2004. Phototrophic bacteria under UV stress. In: J. Seckbach (ed.), *Origins: Genesis, Evolution and Diversity of Life*. Kluwer Academy Publisher, Dordrecht, the Netherlands, pp. 445–461.

Castenholz, R. W. and F. Garcia-Pichel. 2000. Cyanobacterial responses to UV radiation, In: B. A. Whitton and M. Potts (eds.), *Ecology of Cyanobacteria: Their Diversity in Time and Space*. Kluwer Academy Publisher, Dordrecht, the Netherlands, pp. 591–611.

Chakraborty, T. and R. Pal. 2014. An overview of cyanobacterial exopolysaccharides: Features, composition and effects of stress exposure. *International Journal of Life Sciences* 8(4): 1–9.

Cheng, K., Y. Zhao, X. Du, Y. Zhang, S. Lan, and Z. Shi. 2007. Solar radiation driven decay of cyanophage infectivity and photoreactivation of the cyanophage by host cyanobacteria. *Aquatic Microbial Ecology* 48: 13–18.

Crutzen, P. J. 1992. Ultraviolet on the increase. *Nature* 356: 104–105.

Davis, B. H. 1976. Carotenoids. In: T. W. Goodwin (ed.), *Chemistry and Biochemistry of Plant Pigment*. Academic Press, New York/London, pp. 149–154.

De Philippis, R., C. Sili, R. Paperi, and M. Vincenzini. 2001. Exopolysaccharide-producing cyanobacteria and their possible exploitation: A review. *Journal of Applied Phycology* 13: 293–299.

De Philippis, R. and M. Vincenzini. 1998. Exocellular polysaccharides from cyanobacteria and their possible applications. *FEMS Microbiology Reviews* 22: 151–175.

De Philippis, R. and M. Vincenzini. 2003. Outermost polysaccharidic investments of cyanobacteria: Nature, significance and possible applications. *Recent Research Developments in Microbiology* 7: 13–22.

De Vuyst, L. and B. Degeest. 1999. Heteropolysaccharides from lactic acid bacteria. *FEMS Microbiology Reviews* 23: 153–177.

Decho, A. W. 1990. Microbial exopolymer secretions in ocean environments: Their role(s) in food webs and marine processes. *Oceanography and Marine Biology* 28: 73–153.

Dillon, J. G., C. M. Tatsumi, P. G. Tandingum, and R. W. Castenholz. 2002. Effect of environmental factors on the synthesis of scytonemin, a UV screening pigment, in a cyanobacterium (*Chroococcidiopsis* sp.). *Archives of Microbiology* 177: 322–331.

Dudman, W. F. 1977. The role of surface polysaccharides in natural environments. In: I. W. Sutherland (ed.), *Surface Carbohydrates of the Prokaryotic Cell*. Academic Press, New York, pp. 357–414.

Ehling-Schulz, M., W. Bilger, and S. Scherer. 1997. UV-B induced synthesis of photoprotective pigments and extracellular polysaccharides in the terrestrial cyanobacterium *Nostoc commune*. *Journal of Bacteriology* 179: 1940–1945.

Ehling-Schulz, M. and S. Scherer. 1999. UV protection in cyanobacteria. *European Journal of Phycology* 34: 329–338.

Fischer, D., U. G. Schlösser, and P. Pohl. 1997. Exopolysaccharide production by cyanobacteria grown in closed photobioreactors and immobilized using white cotton towelling. *Journal of Applied Phycology* 9: 205–213.

Flaibani, A., Y. Olsen, and T. J. Painter. 1989. Polysaccharides in desert reclamation: Composition of exocellular proteoglycan complexes produced by filamentous blue-green and unicellular green edaphic algae. *Carbohydrate Research* 190: 235–248.

Fleming, E. D. and R. W. Castenholz. 2007. Effects of periodic desiccation on the synthesis of the UV-screening compound, scytonemin, in cyanobacteria. *Environmental Microbiology* 9: 1448–1455.

Fleming, E. D. and R. W. Castenholz. 2008. Effects of nitrogen source on the synthesis of the UV-screening compound, scytonemin, in the cyanobacterium *Nostoc punctiforme* PCC 73102. *FEMS Microbiology Ecology* 63: 301–308.

Garcia-Pichel, F. and R. W. Castenholz. 1991. Characterization and biological implication of scytonemin, a cyanobacterial sheath pigment. *Journal of Phycology* 27: 395–409.

Garcia-Pichel, F., N. D. Sherry, and R. W. Castenholz. 1992. Evidence for a UV sunscreen role of the extracellular pigment scytonemin in the terrestrial cyanobacterium *Chlorogloeopsis* sp. *Photochemistry and Photobiology* 56: 17–23.

Gloaguen, V., H. Morvan, and L. Hoffmann. 1995. Released and capsular polysaccharides Oscillatoriaceae (Cyanophyceae, Cyanobacteria). *Algological Studies* 78: 53–69.

Grant, C. S. and J. W. Louda. 2013. Scytonemin-imine, a mahogany-colored UV/Vis sunscreen of cyanobacteria exposed to intense solar radiation. *Organic Geochemistry* 65: 29–36.

Grilli Caiola, M., D. Billi, and E. I. Friedmann. 1996. Effect of desiccation on envelopes of the cyanobacterium *Chroococcidiopsis* sp. (Chroococcales). *European Journal of Phycology* 31: 97–105.

Grilli Caiola, M., R. Ocampo-Friedmann, and E. I. Friedmann. 1993. Cytology of long-term desiccation in the desert cyanobacterium *Chroococcidiopsis* sp. (Chroococcales). *Phycologia* 32: 315–322.

Gröniger, A., R. P. Sinha, M. Klisch, and D.-P. Häder. 2000. Photoprotective compounds in cyanobacteria, phytoplankton and macroalgae—A database. *Journal of Photochemistry and Photobiology B: Biology* 58: 115–122.

Häder, D.-P., H. D. Kumar, R. C. Smith, and R. C. Worrest. 2007. Effects of solar UV radiation on aquatic ecosystems and interactions with climate change. *Photochemical and Photobiological Sciences* 6: 267–285.

Häder, D.-P. and R. P. Sinha. 2005. Solar ultraviolet radiation-induced DNA damage in aquatic organisms: Potential environmental impact. *Mutation Research* 571: 221–233.

Hargreaves, A., F. A. Taiwo, O. Duggan, S. H. Kirk, and S. I. Ahmad. 2007. Near ultraviolet photolysis of b-phenylpyruvic acid generates free radicals and results in DNA damage. *Journal of Photochemistry and Photobiology B: Biology* 89: 110–116.

Helbling, E. W., K. Gao, H. Ai, Z. Ma, and V. E. Villafañe. 2006. Differential responses of *Nostoc sphaeroides* and *Arthrospira platensis* to solar ultraviolet radiation exposure. *Journal of Applied Phycology* 18: 57–66.

Helms, G. L., R. E. Moore, W. P. Niemczura, G. M. L. Patterson, K. B. Tomer, and M. L. Gross. 1988. Scytonemin A, a novel calcium antagonist from a blue—Green alga. *Journal of Organic Chemistry* 53: 1298–1307.

Hill, D. R., A. Peat, and M. Potts. 1994. Biochemistry and structure of the glycan secreted by desiccation-tolerant *Nostoc commune* (cyanobacteria). *Protoplasma* 182: 126–148.

Hodgson, D. A., W. Vyverman, E. Verleyen, P. R. Leavitt, K. Sabbe, A. H. Squier, and B. J. Keely. 2005. Late Pleistocene record of elevated UV radiation in an Antarctic lake. *Earth and Planetary Science Letters* 236: 765–772.

Hoiczyk, E. 1998. Structural and biochemical analysis of the sheath of *Phormidium uncinatum*. *Journal of Bacteriology* 180: 3923–3932.

Hoiczyck, E. and A. Hansel. 2000. Cyanobacterial cell walls: News from an unusual prokaryotic envelope. *Journal of Bacteriology* 182: 1191–1199.

Hu, C., Y. Liu, B. S. Paulsen, D. Petersen, and D. Klaveness. 2003. Extracellular carbohydrate polymers from five desert soil algae with different cohesion in the stabilization of fine sand grain. *Carbohydrate Polymers* 54: 33–42.

Hunsucker, S. W., B. M. Tissue, M. Potts, and R. F. Helm. 2001. Screening protocol for the ultraviolet-protective pigment scytonemin. *Analytical Biochemistry* 288: 227–230.

Itoh, K., K. I. Tong, and M. Yamamoto. 2004. Molecular mechanism activating Nrf2-Keap1 pathway in regulation of adaptive response to electrophiles. *Free Radical Biology and Medicine* 36: 1208–1213.

Itoh, T., M. Koketsu, N. Yokota, S. Touho, M. Ando, and Y. Tsukamasa. 2014. Reduced scytonemin isolated from Nostoc commune suppresses LPS/IFNγ-induced NO production in murine macrophage RAW264 cells by inducing hemeoxygenase-1 expression via the Nrf2/ARE pathway. *Food and Chemical Toxicology* 69: 330–338.

Itoh, T., R. Tsuzuki, T. Tanaka, M. Ninomiya, Y. Yamaguchi, H. Takenaka, M. Ando, Y. Tsukamasa, and M. Koketsu. 2013. Reduced scytonemin isolated from *Nostoc commune* induces autophagic cell death in human T-lymphoid cell line Jurkat cells. *Food and Chemical Toxicology* 60: 76–82.

Kawaguchi, T. and A. W. Decho. 2000. Biochemical characterization of cyanobacterial extracellular polymers (EPS) from modern marine stromatolites. *Preparative Biochemistry and Biotechnology* 30: 321–330.

Kawaguchi, T. and A. W. Decho. 2002. Isolation and biochemical characterization of extracellular polymeric secretions (eps) from modern soft marine stromatolites (bahamas) and its inhibitory effect on CaCO₃ precipitation. *Preparative Biochemistry and Biotechnology* 32: 51–63.

Kerr, J. B. and C. T. McElroy. 1993. Evidence for large upward trends of ultraviolet-B radiation linked to ozone depletion. *Science* 262: 1032–1034.

Kim, S. T. and A. Sancar. 1995. Photorepair of non adjacent pyrimidine dimers by DNA photolyase. *Photochemistry and Photobiology* 61: 171–174.

Kripke, M. L., P. A. Cox, L. G. Alas, and D. B. Yarosh. 1992. Pyrimidine dimers in DNA initiate systemic immune suppression in UV-irradiated mice. *Proceedings of National Academies of Sciences USA* 89: 7516–7520.

Kylim, H. 1927. Über die Karotinoide Farbstoffe der Algen. Hoppe-Scyler'szeitschr. *Physiological Chemistry* 166: 33–77.

Kylim, H. 1937. Über die Farbstoffe und die Farbe der cyanophyceen. *Fysiogr. Sällsk. Förhandl.* 7: 131–158.

Lee, S. E., S. I. Jeong, H. Yang, S. H. Jeohg, Y. P. Jang, C. S. Park, J. Kim, and Y. S. Park. 2012. Extract of *Salvia miltiorrhiza* (Danshen) induces Nrf2-mediated heme oxygenase-1 expression as a cytoprotective action in RAW264.7 macrophages. *Journal of Ethnopharmacology* 139: 541–548.

Leppard, G. G., A. Heissenberger, and G. J. Herndl. 1996. Ultrastructure of marine snow, I: Transmission electron microscopy methodology. *Marine Ecology Progress Series* 135: 289–298.

Lubin, D. and E. H. Jensen. 1995. Effects of clouds and stratospheric ozone depletion on ultaviolet radiation trends. *Nature* 377: 710–713.

Malla, S. and M. O. A. Sommer. 2014. Sustainable route to produce scytonemin precursor using *Escherichia coli*. *Green chemistry* 16: 3255–3265.

Mancuso Nichols, C. A., J. Guezennec, and J. P. Bowman. 2005. Bacterial exopolysaccharides from extreme marine environments with special consideration of the southern ocean, sea ice, and deep-sea hydrothermal vents: A review. *Marine Biotechnology* 7: 253–271.

Marra, M., A. Palmeri, A. Ballio, A. Segre, and M. E. Slodki. 1990. Structural characterization of the exocellular polysaccharide from *Cyanospira capsulata*. *Carbohydrate Research* 197: 338–344.

Matsui, K., E. Nazifi, Y. Hirai, N. Wada, S. Matsugo, and T. Sakamoto. 2012. The cyanobacterial UV-absorbing pigment scytonemin displays radicals scavenging activity. *Journal of General and Applied Microbiology* 58: 137–144.

Mazur-Marzec, H., J. Meriluoto, and M. Plinśki. 2006. The degradation of the cyanobacterial hepatotoxin nodularin (NOD) by UV radiation. *Chemosphere* 65: 1388–1395.

Nägeli, C. 1849. Gattungeneinzelliger Algen, Physiologisch und Systematisch Bearbeitet. *Neue Denkschrift Allgemeine Schweiz Nature Geselschaft* 10: 1–138.

Nägeli, C. and S. Schwenderer. 1877. *Das Mikroskop*, 2nd edn. Willhelm Engelmann, Leipzig, Germany.

Neu, T. R. and K. C. Marshall. 1990. Bacterial polymers: Physicochemical aspects of their interactions at interfaces. *Journal of Biomaterial Applications* 5: 107–127.

Nguyen, T., C. S. Yang, and C. B. Pickett. 2004. The pathways and molecular mechanisms regulation Nrf2 activation in response to chemical stress. *Free Radical Biology and Medicine* 37: 433–441.

Nicolaus, B., A. Panico, L. Lama, I. Romano, M. C. Manca, A. De Giulio, and A. Gambacorta. 1999. Chemical composition and production of exopolysaccharides from representative members of heterocystous and non-heterocystous cyanobacteria. *Phytochemistry* 52: 639–647.

Pakker, H., C. A. C. Beekman, and A. M. Breeman. 2000. Efficient photoreactivation of UVBR-induced DNA damage in the sublittoral macroalga *Rhodymenia pseudopalmata* (Rhodophyta). *European Journal of Phycology* 35: 109–114.

Parikh, A. and D. Madamwar. 2006. Partial characterization of extracellular polysaccharides from cyanobacteria. *Bioresource Technology* 97: 1822–1827.

Park, E. J., Y. M. Kim, S. W. Park, H. J. Kim, J. H. Lee, D. U. Lee, and K. C. Chang. 2013. Induction of HO-1 through p38 MAPK/Nrf2 signaling pathway by ethanol extract of *Inula helenium* L. reduces inflammation in LPS-activated RAW264.7 cells and CLP-induced septic mice. *Food and Chemical Toxicology* 55: 386–395.

Peak, M. J., J. G. Peak, P. Moehring, and R. B. Webb. 1984. Ultraviolet action spectra for DNA dimer induction, lethality and mutagenesis in *Escherichia coli* with emphasis on the UV-B region. *Photochemistry and Photobiology* 40: 613–620.

Pentecost, A. 1993. Field relationships between scytonemin density, growth and irradiance in cyanobacteria occurring in low illumination regimes. *Microbial Ecology* 26: 101–110.

Pereira, S., A. Zille, E. Micheletti, P. Moradas-Ferreira, R. De Philippis, and P. Tamagnini. 2009. Complexity of cyanobacterial exopolysaccharides: Composition, structures, inducing factors and putative genes involved in their biosynthesis and assembly. *FEMS Microbiology Reviews* 33: 917–941.

Powell, J. T., A. D. Chatziefthimiou, S. A. Banack, P. A. Cox, and J. S. Metcalf. 2015. Desert crust microorganisms, their environment, and human health. *Journal of Arid Environments* 112:127–133.

Proteau, P. J., W. H. Gerwick, F. Garcia-Pichel, and R. W. Castenholz. 1993. The structure of scytonemin, an ultraviolet sunscreen pigment from the sheath of cyanobacteria. *Experimentia* 49: 825–829.

Quaite, F. E., B. M. Sutherland, and J. C. Sutherland. 1992. Action spectrum for DNA damage in alfalfa lowers predicted impact of ozone depletion. *Nature* 358: 577–578.

Quesada, A. and W. F. Vincent. 1997. Strategies of adaptation by Antarctic cyanobacteria to ultraviolet radiation. *European Journal of Phycology* 32: 335–342.

Rastogi, R. P. and R. P. Sinha. 2009. Biotechnological and industrial significance of cyanobacterial secondary metabolites. *Biotechnology Advances* 27: 521–539.

Rastogi, R. P., R. P. Sinha, and A. Incharoensakdi. 2013. Partial characterization, UV-induction and photoprotective function of sunscreen pigment, scytonemin from *Rivularia* sp. HKAR-4. *Chemosphere* 93: 1874–1878.

Rath, J. and S. P. Adhikary. 2007. Response of the estuarine cyanobacterium *Lyngbya aestuarii* to UV-B radiation. *Journal of Applied Phycology* 19: 529–536.

Robbins, R. A., J. Bauld, and D. J. Chapman. 1998. Chemistry of sheath of the cyanobacterium *Lyngbya aestuarii* LIEB. *Cryptogamie Algologie* 19(1–2): 169–178.

Ruas-Madiedo, P., J. Hugenholtz, and P. Zoon. 2002. An overview of the functionality of exopolysaccharides produced by lactic acid bacteria. *International Dairy Journal* 12: 163–171.

Satoh, K., M. Hirai, J. Nishio, T. Yamaji, Y. Kashino, and H. Koike. 2002. Recovery of photosynthetic systems during rewetting is quite rapid in a terrestrial cyanobacterium, *Nostoc commune*. *Plant and Cell Physiology* 43: 170–176.

Scherer, S., T. W. Chen, and P. Böger. 1986. Recovery of adeninenucleotide pools in terrestrial blue-green algae after prolonged drought periods. *Oecologia* 68: 585–588.

Scherer, S., A. Ernst, T. W. Chen, and P. Böger. 1984. Rewetting of drought-resistant blue-green algae: Time course of water uptake and reappearance of respiration, photosynthesis, and nitrogen fixation. *Oecologia* 62: 418–423.

Schmit, T. L. and N. Ahmad. 2007. Regulation of mitosis via mitotic kinases: New opportunities for cancer management. *Molecular Cancer Therapeutics* 6: 1920–1931.

Scott, C., R. L. Fletcher, and G. B. Bremer. 1996. Observations of the mechanisms of attachment of some marine fouling blue-green algae. *Biofouling* 10: 161–173.

Sergeev, V. N., L. M. Gerasimenko, and G. A. Zavarzin. 2002. The proterozoic history and present state of cyanobacteria. *Microbiology* 71: 623–637.

Sinclair, C. and B. A. Whitton. 1977. Influence of nutrient deficiency on hair formation in the Rivulariaceae. *British Phycological Journal* 12: 297–313.

Singh, S. P., D.-P. Häder, and R. P. Sinha. 2010. Cyanobacteria and ultraviolet radiation (UVR) stress: Mitigation strategies. *Ageing Research Reviews* 9: 79–90.

Sinha, R. P., Dautz, M., and D.-P. Häder. 2001. A simple and efficient method for the quantitative analysis of thymine dimers in cyanobacteria, phytoplankton and macroalgae. *Acta Protozoologica* 40: 187–195.

Sinha, R. P. and D.-P. Häder. 1996. Photobiology and ecophysiology of rice field cyanobacteria. *Photochemistry and Photobiology* 64: 887–896.

Sinha, R. P. and D.-P. Häder. 2002. UV-induced DNA damage and repair: A review. *Photochemical and Photobiological Sciences* 1: 225–236.

Sinha, R. P. and D.-P. Häder. 2008. UV protectants in cyanobacteria. *Plant Science* 174: 278–289.

Sinha, R. P., M. Klisch, A. Gröniger, and D.-P. Häder. 1998. Ultraviolet absorbing/screening substances in cyanobacteria, phytoplankton and macroalgae. *Photochemistry and Photobiology* 47: 83–94.

Sinha, R. P., M. Klisch, A. Vaishampayan, and D.-P. Häder. 1999. Biochemical and spectroscopic characterization of the cyanobacterium *Lyngbya* sp. inhabiting Mango (*Mangifera indica*) trees: Presence of an ultraviolet-absorbing pigment, Scytonemin. *Acta Protozoologica* 38: 291–298.

Sinha, R. P., A. Kumar, M. B. Tyagi, and D.-P. Häder. 2005. Ultraviolet-B-induced destruction of phycobiliproteins in cyanobacteria. *Physiology and Molecular Biology of Plants* 11: 313–319.

Smith, R. C., B. B. Prézelin, K. S. Baker, R. R. Bidigare, N. P. Boucher, T. Coley, D. Karentz et al. 1992. Ozone depletion: Ultraviolet radiation and phytoplankton biology in Antarctic waters. *Science* 255: 952–959.

Solheim, B., M. Zielke, J. W. Bjerke, and J. Rozema. 2006. Effects of enhanced UVB radiation on nitrogen fixation in arctic ecosystems. *Plant Ecology* 182: 109–118.

Sorrels, C. M., P. J. Proteau, and W. H. Gerwick. 2009. Organization, evolution, and expression analysis of the biosynthetic gene cluster for scytonemin, a cyanobacterial UV-absorbing pigment. *Applied and Environmental Microbiology* 75: 4861–4869.

Soule, T., F. Garcia-Pichel, and V. Stout. 2009a. Gene expression patterns associated with the biosynthesis of the sunscreen scytonemin in *Nostoc punctiforme* ATCC 29133 in response to UVA radiation. *Journal of Bacteriology* 191: 4639–4646.

Soule, T., K. Palmer, Q. Gao, R. M. Potrafka, V. Stout, and F. Garcia-Pichel. 2009b. A comparative genomics approach to understanding the biosynthesis of the sunscreen scytonemin in cyanobacteria. *BMC Genomics* 10: 336–345.

Soule, T., V. Stout, W. D. Swingley, J. C. Meeks, and F. Garcia-Pichel. 2007. Molecular genetics and genomic analysis of scytonemin biosynthesis in *Nostoc punctiforme* ATCC 29133. *Journal of Bacteriology* 189: 4465–4472.

Stein, B., H. J. Rahmsdorf, A. Steffen, M. Litfin, and P. Herrlich. 1989. UV-induced DNA damage is an intermediate step in UV-induced expression of human immunodeficiency virus type 1 collagenase C-fos and metallothionein. *Molecular and Cellular Biology* 9: 5169–5181.

Stevenson, C. S., E. A. Capper, A. K. Roshak, B. Marquez, C. Eichman, J. R. Jackson, M. Mattern, W. H. Gerwick, R. S. Jacobs, and L. A. Marshall. 2002a. The identification and characterization of the marine natural product scytonemin as a novel antiproliferative pharmacophore. *Journal of Pharmacology and Experimental Therapeutics* 303: 858–866.

Stevenson, C. S., E. A. Capper, A. K. Roshak, B. Marquez, K. Grace, W. H. Gerwick, R. S. Jacobs, and L. A. Marshall. 2002b. Scytonemin—A marine natural product inhibitor of kinases key in hyperproliferative inflammatory diseases. *Inflammation Research* 51: 112–114.

Stewart, I., P. J. Schluter, and G. R. Shaw. 2006. Cyanobacterial lipopolysaccharides and human health—A review. *Environmental Health* 5: 7.

Sutherland, I. W. 1999. Polysaccharases for microbial exopolysaccharides. *Carbohydrate Polymers* 38: 310–328.

Tamaru, Y., Y. Takani, T. Yoshida, and T. Sakamoto. 2005. Crucial role of extracellular polysaccharides in desiccation and freezing tolerance in the terrestrial cyanobacterium *Nostoc commune*. *Applied and Environmental Microbiology* 71: 7327–7333.

Tease, B., U. J. Jurgens, J. R. Golecki, U. R. Heinrich, R. Rippka, and J. Weckesser. 1991. Fine-structural and chemical analyses on inner and outer sheath of the cyanobacterium *Gloeothece* sp. PCC 6909. *Antonie Van Leeuwenhoek* 59: 27–34.

Tease, B. E. and R. W. Walker. 1987. Comparative composition of the sheath of the cyanobacterium *Gloeothece* ATCC 27152 cultured with and without combined nitrogen. *Journal of General Microbiology* 133: 3331–3339.

Thajuddin, N. and G. Subramanian. 2005. Cyanobacterial biodiversity and potential applications in biotechnology. *Current Science* 89: 47–57.

Tirkey, J. and S. P. Adhikary. 2005. Cyanobacteria in biological soil crusts of India. *Current Science* 89: 515–521.

Vaishampayan, A., R. P. Sinha, and D.-P. Häder. 1998. Use of genetically improved nitrogen-fixing cyanobacteria in rice paddy fields: Prospects as a source material for engineering herbicide sensitivity and resistance in plants. *Botanica Acta* 111: 176–190.

Vaishampayan, A., R. P. Sinha, D.-P. Häder, T. Dey, A. K. Gupta, U. Bhan, and A. L. Rao. 2001. Cyanobacterial biofertilizers in rice agriculture. *Botanical Review* 67: 453–516.

Vincent, W. F. and S. Roy. 1993. Solar ultraviolet-B radiation and aquatic primary production: Damage, protection, and recovery. *Environmental Reviews* 1: 1–12.

Von der Gathen, P., M. Rex, N. R. P. Harris, D. Lucic, B. M. Knudsen, G. O. Braathen, H. DeBacker et al. 1995. Observational evidence for chemical ozone depletion over the Arctic in winter 1991–1992. *Nature* 375: 131–134.

Weatherhead, E. C. and S. B. Andersen. 2006. The search for signs of recovery of the ozone layer. *Nature* 441: 39–45.

Wright, D. J., S. C. Smith, V. Joardar, S. Scherer, J. Jervis, A. Warren, R. F. Helm, and M. Potts. 2005. UV irradiation and desiccation modulate the three dimensional extracellular matrix of *Nostoc commune* (cyanobacteria). *The Journal of Biological Chemistry* 280: 40271–40281.

Wulff, A., M. Mohlin, and K. Sundback. 2007. Intraspecific variation in the response of the cyanobacterium *Nodularia spumigena* to moderate UV-B radiation. *Harmful Algae* 6: 388–399.

Xue, L., Y. Zhang, T. Zhang, L. An, and X. Wang. 2005. Effects of enhanced ultraviolet-B radiation on algae and cyanobacteria. *Critical Reviews in Microbiology* 31: 79–89.

*chapter twenty-four*

# Biomedical application of carbohydrates from marine microbes

*Kannan Kamala, Pitchiah Sivaperumal, and Elangovan Dilipan*

## Contents

## 24.1 Introduction

Over the past three centuries, biological resources of terrestrial origin have been discovered and exploited for their bioactive metabolites. However, researchers across the world are in the process of exploring the oceans in search of bioactive compounds. The marine biosphere consists of a large range of ecosystems, such as microbial mats, hypersaline marine environments, shallow regions, deep sea hydrothermal vents, and Arctic and Antarctic sea ice. These distinct environments are characterized by their physical and chemical parameters such as temperature, pH, pressure, and the chemical compounds present. In addition, research on marine biological resources has gathered momentum, and the marine microbes have been proven the best sources for exploring natural products. These marine microbes, such as bacteria, fungi, and some other microalgae, from the marine environment are being intensively investigated for novel bioactive compounds. Generally, the chemical study of marine organisms will likely result in the discovery of new therapeutic compounds. In particular, it is well known that the chemical diversity of marine microbial compounds is very broad (Grabowski et al., 2008). While microorganisms from the sea have been discovered for many years, the development of therapeutic compounds from them has started only recently (Imhoff et al., 2011). Considerably, the majority of substances such as carbohydrates are

present in the form of polysaccharides in marine microbes. Moreover, these microorganisms composed of algae, sponges, mollusks, and echinoderms are rich sources of polysaccharides. In general, polysaccharides are composed of glycosidically linked sugar residues, but these molecules also contain polymeric saccharide structures linked by covalent bonds to amino acids, peptides, proteins lipids, and other structures. These microbial polysaccharides are multifunctional and can be divided into structural polysaccharides, intracellular polysaccharides, and extracellular polysaccharides or exopolysaccharides (EPSs). To date, several polysaccharide-producing marine bacterial strains have been studied, including extremophilic microorganisms, which has led to the discovery and isolation of novel macromolecules (Finore et al., 2014; Sinquin and Colliec-Jouault, 2014). Although polysaccharides are ubiquitous, their chemical structure varies significantly from one another. They have different roles within the cells, from structural attachment to numerous biological activities including interactions. Hence, studying the chemical structure and structure-related activities of polysaccharide remains a challenge for glycochemists and glycobiologists. Moreover, glycobiology has become a major research field, comprising the understanding of human diseases and the discovery of novel therapeutic compounds. This chapter presents an overview of our current knowledge on polysaccharide-producing marine microbes and their bioactivity.

## 24.2 Polysaccharides: A source for biomedical application

Polysaccharides are natural macromolecules composed of osidic monomers and are present in all organisms such as microorganisms, plants, and animals, where they are accessible in the form of glycogen, starch, and cellulose. However, when a polysaccharide contains only one kind of monosaccharide molecule, it is known as a homopolysaccharide, whereas those containing more than one kind of monosaccharide are called heteropolysaccharides. Some monosaccharide derivatives found in polysaccharides include the amino sugars (D-glucosamine and D-galactosamine) as well as their derivatives (N-acetylneuraminic acid and N-acetylmuramic acid), and simple sugar acids (glucuronic and iduronic acids). Polysaccharides have unique characters. Especially, marine microbial polysaccharides synthesized and secreted into the external environment or are synthesized extracellularly by cell-wall-anchored enzymes may be referred to as EPSs. The natural biocompatibility and apparent nontoxic nature of some of these microbial exopolysaccharides have prompted their uses in numerous medical applications, such as scaffolds or matrices in tissue engineering, wound dressing, and drug delivery, thus making them further attractive compared to polysaccharides obtained from plants and microalgae (Sutherland, 1998; Otero and Vincenzini, 2003; Rehm, 2010). Some biopolymers get gradually degraded *in vivo*, making them suitable for use in tissue replacement and controlled drug release (Rehm, 2010). However, the greatest potential of microbial EPSs is their use in pharmaceuticals and biomedicine, wherein traditional polymers fail to comply with the required degree of purity or lack some specific functional properties. Despite the great diversity of molecular structures already described for microbial EPSs, only a few have been industrially developed (Filomena et al., 2011).

Mostly in the form of biofilms, microorganisms occupy only 10% of their dry mass, while the remaining 90% is occupied by the matrix (EPS). This matrix helps the organism to acclimatize with habitat and impart protection against stress such as desiccation, biocides, antibiotics, heavy metals, and UV radiation (Flemming and Wingender, 2010). EPSs mainly consist of polysaccharides, proteins, extracellular DNA, and lipids (Flemming et al., 2007). Polysaccharides are often considered the most abundant component of EPSs (Wingender et al., 1999). EPSs are either homopolysaccharides (made up of one type of sugar) or heteropolysaccharides (with two or more sugars). Most of the bacterial polysaccharides are heteropolysaccharides (Sutherland, 1990).

## 24.3 Bioactivity of polysaccharides from marine microbes

### 24.3.1 Importance of polysaccharides from marine microbes

Marine microorganisms produce novel chemicals to withstand the extreme variations in salinity, temperature, and pressure prevailing in their environment, and the chemicals produced are distinct in terms of diversity, such as structural and functional features (Kathiresan et al., 2008). Secondary metabolites produced by marine bacteria have yielded pharmaceutical products such as novel anti-inflammatory agents (pseudopterosins, topsentins, scytonemin, and manoalide); anticancer agents (bryostatins, discodermolide, sarcodictyin, and eleutherobin); and antibiotics (marinone). The role of probiotic bacteria, such as *Bifidobacteria* and *Lactobacilli*, is mainly in the control of pathogenic microbes through production of the antibacterial protein bacteriocin (DeVugst and Vandamme, 1994; Kathiresan and Thiruneelakandan, 2008) and anticancer substances (Wollowski et al., 2001). Microbial polysaccharides are of growing interest for many sectors of industry. The biggest advantages of microbial polysaccharides over plants polysaccharides are their novel functions and constant chemical and physical properties.

Polysaccharides of bacterial origin are very important in the pharmaceutical industry, and they have been proven to have antioxidant, antitumor, immune stimulatory, antiulcer, and blood-cholesterol-lowering activities (Welman and Maddox, 2003). Moreover, polysaccharides such as EPSs secreted by marine bacteria contain novel combinations, which have potential applications in various biomedical sectors. These EPSs are industrially used as thickeners, stabilizers, and gelling agents in food products also. Previously, they were used as biomedical agents, and there was a growing interest in their biological functions like antitumor, antioxidant, or prebiotic activities (Liu et al., 2010). Recently, Paniagua-Michel et al. (2014) have pointed out that microbial EPSs from marine and deep-sea environments are attracting major interest due to their structural and functional diversity as molecules active on surface and an alternative biomass to replace fossil forms of carbon. These metabolic bioactive products find application in a number of industries and processes, for example, pharmacology, food processing, and bioremediation of oil-polluted environments. Similarly, Finore et al. (2014) also reported that a few EPSs derived from marine microbes have industrial applications in the food, biotechnological, environmental, or health industry.

### 24.3.2   Polysaccharide-producing marine microbes

Polysaccharides derived from microorganisms, including archaea, bacteria, fungi, and yeast, are present in enormous amounts (Sutherland, 2001). In addition, there has been growing interest in isolating new EPS-producing bacteria from marine environments, predominantly from extreme ones (Manusco Nichols et al., 2005a,b). Halophilic strains such as *Halomonas maura* (Arias et al., 2003), *H. ventosae* (Martinez-Canovas et al., 2004), *H. alkaliantarctica* (Poli et al., 2007), *Hahella chejuensis* (Poli et al., 2010), and the archaeal halophilic strain *Haloferax mediterranei* (Anton et al., 1988; Parolis et al., 1996), which were isolated from hyper-saline environments, have also been shown to produce enormous amounts of EPSs; a few of them are sulfated (Poli et al., 2010). The marine thermophilic anaerobe strains *Sulfolobus*, *Thermococcus*, and *Thermotoga* also produce EPSs (Vanfossen et al., 2008). Particularly, *Thermococcus litoralis* produced an EPS that contained sulfate and phosphorus substituents (Rinker and Kelly, 2000), and one of the important archaeal strains, *Sulfolobus solfataricus,* has also been shown to produce a sulfated polysaccharide (Nicolaus et al., 1993). *Bacillus thermodenitrificans*, *Geobacillus* sp., and *B. licheniformis* thermophilic strains have been isolated from shallow marine hydrothermal vents of Vulcano Island (Italy) and their polysaccharides also have been analyzed (Poli et al., 2010).

The marine bacterial strains *Paracoccus zeaxanthinificiens* spp. *Payriae* and *Vibrio* sp. RA29 produce sulfated polysaccharides, and *Vibrio* sp. MO245 produces a polymer very similar to the *Vibrio diabolicus* one (Guezennec et al., 2011). Furthermore, Mancuso Nichols et al. (2004) explained the production of EPS by the marine strains *Pseudoalteromonas* CAM025 and CAM036 isolated from Antarctic seawaters and sea ice. Other *Pseudoalteromonas* strains produce diverse EPSs; SM9913 isolated from deep-sea sediments produces an acetylated EPS (Qin et al., 2007). Recently, some other strains from the Arctic sea ice have been shown to produce EPSs with cryoprotective effect (Liu et al., 2013).

Though marine microbes possess immense diversity and the unique capacity to produce natural products, their importance for marine biotechnology was broadly realized by the scientific community only recently. Marine microorganisms have been studied for various biotechnological applications ranging from bioactive compounds for pharmaceuticals to other high-value products such as enzymes, pigments, nutraceuticals, and cosmetics (Imhoff et al., 2011). The structural diversity, combined with various interesting properties, of EPSs produced by marine bacteria could be explored for biotechnological applications. The EPSs of several marine bacteria have been studied; for example, the hyaluronic acid-like polysachharide produced from the marine bacterium *Vibrio diabolicus* has been commercialized with the trade name "Hyalurift" and known for its restoration of bone integrity (De Morais et al., 2010). Poli et al. (2010) reviewed the bacterial EPSs from extreme marine environments, such as marine hot springs and hydrothermal vents (*Pseudoalteromonas* strain 721, *Alteromonas macleodii* spp. *Fijiensis*, *Thermococcus litoralis*, *Geobacillus* sp. strain 4004, *Bacillus thermodenitrificans* strain B3-72, *Bacillus licheniformis* strain B3-15), as well as cold (*Pseudoalteromonas* strains SM9913, CAM025, and CAM036, *Colwellia psychrerythraea* strain 34H) and hypersaline (*H. mediterranei*, *H. chejuensis*, *Halomonas alkaliantarctica* strain CRSS) marine environments.

The marine biofilm is a good source of extracellular polymeric substances or mainly EPS-producing bacteria, but fewer reports are available from the characterization and properties of these EPSs when compared to bacterial EPS from other sources. Saravanan and Jayachandran (2008) have reported that the EPS of a biofilm-forming marine bacteria *Pseudoalteromonas ruthenica* has pseudoplastic rheology, and preliminary characterization of this EPS was performed using gas chromatography and FTIR (preliminary characterization of EPSs produced by a marine biofilm-forming bacterium *Pseudoalteromonas ruthenica* (SBT 033)). The extracellular materials can be used as stabilizers, gelling agents, adhesives, thickening agents, emulsifying agents, flocculants, and flushing agents (Becker et al., 1998).

### 24.3.3   Biological activities

EPSs secreted from marine bacteria contain novel combinations that have potential applications in various industries. Several Gram-positive and Gram-negative marine bacteria, such as *Bacillus*, *Halomonas*, *Planococcus*, *Alteromonas*, *Enterobacter*, *Zoogloea*, *Rhodococcus*, and *Cyanobacteria*, are well known to produce EPSs (Zukerberg et al., 1979; Nicolaus et al., 2000; Maugeri et al., 2002). Earlier studies have shown that the marine bacteria *Vibrio* sp. producing EPS have effective antitumor and immunostimulant activities, which are very important from a medicinal point of view (Okutani, 1984, 1985). Different properties of EPSs, such as thickening, adhesive, coagulating, stabilizing, and gelling, are used in various industries. EPSs with good viscosity and pseudo-plasticity are resistant to the extremities of temperature, pH, and salinity. Examples of such EPSs are emulsifiers produced by *H. chejuensis* gen. nov., sp. (Lee et al., 2001), *H. mediterranei*, and *Alteromonas* sp. 1545 producing acidic EPSs with thickening property (Talmont et al., 1991) and an

EPS from *Cyanothece* sp. ATCC 51142 with gelling property (Shah et al., 2000).

Moreover, sulfation of this anionic and highly branched heteropolysaccharide performed in an ionic liquid would facilitate the production of new molecules of high specificity for biological targets such as tissue engineering or regenerative medicine. The major metabolite, macrolactin-A, reduces B16-F10 murine melanoma cancer cells and the mammalian herpes simplex virus (types I and II), and protects T lymphocytes against human immunodeficiency virus (HIV) replication (Carte, 1996). Kahalalide F (KF) is a depsipeptide isolated from the mollusk *Elysia rubefescens* from Hawaii, which is believed to be synthesized by microbes associated with the animal. *In vitro*, KF showed activity against solid tumors with an interesting pattern of selectivity in prostate cancer cell lines. Moreover, extensive *in vivo* work has shown that the agent has activity in breast and colon cancers. An interesting patent was filed by Guezennec et al. (2002) on the bacterial species *Vibrio diabolicus* for the production and uses of water-soluble polysaccharides. The molecular weight of the EPS from this marine organism is 800,000 Da and the EPS is similar to heparin, and hence finds important applications in pharmaceutics. Another heparin-like EPS is secreted by the *Vibrio* sp., which has been reported to have anti-thrombotic and anticoagulant properties and can be utilized as an antiviral, antitumor, or antithrombotic agent (Guezennec et al., 2002). In addition, *V. diabolicus* has been reported to secrete HE800 EPS, which possesses significant bone healing property by the formation of extracellular matrix for the direct adhesion of osteoprogenitor cells, osteoblasts, and pericytes, and provides the ideal conditions required for bone growth and healing (Zanchetta et al., 2003). A few reports are available on EPS production by the marine *Bacillus* sp. isolated from a *Bacillus* strain B3-15 (halophilic, thermotolerant) from the shallow marine hot springs at Volcano Island, Italy (Maugeri et al., 2002). Another *Bacillus* strain B3-72 isolated from the same shallow vent produced a structurally different EPS (Nicolaus et al., 2000). It indicates the diversity of the EPSs produced by same genus from similar environments. Matsuda et al. (2003) reported that a sulfated EPS produced by *Pseudomonas* sp. showed cytotoxic effect toward human cancer cell lines such as MT-4, and these findings have resulted in further interest in this polysaccharide as a new anticancer drug suitable for clinical trials.

Arena (2004) also claimed that EPS-2 was successful in partially restoring the immunological disorders caused by by HSV-2. The progress in the field of marine biotechnology has attracted many researchers toward marine microbial surface-active agents and EPSs. The EPS produced by *B. licheniformis* and *Geobacillus thermodenitrificans* can act as strong stimulators for cell-mediated immunity. Such immunomodulatory activity can play a major role in the treatment of immune-compromised patients. In addition, Arena et al. (2006) reported a novel type of EPS-1 polysaccharide-producing thermo-tolerant strain *B. licheniformis* isolated from a shallow marine hot springs in Volcano Island. *Geobacillus thermodenitrificans*, a bacterial isolate recovered from the vent of Vulcano Island, produces EPS-2 that shows immunomodulatory and antiviral effect by hindering HSV-2 replication in human peripheral blood mononuclear cells. The concentration or dose of EPS-2 is important for its immunomodulatory and antiviral activity (Arena et al., 2009). Furthermore, the soil-derived actinobacteria of terrestrial origin have provided a major pharmaceutical resource for the discovery of antibiotics and related bioactive compounds. However, marine actinobacteria have received attention only recently. Recently, Manivasagan et al. (2013) reported EPS production from marine actinobacteria and their biological activities.

## 24.4  Conclusion

Natural bioactive molecules are the focus of interest as new therapeutic drugs. The marine environment holds a variety of natural products and can indeed be a treasure chest for pharmaceutical purposes. Especially, marine microbes can survive on low nutrients and in unfavorable environments due to the EPSs in the surrounding environment. In addition to their diverse marine origins (animals, seaweeds, invertebrates), microorganisms give glycopolymers with original chemical structures and promising biological activities. Basic and applied research efforts in this marine-synthesized biomedical field require close collaboration between biologists and chemists as well as expertise in marine microbiology, biology, biochemistry, chemistry, and computational sciences to fulfill screenings, structural characterization, and bioactivity studies. Owing to the diverse chemical ecology in marine organisms, especially marine microbes hold great promise for providing potent, cheaper, and safer therapeutic drugs, which deserve broad exploration.

## *Acknowledgment*

The authors are grateful to Director CIFE, ICAR—Deemed University, Mumbai, Dr. B. B. Nayak, PHT Department, CIFE, and Dr. R. Rajaram, Department of Marine Science, Bharathidasan University, for giving valuable suggestions.

# References

Anton, J., I. Meseguer, and F. Rodriguez Valera. 1988. Production of an extracellular polysaccharide by *Haloferax mediterranei*. *Applied and Environmental Microbiology* 54: 2381–2386.

Arena, A. 2004. Exopolysaccharides from marine thermophilic bacilli induce a Thl cytokine profile in human PBMC. *Clinical Microbiology and Infection* 10: 366.

Arena, A., C. Gugliandolo, G. Stassi, B. Pavone, D. Iannello, G. Bisignano, and T. L. Maugeri. 2009. An exopolysaccharide produced by *Geobacillus thermodenitrificans* strain B3-72: Antiviral activity on immunocompetent cells. *Immunology Letters* 123: 132–137.

Arena, A., T. L. Maugeri, B. Pavone, D. Iannello, C. Gugliandolo, and G. Bisignano. 2006. Antiviral and immunoregulatory effect of a novel exopolysaccharide from a marine thermotolerant *Bacillus licheniformis*. *International Immunopharmacology* 6: 8–13.

Arias, S., A. Del Moral, M. R. Ferrer, R. Tallon, E. Quesada, and V. Bejar. 2003. Mauran, an exopolysaccharide produced by the halophilic bacterium *Halomonas maura*, with a novel composition and interesting properties for biotechnology. *Extremophiles* 7: 319–326.

Becker, A., F. Katzen, A. Puhler, and L. Ielpi. 1998. Xanthum gum biosynthesis and application: Biochemical/genetic perspective. *Applied Microbiology and Biotechnology* 50: 145–152.

Carte, B. K. 1996. Biomedical potential of marine natural products. *BioScience* 46: 271–286.

De Morais, M. G., C. Stillings, R. Dersch, M. Rudisile, P. Pranke, J. A. V. Costa, and J. Wendorff. 2010. Preparation of nanofibers containing the icroalga *Spirulina* (*Arthrospira*) *Bioresource Technology* 101: 2872–2876.

DeVugst, L. and E. J. Vandamme. 1994. *Bacteriocins of Lactic Acid Bacteria: Microbiology, Genetics, and Application*. Blackie Acadamic & Profession. London, U.K., Vol. 75, pp. 140174–140179.

Finore, I., P. Di Donato, V. Mastascusa, B. Nicolaus, and A. Poli. 2014. Fermentation technologies for the optimization of marine microbial exopolysaccharide production. *Marine Drugs* 12: 3005–3024.

Flemming, H., T. Neu, and D. Wozniak. 2007. The EPS matrix: The "House of Biofilm cells." *Journal of Bacteriology* 189: 7945–7947.

Flemming, H. and J. Wingender. 2010. The biofilm matrix. *Nature Reviews Microbiology* 8: 623–633.

Filmena, F., V. D. Alves, and A. M. Maria. 2011. Advances in bacterial exopolysaccharides: From production to biotechnological applications. *Trends in Biotechnology* 29: 388–398.

Grabowski, K., K. H. Baringhaus, and G. Schneider. 2008. Scaffold diversity of natural products: Inspiration for combinatorial library design. *Natural Product Report* 25: 892–904.

Guezennec, J., X. Moppert, G. Raguénès, L. Richert, B. Costa, and C. Simon-Colin. 2011. Microbial mats in French Polynesia and their biotechnological applications. *Process Biochemistry* 46: 16–22.

Guezennec, J., P. Pignet, G. Raguénès, and H. Rougeaux. 2002. Marine bacterial strain of the genus *Vibrio*, water-soluble polysaccharides produced by said strain and their uses. US patent 6436680.

Imhoff, J. F., A. Labes, and J. Wieses. 2011. Bio-mining the microbial treasures of the ocean: New natural products. *Biotechnology Advances* 29: 468–482.

Kathiresan, K., M. A. Nabeel, and S. Manivannan. 2008. Bioprospecting of marine organisms for novel bioactive compounds. *Scientific Transaction Environmental Technovation* 1: 107–120.

Kathiresan, K. and G. Thiruneelakandan. 2008. Prospects of lactic acid bacteria of marine origin. *Indian Journal of Biotechnology* 7: 170–177.

Lee, H. K., J. Chun, E. Y. Moon, S. H. Ko, D. S. Lee, H. S. Lee et al. 2001. *Hahella chejuensis* gen. nov., sp. nov., an extracellular-polysaccharide-producing marine bacterium. *International Journal of Systematic Evolutionary Microbiology* 2: 661–666.

Liu, C., J. Lu, J. Lu, Y. Liu, F. Wang, and M. Xiao. 2010. Isolation, structural characterization and immunological activity of an exopolysaccharide produced by *Bacillus licheniformis* 8-37-0-1. *Bioresource Technology* 101: 5528–5533.

Liu, S. B., X. L. Chen, H. L. He, X. Y. Zhang, B. B. Xie, Y. Yu et al. 2013. Structure and ecological roles of a novel exopolysaccharide from the Arctic Sea Ice bacterium *Pseudoalteromonas* sp. strain SM20310. *Applied and Environmental Microbiology* 79: 224–230.

Mancuso Nichols, C. A., J. P. Bowman, and J. Guezennec. 2005a. Effects of incubation temperature on growth and production of exopolysaccharides by an Antarctic Sea ice bacterium grown in batch culture. *Applied Environmental Microbiology* 71: 3519–3523.

Mancuso Nichols, C. A., S. Garon, J. P. Bowman, G. Raguenes, and L. Guezennec. 2004. Production of exopolysaccharides by Antarctic marine bacterial isolates. *Journal of Applied Microbiology* 96: 1057–1066.

Mancuso Nichols, C. A., J. Guezennec, and J. P. Bowman. 2005b. Bacterial exopolysaccharides from extreme marine environments with special consideration of the Southern Ocean, sea ice, and deep-sea hydrothermal vents: A review. *Marine Biotechnology* 7: 253–271.

Manivasagan, P., P. Sivasankar, J. Venkatesan, K. Senthilkumar, K. Sivakumar, and S. K. Kim. 2013. Production and characterization of an extracellular polysaccharide from *Streptomyces violaceus* MM72. *International Journal of Biological Macromolecules* 59: 29–38.

Martinez Canovas, M. J., E. Quesada, I. Llamas, and V. Béjar. 2004. *Halomonas ventosae* sp. nov., a moderately halophilic, denitrifying, exopolysaccharide producing bacterium. *International Journal of Systematic Evolutionary Microbiology* 54: 733–737.

Matsuda, M., T. Yamori, M. Naitoh, and K. Okutani. 2003. Structural revision of sulfated polysaccharide B-1 isolated from a marine *Pseudomonas* species and its cytotoxic activity against human cancer cell lines. *Marine Biotechnology* 5: 13–19.

Maugeri, T. L., C. Gugliandolo, D. Caccamo, A. Panico, L. Lama, A. Gambacorta et al. 2002. Halophilic thermo tolerant *Bacillus* isolated from a marine hot spring able to produce a new exopolysaccharide. *Biotechnology Letters* 24: 515–529.

Nicolaus, B., M. C. Manca, I. Ramano, and L. Lama. 1993. Production of an exopolysaccharide from two thermophilic archaea belonging to the genus *Sulfolobus*. *FEMS Microbiology Letters* 109: 203–206.

Nicolaus, B., A. Panico, M. C. Manca, L. Lama, A. Gambacorta, T. L. Maugeri et al. 2000. A thermophilic *Bacillus* isolated from an Eolian shallow hydrothermal vent, able to produce exopolysaccharides. *Systematic and Applied Microbiology* 23: 426–32.

Okutani, K. 1984. Antitumor and immunostimulant activities of polysaccharide produced by a marine bacterium of the genus *Vibrio*. *Bulletin of the Japanese Society for the Science of Fish* 50: 1035–1037.

Okutani, K. 1985. Isolation and fractionation of an extracellular polysaccharide from marine *Vibrio*. *Bulletin of the Japanese Society for the Science of Fish* 51: 493–496.

Otero, A. and M. Vincenzini. 2003. Extracellular polysaccharide synthesis by *Nostoc* strains as affected by N source and light intensity. *Journal of Biotechnology* 102: 143–152.

Paniagua-Michel Jde, J., J. Olmos-Soto, and E. R. Morales-Guerrero. 2014. Algal and microbial exopolysaccharides: New insights as biosurfactants and bioemulsifiers. *Advances in Food and Nutrition Research* 73: 221–257.

Parolis, H., L. A. S. Parolis, I. F. Boan, F. Rodriguez Valera, G. Widmalm, M. C. Manca et al. 1996. The structure of the exopolysaccharide produced by the halophilic archaeon *Haloferax mediterranei* strain R4 (ATCC33500). *Carbohydrate Research* 295: 147–156.

Poli, A., G. Anzelmo, and B. Nicolaus. 2010. Bacterial exopolysaccharides from extreme marine habitats: Production, characterization and biological activities. *Marine Drugs* 8: 1779–1802.

Poli, A., E. Esposito, P. Orlando, L. Lama, A. Giordano, F. De Appolonia et al. 2007. *Halomonas alkaliantarctica* sp. nov., isolated from saline lake Cape Russell in Antarctica, an alkalophilic moderately halophilic, exopolysaccharide producing bacterium. *Systematic and Applied Microbiology* 30: 31–38.

Qin, G., L. Zhu, X. Chen, P. G. Wang, and Y. Zhang. 2007. Structural characterization and ecological roles of a novel exopolysaccharide from the deep-sea psychrotolerant bacterium *Pseudoalteromonas* sp. SM9913. *Microbiology* 153: 1566–1572.

Rehm, B. H. A. 2010. Bacterial polymers: Biosynthesis, modifications and applications. *Nature Review Microbiology* 8: 578–592.

Rinker, K. D. and R. M. Kelly. 2000. Effect of carbon and nitrogen sources on growth dynamics and exopolysaccharide production for the hyper thermophilic archaeon *Thermococcus litoralis* and bacterium *Thermotoga maritima*. *Biotechnology and Bioengineering* 69: 537–547.

Saravanan, P. and S. Jayachandran. 2008. Preliminary characterization of exopolysaccharides produced by a marine biofilm-forming bacterium *Pseudoalteromonas ruthenica* (SBT 033). *Letters in Applied Microbiology* 46: 1–6.

Shah, V., A. Ray, N. Garg, and D. Madamwar. 2000. Characterization of the extracellular polysaccharide produced by a marine cyanobacterium, *Cyanothece* sp. ATCC 51142, and its exploitation toward metal removal from solutions. *Current Microbiology* 40: 274–278.

Sinquin, C. and S. Colliec-Jouault. 2014. Marine polysaccharides and their applications in the field of health— Polysaccharides of macroalgae and their bioactive derivatives. *Chimie Pharmaceutique* BIO6250: 1–5.

Sutherland, I. W. 1990. *Biotechnology of Microbial Exopolysaccharides*. Cambridge University Press, Cambridge, U.K.

Sutherland, I. W. 1998. Novel and established applications of microbial polysaccharides. *Trends in Biotechnology* 16: 41–46.

Sutherland, I. W. 2001. Microbial polysaccharides from Gram negative bacteria. *International Dairy Journal* 11: 663–674.

Talmont, F., P. Vincent, T. F. Ontaine, J. Guezennec, D. Prieur, and B. Fournet. 1991. Structural investigation of an acidic exopolysaccharide from a deep-sea hydrothermal vent marine bacteria. *Food Hydrocolloids* 5: 171–172.

Vanfossen, A. L., D. L. Lewis, J. D. Nichols, and R. M. Kelly. 2008. Polysaccharide degradation and synthesis by extremely thermophilic anaerobes. *Annals of the New York Academic of Sciences* 1125: 322–337.

Welman, A. D. and I. S. Maddox. 2003. Exopolysaccharides from lactic acid bacteria: Perspectives and challenges. *Trends in Biotechnology* 21: 269–274.

Wingender, J., T. R. Neu, and H. C. Flemming. 1999. What are bacterial extracellular polymeric substances? In: Wingender, J., Neu, T. R., and Flemming, H.-C. (eds.), *Microbial Extracellular Polymeric Substances: Characterization, Structure, and Function*. Springer, Berlin, Germany, pp. 1–19.

Wollowski, I., G. Rechkemmer, and B. L. Pool-Zobel. 2001. Protective role of probiotics and prebiotics in colon cancer. *American Journal of Clinical Nutrition* 73: 451–455.

Zanchetta, P., N. Lagarde, and J. Guezennec. 2003. A new bone-healing material: A hyaluronic acid-like bacterial exopolysaccharide. *Calcified Tissue International* 72: 74–79.

Zukerberg, A., A. Diver, Z. Peeri, and D. L. Gutnick. 1979. Rosenberg E. Emulsifier of Arthrobacter RAG-1: Chemical and physical properties. *Applied and Environmental Microbiology* 37: 414–420.

## chapter twenty-five

# Antimicrobial properties of chitosan and chitosan derivatives

*Priyanka Sahariah, Martha Á. Hjálmarsdóttir, and Már Másson*

## Contents

## 25.1  Introduction

Microbial infections remain an important cause of disease and death in modern societies. Microbial contamination is also a serious concern in the areas of medicine, health care, food industry, and agriculture. The emergence of resistance to traditional antibiotics has led to an increased interest in alternatives such as antimicrobial peptides and synthetic antimicrobial polymers that can be used for therapy and disinfection. However, issues such as toxicity and cost of production still remain a significant challenge. Biopolymers and

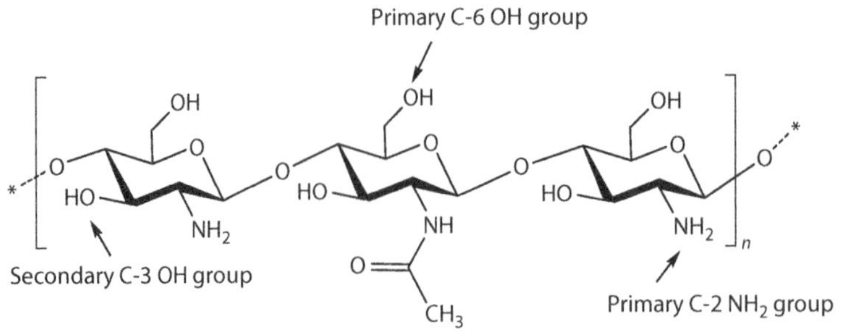

*Figure 25.1* Structure of chitosan showing reactive functional groups.

biopolymer derivatives can be used as effective antimicrobial agents and have therefore also been considered alternatives to traditional antibiotics and disinfectants. Amongst the available biopolymers, chitosan is recognized as having significant antimicrobial properties in addition to other desirable properties such as nontoxicity, biodegradability, and biocompatibility. Chitosan is derived from chitin, which is the most abundant natural polysaccharide of animal origin.[1] It is the basic constituent of the exoskeleton in insects and crustaceans like shrimps, crabs, and lobsters. Chitin differs from cellulose, which is the most abundant polysaccharide derived from plants, in that it contains an acetamido group ($-NHCOCH_3$) at the C-2 position instead of a hydroxyl group. Chemical deacetylation of chitin gives rise to its deacetylated form called chitosan, which occurs naturally only in certain fungi (Mucoraceae).[2] Chitosan is therefore a linear polymer composed of repeating units of glucosamine and N-acetyl glucosamine units that are linked through β-(1–4) glycosidic bonds. The chemical structure of chitosan contains three types of nucleophilic functional groups: a C-2 $NH_2$ group, a C-3 secondary OH group, and a C-6 primary OH group as shown in Figure 25.1. Chitosan can be polycationic, and the density of cationic charge on the polymer will depend on the pH and the number of free amino groups present. Highly deacetylated chitosan is usually insoluble in aqueous medium, but the increased density of such cationic charges at pH below 6.5 improves its water solubility and also contributes to biological properties such as antimicrobial activity. This polymer can be chemically modified to extend the range of conditions where the polymer is soluble and exhibits antimicrobial activity.

## 25.2   Antimicrobial properties of chitosan

Chitosan is available in various forms differing in average chain length and degree of deacetylation. There are three factors that mainly influence the antimicrobial activity of chitosan: molecular weight (Mw), degree of

acetylation, and pH of the medium. The antimicrobial activity of chitosan against various microbes expressed in terms of minimum inhibitory concentration/minimum lethal concentration (MIC/MLC) values are shown in Table 25.1.

### 25.2.1   Molecular weight of chitosan

Several studies have reported that the antimicrobial efficacy of chitosan is linked to the Mw of the polymer. Most studies have reported that chitosan is more effective as an antimicrobial agent than chitooligosaccharides (COS). Chitosan having average Mw > 60 kDa greatly inhibited the growth of several groups of Gram-positive and Gram-negative bacteria, especially the Gram-positive bifidobacteria.[4,8,9] Chitosan (300 kDa) and its (dodec-2-enyl)succinoyl derivatives were hydrolyzed under acidic conditions to obtain low Mw COS (4.6 kDa) and its derivatives, which showed activity against different kinds of bacteria, yeast, and filamentous fungi.[10] The Mw of the COS is also found to be a key factor influencing the inhibition of microorganism growth. COS with a narrow Mw range of 9.6–20 kDa were highly active, whereas materials with 2–4.2 kDa were moderately active and 0.73–1.52 kDa COS were nonactive against methicillin-resistant *Staphylococcus aureus* (MRSA).[11] COS within the range of 5–10 kDa, also displayed inhibitory effect against several strains of bifidobacteria (*B. adolescentis*, *B. bifidum*, *B. breve*, *B. catenulatum*, *B. infantis*, and *B. longum* spp.), while COS of <5 kDa remained inactive against these strains.[12] Additionally, chitosan, low Mw chitosan, and COS have been found to be very effective against several fungi and plant pathogens.[13,14] The dependence of activity on the Mw of the polymer is also found to be related to the cell wall composition of the bacteria tested. Chitosan with Mw of 470 kDa was found to inhibit Gram-positive and Gram-negative bacteria effectively, while chitosan with 1106 kDa was less active against Gram-positive but more effective against Gram-negative strains.[4] Similar observations were made in another study where chitosan showed greater inhibitory effect against *Staphylococcus*

*Table 25.1* MIC/MLC values for chitosan against various microorganisms

| Microorganism strain | MIC/MLC value (µg/mL or ppm) | Molecular weight (kDa) | pH | Degree of acetylation (%) |
|---|---|---|---|---|
| **Gram-positive bacteria** | | | | |
| *Staphylococcus aureus* | | | | |
| [a]NR[3] | 20 | NR | NR | NR |
| ATCC 29737[4] | >1000/800/>1000 | 28/224–740/1106 | 5.9 | NR |
| *Bacillus cereus* | | | | |
| LMG6924[5] | 60 | 43 | 5.5 | 6 |
| ATCC 21366[4] | >1000/500/>1000 | 28/224–470/1106 | 5.9 | NR |
| ATCC 14579[6] | >2000/1000/80/>2000/1330/790 | 2.3/28.4/98.3/11.9/42.5/224 | 6 | 16/16/16/48/48/48 |
| *Listeria monocytogenes* | | | | |
| NR[7] | 150 | 49–1100 | 6 | 2–53 |
| LMG13305[5] | 200 | 43 | 5.5 | 6 |
| Scott A[4] | 800/≥1000 | 470–740/1100–1670 | 5.9 | NR |
| *Bacillus megaterium* | | | | |
| KCTC 3007[4] | 800/500/800/500/800 | 28/220–470/746/1106/1670 | 5.9 | NR |
| *Lactobacillus brevis* | | | | |
| IFO 13109[4] | ≥1000/500 | 224–746/1106 | 5.9 | NR |
| *Lactobacillus bulgaricus* | | | | |
| IFO 3533[4] | ≥1000 | 28–1670 | 5.9 | NR |
| **Gram-negative bacteria** | | | | |
| *Escherichia coli* | | | | |
| CCRC 10674[7] | 100/200/500 | 49.1–310/1080/1100 | 6 | NR |
| ATCC11775[4] | 800/≥1000/≥1000 | 470–746/28–224/1100–1670 | 5.9 | 2–26/47/53 |
| NVH 3793[6] | >2000/30/60/1000/170/130 | 2.3/28.4/98.3/11.9/42.5/224 | 6 | 16/16/16/48/48/48 |
| *Pseudomonas aeruginosa* | | | | |
| CCRC 10944[7] | 150/200 | 49.1/51–1100 | 6 | 2/5–53 |
| *Pseudomonas fluorescens* | | | | |
| LMG 1794[5] | 80 | 43 | 5.5 | 6 |
| ATCC 21541[4] | 800 | 224–746 | 5.9 | NR |
| *Salmonella typhimurium* | | | | |
| CCRC 10746[7] | 1500/>2000 | 49.1–1100/28–1670 | 6 | 2–26/47–53 |
| ATCC 14028[4] | ≥1000 | 746–1671 | 5.9 | NR |
| *Enterobacter aerogenes* | | | | |
| LMG 2094[5] | 60 | 43 | 5.5 | 6 |
| *Vibrio cholera* | | | | |
| CCRC 13860[7] | ≥200 | 49.1–1100 | 6 | 2–53 |
| **Fungi** | | | | |
| *Botrytis cinerea*[3] | 10 | NR | NR | NR |
| *Drechsterasorokiana*[3] | 10 | NR | NR | NR |
| *Fusarium oxysporum* | | | | |
| CCRC 32121[7] | 500–1000/>2000 | 49–310/1080–1100 | 6 | 2–26/43–57 |
| *Microsporum canis*[3] | 1000 | NR | NR | NR |
| *Micronectriella nivalis*[3] | 10 | NR | NR | NR |
| *Trichophyton equinum*[3] | 2500 | NR | NR | NR |
| *Candida lambica* | | | | |
| 194[5] | 400 | 43 | 5.5 | 6 |

[a] NR, not reported.

*aureus* (*S. aureus*), when Mw increased from 5 to 305 kDa, while the reverse trend in activity to Mw relationship was obtained for *Escherichia coli* (*E. coli*).[15] Thus, most studies indicate that there is an optimum number of monomer units in the polymer chain for displaying antimicrobial activity, but they do not agree on the optimal Mw of the polymer needed for such activity, which may depend on the type of microorganism tested.

### 25.2.2 pH of medium

The solubility of chitosan is largely dependent on the number of free amino groups present in the polymer chain. Chitosan ($pK_a$ = 6.3–6.5) is soluble in water at low pH when its amino groups are protonated. Acidic medium imparts high positive charges ($NH_3^+$) on chitosan, which in turn might help in efficient binding of the polymer to the negatively charged components in the bacterial membrane.[16] A loss in activity of chitosan is observed as the pH increases. The activity of acid soluble chitosan with high viscosity is found to have greater inhibitory effect toward various strains of Gram-positive and Gram-negative bacteria as compared to water-soluble chitosan having low viscosity.[17] The antimicrobial effect of chitosan against six different bacterial strains tested within a pH range of 4.5–5.9 revealed that the polymers were more effective in bacterial inhibition at the lowest pH value (4.5).[4] Another study concluded that the water-soluble chitosan did not show activity toward *S. aureus* and *E. coli* and promoted the growth of *Candida albicans*, whereas the water-insoluble chitosan in acidic medium showed significant activity toward all the strains.[18] A study of antimicrobial activity of chitosan at different pH (5.5 and 7.2) showed that chitosan remained inactive toward *S. aureus* at pH 7.2 (MIC ≥ 8192 µg/mL), while its activity at pH 5.5 increased significantly (MIC = 64 µg/mL).[19] Hence, the pH of the solution containing the polymer was found to be crutial for displaying antimicrobial activity.

### 25.2.3 Degree of acetylation of chitosan

Highly water-soluble chitosan can be obtained by partial *N*-acetylation of chitosan.[20] On the other hand, chitosan with a low degree of acetylation (DA) has higher number of free amino groups, which in turn contributes to enhancement of its antimicrobial activity. This has been confirmed in studies where chitosan with less DA was found to have increased antimicrobial activity toward various strains of fungi, Gram-positive and Gram-negative bacteria.[17,21] This effect was especially observed in case of *S. aureus* and *E. coli*.[3,22] However, some studies report different findings. In one study, the optimum activity was reported by chitosan having intermediate DA value (25%) as compared to DA value

of 10% and 50%,[23] but this difference cannot be considered prominent as the variation was only 1–2 dilutions (MIC = 1.25–5 mg/mL), with the MIC values remaining same in most cases. Another study also showed similar results, with very small or no difference in the MIC values with varying DA (2%–53%) of the polymer against various bacteria and fungi.[7] Thus, the results from these studies showed that although some difference in activity might be observed between the very low and the highly acetylated chitosan, no clear relationship between the DA value and antimicrobial activity can be observed when the DA is less than 50%.

## 25.3 Antimicrobial properties of chitosan derivatives

Synthetic modification of chitosan typically target either the C-2 amino group or the C-6 hydroxyl group with the aim of achieving improved antimicrobial properties and sometimes also better aqueous solubility. These modifications have either been carried out on native chitosan or with the use of protected chitosan as a precursor. The most commonly introduced functional groups are the cationic moieties like the quaternary or guanidinium groups and the hydrophobic moieties like alkyl or aromatic groups. This has been achieved by using different synthetic routes as shown in Scheme 25.1. Most of the activity studies have been performed on the quaternary ammonium and acetyl derivatives of chitosan as seen in Table 25.2, with respect to variations in DA, degree of substitution (DS), or Mw. The activity of the derivatives is seen to increase with increasing DS value, while the variation in activity with DA and Mw is found to be less significant. The activity is also found to be dependent on the type of bacterial strain measured. The following categories of derivatives have been observed in the synthesis of antimicrobial chitosan derivatives.

### 25.3.1 Alkyl chitosan derivatives

*N*-Alkyl or *O*-alkyl chitosan is commonly prepared by reaction of chitosan with alkyl halides in the presence of a strong base like NaOH under heating.[24,25] Alkylation with alkyl halides (ethyl, butyl, dodecyl, and cetyl) can also be performed under basic ionic-liquids conditions; however, these conditions gave mixed *N*,*O*-alkylation with a combined DS varying from 35% to 77%. These derivatives showed a decreasing inhibitory effect with increasing chain length against a series of bacterial strains.[26] Aminoalkyl chitosan derivatives are reported to be synthesized either by conjugation with *p*-Benzoquinone in the C-6 OH group, followed by reaction with ethylenediamine or by reacting chitosan with 2-aminoethylchloride in the presence of NaOH. A study of such aminated chitosan

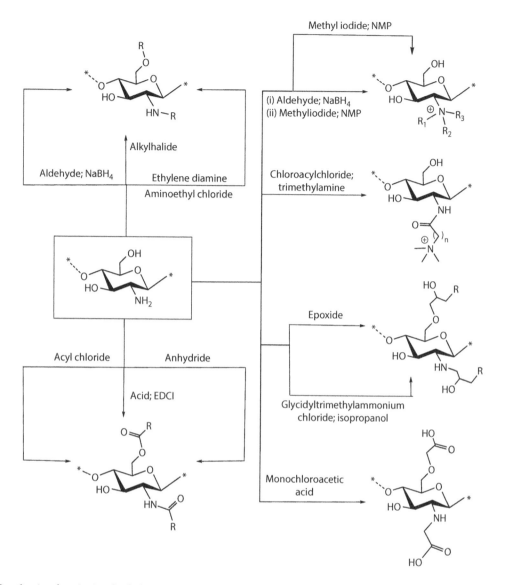

***Scheme 25.1*** Synthesis of antimicrobial chitosan derivatives.

derivatives showed that the increase in the content of amino groups led to an increase in antimicrobial activity, and a decrease in Mw and spacer length also led to an increase in the antibacterial activity against Gram-positive *Bacillus cereus*, *S. aureus* and Gram-negative *E. coli*, *Pseudomonas aeruginosa*.[27] Aminoethyl chitosan (AEC) was found to have higher activity against various strains of MRSA[28] than dimethylaminoethyl chitosan (DMAEC) and diethylaminoethyl chitosan (DEAEC). Another common route to alkylation of chitosan is by reductive amination via Schiff base formation, followed by its reduction with a reducing agent. The antibacterial activity of the Schiff bases of chitosan with citral and cinnamaldehyde against *E. coli* and *S. aureus* and antifungal activity against *Aspergillus niger* was better as compared to chitosan, and maximum efficacy was observed at high concentrations.[29,30] The

reduction of Schiff base of chitosan with substituted cinnamaldehydes produced several analogues having better antibacterial and antifungal activity than chitosan[31] (Table 25.2).

### 25.3.2 *Quaternary chitosan derivatives*

Quaternary chitosan derivatives have been prepared either by quaternizing the C-2 amino group of chitosan by *N,N,N*-alkylation or by introducing quaternary ammonium groups at C-2 or C-6 position by attaching it through spacer groups as seen in Figure 25.2. The simplest and the first example of quaternary chitosan derivative is the *N,N,N*-trimethyl chitosan (TMC).[51] Several works have been reported on the synthesis of a high degree of quaternized TMC; however, only recently the 100% TMC has been reported.[52] The antibacterial

*Table 25.2* Antimicrobial activity of various chitosan derivatives

| Chitosan derivative | DA (%) | DS (%) | Mw (kDa) | pH | Staphylococcus aureus (µg/mL) | Escherichia coli (µg/mL) | Pseudomonas aeruginosa (µg/mL) |
|---|---|---|---|---|---|---|---|
| | | | | | | Bacteria | |
| Cyanoethylchitosan[32] | 21 | [a]NR | NR | NR | 19 | 312 | NR |
| Aminoethylchitosan[33] | 10 | NR | 300c | 5.5 | 125 | 62.5 | 7.8 |
| Dimethylaminoethyl chitosan[33] | 50/10 | NR | 140–310c | 5.5 | 125/125 | 62.5/62.5 | 31.25/15.6 |
| Diethylaminoethyl chitosan[33] | 50/10 | NR | 140–310c | 5.5 | 250/250 | 500/250 | 62.5/125 |
| N-ethyl chitosan[26] | 9.8 | 51.8 | 50c | 7 | 32 | 64 | 128 |
| N-butyl chitosan[26] | 9.8 | 50.7 | 50c | 7 | 64 | 128 | 128 |
| N,N,N-trimethyl chitosan[34–39] | 20/10/10/10/ 10/10/ 10/5/17.3/34.2/ 6/4/4 | –/30/30/65 / 65/65/86/100/ 100/28/64/89.2/90.7 | 9420/281/95.7/ 35.4/84.1/ 42.7/67.3/8/13.2/ 16.5/204.5/ 120.8/19/7.8 | 7.2/7 | 160/1000/2000/ >2000/125/ 500/31/8/4/ 32/500/125 | 40/2000/2000/>2000/ 250/500/31/128/ 64/256/1000/250/1/2 | NR |
| N-(Propyl/furfuryl)-N,N-dimethyl chitosan[39] | 4 | 91.8/89.2/90.7/ 84.1/81.6/89.8 | 214/19/7.8/214 /19/7.8 | 7 | NR | 0.5/1/1/1/1/1 | NR |
| N-(2-(N,N-Trimethyl ammoniumyl) acetyl) chitosan[36,37,40] | 0/0/18/18/18/ 7/17.3/7/17.3 | 83/83/5/5/15/100/ 100/100/100 | 8/42.1/42.1/48/ 48/23.8/16.4/ 18.8/12.2 | 7.2/5.5 | 32/128/512/2048/512 /2048/8/32/8/1024 | 512/>8192/128/2048/ 256/1024/16384/16384/ 16384/16384 | 64 |
| N-(2-(1-Pyridiniumyl)acetyl) chitosan[36,37] | 0/7/6/17.3/19/34.2 | 81/100/100/100/100 | 8/18.8/17.3/16.4/ 16.7/18.9 | 7.2 | 64/8/1024/512/512 | 512/1024/512/ 1024/1024/512 | 512 |
| N-(3-Pyridylmethyl) chitosan[35,38,41] | 10/10/10/10/ 6/6/6/6 | 35/60/80/80/85/ 94/91.2/70/80 | 58/114.15/82.18/ 8.4/276/276/276/ 276/18.8/12.2 | 7.2 | 125/125/125/250/64/ 32/16/8/125/125 | 250/250/250/250/ 64/32/32/32/250/250 | NR |
| N-(Oct/Benz/CH₃-benz/OH-benz/-N-(2-hydroxypropyltrimethylammoniumyl) chitosan[41] | 6 | 93.3/92.5/91.4/90.4 | 276 | 7 | 16/8/16 | 32/32/32 | NR |
| N-(3,4-CH₃O/F/Cl/CF₃/4-Br/NO₂/ COOH)benzyl-N-(2-hydroxypropyltrimethylammoniumyl) chitosan[41] | 6 | 91.4/95/94/94/91/ 91.4/96.5 | 276 | 7 | 32/16/16/16/ 32/16/16 | 32/32/32/32/32/32 | |
| N-dodecyl chitosan[26] | 9.8 | 50.7 | 50c | 7 | 64 | 128 | >256 |
| N-acetyl chitosan[26] | 9.8 | 35.3/51.5/64.4/77.2 | 50c | 7 | 32/64/128/>256 | 128/128/>256/>256 | 256/256/256/>256 |
| (Acetyl/chloroacetyl/benzoyl)thiourea chitosan[42] | 4 | 89.2/91.5/81 | 50c | NR | 62/62/62 | 15/15/15 | 62/15/62 |
| O-biguanidinylbenzoyl chitosan HCl[43,44] | 7.7 | 14.9/24.3/32.7/39.5 | 346/352/375/410 | 7 | 32/16/16/8 | 64/64/32/16 | NR |
| p-biguanidinyl benzoyl chitosan HCl[44] | 7.7 | 16.2/27.1/36.8/44.7 | 349/358/381/424 | 7 | 32/16/16/8 | 64/64/32/16 | NR |
| (Acetyl/chloroacetyl/benzoyl) phenylthiosemicarbazone[45–47] | 4 | 35.6/35.6/35.6 | 7/200 | NR | >250/56/225 | 7/7/7 | 900/225/225 |
| Acetyl(o-Me/p-Me/ p-chloral/p-nitro) phenyl thiosemicarbazone[45] | 4 | 33.7/35.3/32.1/34.5 | 7/200 | NR | 250/250/250/250 | 7/7/7/7 | 450/450/450/450 |
| (O-carboxymethyl/N,O-carboxylethyl) chitosan[48,49] | 20–25/6.7 | NR/72 | NR/347 | NR/7 | 320/31.3 | 320/62.5 | NR |
| Quaternary (Me/Et/Pr/But/Benz) carboxylethylchitosan[49] | 6.7 | 59/57/54/51/55 | 409/436/457/ 475/454/ | 7 | 31.3/15.6/12.5/ 6.3/6.3 | 31.3/31.3/15.6/12.5/6.3 | NR |
| (Acetyl/chloroacetyl/ benzoyl(thioureacarboxymethyl chitosan[50] | 12 | 89.6/88.8/89.6 | 200 | NR | 7.8/31.25/62.5 | 62.5/125/500 | NR |

[a] NR, not reported; DA, degree of acetylation; DS, degree of substitution; Mw, average molecular weight; Mwc, Average molecular weight for starting chitosan.

activity of TMC has been found to be superior to most of the quaternary chitosan derivatives against Gram-positive and Gram-negative bacteria.[34,35] This activity is found to decrease when the C-6 OH group of TMC was modified with carboxymethyl groups.[53] A study of the antibacterial activity of methylated COSs and chitosans showed that while quaternary COS were inactive at neutral pH, quaternary chitosan exhibited high activity with MIC as low as 8 μg/mL (DS = 0.69).[19] When the C-6 OH group of TMC (DS = 0.72) was substituted by quaternary ammonium moiety like 2-hydroxy-3-trialkylammonium propyl (DS = 0.05, 0.10, and 0.20), the

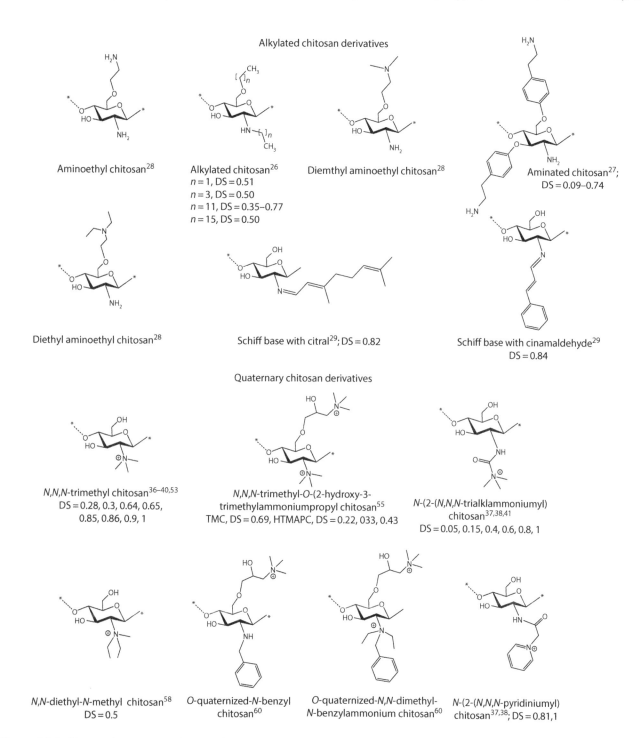

*Figure 25.2* Chemical structure of antimicrobial chitosan derivatives. *Structure and nomenclature have been shown as reported in the studies. The DS for the derivatives were determined by [1]H-NMR spectroscopy in most cases, and by elemental analysis and FT-IR spectroscopy in some cases.

*(Continued)*

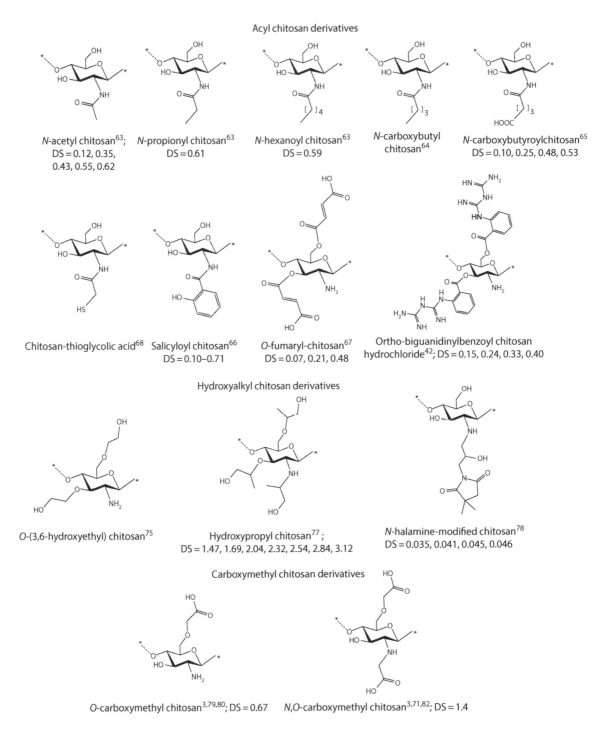

**Figure 25.2 (Continued)** Chemical structure of antimicrobial chitosan derivatives. *Structure and nomenclature have been shown as reported in the studies. The DS for the derivatives were determined by [1]H-NMR spectroscopy in most cases, or by elemental analysis or FT-IR spectroscopy in some cases.

activity of TMC was enhanced up to a maximum of 3-dilutions toward both *S. aureus* and *E. coli*.[54] The presence of carboxyethyl groups in such derivatives was seen to further enhance the antimicrobial activity due to the synergistic effect between the carboxyethyl and the quaternary ammonium groups.[49] The positioning of the

quaternary group in the polymer backbone is another factor influencing the antimicrobial activity of chitosan. Groups like *N*-(2-(*N,N,N*-trialkylammoniumyl) and *N*-(2-pyridiniumyl)acetyl) showed maximum efficacy toward *S. aureus* and *E. coli* as the spacer between the polymer chain and the quaternary group became shorter.[36,37,40] The antimicrobial activity of different N-alkyl chains quaternized using methyl iodide have also been studied and their activity was found to be superior to chitosan against plant pathogenic bacteria and fungi.[55] While an increasing activity was observed with increase in the length of the monoalkyl chain in some studies,[56,57] others have reported TMC to be more active than N,N-diethyl-N-methyl chitosan against *S. aureus*.[58] The quaternary chitosan derivatives were also more effective than chitosan toward fungi *Botrytis cinerea* and *Colletotrichum lagenarium*.[59] Quaternary chitosan derivatives are prepared by reacting chitosan with glycidyl-trimethyl ammoniumchloride in isopropanol. The different derivatives obtained in this study were coated onto fibers that exhibited the following order of activity toward *E. coli* and *S. aureus*: O-quaternized-N,N-diethyl-N-benzylammonium chitosan) ≥ O-quaternized-N-chitosan-Schiff bases > O-quaternized-N-benzyl chitosan.[60] Chitosan modified by cationic dendritic poly(propylene)imine or polyamido amine was also capable of inhibiting the growth of *S. aureus* and *E. coli* when incorporated into wool or cotton fabric.[61,62]

### 25.3.3  Acyl chitosan derivatives

Introduction of the acyl group in chitosan is usually carried out by reacting chitosan with acyl chlorides in the presence of bases, or by heating with anhydrides. A study of N-acetyl, N-propionyl and N-hexanoyl chitosan against two strains of Gram-positive and Gram-negative bacteria showed that higher DA N-acyl compounds formed intermolecular aggregation, which helped in forming bridges to interact with the bacterial cell wall.[63] N-carboxybutyl chitosan has also shown inhibitory effects against several strains of Gram-positive and Gram-negative bacteria and *Candidae* with increasing concentration.[64] N-(4-carboxybutyroyl) chitosan was prepared by reacting 0.1, 0.3, 0.6, and 1.0 mol of glutaric anhydride per glucose amine unit to give derivatives having DS values of 0.10, 0.25, 0.48, and 0.53. Their antibacterial activity against *Agrobacterium tumefaciens* and *Erwinia carotovora* and antifungal activity against *Botrytis cinerea*, *Pythium debaryanum*, and *Rhizoctonia solani* was found to increase with increase in the DS of the derivatives.[65] The enhancement of antibacterial activity with DS was also observed in studies of salicyloyl chitosan,[66] ortho and para-biguanidyl benzoyl chitosan[43,44] and O-fumaryl chitosan.[67] Some of the acyl derivatives were also prepared by coupling chitosan with carboxylic acids through activation by carbodiimide. The synthesis of such derivatives was carried out by this method through activation of various acids like thiogycolic acid,[68] N-acetyl-L-cysteine[69] and 4-CBS[70] with 1-ethyl-3-(3-dimethylaminopropyl)

carbodiimide hydrochloride and then reacting the intermediates with the secondary amino group of chitosan to form the N-acyl products (DS = 4.1–0.13). The activity of chitosan-thioglycolic acid was found to be superior to carboxymethyl chitosan (CMC) and the highly active TMC toward bacteria and fungi.[68] These materials were bactericidal to *E. coli* and *S. aureus*. The investigation was carried out by preparing a dispersion system with oleoyl-chitosan nanoparticles and observing the quick release of the intracellular components and the increased uptake of 1-N-phenylnaphthylamine, as well as the release of cytoplasmic β-galactosidase.[71] The membrane-disrupting ability of *E. coli* and *S. aureus* was also observed in a study of thiol-functionalized chitosan derivative.[69] Thus, the presence of hydrophobic and fatty acid residues in chitosan resulted in greater interaction with the bacterial cell wall.[72,73]

### 25.3.4  Hydroxyalkyl chitosan derivatives

Hydroxyalkyl derivatives of chitosan are usually synthesized by the treatment of epoxides with chitosan under heating.[74] The reported antimicrobial activity of hydroxyalkyl chitosan is variable between different studies. Hydroxyethyl chitosan displayed better activity toward *E. coli* than toward *Enterococcus* and its overall activity was higher than chitosan.[75] Hydroxypropyl chitosan was bactericidal in the case of *E. coli* and *S. aureus* within 30 min of contact with the bacterial cells[76] in one study, while in another study, this derivative having different DS and Mw was found to have no inhibitory effect against *E. coli* and *S. aureus*.[74] O-hydroxy-2,3-propyl-N-methyl-N,N-diallylammonium chitosan methyl sulfate was synthesized via epoxide intermediate, which showed an MIC value of 24 and 155 μg/mL against *S. aureus* and *Klebsiella pneumonia*, respectively.[77] N-halamine-modified chitosan derivatives were synthesized by reacting varying molar equivalents (0.8–1.5) of 3-glycidyl-5,5-dimethylhydantoin with chitosan to get a DS of 3.5%–4.5%. The chlorine containing modified films obtained from the derivatives showed high efficacy against *E. coli* and *S. aureus* following 5 and 10 min of contact time, respectively.[78]

### 25.3.5  Carboxyalkyl chitosan derivatives

CMC have been synthesized by the treatment of chitosan with NaOH followed by reaction with monochloroacetic acid.[48,68,79–81] A study of the antimicrobial properties showed that chitosan was effective against several pathogenic bacteria, while O-CMC showed activity only against a few pathogens[48] and N,O-CMC remained more active than chitosan.[82] The activity of chitosan and CMC against *E. coli* followed the order O-CMC > chitosan > N,O-CMC.[3] O-CMC has also been used as a precursor for a number of synthetic modification of chitosan such as introduction of den-drimer,[83,84] Schiff base,[85] acyl thiourea derivatives,[50] quaternized derivatives,[49,82,86] terephthaloylthiourea cross-linked derivatives,[87] thiosemicarbazone,[88] and poly-*N*-vinyl imidazole.[89] Investigation of several Schiff bases prepared from CMC showed that adding electron-donating benzene substituents increased the activity against *S. aureus* and *E. coli* compared to CMC, while electron-withdrawing benzene substituents reduced the activity.[85] Modifying the amino group of *O*-CMC with functional groups like quaternaryammonium, poly(*N*-vinyl imidazole) and thiosemicarbazone enhances the antibacterial activity toward *S. aureus* and *E. coli*.[86,88,89] This activity was further enhanced when polyamidoamine dendrimers were introduced into the quaternary CMC.[84] Acetyl, chloroacetyl, and benzoyl thiourea derivatives of CMC when tested against *Bacillus subtilis*, *S. aureus*, *E. coli* and the three crop-threatening pathogenic fungi *Aspergillus fumigatus*, *Geotrichum candidum*, and *Candida albicans* showed higher antibacterial and antifungal activity than the parent CMC.[50] CMC hydrogels cross-linked with various terephthaloylthiourea moieties can inhibit the growth of bacteria and pathogenic fungi more effectively than CMC alone.[87] Even the blending of CMC with cellulose and cotton fibers was able to display good antibacterial activity against *E. coli* and *Micrococcus luteus*.[79,81]

## 25.4 Antimicrobial assays

There has been a significant variability in the reported antimicrobial properties of chitosan and its derivatives as observed from the different studies. Apart from the physicochemical factors like pH, DA, and Mw, which affect the antimicrobial properties, the difference in susceptibility of the different strains of the microbial species tested and the use of different methods for assessing the activity against microorganisms also account for this variability. The commonly used methods for determining the antimicrobial activity of chitosan and its derivatives are listed in the following text.

### 25.4.1 Antibacterial assays

The antibacterial activity of chitosan and its derivatives have been reported in terms of the MIC and the MLC values or the rate of growth inhibition of the microorganism. This has been obtained following various standard as well as nonstandard antibacterial assays. The standard method for testing the activity of an antibacterial agent is the Clinical and Laboratory Standards Institute (CLSI) protocol.[90] This institute is responsible for the development of standards and guidelines that address test methods, quality control

procedures, and the interpretation of test results in the clinical microbiology laboratory. Additionally, they provide important performance principles for various culture media and automated procedures. Hence, for assessing the antibacterial activity of chitosan and its derivatives, the use of such standard method is essential to allow reproducibility and a good comparison of the activity of different materials. However, the use of standard assay is found to be limited and different kinds of assays have been reported for determining the antibacterial activity of chitosan and its derivatives as mentioned below.

#### 25.4.1.1 Broth microdilution method

This assay is used for measuring the MIC and the MLC values of chitosan[6,91] and its derivatives according[37] to the standard CLSI methods for antimicrobial dilution susceptibility tests.[92] In the broth microdilution method, the MIC and the MLC values are determined using Mueller–Hinton broth at pH 7.2 and a defined number of bacteria. Samples of the antibacterial agent are prepared in sterile water at an initial concentration, which is then serially diluted by two-fold dilutions in a 96-well plate using Mueller–Hinton broth. The serial dilution is standardized in a way that the concentration of 1 μg/mL is or would be included. Gentamicin or other antibiotic with known MIC against the strains is then used to provide a control for performance during the test. Testing of the Mueller–Hinton broth in absence of bacteria or antimicrobial provides a sterility control and the broth with bacteria provides a growth control. A standard 0.5 McFarland suspension ($1–2 \times 10^8$ CFU/mL) is prepared by direct colony suspension in Mueller–Hinton broth, which is further diluted to achieve a final test concentration of ~$5 \times 10^5$ CFU/mL in the wells in the microtiter plate. The microtiter plates are then incubated at 35°C for 18 h under moistened conditions. The MIC value is defined as the lowest concentration of the antibacterial agent that completely inhibits the visible growth of the microorganism in the microtiter wells. Blood agar or Mueller–Hinton agar is used to measure the MLC. For MLC measurements, 10 μL × 2 (of each of the dilutions that show no visible growth) is placed on an agar plate and incubated at 35°C for 18 h. The MLC is defined as the lowest concentration that achieved a 99.9% decrease in the viable cells.

The broth microdilution method has been used for determining MIC and MLC values for different kinds of chitosan such as chitosan with Mw = 2.3–224 kDa and DA = 16%–48% and chitosan having with Mw = 50–190 kDa and DA = 15%–25%.[6,91,93] The MIC and MLC values of several chitosan derivatives such as TMC (DQ = 67%),[94] TMC (DQ = 100%), quaternary chitosan and chitooligosacharides (DQ = 0%–74%),[19] *N*-(2-(*N,N,N*-trimethylammoniumyl)acetyl) chitosan,[36,37]

*N*-(2-(*N*-pyridiniumyl)acetyl)chitosan,[36,37] *N*-(6-(*N,N,N*-trimethylammoniumyl) hexanoyl) chitosan,[37] *N*-(6-(*N*-pyridiniumyl)hexanoyl)chitosan,[37] quaternary *N*-(2-(*N,N,N*-tri-alkylammoniumyl and 2-pyridiniumyl) acetyl) derivatives of chitosan, chitooligomer, and glucosamine,[36] chitosan-*N*-betainates,[40] methylpiperazine, mono-quaternary dimethylpiperazine, and di-quaternarytrimethylpiperazine derivatives of chitosan (DS = 3%–87%)[37] were determined by this method. In some studies, only the MIC values have been reported for aminoethylchitosan, dimethylaminoethylchitosan, and diethylaminoethylchitosan[33] aminoethyl-modified chitosan derivatives with different Mw[95], *N*-2-hydroxypropyltrimethylammoniumchloridechitosan,[80] quaternaryammonium chitosans containing monosaccharides or disaccharides moieties,[96] quaternary *N*-aryl chitosans,[41] quaternary *N*-substituted CMC,[97] quaternary *N*-(4-*N,N*-dimethylaminobenzyl) chitosan, and quaternary *N*-(4-pyridylmethyl) chitosan.[98] Other studies have followed a modified procedure with changes in pH or medium for chitosan-*N*-hydroxy-2,3-propyl-*N*-methyl-*N,N*-diallylammonium methyl sulfate,[99] *N*-(3-pyridylmethyl) chitosan[35] and quaternary *N*-aryl chitosan.[41]

### 25.4.1.2 Time kill assay

The time kill assay is based on the CLSI (NCCLS) protocol for measurement of concentration- and time-dependent bactericidal activities of antibacterial agents.[100,101] Initially, the MIC value for the antibacterial agent is determined by the broth microdilution method.[92] The sample time points and the test concentrations are then determined, which are usually the MIC, 2 × MIC and 4× MIC. Standard 0.5 McFarland bacterial suspension is prepared and the bacterial suspension is then added to the tubes containing broth and the different concentrations of the antibacterial agent to get a final bacterial concentration of ~5 × 10[5] CFU/mL. The tubes are then incubated at 37°C and 100 µL samples are taken out at time zero and at the time intervals according to the strategy for the test. At each time point, 100 µL of the samples are serially diluted 1:10 in saline, 10 µL of the undiluted and the diluted samples are plated in duplicate on Mueller–Hinton agar and incubated at 37°C overnight. The dilutions that show 20–200 colonies are then counted and the number of colonies for each sample is determined by averaging the counts obtained and CFU/mL is calculated according to the dilution that was counted and volume plated. The number of colonies ($\log_{10}$CFU/mL) versus time intervals is then plotted for the control and the antibacterial agent. Generally, a decrease of 3-$\log_{10}$CFU/mL in bacterial counts in antibacterial solution compared with counts for the growth control indicates an adequate bactericidal response.

The bactericidal activity of different amino derivatized chitosans—monoquaternary dimethylpiperazine and diquaternary trimethylpiperazine derivatives of chitosan (DS = 37%–87%),[102] TMC,[94] aminoethylchitosan, dimethylaminoethylchitosan, diethylaminoethylchitosan (DA = 10%–50%), and ethylenediamine-*p*-benzoquinone derivatized chitosan—have been investigated by such time kill assay.[27,28] *N*-Diazeniumdiolate-functionalized chitosan oligosaccharides were tested for bactericidal activity using the time kill assay with the modification that the MIC values for the compounds were not determined.[103]

### 25.4.1.3 Optical/turbidimetric method

In this method, the inhibitory effect of an antibacterial agent is determined by optically measuring the absorbance at 620 nm. Bacteria is grown in liquid peptone medium and incubated overnight at 37°C. The cultures obtained are diluted with sterilized distilled water to measure the optical absorbance at 620 nm. The cell suspension is added to the liquid peptone medium containing the antibacterial agent that was sterilized at 121°C for 20 min and cultivated at 37°C for 10 h. The number of bacterial cells is measured spectrophotometrically at 620 nm ($A_{620\,nm}$). The inhibition rate is determined by plating the bacterial suspension with the solution of the antibacterial agent on a peptone culture plate and incubating at 37°C for 24 h. Inhibition rate (%) = $N_1 - N_2 / N_1 \times 100$%, where $N_1$ and $N_2$ are the number of colonies on plates before and after inhibition. This is a nonstandard method, but it can be used to assess the reduction of the viable bacterial cells reflected by the spectrophotometric analysis, which can be correlated to the MIC. In some cases, bacterial killing can also confirmed by the inhibition rates.

The optical density of chitosan, chitosan acetate,[104] *O*-CMC chitosan,[3] ethylamine hydroxyethyl chitosan,[25] and several Schiff bases of CMC[85] was determined by measuring their UV absorbance spectrophotometrically. This method was used to determine the MIC values of *N*-acylthiourea derivatives: acetyl thiourea chitosan, chloroacetylthiourea chitosan, and benzylthiourea chitosan.[42] The inhibition rates of ethylenediamine-*p*-benzoquinone derivatized chitosan,[27] *N*-2-hydroxypropyltrimethylammonium chloride chitosan,[80] Schiff base of chitosan, and acylated chitosan were determined by this method.[29,30] The MIC values for chitosan and its thiourea derivatives acetylphenylthiosemicarbazone, acetyl(*o*-methylphenyl)-thiosemicarbazone, and acetyl(*p*-methylphenyl)-thiosemicarbazone, acetyl(*p*-chloralphenyl)-thiosemicarbazone, chloroacetylphenyl-thiosemicarbazone and its substituted derivatives were determined by a using a modified optical method, where Mueller–Hinton broth was used as the medium.[45,46,47] The MIC and MLC values of *N*- and *O*-quaternary chitosan were also reported by this method. MLC was regarded as the lowest concentration that resulted in "100% bacterial killing."[34]

### 25.4.1.4   Agar dilution method

The agar dilution method is a standard method for determining antimicrobial susceptibility.[92] In this method, successive dilutions of the antimicrobial agent (1:2, 1:4, 1:8) is mixed with molten agar, poured into Petri dishes and allowed to cool at room temperature. A standard 0.5 McFarland suspension of the microorganism ($1-2 \times 10^8$ CFU/mL) is prepared by direct colony suspension in Mueller–Hinton broth, which is further diluted to a concentration of $10^7$ CFU/mL using broth or saline. An aliquot of the bacterial suspension is added to the agar plates to get a final concentration of $10^4$ CFU/spot. The agar plates are then incubated at 37°C for 16–20 h. The MIC is recorded as the lowest concentration of the antimicrobial agent that completely inhibits the microorganism growth.

The inhibitory effect of chitosan[90,91] and its derivatives were determined by a nonstandard variation of the agar dilution method. Samples of the antibacterial agent were prepared at an initial concentration and then sterilized at 121°C for 20 min. A two fold serial dilution of each sample was then added to a nutrient broth (at pH 5.5 or pH 7.2) to obtain various concentrations. The pH was then adjusted using 1% AcOH or 1% NaOH solution. The culture of each bacterium was diluted with sterile distilled water to $10^5-10^6$ CFU/mL. A loop of each suspension was inoculated on a nutrient agar with either sample or control and then incubated at 37°C for 48 h. Finally, the colonies were counted to obtain the MIC value (the concentration at which no microorganism colony or less than five colonies were visible). The inhibition rate was calculated by adding the antibacterial solution in broth to the bacterial suspensions ($10^5-10^6$ CFU/mL). After adjusting the pH to 5.5, the mixture was incubated at 37°C for 24 h. Viable cell count was then made by plating on a nutrient agar plate and incubating at 37°C for another 24 h. The number of colonies is then determined and the inhibition rate[105] or antibacterial efficiency[106] calculated using the following equation:

$$\text{Inhibition rate } (\%) = (\text{Initial cell number}$$

$$-\text{Cell number after treatment})/\text{Initial cell number} \times 100$$

The MIC values for *N*-benzylchitosan derivatives,[107] *O*-CMC, quaternized-CMC,[87] *N*-ethylchitosan, *N*-butylchitosan, *N*-dodecylchitosan, *N*-cetylchitosan,[26] quaternary *N*-butyl, *N*-pentyl, *N*-hexyl, *N*-heptyl, and *N*-octyl chitosan,[55] *o*- and *p*-biguanidinylbenzoylchitosan HCl,[43,44] 6-amino-6-deoxychitosan,[108] *N*-cinnamylchitosan derivatives,[31] chitosan, carboxyethylchitosan (CEC), quaternized-CEC,[49] acylthiourea CMC, chloroacylthiourea CMC, benzylthiourea CMC,[50] *N*,*O*-quaternary ammonium chitosan,[54] chitosan, *N*,*O*-CMC, *N*,*O*-CEC, *N*-quaternized chitosan particles,[57] heterochitosan (DA = 10%, 25%, 50%),[23] 2-hydroxypropyldimethylbenzyl ammonium chitosan, 2-hydroxypropyldimethylbenzylammonium CMC, and 2-hydroxypropyldimethylbenzyl ammoniumCEC[82,109] were determined using this plate count method. The antibacterial efficiency of chitin, chitosan, chitosan oligomers, and derivatives like *p*-aminobenzoyl chitosan ester, hydroxypropyltrimethylammoniumchitosan, *N*,*N*-diethylaminoethylchitin, *N*,*N*-diethylamino ethylchitosan, *N*,*N*,*N*-triethylaminoethylchitin, TMC, *N*-butylchitosan, *N*,*N*-dibutylchitosan, *N*-octyl chitosan, and *N*-dodecyl chitosan was determined by this method with modifications such as changes in pH, medium, and incubation time.[4,7,17,56,106,110,111] This method was also used to determine the antibacterial inhibition rate for chitosan acetate[104] chitosan microparticles[112] chitosan derivatives like *N*-acetyl chitosan, *N*-propionyl chitosan, *N*-hexanoyl chitosan,[63] *N*-2-hydroxypropyl trimethylammoniumchitosan,[80] and *N*-carboxybutyl chitosan,[64] TMC and *O*-carboxymethyl TMC,[53] *N*-2-hydroxypropyltrimethylammonium chitosan,[113] arylfuran chitosan derivatives,[114] and *N*,*N*,*N*-trimethyl-*O*-(2-hydroxy-3-trimethylammonium propyl) chitosans.[54]

### 25.4.1.5   Agar diffusion method

This is a CLSI method for qualitative measurement of the susceptibility of bacteria to different antimicrobial agents.[115] In this method, a bacterial suspension of McFarland standard 0.5 is prepared by direct colony suspension and plated all over Mueller–Hinton susceptibility agar plates. The antimicrobial paper discs are then dispensed onto the surface of the inoculated agar plates and incubated for 16–20 h at 37°C. The diameter of the zones of complete inhibition, including the diameter of the discs, are measured and the size of the inhibition zone is then used to assess if the microorganism is susceptible, intermediate, or resistant to the agents that have been tested.

The bacterial growth inhibition of chitosan and its derivatives was determined by a modified agar diffusion assay. Whatman papers were cut into 6 mm diameter discs and sterilized by autoclaving. Each disc was loaded with 50 μL of 1% antibacterial solution. The discs were then placed on nutrient agar in Petri dishes, which had been seeded with 20 μL of bacterial cell suspensions. The Petri dishes were then examined for zone of inhibition after 24 h incubation at 37°C. The area of the whole zone was calculated and then subtracted from the film disc area, and the difference in area was reported as the "zone of inhibition."

This method has been used for determining the bacterial growth inhibition of chitosan, O-CMC,[48] acetyl CMC, chloroacetyl CMC, benzyl CMC,[50] 4-carboxybenzenesulfonamide-chitosan,[70] and quaternized N-substituted carboxymethylchitosan derivatives,[97] aminoethylchitosan, dimethylaminoethylchitosan, diethylaminoethylchitosan having varying DA (10%–50%),[28,116] hydroxypropylchitosan derivatives with different DS and Mw,[74] CMC, thiosemicarbazidecarboxymethyl chitosan, and thiosemicarbazonecarboxymethyl chitosan derivatives.[88] The antibacterial activity of chitosan and chitosan-*graft*-poly(N-vinyl imidazole) was evaluated based on a modified agar diffusion method.[117] The inhibitory effect of salicyloyl chitosan,[66] acylthioureacarboxymethylchitosan,[50] and C-6 quaternaryammonium chitosan derivative[118] was measured by a modified method where stainless steel cylinders of uniform size were placed on the surface of the agar plate, which were filled with 200 μL of sample solutions; these were incubated at 37°C and the zone of inhibition was measured.[12]

## 25.4.2   Antifungal assays

The in vitro antifungal susceptibility testing is influenced by a number of technical variables, including inoculum size and preparation, medium formulation and pH, duration and temperature of incubation, etc. In addition, antifungal susceptibility testing is dependent on the nature of growth of the fungi. Some fungi-like *Rhizopus* grow uncontrollably, while other fungi-like *Aspergillus* grow within a limited number of colonies. Hence, the characteristic of the fungi is likely to be a determining factor while choosing the antifungal assay.[119]

### 25.4.2.1   Radial growth technique

The antifungal activity of most chitosan derivatives has been determined using the radial hyphal growth bioassay.[10,65] This method can only be used for fungi that grow in an uncontrolled manner. In this method, a sample of the agent to be tested was dissolved in 0.25 M HCl and the pH was adjusted to 5.0–5.5 with 1 M NaOH. The serial concentrations were respectively added to a potato dextrose agar (PDA) medium immediately before being poured into Petri dishes. Each concentration was tested in triplicate. The discs of mycelial culture (0.5 cm in diameter) of fungi, taken from 8-day-old cultures on PDA plates, were transferred aseptically to the center of the Petri dishes. The plates were incubated in the dark at 26°C. The colony growth diameter was measured when the fungal growth in the control had completely covered the Petri dishes. The inhibition percentage of mycelial growth was calculated as follows: mycelial growth inhibition (%) = DC − DT/DC × 100, where DC and DT are average diameters of fungal colonies of control and treatment, respectively.

The antifungal activity of N-benzyl chitosan,[107] N-(cinnamyl) chitosan analogs,[31] diethyl dithiocarbamate chitosan,[120] acylthiourea[42] and thiosemicarbazone[47,121] derivatives of chitosan, quaternary N-butyl, N-pentyl, N-hexyl, N-heptyl, and N-octyl derivatives of chitosan[55] were determined using this method.

## 25.4.3   Other methods and general comments

Apart from these assays, methods like drop count method for bactericidal activity,[122] dry cell weight measurement for antifungal activity,[30] and many other non-standard methods have been reported for determining the antimicrobial activity of chitosan and its derivatives. Even the use of standard methods with modifications in pH, incubation time, and medium can bring about significant variability in the results. Also, some studies have been seen to report only the MIC values for the agents. In such cases, the results can sometimes be misleading, as the MICs are determined by visual observation. Hence, measurement of MLC is required to support the MIC data from such studies.

## 25.5   Mechanism of antimicrobial action

Chitosan and its derivatives are known to exhibit broad spectrum activity against a wide variety of target organisms like bacteria, fungi, yeast, and algae. Studies that involve the interaction of the polymer with a bacterial cell have considered such agents to be either bacteriostatic (inhibits the growth of bacteria) or bactericidal (kills the bacteria). The mechanistic studies performed on chitosan and its derivatives have shown that although such interactions usually lead to bacterial cell lysis in most cases,[123–125] at low concentrations they can affect outer membrane permeability without killing the bacterial cell.[126]

### 25.5.1   Bacterial species and cell wall composition

The bacterial species has been classified as Gram-positive and Gram-negative according to their response toward the Gram staining procedure based on the differences in their cell wall. The Gram-positive bacteria possess a cell wall constructed of thick peptidoglycan layer that is anchored to their cytoplasmic membrane by lipoteichoic acid. It also includes teichoic acid, which is a negatively charged polymer that strengthens the cell wall. Most teichoic acids form covalent linkages with *N*-acetylmuramic acid in the peptidoglycan, while the lipopolyteichoic acids are covalently linked to the cell membrane. These teichoic acids provide a high density of charges arranged uniformly in the cell wall, which affects the passage of ions across the outer surface layers.[127] On the other hand, the Gram-negative bacteria possess a thin peptidoglycan layer just above the cytoplasmic membrane, which is covered by an outer membrane (OM). The OM is composed of lipoprotein and lipopolysaccharide (LPS). The protein portion of the lipoprotein is covalently bound to the peptidoglycan, while the hydrophobic lipid portion remains buried in the outer membrane. The LPS of the OM is composed of a hydrophilic O-specific side chain and a hydrophobic component called lipid A. Since the O-specific carbohydrate chain can vary from species to species, these polymers are important in identifying bacteria.[128] The OM in Gram-negative bacteria is highly resistant to macromolecules and hydrophobic compounds; hence, overcoming the OM barrier is a key factor to display activity against Gram-negative bacteria.[126]

### 25.5.2  Models of antimicrobial action

The mode of action of chitosan toward Gram-positive and Gram-negative bacteria is believed to be associated with the interactions at the bacterial membrane in both cases. There are three models that have been proposed so far to explain the mechanism of action.

1. The most commonly accepted mechanism is the electrostatic interaction between the cationic moieties in chitosan and the anionic components of the bacterial membrane. The interaction is initially mediated by the cationic $NH_3^+$ groups of chitosan, which presumably competes with the $Ca^{2+}$ ions for binding to the negative sites of the membrane.[129] Hence, the presence of the amino group ($NH_3^+$) in chitosan is a requirement for the antibacterial activity of chitosan.[22] This interaction operates in two ways. In Gram-negative bacteria, the positive charge of chitosan forms an ionic-type bonding, resulting in changes in the permeability of the membrane wall. This blocks the nutrient flow, which results in the development of internal osmotic pressure and ultimately leads to the death of the cell due to lack of nutrients.[130] In Gram-positive bacteria, the interaction involves the hydrolysis of the peptidoglycans present in the cell wall, leading to leakage of the intracellular components. The release of such electrolytes and proteinaceous substances is confirmed by the increased absorption at 260 nm.[131] Increased electrical interaction involves the hydrolysis of the

peptidoglycans present in the cell wall leading to conductivity of the cell suspension[125] and cytoplasmic β-galactosidase release.[25,132] The interaction between chitooligosaccharides and the bacterial membrane can be seen in Figure 25.3.

The interaction of chitosan with the bacterial membrane has been supported in several studies. It has been observed that at slightly acidic conditions (pH = 5.3), at which chitosan is protonated, the anionic components of the bacterial surface undergo electrostatic binding with chitosan and this in turn opens up the OM.[126] Fluorescence and SEM studies have revealed that in chitosan–arginine derivatives the initial site of action is the OM in Gram-negative *E. coli* and *P. fluorescens*. By affecting the membrane permeability, it eventually leads to leakage of intracellular components and cell lysis.[124]

2. The second mechanism proposes that chitosan targets the intracellular components of the microorganism. According to this mechanism, chitosan penetrates the microbial cell where it interacts with the DNA of fungi and bacteria. It is assumed that chitosan is able to penetrate the multilayered (murein cross-linked) bacterial cell wall and the cytoplasmic membrane.[133] The binding of chitosan with the DNA results in the inhibition of DNA transcription and also interferes with protein and mRNA synthesis.[134,135] The presence of chitooligomers inside *E. coli* cells has been confirmed by the use of confocal laser scanning microscope, and the probable cause of this antibacterial activity seemed to be the inhibition of transcription of DNA.[3] Although this model has been accepted as a possibility, the probability of high Mw chitosan to cross the multilayered cell wall and the cytoplasmic membrane and reach the cell interior seems to be less likely.[92] Hence, chitosan is essentially regarded as a membrane disruptor rather than a penetrating material.[133]

3. The third proposed mechanism involves the chelation of metal ions by chitosan. As the environmental pH is above $pK_a$, the chelation effect seems to be a more prominent factor responsible for antimicrobial activity rather than electrostatic effect.[136] At high pH, the unprotonated amino groups are able to donate its free electron pair to the metal ions, thereby forming complexes. Chitosan is well known for its excellent chelation capability toward metal ions.[137,138] This chelation process takes place in two ways: one is the bridge model, where the metal ions are bound to the amino groups belonging to the same chain or different chains of chitosan; and the other is the pendant model, where the metal ion is bound

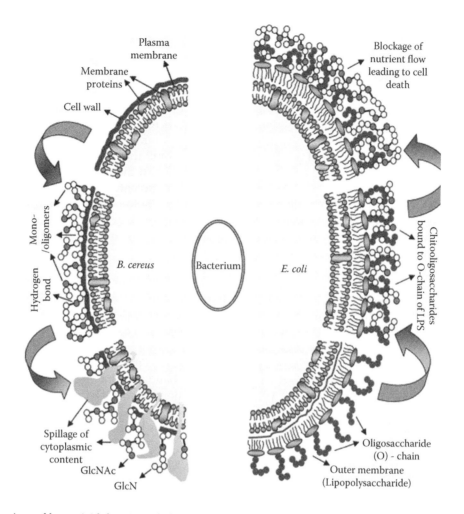

*Figure 25.3* Mechanism of bactericidal action of chito-oligomeric–monomeric mixture towards *B. cereus* and *E. coli*: a hypothetical model. LPS, lipopolysaccharide. The large arrows indicate the sequence of bactericidal actions. Reproduced from Vishu Kumar A.B. et al., *Biochem. J.*, 391, 167, 2005. With permission from the Biochemical Society.

to the amino group in a pendant fashion.[139] The free amino groups in chitosan can thus bind to trace metal ions like $Ca^{2+}$ and $Mg^{2+}$ present in the microorganism cell wall, which then inhibit the production of toxins and prevent the growth of the microorganism.[140] Although this seems to be a plausible mode of action, the chelation capability largely depends on the presence of unprotonated amino groups and hence this effect becomes crucial only after compromising the polycationic structure. Also, in case of chitosan derivatives, where the amino group has mostly been modified by various substituents, this mechanism is unlikely to occur.

## 25.6   Structure–activity relationship

The determination of the structure–activity relationship requires the tested compounds to be well characterized.

A good synthetic procedure for chitosan derivatives would require quantitative reaction and good control of the DS. This would enable reproducible results and help in obtaining uniformity in synthesized derivatives. This should be supported by good analytical data using techniques like $^1$H-NMR, FT-IR, which can help in precise characterization of the derivatives. This is of utmost importance during the assessment of the activity of derivatives since it would highlight the contribution of each of the structural components attached to the polymer backbone. Only a few systematic studies have contributed to the development of structure–activity relationship of such materials. This has mostly been limited to the quaternary derivatives of chitosan. The presence of quaternary groups in chitosan in these derivatives is vital for displaying antimicrobial properties. This has been observed in a study involving methylated chitosan derivatives, where unmethylated, mono-, and dimethylated amino groups of chitosan showed antibacterial properties only at pH 5.5, whereas the *N*-quaternary groups were responsible for high activity at pH 7.2.[19] A study of piperizine derivatives of chitosan showed that methylpiperazine and mono-quaternary dimethylpiperazine substituents did not contribute to activity toward Gram-positive and Gram-negative bacteria, while diquaternary trimethylpiperazine moiety, contributed to antibacterial activity at pH 7.2.[102] In a systematic study of antibacterial activity of analogous derivatives of D-glucosamine monomer, chitooligomer, and chitosan polymer, it was observed that increasing chain length in the quaternary substituent resulted in an increase in the antibacterial activity for monomer and chitooligomer derivatives but decrease in activity for the chitosan derivatives. The following order of activity amongst the chitosan derivatives was observed toward Gram-positive and Gram-negative bacteria: TMC > *N*-(2-(*N,N,N*-trialkylammoniumyl)acetyl) and *N*-(2-pyridiniumyl)acetyl) chitosan > *N*-(2-(*N,N*-dimethyl-*N*-dodecylammoniumyl)acetyl) chitosan.[36] The positioning of the cationic charge was also found to be an important factor in determining the activity of the chitosan derivatives. The highly substituted chitosan *N*-betaines remained low in activity at pH 7.2. When the pH was lowered to 5.5, the activity of this chitosan was found to increase with decrease in the DS of the polymer. This study concluded that the presence of a cationic charge on the polymer backbone is essential for exhibiting high antimicrobial properties.[40] A similar result was reported in a study involving the antibacterial properties of quaternary trimethylammoniumyl and pyridiniumyl chitosan derivatives having varying DA (6%–34.2%) and spacer lengths (C-2–C-6). This study showed that the trimethylammonium group was more active than the pyridinium group toward *S. aureus* and *E. coli*. Also, the antibacterial activity diminished as the quaternary group moved away from the polymer backbone; the activity remaining independent of DA variations.[37] The difference in activity due to the nature of the cationic group was also observed in another study where the trimethylammonium group presented higher bactericidal effect than the *N*-methylpyridinium group toward *S. aureus* and *E. coli* at similar DQ and Mw.[35] Thus, the aforementioned studies lead us to the conclusion that the antimicrobial activity of chitosan derivatives is largely dependent on the nature of the quaternary substituent attached and the positioning of the cationic charge on the polymer backbone.

## 25.7 Applications

The antimicrobial properties exhibited by chitosan and its derivatives make them suitable for a wide range of applications in the field of medicine, pharmaceutical industry, food industry, cosmetics, water treatment,

tissue engineering, and agriculture. Some of the common applications are discussed next.

### 25.7.1 In wound healing

An important application of chitosan and its derivatives is as antibacterial agents in wound healing. Quaternized trimethyl ammonium–modified chitosan fibers were able to inhibit the growth of *S. aureus* and showed low toxicity toward mouse fibroblast cells and hence could be used as potential wound dressing for skin regeneration.[141] Photo-cross-linked quaternary chitosan-containing electro spun mats were highly effective against *S. aureus* and *E. coli* and hence find application as wound dressing.[142,143] A comparative study was conducted on the treatment of open wound in mouse models with HemCom™ bandage, a chitosan acetate bandage, with alginate sponge bandage and silver sulfadiazine cream to compare their antimicrobial activity. It was observed that chitosan acetate was the most efficient in killing bacteria and controlling the growth of *P. aeruginosa*, *Proteus mirabilis*, and *S. aureus* as compared to other bandages that may encourage bacterial growth in the short term.[144] The chitosan dressing was further refined by introducing polyphosphate and silver which improved the hemostatic and antimicrobial efficiency.[145]

### 25.7.2 In food packaging

The contamination of food materials with microbes results in an increasing risk of food-borne illness. The need for storing food for longer time requires the use of chemical food preservatives. Natural and biodegradable antimicrobial polymers like chitosan has found increasing interest since the past decade for its effectiveness in the preservation for food. The potential of chitosan films as edible coatings to reduce microbial growth and enhance the storage life of food products has been investigated in many studies.[146,147] Chitosan forms films having smooth surfaces that can exhibit fat and oil resistance and have selective permeability to gases but fail to resist water transmission.[148] However, this property has been improved by combining chitosan with fatty acids. Chitosan-lauric acid films, for example, showed higher moisture barrier properties compared to pure chitosan films, as well as chitosan-palmitic or stearic acids.[149] The edible coatings prepared from CMC in combination with NaCl provided better preservation of freshly cut pears as compared to chitosan films.[150]

### 25.7.3 Against plant pathogens

Plant pathogens are considered an important group of microorganisms since they cause decay and destroy agricultural crops. The use of synthetic pesticides for controlling these pathogens has been of significant concern due to their ability to contaminate agricultural crops by leaving behind pesticide residues and their increasing resistance toward these pesticides. A number of studies have shown that chitosan and its derivatives have been considered as useful pesticides in the control of plant diseases.[55,107,151] The increasing interest in the antimicrobial properties of chitosan and its derivatives is due to its high antimicrobial activity, lower toxicity, biodegradability, and biocidal activity toward a broad spectrum of pathogenic microorganisms. This makes it potentially superior to synthetic pesticides. Chitosan has been shown to be very effective in controlling diseases in crops such as rice, wheat, barley, oilseed rape, tobacco, etc.[152] Several chitosan derivatives, such as *N*-alkyl-*N,N*-dialkyl quaternary chitosan, *N*-acyl chitosan, Schiff bass of chitosan, and *N*-heterocyclic chitosan, are also very effective against several plant pathogenic bacteria and fungi.[152] Additionally, chitosan showed antiviral activity against plant viruses by inhibiting the infection caused by bacteriophage.[153] Hence chitosan has been widely studied in the control of diseases of various plant systems against bacteria, fungi, and virus.[154]

## 25.7.4   In dentistry

Microorganisms of the *Streptococcus* species, such as, *Streptococcus sanguis, Streptococcus mutans, Streptococcus oralis*, and *Streptococcus mitis*, play a significant role in the formation of dental caries, also caused by mineral imbalances. The microbial community forms the dental biofilm, which is capable of withstanding harsh conditions and, hence, is resistant to many antibacterial agents. Chitosan gels prepared in acetic acid have been found to be a preventative and therapeutic agent for dental caries.[155] The use of chitosan with dental materials such as hydroxyapatite not only increases the biocompatibility of the materials, but also prevents the absorption of oral bacteria to the tooth.[156] Chitosan used in different formulations, such as mouth rinses, toothpaste, and chewing gums,[157] have been effective as antibacterial agents in controlling the *Streptococcus* group of bacteria. They effectively inhibit bacterial plague formation and accelerate the secretion of saliva.[157–159] The order of inhibitory effect of chitosan and its derivatives on the adherence of oral bacteria onto human anterior teeth surfaces are low molecular chitosan > phosphorylated chitosan > amorphous chitosan > CMC.[160] Also, the lactic acid solution of *N*-[1-Hydroxy-3-(trimethylammonium) propyl]chitosan chloride was found to exhibit a better inhibitory effect toward oral pathogens as compared to the lactic acid solution of chitosan.[161]

## 25.8   Conclusion

Chitosan has been seen to exhibit high antimicrobial properties against a wide range of pathogens, its activity being influenced by factors like Mw, pH of the medium, and degree of acetylation. Its antimicrobial properties are only displayed at low pH when the amino group is protonized. Its Mw can have a considerable effect but there is no consensus on the optimal Mw and this may be dependent on the type of microorganism. Also, the DA does not have an effect on the antimicrobial properties when its value is below 50%.

The antimicrobial property of chitosan can be greatly enhanced by derivatization, particularly at the C-2 amino group. Various synthetic modifications of chitosan have been carried out in order to improve its aqueous solubility and its antimicrobial efficacy. The introduction of various functional groups like cationic and hydrophobic moieties has contributed to effective ways to control the antimicrobial property of chitosan, most of the derivatives being able to display better antimicrobial properties than chitosan at normal pH. However, a significant variability has been observed in the reported MIC values for chitosan and its derivatives toward a particular bacterial strain. This discrepancy in the results could be largely due to the inadequate characterization of the derivatives synthesized for these studies. Additional factors like different experimental methods used for determining the activity also contribute to the difference in the results obtained. Hence, use of appropriate synthetic methods, proper characterization of the derivatives, and use of standard assays for activity would contribute to a much better understanding of the structure–activity relationship. Also, more insights into the mode of action of the chitosan derivatives are needed as this could be used to guide the development of new antimicrobial agents based on chitosan.

## Acknowledgment

This study was supported by grants from the Icelandic Centre for Research.

## References

1. Martínez, J. P., Falomir, M. P., and Gozalbo, D. *Chitin: A Structural Biopolysaccharide with Multiple Applications.* eLS: John Wiley & Sons, Ltd.: Chichester, U.K., 2001.
2. Shimahara, K., Tagikushi, Y., Kobayashi, T., Uda, K., and Sannan, T. Screening of mucoraceae strains suitable for chitosan production. In G. Skjak-Brack, T. Anthonsen, and P. Sanford (eds.), *Chitin and Chitosan.* Elsevier Applied Science, London, U.K. pp. 171–178, 1989.
3. Fei Liu, X., Lin Guan, Y., Zhi Yang, D., Li, Z., and De Yao, K. Antibacterial action of chitosan and carboxymethylated chitosan. *Journal of Applied Polymer Science* 79, 7(2001): 1324–1335.

4. No, H. K., Young Park, N., Ho Lee, S., and Meyers, S. P. Antibacterial activity of chitosans and chitosan oligomers with different molecular weights. *International Journal of Food Microbiology* 74, 1–2(2002): 65–72.

5. Devlieghere, F., Vermeulen, A., and Debevere, J. Chitosan: Antimicrobial activity, interactions with food components and applicability as a coating on fruit and vegetables. *Food Microbiology* 21, 6(2004): 703–714.

6. Mellegård, H., Strand, S. P., Christensen, B. E., Granum, P. E., and Hardy, S. P. Antibacterial activity of chemically defined chitosans: Influence of molecular weight, degree of acetylation and test organism. *International Journal of Food Microbiology* 148, 1(2011): 48–54.

7. Tsai, G.-J., Su, W.-H., Chen, H.-C., and Pan, C.-L. Antimicrobial activity of shrimp chitin and chitosan from different treatments and applications of fish preservation. *Fisheries Science* 68, 1(2002): 170–177.

8. Simunek, J., Brandysova, V., Koppova, I., and Simunek, J., Jr. The antimicrobial action of chitosan, low molar mass chitosan, and chitooligosaccharides on human colonic bacteria. *Folia Microbiologica (Praha)* 57, 4(2012): 341–345.

9. Jeon, Y.-J., Park, P.-J., and Kim, S.-K. Antimicrobial effect of chitooligosaccharides produced by bioreactor. *Carbohydrate Polymers* 44, 1(2001): 71–76.

10. Tikhonov, V. E., Stepnova, E. A., Babak, V. G., Yamskov, I. A., Palma-Guerrero, J., Jansson, H.-B., Lopez-Llorca, L. V et al. Bactericidal and antifungal activities of a low molecular weight chitosan and its N-/2(3)-(dodec-2-enyl) succinoyl/-derivatives. *Carbohydrate Polymers* 64, 1(2006): 66–72.

11. Kulikov, S., Tikhonov, V., Blagodatskikh, I., Bezrodnykh, E., Lopatin, S., Khairullin, R., Philippova, Y., and Abramchuk, S. Molecular weight and pH aspects of the efficacy of oligochitosan against methicillin-resistant *Staphylococcus aureus* (MRSA). *Carbohydrate Polymers* 87, 1(2012): 545–550.

12. Simunek, J., Koppova, I., Filip, L., Tishchenko, G., and Belzecki, G. The antimicrobial action of low-molar-mass chitosan, chitosan derivatives and chitooligosaccharides on bifidobacteria. *Folia Microbiologica (Praha)* 55, 4(2010): 379–382.

13. Hirano, S. and Nagao, N. Effects of chitosan, pectic acid, lysozyme, and chitinase on the growth of several phytopathogens. *Agricultural and Biological Chemistry* 53, 11(1989): 3065–3066.

14. Uchida, Y., Izume, M., and Ohtakara, A. *Preparation of Chitosan Oligomers with Purified Chitosanase and its Application.* Elsevier: London, U.K., 1989.

15. Zheng, L.-Y. and Zhu, J.-F. Study on antimicrobial activity of chitosan with different molecular weights. *Carbohydrate Polymers* 54, 4(2003): 527–530.

16. Raafat, D., von Bargen, K., Haas, A., and Sahl, H.-G. Insights into the mode of action of chitosan as an antibacterial compound. *Applied and Environmental Microbiology* 74, 12(2008): 3764–3773.

17. Jung, E. J., Youn, D. K., Lee, S. H., No, H. K., Ha, J. G., and Prinyawiwatkul, W. Antibacterial activity of chitosans with different degrees of deacetylation and viscosities. *International Journal of Food Science & Technology* 45, 4(2010): 676–682.

18. Qin, C., Li, H., Xiao, Q., Liu, Y., Zhu, J., and Du, Y. Water-solubility of chitosan and its antimicrobial activity. *Carbohydrate Polymers* 63, 3(2006): 367–374.

19. Rúnarsson, Ö. V., Holappa, J., Nevalainen, T., Hjálmarsdóttir, M., Järvinen, T., Loftsson, T., Einarsson, J. M., Jónsdóttir, S., Valdimarsdóttir, M., and Másson, M. Antibacterial activity of methylated chitosan and chitooligomer derivatives: Synthesis and structure activity relationships. *European Polymer Journal* 43, 6(2007): 2660–2671.

20. Kurita, K., Kamiya, M., and Nishimura, S.-I. Solubilization of a rigid polysaccharide: Controlled partial N-acetylation of chitosan to develop solubility. *Carbohydrate Polymers* 16, 1(1991): 83–92.

21. Andres, Y., Giraud, L., Gerente, C., and Le Cloirec, P. Antibacterial effects of chitosan powder: Mechanisms of action. *Environmental Technology* 28, 12(2007): 1357–1363.

22. Chung, Y.-C. and Chen, C.-Y. Antibacterial characteristics and activity of acid-soluble chitosan. *Bioresource Technology* 99, 8(2008): 2806–2814.

23. Park, P.-J., Je, J.-Y., Byun, H.-G., Moon, S.-H., and Kim, S.-K. Antimicrobial activity of hetero-chitosans and their oligosaccharides with different molecular weights. *Journal of microbiology and biotechnology* 14, 2(2004): 317–323.

24. Je, J.-Y. and Kim, S.-K. Antimicrobial action of novel chitin derivative. *Biochimica et Biophysica Acta (BBA)— General Subjects* 1760, 1(2006): 104–109.

25. Xie, Y., Liu, X., and Chen, Q. Synthesis and characterization of water-soluble chitosan derivate and its antibacterial activity. *Carbohydrate Polymers* 69, 1(2007): 142–147.

26. Pei, L., Cai, Z., Shang, S., and Song, Z. Synthesis and antibacterial activity of alkylated chitosan under basic ionic liquid conditions. *Journal of Applied Polymer Science* 131, 7(2014): 40052–40059.

27. Mohy Eldin, M. S., Soliman, E. A., Hashem, A. I., and Tamer, T. M. Antimicrobial activity of novel aminated chitosan derivatives for biomedical applications. *Advances in Polymer Technology* 31, 4(2012): 414–428.

28. Lee, D. S., Jeong, S. Y., Kim, Y. M., Lee, M. S., Ahn, C. B., and Je, J. Y. Antibacterial activity of aminoderivatized chitosans against methicillin-resistant *Staphylococcus aureus* (MRSA). *Bioorganic & Medicinal Chemistry* 17, 20(2009): 7108–7112.

29. Jin, X., Wang, J., and Bai, J. Synthesis and antimicrobial activity of the Schiff base from chitosan and citral. *Carbohydrate Research* 344, 6(2009): 825–829.

30. Wang, J., Lian, Z., Wang, H., Jin, X., and Liu, Y. Synthesis and antimicrobial activity of Schiff base of chitosan and acylated chitosan. *Journal of Applied Polymer Science* 123, 6(2012): 3242–3247.

31. Badawy, M. E. and Rabea, E. I. Synthesis and structure-activity relationship of N-(cinnamyl) chitosan analogs as antimicrobial agents. *International Journal of Biological Macromolecules* 57, (2013): 185–192.

32. Abou-Zeid, N. Y., Waly, A. I., Kandile, N. G., Rushdy, A. A., El-Sheikh, M. A., and Ibrahim, H. M. Preparation, characterization and antibacterial properties of cyanoethylchitosan/cellulose acetate polymer blended films. *Carbohydrate Polymers* 84, 1(2011): 223–230.

33. Je, J. Y. and Kim, S. K. Chitosan derivatives killed bacteria by disrupting the outer and inner membrane. *Journal of Agricultural and Food Chemistry* 54, 18(2006): 6629–6633.

34. Huang, J., Jiang, H., Qiu, M., Geng, X., Yang, R., Li, J., and Zhang, C. Antibacterial activity evaluation of quaternary chitin against *Escherichia coli* and *Staphylococcus aureus*. *International Journal of Biological Macromolecules* 52, (2013): 85–91.

35. Sajomsang, W., Ruktanonchai, U. R., Gonil, P., and Warin, C. Quaternization of *N*-(3-pyridylmethyl) chitosan derivatives: Effects of the degree of quaternization, molecular weight and ratio of *N*-methylpyridinium and *N,N,N*-trimethyl ammonium moieties on bactericidal activity. *Carbohydrate Polymers* 82, 4(2010): 1143–1152.

36. Rúnarsson, Ö. V., Holappa, J., Malainer, C., Steinsson, H., Hjálmarsdóttir, M., Nevalainen, T., and Másson, M. Antibacterial activity of N-quaternary chitosan derivatives: Synthesis, characterization and structure activity relationship (SAR) investigations. *European Polymer Journal* 46, 6(2010): 1251–1267.

37. Sahariah, P., Gaware, V., Lieder, R., Jónsdóttir, S., Hjálmarsdóttir, M., Sigurjonsson, O., and Másson, M. The effect of substituent, degree of acetylation and positioning of the cationic charge on the antibacterial activity of quaternary chitosan derivatives. *Marine Drugs* 12, 8(2014): 4635–4658.

38. Sajomsang, W., Gonil, P., and Saesoo, S. Synthesis and antibacterial activity of methylated *N*-(4-*N,N*-dimethylaminocinnamyl) chitosan chloride. *European Polymer Journal* 45, 8(2009): 2319–2328.

39. Jia, Z., Shen, D., and Xu, W. Synthesis and antibacterial activities of quaternary ammonium salt of chitosan. *Carbohydrate Research* 333, 1(2001): 1–6.

40. Holappa, J., Hjálmarsdóttir, M., Másson, M., Rúnarsson, Ö., Asplund, T., Soininen, P., Nevalainen, T., and Järvinen, T. Antimicrobial activity of chitosan *N*-betainates. *Carbohydrate Polymers* 65, 1(2006): 114–118.

41. Sajomsang, W., Tantayanon, S., Tangpasuthadol, V., and Daly, W. H. Quaternization of *N*-aryl chitosan derivatives: Synthesis, characterization, and antibacterial activity. *Carbohydrate Research* 344, 18(2009): 2502–2511.

42. Zhong, Z., Xing, R., Liu, S., Wang, L., Cai, S., and Li, P. Synthesis of acyl thiourea derivatives of chitosan and their antimicrobial activities in vitro. *Carbohydrate Research* 343, 3(2008): 566–570.

43. Cai, Z.-S., Sun, Y.-M., Zhu, X.-M., Zhao, L.-L., and Yue, G.-Y. Preparation and characterization of ortho-biguanidinyl benzoyl chitosan hydrochloride and its antibacterial activities. *Polymer Bulletin* 70, 3(2013): 1085–1096.

44. Zhao-Sheng, C., Yue-Ming, S., Chun-Sheng, Y., and Xue-Mei, Z. Preparation, characterization, and antibacterial activities of para-biguanidinyl benzoyl chitosan hydrochloride. *Journal of Applied Polymer Science* 125, 2(2012): 1146–1151.

45. Zhong, Z., Aotegen, B., and Xu, H. The influence of the different inductivity of acetyl phenyl-thiosemicarbazone-chitosan on antimicrobial activities. *International Journal of Biological Macromolecules* 48, 5(2011): 713–719.

46. Zhong, Z., Aotegen, B., Xu, H., and Zhao, S. The influence of chemical structure on the antimicrobial activities of thiosemicarbazone-chitosan. *Cellulose* 21, 1(2014): 105–114.

47. Zhong, Z., Aotegen, B., Xu, H., and Zhao, S. Structure and antimicrobial activities of benzoyl phenyl-thiosemicarbazone-chitosans. *International Journal of Biological Macromolecules* 50, 4(2012): 1169–1174.

48. Kaya, M., Cakmak, Y., Baran, T., Asan-Ozusaglam, M., Mentes, A,. and Tozak, K. New chitin, chitosan, and O-carboxymethyl chitosan sources from resting eggs of *Daphnia longispina* (Crustacea); with physicochemical characterization, and antimicrobial and antioxidant activities. *Biotechnology and Bioprocess Engineering* 19, 1(2014): 58–69.

49. Cai, Z.-S., Song, Z.-Q., Yang, C.-S., Shang, S.-B., and Yin, Y.-B. Synthesis, characterization and antibacterial activity of quaternized N,O-(2-carboxyethyl) chitosan. *Polymer Bulletin* 62, 4(2009): 445–456.

50. Mohamed, N. A. and Abd El-Ghany, N. A. Preparation and antimicrobial activity of some carboxymethyl chitosan acyl thiourea derivatives. *International Journal of Biological Macromolecules* 50, 5(2012): 1280–1285.

51. Muzzarelli, R. A. A. and Tanfani, F. The N-permethylation of chitosan and the preparation of N-trimethyl chitosan iodide. *Carbohydrate Polymers* 5, 4(1985): 297–307.

52. Benediktsdóttir, B. E., Gaware, V. S., Rúnarsson, Ö. V., Jónsdóttir, S., Jensen, K. J., and Másson, M. Synthesis of *N,N,N*-trimethyl chitosan homopolymer and highly substituted *N*-alkyl-*N,N*-dimethyl chitosan derivatives with the aid of di-tert-butyldimethylsilyl chitosan. *Carbohydrate Polymers* 86, 4(2011): 1451–1460.

53. Xu, T., Xin, M., Li, M., Huang, H., and Zhou, S. Synthesis, characteristic and antibacterial activity of *N,N,N*-trimethyl chitosan and its carboxymethyl derivatives. *Carbohydrate Polymers* 81, 4(2010): 931–936.

54. Xu, T., Xin, M., Li, M., Huang, H., Zhou, S., and Liu, J. Synthesis, characterization, and antibacterial activity of N,O-quaternary ammonium chitosan. *Carbohydrate Research* 346, 15(2011): 2445–2450.

55. Badawy, M. E. I. Structure and antimicrobial activity relationship of quaternary N-alkyl chitosan derivatives against some plant pathogens. *Journal of Applied Polymer Science* 117, 2(2010): 960–969.

56. Kim, C., Choi, J., Chun, H., and Choi, K. Synthesis of chitosan derivatives with quaternary ammonium salt and their antibacterial activity. *Polymer Bulletin* 38, 4(1997): 387–393.

57. Wiarachai, O., Thongchul, N., Kiatkamjornwong, S., and Hoven, V. P. Surface-quaternized chitosan particles as an alternative and effective organic antibacterial material. *Colloids and Surfaces B: Biointerfaces* 92, (2012): 121–129.

58. Sadeghi, A. M. M., Dorkoosh, F. A., Avadi, M. R., Saadat, P., Rafiee-Tehrani, M., and Junginger, H. E. Preparation, characterization and antibacterial activities of chitosan, N-trimethyl chitosan (TMC) and N-diethylmethyl chitosan (DEMC) nanoparticles loaded with insulin using both the ionotropic gelation and polyelectrolyte complexation methods. *International Journal of Pharmaceutics* 355, 1–2(2008): 299–306.

59. Martins, A., Facchi, S., Follmann, H., Pereira, A., Rubira, A., and Muniz, E. Antimicrobial Activity of chitosan derivatives containing Nquaternized moieties in its backbone: A review. *International Journal of Molecular Sciences* 15, 11(2014): 20800–20832.

60. Fu, X., Shen, Y., Jiang, X., Huang, D., and Yan, Y. Chitosan derivatives with dual-antibacterial functional groups for antimicrobial finishing of cotton fabrics. *Carbohydrate Polymers* 85, 1(2011): 221–227.

61. Sadeghi-Kiakhani, M., Arami, M., and Gharanjig, K. Application of a biopolymer chitosan-poly(propylene) imine dendrimer hybrid as an antimicrobial agent on the wool fabrics. *Iranian Polymer Journal* 22, 12(2013): 931–940.

62. Klaykruayat, B., Siralertmukul, K., and Srikulkit, K. Chemical modification of chitosan with cationic hyper-branched dendritic polyamidoamine and its antimicrobial activity on cotton fabric. *Carbohydrate Polymers* 80, 1(2010): 197–207.

63. Hu, Y., Du, Y., Yang, J., Tang, Y., Li, J., and Wang, X. Self-aggregation and antibacterial activity of N-acylated chitosan. *Polymer* 48, 11(2007): 3098–3106.

64. Muzzarelli, R., Tarsi, R., Filippini, O., Giovanetti, E., Biagini, G., and Varaldo, P. Antimicrobial properties of N-carboxybutyl chitosan. *Antimicrobial agents and chemotherapy* 34, 10(1990): 2019–2023.

65. Badawy, M. E. I. and Rabea, E. I. Characterization and antimicrobial activity of water-soluble N-(4-carboxybutyroyl) chitosans against some plant pathogenic bacteria and fungi. *Carbohydrate Polymers* 87, 1(2012): 250–256.

66. He, G., Chen, X., Yin, Y., Zheng, H., Xiong, X., and Du, Y. Synthesis, characterization and antibacterial activity of salicyloyl chitosan. *Carbohydrate Polymers* 83, 3(2011): 1274–1278.

67. Feng, Y. and Xia, W. Preparation, characterization and antibacterial activity of water-soluble O-fumaryl-chitosan. *Carbohydrate Polymers* 83, 3(2011): 1169–1173.

68. Geisberger, G., Gyenge, E. B., Hinger, D., Kach, A., Maake, C., and Patzke, G. R. Chitosan-thioglycolic acid as a versatile antimicrobial agent. *Biomacromolecules* 14, 4(2013): 1010–1017.

69. Fernandes, M. M., Francesko, A., Torrent-Burgués, J., and Tzanov, T. Effect of thiol-functionalisation on chitosan antibacterial activity: Interaction with a bacterial membrane model. *Reactive and Functional Polymers* 73, 10(2013): 1384–1390.

70. Suvannasara, P., Juntapram, K., Praphairaksit, N., Siralertmukul, K., and Muangsin, N. Mucoadhesive 4-carboxybenzenesulfonamide-chitosan with antibacterial properties. *Carbohydrate Polymers* 94, 1(2013): 244–252.

71. Xing, K., Chen, X. G., Kong, M., Liu, C. S., Cha, D. S., and Park, H. J. Effect of oleoyl-chitosan nanoparticles as a novel antibacterial dispersion system on viability, membrane permeability and cell morphology of *Escherichia coli* and *Staphylococcus aureus*. *Carbohydrate Polymers* 76, 1(2009): 17–22.

72. Naberezhnykh, G. A., Bakholdina, S. I., Gorbach, V. I., and Solov'eva, T. F. New chitosan derivatives with potential antimicrobial activity. *Russian Journal of Marine Biology* 35, 6(2009): 498–503.

73. Pavinatto, A., Souza, A. L., Delezuk, J. A., Pavinatto, F. J., Campana-Filho, S. P., and Oliveira, O. N., Jr. Interaction of O-acylated chitosans with biomembrane models: Probing the effects from hydrophobic interactions and hydrogen bonding. *Colloids and Surfaces B: Biointerfaces* 114, (2014): 53–59.

74. Peng, Y., Han, B., Liu, W., and Xu, X. Preparation and antimicrobial activity of hydroxypropyl chitosan. *Carbohydrate Research* 340, 11(2005): 1846–1851.

75. Liu, H., Zhao, Y., Cheng, S., Huang, N., and Leng, Y. Syntheses of novel chitosan derivative with excellent solubility, anticoagulation, and antibacterial property by chemical modification. *Journal of Applied Polymer Science* 124, 4(2012): 2641–2648.

76. Xie, W., Xu, P., Wang, W., and Liu, Q. Preparation and antibacterial activity of a water-soluble chitosan derivative. *Carbohydrate Polymers* 50, 1(2002): 35–40.

77. Jou, C.-H., Yang, M.-C., Suen, M.-C., Yen, C.-K., Hung, C.-C., and Hwang, M.-C. Preparation of O-diallylammonium chitosan with antibacterial activity and cytocompatibility. *Polymer International* 62, 3(2013): 507–514.

78. Li, R., Hu, P., Ren, X., Worley, S. D., and Huang, T. S. Antimicrobial N-halamine modified chitosan films. *Carbohydrate Polymers* 92, 1(2013): 534–539.

79. El-Shafei, A. M., Fouda, M. M. G., Knittel, D., and Schollmeyer, E. Antibacterial activity of cationically modified cotton fabric with carboxymethyl chitosan. *Journal of Applied Polymer Science* 110, 3(2008): 1289–1296.

80. Jin, Z., Li, W., Cao, H., Zhang, X., Chen, G., Wu, H., Guo, C. et al. Antimicrobial activity and cytotoxicity of N-2-HACC and characterization of nanoparticles with N-2-HACC and CMC as a vaccine carrier. *Chemical Engineering Journal* 221, (2013): 331–341.

81. Di, Y., Long, G., Zhang, H., and Li, Q. Preparation and properties of viscose rayon/O-carboxymethyl chitosan antibacterial fibers. *Journal of Engineered Fabrics & Fibers* 6, 3(2011): 39–43.

82. Zhu, X.-M., Sheng, Z., and Yang, G. Synthesis of 2-hydroxypropyl dimethylbenzyl ammonium N, O-carboxymethyl chitosan chloride and its antibacterial activity. *Journal of the Chemical Society of Pakistan* 31, 4(2009): 652–659.

83. Alencar de Queiroz, A. A., Abraham, G. A., Pires Camillo, M. A., Higa, O. Z., Silva, G. S., del Mar Fernandez, M., and San Roman, J. Physicochemical and antimicrobial properties of boron-complexed polyglycerol-chitosan dendrimers. *Journal of Biomaterials Science, Polymer Edition* 17, 6(2006): 689–707.

84. Wen, Y., Tan, Z., Sun, F., Sheng, L., Zhang, X., and Yao, F. Synthesis and characterization of quaternized carboxymethyl chitosan/poly(amidoamine) dendrimer core–shell nanoparticles. *Materials Science and Engineering: C* 32, 7(2012): 2026–2036.

85. Yin, X., Chen, J., Yuan, W., Lin, Q., Ji, L., and Liu, F. Preparation and antibacterial activity of Schiff bases from O-carboxymethyl chitosan and para-substituted benzaldehydes. *Polymer Bulletin* 68, 5(2012): 1215–1226.

86. Sun, L., Du, Y., Fan, L., Chen, X., and Yang, J. Preparation, characterization and antimicrobial activity of quaternized carboxymethyl chitosan and application as pulp-cap. *Polymer* 47, 6(2006): 1796–1804.

87. Mohamed, N. and Abd El-Ghany, N. Synthesis and antimicrobial activity of some novel terephthaloyl thiourea cross-linked carboxymethyl chitosan hydrogels. *Cellulose* 19, 6(2012): 1879–1891.

88. Mohamed, N. A., Mohamed, R. R., and Seoudi, R. S. Synthesis and characterization of some novel antimicrobial thiosemicarbazone O-carboxymethyl chitosan derivatives. *International Journal of Biological Macromolecules* 63, (2014): 163–169.

89. Sabaa, M. W., Mohamed, N. A., Mohamed, R. R., Khalil, N. M., and Abd El Latif, S. M. Synthesis, characterization and antimicrobial activity of poly (N-vinyl imidazole) grafted carboxymethyl chitosan. *Carbohydrate Polymers* 79, 4(2010): 998–1005.

90. CLSI M22-A3, Quality control for commercially prepared microbiological culture media; Approved standard. Clinical Laboratory Standards Institute, Wayne, PA, 2004.

91. Raafat, D., von Bargen, K., Haas, A., and Sahl, H.-G. Insights into the mode of action of chitosan as an antibacterial compound. *Applied and Environmental Microbiology AEM* 74, 12(2008): 3764–3773.

92. CLSI M07-A8, Methods for dilution antimicrobial susceptibility tests for bacteria that grow aerobically. Clinical Laboratory Standards Institute, Wayne, PA, 2009.

93. Jiang, L., Wang, F., Han, F., Prinyawiwatkul, W., No, H. K., and Ge, B. Evaluation of diffusion and dilution methods to determine the antimicrobial activity of water-soluble chitosan derivatives. *Journal of Applied Microbiology* 114, 4(2013): 956–963.

94. Kaur, K. G., Rashmi, K., Manmohan, C., and Moushumi, G. Development of a quaternized chitosan with enhanced antibacterial efficacy. *Journal of Water and Health* 11, 3(2013): 410–418.

95. Meng, X., Xing, R., Liu, S., Yu, H., Li, K., Qin, Y., and Li, P. Molecular weight and pH effects of aminoethyl modified chitosan on antibacterial activity in vitro. *International Journal of Biological Macromolecules* 50, 4(2012): 918–924.

96. Sajomsang, W., Gonil, P., and Tantayanon, S. Antibacterial activity of quaternary ammonium chitosan containing mono or disaccharide moieties: Preparation and characterization. *International Journal of Biological Macromolecules* 44, 5(2009): 419–427.

97. Mohamed, N. A., Sabaa, M. W., El-Ghandour, A. H., Abdel-Aziz, M. M., and Abdel-Gawad, O. F. Quaternized N-substituted carboxymethyl chitosan derivatives as antimicrobial agents. *International Journal of Biological Macromolecules* 60, (2013): 156–164.

98. Sajomsang, W., Tantayanon, S., Tangpasuthadol, V., and Daly, W. H. Synthesis of methylated chitosan containing aromatic moieties: Chemoselectivity and effect on molecular weight. *Carbohydrate Polymers* 72, 4(2008): 740–750.

99. Jou, C.-H. Antibacterial activity and cytocompatibility of chitosan-N-hydroxy-2,3-propyl-N-methyl-N,N-diallylammonium methyl sulfate. *Colloids and Surfaces B: Biointerfaces* 88, 1(2011): 448–454.

100. CLSI M26-A. Methods for determining bactericidal activity of antimicrobial agents. Clinical Laboratory Standards Institute, Wayne, PA, 1999.

101. Garcia, L. S. and Isenberg, H. D. *Clinical Microbiology Procedures Handbook*, Vol. 2. ASM Press, Washington, DC, 2010.

102. Másson, M., Holappa, J., Hjálmarsdóttir, M., Rúnarsson, Ö. V., Nevalainen, T., and Järvinen, T. Antimicrobial activity of piperazine derivatives of chitosan. *Carbohydrate Polymers* 74, 3(2008): 566–571.

103. Lu, Y., Slomberg, D. L., and Schoenfisch, M. H. Nitric oxide-releasing chitosan oligosaccharides as antibacterial agents. *Biomaterials* 35, 5(2014): 1716–1724.

104. Li, Y., Chen, X. G., Liu, N., Liu, C. S., Liu, C. G., Meng, X. H., Yu, L. J., and Kenendy, J. F. Physicochemical characterization and antibacterial property of chitosan acetates. *Carbohydrate Polymers* 67, 2(2007): 227–232.

105. Xu, T., Xin, M., Li, M., Huang, H., Zhou, S., and Liu, J. Synthesis, characterization, and antibacterial activity of N,O-quaternary ammonium chitosan. *Carbohydrate Research* 346, 15(2011): 2445–2450.

106. Peng, Z.-X., Wang, L., Du, L., Guo, S.-R., Wang, X.-Q., and Tang, T.-T. Adjustment of the antibacterial activity and biocompatibility of hydroxypropyltrimethyl ammonium chloride chitosan by varying the degree of substitution of quaternary ammonium. *Carbohydrate Polymers* 81, 2(2010): 275–283.

107. Rabea, E. I., Badawy, M. E. I., Steurbaut, W., and Stevens, C. V. In vitro assessment of N-(benzyl)chitosan derivatives against some plant pathogenic bacteria and fungi. *European Polymer Journal* 45, 1(2009): 237–245.

108. Yang, J., Cai, J., Hu, Y., Li, D., and Du, Y. Preparation, characterization and antimicrobial activity of 6-amino-6-deoxychitosan. *Carbohydrate Polymers* 87, 1(2012): 202–209.

109. Cai, Z. S., Song, Z. Q., Yang, C. S., Shang, S. B., and Yin, Y. B. Synthesis of 2-hydroxypropyl dimethylbenzylammonium N,O-(2-carboxyethyl) chitosan chloride and its antibacterial activity. *Journal of Applied Polymer Science* 111, 6(2009): 3010–3015.

110. Wang, J. and Wang, H. Preparation of soluble p-aminobenzoyl chitosan ester by Schiff's base and antibacterial activity of the derivatives. *International Journal of Biological Macromolecules* 48, 3(2011): 523–529.

111. Kim, C. H., Kim, S. Y., and Choi, K. S. Synthesis and antibacterial activity of water-soluble chitin derivatives. *Polymers for Advanced Technologies* 8, 5(1997): 319–325.

112. Jeon, S. J., Oh, M., Yeo, W.-S., Galvão, K. N., and Jeong, K. C. Underlying mechanism of antimicrobial activity of chitosan microparticles and implications for the treatment of infectious diseases. *PloS One* 9, 3(2014): e927239(1–10).

113. Chi, W., Qin, C., Zeng, L., Li, W., and Wang, W. Microbiocidal activity of chitosan-N-2-hydroxypropyl trimethyl ammonium chloride. *Journal of Applied Polymer Science* 103, 6(2007): 3851–3856.

114. Chethan, P., Vishalakshi, B., Sathish, L., Ananda, K., and Poojary, B. Preparation of substituted quaternized aryl-furan chitosan derivatives and their antimicrobial activity. *International Journal of Biological Macromolecules* 59, (2013): 158–164.

115. CLSI M02-A10. Performance standards for antimicrobial disk susceptibility tests; approved standard. Clinical and Laboratory Standards Institute, Wayne, PA, 2009.

116. Lee, D. S., Kim, Y. M., Lee, M. S., Ahn, C. B., Jung, W. K., and Je, J. Y. Synergistic effects between aminoethyl-chitosans and beta-lactams against methicillin-resistant Staphylococcus aureus (MRSA). *Bioorganic and Medicinal Chemistry Letters* 20, 3(2010): 975–978.

117. Yalinca, Z., Yilmaz, E., Taneri, B., and Bullici, F. T. A comparative study on antibacterial activities of chitosan based products and their combinations with gentamicin against *S. epidermidis* and *E. coli. Polymer Bulletin* 70, 12(2013): 3407–3423.

118. Chen, Y., Wang, F., Yun, D., Guo, Y., Ye, Y., Wang, Y., and Tan, H. Preparation of a C6 quaternary ammonium chitosan derivative through a chitosan schiff base with click chemistry. *Journal of Applied Polymer Science* 129, 6(2013): 3185–3191.

119. Versalovic, J., Carroll, K. C., Funke, G., Jorgensen, J. H., Landry, M. L., and Warnock, D. W. *Manual of Clinical Microbiology, Antifungal Agents and Susceptibility Test Methods.* American Society of Microbiology, 2011.

120. Qin, Y., Xing, R., Liu, S., Li, K., Hu, L., Yu, H., Chen, X., and Li, P. Synthesis of chitosan derivative with diethyldithiocarbamate and its antifungal activity. *International Journal of Biological Macromolecules* 65, (2014): 369–374.

121. Qin, Y., Xing, R., Liu, S., Li, K., Meng, X., Li, R., Cui, J., Li, B., and Li, P. Novel thiosemicarbazone chitosan derivatives: Preparation, characterization, and antifungal activity. *Carbohydrate Polymers* 87, 4(2012): 2664–2670.

122. Da Silva, L. P., de Britto, D., Seleghim, M. H. R., and Assis, O. B. In vitro activity of water-soluble quaternary chitosan chloride salt against *E. coli. World Journal of Microbiology and Biotechnology* 26, 11(2010): 2089–2092.

123. Vishu Kumar, A. B., Varadaraj, M. C., Gowda, L. R., and Tharanathan, R. N. Characterization of chito-oligosaccharides prepared by chitosanolysis with the aid of papain and Pronase, and their bactericidal action against *Bacillus cereus* and *Escherichia coli. Biochemical Journal* 391, 2(2005): 167–175.

124. Tang, H., Zhang, P., Kieft, T. L., Ryan, S. J., Baker, S. M., Wiesmann, W. P., and Rogelj, S. Antibacterial action of a novel functionalized chitosan-arginine against Gram-negative bacteria. *Acta Biomaterialia* 6, 7(2010): 2562–2571.

125. Kong, M., Chen, X. G., Liu, C. S., Liu, C. G., Meng, X. H., and Yu le, J. Antibacterial mechanism of chitosan microspheres in a solid dispersing system against *E. coli. Colloids and Surfaces B: Biointerfaces* 65, 2(2008): 197–202.

126. Helander, I. M., Nurmiaho-Lassila, E. L., Ahvenainen, R., Rhoades, J., and Roller, S. Chitosan disrupts the barrier properties of the outer membrane of Gram-negative bacteria. *International Journal of Food Microbiology* 71, 2–3(2001): 235–244.

127. Jawetz, E., Melnick, J. L., and Adelberg, E. A. *Review of Medical Microbiology.* Lange Medical Publications, Los Altos, CA, 1974.

128. Ketchum, P. A. *Microbiology: Concepts and Applications.* Wiley, New York, 1988.

129. Young, D. H. and Kauss, H. Release of calcium from suspension-cultured glycine max cells by chitosan, other polycations, and polyamines in relation to effects on membrane permeability. *Plant Physiology* 73, 3(1983): 698–702.

130. Kumar, A. B. V., Varadaraj, M. C., Gowda, L. R., and Tharanathan, R. N. Characterization of chito-oligosaccharides prepared by chitosanolysis with the aid of papain and Pronase, and their bactericidal action against *Bacillus cereus* and *Escherichia coli. Biochemical Journal* 391, 2(2005): 167–175.

131. Chen, C. Z. and Cooper, S. L. Interactions between dendrimer biocides and bacterial membranes. *Biomaterials* 23, 16(2002): 3359–3368.

132. Liu, H., Du, Y., Wang, X., and Sun, L. Chitosan kills bacteria through cell membrane damage. *International Journal of Food Microbiology* 95, 2(2004): 147–155.

133. Goy, R. C., Britto, D. D., and Assis, O. B. G. A review of the antimicrobial activity of chitosan. *Polímeros* 19, (2009): 241–247.

134. Hadwiger, L. A., Kendra, D. F., Fristensky, B. W., and Wagoner, W. Chitosan both activates genes in plants and inhibits RNA synthesis in fungi. In *Chitin in Nature and Technology.* Plenum Press, New York, pp. 209–214, 1986.

135. Rabea, E. I., Badawy, M. E. T., Stevens, C. V., Smagghe, G., and Steurbaut, W. Chitosan as antimicrobial agent: Applications and mode of action. *Biomacromolecules* 4, 6(2003): 1457–1465.

136. Kong, M., Chen, X. G., Xing, K., and Park, H. J. Antimicrobial properties of chitosan and mode of action: A state of the art review. *International Journal of Food Microbiology* 144, 1(2010): 51–63.

137. Kołodyńska, D. Adsorption characteristics of chitosan modified by chelating agents of a new generation. *Chemical Engineering Journal* 179, (2012): 33–43.

138. Liu, D., Li, Z., Zhu, Y., Li, Z., and Kumar, R. Recycled chitosan nanofibril as an effective Cu(II), Pb(II) and Cd(II) ionic chelating agent: Adsorption and desorption performance. *Carbohydrate Polymers* 111, (2014): 469–476.

139. Guibal, E. Interactions of metal ions with chitosan-based sorbents: A review. *Separation and Purification Technology* 38, 1(2004): 43–74.

140. Chung, Y.-C., Wang, H.-L., Chen, Y.-M., and Li, S.-L. Effect of abiotic factors on the antibacterial activity of chitosan against waterborne pathogens. *Bioresource Technology* 88, 3(2003): 179–184.

141. Zhou, Y., Yang, H., Liu, X., Mao, J., Gu, S., and Xu, W. Potential of quaternization-functionalized chitosan fiber for wound dressing. *International Journal of Biological Macromolecules* 52, (2013): 327–332.

142. Ignatova, M., Manolova, N., and Rashkov, I. Novel antibacterial fibers of quaternized chitosan and poly(vinyl pyrrolidone) prepared by electrospinning. *European Polymer Journal* 43, 4(2007): 1112–1122.

143. Ignatova, M., Starbova, K., Markova, N., Manolova, N., and Rashkov, I. Electrospun nano-fibre mats with antibacterial properties from quaternised chitosan and poly(vinyl alcohol). *Carbohydrate Research* 341, 12(2006): 2098–2107.

144. Dai, T., Tanaka, M., Huang, Y. Y., and Hamblin, M. R. Chitosan preparations for wounds and burns: Antimicrobial and wound-healing effects. *Expert Review of Anti-Infective Therapy* 9, 7(2011): 857–879.

145. Ong, S. Y., Wu, J., Moochhala, S. M., Tan, M. H., and Lu, J. Development of a chitosan-based wound dressing with improved hemostatic and antimicrobial properties. *Biomaterials* 29, 32(2008): 4323–4332.

146. Coma, V., Deschamps, A., and Martial-Gros, A. Bioactive packaging materials from edible chitosan polymer—Antimicrobial activity assessment on dairy-related contaminants. *Journal of Food Science* 68, 9(2003): 2788–2792.

147. Ribeiro, C., Vicente, A. A., Teixeira, J. A., and Miranda, C. Optimization of edible coating composition to retard strawberry fruit senescence. *Postharvest Biology and Technology* 44, 1(2007): 63–70.

148. Bordenave, N., Grelier, S., Pichavant, F., and Coma, V. Water and moisture susceptibility of chitosan and paper-based materials: Structure-property relationships. *Journal of Agricultural and Food Chemistry* 55, 23(2007): 9479–9488.

149. Wong, D. W. S., Gastineau, F. A., Gregorski, K. S., Tillin, S. J., and Pavlath, A. E. Chitosan-lipid films: Microstructure and surface energy. *Journal of Agricultural and Food Chemistry* 40, 4(1992): 540–544.

150. Xiao, Z., Luo, Y., Luo, Y., and Wang, Q. Combined effects of sodium chlorite dip treatment and chitosan coatings on the quality of fresh-cut d'Anjou pears. *Postharvest Biology and Technology* 62, 3(2011): 319–326.

151. Badawy, M. E. I., Rabea, E. I., Rogge, T. M., Stevens, C. V., Smagghe, G., Steurbaut, W., and Höfte, M. Synthesis and fungicidal activity of new N,O-acyl chitosan derivatives. *Biomacromolecules* 5, 2(2004): 589–595.

152. Badawy, M. E. I. and Rabea, E. I. A biopolymer chitosan and its derivatives as promising antimicrobial agents against plant pathogens and their applications in crop protection. *International Journal of Carbohydrate Chemistry* 1–29, (2011): 460381.

153. Kochkina, Z. M., Pospeshny, G., and Chirkov, S. N. Inhibition by chitosan of productive infection of T-series bacteriophages in the *Escherichia coli* culture. *Mikrobiologiia* 64, 2(1995): 211–215.

154. El, H. A., Adam, L. R., El, H. I., and Daayf, F. Chitosan in plant protection. *Marine Drugs* 8, 4(2010): 968–987.

155. Tarsi, R., Muzzarelli, R. A. A., Guzman, C. A., and Pruzzo, C. Inhibition of streptocccus mutans adsorption to hydroxyapatite by low-molecular-weight chitosans. *Journal of Dental Research* 76, 2(1997): 665–672.

156. Maria Souza Gadelha de Carvalho, T. C. M. S., Emerson, P dos Santos, Tenório, P., and Sampaio, F. In *Science against Microbial Pathogens: Communicating Current Research and Technological Advances*, A. Mendez-Vilas (ed.), Vol. 1. Formatex Research Center, Spain 2011.

157. Miao, D., Blom, D., Zhao, H., Luan, X., Chen, T., Wu, X., and Song, N. The antibacterial effect of CMCTS-containing chewing gum. *Journal of Nanjing Medical University* 23, 1(2009): 69–72.

158. Sano, H., Shibasaki, K., Matsukubo, T., and Takaesu, Y. Effect of chitosan rinsing on reduction of dental plaque formation. *The Bulletin of Tokyo Dental College* 44, 1(2003): 9–16.

159. Hayashi, Y., Ohara, N., Ganno, T., Ishizaki, H., and Yanagiguchi, K. Chitosan-containing gum chewing accelerates antibacterial effect with an increase in salivary secretion. *Journal of Dentistry* 35, 11(2007): 871–874.

160. Sano, H., Shibasaki, K.-I., Matsukubo, T., and Takaesu, Y. Comparison of the activity of four chitosan derivatives in reducing initial adherence of oral bacteria onto tooth surfaces. *The Bulletin of Tokyo Dental College* 42, 4(2001): 243–249.

161. Ji, Q. X., Zhong, D. Y., Lü, R., Zhang, W. Q., Deng, J., and Chen, X. G. In vitro evaluation of the biomedical properties of chitosan and quaternized chitosan for dental applications. *Carbohydrate Research* 344, 11(2009): 1297–1302.

*chapter twenty-six*

# Marine bacterial extracellular polymeric substances
## *Characteristics and applications*

*Vijay Kumar Singh, Avinash Mishra, and Bhavanath Jha*

## Contents

## 26.1 Introduction

The major part of the earth's biosphere is covered by the marine hydrosphere, which covers about 70% of total surface area. Ocean is considered the largest ecosystem, covering an area of ~365 million $km^2$ and a maximum depth of >11,000 m. Ocean is home for the most diverse groups of flora and fauna, and forms a vast source of natural products (Satpute et al., 2010). Marine microorganisms produce unique metabolites to thrive in the extreme conditions of the marine ecosystem, which are unlikely to be found in terrestrial microbes (Fenical, 1993). Bacteria are highly diversified according to the marine environment, and most of the marine bacterial cells are surrounded by extracellular polymeric substances to survive in adverse conditions. Marine bacteria generally exist in two forms: either attached to the substrate and forming biofilms, or free-living as the planktonic cell. Most of the bacteria are reported to be attached to a surface in the marine environment (Zobel, 1943). About 99% of microorganisms inhabiting the earth are found to be present in biofilms or in an aggregate mode of life (Costerton et al., 1987).

Extracellular polymeric substances or exopolysaccharides (EPSs) are carbohydrate polymers produced by bacteria, cyanobacteria, fungi, marine microalgae, and some marine microorganisms. The term *exopolysaccharide* was coined by Sutherland (1972) for high-molecular-weight carbohydrate polymers produced by marine bacteria, and is often used to define extracellular polymeric substances (Wingender et al., 1999). Bacterial cells accumulate a complex organic mixture of EPSs, which include polysaccharides, proteins, nucleic acids, and lipids (McSwain et al., 2005). Bacteria may utilize about 70% of their energy for the production of exopolymers, which is a significant carbon and energy investment for the cell (Harder and Dijkhuizen, 1983; Wolfaardt et al., 1999).

Microbial polysaccharides are classified as (1) cell-wall polysaccharides, (2) intracellular polysaccharides, and (3) EPSs. Cell-wall polysaccharides and intracellular polysaccharides are integral parts of the cell. EPSs are produced by many bacteria in response to environmental stress. EPSs produced by microorganisms are either attached to the cell surface or present as an amorphous slime in the extracellular medium (Sutherland, 1998).

Microorganisms do not grow as dispersed single cells but accumulate at an interface to form microbial aggregates as films, mats, flocs, sludges, or biofilms (Wingender et al., 1999). Generally, more than 90% of the dry mass in a biofilm is occupied by the EPS matrix, and the rest (~10%) is covered by microorganisms. The matrix consists of EPSs, which are mostly produced by

***Table 26.1*** Functions of natural extracellular polymeric substances towards bacterial growth and survival

| Function | Description | EPS components |
| --- | --- | --- |
| Adhesion | Allows bacteria to colonize by providing planktonic cells to long-term attachment to surfaces | Polysaccharides, proteins, and amphiphilic molecules |
| Aggregation | Facilitates bridging between cells and immobilization of cells to develop bacterial populations | Polysaccharides and proteins |
| Cohesion | Construction of hydrated polymer network for cell–cell communication | Neutral and charged polysaccharides, amyloids, and lectins |
| Retention of water | Maintaining a hydrated environment and thus providing tolerance to desiccation in water-deficient environments | Hydrophilic polysaccharides |
| Protection | Provides resistance to nonspecific and specific host defences and tolerance to various antimicrobial agents | Polysaccharides and proteins |
| Sorption of organic compounds | Accumulation of nutrients from the environment | Charged or hydrophobic polysaccharides and proteins |
| Sorption of inorganic ions | Facilitates environmental detoxification by promoting polysaccharide gel formation to accumulate toxic metal ions | Charged polysaccharides (sulfate or/ and phosphate) and proteins |
| Enzymatic activity | Allows digestion of exogenous macromolecules for nutrient acquisition | Proteins |
| Nutrient source | For normal growth and function | All EPS components |
| Genetic information flow | Horizontal gene transfer between cells | DNA |
| Electron donor or acceptor | Allows redox activity | Proteins |
| Export of cell components | Release of cellular material | Membrane vesicles |
| Sink for excess energy | Storage of excess carbon for future uses | Polysaccharides |

*Source:* Adapted from Flemming, H.C. and Wingender, J., *Nature*, 8, 623, 2010.

the cells present in the biofilm. Biofilm formation gives a life style to the cell, which is totally different from that of planktonic cells. EPSs facilitate the biofilm cells to interact, including cell-to-cell communication and the formation of synergistic micro-consortia (Flemming and Wingender, 2010). The function of EPSs in bacterial growth and survival is summarized in Table 26.1.

In the food industry, polysaccharides are used as thickening and gelling agents to improve food quality and texture. Traditionally, such polysaccharides (starch, pectin, alginate, carrageenan, galactomannose, and a variety of exudates and seed gums) are extracted from plants. Bacterial polysaccharides are now considered a potential source of biopolymers for food or general industrial applications because of their novel functions, constant and reproducible chemical and physical properties, and stable cost and supply (MacCormick et al., 1996).

## 26.2 Marine bacterial extracellular polymeric substances

The marine environment is a rich natural source of EPS producers, and several marine microorganisms have been isolated from the seas (Table 26.2) that contain distinct EPSs (Satpute et al., 2010). Many EPS-producing thermotolerant bacterial strains have been isolated from marine hot springs, marine hydrothermal vents (Maugeri et al., 2002; Nicolaus et al., 2002), and deep-sea hydrothermal vents (Raguénès et al., 1996; Rougeaux et al., 1999). Poli et al. (2010) have reported several extreme marine habitats, deep-sea hydrothermal vents, volcanic and hydrothermal marine areas, and shallow submarine thermal springs as new sources of EPS-producing bacteria. Compared to those in thermal marine environments, EPS-producing bacteria from cold marine environments are poorly reported (Mancuso Nichols et al., 2004; Qin et al., 2007; Marx et al., 2009). Bacterial strains of the genus *Pseudoalteromonas* were isolated from melted sea ice, whereas *Colwellia psychrerythraea* strain 34H was isolated from Arctic and Antarctic sea ice. In contrast, there are many reports on EPS-producing bacterial strains from saline environments, and nearly 134 different bacterial strains have been isolated from 18 diverse hypersaline habitats (Martínez-Cánovas et al., 2004). EPSs from the marine microalga *Dunaliella salina* (Mishra and Jha, 2009; Mishra et al., 2011) and biofouling *Vibrio* species were isolated and characterized in detail (Kavita et al., 2011, 2013).

*Table 26.2* Habitat, bacteria, and potential biotechnological application of EPSs

| Source | Bacteria | Possible applications |
| --- | --- | --- |
| Deep-sea hydrothermal vent | *Pseudoalteromonas* sp. | Gelling properties |
| Deep-sea hydrothermal vent | *Alteromonas macleodii* | Thickening agent in food processing industry |
| Shallow submarine thermal spring | *Thermococcus litoralis* | Biofilm formation |
| Sediment in marine hot spring | *Geobacillus* sp. | Pharmaceutical application |
| Water of a shallow hydrothermal vent | *Bacillus thermodenitrificans* | Immunomodulation |
| Water of a shallow marine hot spring | *Bacillus licheniformi* | Antiviral activity |
| Deep-sea sediment | *Pseudoalteromonas* sp. | Flocculation and bio-sorption |
| Melted Antarctic sea | *Pseudoalteromonas* sp. | Cryoprotection |
| Southern Ocean | *Pseudoalteromonas* sp. | Trace-metal binding |
| Arctic marine sediments | *Colwellia psychrerythraea* | Cryoprotection |
| Mediterranean Sea | *Haloferax mediterranei* | In oil recovery |
| Marine sediment | *Hahella chejuensis* | Biosurfactant |
| Salt lake | *Halomonas alkaliantarctica* | High viscosity |
| Hydrothermal vent | *Alteromonas macleodii* | Cosmetics |
| Deep-sea hydrothermal field | *Vibrio diabolicus* | Bone regeneration |
| Hydrothermal vent | *Alteromonas infernus* | Anticoagulant activity |

*Source:* Adapted from Poli, A. et al., *Mar. Drugs*, 8, 1779, 2010.

## 26.3   Characterization of EPSs

Collective study of the physical, chemical, and biological properties of EPSs is called characterization, and it is very important to understand the role of EPSs in different environments. EPSs are difficult to analyze in the natural marine environment due to their low abundance; therefore, growing a single isolated strain under controlled laboratory conditions is the best method for their analysis (Nichols et al., 2005). There is no standard method or culture condition available for the maximum EPS production for all microorganisms because different key factors (carbon and nitrogen source utilization, mineral requirements, temperature, and optimal pH) influence microorganisms for EPS production. Moreover, the quality and quantity of microbial EPS can be affected by nutritional and environmental conditions (Poli et al., 2010). Increased production of EPS was observed when marine bacteria were grown in nutrient-deficient (i.e., nitrogen, phosphorus, sulfur, and potassium) conditions in the laboratory (Sutherland, 1982).

The EPS extraction method is a very important step in the characterization of EPS, because physiochemical properties may change during their isolation and purification. There are different physical and chemical methods available, and the extraction method varies with the source (cell suspension, biofilm, sludge, solid surfaces, and waters). Centrifugation, sonication, heating, and freeze–thawing are different physical methods, while different chemical agents, such as ethylenediamine tetraacetic acid (EDTA), NaOH, and formaldehyde, are used in chemical methods of extraction.

EPSs are generally assayed for total carbohydrates, and quantification is done by the sulfuric acid method (Dubois et al., 1956). The Lowry or Bradford method is used for the determination of the protein content of EPSs (Lowry et al., 1951; Bradford, 1976). Color reaction of methyl pentoses is spectrophotometrically assayed for the determination of uronic acid, with glucuronic acid as standard (Dische and Shettles, 1948). Sulfated sugars are measured after hydrolysis of EPS, where $K_2SO_4$ is used as standard (Terho and Hartiala, 1971).

Various analytical tools such as high-performance liquid chromatography (HPLC), Fourier transform infrared (FTIR) spectroscopy, and nuclear magnetic resonance (NMR) spectroscopy are available for the determination of the physical, chemical, and biological properties of EPSs (Figure 26.1). Functional groups are generally investigated by FTIR and NMR, whereas monosaccharide composition is analyzed by gas chromatography-mass spectrometry (GC-MS) (Mishra and Jha, 2013). Beside these, matrix-assisted laser desorption/ionization time-of-flight mass spectrometry (MALDI-TOF MS), scanning electron microscopy (SEM), energy dispersive X-ray spectroscopy (EDXor EDS), atomic force microscopy (AFM), and X-ray diffraction (XRD) have also been used for the qualitative analyses of EPSs (Kavita et al., 2011; Mishra et al., 2011). IR spectroscopy has been extensively used for the determination of functional groups and structural analysis of polymers

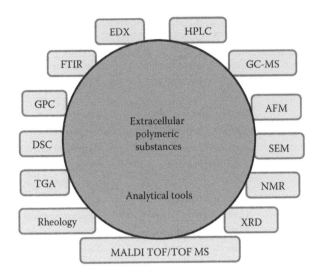

*Figure 26.1* **(See color insert.)** Analytical tools commonly used for EPSs analysis.

(Nyquist, 1961; Zbinden, 1964; Koenig, 1966). Elemental analysis of EPS is carried by EDX, and MALDI MS is an efficient method for the structural analysis of oligosaccharides (Mishra et al., 2013).

Carbohydrates are the main component of EPSs; so for monosaccharide composition analysis, EPS is hydrolyzed followed by alditol acetate derivatization and GC analysis (Freitas et al., 2009). Glucose, galactose, and mannose are the most frequently occurring monomers, whereas rhamnose, fucose, some uronic acids (such as glucuronic and galacturonic acid), and aminosugars (*N*-acetylamino sugars) are also commonly present in EPSs (Freitas et al., 2011). Gel permeation chromatography is one of the traditional methods commonly used for determining the average molecular mass of EPSs. Rheometry is used for the evaluation of the viscosity and viscoelastic properties of EPS solutions. Moreover, the intact, three-dimensional structure of EPSs is studied AFM.

## 26.4 *Properties and applications*

Marine microbes possess immense diversity and the unique capacity to produce natural products, and their importance for marine biotechnology was realized by the scientific community recently (Imhoff et al., 2011). Marine microorganisms have been studied for various biotechnological applications ranging from bioactive compounds for pharmaceuticals to other high-value products such as enzymes, nutraceuticals, and cosmetics (Imhoff et al., 2011). The structural diversity combined with various interesting properties of EPSs produced by marine bacteria could be explored for biotechnological applications.

Bacterial EPSs have a wide range of applications, which depend on their structural composition.

Extracellular polymeric substances play a very important role in the biotechnology and pharmaceutical industries. EPSs have some unique properties such as adhesion, coagulating, gelling, thickening, stabilizing, and pseudoplasticity, which make them widely useful in the food industry as viscosity-increasing, stabilizing, and emulsifying agents (Liu et al., 2010; Mishra and Jha, 2013). Pseudoplasticity of EPSs makes them resistant to extreme temperatures, pH, and salinity (Satpute et al., 2010). As far as industrial applications are concerned, EPSs produced by different marine bacteria are used as antitumor and antiviral agents, immunostimulants (Okutani, 1992), and anticoagulants (Colliec et al., 2001). Some biotechnological applications of marine EPSs are summarized in Table 26.2.

The anti-biofilm activity of polysaccharides was first reported by Valle et al. (2006). The EPS produced by *Pseudomonas aeruginosa* showed anti-biofilm activity of *Staphylococcus epidermis* biofilm without affecting the growth (Qin et al., 2009). EPS from *Bacillus licheniformis* (sponge-associated strain) showed anti-biofilm ability against pathogenic bacteria. Anti-biofilm activity against gram-negative and gram-positive bacteria was also reported in the EPS extracted from *Vibrio* sp. QY10 (Jiang et al., 2011). Recently, Kavita et al. (2014) reported that the EPS isolated from a marine bacterium *Oceanobacillus iheyensis* strain BK6 possesses anti-biofilm properties against the pathogenic bacterium *Staphylococcus aureus*. A unique application of bacterial EPSs was observed when the zoospore settlement on EPS-coated cover slips progressively increased with the incubation time (Singh et al., 2013). Thus, it was established that bacterial EPSs facilitate the primary settlement of spores, which play a crucial role in the macroalgal communities of the coastal environment (Singh et al., 2013).

## 26.5  A case study: Isolation and characterization of EPSs extracted from marine resources

A schematic representation of the commonly used methodology or work plan is given in Figure 26.2. Detailed analysis of EPSs obtained from different marine sources are described in the following.

### 26.5.1  EPS from planktonic marine bacteria

A marine bacterium *Vibrio parahaemolyticus* produced 58.98 mg $L^{-1}$ of EPS with an average particle size of 15.2 μm in a planktonic culture. FTIR analysis of the EPS revealed characteristic functional groups such as aliphatic methyl, primary amine, and halide groups, uronic acid, and saccharides. The NMR spectrum indicated the presence of hexose and pentose, acetyl, and carboxyl groups. Powder XRD analysis revealed the amorphous nature of EPS, with diffraction peaks at 5.98°, 9.15°, and 22.8° with an interplanar spacing of 14.8, 9.3, and 3.9 Å. Among the four monosaccharides detected in the GC-MS analysis, mannose was present in the highest concentration, followed by glucose, galactose, and arabinose. MALDI-TOF analysis showed a series of mass values corresponding to oligosaccharides and polysaccharides with different ratios of pentose and hexose sugars. SEM images revealed the compact nature of the EPS with small pores. Thermogavimetry (TG) and differential scanning calorimetry (DSC) analyses

showed that the EPS degradation took place in two steps (heating and depolymerization). The weight loss of the EPS was 13% at 160°C during the first step. The EPS showed thermal stability up to 250°C. The viscosity of the EPS decreased with the shear rate, showing its pseudoplastic rheological behavior. This effect was more profound up to a shear rate of 250–300 $s^{-1}$. The EPS exhibited shear thinning behavior at both low (3.0) and neutral (7.0) pH. The viscosity of EPS decreased with increasing temperature at both pH values, which may be due to the breakdown of the structural unit by hydrodynamic forces. The EPS also showed 67% emulsifying activity (Kavita et al., 2011).

### 26.5.2  EPS from marine natural biofilm

About 400 mg $L^{-1}$ extracellular polymeric substance with a molecular mass of 2140 kDa was obtained from the marine bacterium *O. iheyensis*, isolated from a marine natural biofilm. Elemental analysis of the EPS showed sulfate and phosphate residues along with common elements such as C, N, O, Mg, and Na. FTIR analysis showed various peaks corresponding to hydroxyl and carboxyl groups, as well as those from uronic acid and the sugar moiety. GC-MS analysis revealed that the major component of EPS was mannose, followed by glucose and arabinose. Powder XRD analysis revealed the amorphous nature of the EPS (Figure 26.3). The SEM image showed the compactness of the EPS (Figure 26.4), while AFM topography suggested tightly packed molecules with rectangular shape, which may be responsible of the pseudoplastic behavior. The thermal properties

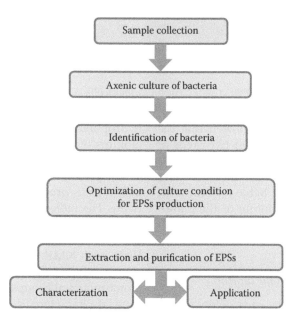

*Figure 26.2* **(See color insert.)** Schematic representation of the commonly used methodology for EPSs studies.

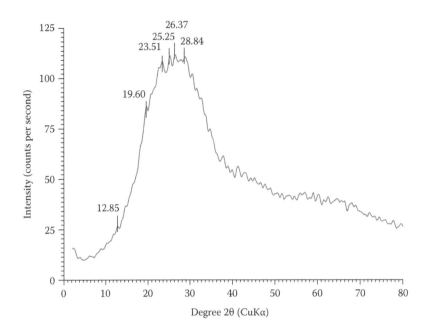

***Figure 26.3*** Powder XRD analysis of EPS obtained from *Oceanobacillus iheyensis*.

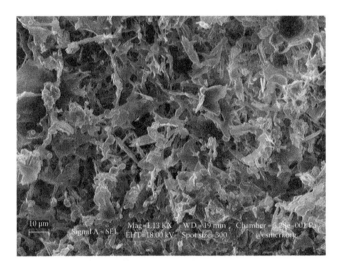

***Figure 26.4*** Representative SEM image of EPS extracted from *Oceanobacillus iheyensis*.

of EPS, as analyzed by TG and DSC, showed that the EPS was stable up to 170°C and followed two steps of degradation (weight loss of 40%). Concerning its potential applications, the EPS exhibited 66% of emulsifying activity and showed pseudoplastic behavior and anti-biofilm activity (62%) against the biofilm of the pathogenic strain of *S. aureus* without inhibiting its growth (Kavita et al., 2014).

### 26.5.3 EPS from marine artificial biofilm

*Vibrio campbellii* and *V. fortis* are the early colonizers of marine artificial biofilms. The cell surface hydrophobicity showed that the cell surface of *V. campbellii* was more

hydrophobic than that of *V. fortis*. The SEM images of the biofilms of these two bacteria revealed that the biofilm of *V. campbellii* was smooth and compact, whereas that of *V. fortis* was irregular in shape and loose. *V. campbellii* and *V. fortis* produced 396 and 134 mg L$^{-1}$ EPS, respectively. The molecular masses of the EPSs from *V. campbellii* and *V. fortis* were 2431 and 2218 kDa, respectively. The functional groups of the EPSs were analyzed using FTIR and NMR. Both EPSs showed almost similar peaks, indicating the presence of the characteristic common functional groups of polysaccharides. Powder XRD showed characteristic diffraction peaks for both EPSs (*V. campbellii* and *V. fortis*) with peaks at 5.6°, 9.4°, and 28.6° with interplanar spacings of 15.07, 9.3, and 3.1 Å. GC-MS

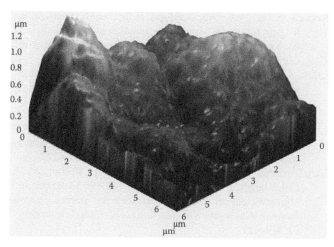 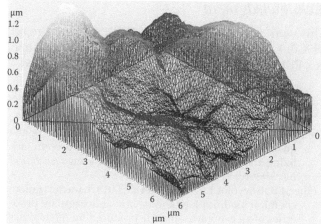

*Figure 26.5* **(See color insert.)** Representative AFM image of EPSs.

analysis revealed the presence of five monosaccharides in the EPS of *V. campbellii;* among these, arabinose was the predominant one, followed by mannose, galactose, glucose, and rhamnose. The EPS of *V. fortis* comprised only three monosaccharides—arabinose, galactose, and mannose—with arabinose as the major component. SEM and AFM images of EPSs showed the exopolymer of *V. campbellii* asymmetric compared to that of *V. fortis.* The EPS of *V. fortis* was porous and compact. AFM showed different morphological parameters and functional indices (Figure 26.5). Thermal study of EPSs by TG and DSC analyses showed weight loss in two steps: EPS from *V. fortis,* with 37% weight loss up to 270°C, was more thermo-stable than the EPS of *V. campbellii,* which showed 57% degradation up to 280°C. DSC showed similar thermograms for both EPSs. Rheological study of these two EPSs showed their pseudoplastic behavior at both pH 3 and 7. The EPS of *V. fortis* was more viscous than that of *V. campbellii* at a constant temperature, while the opposite was true when the EPSs were analyzed at constant shear. The emulsifying activity of EPSs from *V. campbellii* and *V. fortis* was 64% and 72%, respectively (Kavita et al., 2013).

### 26.5.4 Comparative study of EPSs obtained from different sources

The marine bacterium *V. campbellii* produced the largest quantity of EPS among *V. fortis, V. parahaemolyticus,* and *O. iheyensis.* All EPSs obtained in the present study belong to the high molecular weight category. EPSs from *V. fortis* and *V. parahaemolyticus* were found to be highly thermo-stable, followed by those from *V. campbellii* and *O. iheyensis.* FTIR, NMR, and XRD analyses showed common functional groups and chemical shift pattern in all EPSs. All EPSs showed good emulsifying property; however, the EPS of *V. fortis* was the best, with the

highest value. The emulsifying properties of EPSs make them important as food additives. The pseudoplastic rheological behavior was common to the EPS from all four bacteria. This is an important attribute for their utilization as gelling and thickening agents in industries. One of the significant observations was that the EPS produced by *O. iheyensis* inhibited the biofilm formed by the human pathogenic bacterium *S. aureus.* The inhibition was caused by disrupting cell-to-cell attachment and interactions. The polysaccharides present in the EPS mainly blocked biofilm formation by *S aureus.* Thus, the antibiofilm activity of EPS makes them promising agents for therapeutic applications. This may be further explored as a drug for controlling infections caused by *S. aureus.*

### 26.6 Future prospects

Marine bacterial EPSs have several benefits over polysaccharides from other sources because of their stable cost and abundant supply. Furthermore, the EPSs show a wide variety compared to traditional polymers of plant and animal origin. In the marine environment, tidal action, high waves, and irregular exposure to air may cause selection of bacteria with water-holding, adhesive, and protective polysaccharide coating. The marine environment is known to support a large number of unexplored bacterial species, and EPSs produced by marine bacteria possess wide structural diversity, which make them renewable sources of biotechnological importance. Little is known about marine bacterial EPSs, which have enormous structural diversity and physicochemical properties suitable for industrial applications. There is growing demand to explore marine bacteria for their unique EPSs. Nowadays, marine bacterial EPSs are rapidly emerging as new and industrially important sources of polymeric materials.

## Acknowledgment

This work was supported by the CSIR Network Project (BSC0106–BioprosPR; CSIR-CSMCRI Communication No. PRIS-198/2014), which is gratefully acknowledged.

## References

Bradford, M. M. 1976. A rapid and sensitive method for the quantitation of microgram quantities of protein utilizing the principle of protein-dye binding. *Analytical Biochemistry* 72: 248–254.

Colliec, J. S., Chevolot, L., Helley, D. et al. 2001. Characterization, chemical modifications and in vitro anticoagulant properties of an exopolysaccharide produced by *Alteromonas infernos. Biochimica et Biophysica Acta* 1528: 141–151.

Costerton, J. W., Cheng, K. J., Geesey, G. G. et al. 1987. Bacterial biofilms in nature and disease. *Annual Reviews in Microbiology* 41: 435–464.

Dische, Z. and Shettles, L. B. 1948. A specific color reaction of methylpentoses and a spectrophotometric micromethod for their determination. *Journal of Biological Chemistry* 175: 595–603.

Dubois, M., Gilles, K. A., Hamilton, J. K., Rebers, P., and Smith, F. 1956. Colorimetric method for determination of sugars and related substances. *Analytical Chemistry* 28: 350–356.

Fenical, W. 1993. Chemical studies of marine bacteria: Developing a new resource. *Chemical Reviews* 93: 1673–1683.

Flemming, H. C. and Wingender, J. 2010. The biofilm matrix. *Nature Reviews Microbiology* 8: 623–633.

Freitas, F., Alves, V. D., Carvalheira, M., Costa, N., Oliveira, R., and Reis, M. A. 2009. Emulsifying behaviour and rheological properties of the extracellular polysaccharide produced by *Pseudomonas oleovorans* grown on glycerol byproduct. *Carbohydrate Polymers* 78: 549–556.

Freitas, F., Alves, V. D., and Reis, M. A. 2011. Advances in bacterial exopolysaccharides: From production to biotechnological applications. *Trends in Biotechnology* 29: 388–398.

Harder, W. and Dijkhuizen, L. 1983. Physiological responses to nutrient limitation. *Annual Reviews in Microbiology* 37: 1–23.

Imhoff, J. F., Labes, A., and Wiese, J. 2011. Bio-mining the microbial treasures of the ocean: New natural products. *Biotechnology Advances* 29: 468–482.

Jiang, P., Li, J., Han, F. et al. 2011. Antibiofilm activity of an exopolysaccharide from marine bacterium *Vibrio* sp. QY101. *PLoS ONE* 6: e18514.

Kavita, K., Mishra, A., and Jha, B. 2011. Isolation and physicochemical characterisation of extracellular polymeric substances produced by the marine bacterium *Vibrio parahaemolyticus. Biofouling* 27: 309–317.

Kavita, K., Mishra, A., and Jha, B. 2013. Extracellular polymeric substances from two biofilm forming *Vibrio* species: Characterization and applications. *Carbohydrate Polymers* 94: 882–888.

Kavita, K., Singh, V. K., Mishra, A., and Jha, B. 2014. Characterisation and anti-biofilm activity of extracellular polymeric substances from *Oceanobacillus iheyensis. Carbohydrate Polymers* 101: 29–35.

Koenig, S. L. 1966. Applications of infrared spectroscopy to polymers. In *Applied Infrared Spectroscopy*, Ed. D. N. Kendall, pp. 245. New York: Reinhold Publishing Corporation.

Liu, C., Lu, J., Lu, L., Liu, Y., Wang, F., and Xiao, M. 2010. Isolation, structural characterization and immunological activity of an exopolysaccharide produced by *Bacillus licheniformis* 8-37-0-1. *Bioresource Technology* 101: 5528–5533.

Lowry, O. H., Rosebrough, N. J., Farr, A. L., and Randall, R. J. 1951. Protein measurement with the Folin phenol reagent. *Journal of Biological Chemistry* 193: 265–275.

MacCormick, C. A., Harris, J. E., Jay, A. J., Ridout, M. J., Colquhoun, I. J., and Morris, V. J. 1996. Isolation and characterization of a new extracellular polysaccharide from an *Acetobacter* species. *Journal of Applied Bacteriology* 81: 419–424.

Mancuso-Nichols, C. A., Garon, S., Bowman, J. P., Raguénès, G., and Guezennec, J. 2004. Production of exopolysaccharides by Antarctic marine bacterial isolates. *Journal of Applied Microbiology* 96: 1057–1066.

Martínez-Cánovas, M. J., Quesada, E., Martínez-Checa, F., and Béjar, V. 2004. A taxonomic study to establish the relationship between exopolysaccharide-producing bacterial strains living in diverse hypersaline habitats. *Current Microbiology* 48: 348–353.

Marx, J. G., Carpenter, S. D., and Deming, J. W. 2009. Production of cryoprotectant extracellular polysaccharide substances (EPS) by the marine psychrophilic bacterium *Colwellia psychrerythraea* strain 34H under extreme conditions. *Canadian Journal of Microbiology* 55: 63–72.

Maugeri, T. L., Gugliandolo, C., Caccamo, D., Panico, A., Lama, L., Gambacorta, A., and Nicolaus, B. 2002. A halophilic thermotolerant *Bacillus* isolated from a marine hot spring able to produce a new exopolysaccharide. *Biotechnology Letters* 24: 515–519.

McSwain, B. S., Irvine, R. L., Hausner, M., and Wilderer, P. A. 2005. Composition and distribution of extracellular polymeric substances in aerobic flocs and granular sludge. *Applied and Environmental Microbiology* 71: 1051–1057.

Mishra, A. and Jha, B. 2009. Isolation and characterization of extracellular polymeric substances from micro-algae *Dunaliella salina* under salt stress. *Bioresource Technology* 100: 3382–3386.

Mishra, A. and Jha, B. 2013. Microbial exopolysaccharides. In *The Prokaryotes*, Eds. E. Rosenberg, E. F. DeLong, F. Thompson, S. Lory, and E. Stackebrandt, pp. 179–192. Berlin, Germany: Springer.

Mishra, A., Joshi, M., and Jha, B. 2013. Oligosaccharide mass profiling of nutritionally important *Salicornia brachiata*, an extreme halophyte. *Carbohydrate Polymers* 92: 1942–1945.

Mishra, A., Kavita, K., and Jha, B. 2011. Characterization of extracellular polymeric substances produced by micro-algae *Dunaliella salina. Carbohydrate Polymers* 83: 852–857.

Nichols, C. M., Lardière, S. G., Bowman, J. P., Nichols, P. D., Gibson, J. A., and Guézennec, J. 2005. Chemical characterization of exopolysaccharides from Antarctic marine bacteria. *Microbial Ecology* 49: 578–589.

Nicolaus, B., Lama, L., Panico, A., Moriello, V. S., Romano, I., and Gambacorta, A. 2002. Production and characterization of exopolysaccharides excreted by thermophilic bacteria from shallow, marine hydrothermal vents of Flegrean Ares (Italy). *Systematic and Applied Microbiology* 25: 319–325.

Nyquist, R. A. 1961. *Infrared Spectra of Polymers and Resins*, 2nd edn. Midland, MI: The Dow Chemical Company.

Okutani, K. 1992. Antiviral activities of sulfated derivatives of a fucosamine-containing polysaccharide of marine bacterial origin. *Nippon Suisan Gakkaishi* 58: 927–930.

Poli, A., Anzelmo, G., and Nicolaus, B. 2010. Bacterial exopolysaccharides from extreme marine habitats: Production, characterization and biological activities. *Marine Drugs* 8: 1779–1802.

Qin, G., Zhu, L., Chen, X., Wang, P. G., and Zhang, Y. 2007. Structural characterization and ecological roles of a novel exopolysaccharide from the deep-sea psychrotolerant bacterium *Pseudoalteromonas* sp. SM9913. *Microbiology* 153: 1566–1572.

Qin, Z., Yang, L., Qu, D., Molin, S., and Tolker-Nielsen, T. 2009. *Pseudomonas aeruginosa* extracellular products inhibit staphylococcal growth, and disrupt established biofilms produced by *Staphylococcus epidermidis*. *Microbiology* 155: 2148–2156.

Raguenes, G., Pignet, P., Gauthier, G. et al. 1996. Description of a new polymer-secreting bacterium from a deepsea hydrothermal vent, *Alteromonas macleodii* subsp. *fijiensis*, and preliminary characterization of the polymer. *Applied and Environmental Microbiology* 62: 67–73.

Rougeaux, H., Guezennec, J., Carlson, R. W., Kervarec, N., Pichon, R., and Talaga, P. 1999. Structural determination of the exopolysaccharide of *Pseudoalteromonas* strain HYD 721 isolated from a deep-sea hydrothermal vent. *Carbohydrate Research* 315: 273–285.

Satpute, S. K., Banat, I. M., Dhakephalkar, P. K., Banpurkar, A. G., and Chopade, B. A. 2010. Biosurfactants, bioemulsifiers and exopolysaccharides from marine microorganisms. *Biotechnology Advances* 28: 436–450.

Singh, R. P., Shukla, M. K., Mishra, A., Reddy, C. R. K., and Jha, B. 2013. Bacterial extracellular polymeric substances and their effect on settlement of zoospore of *Ulva fasciata*. *Colloids and Surfaces B: Biointerfaces* 103: 223–230.

Sutherland, I. W. 1972. Bacterial exopolysaccharides. *Advances in Microbial Physiology* 8: 143.

Sutherland, I. W. 1982. Biosynthesis of microbial exopolysaccharides. *Advances in Microbial Physiology* 23: 79–150.

Sutherland, I. W. 1998. Novel and established applications of microbial polysaccharides. *Trends in Biotechnology* 16: 41–46.

Terho, T. T. and Hartiala, K. 1971. Method for determination of the sulfate content of glycosaminoglycans. *Analytical Biochemistry* 41: 471–476.

Valle, J., Da Re, S., Henry, N., Fontaine, T., Balestrino, D., Latour-Lambert, P., and Ghigo, J. M. (2006). Broadspectrum biofilm inhibition by a secreted bacterial polysaccharide. *Proceedings of the National Academy of Sciences* 103: 12558–12563.

Wingender, J., Neu, T. R., and Flemming, H. C. 1999. What are bacterial extracellular polymeric substances? In *Microbial Extracellular Polymeric Substances*, Eds. J. Wingender, T. R. Neu, and H. C. Flemming, pp. 1–19. Berlin, Germany: Springer.

Wolfaardt, G. M., Lawrence, J. R., and Korber, D. R. 1999. Function of EPS. In *Microbial Extracellular Polymeric Substances*, Eds. J. Wingender, T. R. Neu, and H. C. Flemming, pp. 171–200. Berlin, Germany: Springer.

Zbinden, R. 1964. *Infrared Spectroscopy of High Polymers*. New York: Academic Press Inc.

Zobell, C. E. 1943. The effect of solid surfaces upon bacterial activity. *Journal of Bacteriology* 46: 39–56.

*chapter twenty-seven*

# Brown algal polysaccharide
## *Alginate and its biotechnological perspectives*

**Vasuki Subramanian, Perumal Anantharaman, and Kandasamy Kathiresan**

## Contents

## 27.1 Introduction

Polysaccharides are polymers of simple sugars (monosaccharides) linked together by glycosidic bonds. Polysaccharides are present in large amounts, especially in marine algae, mainly as cell wall components and as storage products (Murata and Nakazoe 2001; Dettmar et al. 2011), and the polysaccharide content ranges from 4% to 76% of the algal dry weight (Kraan 2012). The carbohydrates of brown seaweed are mainly composed of alginate, laminaran, mannitol, fucoidan, and cellulose in small amounts (Horn et al. 2000). Alginate provides the main structural component of brown algae. It is an important cell wall component, constituting up to 40%–47% of the dry weight of algal biomass (Arasaki and Arasaki 1983; Rasmussen and Morrissey 2007). The chemical composition of alginate varies with species of brown seaweed, different parts of the same plant, place of occurrence, and seasonal changes. Brown seaweed that grow in more turbulent conditions usually have a higher alginate content than those in clear waters. The other biological functions of alginate include prevention of desiccation, maintenance of cell integrity, and provision of mechanical strength and flexibility to the algal tissue, and these properties are adapted as necessary for growth conditions in the sea.

## 27.2   Structure and chemistry

Alginates were discovered in the 1880s by E.C.C. Stanford, a British pharmacist, and industrial production began in California in 1929. The seaweed that produce large amount of alginates belong to the genera *Macrocystis, Laminaria, Ecklonia, Durvillaea Lessonia,* and *Ascophyllum* as well as *Sargassum,* and *Turbinaria.* Alginates are mainly manufactured in the United States, Japan, China, France, Chile, and Norway (Kraan 2012; Hernandez-Carmona et al. 2013).

Alginates are anionic polysaccharides. It is a linear copolymer of $(1 \rightarrow 4)$ linked β-D-mannuronic acid (M) and α-L-guluronic acid (G) residues with widely varying compositions and sequences. The distribution of M and G in alginate chains gives rise to three different block types: blocks of poly-M, blocks of poly-G, and alternating MG blocks. While the M-block segments develop in linear and flexible structures, the G-block residues give rise to fold and rigid structures. Alginates can be identified using IR, Raman, and NIR spectroscopies. More detailed information about the structure is now available with the introduction of high-resolution 1H and 13C NMR spectroscopy in the sequential analysis of alginate (Larsen et al. 2003; Pereira et al. 2003; Leal et al. 2008; Campos-Vallette et al. 2009).

The sequence of mannuronic and guluronic residues significantly affects the physicochemical properties of alginates. The ratio of β-D-mannuronic acid to α-L-guluronic acid residues is usually 2:1, although it may vary with the algal species, the age of the plant as well as the type of tissue used for the extraction of alginates (Milani and Maleki 2012). Most commercial products are of the high M (mannuronic acid) type, for alginate, obtained from giant kelp, *Macrocystis pyrifera,* whereas *Laminaria hyperborea* provides a high G (guluronic acid) alginate. Alginates with a higher M content are preferred for applications where flexible structures are chosen. The different enzymes and genes involved in the alginate biosynthesis are well studied, using alginate-producing bacteria (Gacesa 1998). Detailed work on alginate structure is available in the literature (Moe et al. 1995).

## 27.3   Alginate applications

### 27.3.1   Food

Alginate is widely used in the food industry as a stabilizer, or as thickening or emulsifying agent. It is mainly consumed in southeast Asian countries, whereas in western countries, seaweed phycocolloids—alginate, carrageenan, and agar—are utilized as thickeners and stabilizing or emulsifying agents. As an indigestible polysaccharide, alginate is also viewed as a source of dietary fiber (Brownlee et al. 2005).

Generally, in practice, alginate gels are obtained by using three major methods: diffusion setting, internal setting, and setting by cooling. Diffusion method is effectively utilized to set thin strips of material (e.g., pimiento strips, films, and coatings). This has led to the development of structured fruits and pet foods (Smith and Fischer 1982; Nussinovitch 1997). Alginates can be used as desserts in ice cream products. Alginates give thermostability and desired consistency to food products. Structured onion rings are one of the best known examples of food products obtained by using the algin/calcium reaction. Nowadays, algin-based structures in potatoes reduce oil transfer into the interior of the potato, reduce moisture loss from the potato to the skin, and help to maintain the freshly cooked crispness (Davidson 1980; Fellow 2000).

The gelling applications are utilized in table jellies, aerated desserts, fruit pie fillings, etc. Alginate-based filling has proven exceptionally successful in fruit-filled pie to obtain products ranging from soft cream-type fillings, to soft aerated mousse products, to heavy cheese cake textures. Reduced-calorie products have a wide market demand. With this, the milk-based products can also be formulated with reduced levels of fat, salt, and cholesterol. The chocolate chiffon filling produces extremely rich and tasty chiffon, which can be served as a mousse dessert, frozen dessert, or pie filling. Instant cheesecake and pudding are well accepted today. It offers bakery creams with freeze/thaw stability, improves shelf life, and increases moisture retention in bread and cake mixes. It is commonly used as gelling, thickening, and stabilizing in jams, marmalades, and fruit sauces. In ice-cream, it avoids crystallization and shrinkage in addition to that it allows homogenous melting without whey separation (Sneath 1975; Jaquith and Church 1981; Wotherspoon 1988; Denis et al. 2001).

In meat manufacturing process, alginate is used in meat binding system. It is approved for beef, lamb, pork, and poultry, and also used in fish and other seafood. The products can be presented in a variety of forms: for example, slices for sandwiches or cubes for stews or kebabs. Natural meat flavor is not concealed and color is maintained. Inulinases are enzymes widely used for the production of fructo-oligosaccharide with functional and nutritional properties for use in low-calorie food preparations. Such an enzyme immobilization increases the catalytic properties of enzyme and allows the continuous reuses of costly enzyme to make it economically viable for industrial applications (Rehman et al. 2013). Such an inulinase immobilization has been carried out successfully using alginate-chitosan beads hardened with glutaraldehyde (Missau et al. 2014).

## 27.3.2 Coatings

Edible films are obtained from hydrocolloids, and the edible films provide film cohesion, mechanical properties, and impermeability of aroma compounds and lipids. Such sodium alginate–based edible film is used to limit dehydration of meat, fish, and fruits. Coating the foods with edible materials is an effective method to improve the food quality (Zactiti and Kieckbusch 2006). The transport parameters for aroma compounds through i-carrageenan and sodium alginate-based edible films are reported (Hambleton et al. 2011). But, determination of the transport properties of aroma compounds through biopolymers used in packaging application is challenging, because no standard method is available.

The potential of mucoadhesive materials is used in food application in order to control the delivery of specific ingredients such as salt and flavors. Biopolymers such as pectin, sodium carboxymethylcellulose, sodium alginate, and chitosan are some of the food-grade biopolymers, which possess the adhesive properties on the mucous layer that covers the oral surface. Work is still in progress to study the factors that affect the mucoadhesion such as pH, different molecular weight of same polymers, and new formulation of polymer in order to optimize the performance as delivery system in food engineering (Ali and Bakalis 2011).

Edible coatings are made from alginate to preserve the quality of minimally processed "Gala" apples. The coatings form a good film on the surface of the apple slice, giving the fruit a fresh appearance. It works as a barrier to water vapor by decreasing water loss in apple slices. It also prevents loss of texture and thereby decreases the browning deterioration of apple slices. Such preservation of minimally processed fruit is a promising technology that can improve the quality of fresh products and increase their shelf life and stability (Olives et al. 2007).

At times, delaying the decline of fish quality and extending the shelf life of fish by inhibiting or retarding the growth of microorganisms and reducing the rate of lipid oxidation is necessary. The addition of antioxidants to alginate-calcium-coating solutions preserves good quality and extends the shelf life of bream (*Megalobrama amblycephala*). Coating treatments predominantly reduce chemical spoilage and increase the overall sensory quality of fish, compared to uncoated bream (Song et al. 2011). Similarly, coatings applied to fresh acerolas, alginate-acerola puree film in the form of dispersions, are effective to extend fruit stability by decreasing weight loss, ascorbic acid loss, and decay incidence, and by delaying the ripening process (Azeredo et al. 2012).

## 27.3.3 Food fortification

It plays an important role to compensate for mineral deficiency. Fortification of vitamin A, carotene B, D, K, E, saturated fatty acids, formation of hemoglobin capsules with chitosan and alginate coating and several probiotics with alginate-coating is done by applying the emulsification method (De Vos et al. 2006; Annan et al. 2007; Kassem et al. 2012; Bhujbal et al. 2014; Spasojevic et al. 2014). A ferric-saccharate iron capsulation with alginate calcium coating has been carried out by Khosroyar et al. (2012). The fortification of food rolls with microencapsulated iron sulfate with sodium alginate increases hemoglobin levels and reduces the occurrence of anemia in preschoolers (Barbosa et al. 2012).

Intake of dietary alginate has a number of potentially beneficial physiological effects, such as reduced intestinal absorption, detrimental risks of GI luminal contents, variation in gut microflora, and elevation of colonic barrier function (Sunderland et al. 2000; Brownlee et al. 2005; Paxman et al. 2008; Janczyk et al. 2010).

## 27.3.4 Probiotics

The most important benefits of probiotics are categorized as maintenance of normal intestinal microflora defense against enteropathogen infections and the regulation of immune response and intestinal enzymes (Chavarri et al. 2010; Mortazavian et al. 2012; Nisha et al. 2013; El-Sayed et al. 2014). To get the potential benefits of probiotics, they should safely transit through acidic and enzymatic conditions of gastric tract and colonize and grow on the epithelium of colon (Nazzaro et al. 2009; Fahimdanesh et al. 2012). *Lactobacillus rhamnosus* and *Bifidobacterium animalis* microbial strains are used as probiotic foods. The product developed shows an 82.2% acceptability index of overall characteristics and good market potential (Guimaraes et al. 2013).

Yonekura et al. (2014) evaluated sodium alginate, chitosan, and hydroxypropyl methylcellulose (HPMC) as coencapsulants for spray-dried *Lactobacillus acidophilus*. The impact on cell viability and physicochemical properties of the dried powders shows viability over 35 days of storage at 25°C and survival after simulated digestion. Alginate-gelatinized starch coated with chitosan has considerable effects on probiotic survival as it protects the probiotic bacteria *Lactobacillus casei* and *Bifidobacterium bifidum* from human gastrointestinal ecosystem (Zanjani et al. 2014).

## 27.4 Animal feeds

Generally, enzyme supplementation improves performance and nutrient digestibility of broilers fed with diets containing high levels of grains rich in nonstarch polysaccharides (NSPs) and also plays an important role in efficient utilization of fiber and other nutrients in diet. An *in vitro* study conducted shows the protective effect of sodium alginate and its combinations with other compounds, that is, xanthan gum, carboxy methyl cellulose, and carboxy methyl starch on feed enzymes, namely, amylase and xylanase, against the extreme conditions in GIT of animals. Alginate xanthan and alginate carboxy methyl starch capsules show good tolerance to simulated gastric fluid of pH 2 and intestinal fluid of pH 7.5 (Kumar et al. 2012). In pet food industries, the structuring of meat and fish using algin gel technology is most successfully utilized. It is used to provide either structured meat chunks or a gelled matrix that holds the meat pieces together (Smith and Fischer 1982).

## 27.5 Dental treatments

Alginate is extensively used to study casts, master casts, and working models for the fabrication of intraoral appliances (Pace 2006). It is used to produce dental impressions by using the reaction between calcium salt (usually calcium sulfate dehydrate) and alginate with a resulting irreversible hydrocolloid impression. This dental impression powder is convenient and easy to handle for both dentist and patient with fast setting at room temperature (Onsoyen 1996). The cost-effectiveness of silicone and alginate impressions is well-known for complete dentures (Hulme et al. 2014).

## 27.6 Wound care

Alginate wound dressings have novel hemostatic and antimicrobial properties as well as the ability to promote wound healing. It is commonly used product, especially in the case of exudating wounds. The alginate within the dressing absorbs the exudate, resulting in a hydrophilic gel over the wound, which provides a good moist environment for healing. The gel-forming property of alginate helps in removing the dressing without much trauma and reduces the pain experienced by the patient during the change of dressing. A natural polymer-based complex scaffold of fibrin, chitosan, and sodium alginate with improved mechanical properties is structurally stable due to the strong ionic bonding between the amine groups of chitosan and the carboxyl groups of alginate (Devi et al. 2012). Alginates in the form of films, wool, gauzes, and clots are applied to a wide range of wounds, including burns, lacerations, ulcers, and amputations. In all these cases,

healing is rapid and uneventful. Its use in aural surgery and neurosurgery was reported long back (Passe and Blaine 1948; Thomas 2000).

## 27.7 Tissue engineering

In tissue engineering, numerous alginate applications are currently developed combining the knowledge of biomaterials, cells, signaling molecules, design and engineering, biomechanics, and informatics. Santana et al. (2013) demonstrated the combination of nano/microfibers of titanium dioxide (nfTD) and hydroxyapatite (nfHY) in alginate hydrogel scaffold in mouse fibroblast cell line (NIH/3T3). This alginate hydrogel scaffold retains the chemical characteristics of alginate, and also this association is cytocompatible. Moreover, the combination of nfHY and nfTD with alginate hydrogel favors cell viability in short and long term, respectively.

A three-dimensional follicular growth of intact follicles during *in vitro* growth is established with calcium alginate. The application of tissue engineering principles to the problems of follicle culture *in vitro* may provide an alternative to ovarian transplant for the preservation of fertility in women (Heise et al. 2005). Newer approaches to fiber formation have utilized microfluidic technologies. Many studies have been conducted to test the potentiality of nanofibers in numerous tissue engineering and bioactive factor delivery applications (Andersen et al. 2012).

RGD-conjugated alginate gel is a promising strategy to improve the therapeutic efficacy for the delivery therapeutic cells and angiogenic proteins. Such an injectable multifunctional microgel system can be used for the treatment of vascular diseases (Kim et al. 2014).

## 27.8 Bone regeneration

Alginates are used in the treatment and regeneration of the skin, cartilaginous and bone tissue, the liver, and tissues of the heart muscle (Tan and Marra 2010; Sun and Tan 2013). A hybrid system composed of micrometric fibers (zinc and calcium alginates containing nanoadditive tricalcium phosphate [TCP]) and submicron fibers ([dibutyryl chitin-DBC] and poly-ε-caprolactone [PCL]) is proposed. The biomimetic system of fibers, the presence of a nanometric modifier, and the use of polymers with different resorption times provide a framework with specific properties on which bone cells are able to settle and proliferate (Bogun et al. 2013).

A biomaterial with bioactive nanoparticles (nBG) and $Ca^{2+}$ incorporated into alginate matrix is characterized. The composite biomaterials are attractive candidates for the fabrication of bone tissue engineering scaffolds (Cattalini et al. 2013). The effect of alterations in physical parameters such as oxygen and pH on processes associated with cellular redox system in human osteoarthritic chondrocytes (HOACs) shows that oxygen and pH affect the elements of the redox system in HOAC including cellular antioxidants, mitochondrial membrane potential, and reactive oxygen species levels. The human osteoarthritic chondrocytes isolated from total knee arthroplast samples can be cultured in 3-D alginate beads (Collins et al. 2013).

## 27.9   Drug delivery

A combination of alginate-based hydrogels, porous scaffolds, and microspheres for controlled drug delivery is a promising area of research. Such a porous scaffolding matrix fabricated from alginate and its derivatives has various forms (spheres, sponges, foams, fibers, and rods) for cell culture and response, which makes it particularly suitable for biomedical applications. Likewise numerous reports have been published on the release of proteins from alginate matrices; these have been faithfully summarized by Gombotz and Wee (1998).

### 27.9.1   Sustained drug release

Alginates are used in drug delivery to control/regulate the release of drugs from solid, hydrophilic matrix dosage forms. Following ingestion, the alginate matrix hydrates on contact with gastric fluid, and the surface polymer swells in a predictable manner to form a viscous "gel layer." This gelatinous layer acts as a rate-limiting barrier to drug release as it retards drug efflux and water influx, and prevents matrix disintegration. If the drug is water soluble, it will dissolve as the matrix hydrates and diffuses out of the alginate gel layer. The speed of hydration and diffusion/release of the drug can be controlled by the mannuronic/guluronic acid (M/G) ratio and molecular weight of the alginate (Onsoyen 1996).

Some of the applications of alginate-based biomaterials for sustained drug release are summarized in the Table 27.1.

### 27.9.2   Antireflux drugs

Alginate antireflux preparations are used in the treatment for gastroesophageal reflux disease. These provide physical barrier upon contact with the stomach contents in the form of neutral floating gel or "raft." Products such as Gaviscon contain a high G sodium alginate, calcium carbonate, and sodium (or potassium) bicarbonate, which are formulated to undergo ionic gelation in the stomach. Alginate rafts are formed by the action of gastric acid on a soluble sodium alginate to form an insoluble gel of alginic acid. The insoluble calcium carbonate releases $Ca^{2+}$ on contact with the acid to strengthen the alginic acid gel. The simultaneous action of gastric acid on a bicarbonate source produces $CO_2$ that becomes trapped within the alginic acid gel to provide aeration and buoyancy (Hampson et al. 2005).

### 27.9.3   Herbal drugs

The encapsulation technique enables controlled-delivery and bioavailability of an active compound by tailoring the release mechanism or rate in gastrointestinal tract. For instance, the direct consumption of an herbal active compound sometimes causes inconvenience to the human system. It may promote better product stability by isolating active compounds from the detrimental effects of oxygen, moisture, or incompatible compounds (Chan and Zhang 2005; Chan et al. 2009).

Bioactive compounds of herbal plants such as antioxidants have multiple functional and remedial properties that include antiradial, anticarcinogenic, reduction of oxidative stress, antiinflammatory, and cardioprotection. Chan et al. (2010) have encapsulated the herbal aqueous extract of *Piper sarmentosum* through absorption with calcium-alginate hydrogel beads.

## 27.10   Immunomodulation

Traditional disease control strategies in aquaculture employ antibiotics and chemical disinfectants, but nowadays, these are no longer recommended practices due to the emergence of bacterial resistance, and also owing to concerns over environmental impact (Sakai 1998).

The intraperitoneal injection of sodium alginate increases the phenol peroxidase and respiratory burst activities of white shrimp hemocytes and enhances resistance to *Vibrio alginolyticus* (Cheng et al. 2004). Comparable effects have been observed in a successive study on examining the effects of dietary administration of sodium alginate upon the immune ability of white shrimp (Cheng et al. 2005).

Ergosan is an algal-based product composed of 0.002% unspecified plant extract, 1% alginic acid from *Laminaria digitata*, and 98.998% algal-based carrier. It is used in aquaculture for its immunomodulatory activity. Shrimps fed with ergosan show a significant increase in relative growth (Montero-Rocha et al. 2006). In the same way, immunomodulation in both rainbow trout (*Oncorhynchus mykiss*) and striped snakehead (*Channa striata*) has been observed following intraperitoneal administration of ergosan (Miles et al. 2001; Peddie et al. 2002).

Streptococcus disease is initially described among population of rainbow trout farmed in Japan, South Africa, the United States, Great Britain, and Norway (Austin and Austin 1987). This disease causes huge losses (about 80%–90%) from 1995, but its occurrence

*Table 27.1* Various applications of alginate-based biomaterials

| Drug | Biomaterial | Applications | Study goal/outcome | References |
|---|---|---|---|---|
| Famotidine | Sodium alginate | For gastroretentive drug delivery system | A sustained drug release for more than 8 h. *In vivo* significant antiulcer effect. | Patel et al. (2010) |
| Stavudine | Sodium alginate | For antiretroviral therapy | Sustained release of drug following oral administration. | Revathi et al. (2014) |
| Cisplatin and Rhenium-188 | | Radiochemotherapy of mammary tumors | Possibility of using hydrogel where tumors are inoperable. | Azhdarinia et al. (2005) |
| Metformin HCl | Sodium alginate (with or without pectin) | Gastroretentive multiparticulate drug delivery system of highly water-soluble antihyperglycemic drugs | Alginate and pectin are effective carriers for the drug and management of type 2 diabetes mellitus. | Singh et al. (2013) |
| Insulin | Alginate/chitosan | Oral delivery system for insulin | Encapsulation of insulin into mucoadhesive nanoparticles—a key factor in the improvement of its oral absorption and bioactivity. | Sarmento et al. (2007) |
| Metronidazole | Chitosan/alginate | For colon-specific delivery of drug | Drug release takes place at a highly retarded rate. | Gopinath et al. (2012) |
| Ibuprofen | Sodium alginate | Antirheumatic drug delivery system | Offers a reliable delivering Ibuprofen by the oral route. | Mallick et al. (2013) |
| Theophylline | Sodium alginate | To improve the delivery of TP in bronchial asthma | TP shows extended drug release profile in pH progression medium. | Soni et al. (2010) |
| Paclitaxel | Hydroxyapatite–alginate composites | Drug-delivery matrices for local chemotherapy | Delays the time to progression of subcutaneous breast cancer in rats without any sign of toxicity. | Kobayashi et al. (2012) |
| BSA | L-arginine-grafted alginate | For protein controlled release applications | Grafted alginate improves its release profile behavior particularly in acidic media. | Eldin et al. (2014) |
| Cefalexin | Alginate-g-poly AA/kaolin composite | For site-specific delivery of drugs | The release of drug at pH 7.4 was higher than that at pH 1.6 | Sadeghi et al. (2014) |
| Clarithromycin | Alginate, hydroxypropyl-methylcellulose–sunflower oil | For gastroretentive dosage | To develop an intragastric floating drug delivery system. | Nimase and Vidyasagar (2010) |
| Metronidazole | Guar gum–alginate | For colon-targeted delivery | Natural polymer can be used effectively for the colon targeting. | Mazumder et al. (2010) |
| 5-Fluorouracil | Sodium alginate/chitosan | For colon-specific delivery | Minimal drug release in stimulated stomach and small intestinal pH. | Shabber et al. (2012) |
| Glipizide | Sodium alginate | For treatment of type 2 diabetes mellitus | No interaction between drug and polymers and sustained release. | Allamneni et al. (2012) |
| Diclofenac | Calcium-induced alginate–phosphate composite | For drug delivery | Drug release rate can be controlled by modifying the composition of the matrix with CP (calcium salt of phosphoric acid). | Murata et al. (2009) |
| Trimetazidine dihydrochloride (TMZ) | Sodium alginate | Antianginal agents | The drug concentration exhibits a drug-loading-dependent effect on the release behavior in both sequential and simultaneous methods. | Mandal et al. (2010) |

*(Continued)*

***Table 27.1 (Continued)*** Various applications of alginate-based biomaterials

| Drug | Biomaterial | Applications | Study goal/outcome | References |
|---|---|---|---|---|
| Timolol maleate | Sodium alginate films/ethyl cellulose | Ophthalmic drug delivery systems | The films show promising drug release through diffusion-controlled release process and promising bioerodible material for ocular inserts. | Nazareth et al. (2014) |
| Cefixime trihydrate | Chitosan—alginate | Transdermal drug delivery | The delivery of drug was at a controlled rate across intact skin (*in vitro* skin permeation studies in rat abdomen skin). | Natarajan et al. (2011) |
| Nitrofurazone | Alginate–*Aloe vera* composite membranes | Regeneration of skin | Preliminary *in vitro* drug release tests suggest that the membranes are able to incorporate and release drugs. | Pereira and Bartolo (2013) |
| Amoxicillin | Chitosan–alginate polyelectrolyte complex | Mucopenetrating drug delivery system | Increased mobility of nanoparticles in the gastric mucus by decreasing the surface amino groups of chitosan on ionic interaction with carboxylic groups of alginate. | Arora et al. (2011) |
| Amphotericin B | Sodium alginate | Antifungal therapy | Higher antifungal efficacy than the free drug and sustained *in vitro* release. | Sangeetha et al. (2007) |
| Streptomycin | Chitosan/sodium alginate | Drug delivery | Conc. of chitosan-alginate and calcium chloride had significant synergistic effect on particle size and % encapsulation efficiency. | Chopra et al. (2012) |
| Dust mite allergen, Ag | Chitosan/alginate | For mucosal allergen delivery | Controlled allergen release property. | Suksamran et al. (2011) |
| Nifedipine | Hydroxypropyl methylcellulose/chitosan/alginate | Oral drug delivery | *In vitro* test shows good mucoadhesive properties. | Sherina et al. (2012) |
| Insulin | Alginate/chitosan | Oral insulin delivery | Well-preserved integrity of nanoemulsions in simulated gastric juices and significantly prolonged hypoglycemic effects after oral administration. | Li et al. (2013) |
| Vascular endothelial growth factor (VEGF) ~45 kDa angiogenic protein | Alginate cross-linked by different divalent cations, $Zn^{2+}$ and $Ca^{2+}$ | Protein drug delivery | New methods for controlled, sustained release of VEGF from small alginate microparticles. | Jay and Saltzman (2009) |
| Ovalbumin | Chitosan/alginate | Vaccine delivery systems | Nanoparticles taken up by rat Peyer's patches and slight stimulation of the splenocytes. | Borges et al. (2006) |

gets reduced by using vaccine against this disease (Bachrach et al. 2001). The immunostimulatory effects of alginic acid and antistreptococcus vaccine on some physiological parameters of rainbow trout show that fish resistance increases to environmental stress and pathogen due to the stimulated lymphocyte proliferation (Faghani et al. 2008). Alginic acid supplemented at 4 g kg$^{-1}$ diet has the best effect on growth and immune system parameters in great sturgeon, *Huso huso* juvenile (Ahmadifar et al. 2009). Incorporating the vaccine into food pellets is a favorable oral delivery method for the vaccination of cultured fish. Lyophilized alginate–plasmid complexes added to feed show strong protection with relative survival rates against infections pancreatic necrosis virus in *O. mykiss* (Ballesteros et al. 2014).

Son et al. (2001) describe the stimulation of various function of murine peritoneal macrophages isolated 20 h after intraperitoneal injection with high mannuronic acid–containing alginate (HMA) 25 and 100 mg kg$^{-1}$. It is found that not only HMA increases the number of peritoneal macrophages and phagocytosis but macrophages significantly inhibit the growth of tumor cells. The tumoricidal activity induced by HMA seems to be facilitated by the production of TNF-$\alpha$, NO, and H$_2$O$_2$. At present, the only available vaccine against human tuberculosis (TB) is the Bacillus Calmette-Guerin (BCG). Here, one of the major drawbacks of BCG is its failure to stimulate adequate antimycobacterial T cell responses, required for *M. tuberculosis* control. At present, the research for an improved vaccine is a very active field of investigation. In this regard, the adjuvanticity effect of sodium alginate on subcutaneously injected BCG in BALB/c mice shows proliferative and delayed-type hypersensitivity responses, IFN-g, specific antimycobacterium total IgG, IgG1, and IgG2a production is significantly higher in mice immunized with BCG plus alginate. Also, it has lower mean bacterial count compared to those vaccinated with BCG alone (Dobakhti et al. 2009).

Torres et al. (2007) report the extraction and physicochemical characterization of two *Sargassum vulgare* alginates, denoted as SVLV (*S. vulgare* low viscosity) and SVHV (*S. vulgare* high viscosity). The antitumor activity of low and high viscosity alginate is demonstrated against sarcoma 180 cells transplanted in mice, here SVHV is more active and less toxic. This study suggests that anticancer activity is mediated by a modulation of the immune system (Sousa et al. 2007, 2008). The proinflammatory activity of SVHV in the model of paw edema of rats depends more on the migration of neutrophils than on the activation of mast cells and release of vasoactive substances, such as histamine and nitric oxide, and, also on the activation of resident cells, which have macrophages as the main cell (Lins et al. 2013). Yang and Jones (2009) showed that low viscosity sodium alginates induce activation of macrophages through NF-$\kappa$B pathway, leading to the release of proinflammatory cytokines.

## 27.11  Biological activity

### 27.11.1  Antidiabetic activity

Seaweed polysaccharides are biologically active. The chemical characterization and antihyperlipidemic activity of alginate with a molecular weight of 16,000 Da and a molar ratio of mannuronic acid to guluronic acid of 2.75 from the brown alga *Sargassum fusiforme* markedly decreases the content of total cholesterol, triglyceride and low-density lipoprotein-cholesterol in the serum of experimental hyperlipidemic rats, and significantly increases the level of high-density lipoprotein-cholesterol. The antihyperlipidemic activity is associated with the modes of linkages of pyran rings between polysaccharide chains (Mao et al. 2004). The ultrapure alginate bead is used for encapsulation of pancreatic islets of Langerhans to produce insulin for the treatment of Type I diabetes (King et al. 2003). Such alginate-encapsulated pancreatic islets are currently evaluated in man in early stage of clinical trials (Calafiore et al. 2006).

### 27.11.2  Antioxidant activity

The beneficial effects of calcium alginate in CCL$_4$-induced liver injury model in rats show both healing and preventive effect, which are due to the stabilization of biochemical parameters of oxidative processes in the blood and liver. This ameliorative effect of calcium alginate is related to its antioxidant properties (Khotimchenko and Khotimchenko 2004).

## 27.12  Water treatment

For a long time, industrial wastewater and particularly metal-containing effluent has become a major environmental concern. Sustainable development and better water management enable reuse of the treated water and also offer the possibility of recycling and recovery of the metal ions.

The algae *Lessonia nigrescens* and *Macrocystis integrifolia* are potential biosorbents in the removal of phenol and derivatives from industrial waste. This absorption strongly depends on pH, and a polar adsorption mechanism rather than an electrostatic adsorption is proposed for phenol onto the surface of marine seaweed by the formation of hydrogen bonds with hydroxyl groups of polysaccharides, such as alginates (Navarro et al. 2008). A lot of information about water purification technique using magnetic assistance is available. Magnetic alginate beads are efficiently used to remove heavy metals such as Pb (II) ions from aqueous solutions (Bee et al. 2011).

Chromium occurs in wastewaters of many industries: electroplating, leather tanning, paint, and pigment industries. Numerous studies are made to remove chromium by biosorption from wastewater due to its high toxicity. Calcium alginate gel bead is an effective sorbent for both Cr (III) and Cr (VI) removal from aqueous solutions (Escudero et al. 2013). This kind of procedure is successfully applied to the entrapment of solid by-product from metal surface treatment industry. Two percent of raw material entrapped in calcium alginate gel matrix (2% O-CA) is a promising sorbent candidate for As (V) removal (Garlaschelli et al. 2013). Fe-alginate nanoparticle is an effective adsorbent for the removal of As (V) ions and bacteriological contaminations from aqueous solutions (Singh et al. 2014).

Alginate with a high M/G ratio, extracted from *L. digitata*, is evaluated for heavy metal sorption, such as $Cu^{2+}$, $Cd^{2+}$, and $Pb^{2+}$ in acidic solutions, in the form of calcium cross-linked beads (Papageorgiou et al. 2006). A new method to produce zinc oxide nanocrystals based on the thermal decomposition of zinc alginate gels is used as a potential precursor for the preparation of other technologically important oxides (Baskoutas et al. 2007). Alginate-SBA-15 is synthesized by encapsulation of the nanoporous SBA-15 in the biopolymeric matrix of calcium alginate. It is highly effective in the removal of lead from an aqueous solution (Cheraghali et al. 2013).

The removal of dyes of textile effluents is of significant importance. Dyes are visible in water columns due to their brightness and their presence in industrial effluents affects the photosynthetic process. Positively charged methylene blue and negatively charged methyl orange are used as models of organic pollutants and the magnetic properties of the alginate beads allow their separation from the effluent by a simple magnetic field (Rocher et al. 2008).

## 27.13 Energy

Microbial fuel cells (MFCs) have gained interest among researchers as they hold much potential to provide clean energy with simultaneous waste-water treatment in a more sustainable and cost-effective fashion (Frank and Nevin 2010; Ghangrekar and Shinde 2008). The proton exchange membranes (PEMs) used in an MFC is a limiting factor in determining its overall power performance. MFC with low cost material is one of the components preferred in commercial application (Chai et al. 2010). The cassava starch (a cheaper alternative) is used as a PEM in MFC (Obsai et al. 2012). Further, to increase the commercial application potential, a simple "batch" dual-chambered microbial fuel cell incorporating cassava starch PEM whose proton conductivity is improved with sodium alginate is operated. Here, in addition to the overall cell cost reduction, addition of the alginate improves the performance of the PEM and also promotes cell durability (Obasi et al. 2013).

A low-cost alginate-nano Si-electrode is extracted from seaweed using a straight-forward soda-based process that produces a uniform material. The uniformly distributed carboxylic groups are used to improve the performance of battery electrodes. Such cheap, lightweight, and improved batteries can have a wide range of applications, such as electrical cars, computers, and cell phones (Kovalenko et al. 2011). The mass cultivation of *Botryococcus braunii* in an open raceway pond

for biodiesel production shows the maximum growth and lipid production at 16 mg alginate concentration. It greatly promotes the growth of *B. braunii* for biodiesel production (Chakravarthy et al. 2013).

## 27.14 Miscellaneous uses

### 27.14.1 Surfactants

The polysaccharides of marine algae with innovative structures and functional properties find applications as ingredients for the development of green surfactants and cosmetic actives. Novel biocompatible surfactants based on mannuronate moieties derived from alginates have been developed. Controlled chemical and enzymatic depolymerizations of the polysaccharides give saturated and unsaturated functional oligomannuronates. These ingredients represent a new class of surface-active agents with promising foaming and emulsifying properties (Benvegnu and Sassi 2010).

### 27.14.2 Corrosion

Aluminum alloy 7075 is widely used in aerospace and defense applications. The alloy samples coated with 2000 ppm sodium alginate inhibitor show improved corrosion inhibition. A maximum efficiency of 95% is observed for a 24 h immersion in 3.5% NaCl solution, which is used as the corrosive media (Sriram et al. 2014).

### 27.14.3 Plant growth tonics

Alginates broaden the application for special growth promotion effect on plants (Iwasaki and Mastsubara 2000; Lata et al. 2009). The plant growth and seed yield are high in the plants sprayed with low-molecular-weight sodium alginate (NaAlg) fractions treated with 100 kGy and added potassium persulfate as compared to control plants (El-Mohdy 2013).

## 27.15 Summary

Brown algal polysaccharides have potential for biological and industrial applications. However, a standardized commercial product based on algal polysaccharide is a challenge because of their structural diversity that depends on species and on location and time of harvest. Hence, much information is to be acquired with respect to its structure along with its basic properties which is at this stage possible because of modern instrumentation and technology (Genaro et al. 2014). Successful exploitation based on scientific data reiterates innovative and exciting applications of alginates in the future.

# References

Ahmadifar, E., G. H. A. Takami, and M. Sudagar. 2009. Growth performance, survival and immunostimulation, of Beluga (*Huso huso*) Juvenile following dietary administration of alginic acid (Ergosan). *Pakistan Journal of Nutrition* 8(3): 227–232.

Ali, M. F. and S. Bakalis. 2011. Mucoadhesive polymers for food formulations. *Procedia Food Science* 1: 68–75.

Allamneni, Y., B. V. V. K. Reddy, D. Chary, V. B. Rao, S. C. Kumar, and A. K. Kalekar. 2012. Performance evaluation of mucoadhesive potential of sodium alginate on microspheres containing an anti-diabetic drug: Glipizide. *International Journal of Pharmaceutical Sciences and Drug Research* 4(2): 115–122.

Andersen, T., B. L. Strand, K. Formo, E. Alsberg, and B. E. Christensen. 2012. Alginates as biomaterials in tissue engineering. *Carbohydrate Chemistry* 37: 227–258.

Annan, N. T., A. D. Borza, and H. Truelstrup. 2007. Encapsulation in alginate coated gelatin microspheres improves survival of the probiotic *Bifidobacterium adolescentis* 15703t during exposure to simulated gastro-intestinal conditions. *Journal of Food Science* 41(2): 184–193.

Arasaki, S. and T. Arasaki. 1983. *Low Calorie, High Nutrition Vegetables from the Sea. To Help You Look and Feel Better.* Japan Publication: Tokyo, Japan, 196pp.

Arora, S., S. Gupta, R. K. Narang, and R. D. Budhiraja. 2011. Amoxicillin loaded chitosanalginate polyelectrolyte complex nanoparticles as mucopenetrating delivery system for *H. pylori*. *Scientia Pharmaceutica* 79: 673–694.

Austin, B. and D. A. Austin. 1987. Bacterial fish pathogens: disease in farmed and wild fish. Ellis Horwood Series in Aquaculture and Fisheries Support. University of Aberdeen, Aberdeen, Scotland, pp. 97–111.

Azeredo, H. M. C., K. W. E. Miranda, H. L. Ribeiro, M. F. Rosa, and D. M. Nascimento. 2012. Nanoreinforced alginate-acerola pure coatings on acerola fruits. *Journal of Food Engineering* 113: 505–510.

Azhdarinia, A., D. J. Yang, D.-F. Yu, R. Mendez, C. Oh, S. Kohanim, J. Bryant, and E. E. Kim. 2005. Regional radiochemotherapy using *in situ* hydrogel. *Pharmaceutical Research* 22(5): 776–783.

Bachrach, G., A. Zlotkin, D. Hurvitz, L. Evans, and A. Eldar. 2001. Recovery of *Streptococcus iniae* from diseased fish previously vaccinated with a *Streptococcus* vaccine. *Applied and Environmental Microbiology* 67(8): 3756–3758.

Ballesteros, N. A., S. R. Saint-Jean, and S. I. Perez-Prieto. 2014. Food pellets as an effective delivery method for a DNA vaccine against infectious pancreatic necrosis virus in rainbow trout (*Oncorhynchus mykiss*, Walbaum). *Fish and Shellfish Immunology* 37: 220–228.

Barbosa, T. N. N., J. A. A. C. Taddei, D. Palma, F. Anconalopez, and J. A. P. Braga. 2012. Double-blind randomized controlled trial of rolls fortified with microencapsulation iron. *Revista da Associacao Medica Brasileira* 58(1): 118–124.

Baskoutas, S., P. Giabouranis, S. N. Yannopoulos, V. Dracopoulos, L. Toth, A. Chrissanthopoulos, and N. Bouropoulos. 2007. Preparation of ZnO nanoparticles by thermal decomposition of zinc alginate. *Thin Solid Films* 515: 8461–8464.

Bee, A., D. Talbot, S. Abramson, and V. Dupuis. 2011. Magnetic alginate beads for Pb (II) ions removal from waste water. *Journal of Colloid and Interface Science* 362: 486–492.

Benvegnu, T. and J. F. Sassi. 2010. Oligomannuronates from seaweeds as renewable sources for the development of green surfactants. *Topics in Current Chemistry* 294: 143–164.

Bhujbal, S. V., B. De Haan, S. P. Niclou, and P. De Vos. 2014. A novel multilayer immuno isolating encapsulation system overcoming protrusion of cells. *Scientific Reports* 4: 6856.

Bogun, M., I. Krucinska, A. Kommisarczyk, T. Mikolajczyk, M. Blazewicz, E. Stodolak-Zych, E. Menaszek, and A. Scislowska-Czarnecka. 2013. Fibrous polymeric composites based on alginates fibres and fibres made of poly-ε-caprolactone and dibutyryl chitin for use in regenerative medicine. *Molecules* 18: 3118–3136.

Borges, O., A. Cordeiro-da-Silva, S. G. Romeijn, M. Amidi, A. de Sousa, G. Borchard, and H. E. Junginger. 2006. Uptake studies in rat Peyer's patches, cytotoxicity and release studies of alginate coated chitosan nanoparticles for mucosal vaccination. *Journal of Controlled Release* 114: 348–358.

Brownlee, L. A., A. Allen, J. P. Pearson, P. W. Dettmar, M. E. Havler, M. R. Atherton, and E. Onsoyen. 2005. Alginate as a source of dietary fiber. *Critical Reviews in Food Science and Nutrition* 45: 497–510.

Calafiore, R., G. Basta, G. Luca, A. Lemmi, M. P. Montanucci, G. Calabrese, L. Racanicchi, F. Mancuso, and P. Brunetti. 2006. Microencapsulated pancreatic islet allografts into nonimmunosuppressed patients with type 1 diabetes. *Diabetes Care* 29(1): 137–138.

Campos-Vallette Marcelo, M., N. P. Chandia, E. Clavijo, D. Leal, B. Matsuhiro, I. O. OsorioRomanb, and S. Torres. 2009. Characterization of sodium alginate and its block fractions by surface-enhanced Raman spectroscopy. *Journal of Raman Spectroscopy* 41: 758–763.

Cattalini, J. P., J. Garcia, A. R. Boccaccini, S. Lucangioli, and V. Mourino. 2013. A new calcium releasing nano-composite biomaterial for bone tissue engineering scaffolds. *Procedia Engineering* 59: 78–84.

Chai, L. F., L. C. Chai, N. Suhaimi, and R. Son. 2010. Performance of air-cathode microbial fuel wood charcoal as electrode. *International Food Research Journal* 17: 485–490.

Chakravarthy, K., S. Chinnasamy, S. Bhaskar, and R. Rengasamy. 2013. Effect of sodium alginate on growth and lipid production of *Botryococcus braunii* kutzing for biodiesel production. *Journal of Pharmaceutical and Biomedical Sciences* 35(35): 1802–1807.

Chan, E. S., B. B. Lee, P. Ravindra, and D. Poncelet. 2009. Prediction models for shape and size of Ca-alginate macrobeads produced through extrusion-dripping method. *Journal of Colloids and Interface Science* 338(1): 63–72.

Chan, E.-S., Z.-H. Yim, S.-H. Phan, R. F. Mansa, and P. Ravindra. 2010. Encapsulation of herbal aqueous extract through absorption with Ca-alginate hydrogel beads. *Food and Bioproducts Processing* 88: 195–201.

Chan, E.-S. and Z. Zhang. 2005. Bioencapsulation by compression coating of probiotic bacteria for their protection in an acidic medium. *Process Biochemistry* 40(10): 3346–3351.

Chavarri, M., I. Maranon, R. Ares, F. C. Ibanez, F. Marzo, and M. D. C. Villaran. 2010. Microencapsulation of a probiotic and prebiotic in alginate-chitosan capsules improves survival in simulated gastro-intestinal conditions. *International Journal of Food Microbiology* 142: 185–189.

Cheng, W., C. H. Liu, C. M. Kuo, and J. C. Chen. 2005. Dietary administration of sodium alginate enhances the immune ability of white shrimp *Litopeneaus vannamei* and its resistance against *Vibrio alginolyticus*. *Fish and Shellfish Immunology* 18: 1–12.

Cheng, W., C. H. Liu, S. T. Yeh, and J. C. Chen. 2004. The immune stimulatory effect of sodium alginate on the white shrimp *Litopeneaus vannamei* and its resistance against *Vibrio alginolyticus*. *Fish and Shellfish Immunology* 17: 41–51.

Cheraghali, R., H. Tavakoli, and H. Sepehrian. 2013. Preparation, characterization and lead sorption performance of alginate-SBA-15 composite as a novel adsorbent. *Scientia Iranica* 20(3): 1028–1034.

Chopra, M., P. Kaur, M. Bernela, and R. Thakur. 2012. Synthesis and optimization of streptomycin loaded chitosan-alginate nanoparticles. *International Journal of Scientific and Technology Research* 1(10): 31–34.

Collins, J. A., R. J. Moots, R. Winstanley, P. D. Clegg, and P. I. Milner. 2013. Oxygen and pH sensitivity of human osteoarthritic chondrocytes in 3-D alginate bead culture system. *Osteoarthritis and Cartilage* 21: 1790–1798.

Davidson, R. L. 1980. *Handbook of Water-Soluble Gums and Resins*. McGraw-Hill: New York.

De Vos, P., M. M. Faas, B. Strand, and R. Calafiore. 2006. Alginate-based microcapsules for immuno isolation of pancreatic islets. *Biomaterials* 27: 5603–5617.

Denis, M., E. R. D. Haan, and R. H. Clark. 2001. Particulate natural fruit product and method of making same. US Patent No. 6,251,466 B1.

Dettmar, P. W., V. Strugala, and J. C. Richardson. 2011. The key role alginates play in health. *Food Hydrocolloids* 25: 263–266.

Devi, M. P., M. Sekar, M. Chamundeswari, A. Moorthy, G. Krithiga, N. Selva Murugan, and T. P. Sastry. 2012. A novel wound dressing material-fibrin-chitosan-sodium alginate composite sheet. *Bulletin of Materials Science* 35(7): 1157–1163.

Dobakhti, F., T. Naghibi, M. Taghikhani, S. Ajdary, A. Rafinejad, K. Bayati, S. Rafiei, and M. Rafiee-Tehrani. 2009. Adjuvanticity effect of sodium alginate on subcutaneously injected BCG in BALB/c mice. *Microbes and Infection* 11: 296–301.

Eldin, M. S. M., E. A. Kamoun, M. A. Sofan, and S. M. Elbayomi. 2014. L-Agrinine grafted alginate hydrogel beads: A novel pH-sensitive system for specific protein delivery. *Arabian Journal of Chemistry* 8(3): 355–365.

El-Mohdy, H. L. A. 2013. Radiation-induced degradation of sodium alginate and its plant growth promotion effect. *Arabian Journal of Chemistry*. http://dx.doi.org/10.1016/j. arabjc.2012.10.003.

El-Sayed, H. S., H. H. Salama, and S. M. El-Sayed. 2014. Production of symbiotic ice cream. *International Journal of Chem Tech Research* 7(1): 138–147.

Escudero, C., N. Fiol, I. Villaescusa, and J.-C. Bollinger. 2013. Effect of chromium speciation on its sorption mechanism onto grape stalks entrapped into alginate beads. *Arabian Journal of Chemistry*. http://dx.doi.org/10.1016/j. arabjc.2013.03.011.

Faghani, T., G. H. A. Takami, A. Kousha, and S. Faghani. 2008. Surveying on alginic acid and anti-streptococcus vaccine effects on the growth performance, survival rate, haematological parameters in rainbow trout (*Oncorhynchus mykiss*). *World Journal of Zoology* 3(2): 54–58.

Fahimdanesh, M., N. Mohammadi, H. Ahari, M. A. K. Zanjani, F. Z. Hargalani, and K. Behrouznasab. 2012. Effect of microencapsulation plus resistant starch on survival of *Lactobacillus casei* and *Bifidobacterium bifidum* in mayonnaise sauce. *African Journal of Microbiology Research* 6(40): 6853–6858.

Fellow, P. 2000. *Food Processing Technology: Principles and Practice*. CRC Press: Boca Raton, FL.

Franks, A. E. and K. P. Nevin. 2010. Microbial fuel cells: A current review. *Energies* 3: 899–919.

Gacesa, P. 1998. Bacterial alginate biosynthesis—Recent progress and future prospects. *Microbiology* 144: 1133–1143.

Garlaschelli, F., G. Alberti, N. Fiol, and I. Villaescusa. 2013. Application of anodic stripping voltammetry to assess sorption performance of an industrial waste entrapped in alginate beads to remove As (V). *Arabian Journal of Chemistry*. http://dx.doi.org/10.1016/j. arabjc.2013.01.003.

Genaro, A. P.-J., B. J. D. Haan, M. M. Faas, and P. D. Vos. 2014. A technology platform to test the efficacy of purification of alginate. *Materials* 7: 2087–2103.

Ghangrekar, M. M. and V. B. Shinde. 2008. Simultaneous sewage treatment and electricity generation in membraneless microbial fuel cell. *Water Science and Technology* 58(1): 37–43.

Gombotz, W. R. and S. F. Wee. 1998. Protein release from alginate matrices. *Advanced Drug Delivery Reviews* 31: 267–285.

Gopinath, H., A. Aqther, S. P. Basha, V. K. Siva, and F. Ahmed. 2012. Preparation and *in-vitro* evaluation of chitosan-alginate microcapsules for colon targeted drug delivery of metronidazole. *Journal of Chemical and Pharmaceutical Sciences* 5(3): 117–123.

Guimaraes, R. R., A. L. D. A. Vendramini, A. C. D. Santos, S. G. F. Leite, and M. A. L. Miguel. 2013. Development of probiotic beads similar to fish eggs. *Journal of Functional Foods* 5: 968–973.

Hambleton, A., A. Voilley, and F. Debeaufort. 2011. Transport parameters for aroma compounds through i-carrageenan and sodium alginate-based edible films. *Food Hydrocolloids* 25: 1128–1133.

Hampson, F. C., A. Farndale, V. Strugala, J. Sykes, I. G. Jolliffe, and P. W. Dettmar. 2005. Alginate rafts and their characterization. *International Journal of Pharmaceutics* 294: 137–147.

Heise, M., R. Koepsel, A. J. Russell, and E. A. McGee. 2005. Calcium alginate microencapsulation of ovarian follicles impacts FSH delivery and follicle morphology. *Reproductive Biology and Endocrinology* 3: 47.

Hernandez-Carmona, G., Y. Freile-Pelegrin, and E. Hernandez-Garibay. 2013. Conventional and alternative technologies for the extraction of algal polysaccharides. In *Functional Ingredients from Algae for Foods and Nutraceuticals*; Dominguez, H. Ed. Woodhead Publishing Limited, Cambridge, U.K., pp. 475–516.

Horn, S. J., I. M. Aasen, and K. Ostgaard. 2000. Ethanol production from seaweed extract. *Journal of Industrial Microbiology and Biotechnology* 25: 249–254.

Hulme, C., G. Yu, C. Browne, J. Odwyer, H. Craddock, S. Brown, J. Gray et al. 2014. Cost-effectiveness of silicone and alginate impressions for complete dentures. *Journal of Dentistry* 42: 902–907.

Iwasaki, K. and Y. Matsubara. 2000. Purification of alginate oligosaccharides with root growth promoting activity towards lettuce. *Bioscience Biotechnology and Biochemistry* 64(5): 1067–1070.

Janczyk, P., R. Pieper, C. Wolf, G. Freyer, and W. B. Souffrant. 2010. Alginate fed as a supplement to rats affects growth performance, crude protein digestibility, and caecal bacterial community. *Archiva Zootechnica* 13(2): 5–18.

Jaquith, J. B. and A. R. Church. 1981. Molded gelled pimiento body. US Patent No. 4,296,140.

Jay, S. M. and W. Mark Saltzman. 2009. Controlled delivery of VEGF *via* modulation of alginate microparticle ionic crosslinking. *Journal of Controlled Release* 134(1): 26–34.

Kassem, M. A., M. I. A.-T. E. Assal, and A. A.-S. A. Badrawy. 2012. Preparation and evaluation of certain hydrophilic drug-loaded microspheres. *International Research Journal of Pharmaceuticals* 2(4): 82–90.

Khosroyar, S., A. Akbarzade, M. Arjoman, A. A. Safekordi, and S. A. Mortazavi. 2012. Ferricsaccharate capsulation with alginate coating using the emulsification method. *African Journal of Microbiology Research* 6(10): 2455–2461.

Khotimchenko, Y. S. and M. Y. Khotimchenko. 2004. Healing and preventive effects of calcium alginate on carbon tetrachloride induced liver injury in rats. *Marine Drugs* 2: 108–122.

Kim, P.-H., H.-G. Yim, Y.-J. Choi, B.-J. Kang, J. Kim, S.-M. Kwon, B.-S. Kim, N. S. Hwang, and J.-Y. Cho. 2014. Injectable multifunctional microgel encapsulating outgrowth endothelial cells and growth factors for enhanced neovascularization. *Journal of Controlled Release* 187: 1–13.

King, A., A. Anderson, B. L. Strand, J. Lau, G. Skjak-Braek, and S. Sandler. 2003. The role of capsule composition and biological responses in the function of transplanted microencapsulated islets of Langerhans. *Transplantation* 76: 275–279.

Kobayashi, M., M. Sakane, T. Abe, T. Ikoma, and N. Ochiai. 2012. Anti-tumor effect of a local delivery system; Hydroxyapatite-alginate beads of paclitaxel. *Bioceramics Development and Applications*. doi:10.4303/bda/D110193.

Kovalenko, I., B. Zdyrko, A. Magasinski, B. Hertzberg, Z. Milicev, R. Burtovyy I. Luzinov, and G. Yushin. 2011. A major constituents of brown algae for use in high-capacity Li-ion batteries. *Science* 334: 75–79.

Kraan, S. 2012. Algal polysaccharides, novel applications and outlook. In *Carbohydrates: Comprehensive Studies on Glycobiology and Glycotechnology*; Chang, C.-F., Ed. InTech: Rijeka, Croatia, pp. 489–532.

Kumar, B. N. P., K. K. Reddy, and P. V. R. Mallikarjuna. 2012. Characterization of blended sodium alginate microcapsules for controlled release of animal feed supplements in the GIT. *International Journal of Food, Agriculture and Veterinary Sciences* 2(3): 161–166.

Larsen, B., D. M. S. A. Salem, M. A. E. Sallam, M. M. Mishrikey, and A. I. Beltagy. 2003. Characterization of the alginates from algae harvested at the Egyptian red sea coast. *Carbohydrate Research* 338: 2325–2336.

Lata, H., S. Chandra, I. A. Khan, and M. A. ElSohly. 2009. Propagation through alginate encapsulation of axillary buds of *Cannabis sativa* L.—An important medicinal plant. *Physiology and Molecular Biology of Plants* 15(1): 79–86.

Leal, D., B. Matsuhiro, M. Rossi, and F. Caruso. 2008. FT-IR spectra of alginic acid block fractions in three species of brown seaweeds. *Carbohydrate Research* 343: 308–316.

Li, X., J. Qi, Y. Xie, X. Zhang, S. Hu, Y. Xu, Y. Lu, and W. Wu. 2013. Nanoemulsions coated with alginate/chitosan as oral insulin delivery systems: Preparation, characterization, and hypoglycemic effects in rats. *International Journal of Nanomedicine* 8: 23–32.

Lins, K. O. A. L., M. L. Vale, R. A. Ribeiro, and L. V. Costa-Lotufo. 2013. Proinflammatory activity of an alginate isolated from *Sargassum vulgare*. *Carbohydrate Polymers* 92: 414–420.

Mallick, J., D. Sahoo, D. Madhab Kar, and J. Makwana. 2013. Alginate beads of ibuprofen for oral sustained drug delivery: An *in vitro* evaluation. *International Journal of Pharmaceutical, Chemical and Biological Sciences* 3(3): 595–602.

Mandal, S., S. Senthil Kumar, B. Krishnamoorthy, and S. K. Basu. 2010. Development and evaluation of calcium alginate beads prepared by sequential and simultaneous method. *Brazilian Journal of Pharmaceutical Sciences* 46(4): 785–793.

Mao, W., B. Li, Q. Gu, Y. Fang, and H. Xing. 2004. Preliminary studies on the chemical characterization and antihyperlipidemic activity of polysaccharide from the brown alga *Sargassum fusiforme*. *Hydrobiologia* 512: 263–266.

Mazumder, R., L. K. Nath, A. Haque, T. Maity, P. K. Choudhury, B. Shrestha, M. Chakraborty, and R. N. Pal. 2010. Formulation and *in vitro* evaluation of natural polymers based microspheres for colonic drug delivery. *International Journal of Pharmacy and Pharmaceutical Sciences* 2(1): 211–219.

Milani, J. and G. Maleki. 2012. Hydrocolloids in food industry. In *Food Industrial Processes—Methods and Equipment*; Valdez, B., Ed. InTech: Rijeka, Croatia, pp. 17–38.

Miles, D. J. C., J. Polchana, J. H. Lilley, S. Kanchanakhan, K. D. Thompson, and A. Adams. 2001. Immunostimulation of striped snakehead *Channa striata* against epizootic ulcerative syndrome. *Aquaculture* 195: 1–15.

Missau, J., A. J. Scheid, E. L. Foletto, S. L. Jahn, M. A. Mazutti, and R. C. Kuhn. 2014. Immobilization of commercial inulinase on alginate-chitosan beads. *Sustainable Chemical Processes* 2: 13.

Moe, S., K. Draget, G. Skjak-Braek, and O. Smidsrod. 1995. Alginates. In *Food Polysaccharides and Their Application*; A. M. Stephen, Ed. Marcel Dekker: New York.

Montero-Rocha, A., D. McIntosh, R. Sanchez-Merino, and I. Flores. 2006. Immunostimulation of white shrimp (*Litopeneaus vannamei*) following dietary administration of Ergosan. *Journal of Invertebrate Pathology* 91: 188–194.

Mortazavian, A. M., R. Mohammadi, and S. Sohrabvandi. 2012. Delivery of probiotic microorganisms into gastrointestinal tract by food products. In *New Advances in the Basic and Clinical Gastroenterology*; Brzozowski, T., Ed. InTech: Rijeka, Croatia, pp. 121–146.

Murata, M. and J. Nakazoe. 2001. Production and use of marine algae in Japan. *JARQ-Japan Agricultural Research Quarterly* 35(4): 281–290.

Murata, Y., Y. Kodama, T. Isobe, K. Kofuji, and S. Kawashima. 2009. Drug release profile from calcium-induced alginate-phosphate composite gel beads. *International Journal of Polymer Science*. http://dx.doi.org/10.1155/2009/729057.

Natarajan, R., N. N. Rajendran, and M. R. Priya. 2011. Formulation development and *in vitro* evaluation of cefixime trihydrate loaded chitosan-alginate transdermal patches. *International Journal of Research and Reviews in Pharmacy and Applied Sciences* 1(4): 240–254.

Navarro, A. E., R. F. Portales, M. R. Sun-Kou, and B. P. Llanos. 2008. Effects of pH on phenol biosorption by marine seaweeds. *Journal of Hazardous Materials* 156: 405–411.

Nazareth, C., S. M. Keny, and L. Sawaikar. 2014. Development studies on alginate films for ophthalmic use. *Scholars Academic Journal of Pharmacy* 3(1): 1–5.

Nazzaro, F., F. Fratianni, R. Coppola, A. Sada, and P. Orlando. 2009. Fermentative ability of alginate-prebiotic encapsulated *Lactobacillus acidophilus* and survival under simulated gastrointestinal conditions. *Journal of Functional Foods* 1(3): 319–323.

Nimase, P. K. and G. Vidyasagar. 2010. Preparation and evaluation of floating calcium alginate beads of clarithromycin. *Der Pharmacia Sinica* 1(1): 29–35.

Nisha, Y. R., B. Milind, and K. A. Imran. 2013. Probiotic delivery systems: Applications challenges and prospective. *International Research Journal of Pharmacy* 4(4): 1–9.

Nussinovitch, A. 1997. *Hydrocolloid Applications Gum Technology in the Food and Other Industries*. Blackie Academic and Professional: London, U.K.

Obasi, L. A., C. C. Opara, and A. Oji. 2012. Performance of cassava starch as a proton exchange membrane in a single dual chamber microbial fuel cell. *International Journal of Engineering Science and Technology* 4(1): 227–238.

Obasi, L. A., C. C. Opara, K. Okpala, and A. Oji. 2013. Effect of sodium alginate on proton conductivity of cassava starch in a microbial fuel cell. *Greener Journal of Biological Sciences* 3(2): 074–083.

Olivas, G. I., D. S. Mattinson, and G. V. Barbosa-Canovas. 2007. Alginate coatings for preservation of minimally processed "Gala" apples. *Postharvest Biology and Technology* 45: 89–96.

Onsoyen, E. 1996. Commercial applications of alginates. *Carbohydrates in Europe* 14: 26–31.

Pace, S. L. 2006. Polyvinyl impression material vs. alginate impression materials. *Contemporary Dental Assisting*: 20–23.

Papageorgiou, S. K., F. K. Katsaros, E. P. Kouvelos, J. W. Nolan, H. Le Deit, and N. K. Kanellopoulos. 2006. Heavy metal sorption by calcium alginate beads from *Laminaria digitata*. *Journal of Hazardous Materials* B137: 1765–1772.

Passe, E. R. G. and G. Blaine. 1948. Alginate in endaural wound dressing. *Lancet* 2: 651.

Patel, J. K., J. R. Chavda, and M. K. Modasiya. 2010. Floating *in situ* gel based on alginate as carrier for stomach-specific drug delivery of famotidine. *International Journal of Pharmaceutical Sciences and Nanotechnology* 3(3): 1092–1104.

Paxman, J. R., J. C. Richardson, P. W. Dettmar, and B. A. Corfe. 2008. Alginate reduces the increased uptake of cholesterol and glucose in overweight male subjects: A pilot study. *Nutrition Research* 28(8): 501–505.

Peddie, S., J. Zou, and C. J. Secombes. 2002. Immunostimulation in the rainbow trout (*Oncorhynchus mykiss*) following intraperitoneal administration of ergosan. *Veterinary Immunology and Immunopathology* 86: 101–113.

Pereira, L., A. Sousa, H. Coelho, A. M. Amado, and P. J. A. Ribeiro-Claro. 2003. Use of FTIR, FT-Raman and 13C-NMR spectroscopy for identification of some seaweed phycocolloids. *Biomolecular Engineering* 20: 223–228.

Pereira, R. F. and P. J. Bartolo. 2013. Degradation behavior of biopolymer-based membranes for skin tissue regeneration. *Procedia Engineering* 59: 285–291.

Rasmussen, R. S. and M. T. Morrissey. 2007. *Marine Biotechnology for Production of Food Ingredients*. Reckitt and Colman Product Ltd., Elsevier, New York, pp. 237–292.

Rehman, H. U., A. Aman, A. Silipo, S. A. U. Qader, A. Molinaro, and A. Ansari. 2013. Degradation of complex carbohydrate: Immobilization of pectinase from *Bacillus licheniformis* KIBGE-IB 21 using calcium alginate as a support. *Food Chemistry* 139: 1–4.

Revathi, S., V. Madhulatha, and M. D. Dhanaraju. 2014. Formulation and evaluation of stavudine loaded sodium alginate beads by ionotropic gelation method. *International Research Journal of Pharmacy* 5(9): 706–712.

Rocher, V., J.-M. Siaugue, V. Cabuil, and A. Bee. 2008. Removal of organic dyes by magnetic alginate beads. *Water Research* 42: 1290–1298.

Sadeghi, M., F. Shafiei, and E. Mohammadinasab. 2014. Biosuperabsorbent hydrogel based on alginate-g-poly AA/kaolin composite for releasing Cefalexin drug. *Oriental Journal of Chemistry* 30(1): 285–290.

Sakai, M. 1998. Current research status of fish immunostimulants. *Aquaculture* 172: 63–92.

Sangeetha, S., D. N. Venkatesh, R. Adhiyaman, K. Santhi, and B. Suresh. 2007. Formulation of sodium alginate nanospheres containing amphotericin B for the treatment of systemic candidiasis. *Tropical Journal of Pharmaceutical Research* 6(1): 653–659.

Santana B. P., F. Nedel, E. Piva, R. V. de Carvalho, F. F. Demarco, and N. L. Villarreal Carreno. 2013. Preparation, modification, and characterization of alginate hydrogel with nano/microfibers: A new perspective for tissue engineering. *BioMed Research International* 2013: 1–6.

Sarmento, B., A. Ribeiro, F. Veiga, P. Sampaio, R. Neufeld, and D. Ferreira. 2007. Alginate/chitoson nanoparticles are effective for oral insulin delivery. *Pharmaceutical Research* 24(12): 2198–2206.

Shabber, S., Shaheeda, and K. V. Ramanamurthy. 2012. Formulation and evaluation of chitosan sodium alginate microcapsules of 5-fluorouracil for colorectal cancer. *International Journal of Research in Pharmacy and Chemistry* 2(1): 7–19.

Sherina, V. M., K. Santhi, and C. I. Sajeeth. 2012. Formulation and evaluation of sodium alginate microbeads as a carrier for the controlled release of nifedipine. *International Journal of Pharmaceutical and Chemical Sciences* 1(2): 699–710.

Singh, A. K., A. Mittal, and A. Maiti. 2013. Comparative study of alginate and pectin based sustained release floating beads of metformin hydrochloride. *Pharmaceutical Research* 8(2): 23–30.

Singh, P., S. K. Singh, J. Bajpai, A. K. Bajpai, and R. B. Shrivastava. 2014. Iron crosslinked alginate as novel nanosorbents for removal of arsenic ions and bacteriological contamination from water. *Journal of Materials Research and Technology* 3(3): 195–202.

Smith, J. R. and G. O. J. Fischer. 1982. Simulated food product and method of manufacture therefore. US Patent No. 4348418.

Sneath, M. E. 1975. Simulated soft fruits. US Patent No. 3922360.

Son, E. H., E. Y. Moon, D. K. Rhee, and S. Pyo. 2001. Stimulation of various functions in murine peritoneal macrophages by high mannuronic acid containing alginate (HMA) exposure *in vivo*. *International Immunopharmacology* 1: 147–154.

Song, Y., L. Liu, H. Shen, J. You, and Y. Luo. 2011. Effect of sodium alginate-based edible coating containing different anti-oxidants on quality and shelf life of refrigerated bream (*Megalobrama amblycephala*). *Food Control* 22: 608–615.

Soni, M. L., M. Kumar, and K. P. Namdeo. 2010. Sodium alginate microspheres for extending drug release: Formulation and *in vitro* evaluation. *International Journal of Drug Delivery* 2: 64–68.

Sousa, A. P. A. D., P. S. F. Barbosa, M. R. Torres, A. M. C. Martins, R. D. Martins, R. S. Alves et al. 2008. The renal effects of alginates isolated from brown seaweed *Sargassum vulgare*. *Journal of Applied Toxicology* 28: 364–369.

Sousa, A. P. A. D., M. R. Torres, C. Pessoa, M. O. D. Moraes, F. D. R. Filho, A. P. N. N. Alves, and L. V. Costa-Lotufo. 2007. *In vivo* growth-inhibition of sarcoma 180 tumor by alginates from brown seaweed *Sargassum vulgare*. *Carbohydrate Polymers* 69: 7–13.

Spasojevic, M., S. Bhujbal, G. Paredes, B. J. D. Haan, A. J. Schouten, and P. De Vos. 2014. Considerations in binding diblock copolymers on hydrophilic alginate beads for providing an immunoprotective membrane. *Journal of Biomedical Materials Research. Part A* 102A: 1887–1896.

Sriram, J. G., M. H. Sadhir, M. Saranya, and A. Srinivasan. 2014. Novel corrosion inhibitors based on seaweeds for AA7075 aircraft aluminium alloys. *Chemical Science Review and Letters* 2(5): 402–407.

Suksamran, T., P. Opanasopit, T. Rojanarata, and T. Ngawhirunpat. 2011. Development of alginate/chitosan microparticles for dust mite allergen. *Tropical Journal of Pharmaceutical Research* 10(3): 317–324.

Sun, J. and H. Tan. 2013. Alginate-based biomaterials for regenerative medicine applications. *Materials* 6: 1285–1309.

Sunderland, A. M., P. W. Dettmar, and J. P. Pearson. 2000. Alginates inhibit pepsin activity in vitro; A justification for their use in gastro-oesophageal reflux disease (GORD). *Gastroenterology* 118: 347.

Tan, H. and K. G. Marra. 2010. Injectable, biodegradable hydrogels for tissue engineering application. *Materials* 3: 1746–1767.

Thomas, S. 2000. Alginate dressings in surgery and wound management. *Journal of Wound Care* 9(2): 56–60.

Torres, M. R., A. P. A. Sousa, E. A. T. S. Filho, D. F. Melo, J. P. A. Feitosa, R. C. M. Paula et al. 2007. Extraction and physicochemical characterization of *Sargassum vulgare* alginate from Brazil. *Carbohydrate Research* 342: 2067–2074.

Wotherspoon, C. 1988. Continuous production of gelled chunks of food stuffs. US Patent No. 4,784,862.

Yang, D. and K. S. Jones. 2009. Effect of alginate on innate immune activation of macrophages. *Journal of Biomedical Material Research* 90: 411–418.

Yonekura, L., H. Sun, C. Soukoulis, and I. Fisk. 2014. Microencapsulation of *Lactobacillus acidophilus* NCIMB 701748 in matrices containing soluble fibre by spray drying: Technological characterization, storage stability and survival after *in vitro* digestion. *Journal of Functional Foods* 6: 205–214.

Zactiti, E. M. and T. G. Kieckbusch. 2006. Potassium sorbate permeability in biodegradable alginate films: Effect of the antimicrobial agent concentration and crosslinking degree. *Journal of Food Engineering* 77: 462–467.

Zanjani, M. A. K., B. G. Tarzi, A. Sharifan, and N. Mohammadi. 2014. Microencapsulation of probiotics by calcium alginate-gelatinized starch with chitosan coating and evaluation of survival in stimulated human gastro-intestinal condition. *Iranian Journal of Pharmaceutical Research* 13(3): 843–852.

*chapter twenty-eight*

# Mangroves

## *A potent source of polysaccharides*

*Narayanasamy Rajendran, Kandasamy Saravanakumar, and Kandasamy Kathiresan*

## Contents

## 28.1 Introduction

Mangroves are the only tall tree forest on the Earth, lying between land and sea in the tropical and subtropical coastal regions of the world (Kathiresan and Bingham, 2001). Mangroves are also referred to as "oceanic rain forests," "tidal forests," and "coastal woodlands." Mangroves are found to exist in sheltered shores, estuaries, creeks, backwaters, lagoons, marshes, and mudflats (Tomlinson, 1986). Mangroves are highly productive and enriched with biodiversity. They are well adapted to survive in stressful environmental conditions of salinity, water logging, high wind velocities, high temperatures, and muddy anaerobic soil in which no other vascular plants can thrive. In order to survive in the adverse conditions, mangroves are unique to have arching roots, breathing roots, and salt-vomiting leaves, in addition to biochemical features.

Mangroves possess a number of chemical defense mechanisms against various physical, chemical, and biological stresses. Chemical metabolites of mangroves have been used traditionally by indigenous people for centuries, but, until recently, the chemistry of these natural products has remained poorly defined (Bandaranayake, 2002). Mangroves are well known for containing high phenolic compounds such as tannins, which offer protection to the plant against pest and parasitic damages (Kathiresan and Veera Ravi, 1990; Veera Ravi and Kathiresan, 1990). Mangroves are proven for their potent biological activities against human, animal, and plant pathogens (Premanathan et al., 1992). It is known that the accumulation of low-molecular-weight organic solutes such as sugars, some amino acids, and quaternary ammonium compounds are involved in the adaptation to abiotic stresses (Hibino et al., 2001). Mangroves have also been reported to synthesize and accumulate various chemicals such as aliphatic alcohols and acids, amino acids, alkaloids, carbohydrates, lignins, polysaccharides, carotenoids, hydrocarbons, fatty acids, lipids, pheromones, phorbol esters, phenolics, steroids, troterpenes, tannins and other terpenes, and related compounds (Bandaranayake, 2002). Most mangrove species produce a lot of natural products, such as antioxidants and terpenoids, to modulate physiological activities, including membrane permeability to salt and other solutes (Oku et al., 2003; Parida et al., 2004; Vijayavel et al., 2006) in addition to providing chemical defense.

In order to thrive in stressful environment, mangroves are able to synthesize an array of chemical compounds. However, information about polysaccharides present in mangroves are scattered and, hence, this chapter consolidates such data to improve understanding about mangrove-derived polysaccharides and their biological properties.

## 28.2 Polysaccharides in mangroves

Mangrove fruits are generally rich in polysaccharides and polyphenols (Rubio-Pina and Zapata-Perez, 2011; Table 28.1). The mangrove fruit *Bruguiera gymnorrhiza*

*Table 28.1* Polysaccharides/degrading activity in mangrove environment

| Source of sample | Polysaccharide study | References |
|---|---|---|
| Mangrove fruits | Polysaccharides and polyphenols | Rubio-Pina and Zapata-Perez (2011) |
| Mangrove fruits | Carbohydrate | Hanastiti (2011) |
| Mangrove plants | Sulfated polysaccharides | Kathiresan et al. (2006) |
| Mangrove plants | Starch | Bhosale and Mulik (1992); Desai et al. (2011) |
| Mangrove wood and leaves | Glucose, xylose and arabinose, galactose, rhamnose, xylose | Marchand et al. (2005); Opsahl and Benner (1999) |
| Mangrove wood and leaves | Carbohydrates | Lallier-Vergèsa et al. (2008) |
| Mangrove wood and leaves | Galactose | Benner et al. (1990); Moers et al. (1990) |
| Mangrove wood and leaves | Lignin signature | Hedges and Mann, (1979) |
| Mangrove leaves | Carbon/vanillic units, lignin degradation, cinnamic-to-vanillic ratios | Dittmar and Lara (2001) |
| Red mangrove leaves and wood | Lignocellulose and lignin | Neilson and Richards (1989) |
| Mangrove tissues | Glucose and arabinose | Moers et al. (1993) |
| Mangrove plants | Pinitol and mannitol, proline, and methylated quaternary ammonium compounds | Popp et al. (1985) |
| Mangrove plants | Reduction sugar, soluble sugar, and starch content | Yang et al. (2003) |
| Mangrove plants | Total sugar, carbohydrate, and polyphenols | Das et al. (1997) |
| Mangrove plants | Acid polysaccharide | Premanathan et al. (1992, 1993, 1994, 1995, 1999) |
| Mangrove degradation | Lignin-derived phenols | Sarkanen and Ludwig (1971) |
| Woody and leaf tissues | Lignin | Hedges and Mann (1979); Hedges et al. (1988); Bianchi et al. (1999); Miltner et al. (2005) |
| Bacteria and plankton | Deoxy sugars | Hedges et al. (1988); Hicks et al. (1994) |
| Mangrove leaves | Tannins and sugars | Cundell et al. (1979) |
| Mangrove soil | Acidic amino acids and sugar arabinose | Lacerda et al. (1995) |
| Mangrove sediment | Biodegradation of lignocellulose | Benner and Hodson (1985) |
| | Loss of cinnamic acids | Benner et al. (1990); Opsahl and Benner (1995) |
| Mangrove detritus | Enzymes and proteins | Holguin et al. (2001) |
| Mangrove detritus | Cellulase activity | Nishida et al. (2007) |
| Herbivorous animals | Cellulase activity | Watanabe and Tokuda (2010) |
| Mangrove bacteria | Carboxymethyl cellulose | Behera et al. (2014) |
| Mangrove bacteria | Cellulose degradation | Ramanathan et al. (2008) |
| Mangrove bacteria | Degradation of polysaccharides | Vashist et al. (2013) |
| Mangrove microbes | Degradation of lignin component of lignocelluloses | Crawford (1981) |
| Mangrove environment | Cellulolytic degradation | Thompson et al. (2013) |
| Bacteria, filamentous fungi, basidiomycete, myxomycete, and protozoan | Cellulases | Olson et al. (2010); Trinci et al. (1994); Watanabe et al. (1998); Ronsness (1968); Bera-Maillet et al. (2005) |
| Mangrove microbes | Genes related to cellulose degradation | Sahebi et al. (2013). |
| Fungi | Exopolysaccharide production | Sharmila et al. (2014) |
| Red mangrove | Degradation of structural polysaccharide components | Lee et al. (1990) |
| Mangrove leaf litter | Amino acids and sugars | Kathiresan and Ravikumar (1995); Rajendran and Kathiresan (2000) |
| Thraustochytrids from mangrove detritus | Extracellular polysaccharide production | Devasia and Muraleedharan (2012) |

*(Continued)*

***Table 28.1 (Continued)*** Polysaccharides/degrading activity in mangrove environment

| Source of sample | Polysaccharide study | References |
| --- | --- | --- |
| Mangroves algae | Carbohydrates sorbitol, dulcitol, mannitol, and floridoside | Karsten et al. (1996) |
| Blood cockle | Cellulase activity | Niiyama and Toyohara (2011); Niiyama et al. (2012); Okamura et al. (2010) |
| Brackish water clam | Degradation of cellulose | Sakamoto et al. (2007, 2008, 2009); Sakamoto and Toyohara (2009a,b,c) |
| Termites and nematodes | Occurrence of cellulase | Watanabe et al. (1998); Smant et al. (1998) |
| Abalone, sea urchin, and brackish clam | Occurrence of cellulase | Nishida et al. (2007); Sakamoto et al. (2007); Suzuki et al. (2003) |
| Mangrove sesarmid crabs | Endogenous hepatopancreatic cellulase activity | Adachi et al. (2012); Boon et al. (2008); Thongtham and Kristensen (2005) |
| Mangrove fungi | α-amylase, glucose isomerase, cellulose, protease production | Kathiresan and Manivannan (2006a,b,c; 2007); Manivannan and Kathiresan (2007a,b) |
| Mangrove leaf litter and microbes | Enzyme production—amylase, protease, cellulase, chitinase, and lipase | Kathiresan et al. (2011) |

has a carbohydrate content of 85.1 g per 100 g, which is higher than crop plants: rice 78.9 g per 100 g) and maize (63.6 g per 100 g). The energy content of this mangrove fruit is 371 cal per 100 g. This value is higher than the energy content of rice (360 cal per 100 g) and corn (307 cal per 100 g) (Hanastiti, 2011). Starch plays a vital role in osmoregulatory processes in mangroves to overcome the potentially harmful effects of salts in the substratum (Bhosale and Mulik, 1992; Desai et al., 2011).

Glucose is the most abundant neutral sugar, representing >50% of the total. About 80% of the glucose is present as cellulose, while other neutral sugars are mostly present as hemicellulose. The wood of all mangrove species are almost twice as rich in carbohydrates as their leaves. Glucose is the most abundant neutral sugar in all plants, representing between 26% and 67% of total sugars in leaves and between 47% and 71% in wood. Ribose and deoxy sugars (rhamnose and fucose) are the least abundant monomers, representing generally less than 5% of the total carbohydrates. Leaves are characterized by a higher content in hemicellulosic carbohydrates than wood, for example, up to 81% of total carbohydrates in the leaves of *Rhizophora*. In addition to glucose, mangrove leaves are also characterized by high proportions of arabinose and galactose. Conversely, woody tissues are richer in xylose than leaves, with values ranging from 13% to 22% and from 5% to 13%, respectively (Lallier-Vergèsa et al., 2008). In general, the woods of mangrove plants are characterized by a high content of xylose and cellulosic glucose and low content of galactose and hemicellulosic glucose (Benner et al., 1990; Moers et al., 1990; Opsahl and Benner, 1999). According to Moers et al. (1990) and Benner et al. (1990), in *Rhizophora mangle*, the wood is mainly composed of glucose, xylose, and arabinose, whereas the leaves are composed of glucose > arabinose > galactose

> rhamnose > xylose. Leaves of *Avicennia germinans* are richer in xylose than the leaves of other mangrove species (Opsahl and Benner, 1999; Marchand et al., 2005).

Wood and leaves can be differentiated by their carbohydrate contents. For example, as compared to wood, the leaves are always rich in hemicellulosic carbohydrates especially in galactose in mangroves (Marchand et al., 2005). Similarly, Benner et al. (1990) have reported that *Rhizophora* leaves contain 15% of galactose, and Moers et al. (1990) have found only 5% of galactose in *Rhizophora* wood. Conversely, high xylose content (>13%) is the characteristic feature of wood. For instance, the levels of xylose are 14% and 24% in the woods of *Avicennia* and *Rhizophora*, respectively (Moers et al., 1990; Opshal and Benner, 1999).

According to Marchand et al. (2005), mangrove leaves of three different species (*A. germinans*, *Laguncularia racemosa*, and *R. mangle*) are poor in neutral sugars than woody tissues, with concentrations ranging from 12% to 32%. Mangrove tissues have typical carbohydrate signatures; tend to disappear rapidly in the detritus pool since polysaccharides are highly reactive compounds relative to bulk organic carbon and since microbial communities themselves synthesize new polysaccharides (Kristensen et al., 2008). Carbohydrates can represent up to 65% of organic carbon in mangrove wood (Opsahl and Benner, 1999). However, bacteria and fungi rapidly degrade polysaccharides and synthesize new ones.

The concentrations of lignocellulose and lignin in leaves and wood of living red mangrove *R. mangle* have revealed that approximately half (48.4%) of the dry weight of leaves is of structural, lignocellulosic fiber. The lignocelluloses content of mangrove wood is considerably higher (82.6% dry weight). Leaves and wood have approximately equal percentages of lignin (16%).

Surprisingly, the lignocellulose in leaves is found to be more highly lignified (i.e., a higher lignin: polysaccharide ratio) than that in the woody plant parts; lignocellulose in the leaves is 66% polysaccharide and 34% lignin, whereas, that in the wood is 80% polysaccharide and 20% lignin. In another study, it was found that leaves of the mangrove *Ceriops tagal* consist of 37% aqueous acetone-water-soluble material, 18% water-insoluble polysaccharides, and 50% polyphenols, which include soluble and insoluble tannins and lignin. Pectates are rapidly degraded, while other polysaccharides are reduced proportionately (Neilson and Richards, 1989).

Mangrove leaves contain high amounts of arabinose and glucose and hardly any partially methylated monosaccharides, whereas microbial mats, in general, and lagoonal seagrass sediments show high contributions of fucose, ribose, mannose, galactose, and partially methylated monosaccharides (Moers and Larter, 1993). Galactose, mannose, and fucose are mainly derived from bacteria, whereas glucose and arabinose are largely derived from mangrove tissues, as proved during the analysis of paleosoils of the Abu Dhabi lagoon–sabkha system (Moers et al., 1993).

The organic solutes vary with mangrove species. For instance, leaves of 23 mangrove species have been analyzed from northern Queensland, Australia. The most widely distributed organic solutes are pinitol and mannitol in *C. tagal* (Popp et al., 1985). Other compatible solutes, such as proline and methylated quaternary ammonium compounds get accumulated in *Avicennia eucalyptifolia*, *A. marina*, *Acanthus ilicifolius*, *Heritiera littoralis*, and *Hibiscus tiliaceus*. In all other species, low-molecular-weight carbohydrates are the main organic solutes (Popp et al., 1985). The sugars vary not only within species, but also based on environment stress. For instance, the levels of reduction sugar, soluble sugar, and starch content of two mangrove species (*A. marina* and *Aegiceras corniculatum*) gradually decrease during cold stress, whereas the level of sucrose gradually increases in such conditions (Yang et al., 2003).

## 28.3 Antiviral and anti-blood-clotting polysaccharides of mangroves

Mangrove plants have been used in folklore medicine in India. Our laboratory has reported that the bark extract of *Rhizophora mucronata* has *in vitro* antiviral activity against Newcastle disease virus (Premanathan et al., 1993), vaccinia virus (Premanathan et al., 1994), encephalomyocarditis virus (EMCV) (Premanathan et al., 1994), Semliki forest virus (SFV) (Premanathan et al., 1995), and the inhibitory activity against the binding of hepatitis B virus surface antigen to its antibody (Premanathan

et al., 1992). Our laboratory has further reported that the extract protects Swiss albino mice from lethal infection by EMCV (Premanathan et al., 1994) and SFV (Premanathan et al., 1995). In yet another study, our laboratory extracted the acid polysaccharide from the bark of *Rhizophora mucronata* and proved it to have potent anti-HIV activity *in vitro* (Premanathan et al., 1999). This is due to interference with the adsorption of virus particle to CD4[+] cells. The bark extract has been identified for the presence of a large amount of neutral sugars and uronic acids and these polysaccharides are believed to be acid polysaccharides, since no protein and sulfate can be detected in the active fraction. *Rhizophora mucronata* polysaccharide (RMP) protected MT-4 cells from HIV-induced cytopathogenicity and blocked the expression of HIV antigens. RMP completely inhibited the viral binding to the cell (Premanathan et al., 1999). Thus the polysaccharide (galactose, galactosamine, glucose, and arabinose) from *R. mucronata* seems to be a new class of antiviral agent and a good candidate for therapeutic use against HIV and other viral infections.

Mangrove polysaccharide extracts prolong the time taken for blood clotting. This activity increases proportionally with the concentrations of extracts used (100, 500, and 1000 μg ML[-1]) and also with the levels of sulphate present in the samples. The anticoagulant activity is high in *A. marina* and *Aegiceras corniculatum* when compared to *Excoecaria agallocha* and *Rhizophora* spp. (Kathiresan et al., 2006).

## 28.4 Polysaccharides in mangrove sediment

Neutral carbohydrates show selective degradation patterns in mangrove sediments, which can provide specific details on their sources, despite their low concentrations in sediments. According to Marchand et al. (2005), the debris deriving from mangrove roots strongly contributes to the organic enrichment of sediments and can be discriminated using its content of xylose and cellulosic monosaccharides. In contrast, algal mats developing on the sediment surface during the early stages of mangrove development are responsible for abundant rhamnose content. The carbohydrates that rapidly leach after submersion in water are mostly the nonlignocellulose components (Neilson and Richards, 1989). The polysaccharide (i.e., cellulosic) components of lignocellulose are generally degraded about twice as fast as the lignin component, indicating that mangrove detritus becomes relatively enriched with lignin-derived carbon (Benner and Hodson, 1985; Marchand et al., 2005). While cellulose and lignin can readily be degraded in oxic environments, these compounds are only slowly degraded under anoxic conditions. Lignin, for example, has a half-life of more than 150 years in anoxic mangrove sediments (Dittmar and Lara, 2001). In most mangrove swamps, the upper sediment was rich in arabinose, rhamnose, and fucose, compared to mangrove plants, suggesting that these sugars are produced in sediments by fungi or microbes.

Mangrove sediments are characterized by higher contents in deoxy sugars (rhamnose and fucose) than mangrove plants. Hedges et al. (1988) have shown that deoxy sugars can be derived from bacteria, and later Moers et al. (1990) demonstrated that they can contribute significantly to mangrove peat. In addition, the ratio between xylose and rhamnose decreases in the upper part of the cores beneath all the mangrove stands, thus highlighting increasing bacterial contribution to total carbohydrates with increasing depth. Organic carbon from lignin-derived phenols ranges from 1% to 10% total organic carbon in mangrove tissues and varies between 2% and 10% in mangrove sediments, highlighting the relative stability of these compounds compared to carbohydrates and bulk organic carbon (Benner et al., 1984a,b; Hedges et al., 1985; Cowie et al., 1992; Dittmar and Lara, 2001; Marchand et al., 2005).

Mangrove-derived detritus constitutes a large reservoir of carbon and energy potentially available to the estuarine food web, and bacteria are the major participants in the carbon, sulfur, nitrogen, and phosphorous cycles in mangrove forest (Rojas et al., 2001). Roughly 30%–50% of the organic matter in mangrove leaves is leachable water-soluble compounds such as tannins and sugars (Cundell et al., 1979), the remaining fraction of the organic matter presumably consists of plant structural polymers commonly referred to as cellulosic substance. Organic carbon and nitrogen contents are higher in sediments of *Avicennia* than in *Rhizophora*. The contribution of sugars and amino acids to the total organic carbon pool is constant with depth in *Rhizophora* soils whereas in *Avicennia* soils it increases. Spectral distribution of sugars and amino acids shows dominance of acidic amino acids and sugar arabinose (Lacerda et al., 1995).

The rates of mineralization of the lignocellulosic component of mangrove leaves and wood are 10-fold lower than the mineralization rates of the leachable fraction. The polysaccharide component of the lignocelluloses is mineralized at rates two times higher than rates of mineralization of the lignin component, indicating that mangrove detritus becomes relatively enriched in lignin-derived carbon with time. Anaerobic rates of mineralization of the leachable and lignocellulosic components of mangrove leaves and wood are 10–30 times lower than the respective aerobic mineralization rates, suggesting a very long residence time for mangrove detritus in anaerobic sediments. A comparison of the rates of degradation of mangrove detritus in

sediments from a mangrove swamp and a salt marsh demonstrated that the lignocellulolytic potential in the sediments of the two marine ecosystems is similar, but that lignocelluloses from mangroves is less biodegradable than lignocellulose from the salt-marsh plants *Spartina alterniflora* and *Juncus roemenanus* (Benner and Hodson, 1985).

The proportions of neutral sugars in the sediment varied as follows: glucose, 27%–57%; arabinose, 8%–26%; xylose, 7%–17%; galactose, 6%–15%; rhamnose, 5%–15%; mannose 3%–13%; fucose, 3%–6%; and ribose, <2%. Beneath *R. mangle*, the sediments are characterized by the highest concentrations in xylose and rhamnose. Conversely, on the landward side of the mangrove, the upper sediment is characterized by the highest concentrations in galactose (Marchand et al., 2005). No specific trend with depth can be observed for any neutral sugar. However, the ratios between xylose and rhamnose decrease with depth in the upper core and then increase to reach the highest values in layers that are characterized by higher total organic carbon contents. Along the sea–land transect, xylose and galactose present an opposite trend. The highest xylose concentrations are found beneath *Rhizophora*, the most seaward mangrove plant, whereas the highest galactose content is measured beneath *Pterocarpus*, the most landward plant species. The higher galactose contents determined on the landward side may reflect an accumulation of leaf litter, while on the seaward the root system may be responsible for organic accumulation and its specific sugar composition. Additionally, the high proportion of lignin normalized to total organic carbon recorded in the upper part of the cores collected beneath *Rhizophora* and *Laguncularia* confirms the important contribution of the woody tissues to the organic matter. In the upper cores, total carbohydrate concentrations decrease with depth and their proportions normalize to total organic carbon. Organic carbon from carbohydrates represents between 15% and 50% of total organic carbon in the mangrove plant species but less than 9% in the mangrove sediments, irrespective of the forest zone. This highlights the reactivity of carbohydrates relative to bulk organic carbon as observed in other swamps where degradation of polysaccharides appears to be very efficient even in waterlogged sediments (Cowie and Hedges, 1984; Stout et al., 1988; Moers et al., 1990; Marchand et al., 2005).

Lignocellulose from mangrove leaves is mineralized at different rates. Rates of biodegradation of mangrove lignocellulose are higher in samples from higher salinities and low in the low salinity (Benner and Hodson, 1985). Lignocelluloses derived from mangrove leaves and wood are more resistant to microbial degradation than lignocelluloses derived from other marine macrophytes (Benner and Hodson, 1985). Rates of anaerobic biodegradation of the lignin and polysaccharide components of mangrove leaves and wood are extremely low. The ratio of polysaccharide-to-lignin mineralized under anaerobic conditions is similar to but slightly higher than the ratio of polysaccharide-to-lignin mineralized under aerobic conditions, suggesting that biodegradation of the polysaccharide component of mangrove lignocelluloses is tightly coupled with biodegradation of the lignin component. Low rates of anaerobic biodegradation of mangrove lignocellulose indicate a long residence time for mangrove detritus in anoxic sediments and suggest that a relatively small portion of the carbon and energy is available to the food web under these conditions (Benner and Hodson, 1985).

## 28.5 Polysaccharides in mangrove-associated microbes

Mangrove environments harbor large number of microorganisms and play a major role in the degradation of mangrove litters. Leaves and wood of mangrove plants are degraded on the sediment, primarily by a large variety of microbes which actively participate in the heterotrophic food chain (Alongi et al., 1999). The major product of general recycling of organic matter is detritus, which is rich in enzymes and proteins and contains large microbial population (Holguin et al., 2001). The transfer of carbon and energy from mangrove detritus to animal consumers appears to occur via grazing of easily digestible and highly nutritive microbial biomass resulting from bacterial and fungal transformation of the lignocellulosic detritus. Most bacteria and fungi degrade this cellulosic material by producing an enzyme called cellulase. It is generally accepted that cellulases are exclusively present in microorganisms, such as bacteria and fungi that degrade plant litter. Although cellulase activity is present in herbivorous animals, including ruminants or termites, it has been typically thought to derive from symbiotic organisms (Watanabe and Tokuda, 2010). However, recent data of genomic analysis have revealed that invertebrates contain endogenous cellulolytic enzymes. Mangroves have strong metabolic potential to hydrolyze carbohydrates, including cellulosic compounds (cellulose, hemicellulose, pectin, xylan, among others) by a wide diversity of 58 families of glycosyl hydrolase, 23 families of glycosyltransferases, 5 families of carbohydrate esterases, and 3 families of polysaccharide lyases (Thompson et al., 2013). Specific sequences involved in the cellulolytic degradation, belonging to cellulases, hemicellulases, carbohydrate-binding domains, dockerins, and cohesins have been identified, and thus the mangroves possess all fundamental molecular tools required for building the cellulosome, which is required for the efficient degradation of cellulose material and sugar release (Thompson et al., 2013). Fungal isolates, especially *Trichoderma* species, are more efficient in producing extracellular enzymes than bacterial isolates, revealing the significance of fungi in detritus-based mangrove systems (Kathiresan et al., 2011).

## 28.6 Polysaccharides in mangrove-associated organisms

Mangrove crabs play a crucial role in the carbon cycle in forests by consuming large amounts of mangrove litter, which is mainly composed of cellulose (Adachi et al., 2012). While analyzing the endogenous hepatopancreatic cellulase activity in eight species of crabs, including three mangrove sesarmid crabs (*Episesarma versicolor*, *Perisesarma indiarum*, and *Episesarma palawanense*) native to Thailand, it has been found that Endo-$b$-1,4-glucanase activity is significantly higher in the enzyme extracted from mangrove crabs than in that from Japanese marsh crabs. It is suggested that mangrove crabs efficiently digest cellulose endogenously (Adachi et al., 2012). The sesarmids from mangroves show higher cellulase activity than those from temperate regions and that the cellulase is produced by the sesarmids endogenously (Thongtham and Kristensen, 2005). Recent studies showed hemicellulase, including mannanase, xylanase, xyloglucanase, and licheninase, all of which are required for the complete digestion of hard, degradable plant polysaccharides, are distributed among various aquatic invertebrates in addition to cellulase (Niiyama and Toyohara, 2011). The blood cockle has the ability to digest mangrove-derived cellulose to glucose due to the presence of β-glucosidase (Niiyama and Toyohara, 2011). The blood cockle collected from the river mouth to the coastal area of Malaysia is reported to have lower cellulase activity and this is due to the lower amount of mangrove-derived cellulose in the sediment (Okamura et al., 2010).

## 28.7 Conclusion and future prospects

Mangroves are a rich source of polysaccharides but a focused research is required on mangrove-derived polysaccharides. Generally, mangroves have a higher polysaccharide content than crop plants. Mangrove plant species and plant parts can be differentiated based on carbohydrate content. Even the microbial or algal sources of carbohydrates can be differentiated from mangrove sources. Most of the available work focuses on the degradation of organic matters, microbial enzyme production, and cellulolytic activity in the mangrove environment. Primary litter processing in mangrove environments is by crabs, followed by microbes. This leads to the formation of detritus food that supports many marine detritivorous organisms.

Mangroves have strong metabolic potential to hydrolyze complex polysaccharides by a wide variety of enzymes. The specific sequences involved in the cellulolytic degradation and the carbohydrate-binding domains, dockerins and cohesins, have been identified, and thus the mangroves possess all fundamental molecular tools required for building cellulosome, which is required for the efficient degradation of cellulose material and sugar release. Much more research should be focused on this aspect. In addition, endogenous cellulolytic enzyme production in invertebrates has to be studied in detail.

Acid polysaccharides present in mangroves have antiviral and anti-blood-clotting activities that have potential high commercial utilization and this requires many studies on developing "lead molecules" for drugs.

## Acknowledgments

The authors are thankful to the authorities of Annamalai University, Tamil Nadu, India, and the the principal of Government Arts College, Chidambaram, Tamil Nadu and the authorities of Shanghai Jiao Tong University for the support by the Special Project of Basic Work Project for Science and Technology of China (No. 2014FY120900).

## References

Adachi, K., Kentaro, T., Azekura, T., Morioka, K., Tongnunui, P., and Ikejima, K., 2012. Potent cellulase activity in the hepatopancreas of mangrove crabs. *Fish. Sci.*, 78: 1309–1314.

Alongi, D.M., Tirendi, F., Trott, L.A., and Brunskill, G.J., 1999. Mineralisation of organic matter in intertidal sediments of a tropical semi-enclosed delta. *Estuar. Coast. Shelf Sci.*, 48: 451–467.

Bandaranayake, W.M., 2002. Bioactivities, bioactive compounds and chemical constituents of mangrove plants. *Wet. Ecol. Manag.*, 10: 421–452.

Behera, B.C., Parida, S., Dutta, S.K., and Thatoi, H.N., 2014. Isolation and identification of cellulose degrading bacteria from mangrove soil of Mahanadi River delta and their cellulase production ability. *Am. J. Microbiol. Res.*, 2(1): 41–46.

Benner, R. and Hodson, R.E., 1985. Microbial degradation of the leachable and lignocellulosic components of leaves and wood from *Rhizophora mangle* in a tropical mangrove swamp. *Mar. Ecol. Prog. Ser.*, 23: 221–230.

Benner, R., Maccubbin, A.E., and Hodson, R.E., 1984a. Anaerobic biodegradation of the lignin and polysaccharide components of lignocellulose and synthetic lignin by sediment microflora. *Appl. Environ. Microbiol.*, 47: 998–1004.

Benner, R., Newell, S.Y., Maccubin, A.E., and Hodson, R.E., 1984b. Relative contributions of bacteria and fungi to rates of degradation of lignocellulosic detritus in salt marsh sediments. *Appl. Environ. Microbiol.*, 48: 36–40.

Benner, R., Weliky, K., and Hedges, J.I., 1990. Early diagenesis of mangrove leaves in a tropical estuary: Molecular-level analyses of neutral sugars and lignin derived phenols. *Geochim. Cosmochim. Acta*, 54: 1991–2001.

Bera-Maillet, C., Devillard, E., Cezette, M., Jouany, J.P., and Forano, E., 2005. Xylanases and carboxymethylcellulases of the rumen protozoa *Polyplastron multivesiculatum*, *Eudiplodinium maggii* and *Entodinium* sp. *FEMS Microbiol. Lett.*, 244: 149–156.

Bhosale, L.J. and Mulik, N.G., 1992. Physiology of mangroves. In: Singh, K.P. and Singh, J.S. (eds.), *Tropical Ecosystems: Ecology and Management.* Wiley-Eastern, Delhi, India, pp. 315–320.

Bianchi, T.S., Argyrou, M., and Chippett, H.F., 1999. Contribution of vascular plant carbon to surface sediments across the coastal margin of Cyprus (eastern Mediterranean). *Org. Geochem.*, 30: 287–297.

Boon, P.Y., Daren, C.J.Y., and Todd, P.A., 2008. Feeding ecology of two species of *Perisesarma* (Crustacea: Decapoda: Brachyura: Sesarmidae) in Mandai mangroves, Singapore. *J. Crustac. Biol.*, 28: 480–484.

Cowie, G.L. and Hedges, J.I., 1984. Carbohydrate sources in a coastal marine environment. *Geochim. Cosmochim. Acta*, 48: 2075–2087.

Cowie, G.L., Hedges, J.I., and Calvert, S.E., 1992. Sources and reactivities of amino acids, neutral sugars and lignin in an intermittently anoxic marine environment. *Geochim. Cosmochim. Acta*, 56: 1963–1978.

Crawford, R.L., 1981. *Lignin Biodegradation and Transformation.* Wiley Interscience, New York.

Cundell, A.M., Brown, M.S., Stanford, R., and Mitchell, R., 1979. Microbial degradation of *Rhizophora mangle* leaves immersed in the sea. *Estuar. Coast. Mar. Sci.*, 9: 281–286.

Das, P., Basak, U.C., and Das, A.B., 1997. Metabolic changes during rooting in pre grilled stem cutting and air- layers of Heritiera. *Bot. Bull. Acad. Sin.*, 38: 91–95.

Desai, M.N., P.D. Patil, and Chavan, N.S., 2011. Isolation and characterization of starch from mangroves *Aegiceras corniculatum* (L.) Blanco and *Cynometra iripa* Kostel. *Int. J. Appl. Biol. Pharma. Tech.*, 2(1): 244–249.

Devasia, V.L.A. and Muraleedharan, U.D., 2012. Polysaccharide-degrading enzymes from the marine protists, thraustochytrids. *Biotechnol. Bioinf. Bioeng.*, 2(1): 617–627.

Dittmar, T. and Lara, R.J., 2001. Molecular evidence for lignin degradation in sulfate reducing mangrove sediments (Amazonia, Brazil), *Geochim. Cosmochim. Acta*, 74: 1417–1428.

Hanastiti, W.R., 2011. Potential as an alternative source of mangrove fruit carbohydrates (A. Zani Pitoyo, ed.). doi: http://s4.poltekkes-malang.ac.id/index.php/home/detail/artikel/12/159.

Hedges, J.I. and Mann, D.C., 1979. The characterization of plant tissues by their lignin oxidation products. *Geochim. Cosmochim. Acta*, 43: 1803–1807.

Hedges, J.I., Blanchette, R.A., Weliky K., and Devol, A.H., 1988. Effect of fungal degradation on the CuO oxidation products of lignin: A controlled laboratory study. *Geochim. Cosmochim. Acta*, 52: 2717–2726.

Hedges, J.I., Cowie, G.L., Ertel, J.R., Barbour, R.J., and Hatcher, P.G., 1985. Degradation of carbohydrates and lignins in buried woods. *Geochim. Cosmochim. Acta*, 49: 701–711.

Hibino, T., Meng, Y.L., Kawamitsu, Y., Uehara, N., Matsuda, N., Tanaka, Y., Ishikawa, H., Baba, S., Takabe, T., and Wada, K., 2001. Molecular cloning and functional characterization of two kinds of betaine-aldehyde dehydrogenase in betaine-accumulating mangrove *Avicennia marina* (Forsk.) Vierh. *Plant. Mol. Biol.*, 453: 353–363.

Hicks, R.E., Owen, C.J., and Aas, P., 1994. Deposition, resuspension and decomposition of particulate organic matter in the sediment of Lake Itasca, Minnesota, USA. *Hydrobiol. Int. J. Aquat. Sci.*, 284: 79–91.

Holguin, G., Vazquez, P., and Bashan, Y., 2001. The role of sediment microorganisms in the productivity, conservation, and rehabilitation of mangrove ecosystems: An overview. *Biol. Fertil. Soils*, 33(4): 265–278.

Karsten, U., Mostaert, A.S., King, R.J., Kamiya, M., and Hara, Y., 1996. Osmoprotectors in some species of Japanese mangrove macroalgae Australia. *Phycol. Res.*, 44(2): 109–112.

Kathiresan, K. and Bingham, B.L., 2001. Biology of mangroves and mangrove ecosystems. *Adv. Mar. Biol.*, 40: 81–251.

Kathiresan, K. and Manivannan, S., 2006a. Amylase production by *Penicillium fellutanum* isolated from mangrove rhizosphere soil. *Afr. J. Biotechnol.*, 5(10): 829–832.

Kathiresan, K. and Manivannan, S., 2006b. Cellulase production by *Penicillium fellutanum* isolated from coastal mangrove rhizosphere soil. *Res. J. Microbiol.*, 1(5): 438–442.

Kathiresan, K. and Manivannan, S., 2006c. Glucose isomerase production by *Penicillium fellutanum* isolated from mangrove rhizosphere sediment. *Trends Appl. Sci. Res.*, 1(5): 524–528.

Kathiresan, K. and Manivannan, S., 2007. Production of alkaline protease by *Strptomycetes* sp. isolated from coastal mangrove sediment. *Int. J. Biol. Chem.*, 1(2): 98–103.

Kathiresan, K. and Ravikumar, S., 1995. Influence of tannins, sugars and amino acids on bacterial load of marine halophytes. *Environ. Ecol.*, 13(1): 94–96.

Kathiresan, K., Ravindran, V.S., and Muruganandham, A., 2006. Mangrove extracts prevent the blood coagulate!. *Indian J. Biotech.*, 5: 252–254.

Kathiresan, K., Saravanakumar, K., Anburaj, R., Gomathi, V., Abirami, G., Sahu, S.K., and Anandhan, S., 2011. Microbial enzyme activity in decomposing leaves of mangroves. *Int. J. Adv. Biotechnol. Res.*, 2(3): 382–389.

Kathiresan, K. and Veera Ravi, A., 1990. Seasonal changes in tannin content of mangrove leaves. *Indian For.*, 116(5): 390–392.

Kristensen, E., Bouillon, S., Dittmar, T., and Marchand, C., 2008. Organic carbon dynamics in mangrove ecosystem. *Aquat. Bot.*, 89(2): 210–219.

Lacerda, L.D., Ittekkot, V., and Patchineelam, S.R., 1995. Biogeochemistry of mangrove soil organic matter: A comparison between Rhizophora and Avicennia soils in south-eastern Brazil. *Estuar. Coast. Shelf Sci.*, 40(6): 713–720.

Lallier-Vergès, E., Marchand, C., Disnar, J.R., and Lottier, N., 2008. Origin and diagenesis of lignin and carbohydrates in mangrove sediments of Guadeloupe (French West Indies): Evidence for a two-step evolution of organic deposits. *Chem. Geol.*, 255: 388–398.

Lee, K.H., Moran, M.A., Benner, R., and Hodson, R.E., 1990. Influence of soluble components of red mangrove (Rhizophora mangle) leaves on microbial decomposition of structural (lignocellulosic) leaf components in seawater. *Bull. Mar. Sci.*, 46(2): 374–386.

Manivannan, S. and Kathiresan, K., 2007a. Alkaline protease production by *Penicillium fellutanum* isolated from mangrove sediment. *Int. J. Biol. Chem.*, 1(2): 98–103.

Manivannan, S. and Kathiresan, K., 2007b. Effect of medium composition on glucose oxidase production by *Penicillium fellutanum* isolated from mangrove rhizosphere soil. *Res. J. Microbiol.*, 2(3): 294–298.

Marchand, C., Disnar, J.R., Lallier-Vergès, E, and Lottier, N., 2005. Early diagenesis of carbohydrates and lignin in mangrove sediments subject to variable redox conditions (French Guiana). *Geochim. Cosmochim. Acta*, 69: 131–142.

Miltner, A., Emeis, K.C., Struck, U., Leipe, T., and Voss, M., 2005. Terrigenous organic matter in Holocene sediments from the central Baltic Sea, NW Europe. *Chem. Geol.*, 216: 313–328.

Moers, M.E.C., Baas, M., De Leuw, J.W., Boon, J.J., and Schenck, P.A., 1990. Occurrence and origin of carbohydrates in peat samples from a red mangrove environment as reflected by abundances of neutral monosaccharides. *Geochim. Cosmochim. Acta*, 54: 2463–2472.

Moers, M.E.C., Jones, D.M., Eakin, P.A., Fallick, A.E., Griffiths, H., and Larter, S.R., 1993. Carbohydrate diagenesis in hypersaline environments: Application of GC-IRMS to the stable isotope analysis of derivatized saccharides from surficial and buried sediments. *Org. Geochem.*, 20(7): 927–933.

Moers, M.E.C. and Larter, S.R., 1993. Neutral monosaccharides from a hypersaline tropical environment: Applications to the characterization of modern and ancient ecosystems. *Geochim. Cosmochim. Acta*, 57(13): 3063–3071.

Neilson, M.J. and Richards, G.N., 1989. Chemical-composition of degrading mangrove leaf litter and changes produced after consumption by mangrove crab *Neosarmatium smithi* (Crustacea, Decapoda, Sesarmidae). *J. Chem. Ecol.*, 15: 1267–1283.

Niiyama, T. and Toyohara, H., 2011. Widespread distribution of cellulase and hemi-cellulase activities among aquatic invertebrates. *Fish. Sci.*, 77: 649–655.

Niiyama, T., Toyohara, H., and Tanaka, K., 2012. Cellulase activity in blood cockle (*Anadara granosa*) in the matang mangrove forest reserve, Malaysia. *JARQ*, 46(4): 355–359.

Nishida, Y., Suzuki, K., Kumagai, Y., Tanaka, H., Inoue, A., and Ojima, T., 2007. Isolation and primary structure of a cellulase from the Japanese sea urchin *Strongylocentrotus nudus. Biochimie*, 89: 1002–1011.

Okamura, K., Tanaka, K., Siow, R., Man, A., Kodama, M., and Ichikawa, T., 2010. Spring tide hypoxia with relation to chemical properties of the sediments in the Matang Mangrove Estuary, Malaysia. *JARQ*, 44: 325–333.

Olson, D.G., Tripathi, S.A., Giannone, R.J., Lo, J., and Caiazza, N.C., 2010. Deletion of the Cel48S cellulose from *Clostridium thermocellum. Proc. Natl. Acad. Sci. U.S.A.*, 107: 17727–17732.

Ronsness, P.A., 1968. Cellulolytic enzymes during morphogenesis in *Dictyostelium discoideum. J. Bacteriol.*, 96: 639–645.

Oku, H., Baba, S., Koga, H., Takara, K., and Iwasaki, H., 2003. Lipid composition of mangrove and its relevance to salt tolerance. *J. Plant Res.*, 116: 37–45.

Opsahl, S. and Benner, R., 1995. Early diagenesis of vascular plant tissues: Lignin and cutin decomposition and biogeochemical implications. *Geochim. Cosmochim. Acta*, 59: 4889–4904.

Opsahl, S. and Benner, R., 1999. Characterization of carbohydrates during early diagenesis of five vascular plant tissues. *Org. Geochem.*, 30: 83–94.

Parida, A.K., Das, A.B., and Mohanty, P., 2004. Defense potentials to NaCl in a mangrove, *Bruguiera parviflora*: Differential changes of isoforms of some antioxidative enzymes. *J. Plant Physiol.*, 161: 531–542.

Popp, M., Larher, F., and Weigel, P., 1985. Osmotic adaption in Australian mangroves. *Symp. Coastal Vegetation*, 61(1–3): 247–253.

Premanathan, M., Arakaki, R., Izumi, H., Kathiresan, K., Nakano, M., Yamamoto, N., Nakashima, H., and Nakashima, H., 1999. Antiviral properties of a mangrove plant, *Rhizophora apiculata* Blume. against human immunodeficiency virus. *Antiviral Res.*, 44: 113–122.

Premanathan, M., Chandra, K., Bajpai, S.K., and Kathiresan, K., 1992. A survey of some Indian marine plants of antiviral activity. *Bot. Mar.*, 35(4): 321–324.

Premanathan, M., Kathiresan, K., and Chandra, K., 1995. Antiviral evaluation of some marine plants against Semliki forest virus. *Int. J. Pharm.*, 33(1): 75–77.

Premanathan, M., Kathiresan, K., Chandra, K., and Bajpai, S.K., 1993. Antiviral activity of marine plants against New castle disease virus. *Trop. Biomed.*, 10: 31–33.

Premanathan, M., Kathiresan, K., Chandra, K., and Bajpai, S.K., 1994. *In vitro* anti-vaccinia virus activity of some marine plants. *Indian J. Med. Res.*, 99: 236–238.

Rajendran, N. and Kathiresan, K., 2000. Biochemical changes in decomposing leaves of mangroves. *Chem. Ecol.*, 17: 91–102.

Ramanathan, A.L., Singh, G., Majunder, J., Samal, A.C., Chahuan, R., Ranjan, R.K., Rajkumar, K., and Santra, S.C., 2008. A study of microbial diversity and its interactions with nutrients in the sediment of Sunderban mangrove. *Indian J. Mar. Sci.*, 37: 159–165.

Rojas, A., Holguin, G., Glick, B.R., and Bashan, Y., 2001. Synergism between *Phyllobacterium* sp. (N₂ fixer) and *Bacillus licheniformis* (P-solubilizer), both from the semi-arid mangrove rhizosphere. *FEMS Micro. Ecol.*, 35: 181–187.

Rubio-Pina, J.A. and Zapata-Perez, O., 2011. Isolation of total RNA from tissues rich in polyphenols and polysaccharides of mangrove plants. *Electron. J. Biotechnol.*, 14: 1–8.

Sahebi, M., Hanafi, M.M., Abdullah, S.N.A., Nejati, N., Rafii, M.Y., and Azizi, P., 2013. Extraction of total RNA from mangrove plants to identify different genes involved in its adaptability to the variety of stresses. *Pak. J. Agric. Sci.*, 50(4): 529–536.

Sakamoto, K., Touhata, K., Yamashita, M., Kasai, A., and Toyohara, H., 2007. Cellulose digestion by common Japanese freshwater clam *Corbicula japonica*. *Fish. Sci.*, 73: 675–683.

Sakamoto, K. and Toyohara, H., 2009a. Molecular cloning of glycoside hydrolase family 45 cellulase genes from brackish water clam *Corbicula japonica*. *Comp. Biochem. Physiol. B*, 152: 390–396.

Sakamoto, K. and Toyohara, H., 2009b. Putative endogenous xylanase from brackish-water clam *Corbicula japonica*. *Comp. Biochem. Physiol. B*, 154: 85–92.

Sakamoto, K. and Toyohara, H., 2009c. A comparative study of cellulase and hemicellulase activities of brackish water clam *Corbicula japonica* with those of other marine Veneroida bivalves. *J. Exp. Biol.*, 212: 2812–2818.

Sakamoto, K., Uji, S., Kurokawa, T., and Toyohara, H., 2008. Immunohistochemical, in situ hybridization and biochemical studies on endogenous cellulose of *Corbicula japonica*. *Comp. Biochem. Physiol. B*, 150: 216–221.

Sakamoto, K., Uji, S., Kurokawa, T., and Toyohara, H., 2009. Molecular cloning of endogenous β-glucosidase from common Japanese Brackish water clam *Corbicula japonica*. *Gene*, 435: 72–79.

Sarkanen, K.V. and Ludwig, C.H., 1971. *Lignins: Occurrence, Formation, Structure and Reactions.* Wiley-Intersciences, New York.

Sharmila, K., Thillaimaharani, K.A., Durairaj, R., and Kalaiselvan, M., 2014. Production and characterization of exopolysaccharides (EPS) from mangrove filamentous fungus, *Syncephalastrum* sp. *Afr. J. Microbiol. Res.*, 8(21): 2155–2161.

Smant, G., Stokkermans, J.P., Yan, Y., de Boer, J.M., Baum, T.J., Wang, X. Hussey, R.S. et al., 1998. Endogenous cellulases in animals: Isolation of b-1,4-endoglucanase genes from two species of plant parasitic cyst nematodes. *Proc. Natl. Acad. Sci. U.S.A.*, 95: 4906–4911.

Stout, S.A., Boon J.J., and Spackman, W., 1988. Molecular aspects of the peatification and early coalification of angiosperm and gymnosperm woods, *Geochim. Cosmochim. Acta*, 52: 405–414.

Suzuki, K., Ojima, T., and Nishita, K., 2003. Purification and cDNA cloning of a cellulase from abalone *Haliotis discus hannai*. *Eur. J. Biochem.*, 270: 771–778.

Thompson, C.E., Beys-da-Silva, W.O., Santi, L., Berger, M., Vainstein, M.H., Guima Raes, J.A., and Vasconcelos, A.T., 2013. A potential source for cellulolytic enzyme discovery and environmental aspects revealed through metagenomics of Brazilian mangroves. *AMB Express*, 3: 65.

Thongtham, N. and Kristensen, E., 2005. Carbon and nitrogen balance of leaf-eating sesarmid crabs (*Neoepisesarma versicolor*) offered different food sources. *Estuar. Coast. Shelf Sci.*, 65: 213–222.

Tomlinson, P.B., 1986. *The Botany of Mangroves.* Cambridge University Press, Cambridge, U.K., 413pp.

Trinci, A.P.J., Davies, D.R., Gull, K., and Lawrence, M.I., 1994. Anaerobic fungi in herbivorous animals. *Mycol. Res.*, 98: 129–152.

Vashist, P., Ghadi, S.C., Verma, P., and Shouche, Y.S., 2013. *Microbulbifer mangrovi* sp. nov., a polysaccharide degrading bacterium isolated from the mangroves of India. *Int. J. Sys. Evol. Microbiol.*, 63, 2532–2537.

Veera Ravi, A. and Kathiresan, K., 1990. Seasonal variation in gallotannin from mangroves. *Indian J. Mar. Sci.*, 19(3): 224–225.

Vijayavel, K., Anbuselvam, C., and Balasubramanian, M.P., 2006. Free radical scavenging activity of the marine mangrove *Rhizophora apiculata* bark extract with reference to naphthalene induced mitochondrial dysfunction. *Chem. Biol. Interact.*, 165: 170–175.

Watanabe, H., Noda, H., Tokuda, G., and Lo, N., 1998. A cellulase gene of termite origin. *Nature*, 394: 330–331.

Watanabe, H. and Tokuda, G., 2010. Cellulolytic systems in insects. *Annu. Rev. Entomol.*, 55: 609–632.

Yang, S., Li, Y., and Lin, P., 2003.Change of leaf caloric value from *Avicennia marina* and *Agiceras corniculatum* mangrove plants under cold stress. *J. Oceanogra. Taiwan Strait; Taiwan-Haixia*, 22(1): 46–52.

# chapter twenty-nine

# It's all about the marine carbohydrates

*Thangapandi Marudhupandi, Radhika Rajasree Santha Ravindranath,*
*and Dhinakarasamy Inbakandan*

## Contents

## 29.1 Introduction

Marine environments offer plenty of resources and a variety of natural products for the benefit of humans; research on marine carbohydrates is constantly discovering their novel properties to utilize them in various industrial applications (Figure 29.1). A carbohydrate is a biological molecule consisting of carbon, hydrogen, and oxygen atoms with the empirical formula $C_nH_{2n}O_n$. In general, based on the degree of polymerization, they are classified as mono and disaccharides, oligosaccharides, and polysaccharides (Edwards and Garcia 2009). Marine carbohydrates can be classified based on their source of isolation (marine animals, plants, and microbes), and they are considered as an extraordinary source of chemical diversity for drug discovery (Laurienzo 2010). In addition, research on marine carbohydrates has increasingly shown the utilization of their valuable properties to various industrial applications including cosmetics, food and agriculture, pharmaceuticals, nanotechnology, biotechnology and microbiology, industrial effluent treatment, and so on. In the context of this chapter we will discuss the recent advances and applications of carbohydrates derived from the marine sources (microbes, plants, and animals).

## 29.2 Chitin and chitosan polymers

Chitin is a long-chain biopolymer consisting of 2-acetamido-2-deoxy-(1-4)-β-D-glucopyranose residues (*N*-acetyl-D-glucosamine units) and is found in the cell walls of fungi and microorganisms, exoskeletons of arthropods, cuticle of insects, and internal shells of cephalopods (Je and Kim 2012). Interestingly, chitin isomorphs' occurrence was observed in fungi, arthropods, and mollusks in the form of granules, sheets, or powders (Khor 2001), whereas chitin-based scaffolds of Poriferan (marine sponge) exhibited three-dimensional networks of tube-like interconnected fibers in *Aplysina aerophoba* (Brunner et al. 2009). Chitin is insoluble in water because of its strong intra and intermolecular hydrogen bonding (Je and Kim 2012). In contrast, chitosan (β-(1,4)-2-amino-2-deoxy-D-glucose), which is a cationic polymer obtained by the deacetylation of chitin, can be dissolved in water by converting it into salts with different acids on the amino group of D-glucosamine units. Moreover, partially acetylated chitosan can be dissolved in water alone (Aiba 1989). Chitin and chitosan have been widely used in various fields due to their unique features such as, low cost, eco-friendliness, and biocompatibility (Cao et al. 2014).

### 29.2.1 Bioadsorbent potential of chitin and chitosan

Growing human population and industrialization throughout the world have been increasing environmental pollution. Of the pollutants, heavy metals are considered the most problematic (Kandile et al. 2015). Currently, chitin and its derivatives are receiving

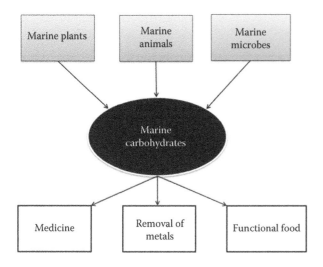

**Figure 29.1** Utilization of marine carbohydrates for industrial applications.

attention as natural bioadsorbents. Studies with chitin as a biosorbent are fewer than with chitosan, and it is believed that the concentration of primary amine groups of chitosan is the main reason for its metal adsorption capacity, which is characterized through different processes such as cationic (chelation), anionic (electrostatic attraction), or polar interactions (ion–dipole, dipole–dipole, van der Waals' interactions, etc.) (Jose et al. 2010). However, chitin also possesses considerable concentration of primary amine groups even at high degrees of acetylation; this makes chitin a competitive absorbent to other materials (Jose et al. 2010). For example, chitin from the marine sponge *A. aerophoba* efficiently adsorbed uranium content up to 288 mg/g compared to many other types of chitinous sorbents (Schleuter et al. 2013). Similarly, Liu et al. (2013b) investigated the heavy metal removing efficiency of chitin microparticles and chitin nanofibrils. Their results indicated that chitin nanofibrils could more efficiently remove heavy metals such as Cd(II), Ni(II), Cu(II), Zn(II), Pb(II), and Cr(III) than chitin microparticles, and they have suggested that the higher adsorption was due to the higher specific surface area and widely distributed pores of the nanofibrils.

Chitosan polymer has been extensively studied for the adsorption of organic and inorganic compounds, dyes, and anionic compounds, such as amoxicillin (Adriano et al. 2005), chromium (Liu et al. 2010), copper(II), cadmium(II) (Kolodynska 2012), fluoride (Viswanathan et al. 2009), perchlorate (Xie et al. 2010), cobalt(II), lead(II) (Gotoh et al. 2004), arsenic (Boddu et al. 2008), nitrate (Chatterjee et al. 2009), molybdate (Guibal et al. 1998), mercury (Merrifield et al. 2004), and tungsten (Qian et al. 2004) from wastewater. However, to improve the heavy metal absorption efficiency of chitosan, a novel carbonaceous sulfur-containing

chitosan–Fe(III) complex was prepared for removing copper(II) from water by Wen et al. (2015). Interestingly, they found the adsorption capacity of carbonaceous sulfur-containing chitosan–Fe(III) for $Cu^{2+}$ up to 413.2 mg/g, which was higher than that in a previous study on copper removal in water. Similarly, Mi et al. (2015) investigated the copper(II) ions adsorption efficiency of the chitosan–oxalate complex biosorbent (COCB) synthesized by an ionotropic cross-linking method. The maximum adsorption capacity of this porous COCB bead was 227.27 mg/g and for nonporous COCB beads 175.44 mg/g at pH 5.0. Their results showed that COCB bioadsorbent has fast and high adsorption capacity for Cu(II) uptake.

## 29.2.2 Biomedical application of chitin and chitosan

Evaluation of the antimicrobial activity of chitin, chitosan, and their derivatives has received significant interest in recent years. It is believed that the antibacterial activity of chitin and chitosan depends on a number of factors including the degree of acetylation, molecular weight, physicochemical parameters, and the type of bacteria. Modification of these polymers using nanotechnology for enhancing their antimicrobial activity has been attempted by many researchers. More recently, Shankar et al. (2015) studied the antibacterial capacity of chitin nanofibrils reinforced with carrageenan nanocomposite films prepared by solution-casting against the gram-positive food-borne pathogen *Listeria monocytogenes* and the gram-negative bacteria *Escherichia coli*. In their study, the neat carrageenan film did not show antibacterial activity against both organisms, but chitin nanofibrils including carrageenan films showed notable antimicrobial activity against *L. monocytogenes* and minimal activity against *E. coli*. However, the mechanism of the antibacterial activity of chitin nanofibrils has not been explained yet. The antimicrobial action of positively charged chitin against gram-negative bacteria is due to its interactions with the negatively charged bacterial cell membranes, which provide enhanced membrane permeability and eventually cause the rupture and leakage of the intracellular material (De Azeredo 2009). Moreover, chitosan nanoparticles have exhibited potent antibacterial activity against the *E. coli* K88, *E. coli* ATCC 25922, *S. choleraesuis* ATCC 50020, *S. typhimurium* ATCC 50013, and *S. aureus* ATCC 25923 strains when compared with native chitosan (Qi et al. 2004). A comparative study of the antibacterial activity of chitin and chitosan from the shrimp cell waste has clearly indicated that chitosan possessed high antibacterial activity against gram-negative and gram-positive bacteria, while chitin showed very low activity (Tareq et al. 2013). In addition, the molecular weight of chitosan, rather than the degree of acetylation, has a strong influence on the antimicrobial activity (Sekiguchi et al. 1994). Several authors have shown that the low-molecular-weight (LMW) chitosan shows tremendous antibacterial activity against a range of bacterial pathogens compared to high MW (HMW) chitosan (Tikhonov et al. 2006; Jing et al. 2007).

Better biodistribution, improved specificity and sensitivity, and less toxicity can achieved by combining chitosan with nanoparticles, which is an ideal polymeric platform for the development of new pharmacological and therapeutic drug-release systems (Peniche and Peniche 2011). Insulin-loaded chitosan nanoparticles showed enhanced intestinal absorption of insulin and also enhanced bioavailability (Pan et al. 2002). Chitosan nanoparticles have been used as a gene carrier to enhance the gene transfer efficiency in cells (Kim et al. 2004). Reacetylated chitosan microspheres were prepared for the controlled release of antimicrobial agents, such as amoxicillin and metronidazole, in the gastric cavity (Portero et al. 2002). However, some studies have reported that chitosan caused damage in the cells, including lysis of red blood cells (Lee et al. 1995; Gomez and Duncan 1997). However, carboxymethyl chitin nanoparticles (CMCs) showed no toxic effect against mouse L929 cells, and 5-Fu loaded CMC nanoparticles enhanced the toxicity against oral epithelial carcinoma KB cells (Dev et al. 2010).

Adjuvants are compounds that activate the immune system and thereby enhance the response to a vaccine, without having a specific antigenic effect (Montomoli et al. 2011). Thirty percent deacetylated chitin effectively induced *in vivo* cytotoxic effect in macrophages compared to chitosan and chitin derivatives (Nishimura et al. 1984). In another study, Nishimura et al. (1985) demonstrated that 70% of dihydroxypropyl-chitosan and deacetylated chitin effectively suppressed Meth-A tumor growth in mice. In addition, both 30% and 70% of deacetylated-chitin-supplemented host enhanced the nonspecific immune response against *E. coli* infection (Nishimura et al. 1986).

The poor solubility of chitosan has limited its biological applications. Notably, in the area of immunology, aqueous solution is essential for using an immunostimulant in clinical applications. To resolve this issue, glycated chitosan (GC), a water-soluble compound prepared by attaching galactose molecules to the chitosan molecules, was used (Chen et al. 2003). The immunoadjuvant effect of GC has been proven in the treatment of metastatic cancers (Chen et al. 2005). The low toxicity of GC makes it suitable for late-stage cancer patients who cannot tolerate chemotherapy and radiation therapy (Li et al. 2011).

Chitin and chitosan in the form of composites, gels, nanofibers, films, nonwovens, and scaffolds have been used for regenerating wounded tissues, taking advantage of their tremendous biological and physicochemical properties (Muzzarelli 2012). However, the antibacterial resistance of microorganisms is a major concern in the area of wound care and management, which results in infection and delayed wound healing. Several studies have indicated that the use of chitosan scaffolds and membranes is promising to treat patients with deep burns, wounds, and so on. In this context, Madhumathi et al. (2010) have reported that α-chitin/nanosilver composite scaffolds possess excellent antibacterial activity against the *S. aureus* and *E. coli* pathogens combined with fine blood clotting capability. In addition, a β-chitin hydrogel containing silver nanoparticles also exhibited potential antibacterial activity. Moreover, β-chitin/nanosilver composite scaffolds showed promising cell adhesion properties on epithelial cells (vero cells), which is ideal for wound-healing applications (Sudheesh Kumar et al. 2010).

Overproduction of oxygen free radicals causes membrane lipid peroxidation, DNA alteration, and enzyme inactivation. It is also responsible for aging, cancer, and some other human diseases. The use of synthetic antioxidants has been restricted due to their possible toxic and carcinogenic effects. To avoid this problem, recent research has focused on the extraction, identification, modification, and utilization of natural antioxidants in foods or medical materials (Wu et al. 2011). Several studies have suggested that chitin and chitosan derivatives possess antioxidant properties and they could be used as natural antioxidants, potential food supplements, or ingredients in the pharmaceutical industry (Lin and Chou 2004; Xing et al. 2005; Anraku et al. 2011).

## 29.3 Carbohydrates of marine mollusks and their biological applications

Heparin is a linear polysaccharide consisting of sulfated $1 \rightarrow 4$ linked uronic acid-$(1 \rightarrow 4)$-D-glucosamine repeating disaccharide units and has been widely used as an anticoagulant for more than half a century. Commercial production of heparin depends on either porcine (or) bovine sources as raw material. However, bovine spongiform encephalopathy disease, commonly called "mad cow disease," has limited the commercial production of heparin from bovine sources. Moreover, differentiating bovine and porcine heparin is difficult to ensure the source of species (Saravanan and Shanmugam 2011). This issue has prompted the development of nonmammalian sources for heparin. In this context, heparin from mollusks has been shown to be very similar to those present in mammalian tissues in terms of both physicochemical and biological properties (Saravanan and Shanmugam 2011). Dietrich et al. (1985) have shown that heparin from the marine mollusk *Anomalocardia brasiliana* exhibited a higher degree of binding capacity with antithrombin III (45%), molecular weight (27–43 kDa), and anticoagulant activity (320 IU/mg) than mammalian heparin. In addition, Saravanan and Shanmugam (2011) have confirmed that the isolated polymer glycosaminoglycan from the marine mollusk *Amussium pleuronectus* is heparin sulfate, by $^1$H NMR analysis, but it showed lower activity against partial thromboblastin and prothrombin when compared to standard heparin sulfate. Moreover, Saravanan et al. (2010) have reported the activated partial thromboplastin time at 72 IU/mg with a molecular mass of 15,000 for heparin-like glycosminoglycan from the marine clam *Meretrix meretrix*. Further, they suggested that glycosminoglycan could be used as an alternate source of heparin.

Recently, Liu et al. (2013) investigated the antioxidant capacity (superoxide radicals and reducing power) of a novel mannoglucan from the marine bubble snail *Bullacta exarata* (Philippi). The isolated mannoglucan consisted of Glcp6Manp heptasaccharide repeating unit, with infrequent branch chains of mannose residues (14%), indicating the backbone of mannose. In another study, the polysaccharide isolated from the mollusk *Sepia aculeata* has shown more potent antioxidant activity (1,1-diphenyl-2-picrylhydrazyl (DPPH) scavenging assay, hydroxyl radical scavenging, superoxide radical scavenging, and chelating ability on ferrous ions) than standard commercial antioxidants such as butylated hydroxyanisole and ascorbic acid (Subhapradha et al. 2014). In addition, glycosaminoglycans from *Sepia brevimana* have also shown potential antioxidant and cardioprotective activities in male Wistar rats (Barwinvino 2010). Similarly, polysaccharides isolated from the marine mussels *Mytilus coruscus* and *Mytilus edulis* also have potent antioxidant activity (Xu et al. 2008; Cheng et al. 2010). Interestingly, Zhang and his colleagues (2008) have shown that the polysaccharide from the marine mollusk *Ruditapes philippinarum* displayed potential antitumor and immunomodulatory activities. The authors have stated that the polysaccharide was mainly composed of homoglucan–protein complexes and the glucan moiety was $(1 \rightarrow 6)$-branched $(1 \rightarrow 4)$-$\alpha$-D-glucan, $(1 \rightarrow 4)$-$\alpha$-D-glucan, and $(1 \rightarrow 6)$-$\beta$-D-glucan.

## 29.4 Polymers of sea cucumber and their medicinal property

Sea cucumber has been used as a fundamental source of functional food and drug in traditional Chinese medicine. A novel fucosylated chondroitin sulfate or glycosaminoglycan had been isolated from the body wall of sea cucumber in the past decades. The glycosaminoglycans

isolated from the sea cucumbers *Stichopus japonicas* and *Ludwigothurea grisea* (Yoshida et al. 1992; Mourao et al. 1996) have shown various biological activities, including antiviral, antitumor, antithrombotic, and anticoagulant (Hoshino and Heiwamachi 1990; Sheehan and Walke 2006; Buyue and Sheehan 2009). In addition, a polysaccharide isolated from the sea cucumber *Apostichopus japonicus* also displayed anticoagulant (Gao et al. 1996), antitumor, and immunomodulating activities (Guo et al. 2009), osteoclastogenesis inhibitory activity (Kariya et al. 2004), and neurosphere formation enhancing activity (Zhang et al. 2010). Moreover, polysaccharides from the sea cucumber *A. japonicus* showed potential free-radical scavenging effects in various antioxidant assays. The polysaccharides significantly reduced the serum total cholesterol, low-density lipoprotein, and triglyceride and increased high-density lipoprotein cholesterol after hyperlipidemic treatment in Wistar rats (Liu et al. 2012). Recently, Wu et al. (2012) reported that the presence of sulfated fucose branches played an important role in the anticoagulant activity of isolated fucosylated chondroitin sulfate from the sea cucumber *Thelenata ananas*. They also stated that this activity is probably due to the occurrence of 2,4-di-*O*-sulfated fucose units in fucosylated chondroitin sulfate. Fucoidan, a sulfated polysaccharide isolated from the sea cucumber *Acaudina molpadioides*, has effectively prevented the formation of ethanol-induced gastric ulcer in rat (Wang et al. 2012). It is believed that oxidative stress and reduction of antioxidants are the major steps in ethanol-induced mucosal damage (La Casa et al. 2000). When compared to algal polysaccharides, sulfated fucans from sea urchins (Echinodermata) can be easily purified, and they have the simple, linear chain structure of $\alpha$-L-fucose with clear repetitive patterns. In addition, their specific sulfating patterns and glycosidic bond positions vary depending on the species. The simplest form of sulfated fucan ($[3$-$\alpha$-L-Fuc-2(OSO3)-1$]_n$) was observed in the eggs of the sea cucumber *Strongylocentrotus franciscanus* (Vilela-Silva et al. 1999). However, the sulfated fucan structure varies with the species depending on the position of glycosidic linkages $[\alpha (1 \rightarrow 3)$ or $\alpha (1 \rightarrow 4)]$ and the sulfating position (2-*O* or 4-*O*) (Pomin and Mourao, 2008).

## 29.5 Marine microbial polysaccharides and their biological activities

Microorganisms from oceanic environments often produce extracellular polysaccharides (EPSs) with novel structures and varied biological activities (Arena et al. 2009). The osmo-tolerant marine bacterium *Alteromonas macleodii* produced 23.4 g/L of EPS while supplementing the medium with 15% (w/v) lactose, with a productivity rate of 7.8 g/L/day. Further, the isolated EPS produced spherical silver nanoparticles of ~70 nm, which clearly indicated the capability of EPS for synthesizing biocompatible metal nanoparticles (Mehta et al. 2014). Interestingly, *Pseudomonas* sp. WAK-1 isolated from the brown seaweed *Undaria pinnatifida* simultaneously produced the extracellular glycosaminoglycan and sulfated polysaccharide. These polysaccharides showed antiviral activity ($EC_{50}$ at 1.4 µg/mL) against anti-HSV-1 in RPMI 8226 cells. In addition, the oversulfated derivatives of these polysaccharides displayed antiviral activities against influenza virus type A, whereas no activity was detected for glycosaminoglycan, sulfated polysaccharide, and their oversulfated derivatives against influenza virus type B and HIV type 1 (Matsuda et al. 1999). Similarly, Sun et al. (2011) investigated the antioxidant effect of the EPS fractions (ENP1 and ENP2) obtained by anion-exchange chromatography and gel-filtration chromatography from the marine fungus *Epicoccum nigrum* JJY40. The structural relationship between these two polysaccharides was studied by chemical analysis (methylation) and NMR spectroscopy. The fraction ENP2 showed good antioxidant effect against the hydroxyl, superoxide, and DPPH radical scavenging activities and lipid peroxidation inhibition under *in vitro* condition, when compared to the ENP1 fraction. It was concluded that the higher antioxidant activity of ENP2 depended on the uronic acid content, molecular weight, and glycosidic linkage patterns of EPS. Research on the EPS produced by symbiotic and epiphytic microorganisms is increasing, and it could be explored as a source of functional biopolymers. In this context, the EPS isolated from the coral-associated fungus *Aspergillus versicolor* LCJ-5-4 was identified as mannoglucan. One-dimensional (1D) and 2D NMR spectroscopic results revealed that the EPS was composed of (1 → 6)-linked α-D-glucopyranose and (1 → 2)-linked α-D-mannopyranose units; it also exhibited potential antioxidant activity, notably scavenging effects on superoxide radicals assay (Chen et al. 2012). In addition, Chen et al. (2013) demonstrated the preparation and structural characterization of EPS from the coral-associated fungus *Penicillium commune*. They found a novel galactofuranose-containing EPS, which consisted of branches of four (1 → 6)-linked β-galacto furanose residues with a terminal of α-glucopyranose residue attached to the galactofuranose residue at the 2-position.

Research on microalgae-derived products or biologically active compounds has been increasing in recent times because of their unique characteristic features such as photosynthetic, fast growth rate, $CO_2$ reduction, and so on. For the first time, Sun et al. (2012) identified the intra and extracellular polysaccharides from the marine microalga *Isochrysis galbana*. Furthermore, a crude polysaccharide from this species also showed enhanced immunomodulatory properties in murine macrophages by inducing interleukin-1 (Yu et al. 2010). Moreover, a polysaccharide, consisting of highly branched (1 → 3, 1 → 6)-β-D-glucan of *I. galbana*, showed the direct inhibition against the proliferation of U937 human leukemic monocyte lymphoma cells. Thus, it could be used as a potential anticancer agent (Sadovskaya et al. 2014). Similarly, Yingying et al. (2014) investigated the antioxidant effect of three different polysaccharides (IPSI-A, IPSI-B, and IPSII) isolated from *I. galbana* by using anion exchange and repeated gel chromatography. Of these, the polysaccharide fraction IPSII showed more effective antioxidant activity than IPSI-A and IPSI-B. IPSII has a molecular weight of 15.934 kDa and belongs to β-type heteropolysaccharides with a pyran group. Additionally, the sulfated exopolysaccharide isolated from the marine microalga *Gyrodinium impudicum* strain KG03 showed notable antiviral activity under *in vitro* condition against the encephalomyocarditis virus. Impressively, no cytotoxic effect was observed in HeLa cells, even at higher concentrations (1000 µg/mL). Further, this polysaccharide was typified as a homopolysaccharide of galactose consisting of uronic acid (2.96% wt/wt) and sulfate (10.32% wt/wt) content with a molecular weight of $1.87 \times 10^7$ (Yim et al. 2004).

## 29.6 Seaweed polysaccharides and their applications

Alginates are linear acidic polysaccharides found in the cell wall and matrix of brown seaweeds; they also produce by some bacterial species (Rinaudo 2007). Alginates from *Sargassum* sp. (Phaeophyta) demonstrated significant anticancer activity against the various murine tumors, such as Sarcoma 180, Ehrlich ascites carcinoma, and IMC carcinoma (Fujiihara et al. 1984; Fujiihara and Nagumo 1993; De Sousa et al. 2007). The stronger antitumor activity of alginates was due to the composition of the homopolymer, which has a higher MM-block content (Ooi and Liu 2000). Similarly, Paxman et al. (2008) investigated the appetite-controlling effect of sodium alginate in free-living adults (68 males and females). Their results revealed that the intake of the sodium alginate formulation resulted in a significant reduction of 7% (134.8 kcal) in terms of daily energy intake, and suggested that the strong gelling property of this sodium alginate formulation could be used for controlling overweight and obesity in future. In addition, alginates play an important role in the food and beverages and pharmaceutical industries (Goh et al. 2012). Apart from these, alginates have been used for ceramic shaping in gel casting, as biocatalysts in the water treatment (Wang et al. 2002), and also as a noble metal sequestrants (Aderhold et al. 1996).

Fucoidans are a group of polysaccharides mainly composed of sulfated L-fucose with less than 10% other monosaccharides. They are generally found in the cell walls of brown seaweeds (Berteau and Mulloy 2003). Major physiological roles of fucans in algae are not well understood, but it is well known that the bioactivity of fucoidan in human health is varied and it has been used as a functional food for human health (Berteau and Mulloy 2003).

Synytsya and his colleagues (2014) have evaluated the structural features and the protective effect of fucoidan isolated from the sporophyll (Mekabu) of the brown seaweed *U. pinnatifida* against avian influenza A viruses. Structural characterization by FTIR, FT Raman, and NMR indicated that the polysaccharide is *O*-acetylated sulfated fucogalactan. Oral administration of this polysaccharide suppressed the virus yield by increasing antibody production in mice. In addition, Lim et al. (2014) have reported that the isolated fucoidan with a yield of 6.16% ± 0.08% from *Sargassum binderi* possessed excellent antioxidant activity to the superoxide anion and hydroxyl radical scavenging assays than synthetic antioxidants. Marudhupandi et al. (2014) have reported that the antioxidant capacity of fucoidan depends on the percentage of sulfate content in it.

Many studies have suggested that fucoidan can be used as a potential anti-inflammatory agent. A study showed that fucoidan treatment reduced the severe effects in the early stages of *Staphylococcus aureus*-caused arthritis in mice, but it delayed phagocyte recruitment and decreased clearance of the bacterium (Verdrengh et al. 2000). Additionally, injection of fucoidan into sensitized mice before hapten challenge could reduce the contact hypersensitivity reactions (Nasu et al. 1997). Furthermore, recruitment of leukocytes into the cerebrospinal fluid in a meningitis model was reduced by fucoidan (Granert et al. 1999), as was interleukin-1 (IL-1) produced in a similar model (Ostergaard et al. 2000).

Fucoidan is known to have antitumor effects, but its mode of action is not fully understood. A study by Alekseyenko et al. (2007) demonstrated that when 10 mg/kg of fucoidan was administered in mice with transplanted Lewis lung adenocarcinoma, it produced moderate antitumor and anti-metastatic effects (Li et al. 2008). Further, Marudhupandi et al. (2015) have reported that fucoidan isolated from *Turbinaria conoides* showed prominent anticancer activity against A549 lung cancer cells.

In diabetes research, fucoidan has gained importance in recent times. Particularly, in type 2 diabetes, which is called non-insulin-dependent diabetes mellitus, fucoidan plays an important role in enzyme inhibition (α-amylase and α-glucosidase). Lakshmanasenthil

et al. (2014) have reported that fucoidan isolated from *Turbinaria ornata* inhibits α-amylase by 86% with the $IC_{50}$ value of 33.6 µg. Vinothkumar et al. (2015) have reported the inhibition effect of fucoidan isolated from *Sargassum wigthii* against α-glucosidase.

More recently, Thuy et al. (2015) have investigated the anti-HIV activity of fucoidan from brown seaweeds such as *Sargassum mcclurei*, *S. polycystum*, and *T. ornata*. The isolated fucoidans exhibited similar activity, with a mean $IC_{50}$ value ranging from 0.33 to 0.7 µg/mL, with no cytotoxic effect. The authors suggested that the antiviral activity of these fucoidans is due to their binding capacity to HIV-1, blocking the early steps of HIV entry.

Carrageenan is the common name for a family of HMW sulfated polysaccharides isolated from red seaweeds, and it consists of galactose and anhydrogalactose units linked by glycosidic units (Jiao et al. 2011). In general, carrageenans are classified as $k$, $l$, and $\lambda$ forms based on their sulfating patterns and existence of 3,6-anhydro-α-D-galactopyranose on the D units (Shanmugam and Mody 2000). Carrageenan has been widely utilized in the food and cosmetic industries owing to its excellent physical and functional properties, including gelling, thickening, emulsifying, and stabilizing capacity (Necas and Bartosikova 2013). Liang et al. (2014) investigated the sulfate group position of red seaweed polysaccharides (carrageenan and agarose) and its influence on the anticoagulant and cytotoxicity activity. For this, they used the pyridine-chlorosulfonic acid method for sulfating agarose and $k$-carrageenan. The degree of sulfate substitution was analyzed by a CHNS elemental analyzer and the barium chloride method, and the structural feature was determined by NMR spectroscopy. Their results demonstrated that C-2 of 3,6-anhydro-α-D-Galp is the most favorable position for substituting the sulfate group of this polysaccharide, which also played a major role in both the anticoagulant and cell proliferation activity. In addition, the lambda-carrageenan isolated from the red seaweed *Gigartina skottsbergii* exhibited activity against the animal viruses BoHV-1 and SuHV-1 (Diogo et al. 2014); also, lambda-carrageenan has the ability to bind viral envelope glycoproteins and prevent virus binding to cell surface receptors (Carlucci et al. 1997). Likewise, Ordonez et al. (2014) isolated soluble and insoluble polysaccharide fractions from the red seaweed *Mastocarpus stellatus* using different extraction methods, including cold water, hot water, acid, and alkali. They tested the antioxidant and anticoagulant activity of these polysaccharide fractions. Out of these, the polysaccharide extracted with cold water exhibited potential antioxidant and anticoagulant activity, and the authors suggested that this enhanced activity was due to its higher sulfate content and higher molecular weight. Recently, Sokolova et al. (2014) showed that carrageenan-supplemented food did not stimulate hyperactivation of the immune system in hypercholesterolemia patients with cardiovascular disease. However, it moderately altered the biomarkers of chronic inflammation and also decreased the cholesterol level by 16.5% and low-density lipoprotein (LDL) cholesterol by 33.5%.

More recently, Ropellato et al. (2015) investigated the antitumor effect of sulfated heterorhamnans from the green seaweed *Gayralia oxysperma* against the U87MG cells. They prepared two homogeneous and highly sulfated polymers by Smith degradation, with molecular weights of 109 and 251 kDa, and these polymers contained 3-linked α-L-rhamnosyl units 2- or 4-sulfate and 2-linked α-L-rhamnosyl units 4 at different percentages. Besides, Arata et al. (2015) isolated the sulfated and pyruvylated galactans from tropical green seaweeds such as *Penicillus capitatus*, *Udotea flabellum*, and *Halimeda opuntia*. They found important sulfated polysaccharides from the suborder Halimedineae, whose backbone consists of 3,6-linkages, comprising major amounts of 3-linked 4,6-*O*-(1-carboxy)ethylidene-D-galactopyranose units in the part sulfated on C-2 position, and this polysaccharide exhibited anticoagulant activity.

## 29.7 Summary

In this chapter, we discussed the recent research on marine carbohydrates and their utilization in various industrial applications for the betterment of humankind. We addressed the nature of carbohydrates and their structural and functional properties from marine sources, including marine microbes (bacteria, fungi, and microalgae), animals (mollusks and sea cucumbers), and seaweeds (brown, red, and green).

## Acknowledgment

The authors are grateful to the authorities of Sathyabama University.

## References

Aderhold, D., C.J. Williams, and R.G.J. Edyvean. 1996. The removal of heavy-metal ions by seaweeds and their derivatives. *Bioresource Technology* 58(1): 1–6.

Adriano, W., V. Veredas, C. Santana, and L. Gonçalves. 2005. Adsorption of amoxicillin on chitosan beads: Kinetics, equilibrium and validation of finite bath models. *Biochemical Engineering Journal* 27: 132–137.

Aiba, S. 1989. Studies on chitosan: 2. Solution stability and reactivity of partially N-acetylated chitosan derivatives in aqueous media. *International Journal of Biological Macromolecules* 11: 249–252.

Alekseyenko, T.V., S.Y. Zhanayeva, A.A. Venediktova et al. 2007. Antitumor and antimetastatic activity of fucoidan, a sulfated polysaccharide isolated from the Okhotsk Sea *Fucus evanescens* brown alga. *Bulletin of Experimental Biology and Medicine* 143: 730–773.

Arata, P.X., I. Quintana, D.J. Canelon, E.B. Vera, R.S. Compagnone, and M. Ciancia. 2015. Chemical structure and anticoagulant activity of highly pyruvylated sulfated galactans from tropical green seaweeds of the order Bryopsidales. *Carbohydrate Polymers* 122: 376–386. http://dx.doi.org/doi:10.1016/j.carbpol.2014.10.030.

Arena, A., C. Gugliandolo, G. Stassi, B. Pavone, D. Iannello, and G. Bisignano. 2009. An exopolysaccharide produced by *Geobacillus thermodenitrificans* strain B3-72: Antiviral activity on immunocompetent cells. *Immunology Letters* 123: 132–137.

Barwinvino, A. 2010. Bioactive compounds from cephalopod mollusk: Isolation, characterization and in vitro antioxidant activity of glycosa-minoglycans from cuttlefish *Sepia brevimana* Steenstrup, 1875 and its cardioprotective effect on isoproterenol-induced myocardial infarction in male wistar rats. PhD thesis, Annamalai University, Tamil Nadu, India, pp. 1–230.

Berteau, O. and B. Mulloy. 2003. Sulfated fucans, fresh perspectives: Structures, functions, and biological properties of sulfated fucans and an overview of enzymes active toward this class of polysaccharide. *Glycobiology* 13(6):29R–40R.

Boddu, K., J.L. Abburi, E.D. Talbott, and R.H. Smith. 2008. Removal of arsenic (III) and arsenic(V) from aqueous medium using chitosan-coated biosorbent. *Water Research* 42: 633–642.

Brunner, E., H. Ehrlich, P. Schupp, R. Hedrich, S. Hunoldt, M. Kammer, S. Machill, S Paasch, V.V. Bazhenov, D.V. Kurek, et al. 2009. Chitin-based scaffolds are an integral part of the skeleton of the marine demosponge *Ianthella basta*. *Journal of Structural Biology* 168: 539–547.

Buyue, Y. and J.P. Sheehan. 2009. Fucosylated chondroitin sulfate inhibits plasma thrombin generation via targeting of the factor IXa heparin-binding exosite. *Blood* 114: 3092–3100.

Cao, C., L. Xiao, C. Chen, X. Shi, Q. Cao, and L. Gao. 2014. In situ preparation of magnetic $Fe_3O_4$/chitosan nanoparticles via a novel reduction–precipitation method and their application in adsorption of reactive azo dye. *Powder Technology* 260: 90–97.

Carlucci, M.J., C.A. Pujol, M. Ciancia, M.D. Noseda, M.C. Matulewicz, E.B. Damonte, and A.S. Cerezo. 1997. Antiherpetic and anticoagulant properties of carrageenans from the red seaweed *Gigartina skottsbergii* and their cyclized derivatives: Correlation between structure and biological activity. *International Journal of Biological Macromolecules* 20: 97–105.

Chatterjee, S., D.S. Lee, M.W. Lee, and S.H. Woo. 2009. Nitrate removal from aqueous solutions by cross-linked chitosan beads conditioned with sodium bisulfate. *Journal of Hazardous Materials* 166: 508–513.

Chen, W.R., S.W. Jeong, M.D. Lucroy et al. 2003. Induced antitumor immunity against DMBA-4 metastatic mammary tumors in rats using a novel approach. *International Journal of Cancer* 107(6): 1053–1057.

Chen, W.R., M. Korbelik, K.E. Bartels, H. Liu, J. Sun, and R.E. Nordquist. 2005. Enhancement of laser cancer treatment by a chitosan-derived immunoadjuvant. *Photochemistry and Photobiology* 81(1): 190–195.

Chen, Y., W. Mao, J. Wang, W. Zhu, C. Zhao, N. Li, C. Wang, M. Yan, T. Guo, and X. Liu. 2013. Preparation and structural elucidation of a glucomannogalactan from marine fungus *Penicillium commune. Carbohydrate Polymers* 97: 293–299.

Chen, Y., W. Mao, Y. Yang et al. 2012. Structure and antioxidant activity of an extracellular polysaccharide from coral-associated fungus, *Aspergillus versicolor* LCJ-5-4. *Carbohydrate Polymers* 87: 218–226.

Cheng, S., X. Yu, and Y. Zhang. 2010. Extraction of polysaccharides from *Mytilus edulis* and their antioxidant activity in vitro. *Shipin Gongye Keji* 31: 132–134.

De Azeredo, H.M.C. 2009. Nanocomposites for food packaging applications. *Food Research International* 42: 1240–1253.

De Sousa, A.P.A., M.R. Torres, C. Pessoa, M.O.D. Moraes, F.D.R. Filho, A.P.N.N. Alves, and L.V.C. Lotufo. 2007. In vivo growth-inhibition of Sarcoma 180 tumor by alginates from brown seaweed *Sargassum vulgare. Carbohydrate Polymers* 69: 7–13.

Dev, A., C.J. Mohan, V. Sreeja, H. Tamura, G.R. Patzke, F. Hussain, S. Weyeneth, S.V. Nair, and R. Jayakumar. 2010. Novel carboxymethyl chitin nanoparticles for cancer drug delivery applications. *Carbohydrate Polymers* 79: 1073–1079.

Dietrich, C.P., J.F. de-Paiva, C.T. Moraes, H.K. Takahashi, M.A. Porcionatto, and H.B. Nader. 1985. Isolation and characterization of a heparin with high anticoagulant activity from *Anomalocardia brasiliana. Biochimica Biophysica Acta* 843: 1–7.

Diogo, J.V., S.G. Novo, M.J. Gonzalez, M. Ciancia, and A.C. Bratanich. 2015. Antiviral activity of lambda-carrageenan prepared from red seaweed (*Gigartina skottsbergii*) against BoHV-1 and SuHV-1. *Research in Veterinary Science* 98: 142–144. doi:10.1016/j.rvsc.2014.11.010.

Edwards, C.A. and A.L. Garcia. 2009. The health aspects of hydrocolloids. In *Handbook of Hydrocolloids*, 2nd edn., Phillips, G.O. and Williams, P.A. (eds.). Boca Raton, FL: CRC Press, p. 50.

Fujiihara, M., N. Izimma, I. Yamamoto, and T. Nagumo. 1984. Purification and chemical and physical characterization of an antitumor polysaccharide from the brown seaweed *Sargassum fulvellum. Carbohydrate Research* 125: 97–106.

Fujiihara, M. and T. Nagumo. 1993. An influence of the structure of alginate on the chemotactic activity of macrophage and the antitumor activity. *Carbohydrate Research* 243: 211–216.

Gao, C.J., J. Li, Z. Peng, L. Wu, H.Z. Ma, and J.W. Liu. 1996. Effect of acidic mucopolysaccharide extracted from *Stichopus japonicus* selenka on the structure and dissolubility of fibrin gels. *Chinese Journal of Hematology* 17: 458–461.

Goh, C.H., P.W.S. Heng, and L.W. Chan. 2012. Alginates as a useful natural polymer for microencapsulation and therapeutic applications. *Carbohydrate Polymers* 88: 1–12.

Gomez, C.B. and R. Duncan. 1997. Evaluation of the biological properties of soluble chitosan and chitosan microspheres. *International Journal of Pharmaceutics* 148: 231–240.

Gotoh, T., K. Matsushima, and K.I. Kikuchi. 2004. Preparation of alginate–chitosan hybrid gel beads and adsorption of divalent metal ions. *Chemosphere* 55: 135–140.

Granert, C., J. Raud, A. Waage, and L. Lindquist. 1999. Effects of polysaccharide fucoidin on cerebrospinal fluid interleukin-1 and tumor necrosis factor alpha in pneumococcal meningitis in the rabbit. *Infection and Immunity* 67(5): 2071–2074.

Guibal, E., C. Milot, and J.M. Tobin. 1998. Metal-anion sorption by chitosan beads: Equilibrium and kinetic studies. *Industrial & Engineering Chemistry Research* 37: 1454–1463.

Guo, L.Y., L. Wang, X.Y. Zou et al. 2009. Antitumor and immunomodulating effects of *Apostichopus japonicus* polysaccharides (AJPS) in tumor-bearing mice. *Chinese Journal of Microecology* 21: 806–808.

Hoshino, H. and M. Heiwamachi. 1991. Anti-HIV drug. European Patent. EP 0410002A1.

Je, J.Y. and S.K. Kim. 2012. Chitosan as potential marine nutraceutical. *Advances in Food and Nutrition Research*, 65: 121–135.

Jiao, G., G. Yu, J. Zhang, and H.S. Ewart. 2011. Chemical structures and bioactivities of sulfated polysaccharides from marine algae. *Marine Drugs* 9(2): 196–223.

Jing, Y.J., Y.J. Hao, H. Qu, Y. Shan, D.S. Li, and R.Q. Du. 2007. Studies on the antibacterial activities and mechanisms of chitosan obtained from cuticles of housefly larvae. *Acta Biologica Hungarica* 58: 75–86.

Jose, R., R. Mendez, A. Vladimir, E. Barrios, and J.L.D. Rodriguez. 2010. Chitin based biocomposites for removal of contaminants from water: A case study of fluoride adsorption. In *Biopolymers*, Elnashar, M. (ed.). InTech, Croatia - European Union. http://www.intechopen.com/books/biopolymers/chitin-based-biocomposites-for-removal-of-contaminants-fromwater-a-case-study-of-fluorideadsorptio. Accessed August 2014.

Kandile, N.G., M.H. Mohamed, and M.I. Mohamed. 2015. New heterocycle modified chitosan adsorbent for metal ions (II) removal from aqueous systems. *International Journal of Biological Macromolecules* 72: 110–116.

Kariya, Y., B. Mulloy, K. Imai, A. Tominaga, T. Kaneko, and A. Asari. 2004. Isolation and partial characterization of fucan sulfates from the body wall of sea cucumber *Stichopus japonicus* and their ability to inhibit osteoclastogenesis. *Carbohydrate Research* 339: 1339–1346.

Khor, E. 2001. *Chitin: Fulfilling a Biomaterials Promise*. Amsterdam, the Netherlands: Elsevier, p. 136.

Kim, T.H., I.K. Park, J.W. Nah, Y.J. Choi, and C.S. Cho. 2004. Galactosylated chitosan/DNA nanoparticles prepared using water-soluble chitosan as a gene carrier. *Biomaterials* 25: 3783–3792.

Kolodynska, D. 2012. Adsorption characteristics of chitosan modified by chelating agents of a new generation. *Chemical Engineering Journal* 179: 33–43.

La Casa, C., I. Villegas, C.A.D.L. Lastra, V. Motilva, and M.M. Calero. 2000. Evidence for protective and antioxidant properties of rutin, a natural flavone, against ethanol induced gastric lesions. *Journal of Ethnopharmacology* 71: 45–53.

Lakshmanasenthil, S., T. Vinothkumar, D. Geetharamani, T. Marudhupandi, G. Suja, and N.S. Sindhu. 2014. Fucoidan—A novel α-amylase inhibitor from *Turbinaria ornata* with relevance to NIDDM therapy. *Biocatalysis and Agricultural Biotechnology* 3(3): 66–70.

Laurienzo, P. 2010. Marine polysaccharides in pharmaceutical applications: An overview. *Marine Drugs* 8: 2435–2465.

Lee, K.Y., W.S. Ha, and W.H. Park. 1995. Blood compatibility and biodegradability of partially N-acetylated chitosan derivatives. *Biomaterials* 16: 1211–1216.

Li, B., F. Lu, X. Wei, and R. Zhao. 2008. Fucoidan: Structure and bioactivity. *Molecules* 13: 1671–1695.

Li, X., G.L. Ferrel, M.C. Guerra et al. 2011. Preliminary safety and efficacy results of laser immunotherapy for the treatment of metastatic breast cancer patients. *Photochemical and Photobiological Sciences* 10(5): 817–821.

Liang, W., X. Mao, X. Peng, and S. Tang. 2014. Effects of sulfate group in red seaweed polysaccharides on anticoagulant activity and cytotoxicity. *Carbohydrate Polymers* 30(101): 776–785.

Lim, S.J., W.M.W. Aida, M.Y. Maskat, S. Mamot, J. Ropien, and D.M. Mohd. 2014. Isolation and antioxidant capacity of fucoidan from selected Malaysian seaweeds. *Food Hydrocolloids* 42: 280–288.

Lin, H.Y. and C.C. Chou. 2004. Antioxidant activities of water-soluble disaccharide chitosan derivatives. *Food Research International* 37: 883–889.

Liu, D., N. Liao, X. Ye, Y. Hu, D. Wu, X. Guo, J. Zhong, J. Wu, and S. Chen. 2013a. Isolation and structural characterization of a novel antioxidant mannoglucan from a marine bubble snail, *Bullacta exarata* (Philippi). *Marine Drugs* 11: 4464–4477.

Liu, D., Y. Zhu, Z. Li, D. Tian, L. Chen, and P. Chen. 2013b. Chitin nanofibrils for rapid and efficient removal of metal ions from water system. *Carbohydrate Polymers* 98: 483–489.

Liu, T., L. Zhao, D. Sun, and X. Tan. 2010. Entrapment of nanoscale zero-valent iron in chitosan beads for hexavalent chromium removal from wastewater. *Journal of Hazardous Materials* 184: 724–730.

Liu, X., S. Zhenliang, M. Zhang, X. Meng, X. Xia, W. Yuan, F. Xue, and C. Liu. 2012. Antioxidant and antihyperlipidemic activities of polysaccharides from sea cucumber *Apostichopus japonicas. Carbohydrate Polymers* 90: 1664–1670.

Madhumathi, K., P.T. Sudhessh Kumar, S. Abhilash, V. Sreeja, H. Tamura, K. Manzoor, S.V. Nair, and R. Jayakumar. 2010. Development of novel chitin/nanosilver composite scaffolds for wound dressing applications. *Journal of Materials Science: Materials in Medicine* 21: 807–813.

Marudhupandi, T., T.T. Ajith kumar, S. Lakshmana Senthil, and K. Nandini Devi. 2014. In vitro antioxidant properties of fucoidan fractions from *Sargassum tenerrimum. Pakistan Journal of Biological Science* 17(3): 402–407.

Marudhupandi, T., T.T. Ajith Kumar, S. Lakshmana Senthil, G. Suja, and T. Vinothkumar. 2015. In vitro anticancer activity of fucoidan from *Turbinaria conoides* against A549 cell lines. *International Journal of Biological Macromolecules* 72: 919–923.

Matsuda, M., S. Shigeta, and K. Okutani. 1999. Antiviral activities of marine Pseudomonas polysaccharides and their oversulfated derivatives. *Marine Biotechnology* 1(1): 68–73.

Mehta, A., C. Sidhu, A.K. Pinnaka, and A.R. Choudhury. 2014. Extracellular polysaccharide production by a novel osmotolerant marine strain of *Alteromonas macleodii* and its application towards biomineralization of silver. *PLoS ONE* 9(6): e98798. doi:10.1371/journal.pone.0098798.

Merrifield, J.D., W.G. Davids, J.D. MacRae, and A. Amirbahman. 2004. Uptake of mercury by thiol-grafted chitosan gel beads. *Water Research* 38: 3132–3138.

Mi, F.L., S.J. Wu, and F.M. Lin. 2015. Adsorption of copper (II) ions by a chitosan–oxalate complex biosorbent. *International Journal of Biological Macromolecules* 72: 136–144.

Montomoli, E., S. Piccirella, B. Khadang et al. 2011. Current adjuvants and new perspectives in vaccine formulation. *Expert Review of Vaccines* 10(7): 1053–1061.

Mourao, P.A., M.S. Pereira, M.S. Pavao, B. Mulloy, D.M. Tollefsen, M.C. Mowinckel, and U. Abildgaard. 1996. Structure and anticoagulant activity of a fucosylated chondroitin sulfate from echinoderm. Sulfated fucose branches on the polysaccharide account for its high anticoagulant action. *The Journal of Biological Chemistry* 271(390): 23973–23984.

Muzzarelli, R.A.A. 2012. Nanochitins and nanochitosans, paving the way to eco-friendly and energy-saving exploitation of marine resources. In *Polymer Science: A Comprehensive Reference*, Maty-jaszewski, K. and Moller, M. (eds.). Amsterdam, the Netherlands: Elsevier, pp. 153–164.

Nasu, T., Y. Fukuda, K. Nagahira, H. Kawashima, C. Noguchi, and T. Nakanishi. 1997. Fucoidin, a potent inhibitor of L-selection function, reduces contact hypersensitivity reaction in mice. *Immunology Letters* 59(1): 47–51.

Necas, J. and L. Bartosikova. 2013. Carrageenan: A review. *Veterinarni Medicina* 58: 187–205.

Nishimura, K., C. Ishihara, and S. Ukei. 1986. Stimulation of cytokine production in mice using deacetylated chitin. *Vaccine* 4(3): 151–156.

Nishimura, K., S. Nishimura, and N. Nishi. 1984. Immunological activity of chitin and its derivatives. *Vaccine* 2(1): 93–99.

Nishimura, K., S.I. Nishimura, and N. Nishi. 1985. Adjuvant activity of chitin derivatives in mice and guinea-pigs. *Vaccine* 3(5): 379–384.

Ooi, V.E.C. and F. Liu. 2000. Immunomodulation and anticancer activity of polysaccharide-protein complexes. *Current Medicinal Chemistry* 7: 715–729.

Ordonez, E.G., A.J. Escrig, and P. Ruperez. 2014. Bioactivity of sulfated polysaccharides from the edible red seaweed *Mastocarpus stellatus. Bioactive Carbohydrates and Dietary Fiber* 3: 29–40.

Ostergaard, C., R.V.Y. Kow, T. Benfield, N.F. Moller, F. Espersen, and J.D. Lundgren. 2000.Inhibition of leukocyte entry into the brain by the selection blocker fucoidin decreases interleukin-1 (IL-1) levels but increases IL-8 level in cerebrospinal fluid during experimental pneumococcal meningitis in rabbits. *Infection and Immunity* 68(6): 3153–3157.

Pan, Y., L. Ying-jian, Z. Hui-ying , Z. Jun-min, X. Hui, W. Gang, H. Jin-song, C. Fu-de. 2002. Bioadhesive polysaccharide in protein delivery system: chitosan nanoparticles improve the intestinal absorption of insulin in vivo. *International Journal of Pharmaceutics* 249: 139–147.

Paxman, J.R., J.C. Richardson, P.W. Dettmar, and B.M. Corfe. 2008. Daily ingestion of alginate reduces energy intake in free-living subjects. *Appetite* 51: 713–719.

Peniche, H. and C. Peniche. 2011. Chitosan nanoparticles: A contribution to nanomedicine. *Polymer International* 60(6): 883–889.

Pomin, V.H. and P.A.S. Mourao. 2008. Structure, biology, evolution, and medical importance of sulfated fucans and galactans. *Glycobiology* 18(12): 1016–1027.

Portero, A., C. Remunan-Lopez, M.T. Criado, and M.J.J. Alonso. 2002. Reacetylated chitosan microspheres for controlled delivery of anti-microbial agents to the gastric mucosa. *Journal of Microencapsulation* 19: 797–809.

Qi, L., Z. Xu, X. Jiang, C. Hu, and X. Zou. 2004. Preparation and antibacterial activity of chitosan nanoparticles. *Carbohydrate Research* 339: 2693–2700.

Qian, S., H. Wang, G. Huang, S. Mo, and W. Wei. 2004. Studies of adsorption properties of cross-linked chitosan for vanadium (V), tungsten (VI). *Journal of Applied Polymer Science* 92: 1584–1588.

Rinaudo, M. 2007. Seaweed polysaccharides. Comprehensive glycoscience, 2.2. In *Polysaccharide Functional Properties*, Kamerling, J.P. (ed.). Amsterdam, the Netherlands: Elsevier, pp. 691–773.

Ropellato, J., M.M. Carvalho, L.G. Ferreira et al. 2015. Sulfated heterorhamnans from the green seaweed *Gayralia oxysperma*: Partial depolymerization, chemical structure and antitumor activity. *Carbohydrate Polymers* 117: 476–485.

Sadovskaya, I., A. Souissi, S. Souissi et al. 2014. Chemical structure and biological activity of a highly branched (1 → 3, 1 → 6)-β-D-glucan from *Isochrysis galbana*. *Carbohydrate Polymers* 13(111): 139–148.

Saravanan, R. and A. Shanmugam. 2011. Is isolation and characterization of heparan sulfate from marine scallop *Amussium pleuronectus* (Linne.) an alternative source of heparin? *Carbohydrate Polymers* 86: 1082–1084.

Saravanan, R., S. Vairamani, and A. Shanmugam. 2010. Glycosaminoglycans from marine clam *Meretrix meretrix* (Linne.) are an anticoagulant. *Preparative Biochemistry and Biotechnology* 40(4): 305–315.

Schleuter, D., A. Guntherb, S. Paasch, H. Ehrlich, Z. Kljaji, T. Hanke, G. Bernhard, and E. Brunner. 2013. Chitin-based renewable materials from marine sponges for uranium adsorption. *Carbohydrate Polymers* 92: 712–718.

Sekiguchi, S., Y. Miura, H. Kaneko, S.I. Nishimura, N. Nishi, M. Iwase, and S. Tokura. 1994. Molecular weight dependency of antimicrobial activity by chitosan oligomers. In *Food Hydrocolloids: Structures, Properties and Functions*, Nishinari, K. and Doi, E. (eds.). New York: Plenum Press.

Shankar, S., J.P. Reddy, J.W. Rhim, and H.Y. Kim. 2015. Preparation, characterization, and antimicrobial activity of chitin nanofibrils reinforced carrageenan nanocomposite films. *Carbohydrate Polymers* 117: 468–475.

Shanmugam, M. and K. Mody. 2000. Heparinoid-active sulfated polysaccharides from marine algae as potential blood anticoagulant agents. *Current Science* 79(12): 1672–1683.

Sheehan, J.P. and E.N. Walke. 2006. Depolymerized holothurian glycosaminoglycan and heparin inhibit the intrinsic tenase complex by a common antithrombin independent mechanism. *Blood* 107: 3876–3882.

Sokolova, E.V., L.N. Bogdanovich, T.B. Ivanova, A.O. Byankina, S.P. Kryzhanovskiy, and I.M. Yermak. 2014. Effect of carrageenan food supplement on patients with cardiovascular disease results in normalization of lipid profile and moderate modulation of immunity system markers. *Pharma Nutrition* 2: 33–37.

Subhapradha, N., P. Ramasamy, P. Seedevi, V. Shanmugam, A. Srinivasan, and A. Shanmugam. 2014. Extraction, characterization and its antioxidant efficacy of polysaccharides from *Sepia aculeata* (Orbigny, 1848) cuttlebone. *African Journal of Biotechnology* 13(1): 138–144.

Sudheesh Kumar, P.T.S., S. Abilash, K. Manzoor, S.V. Nair, H. Tamura, and R. Jayakumar. 2010. Preparation and characterization of novel β-chitin/nano silver composite scaffolds for wound dressing applications. *Carbohydrate Polymers* 80: 761–767.

Sun, H.H., W.J. Mao, J.Y. Jiao et al. 2011. Structural characterization of extracellular polysaccharides produced by the marine fungus *Epicoccum nigrum* JJY-40 and their antioxidant activities. *Marine Biotechnology* 13(5): 1048–1055.

Sun, Y., B. Zhou, S. Xu, W. Li, and B. Yan. 2012. Separation and purification of intracellular and extracellular polysaccharides from *Isochrysis galbana* and their antimicrobial activity. *Food Science* 33(11): 137–141.

Synytsya, A., R. Bleha, A. Synytsya, R. Pohl, K. Hayashi, K. Yoshinaga, T. Nakano, and T. Hayashif. 2014. Mekabu fucoidan: Structural complexity and defensive effects against avian influenza A viruses. *Carbohydrate Polymers* 111: 633–644.

Tareq, A., M. Alam, S. Raza, T. Sarwar, Z. Fardous, A.Z. Chowdhury, and S. Hossain. 2013. Comparative study of antibacterial activity of chitin and chemically treated chitosan prepared from shrimp (*Macrobrachium rosenbergii*) shell waste. *Journal of Virology and Microbiology* 2013: 9.

Thuy, T.T.T., B.M. Ly, T.T.T. Van, N.V. Quang, H.C. Tu, Y. Zheng, C. Seguin-Devaux, B. Mid, and U. Ai. 2015. Anti-HIV activity of fucoidans from three brown seaweed species. *Carbohydrate Polymers* 115: 122–128.

Tikhonov, V.E., E.A. Stepnova, V.G. Babak et al. 2006. Bactericidal and antifungal activities of a low molecular weight chitosan and its *N*-/2(3)-(dodec-2-enyl) succinoyl/-derivatives. *Carbohydrate Polymers* 64: 66–72.

Verdrengh, M., H.E. Harris, and A. Tarkowski. 2000. Role of selections in experimental *Staphylococcus aureus* induced arthritis. *European Journal of Immunology* 30(6):1606–1613.

Vilela-Silva, A.C.E.S., A.P. Alves, A.P. Valente, V.D. Vacquier, and P.A.S. Mourão. 1999. Structure of the sulfated α-L-fucan from the egg jelly coat of the sea urchin *Strongylocentrotus franciscanus*: Patterns of preferential 2-O- and 4-O-sulfation determine sperm cell recognition. *Glycobiology* 9(9): 927–933.

Vinoth Kumar, T., S. Lakshmanasenthil, D. Geetharamani, T. Marudhupandi, G. Suja, and P. Sugany. 2015. Fucoidan—A α-D-glucosidase inhibitor from *Sargassum wightii* with relevance to type 2 diabetes mellitus therapy. *International Journal of Biological Macromolecules* 72: 1044–1047.

Viswanathan, N., C.S. Sundaram, and S. Meenakshi. 2009. Removal of fluoride from aqueous solution using protonated chitosan beads. *Journal of Hazardous Materials* 161: 423–430.

Wang, X., Z.P. Xie, Y. Huang, and Y.B Cheng. 2002. Gelcasting of silicon carbide based on gelation of sodium alginate. *Ceramics International* 28(8): 865–871.

Wang, Y., W. Su, C. Zhang, C. Xue, Y. Chang, X. Wu, Q. Tang, and J. Wang. 2012. Protective effect of sea cucumber (*Acaudina molpadioides*) fucoidan against ethanol-induced gastric damage. *Food Chemistry* 133: 1414–1419.

Wen, Y., J. Ma, J. Chen, C. Shen, H. Li, and W. Liu. 2015. Carbonaceous sulfur-containing chitosan–Fe (III): A novel adsorbent for efficient removal of copper (II) from water. *Chemical Engineering Journal* 259: 372–380.

Wu, M., R. Huang, D. Wena, N. Gao, J. He, Z. Li, and J. Zhao. 2012. Structure and effect of sulfated fucose branches on anticoagulant activity of the fucosylated chondroitin sulfate from sea cucumber *Thelenata ananas*. *Carbohydrate Polymers* 87: 862–868.

Wu, Z., J. Ming, R. Gao et al. 2011. Characterization and antioxidant activity of the complex of tea polyphenols and oat β-glucan. *Journal of Agricultural and Food Chemistry* 59(19): 10737–10746.

Xie, Y., S. Li, F. Wang, and G. Liu. 2010. Removal of perchlorate from aqueous solution using protonated cross-linked chitosan. *Chemical Engineering Journal* 156: 56–63.

Xing, R., H. Yu, S. Liu, W. Zhang, Q. Zhang, and Z. Li. 2005. Antioxidative activity of differently region selective chitosan sulfates in vitro. *Bioorganic and Medicinal Chemistry* 13: 1387–1392.

Xu, H., T. Guo, Y.F. Guo, J. Zhang, Y. Li, and W. Feng. 2008. Characterisation and protection on acute liver injury of a polysaccharide MP-I from *Mytilus coruscus*. *Glycobiology* 18(1): 97–103.

Yim, J.H., S.J. Kim, S.H. Ahn, C.K. Lee, K.T. Rhie, and H.K. Lee. 2004. Antiviral effects of sulfated exopolysaccharide from the marine microalga *Gyrodinium impudicum* strain KG03. *Marine Biotechnology* 6: 17–25.

Yingyinga, S., W. Hui, G. Ganlin, P. Yinfang, and Y. Binlun. 2014. The isolation and antioxidant activity of polysaccharides from the marine microalgae *Isochrysis galbana*. *Carbohydrate Polymers* 113: 22–31.

Yoshida, K., Y. Minami, H. Nemoto, K. Numata, and E. Yamanaka. 1992. Structure of DHG, a depolymerized glycosaminoglycan from sea cucumber *Stichopus japonicus*. *Tetrahedron Letters* 33: 4959–4962.

Yu, C.C., H.W. Chen, M.J. Chen, Y.C. Chang, S.C. Chien, and Y.H. Kuo. 2010. Chemical composition and bioactivities of the marine alga *Isochrysis galbana* from Taiwan. *Natural Product Communications* 5: 1941–1944.

Zhang, L., W. Liu, B. Han, J. Sun, and D. Wang. 2008. Isolation and characterization of antitumor polysaccharides from the marine mollusk *Ruditapes philippinarum*. *European Food Research and Technology* 227: 103–110.

Zhang, Y., S. Song, H. Liang, Y. Wang, W. Wang, and A. Ji. 2010. Enhancing effect of a sea cucumber *Stichopus japonicus* sulfated polysaccharide on neurosphere formation in vitro. *Journal of Bioscience and Bioengineering* 110(4): 479–486.

# Bioinformatics of glycobiology

*chapter thirty*

# Glycans predictive modeling using modern algorithms

*Imran Ahmad and Athapol Noomhorm*

## Contents

## 30.1 Introduction

The increased interest in carbohydrates in drug development and their therapeutic uses has led to the rapid evolution of glycomics, resulting in large sets of data that need to be analyzed and possibly used for the prediction of their structure, biopolymer interaction, and data visualization in general.

The science of bioinformatics related to glycans is constantly evolving. Unlike nucleotide and protein sequences, glycan and lipid structures have not received comparable attention of researchers working on bioinformatics. Glycans are generally tree structures, which are more complex than linear DNA or protein sequences, because of their complex nature.

Glycans are abundant in nature, and a higher degree of diverse biopolymer structures require sophisticated algorithms to classify them into two or more classes. The main research areas of interest are, therefore, glycosylation (prediction of binding sites with protein), biomarker prediction (disease-related), and structural characterization.

Although various aspects of glycan informatics research have been covered in earlier publications (von der Lieth et al. 2009), this chapter attempts to combine these aspects by outlining new advances in data science and algorithms.

In this chapter, recent advances in data-driven glycan bioinformatics are given, which are related to pattern recognition. While a general introduction is given to data-driven prediction methods for glycosylation and structural characterization, including classification trees, artificial neural networks (ANNs), and support vector machines (SVMs), it is by no mean an exhaustive discussion on general artificial intelligence methods that are available in the literature. One the most important issues concerning prediction procedures is the performance evaluation of the developed schemes and cross-validation, which have been adequately discussed.

## 30.2 Prerequisite for glycan bioinformatics

General information on glycan bioinformatics is covered elsewhere in this book; however, we intend to provide a brief note on how experimental data is obtained and used in predictive modeling, as well as compatibility of analytical methods with the algorithms. Care must be taken in utilizing analytical data stored in databases, as one may find variations in data from different sources despite having been obtained using the same experimental procedure. Such variations are common due to the variation in reagent purity, detection method, and, above all, sample size of biological materials. It is suggested that proper experimental design procedures be followed and comparisons with other studies made (von der Lieth et al. 2009).

Glycan structure profiles from various biological samples can be obtained by mass spectrometry (MS), high-performance liquid chromatography (HPLC), and nuclear magnetic resonance (NMR).

### 30.2.1   Mass spectrometry

Traditionally, carbohydrate structures are determined by using mass spectral data, in which each mass peak is manually marked to develop one mass spectrum. This is understandably a tedious task, but a database of theoretical mass spectra helps in finding new structures having similar spectra. A more advanced form of MS is MS/MS or tandem MS, which detects the molecules' structural information by measuring their weight (mass) by fragmenting them into ions in multiple rounds of spectrometry. Tang et al. (2005b) have reported an automated interpretation of MS/MS spectra of oligosaccharides. A commercial tool called Thermo Novel Bioinformatics is available for the interpretation of glycan mass spectra with metal adducts and multiple adduct combinations by Waters® (Saba et al. 2012).

### 30.2.2   High-performance liquid chromatography

To ensure integrity of glycan analysis and reduce uncertainty, MS is often used with HPLC simultaneously. Reliable results were obtained with structural analysis of N-glycans released from glycoproteins in sodium dodecyl sulfate-polyacrylamide gel electrophoresis (SDS-PAGE) gel bands using HPLC (Royle et al. 2006). HPLC techniques are used for mapping oligosaccharides and have been demonstrated for the separation and detection of mixtures of N- and O-glycans (Royle 2006). The availability of internal standards makes HPLC a very handy tool for fast identification of the eluted peaks and subsequent analysis.

### 30.2.3   Nuclear magnetic resonance (NMR/¹³C NMR)

NMR spectroscopy is used to study the interaction of electromagnetic radiation with molecules, which allows interpreting the physical, chemical, and biological properties of the glycan molecule under investigation. The main areas of NMR-based structural determination of carbohydrate structures are those associated with microorganisms (Feng et al. 2005) and polysaccharides such as those found in marine products (Ahrazem et al. 2005). Typically, $^{13}C$ NMR chemical shifts of carbohydrate ring carbons (60–110 ppm) have been used (Mathias 2006).

### 30.3   Databases

The major databases for complex carbohydrates are Glycosciences, KEGG GLYCAN, and the database

**Table 30.1** Summary of tools and utilities available for glycan bioinformatics study

| | | |
|---|---|---|
| GlycoMod | Prediction of glycan structure from mass spectra | Cooper et al. (2001) |
| GLYCH | Prediction of glycan structure from mass spectra | Tang et al. (2005a) |
| GlycoPep ID | Prediction of glycan structure from mass spectra | Irungu et al. (2007) |
| GlycanBuilder | Building and displaying glycan structures | Ceroni et al. (2007) |
| DrawRINGS | Glycan structure drawing tool | Akune et al. (2010) |
| Glycan Kernel Tool | Kernel-based classification | Jiang et al. (2007) |

developed by the Consortium for Functional Glycomics (CFG) linked to the CarbBank database. The information on databases has been documented in various publications (Doubet and Albersheim 1992) and need not be repeated here. However, as rightly pointed out by the authors of *Bioinformatics for Glycobiology and Glycomics: An Introduction* (Lieth et al. 2010), the absence of a universal system of storing new developments and retrieving this information would make the database description obsolete.

Efforts to establish a centralized database linked to other molecular databases are well under way. This development will allow making full use of the existing body of knowledge, similar to the Protein Data Bank and GenBank.

Some of the handy tools and utilities freely available on the Internet are listed in Table 30.1, which readers may find useful for glycan bioinformatics study.

## 30.4   Classical mathematical modeling

Glycosylation, as the foremost area of interest, has been statistically analyzed for amino acids surrounding glycosylation-binding sites based on protein sequences (Lutteke et al. 2005). Krambeck and Betenbaugh (2005) and Umana and Bailey (1997) approached glycosylation prediction problem using nonlinear algebra models. The developed models of N-linked glycosylation are substrate specific and predicted possible synthesis of glycans.

*Table 30.2* Innovative modeling techniques used in bioinformatics as applied to glycans

| Application | Problem | References |
|---|---|---|
| Glycosylation (GlySeq) | Prediction of protein binding sites using statistics | The German Cancer Research Center |
| Glycosylation (GlyVicinit) | Distance of amino acids in the spatial vicinity of carbohydrate residues | Lutteke et al. (2005) |
| Phylogenetic tree estimation | Tree structure | Csürös (2002); |
| Glycan structural studies | Structure matching | Aoki et al. (2003); |
| Prediction | Stochastic modeling | Knudsen and Hein (1999) |
| Probabilistic models such as (hidden) Markov models | Biological sequence analysis | Durbin et al. (1998) |
| Bayesian networks | Classification | Friedman (1998) |
| Stochastic context-free grammars (SCFGs) | Prediction (transfer RNA modeling) | Sakakibara et al. (1994) |

Table 30.2 lists some of the significant studies based on innovative techniques using modern algorithms, which will shape the future of the study of glycans.

## 30.5  Data mining approach

Glycan structure mining was used for capturing significant hidden tree structures, where binding events generate signaling processes due to specific proteins, resulting in unconnected tree structures. Because of this difficulty, models discussed before are unable to determine such structures.

Hashimoto et al. (2008) developed an efficient method for mining motifs or significant subtrees from glycans. They proposed a new concept called "α-closed frequent subtrees" and an efficient method for mining all these subtrees from given trees using a statistical hypothesis testing procedure to reorganize the frequent subtrees in the order of their significance.

Moreover, the general classification tree algorithms are limited to forming tree structure in which the parent branch is connected through nodes with the siblings and offspring. However, in order to explain the relationships between nodes and trees, as is the case for glycans, one has to depend on glycan biological structures.

The classification tree algorithms have not been fully explored for explaining such relationships.

Diligenti et al. (2003) reported the application of hidden tree Markov models for image classification to show the interdependence between the parent and offspring. Furthermore, Aoki et al. (2004) applied a new probabilistic model, called probabilistic sibling-dependent tree Markov model (PSTMM), which is able to inherently capture such complex patterns of glycans by including siblings. The PSTMM approach can be utilized to align glycan structures, and it offers deeper insight into glycan functions.

## 30.6  Artificial neural networks

ANNs are capable of arbitrarily classifying complex motifs containing correlations between different positions in the sequence. A typical ANN contains two-layered, feed-forward neural networks. Traditionally, ANNs are trained with backpropagation; however, due to concerns of overfitting and low generalization ability, other training procedures are also employed, such as the Levenberg–Marquard (LM) algorithm. The selection of an appropriate NN topology to predict the glycosylation sites is important in terms of model accuracy and model simplicity. The architecture of an NN greatly influences its performance. Many algorithms for finding the optimal NN structure are derived based on specific data in a specific area of application, but predicting the optimal NN topology is a difficult task since choosing the neural architecture requires expertise and substantial biochemistry knowledge. The network operating parameter optimization schemes are, but not limited to, random-centroid optimization (RCO) (Nakai 1990), genetic algorithm (GA) (Goldberg 1989), and trial and error with some previous experience of network optimization. The parameters in question are the step size, weight change momentum (1, –1), learning rate, and the number of neurons in the hidden layer. In addition, selection of activation function depends upon the geometrical trend present in the data. For nonlinear arrangements, a hyperbolic tangent activation function has been recommended ((Julenius et al. 2007).

In a glycosylation problem (e.g., to predict glycosylation sites based on the amino acid sequence), a fairly large number of input variables (predictors) would include both protein binding and nonbinding sites to be represented by binary (1, 0) classification, respectively. The dataset must be randomized to the maximum possible ways to avoid accidental bias during learning.

The higher the number of patterns on which a network is trained, a more generalized the model obtained. Unfortunately, however, experimentally obtained sequence data is scarce especially in the case of glycans. Therefore, a simpler network with fewer layers and fewer neurons in the hidden (middle) layers of the network is recommended (Julenius et al. 2007).

ANN, however, has a drawback in terms of its inability to explain the biological relevance of the output results with the features it correlates despite having very good post-modeling performance indications in the test and cross-validation sets.

Despite the fact that biological processes and ANNs are in harmony, their application in biological systems has not been fully explored. One of the major reasons, as described earlier, is that the technique follows a "black box" approach and that there is lack of interpretability of the network weights obtained during the model-building process. The set of numbers obtained in a training ANN session is hard to be related to the application in a meaningful manner. A few researchers have addressed the issue of interpretation of trained network results in terms of connecting weights and biases (Duh et al. 1998; Olden and Jackson 2002). The approach here is to establish the statistical significance of the contributions in terms of individual connection weights and overall influence of each of the input variables. To establish the relationship between the input variables and the network response, sensitivity analysis of the validation set is carried out by setting all predictor values to their mean, and then obtaining the network prediction. The process is repeated by setting each predictor consecutively to its minimum and maximum value by comparing the network response of the mean, minimum, and maximum levels of predictors, and exploring an indicative effect of the input variables. Finally, the product sums of the connecting weights are used to determine the significance of the input variables. In order to explain the contribution of each variable for the prediction of a dependent variable, the relative importance of the input variables is determined by the product sum of the connecting weights of the network, as suggested by Olden and Jackson (2002). It was demonstrated that, by using Garson's algorithm (Garson 1991) for partitioning and quantifying the NN connection weights, the relationship with predictor-dependent variables can be explained; the larger the sum for a given input node, the greater the importance of the corresponding input variable (Ahmad and Jeenanunta 2014).

NeuroCarb utilizes NMR and ANN for the structure interpretation of glycans (Matthias 2006). In this work, various spectral data (MS, HPLC, and NMR) were evaluated for rapid and efficient determination of sugars. It was concluded that ANNs should be trained with $^{13}C$ NMR spectra. The author first normalized entire dataset, and then used backpropagation network layouts and training parameters with Kohonen networks for training the network. A fairly large dataset (1000 oligosaccharides and ~2500 monosaccharide) yielded satisfactory results in recognizing the type of sugar.

## 30.7   Support vector machines

Among the other high-level machine learning algorithms, SVMs are increasingly used by researchers for biological classification including gene expression microarrays. SVMs have shown powerful classification ability with excellent performance in many practical classification problems (Vapnik 1998; Ahmad et al. 2013).

The theoretical advantage of SVMs is that the idea of margin or stability mitigates the problem of overfitting the training data, which is one of the inherent drawbacks of ANNs. SVM, on the other hand, works on the principle of learning a classification boundary separating positive and negative examples. Unlike the ANNs, which are regarded as "black boxes" due to their inability to express the underlying mathematical basis, SVMs are based on a strong mathematical background with a clearly apprehendable geometric insight. Readers are encouraged to use SVMs and then give a more general mathematical formulation.

The application of SVM for classifying microarray data has been presented in detail by Mukherjee (1999). The theoretical background, in terms of geometric and algorithmic, of SVMs as applicable to biological data classification, gene selection, cancer morphology, treatment outcome, and multiclass classification has been described by Mukherjee (2000), which is a good source for beginners. Mukherjee (1999) have reported that SVM performed very well even without feature selection using all 7,000–16,000 expression values in DNA microarrays.

Glycoproteins are increasingly required to be analyzed and interpreted due to their role in diseases including cancer (Yamanishi et al. 2007). SVM appeared to be a strong contender for glycan biomarker predictions. Kernel methods are now well known in the classification of large datasets utilizing features as vectors to extract the useful classes present. For the successful application of kernel-based classification, the most important task is "feature selection," which means the characteristics of the dataset that may have higher influence potential, which may be two or more. Other prerequisites are data clean-up, normalization, scaling, and selection of the right type of kernel. There are other operating parameters and evaluation criteria that need to be adjusted.

Application of kernels in glycomics involves labeling a rooted tree as a function of vertices (monosaccharides) and edges (sugar bounds) of the tree (Yamanishi et al. 2007). Selecting the best feature that would generalize into target classes is an arduous task. Researchers have attempted to find an optimized feature selection method, for example, as reported by Golub et al. (1999) and subsequently employed by Yamanishi et al. (2007), in which a ranking procedure based on the features' (both positives and negatives) statistical value (standard deviation) was used.

A new class of tree kernels to measure the similarity between glycans based on the comparison of tree substructures was reported by Hashimoto et al. (2008). They used glycan features such as the sugar type and the sugar bound type. The proposed scheme was independently verified by classifying human glycans into four blood components: leukemia cells, erythrocytes, plasma, and serum. Hashimoto et al. (2008) further examined the top 10 subtrees obtained by the proposed method and found that all subtrees were significant motifs in glycobiology.

Despite having the added advantage of machine learning approaches such as SVMs, the results are largely regarded as noncomprehensible in nature. Hashimoto et al. (2008) have suggested that developing computational methods that can find biologically understandable patterns from the collected data would be more beneficial.

## 30.8 Conclusion

Glycans are regarded as biological materials with high information density per fundamental unit, which have structural diversity, heterogeneity, and complex non-template biosynthesis. Because of these characteristics, glycans remain a challenging part of molecular biology. With recent advances, much work has been reported on the use of modern algorithms, which has made possible the characterization molecular information of complex biological systems. Machine learning algorithms are applied to glycan bioinformatics in the structural analysis, biomarkers prediction, and functionality of carbohydrates.

## References

Ahmad, I. and Jeenanunta, C. (2014). Suitability of classification algorithms for predictive modeling of quality parameters of frozen shrimps (*Litopenaeus vannamei*) through sensor based time-temperature monitoring. *Food and Bioprocess Technology* 8(1): 134–147.

Ahmad, I., Jeenanunta, C., Chanvarasuth, P., and Komolavanij, S. (2013). Prediction of physical quality parameters of frozen shrimp (*Litopenaeus vannamei*) through an artificial neural networks and genetic algorithm approach. *Food and Bioprocess Technology*, 7(5): 1433–1444.

Ahrazem, O., Prieto, A., Giménez-Abián, M., Leal, J., Jiménez-Barbero, J., and Bernabé, M. (2005). Structural elucidation of fungal polysaccharides isolated from the cell wall of *Plectosphaerella cucumerina* and *Verticillium* spp. *Carbohydrate Research*, 341: 246–252.

Akune, Y., Hosoda, M., Kaiya, S., Shinmachi, D., and Aoki-Kinoshita, K.F. (2010). The RINGS resource for glycome informatics analysis and data mining on the Web. *OMICS*, 14(4): 475–486.

Aoki, K.F. et al. (2003). Efficient tree-matching methods for accurate carbohydrate database queries. *Genome Informatics*, 14: 134–143.

Aoki, K.F., Ueda, N., Yamaguchi, A., Kanehisa, M., Akutsu, T., and Mamitsuka, H. (2004). Application of a new probabilistic model for recognizing complex patterns in glycans. *Bioinformatics*, 20(1): i6–i14.

Ceroni, A., Dell, A., and Haslam, S.M. (August 7, 2007). The GlycanBuilder: A fast, intuitive and flexible software tool for building and displaying glycan structures. *Source Code for Biology and Medicine*, 2: 3.

Cooper, C.A., Gasteiger, E., and Packer, N.H. (2001). GlycoMod: A software tool for determining glycosylation compositions from mass spectrometric data. *Proteomics*, 1: 340–349.

Csürös, M. (2002). Fast recovery of evolutionary trees with thousands of nodes. *Journal of Computational Biology*, 9: 277–297.

Diligenti, M., Frasconi, P., and Gori, M. (2003). Hidden tree Markov models for document image classification. *Transactions on Pattern Analysis Machine Intelligence* 25: 519–523.

Doubet, S. and Albersheim, P. (1992). CarbBank. *Glycobiology* 2: 505.

Duh, M.S., Walker, A.M., and Ayanian, J.Z. (1988). Epidemiologic interpretation of artificial neural networks. *American Journal of Epidemiology*, 147: 1112–1122.

Durbin, R., Eddy, S., Krogh, A., and Mitchison, G. (1998). *Biological Sequence Analysis*. Cambridge, U.K.: Cambridge University Press.

Feng, L., Senchenkova, S., Wang, W., Shashkov, A., Liu, B., Shevelev, S., Liu, D., Knirel, Y., and Wang, L. (2005). Structural and genetic characterization of the *Shigella boydii* type 18 O antigen. *Gene*, 355: 79–86.

Friedman, N. (1998). The Bayesian structural EM algorithm. In *Proceedings of UAI-98*. San Francisco, CA: Morgan Kaufmann Publishers, pp. 129–138.

Garson, G.D. (1991). Interpreting neural-network connection weights. *Artificial Intelligence Expert*, 6: 47–51.

Goldberg, D.E. (1989). *Genetic Algorithm Search, Optimization and Machine Learning*. New York: Addison-Wesley.

Golub, T.R. et al. (1999). Molecular classification of cancer: Class discovery and class prediction by gene expression monitoring. *Science*, 286: 531–537.

Hashimoto, K., Takigawa, I., Shiga, M., Kanehisa, M., and Mamitsuka, H. (2008). Mining significant tree patterns in carbohydrate sugar chains. *Bioinformatics*, 24 ECCB: i167–i173.

Irungu, J., Go, E.P., Dalpathado, D.S., and Desaire, H. (2007). Simplification of mass spectral analysis of acidic glycopeptides using GlycoPep ID. *Analytical Chemistry*, 79: 3065–3074.

Jiang, H., Aoki-Kinoshita, K.F., and Ching, W.K. (2007). Extracting glycan motifs using a biochemically weighted kernel. *Bioinformation*, 7(8): 405–412.

Julenius, K., Johansen, M.B., Zhang, Y., Brunak, S., and Gupta, R. (2007). Prediction of glycosylation sites in proteins. In *Bioinformatics for Glycobiology and Glycomics: An Introduction*, C.-W. von der Lieth, T. Lutteke, and M. Frank (Eds.). Wiley-Blackwell, New York.

Knudsen, B. and Hein, J. (1999). RNA secondary structure prediction using stochastic context-free grammars and evolutionary history. *Bioinformatics*, 15: 446–454.

Krambeck, F.J., and Betenbaugh, M.J. (2005). A mathematical model of N-linked glycosylation. *Biotechnology and Bioengineering*, 92: 711–728.

Lieth, C.-W. von der, T. Lutteke, and M. Frank (Eds.). (2010). Bioinformatics for glycobiology and glycomics: An introduction. Wiley-Blackwell, Chichester, U.K.

Lutteke, T., Frank, M., and von der Lieth, C.W. (2005). Carbohydrate Structure Suite (CSS): analysis of carbohydrate 3-D structures derived from the PDB. *Nucleic Acids Research*, 33: D242–D246.

Matthias, S.-I. (2006). NeuroCarb: Artificial neural networks for NMR structure elucidation of oligosaccharides. PhD thesis, Faculty of Science, University of Basel, Basel, Switzerland.

Mukherjee, S. (1999). *Classifying Microarray Data Using Support Vector Machines Classifying Microarray Data Using Support Vector Machines*. MIT Press, Cambridge, MA.

Mukherjee S., Tamayo P., Slonim D., Verri A., Golub T., Mesirov J.P., and Poggio T. (2000). Support vector machine classification of microarray data. Technical Report, Artificial Intelligence Laboratory, Massachusetts Institute of Technology, Cambridge, MA.

Nakai, S. (1990). Computer-aided optimization with potential application in biorheology. *Journal of Japanese Biorheology Society*, 4: 143.

Olden, J.D. and Jackson, D.A. (2002). Illuminating the "black box": A randomization approach for understanding variable contributions in artificial neural networks. *Ecological Modeling*, 154: 135–150.

Royle, L., Radcliffe, C., Dwek, R., and Rudd, P. (2006). Detailed structural analysis of N-glycans released from glycoproteins in SDS-PAGE gel bands using HPLC combined with exoglycosidase array digestions. *Methods in Molecular Biology*, 347: 125–143.

Saba, J.U., Meitei, N.S., and A. Apte (2012). Novel bioinformatics tool: Interpretation of glycan mass spectra with metal adducts and multiple adduct combinations. San Jose, CA: Thermo Fisher Scientific.

Sakakibara, Y., Brown, M., Hughey, R., Mian, I.S., Sjolander, K., Underwood, R.C., and Haussler, D. (1994). Stochastic context-free grammars for tRNA modeling. *Nucleic Acids Research*, 22: 5112–5120.

Tang, H. et al. (2005a). Reported an automated interpretation of MS/MS spectra of oligosaccharides. *Bioinformatics*, 21: i431–i439.

Tang, H., Mechref, Y., and Novotny, M. (2005b). Automatic interpretation of MS/MS spectra of oligosaccharides. *Bioinformatics*, 21(Suppl. 1): i431–i439.

Umana, P. and Bailey, J.E. (1997). A mathematical model of N-linked glycoform biosynthesis. *Biotechnology Bioengineering*, 55: 890–908.

Vapnik V.N. (1998). Statistical Learning Theory. John Wiley & Sons, New York.

von der Lieth, C.-W. (2009). Chapter 10: Experimental methods for the analysis of glycans and their bioinformatics requirements. In *Bioinformatics for Glycobiology and Glycomics: An Introduction*, C.-W. von der Lieth, T. Lutteke, and M. Frank (Eds.). Chichester, U.K.: Wiley-Blackwell.

Yamanishi, Y. et al. (2007). Glycan classification with tree kernels. *Bioinformatics*, 23: 1211–1216.

# *Biological role of glycoconjugates*

*chapter thirty-one*

# Glyco-conjugated bioactive compounds derived from brown algae and its biological applications

*Bakrudeen Ali Ahmed, Abdul Shirin Alijani, and Reza Farzinebrahimi*

## Contents

## 31.1 Introduction

Brown algae constitute a rich source of biologically active polysaccharides. Sulfated fucans are one of those polysaccharides isolated from brown algae ~90 years ago. These polysaccharides, which mainly consist of sulfated L-fucose, is extracted from the cell walls of brown algae (Killing, 1913) and constitute 40% of dry weight of the cell wall (Kloareg, 1984). In this chapter, we investigate the biomedical effects of these polysaccharides as glyco-conjugated bioactive compounds and their biological applications. Some aquatic animals such as shrimps, juvenile crabs, fish, and phytoplanktones, especially diatoms, nourish and build their shelters from materials provided by these seaweeds, such as laminaran, a β(1 → 3)glucan with β(1 → 6) branches, which is generated by photosynthesis (Beattle et al., 1961). It means that brown algae are able to maintain the condition of their environment. Some bioactive primary and secondary metabolites such as terpenoids, phlorotanins, fucoidans, sterols, and glycolipids, as well as iodine, minerals, fats, and sometimes vitamins, have also been found in brown algae (Iwamoto et al., 1999, 2001). Numerous novel compounds have been extracted during the past four decades that play important roles in biological activities. The extracts or isolated pure components from seaweeds possess a wide range of pharmacological properties such as anticancer, antibacterial,

antifungal, antiviral, anti-inflammatory, anticoagulant, antioxidant, hypoglycemic, hypolipidemic, antimelanogenic, anti-boneloss, hepatoprotective, and neuroprotective (Faulkner, 1999, 2000, 2001, 2002). Phaeophyta is easy to find mostly in the coastal areas, preferably in cold water bodies, because this area is its natural habitat. Kelps, macrocystis, nereocystis, and sargassum are the most common brown seaweeds whose height can be up to 35 m (100 ft) (van den Hoek et al., 1995). To produce algin, carrageenan, and agar, dried kelp or brown seaweed is used. These products are extensively used in ice creams, as suspended antibiotics in solution, as pigments in paints, in canning meat or fish, and as glue (Dhargalkar and Pereira, 2005).

### 31.1.1 Mechanism of action

Glyco-conjugates are generated when carbohydrates covalently join with other chemical materials to produce the simplest chemical modifications of biomolecules such as proteins, peptides, lipids, and saccharides. This happens in a process named "glycosylation" to regulate the biological activities and is recognizable by microarray techniques. These molecular and supermolecular scaffolds are very significant bioactive compounds. They are of different types, including glycoproteins, glycopeptides, peptidoglycans, glycolipids, glycosides, and lipopolysaccharides. They are involved in cell–matrix interactions including cell–cell interactions, detoxification processes, and cell–cell recognitions. They also have an important role in the evolution of vaccines against cancer, viral and bacterial infections, and many other diseases (Kwak et al., 2013).

### 31.1.2  Marine resources

All organisms that live in a marine environment have the potential to generate sources of secondary metabolites that might display useful roles in the production new pharmaceutical agents (Iwamoto et al., 1998, 1999, 2001). Algae are a very simple heterogeneous group of marine organisms that contain chlorophyll in their structure (Bold and Wynne, 1985) and can be found in the form of unicellular, as a group in colonies, or as a multicellular organisms, or even some cells combining together as a simple tissue (Hillison, 1977). Two main types of algae can be described: macroalgae and microalgae. Seaweeds or macroalgae are mostly found in coastal areas and included in them are three groups: green algae, brown algae, and red algae. In contrast, microalgae occupy both benthic and littoral habitats and throughout of the ocean waters as phytoplanktons (Garson, 1989). The supremacy of the xanthophyll pigments and fucoxanthin gives rise to the brown color of this kind of algae, as well as the coating of other pigments like chlorophyll a and c, β carotenes, and other xanthophylls (Bold and Wynne, 1985). The cell wall of brown algae is made of cellulose and alginic acid, which bind the food reserve laminaran. There are three classes of compounds in brown algae that have important biological functions and advantageous in industrial production: (1) polysaccharides, (2) phenolic compounds, and (3) terpenes (Blunt et al., 2006, 2007).

## 31.2  Polysaccharides

The cell wall of brown algae is composed of an inner cellulose layer and an outer gelatinous coating like microfibril sections. These microfibrils are placed in a formless matrix of acid polysaccharides (alginic acid and sulfated fucoidan) linked to each other by proteins (Kloareg et al., 1986). Alginic acid is composed of two linear polysaccharides, 1,4-linked β-D-mannuronic (M) and α-L-guluronic acid (G) residues, which are ordered in a anticlockwise arrangement along the chain (Andrade et al., 2004). Poly β-(1-4)-linked D-mannuronate forms a threefold left-handed helix with (weak) intramolecular hydrogen bonding, whereas poly α-(1-4)-linked L-guluronate forms stiffer (and more acid-stable) twofold twisted helical chains with two kinds of intramolecular hydrogen bonding. Polyguluronate has specific binding sites for calcium, consisting of five oxygen ligands. The

alginate-containing cell wall of the algae attach together and provide certain mechanical character to the algae (Caroline et al., 2014). The mechanical character of alginate depends on amount of three substances (M, G, and GMGM blocks) in the form of a copolymer (Kloareg and Quatrano, 1988; Indegaard and Minsaas, 1991). Generally, alginic acid with the low M/G ratio generates strong and rigid gels with good heat stability in contrast to the high proportion of M/G; in other words, a small number of guluronic acids produce soft and elastic gels with good freeze–thaw characteristics. The high GMGM content zips with $Ca^{2+}$ ions to decrease shearing. The other soluble and branched oligosaccharides are called fucans. They have repeating units and consist of a large range of sugars, such as L-fucose and D-galactose, and are distinguished by the specific pattern of sulfation among different species (Berteau and Mulloy, 2003). Fucan structure consists of linear and branched forms, and each one shows a different function. Fucoidans play a role in the cell wall structure and is involved in connecting alginate and cellulose (Kloareg and Quatrano, 1988; Bisgrove and Kropf, 2001). It increases the immune response, suppresses angiogenesis, reduces adhesion of tumor cells to human platelets, and activates the intrinsic and extrinsic pathways of apoptosis.

Polysaccharides extracted from brown seaweeds have shown several inhibitory effects on tumor cell growth. Fucoidan extracted from *Ascophyllum nodosum* exerts *in vitro* a reversible anti-proliferative activity on a cell line derived from a non-small-cell human bronchopulmonary carcinoma (NSCLC-N6) with a block in the G1 phase of the cell cycle. Peruses accomplished with NSCLC-bearing nude mice show antitumor activity at subtoxic doses. These primary results demonstrate that HF (fucoidan extract) exerts an inhibitory effect both *in vitro* and *in vivo* and is a very powerful antitumor agent in cancer therapy (Riou et al., 1996). The antitumor activity of fucoidans isolated from brown algae *Saccharina cichorioides*, *Fucus evanescens*, and *Undaria pinnatifida* significantly suppressed the proliferation of human cancer cells. The highly sulfated $(1 \rightarrow 3)$-α-L-fucan from *S. cichorioides* was found effective against colon cancer, and the sulfated and acetylated $(1 \rightarrow 3)$, $(1 \rightarrow 4)$-α-L-fucan suppressed the development of melanoma (Olesya et al., 2013).

## 31.3  Phenolic compounds

Brown algae habitats in tropical regions have low levels of polyphenolic compounds and are consumed by herbivorous fish. In contrast, brown algae grown in temperate regions often produce large quantities of phenolic compounds. These compounds make them distasteful to herbivorous gastropods and sea urchins. Phlorotannins, which are aromatic phenolic compounds in brown algae, are specific compounds that show natural anti-toxicity and are used in the treatment and prevention of cancer as well as in inflammatory, cardiovascular, and neurodegenerative diseases. Phenolic compounds were extracted in high levels of up to nearly 20% DW (dry weight) in fucales and 30% DW in dictyotales (Ragan and Glombitza, 1986; Targett et al., 1995; Amsler and Fairhead, 2006). This class of molecules is divided into phloroglucinols (mono-, di-, tri-, tetra-, and oligomeric) and phlorotanins (Singh and Bharate, 2006). They are mostly found in red and brown algae, but halogenated monomeric phenolics are also found in their chemical structure. Phenolic compounds have a considerable chemical structure, consisting of always one or several aromatic cycle(s) and one or several hydroxyl radicals (Singh and Bharate, 2006). Small phenolic molecules are water soluble, while some of polymeric forms are lipid soluble. In addition, brown algae can produce these compounds, of which the most important are phlorotannins, which are derived from phloroglucinol in high molecular weight (Singh and Bharate, 2006). Six types of phlorotannins produced by brown algae are fuhalos, phlorethols, fuculs, fucophlorethols, eckols, and carmalols.

The total amount of phenolic compounds in the brown seaweed *Sargassum muticum* has been evaluated to determine its cytotoxic effect against a breast cancer cell line and its anti-angiogenic activity in the chorioallantoic membrane (CAM). These works have indicated that the polyphenol, which is rich in the seaweed *S. muticum*, has a significant effect on the induction of apoptosis as well as antioxidant and anti-angiogenesis effects. The phenolic compounds in brown algae can act as antioxidants to treat serious human diseases including melanoma, cardiac disorders, diabetes mellitus, as well as inflammatory and neurodegenerative diseases. Free-radical-induced oxidation is one of the major reasons for the deterioration of nutritional quality. *Turbinaria* spp. were found to have antioxidant and anti-inflammatory activities, being a source of phenolic compounds (Farideh et al., 2013).

## 31.4  Terpene compounds

In brown algae, terpenes are known in two main orders: fucales contain linear diterpenes, and dictyotales contain cyclic diterpenes and sesquiterpenes (both are large orders in the class Phaeophyceae). Carotenoids are tetraterpenoids with a characteristic linear C40 molecular backbone containing up to 11 conjugated double bonds, and are produced in photosynthetic organisms. They form a group of natural pigments with more than 600 members, and possess a variety of biological activities including radical scavenging and immunomodulation, and as potential chemotherapeutic (or)

chemo-preventive agents, singlet oxygen-quenchers, and other pharmacological agents. Carotenoids include two main subclasses: nonpolar hydrocarbon carotenes and polar compounds called "xanthophylls" (Koji and Masashi, 2013). One well-known example of a xanthophyll with anticancer activity is fucoxanthin. For example, fucoxanthin isolated from *U. pinnatifida* has shown remarkable inhibitory effect against human leukemic HL-60 and three lines of human prostate cancer cells (Soheil et al., 2014).

Inflammation is part of a collection of biological responses of the vascular system to the presence of foreign pathogens such as parasites, bacteria, viruses, and damaged cells. Chronic inflammation can result in various pathological conditions by activated immune cells, such as macrophages that secrete large amounts of pro-inflammatory cytokines, nitric oxide (NO), and prostaglandin E2 (PGE2). High levels of NO and PGE2 can result in various pathological conditions. *S. muticum* is one of the brown algae containing apo-9′-fucoxanthinone, which has effects on inflammatory diseases. In fact, apo-9′-fucoxanthinone inhibits the production of NO and PGE2 and expression of inducible nitric oxide synthase (iNOS) and cyclooxygenase (COX)-2 in LPS-stimulated macrophages (Young et al., 2013).

## 31.5  Bioactive compounds' pharmacological activity

### 31.5.1  Cytotoxic and antitumor activity

Cancer cells of diverse origins exhibit dramatic consumption of glucose and high rates of aerobic glycolysis. Tumors are the result an overactive metabolism, and cancer cells need both energy and bioactive intermediates to grow. One strategy for destroying tumor cells is glycol-conjugation, the linking of a drug to glucose or other sugars on the tumor cells. Glyco-conjugates bioactive compounds are attached to the tumor cells and enter them by receptor-mediated endocytosis and traffic at the receptors of glucose and endosomes where degradative enzymes are accumulated, thereby increasing the acidity, which leads to further change in the cell situation and damaging the tumor cells. In 1999, Guardia and colleagues isolated a linear cytotoxic diterpene from the brown alga *Bifurcaria bifucata*, known as bifurcadiol, which showed toxicity against cell culture containing the human tumor cell lines A549, SK-OV-3, SKL-2, XF 498, and HCT (Figure 31.1) (Guardia et al., 1999). Meroterpenoids (Figure 31.2), sargol I (Figure 31.3), and sargol II (Figure 31.4) were extracted from the brown alga *Sargassum tortile*, which exhibited cytotoxic activity (Numata et al., 1991).

Leptosin or methyl syringate 4-*O*-β-D-gentiobiose is a glycosidic compound found in manuka honey

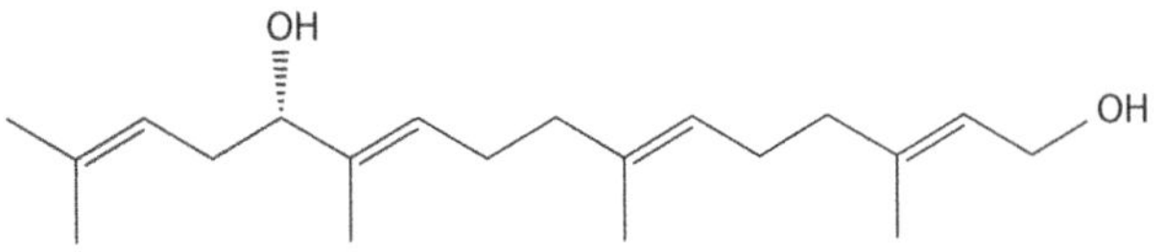

**Figure 31.1** Biforcadiol.

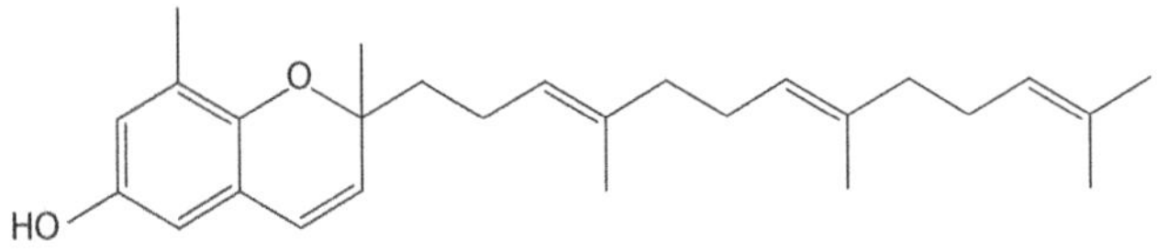

**Figure 31.2** Meroterpenoids.

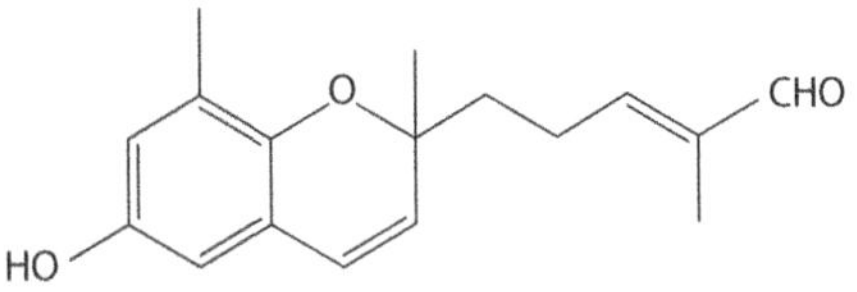

**Figure 31.3** Sargol I.

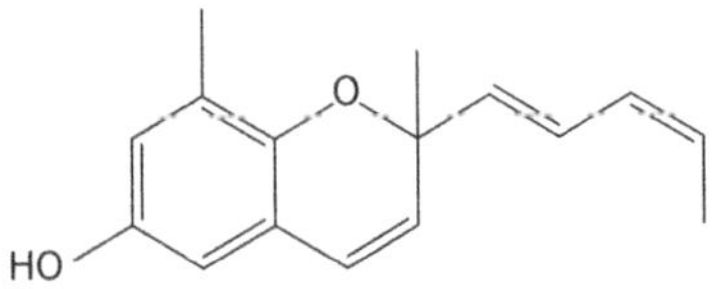

**Figure 31.4** Sargol II.

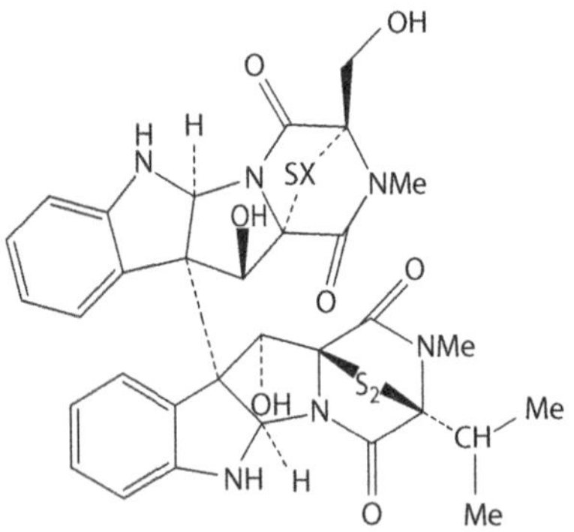

**Figure 31.5** Leptosin A, B, and C.

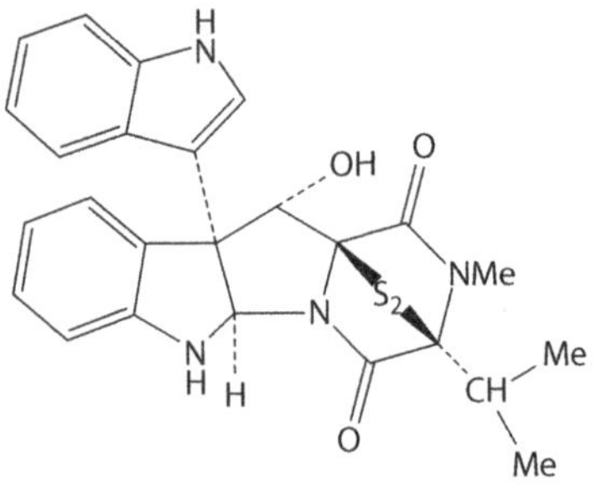

**Figure 31.6** Leptosin D, E, and F.

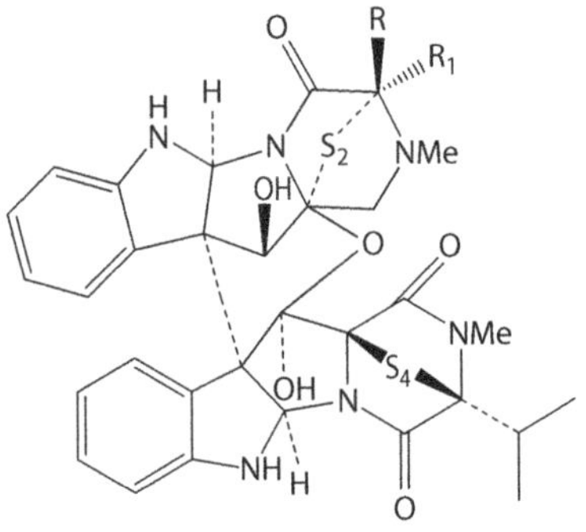

**Figure 31.7** Leptosin I: $R=CH_2OH$, $R_1=H$. J: $R=H$, $R_1=CH_2OH$.

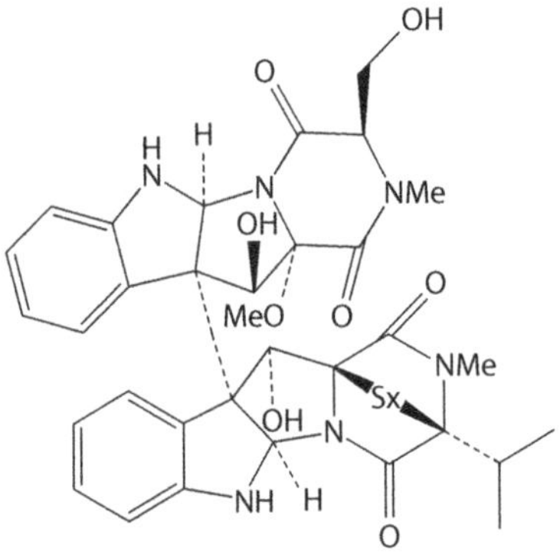

**Figure 31.8** Leptosin M.

and has both antitumor and antibacterial activity (Kato et al., 2012). Leptosin has different structures such as A, B, C (Figure 31.5), D, E, and F (Figure 31.6). Epipolythiodioxopiperazine (ETP) extracted from the mycelia of a strain of *Leptosphaeria* species was linked to the marine alga *S. tortile*. All these combinations showed cytotoxicity against cultured P388 cells, except leptosins A and C, which showed considerable antitumor activity against sarcoma 180 ascites (Takahashi et al., 1994a,b). Leptosin has two other structures,

I and J, which show cytotoxicity against cultured P388 cells as well (Figure 31.7).

Leptosin M (Figure 31.8), also isolated from a strain of *Leptosphaeria* species, is separated from the strain *S. tortile*. It shows significant cytotoxicity against of P388; especially, leptosin M has proven cytotoxicity against of human cancer cell line, and was found to inhibit particularly human topoisomerase II and two protein kinases CaMKIII (Ca$^{2+}$/calmodulin-dependent protein kinase II, a serine/threonine-specific protein kinase involved in signaling cascades) and PTK (protein tyrosine kinase; tyrosine kinases attach phosphate groups to other amino acids such as serine and threonine to

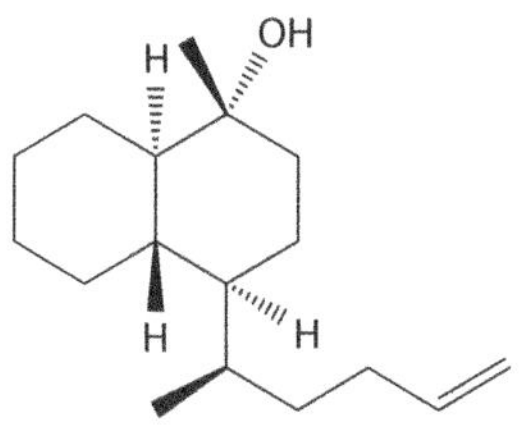

*Figure 31.9* Dictyotin A.

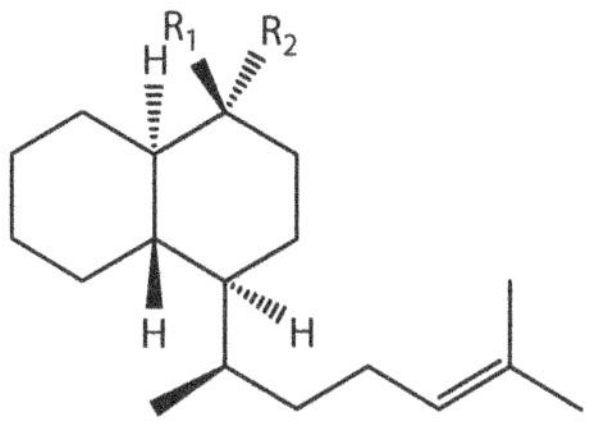

*Figure 31.10* Dictyotin B, C. B: $R_1$ = Me, $R_2$ = OH. C: $R_1$ = OH, $R_2$ = Me.

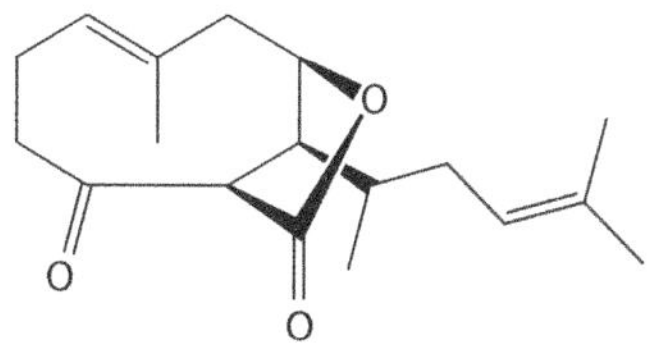

*Figure 31.12* Dictyota with xencicane and norxenicane.

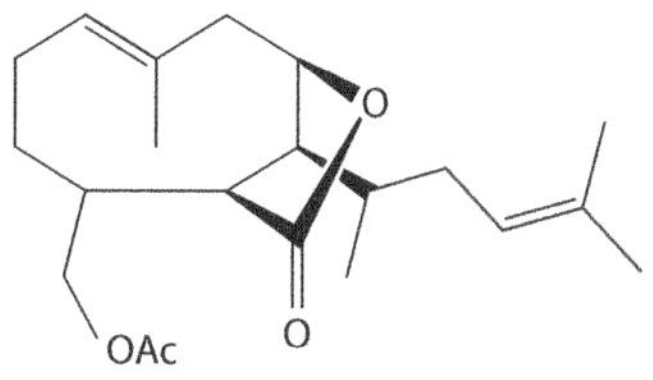

*Figure 31.13* Dictyota with xencicane and norxenicane.

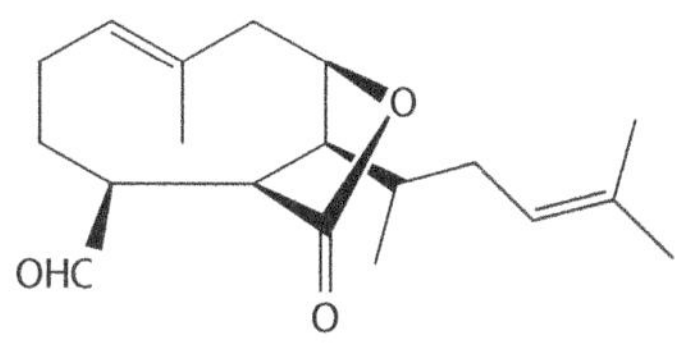

*Figure 31.14* Dictyota with xencicane and norxenicane.

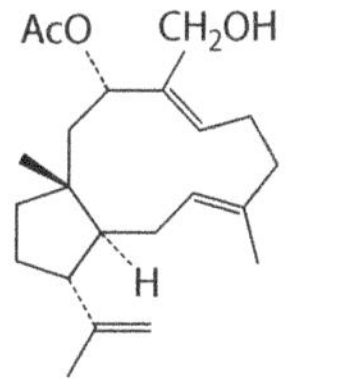

*Figure 31.15* Diterpene derivative.

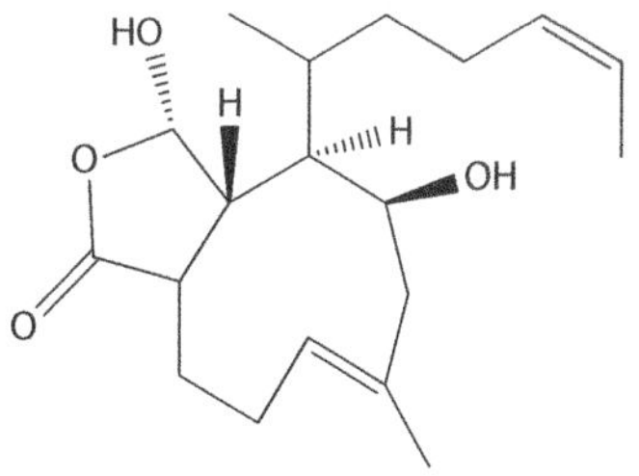

*Figure 31.16* 4,18-Dihydroxydictyolactone.

communicate signals within a cell and regulate cellular activity, such as cell division).

The brown alga *Dictyota dichotoma* contains three cytotoxic diterpene dictyotins A, B, and C, isolated by Wu and coworkers in 1990 (Wu et al. 1990; Figures 31.9 and 31.10). *D. dichotoma* from THE Okinawa Islands has four diterpenes with xenicane and norxenicane derivatives, which have cytotoxic activity and antitumor activity (Figures 31.11 through 31.14). *Dictyota* has an unidentified species named *dolabellane*, which contains also specific diterpene derivative  showing cytotoxic activity (Tringali et al., 1984; Figure 31.15). The brown alga *Dictyota* sp. has two diterpenes, 4,18-dihydroxydictyolactone (Figure 31.16) and 8α,11-dihydroxypachydictyol A (Figure 31.17), which have shown cytotoxic activity; specially, 4,18-dihydroxydictyolactone exhibited strong effects on NCI-H187 (Jongaramruong and Kongkam, 2007). Another type of brown algae is *Turbinaria ornate*, which has a cytotoxic compound known as turbinaric (Figure 31.18; Asari et al., 1989).

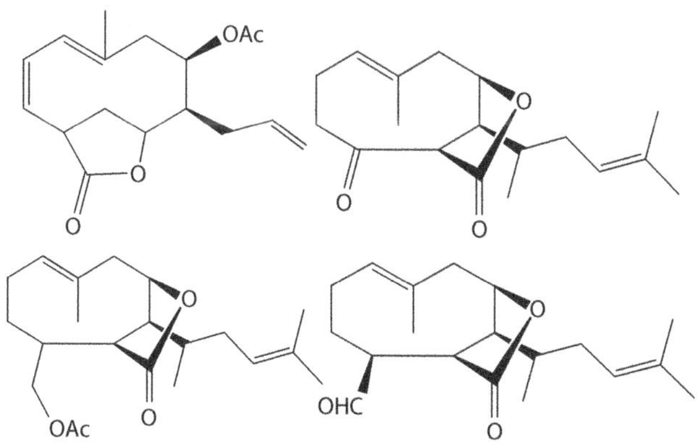

The brown alga *Turbinaria confides* has oxygenated fucosterol components that show cytotoxic activity against of P-388, KB, A-549, and HT-29 cell lines. Some of these fucosterols were designated as 24-ethylcholesta-4,24(28)-diene 3-one (Figure 31.19), 24-ethylcholesta-4,28(29)-diene-3-one (Figure 31.20), 24-ethylcholesta-4,24(28)-diene-3,6-dione (Figure 31.21),

*Figure 31.11* Dictyota with xencicane and norxenicane.

**Figure 31.17**  8α,11-Dihydroxypachydictyol A.

**Figure 31.18**  Turbinaric.

**Figure 31.19**  24-Ethylcholesta-4,24(28)-diene 3-one.

**Figure 31.20**  24-Ethylcholesta-4, 28(29)-diene-3-one.

**Figure 31.21**  24-Ethylcholesta-4, 24(28)-diene-3, 6-dione.

**Figure   31.22**  24β-Hydroperoxy-24-ethylcholesta-4,   28(29)-diene-3, 6-dione.

**Figure   31.23**  60-Hydroxy-24-ethylcholesta-4,24(28)-diene-3-one.

**Figure   31.24**  24-Hydroperoxy-6β-hydroxy-24-ethylcholesta-4,28(29)-diene-3-one.

24β-hydroperoxy-24-ethylcholesta-4,28(29)-diene-3,6-dione (Figure 31.22), 60-hydroxy-24-ethylcholesta-4,24(28)-diene-3-one (Figure 31.23), and 24-hydroperoxy-6β-hydroxy-24-ethylcholesta-4,28(29)-diene-3-one (Figure 31.24) (Sheu et al., 1999).

The brown alga *Hizikia fusiforme*, which is an edible brown alga in Japan, has four arsenic-containing ribofuranosides together with inorganic arsenic (Edmonds et al., 1987; Figure 31.25). The brown alga *Stypopodium*

**Figure   31.25**  Ribofuranoside.   R$_1$=OH,   R$_2$=OSO$_3$H. R$_1$=OH,   R$_2$=SO$_3$H.   R$_1$=OH,   R$_2$=O-POO-OCH$_3$-CH(OH) CH$_2$OH. R$_1$=NH$_2$, R$_2$ = SO$_3$H.

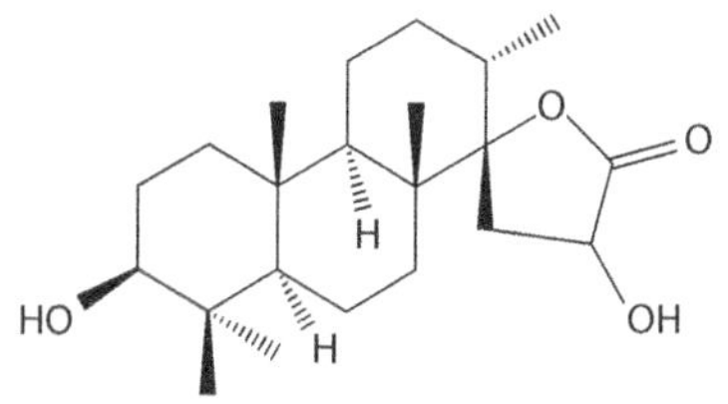

**Figure 31.26** Stypolactone.

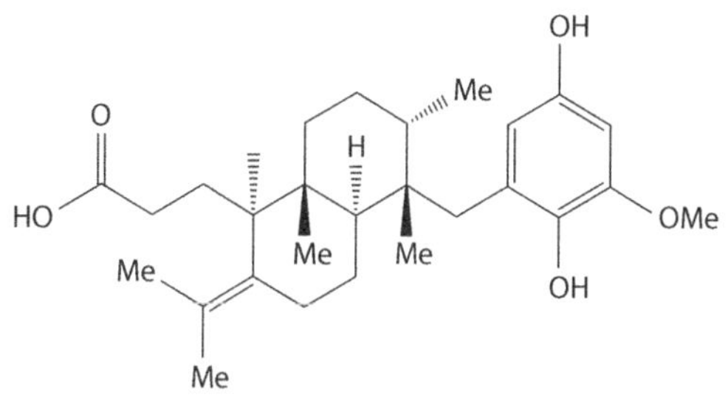

**Figure 31.30** Terpenoid C.

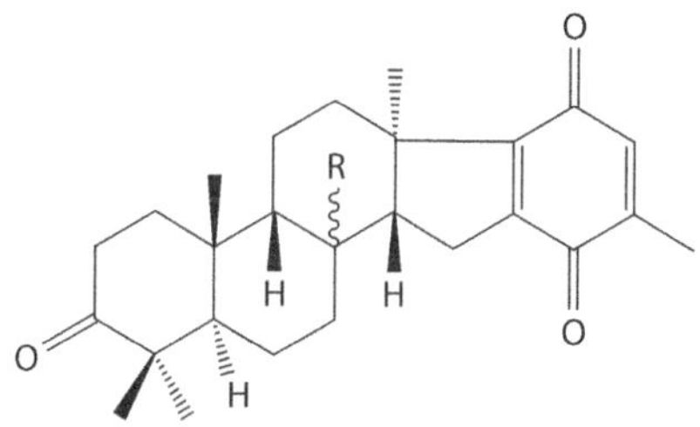

*zonale*, by a diterpenoid mixed biogenesis, stypolactone (Figure 31.26), has shown weak cytotoxic reactions to the H-116 and A-549 cell lines (Dorta et al., 2002). The brown alga *Cystoseira myrica*, found and gathered in the Gulf of Suez, showed moderate cytotoxicity against the murine cancer cell line KA31T, but reduced cytotoxicity against normal NIH3T3 by the four hydroazulene diterpen compounds dictyone acetate, dictyol F monoacetate, isodictytiol monoacetate, and cystoseirol monoacetate (Ayyad et al., 2003). A brown alga named *Stypopodium carpophyllumm* exhibited cytotoxic activity against several cultured cancer cell lines through its strol B compound (Figure 31.27; Tang et al., 2002). A brown alga named *Bifurcaria bifurcarate* has shown cytotoxic activity through two trihydroxylated diterpenes based on 12-hydroxygeranylgeraniol compounds (Figures 31.28 and 31.29). A terpenoid C containing a methyl ester of C in the tropical brown alga *S. zonale* has shown cytotoxic activity *in vitro* against HT-29, A-549, and H-116 (Dorta

et al., 2002; Figure 31.30). There are terpenes named meroditerpenes atomarianones A and B in the brown alga *Taonia atomaria*, which show cytotoxic activity against of NSCLC-N6 and A-549 cell lines (Abatis et al., 2005; Figure 31.31). There are some compounds isolated from brown algae with cytotoxic activity against several human tumor cell lines, known as yahazunol (Figure 31.32), cyclozonarone (Figure 31.33) (Ochi et al., 1979)

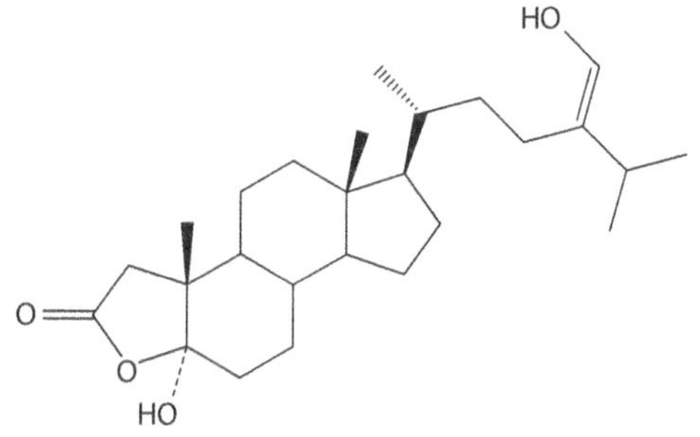

**Figure 31.31** Meroditerpenes atomarianones A and B. R = βMe. R = αMe.

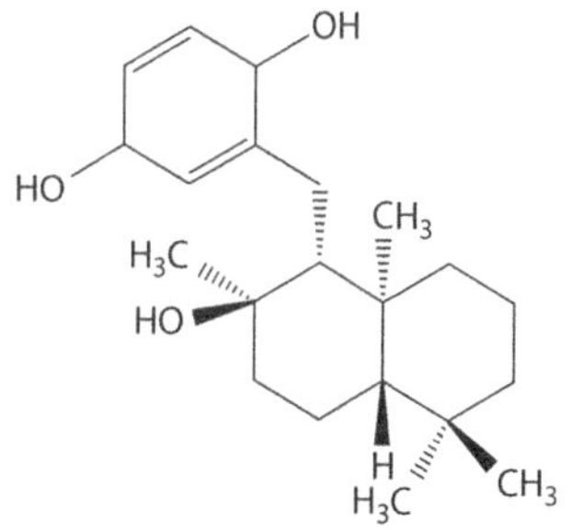

**Figure 31.27** Strol B.

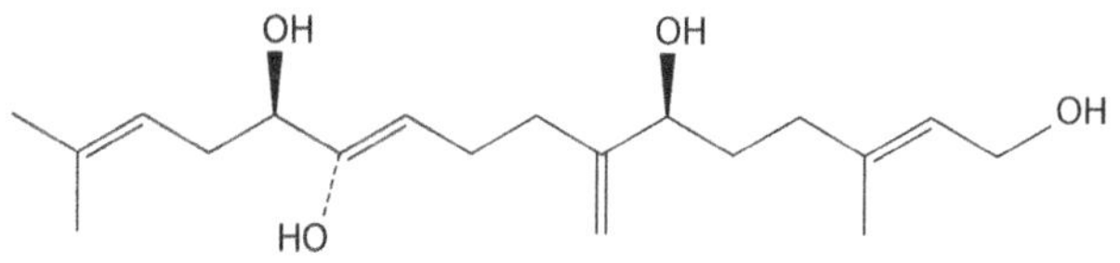

**Figure 31.28** Tridoxylated diterpenes.

**Figure 31.32** Yahazunol.

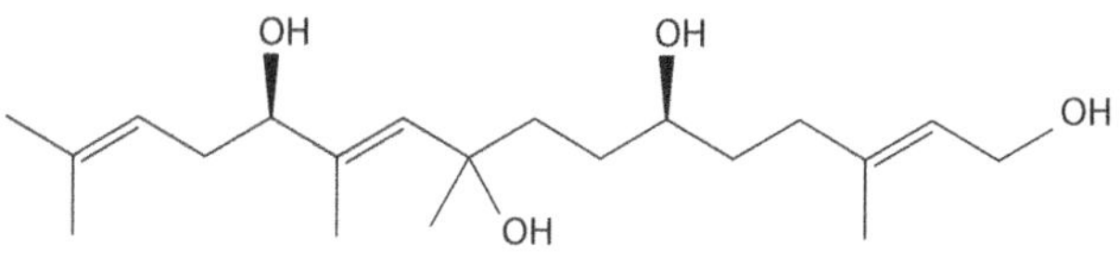

**Figure 31.29** Tridoxylated diterpenes.

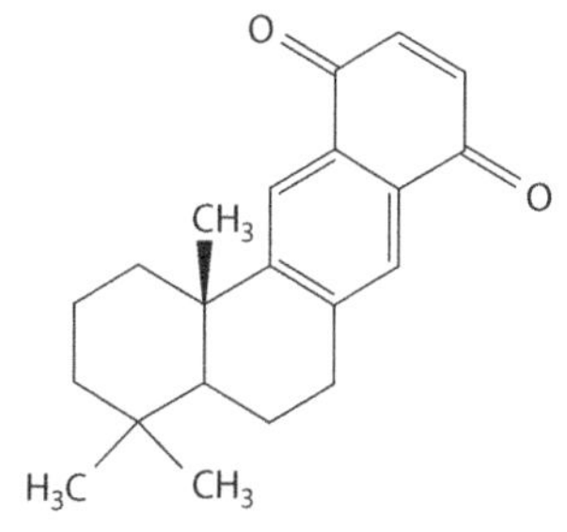

**Figure 31.33** Cyclozonarone.

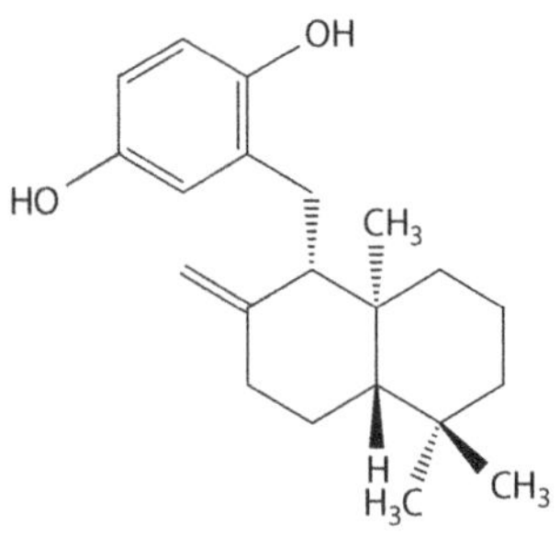

**Figure 31.34** Zonarol.

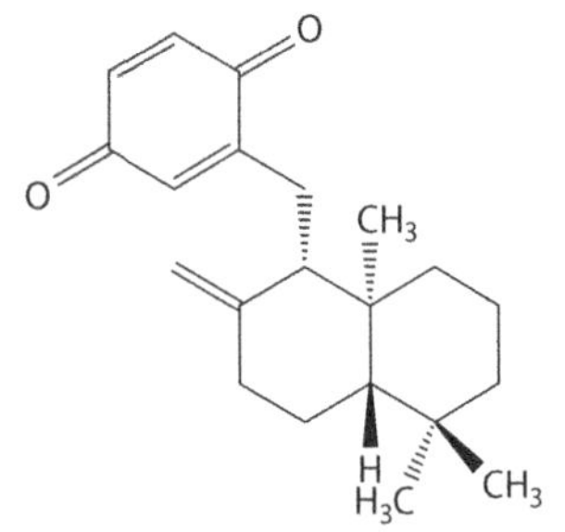

**Figure 31.35** Zonarone.

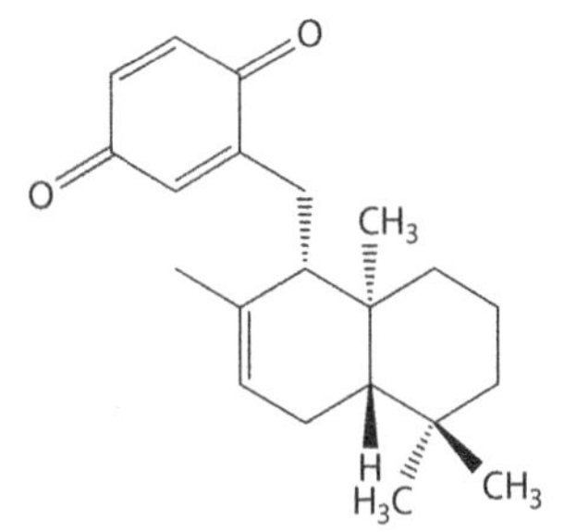

**Figure 31.36** Isozonarol.

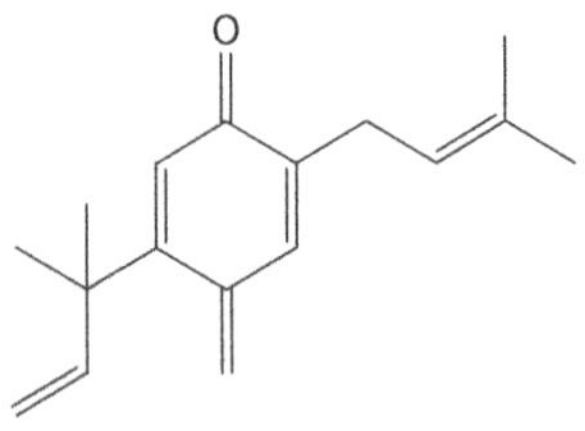

**Figure 31.37** Bisprenylated quionones.

zonarol (Figure 31.34), zonarone (Figure 31.35), and isozonarol (Figure 31.36; Laube et al., 2005). A bisprenylated quinone (Figures 31.37 and 31.38), which is produced by the brown alga *Perithalia capillaris*, inhibits superoxide generation in human neutrophils *in vitro* and proliferation of HL-60 cells (Blackman et al., 1979).

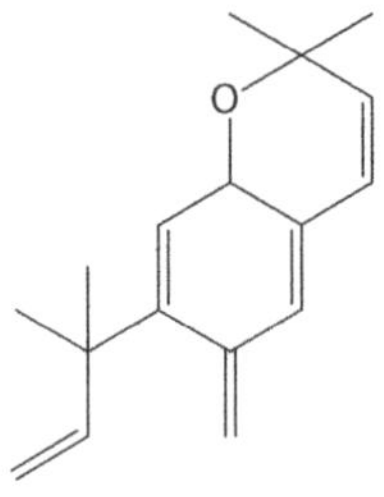

**Figure 31.38** Bisprenylated quionones.

## 31.5.2 Antiviral activity

As pathogens, viruses pose one of the crucial challenges in current basic medical research. Glyco-conjugate compounds have been used as tools to interfere with the infectious process. For example, galectins are lectins with affinity for β-galactosides, which are involved in self-/non-self-recognition. Galectins from both invertebrates and vertebrates recognize a variety of viral and bacterial pathogens and protozoan parasites. From the brown alga *Dictyota pfaffi*, a new dollabelladiene (Figure 31.39), 10,18-diacetoxy-8-hydroxy 2,6-dollabelladiene (Figure 31.40), was isolated (Ireland and Faulkner, 1977). Dollabelladiene and its derivatives are compounds showing massive anti-human (HSV)-1 activity *in vitro*. Unfortunately, they show only weak inhibition against the human immunodeficiency virus (HIV)-1 reverse transcriptase. The brown alga has two diterpenes, (6*R*)-6-hydroxy dichototomo-3,14-diene-1,17-dial (Figure 31.41) and the 6-acetate derivative (Figure 31.42), which show antiretroviral activity *in vitro* (Pereira et al., 2004). The brown alga *Ecklonia cava* produces a specific compound named phlorotannin that has derivatives such as 8,8′-bieckol (Figure 31.43) and 8,4″-bieckol (Figure 31.44)

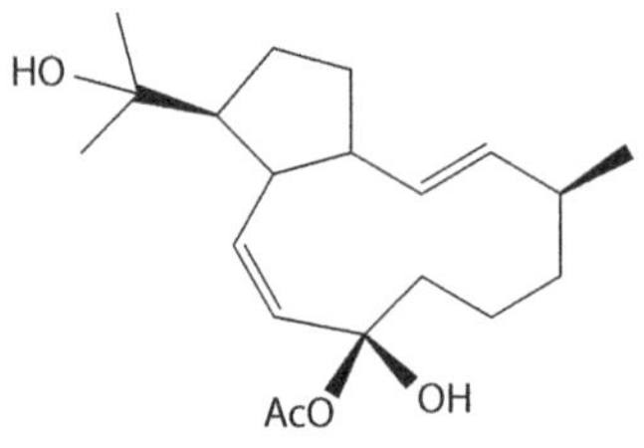

**Figure 31.39** Dollabelladiene.

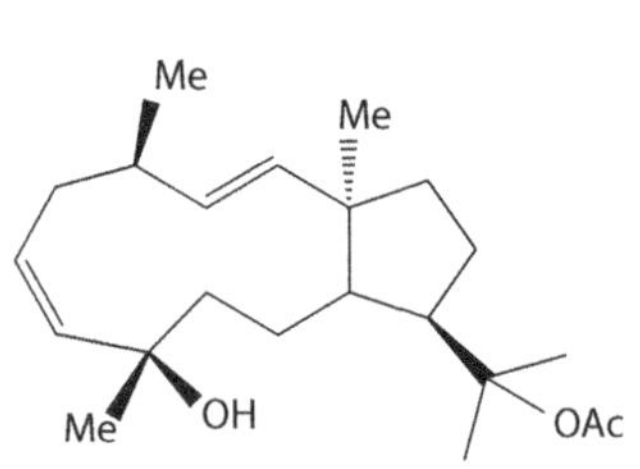

**Figure 31.40** 10, 18-Diacetoxy-8-hydroxy 2,6dollabelladiene.

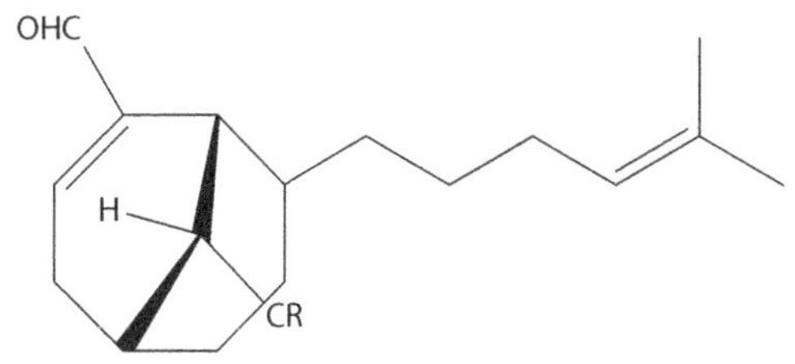

***Figure   31.41*** R=H   two   diterpenes   (6R)-6-hydroxy dichototomo-3,14-diene-1,17-dial.

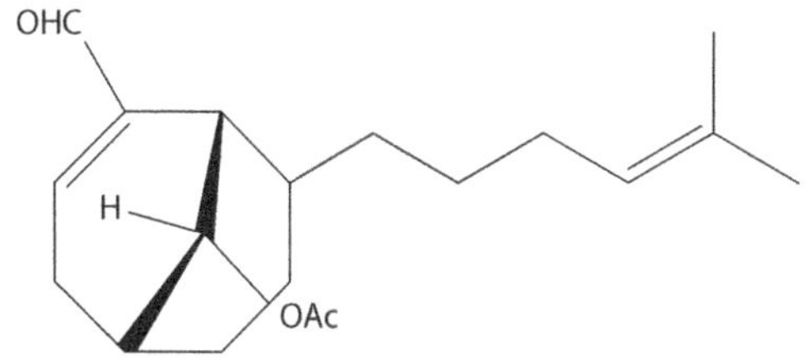

***Figure 31.42*** R = 6 acetate derivative.

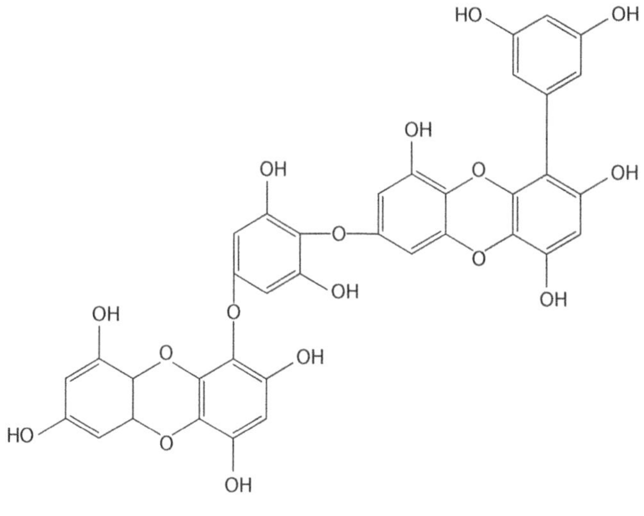

***Figure 31.43*** 8,8″-Bieckol.

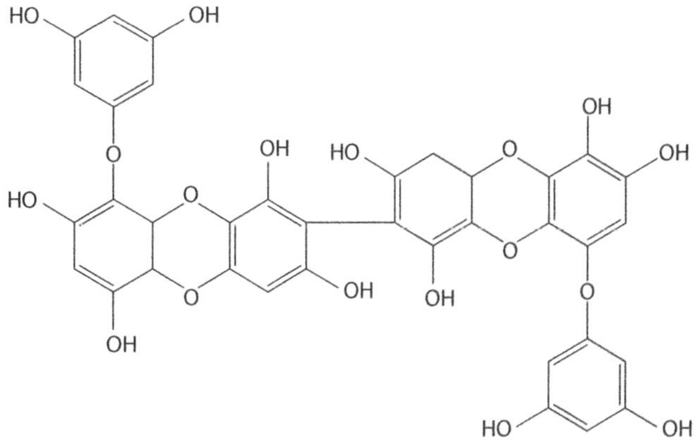

***Figure 31.44*** 8,4′-bieckol.

(Fukuyama et al., 1989), which are inhibitors of HIV-1 reverse transcriptase (RT) and protease. Although their effect on RT is more than on the protease, the inhibitory activity of 8,8′-bieckol against HIV-I was comparable to that of a reference compound nevirapine (a non-nucleoside RT inhibitor (NNRTI) used to treat HIV-1 infection and AIDS).

### 31.5.3   Antidiabetic activity

Phlorotannins are a type of tannins and one of the most important components produced by brown algae, and show active biological effects on diabetes. Phlorotannins in brown algae have shown some mechanisms against diabetes, such as α-glucosidase and α-amylase inhibitory effect, by glucose uptake in skeletal muscles and improvement of insulin sensitivity in type 2 diabetic db/db mice, protein tyrosine phosphatase 1B (PTP 1B), enzyme inhibition, and so on. Fucosterol is a sterol extracted from the brown alga *Pelvetia siliquosa*, and it is the main antidiabetic compound was found (Lee et al., 2004). The major sites of glucose consumption are the skeletal muscles. One way to decrease insulin resistance is by making glucose to be taken up into this tissue in type 2 diabetes. *Ishige foliacea* contains octaphlorethol A (OPA), a novel phenolic compound that has effect on glucose uptake in skeletal muscle cells (Figure 31.45; Seung et al., 2012).

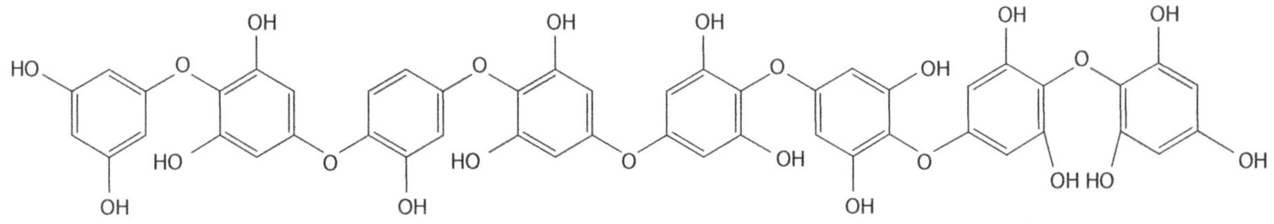

***Figure 31.45*** Octaphlorethol A.

## 31.6  Conclusion

All bioactive compounds introduced and discussed in this chapter are collected from brown seaweeds, and represent different chemical structures and display a variety of biological effects on cancer, viral infections, and diabetes. The role of glycol-conjugates, such as polysaccharides, terpenoids, and phenolic compounds such as highly sulfated fucans, dollabelladiene, phlorotannins, fucostrol, xanthophyll, focuxanthin, and so on, is highly significant. These metabolites in various species of brown algae have shown vast efficacy in regulating bioprocesses such as inhibition effects on the growth of tumor cells and viral particles, antiproliferative activity, cytotoxicity, induction of apoptosis, and as potential chemotherapeutic, anti-inflammatory, and antidiabetic agents. Although the last activity (antidiabetic), it is, stimulation of glucose consumption and metabolism in the skeletal muscle, has not been understood very well, there are promising ways for further achievements. In addition, brown algae that produce underwater forests contain laminaran, mannitol, saccharides, and glycerol, which are very useful in the production of agar, carrageenan, alginate, and glue. Nowadays, the utilization of brown seaweed is increasing, and more investigations and more finding are predictable in future.

## References

Abatis, D., Vigias, C., Galanakis, D. et al. (2005) Atomarianones A and B: Two cytotoxic meroditerpenes from the brown alga *Taonia atomaria*. *Tetrahedron Lett.*, 46, 8525–8529.

Amsler, C.D. and Fairhead, V.A. (2006) Defensive and sensory chemical ecology of brown algae. *Adv. Bot. Res.*, 43, 1–91.

Andrade, L.R., Salgado, L.T., Farina, M., Pereira, M.S., Mourão, P.A.S., and Amado Filho, G.M. (2004) Ultrastructure of acidic polysaccharides from the cell walls of brown algae. *J. Struct. Biol.*, 145, 216–225.

Asari, F., Kusumi, T., and Kakisawa, H. (1989) Turbinaric acid, a cytotoxic secosqualene carboxylic acid from the brown alga *Turbinaria ornate*. *J. Nat. Prod.*, 52, 1167–1169.

Ayyad, S.-E.N., Abdel-Halim, O.B., Shier, W.T., and Hoye, T.R. (2003) Cytotoxic hydroazulene diterpenes from the brown alga *Cystoseira myrica*. *Z. Natuforsch., C., Biosci.*, 58, 33–38.

Beattie, A., Hirst, E.L., and Percival, E. (1961) Studies on the metabolism of the *Chrysophyceae*. *Biochem. J.* (*England*), 79, 531–537.

Berteau, O. and Mulloy, B. (2003) Sulfated fucans, fresh perspectives: Structures, functions, and biological properties of sulfated fucans and an overview of enzymes active toward this class of polysaccharide. *Glycobiology*, 13, 29–40.

Bisgrove, S.R. and Kropf, D.L. (2001) Asymmetric cell division in fucoid algae: A role for cortical adhesions in alignment of the mitotic apparatus. *J. Cell Sci.*, 114, 4319–4328.

Blackman, A.J., Dragar, C., and Wells, R.J. (1979) A new phenol from the brown alga *Perithalia caudata* containing a "reverse" isoprene unit at the 4-position. *J. Aust. J. Chem.*, 32, 2783–2786.

Blunt, J.W., Copp, B.R., Munro, M.H., Northcote, P.T., and Prinsep, M. R. (2006). Marine Natural Products. *Nat. Prod. Rep.* 23, 26–78.

Blunt, J.W., Copp, B.R., Munro, M.H., Northcote, P.T., and Prinsep, M. R. (2007). Marine Natural Products. *Nat. Prod. Rep.* 24, 31–86.

Bold, H.C. and Wynne, M.J. (1985) *Introduction to the Algae: Structure and Reproduction*, 2nd edn. Prentice-Hall Inc., Englewood Cliffs, NJ, pp. 1–33.

Caroline, B., Ana, P., Espindolab, D.M., Sirlei, J.K., Ljubica, T., Meuris, G., and Carlos, D.S. (2014) *Sargassum filipendula* alginate from Brazil: Seasonal influence and characteristics. *Carbohydr. Polym.*, 111, 619–623. doi:10.1016/j.carbpol.2014.05.024.

Dhargalkar, V.K. and Pereira, N. (2005) Seaweed: Promising plant of the millennium. *Sci. Cult.*, 71, 3, 60–66.

Dorta, E., Cueto, M., Bito, I., and Darias, J. (2002) New terpenoids from the brown alga *Stypopodium zonale*. *J. Prod.*, 65, 1727–1730.

Edmonds, S.L., Morita, M., and Shibata, Y. (1987) Isolation and identification of arsenic-containing ribfurnaoside and inorganic arsenic from Japanese edible seaweed *Hizikia fusiforme*. *J. Chem. Soc. Perkin. Trans.*, I, 577–580.

Faridch, N., Rosfarizan, M., Javad, B., Saeedeh, Z.B., Fahimeh, F., and Heshu, S.R. (2013) Antioxidant, antiproliferative, and antiangiogenesis effects of polyphenol-rich seaweed *Sargassum muticum*. *BioMed Res. Int.*, 2013, Article ID 604787, 9pp.

Faulkner, D.J. (1999) Marine natural products. *Nat. Prod. Rep.*, 16, 33–43.

Faulkner, D.J. (2000) Marine natural products. *Nat. Prod. Rep.*, 17, 7–55.

Faulkner, D.J. (2001) Marine natural products. *Nat. Prod. Rep.*, 18, 1–49.

Faulkner, D.J. (2002) Marine natural products. *Nat. Prod. Rep.*, 19, 1–48.

Fukuyama, Y., Kodaama, M., Miura, I. et al. (1989) Antiplasmin inhibitor. V. Structures of novel dimeric eckols isolated from the brown alga *Ecklonia kurome* Okamura. *Chem. Pharm. Bull.*, 37, 2438–2440.

Garson, J. (1989) Marine natural products. *Nat. Prod. Rep.*, 6, 143–170.

Guardia, S.D., Valls, R., Mesguiche, V., Brunei, J.-M., and Gulioli, G. (1999) Enantioselective synthesis of bifuracadiol: A natural antitumor marine product. *Tetrahedron Lett.*, 40, 8359–8360.

Hillison, C.I. (1977) *Seaweeds, a Color-Coded, Illustrated Guide to Common Marine Plants of East Coast of the United States*. Keystone Books, The Pennsylvania State University Press, University Park, PA, p. 15.

Indegaard, M. and Minsaas, J. (1991) Animal and human nutrition. In *Seaweed Resources in Europe—Uses and Potential*, M.D. Guiry and G. Blunden (eds.). John Wiley & Sons, Chichester, U.K., pp. 21–64.

Ireland, C. and Faulkner, D.J. (1977) Diterpenes from *Dolabella californica*. *J. Org. Chem.*, 42, 3157–3162.

Iwamoto, C., Minoura, K., Hagishita, S., Nomoto, K., and Numata, A. (1998) Penostatins F–I, novel cytotoxic metabolites from a *Penicillium* species from an Enteromorpha marine alga. *J. Chem. Soc. Perkin Trans.*, 1, 3, 449–456.

Iwamoto, C., Minoura, K., Hagishita, S. et al. (1999) Absolute sterostructures of novel penostatins. A–E from a *Penicillium* species from an Enteromorpha marine alga. *Tetrahedron*, 55, 14353–14368.

Iwamoto, C., Yamada, T., Ito, Y., Minoura, K., and Numata, A. (2001) Cytotoxic cytochalasans from a *Penicillium* species separated from a marine alga. *Tetrahedron*, 57, 2997–3004.

Jongaramruong, J. and Kongkam, N. (2007) Novel diterpenes with cytotoxic, anti-malarial and anti-tuberculosis activities from a brown alga *Dictyota* sp. *J. Asian Nat. Prod. Res.*, 9, 743–751.

Kato, Y., Umeda, N., Maeda, A., Matsumoto, D., Kitamoto, N., and Kikuzaki, H. (2012). Identification of a novel glycoside, leptosin, as a chemical marker of manuka honey. *J. Agric. Food. Chem.* 60:3418–3423.

Killing, H. (1913) Zur biochemie der Meersalgen. *Z. Physiol. Chem.*, 83, 171–197.

Kloareg, B. (1984) Isolation and analysis of cell walls of the brown marine algae *Pelvetia canaliculata* and *Ascophyllum nodosum*. *Physiol. Vég.*, 22, 47–56.

Kloareg, B., Demarty, M., and Mabeau, S. (1986) Polyanionic characteristics of purified sulfated homofucans from brown algae. *Int. J. Biol. Macromol.*, 8, 380–386.

Koji, M. and Masashi, H. (2013) Biosynthetic pathway and health benefits of fucoxanthin, an algae-specific xanthophyll in brown seaweeds. *Int. J. Mol. Sci.*, 14(7), 13763–13781.

Kwak, S.Y., Yang, J.K., Kim, J.H., and Lee, Y.S. (2013) Chemical modulation of bioactive compounds via oligopeptide or amino acid conjugation. *Pep. Sci.*, 100(6), 584–591. doi:10.1002/bip.22307.

Laube, T., Beil, W., and Seifert, K. (2005) Total synthesis of two 12-nordrimanes and the pharmacological active sesquiterpene hydroquinone yahazunol. *Tetrahedron*, 61, 1141–1148.

Lee, Y.S., Shin, K.H., Kim, B.K., and Lee, S. (2004) Antidiabetic activities of fucosterol from *Pelvetia siliquosa*. *Arch. Pharmacol. Res.*, 27, 1120–1122.

Numata, A., Kambara, S., Takahashi, C. et al. (1991) Cytotoxic activity of marine algae and a cytotoxic principle of the brown alga *Sargassum tortile*. *Chem. Pharm. Bull.*, 39, 2129–2131.

Pereira, H.S., Leao-Ferreira, L.R., Moussatche, N. et al. (2004) Antiviral activity of diterpenes isolated from the Brazilian marine alga *Dictyota menstrualis* against human immunodeficiency virus type 1 (HIV-1). *Antiviral Res.*, 64, 69–76.

Ragan, M.A. and Glombitza K.W (1986). Phlorotannins, brown algal polyphenols. *Prog Phycol Res* 4, 130–230.

Riou, D., Colliec-Jouault, S., Pinczon, D.S.D., and Bosch S. (1996) Antitumor and antiproliferative effects of a fucan extracted from *Ascophyllum nodosum* against a non-small-cell bronchopulmonary carcinoma line. *Anticancer Res.*, 3A, 1213–1218.

Seung, H.L., Sung, M.K., Seok-Chun, K., Dae-Ho, L., and You-Jin J. (2012). Octaphlorethol A, a novel phenolic compound isolated from a brown alga, Ishige foliacea, increases glucose transporter 4-mediated glucose uptake in skeletal muscle cells. *Biochem. Biophys. Res. Commun.* 420, 576–581

Sheu, J.-H., Wang, G.-H., Sung, P.-J., and Duh, C.-Y. (1999) New cytotoxic oxygenated fucosterols from the brown alga *Turbinaria conoides*. *J. Nat. Prod.*, 62, 224–227.

Singh, I.P. and Bharate, S.B. (2006) Phloroglucinol compounds of natural origin. *Nat. Prod. Rep.*, 23, 558–591.

Soheil, Z.M., Hamed, K., Ramin, K., Mahboubeh, R., Mohammad, F., Keivan, Z., and Habsah, A. (2014) Anticancer and antitumor potential of fucoidan and fucoxanthin, Two main metabolites isolated from brown algae. *ScientificWorldJournal*, 2014, Article ID 768323, 10pp.

Takahashi, C., Numata, A., Ito, Y. et al. (1994a) Leptosins, antitumor metabolites of a fungus isolated from a marine alga. *J. Chem. Soc. Perkin Trans. 1*, 13, 1859–1864.

Takahashi, C., Numata, A., Matsumura, E. et al. (1994b) Leptosins I and J, cytotoxic substances produced by a *Leptosphaeria* species physico-chemical properties and cytotoxic substances structures. *J. Antibiot.*, 47, 1242–1249.

Tang, H.-F., Yi, Y.-H., Yao, X.-S., Xu, Q.-Z., Zhang, S.-Y., and Lin, H.-W. (2002) Bioactive steroids from the brown alga *Sargassum carpophyllum*. *J. Asian Nat. Prod. Res.*, 4, 95–101.

Targett, N., Boettcher, A., Targett T., and Vrolijk, N. (1995) Tropical marine herbivore assimilation of phenolic-rich plants. *Oecologia*, 103, 170–179.

Tringali, C., Prattellia, M., and Nicols, G. (1984) Structure and conformation of new diterpenes based on the *dolabellane* skeleton from *Dictyota* species. *Tetrahedron*, 40, 799–803.

Vishchuk, O.S., Svetlana, P.E., and Tatyana, N.Z. (2013). The fucoidans from brown algae of Far-Eastern seas: Antitumor activity and structure–function relationship. *Food Chem.*, 141(2), 1211–1217. doi:10.1016/j.foodchem.03.065.

Van den Hoek, C., Mann, D. G., and Jahns, H. M. (1995) *Algae: An Introduction to Phycology*. Cambridge University Press, Cambridge, U.K., pp. 165–218. ISBN 0-521-31687-1.

Wu, C.X., Li, Z.G., and Li, H.W. (1990) Effect of berbamine on action potential in isolated human atrial tissues. *Asia Pac. J. Pharmacol.*, 5, 191–193.

Yang, E.J., Ham, Y.M., Lee, W.J., Lee, N.H., and Hyun, C.G. (2013) Anti-inflammatory effects of apo-9′ fucoxanthinone from the brown alga, *Sargassum muticum*. *J. Fac. Pharm.*, 21(1), 62. doi:10.1186/2008-2231-21-62.

# Glycoconjugates in biomedicine and biotechnology

## chapter thirty-two

# D-Glucosamine contributes to cell membrane stability

*Yoshihiko Hayashi, Kei Kaida, Kazunari Igawa, Shizuka Yamada,*
*Takeshi Ikeda, and Kajiro Yanagiguchi*

### Contents

## 32.1 Introduction

D-glucosamine (GlcN), a natural amino monosaccharide, universally distributes in the connective and cartilage tissues as a component of glycosaminoglycana and contributes to maintaining the strength, flexibility, and elasticity of these tissues (Setnikar 1992). GlcN has a variety of biological activities. It exhibits anti-inflammatory reactions (Hua et al. 2002) and can suppress adjuvant arthritis (Hua et al. 2005) and platelet aggregation (Hua et al. 2004). GlcN also has an analgesic effect in patients with osteoarthriris. Therefore, it is widely used in attempts to suppress the pain associated with the disability caused by osteoarthritis (Crolle and D'Este 1980).

Recently, our group reported that bradykinin and 5-hydroxytryptamine-induced nociceptive responses were significantly suppressed by the direct application of GlcN (Kaida et al. 2013, 2014). It has been reported that GlcN reduces the elementary current amplitude and increases the mean channel open time (Marchais and Marty 1980). Because GlcN has a weak binding site in the channel itself, the channel cannot be closed (Marchais and Marty 1980). Voltage-gated sodium channels are necessary for electrogenesis and nerve impulse conduction (Cummins et al. 2007). GlcN may, thus, exhibit an antinociceptive effect by binding to these sodium channels, resulting in a longer open time, which produces the hyperpolarization of nerves in relation to cell membrane stability.

This chapter focuses on and reviews GlcN's contribution to cell membrane stability. Only limited data have been presented in this field. Cell biological approaches using electroporation and quantum dots (QDs) have been applied as the experimental materials and methods to understand the relationship between GlcN and membrane stability, which support the discussion of the electrophysiological data.

## 32.2 Electroporation technology

Exposure of biological cells to a sufficiently strong external electric field results in transiently or permanently increased permeability of cell membranes, referred to as electroporation (Neumann et al. 1982). Electroporation is currently used for food and biomass processing (Toepfl et al. 2007; Sack et al. 2010; Martin-Belloso and Sobrino-Lopez 2011) and as a local treatment of cancer (Sersa et al. 2012; Testori et al. 2012). Nonthermal irreversible electroporation for the ablation of solid tumors has recently emerged as a new application of electroporation technology (Golberg and Yarmush 2013). Since all types of cells (animal, plant, and microorganism) can be effectively electroporated, without addition of any viral or chemical compounds, electroporation is considered to be a universal method and a platform technology (Miklavcic 2012).

According to the theory of aqueous pore formation, which is largely based on thermodynamic consideration, the formation of aqueous pores is initiated by the penetration of water molecules into the lipid bilayer of the membrane. This phenomenon leads to reorientation of the adjacent lipids with their polar head groups pointing toward these water molecules (Kotnik et al. 2012; Kotnik 2013). It is interesting whether the GlcN promotes to repair the penetration of lipid bilayer by the aqueous pore.

### 32.2.1 Electroporation with GlcN

The formation of electropores takes nano- to microseconds, whereas their resealing—as revealed by the return of the membrane's electric conductivity to its preparation value and by termination of detectable transmembrane transport—takes seconds or even minutes after the end of the exposure (Saulis et al. 1991). More detailed measurements reveal that the resealing proceeds in several stages with time constants ranging from micro- and/or milliseconds up to tens of seconds (Hibino et al. 1993; Puchihar et al. 2008). However, neither the existing theory nor the experiments can provide a reliable picture of specific events characterizing each of these distinctive stages, and reliable molecular dynamics cannot be yet demonstrated.

Easy attachment of GlcN to a negative charged cell membrane is induced by a positive charge and biocompatibility of GlcN. This electrostatic interaction is thought to bring the early repair of electropores in cell membrane and indirectly demonstrates cell membrane protection and stability.

### 32.2.2 Transfection efficiency

Recent work presents that the transfection efficiency increased over 30% in electroporation samples treated with GlcN-supplemented buffer after 1 day (Figure 32.1) compared to the controls (Igawa et al. 2014a). The proton sponge hypothesis/effect (Boussif et al. 1995; Sonawane et al. 2003; Yang and May 2008), while not definitively proven, has been invoked to explain the relatively high transfection efficiency of other proton-sponge-type materials, such as lipopolyamines (Remy et al. 1994; Behr 1997), polyamidoamine (PAMAM) dendrimers (Tang and Szoka 1997), and various imidazole-containing polymers (Boussif et al. 1995; Midoux and Monsigny 1999; Pack et al. 2000). The applied GlcN was used after neutralization. The concentration of endosomal chloride ions originated from GlcN hydrochloride leads to osmotic rupture in endosomes (Boussif et al. 1995; Behr 1997), which involves the escape of plasmid vectors from endosomes and lysosomes. The adsorption process to the cell membrane of GlcN is first indispensable for the appearance of newly proven polycationic action and, second, the accumulation into cytoplasm occurs for the proton sponge effect (Igawa et al. 2014a,b).

### 32.2.3 Application for biomedical fields

The electrical pulses induce nanoscale pores within the phospholipid bilayer, thus changing cell permeability (Rubinsky 2007). There are two types of electroporations: reversible (RE) and irreversible (IRE).

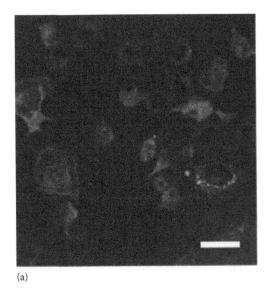
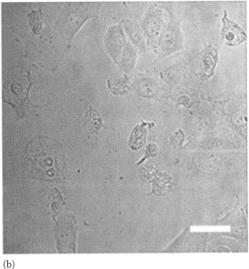

(a)                                          (b)

*Figure 32.1* (a) GFP-positive cells after electroporation with 0.005% GlcN-containing buffer. (b) Phase contrast image of area *A*. Scale bar = 20 μm. (From Igawa, K. et al., *BioMed Res. Int.*, 2014, Article ID 485867, 2014a. With permission.)

The lesser electrical strength and duration of the applied pulses during RE allow pores in the membrane to spontaneously seal by cells themselves, and GlcN may promote this sealing activity. This unique effect of a transient increase in cell permeability brings foreign materials (gene and low-molecular-weight protein) to the cytoplasm. Several clinical trials (Benevento et al. 2012; Campana et al. 2012; Curatolo et al. 2012) show that RE applied for this purpose enhances the delivery of chemotherapy to the desired tissues, such as skin and breast.

IRE depends on delivering electrical pulses whose strength and duration exceeds the threshold of spontaneous cell membrane repair. IRE brings the permanent permeability to cell membranes and disrupts homeostasis of cells, leading to cell death without thermal energy (Lee et al. 2010; Narayanan 2011). The first experiment of IRE was performed on swine livers (Rubinsky et al. 2007). Furthermore, there was histologically haemorrhagic necrosis in the liver, vessels and bile ducts within the zone of ablation were interestingly preserved (Charpentire 2012). The demand of clinical application for tumor tissues is increasing, especially for use in locations where thermal damage is a substantial concern, such as the pancreas (Charpentier et al. 2010; Martin et al. 2012), or near vital structures, such as the bile ducts in the liver (Lee et al. 2012). Electrode location, orientation, and heterogeneities in local environment must be considered in planning ablation treatments (Ben-David et al. 2013). The protective effects of GlcN against many types of wounds (Sal'nikova et al. 1990; McCarty 1996; Matsunaga et al. 2006), including its antioxidative and immunostimulating properties (Yan et al. 2007), are extensively expected after the application of IRE.

## 32.3 QD technology

Luminescent semiconducting nanocrystals called QDs are mainly based on cadmium selenide (core size: approximately 3 nm) (Murray et al. 1993; Park et al. 2007; Reiss et al. 2009), which provide a very large absorption spectrum with a size-dependent narrow emission spectrum. Then, QDs are useful imaging tools for cellular labeling and monitoring due to their optical properties, such as a high fluorescence intensity, remarkable resistance to photobleaching, broad absorption spectra, and narrow emission spectra (Medintz et al. 2005; Michalet wt al. 2005).

### 32.3.1 QDs conjugated with GlcN

Cationic polymers with a large number of primary amine groups, such as PAMAM dendrimer, grafted polyethylene imine (PEI), and peptides, have been widely used to promote cellular uptake through the electrostatic interaction between their positive and negative charges on the cell membrane (Kawakami et al. 2008; Higuchi et al. 2010; Algarra et al. 2011). QDs were modified with PAMAM dendrimer to increase the uptake efficiency and cytosolic distribution of QDs in primary cultured mesenchymal stem cells (Higuchi et al. 2011). A similar approach was recently made using conjugation of carboxylated QDs with GlcN (final modified size: 20–30 nm) to directly investigate QDs uptake process into osteoblastic cells (Igawa et al. 2014b). The observation of these processes can be expected to verify the electrophysiological data using a fluorescent nano-imaging technique.

### 32.3.2 Attachment of conjugated GlcN to the cell membrane

Previously, QDs were modified with several types of ligands, such as antibodies (Ohyabu et al. 2009) and sugars (Higuchi et al. 2008), to enhance the cellular uptake via specific interactions with target cell membranes. Cationic polymers with a large number of primary amine groups have been recently applied to modify the surface of QDs. This type of modification is expected to enhance cellular uptake through the nonspecific electrostatic interaction between their positive and negative charges on the cell membrane.

Although GlcN is not a polymer but monomer type of chitosan, the attachment of GlcN to odontoblastic cell membrane occurs similarly due to a positive charge and the biocompatibility of conjugated GlcN (Figure 32.2). This phenomenon is supported by the finding that nonconjugated GlcN have difficulty entering into cells. It has been recently reported that within three days, GlcN gets attached to the cell membrane and entered into cytoplasm (Igawa et al. 2014b). Furthermore, the interesting findings that the cell membrane was intact even after conjugated QDs entered into cells and released extracellularly indicates GlcN's potential toward membrane stability.

### 32.3.3 Distribution in cytoplasm

After 1 day of culture, many QDs were taken up into osteoblastic cells (Figure 32.3). After 7 days of culture, the distribution of QDs dramatically decreased inside the cells. The more interesting finding was that no overlap was recognized between QDs and lysosomes after merging the fluorescent images of these two structures (Figure 32.2). This is the first experiment to observe the stability and protection of intracellularly distributed QDs following the conjugation with GlcN.

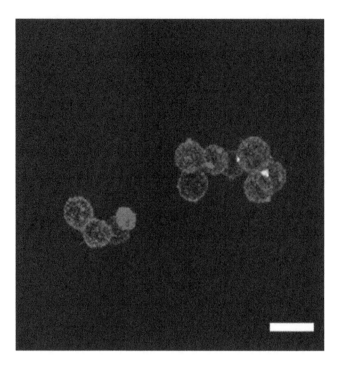

**Figure 32.2** QD (white) inside the cell membrane (gray) after 3 h of culture. Scale bar = 15 µm. (From Igawa, K. et al., *BioMed Res. Int.*, 2014, Article ID 821607, 2014b. With permission.)

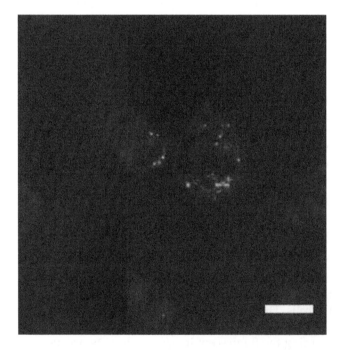

**Figure 32.3** Many QDs (large white dots) are labeled inside the cells after 1 day of culture with GlcN. Note the lack of overlap between the QDs and lysosomes (small white dots). Scale bar = 15 µm. (From Igawa, K. et al., *BioMed Res. Int.*, 2014, Article ID 821607, 2014b. With permission.)

QDs are generally trapped in endosomes/lysosomes after endocytosis. The fluorescence intensity of QDs would be decreased irreversibly after QDs move into endosomes/lysosomes with a pH of around 4.0–5.0 in the cells (Silver and Ou 2005; Liu et al. 2007; Manda and Tamai 2008). Cationic polymers such as PAMAM and PEI have a strong pH-buffering capacity that could enhance proton absorption in acid organelles and an osmotic pressure buildup across the organelle membrane, which could promote endosomal escape and release into the cytoplasm (Sonawane et al. 2003; Akinc et al. 2005). The pH of D-glucosamine hydrochloride is acidic (3.5–4.5). This acidic condition supports the proton sponge hypothesis (Remy et al. 1994; Behr 1997; Tang and Szoka 1997) involving escape from endosomes/lysosomes due to QDs labeling for long periods (over 30 min; personal communication). The polycationic behavior of GlcN together with the action for cell membrane stability is useful and meaningful for cell biological studies (Igawa et al. 2014b), such as the investigation of the fate of small molecular protein in cytoplasm.

## 32.4   Summary

The membrane stabilizing activity of GlcN was addressed in our recent studies, QDs, and electroporation applications into cultured osteoblastic cells. Although many other biological activities of GlcN have not been fully reviewed in this chapter, cell membrane stability by GlcN application may be related to the basic and fundamental mechanism for promoting cell activity, probably through its antioxidative and immunostimulating properties. GlcN can function as a biological response modifier and can boost natural protective responses to both cells and tissues.

## References

Akinc, A., M. Thomas, A. M. Klibanov, and R. Langer. 2005. Exploring polythylenimine-mediated DNA transfection and the proton sponge hypothesis. *Journal of Gene Medicine* 7:657–663.

Algarra, M., B. B. Campos, M. S. Miranda, and J. C. G. Esteves da Silva. 2011. CdSe quantum dots capped PAMAM dendrimer nanocomposites for sensing nitroaromatic compounds. *Talanta* 83:1335–1340.

Behr, J.-P. 1997. Theproton sponge: A trick to enter cells the viruses did not exploit. *Chimia* 51:34–36.

Ben-David, E., L. Appelbaum, J. Sosna, I. Nissenbaum, and S. N. Goldberg. 2012. Characterization of irreversible electroporation ablation in *in vivo* porcine liver. *American Journal of Roentgenology* 198:W62–W68.

Benevento, R., A. Santoriello, G. Perna, and S. Canonico. 2012. Electrochemotherapy of cutaneous metastases from breast cancer in elder patients: A preliminary report. *BMC Surgery* 12(Suppl. 1):S6.

Boussif, O., F. Lezoualc'h, M. A. Zanta, M. D. Mergny, D. Scherman, B. Demenelx, and J.-P. Behr. 1995. A versatile vector for gene and oligonucleotide transfer into cells in culture and *in vivo*: Polyethylenimine. *Proceedings of the National Academy of Sciences of the United States of America* 92:7297–7301.

Campana, L. G., S. Valpione, C. Falci, S. Mocellin, M. Basso, L. Corti, N. Balestrieri et al. 2012. The activity and safety of electrochemotherapy in persistent chest wall recurrence from breast cancer after mastectomy: A phase-II study. *Breast Cancer and Treatment* 134:1169–1178.

Charpentier, K. P. 2012. Irreversible electroporation for the ablation of liver tumors: Are we there yet? *Archives of Surgery* 147:1053–1061.

Charpentier, K. P., F. Wolf, L. Noble, B. Winn, M. Resnick, and D. E. Dupuy. 2010. Irreversible epectroporation of the pancreas in swine: A study. *HPB (Oxford)* 12:348–351.

Crolle, G. and E. D'Este. 1980. Glucosamine sulphate for the management of arthrosis: A controlled clinical investigation. *Current Medical Research and Opinion* 7:104–109.

Cummins, T. R., P. L. Sheets, and S. G. Waxman. 2007. The roles of sodium channels in nociception: Implications for mechanisms of pain. *Pain* 131:243–257.

Curatolo, P., P. Quaglino, F. Marenco, M. Mancini, T. Nardo, C. Mortera, R. Rotunno et al. 2012. Electrochemotherapy in the treatment of Kaposi sarcoma cutaneous lesions: A two-center prospective phase II trial. *Annals of Surgical Oncology* 19:192–198.

Goldberg, A. and M. L. Yarmush. 2013. Nonthermal irreversible electroporation: Fundamentals, applications, and challenges. *IEEE Transactions on Biomedical Engineering* 60:707–714.

Hibino, M., H. Itoh, and K. Jr. Kinoshita. 1993. Time courses of cell electroporation as revealed by submicrosecond imaging of transmembrane potential. *Biophysical Journal* 64:1789–1800.

Higuchi, Y., S. Kawakami, and M. Hashida. 2010. Strategies for *in vivo* delivery of siRNAs: Recent progress. *BioDrugs* 24:195–205.

Higuchi, Y., M. Oka, S. Kawakami, and M. Hashida. 2008. Mannosylated semiconductor quantum dots for the labeling of macrophages. *Journal of Controlled Release* 125:131–136.

Higuchi, Y., C. Wu, K.-L. Chang, K. Irue, S. Kawakami, F. Yamashita, and M. Hashida. 2011. Polyamidoamine dendrimer-conjugated quantum dots for efficient labeling of primary cultured mesenchymal stem cells. *Biomaterials* 32:6676–6682.

Hua, J., K. Sakamoto, and I. Nagaoka. 2002. Inhibitory actions of glucosamine, a therapeutic agent for osteoarthritis, on the functions of neutrophils. *Journal of Leukocyte Biology* 71:632–640.

Hua, J., S. Suguro, S. Hirano, K. Sakamoto, and I. Nagaoka. 2005. Preventive actions of a high dose of glucosamine on adjuvant arthritis in rats. *Inflammation Research* 54:127–132.

Hua, J., S. Suguro, K. Iwabuchi, Y. Tsutsumi-Ishii, K. Sakamoto, and I. Nagaoka. 2004. Glucosamine, a naturally occurring amino monosaccharide, suppresses the ADP-mediated platelet activation in humans. *Inflammation Research* 53:680–688.

Igawa, K., N. Ohara, A. Kawakubo, K. Sugimoto, K. Yanagiguchi, T. Ikeda, S. Yamada et al. 2014a. D-Glucosamine promotes transfection efficiency during electroporation. 2014: Article ID 485867, 4pp.

Igawa, K., M.-F. Xie, H. Ohba, S. Yamada, and Y. Hayashi. 2014b. D-Glucosamine conjugation accelerates the labeling efficiency of quantum dots in osteoblastic cells. *BioMed research International* 2014: Article ID 821607, 5pp.

Kaida, K., H. Yamashita, K. Toda, and Y. Hayashi. 2013. Effects of glucosamine on tooth pulpal nociceptive responses in the rat. *Journal of Dental Sciences* 8:68–73.

Kaida, K., H. Yamashita, K. Toda, and Y. Hayashi. 2014. Suppressive effects of D-glucosamine on the 5-HT sensitive nociceptive units in the rat tooth pulpal nerve. *BioMed Research International* 2014: Article ID 187989, 6pp.

Kawakami, S., Y. Higuchi, and M. Hashida. 2008. Nonviral approaches for targeted delivery of plasmid DNA and oligonucleotide. *Journal of Pharmaceutical Sciences* 97:726–745.

Kotnik, T. 2013. Lightning-triggered electroporation and electrofusion as possible contributors to natural horizontal gene transfer. *Physics of Life Reviews* 10:351–370.

Kotnik, T., P. Kramar, G. Pucihar, D. Miklavcic, and M. Tarek. 2012. Cell membrane electroporation-part 1: The phenomenon. *IEEE Electrical Insulation Magazine* 28:14–23.

Lee, E. W., C. Chen, V. E. Prieto, S. M. Dry, C. T. Loh, and S. T. Kee. 2010. Advanced hepatic ablation technique for creating complete cell death: Irreversible electroporation. *Radiology* 255:426–433.

Lee, Y. J., D. S. Lu, F. Osuagwu, and C. Lassman. 2012. Irreversible electroporation in porcine liver: Short- and long-term effect on the hepatic veins and adjacent tissue by CT with pathological correlation. *Investigative Radiology* 47:671–675.

Liu, Y. S., Y. H. Sun, P. T. Vernier, C. H. Liang, S. Y. C. Chong, and M. A. Gundersen. 2007. pH-sensitive photoluminescence of CdSe/ZnSe/ZnS quantum dots in human ovarian cancer cells. *Journal of Physical Chemistry C* 111:2872–2878.

Mandal, A. and N. Tamai. 2008. Influence of acid on luminescence properties of thioglycolic acid-capped CdTe quantum dots. *Journal of Physical Chemistry C* 112:8244–8250.

Marchais, D. and A. Marty. 1980. Action of glucosamine on acetylcholine-sensitive channels. *Journal of Membrane Biology* 56:43–48.

Martin, R. C. 2nd, K. McFarland, S. Ellis, and V. Velanovich. 2012. Irreversible electroporation therapy in the management of locally advanced pancreatic adenocarcinoma. *Journal of American College of Surgery* 215:361–369.

Martin-Belloso, O. and A. Sorbrino-Lopez. 2011. Combination of pulsed electric fields with other preservation techniques. *Food Bioprocess Technology* 4:954–968.

Matsunaga, T., K. Yanaguchi, S. Yamada, N. Ohara, T. Ikeda, and Y. Hayashi. 2006. Chitosan monomer promotes tissue regeneration on dental pulp wounds. *Journal of Biomedical Materials Research A* 15:711–720.

McCarty, M. F. 1996. Glucosamine for wound healing. *Medical Hypotheses* 47:273–275.

Medints, I. L., H. T. Uyeda, E. R. Goldman, and H. Mattoussi. 2005. Quantum dot bioconjugates for imaging, labeling and sensing. *Nature Materials* 4:435–446.

Michalet, X., F. F. Pinaud, L. A. Bentolila, J. M. Tsay, S. Doose, J. J. Li, G. Sundaresan et al. 2005. Quantum dots for liver cells, *in vivo* imaging, and diagnosis. *Science* 307:538–544.

Midoux, P. and M. Monsigny. 1999. Efficient gene transfer by histidylated polylysine/pDNA complexes. *Bioconjugate Chemistry* 10:406–411.

Miklavcic, D. 2012. Network for development of electroporation-based technologies and treatment: COST TD1104. *Journal of Membrane Biology* 25:591–598.

Murray, C. B., D. J. Norris, and M. G. Bewendi. 1993. Synthesis and characterization of nearly monodisperse CdE (E=sulfur, selenium, tellurium) semiconductor nanocrystallites. *Journal of American Chemical Association* 115:8706–8715.

Narayanan, G. 2011. Irreversible electroporation for treatment of liver cancer. *Gastroenterology and Hepatology (NY)* 7:313–316.

Neumann, E., M. Schaefer-Ridder, Y. Wang, and P. H. Hofschneider. 1982. Gene transfer into mouse lyoma cells by electroporation in high electric fields. *EMBO Journal* 1:841–845.

Ohyabu, Y., Z. Kaul, T. Yoshioka, K. Inoue, S. Sasaki, H. Mishima, T. Uemura et al. 2009. Stable and nondisruptive in vitro/in vivo labeling of mesenchymal stem cells by internalizing quantum dots. *Human Gene Therapy* 20:217–224.

Pack, D. W., D. Putnam, and R. Langer. 2000. Design of imidazole-containing endosomolytic biopolymers for gene delivery. *Biotechnology and Bioengineering* 67:217–223.

Park, J., J. Joo, S. G. Kwon, Y. Jang, and T. Hyeon. 2007. Synthesis of monodisperse spherical nanocrystals. *Angewandte Chemie International Edition in English* 46:4630–4660.

Pucihar, G., T. Kotnik, D. Miklavcic, and J. Teissie. 2008. Kinetics of transmembrane transport of small molecules into electropermeabilized cells. *Biophysical Journal* 95:2837–2848.

Reiss, P., M. Protiere, and L. Li. 2009. Core/shell semiconductor nanocrystals. *Small* 5:154–168.

Remy, J.-S., C. Sirlin, P. Vierling, and J.-P. Behr. 1994. Gene transfer with a series of lipophilic DNA-binding molecules. *Bioconjugate Chemistry* 5:647–654.

Rubinsky, B. 2007. Irreversible electroporation in medicine. *Technology in Cancer Research and Treatment* 6:255–260.

Rubinsky, B., G. Onik, and P. Mikus. 2007. Irreversible electroporation: A new ablation modality-clinical implications. *Technology in Cancer Research and Treatment* 6:37–48.

Sack, M., J. Sigler, S. Frenzel, C. Eling, J. Arnold, T. Michelberger, W. Frey et al. 2010. Research on industrial-scale electroporation devices fostering the extraction of substances from biological tissue. *Food Engineering Review* 2:147–156.

Sal'nikova, S. I., S. M. Droqovoz, and A. Zupanets. 1990. The liver-protective properties of D-glucosamine. *Famakol Toksikol* 53:33–35.

Saulis, G., M. S. Venslauskas, and J. Naktinis. 1991. Kinetics of pore resealing in cell membranes after electroporation. *Bioelectrochemistry and Bioenergetics* 26:1–13.

Sersa, G., T. Cufer, S. M. Paulin, M. Cemazar, and M. Snoj. 2012. Electrochemothapy of chest wall breast cancer recurrence. *Cancer Treatment Review* 38:379–386.

Setnikar, I. 1992. Antireactive properties of "chondroprotective" drugs. *International Journal of Tissue Reaction* 14:253–261.

Silver, J. and W. Ou. 2005. Photoactivation of quantum dot fluorescence following endocytosis. *Nano Letters* 5:1445–1449.

Sonawane, N. D., F. C. Szoka, and A. S. Verkman. 2003. Chloride accumulation and swelling in endosomes enhances DNA transfer by polyamine-DNA polyplexes. *Journal of Biological Chemistry* 278:44826–44831.

Tang, M. X. and F. C. Szoka. 1997. The influence of polymer structure on the interaction of cationic polymers with DNA and morphology of the resulting complexes. *Gene Therapy* 4:823–832.

Testori, A., C. R. Rossi, and G. Tosti. 2012. Utility of electrochemotherapy in melanoma treatment. *Current Opinion in Oncology* 24:155–161.

Toepfl, S., V. Heinz, and D. Knorr. 2007. High intensity pulsed electric fields applied for food preservation. *Chemical Engineering and Processing* 46:537–546.

Yan, Y., L. Wanshun, H. Baoqin, W. Changhong, F. Chenwei, L. Bing, and C. Liehuan. 2007. The antioxidative and immunostimulating properties of D-glucosamine. *International Immunopharmacology* 7:29–35.

Yang, S. and S. May. 2008. Release of cationic polymer-DNA complexes from the endosomes: A theoretical investigation of the proton sponge hypothesis. *Journal of Chemical Physics* 129:185105.

*chapter thirty-three*

# Structural glycobiology for lectin to promote advanced biomedical research

*Imtiaj Hasan, Yuki Fujii, Sarkar M.A. Kawsar, Sultana Rajia,*
*Shigeki Sugawara, Masahiro Hosono, Yukiko Ogawa, Yasushi Kawakami,*
*Yasuhiro Koide, Daiki Yamamoto, Robert A. Kanaly, and Yasuhiro Ozeki*

## Contents

## 33.1 Introduction

### 33.1.1 The appearance of glycans and their recognition molecules in the primitive sea

Deoxyribose is a key component of nucleotides in adenosine triphosphate (ATP) that transfer energy from glucose to chemical bonds through the glycolysis pathway in all present-day organisms. Therefore, saccharides have clearly been selected as the primary essential molecules for energy transfer reactions during the course of evolution. During chemical evolution in the primitive sea, the first monosaccharides were presumably generated by processes similar to the "formose reaction" which is autopolymerization of formaldehydes to make glyceraldehyde and dihydroxyacetone (nucleophilic) in the presence of basic catalysts (e.g., clay) under the prevailing weak reducing conditions. These two carbonyl compounds were able to bind to ketose sugars such as fructose (Frc) through the aldol condensation reaction. Further, catalytic reactions generated hexoses such as glucose (Glc). Following the appearance of Frc, both enolization (e.g., from Frc [ketose] to Glc [aldose]) and

epimerization of an hydroxyl group at the contiguous position (e.g., from Glc to mannose [Man; the C2 epimer of Glc]) led to the widespread production of Frc, Glc, and Man by nonenzymatic processes, through the so-called Lobry de Bruyn transformation (Hirabayashi 1996). In monosaccharides and other cyclohexane compounds, chair forms are more stable than boat forms, according to the concept of 1,3-diaxial interaction. The chair form of β-Glc does not involve 1,3-diaxial interaction because all hydroxyl groups at the C1–C4 positions take an equatorial configuration, which is the most stable thermodynamically. In contrast, Man and galactose (Gal) have a single axial hydroxyl group at the 2- and 4-positions, respectively (Figure 33.1; Glc vs. Man, Gal). Thus, each monosaccharide derived from a Glc epimer has one 1,3-diaxial interaction between C2-OH (axial) and C4-H Man, whereas Gal has an interaction between C2-H and C4-OH (axial).

As living organisms evolved, monosaccharides were used for various purposes, including energy reserves (Glc), catalytic nucleotides (ribose in ribozyme), frameworks of coenzymes (deoxyribose for ATP and

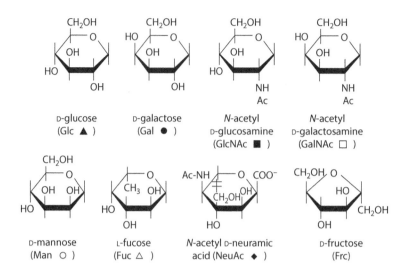

***Figure 33.1*** Monosaccharides that serve as building blocks for oligosaccharides (glycans). Symbol nomenclature follows the recommendations of the Human Disease Glycomics/Proteome Initiative (Wada et al. 2007).

NAD) and genes (deoxyribose and ribose). In contrast to Man (C2-epi-Glc), which could be synthesized by nonenzymatic processes during early chemical evolution, synthesis of Gal (C4-epi-Glc) required enzymes produced by organisms (Hirabayashi 1996, 2004). The many monosaccharide epimers resulting from the activity of various enzymes gave rise to great diversification of monosaccharide structural and functional properties (Figure 33.1). Monosaccharides are covalently bonded together by glycosyltransferases as molecular chains termed "oligosaccharides" or "glycans." Specific glycosyltransferases and their substrates (nucleotide sugars) give rise to many types of oligosaccharide structures whose diversity is based on monosaccharide sequence, glycosidic bonds (α- or β-), and the number of branches. Oligosaccharide chains are involved in formation of advanced biomolecules through covalent conjugation of proteins and lipids into glycoproteins and glycosphingolipids (GSLs).

Glycan structures play crucial roles in nearly all biological phenomena. Recent "omics" studies focused on glycans (glycomics) provide new ideas regarding the diversification of glycan structures in experimental model organisms and the dynamic changes of glycan structures observed in normal (e.g., differentiation) and abnormal (e.g., disease) processes. Studies using glycan-based microarray technology (Tateno et al. 2010a) demonstrated that insertion of four essential genes into somatic cells generated "iPS cells" with pluripotency similar to that of stem cells. The iPS cells displayed not only regenerative ability but also a pattern of surface glycan structures similar to that of natural stem cells, even though the transfected genes were unrelated to

these glycan structures (Figure 33.2) (Tateno et al. 2010b, 2013).

Proteins termed "lectins" that bind to oligosaccharides are present in all present-day organisms and function as glycan receptors. The biological roles of lectins were unclear for many decades, but recent studies have demonstrated the following functions: intracellular trafficking of glycoproteins in Golgi apparatus by the calnexin/calreticulin system (Williams 2006),

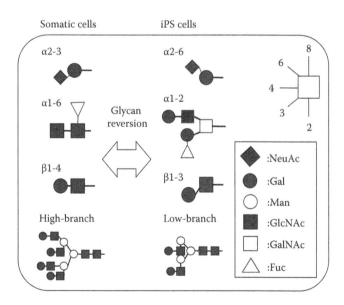

***Figure 33.2*** Alteration of glycosylation pattern during transformation of somatic to iPS cells. (Modified from Tateno, H. et al., *J. Biol. Chem.*, 286, 20345, 2010b; Tateno, H. et al., *Stem Cells Transl. Med.*, 2, 265, 2013.)

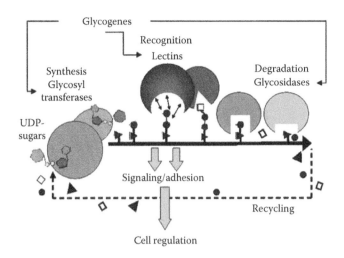

*Figure 33.3* Glycan-dependent cell regulation based on biosynthesis, recognition, and degradation controlled by glycogenes.

trafficking of extracellular matrix components by galectins (Berois and Osinaga 2014), protection from invading microorganisms (Grewal et al. 2013), host infection by influenza virus hemagglutinin (Sauer et al. 2014), bacterial and plant toxins (Tesh 2012), quality control of serum glycoproteins in blood by asialoglycoproteins (Christie et al. 2014; Mi et al. 2014), control of cancer cell metastasis by selectins (Shirure et al. 2012), and cell signaling by certain lectins, including siglecs (Müller et al. 2013). Lectins have the advantage of being able to recognize specific glycan structures through weak and reversible interactions to promote these processes. Glycans synthesized by transferases in the endoplasmic reticulum and Golgi apparatus are recognized by lectins with a low dissociation constant ($10^{-4}$ to $10^{-6}$ M), leading to transduction of cellular signals. Following the recognition process, glycans undergo a recycling process whereby they are hydrolyzed by glycosidases to supply UDP-sugars as substrates for glycosyltransferases. These biosynthesis, recognition, and degradation processes contributing to glycan-dependent cell regulation (signaling, adhesion) are under the control of glycogenes (Figure 33.3).

### 33.1.2   Galactose-binding lectins in marine invertebrates and their usefulness in biomedical studies

The ancestors of present-day marine invertebrates first appeared during the Cambrian era, as part of the "Cambrian explosion" (rapid diversification of organisms in the ocean). Many of the essential or important genes (including lectin genes) found in present-day species most likely arose during the Cambrian and were

maintained during the course of evolution. Even in invertebrates whose immune systems do not include antibodies, certain lectins may help prevent pathogen invasion (Hoving et al. 2014). Many lectins that recognize D-Gal and its derivatives have been obtained experimentally from organs of marine invertebrates (Ozeki et al. 1991; Kawsar et al. 2008, 2009a,b, 2011; Fujii et al. 2009, 2011, 2012; Matsumoto et al. 2011), even though Gal residues may not be expressed in the animal's glycans. Some Gal-binding lectins of marine invertebrates have been found to kill *Artemia* (brine shrimp) (Kawsar et al. 2010) and human cancer cells (Kawsar et al. 2011; Fujii et al. 2012). Lectins in these animals have evidently evolved over time for protection from predators, similarly to the evolution of lectins and other toxins in plant seeds. Many marine invertebrate lectins affect mammalian (including human) cells, and therefore have great potential for application in biomedical research.

Mammals express both Gal and sialic acid residues at nonreducing termini of glycans. Galactosyltransferases are located in the lumen of the trans-Golgi network. Gal residues in glycans play important roles in biological recognition related to cell adhesion and signal transduction (Steelant et al. 2002; Malagolini et al. 2009; Boggs et al. 2010). The application of Gal-binding lectins with diverse biochemical characteristics obtained from marine invertebrates in life science and biomedical research is likely to reveal novel properties of mammalian cells.

We describe in this chapter glycome analytical studies of an α-Gal-binding lectin isolated from *M. galloprovincialis* (Mediterranean mussel) and the potential application of this lectin in cancer-related biomedical research.

## 33.2   Diversity of glycan-binding patterns of Gal-binding lectins as revealed by glycome analysis

### 33.2.1   Frontal affinity chromatography technology: An excellent glycome analytical system for lectins

Frontal affinity chromatography technology (FACT) was originally established as a procedure to evaluate the weak and reversible molecular interactions ($KD \approx 10^{-4}$ to $10^{-6}$ M) between proteases and their inhibitors (Kasai and Ishii 1978a,b). It was subsequently refined as a sophisticated glycome technology to evaluate the glycan-binding profiles of lectins (Hirabayashi et al. 2002, 2003). The instrumentation consists of a small column with lectin-conjugated affinity gel between a high-performance liquid chromatography (HPLC) pump

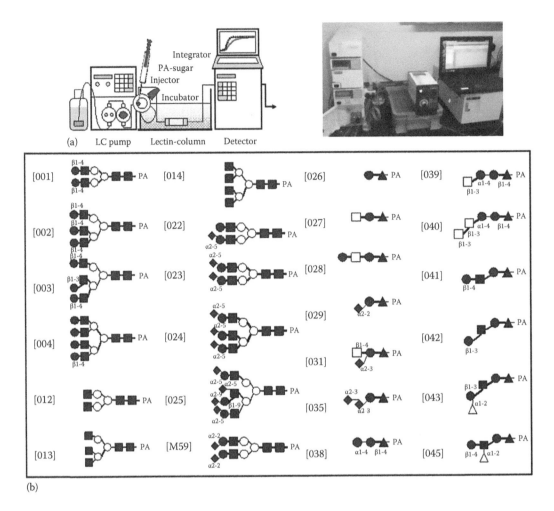

**Figure 33.4** FACT instrumentation and the 25 PA-labeled oligosaccharides discussed in this chapter. (a) The FACT system consists of a liquid chromatography pump, immobilized-lectin affinity column, fluorescence detector, and integrator. Twenty-five PA-oligosaccharides and inert control PA-rhamnose are injected manually into the column at a flow rate of 250 μL/min. The elution volume of PA-oligosaccharides is detected by fluorescence (Ex. 310 nm/Em. 380 nm). The retardation volume ($V$–$V_0$) of each PA-oligosaccharide ($V$) in comparison with control PA-rhamnose ($V_0$) is estimated to give a glycan-binding profile. (b) Twenty-five PA-oligosaccharides. Symbol nomenclature as in Figure 33.1. Inserted α- and β-numbers indicate the nature of glycosidic bond connections between monosaccharides. 001-025 and M059: oligosaccharides derived from glycoproteins. 026-045: oligosaccharides derived from GSLs.

and a fluorescence detector (Figure 33.4). Several pmol of a pyridylamino(PA)-group-labeled oligosaccharide library consisting of 25 kinds of PA-oligosaccharide including serum glycoprotein-type oligosaccharides (001-025, M59) and GSL-type oligosaccharides (026-045) are injected into the column. If the PA-oligosaccharide is recognized by the lectin in the column, delays in the elution of front peak detected by fluorescence (emission and excitation wavelengths 380 and 310 nm, respectively) are comparable to those of negative control PA-sugar, and glycan-binding properties of the lectins are compared. This system has been used to document characteristic, distinct glycan-binding profiles of D-Gal-binding lectins purified from various marine invertebrates.

### 33.2.2  Gal-binding lectins in marine animals

Various groups were searched during 2008–2011 for novel Gal-binding lectins from marine invertebrates (phyla Porifera, Annelida, Mollusca, Echinodermata) and vertebrates (fish, frog) using lactosyl- (Galβ1-4Glc) or melibiosyl- (Galα1-6Glc) agarose affinity columns, resulting in the purification of lectins with molecular masses ranging from 11.5 to 63 kDa, from 10 species (Figure 33.5). Based on a primary structural analysis, "sea urchin egg lectin (SUEL)/rhamnose-binding lectin" (SUEL) and "Ricin B-chain like (R-type) lectin" (PnL) families were purified from unfertilized sea urchin eggs and lugworm body wall, respectively. A "galectin"

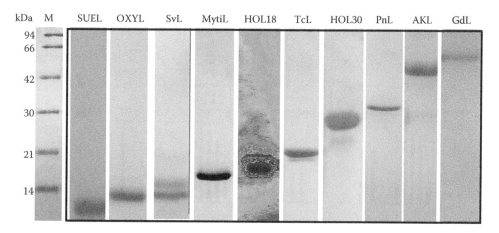

**Figure 33.5** Molecular diversity of Gal-binding lectins isolated from marine invertebrates. SUEL: sea urchin egg lectin from *Anthocidaris crassispina* (Ozeki et al. 1991). OXYL: feather star lectin from *Oxycomanthus japonicus* (Matsumoto et al. 2011). SvL: mytilid lectin from *Septifer virgatus*. MytiL (MytiLec): 17 kDa lectin from mussel *Mytilus galloprovincialis* (Fujii et al. 2012). HOL18: 18 kDa GalNAc/GlcNAc-binding lectin from *Halichondria okadai* (Matsumoto et al. 2012). TcL: coronate moon turban shell lectin from *Turbo* (Lunella) *coreensis* (Fujii et al. 2011). HOL30: 30 kDa Gal-binding lectin from *Halichondria okadai* (Japanese sponge) (Kawsar et al. 2008). PnL: lugworm lectin from *Perinereis nuntia* (Kawsar et al. 2009a). AKL: sea hare egg lectin from *Aplysia kurodai* (Kawsar et al. 2009b, 2011). GdL: bladder moon shell lectin from *Glossaulax didyma* (Fujii et al. 2009).

representative of β-galactoside-specific lectins was found as a sponge lectin (HOL30).

FACT analysis revealed great diversity in these Gal-binding lectins purified and marine invertebrates, even though the purification steps in each case were nearly identical (Figure 33.6).

Each of these Gal-binding lectins from a variety of invertebrates and vertebrates had a characteristic glycan-binding profile. The target glycans can be classified as (1) complex-type glycans (for HOL30, PnL), (2) complex-type glycans with type-2 lactosamine (Galβ1-4GlcNAc) structure (OXYL), (3) complex-type glycans and ABH blood type lacto-series (TcL), (4) complex-type glycans and α-galactoside lacto-series (Gb3 [globotriose]) (AKL), (5) α-galactoside (MytiLec, SAL), and (6) α-N-acetylgalactosamine containing GSL oligosaccharides (RCL). Some of the lectins recognized a narrow range of specific glycans (Gb3 [MytiLec, SAL], complex-type glycan [HOL30, PnL], type-2 lactosamine [OXYL], α-GalNAc [RCL]). Other lectins recognized a mixture or wide range of glycan structures (complex type glycan + lacto-series [TcL and AKL]).

The comparative glycomic approach has given rise to a new concept that recognition properties of Gal-binding lectins isolated from marine animals are considerably more diverse than previously thought. Some lectins (MytiLec, SAL, AKL) recognize α-galactoside. These characteristics are advantageous for studies of glycan-dependent cell signaling in a biomedical context.

## 33.3    Structure and function of MytiLec, a Gb3-specific lectin *from* Mytilus galloprovincialis

### 33.3.1    α-GAL in Gb3 as a recognition sugar

A lectin glycome (FACT) analysis showed that MytiLec, a Gal-binding lectin purified from the mussel *M. galloprovincialis*, specifically recognizes the GSL globotriaosylceramide (Gb3; Galα1-4Galβ1-4Glc-ceramide) (Figure 33.6). Gb3, also known as CD77 or P$^k$ blood-group antigen, is highly expressed in GSL-enriched microdomain (GEM) of Burkitt's lymphoma cells that result from herpes virus infection (Wiels et al. 1981, 1984; Hakomori and Handa 2002). Gb3 is also a target of verotoxins and the HIV adhesion molecule gp120 (Lingwood 2011; Lingwood and Branch 2011). It is synthesized in vertebrate cells by the transfer of UDP-Gal (donor) to lactosylceramide (Galβ1-4Glc-Cer) (acceptor) by Gb3 synthase (α1-4 galactosyltransferase) (Shin et al. 2009). Interestingly, homologous genes encoding Gb3 synthase are conserved in invertebrates and plants (Keusch et al. 2000). Lactosylceramide is expressed on vertebrate cell surfaces, but is also found in the sea star *Asterias amurensis* (Irie et al. 1990). α1-4 galactosyltransferase from cultured butterfly cells was found to transfer UDP-Gal to Galβ1-4GalNAc-Ser/Thr of mucin-type glycans (Lopez et al. 1998). α-Gal residues are located at the nonreducing terminus of glycans and function as xenograft antigens, toxin

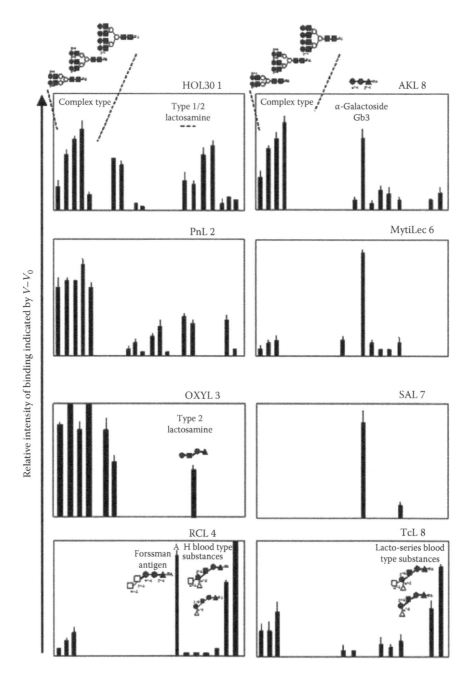

***Figure 33.6*** Glycan-binding profiles of Gal-binding lectins from aquatic animals, analyzed by FACT. *Y*-axis: relative intensity of the difference between volume elution of each PA-oligosaccharide (*V*) and PA-rhamnose ($V_0$). Height of bars was calculated as $V–V_0$. *X*-axis: library of PA-oligosaccharides as in Figure 33.4. Lectins as described in Figure 33.5 except as noted. 1: HOL30. 2: PnL. 3: OXYL. 4: RCL from *Rana catesbeiana* (bullfrog) eggs (Kawsar et al. 2009c). 5: AKL. 6: MytiLec. 7: SAL from catfish egg *Silurus asotus* (Hosono et al. 2013). 8: TcL.

receptors, and histo-blood group type carriers for various microbial diseases with differing susceptibility. Gal is generally the sugar with a role in recognition (Figure 33.7), whereas Glc and Man have structural roles such as components of glycogen and cell walls. Therefore, Gal residues transferred to glycans by galactosyltransferases at the termini of glycans, have an advantage in recognition processes (Hirabayashi 1996, 2004). We have studied in detail and describe, in the following sections, the structural and functional properties of MytiLec, a lectin that specifically recognizes the Gb3 structure.

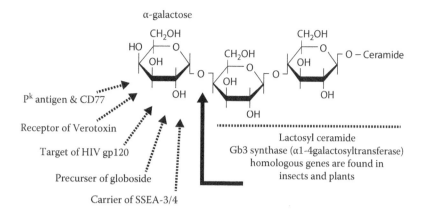

*Figure 33.7* Reported roles as recognition sugar of α-galactose in Gb3.

### 33.3.2   Novel primary structure of MytiLec

We isolated a 17 kDa Gb3-binding lectin (termed MytiLec) from the marine bivalve mussel *M. galloprovincialis* (family Mytilidae). *M. galloprovincialis* is native to the Mediterranean, but is now found worldwide as a result of human activities. Studies using genome databases led to the establishment of an expressed sequence tag (EST) library, "MytiBase," for *M. galloprovincialis* (Venier et al. 2011a,b). This EST database is useful for research on physiology, development, and defensive systems. Although MytiLec was described independently based on biochemical procedures, a focus on the animal model utilizing the EST database will promote future discovery of novel genes regulated by the lectin.

The lectin was purified using a column packed with melibiosyl (Galα1-6Glc)-agarose affinity gel, a technique useful for providing new information on primary structure (Fujii et al. 2012). MytiLec is composed of 149 amino acids, including a 50-amino acid. triple-tandem structure. The *N*-terminus is acetyl threonine. The complete primary structure of MytiLec is novel in comparison with other lectins having known primary structure. Its highest sequence similarity (88%) is with a Gal/GalNAc-binding lectin isolated from the related marine mussel *Crenomytilus grayanus* (Kovalchuk et al. 2013). In each domain of MytiLec, the basic amino acid. residues Lys, His, and Arg (*) are highly conserved around the polypeptide, whereas acidic amino acid residues such as Asp and Glu (#) are conserved on the C-terminal side.

MytiLec treatment of Burkitt's lymphoma Raji cells (0–50 μg/mL) resulted in dose-dependent reduction of cell viability, based on trypan blue exclusion assay (to detect cytotoxicity) and water-soluble tetrazolium salt assay (to detect cell growth). MytiLec treatment had no effect on viability of erythroleukemia K562 cells, which do not express Gb3. Lectin-dependent cytotoxicity

against Raji was more strongly inhibited by the co-presence of an α-galactoside (melibiose), which mimics Gb3 on the cell surface, than by the co-presence of a β-galactoside (lactose).

MytiLec treatment of Raji cells induced inversion of phosphatidylserine and a loss of cell membrane integrity, as assessed by anti-annexin V antibody binding and propidium iodate incorporation measured by fluorescence-activated cell sorting (FACS). MytiLec caused late apoptosis or necrosis of cells in a dose-dependent manner. This effect was inhibited by the co-presence of melibiose. The results of primary structure analysis and cytotoxicity and FACS assays, taken together, indicate that MytiLec has a unique primary structure with short motifs and promotes glycan-dependent cytotoxicity with late apoptosis/necrosis. Both the primary structure and cell regulatory activity of MytiLec are unique among known animal lectins. To clarify the relationship between the conformation of MytiLec and mechanism of its regulatory effect on cell growth, we determined its 3-D structure.

### 33.3.3   Conformational identity of MytiLec and ricin B-chain

MytiLec was crystallized using gene recombinant technology (Figure 33.8, inset), and its 3-D structure was resolved to 1.1 Å by X-ray diffraction analysis. The five amino acids in each of the three domains in the polypeptide essential for sugar binding were identified (asterisks in panel a). Conformational analysis demonstrated why both MytiLec and the Gal/GalNAc-binding lectin from *C. grayanus* are able to bind Gal and GalNAc residues (Belogortseva et al. 1998): these five essential amino acids are conserved in the two lectins (Kovalchuk et al. 2013), and the hydroxyl groups at positions C3, C4, and C6 of the sugars bind to the polypeptide. Neither the hydroxyl

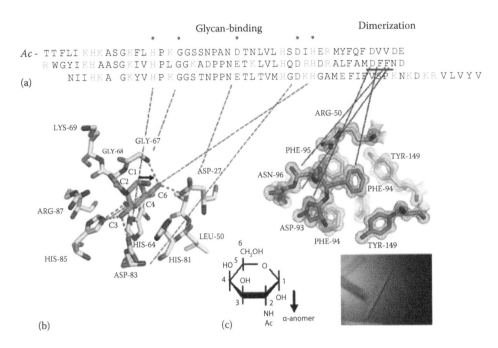

(a)

(b)                                                    (c)

***Figure 33.8*** Binding sites of MytiLec. (a) Five essential amino acids in each subdomain are indicated by asterisks (*). (b) Crystallographic analysis indicating that MytiLec binds to hydroxyl groups at positions C3, C4, and C6 of GalNAc. The α-anomer direction of the C1 hydroxyl group is compatible with the site within the lectin (arrow at C1 of α-GalNAc directs to α-anomer). (c) Asp[93]-Asn[96] is the binding site for the dimerization of MytiLec. *Inset*: lectin crystal used for structural analysis.

nor the *N*-acetyl group at position C2 is involved in lectin binding (C3, C4, and C6 in Panel b). Crystallographic analysis also showed that the lectin preferentially recognizes the α-anomer of Gal/GalNAc over the β-anomer (arrow direction in Panel b). MytiLec undergoes dimerization through hydrophobic interaction at the Asp[93]-Asn[96] region in each polypeptide (Panel c, and bar in Panel a).

Although the primary structure of MytiLec is novel in comparison with other known lectins, crystallographic study showed that its conformation is identical with that of Ricin B-chain (β-trefoil structure) (Figure 33.9: MytiLec vs. Ricin B). Inside each triple tandem repeat in the polypeptide, all four β-sheets with one α-helix are conserved in MytiLec, as in Ricin B-chain (panel b; secondary structure prediction).

Ricin, a typical plant toxin found in castor beans (*Ricinus communis*), has been studied as a model lectin since 1888 (Sharon and Lis 2004). It has two subunits for the toxic *N*-glycosidase (A subunit) and the β-galactoside-binding lectin (B subunit) (Roberts et al. 1985; Lord et al. 1994). The Ricin B-chain type lectin family has many members found in a wide variety of organisms, including microorganisms (plant pathogenic fungus: Sulzenbacher et al. 2010; mushroom: Bovi et al. 2011), plants (Fouquaert and Van Damme 2012), invertebrates (butterfly larvae: Matsushima-Hibiya et al. 2003; earthworm: Hemmi et al. 2013; lugworm: Kawsar et al. 2009a; sea cucumber: Uchida et al. 2004), and vertebrates (macrophage Man

receptor: Fiete and Baenziger 1997). β-trefoil structures are also found in glycosyltransferases, interleukin-1, and fibroblast growth factors (Murzin et al. 1992; Gerken et al. 2013). The β-trefoil framework thus appears to have essential importance in living systems. Although we did not find a common ancestral primary structure consisting of typical R-type lectins in genes of MytiLec and other R-type lectins with new sequences, it should be possible to construct a β-trefoil conformation through preparation of 12 β-sheets in the polypeptide and assemble a framework that represents synchronic evolution. A comparison of properties between MytiLec and pierisin (a cytotoxic protein from butterfly larvae) would be particularly interesting. Pierisin has a domain construction similar to that of Ricin (two R-type lectin domains and one cytotoxic ADP-ribosyltransferase domain) and Gb3-binding specificity similar to that of MytiLec. In HeLa cells, pierisin combined with surface Gb3 glycans and caused apoptosis through R-type lectin domains and the ADP ribosyltransferase domain (Matsushima-Hibiya et al. 2003). Studies of the cell suicidal mechanism by MytiLec (which has no cytotoxic domain) will help to clarify apoptotic processes more.

The fact that an identical β-trefoil structure conformation is found in MytiLec and Ricin B-chain suggests that MytiLec could be a novel research target in cell regulation studies through retrograde trafficking of the lectin. Some R-type lectins have been reported to be incorporated into cultured cells through the activation

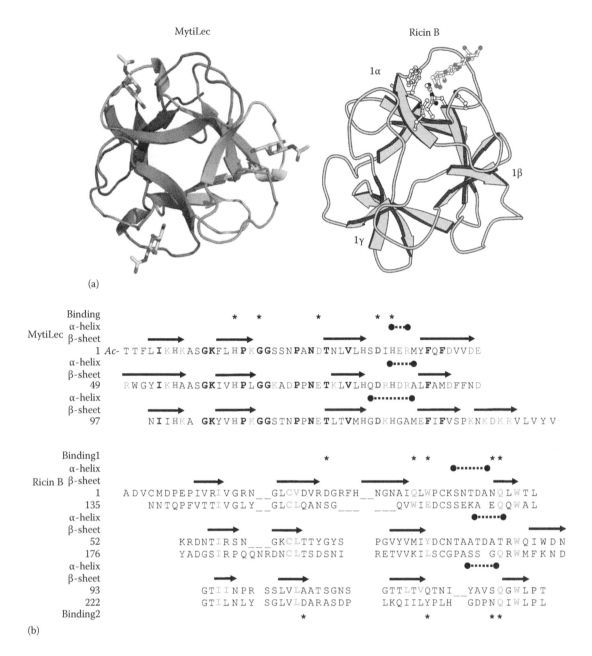

*Figure 33.9* Conformational identity of MytiLec and Ricin B-chain. (a) 3-D structures. (b) Secondary structure predictions (dotted lines: α-helices; arrows: β-sheets). Although no primary structure similarity was found for the two lectins, 12 β-sheets and 3 α-helices were highly conserved. The structure data for Ricin B-chain is from the Swiss-Prot database (P02879). Asterisks indicate essential amino acids in a polypeptide that bind to two lactose molecules. (From Lord, J.M. et al., *FASEB J.*, 8, 201, 1994.)

of signal transduction molecules, such as caspases (Hasegawa et al. 2000; Tesh 2012).

## 33.4   Intracellular trafficking of MytiLec and its cytotoxic effects

Crystallographic studies, as described earlier, showed that MytiLec protein has a β-trefoil conformation identical to that of Ricin B-chain. MytiLec also displays cytotoxic activity; it kills Gb3-expressing cells by binding to cell surface glycans. Thus, MytiLec resembles Ricin in its ability to be incorporated by and kill cells. In contrast to Ricin, however, MytiLec lacks a cytotoxic subunit analogous to the Ricin A-chain whose *N*-glycosidase activity causes ribosomal inactivation. We hypothesized that MytiLec itself activates cytotoxic signaling molecules during retrograde intracellular trafficking after being incorporated into cells through binding to Gb3

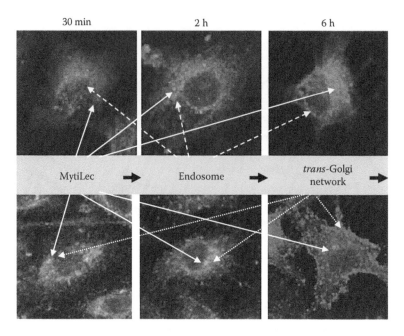

*Figure 33.10* Translocation of MytiLec from endosome to trans-Golgi network of HeLa cells. Cells were observed at 30 min, 2 h, and 6 h after administration of FITC-labeled MytiLec. Images of the distribution pattern were merged with those of endosome marker (transferrin receptor) and trans-Golgi network marker (TGN46) (solid lines: location of MytiLec; dotted lines: location of the organelle markers).

glycans. Fluorescent probe (FITC)-labeled MytiLec was introduced and tracked inside Gb3-expressing HeLa cells to observe how the lectin migrates and becomes quickly incorporated into the cells (Figure 33.10). The applied MytiLec became attached to the cells and was retro-translocated to the endosomes within 2 h to sort cell surface molecules, as evidenced by the co-presence of transferrin receptors, a molecular marker indicating endosome location. MytiLec also migrated into the trans-Golgi network after a few hours, but was then released gradually from this network to the cytoplasm. It remains to be determined precisely where MytiLec was localized throughout the 12 h period following its administration to the cells. Further studies will clarify the relationship between the cytotoxic activity of MytiLec and its retrograde intercellular trafficking from the cell surface to cytosol through the endosome/trans-Golgi network.

We found that MytiLec is incorporated into cytoplasm of other Gb3-expressing cell types. The cytotoxic activity of MytiLec varied among the cell types (Figure 33.11). The mechanisms of Gb3-dependent cell signaling and cytotoxicity will be clarified by further studies on apoptotic effects of MytiLec. Molecules that activate cell regulatory events through intracellular trafficking can be identified by comparison of various Gb3-expressing cell types, using MytiLec as a tool.

## 33.5 Future perspective: "blue glycomics" and biomedical studies

We use the term "blue glycomics" to refer to glycomic studies of lectins and other glycan-related molecules isolated from marine animals and the application of these molecules in biomedical research. For example, the lectin MytiLec was isolated from a mussel species, specifically recognizes Gb3 glycans, and was found to trigger Gb3-dependent late apoptosis/necrosis in cancer cells. The primary structure of MytiLec was distinct from those of other known lectins. Crystallographic studies showed that MytiLec has a β-trefoil structure identical to that of Ricin B-chain, although again the primary structures were different. We therefore hypothesized that MytiLec is incorporated into cells similarly to Ricin, and is translocated inside Gb3-expressing cells. Despite having only a glycan-binding subunit, MytiLec displays cytotoxicity similar to that of Ricin, which also has an N-glycosidase subunit and kills cells through inactivation of ribosomes and lectin subunits. MytiLec may activate apoptosis-inducing molecules during retrograde intracellular trafficking. In U937 cells, apoptosis was induced by a separated Ricin B-chain that had only lectin activity (Hasegawa et al. 2000). We found recently that a sialic

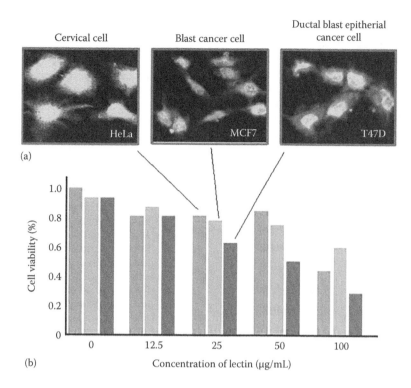

**Figure 33.11** Incorporation and cytotoxic effects of MytiLec in three cell lines. (a) Incorporation of FITC-labeled MytiLec into Gb3-expressing cells (HeLa, MCF7, T47D). Cells were observed by fluorescence microscopy following DAPI staining of nuclei. (b) Cell viability 48 h after administration of MytiLec. Three bar colors correspond to the three cell types as indicated.

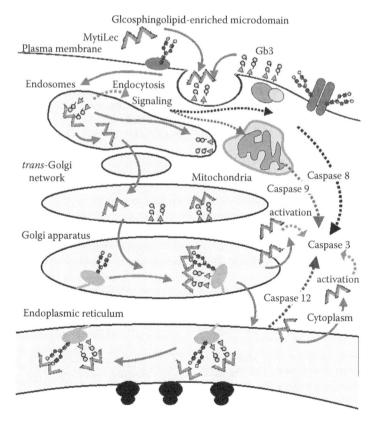

**Figure 33.12** Schematic representation of retrograde intracellular trafficking of MytiLec, with cytotoxic signals, from the endosome to cytoplasm via the *trans*-Golgi network and endoplasmic reticulum.

acid–binding lectin with primary structure identical to that of ribonuclease in frog eggs triggered apoptosis of cancer cells through the activation of caspase-3 and -8 (Ogawa et al. 2014).

The presence of apoptosis-inducing molecules in cytoplasm and our molecular morphological findings (Section 33.4.1) suggest that MytiLec is released into cytoplasm after traveling through a cell surface/endosome/trans-Golgi network pathway (Figure 33.12). Detailed studies on the direction of MytiLec trafficking in cells and on the upregulation of caspase levels will provide new insights in glycobiology and design of novel anticancer drugs that modify caspase activation. More generally, "blue glycomics" approaches will lead to valuable new biomedical research strategies in the future.

## Acknowledgment

The authors are grateful to Dr. S. Anderson for English editing of the manuscript.

## References

Belogortseva, N. I., V. I. Molchanova, A. V. Kurika, A. S. Skobun, and V. E. Glazkova. 1998. Isolation and characterization of new GalNAc/Gal-specific lectin from the sea mussel *Crenomytilus grayanus*. *Comparative Biochemistry and Physiology Part C: Pharmacology, Toxicology and Endocrinology* 119:45–50.

Berois, N. and E. Osinaga. 2014. Glycobiology of neuroblastoma: Impact on tumor behavior, prognosis, and therapeutic strategies. *Frontiers Oncology* 4:114.

Boggs, J. M., W. Gao, J. Zhao, H. J. Park, Y. Liu, and A. Basu. 2010. Participation of galactosylceramide and sulfatide in glycosynapses between oligodendrocyte or myelin membranes. *FEBS Letter* 584:1771–1778.

Bovi, M., M. E. Carrizo, S. Capaldi, M. Perduca, L. R. Chiarelli, M. Galliano, and H. L. Monaco. 2011. Structure of a lectin with antitumoral properties in king bolete (*Boletus edulis*) mushrooms. *Glycobiology* 21:1000–1009.

Christie, M. P., P. Simerská, F. E. Jen, W. M. Hussein, M. F. Rawi, L. E. Hartley-Tassell, C. J. Day, M. P. Jennings, and I. Toth. 2014. A drug delivery strategy: Binding enkephalin to asialoglycoprotein receptor by enzymatic galactosylation. *PLoS One* 9:e95024.

Fiete, D. and J. U. Baenziger. 1997. Isolation of the SO₄–4-GalNAcβ1,4GlcNAcβ 1,2Manα-specific receptor from rat liver. *Journal of Biological Chemistry* 272: 14629–14637.

Fouquaert, E. and E. J. Van Damme. 2012. Promiscuity of the euonymus carbohydrate-binding domain. *Biomolecules* 2:415–434.

Fujii, Y., N. Dohmae, K. Takio, S. M. A. Kawsar, R. Matsumoto, I. Hasan, Y. Koide et al. 2012. A lectin from the mussel *Mytilus galloprovincialis* has a highly novel primary structure and induces glycan-mediated cytotoxicity of globotriaosylceramide-expressing lymphoma cells. *Journal of Biological Chemistry* 287:44772–44783.

Fujii, Y., S. M. A. Kawsar, R. Matsumoto, H. Yasumitsu, N. Ishizaki, C. Dogasaki, M. Hosono et al. 2011. A D-galactose-binding lectin purified from coronate moon turban, *Turbo (Lunella) coreensis*, with a unique amino acid sequence and the ability to recognize lacto-series glycosphingolipids. *Comparative Biochemistry and Physiology Part B: Biochemistry and Molecular Biology* 158:30–37.

Fujii, Y., S. M. A. Kawsar, R. Matsumoto, H. Yasumitsu, N. Kojima, and Y. Ozeki. 2009. Purification and characterization of a D-galactoside-binding lectin purified from bladder moon shell (*Glossaulax didyma* Roding). *Journal of Biological Sciences* 9:319–325.

Gerken, T. A., L. Revoredo, J. J. Thome, L. A. Tabak, M. B. Vester-Christensen, H. Clausen, G. K. Gahlay et al. 2013. The lectin domain of the polypeptide GalNAc transferase family of glycosyltransferases (ppGalNAc Ts) acts as a switch directing glycopeptide substrate glycosylation in an N- or C-terminal direction, further controlling mucin type O-glycosylation. *Journal of Biological Chemistry* 288:19900–19914.

Grewal, P. K., P. V. Aziz, S. Uchiyama, G. R. Rubio, R. D. Lardone, D. Le, N. M. Varki, V. Nizet, and J. D. Marth. 2013. Inducing host protection in *Pneumococcal sepsis* by preactivation of the Ashwell-Morell receptor. *Proceedings of the National Academy of Sciences USA* 110:20218–20223.

Hakomori, S. and K. Handa. 2002. Glycosphingolipid-dependent cross-talk between glycosynapses interfacing tumor cells with their host cells: Essential basis to define tumor malignancy. *FEBS Letter* 531:88–92.

Hasegawa, N., Y. Kimura, T. Oda, N. Komatsu, and T. Muramatsu. 2000. Isolated ricin B-chain-mediated apoptosis in U937 cells. *Bioscience, Biotechnology, and Biochemistry* 64:1422–1429.

Hemmi, H., A. Kuno, and J. Hirabayashi. 2013. NMR structure and dynamics of the C-terminal domain of R-type lectin from the earthworm *Lumbricus terrestris*. *FEBS Journal* 280:70–82.

Hirabayashi, J. 1996. On the origin of elementary hexoses. *The Quarterly Review of Biology* 71:365–380.

Hirabayashi, J. 2004. On the origin of glycome and saccharide recognition. *Trends in Glycoscience and Glycotechnology* 16:63–85.

Hirabayashi, J., Y. Arata, and K. Kasai. 2003. Frontal affinity chromatography as a tool for elucidation of sugar recognition properties of lectins. *Methods in Enzymology* 362:353–368.

Hirabayashi, J., T. Hashidate, Y. Arata, N. Nishi, T. Nakamura, M. Hirashima, T. Urashima et al. 2002. Oligosaccharide specificity of galectins: A search by frontal affinity chromatography. *Biochimica et Biophysica Acta* 1572:232–254.

Hosono, M., S. Sugawara, T. Tatsuta, T. Hikita, J. Kominami, S. Nakamura-Tsuruta, J. Hirabayashi et al. 2013. Domain composition of rhamnose-binding lectin from shishamo smelt eggs and its carbohydrate-binding profiles. *Fish Physiology and Biochemistry* 39:1619–1630.

Hoving, J. C., G. J. Wilson, and G. D. Brown. 2014. Signalling C-type lectin receptors, microbial recognition and immunity. *Cell Microbiology* 16:185–194.

Irie, A., H. Kubo, F. Inagaki, and M. Hoshi. 1990. Ceramide dihexosides from the spermatozoa of the starfish, *Asterias amurensis*, consist of gentiobiosyl- cellobiosyl-, and lactosylceramide. *Journal of Biochemistry (Tokyo)* 108:531–536.

Kasai, K. and S. Ishii. 1978a. Affinity chromatography of trypsin and related enzymes. V. Basic studies of quantitative affinity chromatography. *Journal of Biochemistry (Tokyo)* 84:1051–1060.

Kasai, K. and S. Ishii. 1978b. Studies on the interaction of immobilized trypsin and specific ligands by quantitative affinity chromatography. *Journal of Biochemistry (Tokyo)* 84:1061–1069.

Kawsar, S. M. A., S. Aftabuddin, H. Yasumitsu, and Y. Ozeki. 2010. The cyotoxic activity of two D-galactose-binding lectins purified from marine invertebrates. *Archives of Biological Science Belgrade* 62:1027–1034.

Kawsar, S. M. A., Y. Fujii, R. Matsumoto, T. Ichikawa, H. Tateno, J. Hirabayashi, H. Yasumitusu et al. 2008. Isolation, purification, characterization and glycan-binding profile of a D-galactoside specific lectin from the marine sponge, *Halichondria okadai*. *Comparative Biochemistry and Physiology Part B: Biochemistry and Molecular Biology* 150:349–357.

Kawsar, S. M. A., R. Matsumoto, Y. Fujii, H. Matsuoka, N. Masuda, C. Iwahara, H. Yasumitsu et al. 2011. Cytotoxicity and glycan-binding profile of a D-galactose-binding lectin from the eggs of a Japanese sea hare (*Aplysia kurodai*). *Protein Journal* 30:509–519.

Kawsar, S. M. A., R. Matsumoto, Y. Fujii, H. Yasumitsu, C. Dogasaki, M. Hosono, K. Nitta et al. 2009b. Purification and biochemical characterization of D-galactose binding lectin from Japanese sea hare (*Aplysia kurodai*) eggs. *Biochemistry (Moscow)* 74:709–716.

Kawsar, S. M. A., R. Matsumoto, Y. Fujii, Y. Yasumitsu, H. Uchiyama, M. Hosono, K. Nitta et al. 2009c. Glycan-binding profile and cell adhesion activity of American bullfrog (*Rana catesbeiana*) oocyte galectin-1. *Protein and Peptide Letter* 16:677–684.

Kawsar, S. M. A., T. Takeuchi, K.-I. Kasai, Y. Fujii, R. Matsumoto, H. Yasumitsu, and Y. Ozeki. 2009a. Glycan-binding profile of a D-galactose binding lectin purified from the annelid, *Perinereis nuntia* ver. *vallata*. *Comparative Biochemistry and Physiology Part B: Biochemistry and Molecular Biology* 152:382–389.

Keusch, J. J., S. M. Manzella, K. A. Nyame, R. D. Cummings, and J. U. Baenziger. 2000. Cloning of Gb3 synthase, the key enzyme in globo-series glycosphingolipid synthesis, predicts a family of α1, 4-glycosyltransferases conserved in plants, insects, and mammals. *Journal of Biological Chemistry* 275:25315–25321.

Kovalchuk, S. N., I. V. Chikalovets, O. V. Chernikov, V. I. Molchanova, W. Li, V. A. Rasskazov, and P. A. Lukyanov. 2013. cDNA cloning and structural characterization of a lectin from the mussel *Crenomytilus grayanus* with a unique amino acid sequence and antibacterial activity. *Fish and Shellfish Immunology* 35:1320–1324.

Lingwood, C. A. 2011. Glycosphingolipid functions. *Cold Spring Harbor Perspectives in Biology* 3:a004788.

Lingwood, C. A. and D. R. Branch. 2011. The role of glycosphingolipids in HIV/AIDS. *Discovery Medicine* 11:303–313.

Lord, J. M., L. M. Roberts, and J. D. Robertus. 1994. Ricin: Structure, mode of action, and some current applications. *FASEB Journal* 8:201–208.

Lopez, M., M. Gazon, S. Juliant, Y. Plancke, Y. Leroy, G. Strecker, J. P. Cartron et al. 1998. Characterization of a UDP-Gal:Galβ1–3GalNAc α1, 4-galactosyltransferase activity in a *Mamestra brassicae* cell line. *Journal of Biological Chemistry* 273:33644–33651.

Malagolini, N., M. Chiricolo, M. Marini, and F. Dall'Olio. 2009. Exposure of α2,6-sialylated lactosaminic chains marks apoptotic and necrotic death in different cell types. *Glycobiology* 19:172–181.

Matsumoto, R., Y. Fujii, S. M. A. Kawsar, R. A. Kanaly, H. Yasumitsu, Y. Koide, I. Hasan et al. 2012. Cytotoxicity and glycan-binding properties of an 18 kDa lectin isolated from the marine sponge *Halichondria okadai*. *Toxins (Basel)* 4:323–338.

Matsumoto, R., T. F. Shibata, H. Kohtsuka, M. Sekifuji, N. Sugii, N. Nakajima, N. Kojima et al. 2011. Glycomics of a novel type-2 *N*-acetyllactosamine-specific lectin purified from the feather star, *Oxycomanthus japonicus* (*Pelmatozoa: Crinoidea*). *Comparative Biochemistry and Physiology Part B: Biochemistry and Molecular Biology* 158:266–273.

Matsushima-Hibiya, Y., M. Watanabe, K. I. Hidari, D. Miyamoto, Y. Suzuki, T. Kasama, K. Koyama, T. Sugimura, and K. Wakabayashi. 2003. Identification of glycosphingolipid receptors for pierisin-1, a guanine-specific ADP-ribosylating toxin from the cabbage butterfly. *Journal of Biological Chemistry* 278:9972–9978.

Mi, Y., A. Lin, D. Fiete, L. Steirer, and J. U. Baenziger. 2014. Modulation of mannose and asialoglycoprotein receptor expression determines glycoprotein hormone half-life at critical points in the reproductive cycle. *Journal of Biological Chemistry* 289:12157–12167.

Müller, J., I. Obermeier, M. Wöhner, C. Brandl, S. Mrotzek, S. Angermüller, P. C. Maity, M. Reth, and L. Nitschke. 2013. CD22 ligand-binding and signaling domains reciprocally regulate B-cell Ca$^{2+}$ signaling. *Proceedings of the National Academy of Sciences USA* 110:12402–12407.

Murzin, A. G., A. M. Lesk, and C. Chothia. 1992. β-Trefoil fold. Patterns of structure and sequence in the Kunitz inhibitors interleukins-1β and 1α and fibroblast growth factors. *Journal of Molecular Biology* 223:531–543.

Ogawa, Y., S. Sugawara, T. Tatsuta, M. Hosono, N. Nitta, Y. Fujii, H. Kobayashi et al. 2014. Sialyl-glycoconjugates in cholesterol-rich microdomains of P388 cells are the triggers for apoptosis induced by *Rana catesbeiana* oocyte ribonuclease. *Glycoconjugate Journal* 31:171–184.

Ozeki, Y., T. Matsui, M. Suzuki, and K. Titani. 1991. Amino acid sequence and molecular characterization of a D-galactoside-specific lectin purified from sea urchin (*Anthocidaris crassispina*) eggs. *Biochemistry* 30:2391–2394.

Roberts, L. M., F. I. Lamb, D. J. Pappin, and J. M. Lord. 1985. The primary sequence of *Ricinus communis* agglutinin. Comparison with ricin. *Journal of Biological Chemistry* 260:15682–15686.

Sauer, A. K., C. H. Liang, J. Stech, B. Peeters, P. Quéré, C. Schwegmann-Wessels, C. Y. Wu, C. H. Wong, and G. Herrler. 2014. Characterization of the sialic acid binding activity of influenza A viruses using soluble variants of the H7 and H9 hemagglutinins. *PLoS One* 9:e89529.

Sharon, N. and H. Lis. 2004. History of lectins: From hemagglutinins to biological recognition molecules. *Glycobiology* 14:53–62.

Shin, I. S., S. Ishii, J. S. Shin, K. I. Sung, B. S. Park, H. Y. Jang, and B. W. Kim. 2009. Globotriaosylceramide (Gb3) content in HeLa cells is correlated to Shiga toxin-induced cytotoxicity and Gb3 synthase expression. *BMB Reports* 42:310–314.

Shirure, V. S., N. M. Reynolds, and M. M. Burdick. 2012. Mac-2 binding protein is a novel E-selectin ligand expressed by breast cancer cells. *PLoS One* 7:e44529.

Steelant, W. F., Y. Kawakami, A. Ito, K. Handa, E. A. Bruyneel, M. Mareel, and S. Hakomori. 2002. Monosialyl-Gb5 organized with cSrc and FAK in GEM of human breast carcinoma MCF-7 cells defines their invasive properties. *FEBS Letter* 531:93–98.

Sulzenbacher, G., V. Roig-Zamboni, W. J. Peumans, P. Rouge, E. J. Van Damme, and Y. Bourne. 2010. Crystal structure of the GalNAc/Gal-specific agglutinin from the phytopathogenic ascomycete *Sclerotinia sclerotiorum* reveals novel adaptation of a β-trefoil domain. *Journal of Molecular Biology* 400:715–723.

Tateno, H., A. Kuno, Y. Itakura, and J. Hirabayashi. 2010a. A versatile technology for cellular glycomics using lectin microarray. *Methods Enzymol.* 478:181–195.

Tateno, H., A. Matsushima, K. Hiemori, Y. Onuma, Y. Ito, K. Hasehira, K. Nishimura et al. 2013. Podocalyxin is a glycoprotein ligand of the human pluripotent stem cell-specific probe rBC2LCN. *Stem Cells Translational Medicine* 2013:2:265–273.

Tateno, H., M. Toyota, S. Saito, Y. Onuma, Y. Ito, K. Hiemori, M. Fukumura et al. 2010b. Glycome diagnosis of human induced pluripotent stem cells using lectin microarray. *Journal of Biological Chemistry* 286:20345–20353.

Tesh, V. L. 2012. The induction of apoptosis by Shiga toxins and ricin. *Current Topics in Microbiology and Immunology* 357:137–178.

Uchida, T., T. Yamasaki, S. Eto, H. Sugawara, G. Kurisu, N. Nakagawa, M. Kusunoki, and T. Hatakeyama. 2004. Crystal structure of the hemolytic lectin CEL-III isolated from the marine invertebrate *Cucumaria echinata*: Implications of domain structure for its membrane pore-formation mechanism. *Journal of Biological Chemistry* 279:37133–37141.

Venier, P., C. De Pittà, F. Bernante, L. Varotto, B. De Nardi, G. Bovo, P. Roch et al. 2011a. MytiBase: A knowledgebase of mussel (*M. galloprovincialis*) transcribed sequences. *BMC Genomics* 10:72.

Venier, P., L. Varotto, U. Rosani, C. Millino, B. Celegato, F. Bernante, G. Lanfranchi et al. 2011b. Insights into the innate immunity of the Mediterranean mussel *Mytilus galloprovincialis*. *BMC Genomics* 12:69.

Wada, Y., P. Azadi, C. E. Costello, A. Dell, R. A. Dwek, H. Geyer, R. Geyer et al. 2007. Comparison of the methods for profiling glycoprotein glycans—HUPO Human Disease Glycomics/Proteome Initiative multi-institutional study. *Glycobiology* 17:411–422.

Wiels, J., M. Fellous, and T. Tursz. 1981. Monoclonal antibody against a Burkitt lymphoma-associated antigen. *Proceedings of the National Academy of Sciences USA* 78:6485–6488.

Wiels, J., E. H. Holmes, N. Cochran, T. Tursz, and S. Hakomori. 1984. Enzymatic and organization difference in expression of a Burkitt lymphoma-associated antigen (globotriaosylceramide) in Burkitt lymphoma and lymphoblastoid cell lines. *Journal of Biological Chemistry* 259:14783–14787.

Williams, D. B. 2006. Beyond lectins: The calnexin/calreticulin chaperone system of the endoplasmic reticulum. *Journal of Cell Science* 119: 615–623.

*chapter thirty-four*

# Chitin derivatives as functional foods

*Kazuo Azuma, Shinsuke Ifuku, Tomohiro Osaki,*
*Hiroyuki Saimoto, and Yoshiharu Okamoto*

## Contents

## 34.1 Introduction

Chitin (β-(1-4)-poly-N-acetyl-D-glucosamine (Figure 34.1) is widely distributed in nature and is the second most abundant polysaccharide after cellulose (Muzzarelli, 2011). Chitin occurs in nature as ordered macrofibrils and is the major structural component in the exoskeletons of crabs and shrimp and the cell walls of fungi and yeast. As chitin is not readily dissolved in common solvents, it is often converted to its more deacetylated derivative, chitosan (Kurita, 2001; Rinaudo, 2006; Pillai et al., 2009).

Chitin and chitosan are biocompatible, biodegradable, and nontoxic and act as antimicrobial and hydrating agents (Jayakumar et al., 2010, 2011). Chitosan and its derivatives, including N-acetyl-D-glucosamine (GlcNAc), chitin oligosaccharide (NACOS), glucosamine (GlcN), and chitosan oligosaccharide (COS), are used as functional foods because evidence indicates that oral intake of these compounds produces beneficial health effects (De Silva et al., 2011; Je and Kim, 2012a,b; Reginster et al., 2012).

Recently, a method for the preparation of chitin nanofibers (CNFs) was reported (Ifuku et al., 2009). CNFs are considered to have several potential applications because they have useful properties such as high specific surface area and porosity. CNFs are expected to be used in applications such as tissue engineering, drug delivery, dental materials, wound healing, cosmetics, and medical implants (Zhang et al., 2005a,b; Muzzarelli et al., 2007; Fang et al., 2008).

The beneficial effects of GlcNAc, NACOS, and CNFs have been evaluated in experimental disease models. In this chapter, recent reports concerning GlcNAc, NACOS, COS, and CNFs are discussed.

## 34.2 GlcNAc as functional foods

### 34.2.1 Absorption of GlcNAc by oral administration

Pharmacokinetic studies indicate that orally administered GlcN is rapidly absorbed from the GI tract (Barclay et al., 1998; Adebowale et al., 2002; Aghazadeh-Habashi et al., 2002; Du et al., 2004; Laverty et al., 2005; Meulyzer et al., 2008). A comparison study of GlcN and N-acetyl-D-glucosamine (GlcNAc) absorption was performed in healthy dogs (Azuma et al., 2011, 2014a). Changes in plasma GlcN or GlcNAc concentration after oral GlcN or GlcNAc administration (both at a dose of 300 mg/kg) are shown in Figure 34.2. The maximum plasma GlcN concentration was 100 μM and the maximum plasma GlcNAc concentration was about 20 μM. These results indicate that absorption of GlcNAc by the canine GI tract may be inferior to that of GlcN. Differences in GlcN and GlcNAc uptake and subsequent effects on glucose transport, glucose transporter (GLUT) expression, and

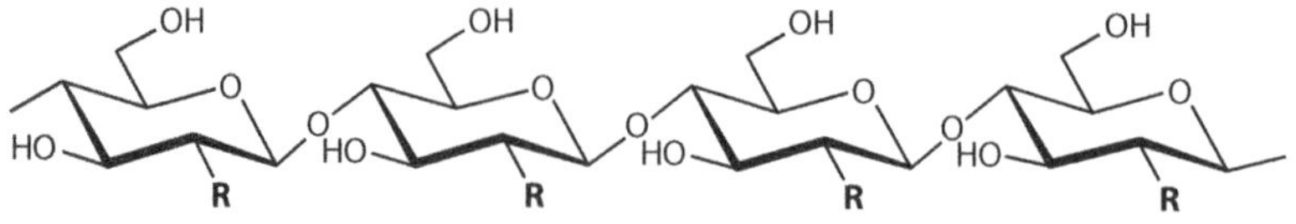

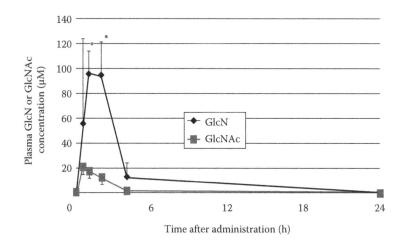

Cellulose: **R** = OH
Chitin: **R** = NHCOCH$_3$
Chitosan: **R** = NH$_2$

**Figure 34.1** Chemical structure of linear polysaccharides: cellulose, chitin, and chitosan. (Reprinted from Azuma, K. et al., *J. Biomed. Nanotechnol.*, 10, 2891, 2014. Copyright American Scientific Publishers. With permission.)

**Figure 34.2** Changes in plasma GlcN or GlcNAc concentration after oral GlcN or GlcNAc administration. All data are presented as mean ± S.D. *$P < 0.05$ vs. the GlcN group. (Reprinted from Azuma, K. et al., *Mar. Drugs*, 9, 712, 2011. Copyright MPDI. With permission.)

sulfated glycosaminoglycans (sGAG) and hyaluronan synthesis in vitro have been reported (Shickman et al., 2009). However, the mechanism of GlcNAc absorption in vivo is unclear; further investigation into the absorption and metabolism of GlcNAc is necessary.

### 34.2.2 Anti-inflammatory effects of GlcNAc

Previous reports indicate that GlcN and GlcNAc produce different effects in vitro. For example, GlcN suppressed interleukin (IL)-8, prostaglandin E2, and nitric oxide synthesis in Il-1β-treated synoviocytes; however, GlcNAc did not produce such an effect (Hua et al., 2007). Moreover, GlcN has been shown to produce anti-inflammatory effects in experimental models of rheumatoid arthritis (RA) (Hua et al., 2005). A recent report showed similar suppressive effects of oral administration of GlcNAc and GlcN on arthritis scores and histopathological scores in an experimental RA model (Azuma et al., 2012c). In the GlcN-treated group, the levels of serum tumor necrosis factor-α and IL-6 concentrations were significantly decreased in comparison with those of the control group (Table 34.1). In the GlcNAc group, the levels of serum IL-10, transforming

growth factor-β, and IL-2 abundance were significantly increased in comparison with those of the control group. These results suggest that, in comparison with GlcN, GlcNAc acts via a different suppressive mechanism in experimental models of RA, and this difference manifests as increases in IL-10, TGF-β1, and IL-2 abundance (Table 34.1). These results indicate that GlcNAc may be a useful supplement for patients with RA.

## 34.3 NACOS and COS as a functional food

### 34.3.1 Antitumor effects of NACOS and COS

Previous reports have indicated that chitin oligosaccharide (NACOS) and chitosan oigosaccharide (COS) possess antitumor properties (Suzuki et al., 1986; Tokoro et al., 1989; Harish-Prashanth and Tharanathan, 2005; Huang et al., 2006; Wang et al., 2008; Shen et al., 2009). Furthermore, it was revealed that the tumor inhibitory effects of NACOS and COS are likely related to their ability to induce lymphocyte cytokine release through increased T-cell proliferation. Essentially, the antitumor mechanisms of NACOS and COS are enhanced by

*Table 34.1* Effects of GlcNAc on histopathological score and serum cytokine concentration in SKG/jcl mice

|  | Control | GlcNAc | GlcN |
|---|---|---|---|
| Histopathological index | 6.5 ± 1.3 | 4.5 ± 1.0* | 4.3 ± 0.6* |
| Serum TNF-α (pg/ml) | 13.5 ± 1.7 | 4.5 ± 4.4 | 3.8 ± 0.9** |
| IL-6 (pg/ml) | 73.3 ± 18.6 | 44.9 ± 30.8 | 25.2 ± 3.2* |
| IL-2 (pg/ml) | 13.9 ± 9.4 | 46.0 ± 17.8* | 19.6 ± 2.5 |
| IL-10 (pg/ml) | 18.7 ± 12.4 | 104.5 ± 39.3**† | 25.2 ± 4.2 |
| TGF-β1 (pg/ml) | 230.4 ± 33.6 | 334.6 ± 52.1* | 282.9 ± 159.7 |

*Source:* With kind permission from Springer Science+Business Media: *Inflammation*, Suppressive effects of N-acetyl-D-glucosamine on rheumatoid arthritis mouse models, Azuma, K. et al., Inflammation, 35, 2012, 1462, Azuma, K., Osaki, T., Wakuda, T., Tsuka, T., Imagawa, T., Okamoto, Y., and Minami, S.

*Note:* Each value represents the mean ± SD of 3–4 mice in each group. **: $p < 0.01$ compared to control group. *: $p < 0.05$ compared to control group. †: $p < 0.05$ compared to GlcN group.

acquired immunity via acceleration of T-cell differentiation, which in turn increases cytotoxicity and maintains T-cell activity (Suzuki et al., 1986). However, most reports that describe the *in vivo* antitumor effects of NACOS have evaluated intraperitoneal or intravenous administration of NACOS and COS.

Recently, the antitumor effect of oral administration of NACOS was evaluated using a tumor bearing mouse model (Masuda et al., 2014). NACOS and COS significantly reduced tumor growth, and induced apoptosis in tumor tissues. The effects of NACOS and COS on tumor growth are shown in Figure 34.3. Furthermore, NACOS and COS significantly increased serum IL-12p70 and interferon-γ levels in comparison with those of the control group. Results of experiments

in myeloid differentiation primary response 88 (Myd-88) knockout mice revealed that the effects of NACOS and COS were related to both Myd-88-dependent and Myd-88-independent signaling pathways. These results indicated that oral administration of NACOS and COS produced antitumor effects by inducing apoptosis and stimulating the immune system.

### 34.3.2   Anti-inflammatory effects of COS

IBD is a common group of conditions characterized by inflammation of the intestinal tract. Crohn's disease and ulcerative colitis (UC) account for the majority of cases of IBD (Morrison et al., 2009). Intraperitoneal injection of COS suppressed inflammation of the colon

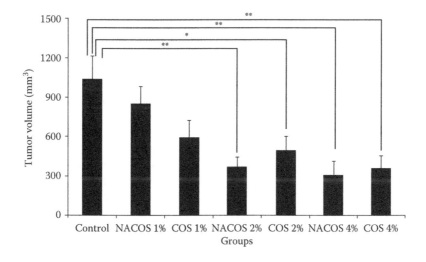

*Figure 34.3* Effects of oral administration of NACOS and COS on tumor growth. Data are presented as mean ± S.E.M. There were eight subjects in the NACOS 1%, COS 1%, NACOS 2%, COS 2%, NACOS 4%, and COS 4% groups, and there were 10 subjects in the control group. **$P < 0.01$, *$P < 0.05$ vs. the control group (Tukey–Kramer test). NACOS, N-acetyl-D-glucosamine oligomer; COS, glucosamine oligomer. (Reprinted from *Carbohydr. Polym.*, 111, Masuda, S., Azuma, K., Kurozumi, S., Kiyose, M., Osaki, T., Tsuka, T., Itoh, N., Imagawa, T., Minami, S., Sato, K., and Okamoto, Y., Anti-tumor properties of orally administered glucosamine and N-acetyl-D-glucosamine oligomers in a mouse model, 783–787, Copyright 2014, with permission from Elsevier.)

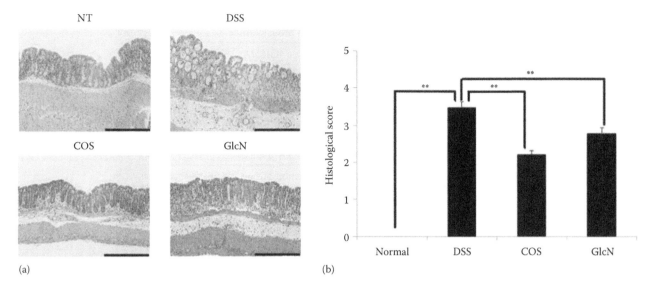

*Figure 34.4* Effect of oral COS administration on histopathological changes in an experimental model of IBD. (a) Colon tissue sections were stained with hematoxylin and eosin. Representative images from the NT, DSS, COS, and GlcN groups are shown. Bar = 200 μm. (b) Data are presented as mean ± S.E.M. of 30 fields (100× magnification) for each group (Steel–Dwass test). \*\**P* < 0.01. (Reprinted from *Carbohydr. Polym.*, 115, Azuma, K., Osaki, T., Kurozumi, S., Kiyose, M., Tsuka, T., Murahata, Y., Imagawa, T., Itoh, N., Minami, S., Sato, K., and Okamoto, Y., Anti-inflammatory effects of orally administered glucosamine oligomer in an experimental model of inflammatory bowel disease, 448, Copyright 2015, with permission from Elsevier.)

in an experimental model of IBD (Yousef et al., 2012). Moreover, oral administration of COS inhibited colon shortening and tissue injury (as assessed by histology) in an experimental model of IBD in mice (Azuma et al., 2015; Figure 34.4). Oral administration of COS inhibited inflammation of the colonic mucosa by suppressing myeloperoxidase activation in inflammatory cells, as well as inhibiting activation of nuclear factor-kappa B, cyclooxygenase-2, and inducible nitric oxide synthase, suggesting that these factors play important roles in colon inflammation. Oral administration of COS also reduced serum levels of pro-inflammatory cytokines tumor necrosis factor-α and IL-6, and prolonged survival time in mice. These data suggest that COS could produce anti-inflammatory effects in IBD, and thus it could be a promising new functional food for patients with IBD.

## 34.4   CNFs as functional foods

### 34.4.1   Preparation of CNFs

Chitin polymers form a nanofiber with an extended crystalline structure that is coated with proteins (Figure 34.5). Chitin polymer nanofibers form bundles of chitin/protein composite fibers that make up a planar, woven, branched network of bundles of nanofibers in which minerals, mainly calcium carbonate, are embedded. The woven network of chitin polymer nanofiber bundles form a twisted pattern with a helicoidal stacked structure.

A simple method for the preparation of chitin nanofibers (CNFs) has been reported, in which crab shell

powder was used as the starting material for nanofiber preparation. To extract CNFs from crab shells, proteins and minerals were removed using aqueous solutions of KOH and HCl, respectively (Gopalan and Dufresne, 2003; Zeng et al., 2012). Most protein and mineral impurities are removed by alkali and acid treatments (Shimahara and Takiguchi, 1998). The strong hydrogen bonds that result from the drying process make it difficult to obtain thin and uniform nanofibers; thus, the substance should be kept wet after the removal of proteins and minerals (Abe et al., 2007).

Figure 34.6 shows field emission scanning electron microscopic (FE-SEM) images of the crab shell surface after removal of proteins and calcium carbonate. CNFs of approximately 10 nm thickness were observed. Thicker chitin-protein fibers with a diameter of approximately 100 nm were also observed, and confirmed to be bundles of nanofibers with a width of approximately 10 nm. The slurry of chitin and neutral water derived from crab shells was passed through a grinder, after which the width of the fibers was distributed over a wide range of, for example, 10–100 nm (Figure 34.7a). The twisted structure seemed to disintegrate after a single application of the grinder treatment. As a result, 100 nm thick fibers were isolated from the slurry of chitin-protein fibers. However, thicker fibers were not successfully disintegrated by the grinder treatment, even though the protein layers were removed before the drying process.

Isogai et al. reported a method for the preparation of CNFs from squid pen β-chitin in acidic water

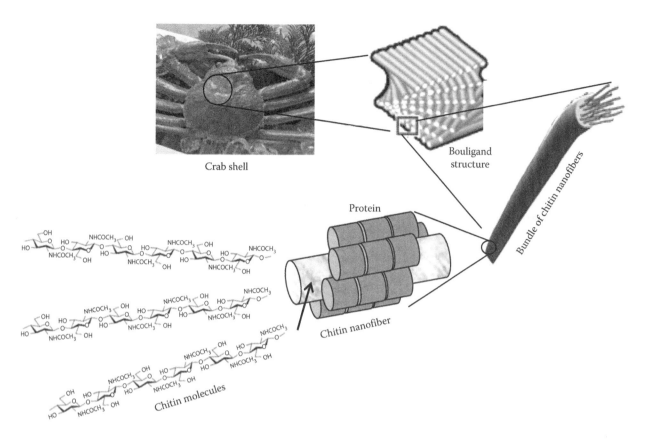

***Figure 34.5*** Schematic presentation of the structure of the crab shell exoskeleton. (From Ifuku, S. and Saimoto, H., Chitin nano-fibers: Preparations, modifications, and applications, *Nanoscale*, 4, 3308, 2012. Reproduced by permission of The Royal Society of Chemistry.)

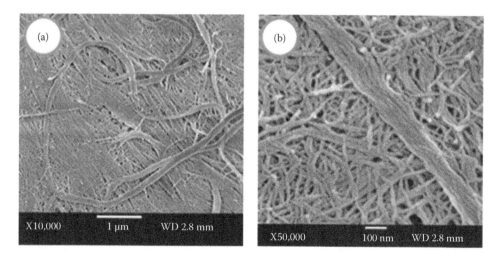

***Figure 34.6*** FE-SEM micrographs of the crab shell surface after removal of the matrix. The lengths of the scale bars are (a) 1000 nm and (b) 100 nm, respectively. (Reprinted with permission from Ifuku, S., Nogi, M., Abe, K., Yoshioka, M., Morimoto, M., Saimoto, H., and Yano, H., Preparation of chitin nanofibers with a uniform width as alpha-chitin from crab shells, *Biomacromolecules*, 10, 1584. Copyright 2009 American Chemical Society.)

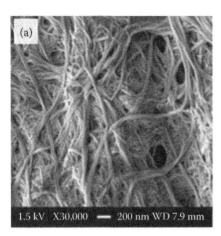 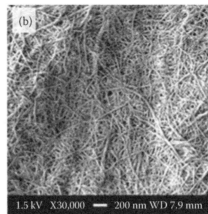 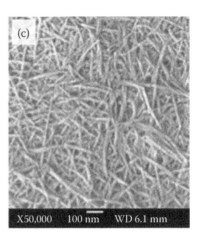

*Figure 34.7* FE-SEM micrographs of chitin nanofibers from crab shells after a single pass through the grinder: (a) without acetic acid (pH 7); (b, c) with acetic acid (pH 3). The lengths of the scale bars are (a) 200 nm, (b) 200 nm, and (c) 100 nm, respectively. (Reprinted with permission from Ifuku, S., Nogi, M., Abe, K., Yoshioka, M., Morimoto, M., Saimoto, H., and Yano, H., Preparation of chitin nanofibers with a uniform width as alpha-chitin from crab shells, *Biomacromolecules*, 10, 1584. Copyright 2009 American Chemical Society.)

(Fan et al., 2009). Cationization of the amino groups in β-chitin under acidic conditions is necessary to maintain a stable dispersion state by electrostatic repulsion in water. Therefore, purified β-chitin from crab shells would also be homogeneously dispersed under acidic conditions by the cationization of the amino groups on the fiber surface, which would facilitate nanofibrillation. Thus, purified chitin was dispersed in aqueous acetic acid and subjected to grinder treatment. The chitin slurry thus obtained formed a gel after a single grinder treatment, suggesting that nanofibrillation was accomplished because of the high dispersion property in acidic water and the high surface-to-volume ratio of the nanofiber. The disintegrated chitin was observed as uniform nanofibers with a width of 10 nm, suggesting that the fibrillation process was facilitated in acidic water (Figure 34.7b and c). Broken fibers were not observed over a wide observation area. The aspect ratios of the nanofibers were very high, indicating that CNFs were successfully isolated from crab shells while maintaining their original structure.

### 34.4.2 Anti-inflammatory effects of CNFs

Some biological activities of CNFs have been studied. The anti-inflammatory effect of oral administration of chitin nanofibrils was demonstrated in a mouse model of IBD (Azuma et al., 2012a,b). We compared the anti-inflammatory effects of chitin nanofibrils with those of chitin suspensions (chitin-PS) using the clinical score (DAI, disease activity index), colon length, colon weight/length ratio, and histological observations, and found that CNFs improved clinical symptoms and suppressed ulcerative colitis. Furthermore, chitin

nanofibrils suppressed myeloperoxidase activation in the colon and decreased serum IL-6 concentrations. Conversely, chitin-PS did not suppress experimental colitis. CNFs decreased the sizes of areas with nuclear factor-κB (NF-κB) staining in colon tissue samples, and decreased serum monocyte chemotactic protein-1 concentration in a colitis model. The NF-κB immunohistochemistry results are shown in Figure 34.8. We also found that CNFs suppressed the increase in Masson's trichrome stained area in colon tissue samples in the colitis model. In contrast, chitin-PS did not show these effects in the experimental model of colitis in mice. Our results suggest that chitin nanofibrils are a promising functional food for patients with inflammatory bowel disease.

We tested α-CNFs in a mouse model of experimental IBD (Azuma et al., 2012a,b). Chitin exists in different allomorphic forms in nature, which vary in terms of their polymer chain structure and crystallinity (Jang et al., 2004; Khoushab and Yamabhai, 2010). Most chitins, including insect and crustacean chitins, are composed of α-chitin in its native state, whereas the rarer β-chitin allomorph is found in squid pen and some diatoms. α-Chitin exhibits a two-chain, antiparallel structure, whereas β-chitin has a one-chain unit cell with a parallel-chain structure and intramolecular hydrogen bonding (Blackwell, 1969; Rudall and Kenchington, 1973; Minke and Blackwell, 1978). The comparatively weaker hydrogen bonds in the parallel-chain structure of β-chitin may account for its higher chemical reactivity in comparison with α-chitin (Atkins, 1985; Kurita et al., 2005). In addition, β-chitin can incorporate small molecules, including water, into its crystal lattice to form crystalline complexes, but α-chitin does not share

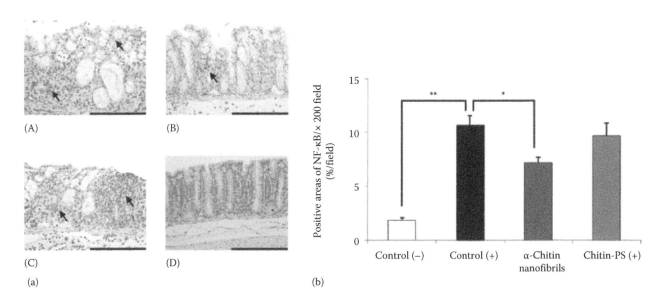

*Figure 34.8* Effects of CNF on colon NF-κB activation in a mouse model of DSS-induced acute ulcerative colitis. (a) NF-κB-positive areas are indicated by the arrows. Representative images from the control (+) (A), chitin nanofibrils (+) (B), chitin-PS (C), and control (−) (D) groups are shown. Bars = 100 μm. (b) Data are presented as mean ± S.E.M. of 30 fields (100× magnification) for each group. Differences were analyzed with the Steel–Dwass test. *$P < 0.05$, **$P < 0.01$. (Reprinted from *Carbohydr. Polym.*, 90, Azuma, K. Osaki, T., Ifuku, S., Saimoto, H., Tsuka, T., Imagawa, T., Okamoto, Y., and Minami, S., α-Chitin nanofibrils improve inflammatory and fibrosis responses in mice with inflammatory bowel disease, 197, Copyright 2012, with permission from Elsevier.)

this characteristic (Saito et al., 1997; Sikorski et al., 2009). α-Chitin chains give rise to strong hydrogen bonds that confer high stability (Jang et al., 2004).

We evaluated differences in the biological effects of α-CNFs and β-CNFs using a mouse model of IBD. α-CNFs significantly suppressed increases in histological scores and MPO-positive cell counts. Furthermore, α-chitin nanofibrils significantly decreased NF-κB-positive areas and areas of fibrosis. In contrast, no anti-inflammatory effects were observed in the group of IBD mice treated with β-CNFs. These results suggest that the α-crystal structure of chitin nanofibrils confers the ability to exert an anti-inflammatory effect (Azuma et al., 2013). Therefore, chitin nanofibrils prepared from α-chitin are a promising functional food for patients with IBD.

## 34.4.3  Anti-obesity effects of CNFs

Obesity is a growing health problem worldwide that has been associated with the metabolic syndrome (Vucenik and Stains, 2012). The metabolic syndrome is a constellation of risk factors, including atherogenic dyslipidemia, impaired fasting glucose, hypertension, and central adiposity, which predispose affected individuals to a higher risk of type 2 diabetes, cardiovascular diseases, and cancer (Park et al., 2007; Chen et al., 2008;

Grattagliano et al., 2008; Ishizaka et al., 2009; Vucenik and Stains, 2012).

The increasing incidence of obesity suggests that this epidemic will worsen in the future (Haslam and James, 2005). Some dietary modifications have been found to prevent or suppress the metabolic syndrome, including catechins, flavonoids, and CS (Ahn et al., 2008; Thielecke and Boschmann, 2009; Seiva et al., 2012).

We evaluated the anti-obesity effects of CNFs and surface-deacetylated chitin nanofibrils (sda-CNs) in a mouse model of high-fat diet-induced obesity (Azuma et al., 2014b). SDACNFs suppressed the increase in body weight produced by the high-fat diet; however, CNFs did not suppress such weight gain (Figure 34.9). sda-CNs also suppressed the increase in epididymal tissue weight produced by the high-fat diet. Moreover, SDACNFs decreased serum levels of leptin and TNF-α. These results could be explained by the high surface area of the sda-CNs. Surface-deacetylation of CNFs resulted in a compound that produced anti-obesity effects in experimental models. Adipocytes in adipose tissue secrete proteins known as adipocytokines, including TNF-α, IL-6, resistin, leptin, and adiponectin (Gnacińska et al., 2009). Plasma leptin concentrations are positively correlated with adiposity (excessive body fat) and body weight changes in humans and rodents (Rosen and Spiegelman, 2006).

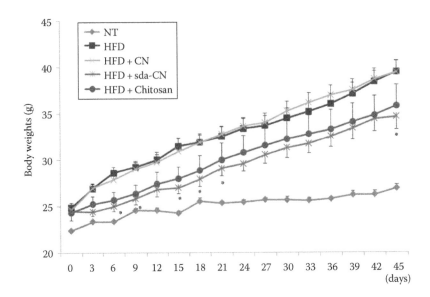

*Figure 34.9* Effects of surface-deacetylated chitin nanofibrils (sda-CN) on body weight changes in a mouse model of high-fat-diet-induced obesity. The body weight changes of the mice are shown. The body weight of the high-fat diet (HFD) + sda-CN group was significantly decreased in comparison with those of the HFD and HFD + sda-CN groups on days 6, 9, 15, 18, 21, and 45 ($P < 0.05$). (Reprinted from Azuma, K. et al., *J. Biomed. Nanotechnol.*, 10, 2891, 2012. Copyright American Scientific Publishers. With permission.)

Adiponectin contributes to insulin sensitivity and fatty acid oxidation, and circulating concentrations of adiponectin are inversely correlated with body mass (Arita et al., 1999). The results of our study indicated that sda-CN suppressed the secretion of leptin from adipocytes. Liver steatohepatitis was markedly suppressed by oral administration of SDACNFs.

## 34.5   Summary

Reports indicate that GlcNAc, NACOS, and CNFs have many beneficial health effects, including anti-inflammatory, antitumor, and anti-obesity effects. These data strongly suggest that GlcNAc, NACOS, and CNFs are potent functional foods. However, their mechanisms of action are unknown and further study is needed.

## References

Abe, K., S. Iwamoto, and H. Yano. 2007. Obtaining cellulose nanofibers with a uniform width of 15 nm from wood. *Biomacromolecules* 8: 3276–3278.

Adebowale, A., J. Du, Z. Liang, J. L. Leslie, and N. D. Eddington. 2002. The bioavailability and pharmacokinetics of glucosamine hydrochloride and low molecular weight chondroitin sulfate after single and multiple doses to beagle dogs. *Biopharmaceutics and Drug Disposition* 23: 217–225.

Aghazadeh-Habashi, A., S. Sattari, F. Pasutto, and F. Jamali. 2002. Single dose pharmacokinetics and bioavailability of glucosamine in the rat. *Journal of Pharmacy and Pharmaceutical Sciences* 5: 181–184.

Ahn, J., H. Lee, S. Kim, J. Park, and T. Ha. 2008. The anti-obesity effect of quercetin is mediated by the AMPK and MAPK signaling pathways. *Biochemical and Biophysical Research Communication* 373: 545–549.

Arita, Y., S. Kihara, N. Ouchi, M. Takahashi, K. Maeda, J. Miyagawa, K. Hotta et al. 1999. Paradoxical decrease of an adipose-specific protein, adiponectin, in obesity. *Biochemical and Biophysical Research Communication* 257: 79–83.

Atkins, E. 1985. Conformations in polysaccharides and complex carbohydrates. *Journal of Bioscience* 8: 375–387.

Azuma, K., S. Ifuku, T. Osaki, Y. Okamoto, and S. Minami. 2014b. Preparation and biomedical applications of chitin and chitosan nanofibers. *Journal of Biomedical Nanotechnology* 10: 2891–2920.

Azuma, K., T. Osaki, S. Ifuku, H. Saimoto, T. Tsuka, T. Imagawa, Y. Okamoto, and S. Minami. 2012a. α-Chitin nanofibrils improve inflammatory and fibrosis responses in mice with inflammatory bowel disease. *Carbohydrate Polymers* 90: 197–200.

Azuma, K., T. Osaki, S. Ifuku, H. Saimoto, T. Tsuka, T. Imagawa, Y. Okamoto, and S. Minami. 2013. A comparative study analysis of α-chitin and β-chitin nanofibrils by using an inflammatory-bowel disease mouse model. *Journal of Chitin and Chitosan Science* 1: 144–149.

Azuma, K., T. Osaki, S. Kurozumi, M. Kiyose, T. Tsuka, Y. Murahata, T. Imagawa et al. 2015. Anti-inflammatory effects of orally administered glucosamine oligomer in an experimental model of inflammatory bowel disease. *Carbohydrate Polymers* 115: 448.

Azuma, K., T. Osaki, T. Tsuka, T. Imagawa, Y. Okamoto, Y. Takamori, and S. Minami. 2011. Effects of oral glucosamine hydrochloride administration on plasma free amino acid concentrations in dogs. *Marine Drugs* 9: 712–718.

Azuma, K., T. Osaki, T. Wakuda, S. Ifuku, H. Saimoto, T. Tsuka, T. Imagawa, Y. Okamoto, and S. Minami. 2012b. Beneficial and preventive effect of chitin nanofibrils in a dextran sulfate sodium-induced acute ulcerative colitis model. *Carbohydrate Polymers* 87: 1399–1403.

Azuma, K., T. Osaki, T. Wakuda, T. Tsuka, T. Imagawa, Y. Okamoto, and S. Minami. 2012c. Suppressive effects of N-acetyl-D-glucosamine on rheumatoid arthritis mouse models. *Inflammation* 35: 1462–1465.

Azuma, K., S. Suguro, Y. Yamagishi, M. Yamashita, T. Osaki, T. Tsuka, T. Imagawa, I. Arifuku, and Y. Okamoto. 2014a. Pharmacokinetic study of D-glucosamine hydrochloride produced by microbial fermentation and N-acetyl-D-glucosamine synthesized from D-glucosamine hydrochloride after oral administration to dogs. *Journal of Chitin and Chitosan Science* 2: 89–92.

Barclay, T. S., C. Tsourounis, and G. M. McCart. 1998. Glucosamine. *Annals of Pharmacotherapy* 32: 574–579.

Blackwell, J. 1969. Structure of β-chitin or parallel chain systems of poly-β-(1→4)-N-acetyl-D-glucosamine. *Biopolymers* 7: 281–298.

Chen, K., K. J. B. Lindsey, A. Khera, J. A. De Lemos, C. R. Ayers, A. Goyal, G. L. Vega, S. A. Murphy, S. M. Grundy, and D. K. McGuire. 2008. Independent associations between metabolic syndrome, diabetes mellitus and atherosclerosis: Observations from the Dallas Heart Study. *Diabetes and Vascular Disease Research* 5: 96–101.

De Silva, V., A. El-Metwally, E. Ernst, G. Lewith, and G. J. Macfarlane. 2011. Evidence for the efficacy of complementary and alternative medicines in the management of osteoarthritis: A systematic review. *Rheumatology (Oxford)* 50: 911–920.

Du, J., N. White, and N. D. Eddington. 2004. The bioavailability and pharmacokinetics of glucosamine hydrochloride and chondroitin sulfate after oral and intravenous single dose administration in the horse. *Biopharmaceutics and Drug Disposition* 25: 109–116.

Fan, Y., T. Saito, and A. Isogai. 2008. Preparation of chitin nanofibers from squid pen β-chitin by simple mechanical treatment under acid conditions *Biomacromolecules* 9: 1919–1923.

Fan, Y., T. Saito, and A. Isogai. 2010. Individual chitin nanowhiskers prepared from partially deacetylated α-chitin by fibril surface cationization. *Carbohydrate Polymers* 79: 1046–1051.

Fang, J., H. Niu, T. Lin, and X. Wang. 2008. Applications of electrospun nanofibers. *Chinese Science Bulletin* 53: 2265–2286.

Gnacińska, M., S. Małgorzewicz, M. Stojek, W. Łysiak-Szydłowska, and K. Sworczak. 2009. Role of adipokines in complications related to obesity: A review. *Advance in Medical Science* 54: 150–157.

Gopalan, N. K. and A. Dufresne. 2003. Crab shell chitin whisker reinforced natural rubber nanocomposites. 1. Processing and swelling behavior. *Biomacromolecules* 4: 657–665.

Grattagliano, I., V. O. Palmieri, P. Portincasa, A. Moschetta, and G. Palasciano. 2008. Oxidative stress-induced risk factors associated with the metabolic syndrome: A unifying hypothesis. *Journal of Nutritional Biochemistry* 19: 491–504.

Harish-Prashanth, K. V. and K. V. Tharanathan. 2005. Depolymerized products of chitosan as potent inhibitors of tumor-induced angiogenesis. *Biochimica et Biophysica Acta* 1722: 22–29.

Haslam, D. W. and W. P. James. 2005. Obesity. *Lancet* 366: 1197–1209.

Hua, J., K. Sakamoto, T. Kikukawa, C. Abe, H. Kurosawa, and I. Nagaoka. 2007. Evaluation of the suppressive actions of glucosamine on the interleukin-1β-mediated activation of synoviocytes. *Inflammation Research* 56: 432–438.

Hua, J., S. Suguro, S. Hirano, K. Sakamoto, and I. Nagaoka. 2005. Preventive actions of a high dose of glucosamine on adjuvant arthritis in rats. *Inflammation Research* 54: 127–132.

Huang, R., E. Mendis, N. Rajapakse, and S. K. Kim. 2006. Strong electronic charge asan important factor for anticancer activity of chitooligosaccharides (COS). *Life Science* 78: 2399–2408.

Ifuku, S., M. Nogi, K. Abe, M. Yoshioka, M. Morimoto, H. Saimoto, and H. Yano. 2009. Preparation of chitin nanofibers with a uniform width as alpha-chitin from crab shells. *Biomacromolecules* 10: 1584–1588.

Ifuku, S. and H. Saimoto. 2012. Chitin nanofibers: Preparations, modifications, and applications. *Nanoscale* 4: 3308.

Ishizaka, N., Y. Ishizaka, M. Yamakado, E. Toda, K. Koike, and R. Nagai. 2009. Association between metabolic syndrome and carotid atherosclerosis in individuals without diabetes based on the oral glucose tolerance test. *Atherosclerosis* 204: 619–623.

Jang, M. K., B. G. Kong, Y. I. Jeong, C. H. Lee, and J. W. Nah. 2004. Physicochemical characterization of α-chitin, β-chitin, and γ-chitin separated from natural resources. *Journal of Polymer Science Part A: Polymer Chemistry* 42: 3423–3432.

Jayakumar, R., M. Prabaharan, S. V. Nair, and H. Tamura. 2010. Novel chitin and chitosan nanofibers in biomedical applications. *Biotechnology Advances* 28: 142–150.

Jayakumar, R., M. Prabaharan, P. T. Sudheesh Kumar, S. V. Nair, and H. Tamura. 2011. Biomaterials based on chitin and chitosan in wound dressing applications. *Biotechnology Advances* 29: 322–327.

Je, J. Y. and S. K. Kim. 2012a. Chitosan as potential marine nutraceutical. *Advances in Food and Nutrition Research* 65: 121–135.

Je J. Y. and S. K. Kim. 2012b. Chitooligosaccharides as potential nutraceuticals: Production and bioactivities. *Advances in Food and Nutrition Research* 65: 321–336.

Khoushab, F. and M. Yamabhai. 2010. Chitin research revisited. *Marine Drugs* 8: 1988–2012.

Kurita, K. 2001. Controlled functionalization of the polysaccharide chitin. *Progress in Polymer Science* 26: 1921–1971.

Kurita, K., K. Sugita, N. Kodaira, M. Hirakawa, and J. Yang. 2005. Preparation and evaluation of trimethylsilylated chitin as a versatile precursor for facile chemical modifications. *Biomacromolecules* 6: 1414–1418.

Laverty, S., J. D. Sandy, C. Celeste, P. Vachon, J. F. Marier, and A. H. Plaas. 2005. Synovial fluid levels and serum pharmacokinetics in a large animal model following treatment with oral glucosamine at clinically relevant doses. *Arthritis and Rheumatism* 52: 181–191.

Masuda, S., K. Azuma, S. Kurozumi, M. Kiyose, T. Osaki, T. Tsuka, N. Itoh et al. 2014. Anti-tumor properties of orally administered glucosamine and N-acetyl-D-glucosamine oligomers in a mouse model. *Carbohydrate Polymers* 111: 783–787.

Meulyzer, M., P. Vachon, F. Beaudry, T. Vinardell, H. Richard, G. Beauchamp, and S. Laverty. 2008. Comparison of pharmacokinetics of glucosamine and synovial fluid levels following administration of glucosamine sulphate or glucosamine hydrochloride. *Osteoarthritis and Cartilage* 16: 973–979.

Minke, R. and J. Blackwell. 1978. The structure of alpha-chitin. *Journal of Molecular Biology* 120: 167–181.

Morrison, G., B. Headon, and P. Gibson. 2009. Update in inflammatory bowel disease. *Australian Family Physician* 38: 956–961.

Muzzarelli, R. A. A. 2011. Chitin nanostructures in living organisms. In *Chitin: Formation and Diagenesis*, N. Gupta (ed.). Springer, Dordrecht, the Netherlands, Vol. 34, pp. 1–34.

Muzzarelli, R. A. A., P. Morganti, G. Morganti, P. Palombo, M. Palombo, G. Biagini, M. M. Belmonte, F. Giantomassi, F. Orlandi, and C. Muzzarelli. 2007. Chitin nanofibrils/chitosan glycolate composites as wound medicaments. *Carbohydrate Polymers* 70: 274–284.

Park, S. H., B. I. Kim, S. H. Kim, H. J. Kim, D. I. Park, Y. K. Cho, I. K. Sung et al. 2007. Body fat distribution and insulin resistance: Beyond obesity in nonalcoholic fatty liver disease among overweight men. *Journal of American College of Nutrition* 26: 321–326.

Pillai, K. S., W. Paul, and C. P. Sharma. 2009. Chitin and chitosan polymers: Chemistry, solubility and fiber formation. *Progress in Polymer Science* 34: 641–678.

Reginster, J. Y., A. Neuprez, M. P. Lecart, N. Sarlet, and O. Bruyere. 2012. Role of glucosamine in the treatment for osteoarthritis. *Rheumatology International* 32: 2959–2967.

Rinaudo, M. 2006. Chitin and chitosan: Properties and applications. *Progress in Polymer Science* 31: 603–632.

Rosen, E. D. and B. M. Spiegelman. 2006. Adipocytes as regulators of energy balance and glucose homeostasis. *Nature* 444: 847–853.

Rudall, K. M. and W. Kenchington. 1973. The chitin system. *Biological Reviews* 48: 597–633.

Saito, Y., J. L. Putaux, T. Okano, F. Gaill, and H. Chanzy. 1997. Structural aspects of the swelling of beta-chitin in HCl and its conversion into alpha-chitin. *Macromolecules* 30: 3867–3873.

Seiva, F. R., L. G. Chuffa, C. P. Braga, J. P. Amorim, and A. A. Fernandes. 2012. Quercetin ameliorates glucose and lipid metabolism and improves antioxidant status in postnatally monosodium glutamate-induced metabolic alterations. *Food Chemistry and Toxicology* 50: 3726–3561.

Shen, K. T., M. H. Chen, H. Y. Chan, J. H. Jeng, and Y. J. Wang. 2009. Inhibitory effects of chitooligosaccharides on tumor growth and metastasis. *Food and Chemical Toxicology* 47: 1864–1871.

Shikhman, A. R., D. C. Brinson, J. Valbracht, and M. K. Lotz. 2009. Differential metabolic effects of glucosamine and N-acetylglucosamine in human articular chondrocytes. *Osteoarthritis Cartilage* 17: 1022–1028.

Shimahara, K. and Y. Takiguchi. 1998. Preparation of crustacean chitin. In *Methods in Enzymology*, W. A. Wood and S. T. Kellogg (eds.). Academic Press, New York, Vol. 161, pp. 417–423.

Sikorski, P., R. Hori, and M. Wada. 2009. Revisit of alpha-chitin crystal structure using high-resolution X-ray diffraction data. *Biomacromolecules* 10: 1100–1105.

Suzuki, K., T. Mikami, Y. Okawa, A. Tokoro, S. Suzuki, and M. Suzuki. 1986. Antitu-mor effect of hexa-N-acetylchitohexaose and chitohexaose. *Carbohydrate Research* 151: 403–408.

Thielecke, F. and M. Boschmann. 2009. The potential role of green tea catechins in the prevention of the metabolic syndrome—A review. *Phytochemistry* 70: 11–24.

Tokoro, A., N. Tatewaki, T. Mikami, S. Suzuki, and M. Suzuki. 1989. Effect of NACOS-6 on lymphokine-activated killer cell (LAK) activity. *Biotherapy* 3: 51–54.

Vucenik, I. and J. P. Stains. 2012. Obesity and cancer risk: Evidence, mechanisms, and recommendations. *Annals of the New York Academy of Sciences* 1271: 37–43.

Wang, S. L., H. T. Lin, T. W. Liang, Y. J. Chen, Y. H. Yen, and S. P. Guo. 2008. Reclamation of chitinous materials by bromelain for the preparation of antitumor and antifungal materials. *Bioresource Technology* 99: 4386–4393.

Yousef, M., R. Pichyangkura, S. Soodvilai, V. Chatsudthipong, and C. Muanprasat. 2012. Chitosan oligosaccharide as potential therapy of inflammatory bowel disease: Therapeutic efficacy and possible mechanisms of action. *Pharmacological Research* 66: 66–79.

Zeng, J. B., Y. S. He, S. L. Li, and Y. Z. Wang. 2012. Chitin whiskers: An overview. *Biomacromolecules* 13: 1–11.

Zhang, C., X. Yuan, L. Wu, Y. Han, and J. Sheng. 2005a. Study on morphology of electrospun poly(vinyl alcohol) mats. *European Polymer Journal* 41: 423–432.

Zhang, Y., C. T. Lim, S. Ramakrishna, and Z. M. Huang. 2005b. Recent development of polymer nanofibers for biomedical and biotechnological applications. *Journal of Materials Science: Materials in Medicine* 16: 933–946.

## chapter thirty-five

# Marine carbohydrates and their applications

*Bishnu Pada Chatterjee and Partha Pratim Bose*

## Contents

## 35.1 Introduction

Marine carbohydrates are one of the most important naturally occurring polymers that are produced by photosynthesis in marine living organisms. They are considered as the biggest store-houses of polysaccharides with uncommon structures. Carbohydrates are important due to their activity in the immune system. Among the many polysaccharides produced by aquatic organisms, chitin is the second abundant polysaccharide after cellulose. Carbohydrate contents and carbohydrate production in marine algae are found to be proportional to the extent of respiration and light. Marine polysaccharides from seaweed find applications in various fields such as pharmaceuticals, food production, cosmeceuticals, and so on. Marine organisms are good sources of nutrients, since they are rich in sulfated polysaccharides.

Over the past few decades, a diverse array of marine organisms (flora and fauna) containing carbohydrates have gained tremendous interest in medical and pharmaceutical industries and research institutions for their wide potential applications in several biological, biomedical, and nutritional fields such as antiproliferative, antitumor, antiviral, anticoagulant, antioxidant, anti-inflammatory, and many others. Fucosylated chondroitin sulfate (FucCS) isolated from the sea cucumber *Ludwigothurea grisea* and sulfated galactan (SG) from the red alga *Botryocladia occidentalis* are found to be rare and

exhibit high anticoagulant and antithrombotic properties and could promote the possibility of developing new antithrombotic drugs.

## 35.2 Marine polysaccharides

A great diversity of polysaccharides are present in the cell walls of marine algae, which are not found in terrestrial plants. The three types of marine algae—such as green algae (Chlorophyta), red algae (Rhodophyta), and brown algae (Phaeophyta)—are sources of sulfated polysaccharides (SPs), namely, sulfated galactans (SGs), sulfated fucans (SFs), xylans, alginic acid, fucoidan (sulfated fucose), laminarin, sargassan, agar, carrageenans, xylans, water-soluble sulfated galactan, and mucopolysaccharides. SFs are composed of repeating units of α-L-fucopyranosyl (Fuc*p*), and SGs are homopolymers of α-L-, α-D-, β-D-galactopyranosyl (Gal*p*) residues and are different from the other glycans, because they are mostly derived from marine organisms in which the structures are taxonomically related, exhibit very uncommon structural patterns. SFs have been synthesized in the cell wall of brown seaweeds (Phaeophyceae), which are commonly known as fucoidans, whereas green (Chlorophyceae) and red seaweeds (Rhodophyceae) contain only SGs. Besides seaweeds SFs and SGs can both be found in marine angiosperm and invertebrates, such as ascidians, also known as sea squirts or tunicates (Urochordata, Ascidiaceae), sea cucumber (Echinodermata, Holothuroidea), and sea urchins (Echinodermata, Echinoidea) (Santos et al. 1992). However, SPs play different roles in these animals. For example, in ascidians and sea cucumbers, these SPs are involved in making their body walls, whereas in sea urchins, these polysaccharides have been known to form the jelly coat, which surrounds the female gametes for selective participation in species recognition during the fertilization process of these animals. SGs also help to build up different tissues in cell walls of marine angiosperms. SFs and SGs have been known to be present as essential components of the extracellular matrices in these marine organisms. Brown algal SFs are the most abundant SPs in the sea world, because brown seaweeds are believed to dominate the sea environment in both number of species (1.5–2000) and biomass. Marine organisms such as giant squid, skate, and salmon cartilage are the potential sources of chondroitin sulfate and contain no other glycosaminoglycans, such as heparin, hyaluronic acid, keratan sulfate, or dermatan sulfate. Giant squid and skate cartilage contain 21.6% and 12.48% of chondroitin sulfate, respectively, much higher than that of the shark cartilage (9.71%) (Im et al. 2008).

In contrast to most algal SPs, the invertebrate polysaccharides exhibit highly regular chemical structures (Figures 35.1 and 35.2), which makes it easier to correlate their biological functions with their respective structural features (Pomin and Mourao 2008; Pomin 2009, 2010). This gives advantage to the determination of structure activity relationship of these important marine carbohydrates. The extracellular matrices of certain marine invertebrates, however, have been found as newer sources of these compounds.

The chemical structures of repeating units of the SFs isolated from the cell wall of the sea cucumber (Figure 35.1a) and from the egg jelly coat of sea urchins (Figure 35.1b through h) are composed of α-L-fucopyranosyl residues containing sulfate groups exclusively at 2- and 4-positions. The predominant glycosidic linkages, α(1-3) and α(1-4) are all linear. Some of the repeating units are tetrasaccharides (Figure 35.1a through d) or trisaccharides (Figure 35.1f), or monosaccharides (Figure 35.1e, g, and h).

The chemical structures of the repeating oligosaccharide units of SGs from the egg jelly coat of sea urchins (Figure 35.2a, b), from the tunic of ascidians (Figure 35.2c, d), from red algae (Figure 35.2e), and the repeating units of glycosaminoglycans (GAGs) and dermatan sulfates (DSs) from ascidians (Figure 35.2f) and of FucCS from sea cucumber (Figure 35.2g) are shown below. These polysaccharides are composed of L-galactopyranose, D-galactopyranose, L-iduronic acid, D-glucuronic acids, and D-$N$-acetylgalactosamines.

The backbone of the polysaccharide from *L. grisea* is made up of repeating disaccharide units of alternating 4-linked β-D-glucuronic acid (GlcA) units and 3-linked $N$-acetyl-β-D-galactosamine (GalNAc) (Figure 35.2g). However, the β-D-GlcA residues contain sulfated Fuc$p$ branches bound at 3-position of GlcA and the GalNAc units from the central core with a complex sulfation pattern (Glauser et al. 2013).

[-3)-α-L-Fuc$p$-2,4(OSO$_3^-$)-(1-3)-α-L-Fuc$p$-(1-3)-α-L-Fuc$p$-2(OSO$_3^-$)-(1-3)- α-L-Fuc$p$-2(OSO$_3^-$)-(1-]$_n$

**(a)** *Ludwigothurea grisea*

[-3)-α-L-Fuc$p$-2,4(OSO$_3^-$)-(1-3)-α-L-Fuc$p$-2(OSO$_3^-$)-(1-3)-α-L-Fuc$p$-2(OSO$_3^-$)-(1-3)-α-L-Fuc$p$-4(OSO$_3^-$)-(1-]$_n$

**(b)** *Lytechinus variegatus*

[-3)-α-L-Fuc$p$-4(OSO$_3^-$)-(1-3)-α-L-Fuc$p$-4(OSO$_3^-$)-(1-3)-α-L-Fuc$p$-2(OSO$_3^-$)-(1-3)-α-L-Fuc$p$-2(OSO$_3^-$)-(1-]$_n$

**(c)** *Strongylocentrotus pallidus*

[-4)-α-L-Fuc$p$-2(OSO$_3^-$)-(1-4)- α-L-Fuc$p$-2(OSO$_3^-$)-(1-4)-α-L-Fuc$p$-(1-4)-α-L- Fuc$p$-(1-]$_n$

**(d)** *Arboaia lixula*

[-3)-α-L-Fuc$p$-2,4(OSO$_3^-$)-(1-]$_n$ ~ 80% [-3)-α-L-Fuc$p$-2(OSO$_3^-$)-(1-]$_n$ ~ 20%

**(e)** *Strongylocentrotus purpuratus* I

[-3)-α-L-Fuc$p$-2,4(OSO$_3^-$)-(1-3)-α-L-Fuc$p$-4(OSO$_3^-$)-(1-3)-α-L-Fuc$p$-4(OSO$_3^-$)-(1-]$_n$

**(f)** *Strongylucentrotus purpuratus* II

[-4)-α-L-Fuc$p$-2(OSO$_3^-$)-(1-]$_n$

**(g)** *Strongylocentrotus droebachiensis*

[-3)-α-L-Fuc$p$-2(OSO$_3^-$)-(1-]$_n$

**(h)** *Strongylocentrotus franoiscanus*

*Figure 35.1* Repeating units of the sulfated fucans from the cell wall of the sea cucumber (a) and from the egg jelly coat of sea urchins (b–h).

[3]-α-L-Gal*p*-2(OSO₃⁻)-(1-]ₙ

**(a)** *Echinometra lucunter*

[-3)-β-D-Gal*p*-2(OSO₃⁻)-(1-3)-β-D-Gal*p*-1-)ₙ

**(b)** *Glyptocidans crenularis*

[-4)-β-L-Gal*p*-2[-1)-α-L-Gal*p*-3(OSO₃⁻)]-3(SO₃⁻)-(1-)ₙ

**(c)** *Styela plicata*

[-4)-β-L-Gal*p*-2[-1)-α-L-Gal*p*-3(OSO₃⁻)]-3(SO₃⁻)-(1-)ₙ

**(d)** *Herdm aniamonus*

[-3)-β-D-Gal*p*-(1-4)-α-D-Gal*p*-1-)ₙ

**(e)** *Botryododia occidentalis* and *Gelidium crinale* with different sulfation contents

[-4)-α-L-IdoA-(1-3)-β-D-GalNAc-(-1]ₙ

**(f)** *Styela plicata, Halocynthia pyriformis,* and *Ascidia nigra* with different sulfation patterns

[-4)-β-D-GlcA-3[-1)-α-L-Fuc*p*-2,4-di(OSO₃⁻)]-(1-3)-β-D-GalNAc-(-1-)ₙ

**(g)** *Ludwigothgurea grisee*

*Figure 35.2* Repeating units of sulfated galactans from structures of egg jelly coat of sea urchin eggs (a, b), tunic of ascidians (c, d), red algae (e), glycosaminoglycans and dermatan sulfates from ascidians (f), and fucosylated chondroitan sulfate from sea cucumber (g).

## 35.3   Applications

### 35.3.1   Anticoagulant effect

Though FucCS from *L. grisea* displays antimetastatic and anti-inflammatory activities in mammalian system (Borsig et al. 2007), it is mostly recognized as an oral antithrombotic agent (Fonseca and Mourao 2006). FucCS can substitute heparin as an alternative drug for the treatment of thrombosis, with many advantages over heparin. Like heparin, FucCS has serine protease inhibitor (serpin)-dependent anticoagulant activity for its increasing action of thrombin (IIa) inhibition by heparin cofactor II (HCII) and by antithrombin (AT), which is more prominent in FucCS than heparin. However, unlike heparin, FucCS also shows a serpin-independent anticoagulant activity by inhibiting tenase and prothrombinase procoagulant complexes (Glauser et al. 2008).

Removal of sulfated Fuc*p* units in FucCS (Figure 35.3) by mild acid hydrolysis as well as desulfation reaction considerably reduces its anticoagulant and antithrombotic activities, which proves its crucial role in anticoagulant activity (Glauser et al. 2008). Moreover, sulfated Fuc*p* branches in FucCS molecules prevent its rapid digestion by hyaluronidase that normally degrades GAGs in the gastrointestinal tract of vertebrates; this proves its effectiveness by oral administration, which is different from mode of heparin treatment (Fonseca and Mourao 2006). Sulfated fucans containing 2,4-disulfated Fuc*p* units (Figure 35.1a, b, e, f) augment the effect of the antithrombin-mediated

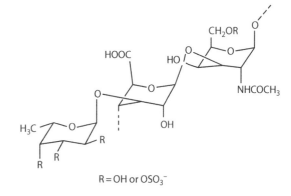

*Figure 35.3* Fucosylated chondroitin sulfate (FucCS) from the sea cucumber.

anticoagulant activity (Pereira et al. 2002a,b; Fonseca et al. 2009). However, when these 2,4-disulfated Fuc*p* units are present as branches, they did not favor interaction with antithrombin in a similar way. This clearly suggests the anticoagulant effect of FucCS is mostly via HCII (Glauser et al. 2008). The reduction of the carboxyl group of the GlcA residue to glucose (Figure 35.3) does not affect its anticoagulant activity, indicating that negatively charged carboxyl group has no influence on interaction with the coagulation cofactors. In contrast, carboxyl-reduced FucCS is still able to retain the anticoagulant activity. FucCS and SG present at high concentrations in sea cucumber and in red algae, respectively, are used as new therapeutics.

## 35.3.2 Nutraceuticals and pharmaceuticals

*Ascophyllum nodosum, Caulerpa racemosa, Codium fragile, Codium pugniformis, Cryptonemia crenulata, Ecklonia cava, Ecklonia kurome, Gayralia oxysperma, Gigartina skottsbergii, Grateloupia indica, Laminaria digitata, Laminaria japonica, Lessonia vadosa, Monostroma latissimum, Nemalion helminthoides, Nothogenia fastigiata, Porphyra haitanensis, Sargassum horneri, Schizymenia binderi, Ulva conglobata, Ulva pertusua, Ulva lactuca, Undaria pinnatifida* are among some of the important species of algae that have gained interest for the nutraceuticals and pharmaceuticals in their sulfated polysaccharides as a medicine against various diseases (Jiménez-Escrig et al. 2011; Wijesekara et al. 2011). Biological activity of SPs is widely dependent on chain length, molecular weight, and structure of polysaccharide in the marine organism (Ye et al. 2008). Besides the carbohydrates, other components such as enzymes, antioxidants, vitamins, and bioactive peptides are available in most marine organisms (Wijesekara et al. 2011). Most of the marine species because of their edibility and higher biomass content are considered as diet for medicinal value. Biomass from marine microalgal species is rich in lipids, carbohydrates, and proteins. The lipid-extracted microalgal debris consisting of mainly proteins and carbohydrates are used for animal feed. After extraction the residual carbohydrate material can be utilized as nutraceutical (Pleissner and Lin 2013). Marine microbial carbohydrates have been screened with respect to their structure and function and found largely in pharmaceuticals industries (Kim 2013). The application of nanoparticles in medicine is very promising since nanoparticles exhibit high stability and loading capacity, as well as responsiveness to environmental factors such as pH, ionic strength, and temperature, making them viable candidates for drug transport and drug release. However, when nanoparticles enter the bloodstream, they are often opsonized by plasma proteins and/or rapidly removed from the blood by the mononuclear phagocytic system. Recently, self-assembled nanoparticles based on natural polysaccharides have been of particular interest in light of their good biocompatibility, biodegradability, reduced toxic side effects, and improved therapeutic effects.

## 35.3.3 Industrial applications

Marine animal wastes like shellfish wastes from scallops (*Chlamys hastate*), cockles (*Cerastoderma edule, Clinocardium nuttallii*), clams and mussels (*Mercenaria mercenaria, Mytilus galloprovincialis, Mytilus edulis*), oysters (*Crassostrea gryphoides, Crassostrea gigas*), and crustaceans (crab—*Cancer pagurus*; lobster—*Homarus americanus*; shrimp—*Crangon crangon*) have been chemically and enzymatically processed to develop various important biomacromolecules, which can be used as nutritional substances, animal feed, biomedical goods, and so on. Chitin, a β-1,4 linked linear polymer of N-acetyl-D-glucosamine, is the main structural component of crustacean and molluscs and is extracted mainly as a by-product of the fishing industry. It is considered one of the most abundant polysaccharides in nature after cellulose and its production is estimated at $10^9$–$10^{11}$ tons per year (Kurita 2006; Dash et al. 2011). Chitin occurs in three polymorphic forms, α, β, and γ chitin. α chitin is the most common and is usually isolated from the exoskeletons of crustaceans and mollusks where it is the major structural element (Al Sagheer et al. 2009; Hayes et al. 2008). β chitin is less widespread and occurs in squid pens, and in the spines of some diatoms (Gaill et al. 1992). There is not much research on γ chitin, which is found in fungi and yeast and is believed to be a combination of α and β chitins (Al Sagheer et al. 2009). Chitin is insoluble in most solvents due to its rigid structure; however, it could have industrial applications when chemically modified. Chitin is normally converted by deacetylation through alkaline hydrolysis and due to the presence of amino groups becomes soluble in certain media. Chitosan is a linear polymer consisting of β-1,4-linked D-glucosamine obtained by stepwise deproteination, demineralization, decoloration, and deacetylation of chitin (Chatterjee et al. 2003). Partial deacetylation of chitin leads to chitosan, a polysaccharide composed of glucosamine and N-acetyl glucosamine units linked by β(1 → 4) bonds. Chitosan is the only natural polysaccharide that presents cationic character due to its amino groups which at low pH are protonated and can interact with negatively charged compounds such as proteins, anionic polysaccharides (e.g., alginates, carragenates, and pectins) fatty acids, bile acids, and phospholipids (Ramos et al. 2003). Production of chito-oligosaccharides by enzymatic hydrolysis besides acidic hydrolysis is also employed to produce chito-oligosaccharides. Macroalgal polysaccharides such as carrageenans, alginates, and agars are of much industrial use. The extent of the deacetylation of the chitin determines its physical, chemical, and biological properties. Chitin and chitosan have applications in numerous industries. Commercially produced chitosan has antimicrobial properties as well as increased biocompatibility and biodegradability (Abdou et al. 2008). This behavior along with its lack of toxicity has led to the usage of chitosan in diverse fields as agriculture, food, pharmaceutical, biomedical, cosmetics, medicine, biotechnology, paper industry, wastewater, and the paper industry (Kurita 2006; Kittur et al. 2003; Chatterjee et al. 2004, 2005, 2006, 2007; Rashidova et al. 2004; Sashiwa and Aiba 2004; Kim and Rajapakse 2005; Bautista-Baños et al. 2006). Recently, chitosan has been formulated in mouthwash, which is capable of inhibiting microbial adhesion and biofilm formation and at the same time being capable of promoting the dissolution of already formed biofilms. Additionally, in comparison with two commercially available mouthwashes, the chitosan-containing mouthwash presented significantly superior activity (Costa et al. 2014a). Chitosan inhibited *Candida albicans* adhesion (95%), biofilm formation (90%), and reduced mature biofilms by 65% and dual species biofilms (*C. albicans* and *S. mutans*) by 70%. The potentiality of this molecule to be used as an effective anti-*Candida* agent against *C. albicans* infections has been documented (Costa 2014b).

## 35.3.4 Antiviral effect

Sulfated homopolysaccharides and heteropolysaccharides isolated from a number of algae and cyanobacteria have demonstrated potent antiviral activity against retroviruses, for example, human immunodeficiency virus and herpes simplex virus (Schaeffer and Krylov 2000). HIV-induced syncytium or multinucleated cell formation as well as cytopathogenicity has been found to be prevented by these polysaccharides, The green seaweed, *Codium fragile*, contains many polysaccharides, for example, mannans, pyruvated arabinogalactan sulfates, and hydroxyproline-rich glycoprotein epitopes (HRGP; Estevez et al. 2009). HRGPs are believed to be involved in cell biology such as cell–cell recognition, cell expansion, and cell proliferation as well as cell wall structure (Seifert and Robert 2007).

## 35.3.5 Antioxidant activity

Antioxidants such as vitamins, minerals, polyphenols, flavanol, and other nutrients inhibit the oxidation process by protecting the cells from damaging by oxidation and play a vital role against various diseases such as aging processes, chronic inflammation, atherosclerosis, and cardiovascular disorders (Kohen and Nyska 2002). Marine polysaccharides having antimicrobial, antiviral, and anticancer properties are supposed to have a high antioxidative effect. Most of the organisms have antioxidant activity to defend themselves against oxidative damages. The bioactive compounds and antioxidants that marine organisms produce are important in the pharmaceutical industry. *Keissleriella* sp. is a marine fungus that has antioxidant activity (Wang et al. 2007). Fucoidan and lambda carrageenan have the highest antioxidant activity among the SP from brown and red seaweeds (De Souza et al. 2007).

## 35.3.6 Anticancer activity

It is well documented that SP showed *in vitro* inhibition of tumor, antiproliferative activity, and antimetastatic activity. SP from *Porphyra yezoensis* (Rhodophyceae) showed an apoptosis *in vitro* when the normal cells were not affected (Kwon and Nam 2006). Phosphated polysaccharide, GA3P, isolated from *Gymnodinium* sp. is an extracellular acidic polysaccharide, which is composed of D-galactose sulfate and lactic acid inhibits the growth of human leukemic cell lines. It is shown that GA3P has inhibitory effects on various cancer cell lines (Umemura et al. 2003) It is reported that many polysaccharides isolated from marine organisms can inhibit the growth of human cancer cells (Ahmed et al. 2013).

## 35.3.7 Biomedical and biotechnological applications

There has been growing interest in developing and exploiting glycoconjugates and their related molecules for biomedical and industrial applications. Glycans and glycoconjugates are involved in malignancies, cystic fibrosis, and inflammatory disease. The importance of the glycosidic residue in pharmaceuticals for the biological activity of a number of compounds including antibiotics, steroids, and vitamins has been reviewed by Kren and Martinkova (2001). In marine organisms, the majority of research has focused on compounds derived from molluscs and sponges. Pachymatismin, a glycoprotein isolated from the sponge *Pachymatisma johnstonii*, exhibits antileishmanial activity (Le Pape et al. 2000) and is also strongly antineoplastic (Sangrajrang et al. 2000). Glycoproteins isolated from *Aplysia kurodai*, *Aplysia juliana*, and *Dolabella auricularia* designated as aplysianins, julianins, and dolabellanins, respectively, show pronounced antineoplastic and antimicrobial activities. Aplysianin-A is a strongly antibacterial glycoprotein present in the albumen gland, which inhibits growth of both Gram-positive and Gram-negative bacteria. Aplysianin-E is not only antineoplastic and antibacterial but also possesses fungicidal activity (Iijima et al. 1995). The antineoplastic activity is via cytolysis whereas the antimicrobial activity is cytostatic (Yamazaki 1993). Dolabellanin A, also located in the albumen gland, has antineoplastic, antibacterial, and fungicidal activities (Iijima et al. 1994). Its mode of action is not bactericidal but bacteriostatic (Kisugi et al. 1992). A limpet hemocyanin, from a circulating glycoprotein from *Megathura crenulata*, is a novel immunostimulant in humans (McFadden et al. 2003). Fucan sulfates from the sea cucumber, *Stichopus japonicus*, are potent inhibitors of ostoclastogenesis in mice. These inhibitors might possess the potential to alleviate bone disorders like rheumatoid arthritis, and osteoporosis which could be used as future drug for the treatment of such diseases (Kariya et al. 2004). In osteoarthritis, chondroitin sulfate has been promoted as a potential drug (Kim et al. 2008) for improvement of the disease.

Sulfated homopolysaccharides and heteropolysaccharides isolated from algae and cyanobacteria have shown potent activity against retroviruses including human immunodeficiency virus and herpes simplex virus (Schaeffer and Krylov 2000).

They exhibit a broad range of medicinal benefits in inflammation, coagulation, thrombosis, angiogenesis, cancer, and infections. They are distinct from the majority of other glycans, because they are mostly derived from marine organisms in which the structures are taxonomically related, exhibit very rare patterns of structural regularity, and have a broad range of therapeutic actions.

## 35.4   Summary

Marine organisms are regarded as sources of sulfated polysaccharides. Over the past few decades, marine algal polysaccharides from green algae, red algae, and brown algae have gained great interest in medical and pharmaceutical industries for their wide potential applications in several biological, biomedical, and nutritional fields. These polysaccharides exhibit antiproliferative, anticancer, antiviral, anticoagulant, antioxidant, and anti-inflammatory activities. Sulfated galactans (SGs), sulfated fucans (SFs), fucoidans, and carrageenans are example of such marine polysaccharides. SFs are composed of repeating units of α-L-Fuc*p* and SGs are homopolymers of α-L, α-D, β-D Gal*p* residues. Fucoidans are present in the cell walls of brown seaweeds, whereas the green and red sea weeds are the sources of only SGs. SFs and SGs are also found to be present in tunicates, sea cucumber and sea urchins. It is suggested that the anticoagulating effect of FucCS is mostly via HCII. FucCS and SGs are used as new therapeutics. Marine algae, because of the edibility and higher biomass contain, are considered as diet of medicinal value, which is rich in lipid, carbohydrate, and protein. Chitosan obtained by deacetylation of chitin, which is derived from exoskeleton of crustacean and molluscs, has found tremendous applications in diverse fields, namely agriculture, food, paper industries, waste water management, and so on. Sulfated polysaccharide from a number of algae and cyanobacteria shows potent antiviral activity against retrovirus. GA3P from *Gymnodinium* sp. has inhibitory effects on various cancer cell lines. Fucoidan and lambda carrageenan are found to be the highest antioxidant.

## References

Abdou, E. S., S. S. Elkholy, M. Z. Elsabee, and E. Mohamed. 2008. Improved antimicrobial activity of polypropylene and cotton nonwoven fabrics by surface treatment and modification with chitosan. *Journal of Applied Polymer Science* 108: 2290–2296.

Ahmed, A. B. A., M. Vijayakumar, R. Pallela, N. Abdullah, and R. M. Taha. 2013. Marine derived bioactive peptides and their cardioprotective activities with potential applications. In *Marine Proteins and Peptides: Biological Activities and Applications*, S. Kim (Ed.), pp. 631–638. Chichester, U.K.: John Wiley & Sons, Inc.

Al Sagheer, F. A., M. A. Al-Sughayer, S. Muslim, and M. Z. Elsabee. 2009. Extraction and characterization of chitin and chitosan from marine sources in Arabian Gulf. *Carbohydrate Polymer* 77: 410–419.

Bautista-Baños, S., A. N. Hernández-Lauzardo, M. G. Velázquez-del Valle, M. Hernández-López, E. Ait Barka, E. Bosquez-Molina, and C. L. Wilson. 2006. Chitosan as a potential natural compound to control pre and postharvest disease of horticultural commodities. *Crop Protection* 25: 108–118.

Borsig, L., L. Wang, and M. C. M. Cavalcante. 2007. Selectin blocking activity of a fucosylated chondroitin sulfate glycosaminoglycan from sea cucumber. *Journal of Biological Chemistry* 282: 14984–14991.

Chatterjee, S., S. Chatterjee, B. P. Chatterjee, A. R. Das, and A. K. Guha. 2005. Adsorption of a model anionic dye, eosin Y, from aqueous solution by chitosan hydrobeads. *Journal Colloid and Interface Science* 288: 30–35.

Chatterjee, S., S. Chatterjee, B. P. Chatterjee, and A. K. Guha. 2004. Clarification of fruit juice with chitosan. *Process Biochemistry* 39: 2229–2232.

Chatterjee, S., S. Chatterjee, B. P. Chatterjee, and A. K. Guha. 2007. Adsorptive removal of Congo red, a carcinogenic textile dye by chitosan hydrobeads: Binding mechanism, equilibrium and kinetics. *Colloid and Surface A: Physicochemical and Engineering Aspects* 299: 146–152.

Chatterjee, S., A. K. Guha, and B. P. Chatterjee. 2003. Preparation and characterization of lobster shell chitosan: Modification of traditional method. *Indian Journal of Chemical Technology* 10: 350–354.

Costa, E. M., S. Silva, A. R. Madureira, A. Cardelle-Cobas, F. K. Tavaria, and M. M. Pintado. 2014. A comprehensive study into the impact of a chitosan mouthwash upon oral microorganism's biofilm formation in vitro. *Carbohydrate Polymers* 101: 1081–1086.

Costa, E. M., S. Silva, F. K. Tavaria, and M. M. Pintado. 2014. Antimicrobial and antibiofilm activity of chitosan on the oral pathogen *Candida albicans*. *Pathogens* 3: 908–919.

Dash, M., F. Chiellini, R. M. Ottenbrite, and E. Chiellini. 2011. Chitosan—A versatile semi-synthetic polymer in biomedical applications. *Progress in Polymer Science* 36: 981–1014.

De Souza, M., C. T. Marques, C. M. Dore, F. R. Silva, H. A. Rocha, and E. L. Leite. 2007. Antioxidant activities of sulfated polysaccharides from brown and red seaweeds. *Journal of Applied Phycology* 19: 153–160.

Estevez, J. M., P. V. Fernández, L. Kasulin, P. Dupree, and M. Ciancia. 2009. Chemical and in situ characterization of macromolecular components of the cell walls from the green seaweed *Codium fragile*. *Glycobiology* 19: 212–228.

Fonseca, R. J. and P. A. Mourão. 2006. Fucosylated chondroitin sulfate as a new oral antithrombotic agent. *Thrombosis and Haemostasis* 96: 822–829.

Fonseca, R. J. C., G. R. C. Santos, and P. A. Mourão. 2009. Effects of polysaccharides enriched in 2,4-disulfated fucose units on coagulation, thrombosis and bleeding. Practical and conceptual implications. *Thrombosis and Haemostasis* 102: 829–836.

Gaill, F., J. Persson, J. Sugiyama, R. Vurong, and H. Chanzy. 1992. The chitin system in the tubes of deep sea hydrothermal vent worms. *Journal of Structural Biology* 109: 116–128.

Glauser, B. F., P. A. S. Mourão, and V. H. Pomin. 2013. Marine sulfated glycans with serpin-unrelated anticoagulant properties. *Advance in Clinical Chemistry* 62: 269–303.

Glauser, B. F., M. S. Pereira, and R. Q. Monteiro. 2008. Serpin-independent anticoagulant activity of a fucosylated chondroitin sulfate. *Thrombosis and Haemostasis* 100: 420–428.

Hayes, M., B. Carney, J. Slater, and W. Brück. 2008. Mining marine shellfish wastes for bioactive molecules: Chitin and chitosan—Part A: Extraction methods. *Biotechnology Journal* 3: 871–877.

Iijima, R., J. Kisugi, and M. Yamazaki. 1994. Biopolymers from marine invertebrates. XIV. Antifungal property of Dolabellanin-A, a putative self-defense molecule of the sea hare, *Dolabella auricularia*. *Biological Pharmaceutical Bulletin* 17: 1144–1146.

Iijima, R., J. Kisugi, and M. Yamazaki. 1995. Antifungal activity of Aplysianin-E, a cytotoxic protein of sea hare (*Aplysia kurodai*) eggs. *Developmental Comparative Immunology* 19: 13–19.

Im, A. R., J. S. Sim, Y. Park, Y. S. Kim, and T. Toshihiko. 2008. Isolation and characterization of chondroitin sulfate from marine organisms. *Glycobiology* 18: 995–996.

Jiménez-Escrig, A., E. Gómez-Ordóñez, and P. Rupérez. 2011. Seaweed as a source of novel nutraceuticals: Sulfated polysaccharides and peptides. In *Advances in Food and Nutrition Research*, S. K. Kim (Ed.), Vol. 64, pp. 325–337. Burlington, U.K.: Academic Press.

Kariya, Y., B. Mulloy, K. Imai, A. Tominaga, T. Kaneko, A. Asari, K. Suzuki, H. Masuda, and I. T. Kyogashima. 2004. Isolation and partial characterization of fucan sulfates from the body wall of sea cucumber *Stichopus japonicas* and their ability to inhibit osteoclastogenesis. *Carbohydrate Research* 339: 1339–1346.

Kim, S. K. 2013. *Marine Nutraceuticals: Prospects and Perspectives*, Vol. 1, p. 464. Boca Raton, FL: CRC Press.

Kim, S.-K. and N. Rajapakse. 2005. Enzymatic production and biological activities of chitosan oligosaccharides (COS): A review. *Carbohydrate Polymers* 62: 357–368.

Kim, S. K., Y. D. Ravichandran, S. B. Khan, and Y. T. Kim. 2008. Prospective of the cosmeceuticals derived from marine organisms. *Biotechnology and Bioprocess Engineering* 13: 511–523.

Kisugi, J., H. Ohye, H. Kamiya, and M. Yamazaki. 1992. Biopolymers from marine invertebrates. 13. Characterization of an antibacterial protein, Dolabellanin-A, from the albumin gland of the sea hare, *Dolabella auricularia*. *Chemical and Pharmaceutical Bulletin* 40: 1537–1539.

Kittur, F. S., A. B. V. Kumar, L. R. Gowda, and R. N. Tharanathan. 2003. Chitosanolysis by a pectinase isozyme of *Aspergillus niger*—A non-specific activity. *Carbohydrate Polymers* 53: 191–196.

Kohen, R. and Nyska, A. 2002. Oxidation of biological systems: Oxidative stress phenomena, antioxidants,

redox reactions, and methods for their quantification. *Toxicologic Pathology* 30(6): 620–650.

Kren, V. and L. Martinkova. 2001. Glycosides in medicine: The role of glycosidic residue in biological activity. *Current Medicinal Chemistry* 8: 1303–1328.

Kurita, K. 2006. Chitin and chitosan: Functional biopolymers from marine crustaceans. *Marine Biotechnology*. 8: 203–226.

Kwon, M. J. and T. J. Nam. 2006. Porphyran induces apoptosis related signal pathway in AGS gastric cancer cell lines. *Life Sciences* 79: 1956–1962.

Le Pape, P., M. Zidane, H. Abdala, and M. T. More. 2000. A glycoprotein isolated from the sponge, *Pachymatisma johnstonii*, has antileishmanial activity. *Cell Biology International* 24: 51–56.

McFadden, D. W., D. R. Riggs, B. J. Jackson, and L. Vona-Davis. 2003. Keyhole limpet hemocyanin, a novel immune stimulant with promising anticancer activity in Barrett's esophageal adenocarcinoma. *American Journal of Surgery* 186: 552–555.

Pereira, M. S., F. R. Melo, and P. A. S. Mourão. 2002. Is there a correlation between structure and anticoagulant action of sulfated galactans and sulfated fucans? *Glycobiology* 12: 573–580.

Pereira, M. S., A.-C. E. S. Vilela-Silva, and A.-P. Valente. 2002. A 2-sulfated, 3-linked α-l-galactan is an anticoagulant polysaccharide. *Carbohydrate Research* 337: 2231–2238.

Pleissner, D. and C. S. Lin. 2013. Valorisation of food waste in biotechnological processes. *Sustainable Chemical Processes* 1: 21.

Pomin, V. 2010. Structural and functional insights into sulfated galactans: A systematic review. *Glycoconjugate Journal* 27: 1–12.

Pomin, V. H. 2009. Review: An overview about the structure–function relationship of marine sulfated homopolysaccharides with regular chemical structures. *Biopolymers* 91: 601–609.

Pomin, V. H. and P. A. S. Mourão. 2008. Structure, biology, evolution, and medical importance of sulfated fucans and galactans. *Glycobiology* 18: 1016–1027.

Rademacher, T., R. Parekh, and R. Dwek. 1988. Glycobiology. *Annual Review of Biochemistry* 57: 785–838.

Ramos, V. M., N. M. Rodriguez, M. S. Rodriguez, A. Heras, and E. Agulló. 2003. Modified chitosan carrying phosphonic and alkyl groups. *Carbohydrate Polymers* 51: 425–429.

Rashidova, S. S., R. Y. Milusheva, N. L. Voropaeva, S. R. Pulatova, G. V. Nikonovich, and I. N. Ruban. 2004. Isolation of chitin from a variety of raw materials, modification of the material, and interaction its derivatives with metal ions. *Chromatographia* 59: 783–786.

Sangrajrang, S., M. Zidane, P. Berda, M. T. More, F. Calvo, and A. Fellous. 2000. Different microtubule network alterations induced by pachymatismin, a new marine glycoprotein, on two prostatic cell lines. *Cancer Chemotherapy and Pharmacology* 45: 120–126.

Santos, J. A., B. Mulloy, and P. A. S. Mourao. 1992. Structural diversity among sulfated α-L-galactans from ascidians (tunicates)—Studies on the species *Ciona intestinalis* and *Herdmania monus*. *European Journal of Biochemistry* 204: 669–677.

Sashiwa, H. and S. Aiba. 2004. Chemistry modified chitin and chitosan as biomaterials. *Progress in Polymer Science* 29: 887–908.

Schaeffer, D. J. and V. S. Krylov. 2000. Anti-HIV activity of extracts and compounds from algae and cyanobacteria. *Ecotoxicology Environmental Safety* 45: 208–227.

Seifert, G. J. and K. Robert. 2007. The biology of arabinogalactan proteins. *Annual Review of Plant Biology* 58: 137–161.

Umemura, K., K. Yanase, M. Suzuki, K. Okutani, T. Yamori, and T. Andoh. 2003. Inhibition of DNA topoisomerases I and II, and growth inhibition of human cancer cell lines by a marine microalgal polysaccharide. *Biochemical Pharmacology* 66: 481–487.

Wang, S., Bligh, S., Shi, S., Wang, Z., Hu, Z., Crowder, J. et al. (2007). Structural features and anti-HIV-1 activity of novel polysaccharides from red algae *Grateloupia longifolia* and *Grateloupia filicina*. *International Journal of Biological Macromolecules* 41(4): 369–375.

Wijesekara, I., R. Pangestuti, and S. K. Kim. 2011. Biological activities and potential health benefits of sulfated polysaccharides derived from marine algae. *Carbohydrate Polymers* 84: 14–21.

Yamazaki, M. 1993. Antitumor and antimicrobial glycoproteins from sea hares. *Comparative Biochemistry Physiology C* 105: 141–146.

Ye, H., K. Wang, C. Zhou, J. Liu, and X. Zeng. 2008. Purification, antitumor and antioxidant activities in vitro of polysaccharides from the brown seaweed *Sargassum pallidum*. *Food Chemistry* 111: 428–432.

*chapter thirty-six*

# Glycobiotechnology

*Ramachandran Sarojini Santhosh, Visamsetti Amarendra,*
*Dharmalingam Gowdhaman, and Nallthambi Tamilkumar Varsha*

## Contents

## 36.1 Introduction

Glycobiotechnology is an interdisciplinary science that deals with highly efficient synthesis strategies using chemical or biological methods, to produce glycans for different applications to bring about human welfare. Generally, carbohydrates in the form of starch are used as food/storage material in the form of cellulose as structural material. But these polymers and simple sugars are also exploited as drugs, biomarkers, scaffolds, sensory molecules, and in environmental applications. Starch, generally produced through photosynthesis, is later converted into different structural components and sensory molecules of plant and animals. Other than an energy source, it is chemically linked with different biological molecules and by itself in different combinations and configurations for a variety of functions in organisms. Glycans are generally the underappreciated building blocks of life. The main source of glycans are terrestrial plants, but a large biomass is available from ocean; hence, recently, marine plants and animals were explored, and this field of work is getting separated from marine biotechnology; many carbohydrates separated from marine sources are industrially produced with different trademarks. This chapter mainly highlights the different applications of marine carbohydrates.

## 36.2 Applications of carbohydrates

### 36.2.1 Glycan drugs

In many antibiotics, the presence of glycan moiety in their structure component enhances the potential of the drug. For example, less cytotoxicity is exhibited when a polymeric galactomannan is conjugated to doxorubicin. Glycan drugs are classified based on the attached glycan structure, complexity, and use.

#### 36.2.1.1 Monosaccharides

The monosaccharides are glycan drugs (Figure 36.1a), is used in intravenous fluid and in oral rehydrating agents and mannitol to reduce intracranial tension. A metabolic product of fluorinated arabinofuranosyladenine 5'-monophosphate (1) was found to be very effective against chronic B-cell lymphocyte leukemia; other such monosaccharides in use are stavudine (2) (HIV); adenosine (3) (paroxysmal supraventricular tachycardia); ribavirin (4) (respiratory syncytial virus); and acadesine (5) (coronary artery bypass graft surgery). Derivatized monosaccharides such as amphotericin B (6) (derivative of an octahydroxypolyene) and etoposide (7) (derivative of podophyllotoxin) possess inhibitory action against fungal infection and cancer cells, respectively. Lincomycin (8, 1-thio-D-*erythro*-α-D-*galacto*-octopyranoside) and clindamycin

(9, 1-thio-L-*threo*-α-D-*galacto*-octopyranoside) have antibacterial activity. Pentostatin (10), an antitumor agent, mimics adenosine and inhibits adenosine deaminase of lymphoid cells (Mahajan et al. 2014).

#### 36.2.1.2 Disaccharides and disaccharide conjugates

Disaccharides and disaccharide conjugates are also used as glycan drugs (Figure 36.1b). Sucralfate (11) (basic aluminum sucrose sulfate complex) and lactulose (12) (4-*O*-β-D galactosyl-D-fructose) were used as ulcer-protective and laxative agents, respectively. Antibacterial vancomycin (13) (amphoteric tricyclic glycopeptides) and tobramycin (14) (aminoglucopyranosyl-ribohexopyranosyl-L-streptamine) were used against cystic fibrosis. Cardiac drug, Digoxin (15), contains a sugar molecule—*O*-2,6-dideoxy-β-D-*ribo*-hexapyranosyl. Other than the naturally existing, designed glycan drugs are Tamiflu (16) and Relenza (17). A guanidino side chain was added to the C4 position of 2-deoxy-2, 3-dehydro-*N*-acetyl-neuraminic acid (18) for increasing its affinity for influenza neuraminidase, without any affinity for host-cell neuraminidases by optimized synthesis. Li et al. synthesized 32 analogues using a mangrove-derived fungus based on the natural isoprenyl phenyl ether against influenza H1N1 neuraminidase (Li et al. 2011).

#### 36.2.1.3 Oligosaccharide and polysaccharides drugs

Erythromycin semisynthetic derivatives like dirithromycin (19), clarithromycin (20), and azithromycin (21) are available for antimicrobial therapy (Kirst 1993). Oligosaccharides and polysaccharide drugs (Figure 36.1c) include glycosaminoglycans (GAGs), heparin-like saccharides, and other complex oligosaccharides. Heparin (22) and heparan sulfate (23) differ by ratios of *N*-acetyl to *O*-sulfo groups and different sensitivity to heparin lyases. Enoxaparin (24) and Tinzaparin (25) (Dalteparin) are prepared from heparin by alkaline degradation, enzymatic hydrolysis, and nitrous acid fragmentation, respectively, and are effectively used against thrombotic diseases (Islam and Linhardt 2003). Danaparoid (26) is a complex glycosaminoglycuronan having active components such as heparan sulfate (23), dermatan sulfate (27), and chondroitin sulfate (28) in its structure. Synthetic heparin pentasaccharide (Arixtra, 29) binds to antithrombin and is used to prevent deep vein thrombosis and pulmonary embolism. Pentosan polysulfate (30) (β-D-xylopyranose) has both anticoagulant and fibrinolytic effects. The concentration of heparan sulfate (23) and dermatan sulfate (27) proteoglycans change during atherogenesis (Edwards et al. 1990). Dermatan sulfate (27) is a weak anticoagulant used in medical devices and tissue engineering. A microporous polyurethane tube coated with a mixture of type I

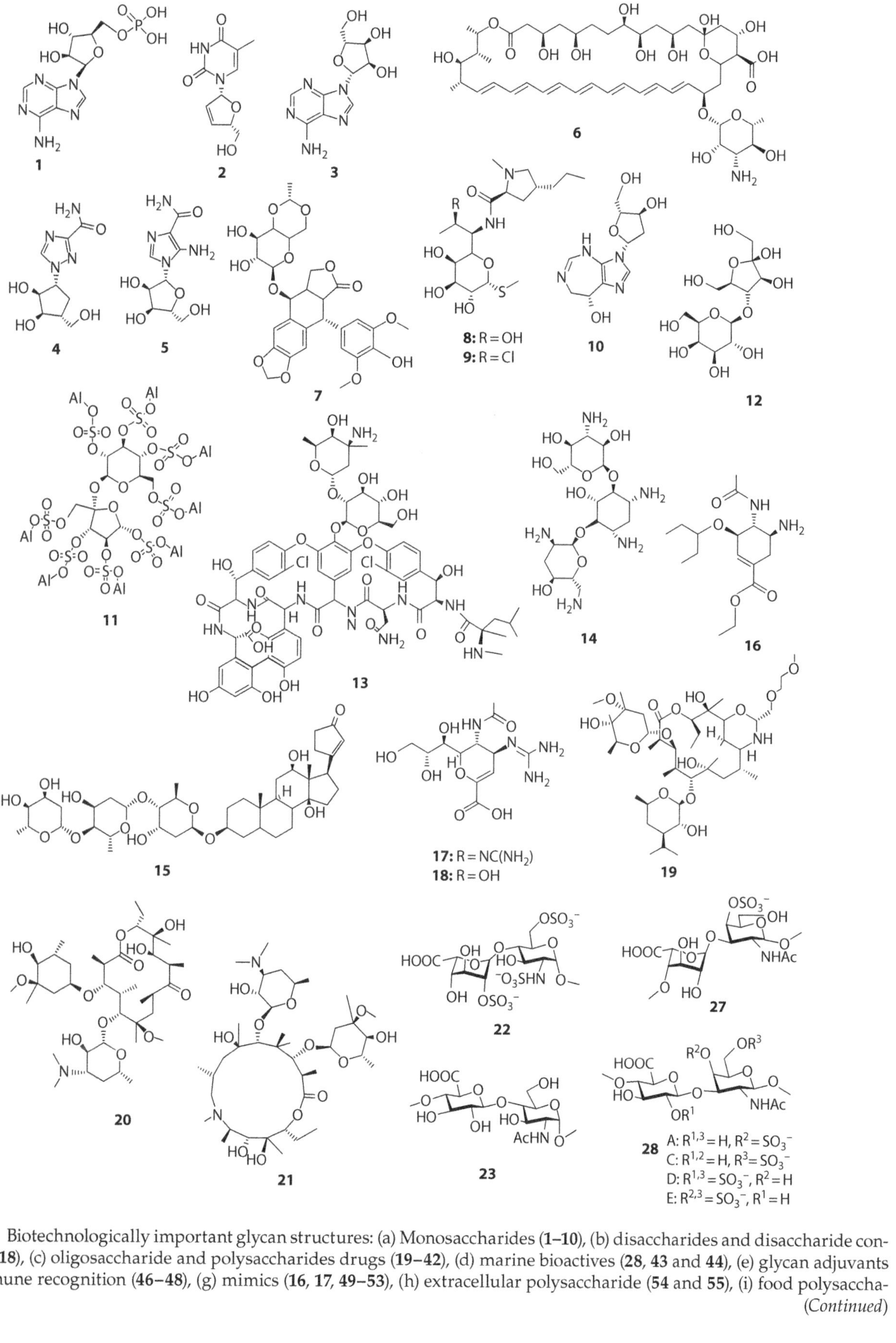

*Figure 36.1* Biotechnologically important glycan structures: (a) Monosaccharides (**1–10**), (b) disaccharides and disaccharide conjugates (**11–18**), (c) oligosaccharide and polysaccharides drugs (**19–42**), (d) marine bioactives (**28**, **43** and **44**), (e) glycan adjuvants (**45**), (f) immune recognition (**46–48**), (g) mimics (**16**, **17**, **49–53**), (h) extracellular polysaccharide (**54** and **55**), (i) food polysaccharide (**56**).

(*Continued*)

**Figure 36.1 (Continued)** Biotechnologically important glycan structures: (a) Monosaccharides (**1–10**), (b) disaccharides and disaccharide conjugates (**11–18**), (c) oligosaccharide and polysaccharides drugs (**19–42**), (d) marine bioactives (**28, 43 and 44**), (e) glycan adjuvants (**45**), (f) immune recognition (**46–48**), (g) mimics (**16, 17, 49–53**), (h) extracellular polysaccharide (**54 and 55**), (i) food polysaccharide (**56**).                                                                                           *(Continued)*

***Figure 36.1 (Continued)*** Biotechnologically important glycan structures: (a) Monosaccharides (**1–10**), (b) disaccharides and disaccharide conjugates (**11–18**), (c) oligosaccharide and polysaccharides drugs (**19–42**), (d) marine bioactives (**28, 43** and **44**), (e) glycan adjuvants (**45**), (f) immune recognition (**46–48**), (g) mimics (**16, 17, 49–53**), (h) extracellular polysaccharide (**54** and **55**), (i) food polysaccharide (**56**).                    (*Continued*)

collagen and dermatan sulfate (**27**) is used for arterial prostheses. Chondroitin sulfate (**20**), a nutriceutical (osteoarthritis), from human milk binds to host cell CD4 receptor and inhibits interaction of HIV glycoprotein gp120. Hyaluronan (**31**) are cell immobilizers that act by binding to agrecan, versican, neurocan, brevican, and CD44 (Tammi et al. 2002). *N*-sulfoacharan sulfate (**32**) is a moderate inhibitor of thrombin and shows reduced heparin-like effect on fibroblast growth factor-2 (Wang et al. 1997; Wu et al. 1998). *Fucus* fucoidan (**33**) is an anticoagulant, antiviral, anti-inflammatory, antigastriculcer, and contraceptive (Islam and Linhardt 2003). Irish

moss (*Chondrus crispus*) carrageenan is a cough medicine (Prajapati et al. 2014), degraded ɩ-carrageenan is antiulcer (Pittman et al. 1976), and both κ- and ɩ-carrageenan have antitumor activity (Yao et al. 2014). Sulfated galactans such as carrageenans produced from red algae and their depolymerized products possess immunomodulation and antitumor activity. Chitins (**34**) have anticoagulant activity and a carboxymethylated sulfochitosan inhibits thrombin activity similar to heparin (Chaudhari et al. 2014). In nature, three polymorphic forms of chitin are available. α-chitin is obtained from exoskeletons of crustaceans, mollusks, and shrimp. These α-chitin

**Figure 36.1 (Continued)** Biotechnologically important glycan structures: (a) Monosaccharides (**1–10**), (b) disaccharides and disaccharide conjugates (**11–18**), (c) oligosaccharide and polysaccharides drugs (**19–42**), (d) marine bioactives (**28, 43** and **44**), (e) glycan adjuvants (**45**), (f) immune recognition (**46–48**), (g) mimics (**16, 17, 49–53**), (h) extracellular polysaccharide (**54** and **55**), (i) food polysaccharide (**56**).

microparticles in living system convert allergic response into inflammatory response (Muzzarelli 2010). β-chitin is found in squid pens, pogonophoran tubes, and in the spines of some diatoms (Gaill et al. 1992). γ-chitin, a combination of α- and β-chitins, is found in fungi and yeast (Sagheer et al. 2009). Chitosan can lower the total cholesterol, plasma, and liver triacylgycerol levels and also used to stop bleeding of the liver, aorta, lung, kidney, and cardiac ventricle as an internal wound dressing agent. Alginates (**35**), used to treat wounds, gastric ulcers, reflux esophagitis, plasma cholesterol level, and allergic reactions, also strongly inhibit hyaluronidase and mast cell degranulation (Mahajan 2014). Suramin (**36**) is an antihelminthic, antiprotozoal, antineoplastic, and antiviral agent. Simple aliphatic disulfates and disulfonates are used in the treatment of Alzheimer's disease (Ferrao-Gonzales et al. 2005). Dextran sulfate (**37**) inhibits the binding of HIV to T-lymphocytes (Jaques et al. 1991). β-cyclodextrin tetradecasulfate, having anticancer activity, can also inhibit complement activation system (Gerloczy et al. 1994; Watson et al. 2013). Phosphosulfomannan (PI-88) (sulfonated phosphomannan) from *Pichia holstii* is in Phase I clinical trials as an antitumor agent (Islam and Linhardt 2005). Pentosan from the bark of *Fagus sylvantica* is an anticoagulant (Scully et al. 1983). Trestatin A (**38**) (pseudo-nonasaccharide) is an α-amylase inhibitor from *Streptomyces*

*dimorphogenes*. A highly sulfated maltotriosyl trehalose pentasaccharide, a modification of Trestatin, has antiproliferative activity (Xie et al. 2000) and sulfated lactobionic acid (**39**) is an antithrombotic agent (Giedrojc et al. 1996). Aminoglycoside antibiotics such as streptomycin (**40**) (antituberculosis) and neomycin (**41**) interfere with protein synthesis. Another complex oligosaccharide, acarbose (**42**), delays the digestion of ingested carbohydrates by inhibiting α-glucosidase (beneficial in type 2 diabetes mellitus) (Anatole 2012).

### 36.2.2 Marine bioactives

A detailed description about marine bioactive glycans (Figure 36.1d) is written by the same authors as a chapter in this book; however, applications of this group of glycans are explained briefly. Many glycoproteins like Pachymatismin (*Pachymatisma johnstonii*) have antileishmanial (Le Papeet al. 2000) and antineoplastic (Sangrajrang et al. 2000) effects. Aplysianins, julianins, and dolabellanins possessing antineoplastic and antimicrobial activities were isolated from *Aplysia kurodai*, *Aplysia juliana*, and *Dolabella auricularia*, respectively (Kisugi et al. 1992; Iijima et al. 1994; Lijima et al. 1995). Keyhole limpet hemocyanin derived from mollusk *Megathura crenulata* is a novel immunostimulant in humans (McFadden et al. 2003). Crambescidin 826 (**43**) and Dehydrocrambine A (**44**) prevent membrane fusion of HIV-1 virus (Chang et al. 2003). Fucan sulfates from *Stichopus japonicus* are potent inhibitors of osteoclastogenesis and used as an accelerant for bone formation (Kariya et al. 2004). Chondroitin sulfate (28), widely used in cosmoceutical products (Kim et al. 2008), is naturally present in giant squid, skate, and salmon cartilage. Algae and cyanobacteria produces sulfated homopolysaccharides and heteropolysaccharides that are effective against retroviruses (Schaeffer and Krylov 2000). In diatoms, silica deposition takes place with the help of glycoproteins having high affinity for calcium (Kroger et al. 1994).

## 36.2.3 Fouling and antifouling agents

There are fouling and antifouling marine carbohydrates. In barnacles and mussels, glycoconjugates that help in the settlement and attachment are fouling agents. "Foot protein 1" present in the byssus thread of freshwater quagga mussel (*Dreissena bugensis*) is a fouling agent (Anderson and Waite 2000). Mucin-type glycoproteins produced by starfish and brittlestars is an antifouling agent (Bavington et al. 2004).

## 36.2.4 Vaccines

Natural polysaccharides and synthetic oligosaccharide epitopes are commercially used as vaccines. Natural polysaccharides conjugated to carrier proteins are used against *Neisseria meningitidis*, *Streptococcus pneumoniae*, *Haemophilus influenza type b* (Hib) and synthetic oligosaccharide epitopes against HIV, *Plasmodium falciparum*, *Vibrio cholerae*, *Cryptococcus neoformans*, *S. pneumoniae*, Shiga toxin, *Bacillus anthracis*, and *Candida albicans*. TLR2 agonist is a three component anticancer vaccine having a promiscuous peptide, T-helper epitope, and a tumor-associated glycopeptide (Mahajan 2014). Gangliosides (GM2 and GD2 of melanomas) and globo H (breast cancer antigen—sialylα2-6GalNAcα-) are utilized in the development of cancer vaccines (Wang et al. 2000).

## 36.2.5 Glycan adjuvants

The adjuvanticity of *Sargassum pallidum* polysaccharides in Newcastle disease, infectious bronchitis, and avian influenza was investigated (Li et al. 2012). The α-Galactosylceramide (45) (Figure 36.1e) (α-GalCer or C1) isolated from marine sponges is considered a possible adjuvant candidate. CD1d on dendritic cells is the receptor of C1 (Kobayashi et al. 1995; Brossay et al. 1997; Kawano et al. 1997). The high avidity of CD1d–glycolipid–T-cell receptor complex selectively enhanced the Th1 pathway and, thus, the adjuvant activity (Wu et al. 2006).

## 36.2.6 Diagnosis

### 36.2.6.1 Biomarkers

Glycoconjugates were characterized as pluripotency biomarkers, stage-specific embryonic antigens, and tumor rejection antigens. Mass spectrometry–based analysis of total glycome is needed for identifying more versatile glycobiomarkers (Parry et al. 2007; Babu et al. 2009; Fujitani et al. 2013). Often dietary glycemic index can predict possible metabolic syndrome (Danielsen et al. 2013). Main cancer biomarkers are carbohydrate antigens (CA19-9, CA125, DUPAN-II, AFP-L3) (Ueda 2013). CA-125 and CA 19-9 are potential biomarkers in heart failure and pancreatic cancer, respectively (Bauer et al. 2013; Vizzardi et al. 2013). Glycoprofiling of serum, urine, saliva, or various cell or tissue extracts observed alterations in glycostructures in leukemia, hematopoietic diseases, carbohydrate deficiency diseases, leukocyte adhesion deficiencies, cystic fibrosis, diabetes, intestinal inflammatory and liver diseases, osteoarthritis, rheumatoid arthritis, thrombosis, etc. (Janković 2011); Lewis antigen (Le) present on the plasma membrane of red blood cells is an exception to this. *Helicobacter pylori* Lex and Ley antigens responsible for the adhesion are cell-surface glycoconjugate molecules, which lead to gastric cancer (Moran and Prendergast 2001). Glycan structure variation in Ig is the leading cause of rheumatoid arthritis, systemic lupus erythematosus (Arnold et al. 2007; Wuhrer 2007), and IgA nephropathy (Giannakakis et al. 2007). Another example is the higher fucosylation observed in serum glycoproteins isolated from liver cirrhosis patients.

The glycosylation pattern is different in malignant cells and diseased tissues (Ghazarian et al. 2011) due to the altered activity of glycosyltransferases and glycosidases (Powlesland et al. 2009), which result in the increased exposure of terminal galactose residues, such as in the T antigen (Galβ1–3GalNAc) and the Lewis trisaccharide (Galβ1–4(Fucα1–3)GlcNAc) (Kim and Varki 1997). These alterations can be biomarkers for diagnosis or treatment.

Absence of a carbohydrate molecule on transferrin acts as a biomarker in alcohol abuse (Stibler 1991). In tumor progression, cell–cell interaction during uncontrolled cell division is determined by alternations in glycosylation. In invasive human bladder transitional cell, beta-1,3-*N*-acetylglucosaminyl-transferase-T2 expression is downregulated. In colorectal cancers, expression of various glycosyltransferases is altered. Tumor-specific fucosylated alpha-fetoprotein (AFP) glycoform is documented for hepatocellular carcinoma (HCC) and chronic liver diseases. Using lectin affinity electrophoresis, the glycoform between chronic liver diseases and HCC can be differentiated. A monosialylated glycoform of AFP can be used for detecting early stage HCC. Fucosylated haptoglobin associated with tumor progression is a potential HCC biomarker. Carbohydrate pattern can be used for studying land use pattern, coastal pollution monitoring by measuring glyconitrile levels. Composition and distribution of dissolved carbohydrates in the Beaufort Sea Mackenzie margin (Arctic Ocean) gave an idea about the pollution in that region (Panagiotopoulos et al. 2014). Glycan-based biomarkers for mucopolysaccharidosis (Lawrence et al. 2014).

### 36.2.7 Physiology

#### 36.2.7.1 Metabolic diseases

The glycan salvage pathway was utilized for the treatment of metabolic disorders. In patients who are deficient in phosphomannose isomerase, oral mannose supplementation was found to be very effective (Chu et al. 2013). In CDG-IIc patients, treatment with fucose restored the synthesis of sialyl Lewis on leukocytes (Moriwaki and Miyoshi 2010). In patients lacking an enzyme involved in polysaccharide degradation, an alternate sugar molecule was suggested for meeting the energy requirements. Lysosomal enzyme replacement therapy (LERT) and substrate reduction therapy (SRT) were used for patients with lysosomal storage disorders that arise due to the failure in the turnover of glycans in lysosome (Futerman and van Meer 2004).

### 36.2.8 Pathology

In the place of an adaptive immune system, sugar components of bacterial cell walls are recognized to induce bacterial agglutination in invertebrate aquatic animals (Alpuche et al. 2005). Lectins can recognize specific sugars and may prevent the invasion of pathogens. The marine sponge, *Halichondria okadai*, is protected from infection by three lectins HOL-I, HOL-II, and HOL-30 (Kawsar et al. 2008) or help in the symbiosis with microbes (Jimbo et al. 2000). A novel lectin present in the mantle of the penguin wing oyster inhibits the growth of *Escherichia coli* (Naganuma et al. 2006). The multixenobiotic resistance phenotype (MXR) of filter feeding mollusks is mediated by permeability glycoproteins (Smital et al. 2000, 2003; Eufemia et al. 2002). Kendarimide A (**46**) from *Haliclona* sp. (Aoki et al. 2004), Dictyostatin-1 (**47**) from *Spongia* sp. (Isbrucker et al. 2003), and irciniasulfonic acid (**48**) from *Ircinia* sp. (Kawakami et al. 2001) inhibit the P-gp efflux pump associated with human pathogens (Lopez and Martinez-Luis 2014) (Figure 36.1f). Oligosaccharides also have a role in the *Chondrus crispus*–*Acrochaete operculata* and *Paralichthys olivaceus*–*Neobenedenia girellae* pathosystem. Heterogenic N-linked glycans of IgM present in Atlantic cod (*Gadus morhua*) and *Salmo salar* are similar to mammalian but have lower levels of oligomannose glycans (Magnadóttir et al. 1997).

### 36.2.9 Pheromones and chemoattractant

Pheromones and chemoattractants transmit information to cospecies and heterospecies, respectively. Barnacles use a glycoprotein, a settlement pheromone secreted by females (Clare and Matsumura 2000), which is detected by a protein receptor on the males' antennules (Kelly and Snell 1998). Glycoproteins are also important for mate recognition in calanoid copepods (Snell and Carmona 1994), and act as potent chemoattractants in eggs (Ferrari and Targett 2003).

### 36.2.10 Glycoprotein hormones

Stanniocalcin (hypocalcin or teleocalcin) is a homodimeric glycoprotein from Stannius endocrine glands and plays an integral role in calcium and phosphate homeostasis and functions to prevent hypercalcemia in salmon (*S. salar*) (Wagner et al. 1998).

### 36.2.11 Reproduction

Glycoconjugates are responsible for the sperm-egg binding in marine organisms (Caldwell and Pagett 2010). Using marine organisms as a model, studying the role of glycans in reproduction helps clinicians and scientists to know more about the molecular details of human and animal fertility (Gallo and Costantini 2012) as gender-specific glycosylation in human reproduction has been reported (Morris et al. 1996).

### 36.2.12 MIMICS

Monosaccharides like voglibose (**49**) and miglitol (**50**) inhibit intestinal α-glycosidases and prevent the hydrolysis of polysaccharides and oligosaccharides in diabetes treatment (Dabhi et al. 2013). Zanamivir (**17**) and oseltamivir (**16**) prevent influenza virus infection by mimicking the sugar structure to which the virus binds by receptors (Shtyrya et al. 2009). Other examples are miglustat (**51**) (cystic fibrosis) and topiramate (**52**) (anticonvulsant) drugs. Validamycin A (**53**) is an oligosaccharide from *Streptomyces hygroscopicus* subspecies *limoneus* (Bai et al. 2006). Carbohydrate-based mimetic drugs (Figure 36.1g) currently in the market are BGM Galectin-3, GSC100, GM-CT-01, and GM-MD-02. Glycan mimetics prevent the attachment of microbes to cell wall and others lead to the inhibition of selectin-mediated leukocyte trafficking to prevent inflammation. Transfusion and transplantation rejection are carried out using anti-glycan antibodies (Varki and Chrispeels 1999).

### 36.2.13 Tissue engineering

In *ex vivo* and *in situ* tissue engineering to improve or replace biological functions, biomaterials used in tissue scaffolds are important. The scaffolds should posses three-dimensional (3D) porous structure with a high surface to volume ratio for seeding maximum number of cells, physicochemical structures to support cell adhesion, proliferation, differentiation and extracellular matrix (ECM) production to organize cells into a 3D structure, capable of good nutrient and waste exchange, nontoxic, biological property for vasculature formation, capacity to retain mechanical properties and mechanical architecture until cells produce ECM, slow delivery of drug and growth factors, clinically relevant size and shape, sterile and stable for long-time storage, should not have cell adhesive motifs, and should be economically viable. Using polysaccharides, fiber mesh, sponge type, fine filament mesh, fibrin gel, injectable scaffolds, and soft elastic scaffolds were prepared, but marine polysaccharides are still underexploited. GAGs, hyaluronic acid, chondroitin sulfate, dermatan sulfate, and heparan sulfate are used in scaffold preparations.

### 36.2.13.1  Glycosaminoglycans

Marine invertebrates like marine mollusks (*Amussium pleuronectus* (Linne)) or echinoderms such as sea urchins or sea cucumbers (ascidians) are rich in GAGs. GAGs from *A. pleuronectus* (Linne). contain uronic acid and hexosamine (Senni et al. 2011) in equivalent amounts. The bioactivity of marine polysaccharides makes them suitable for preparing tissue scaffolds. Fucosylated chondroitin sulfate obtained from sea cucumber is effective against thrombosis and ischemia. Chondroitin/dermatan sulfate hybrid chains extracted from sharkskin showed a high affinity for growth factors and neurotrophic factors. Chondroitin sulfate forms the arms of the aggrecan molecules in cartilage. Aggrecan, proteoglycans, and type II collagen present in the cartilage ECM is responsible for the tissues' compressive and tensile strength (Fox et al. 2009).

### 36.2.13.2  Hyaluronic acid

The hyaluronic acid, which helps in wound healing, is a major component of connective tissues and ECM. Hyaluronic acid is present in vitreous and synovial fluid, cartilage, and serum. It plays an important role in tissue morphology, extracellular space, and also in the transport of nutrients and ions. Low-molecular-weight hyaluronic acid helps in angiogenesis (Ke et al. 2013). Hyaluronan is used as a gel in nerve, cartilage, and skin tissue engineering, because it can be removed by hyaluronidase present in these tissues.

### 36.2.13.3  Alginate

Alginates from brown seaweed and algae are used in preparing gels. Alginate having chemical modifications is also a useful biomaterial for cell immobilization, controlled drug delivery, and enhances neocartilage or neobone formation while transplanting chondrocytes and osteoblasts (Senni et al. 2011). Transplantation rejection can be prevented by encapsulating the cells in calcium alginate to escape from the immune system. Encapsulated pancreatic islet cells produced insulin according to the needs of the host as the encapsulation prevented them from immune reaction against the grafted cells.

### 36.2.13.4  Chitosan

Chitosan derived by deacetylation of chitin is a strong cationic polysaccharide used for bandages in wound dressing and blood clotting. Composite chitosan/nano-hydroxyapatite scaffolds support osteoblast differentiation (Chesnutt et al. 2009).

### 36.2.13.5  Carrageenans

Red seaweeds (Rhodophyceae), *Chondrus crispus*, *Gigartina*, *Eucheuma cottonii*, and *Spinosum* are main sources of sulfated polysaccharide and carrageenans. Human-adipose-derived stem cells and human nasal chondrocytes encapsulated with carrageenan hydrogel provide support for cell culture, viability, cartilage matrix extracellular formation, and chondrogenic differentiation. Some of the polysaccharide producing green algae are *Bryopsis plumosa* (rhamnan sulfate), *Chaetomorpha aerea* (sulfated galactan), *Codium dwarkense* (ulfated arabinan, sulfated arabinogalactan), *Codium fragile* (sulfated arabinogalactans, pyruvylated galactan sulfate), *Codium vermilara* (sulfated mannan), *Codium yezoense* (pyruvylated galactan sulfate), *Monostroma latissimum* and *Monostroma nitidum* (rhamnan sulfate), *Ulva lactuca, Ulva rotundata, Enteromorpha compressa*, and *Ulva rigida* (sulfated rhaman) (Silva et al. 2012).

### 36.2.13.6  Fucoidans

The depolymerized low-molecular-weight fractions of simple (marine echinoderms and sea cucumbers) and more complex fucoidans (sea weeds) were used for angiogenesis in tissue engineering (Silva et al. 2012).

### 36.2.13.7  Extracellular polysaccharide

*Arthrospira platensis* (Spirulina) produces a highly porous scaffold of spirulan, which is used for stem cell culture and treatment of spinal cord injury. *Aphanocapsa, Cyanothece, Gloeothece, Synechocystis, Phormidium*, Anabaena, and *Nostoc* produce EPS having potential use in tissue engineering (Figure 36.1h). There are three bacteria in ocean producing exopolysaccharides, namely, *Pseudoalteromonas* sp., *Alteromonas* sp., and *Vibrio* sp. *Vibrio diabolicus* bacterium isolated from a Pompeii worm tube (*Alvinella pompejana*) produces a hyaluronic acid-like exopolysaccharide (HE800 EPS)-Hyalurift® **(54)**. It enhanced the collagen structuring of the engineered connective tissue and promoted fibroblast settlement in extracellular matrix. Bone healing using Hyalurift was quick compared to collagen in treated animals and it is suggested for skin or cartilage grafting (Nwodo et al. 2012). *Alteromonas infernus* isolated from a dense population of *Riftia pachyptila* produces EPS GY 785 (**55**), a highly branched acidic heteropolysaccharide with a high molecular weight and low sulfate content. It is an injectable sialylated hydroxypropyl methylcellulose-based hydrogel (Si-HPMC). It improved mechanical properties of the scaffold and attachment of chondrocytes and osteoblasts (Courtois et al. 2014).

## 36.2.14  Electronics

### 36.2.14.1  Biosensor

Biosensors based on carbohydrates, antibodies, and DNA are available. But due to the instability of antibodies in extreme environments and the purification steps required for long nucleic acid sequences, the carbohydrate–protein recognition mechanism gained importance as an alternative detection and identification principle and the first glucose biosensor was developed in 1962. The carbohydrate biosensors are based on the isothermal titration calorimetry (ITC), surface plasmon resonance (SPR), capillary electrophoresis (CE), nuclear magnetic resonance (NMR), quartz crystal microbalance (QCM), fluorescence spectroscopy, and atomic force microscopy (AFM). Carbohydrate sensors detect carbohydrates in biological samples by immobilized glycoenzymes using any one of the principles mentioned earlier. In the sensors, carbohydrate recognition elements are immobilized on polymers or glass or gold nanoparticles (NPs), or quantum dots by physisorption, covalent immobilization, and bioaffinity-based interaction. For immobilization, lectin or receptor binding method is also utilized. For heparin detection, SPR is used (Liljeblad et al. 1998). *Neoglycolipids*–oligosaccharides coupled to lipid residues are also used in biosensor design (Feizi et al. 1994). Carbohydrate components in biosensors are either carbohydrate recognition elements or a scaffold. Structural profiles of glycans and their recognition by lectins can be used as biomarker as a diagnostic tool (Kim et al. 1990). Chitin is also an immobilizing agent of recognition elements. Lectins from algae (Pinto et al. 2009), sponges (Bretting et al. 1981), mollusks (Bulgakov et al. 2004), and echinoderms (Hatakeyama et al. 1994) are used in sensors for identifying pathogen and viral species (Mahmood and Hay 1992) and these are capable of detecting subnanomolar concentrations of glycogen from sample (Krull et al. 1989). Antibody-tagged, glycoprotein-based amperometric impedance biosensor was used for the detection of a human mammary tumor. Bertucci et al. exploited a recombinant of glycoproteins of herpes simplex virus

(HSV) and human cellular receptor for HSV for constructing an optical biosensor (Bertucci et al. 2003). Mostly, carbohydrate-based pathogen biosensors rely on LPS moieties on the pathogen. Sialylated Lewis antigens (SLeAs) (Davidson et al. 2000; Nakagoe et al. 2002) and mucins were developed for cancer and tumor detection (Devine et al. 1992). Carbohydrate nanobiosensors are the contribution of nanotechnology. Nickel NPs within a graphite film electrode confer high-sensitivity detection of sugar moieties, a contribution of nanotechnology for making carbohydrate nanobiosensors (You et al. 2003). In the fabrication of nanomaterial-based electrochemical biosensors, the immobilization techniques improved the efficiency of detection of glucose (Putzbach and Ronkainen 2013).

## 36.2.15  Polysaccharides are used as prebiotic, probiotic, dietary fiber, or a functional food

### 36.2.15.1  Food polysaccharides

Probiotics and prebiotics bring out healthier effects in human by equilibrating intestinal microflora either present or therapeutically introduced in the intestine. Diet polysaccharides can be both starch (amylose and amylopectin) and nonstarch. Resistant starches (RS) formed upon food processing are indigestible and have beneficial nutritional effects. Marine algal polysaccharides possess hypocholesterolemic activities (Lordan et al. 2011). Red algae and brown algae are rich sources of dietary polysaccharides. These indigestible polysaccharides are regarded as dietary fibers (Champ et al. 2003). Storage polysaccharides, like agar, carrageenans and alginates, are commercially exploited for food applications. Porphyran (**56**) (Figure 36.1i) (sulfated polysaccharide) from *Porphyra* is a nutritional supplement as gelling and antioxidant agent (Rocha de Souza et al. 2007). Laminarin, from brown algae (Pheophyta), modulates intestinal metabolism. Nondigestible xylo- and fructo-oligosaccharides are preferential substrate for *Bifidobacteria* and *Lactobacilli*. Even though algal polysaccharides have the potential to serve as functional foods, they are underexploited (O'Sullivan et al. 2010).

*Ulvan* from *Ulva rigida* (*chlorophyta*), which is a water-soluble dietary fiber, is composed of β-(1,4)-xyloglucan, glucuronan, and cellulose. Neoagaro-oligosaccharides (NAOS) are resistant to amylolytic enzymes (Hu et al. 2006). Glycerol galactoside, from *Porphyra yezoensis*, is also resistant to intestinal enzymes. Colon bacteria can degrade laminarin and alginate; red seaweed (*Porphyra yezoensis*) polysaccharides and NAOS were fermentable by *Lactobacillus* and *Bifidobacterium* but not *E. coli* or *Enterococcus*. Algal fibers from *Himanthalia elongata*,

*Laminaria digitata*, and *Undaria pinnatifida* were fermented by human intestinal microflora (Muraoka et al. 2008). *Ascophyllum nodosum* dried to powder decreased the growth of *E. coli* and *Streptococci* in a mixed culture in porcine intestine (Dierick et al. 2009). Polysaccharides from plant and *Undaria pinnatifida* fucoidans increased the *Bifidobacterium* population (Marzorati et al. 2010). Diet supplemented with alginate oligosaccharides increased *Bifidobacterium* and *Lactobacillus* population in intestine and in pigs, it helped in weight gain. Inulin in diet reduced fecal ammonia concentrations in pigs. Animal diet containing inulin reduced skatole in plasma, cecal, and fat (O'Sullivan et al. 2010).

### 36.2.16  Derivatized starches

Chemically derivatized starch granules like phosphorylated and citrate starches are resistant to digestion. Oxidized starches are useful as dietic food. Modified starch (Fat replacers) have same organoleptic characteristics of fat. Soft texture, juiciness, and flavor are due to the modified starch and carrageenan present in them.

### 36.2.17  Nonstarch polysaccharides

#### 36.2.17.1  Algal source

Alginate (brown seaweeds, *Azotobacter*), carrageenan (red seaweeds), and ulvan (green seaweeds) are rich in soluble fibers. Seaweeds are eaten in Southeast Asia, but the NSPs are used mainly for food applications in Western world. Alginate can control plasma cholesterol by increasing fecal bile and cholesterol excretions. Na-alginate was shown to have less hypocholesteremic effect than carrageenan. Alginates repair the mucosal damage in the gastrointestinal tract. Alginates stimulate interleukins (1 and 6) and TNF-α production by human monocytes (Otterlei et al. 1993). Ulvan augments the binding of growth factors to receptors of intestinal membrane for growth and repair (Lahaye and Kaeffer 1997). Carrageenans, with ulvan, help colonocytes (Michel and Macfarlane 1996).

#### 36.2.17.2  Bacterial source

Bacterial exopolysaccharides (EPS), curdlan (*Agrobacterium*), xanthan (*Xanthomonas campestris*), and gellan (*Sphingomonas paucimobilis*) are approved as food adjuncts by the United States Food and Drug Administration (McIntosh et al. 2005). Curdlan is used as a thickening and gelling agent in foods. Curdlan is used in tofu noodles as calorie-reduced food, fat-substitute and for lowering cholesterol concentration in the liver (Warrand 2006). Xanthan is used in food applications as a thickener and emulsifier (Sutherland 2001). Gellan is produced by *Sphingomonas paucimobilis*, and is marketed as Gelrite, Phytagel, and Kelcogel.

#### 36.2.17.3  Plant source

*36.2.17.3.1 Cereal grains and β-glucans* β-glucan from oats reduces the risk of heart disease. Oatrim™ is an oat β-glucan concentrate and Nutrim™ is a barley β-glucan, which regulates glucose and insulin response (Brennan and Cleary 2005). β-glucans also stimulate immune responses by the alteration they cause in the colonic microflora or the alteration in SCFA production by the microflora (Brownlee et al. 2005).

*36.2.17.3.2 Galactomannans* Galactomannan gums are available from fenugreek (*Trigonella foenumgraecum* L.), guar (*Cyanopsis tetragonoloba*), and locust bean (*Ceratonia siliqua*) (Srivastava and Kapoor 2005). Ground fenugreek lowers blood glucose level and guar gum exerts hypocholesterolemic effect.

*36.2.17.3.3 Inulin* The fructans present in chicory modulates microflora and physiology of gastrointestinal tract to reduce the risk of colonic diseases. Inulin improves colonic action by facilitating defecation. Pectin had greater effect in lowering total cholesterol levels per gram when compared to oats, psyllium, and guar gum (Brownlee et al. 2005).

*36.2.17.3.4 Cellulose* Microcrystalline cellulose (MC) enhances sensory properties and provides the creamy mouth-feel of fat in oil-in-water.

#### 36.2.17.4  Animal source

Chitin is the second most abundant natural biopolymer. It can decrease the availability of bile acids. The USFDA has approved chitosan as a food additive, and a natural food preservative. It lowered cholesterol and triacylglycerol values in animal fed with chitin or chitosan and also enhanced the growth of *Bifidobacteria* in the guts of chicken, which prevent the growth of other microorganisms (Synowiecki and Al-Khateeb 2003). An et al. (2009) described the health effects induced by different polysaccharides.

## 36.3  Conclusion

Marine glycobiotechnology is a new branch of science emerging in marine biotechnology. Glycans find a wide range of applications; besides being an energy source, they are commonly encountered as drugs, vaccines, biomarkers, and potential biomaterials. Antibiotics conjugated with glycan moiety have been shown to have a better activity than the antibiotics alone. These molecules are also exploited for their adjuvanticity. Often the glycosylation pattern is found to be altered in any kind of abnormality or even as a part of normal developmental stages in an organism; hence, the presence or absence of a glycan is utilized to understand the nature of the abnormality. They are also given as dietary supplement to treat a few metabolic disorders. They play a vital role in the reproduction process. More recently, they have been exploited in the biomedical industry for scaffolds and biosensors. Glycans are also used as prebiotics, probiotics, and food substituent for their good health-imparting effects besides possessing an organoleptic properties. Despite the wide range of beneficial properties, it is still an underexploited field.

## Acknowledgments

The authors thank the funding agencies, Department of Science and Technology, Department of Biotechnology, Indian Council of Medical Research, Government of India, New Delhi, and Vice Chancellor, SASTRA University, Thanjavur, India, for financial and infrastructural facilities.

## References

Al Sagheer, F. A., M. A. Al-Sughayer, S. Muslim, and M. Z. Elsabee. 2009. Extraction and characterization of chitin and chitosan from marine sources in Arabian Gulf. *Carbohydr Polym* 77(2):410–419.

Alpuche, J., A. Pereyra, C. Agundis et al. 2005. Purification and characterization of a lectin from the white shrimp *Litopenaeus setiferus* (Crustacea decapoda) hemolymph. *Biochim Biophys Acta* 1724(1–2):86–93.

An, H. J., S. R. Kronewitter, M. L. de Leoz, and C. B. Lebrilla. 2009. Glycomics and disease markers. *Curr Opin Chem Biol* 13(5–6):601–607.

Anatole, A. K. 2012. Carbohydrates and drug design. In *Glycobiology and Drug Design*. American Chemical Society: Washington, DC.

Anderson, K. E. and J. H. Waite. 2000. Immunolocalization of Dpfp1, a byssal protein of the zebra mussel Dreissena polymorpha. *J Exp Biol* 203(Pt 20):3065–3076.

Aoki, S., L. Cao, K. Matsui, R. Rachmat, S.-I. Akiyama, and M. Kobayashi. 2004. Kendarimide A, a novel peptide reversing P-glycoprotein-mediated multidrug resistance in tumor cells, from a marine sponge of Haliclona sp. *Tetrahedron* 60(33):7053–7059.

Arnold, J. N., M. R. Wormald, R. B. Sim, P. M. Rudd, and R. A. Dwek. 2007. The impact of glycosylation on the biological function and structure of human immunoglobulins. *Annu Rev Immunol* 25:21–50.

Babu, P., S. J. North, J. Jang-Lee et al. 2009. Structural characterisation of neutrophil glycans by ultra sensitive mass spectrometric glycomics methodology. *Glycoconj J* 26(8):975–986.

Bai, L., L. Li, H. Xu et al. 2006. Functional analysis of the Validamycin biosynthetic gene cluster and engineered production of Validoxylamine A. *Chem Biol* 13(4):387–397.

Bauer, T. M., B. F. El-Rayes, X. Li et al. 2013. Carbohydrate antigen 19-9 is a prognostic and predictive biomarker in patients with advanced pancreatic cancer who receive gemcitabine-containing chemotherapy. *Cancer* 119(2):285–292.

Bavington, C. D., R. Lever, B. Mulloy et al. 2004. Anti-adhesive glycoproteins in echinoderm mucus secretions. *Comp Biochem Physiol B Biochem Mol Biol* 139(4):607–617.

Bertucci, C., S. Cimitan, and L. Menotti. 2003. Optical biosensor analysis in studying herpes simplex virus glycoprotein D binding to target nectin1 receptor. *J Pharm Biomed Anal* 32(4–5):697–706.

Brennan, C. S. and L. J. Cleary. 2005. The potential use of cereal $(1 \rightarrow 3, 1 \rightarrow 4)$-β-D-glucans as functional food ingredients. *J Cereal Sci* 42(1):1–13.

Bretting, H., S. G. Phillips, H. J. Klumpart, and E. A. Kabat. 1981. A mitogenic lactose-binding lectin from the sponge *Geodia cydonium*. *J Immunol* 127(4):1652–1658.

Brossay, L., D. Jullien, S. Cardell et al. 1997. Mouse CD1 is mainly expressed on hemopoietic-derived cells. *J Immunol* 159(3):1216–1224.

Brownlee, I. A., A. Allen, J. P. Pearson et al. 2005. Alginate as a source of dietary fiber. *Crit Rev Food Sci Nutr* 45(6):497–510.

Bulgakov, A. A., K. I. Park, K. S. Choi, H. K. Lim, and M. Cho. 2004. Purification and characterisation of a lectin isolated from the Manila clam *Ruditapes philippinarum* in Korea. *Fish Shellfish Immunol* 16(4):487–499.

Caldwell, G. S. and H. E. Pagett. 2010. Marine glycobiology: Current status and future perspectives. *Mar Biotechnol (NY)* 12(3):241–252.

Champ, M., A.-M. Langkilde, F. Brouns, B. Kettlitz, and Y. Le Bail Collet. 2003. Advances in dietary fibre characterisation. 1. Definition of dietary fibre, physiological relevance, health benefits and analytical aspects. *Nutr Res Rev* 16(1):71–82.

Chang, L., N. F. Whittaker, and C. A. Bewley. 2003. Crambescidin 826 and dehydrocrambine A: New polycyclic guanidine alkaloids from the marine sponge *Monanchora* sp. that inhibit HIV-1 fusion. *J Nat Prod* 66(11):1490–1494.

Chaudhari, S. S., B. Moussian, C. A. Specht et al. 2014. Functional specialization among members of Knickkopf family of proteins in insect cuticle organization. *PLoS Genet* 10(8):e1004537.

Chesnutt, B. M., Y. Yuan, K. Buddington, W. O. Haggard, and J. D. Bumgardner. 2009. Composite chitosan/nanohydroxyapatite scaffolds induce osteocalcin production by osteoblasts in vitro and support bone formation in vivo. *Tissue Eng Part A* 15(9):2571–2579.

Chu, J., A. Mir, N. Gao et al. 2013. A zebrafish model of congenital disorders of glycosylation with phosphomannose isomerase deficiency reveals an early opportunity for corrective mannose supplementation. *Dis Model Mech* 6(1):95–105.

Clare, A. S. and K. Matsumura. 2000. Nature and perception of barnacle settlement pheromones. *Biofouling* 15(1–3):57–71.

Courtois, A., C. Berthou, J. Guézennec, C. Boisset, and A. Bordron. 2014. Exopolysaccharides isolated from hydrothermal vent bacteria can modulate the complement system. *PLoS One* 9(4):e94965.

Dabhi, A. S., N. R. Bhatt, and M. J. Shah. 2013. Voglibose: An alpha glucosidase inhibitor. *J Clinic Diagn Res* 7(12):3023–3027.

Danielsen, I., C. Granström, T. Haldorsson et al. 2013. Dietary glycemic index during pregnancy is associated with biomarkers of the metabolic syndrome in offspring at age 20 years. *PLoS One* 8(5):e64887.

Davidson, B., A. Berner, J. M. Nesland et al. 2000. Carbohydrate antigen expression in primary tumors, metastatic lesions, and serous effusions from patients diagnosed with epithelial ovarian carcinoma: Evidence of up-regulated Tn and Sialyl Tn antigen expression in effusions. *Hum Pathol* 31(9):1081–1087.

Devine, P. L., M. A. McGuckin, and B. G. Ward. 1992. Circulating mucins as tumor markers in ovarian cancer (review). *Anticancer Res* 12(3):709–717.

Dierick, N., A. Ovyn, and S. De Smet. 2009. Effect of feeding intact brown seaweed *Ascophyllum nodosum* on some digestive parameters and on iodine content in edible tissues in pigs. *J Sci Food Agric* 89(4):584–594.

Edwards, I. J., W. D. Wagner, and R. T. Owens. 1990. Macrophage secretory products selectively stimulate dermatan sulfate proteoglycan production in cultured arterial smooth muscle cells. *Am J Pathol* 136(3):609–621.

Eufemia, N., S. Clerte, S. Girshick, and D. Epel. 2002. Algal products as naturally occurring substrates for *p*-glycoprotein in *Mytilus californianus*. *Mar Biol* 140:343–353.

Feizi, T., M. S. Stoll, C. T. Yuen, W. Chai, and A. M. Lawson. 1994. Neoglycolipids: Probes of oligosaccharide structure, antigenicity, and function. *Methods Enzymol* 230:484–519.

Ferrao-Gonzales, A. D., B. K. Robbs, V. H. Moreau et al. 2005. Controlling {beta}-amyloid oligomerization by the use of naphthalene sulfonates: Trapping low molecular weight oligomeric species. *J Biol Chem* 280(41):34747–34754.

Ferrari, K. M. and N. M. Targett. 2003. Chemical attractants in horseshoe crab, *Limulus polyphemus*, eggs: The potential for an artificial bait. *J Chem Ecol* 29(2):477–496.

Fox, S., J. Alice, A. Bedi, and S. A. Rodeo. 2009. The basic science of articular cartilage: Structure, composition, and function. *Sports Health* 1(6):461–468.

Fujitani, N., J. Furukawa, K. Araki et al. 2013. Total cellular glycomics allows characterizing cells and streamlining the discovery process for cellular biomarkers. *Proc Natl Acad Sci USA* 110(6):2105–2510.

Futerman, A. H. and G. van Meer. 2004. The cell biology of lysosomal storage disorders. *Nat Rev Mol Cell Biol* 5(7):554–565.

Gaill, F., J. Persson, J. Sugiyama, R. Vuong, and H. Chanzy. 1992. The chitin system in the tubes of deep sea hydrothermal vent worms. *J Struct Biol* 109(2):116–128.

Gallo, A. and M. Costantini. 2012. Glycobiology of reproductive processes in marine animals: The state of the art. *Mar Drugs* 10(12):2861–2892.

Gerloczy, A., T. Hoshino, and J. Pitha. 1994. Safety of oral cyclodextrins: Effects of hydroxypropyl cyclodextrins, cyclodextrin sulfates and cationic cyclodextrins on steroid balance in rats. *J Pharm Sci* 83(2):193–196.

Ghazarian, H., B. Idoni, and S. B. Oppenheimer. 2011. A glycobiology review: Carbohydrates, lectins and implications in cancer therapeutics. *Acta Histochem* 113(3):236–247.

Giannakakis, K., S. Feriozzi, M. Perez, T. Faraggiana, and A. O. Muda. 2007. Aberrantly glycosylated IgA1 in glomerular immune deposits of IgA nephropathy. *J Am Soc Nephrol* 18(12):3139–3146.

Giedrojc, J., K. Krupinski, H. K. Breddin, and M. Bielawiec. 1996. Interaction between the sulfated lactobionic acid (LW 10082) and other antithrombotic agents in animal thrombosis model. *Pol J Pharmacol* 48(3):317–322.

Hatakeyama, T., H. Kohzaki, H. Nagatomo, and N. Yamasaki. 1994. Purification and characterization of four Ca$^{2+}$-dependent lectins from the marine invertebrate, *Cucumaria echinata*. *J Biochem* 116(1):209–214.

Hu, B., Q. Gong, Y. Wang, Y. Ma, J. Li, and W. Yu. 2006. Prebiotic effects of neoagaro-oligosaccharides prepared by enzymatic hydrolysis of agarose. *Anaerobe* 12(5–6):260–266.

Iijima, R., J. Kisugi, and M. Yamazaki. 1994. Biopolymers from marine invertebrates. XIV. Antifungal property of Dolabellanin A, a putative self-defense molecule of the sea hare, *Dolabella auricularia*. *Biol Pharm Bull* 17(8):1144–1146.

Isbrucker, R. A., J. Cummins, S. A. Pomponi, R. E. Longley, and A. E. Wright. 2003. Tubulin polymerizing activity of dictyostatin-1, a polyketide of marine sponge origin. *Biochem Pharmacol* 66(1):75–82.

Islam, T. and R. J. Linhardt. 2003. Chemistry, biochemistry, and pharmaceutical potentials of glycosaminoglycans and related saccharides. *Carbohydr Based Drug Discov* 1:407–439.

Islam, T. and R. J. Linhardt. 2005. Chemistry, biochemistry, and pharmaceutical potentials of glycosaminoglycans and related saccharides. In *Carbohydrate-Based Drug Discovery*, C. -H. Wong (Ed.): Wiley-VCH Verlag GmbH & Co. KGaA: Weinheim, Germany.

Janković, M. 2011. Glycans as biomarkers: Status and perspectives. *J Med Biochem* 30:213–223.

Jaques, L. B., L. M. Hiebert, and S. M. Wice. 1991. Evidence from endothelium of gastric absorption of heparin and of dextran sulfates 8000. *J Lab Clin Med* 117(2):122–130.

Jimbo, M., T. Yanohara, K. Koike et al. 2000. The D-galactose-binding lectin of the octocoral *Sinularia lochmodes*: Characterization and possible relationship to the symbiotic dinoflagellates. *Comp Biochem Physiol B Biochem Mol Biol* 125(2):227–236.

Kariya, Y., B. Mulloy, K. Imai et al. 2004. Isolation and partial characterization of fucan sulfates from the body wall of sea cucumber *Stichopus japonicus* and their ability to inhibit osteoclastogenesis. *Carbohydr Res* 339(7):1339–1346.

Kawakami, A., T. Miyamoto, R. Higuchi, T. Uchiumi, M. Kuwano, and R. W. M. Van Soest. 2001. Structure of a novel multidrug resistance modulator, irciniasulfonic acid, isolated from a marine sponge, *Ircinia* sp. *Tetrahedron Lett* 42(19):3335–3337.

Kawano, T., J. Cui, Y. Koezuka et al. 1997. CD1d-restricted and TCR-mediated activation of valpha14 NKT cells by glycosylceramides. *Science* 278(5343):1626–1629.

Kawsar, S. M., Y. Fujii, R. Matsumoto et al. 2008. Isolation, purification, characterization and glycan-binding profile of a D-galactoside specific lectin from the marine sponge, *Halichondria okadai*. *Comp Biochem Physiol B Biochem Mol Biol* 150(4):349–357.

Ke, C., D. Wang, Y. Sun, D. Qiao, H. Ye, and X. Zeng. 2013. Immunostimulatory and antiangiogenic activities of low molecular weight hyaluronic acid. *Food Chem Toxicol* 58:401–407.

Kelly, L. S. and T. W. Snell. 1998. Role of surface glycoproteins in mate guarding of the marine harpacticoid. *Mar Biol* 130:605–612.

Kim, B., G. S. Cha, and M. E. Meyerhoff. 1990. Homogeneous enzyme-linked binding assay for studying the interaction of lectins with carbohydrates and glycoproteins. *Anal Chem* 62(24):2663–2668.

Kim, Se-K., Y. Dominic Ravichandran, S. Khan, and Y. T. Kim. 2008. Prospective of the cosmeceuticals derived from marine organisms. *Biotechnol Bioprocess Eng* 13(5):511–523.

Kim, Y. J. and A. Varki. 1997. Perspectives on the significance of altered glycosylation of glycoproteins in cancer. *Glycoconj J* 14(5):569–576.

Kirst, H. A. 1993. Semi-synthetic derivatives of erythromycin. *Prog Med Chem* 30:57–88.

Kisugi, J., H. Ohye, H. Kamiya, and M. Yamazaki. 1992. Biopolymers from marine invertebrates. XIII. Characterization of an antibacterial protein, dolabellanin A, from the albumen gland of the sea hare, *Dolabella auricularia*. *Chem Pharm Bull* 40(6):1537–1539.

Kobayashi, E., K. Motoki, T. Uchida, H. Fukushima, and Y. Koezuka. 1995. KRN7000, a novel immunomodulator, and its antitumor activities. *Oncol Res* 7(10–11): 529–534.

Kroger, N., C. Bergsdorf, and M. Sumper. 1994. A new calcium binding glycoprotein family constitutes a major diatom cell wall component. *Embo J* 13(19):4676–4683.

Krull, U. J., R. S. Brown, R. N. Koilpillai, R. Nespolo, A. Safarzadeh-Amiri, and E. T. Vandenberg. 1989. Selective electrochemical biosensors from state-switching of bilayer and monolayer lipid membranes by lectin-polysaccharide complexes. *Analyst* 114(1):33–40.

Lahaye, M. and B. Kaeffer. 1997. Seaweed dietary fibres: Structure, physico-chemical and biological properties relevant to intestinal physiology. *Sciences des aliments* 17(6):563–584.

Lawrence, R., J. R. Brown, F. Lorey, P. I. Dickson, B. E. Crawford, and J. D. Esko. 2014. Glycan-based biomarkers for mucopolysaccharidoses. *Mol Genet Metab* 111(2):73–83.

Le Pape, P., M. Zidane, H. Abdala, and M. T. More. 2000. A glycoprotein isolated from the sponge, *Pachymatisma johnstonii*, has anti-leishmanial activity. *Cell Biol Int* 24(1):51–56.

Li, J., D. Zhang, X. Zhu et al. 2011. Studies on synthesis and structure-activity relationship (SAR) of derivatives of a new natural product from marine fungi as inhibitors of influenza virus neuraminidase. *Mar Drugs* 9(10):1887–1901.

Li, L. J., M. Y. Li, Y. T. Li, J. J. Feng, F. Q. Hao, and Z. Lun. 2012. Adjuvant activity of *Sargassum pallidum* polysaccharides against combined Newcastle disease, infectious bronchitis and avian influenza inactivated vaccines. *Mar Drugs* 10(12):2648–2660.

Lijima, R., J. Kisugi, and M. Yamazaki. 1995. Antifungal activity of Aplysianin E, A cytotoxic protein of sea hare (*Aplysia kurodai*) eggs. *Dev Comp Immunol* 19(1):13–19.

Liljeblad, M., A. Lundblad, S. Ohlson, and P. Pahlsson. 1998. Detection of low-molecular-weight heparin oligosaccharides (Fragmin) using surface plasmon resonance. *J Mol Recognit* 11(1–6):191–193.

Lopez, D. and S. Martinez-Luis. 2014. Marine natural products with P-glycoprotein inhibitor properties. *Mar Drugs* 12(1):525–546.

Lordan, S., R. P. Ross, and C. Stanton. 2011. Marine bioactives as functional food ingredients: Potential to reduce the incidence of chronic diseases. *Mar Drugs* 9(6):1056–1100.

Magnadóttir, B., S. Gudmundsdóttir, and B. K. Gudmundsdóttir. 1997. The carbohydrate moiety of IgM from salmon (*Salmo salar* L.). *Comp Biochem Physiol B* 116:423–430.

Mahajan, P., K. S. Sodhi, R. Pandey, and J. Singh. 2014. Glycotherapeutics: Clinical implications. *Int J Adv Pharm Biol Chem* 3(3):786–795.

Mahmood, N. and A. J. Hay. 1992. An ELISA utilizing immobilised snowdrop lectin GNA for the detection of envelope glycoproteins of HIV and SIV. *J Immunol Methods* 151(1–2):9–13.

Marzorati, M., A. Verhelst, G. Luta et al. 2010. In vitro modulation of the human gastrointestinal microbial community by plant-derived polysaccharide-rich dietary supplements. *Int J Food Microbiol* 139(3):168–176.

McFadden, D. W., D. R. Riggs, B. J. Jackson, and L. Vona-Davis. 2003. Keyhole limpet hemocyanin, a novel immune stimulant with promising anticancer activity in Barrett's esophageal adenocarcinoma. *Am J Surg* 186(5):552–555.

McIntosh, M., B. A. Stone, and V. A. Stanisich. 2005. Curdlan and other bacterial (1 → 3)-beta-D-glucans. *Appl Microbiol Biotechnol* 68(2):163–173.

Michel, C. and G. T. Macfarlane. 1996. Digestive fates of soluble polysaccharides from marine macroalgae: Involvement of the colonic microflora and physiological consequences for the host. *J Appl Bacteriol* 80(4):349–369.

Moran, A. P. and M. M. Prendergast. 2001. Molecular mimicry in *Campylobacter jejuni* and *Helicobacter pylori* lipopolysaccharides: Contribution of gastrointestinal infections to autoimmunity. *J Autoimmun* 16(3):241–256.

Moriwaki, K. and E. Miyoshi. 2010. Fucosylation and gastrointestinal cancer. *World J Hepatol* 2(4):151–161.

Morris, H. R., A. Dell, R. L. Easton et al. 1996. Gender-specific glycosylation of human glycodelin affects its contraceptive activity. *J Biol Chem* 271(50):32159–32167.

Muraoka, T., K. Ishihara, C. Oyamada, H. Kunitake, I. Hirayama, and T. Kimura. 2008. Fermentation properties of low-quality red alga Susabinori *Porphyra yezoensis* by intestinal bacteria. *Biosci Biotechnol Biochem* 72(7):1731–1739.

Muzzarelli, R. A. 2010. Chitins and chitosans as immunoadjuvants and non-allergenic drug carriers. *Mar Drugs* 8(2):292–312.

Naganuma, T., T. Ogawa, J. Hirabayashi, K. Kasai, H. Kamiya, and K. Muramoto. 2006. Isolation, characterization and molecular evolution of a novel pearl shell lectin from a marine bivalve, *Pteria penguin*. *Mol Divers* 10(4):607–618.

Nakagoe, T., K. Fukushima, T. Sawai et al. 2002. Increased expression of sialyl Lewis(x) antigen in penetrating growth type A early gastric cancer. *J Exp Clin Cancer Res* 21(3):363–369.

Nwodo, U. U., E. Green, and A. I. Okoh. 2012. Bacterial exopolysaccharides: Functionality and prospects. *Int J Mol Sci* 13(11):14002–14015.

O'Sullivan, L., B. Murphy, P. McLoughlin et al. 2010. Prebiotics from marine macroalgae for human and animal health applications. *Mar Drug* 8(7):2038–2064.

Otterlei, M., A. Sundan, G. Skjåk-Braek, L. Ryan, O. Smidsrød, and T. Espevik. 1993. Similar mechanisms of action of defined polysaccharides and lipopolysaccharides: Characterization of binding and tumor necrosis factor alpha induction. *Infect Immun* 61(5):1917–1925.

Panagiotopoulos, C., R. Sempéré, V. Jacq, and B. Charriere. 2014. Composition and distribution of dissolved carbohydrates in the Beaufort Sea Mackenzie margin (Arctic Ocean). *Mar Chem* 166:92–102.

Parry, S., V. Ledger, B. Tissot et al. 2007. Integrated mass spectrometric strategy for characterizing the glycans from glycosphingolipids and glycoproteins: Direct identification of sialyl Lex in mice. *Glycobiology* 17(6):646–654.

Pinto, V. P., H. Debray, D. Dus et al. 2009. Lectins from the red marine algal species *Bryothamnion seaforthii* and *Bryothamnion triquetrum* as tools to differentiate human colon carcinoma cells. *Adv Pharmacol Sci* 2009:862162.

Pittman, K. A., L. Goldberg, and F. Coulston. 1976. Carrageenan: The effect of molecular weight and polymer type on its uptake, excretion and degradation in animals. *Food Cosmet Toxicol* 14(2):85–93.

Powlesland, A. S., P. G. Hitchen, S. Parry et al. 2009. Targeted glycoproteomic identification of cancer cell glycosylation. *Glycobiology* 19(8):899–909.

Prajapati, V. D., P. M. Maheriya, G. K. Jani, and H. K. Solanki. 2014. Carrageenan: A natural seaweed polysaccharide and its applications. *Carbohydr Polym* 105:97–112.

Putzbach, W. and N. J. Ronkainen. 2013. Immobilization techniques in the fabrication of nanomaterial-based electrochemical biosensors: A review. *Sensors (Basel)* 13(4):4811–4840.

Rocha de Souza, M. C., C. T. Marques, C. M. Guerra Dore, F. R. Ferreira da Silva, H. A. Oliveira Rocha, and E. L. Leite. 2007. Antioxidant activities of sulfated polysaccharides from brown and red seaweeds. *J Appl Phycol* 19(2):153–160.

Sangrajrang, S., M. Zidane, P. Berda, M. T. More, F. Calvo, and A. Fellous. 2000. Different microtubule network alterations induced by pachymatismin, a new marine glycoprotein, on two prostatic cell lines. *Cancer Chemother Pharmacol* 45(2):120–126.

Schaeffer, D. J. and V. S. Krylov. 2000. Anti-HIV activity of extracts and compounds from algae and cyanobacteria. *Ecotoxicol Environ Saf* 45(3):208–227.

Scully, M. F., K. M. Weerasinghe, V. Ellis, B. Djazaeri, and V. V. Kakkar. 1983. Anticoagulant and antiheparin activities of a pentosan polysulphate. *Thromb Res* 31(1):87–97.

Senni, K., J. Pereira, F. Gueniche et al. 2011. Marine polysaccharides: A source of bioactive molecules for cell therapy and tissue engineering. *Mar Drugs* 9(9):1664–1681.

Shtyrya, Y. A., L. V. Mochalova, and N. V. Bovin. 2009. Influenza virus neuraminidase: Structure and function. *Acta Naturae* 1(2):26–32.

Silva, T. H., A. Alves, E. G. Popa et al. 2012. Marine algae sulfated polysaccharides for tissue engineering and drug delivery approaches. *Biomatter* 2(4):278–289.

Smital, T., R. Sauerborn, and B. K. Hackenberger. 2003. Inducibility of the P-glycoprotein transport activity in the marine mussel *Mytilus galloprovincialis* and the freshwater mussel *Dreissena polymorpha*. *Aquat Toxicol* 65(4):443–465.

Smital, T., R. Sauerborn, B. Pivcevic, S. Krca, and B. Kurelec. 2000. Interspecies differences in P-glycoprotein mediated activity of multixenobiotic resistance mechanism in several marine and freshwater invertebrates. *Comp Biochem Physiol C: Toxicol Pharmacol* 126(2):175–186.

Snell, T. W. and M. J. Carmona 1994. Surface glycoproteins in copepods—Potential signals for mate recognition. *Hydrobiologia* 293:255–264.

Srivastava, M. and V. P. Kapoor. 2005. Seed galactomannans: An overview. *Chem Biodiversity* 2(3):295–317.

Stibler, H. 1991. Carbohydrate-deficient transferrin in serum: A new marker of potentially harmful alcohol consumption reviewed. *Clin Chem* 37(12):2029–2037.

Sutherland, I. W. 2001. Microbial polysaccharides from Gram-negative bacteria. *Int Dairy J* 11(9):663–674.

Synowiecki, J. and N. A Al-Khateeb. 2003. Production, properties, and some new applications of chitin and its derivatives. *Crit Rev Food Sci Nutr* 43(2):145–171.

Tammi, M. I., A. J. Day, and E. A. Turley. 2002. Hyaluronan and homeostasis: A balancing act. *J Biol Chem* 277(7):4581–4884.

Ueda, K. 2013. Glycoproteomic strategies: From discovery to clinical application of cancer carbohydrate biomarkers. *Proteomics—Clin Appl* 7(9–10):607–617.

Varki, A. and M. J. Chrispeels. 1999. *Essentials of Glycobiology*. Cold Spring Harbor Laboratory Press: New York.

Vizzardi, E., A. D'Aloia, A. Curnis, and L. Dei Cas. 2013. Carbohydrate antigen 125: A new biomarker in heart failure. *Cardiol Rev* 21(1):23–26. doi:10.1097/CRD.0b013e318265f58f.

Wagner, G. F., E. M. Jaworski, and M. Haddad. 1998. Stanniocalcin in the seawater salmon: Structure, function, and regulation. *Am J Physiol* 274(4 Pt 2):R1177–R1185.

Wang, H., T. Toida, Y. S. Kim et al. 1997. Glycosaminoglycans can influence fibroblast growth factor-2 mitogenicity without significant growth factor binding. *Biochem Biophys Res Commun* 235(2):369–373.

Wang, Z. G., L. J. Williams, X. F. Zhang et al. 2000. Polyclonal antibodies from patients immunized with a globo H-keyhole limpet hemocyanin vaccine: Isolation, quantification, and characterization of immune responses by using totally synthetic immobilized tumor antigens. *Proc Natil Acad Sci USA* 97(6):2719–2724.

Warrand, J. 2006. Healthy polysaccharides. *Food Technol Biotechnol* 44(3):355–370.

Watson, C. A., K. L. Vine, J. M. Locke, A. Bezos, C. R. Parish, and M. Ranson. 2013. The antiangiogenic properties of sulfated beta-cyclodextrins in anticancer formulations incorporating 5-fluorouracil. *Anticancer Drugs* 24(7):704–714.

Wu, D., D. M. Zajonc, M. Fujio et al. 2006. Design of natural killer T cell activators: Structure and function of a microbial glycosphingolipid bound to mouse CD1d. *Proc Natl Acad Sci USA* 103(11):3972–3977.

Wu, S. J., M. W. Chun, K. H. Shin et al. 1998. Chemical sulfonation and anticoagulant activity of acharan sulfate. *Thromb Res* 92(6):273–281.

Wuhrer, M. 2007. Glycosylation profiling in clinical proteomics: Heading for glycan biomarkers. *Expert Rev Prot* 4(2):135–136.

Xie, X., A. S. Rivier, A. Zakrzewicz et al. 2000. Inhibition of selectin-mediated cell adhesion and prevention of acute inflammation by nonanticoagulant sulfated saccharides. Studies with carboxyl-reduced and sulfated heparin and with trestatin a sulfate. *J Biol Chem* 275(44):34818–34825.

Yao, Z., H. Wu, S. Zhang, and Y. Du. 2014. Enzymatic preparation of kappa-carrageenan oligosaccharides and their anti-angiogenic activity. *Carbohydr Polym* 101:359–367.

You, T., O. Niwa, Z. Chen, K. Hayashi, M. Tomita, and S. Hirono. 2003. An amperometric detector formed of highly dispersed Ni nanoparticles embedded in a graphite-like carbon film electrode for sugar determination. *Anal Chem* 75(19):5191–5196.

## chapter thirty-seven

# Application of bacterial chitinase

*Siswa Setyahadi*

## Contents

## 37.1 Introduction

As one of the marine by-products having large amount of chitin, proteins, and minerals, shrimp can be used for producing chitin and chitosan. Chitosan and its derivatives are examples of value-added materials. They are produced from chitin, which is a natural carbohydrate polymer found in the skeleton of crustaceans, such as crabs, shrimps, and lobsters, as well as in the exoskeleton of marine zooplankton species, including corals and jellyfish. Insects, such as butterflies and ladybugs, also have chitin in their wings, and the cell walls of yeast, mushrooms, and other fungi also contain this substance. Industrial-scale chitosan production involves four steps: (1) demineralization, (2) deproteinization, (3) decoloration, and (4) deacetylation. Despite the widespread occurrence of chitin in nature, presently crab and shrimp shells remain its primary commercial sources.

Chitin is the second most abundant natural polymer in the world after cellulose. Upon deacetylation, it yields the novel biomaterial chitosan, which upon further hydrolysis yields an oligosaccharide of extremely low molecular weight. Chitosan possesses a wide range of useful properties. Specifically, it is a biocompatible, antibacterial, and environmentally friendly polyelectrolyte, thus lending itself to a variety of applications, including water treatment, chromatography, additives for cosmetics, in textile treatment for antimicrobial activity, as novel fibers for textiles, photographic papers, biodegradable films, biomedical devices, and microcapsule implants for controlled release in drug delivery (Aye and Stevens 2004, Gossen 1997).

Chitin is a polysaccharide. A polysaccharide is a giant polymer molecule consisting of smaller molecules of sugar strung together. Chitin can be described as a biopolymer composed of N-acetyl-D-glucosamine (NAG), a chemical structure very close to that of cellulose except that the hydroxyl group in C2 of cellulose is replaced by an acetamido group in chitin. One can associate this chemical similarity between cellulose and chitin as serving similar structural and defensive functions (Rinaudo 2006). In most organisms, chitin is modified by forming linkages with other polymers, such as glucans, proteins, and so on. However, some centric diatoms, such as *Thalassiosira fluviatilis*, produce radiating spines that are the purest form of chitin known in nature, as they are fully acetylated and unlinked with other extracellular components (Gooday et al. 1985). The complete enzymatic hydrolysis of chitin to free N-acetylglucosamine (GlcNAc) is performed by a chitinolytic system, the action of which is known to be synergistic and consecutive (Deshpande 1986, Shaikh and Deshpande 1993). The studies of chitnolytic enzymes from plants, insects, and microorganisms with respect to their role and applications have been extensively reviewed (Kramer and Muthukrishnan 1997, Zikakis 1989).

Chitinolytic enzymes are produced by a wide range of organisms including bacteria, fungi, insects, plants, and animals for different purposes such as nutrition, morphogenesis, and defense against chitin-containing pathogens (Adrangi et al. 2010). Many of these organisms possess several genes that encode chitinolytic enzymes. For example, most filamentous fungi have 10–20 different chitinolytic genes, while in mycoparasitic species the number of such genes may reach 30 or even higher (Hartl et al. 2012). These enzymes act in a synergetic or successive manner to degrade chitin (Patil et al. 2000). Higher organisms such as *Arabidopsis* have also been reported to contain a large number of chitinolytic genes (Hossain et al. 2010). However, not all of these gene codes are for active enzymes. Many organisms including plants, invertebrates, and higher animals express genes encoding the so-called chitinase-like lectins (chi-lectins) that are devoid of chitinolytic activity due to substitutions in their key catalytic residues (Arakane and Muthukrishnan 2010, Vega and Kalkum 2012).

Chitinolytic enzymes have found several applications, such as the production of single-cell proteins, isolation of fungal protoplasts, estimation of fungal biomass, development of three-dimensional (3D) cell culture scaffolds, biocontrol of plant pathogenic fungi and insect vectors, and the production of chito-oligosaccharides, glucosamine, GlcNAc, neoglycoproteins, and artificial polysaccharides (Adrangi et al. 2010, Jamialahmadi et al. 2011, Li et al. 2008, Lu et al. 2012, Ortiz-Rodriguez et al. 2010, Tajdini et al. 2010, Zakariassen et al. 2011).

## 37.2 Chitinases

Glycoside hydrolases as chitinases are enzymes that cleave the glycosidic bonds of glycans. Chitinases, which hydrolyze chitin, occur in a wide range of organisms including viruses, bacteria, fungi, insects, higher plants, and animals (Park et al. 1997). The roles of chitinases in these organisms are diverse. In vertebrates, chitinases are usually part of the digestive tract. In insects and crustaceans, chitinases are associated with the need for partial degradation of old cuticle. Chitinases have been implicated in plant resistance against fungal pathogens because of their inducible nature and antifungal activities in vitro (Taira et al. 2002). Chitinase in fungi is thought to have autolytic, nutritional, and morphogenetic roles. In viruses, chitinases are involved in pathogenesis (Patil et al. 2000).

In bacteria, chitinases play a role in nutrition and parasitism. In addition to these potential applications, they can be used for the production of chito-oligosaccharides, which have been found to function as antibacterial agents, elicitors of lysozyme inducers, and immunoenhancers (Wen et al. 2002). Chitinases can also be used in agriculture to control plant pathogens (Dahiya et al. 2005a, Karasuda et al. 2003).

The findings in the catalytic and substrate-binding mechanisms of chitinolytic enzymes, as well as their sequence homology and applications to plant protection against fungal pathogens and insect pests, have been reported by Fukamizo (2000). Various molecular and biotechnological aspects such as regulation strategies, gene cloning of chitinase from microorganisms and plants, and various industrial and agricultural applications of chitinases have been described by Patil et al. (2000).

Based on amino acid sequence similarity, chitinolytic enzymes are grouped into families 18, 19, and 20 of glycosyl hydrolases (Henrissat and Bairoch 1993). Family 18 is diverse in evolutionary terms and contains chitinases from bacteria, fungi, viruses, animals, and some plant chitinases. Family 19 consists of plant chitinases (classes I, II, and IV) and some Streptomyces chitinases (Hart et al. 1995). The chitinases of the families 18 and 19 do not share amino acid sequence similarity. They have completely different 3D structures and molecular mechanisms and are therefore likely to have evolved from different ancestors (Suzuki et al. 1999). Chitinases as glycoside hydrolase family 18 contain hydrolytic enzymes with chitinase or endo-N-acetyl-β-D-glucosaminidase (ENGase) activity, while glycoside hydrolase family 20 contains enzymes with β-N-acetylhexosaminidase (NAGase) activity. Chitinases and NAGases are involved in chitin degradation. Enzymes acting on somewhat similar substrates (N-acetylglucosaminidases (NAGase), lysozymes, and chitosanases) can be found in other glycoside hydrolase families. For instance, NAGase can be found in families 3 and 20, lysozymes in families 22–25, and chitosanases in family 46. Family 20 includes the β-N-acetylhexosaminidases from bacteria, Streptomycetes, and humans.

Bacterial chitinases are clearly separated into three major subfamilies, A, B, and C, based on the amino acid sequence of individual catalytic domains (Watanabe et al. 1993). Subfamily A chitinases have the presence of a third domain corresponding to the insertion of an $\alpha + \beta$ fold region between the seventh and eighth $(\alpha/\beta)_8$ barrel. On the other hand, none of the chitinases in subfamilies A and B has this insertion. Several chitinolytic bacteria that possess chitinases belonging to different subfamilies like Serratia marcescens (Suzuki et al. 1999), Bacillus circulans WL-12 (Alam et al. 1995), and Streptomyces coelicolor A3(2) (Saito et al. 1999) have been reported.

Chitinases can be classified into two major categories. Endochitinases (EC 3.2.1.14) cleave chitin randomly at internal sites, generating low molecular mass multimers of GlcNAc, such as chitotetraose, chitotriose, and diacetylchitobiose. Exochitinases can be divided into two subcategories: (1) chitobiosidases (EC 3.2.1.29), which catalyze the progressive release of diacetylchitobiose starting at the nonreducing end of chitin microfibril and (2) β-(1,4) N-acetyl glucosaminidases (EC 3.2.1.30), which cleave the oligomeric products of endochitinases and chitobiosidases, generating monomers of GlcNAc (Sahai and Manocha 1993). An alternative pathway involves the deacetylation of chitin to chitosan, which is finally converted to glucosamine residues by the action of chitosanase (EC 3.2.1.132). Characteristics of some chitinases and their sources are summarized in Table 37.1.

**Table 37.1** Characteristics of some chitinases and their applications

| Microorganism | Optimum pH | Optimum temperature | Application | References |
|---|---|---|---|---|
| *Enterobacter* sp. NRG4 | 5.5 | 45 | Release of fungal protoplasts, production of GlcNAc, antifungal potential | Dahiya et al. (2005b) |
| *Alcaligenes xylosoxidans* | 5.0 | 50 | Antifungal potential | Vaidya et al. (2003) |
| *Bacillus* sp. BG-11 | 7.5–9.0 | 45–55 | Antifungal potential | Bhushan and Hoondal (1998) |
| *Bacillus* sp. NCTU2 | 7.0 | 60 | | Wen et al. (2002) |
| *Bacillus* sp. 13.26 | 7.0–8.0 | 60 | | Yuli et al. (2004) |
| *Serratia marcescens* QMB1466 | 4.0–7.0 | 30 | | Roberts and Cabib (1982) |
| *Bacillus cereus* 65 | 4.5–7.5 | | Protected root disease | Pleban et al. (1997) |

## 37.3  Bacterial chitinase inhibits growth of Ganoderma boninense

The use of fungicides intensively results in the accumulation of toxic compounds potentially hazardous to humans and the environment (Cook and Baker 1983) and also in the buildup of resistance of the pathogens (Dekker and Georgopolous 1982). In order to solve these problems, alternatives to chemical control are being investigated, and the use of antagonistic microbes seems to be one of the promising approaches. Antagonism may be accomplished by competition, parasitism, or antibiotics, or by a combination of these modes of action (Deacon and Barry 1992, Whipps 1992). Parasitism involves the production of several hydrolytic enzymes that degrade the cell walls of pathogenic fungi (Elad et al. 1985). The importance of β-1,3-glucanase and chitinase as key enzymes responsible for fungal cell and sclerotial wall lysis and degradation has been reported (Cook and Baker 1983). These enzymes have been shown to be produced by several fungi and bacteria and may be an important factor in biological control (Artigues and Davet 1984, Elad et al. 1982, Ordentlich et al. 1988). Fungal cell walls contain chitin composed of N-acetyl-D-glucosamine (NAG) molecules cross-linked with other by β(1-4) glycosidic linkages, which results in a highly insoluble crystalline structure organized into microfibrils (Adams 2004). Most of the fungal cell wall hydrolyses have chitinolytic activity. Potential application of chitinases in the biocontrol of unwanted fungi is promising.

Enzymes of chitin and glucan biosynthesis manufacture long, linear chains of β1,4-linked N-acetylglucosamine and β1,3-linked glucose, respectively. However, the fungal cell wall contains abundant quantities of branched 1,3-β-, 1,6-β-glucan, and there is evidence of extensive cross-linking between chitin, glucan, and other wall components

(Cabib et al. 2001, Klis et al. 2002). Furthermore, the wall is a highly dynamic structure subject to constant change, for example, during cell expansion and division in yeasts, and during spore germination, hyphal branching, and septum formation in filamentous fungi. Cell-wall polymer branching and cross-linking, and the maintenance of wall plasticity during morphogenesis, may depend upon the activities of a range of hydrolytic enzymes found intimately associated with the fungal cell wall. Most of the fungal cell wall hydrolases characterized to date have chitinase or glucanase activity, and a number of these enzymes also exhibit transglycosylase activity. They may therefore contribute to the breakage and re-formation of bonds within and between polymers, leading to remodeling of the cell wall during growth and morphogenesis.

Fontaine (Fontaine et al. 1997) detected monomeric and dimeric exo-1,3-β-glucanases of molecular mass 82 and 230 kDa, respectively, in the cell wall autolysates of *Aspergillus fumigatus*. An endo-1,3-β-glucanase of molecular mass 74 kDa was also detected. The cell wall of this species also contains a number of glucanases that exhibit 1,3-β-glucanosyltransferase activity. These include a 49 kDa endoglucanase that cleaves 1,3-β-glucan, and then transfers the newly generated reducing end to the nonreducing end of another 1,3-β-glucan molecule. With the formation of a new β-1,3 linkage, the 1,3-β-glucan chain is elongated, and the gene encoding this enzyme is described as GEL1 for glucan elongating glucanosyltransferase (Mouyna et al. 2000). GEL1 encodes a GPI (glycosylphosphatidylinositol)-anchored protein homologous to the GAS1 (Gas1p is a glycoprotein that is GPI-anchored to the plasma membrane of *Streptococcus cerevisiae*) and PHR gene products of *S. cerevisiae* and *Candida albicans*.

Disruption of the gene encoding the ChiB1p chitinase of *A. fumigatus* or the Cts1p chitinase of *Coccidioides immitis* had no effect on the growth or morphogenesis in these organisms (Jaques et al. 2003, Reichard et al.

2000). It is possible that the related enzymes in *A. fumigatus* and *C. immitis* compensate for the loss of ChiB1p or Cts1p. Alternatively, these secreted enzymes may not have morphogenetic roles. Instead, they may contribute to the digestion and utilization of exogenous chitin as a source of organic nutrients for energy and biosynthesis. The fungal/plant chitinases of Aspergillus species and *C. immitis* (deduced molecular mass 83–97 kDa (Jaques et al. 2003, Reichard et al. 2000, Takaya et al. 1998) are much larger than the enzymes of the fungal/bacterial class. Furthermore, each fungal/plant chitinase contains a serine/threonine-rich domain and a putative GPI anchor and cleavage site (Figure 37.1).

Many chitinolytic bacteria can play an important role in the biocontrol of fungi. The major genera containing chitinolytic species are *Pseudomonas aeroginosa* (Wang and Chang 1997), *Aeromonas* (Lan et al. 2006), *Xanthomonas*, *Serratia* (Kim et al. 2007), and *Bacteria* (Chuang et al. 2008).

Fortunately, chitin is not a constituent biological part or metabolite of vertebrates. Thus bacteria producing chitinases have great potential in controlling fungi and insects under select conditions (Bhattacharya et al. 2007).

*Ganoderma boninense* is a serious fungal disease called basal stem rot (BSR) in oil palm plantation where the microbial ecosystem has been destroyed, and it could thrive and occupy the space that was created by sterilizing the soil (Kandan et al. 2010). Infection by *Ganoderma* spp. begins by their attack in the palm roots, which gradually spreads to the bole of the stem where they cause dry rot, which prevents absorption and

transport of nutrients (Sanderson et al. 2002). Infected oil palms gradually lose their ability to produce fruits and eventually collapse. By the time *Ganoderma* fruiting bodies are detectable on the oil palm, about 50% of the internal tissues would have already rotted (Kandan et al. 2010). *G. boninense* also has many forms of resting stages, including resistant mycelium, basidiospores, chlamydospores, and pseudosclerotia, and these are difficult to control (Susanto et al. 2005).

In invertebrates, fungi have a coating of chitin in the cell wall, which gives them their shape (Donderski and Swiontek 2001). Chitinoloytic microbes when in contact with chitin secrete an enzyme called chitinase to mainly degrade the chitin and utilize it as an energy source. The enzyme chitinase inhibits fungal growth by hydrolyzing the chitin present in the fungal cell wall, resulting in inhibition of further progress of the fungi. There has been a lot of interest in this biological process because of its potential to be an agent for the biological control of plant disease and for engineering plants for resistance to phytopathogenic fungi by inducing the systemic resistance (Kamil et al. 2007).

One of the best studied responses to chitin addition is the effect on the microbial species that act as antagonists of crop pathogens. Antagonistic microbes employ a number of methods to attack plant pests and pathogens. This includes, but is not limited to, the production of chitinases (Maksimov et al. 2011), the production of toxins (e.g., antibiotics and toxins), direct parasitism, competition for nutriment, and the induction of defense responses in the plant. Therefore, adding chitin-based products to the growing environment may aid beneficial antagonists by stimulating the production and activation of chitinases, which can then be used to attack pests and pathogens, or be used as a stable nitrogen-rich polysaccharide food source that boosts the population to the level where other mechanisms control the plant pathogens.

The bacterium *Bacillus subtilis* is a pathogen of fungi and is one of the most widely used biopesticide in agriculture (Chandler et al. 2011). *B. subtilis* is known to secrete chitinases into the medium in which it is growing (Chen et al. 2010). Manjula and Podile (2001) showed that the addition of chitin to the carrier material improved the multiplication of *B. subtilis*, improved the bacteria's fungicidal action, and improved the control of Fusarium wilt in pigeon pea and crown rot in peanut caused by *Aspergillus niger*. Chitosan addition also improved the action of *B. subtilis* against powdery mildew in strawberry (Lowe et al. 2012).

The beneficial effect of chitin-based treatments to antagonistic bacteria is not restricted to *B. subtilis*, with both chitin and chitosan improving the control of Fusarium wilt in both tomato (Toyoda et al. 1996) and

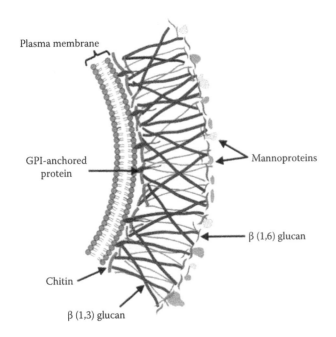

*Figure 37.1* Structure of fungal membrane.

cucumber (Singh et al. 1999) when applied to the soil with a range of different species of chitinolytic microbes.

Like all invertebrates, fungi have a coating of chitin in the cell wall that gives them their shape (Donderski and Swiontek 2001). Chitinoloytic microbes when in contact with chitin secrete an enzyme called chitinase to mainly degrade the chitin and utilize it as an energy source. The enzyme chitinase inhibits fungal growth by hydrolyzing the chitin present in the fungal cell wall, resulting in the inhibition of further progress of the fungi. There has been a lot of interest in this biological process because of its potential to be an agent for the biological control of plant disease and for engineering plants for resistance to phytopathogenic fungi by inducing the systemic resistance (Kamil et al. 2007).

The hydrolyzed chitin of fungi cell wall will become chitosan, which itself helps or induces the plants to secrete their own growth hormones. This will help build a much stronger root system and a much more vigorous vegetative growth.

Screening of potential chitinolytic bacteria has been done in Bioindustrial Technology Laboratory, Serpong, Indonesia. Based on the antifungal test, chitinolytic *Bacillus* sp. BPPTCC2 was found to have the ability to inhibit *G. boninense*. Inhibition of the growth of *G. boninense* can be seen from the clear zone (Figure 37.2).

No hyphae were grown on potato dextrose agar (PDA) taken from the clear zone that has been overgrown by *G. boninense* for 33 days. Chitin degradation by bacteria occurs in two phases: the process involves the hydrolysis of β-1,4-glycoside bond linking the

subunits GlcNAc. First, binding endochitinase tetramer and GlcNAc polymer produces a disaccharide chitobiose. Second, chitobiase hydrolyzes chitobiose to monomeric GlcNAc (Connell et al. 1998). Not all chitinolytic bacteria are capable of inhibiting *G. boninense*. *G. boninense* inhibition by chitinolytic bacteria is determined not only by the chitinase activity but also by the ability of the chitinase to degrade chitin in *G. boninense*.

## 37.4   Production of NAG

*N*-Acetyl-D-glucosamine (D-GlcNAc) is a derivative glucose monomer found in the polymers of bacterial cell walls, chitin, hyaluronic acids, and various glycans. D-GlcNAc is used to identify, differentiate, and characterize *N*-acetyl-β-D-hexoaminidase(s). In commercial production, GlcNAc is produced by acid hydrolysis of chitin with concentrated HCl at high temperature (above 80°C) (Rupley 1964), but this chemical process has problems, including production of acidic wastes, low yield, and high cost. For solving these problems, enzymatic hydrolysis was attempted for the production of GlcNAc from chitin (Il'ina et al. 2004, Sashiwa et al. 2001). Many chitinases produced from *Aspergillus* sp. (Rattanakit et al. 2003), *Bacillus* sp. (Thamthiankul et al. 2001, Wang et al. 2001), *Clostridium* sp. (Li et al. 2002), *Serratia* sp. (Brurberg et al. 1994), and *Aeromonas* sp. (Choi et al. 2003, Ueda et al. 2003) have been studied and applied to the production of GlcNAc from chitin. However, only a few industrial applications of enzyme preparations for the production of GlcNAc are currently available. Therefore, there is a need for research into enzymatic production of GlcNAc to find a more effective method.

NAG can be produced by the acid hydrolysis of chitin, but it is still expensive owing to the low yield. The product manufactured by *N*-acetylation of glucosamine is not approved as a natural material due to the chemical process (Sashiwa et al. 2003). Several natural methods have been proposed to produce NAG. Using these methods, NAG is obtained by chitin hydrolysis using chitin-degrading enzymes, containing endochitinases and NAGase. The enzymes were usually isolated from microorganisms having strong chitinolytic activities (Kuk et al. 2005, Li et al. 2005). Sashiwa et al. (2001) performed selective and efficient methods to produce NAG using a mixed enzyme preparation with high NAGase/endochitinase ratio (Sashiwa et al. 2001, 2002).

NAG was produced from chitins with various particle size by crude enzymes from *A. hydrophila* H-2330 to yields of 64%–77% within 10 days (Sashiwa et al. 2002). Another researcher reported that *Aeromonas* sp. GJ-18 crude enzymes produced NAG as the main enzymatic product, and swollen chitin was hydrolyzed to 83.0% and 94.9% yield of NAG within 5 and 9 days. Crude chitinase from *Bacillus* sp. BPPTCC-2, an indigenous strain,

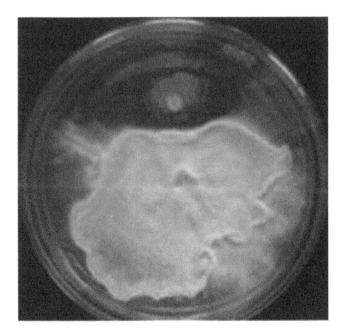

*Figure 37.2* Inhibition of *Ganoderma boninense* growth by *Bacillus* sp. BPPTCC 2.

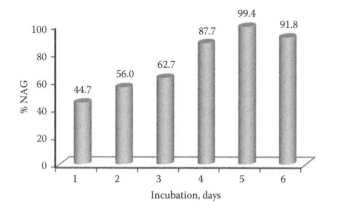

*Figure 37.3* Effects of incubation time on the percentage yield of NAG with 0.2 U/mL chitinase from *Bacillus* sp. BPPTCC 2 and substrate 3% dw/v at pH 6.0 and temperature 50°C.

can produce NAG as a the result of hydrolysis of chitin with a maximum yield of 99.4% in 5 days (Figure 37.3).

## 37.5  Conclusion

Discovering and implementing microorganisms as biological control agents or in biological processes have challenges and opportunities in the future, whether managed as resident communities or introduced as individual or a mixture of strains. Biological agents are treated on par with commercial fungicides for the purpose registration. Toxicological and biosafety tests have to be obtained and furnished to regulatory agencies, which ultimately leads to a long waiting period before registration and a high cost for the developed product.

The utilization of bacterial chitinases for various applications depends on the supply of highly active preparations at a reasonable cost. Most of the suppliers use either natural microbial biodiversity or genetically engineered chitinase overproducing microbial strains to obtain efficient preparations. The use of chitinases for the biocontrol of plant pathogens, and for developing transgenic plants is one of the major applications. Chitinases can also be employed in human health care, such as making opthalmic preparations with chitinases and microbicides. The understanding of the biochemistry of chitinolytic enzymes will make them more useful in a variety of processes in the near future.

## References

Adams D.J. Fungal cell wall chitinases and glucanases. *Microbiology* 150 (2004): 2029–2035.

Adrangi S., Faramarzi M.A., Shahverdi A.R., and Sepehrizadeh Z. Purification and characterization of two extracellular endochitinases from *Massilia timonae*. *Carbohydr Res* 345 (2010): 402–407.

Alam M.M., Nikaidou N., Tanaka H., and Watanabe T. Cloning and sequencing of chic gene of *Bacillus circulans* WL-12 and relationship of its product to some other chitinases and chitinase like proteins. *J Ferment Bioeng* 80 (1995): 454–461.

Arakane Y. and Muthukrishnan S. Insect chitinase and chitinase-like proteins. *Cellular Molecular Life Sciences* 67 (2010): 201–216.

Artigues M. and Davet P. Activites β(1–3) glucanasique et chitinasique de quelques champignons, en relation avec leur aptitude a detruire les sclerotes de *Corticium rolfsiidans* de la terre sterile. *Soil Biol Biochem* 16(5) (1984): 527–528.

Aye K.N. and Stevens W.F. Improved chitin production by pretreatment of shrimp shells. *J Chem Technol Biotechnol* 79 (2004): 421–425.

Bhattacharya D., Nagpur A., and Gupta R.K. Bacterial chitinases: Properties and potential. *Crit Rev Biotechnol* 27 (2007): 21–28.

Bhushan B. and Hoondal G.S. Isolation, purification and properties of a thermostable chitinase from an alkalophilic *Bacillus* sp. BG-11. *Biotechnol Lett* 20 (1998): 157–159.

Brurberg M.B., Eijsink V.G., and Nes I.F. Characterization of a chitinase gene (chiA) from *Serratia marcescens* BJL200 and one-step purification of the gene product. *FEMS Microbiol Lett* 124 (1994): 399–404.

Cabib E., Dong-Hyun R., Schmidt M., Crotti L.B., and Varma A. The yeast cell wall and septum as paradigms of cell growth and morphogenesis. *J Biol Chem* 276 (2001): 19679–19682.

Chandler D., Bailey A.S., Tatchell G.M., Davidson G., Greaves J., and Grant W.P. The development, regulation and use of biopesticides for integrated pest management. *Philos Trans R Soc B* 366 (2011): 1987–1998.

Chen F., Wang M., Zheng Y., Luo J., Yang X., and Wang X. Quantitative changes of plant defense enzymes and phytohormone in biocontrol of cucumber Fusarium wilt by *Bacillus subtilis* B579. *World J Microbiol Biotechnol* 26 (2010): 675–684.

Choi Y.U., Kang J.H., Lee M.S., and Lee W.J. Isolation and characterization of a chitinolytic enzyme producing marine bacterium, *Aeromonas* sp. J-5003. *J Fish Sci Technol* 6 (2003): 1–6.

Chuang H.H., Lin H.Y., and Lin F.P. Biochemical characteristics of C-terminal region of recombinant chitinase from *Bacillus licheniformis*: Implication of necessity for enzyme properties. *FEBS J* 275 (2008): 2240–2254.

Connell T.D., Metzger D.J., Lynch J., and Folster J.P. Endochitinase is transported to the extracellular milieu by the Eps-encoded general secretory pathway of vibrio cholera. *J Bacteriol* 180 (1998): 5591–5600.

Cook R.J. and Baker K.F. The nature and practice of biological control of plant pathogens. *Am Phytopathol Soc. Minnesota* 2 (1983): 539.

Dahiya N., Tewari R., Tiwari R.P., and Hoondal G.S. Chitinase from *Enterobacter* sp. NRG4: Its purification, characterization and reaction pattern. *Electron J Biotechnol* 8(2) (2005a): 134–145.

Dahiya N., Tewari R., Tiwari R.P., and Hoondal G.S. Production of an antifungal chitinase from *Enterobacter* sp. NRG4 and its application in protoplast production. *World J Microbiol Biotechnol* 21 (2005b): 1611–1616.

Deacon J.W. and Barry L.A. Modes of action of mycoparasites in relation to biocontrol of soilborne plant pathogens. In: *Biological Control of Plant Diseases*, E.C. Tjamos, G.C. Papavizas, and R.J. Cook (Eds.). Plenum Press: New York, pp. 157–167 (1992).

Dekker J. and Georgopolous S.G. *Fungicide Resistance in Crop Protection*. Centre for Agricultural Publication and Documentation: Wageningen, the Netherlands, p. 265 (1982).

Deshpande M.V. Enzymatic degradation of chitin & its biological applications. *J Sci Ind Res* 45 (1986): 273–281.

Donderski W. and Swiontek M.B. Occurrence of chitionlytic bacteria in water and bottom sediment of eutrophic lakes in Ilawskie Lake District. *Pol J Environ Stud* 10(5) (2001): 331–336.

Elad Y., Chet I., and Henis Y. Degradation of plant pathogenic fungi by *Trichoderma harzianum*. *Can J Microbiol* 28(7) (1982): 719–725.

Elad Y., Lifshitz R., and Baker R. Enzymatic activity of the mycoparasite *Pythium nunn* during interaction with host and non-host fungi. *Physiol Plant Pathol* 2 (1985): 131.

Fontaine T., Hartland R.P., Diaquin M., Simenel C., and Latgé J.-P. Differential patterns of activity displayed by two exo-β-1,3-glucanases associated with the *Aspergillus fumigatus* cell wall. *J Bacteriol* 179 (1997): 3154–3163.

Fukamizo T. Chitinolytic enzymes: Catalysis, substrate binding and their application. *Curr Protein Pept Sci* 1 (2000): 105–124.

Gooday G.W., Woodman J., Casson E.A., and Browne C.A. Effect of nikkomycin on chitin spine formation in the diatom *Thalassiosira fluviatilis*, and observations on its peptide uptake. *FEMS Microbiol Lett* 28 (1985): 335–340.

Gossen M.F.A. *Applications of Chitin and Chitosan*. Technomic Publishing Company Book: Lancaster, PA (1997).

Hart P.J., Pfluger H.D., Monzingo A.F., Hoihi T., and Robertus J.D. The refined crystal structure of an endochitinase from *Hordeum vulgare* L. seeds at 1.8 Å resolution. *J Mol Biol* 248 (1995): 402–413.

Hartl L., Zach S., and Seidl-Seiboth V. Fungal chitinases: Diversity, mechanistic properties and biotechnological potential. *Appl Microbiol Biotechnol* 93 (2012): 533–543.

Henrissat B. and Bairoch A. New families in the classification of glycosylhydrolases based on amino acid sequence similarities. *Biochem J* 293 (1993): 781–788.

Hossain M.A., Noh H.N., Kim K.I., Koh E.J., Wi S.G., Bae H.J., Lee H., and Hong S.W. Mutation of the chitinase-protein-encoding AtCTL2 gene enhances lignin accumulation in dark-grown *Arabidopsis* seedlings. *J Plant Physiol* 167(8) (2010): 650–658.

Il'ina A.V., Zueva O.Y., Lopatin S.A., and Varlamov V.P. Enzymatic hydrolysis of α-chitin. *Appl Biochem Microbiol* 40 (2004): 42–45.

Jamialahmadi K., Behravan J., Najafi M.F., Yazdi M.T., Shahverdi A.R., and Faramarzi M.A. Enzymatic production of N-acetyl-D-glucosamine from chitin using crude enzyme preparation of *Aeromonas* sp. PTCC1691. *Biotechnology* 10 (2011): 292–297.

Jaques A.K., Fukamizo T., Hall D., Barton R.C., Escott G.M., Parkinson T., Hitchcock C.A., and Adams D.J. Disruption of the gene encoding the ChiB1 chitinase of *Aspergillus fumigatus* and characterization of a recombinant gene product. *Microbiology* 149 (2003): 2931–2939.

Kandan A., Bhaskaran R., and Samiyappan R. Ganoderma a basal stem rot disease of coconut palm in South Asia and Asia Pacific regions. *Arch Phytopathol Plant Protect* 43 (2010): 1445–1149.

Kamil Z., Rizk M.S., and Moustafa, S.A. Isolation and identification of rhizosphere soil chitinolytic bacteria and their potential in antifungal bio-control. *Glob J Mol Sci* 2(2) (2007): 57–66.

Karasuda S., Tanaka S., Kajihara H., Yamamoto Y., and Koga D. Plant chitinase as a possible biocontrol agent for use instead of chemical fungicides. *Biosci Biotechnol Biochem* 67 (2003): 221–224.

Kim H.S., Timmis K.N., and Golyshin P.N. Characteristics of a chitinolytic enzyme from *Serratia* sp. KCK isolated from kimchi juice. *Appl Microbiol Biotechnol* 75 (2007): 1275–1283.

Klis F.M., Mol P., Hellingwerf K., and Brul S. Dynamics of cell wall structure in *Saccharomyces cerevisiae*. *FEMS Microbiol Rev* 26 (2002): 239–256.

Kramer K.J. and Muthukrishnan S. Insect chitinases: Molecular biology and potential use as biopesticides. *Insect Biochem Mol Biol* 27 (1997): 887–900.

Kuk J.H., Jung W.J., Hyun J.G., Ahn J.S., Kim K.Y., and Park R.D. Selective preparation of N-acetyl-D-glucosamine and N,N'-diacetylchitobiose from chitin using a crude enzyme preparation from *Aeromonas* sp. *Biotechnol Lett* 27 (2005): 7–11.

Lan X., Zhang X., Hu J., and Shimsaka M. Cloning, expression and characterization of a chitinase from the chitinolytic bacterium *Aeromonas hydrophila* strain SUWA-9. *Biosci Biotechnol Biochem* 70 (2006): 2437–2442.

Li C., Huang W., and Wang L.X. Chemoenzymatic synthesis of N-linked neoglycoproteins through a chitinase-catalyzed transglycosylation. *Bioorg Med Chem* 16 (2008): 8366–8372.

Li H., Morimoto K., Katagiri N., Kimura T., Sakka K., Lun S., and Ohmiya K. A novel β-N-acetylglucosaminidase of *Clostridium paraputrificum* M—21 with high activity on chitobiose. *Appl Microbiol Biotechnol* 60 (2002): 420–427.

Li Y.L., Wu S.T., Yu S.T., and Too J.R. Screening of a microbe to degrade chitin. *Taiwan J Agric Chem Food Sci* 43 (2005): 410–418.

Lowe A., Rafferty-McArdle S.M., and Cassells A.C. Effects of AMF- and PGPR-root inoculation and a foliar chitosan spray in single and combined treatments on powdery mildew disease in strawberry. *Agric Food Sci* 21 (2012): 28–38.

Lu H.F., Narayanan K., Lim S.X., Gao S., Leong M.F., and Wan A.C. A 3D microfibrous scaffold for long-term human pluripotent stem cell self-renewal under chemically defined conditions. *Biomaterials* 33 (2012): 2419–2430.

Maksimov I.V., Abizgil'dina R.R., and Pusenkova L.I. Plant growth promoting rhizobacteria as alternative to chemical crop protectors from pathogens. *Appl Biochem Microbiol* 47 (2011): 333–345.

Manjula K. and Podile A.R. Chitin-supplemented formulations improve biocontrol and plant growth promoting efficiency of *Bacillus subtilis* AF1. *Can J Microbiol* 47 (2001): 618–625.

Mouyna I., Fontaine T., Vai M., Monod M., Fonzi W.A., Diaquin M., Popolo L., Hartland R.P., and Latgé J.-P. Glycosylphosphatidylinositol-anchored glucanosyltransferases play an active role in the biosynthesis of the fungal cell wall. *J Biol Chem* 275 (2000): 14882–14889.

Ordentlich A., Elad Y., and Chet I. The role of chitinase of *Serratia marcescens* in biocontrol of *Sclerotium rolfsii*. *Phytopathology* 78 (1988): 84–88.

Ortiz-Rodriguez T., de la Fuente-Salcido N., Bideshi D.K., Salcedo-Hernandez R., and Barboza-Corona J.E. Generation of chitin-derived oligosaccharides toxic to pathogenic bacteria using ChiA74, an endochitinase native to *Bacillus thuringiensis*. *Lett Appl Microbiol* 51 (2010): 184–190.

Park J.K., Morita K., Fukumoto I., Yamasaki Y., Nakagawa T., Kawamukai M., and Matsuda H. Purification and characterization of the Chitinase (ChiA) from *Enterobacter* sp. G-1. *Biosci Biotechnol Biochem* 61 (1997): 684–689.

Patil S.R., Ghormade V., and Deshpande M.V. Chitinolytic enzymes: An exploration. *Enzyme Microb Technol* 26 (2000): 473–483.

Pleban S., Chernin L., and Chet I. Chitinolytic activity of an endophytic strain of *Bacillus cereus*. *Lett Appl Microbiol* 25 (1997): 284–288.

Rattanakit N., Yano S., Wakayama M., Plikomol A., and Tachiki T. Saccharification of chitin using solid-state culture of *Aspergillus* sp. S1–13 with shellfish waste as a substrate. *J Biosci Bioeng* 95 (2003): 391–396.

Rinaudo M. Chitin and chitosan: Properties and applications. *Prog Polym Sci* 31 (2006): 603–632.

Reichard U., Hung C.-Y., Thomas P.W., and Cole G.T. Disruption of the gene which encodes a serodiagnostic antigen and chitinase of the human fungal pathogen *Coccidioides immitis*. *Infect Immun* 68 (2000): 5830–5838.

Roberts R.L. and Cabib E. *Serratia marcescens* chitinase: One step purification and use for the determination of chitin. *Anal Biochem* 127 (1982): 402–412.

Rupley J.A. The hydrolysis of chitin by concentrated hydrochloric acid, and the preparation of low-molecular-weight substrates for lysozyme. *Biochim Biophys Acta* 83 (1964): 445–455.

Sahai A.S. and Manocha M.S. Chitinases of fungi and plants: Their involvement in morphogenesis and host parasite interaction. *FEMS Microbiol Rev* 11 (1993): 317–338.

Saito A., Fujii T., Yoneyama T., Redenbach M., Ohno T., Watanabe T., and Miyashita K. High multiplicity of chitinase genes in *Streptomyces coelicolor* A3(2). *Biosci Biotechnol Biochem* 63 (1999): 710–718.

Sanderson F.R., Pilotti C.A., and Bridge P.D. Basidiospores: Their influence on our thinking regarding a control strategy for basal stem rot. In: *Ganoderma Diseases of Perennial Crops*, J. Flood, P.D. Bridge, and M. Holderness (Eds.). CABI Publishing: Wallingford, U.K. (2002).

Sashiwa H., Fujishima S., Yamano N., Kawasaki N., Nakayama A., Muraki E., and Aiba S. Production of N-acetyl-D-glucosamine from β-chitin by enzymatic hydrolysis. *Chem Lett* 30 (2001): 308–309.

Sashiwa H., Fujishima S., Yamano N., Kawasaki N., Nakayama A., Muraki E., Sukwattanasinitt M., Pichyangkura R., and Aiba S.I. Enzymatic production of N-acetyl-D-glucosamine from chitin. Degradation study of N-acetylchitooligosaccharide and the effect of mixing of crude enzymes. *Carbohydr Polym* 51 (2003): 391–395.

Sashiwa H., Fujishima S., Yamano N., Kawasaki N., Nakayama A., Muraki E., Hiraga K., Oda K., and Aiba S. Production of N-acetyl-D-glucosamine from alpha-chitin by crude enzymes from *Aeromonas hydrophila* H-2330. *Carbohydr Res* 337 (2002): 761–763.

Shaikh S.A. and Deshpande M.V. Chitinolytic enzymes: Their contribution to basic and applied research. *World J Microbiol Biotechnol* 9 (1993): 468–475.

Singh P.P., Shin Y.C., Park C.S., and Chung Y.R. Biological control of Fusarium wilt of cucumber by chitinolytic bacteria. *Phytopathology* 89 (1999): 92–99.

Susanto A., Sudharto P.S., and Purba R.Y. Enhancing biological control of basal stem rot diseases (*Ganoderma boninense*) in oil palm plantations. *Mycopathology* 159 (2005): 153–157.

Suzuki K., Taiyoji M., Sugawara N., Nikaidou N., Henrissat B., and Watanabe T. The third chitinase gene (chiC) of *Serratia marcescens* 2170 and the relationship of its product to other bacterial chitinases. *Biochem J* 343 (1999): 587–596.

Taira T., Ohnuma T., Yamagami T., Aso Y., Ishiguro M., and Ishihara M. Antifungal activity of rye (*Secale cereale*) seed chitinases: The different binding manner of class I and class II chitinases to the fungal cell wall. *Biosci Biotechnol Biochem* 66 (2002): 970–977.

Tajdini F., Amini M.A., Nafissi-Varcheh N., and Faramarzi M.A. Production, physiochemical an antimicrobial properties of fungal chitosan from *Rhizomucor miehei* and *Mucor racemosus*. *Int J Biol Macromol* 47 (2010): 180–183.

Takaya N., Yamazaki D., Horiuchi H., Ohta A., and Takagi M. Cloning and characterization of a chitinase-encoding gene (chiA) from *Aspergillus nidulans*, disruption of which decreases germination frequency and hyphal growth. *Biosci Biotechnol Biochem* 62 (1998): 60–65.

Toyoda H., Matsud Y., Fukamizo T., Nonomura T., Kukutani K., and Ouchi S. Application of chitin and chitosan degrading microbes to comprehensive biocontrol of fungal wilt pathogen, *Fusarium oxysporum*. In: *Chitin Handbook*, R.A.A. Muzzarelli and M.G. Peter (Eds.). European Chitin Society, Atec: Grottammare, Italy, pp. 359–370 (1996).

Thamthiankul S., Suan-Ngay S., Tantimavanich S., and Panbangred W. Chitinase from *Bacillus thuringiensis* subsp. pakistani. *Appl Microbiol Biotechnol* 56 (2001): 395–401.

Ueda M., Kojima M., Yoshikawa T., Mitsuda N., Araki K., Kawaguchi T., Miyatake K., Arai M., and Fukamizo T. A novel type of family 19 chitinase from *Aeromonas* sp. No. 10S-24. Cloning, sequence, expression, and the enzymatic properties. *Eur J Biochem* 270 (2003): 2513–2520.

Vaidya R., Roy S., Macmil S., Gandhi S., Vyas P., and Chhatpar H.S. Purification and characterization of chitinase from *Alcaligenes xylosoxydans*. *Biotechnol Lett* 25 (2003): 715–717.

Vega K. and Kalkum M. Chitin, chitinase responses, and invasive fungal infections. *Int J Microbiol* 2012 (2012): 920459. doi:10.1155/2012/920459.

Wang S.L. and Chang W.T. Purification and characterization of two bifunctional chitinase/lysozyme extracellularly produced by *Pseudomonas aeruginosa* K-187 in a shrimp and crab shell powder medium. *Appl Environ Microbiol* 63 (1997): 380–386.

Wang S.Y., Moyne A.L., Thottappilly G., Wu S.J., Locy R.D., and Singh N.K. Purification and characterization of a *Bacillus cereus* exochitinase. *Enzyme Microb Technol* 28 (2001): 492–498.

Watanabe T., Kobori K., Miyashita K., Fujii T., Sakai H., Uschida M., and Tanaka H. Identification of glutamic acid 204 and aspartic acid 200 in chitinase A1 of *Bacillus circulans* WL-12 as essential residues for chitinase activity. *J Biol Chem* 268 (1993): 18567–18572.

Wen C.M., Tseng C.S., Cheng C.Y., and Li Y.K. Purification, characterization and cloning of a chitinase from *Bacillus* sp. NCTU2. *Biotechnol Appl Biochem* 35 (2002): 213–219.

Whipps J.M. Status of biological disease control in horticulture. *Biocontrol Sci Technol* 2 (1992): 3–24.

Yuli P.E., Suhartono M.T., Rukayadi Y., Hwang J.K., and Pyun Y.R. Characteristics of thermostable chitinase enzymes from the Indonesian *Bacillus* sp. 13.26. *Enzyme Microb Technol* 35 (2004): 147–153.

Zakariassen H., Hansen M.C., Joranli M., Eijsink V.G., and Sørlie M. Mutational effects on transglycosylating activity of family 18 chitinases and construction of a hypertransglycosylating mutant. *Biochemistry* 50 (2011): 5693–5703.

Zikakis J.P. Chitinolytic enzymes and their applications. In: *Biocatalysts in Agricultural Biotechnology*, ACS Symposium Series, J.R. Whitaker and P.E. Sonnet (Eds.), Vol. 389. American Chemical Society: Washington, DC, pp. 116–126 (1989).

*chapter thirty-eight*

# Biological activity of marine sponge lectins

*Partha Pratim Bose, Urmimala Chatterjee, and Bishnu Pada Chatterjee*

## Contents

## 38.1 Introduction

Nature has endowed biomolecules with structural diversity to meet the high demand for information, and for this nature harnesses some building blocks such as nucleotides, amino acids, and sugars. Of these, sugars are the best choice for coding biological information. A simple fact can be put forward as evidence of the superiority of sugar building blocks as a natural coding language. To code two different pieces of information, both nucleotides and amino acids require their sequence to be modified, as both these units have only one exclusive way of two-end linkage formation (Figure 38.1), whereas with a same sequence of sugar, coding of diverse types of information is achievable by employing only the different available linkage points in a single sugar unit (1,2 or 1,3 or 1,4 or 1,6, etc.) or utilizing the ring size of the same sugar unit (pyranose or furanose). Thus, nature adopts sugars as coding language in the highly space-saving situation of biological membranes, where an enormous amount of flow of information occurs between the inner and outer environments of cells, which demands immense structural diversity (Gabius and Kayser 2014).

To utilize the structural diversity of sugar structures in the biological information transfer network, another type of biomolecules—lectins—was evolved by nature of nonimmune origin. Lectins are proteins or glycoproteins having specificity for sugars, but unlike the sugar binding enzymes, they retain the structural integrity of the sugar even after binding. Thus, lectin has a carbohydrate recognition domain (CRD) without any catalytic activity. Since its first historical identification as hemagglutinin, new findings of lectin–sugar interplay in glycobiology have upgraded the knowledge base of a range of complex biological mechanisms associated with molecular and cellular crosstalks relevant in immunology, developmental biology, host–pathogen interaction, and so on. Lectins are ubiquitous in nature and much research endeavor has gone into their identification and biological characterization. However, only a comparatively small volume of literature is available for lectin–sugar interactions evolved in marine invertebrates. Many of these lectins are protective molecules against the predators or pathogens and can be promising sources of new biologically active molecules. Among the lectins from marine invertebrates, sponge-derived lectins and their biological and biopharmaceutical activities will be discussed in this chapter. The natural habitat of marine invertebrates often has a very high microbial load of $10^6$ bacteria and $10^9$ viruses per milliliter of seawater. Lacking the adaptive immunological machinery, marine invertebrates such as sponges have to solely depend on their innate immunity. Invertebrate lectins are found to play an essential role in their innate immune system. Sponges, the primitive metazoans, are filter feeders and do not have any true circulatory system other than water

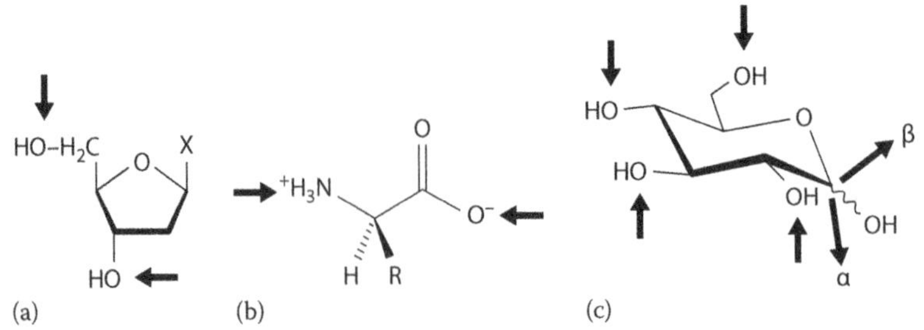

**Figure 38.1** (a) - (c) Diversity of linkages only possible across sugar moieties as compared to amino acids and nucleotides. (Data from Gabius, H.J. and Kayser, K., *Diagn. Pathol.*, 9, 4, 2014.)

**Table 38.1** Marine sponge lectins: Their structural features and carbohydrate specificities

| Name of marine sponges | Subunit molecular weight (kDa) | Monomeric arrangements | Disulfide linkage | Carbohydrate specificity |
|---|---|---|---|---|
| *Aplysina archeri* | 16 | Tetramer | N | β-D-Gal |
| *Aplysina lawnosa* | 16 | Tetramer | N | β-D-Gal |
| *Axinella polypoides* I | 21 | Dimer | N.D. | D-Gal and D-Fuc |
| *Axinella polypoides* II | 15 | Monomer | N | D-Gal and D-Fuc |
| *Axinella corrugata* I | 13.9 | Hexamer | Y | *N,N′,N″*-Triacetylchitotriose |
| *Axinella corrugata* II | 13.9 | N.D. | N.D. | *N,N′,N″*-Triacetylchitotriose |
| *Cliona varians* | 28.5 | Tetramer | Y | Galactose |
| *Craniella australiensis* | 17.8 | Trimer | Y | Asialo-PSM |
| *Cinachyrella alloclada* | 17 | Mono/Dimer | Y | Lactose |
| *Cinachyrella apion* | 15.5 | Octamer | N.D. | Lactose |
| *Cinachyrella* sp. | 16 | Tetramer | N.D. | Lactose |
| *Geodia cydonium* | 13 | Trimer | Y | Lactose |
| *Halichondria okadai* | 30 | Dimer | N | Galactose |
| *Haliclona cratera* | 29 | Monmer | N | D-Galactose and *N*-acetyl-D-galactosamine |
| *Haliclona caerulea* I | 14 | Monomer | N.D. | N.D. |
| *Haliclona caerulea* II | 15 | Dimer | N.D. | N.D. |
| *Pellina semitubulosa* | 34 | Hexamer | N.D. | Lactose |

*Source:* Adapted from Gomes Filho et al., *Molecules*, 20, 348, 2014.

*Notes:* Y, present; N, absent; N.D., not determined.

circulatory systems. Therefore, they have to circulate enormous amounts of water through their aquiferous system and thus are exposed to a high risk of infection. Therefore, they have evolved a chemical defense system to protect themselves. It has been found that lectins are the major components of this chemical defense shield (Proksch 1994). Sponges are also equipped with efficient humoral and cellular immune mechanisms that show astoundingly high sequence similarity with human immune molecules, and this acts as an impetus for research in these areas with the hope that sponge lectins might have a similar type of biological response and relevance in humans (Muller et al. 1999). Table 38.1 summarizes some of the lectins isolated from marine sponge species along with their carbohydrate specificity and structural units.

## 38.2 Biological activity of the sponge lectins

### 38.2.1 Sponge lectins and their immunological aspects

Sponges like other invertebrates with no adaptive immunity have developed a chemical defense mechanism that responds to the surface antigens of the potential pathogens. The sponge defense systems depend mostly on the innate immune network that is activated when pathogenic molecular patterns are recognized by cell surface sponge proteins. A major part of these proteins are lectins with diverse modes of action, such as antimicrobial, microbial agglutinin, and pattern recognition for pathogenic proteins. Any of these molecular reactions activates cellular or humoral effector molecules to act against the invading pathogens. These lectins participate in immune defense by specific recognition of various pathogen-associated carbohydrate containing molecules, that is, glycans, glycoproteins, peptidoglycans, glycolipids, or lipopolysaccharides, from Gram-negative and Gram-positive bacteria, viruses, or fungi. Among the immune active sponge lectins, a D-galactose-, D-arabinose-, and D-fucose-binding lectin was isolated from *Pellina semitubulosa* having a putative hexameric structure with a monomeric size 34 kDa. This lectin was found to be strongly mitogenic toward mice spleen lymphocyte, and showed a similar character in a mixed culture of T and B lymphocytes and macrophages. This had a pronounced immunomodulatory effect on mouse. An increase in $^3$H-thymidine incorporation into mouse thymocyte DNA, upon addition of the lectin, showed enhanced production of IL-1. Similarly, the lectin was also found to be enhancing the level of IL-2 production in a mouse mixed lymphocyte culture (Engel et al. 1992). *Axinella corrugata* is a marine sponge found in the South Atlantic coast of Brazil, Panama, and Colombia. A hexameric glycoprotein lectin of molecular mass 82.3 kDa was isolated from *A. corrugata* and named as ACL-1 (Roger et al. 2008). Hemagglutination activity of ACL-I was found to be inhibited specifically by GlcNAc, GalNAc, and ManNAc with the same intensity but not by D-galactose. Another very interesting observation about ACL-1 was its high resistance to proteolytic enzymes such as trypsin and chymotrypsin, which might be attributed to the high compactness of its hexameric structure. Rat neutrophils were shown to undergo chemotaxis upon treatment with ACL-I at concentrations of 0.8–12.8 µg/mL, as compared to the lipopolysaccharide control in an *in vitro* neutrophil migration study. However, the effect was found to be independent of the ACL-I concentration. The chemotaxis of neutrophils suggested that ACL-I might present a pro-inflammatory role, and it might be used as a molecular tool in the study of cellular events related to inflammation (Dresch et al. 2008). ACL-1 was also found to be mitogenic to human mononuclear cells, which were inhibited by GlcNAc (Dresch et al. 2012).

The presence of allo- and auto-immune recognition systems in sponge has been demonstrated in last decades (Kruse et al. 1999; Müller et al. 1999). It has been found that the glass sponges *Hexactinellida* are endowed with cell-surface aggregation factor (lectin) and complex signal transduction systems such as the tyrosine kinase receptor (Skorokhod et al. 1999). In an experiment, *Rhabdocalyptus dawsoni* was subjected to allo- and auto-grafts separately, wherein successful aggregation and fusion with the host syncytia was observed only when the grafts from the same individuals were placed on a slab of the sponge body wall; for other different individuals, the grafting was rejected. This is evidence for the allo-immune recognition in sponge probably acting via the involvement of the lectin–carbohydrate recognition system (Leys et al. 1999).

### 38.2.2 Sponge lectin in protection against pathogens

Sponges have developed a highly efficient molecular recognition system against invasion of pathogenic bacteria and fungi. The sponge cell-surface-associated lipopolysaccharide (LPS)-recognizing protein is a potential sensor for Gram-negative bacteria. Upon binding to bacterial LPS, some cascade pathways are found to be initiated in sponge, such that the sensor protein dimerizes and interacts with MyD88-like protein (myeloid differentiation response associated protein) to start an immunological cascade (Dzik 2010). The marine sponge *Suberites domuncula* reacts to LPS and subsequently upregulates a D-GlcNAc-binding lectin, the tachylectin related protein (Schröder et al. 2003). Three key components have been identified that are associated with the sponge immune network against bacterial lipopeptides: Toll-like receptor (TLR), IL-1 receptor associated kinase-4 like protein (IRAK-4l), and an effector caspase protein isolated from *S. domuncula* (Dzik 2010). Recognition of Gram-positive bacteria by *Suberites* occurs through sensing the membrane-associated proteoglycan of bacteria (Wiens et al. 2005) and a subsequent enhancement in the level of lysozyme and some endocytic activity have been observed. Upon LPS binding, sometimes sponge cells are found to activate protein kinase pathways such as p38 kinase and JNK (c-Jun N-terminal kinase) similar to those in mammals (Böhm et al. 2000; Wiens et al. 2005). Therefore, lectin response of sponge triggered by pathogen binding has been an active area of research.

A 27 kDa, D-GlcNAc-specific demosponge lectin, Suberites lectin (LEC_SUBDO), was isolated from *Suberites domuncula*, which showed its potential against both Gram-positive *Staphylococcus aureus* and Gram-negative *Escherichia coli* (Schroder et al. 2003). The identification of this lectin and its striking similarity in bactericidal effect of tachylectin-1 from *T. trunculus* (sea snail) give insights into the evolutionary history of the molecular immunology in these sea animals (Kawabata et al. 2001). It has been hypothesized that these sponge immune lectins have been subsequently lost in the evolution route to the crown taxa of Deuterostomia (animals where the blastopore becomes the anus in the adult) and of Protostomia (animals where the mouth results from the blastopore). This sponge immune lectin is also reminiscent of some protecting molecules of multicellular ancestors, such as the tectonins of the slime mold *P. polycephalum* (Schroder et al. 2003). Kawser et al. (2011) have isolated and characterized a 30 kDa D-galactose-specific lectin, HOL-30, from the demosponge *Halichondria oka-dai*, which showed potent antibacterial effect against Gram-positive bacteria.

The marine sponge species *Cinachyrella apion* and *Cliona varians* share the same habitat and the lectin isolated from these two specimens of sponge (CaL and CvL, respectively) were found to be leactose- and D-galactose-binding lectins, respectively. Interestingly, the agglutinating activity of both CaL and CvL has been recorded on *Leishmania chagasi* promastigotes, revealing the presence of both D-galactose and lactose receptors in the glycoclyx (Moura et al. 2006; Mederios et al. 2010). Another lectin, ACL-II, was isolated from *A. corrugata* with diverse binding specificities for GlcNAc, ManNAc, and triacetylchitotriose, with highest affinity toward triacetylchitotriose. A binding event with chitosan may be indicative of the role of ACL-II in the defense mechanism of the sponge since chitin is an essential structural component of the fungal cell wall. ACL-II was also shown to be agglutinating promastigotes of *L. chagasi* as CaL or CvL (Dresch et al. 2012).

### 38.2.3 Sponge lectin in anti-predation

Evading predators is one of the most critical means for survival especially for lower organisms. To fight predators, organisms have developed an array of chemical and physical defense shields. From the battle of prey defense and predator counter-defense, the latter also have developed some adaptive strategies to dodge these prey defenses shields. Sponges are perhaps one of the most endangered and vulnerable marine animals due to predation because of their motionless and often prominent physical appearance. Their nutritional value also makes them a common target for different predators. Consequently, many sponges defend themselves by utilizing various chemicals, and lectins have also been found to carry out this task. Information on feeding biology has demonstrated some of very interesting facts in marine ecology and evolution, where chemicals synthesized by one animal is utilized by others for aposematic purpose. Pawlik et al. showed that the sponge feeder mollusk *Hexabranchus sanguineus*, the Spanish dancer nudibranch, seized the chemical defense materials from its food sponge *Halichondria* sp., and distributed and concentrated the chemical in their mantle, mucus secretion, and the egg masses (Pawlik et al. 1988).

Mebs et al. isolated hemagglutinins from four species of marine sponges, *Haliclona* sp., *Cinachyra tenuifolia*, *Callyspongia viridis*, *Terpios zeteki*, and found that their hemagglutination to be inhibited by lactose but not melibiose or other oligosaccharides. It indicated that those lectins might react with terminal D-Galβl-4-residues. Among the above four marine sponges,

*Callyspongia viridis* hemagglutinin showed hemolysis along with ichthyotoxic activity. Ichthyotoxic activity of marine sponge lectins is evidence for the typical evolution of these molecules for defense mechanism of sedentary coral reef animals like sponges against predation (Mebs et al. 1985). Halilectin-1 and 2 (H-1 and H-2) from *Haliclona caerulea* demonstrated their specificity for orosomucoid and porcine stomach mucin. Both H-1 and H-2 were shown to be toxic against *Artemia nauplii*, the brine shrimp, with lethal effect of H-1, indicating their role in protection against predation (Carneiro et al. 2013). Another very interesting anti-predation mechanism adopted by sponges is the touch sensitivity or the stiffening of the whole sponge (*Chondrosia reniformis*) body when touched (Wilkie et al. 2006). It has been found that the stiffening of mesohyl, the collagenous tissue of sponges, could be manipulated by the modification of $Ca^{+2}$ ion concentration, using calcium channel blockers, cell membrane disrupters, or even by the addition of self-tissue extract that had been repeatedly frozen and thawed. Therefore, it is evident that the mechanical stiffening of sponge mesohyl ECM is under cellular control. Though the putative effector molecule has not been identified to date, it has been hypothesized that this has to be a lectin-type molecule as such molecules were found to play a crucial role in mammalian collagenous structure (Ozeki et al. 1995). The lectins might be stored in the spherulous cells abundantly present in *C. reniformis* ectosome, because the ectosome extract produced most pronounced stiffening response in intact sponge (Bretting et al. 1983).

### 38.2.4 Sponge lectin for symbiosis

Most sponge habitats contain a very high load of different bacteria and other unicellular organisms. Therefore, very interesting types of association and symbiosis are often observed between sponge and bacteria. In the mesohyl of some of the *Verongia* species of sponges, a huge bacteria population of genera *Pseudomonas* and *Aeromonas* has been found to be present, which is about 38% of the total sponge tissue volume (Bertrand and Vacelet 1971). In the symbiosis between bacteria and sponge, the latter provides bacteria with growth medium and amino acids, and in turn bacteria help the host to meet their requirement for sugars. The bacteria also help the sponge in their growth by involving in the controlled denaturation of collagen bundles (Garrone 1978).

Lectins mediating the cell-to-cell symbiosis between marine sponges and their bacterial flora have been an interesting area of research. A 78 kDa lectin having specificity for D-galacturonic acid, D-glucuronic acid, and L-fucose was isolated from *Halicondria panacea*, and the lectin molecule was found to be present in very high copy number of $6 \times 10^3$ per cell. Under culture condition, it was found that *Halicondria* lectin was essential for the growth of the bacterial species *Pseudomonas insolita*, and the latter was found to grow only in presence of the *Halicondria* lectin. However, *P. insolita* did not use the lectin as a nutrient source because no growth was found when heat-denatured lectin was added to the culture medium. It was envisaged that big molecules such as *Halicondria* lectin would undergo irreversible structural change during thermal denaturation with subsequent loss of secondary-tertiary structural integrity, which prevented the lectin from binding with the growth receptor on *P. insolita* to stimulate growth. The fact that the bacteria did not directly assimilate the lectin either as growth factor or as carbon source was demonstrated by the same hemagglutination titer values obtained from noninoculated culture broth and medium under the growth of bacteria. Binding studies indicated that the lectin had two different binding sites; one site recognized the sponge cell membrane receptors, and the second type bound the *P. insolita* surface antigen (Muller et al. 1981). Therefore, it is obvious that to run symbiosis and pathogen shielding in parallel, sponges have developed two chemical recognition systems to identify xenogeneic bacteria species.

## 38.2.5 Sponge lectin as antioxidant or protection from ultraviolet damage

A few organic chromophore-containing colored lectins have been isolated from some sponge species, but their activity has not been properly elucidated to date.

A lectin having specificity toward *N*-acetyl-galactosamine, H-3, has been isolated from *H. caerulea*. This lectin has blue color because one of its domains is linked with a hydrophobic organic chromophore of mass 597 Da (Carneiro et al. 2013). Another new chromophore-containing lectin, the *Haliclona manglaris* agglutinin (HMA), has been isolated from tropical sponge *H. manglaris* (Carneiro et al. 2015). This is a heterotrimeric glycoprotein consisting of two 15 kDa β-chains and one 22 kDa α-chain. It showed specificity for thyroglobulin. A 581 Da chromophore was found to be linked with the α-chain of HMA, probably by weak interactions. Neither H-3 nor HMA showed any toxicity against *A. nauplii*. They did not even have any agglutinating effect on bacteria. Thus, these lectins did not evolve as protecting molecules in those marine species. However, both H-3 and HMA were endowed a very interesting type of protection to β-carotene against oxidative damage mediated by the chromophores present, although neither lectin showed any potential for free-radical scavenging. Both *H. caerulea* and *H. manglaris*

are tropical sponges, and therefore in their habitat a high degree of ultraviolet assault from solar radiation is very common. Therefore, probably in those species chromophore-attached lectins were evolved to ensure antioxidant protection for their vital metabolites (Harrison and Cowden 1976).

## 38.2.6 Sponge lectin having glutamate receptor activity

A set of novel galectins were isolated from the marine sponge *Cinachyrella* sp. and was named *Cinachyrella* galectins (CchGs). These galectins showed very high thermostability with only a negligible drop in hemagglutination titer at 100°C. CchG-1 and 2 showed unprecedented activity on ionotropic glutamate receptors (iGluRs). Glutamate, one of the most important excitory neurotransmitters, is abundant in human brain, and iGluRs are mostly present in the neuronal membranes. To gain insight into the mechanistic basis of the binding activity with iGluR, the X-ray structure of one member of this important galectin family, CchG-1, was studied. In the high-resolution X-ray structure, it was found that the tetrameric structure of CchG-1 has been stabilized as a rigid toroidal-shaped "donut" formation, partly stabilized by packed pairs of vicinal disulfide linkages (Figure 38.2). The pair of binding sites were found to be exchanged by twofold symmetry, which is an indication for the probable model of interaction with ionotropic glutamate receptors. In this proposed model, the organization of the ligand-binding sites within the tetramericCchG-1 stretches the structure in a plane parallel to the membrane. This extended planar structure

*Figure 38.2* Structure of tetrameric galectin (CchG-1) from *Cinachyrella* sp. (Data from Freymann, D.M. et al., Acta Crystallogr. Sect. D, 68, 1163, 2012). PDB entry: 4AGV.

can support two crucial mechanism: first, the structure can stay as a mediator across extracellular domains of neighboring membrane receptors, which can act as a redox regulator; second, on the rigid planar orientation binding-site pairs within the CchG tetramer across the CC-loop, the protein could readily act as an allosteric liaison from receptor to receptor across the cell surface (Freymann et al. 2012). Thus, considering this strong structural evidence of CchG-1 upon putative interaction with iGluR, neuromodulatory effect of these galectins has been envisaged and studies are under way for their application (Ueda et al. 2013).

### 38.2.7   Sponge lectins with antitumor activity

Due to their outstanding carbohydrate specificity, lectins are used in various applications in biomedical research including cancer. Many lectins have been isolated and characterized as well-established and unprecedented sources of anticancer compounds. In the past decade, lectins from various sources have become a regular means for understanding the various glycobiological features of cancer and metastasis. The crucial role of lectins has already been justified in various cellular recognition processes known to be key steps in cancer biology such as recognition of surface markers of tumor cell, cell adhesion, trans-membrane signaling, mitogenic signaling, and apoptosis. To further the level of knowledge of cancer glycobiology and related lectin-associated processes, efforts are in progress to discover new lectins with newer specificities. This will enable us to develop analytical tools based on lectin–glycoconjugate recognition for the sensitive detection and monitoring of cancer. Sponge lectins are one of the most scarcely studied lectin families for their molecular basis of the protein–carbohydrate interactions relevant in cancer. In recent years, one of the central interests in this lectin family has been to reveal their potential role as antitumor agents and specific recognition of surface glycoconjugates of various cancer cells.

*Heliclona cratera* is an Adriatic sponge contains a 29 kDa lectin which shows moderate binding specificity for D-Gal and *N*-acetyl D-GalNAc. *H. crater* is endowed antitumor effect on HeLa and FemX cells, and it also has a weak mitogenic effect on human T lymphocytes (Pajic et al. 2002). Another lectin, HOL-18 from the demosponge *Halichondria okadai*, has been isolated, which is a 72 kDa tetrameric lectin having four 18 kDa noncovalently bonded monomers (Matsumoto et al. 2012). This lectin has specificity for binding chitotriose and fetuin. Frontal affinity chromatography showed that HOL-18 had affinity for N-linked sphingolipid-associated oligosaccharides with *N*-acetylated hexosamines and neuraminic acid at the nonreducing end. HOL-18 showed its

cytotoxicity toward both T-cell leukemia (Jurkat) and erythroleukemia (K562) cells expressing D-GlcNAc, D-GalNAc, and D-NeuAc at the nonreducing ends of their surface glycoprotein (Piller et al. 1990). Queiroz et al. isolated a D-galactose-specific lectin from the marine sponge *Cliona varians*, CvL, which showed specific antitumor activity against the human erythroleukemia cell K562 and Jurkat but not against mammalian solid tumor cell lines (B16 melanoma and PC3 prostate tumor cell) or against human peripheral blood lymphocytes (Queiroz et al. 2009). Mechanistic studies proved that CvL did not induce caspase-dependent cell death pathways; rather, the effect of cell death by CvL was blocked by the cathepsin inhibitor E-64, and thus CvL promotes cathepsin-dependent cell death mechanism in K562 and Jurkat cells.

The previously described CaL (*Cinachyrella apion* lectin) was also found to have antiproliferative activity against HeLa cells, whereas it has no toxicity at all for human blood mononuclear cells (Figure 38.3). Studies suggested that CaL might act as a Bax promoter to induce apoptosis. These results indicate the pharmacological potential of CaL as a medicine for cervical cancer (Rabelo et al. 2012). ACL-1 (*A. corrugate* lectin) was also potent in binding transformed malignant cells such as those of lung (H460), colon (HT-29), ovary (OVCAR-3), bladder (T24) and breast (T-47D and MCF7).

During malignant transformation, transformed cells show some aberrant glycosylation pattern in the cell surface glycoclyx, which are different from those of a normal cell surface. Therefore, these results may find important use in the determination of malignancy level, metastatic capacity of tumor cells, and progression or staging of cancers from tissue samples (Dresch et al. 2013).

### 38.3   Summary

In this chapter, we reviewed lectins from diverse sources of marine sponges and their biological activities. These lectins are found to be diverse with respect to their composition (glycosylated or non-glycosylated), specificities (mono- to oligo-saccharide binding, some binding to complex sugars), biological function (defense, symbiosis, anti-predation, etc.), and application (antitumor, antimicrobial, etc.). These sponge lectins have already demonstrated their applicability in different fields of biological science, such as a modulator of action between neurotransmitter and receptor and as antimicrobial or antitumor agents, and so on. All these interesting activities hold great prospects for these sponge-derived lectins in pharmaceutical science. However, further studies on the toxicology, delivery, and animal trials are needed before their practical utilization.

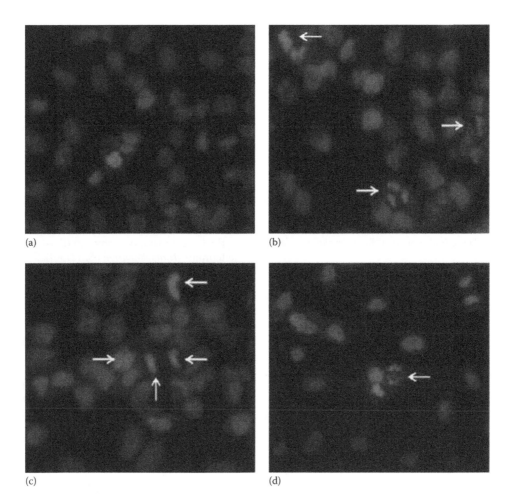

*Figure 38.3* Micrograph of HeLa cells treated with CaL. HeLa cells were incubated with 10 µg/mL (80.65 nM) CaL for 24 h and labeled with DAPI to show nuclear morphology. (a) Control HeLa cells, without CaL; (b–d) HeLa cells treated with CaL, showing nuclear morphological changes such as pyknosis and fragmentation (arrows). (Courtesy of Rabelo, L. et al., *Mar. Drugs*, 10, 727, 2012.)

# References

Atta, A. M., Barral-Netto, M. Peixinho, S., and Sousa-Atta, M. L. B. 1989. Isolation and functional characterization of a mitogenic lectin from the marine sponge *Cinachyrella alloclada*. *J. Med. Biol. Res.* 22: 379–385.

Bertrand, J. C. and Vacelet, J. 1971. L'association entre éponge-scornéesetbactéries. C. R. Hebd. *Seances Acad. Sci. Paris* 273: 638–641.

Böhm, M., Schröder, H. C., Müller, I. M., Müller, W. E., and Gamulin, V. 2000. The mitogen-activated protein kinase p38 pathway is conserved in metazoans: Cloning and activation of p38 of the SAPK2 subfamily from the sponge *Suberites domuncula*. *Biol. Cell* 92: 95–104.

Bretting, H., Jacobs, G., Donadey, C., and Vacelet, J. 1983. Immunohistochemical studies on the distribution and the function of the D-galactose-specific lectins in the sponge *Axinella polypoides* (Schmidt). *Cell Tissue Res.* 229: 551–571.

Carneiro, R. F., de Almeida, A. S., de Melo, A. A., de Alencar, D. B., de Sousa, O. V., Delatorre, P. K. S., Sampaio, S., Cavada, B. S., Nagano, C. S., and Sampaio, A. H. 2015. A chromophore-containing agglutinin from *Haliclona manglaris*: Purification and biochemical characterization. *Int. J. Biol. Macromol.* 72: 1368–1375.

Carneiro, R. F., de Melo, A. A., Nascimento, F. E. P., Simplicio, C. A., do Nascimento, K. S., da Rocha, B. A. M., Saker-Sampaio, S. et al. 2013. Halilectin 1 (H-1) and Halilectin 2 (H-2): Two new lectins isolated from the marine sponge *Haliclona caerulea*. *J. Mol. Recogn.* 26: 51–58.

Carneiro, R. F., Melo, A. A., Almeida, A. S., Moura, R. M., Chaves, R. P., Sousa, B. L., Nascimento, K. S. et al. 2013. H-3, a new lectin from the marine sponge *Haliclona caerulea*: purification and mass spectrometric characterization. *Int. J. Biochem. Cell. Biol.* 45: 2864–2873.

Dresch, R. R., Zanetti, G. D., Lerner, C. B., Mothes, B., Trindade, V. M. T., Henriques, A. T., and Voz.ri-Hampe, M. M. 2008. ACL-1, a lectin from the marine sponge *Axinella corrugata*: Isolation, characterization and chemotactic activity. *Comp. Biochem. Physiol. Part C* 148: 23–30.

Dresch, R. R., Lerner, C. B., Mothes, B., Trindade, V. M. T., Henriques, A. T., and Vozári-Hampe, M. M. 2012. Biological activities of ACL-I and physicochemical properties of ACL-II, lectins isolated from the marine sponge *Axinella corrugata*. *Comp. Biochem. Physiol. Part B* 161: 365–370.

Dresch, R. R., Zanetti, G. D., Irazoqui, F. J., Sendra, V. G., Zlocowski, N., Bernardi, A., Rosa, R. M., Battastini, A. M. O., Henriques, A. T., and Vozári-Hampe, M. M. 2013. Staining tumor cells with biotinylated ACL-I, a lectin isolated from the marine sponge *Axinella corrugate*. *Biotech. Histochem.* 88(1): 1–9.

Dzik, J. M. 2010. The ancestry and cumulative evolution of immune reactions. *Acta Biochim. Polonica* 57(4): 443–466.

Engel, M., Bachmann, M., Schröder, H. C., Rinkevich, B., Kljajic, Z., Uhlenbruck, G., and Müller, W. E. G. 1992. A novel galactose- and arabinose-specific lectin from the sponge *Pellina semitubulosa*: Isolation, characterization and immunobiological properties. *Biochimie* 74(6): 527–537.

Freymann, D. M., Nakamura, Y., Focia, P. J., Sakai, R., and Swanson, G. T. 2012. Structure of a tetrameric galectin from *Cinachyrella* sp. (ball sponge). *Acta Crystallogr. Sect. D* 68: 1163.

Gabius, H. J. and K. Kayser. 2014. Introduction to glycopathology: The concept, the tools and the perspectives. *Diagn. Pathol.* 9: 4.

Garrone, R. 1978. Phylogenesis of connective tissue: Morphological aspects and biosynthesis of sponge intercellular matrix. In *Frontiers of Matrix Biology*, Robert, L. (Ed.). Karger Press: Basel, Switzerland, Vol. 5, pp. 1–250.

Gomes Filho, S. M., Cardoso, J. D., Anaya, K., Silva do Nascimento, E., de Lacerda, J. T., Mioso, R., and Santi Gadelha, T., and de Almeida Gadelha, C. A. 2014. Marine sponge lectins: Actual status on properties and biological activities. *Molecules* 20: 348–357.

Harrison, F. W. and Cowden, R. R. 1976. *Aspects of Sponge Biology*. Academic Press: New York.

Kawabata, S., Beisel, H. G., Huber, R., Bode, W., Gokudan, S., Muta, T. Tsuda, R. et al. 2001. Role of tachylectins in host defense of the Japanese horseshoe crab *Tachypleus tridentatus*. In *Phylogenetic Perspectives on the Vertebrate Immune System*, Beck, G., Sugumaran, M., and Cooper, E. L., (Eds.). Kluwer Academic Publishers Group: Dordrecht, the Netherlands, pp. 195–202.

Kawsar, S. M. A., Mamun, S. M. A., Rahman, M. D. S., Hidetaro, Y and Ozeki, Y. 2011. In vitro antibacterial and antifungal effects of a 30 kDa D-galactoside-specific lectin from the desmosponge *Helichondria okadai*. *Int. J. Biol. Life Sci.* 7: 31–37.

Kruse, M., Steffen, R., Batel, R., Müller, I. M., and Müller, W. E. G. 1999. Differential expression of allograft inflammatory factor 1 and of glutathione peroxidase during auto- and allograft response in marine sponges. *J. Cell Sci.* 112: 4305–4313.

Leys, S. P., Mackie, G. O., and Meech, R. W. 1999. Impulse conduction in a sponge. *J. Exp. Biol.* 202: 1139–1150.

Matsumoto, R., Fujii, R. M. Y., Sarkar, M. A. K., Robert, A. K., Yasumitsu, H., Koide, Y., Hasan, I., Iwahara, C. et al. 2012. Cytotoxicity and glycan-binding properties of an 18 kDa lectin isolated from the marine sponge *Halichondria okadai*. *Toxins* 4: 323–338.

Mebs, D., Weiler, I., and Heinke, H. F. 1985. Bioactive proteins from marine sponges: Screening of sponge extracts for hemagglutinating, hemolytic, ichthyotoxic and lethal properties and isolation and characterization of hemagglutinins. *Toxicon* 23: 955–962.

Medeiros, D. S., Medeiros, T. L., Ribeiro, J. K. C., Monteiro, N. K. V., Migliolo, L., Uchoa, A. F., Vasconcelos, I. M., Oliveira, A. S.,

de Sales, M. P., and Santos, E. A. 2010. A lactose specific lectin from the sponge *Cinachyrella apion*: Purification, characterization, N-terminal sequences alignment and agglutinating activity on *Leishmania promastigotes*. *Comp. Biochem. Physiol. Biochem. Mol. Biol.* 155: 211–216.

Moura, R. M., Queiroz, A. F., Fook, J. M., Dias, A. S., Monteiro, N. K., Ribeiro, J. K., Moura, G. E., Macedo, L. L., Santos, E. A., and Sales, M. P. 2006. CvL, a lectin from the marine sponge *Cliona varians*: Isolation, characterization and its effects on pathogenic bacteria and *Leishmania promastigotes*. *Comp. Biochem. Physiol. A Mol. Integr. Physiol.* 145: 517–523.

Müller, W. E. G., Blumbach, B., and Müller, I. M. 1999. Evolution of the innate and adaptive immune systems: Relationships between potential immune molecules in the lowest metazoan phylum (Porifera) and those invertebrates. *Transplantation* 68: 1215–1227.

Muller, W. E. G., Zhan, R. K., Kurelec, B., Lucu, C., Muller, I., and Uhlenbruck, G. 1981. Lectin, a possible basis for symbiosis between bacteria and sponges. *J. Bacteriol.* 145: 548–558.

Ozeki, Y., Matsui, T., Yamamoto, Y., Funahashi, M., Hamako, J., and Titani, K. 1995. Tissue fibronectin is an endogenous ligand for galectin-1. *Glycobiology* 5: 255–261.

Pajic, I., Kljajic, Z., Sladic, D., Juranic, Z., and Gasic, M. J. 2002. A novel lectin from the sponge *Haliclona cratera*: Isolation, characterization and biological activity. *Comp. Biochem. Physiol. Part C: Toxicol. Pharmacol.* 132: 213–221.

Pawlik, J. R., Kernan, M. R., Molinski, T. F., Harper, M. K., and Faulkner, D. J. 1988. Defense chemicals of the Spanish dancer nudibranch *Hexabranchus sanguineus* and its egg ribbons: Macrolides derived from a sponge diet. *J. Exp. Mar. Biol. Ecol.* 119: 99–109.

Piller, V., Piller, F., and Fukuda, M. 1990. Biosysthesis of truncated O-glycans in the T cell line Jurkat. Localization of O-glycan initiation. *J. Biol. Chem.* 265: 9264–9271.

Proksch, P. 1994. Defensive roles for secondary metabolites from marine sponges and sponge-feeding nudibranchs. *Toxicon* 32: 639–655.

Queiroz, A. F., Silva, R. A., Moura, R. M., Dreyfuss, J. L., Paredes-Gamero, E. J., Souza, A. C., Tersariol, I. L. et al. 2009. Growth inhibitory activity of a novel lectin from Cliona varians against K562 human erythroleukemia cells. *Cancer Chemother. Pharmacol.* 63: 1023–1033.

Rabelo, L., Monteiro, N., Serquiz, R., Santos, P., Oliveira, R., Oliveira, A., Rocha, H., Morais, A. H., Uchoa, A., and Santos, A. 2012. A lactose-binding lectin from the marine sponge *Cinachyrella apion* (CaL) induces cell death in human cervical adenocarcinoma. *Cells Mar. Drugs* 10: 727–743.

Roger, R. D., Zanetti, G. D., Lerner, C. B., Mothes, B., Trindade, V. M. T., Henriques, A. T., and Vozári-Hampe, M. M. 2008. ACL-I, a lectin from the marine sponge *Axinella corrugata*: Isolation, characterization and chemotactic activity. *Comp. Biochem. Physiol.* 148: 23–30.

Schroder, H. C., Ushijima, H., Krasko, A., Gamulin, V., Thakur, N. L., Diehl-Seifert, B. I., Muller, M., and Muller, W. E. G. 2003. Emergence and disappearance of an immune molecule, an antibacterial lectin, in basal metazoan; A tachylectin-related protein in the sponge *Suberites domuncula*. *J. Biol. Chem.* 278: 32810–32817.

Skorokhod, A., Gamulin, V., Gundacker, D., Kavsan, V., Müller, I. M., and Müller., W. E. G. 1999. Origin of insulin receptor tyrosine kinases in marine sponges. *Biol. Bull.* 197: 198–206.

Ueda, T., Nakamura, Y., Smith, C., Copits, M. B. A., Inoue, A., Ojima, T., Matsunaga, S., Swanson, G. T., and Sakai, R. 2013. Isolation of novel prototype galectins from the marine ball sponge *Cinachyrella* sp. guided by their modulatory activity on mammalian glutamate-gated ion channels. *Glycobiology* 23: 412–425.

Wiens, M., Korzhev, M., Krasko, A., Thakur, N. L., Perović-Ottstadt, S., Breter, H. J., Ushijima, H., Diehl-Seifert, B., Müller, I. M., and Müller, W. E. 2005. Innate immune defense of the sponge *Suberites domuncula* against bacteria involves a MyD88-dependent signaling pathway. Induction of a perforin-like molecule. *J. Biol. Chem.* 280: 27949–27959.

Wilkie, I. C., Parma, L., Bavestrello, G., Cerrano, C., and Candia Carnevali., M. D. 2006. Mechanical adaptability of a sponge extracellular matrix: Evidence for cellular control of mesohyl stiffness in *Chondrosia reniformis* Nardo. *J. Exp. Biol.* 209: 4436–4443.

# Index

Printed and bound by CPI Group (UK) Ltd, Croydon, CR0 4YY

01/11/2024

01782602-0005